# FORMULAS

**Universal law of gravitation**  The force of attraction in Newtons between two objects with masses $m_1$ and $m_2$ kilograms that are a distance $r$ meters apart is given by

$$F = G\frac{m_1 m_2}{r^2}, \qquad \text{where} \quad G = 6.672 \times 10^{-11}\ \text{Nm}^2/\text{kg}^2$$

**Height of a falling object**  If an object is thrown into the air from an initial height $s_0$ feet with an initial velocity $v_0$ feet per second, its height in feet after $t$ seconds is given by $h(t) = -16t^2 + v_0 t + s_0$.

**Time dilation formula**  If a person is traveling at a velocity of $v$ feet per second for a total elapsed time of $T_0$ seconds, then the elapsed time in seconds on Earth is given by

$$T = \frac{T_0}{\sqrt{1 - \dfrac{v^2}{186,000^2}}}$$

**Geometric series**  If $a_1, a_2, a_3, \ldots, a_n$ is a geometric sequence with common ratio $r \neq 1$, then

$$\sum_{k=1}^{n} a_k = \frac{a_1(1 - r^n)}{1 - r}$$

**Arithmetic series**  If $a_1, a_2, a_3, \ldots, a_n$ is an arithmetic sequence, then $\displaystyle\sum_{k=1}^{n} a_k = \frac{n}{2}(a_1 + a_n)$.

**Quadratic formula**  If $a \neq 0$, then the solution(s) of $ax^2 + bx + c = 0$ are $x = \dfrac{-b \pm \sqrt{b^2 - 4ac}}{2a}$.

**Simple interest**  If a principal of $P$ dollars is deposited in an account paying simple interest at a rate $r$, then the balance $B$ after $t$ years is given by $B = P(1 + rt)$.

**Compound interest**  If a principal of $P$ dollars is deposited in an account paying interest at a rate $r$ compounded $n$ times per year, then the balance $B$ after $t$ years is given by $B = P\left(1 + \frac{r}{n}\right)^{nt}$.

**Binomial theorem**  $(a + b)^n = \binom{n}{n}a^n + \binom{n}{n-1}a^{n-1}b + \binom{n}{n-2}a^{n-2}b^2 + \cdots + \binom{n}{1}ab^{n-1} + \binom{n}{0}b^n$

**Distance formula**  The distance between two points $(x_1, y_1)$ and $(x_2, y_2)$ is $d = \sqrt{(x_2 - x_1)^2 + (y_2 - y_1)^2}$.

**Midpoint formula**  The midpoint of the line segment connecting two points $(x_1, y_1)$ and $(x_2, y_2)$ is the point $(\bar{x}, \bar{y})$ where

$$\bar{x} = \frac{x_1 + x_2}{2} \qquad \text{and} \qquad \bar{y} = \frac{y_1 + y_2}{2}$$

# ALGEBRAIC FORMULAS

**Sum of cubes**
$a^3 + b^3 = (a + b)(a^2 - ab + b^2)$

**Difference of cubes**
$a^3 - b^3 = (a - b)(a^2 + ab + b^2)$

**Difference of squares**
$a^2 - b^2 = (a + b)(a - b)$

# COLLEGE ALGEBRA AND TRIGONOMETRY

**David Dwyer**

*University of Evansville*

**Mark Gruenwald**

*University of Evansville*

# COLLEGE ALGEBRA AND TRIGONOMETRY

**David Dwyer**

*University of Evansville*

**Mark Gruenwald**

*University of Evansville*

**BROOKS / COLE PUBLISHING COMPANY**

 An International Thomson Publishing Company

*Pacific Grove* ▪ *Albany* ▪ *Bonn* ▪ *Boston* ▪ *Cincinnati* ▪
*Detroit* ▪ *London* ▪ *Madrid* ▪ *Melbourne* ▪ *Mexico City* ▪
*New York* ▪ *Paris* ▪ *San Francisco* ▪ *Singapore* ▪
*Tokyo* ▪ *Toronto* ▪ *Washington*

PRODUCTION CREDITS

*Copy editing: Luana Richards*
*Design: Techarts*
*Composition: University Graphics, Inc.*
*Art: University Graphics, Inc. and Randy Miyake*
*Cover illustration: © David Rickerd/The Image Bank*
*Photo credits follow index.*

*For more information, contact:*

BROOKS/COLE PUBLISHING COMPANY
511 Forest Lodge Road
Pacific Grove, CA 93950
USA

International Thomson Publishing Europe
Berkshire House 168-173
High Holborn
London WCIV 7AA
England

Thomas Nelson Australia
102 Dodds Street
South Melbourne, 3205
Victoria, Australia

Nelson Canada
1120 Birchmount Road
Scarborough, Ontario
Canada MIK 5G4

International Thomson Editores
Campos Eliseos 385, Piso 7
Col. Polanco
11560 México D. F. México

International Thomson Publishing GmbH
Königswinterer Strass 418
53227 Bonn
Germany

International Thomson Publishing Asia
221 Henderson Road
#05–10 Henderson Building
Singapore 0315

International Thomson Publishing Japan
Hirakawacho Kyowa Building, 3F
2-2-1 Hirakawacho
Chiyoda-ku, Tokyo 102
Japan

Beginning February 22, 1999, you can request permission to use material from this text through the following phone and fax numbers: Phone: 1-800-730-2214; Fax: 1-800-730-2215.

Printed in the United States of America.

10  9  8  7  6  5  4  3

A catalogue record for this book is available from the British Library.

LIBRARY OF CONGRESS
CATALOGING-IN-PUBLICATION
DATA

Dwyer, David.
College algebra and trigonometry / David Dwyer, Mark Gruenwald.
  p.   cm.
Includes index.
ISBN 0-314-09970-0 (hc : alk. paper)
1. Algebra.  2. Trigonometry  I. Gruenwald, Mark (Mark Edward)  II. Title.
QA152.2.D894   1997
512′.13--dc 20                                                                 96-45932
                                                                                      CIP

*Dedicated to the many brilliant and dedicated teachers who instilled in us a love for learning, including Fred Boos, Michael Golomb, Janet Koepp, Jerry Miller, Richard Penney, and Bob Wheeler.*

# CONTENTS

Mathematics is too often viewed as a collection of formulas and rules chiseled on stone tablets by ancient scribes, to be memorized and quickly forgotten. Surely some of the blame for this point of view rests with textbooks which present mathematics in a dust-dry fashion, carefully sanitizing the material of relevance and stripping it of all cultural context. Mathematics is not divorced from the rest of the human experience: it has developed from and shaped art, astronomy, chemistry, cinema, law, literature, music, religion, philosophy, physics, politics, sports, and technology. Moveover, mathematics is a work in progress, its frontiers ever expanding, new areas of application arising daily. Ozone depletion, destruction of tropical rain forest, the spread of AIDS, nuclear proliferation, and the population explosion are just a few of the issues confronting society which are illuminated by mathematical techniques and which are considered in this text. One could argue that the destiny of planet Earth hinges upon our ability to make intelligent decisions based in part upon mathematical models of these real world phenomena.

Mathematics is much more than a game in which symbols are manipulated on a piece of paper. It is the universal language of quantification, the mode of expressing and answering questions such as, how much, how many, how long and how likely. Just as truly literate people have the ability to not only read the written word but also to interpret it, so too are the mathematically numerate able to intertpret numerical data, to make judgments regarding its validity, and to quickly estimate unknown quantities based on a sense of scale, intuition and experience. In order to improve the student's sense of scale, numerous examples and exercises are included in this text which involve either extremely large or extremely small numbers. Mathematical numeracy is further enhanced by including dozens of real world applications in which actual data is used, mathematical models are constructed, and predictions are made and analyzed.

Not only are the areas of application of precalculus mathematics changing rapidly, but also the tools with which such investigations are conducted. In recent years inexpensive hand-held calculators capable of graphing complex mathematical functions have become readily available. The graphics calculator is an invaluable tool for solving equations, exploring the connection between graphical and algebraic representations of functions, and solving optimization problems. For these reasons the graphics calculator has been integrated within the text, not treated as an add-on or relegated to a supplement. It is thus recommended that the student have access to a graphics calculator and keep it handy while reading the text and doing the exercises.

The ability to produce virtually instantaneous graphical depictions of mathematical functions not only provides students with new ways of solving old problems, but also allows them to tackle problems that would have at one time been only the province of calculus and other branches of higher mathematics. But the graphics calculator is in no way a substitute for understanding the underlying concepts. On the contrary, there is a synergistic relationship between the two: facility with the calculator can be a tremendous help in understanding, and students with a thorough understanding will more easily gain dexterity with the graphics calculator. In order to take full advantage of the graphics calculator it is essential that the student understand its limitations as well as its capabilities. For this reason, you will occasionally find examples and exercises in which naive use of the calculator, without understanding underlying concepts, can lead to erroneous conclusions.

## FEATURES OF THE TEXT

**Chapter opening photographs**

Each chapter begins with a photograph and a brief description of how the photo relates to the subject matter of the chapter. For example, at the beginning of Chapter 2 we have included a striking photograph of a denuded tropical rain forest. We explain that the primary causes of deforestation are slash and burn clearing by ranchers and irresponsible timber harvesting by loggers. The reader is then referred to an exercise in the text that relates to this subject.

**Motivating questions**

Each section begins with a list of questions that can be answered by the student upon completion of the section. For example, the following motivating questions are asked at the beginning of Section 3 of Chapter 6. How powerful would the ''bombquake'' be that would result from simultaneously detonating all of the world's nuclear warheads? How can multiplication and division be performed by simply sliding pieces of paper? How many jam boxes would it take to cause a fatal musical overdose? If the entire population of China simultaneously jumped a foot off the ground, how strong would the resulting earthquake be?

**Examples**

Examples are used to illustrate the techniques and theory discussed in the text, or in some cases to introduce a topic or motivate a theoretical concept. The primary focus of application examples is on problem-solving strategy, while numerous standard examples illustrate the mechanics of algebraic, trigonometric and graphical techniques. Multi-step examples often include explanatory notes beside each step, or detailed explanations between important steps.

**Graphics calculators**

Although there are significant differences in the keystrokes required to perform tasks among leading brands of graphics calculators, there is a surprising degree of uniformity in the features that are available and the terminology used to describe them. For example, all leading graphics calculators have the same zoom features (zoom in, zoom out, and zoom box) and even use the same terminology for the range variables (xmin, xmax, ymin, ymax). For this reason, all of the most important graphics calculator functions can be defined and described using a ''generic'' graphics calculator language that is independent of the particular model of graphics calculator chosen. Detailed model-specific keystroke instructions can be found in the graphics calculator supplement.

**Art and graphics**

The text includes hundreds of graphs, all of which are computer generated for accuracy. In addition, dozens of computer generated graphics calculator screen images have been incorporated wherever students will be performing a task with the graphics calculator. Many applied examples and exercises are accompanied by color illustrations or bar graphs. Occasionally cartoon art has been used to illustrate examples or exercises in which there is a whimsical premise. The text also includes dozens of four color photographs, selected for their relevance, novelty, and beauty. In the text you will find photographs of the world's tallest human, tropical rain forests, radio telescopes, a solar flare, a human powered airplane, and the devastation arising from the possible cometary collision in Siberia in 1908.

**Mathematical notes**

In order to underscore the vitality of the subject, the text includes brief accounts of recent developments in mathematics, such as Andrew Wiles' proof of Fermat's Last Theorem and recent investigations of dynamical systems. In addition, we explore several intriguing applications of mathematics, such as the use of mathematical morphing in movies and television, radiocarbon dating to test the authenticity of the Shroud of Turin, and the spectral analysis of light and sound.

**Warnings and rules of thumb**

Warning boxes are provided to point out common mathematical pitfalls and how they can be avoided. Rule of thumb boxes describe helpful hints and tricks of the trade. For example, when solving work problems, students are encouraged to first express the *rate* at which work is done.

**Exercise sets**

There are some tasks for which pencil and paper techniques are superior, others where the graphics calculator is much more efficient, and still others that are best solved using some combination of algebraic, trigonometric and graphical tools. In order to develop the ability to select the mathematical tool most appropriate for a given task, many exercises have been included in which the student must first decide whether or not to use the graphics calculator. Thus, exercises which require the use of a graphics calculator are integrated within all of the following categories of exercises.

Numerous **standard exercises**, graded in difficulty, are included to reinforce the development of important skills and concepts. Many of the standard exercises are similar to examples and are intended primarily to test the student's manipulative ability. Others are unlike any example in the text in order to ensure that the student has synthesized the underlying concepts and isn't just blindly using examples as templates.

**Applied problems** vary in scope from explorations of topical issues with important social, environmental, and economic ramifications, to problems with whimsical, thought provoking, and hopefully entertaining premises. Many involve modeling with real world data from such diverse sources as governmental agencies, the World Almanac, the Environmental Almanac and the Guinness Book of World Records. Some involve important notions from the physical and social sciences, such as gravity, projectile motion, earthquakes, the speed of light, time dilation, chemical equations, nuclear power, surveying, navigation, population dynamics, and the spread of disease. On the lighter side are problems involving whimsical premises such as two cows tormenting a train engineer, a cyclist powering a small city, and a hot-air balloon raising the Statue of Liberty.

**Projects for Enrichment** are nontrivial multi-step problems designed for either group work or as especially challenging problems, possibly for extra credit. Some explore related mathematical topics that go beyond the classical boundaries of college algebra and trigonometry. Examples of such topics include the cubic formula, fractal geometry, partial fraction decomposition, and functions of several variables. Other projects are in-depth application problems. Examples here include an investigation of the relationship between alcohol consumption and blood alcohol level, a development of Kepler's third law for planetary motion, an application of systems of equations to balancing chemical equations and construction of a surveying tool based on trigonometric notions.

**Questions for Discussion or Essay** are intended for use as thought provoking discussion questions, or as subjects for short essays. Some are *critical thinking problems* that involve a mathematical notion presented in the text. Others ask the students to analyze the premises of application problems. Still others require that students connect the material with their experiences in other math classes, other disciplines, or even life in general.

**Chapter reviews and tests**

Each chapter concludes with a chapter review and a chapter test. The chapter reviews consist of comprehensive selections of exercises representative of those previously encountered in the chapter. The chapter tests include a sampling of representative problems but are not intended to be comprehensive. Each includes a section of true and false and ''give an example of'' problems that are designed to test the student's mastery of important concepts and key definitions.

## ACKNOWLEDGMENTS

The candid comments and insightful suggestions of our colleagues have been invaluable. Without their encouragement this text would never have been completed.

*Reviewers*

Dennis Alber, *Palm Beach Community College*
Judith Barclay, *Cuesta College*
Kathy Bavelas, *Manchester Community College*
Jim Blackburn, *Tulsa Junior College*
Louise M. Boyd, *Livingston University*
James Brasel, *Phillips County Community College*
Mary Cabral, *Middlesex Community College*
Terry Czerwinski, *University of Illinois at Chicago*
Bettyann Daley, *University of Delaware*
Gregory Dotseth, *Iowa State College*
Sharon Edgmon, *Bakersfield College*
Iris B. Fetta, *Clemson University*
Virginia Fisher, *Albuquerque T-VI*
Richard Faulkenberry, *University of Massachusetts, Dartmouth*
Nina Girard, *University of Pittsburgh at Johnstown*
Kalin Godev, *Penn State University/College of Charlestown*
Marvin Goodman, *Monmouth College*
Michael Gutowski, *Johnson County Community College*
Nancy Henry, *Indiana University at Kokomo*
Randal Hoppens, *Blinn College*
Nancy G. Hyde, *Broward Community College*
Maureen Kelley, *Northern Essex Community College*
Sylvia Kennedy, *Broome Community College*
Gary King, *New Mexico State University at Alamogordo*
C. J. Knickerbocker, *St. Lawrence University*
David Lawrence, *Southwestern Oklahoma State University*
James W. Lea, Jr., *Middle Tennessee State University*
Peter Livorsi, *Oakton Community College*
Debra Madrid-Doyle, *Santa Fe Community College*
P. E. Marion, *North Harris Community College*
Gael Mericle, *Mankato State University*
Carol Page, *St. Charles County Community College*
Beverly Phillips, *Thomas Nelson Community College*
Geraldine Prange, *University of Delaware*
William Radulovich, *Florida Community College*
Jane Ringwald, *Iowa State*
Stephen Rodi, *Austin Community College*
Ned W. Schillow, *Lehigh Carbon Community College*
Karl Schroeder, *Canisius College*
Burla Sims, *University of Arkansas–Little Rock*
Michael Smith, *Auburn University*
Mary Jane Sterling, *Bradley University*
Ara Sullenberger, *Tarrant County Junior College*
Lana Taylor, *Siena Heights College*
Kathy Wetzel, *Amarillo College*
Jamie Whitehead, *Texarkana College*
Mary Wilbur, *Pasco-Hernando Community College*
Francher Wolfe, *Metropolitan State University*

***Special Thanks***

We are appreciative to the staff at West Publishing for their assistance, with special thanks to Acquisitions Editor Nancy Hill-Whilton, Developmental Editor Denis Bayko, Production Editor Tamborah Moore, Copy Editor Luana Richards, and Promotion Managers Doug Abbott and Stacie Falvey.

Photographical acquisitions researcher Dallas Chang provided us with eye-catching photographs from a variety of sources. Thanks also to Judy Barclay for checking the manuscript for accuracy.

Finally, we would like to thank those who prepared supplements, including Kathryn Wetzel, Kalin Godev and David Lawrence.

## SUPPLEMENTS FOR THE TEXT

*Instructor's Solutions Manual* by Kathryn C. Wetzel of Amarillo College includes worked-out solutions to all of the even-numbered text exercises.

*Student's Solutions Manual* also by Kathryn C. Wetzel includes worked-out solutions to all of the odd-numbered text exercises.

*Instructor's Manual with Test Bank* by Kalin Godev of the College of Charleston and James Brasel of Phillips Community College, includes suggested course schedules, chapter outlines, suggested homework assignments and a test bank of multiple choice and open-ended questions. The manual also includes instructions on how to use the graphing calculator in the classroom as well as how to integrate the features of the text including collaborative learning and projects for enrichment. Special attention is given to identifying the most challenging exercises and problems.

*Graphing Calculator and Software Laboratory Manual* by David P. Lawrence of Southwestern Oklahoma State University includes keystroke instructions for the TI-81, TI-82, and TI-85, Casio 7700GE and 9700GE, Sharp EL9200C and EL9300C, and HP48G/GX as well as DERIVE software.

*50 Transparency Masters* contain important text illustrations and examples.

*Westest 3.0* computerized testing program includes algorithmically-generated questions. Versions are available for IBM PC and Macintosh systems to qualified adopters.

*West Math Tutor Software* by Mathens is an algorithmically-based tutorial for Macintosh, IBM PCs or compatibles. Available to qualified adopters.

Two video tapes provide a general overview of the TI-81, TI-82, TI-85, Casio fx7700, HP48G, and Sharp EL9200. The videos also provide detailed information and examples appropriate for precalculus-level courses. Available to qualified adopters.

Video series - ''In the Simplest Terms'' available from PBS/Annenberg, a series of 26 half-hour programs, provides a comprehensive overview of algebra.

# FOUNDATIONS AND FUNDAMENTALS

■ The Nashua River, desecrated by human waste and bleeding red from industrial dyes, had become a lifeless cocktail of toxins by the 1960s. Today the river teems with life, its pristine beauty restored, thanks to a concerted community campaign. Mathematical models, taking into account such factors as water temperature, levels of pollutants, and flow rates, help ecologists design such clean-up efforts. In Example 11 of Section 7 we will consider a simplified model for predicting the expense associated with removing pollution.

---

### SECTION 1

## NUMBER SYSTEMS AND ESTIMATION

- Which is larger, 0.999… or 1?
- In which state could the entire population of Earth fit if each person were allotted a square yard?
- How long a convoy of trucks would be required to haul away Mt. Everest?
- What fraction when multiplied by itself gives 2?
- How far will you walk in your lifetime?
- How much gasoline would be saved if every car in the nation got 50 miles per gallon?

---

### NUMBER SYSTEMS

Number systems were developed as a language for expressing how many, how much, and how far. As with all languages, number systems evolved so that more sophisticated ideas could be expressed. In particular, new number systems arose whenever an old number system was found incomplete or inadequate.

The **natural numbers** (sometimes called the counting numbers) are the numbers 1, 2, 3, …. They arose out of our need to count—livestock, furs, days, tribe members, and so on. Millennia elapsed before the number zero was introduced as a notational convention for writing numbers. The **whole numbers** are simply the natural numbers together with 0—that is, 0, 1, 2, 3, …. The whole numbers, however, are incomplete. We would like to be able to subtract any two numbers and get a third number, but what whole number is $1 - 2$? In order to be able to perform such subtractions, we must extend our number system to the **integers**, the numbers …, $-3$, $-2$, $-1$, 0, 1, 2, 3, …. The integers are complete with respect to subtraction in the sense that the difference between any two integers is an integer.

### EXAMPLE 1    *Recognizing natural numbers and integers*

For each of the following numbers, determine whether it is an integer. If so, determine whether it is a natural number.
a. $1,234,000,031 - 23,886,986,102$
b. The number of atoms that constitute the planet Jupiter
c. The number of complete minutes from the birth of Jesus until now minus the number of complete seconds from the birth of Jesus until now
d. Your height in feet

#### SOLUTION

a. Since $1,234,000,031 - 23,886,986,102$ is formed by subtracting two integers, it too is an integer. However, since the second number is larger than the first, the result is negative and hence not a natural number.

b. Although the number of atoms constituting the planet Jupiter is unknown (and no doubt enormous), this number will be a natural number and hence an integer as well.

c. Since there are 60 times as many seconds in this time interval as there are minutes, the difference between the number of complete minutes and the number of complete seconds will be negative and hence not a natural number.

**d.** If you happen to be *exactly* 3, 4, 5, 6, 7, or 8 feet tall, then your height in feet will be a natural number (and, of course, an integer also). Otherwise, your height in feet will not be an integer. For example, if you are 5 feet 5 inches or 65 inches tall, then your height in feet would be $65 \div 12$, which is not an integer.

As part (d) of the previous example illustrates, the integers are also incomplete. The expression $65 \div 12$ has no meaning if we are confined to the set of integers. To give such expressions meaning, we must extend our number system to the **rational-number system**, the set of fractions. In the rational-number system, $65 \div 12$ is simply the fraction $\frac{65}{12}$. In fact, whenever we divide any integer by a nonzero integer, the result is a rational number. A rational number can be expressed as the quotient of two integers, or it can be written as a decimal. It will be shown in Exercise 59 that whenever a rational number is expressed as a decimal, the expansion either terminates, like $\frac{1}{4} = 0.25$, or repeats, like $\frac{1}{3} = 0.333\ldots$.

**EXAMPLE 2**    *Recognizing rational numbers*

Which of the following numbers are rational?
**a.** 22.47
**b.** $-2$
**c.** 0.24343434…
**d.** The price in dollars of a compact disc
**e.** 0.101001000100001000001…

**SOLUTION**

**a.** 22.47 is a terminating decimal, so it is a rational number. Written as a fraction, $22.47 = \frac{2247}{100}$.

**b.** $-2$ is a rational number since it can be expressed as $-2 = \dfrac{-2}{1}$. (In fact, all integers are rational numbers.)

**c.** 0.24343434… is a repeating decimal, so it is a rational number. (Using the technique outlined in Exercise 59, we find that $0.24343434\ldots = \frac{241}{990}$.)

**d.** In the United States, all retail prices are rounded to the nearest cent, which means that the price of the compact disc in dollars could be expressed as a decimal with two digits to the right of the decimal point, like 12.95. Thus, all prices are rational numbers.

**e.** Since 0.101001000100001000001… neither repeats nor terminates, it is not a rational number.

The rational numbers are complete with respect to the four basic operations of addition, subtraction, multiplication, and division. That is, if we perform any of these operations on a pair of rational numbers (with the exception of division by zero), the result will be a rational number. Numbers that are not rational, such as 0.1010010001…, might seem unnecessary. However, as the ancient Greeks discovered, the rational-number system is inadequate for expressing distances. According to the Pythagorean Theorem, the square of the length of the hypotenuse of a right triangle is the sum of the squares of the legs, $c^2 = a^2 + b^2$ (see Figure 1). A right triangle with legs of length 1 has a hypotenuse with length $c$, where $c^2 = 1^2 + 1^2 = 2$; thus, $c = \sqrt{2}$. But it can be shown (see Exercise 60) that $\sqrt{2}$ is not a rational number. This means that there is at least one distance that is not a rational number. In fact, there are many other distances

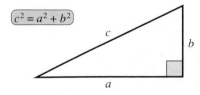

$c^2 = a^2 + b^2$

**FIGURE 1**

that are not rational numbers. Every number of the form $\sqrt{n}$, where $n$ is not a perfect square, is not rational but corresponds to a distance. The number $\pi \approx 3.14159\ldots$ is not a rational number, yet it corresponds to the distance around a circle with diameter 1.

To remedy the inadequacy of the rational numbers for expressing distance, we extend to the **real-number system**. The real-number system can be described as the set of all numbers with decimal expansions. Since the rational numbers have terminating or repeating decimal expansions, the nonrational real numbers, or **irrationals** as they are called, consist of those numbers with nonterminating, nonrepeating decimal expansions. Examples of irrational numbers include $\sqrt{2}$, $\pi$, and $0.1010010001\ldots$.

## THE REAL-NUMBER LINE

**FIGURE 2**

The real-number system can be represented geometrically by the **real-number line** depicted in Figure 2. The point labeled zero is called the **origin**. The points to the left of the origin correspond to the negative real numbers, and the points to the right of the origin correspond to the positive real numbers.

The number line establishes a **one-to-one correspondence** with the real numbers in the sense that every point on the number line corresponds to a real number, and every real number corresponds to a point on the number line. As a consequence, we can order the real numbers according to their position on the number line.

**Ordering the real-number system**

> For two real numbers $a$ and $b$, we say that $a$ is **less than** $b$ (denoted $a < b$) if $a$ lies to the left of $b$ on the number line, and $a$ is **greater than** $b$ (denoted $a > b$) if $a$ lies to the right of $b$.

The symbols "$<$" and "$>$" are called **inequality symbols**. Two other inequality symbols are "$\leq$," which denotes **less than or equal to**, and "$\geq$," which denotes **greater than or equal to**.

**EXAMPLE 3**

*Ordering real numbers*

Place the appropriate inequality symbol in the box between the numbers given below.
a. $2 \ \square \ 5$
b. $-\frac{2}{3} \ \square \ -\frac{5}{6}$
c. $0 \ \square$ The weight of a bacterium in pounds
d. $-2.34 \ \square \ 1.03$
e. $0 \ \square$ The number of dimes a man has in his pocket
f. $100 \ \square$ Your score as a percentage on your next mathematics exam

**SOLUTION**

a. Since 2 lies to the left of 5 on the number line, we write $2 < 5$.

b. Since $-\frac{2}{3} = -\frac{4}{6}$ and $-\frac{4}{6}$ lies to the right of $-\frac{5}{6}$ on the number line, we write $-\frac{2}{3} > -\frac{5}{6}$.

c. The weight of a bacterium, no matter how small, is positive, so we have

$$0 < \text{The weight of a bacterium in pounds}$$

d. Every negative number is less than every positive number; thus, we have $-2.34 < 1.03$.

**e.** It is possible that a man might have no dimes in his pocket, so we write

$0 \leq$ The number of dimes a man has in his pocket

**f.** Since your score will be a number that is at most 100, we have

$100 \geq$ Your score as a percentage on your next mathematics exam

## SETS OF REAL NUMBERS

The most direct way to write a set is to simply list all of its elements within braces. For example, the set consisting of the positive even numbers less than 10 could be written as {2, 4, 6, 8}. The set of all prime numbers less than 30 is {2, 3, 5, 7, 11, 13, 17, 19, 23, 29}. The **empty set**, the set with no elements, may be written as { } or ∅.

Although listing is convenient for small sets, it is impractical for large sets and impossible for sets with infinitely many elements. **Set-builder notation** is useful for describing large or infinite sets that are defined by a property. For example, consider the set of real numbers less than 2. In set-builder notation we write $\{x|x < 2\}$. The set of real numbers between and including $-3$ and 3 can be written $\{x|-3 \leq x \leq 3\}$.

## THE DIGITS OF PI

The number $\pi$, the ratio of the circumference of a circle to its diameter, is irrational, and hence its decimal expansion neither repeats nor ends. It begins

3.14159265358979323846264338327950288419716939937510582097494459 23...

For most practical purposes, the first few digits of $\pi$ provide us with a sufficiently accurate approximation. Indeed, the 65 digits given above provide us with sufficient accuracy to estimate the circumference of a circle with a radius equal to the distance to the most distant known object (a quasar some $7.8 \times 10^{22}$ miles away) to within the length of the smallest hypothesized object in the universe (a "string" approximately $10^{-33}$ centimeter in length). But mathematicians are not constrained by the boundaries of the physical universe and have computed $\pi$ to what at first glance seems a ridiculous number of decimal places. For example, the Chudnovsky brothers, David and Gregory, have computed $\pi$ to over 2 billion decimal places; if their decimal expansion were written on a single line with the same size type that is used in this sentence, it would extend from New York to San Francisco!

Mathematicians have been competing to compute the most accurate version of $\pi$ since ancient times. Although the need for increased accuracy may have been initially dictated by practical concerns, the driving force behind the quests of many modern

"digit hunters" is clearly the competition itself. Also, computation of $\pi$ to millions of decimal places provides an excellent means for testing the speed and accuracy of supercomputers, such as the massively parallel computer shown here.

But there is more here than meets the eye. Although we *define* $\pi$ to be the ratio of the circumference of a circle to its diameter, it arises in many settings having nothing to do with circles. It appears in probability theory, calculus, number theory, and physics. To many mathematicians, $\pi$ is not simply a human-made convention; $\pi$ is an integral part of our universe. If we were to one day encounter intelligent beings from another galaxy, then surely they would know about $\pi$. To a digit hunter, the digits of $\pi$ represent a piece of the cosmic code, a divine communiqué containing, perhaps, revelations of nature. And so they compute digits of $\pi$, looking for patterns, trying to detect structures. Are the digits truly random? Are there as many strings of consecutive digits as we would expect? These are the sorts of questions that have not yet been answered, but perhaps we will know more once we see the next 2 billion digits.

**FIGURE 3**

It is often helpful to graph sets of real numbers on a number line. To do this, we simply draw a number line and then draw a heavy line through all values that are included in the set. For example, the set $\{x | -2 < x \leq 4\}$ is graphed as shown in Figure 3. Note that a parenthesis, "(", is used at $-2$ to indicate that $-2$ is *not* included, whereas a bracket, "]", is used at 4 to indicate that 4 *is* included.

**EXAMPLE 4**    *Graphing sets of real numbers*

Graph the following sets.

**a.** $\{x | x \geq 3\}$     **b.** $\{x | -\frac{3}{2} \leq x < 3\}$

**SOLUTION**

**a.**   **b.**

   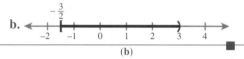

(a)   (b)

Sets of real numbers such as those graphed in Example 4 can be conveniently described using **interval notation**. For example, instead of writing $\{x | -\frac{3}{2} \leq x < 3\}$, we can simply write $[-\frac{3}{2}, 3)$. Notice that we are following our graphing convention of using a parenthesis to indicate that an endpoint of the interval is not included and a bracket to indicate that an endpoint is included. The set $\{x | x \geq 3\}$ can be written as $[3, \infty)$. Note that the symbol "$\infty$" (infinity) does not represent a real number but rather indicates that the interval includes all numbers to the right of 3. Table 1 summarizes the nine types of intervals.

**TABLE 1**

*Interval notation*

| Interval | Set-builder notation | Graph |
|----------|----------------------|-------|
| $(a, b)$ | $\{x | a < x < b\}$ | |
| $[a, b]$ | $\{x | a \leq x \leq b\}$ | |
| $(a, b]$ | $\{x | a < x \leq b\}$ | |
| $[a, b)$ | $\{x | a \leq x < b\}$ | |
| $(a, \infty)$ | $\{x | x > a\}$ | |
| $[a, \infty)$ | $\{x | x \geq a\}$ | |
| $(-\infty, b)$ | $\{x | x < b\}$ | |
| $(-\infty, b]$ | $\{x | x \leq b\}$ | |
| $(-\infty, \infty)$ | $\{x | -\infty < x < \infty\}$ | |

**EXAMPLE 5**    *Using interval notation*

Write the following sets using interval notation and sketch their graph.

**a.** $\{x | -7 \leq x < 2\}$     **b.** $\{s | s < \sqrt{5}\}$

**SOLUTION**

**a.** $[-7, 2)$

**b.** $(-\infty, \sqrt{5})$

**EXAMPLE 6**  *Converting from interval notation to set-builder notation*

Express each of the following sets using set-builder notation.
**a.** $(2, 5]$      **b.** $(-\infty, 3)$      **c.** $[-4, \infty)$

**SOLUTION**

**a.** $\{x | 2 < x \le 5\}$      **b.** $\{x | x < 3\}$      **c.** $\{x | x \ge -4\}$

---

## ESTIMATION

In many real-life situations (and problems that you will encounter in this text), precise information about real-number quantities is unavailable. Usually, however, the question at hand can be answered just by knowing a range of values for the quantities.

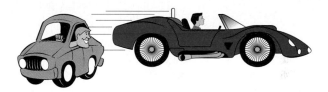

For example, suppose that you drive by a Maserati (an Italian sports car) with a for-sale sign in the rear window. You know that you have some change in your pocket left over from the grilled chicken sandwich that you just purchased, but you are not sure exactly how much. You wonder, "Do I have enough money to buy this car?" In spite of your lack of precise information, either concerning the cost of the car or the amount of change in your pocket, you deduce that you cannot buy the car. How did you do it? Although you were probably not conscious of this process, your thinking probably went something like this. "I know I have less than $20 in my pocket, and the Maserati must cost way more than $10,000. Guess I'll be driving this Yugo for a while." Although the approximations used in this example were trivial, the same sort of reasoning can be used to answer much more difficult questions.

**EXAMPLE 7**  *Lifetime walking—an estimation problem*

Estimate the distance you will walk in your lifetime.

**SOLUTION**   We will begin by making a rough estimate of the distance you might walk in a single day. It's not unreasonable to assume that if we include walking from room to room, walking to and from the parking lot, and all the other walks that occur in the course of a day, then most people walk about 40 minutes per day. Now a pace of 20 minutes per mile is typical, so we'll assume that you walk 2 miles in the course of a day.

Next we compute the number of days in your lifetime. If you live 75 years, then (ignoring leap years) you will live $75 \times 365 = 27,375$ days. Multiplying the 2 miles that you walk every day by the number of days in your lifetime, we have $2 \times 27,375 = 54,750$ miles—far enough to circle the Earth twice!

A team of scientists headed by B. Chamoux and A. de Polenza measures Mount Everest.

| EXAMPLE 8 |

*Leveling Everest—an estimation problem*

If it were possible to grind up Mt. Everest, fill semitrailers with the contents, and then form the semitrailers into a convoy, how long would the convoy be?

**SOLUTION**    First we will estimate the volume of Mt. Everest. We look in a reference book (such as an almanac or *The Guinness Book Of World Records*) and find that Mt. Everest is approximately 29,000 feet tall. Now unless we are lucky enough to have a reference that gives the volume of Mt. Everest, we are going to have to make an educated guess as to its shape. Let us assume that the mountain is roughly the shape of a cone with a height of 29,000 feet. We look up the formula for the volume of a cone, and find that $V = \frac{1}{3}\pi r^2 h$, where $r$ is the radius of the circle that forms the base of the cone and $h$ is the height. We are taking $h$ to be 29,000, but we don't know $r$. Here again, we will make an educated guess that the diameter of the base is roughly twice the height, so that the radius equals the height. Thus, we take $r$ to be 29,000 feet. Upon substituting the values of $r$ and $h$ into the formula for $V$, we find that

$$V = \frac{1}{3}\pi(29{,}000 \text{ ft})^2(29{,}000 \text{ ft}) \approx 25{,}540{,}000{,}000{,}000 \text{ ft}^3$$

Next we estimate the capacity of a semitrailer. They appear to be about 40 feet long, 8 feet wide, and 9 feet tall. This would make the volume of a semitrailer equal to 40 ft × 8 ft × 9 ft = 2880 cubic feet. Hence it would take 25,540,000,000,000 ÷ 2880 ≈ 8,868,000,000 semitrailers to haul off Mt. Everest. Now we must figure out how long a convoy this would make. The cab of a semitruck is about 10 feet in length, so that the entire truck would be about 10 ft + 40 ft = 50 feet long. Assuming two truck lengths (or 100 feet) between trucks, the convoy would be about 8,868,000,000 × 150 ft = 1,330,200,000,000 feet long. Using the fact that a mile is 5280 feet, we determine that the convoy is 1,330,200,000,000 ÷ 5280 ≈ 251,900,000 miles long. This, by the way, is long enough to reach to the Sun

and back, and then make over 137 round-trips to the Moon! Although our estimates were crude, even if we are off by a factor of 10 or even 100, it is clear that such a convoy would be impossible to form.

## COMPLEX NUMBERS

If a mathematician of only a few hundred years ago were asked to produce the square root of $-1$ (a number whose square is $-1$), the response would have been, "There is no such number." And indeed, there is no such *real* number. But strangely, square roots of negative numbers reared their heads in problems in number theory, geometry, and even physics. Although such solutions were at first discarded, it soon became evident that at the very least, the square roots of negative numbers could be used as a sort of "trick" to solve many problems. Mathematicians found that by temporarily pretending such numbers existed, they were able to produce correct results. Since the square roots of negative numbers were not seen as being full-fledged numbers, they were called **imaginary numbers**. The square root of $-1$ was called $i$ (i.e., $i^2 = -1$), and any other imaginary number could be expressed as a real number times $i$, like $2i$, $5i$, $-1.23i$, and so on.

As mathematics has evolved, the status of the so-called imaginary numbers has been elevated; modern mathematicians view imaginary numbers as being no less "real" than the real numbers themselves. The **complex-number system** is an extension of the real-number system formed by including all possible sums of real numbers with imaginary numbers, like $2 + 3i$ and $3 - 4i$.

**EXAMPLE 9**   *Classification of complex numbers*

Determine whether or not each of the following numbers is real, imaginary, or neither.
a. A number that when multiplied by itself gives $-25$
b. The distance that light travels in a year expressed in millimeters
c. A number that when squared gives $2i$

**SOLUTION**

a. There is no real number whose square is negative; only imaginary numbers have negative squares. Thus, any number that when multiplied by itself gives $-25$ is imaginary.

b. All distances can be expressed as real numbers, so this number is real. In fact, it is approximately 9,439,922,663,424,000,000.

c. When imaginary numbers are squared, the result is a negative real number. When real numbers are squared, the result is a positive real number. Thus, any number whose square is $2i$ is neither real nor imaginary. However, it can be shown that $2i$ is the square of two complex numbers, $1 + i$ and $-1 - i$.

We now summarize the number systems that have been discussed. The relationships among the systems are diagrammed in Figure 4.

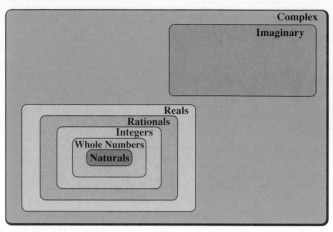

**FIGURE 4**

---

**Number systems**

> **Natural numbers** are the numbers 1, 2, 3, ….
>
> **Whole numbers** are the numbers 0, 1, 2, 3, ….
>
> **Integers** are the numbers …, $-2$, $-1$, 0, 1, 2, 3, ….
>
> **Rational numbers** are the fractions—that is, the numbers of the form $p/q$, where $p$ and $q$ are integers and $q \neq 0$. Rational numbers have terminating or repeating decimal expansions.
>
> **Real numbers** are the numbers that may be written as decimals. The real numbers that are not rational are called **irrational**, and they have nonterminating, nonrepeating decimal expansions.
>
> **Complex numbers** are numbers of the form $a + bi$, where $a$ and $b$ are real numbers and $i = \sqrt{-1}$. The term **imaginary** is used to describe complex numbers of the form $bi$.

---

## EXERCISES 1

**EXERCISES 1–18** ☐ *List all the terms that describe the given number. Choose from natural, whole, integer, rational, irrational, real, and complex.*

1. $\dfrac{6}{3}$

2. $-47.134$

3. $\sqrt{7}$

4. $\dfrac{12}{5}$

5. $3 - i$

6. $-\sqrt{3}$

7. The number of goats on Mars

8. $0.8888888\ldots$

9. $0.8888888$

10. $\sqrt{-4}$

11. 4

12. The ratio of the circumference of a circle to its diameter

13. $-\sqrt{25}$

14. $34{,}343{,}434{,}343 - 35{,}343{,}434{,}343$

15. A number whose square is $-7$

16. The number of men named Bob living as of 9:43:26.24 A.M. EST on June 17, 1994

17. The height in miles of a 1-foot-tall rooster

18. The length of the hypotenuse of an isosceles right triangle with legs of length 1

**EXERCISES 19–24** □ *Answer true or false.*

19. Every natural number is a whole number.

20. Some integers are irrational numbers.

21. Some complex numbers are real.

22. All rational numbers have a finite number of decimal places.

23. Negative integers are rational numbers.

24. If the square of a number is negative, then the number itself cannot be rational.

**EXERCISES 25–30** □ *Place the most appropriate order symbol between each pair of quantities given below.*

25. 3 □ 0

26. $-4$ □ $-5$

27. $-\dfrac{6}{5}$ □ $-\dfrac{4}{5}$

28. $\dfrac{5}{6}$ □ $\dfrac{2}{3}$

29. The number of "Ben Franklins" in your instructor's wallet □ 0

30. A person's age in years on or after their twentieth birthday □ 20

**EXERCISES 31–36** □ *Place the given list of numbers in ascending order (the smallest first, the largest last). Estimate as needed.*

31. 23, $-22.9$, $-23$, 22.9

32. $-\dfrac{1}{4}$, $-0.26$, $-\dfrac{3}{16}$, $-0.2$

33. $\dfrac{22}{7}$, $\pi$, 3.15, $3 + \dfrac{1}{10} + \dfrac{4}{100}$

34. $\dfrac{1}{3}$, 0.33333, 0.3334, $\dfrac{1001}{3000}$

35. The number of miles from the Earth to the Sun, the gross national product of the United States in dollars, 1,000,000,000, the number of living organisms on Earth

36. The diameter of the period "." in feet, 0.0001, the weight of this page in tons, Shaquille O'Neal's height in light-years

**EXERCISES 37–44** □ *Write each set using interval notation and sketch the graph.*

37. $\{x | -2 < x \le 4\}$

38. $\{x | 3 \le x < 8\}$

39. $\left\{x \middle| -7 < x < -\dfrac{1}{2}\right\}$

40. $\{x | -2 \le x \le \sqrt{3}\}$

41. $\{x | x < -3\}$

42. $\{x | x \ge -1\}$

43. $\{x | x \ge \sqrt{2}\}$

44. $\left\{x \middle| x \le \dfrac{3}{2}\right\}$

**EXERCISES 45–52** □ *Write each set using set-builder notation.*

45. (3, 5)

46. $(-\infty, 2)$

47. The real numbers from 1 to $10^{100}$

48. $(0, \infty)$

49. [1, 2]

50. The real numbers from 0 to $-100$

51. $(100, \infty)$

52. (1, 2)

■ *Applications*

53. *Cow Volume* Estimate the volume of a cow in liters.

54. *Cigarette Expenses* Estimate the amount of money that a 3-pack-a-day smoker will spend on cigarettes in 30 years.

55. *Lifting a Van* Estimate the number of kindergartners required to lift a mini-van.

56. *Earth Population* Estimate the number of square yards of living space available to each person if the entire population of Earth was crammed into the state of Delaware.

57. *Garbage Weight* Estimate the weight of all the garbage produced in residences in the United States in a year's time.

58. *Sharing Wealth* Estimate the number of families that the world's richest man could feed for a year if he were willing to spend his entire fortune.

### ◼ *Projects for Enrichment*

**59.** *Decimal Expansions* The text states that rational numbers have terminating or repeating decimal expansions. The converse of this statement is also true; that is, if a decimal expansion terminates or repeats, then it represents a rational number. In this project we will learn how to express terminating and repeating decimals as fractions.

Terminating decimals are converted to fractions as follows. The numerator of the fraction is the integer obtained by removing the decimal point; the denominator is an appropriate power of 10 (10, 100, 1000, etc.). For example, 0.123 may be written as $\frac{123}{1000}$, 1.03 as $\frac{103}{100}$, and so on.

**a.** Express each of the terminating decimals as a fraction.

    **i.** 87.096            **ii.** 232.1

    **iii.** 0.0034         **iv.** 0.000048

The following examples illustrate a method for converting *repeating* decimals to fractions.

Example: Convert 13.454545... to a fraction.
*Solution:* Let $x = 13.454545...$. Then $100x = 1345.454545...$. We subtract $x$ from $100x$ as follows.

$$\begin{aligned} 100x &= 1345.454545... \\ -x &= -13.454545... \\ \hline 99x &= 1332.0 \end{aligned}$$

Dividing both sides by 99, we obtain $x = \frac{1332}{99}$.

Example: Convert 2.512012012... to a fraction.
*Solution:* Let $x = 2.512012012...$. Then $10,000x = 25,120.12012012...$ and $10x = 25.12012012...$. Subtracting, we obtain

$$\begin{aligned} 10,000x &= 25,120.120120120... \\ -10x &= -25.120120120... \\ \hline 9,9990x &= 25,095.0 \end{aligned}$$

Thus, $x = \dfrac{25,095}{9990} = \dfrac{1673}{666}$.

**b.** Express each of the following repeating decimals as a fraction.

    **i.** 0.252525...         **ii.** 0.999...

    **iii.** 12.0010101...     **iv.** 344.123123123...

**60.** *The Irrationality of* $\sqrt{2}$ In this project we will show that $\sqrt{2}$ is irrational. We will use the method called *reductio ad absurdum* or reduction to an absurdity. That is, we will assume the opposite of what we think is true and show that this leads to a contradiction.

**a.** Suppose that $\sqrt{2}$ is rational—that is, that $\sqrt{2} = p/q$ for a pair of integers $p$ and $q$ with no common factors. (If $p$ and $q$ have a common factor, then $p/q$ is not in lowest terms.) Show that $p^2 = 2q^2$.

**b.** Use part (a) to show that 2 must be a factor of $p^2$. If 2 is a factor of $p^2$, does it follow that 2 is a factor of $p$?

**c.** Suppose that 2 divides evenly into $p$. Then $p = 2 \cdot m$ for some integer $m$. Show that $4m^2 = 2q^2$.

**d.** Show that $q^2 = 2m^2$ and hence that 2 divides evenly into $q$.

**e.** The result to part (d) is an absurdity. Why?

Since the assumption that $\sqrt{2}$ is rational leads to an absurdity, it follows that $\sqrt{2}$ is irrational.

**61.** *Estimating Gasoline Usage* Estimate the number of gallons of gasoline that would be saved each year if every passenger car in the United States got 50 miles per gallon under all conditions (highway, city, etc.).

**62.** *Estimating Hair* Estimate the number of hairs on your head.

**63.** *Estimating Words* Estimate the number of words that the average American speaks in a day. About how many words are spoken every day in the United States?

### ◼ *Questions for Discussion or Essay*

**64.** It is a fact (see Exercise 59) that 0.999... = 1. Although mathematicians are in universal agreement, many other people refuse to believe it. What makes this fact so difficult to believe? Does the statement that 0.999... = 1 seem to defy common sense? Why?

**65.** People often complain about being treated as if they were "just a number." How many numbers can you think of that are used for identification? Are we dehumanized by being identified by numbers?

**66.** When solving estimation problems, it is important to bear in mind the desired degree of accuracy. For example, if someone asks you for the time, he is really asking you for an estimation of the time. If you were to answer either "the twentieth century" or "at the instant that you perceive that I have blinked my left eye, the time will be 3:25.62859 P.M. September the twentieth A.D. nineteen hundred and ninety-seven," you would no doubt be institutionalized. How does one determine the degree of precision required of an estimation?

**67.** Many of the estimation problems in this text have unrealistic premises (Mt. Everest being "ground up" and hauled away in semitrailers, for example). Do such problems have any importance? If algebra can be used to solve important real-world problems, why would the authors include unrealistic problems?

**SECTION 2**

# INTRODUCTION TO COORDINATE GEOMETRY

- What can coordinate geometry reveal about the national debt or the concentration of $CO_2$ in the atmosphere?

- In what way does a city's system of streets and avenues suggest a link between algebra and geometry?

- How can the ideas of a Greek mathematician from several hundred years B.C. and a French philosopher from the seventeenth century A.D. be combined to determine distances without a tape measure?

- When is a 45-yard punt not 45 yards?

## THE CARTESIAN PLANE

In Section 1 we saw that real numbers can be represented by points on the number line. This correspondence between real numbers and points on a line establishes a connection between algebra and geometry, and it was this connection that led the brilliant French mathematician and philosopher René Descartes to the discovery of coordinate geometry. He formulated a simple yet surprisingly useful way to associate a pair of numbers with each point in the **Cartesian plane**. To do this, he placed a horizontal number line, called the *x*-**axis**, and a vertical number line, called the *y*-**axis**, so that they intersected at the zero of each line. The point of intersection of the two axes is called the **origin**, and the axes divide the plane into four regions, called **quadrants**, as shown in Figure 5.

Each point in the Cartesian plane corresponds to an ordered pair of numbers $(x, y)$, where $x$ is the number on the *x*-axis directly above or below the point and $y$ is the number on the *y*-axis directly to the right or left of the point, as shown in Figure 6. In other words, the number $x$ indicates how many units to move to the left or right of the origin, while the number $y$ indicates how many units to move up or down. The numbers $x$ and $y$ are called the **coordinates** of the point.

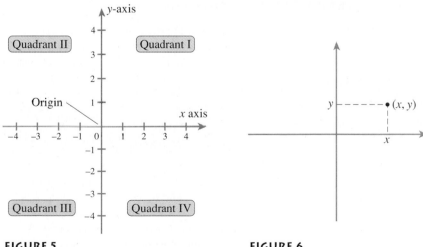

**FIGURE 5**                                    **FIGURE 6**

**EXAMPLE 1**  *Plotting points*

Plot the points $(2, 3)$, $(-1, -4)$, $(-3, 0)$, and $(\pi, -\sqrt{2})$.

**SOLUTION**   To plot the point $(2, 3)$, we move 2 units to the right of the origin and 3 units up. For the point $(-1, -4)$, we move 1 unit to the *left* of the origin and 4 units *down*. The point $(-3, 0)$ is on the $x$-axis, 3 units to the left of the origin. Finally, the point $(\pi, -\sqrt{2})$ is $\pi$ units to the right of the origin and $\sqrt{2}$ units *down*.

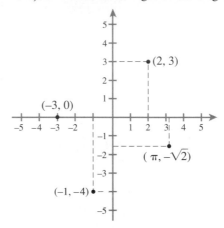

The principles of coordinate geometry are involved whenever a grid system is used to identify locations of objects. Two examples of this are the rectangular grids formed by streets and avenues in cities and the yard lines and hash marks on football fields. Nonrectangular grids are used on radar screens and for marking latitude and longitude on the surface of the Earth. Some of these examples will be considered in the exercises.

Another common application of coordinate geometry is in the area of *data analysis*. Plots of points can be used to organize and display data so that trends can be easily recognized.

**EXAMPLE 2**  *Plotting real-world data*

The U.S. Department of the Treasury records the following data on the national debt since 1950. Using the year as the $x$-coordinate and the debt as the $y$-coordinate, plot the data on a rectangular coordinate system.

| *Year* | 1950 | 1955 | 1960 | 1965 | 1970 | 1975 | 1980 | 1985 | 1990 |
|---|---|---|---|---|---|---|---|---|---|
| *Debt (in billions of dollars)* | 256 | 273 | 284 | 314 | 370 | 533 | 910 | 1,823 | 3,233 |

**SOLUTION**   Two possible choices for the scale on the horizontal axis are shown in Figures 7 and 8. The axis in Figure 7 is broken between the origin and the year 1950 to indicate that the years between those dates have been omitted. Figure 8 illustrates the loss of clarity if this is not done. As you will see in Exercise 27, the broken-axis technique must be used carefully because it can result in misleading conclusions.

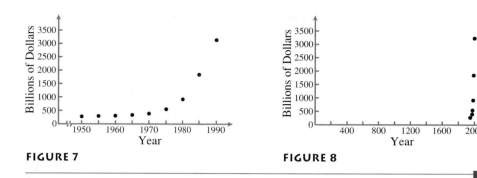

**FIGURE 7**                                             **FIGURE 8**

## PLOTTING POINTS WITH A GRAPHICS CALCULATOR

Most graphics calculators are capable of plotting individual points. However, if more than just one or two points are to be plotted, it is much more convenient to use a **scatter plot**. This feature is particularly useful when one is looking for trends in data.

A **scatter plot** is simply a graph of a collection of data points $(x, y)$. The creation of such a graph involves the following four steps.

1. Clear any existing graphs.
2. Clear any existing statistical data.
3. Enter the points $(x, y)$ as statistical data.
4. Select the scatter plot option from the list of available statistical plots.

For a detailed description of how these steps are done on your calculator, refer to the graphics calculator supplement.

**EXAMPLE 3**   *Using a scatter plot to identify trends in data*

Plot the following points with the scatter plot option on a graphics calculator. Describe any trend you see in the data.

| $x$ | 1.4 | 2.0 | 2.9 | 3.5 | 4.4 |
|-----|-----|-----|-----|-----|-----|
| $y$ | 1.0 | 1.5 | 1.7 | 2.0 | 2.4 |

**SOLUTION**   Before creating a new scatter plot, it is usually necessary to clear any existing graphs and any existing statistical data. Next, we enter the new data points as statistical data. Finally, we select the scatter plot option from the list of available statistical plots. The display should look similar to Figure 9. The plot suggests that the points lie on a straight line and that the line is rising from left to right.

**FIGURE 9**

Often it is necessary to adjust the *viewing rectangle* of the graphics calculator so that an appropriate portion of the graph can be seen on the display. This is done by modifying the *range variables*.

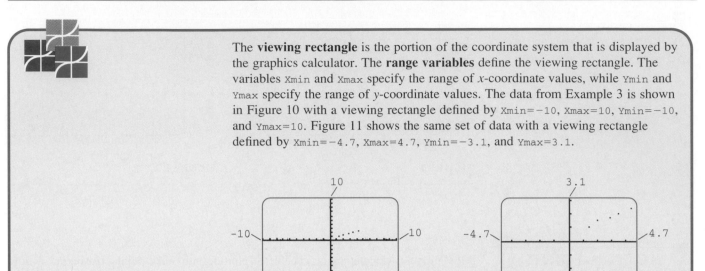

The **viewing rectangle** is the portion of the coordinate system that is displayed by the graphics calculator. The **range variables** define the viewing rectangle. The variables `Xmin` and `Xmax` specify the range of $x$-coordinate values, while `Ymin` and `Ymax` specify the range of $y$-coordinate values. The data from Example 3 is shown in Figure 10 with a viewing rectangle defined by `Xmin=-10`, `Xmax=10`, `Ymin=-10`, and `Ymax=10`. Figure 11 shows the same set of data with a viewing rectangle defined by `Xmin=-4.7`, `Xmax=4.7`, `Ymin=-3.1`, and `Ymax=3.1`.

**FIGURE 10**          **FIGURE 11**

**EXAMPLE 4**     *Using a scatter plot to identify trends in real-world data*

The atmospheric concentration of $CO_2$ over a recent time period is given in the table below. (Data Source: *The Challenge of Global Warming*, Dean Abrahamson, Ed., Island Press, 1989.) Describe any trends in the data.

| *Month* | 0 | 3 | 6 | 9 | 12 | 15 | 18 | 21 | 24 |
|---|---|---|---|---|---|---|---|---|---|
| *$CO_2$ concentration (parts per million)* | 341 | 343 | 341 | 338 | 342 | 345 | 343 | 340 | 343 |

**SOLUTION**   We enter the data points as statistical data with $x$ representing the month and $y$ representing the corresponding $CO_2$ concentration. The scatter plot is shown with two different viewing rectangles in Figures 12 and 13. The view in Figure 13 suggests that there are regular fluctuations in the $CO_2$ level and also that there is a slight upward shift in the second year (see Exercise 29).

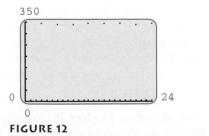

**FIGURE 12**

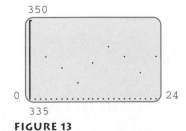

**FIGURE 13**

## DISTANCES BETWEEN POINTS

The distance between two points on the number line can be found by subtracting the smaller number from the larger. In a similar way, we can find the distance between points that lie on the same horizontal or vertical line. For example, to find the distance between the points $(2, -1)$ and $(2, 5)$ shown in Figure 14, we can compute the distance between the $y$-coordinates to obtain $5 - (-1) = 6$. Likewise, the distance between $(-6, 1)$ and $(-3, 1)$ is the distance between the $x$-coordinates, namely $-3 - (-6) = 3$.

Suppose now that we are interested in the distance between two points that are *not* on the same horizontal or vertical line, say $(2, 5)$ and $(6, 8)$. From Figure 15 we see that the horizontal distance between the points is $6 - 2 = 4$ and the vertical distance is $8 - 5 = 3$. Thus, the line segment joining the points is the hypotenuse of a right triangle with side lengths of 3 and 4. By the Pythagorean Theorem, the distance between the points is given by $d = \sqrt{3^2 + 4^2} = \sqrt{25} = 5$.

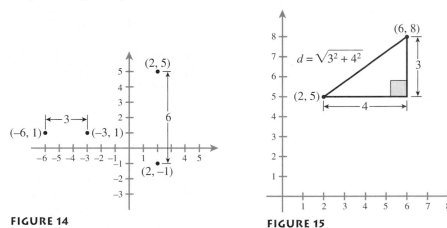

**FIGURE 14**                                   **FIGURE 15**

By applying this process more generally to the points $(x_1, y_1)$ and $(x_2, y_2)$, we obtain the following formula for the distance between two points.

### *The distance formula*

The distance $d$ between the two points $(x_1, y_1)$ and $(x_2, y_2)$ is given by

$$d = \sqrt{(x_2 - x_1)^2 + (y_2 - y_1)^2}$$

**EXAMPLE 5**

### *Finding the distance between points*

Find the distance between the points $(3, 5)$ and $(-1, -2)$.

**SOLUTION**   According to the distance formula, we compute the square of the differences between the $x$- and $y$-coordinates, add them, and take the square root of the sum.

$$d = \sqrt{(-1 - 3)^2 + (-2 - 5)^2}$$
$$= \sqrt{(-4)^2 + (-7)^2}$$
$$= \sqrt{16 + 49}$$
$$= \sqrt{65} \approx 8.06$$

| EXAMPLE 6 | *Identifying a polygon* |

Show that the points $(-4, 2)$, $(0, 0)$, $(-1, 5)$ and $(3, 3)$ form the vertices of a parallelogram. Is the parallelogram in fact a rhombus? That is, are all four sides of equal length?

**SOLUTION**    The four points are plotted in Figure 16. Since a quadrilateral is a parallelogram if and only if opposite sides have equal length, we must show that $a = c$ and $b = d$.

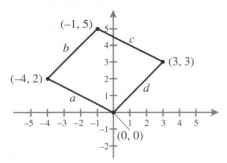

**FIGURE 16**

$$a = \sqrt{(-4 - 0)^2 + (2 - 0)^2} = \sqrt{20}$$
$$c = \sqrt{(-1 - 3)^2 + (5 - 3)^2} = \sqrt{20}$$
$$b = \sqrt{[-1 - (-4)]^2 + (5 - 2)^2} = \sqrt{18}$$
$$d = \sqrt{(3 - 0)^2 + (3 - 0)^2} = \sqrt{18}$$

Since the opposite sides have the same length, the points are the vertices of a parallelogram. However, they are not the vertices of a rhombus since the four sides do not have the same length.

---

## THE MIDPOINT FORMULA

Suppose that we are interested in finding the coordinates $(\bar{x}, \bar{y})$ of the midpoint $M$ of the line segment joining two arbitrary points $P(x_1, y_1)$ and $Q(x_2, y_2)$ as in Figure 17. First, we note that $\triangle PMR$ and $\triangle MQS$ are congruent. Thus, $PR = MS$ and $MR = QS$. Using the coordinates of the points involved, we can compute $\bar{x}$ and $\bar{y}$.

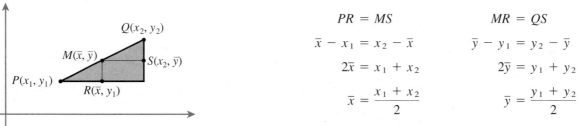

**FIGURE 17**

$$PR = MS \qquad\qquad MR = QS$$
$$\bar{x} - x_1 = x_2 - \bar{x} \qquad\qquad \bar{y} - y_1 = y_2 - \bar{y}$$
$$2\bar{x} = x_1 + x_2 \qquad\qquad 2\bar{y} = y_1 + y_2$$
$$\bar{x} = \frac{x_1 + x_2}{2} \qquad\qquad \bar{y} = \frac{y_1 + y_2}{2}$$

We have thus derived the midpoint formula.

---

### *The midpoint formula*

The midpoint of the line segment joining the points $(x_1, y_1)$ and $(x_2, y_2)$ has coordinates $(\bar{x}, \bar{y})$, where

$$\bar{x} = \frac{x_1 + x_2}{2} \quad \text{and} \quad \bar{y} = \frac{y_1 + y_2}{2}$$

Note that $\bar{x}$ is the average of the $x$-coordinates and $\bar{y}$ is the average of the $y$-coordinates.

**EXAMPLE 7**   *Using the midpoint formula*

Find the coordinates of the midpoint of the line segment connecting (3, 5) and (−1, −2).

**SOLUTION**   According to the midpoint formula, we have

$$\bar{x} = \frac{3 + (-1)}{2} = 1 \quad \text{and} \quad \bar{y} = \frac{5 + (-2)}{2} = \frac{3}{2}$$

Thus, the midpoint has coordinates $(1, \frac{3}{2})$.

**EXAMPLE 8**   *An application involving the midpoint formula*

A subway is to be built that connects points *A* and *B* on the city map in Figure 18. Find the location of a subway stop that is exactly half-way between *A* and *B*.

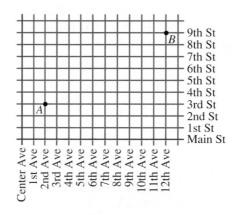

**FIGURE 18**

**SOLUTION**   By placing a coordinate system with the origin at the intersection of Center Avenue and Main Street, we can express point *A* as (2, 3) and point *B* as (12, 9). The midpoint of $\overline{AB}$ has coordinates

$$\bar{x} = \frac{2 + 12}{2} = \frac{14}{2} = 7 \quad \text{and} \quad \bar{y} = \frac{3 + 9}{2} = \frac{12}{2} = 6$$

So the subway stop should be located at the intersection of 7th Avenue and 6th Street.

---

**EXERCISES 2**

**EXERCISES 1–2** □ *Plot the given points.*

1. $(-2, 1)$, $(\sqrt{3}, -5)$, $\left(1, \frac{3}{2}\right)$, $(0, -4)$

2. $(1, -1)$, $\left(\frac{2}{3}, 0\right)$, $(-1, 3)$, $(-2, -\sqrt{5})$

**EXERCISES 3–6** □ *Plot the pair of points and find the distance between them.*

3. $(3, 5)$ and $(3, -2)$       4. $(-6, -1)$ and $(-2, -1)$

5. $(-3, -2)$ and $(5, 4)$      6. $(-8, 6)$ and $(4, 1)$

**EXERCISES 7–10** ☐ *Plot the pair of points, find the distance between them, and find the coordinates of the midpoint of the line segment connecting them.*

**7.** (1, 2) and (5, 4)

**8.** (−2, 3) and (4, −1)

**9.** $\left(-\dfrac{4}{3}, 2\right)$ and $\left(\dfrac{8}{3}, 1\right)$

**10.** $\left(-2, -\dfrac{7}{2}\right)$ and $\left(5, \dfrac{1}{2}\right)$

**EXERCISES 11–14** ☐ *Show that the given points form the vertices of the indicated polygon.*

**11.** Parallelogram: (2, 3), (4, 5), (6, 4), (8, 6)   (*Hint:* Show opposite sides are of equal length.)

**12.** Rhombus: (−1, −3), (1, 5), (8, −1), (−8, 3)   (*Hint:* Show all four sides are of equal length.)

**13.** Right triangle: (1, 0), (2, 3), (4, −1)   (*Hint:* Use the Pythagorean Theorem.)

**14.** Square: (1, 10), (−5, 2), (3, −4), (9, 4)   (*Hint:* The diagonals must be the same length, as must the adjacent sides.)

**EXERCISES 15–18** ☐ *Use a scatter plot to graph the given data points and identify any trends.*

**15.**

| $x$ | 0.7 | 1.2 | 1.6 | 2.0 | 2.6 |
|-----|-----|-----|-----|-----|-----|
| $y$ | 2.7 | 2.5 | 2.4 | 2.2 | 1.9 |

**16.**

| $x$ | 1.4 | 1.8 | 2.2 | 2.6 | 3.0 | 3.4 |
|-----|-----|-----|-----|-----|-----|-----|
| $y$ | 1.0 | 1.7 | 2.1 | 2.2 | 2.0 | 1.4 |

**17.**

| $x$ | 0.7 | 1.6 | 2.4 | 3.0 | 3.9 | 4.9 | 5.3 | 6.0 | 6.6 |
|-----|-----|-----|-----|-----|-----|-----|-----|-----|-----|
| $y$ | 2.5 | 2.0 | 1.5 | 1.9 | 2.5 | 1.8 | 1.5 | 1.7 | 2.3 |

**18.**

| $x$ | 1.4 | 2.2 | 2.6 | 3.3 | 3.5 | 4.1 | 4.7 | 5.2 | 5.6 |
|-----|-----|-----|-----|-----|-----|-----|-----|-----|-----|
| $y$ | 1.0 | 3.0 | 4.1 | 3.3 | 3.1 | 4.3 | 6.2 | 5.8 | 5.2 |

■ *Applications*

**19.** *Crawling Ant* Suppose that an ant is initially at the point (−2, 4) and that it makes an "antline" for the point (−7, −4). From there it crawls to the point (3, −2) and then returns to (−2, 4). Is the ant traveling clockwise or counterclockwise? What is the total distance traveled by the ant?

**20.** *Punctuation Mark* Plot the points (−2, 4), (−1, 5), (0, 5), (1, 4), (1, 3), (0, 2), (0, 1), (0, 0), and (0, −2). What punctuation mark does this set of points suggest?

**21.** *Minimum Wage* The U.S. Department of Labor records the following data on the minimum hourly wage from 1975 to 1981.

| Year | 1975 | 1976 | 1977 | 1978 | 1979 | 1980 | 1981 |
|------|------|------|------|------|------|------|------|
| Wage | 2.10 | 2.30 | 2.30 | 2.65 | 2.90 | 3.10 | 3.35 |

Use the year as the $x$-coordinate and the wage as the $y$-coordinate to plot the data on a rectangular coordinate system. Describe the trend you see. On the basis of the data, what would you predict for the minimum wage in 1990? The actual value was $3.80. How far off is your estimate?

**22.** *Oil Imports* The U.S. Department of Energy records the following data on oil imports from 1983 to 1991.

| Year | 1983 | 1984 | 1985 | 1986 | 1987 | 1988 | 1989 | 1990 | 1991 |
|------|------|------|------|------|------|------|------|------|------|
| Oil (thousands of barrels per day) | 4,312 | 4,715 | 4,286 | 5,439 | 5,914 | 6,587 | 7,202 | 7,161 | 6,575 |

Use the year as the $x$-coordinate and the amount of oil as the $y$-coordinate to plot the data on a rectangular coordinate system.

**23.** *City Street Map* A coordinate system is placed on a city street map with the origin located at the intersection of Main Street and Center Avenue, as shown in Figure 19. All streets and avenues, except Lloyd Avenue, run north/south or east/west. What is the shortest route from point $A(-6, -2)$ to point $B(5, 4)$? Compute the total distance for this shortest route (using a standard city block as 1 unit), and compare it to the distance one would have to travel if Lloyd Avenue were closed for repairs.

**24.** *City Street Map* Refer to the city street map in Figure 19. A bicycle messenger, currently at point $A(-6, -2)$, receives a message to pick up a package from an office building at the intersection of Main and Center and deliver it to a building at point $B(5, 4)$. Find the shortest route for her to take, and compute the total distance (using a standard city block as 1 unit).

**25.** *Suspended Ceiling* The suspended ceiling shown in Figure 20 consists of $2' \times 2'$ square tiles. An electrical junction box is to be placed half-way between points $A$ and $B$. Use the midpoint formula to locate the half-way point.

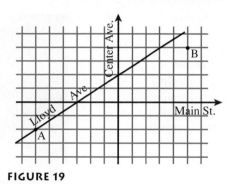

**FIGURE 19**

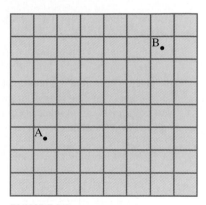

**FIGURE 20**

---

### Projects for Enrichment

**26.** *Polar Coordinate Systems and Radar Screens* The rectangular coordinate system discovered by Descartes is not the only way of identifying locations of points. About 40 years after Descartes' discovery, Sir Isaac Newton developed a system of *polar coordinates* that references points by their distance and "compass direction" from the origin. (See Section 9.5 for a more detailed treatment of polar coordinates.) To define polar coordinates, we fix the origin at a point $O$ and construct a horizontal ray with $O$ as the initial point as shown in Figure 21. Each point $P$ is then associated with a pair $(r, \theta)$, where $r$ is the *directed distance* from $O$ to $P$ and $\theta$ is the *directed angle measure* from the ray to the segment $\overline{OP}$.

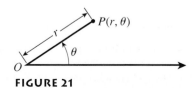

**FIGURE 21**

In Figure 22 we have plotted the four points $A(2, 45°)$, $B(3, 120°)$, $C(-1, 30°)$, and $D(4, -30°)$. Notice that positive angle measures are measured counterclockwise from the hori-

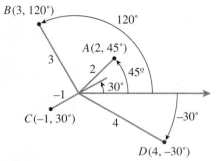

**FIGURE 22**

zontal ray while negative angles are measured clockwise. Notice also that because $r$ refers to a directed distance, the point $C$ with coordinates $(-1, 30°)$ is 1 unit from the origin in the exact opposite direction as that of the point $(1, 30°)$.

Points with polar coordinates are much easier to plot on *polar graph paper*, which consists of concentric circles centered at the origin and rays coming out of the origin that correspond to common angle measures as shown in Figure 23.

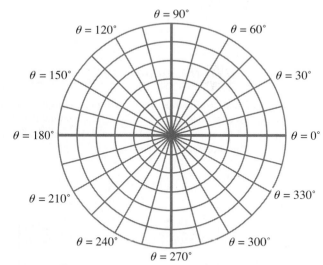

**FIGURE 23**

**a.** Photocopy the polar graph paper in Figure 23 and plot the points $(3, 60°)$, $(2, 135°)$, $(4, -90°)$, and $(-5, 30°)$.

One significant difference between rectangular and polar coordinates is that the polar coordinates of a point are not unique. For example, the point with polar coordinates $(2, -30°)$ can

also be identified using the coordinates (2, 330°) or even
(−2, 150°).

**b.** Find a second representation for each of the points in part
(a). Use a degree measure that is between −360° and 360°.

By placing a rectangular coordinate system on top of a polar
coordinate system, as shown in Figure 24, it is possible to convert coordinates from one system to the other. For example, the
point with rectangular coordinates (2, 2) has approximate polar
coordinates of (2.8, 45°). The point with polar coordinates
(3, 30°) has approximate rectangular coordinates of (2.6, 1.5).

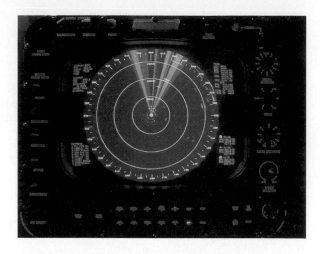

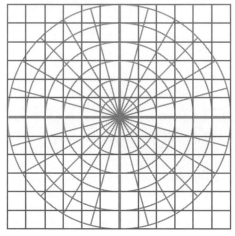

**FIGURE 24**

**c.** Photocopy Figure 24 and use it to convert the points given
in part (a) to rectangular coordinates.

**d.** Convert the points with rectangular coordinates (2, 0),
(3, 5), (−4, 4), and (−3, −2) to polar coordinates.

**e.** A radar screen reports the *bearing* (compass direction) and
*range* (distance) of an object as it relates to the source of
the radar signal. Explain why a polar coordinate system
would be the system of choice for recording the readings
from a radar screen.

## ▪▪ *Questions for Discussion or Essay*

**27.** Figures 25 and 26 illustrate the growth in newspaper circulation
for a certain city over a 3-year period. Discuss how the choice
of scale and the use of a broken vertical axis could lead to differences in interpretation.

**28.** A punter kicks a football from a point on the 2-yard line
approximately 18 yards from the left edge of the field (point *A*
in Figure 27). The ball is picked up on the 47-yard line approximately 5 yards from the right edge of the field (point *B* in Figure 27). From yard line to yard line, the punt would appear to

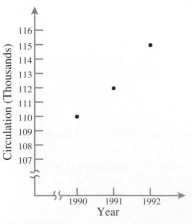

**FIGURE 25**

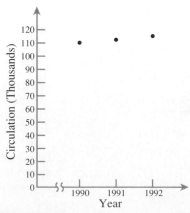

**FIGURE 26**

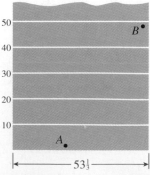

**FIGURE 27**

be $47 - 2 = 45$ yards. Explain why this is not technically correct. Given that the width of a football field is $53\frac{1}{3}$ yards, explain how the principles of coordinate geometry can be used to calculate the actual distance of the punt.

**29.** In Example 4 we observed a regular fluctuation in the atmospheric concentration of $CO_2$. Study the pattern more closely and determine the length of time between the "peaks" and "valleys." What explanation can you give for the fluctuations?

---

## SECTION 3

# PROPERTIES AND OPERATIONS

■ What property do addition, multiplication, and marriage have, that subtraction, division, and love do not?

■ Why can't you divide by zero?

■ Is 7 really equal to 2 and is 0 the same as $-4$?

■ In what sense is $-17{,}000$ larger than $0.0000001$?

### ADDITION AND MULTIPLICATION OF REAL NUMBERS

The operations of addition and multiplication have properties that can be used to simplify many mathematical problems. For example, addition is *commutative*; that is, the order in which numbers are added is immaterial—we get the same answer either way. We can state the commutative property of addition algebraically as $a + b = b + a$.

Table 2 lists the most important properties involving the operations of multiplication

**TABLE 2**

*Properties of the real-number system*

| Name of property | Verbal description | Algebraic description |
|---|---|---|
| Commutative property of addition | The order doesn't matter when adding. | $a + b = b + a$ |
| Commutative property of multiplication | The order doesn't matter when multiplying. | _____ |
| Associative property of addition | The grouping doesn't matter when adding. | _____ |
| Associative property of multiplication | _____ | $a(bc) = (ab)c$ |
| Distributive property | _____ | $a(b + c) = ab + ac$ |
| Existence of the additive identity | There is a number (called zero) whose sum with any number is that number. | $a + 0 = a$ |
| Existence of the multiplicative identity | There is a number (called 1) whose product with any number is that number. | _____ |
| Existence of the additive inverse | Every real number has an additive inverse; that is, for each number there is a second number that when added to the first gives zero. | $a + (-a) = 0$ |
| Existence of the multiplicative inverse | _____ | $a\left(\dfrac{1}{a}\right) = 1$ |

and addition of real numbers, along with a verbal and algebraic description of each property. Some of the descriptions have been left for the exercises.

As you can see, many of these properties are very awkward to state in words. It is much more efficient to use the language of *algebra*. Whenever we use letters to represent numbers, we are using algebra.

**EXAMPLE 1**    *Translating English into algebra*

Translate the following statement algebraically: Whenever a number is multiplied by the additive identity, the result is the additive identity.

**SOLUTION**    Using $x$ to represent "a number" and writing zero for the additive identity, we obtain

$$0 \cdot x = 0$$

**EXAMPLE 2**    *Computing a tip with the distributive property*

Use the distributive property to compute a 15% tip on a restaurant bill of $42.00.

**SOLUTION**    We wish to compute $15\% \cdot 42.00$.

$$
\begin{aligned}
0.15 \cdot 42.00 &= (0.1 + 0.05) \cdot 42.00 \\
&= (0.1 \cdot 42.00) + (0.05 \cdot 42.00) \\
&= 4.20 + 2.10 \\
&= 6.30
\end{aligned}
$$

Thus, the tip should be $6.30. In words, since 15% is 10% + 5%, we compute 10% of $42.00, obtaining $4.20, and add half of that, namely $2.10, to obtain $6.30.

**EXAMPLE 3**    *The property of commutativity in other contexts*

For each operation given below, indicate whether or not it is commutative.
a. Taking the minimum of two real numbers
b. Forming a hyphenated surname from a couple's surnames prior to their marriage

**SOLUTION**

a. Taking the minimum of two numbers is commutative: the minimum of $x$ and $y$ is the same as the minimum of $y$ and $x$.

b. The formation of a hyphenated surname is not commutative; Olivia John-Newton is not the same as Olivia Newton-John.

**ADDITION AND MULTIPLICATION OF COMPLEX NUMBERS**

In Section 1 we defined a complex number as one of the form $a + bi$ where $a$ and $b$ are real numbers and $i = \sqrt{-1}$ (i.e., $i^2 = -1$). Two complex numbers $a + bi$ and $c + di$ can be added or multiplied according to the following definitions.

**Addition and
multiplication
of complex numbers**

$$(a + bi) + (c + di) = (a + c) + (b + d)i$$

$$(a + bi)(c + di) = (ac - bd) + (ad + bc)i$$

Using these definitions, it can be shown (see Exercise 58) that the commutative, associative, and distributive properties hold for the addition and multiplication of complex numbers. Thus, it is *not* necessary to memorize the definitions. Instead, we can simply apply the appropriate properties to compute the required sum or product.

**EXAMPLE 4**    *Using the properties of complex numbers to add nonreal
numbers*

Perform the operation $(2 + i) + (13 + 4i)$.

**SOLUTION**    By combining the real and imaginary parts, one can probably tell at a glance that the answer is $15 + 5i$. But let us analyze how the properties of complex numbers are used to obtain this answer.

$$(2 + i) + (13 + 4i) = (2 + 13) + (i + 4i) \qquad \text{Commutative and associative}$$
$$\text{properties of addition}$$

$$= 15 + (1 + 4)i \qquad \text{Distributive property}$$

$$= 15 + 5i$$

**EXAMPLE 5**    *Using the properties of complex numbers to multiply
nonreal numbers*

Perform the operation $(5 - 2i)(8 + 3i)$.

**SOLUTION**    We apply the distributive property and the fact that $i^2 = -1$.

$$(5 - 2i)(8 + 3i) = 5(8 + 3i) + (-2i)(8 + 3i) \qquad \text{Distributive property}$$

$$= (40 + 15i) + (-16i - 6i^2) \qquad \text{Distributive property}$$

$$= (40 + 15i) + (-16i - 6(-1)) \qquad \text{Definition of } i \ (i^2 = -1)$$

$$= (40 + 15i) + (6 - 16i)$$

$$= 46 - i \qquad \text{Addition of complex numbers}$$

The other properties listed in Table 2 also hold for complex numbers. The number $0 + 0i = 0$ is the additive identity, the number $1 + 0i = 1$ is the multiplicative identity, and the additive inverse of a complex number $a + bi$ is the complex number $-a - bi$. The multiplicative inverse of a complex number $a + bi$ should naturally be $\dfrac{1}{(a + bi)}$, but it is preferable to write it in the form $c + di$. To do this, we multiply the numerator and denominator of $\dfrac{1}{(a + bi)}$ by the **complex conjugate** of $a + bi$, namely $a - bi$, as shown in the following example.

**EXAMPLE 6**    *Finding the multiplicative inverse of a complex number*

Write the multiplicative inverse of $4 - 2i$ in the standard form $a + bi$.

**SOLUTION**    We multiply the numerator and denominator of $\dfrac{1}{4 - 2i}$ by the complex conjugate $4 + 2i$.

$$\frac{1}{4 - 2i} = \frac{1}{4 - 2i} \cdot \frac{4 + 2i}{4 + 2i} \qquad \text{Multiply by } 1 = \frac{4 + 2i}{4 + 2i}$$

$$= \frac{4 + 2i}{16 - 8i + 8i - 4i^2} \qquad \text{Multiplying numerators and denominators}$$

$$= \frac{4 + 2i}{16 - 4(-1)} \qquad i^2 = -1$$

$$= \frac{4 + 2i}{20}$$

$$= \frac{4}{20} + \frac{2}{20}i$$

$$= \frac{1}{5} + \frac{1}{10}i$$

**CHECK**    We can check our answer by multiplying $4 - 2i$ by $\frac{1}{5} + \frac{1}{10}i$ to see if we get 1 as the answer.

$$(4 - 2i)\left(\frac{1}{5} + \frac{1}{10}i\right) = \frac{4}{5} - \frac{2}{5}i + \frac{4}{10}i - \frac{2}{10}i^2$$

$$= \frac{4}{5} - \frac{1}{5}(-1) = 1$$

---

**SUBTRACTION AND DIVISION**

Subtraction and division can be defined in terms of addition and multiplication. Subtraction is just addition of the additive inverse, and division is nothing more than multiplication by the multiplicative inverse. More formally, we have

$$x - y = x + (-y)$$

$$x \div y = x \cdot \frac{1}{y} \qquad \text{for } y \neq 0$$

Note that division by zero is not defined. But why is this? Let us consider the expression $\frac{1}{0}$. According to the definition, $\frac{1}{0}$ would be the multiplicative inverse of zero. In other words, if we let $c = \frac{1}{0}$, then $0 \cdot c = 1$. But we know that $0 \cdot c = 0$. This is a contradiction, so there is no such number as $\frac{1}{0}$.

The fact that division by zero is not defined is much more than just a novelty; it is at the heart of many mathematical phenomena. If we inadvertently allow a division by zero, an absurdity can result.

**EXAMPLE 7**   *A false demonstration that* $7 = 2$

What is wrong with the following proof that $7 = 2$?

Suppose that $x = 1$. Then $5x - 8 = -3$. We perform the following operations:

$$5x - 8 = -3$$

$$5x - 8 + 2x + 1 = -3 + 2x + 1 \qquad \text{Adding } 2x + 1 \text{ to both sides}$$

$$7x - 7 = 2x - 2 \qquad \text{Simplifying}$$

$$7(x - 1) = 2(x - 1) \qquad \text{Distributive property}$$

$$7 = 2 \qquad \text{Dividing by } (x - 1)$$

**SOLUTION**   The error is in the last step where we divided by $(x - 1)$. We began the proof by supposing that $x = 1$, so that $x - 1 = 0$. Thus, the last step of the proof was a division by zero, which is not allowed.

When dividing complex numbers, we use the complex conjugate of the denominator to write the quotient in the standard form $a + bi$.

**EXAMPLE 8**   *Dividing complex numbers*

Write the quotient $\dfrac{3 - 2i}{1 + 4i}$ in the standard form $a + bi$.

**SOLUTION**   Since the conjugate of $1 + 4i$ is $1 - 4i$, we proceed as follows.

$$\frac{3 - 2i}{1 + 4i} = \frac{3 - 2i}{1 + 4i} \cdot \frac{1 - 4i}{1 - 4i} \qquad \text{Multiplying by } 1 = \frac{1 - 4i}{1 - 4i}$$

$$= \frac{3 - 2i - 12i + 8i^2}{1 + 4i - 4i - 16i^2} \qquad \text{Multiplying numerators and denominators}$$

$$= \frac{3 - 14i + 8(-1)}{1 - 16(-1)} \qquad \text{Using } i^2 = -1$$

$$= \frac{-5 - 14i}{17}$$

$$= -\frac{5}{17} - \frac{14}{17}i$$

## ORDER OF OPERATIONS

The associative properties of addition and multiplication tell us that, when either adding or multiplying a string of numbers together, we may group them in any way that we wish. For example, we could evaluate $1 + 2 + 3$ in either of the following ways.

$$(1 + 2) + 3 = 3 + 3 = 6, \quad \text{or}$$

$$1 + (2 + 3) = 1 + 5 = 6$$

But suppose that we need to evaluate $1 - 2 - 3$. We could obtain either

$$(1 - 2) - 3 = -1 - 3 = -4, \quad \text{or}$$

$$1 - (2 - 3) = 1 - (-1) = 2$$

The difficulty is that there is no associative property for subtraction, and so grouping is important when subtracting. Similarly, there is no associative property for division. In fact, the grouping matters if we have both addition and multiplication in the same expression. For example, if we wish to evaluate $2 \times 3 + 5$, we might obtain either

$$(2 \times 3) + 5 = 6 + 5 = 11, \quad \text{or}$$

$$2 \times (3 + 5) = 2 \times 8 = 16$$

The question is, Which result is correct? The answer is simple. We agree that unless grouping symbols (parentheses, brackets, etc.) indicate otherwise, we will evaluate expressions involving the four fundamental operations of addition subtraction, multiplication, and division according to the following rule.

*Order of arithmetic operations*

> If no grouping symbols are used, first perform all multiplications and divisions from left to right, and then perform any additions or subtractions from left to right. If grouping symbols (such as parentheses) are present, the operations within the grouping symbols are performed first, again following the rules for order of operations.

**EXAMPLE 9** *Order of operations*

Evaluate the following expressions.
**a.** $10 - 6 \div 2 \times 5$ **b.** $(10 - 6) \div (2 \times 5)$

**SOLUTION**

**a.** We first perform divisions and multiplications in left-to-right order.

$$10 - 6 \div 2 \times 5 = 10 - 3 \times 5$$
$$= 10 - 15$$
$$= -5$$

**b.** We perform the operations in parentheses first.

$$(10 - 6) \div (2 \times 5) = 4 \div 10$$
$$= 0.4$$

**EXAMPLE 10** *A false proof that $0 = -4$*

What is wrong with the following proof that $0 = -4$?
Let $x = 2 \times 1 + 2 \times (-1)$.

Then we have

$$x = \underbrace{2 \times 1} + 2 \times (-1)$$
$$= \underbrace{2 + 2} \times - 1$$
$$= 4 \times - 1$$
$$= -4$$

On the other hand, we have

$$x = \underbrace{2 \times 1} + \underbrace{2 \times (-1)}$$
$$= 2 + -2$$
$$= 0$$

Thus, $-4 = 0$.

**SOLUTION**   The second step of the first evaluation of $x$ is in error since multiplications are to be performed before additions.

## ABSOLUTE VALUE

It is hard for some students to accept the fact that $-17{,}000$ is less than $0.0000001$. Although $-17{,}000$ is further to the left on the number line than is $0.0000001$, it somehow *seems* like a larger quantity. This is because it is so much further from zero than is $0.0000001$. A number's distance from zero is known as the *absolute value* of the number. For example, since $-17{,}000$ is $17{,}000$ units distant from zero, the absolute value of $-17{,}000$ is $17{,}000$. We write $|-17{,}000| = 17{,}000$. Since the absolute value of a number represents a distance, it is never negative. In general, we make the following definition.

***Definition of***
***absolute value***

> The **absolute value** of a real number $x$, denoted $|x|$, is its distance from zero. In symbols,
>
> $$|x| = \begin{cases} x & \text{if } x \ge 0 \\ -x & \text{if } x < 0 \end{cases}$$

**RULE OF THUMB**   When finding the absolute value of a negative number, take its opposite to make it positive. When finding the absolute value of a positive number, leave it alone.

**EXAMPLE 11**

*Finding absolute value*

Evaluate each of the following.
a. $|14.5|$          b. $|-\pi|$          c. $|0|$
d. $-|-4|$          e. $|x|$, where $x$ is 27 units from the origin

**SOLUTION**

a. Since $14.5 \ge 0$, $|14.5| = 14.5$.

b. Since $-\pi < 0$, $|-\pi| = -(-\pi) = \pi$.

c. Since $0 \geq 0$, $|0| = 0$.

d. $-|-4|$ means $-(|-4|)$ or "the opposite of $|-4|$." Since $-4 < 0$, $|-4| = 4$. Thus, $-|-4| = -4$. Note that this does not contradict the rule of thumb since it is the negative sign *outside* of the absolute value that makes the result negative.

e. Since the absolute value of a number is its distance from the origin, $|x| = 27$. Of course, this means that either $x = 27$ or $x = -27$.

---

**EXAMPLE 12**    *Rewriting an absolute-value expression*

If $x < 2$, write the expression $|x - 2|$ without using absolute value.

**SOLUTION**    Since $x < 2$, $x - 2 < 0$. Thus, $|x - 2| = -(x - 2) = -x + 2$.

---

We have seen that the distance between two real numbers can be found by subtracting the smaller from the larger. Using absolute value, the distance between two numbers can be defined without regard to which of the two is larger. For example, the distance between 2 and 6 can be found by computing $|2 - 6| = 4$. More generally, we have the following definition.

**Distance**

> The **distance** between two real numbers $a$ and $b$ is given by $|a - b|$.

**EXAMPLE 13**    *Finding distance*

Find the distance between $\frac{4}{5}$ and $\frac{9}{11}$.

**SOLUTION**    The distance is $\left|\frac{4}{5} - \frac{9}{11}\right| = \left|\frac{44}{55} - \frac{45}{55}\right| = \left|-\frac{1}{55}\right| = \frac{1}{55}$.

---

# EXERCISES 3

**EXERCISES 1–4** ☐ *See Table* 2.

1. Write an algebraic description of the commutative property of multiplication.

2. Write an algebraic description of the associative property of addition.

3. Write a verbal description of the distributive property.

4. Write a verbal description of the existence of the multiplicative inverse.

5. An operation is said to be commutative if "the order doesn't matter." For example, addition is commutative because $a + b = b + a$. Subtraction, on the other hand, can be shown to be noncommutative by considering, for example, $a = 0$ and $b = 1$. Here $a - b = 0 - 1 = -1$, whereas $b - a = 1 - 0 = 1$. The pair of numbers $a = 0$, $b = 1$, are a *counterexample* to the statement that subtraction is commutative. State whether each of the operations given below is commutative; if not, give a counterexample.

a. Division

b. Multiplication

c. Marriage (If Bob marries Sue, does Sue marry Bob?)

d. Love (If Dirk loves Camille, does Camille love Dirk?)

e. Taking the maximum of two numbers

f. Addressing (Is "Jamal speaks to Ann" the same as "Ann speaks to Jamal"?)

g. Implication (Is "If $A$, then $B$" the same as "If $B$, then $A$"?)

h. Finding the greatest common divisor of two numbers

6. An operation is said to be associative if the grouping doesn't matter. For example, addition is associative because $a + (b + c) = (a + b) + c$. For each operation given below, indicate if it is associative. If it isn't, give a counterexample.

   a. Multiplication

   b. Division

   c. Subtraction

   d. Linking chains

   e. Connecting Lego blocks together

7. Place a one dollar bill in front of you right-side-up. Rotate the front of the bill 90° clockwise so that George Washington's nose points towards you. Now flip the bill over (top-to-bottom) so that you now see the reverse side of the dollar bill with the seal of the United States nearest you. Now, return the dollar bill to its original position, and this time, flip the bill first and then rotate it 90° clockwise. Do rotations and flips commute with one another?

8. The distributive property was given in Table 2 as $a(b + c) = ab + ac$. Show that it is also true that $(a + b)c = ac + bc$. (*Hint:* Use the commutative property of multiplication and the distributive property.)

9. Use the distributive property to compute a 15% tip on a bill of $64.00.

10. Use the distributive property to compute a 15% tip on a bill of $52.00.

11. Compute $999,999 \times 75$ in your head. (*Hint:* Think of 999,999 as $1,000,000 - 1$ and use the distributive property.)

12. Compute $1,000,003 \times 90$ in your head.

13. Use the properties of real numbers to show that $a \cdot 0 = 0$. (*Hint:* Use $0 = 1 + (-1)$ and the distributive property.)

14. Suppose that we were unaware of the rules regarding order of operations. List all possible results of the computation $3 \cdot 5 + 6 \div 2$. (*Hint:* There are four in all.)

**EXERCISES 15–24** □ *Perform the indicated operation.*

15. $(3 + 4i) + (1 - 2i)$

16. $(2 - i) - (5 - 9i)$

17. $7 - (3 + 2i)$

18. $(4 - 8i) + 2i$

19. $2 \cdot (3 + i)$

20. $(4 + i) \cdot 5$

21. $(1 + i) \cdot i$

22. $6i \cdot (2 - 3i)$

23. $(2 + 3i) \cdot (1 - i)$

24. $(3 - 4i) \cdot (-5 + i)$

**EXERCISES 25–28** □ *Write the multiplicative inverse of the given number in the form $a + bi$. Check your answer by showing that the product of the number with its multiplicative inverse is equal to 1.*

25. $2i$

26. $-3i$

27. $1 + i$

28. $2 - 6i$

**EXERCISES 29–34** □ *Write the quotient in the form $a + bi$.*

29. $\dfrac{5}{2 + i}$

30. $\dfrac{-13i}{2 - 3i}$

31. $\dfrac{-18 + 13i}{5 - 2i}$

32. $\dfrac{29 - 3i}{3 + 4i}$

33. $\dfrac{8 + 3i}{-4i}$

34. $\dfrac{3 - 2i}{5i}$

**EXERCISES 35–40** □ *Evaluate the given expression.*

35. $5 - 3 \times 4$

36. $6 + 3 \div 2$

37. $1.2 - 2.5 - 3.4 - 1.8$

38. $3 \div 1 + 2 - 4 + 2 \cdot 3$

39. $4 \times [3 - (4 - 5)]$

40. $12 \div 2 \times 3 \div 2 \times 3$

**EXERCISES 41–48** □ *Evaluate the given expression.*

41. $|-6|$

42. $-|4|$

43. $|0|$

44. $|10.61|$

45. $|3 - 1|$

46. $|-1 - 4|$

47. $\dfrac{|-3.7|}{-3.7}$

48. $\dfrac{|2 - 5|}{2 - 5}$

**EXERCISES 49–52** □ *Write the expression without using absolute value symbols.*

49. $|x - 1|$  if $x > 1$

50. $|t + 4|$  if $t < -4$

51. $|y + 2|$  if $y < -2$

52. $|x - 7|$  if $x > 7$

**EXERCISES 53–56** □ *Find the distance between the given real numbers.*

53. $-3$ and $7$

54. $-5$ and $-12$

55. $-\dfrac{1}{3}$ and $-\dfrac{2}{5}$

56. $\dfrac{3}{4}$ and $\dfrac{5}{6}$

■ *Projects for Enrichment*

57. *False Proof* Using Example 7 as a model, construct a false proof that $1 = 1,000,000$.

58. *Verifying Properties* Using the properties of addition and multiplication of real numbers, we can prove that similar prop-

erties hold for the addition and multiplication of complex numbers. For example, we may establish the commutative property of addition of complex numbers as follows:

$$(a + bi) + (c + di) = (a + c) + (b + d)i$$

Definition of complex addition

$$= (c + a) + (d + b)i$$

Commutative property of addition of reals

$$= (c + di) + (a + bi)$$

Definition of complex addition

Use the properties of real-number addition and multiplication to establish each of the following properties for complex-number addition and multiplication.

a. The commutative property of multiplication

b. The associative property of addition

c. The associative property of multiplication

d. The distributive property

59. *The Complex Plane* For every operation with real numbers, there is a corresponding operation with complex numbers. We have seen that complex numbers can be added, subtracted, multiplied, and divided just like real numbers. We will learn in later sections that complex numbers can be raised to powers. However, what in the world would the absolute value of a complex number be? In the case of real numbers, the absolute value of a number was defined as being the distance between that number and zero. But how far is $3 + 4i$ from 0? In fact, *where* on a number line is $3 + 4i$ anyway?

Let us answer the last question first: $3 + 4i$ isn't anywhere on a number line; only real numbers are on the number line. But $3 + 4i$ is in the *complex plane*. The complex plane looks a lot like the rectangular coordinate plane, with a few important differences. Instead of the $x$-axis, we have the real axis; instead of the $y$-axis, we have the imaginary axis; and in place of the origin, we have the number 0. The complex number $3 + 4i$, for

example, is located in the first quadrant 3 units over and 4 units up. In general, the complex number $x + yi$ corresponds to the point $(x, y)$ in the complex plane. The complex plane is shown in Figure 28 with several complex numbers plotted. (See Section 9.3 for a more detailed treatment of the complex plane.)

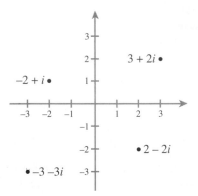

**FIGURE 28**

a. Indicate the quadrant in which each of the following complex numbers lies.

   i. $-2 + 4i$       ii. $3 - 7i$

   iii. $1 + i$        iv. $-3 - \dfrac{8}{3}i$

The absolute value of a complex number is defined as the distance of the number from zero, just as in the case of real numbers. For example, $|3i| = 3$ and $|-4i| = 4$.

b. Find the absolute value of the following complex numbers. (*Hint:* It is helpful to first plot the given number in the complex plane. The Pythagorean Theorem may be useful.)

   i. $1.5i$       ii. $3 + 4i$

   iii. $-4 - 3i$     iv. $1 + i$

c. Describe the set of points in the complex plane with absolute value equal to 1.

---

■ *Questions for Discussion or Essay*

---

60. How would you respond to someone who states, "Just because we don't know how to divide by zero now doesn't mean that some day someone might not be able to figure out how to do it."

61. The definition of absolute value as stated in the text is

$$|x| = \begin{cases} x & \text{if } x \geq 0 \\ -x & \text{if } x < 0 \end{cases}$$

Thus, under some circumstances, $|x| = -x$. On the other hand, absolute values are never negative. Explain this apparent contradiction.

62. Are the properties of addition and multiplication discovered or created? Support your claim.

63. Since most of us are able to successfully add and multiply numbers without being aware of the properties that we are employing, why should algebra students be required to learn these properties? Discuss the differences between doing mathematics and understanding mathematics.

64. To prove that subtraction is *not* commutative, it is sufficient to produce a single pair of numbers (such as 0 and 1) for which we obtain a different answer depending on the order in which

we perform the subtraction. Why then *can't* we establish the commutative property of addition by simply producing a single pair of numbers such that the order in which we add these numbers doesn't matter? For example, what is wrong with this "proof" of the commutativity of addition? "Consider the numbers 2 and 3. Since 2 + 3 = 5, and also 3 + 2 = 5, addition is commutative."

65. How is algebra like a language? What mathematical entities correspond to letters, words, and sentences?

---

## SECTION 4

# INTEGER EXPONENTS

- How thick would a piece of paper be if you folded it in half 52 times?
- If $2^3$ is 2 multiplied by itself 3 times, then what is $2^{-3}$?
- Which would you rather have, the money accumulated from Julius Caesar's deposit of $100 at 10% simple interest, or from Ben Franklin's deposit of $1 at 6% interest compounded annually?
- An individual doubles in weight between the ages of birth and 6 months. If he continued to double in weight every 6 months, how much would he weigh on his forty-first birthday?
- What is so scientific about scientific notation?

## NATURAL-NUMBER EXPONENTS

Suppose that you wish to compute the number of people in an auditorium consisting of 50 rows with 40 seats in each row. You *could* add the number of seats in each row 50 times:

$$40 + 40 + 40 + 40 + 40 + 40 + 40 + 40 + 40 + 40 +$$
$$40 + 40 + 40 + 40 + 40 + 40 + 40 + 40 + 40 + 40 +$$
$$40 + 40 + 40 + 40 + 40 + 40 + 40 + 40 + 40 + 40 +$$
$$40 + 40 + 40 + 40 + 40 + 40 + 40 + 40 + 40 + 40 +$$
$$40 + 40 + 40 + 40 + 40 + 40 + 40 + 40 + 40 + 40 = 2000$$

However, this is incredibly tedious. Instead, we could express this sum as the product $40 \times 50 = 2000$. Multiplication, then, is a shorthand notation for repeated addition.

Now suppose that we want to find the thickness of a piece of paper that is folded in half 52 times. Since there are 500 sheets of paper in a ream and a ream is approximately 1.5 inches thick, we'll assume that a single piece of paper is approximately 0.003 inch thick. After one folding, it will be $2 \times 0.003$ inch thick, after two foldings it will be $2 \times 2 \times 0.003$ inch thick, and so on. Thus, after 52 foldings, it would be $2 \times 2 \times 2 \times 2 \times 2 \times 2 \times 2 \times 2 \times 2 \times 2 \times 2 \times 2 \times 2 \times 2 \times 2 \times 2 \times 2 \times 2 \times 2 \times 2 \times 2 \times 2 \times 2 \times 2 \times 2 \times 2 \times 2 \times 2 \times 2 \times 2 \times 2 \times 2 \times 2 \times 2 \times 2 \times 2 \times 2 \times 2 \times 0.003$ inches thick. Again, it is incredibly tedious to write out all the multiplications. Instead, we could express the thickness in exponential form as $2^{52} \times 0.003$ inches. (This is over 200,000,000 miles, which is more than twice the distance to the Sun!) Exponentiation, then, is a shorthand notation for repeated multiplication.

| *Definition of natural-*<br>*number exponents* | If $n$ is a natural number, then $a^n = a \times a \times \cdots \times a$, where there are $n$ factors of $a$. The number $a$ is called the **base**, and the number $n$ is called the **exponent**. |

**EXAMPLE 1**    *Evaluating expressions involving natural-number exponents*

Evaluate each of the following.

**a.** $2^5$        **b.** $(-3)^3$        **c.** $\left(\dfrac{3}{4}\right)^2$

**SOLUTION**

**a.** $2^5 = 2 \cdot 2 \cdot 2 \cdot 2 \cdot 2 = 32$        **b.** $(-3)^3 = (-3) \cdot (-3) \cdot (-3) = -27$

**c.** $\left(\dfrac{3}{4}\right)^2 = \dfrac{3}{4} \cdot \dfrac{3}{4} = \dfrac{9}{16}$

Note that the parentheses in parts (b) and (c) of Example 1 indicate that the base is the entire quantity inside the parentheses. What about expressions such as $-3^2$ where no parentheses are given? Should we assume that $-3^2$ means $(-3)^2 = 9$ or that $-(3)^2 = -9$? In Section 3 we discussed the fact that whenever an algebraic expression could have more than one meaning, conventions determine the order in which operations are to be performed. For example, if there are no grouping symbols, then multiplication and division are performed before addition and subtraction. In the case of exponents, the rule is that we perform all exponentiations before we perform multiplications, divisions, additions, subtractions, and negations (taking the negative). Thus, $-3^2 = -(3^2) = -9$.

| *Order of operations* | If no grouping symbols are used, first perform all exponentiations, then perform all multiplications and divisions from left to right, and finally perform any additions or subtractions from left to right. If grouping symbols are present, the operations within the grouping symbols are performed first, again using these rules for order of operations. |

**EXAMPLE 2**    *Evaluating expressions involving natural-number exponents*

Evaluate each of the following.

**a.** $\dfrac{9^2}{3}$        **b.** $[(2^3)^2]^2$        **c.** $-(-4)^2$

**SOLUTION**

**a.** We evaluate the exponent first to obtain $\dfrac{9^2}{3} = \dfrac{81}{3} = 27$.

**b.** We evaluate the innermost exponent first to obtain $[(2^3)^2]^2 = [(8)^2]^2 = [64]^2 = 4096$.

**c.** $-(-4)^2 = -[(-4)^2] = -[16] = -16$

| EXAMPLE 3 | *A doubling problem* |

Suppose that a child doubles in weight in the first six months of his life, and continues to double in weight every six months. How much would he weigh on his forty-first birthday? His fiftieth birthday?

**SOLUTION**   We will assume that he weighs 7 pounds at birth. Then after 1 six-month period he will weigh $2 \cdot 7$ pounds; after 2 six-month periods he will weigh $2 \cdot 2 \cdot 7 = 2^2 \cdot 7$ pounds, and so forth. In a similar fashion, if $n$ represents the number of six-month periods that have elapsed, he will weigh $2^n \cdot 7$ pounds after $n$ six-month periods. On his forty-first birthday, $2 \cdot 41 = 82$ six-month periods will have elapsed. Thus, he will weigh

$$2^{82} \cdot 7 = 33,849,922,949,209,616,891,772,928 \text{ pounds}$$

more than double the weight of Earth. On his fiftieth birthday he would weigh

$$2^{100} \cdot 7 = 8,873,554,201,597,605,810,476,922,437,632 \text{ pounds}$$

more than double the weight of the Sun!

---

The following properties are used for simplifying expressions involving exponents. Each can be proven using the definition of natural-number exponents, as suggested by the accompanying illustrations.

**Properties of exponents**

Assume that $m$ and $n$ are natural numbers.

| Property | Example | Illustration |
|---|---|---|
| 1. $a^m \cdot a^n = a^{m+n}$ | $2^3 2^4 = 2^{3+4} = 2^7$ | $(2 \cdot 2 \cdot 2) \cdot (2 \cdot 2 \cdot 2 \cdot 2) =$ $2 \cdot 2 \cdot 2 \cdot 2 \cdot 2 \cdot 2 \cdot 2 = 2^7$ |
| 2. $\dfrac{a^m}{a^n} = a^{m-n}$ | $\dfrac{3^5}{3^3} = 3^{5-3} = 3^2$ | $\dfrac{3 \cdot 3 \cdot 3 \cdot 3 \cdot 3}{3 \cdot 3 \cdot 3} = \dfrac{\not{3} \cdot \not{3} \cdot \not{3} \cdot 3 \cdot 3}{\not{3} \cdot \not{3} \cdot \not{3}}$ $= 3^2$ |
| 3. $(a^m)^n = a^{mn}$ | $(5^2)^3 = 5^{2 \cdot 3} = 5^6$ | $(5 \cdot 5) \cdot (5 \cdot 5) \cdot (5 \cdot 5) = 5^6$ |
| 4. $(ab)^n = a^n b^n$ | $(3 \cdot 5)^2 = 3^2 \cdot 5^2$ | $(3 \cdot 5) \cdot (3 \cdot 5) = (3 \cdot 3) \cdot (5 \cdot 5)$ $= 3^2 \cdot 5^2$ |
| 5. $\left(\dfrac{a}{b}\right)^n = \dfrac{a^n}{b^n}$ | $\left(\dfrac{2}{5}\right)^3 = \dfrac{2^3}{5^3}$ | $\dfrac{2}{5} \cdot \dfrac{2}{5} \cdot \dfrac{2}{5} = \dfrac{2 \cdot 2 \cdot 2}{5 \cdot 5 \cdot 5} = \dfrac{2^3}{5^3}$ |

| EXAMPLE 4 | *Simplifying expressions involving natural-number exponents* |

Simplify the following expressions using the rules of exponents.

**a.** $3^4 \cdot 3^{17}$        **b.** $(x^3)^5$

**SOLUTION**

**a.** $3^4 \cdot 3^{17} = 3^{(4+17)}$   Using Property 1   **b.** $(x^3)^5 = x^{(3 \cdot 5)}$   Using Property 3

$\qquad\qquad\quad = 3^{21}$ $\qquad\qquad\qquad\qquad\qquad\qquad\qquad = x^{15}$

**EXAMPLE 5**    *Simplifying an expression involving natural-number exponents*

Simplify the expression $(2^5 x^3 y^2 z)(2^3 xy^4 z^2)$.

**SOLUTION**

$$(2^5 x^3 y^2 z)(2^3 xy^4 z^2) = 2^{5+3} x^{3+1} y^{2+4} z^{1+2} \qquad \text{Using Property 1}$$
$$= 2^8 x^4 y^6 z^3$$

**EXAMPLE 6**    *Simplifying an expression involving natural-number exponents*

Simplify the expression $\dfrac{7^8 a^5}{7^3 a^3}$.

**SOLUTION**

$$\frac{7^8 a^5}{7^3 a^3} = \frac{7^8}{7^3} \cdot \frac{a^5}{a^3} \qquad \text{Separating factors}$$
$$= 7^5 a^2 \qquad \text{Using Property 2}$$

**EXAMPLE 7**    *Simplifying an expression involving natural-number exponents*

Simplify the expression $\dfrac{3^4 x^6 y^2}{3^2 x^3 y} \cdot \dfrac{3x^3 y^3}{(3x^2 y)^2}$.

**SOLUTION**

$$\frac{3^4 x^6 y^2}{3^2 x^3 y} \cdot \frac{3x^3 y^3}{(3x^2 y)^2} = \frac{3^4 x^6 y^2 \cdot 3x^3 y^3}{3^2 x^3 y \cdot 3^2 x^4 y^2} \qquad \text{Multiplying fractions and using Property 4 to remove parentheses}$$
$$= \frac{3^5 x^9 y^5}{3^4 x^7 y^3} \qquad \text{Using Property 1 in the numerator and denominator}$$
$$= 3x^2 y^2 \qquad \text{Using Property 2}$$

## NEGATIVE AND ZERO EXPONENTS

If exponentiation is repeated multiplication, then what is $2^0$? That is, how do you multiply 2 by itself 0 times? The answer is that although "multiplying 2 by itself 0 times" *doesn't* have any meaning, we can still define $2^0$ in a sensible way using the properties of exponents given above. For example, according to Property 2 for exponents, $2^0$ should have the same value as $2^1/2^1 = 1$; therefore we *define* $2^0$ to be 1. Similarly, we can define $2^{-3}$ to be $2^0/2^3 = 1/2^3$. More generally, we give the following definition.

**Definition of negative and zero exponents**

> For $a \neq 0$, $a^{-n} = \dfrac{1}{a^n}$ for positive integers $n$, and $a^0 = 1$.

**EXAMPLE 8** *Evaluating expressions with nonpositive exponents*

Evaluate each of the following expressions.

**a.** $3^{-2}$     **b.** $10^{-1}$     **c.** $(23457892.09845987234857)^0$

**SOLUTION**

**a.** $3^{-2} = \dfrac{1}{3^2} = \dfrac{1}{9}$

**b.** $10^{-1} = \dfrac{1}{10^1} = \dfrac{1}{10}$

**c.** $(23457892.09845987234857)^0 = 1$, since anything (except 0) raised to the zero power is 1.

Exponentiation is now defined for all integer exponents in such a way that all the familiar properties of natural-number exponents are valid.

**EXAMPLE 9** *Simplifying an expression with a negative exponent*

Simplify the expression $x^3 \cdot x^{-5}$. Express your answer using only positive exponents.

**SOLUTION**

$$
\begin{aligned}
x^3 \cdot x^{-5} &= x^{3+(-5)} & \text{Using Property 1} \\
&= x^{-2} & \text{Simplifying} \\
&= \frac{1}{x^2} & \text{Using the definition of negative exponents}
\end{aligned}
$$

**EXAMPLE 10** *Simplifying an expression with negative exponents*

Simplify the expression $\dfrac{2^3 a^{-4} b^2}{2^{-2} a^{-2} b^5}$. Express your answer using only positive exponents.

**SOLUTION**

$$
\begin{aligned}
\frac{2^3 a^{-4} b^2}{2^{-2} a^{-2} b^5} &= \frac{2^3}{2^{-2}} \cdot \frac{a^{-4}}{a^{-2}} \cdot \frac{b^2}{b^5} & \text{Grouping like factors} \\
&= 2^{3-(-2)} \cdot a^{-4-(-2)} \cdot b^{2-5} & \text{Using Property 2} \\
&= 2^5 a^{-2} b^{-3} & \text{Simplifying} \\
&= 2^5 \cdot \frac{1}{a^2} \cdot \frac{1}{b^3} & \text{Using the definition of negative exponents} \\
&= \frac{32}{a^2 b^3} & \text{Simplifying}
\end{aligned}
$$

EXAMPLE 11    *Simplifying an expression with negative exponents*

Simplify the expression $\left(\dfrac{x^3 y^{-2} z^{-3}}{x^{-5} y z^{-1}}\right)^{-1}$. Express your answer using only positive exponents.

**SOLUTION**

$$\left(\frac{x^3 y^{-2} z^{-3}}{x^{-5} y z^{-1}}\right)^{-1} = \left(\frac{x^3}{x^{-5}} \cdot \frac{y^{-2}}{y} \cdot \frac{z^{-3}}{z^{-1}}\right)^{-1} \qquad \text{Grouping like factors}$$

$$= (x^8 y^{-3} z^{-2})^{-1} \qquad \text{Using Property 2}$$

$$= x^{-8} y^3 z^2 \qquad \text{Using Property 4}$$

$$= \frac{y^3 z^2}{x^8} \qquad \text{Using the definition of negative exponents}$$

## COMPOUND INTEREST

If you invest $1000 at 5% *simple interest*, then each year you will earn $5\% \cdot \$1000 = 0.05 \cdot \$1000 = \$50$. Thus, after 1 year the balance will be $1000 + $50, after 2 years it will be $\$1000 + 2 \cdot \$50$, and so on. Thus, after $t$ years the balance will be $\$1000 + t \cdot \$50$. More generally, if $P$ is the **principal** (the initial amount invested) in dollars and if $r$ is the annual interest rate, then the account will earn $Pr$ dollars each year. After $t$ years, the account will have earned $Prt$ dollars. Thus, after $t$ years the balance will be $B = \text{Principal} + \text{Interest} = P + Prt = P(1 + rt)$.

With a compound interest account, interest is earned on interest as well as on principal. For example, if interest is 5% compounded annually, then after the first year the balance would be $1050, just as with simple interest. But in the second year, the account will earn $5\% \cdot \$1050 = \$52.50$. Thus, after 2 years the balance would be $1102.5. In the third year, the account would earn $5\% \cdot \$1102.50 = \$55.125$, so that the balance would be $\$1102.50 + \$55.125 = \$1157.625$ (which would probably be rounded to $1157.63). In general, if the principal is $P$ and the interest rate is $r$, then the balance at

**TABLE 3**

*Simple versus compound interest*

| Year | Balance (5% simple interest) | Balance (5% interest compounded annually) |
|---|---|---|
| 0 | $1,000.00 | $1,000.00 |
| 1 | $1,050.00 | $1,050.00 |
| 2 | $1,100.00 | $1,102.50 |
| 3 | $1,150.00 | $1,157.63 |
| 4 | $1,200.00 | $1,215.51 |
| 5 | $1,250.00 | $1,276.28 |
| 10 | $1,500.00 | $1,628.89 |
| 15 | $1,750.00 | $2,078.93 |
| 20 | $2,000.00 | $2,653.30 |
| 30 | $2,500.00 | $4,321.94 |
| 50 | $3,500.00 | $11,467.40 |
| 100 | $6,000.00 | $131,501.26 |
| 1,000 | $51,000.00 | $1,546,318,920,731,927,238,984.57 |

the end of the first year is $P + Pr = P(1 + r)$. The balance at the end of the second year will then be $(1 + r)$ times the balance at the beginning of the second year or $P(1 + r)^2$. Similarly, the balance at the end of the third year will be $P(1 + r)^3$, and the balance at the end of the $t$ years will be $P(1 + r)^t$. Table 3 illustrates the advantage of compound interest over simple interest when $1000 is invested at 5%.

We summarize the formulas for simple interest and compound interest below.

***Simple and annually compounded interest formulas***

> If a principal of $P$ dollars is deposited in an account paying an annual rate of interest $r$, then the balance $B$ in the account after $t$ years can be computed using the following formulas.
>
> $$\text{Simple Interest:} \quad B = P(1 + rt)$$
>
> $$\text{Interest Compounded Annually:} \quad B = P(1 + r)^t$$

**EXAMPLE 12**

*Compound interest*

Find the balance that would result if $500 were invested for 20 years with 6% interest compounded annually.

**SOLUTION**    Using the exponent key on a calculator we obtain $B = P(1 + r)^t = 500(1.06)^{20} \approx 1603.57$. Thus, the balance after 20 years would be $1603.57.

## SCIENTIFIC NOTATION

Many numbers that arise in physics, biology, astronomy, mathematics, economics, and other sciences are either very large or very small. For example, a bloodsucking banded louse weighs 0.00000001101 pound. Light travels 5,878,000,000,000 miles in a year. There are about 10,000,000,000,000,000,000,000,000,000,000,000,000,000,000,000, 000,000,000,000,000,000,000,000,000,000,000 atoms in the observable universe. An electron tips the scales at 0.000000000000000000000000000910953 gram. There are about 80,660,000,000,000,000,000,000,000,000,000,000,000,000,000,000, 000,000,000,000,000,000,000 different orderings for a deck of cards. As these examples illustrate, standard decimal notation is cumbersome for extremely large and extremely small numbers. For this reason, *scientific notation* was invented. Using scientific notation, we would say that the louse weighs $1.1101 \times 10^{-8}$ pound, there are $5.878 \times 10^{12}$ miles in a light-year, there are about $1 \times 10^{85}$ atoms in the observable universe, an electron weighs $9.10953 \times 10^{-28}$ gram, and there are about $8.066 \times 10^{67}$ different orderings of a deck of cards.

Numbers written in scientific notation are expressed as a real number between 1 and 10 multiplied by 10 to an integer power. Numbers that are less than 1 will have a negative exponent when written in scientific notation.

***Scientific notation***

> A number is said to be expressed in scientific notation if it is written in the form $a \times 10^n$ where $1 \le a < 10$ and $n$ is an integer.

> **RULE OF THUMB**   When **converting from decimal notation to scientific notation**, the number between 1 and 10 is obtained by shifting the decimal point to the position immediately after the first nonzero number. The power of 10 is obtained by counting the number of places that the decimal point has been shifted. If the original number is larger than 1, the exponent is nonnegative. If the original number is smaller than 1, then the exponent will be negative.

**EXAMPLE 13**   *Converting to scientific notation*

Express the following numbers in scientific notation.

a. 234.0023                                          b. One billion
c. 0.00012                                           d. 123,456,789.01

**SOLUTION**

a. $234.0023 = 2.340023 \times 10^2$

b. One billion $= 1,000,000,000 = 1 \times 10^9$

c. $0.00012 = 1.2 \times 10^{-4}$

d. $123,456,789.01 = 1.2345678901 \times 10^8$

---

**EXAMPLE 14**   *Converting from scientific notation to decimal notation*

Express the following numbers in standard decimal notation.

a. $7.03 \times 10^{-3}$          b. $9.05 \times 10^5$          c. $1.47 \times 10^0$

**SOLUTION**

a. $7.03 \times 10^{-3} = 0.00703$          The decimal point is moved 3 units to the left.

b. $9.05 \times 10^5 = 905,000$          The decimal point is moved 5 units to the right.

c. $1.47 \times 10^0 = 1.47$          Since $10^0 = 1$, the decimal point isn't moved at all.

---

Numbers expressed in scientific notation can be entered into a calculator using a key that is usually labeled either EE or EXP. For example, to enter the number $2.04 \times 10^{32}$, we first enter the number 2.04, then press EE (or EXP), and finally enter the number 32.

## EXERCISES 4

**EXERCISES 1–4** □ *Express each product in exponential form.*

1. $2 \cdot 2 \cdot 2 \cdot 2 \cdot 2 \cdot 2$

2. $(-1) \cdot (-1) \cdot (-1)$

3. $x \cdot x \cdot x \cdot x \cdot x \cdot x \cdot x$

4. $a \cdot b \cdot b \cdot a \cdot a \cdot b \cdot a \cdot b \cdot b$

**EXERCISES 5–28** □ *Evaluate the given expression.*

5. $(-1)^5$

6. $\left(\dfrac{2}{5}\right)^3$

7. $(-2)^4$

8. $\left(-\dfrac{1}{10}\right)^3$

9. $3^{-2}$

10. $4^{-3}$

11. $(-2)^{-1}$

12. $(-1)^{-4}$

13. $\left(\dfrac{1}{3}\right)^{-1}$

14. $\left(\dfrac{4}{3}\right)^{-2}$

15. $0.1^5$

16. $0.02^3$

17. $1.2^{-2}$

18. $0.1^{-3}$

19. $-5^2$

20. $-3^{-2}$

21. $\left(-\dfrac{3}{7}\right)^{-2}$

22. $(2^3)^2$

23. $(4^{-2})^3$

24. $2^{(3^2)}$

25. $[(2^{-3})^2]^{-2}$

26. $[(10^{-1})^2]^{-1}$

27. (The number of ozone molecules in the atmosphere)$^{\text{(The number of talking rhinos)}}$

28. (The number of talking rhinos)$^{\text{(The number of ozone molecules in the atmosphere)}}$

**EXERCISES 29–52** □ *Simplify the given expression. Write your final answer without using negative exponents.*

29. $a^3a^6$

30. $x^5x^{-3}$

31. $\dfrac{2x^8}{x^3}$

32. $\dfrac{b^3}{b^1}$

33. $\dfrac{r^5}{r^7}$

34. $\dfrac{x^{-2}}{x^3}$

35. $\dfrac{y^{-1}}{y^{-2}}$

36. $x^2y^4x^{-3}$

37. $(a^3b^{-2})(a^{-2}b^{-4})$

38. $(x^{-1})^2$

39. $(x^2y^3)^2$

40. $(s^{-4}t^3)^{-2}$

41. $\dfrac{p^{10}q^6}{p^4q^2}$

42. $\dfrac{2^3s^5t^2}{2^2s^6t^5}$

43. $(a^3b^{-1}c^2)\cdot(a^{-2}b^{-3}c^3)$

44. $\dfrac{(a^2+b^2)^3}{(a^2+b^2)^2}$

45. $\dfrac{(x-y)(x+y)}{(x-y)^2}$

46. $\dfrac{p^{-2}}{q^{-3}}\cdot\dfrac{q^{-1}}{p^4}$

47. $\dfrac{3^{-1}x^4y^{-2}}{3^{-2}x^{-2}y}$

48. $\left(\dfrac{p^2}{q^3}\right)^{-1}$

49. $(y^{-1}z^2)\cdot(yz^{-1})^{-1}$

50. $\left(\dfrac{a^2b^3c}{a^2b^{-1}c^2}\right)^2$

51. $\left(\dfrac{xy^{-2}z^{-3}}{2x^{-2}y^1z^2}\right)^{-2}$

52. $\left(\dfrac{p^2q^{-1}}{p^4q^{-2}r^3}\right)^{-3}$

**EXERCISES 53–58** □ *Convert the given number to standard decimal notation.*

53. $3.04\times10^{-3}$

54. $8.38\times10^3$

55. $3.001\times10^5$

56. $2.7\times10^{-1}$

57. $7.5\times10^0$

58. $3.2\times10^1$

**EXERCISES 59–70** □ *Convert the given number to scientific notation.*

59. 4,000,000

60. 10,000.001

61. 0.0943

62. 0.0008456

63. Fifty-two trillion

64. One-billionth

65. $20.31\times10^3$

66. $0.00737\times10^{-4}$

67. $(2.5\times10^{-45})\cdot(4\times10^{52})$

68. $(8.1\times10^{12})\div(2.7\times10^{-5})$

69. The weight of the Earth in pounds

70. The distance from the Earth to the Sun in miles

■ **Applications**

71. **Simple Interest** What would the balance be on an account paying 8% simple interest if $100 were deposited for 6 years?

72. **Simple Interest** Suppose that $750.00 is deposited in an account paying 4% simple interest. How much interest will be earned in the first 10 years?

73. **Compound Interest** If the dollar that George Washington threw across the Potomac in 1776 had instead been deposited in an account paying 5% interest compounded annually, what would the balance be in the year 2001?

74. **Compound Interest** What is the balance after 50 years in an account paying 10% interest compounded annually with a principal of 1¢?

75. **Salary Options** You are hired as a temporary worker to begin work on the first of May; the job is scheduled to be completed on May 31. Your employer gives you the choice either of receiving $300,000 a day for a month or of receiving 1¢ on the first day of the month, 2¢ on the second day, 4¢ on the third, doubling your pay on each successive day of the month. Which deal should you take? Justify your answer.

76. **Faded Jeans** Suppose that a pair of jeans fades in such a way that only $\frac{3}{4}$ of the dye remains after every washing. Assuming that the jeans are washed once a week for a year, how much of the dye remains after 1 year?

■ *Projects for Enrichment*

**77.** *Julius Caesar Versus Ben Franklin* In this project we will see the enormous advantage of compound interest over simple interest.

  **a.** Which would be worth more today, Julius Caesar's deposit of $100 at 10% simple interest or Ben Franklin's deposit of $1 at 6% interest compounded annually?

  **b.** What approximations did you make in order to solve part (a)?

  **c.** What aspects of the premise of part (a) are unrealistic?

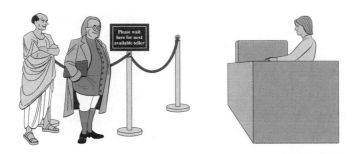

**78.** *Exponential Growth* In this project we will attempt to fathom the incredible consequences of exponential growth over long periods of time.

  **a.** The Earth is approximately spherical with a radius of about 4000 miles. The volume of a sphere is given by the formula $V = \frac{4}{3}\pi r^3$, where $r$ represents the radius and $V$ is the volume. Estimate the volume of the Earth. Express the answer in scientific notation.

  **b.** Estimate the volume of a dollar bill. (*Hint:* A dollar bill can be viewed as a box with a very small thickness. The formula for the volume of a box is $V = l \cdot w \cdot t$, where $l$ is the length, $w$ is the width, and $t$ is the thickness.)

  **c.** Using parts (a) and (b), estimate the number of dollar bills that could fit inside the Earth, if the Earth were hollow. Express the answer in scientific notation.

**d.** How much would $1 deposited in A.D. 1 be worth in 1997? Assume interest is 6% compounded annually. Express the answer in scientific notation.

**e.** Using the results to parts (b) and (c) compute the number of "Earth banks" that could be filled if the balance from a deposit of $1 deposited in A.D. 1 were paid in one dollar bills in 1997.

**79.** *Repeated Exponentiation* As was mentioned in the text, multiplication represents repeated addition, and exponentiation represents repeated multiplication. Suppose that we take this one step further and define "hyperexponentiation" to be repeated exponentiation. For example, just as $2 \cdot 3$ represents 2 added to itself 3 times, and $2^3$ represent 2 multiplied by itself 3 times, we could define $2 \boxed{h} 3$ to be 2 "exponentiated 3 times," or $2^{(2^2)} = 2^4 = 16$. More generally, we could define $m \boxed{h} n$ to be

$$m^{\left(m^{\left(m^{\left(\cdot^{\cdot^{\cdot^m}}\right)}\right)}\right)}$$

where $m$ appears $n$ times in all.

Compute the following quantities and express the answer in scientific notation.

  **a.** $5 \boxed{h} 2$

  **b.** $2 \boxed{h} 4$

  **c.** $1 \boxed{h} 10$

  **d.** $10 \boxed{h} 3$

**80.** *The Bouncing Ball* Suppose that whenever a certain ball is dropped, it bounces to two-thirds of its original height. Suppose the ball is dropped off the Sears Tower in Chicago (height 1454 feet).

  **a.** How high would the ball rise after one bounce?

  **b.** How high would the ball rise after two bounces?

Dominique Wilkins soars for a slam dunk.

c. How high would the ball rise after $n$ bounces?

d. It has been estimated that the basketball player Dominique Wilkins could jump and extend his hand to a height of 12 feet 6 inches. How many times must the ball bounce before Dominique can catch the ball at the peak of a bounce?

e. Suppose that a film of the bouncing ball is run backwards. The first visible bounce appears to be to a height of about 1 inch. How many additional bounces will occur before the ball is seen leaping over the edge of the Sears Tower?

---

![] ***Questions for Discussion or Essay***

81. In this section we have dealt with many very large numbers. But do such numbers have any significance? Can you think of any context (other than those mentioned in the text) in which numbers larger than a million arise? In particular, what topics discussed on the nightly news might involve very large numbers?

82. In the text we define $a^{-n} = 1/a^n$ for positive integers $n$ and $a \neq 0$. Why do we exclude $a = 0$? Also, we define $a^0 = 1$ for $a \neq 0$. Why can't we define $0^0 = 1$? (*Hint:* If we define $0^0 = 1$, then $1^0/0^0 = 1$. Yet if the properties of exponents are to apply, then $1^0/0^0$ can be rewritten.)

83. Suppose that the Board of Directors of Simpleton Bank and Trust decides to offer only simple interest in order to cut expenses. How could you foil their efforts by obtaining compound interest from their bank?

84. One of the examples in the text involved the premise that a child continues to double in weight every 6 months for the rest of his life. This, of course, led to ridiculous results. What factors prevent a child from continuing to double in weight every 6 months? The hypothesis that a child's weight doubles every 6 months is an example of a *mathematical model*. This particular mathematical model was a complete failure. What characteristics should a good mathematical model have? What information would be helpful in constructing a good mathematical model of weight gain?

85. In the text we discussed folding a piece of paper in half 52 times creating a piece of paper more than 200,000,000 miles thick. In practice, about how many times can a piece of paper be folded? What factors prevent one from folding a piece of paper 52 times?

---

**SECTION 5**

## POLYNOMIAL EXPRESSIONS

- How can you estimate the height of a bridge with only a stopwatch and a rock?
- How can you estimate the number of board feet in a tree with a circumference of 60 inches?
- If you are traveling at a rate of 65 mph and you suddenly notice an obstruction in the road 200 feet ahead of you, will you be able to stop in time?
- How is it possible for a mind reader to determine something only you could know?

---

**DEFINITION OF POLYNOMIAL**

Expressions involving variables raised to natural-number powers arise frequently when studying real-world phenomena. For example, the laws of physics tell us that if an object is dropped near the surface of the Earth, it will fall $16t^2$ feet in $t$ seconds. The expression $16t^2$ is an example of a *monomial*.

**Definition of monomial**

> A **monomial** is the product of a number and one or more variables raised to non-negative integer powers. The **coefficient** of a monomial is the number preceding the variable(s). The **degree** of a monomial is the sum of the powers.

We list some examples of expressions that are monomials and some that are not. For those that are monomials, we identify the coefficient and the degree.

| *Expression* | *Monomial?* | *Coefficient* | *Degree* | *Comment* |
|---|---|---|---|---|
| $2x^3$ | Yes | $2$ | $3$ | |
| $\frac{1}{3}p^2q^5$ | Yes | $\frac{1}{3}$ | $7$ | The degree is the sum of the powers. |
| $5y^{-2}$ | No | | | The exponent is negative. |
| $-4$ | Yes | $-4$ | $0$ | Numbers can be considered to be monomials since, for example, $-4 = -4x^0$. |
| $1.5t^{1/3}$ | No | | | The exponent is not an integer. |

**Definition of polynomial**

> A **polynomial** is a finite sum of monomials. The monomials that make up a polynomial are called the **terms** of the polynomial. The **degree** of a polynomial is the highest degree of its terms.

A monomial can be viewed as a polynomial with only one term. Two other categories of polynomials are *binomials* and *trinomials*. As the prefixes *bi* and *tri* suggest, a **binomial** is a polynomial with *two* terms while a **trinomial** is a polynomial with *three* terms.

**EXAMPLE 1**     *Identifying polynomials*

Identify each of the following as a monomial, binomial, trinomial, polynomial, or none of these, and determine the degree of any that are polynomials.

a. $-2a^5 + 4a^3 - 3a + 4$          b. $3x - 5$

c. $3x^2y + 4y^{1/3} + 2xy^3$          d. $\frac{1}{2}p^2q^3 + \frac{2}{3}p^3q^5 - pq^6$

**SOLUTION**

a. Polynomial of degree 5.

b. Binomial (polynomial also) of degree 1.

c. This is not a polynomial because the exponent in $y^{1/3}$ is not an integer.

d. Trinomial (polynomial also) of degree 8.

## EVALUATING POLYNOMIAL EXPRESSIONS

The variables in a polynomial expression represent numbers. In many instances, we will be interested in evaluating such expressions with specific values for the variables.

**EXAMPLE 2**    *Estimating the height of a bridge*

An object dropped near the surface of the Earth will fall $16t^2$ feet in $t$ seconds. If a rock that is dropped off the edge of a bridge hits the water after 1.2 seconds, how high above the water is the bridge?

**SOLUTION**    We evaluate the monomial $16t^2$ using $t = 1.2$ to obtain $16(1.2)^2 \approx 23$ feet.

**EXAMPLE 3**    *Evaluating a polynomial in one variable*

Evaluate the polynomial $-2a^4 + 4a^3 - a^2 - 3a + 4$ for $a = 2$.

**SOLUTION**    Substituting $a = 2$, we have

$$-2(2)^4 + 4(2)^3 - (2)^2 - 3(2) + 4 = -32 + 32 - 4 - 6 + 4 = -6$$

**EXAMPLE 4**    *Evaluating a polynomial in two variables*

Evaluate the polynomial $\frac{1}{2}p^2q^3 + \frac{2}{3}p^3q^5 - pq^6$ for $p = 3$ and $q = -1$.

**SOLUTION**    Substituting $p = 3$ and $q = -1$, we obtain

$$\frac{1}{2}(3)^2(-1)^3 + \frac{2}{3}(3)^3(-1)^5 - (3)(-1)^6 = -\frac{9}{2} - 18 - 3 = -\frac{9}{2} - 21 = -\frac{51}{2}$$

The evaluation of a polynomial can be conveniently denoted with the *function* notation that will be developed in more detail in Chapter 4. To illustrate this notation, consider the polynomials in Examples 3 and 4. If we write $f(a) = -2a^4 + 4a^3 - a^2 - 3a + 4$ to indicate the polynomial in the single variable $a$ from Example 3, then $f(2) = -2(2)^4 + 4(2)^3 - (2)^2 - 3(2) + 4$ gives the value of the polynomial when $a = 2$. Similarly, if we write $h(p, q) = \frac{1}{2}p^2q^3 + \frac{2}{3}p^3q^5 - pq^6$ to indicate the polynomial of two variables $p$ and $q$ from Example 4, then $h(3, -1)$ represents the value of the polynomial $h$ when $p = 3$ and $q = -1$, namely $\frac{1}{2}(3)^2(-1)^3 + \frac{2}{3}(3)^3(-1)^5 - (3)(-1)^6$. Notice that since the polynomial was defined as $h(p, q)$, the first number inside the parentheses must be substituted for $p$, and the second must be substituted for $q$.

**EXAMPLE 5**    *Evaluating a polynomial in one variable using function notation*

Given $p(x) = x^2 - 3x + 7$, find $p(-2)$.

**SOLUTION**

$$p(-2) = (-2)^2 - 3(-2) + 7 = 4 + 6 + 7 = 17$$

| EXAMPLE 6 | *Evaluating a polynomial in two variables using function notation* |

Given $g(u, v) = u^2v + 5uv - 4v^3$, find $g(-3, 2)$.

**SOLUTION**   For the polynomial $g(u, v)$, the notation $g(-3, 2)$ means we let $u = -3$ and $v = 2$ to obtain

$$g(-3, 2) = (-3)^2(2) + 5(-3)(2) - 4(2)^3 = 18 - 30 - 32 = -44$$

## ADDITION AND SUBTRACTION OF POLYNOMIALS

The operations of addition and subtraction of polynomials are performed by *combining like terms*. The mechanics of this process involve the associative, commutative, and distributive properties.

| EXAMPLE 7 | *Adding polynomials* |

Find the sum $(2x^3 - 5x^2 + x - 4) + (8x^2 - 3x + 4)$.

**SOLUTION**

$$\begin{aligned}
(2x^3 - 5x^2 + x - 4) + (8x^2 - 3x + 4) &= 2x^3 + (-5x^2 + 8x^2) \\
&\quad + (x - 3x) + (-4 + 4) \\
&= 2x^3 + 3x^2 - 2x
\end{aligned}$$

| EXAMPLE 8 | *Subtracting polynomials* |

Find the difference $(7m^2 - 4mn + n^2 + 5n) - (3m^2 + 2mn - 2n^2)$.

**SOLUTION**

$$\begin{aligned}
(7m^2 &- 4mn + n^2 + 5n) - (3m^2 + 2mn - 2n^2) \\
&= 7m^2 - 4mn + n^2 + 5n - 3m^2 - 2mn + 2n^2 \\
&= (7m^2 - 3m^2) + (-4mn - 2mn) + (n^2 + 2n^2) + 5n \\
&= 4m^2 - 6mn + 3n^2 + 5n
\end{aligned}$$

## MULTIPLICATION OF POLYNOMIALS

Polynomials are multiplied by repeated application of the distributive property, as the following examples demonstrate.

| EXAMPLE 9 | *Multiplying polynomials* |

Find the product $3mn(8m^3 - 12m^2n + 6mn^2 - n^3)$.

**SOLUTION**

$$\begin{aligned}
3mn(8m^3 &- 12m^2n + 6mn^2 - n^3) \\
&= 3mn(8m^3) - 3mn(12m^2n) + 3mn(6mn^2) - 3mn(n^3) \\
&= 24m^4n - 36m^3n^2 + 18m^2n^3 - 3mn^4
\end{aligned}$$

**EXAMPLE 10**   *Multiplying polynomials*

Find the product $(x - 1)(2x + 3)$.

**SOLUTION**

$$
\begin{aligned}
(x - 1)(2x + 3) &= x(2x + 3) + (-1)(2x + 3) \\
&= 2x^2 + 3x - 2x - 3 \\
&= 2x^2 + x - 3
\end{aligned}
$$

---

**EXAMPLE 11**   *Multiplying polynomials*

Find the product $(y^2 + 2y - 5)(3y^2 - y + 2)$.

**SOLUTION**

$$
\begin{aligned}
(y^2 &+ 2y - 5)(3y^2 - y + 2) \\
&= y^2(3y^2 - y + 2) + 2y(3y^2 - y + 2) + (-5)(3y^2 - y + 2) \\
&= (3y^4 - y^3 + 2y^2) + (6y^3 - 2y^2 + 4y) + (-15y^2 + 5y - 10) \\
&= 3y^4 + (-y^3 + 6y^3) + (2y^2 - 2y^2 - 15y^2) + (4y + 5y) - 10 \\
&= 3y^4 + 5y^3 - 15y^2 + 9y - 10
\end{aligned}
$$

---

Two shortcuts are often used for multiplying polynomials. The first, the column method, is especially useful when multiplying polynomials with three or more terms. We demonstrate this method below for the polynomials from Example 11.

$$
\begin{array}{r}
y^2 + 2y - 5 \\
3y^2 - \; y + 2 \\
\hline
2y^2 + 4y - 10 \\
-y^3 - \; 2y^2 + 5y \\
3y^4 + 6y^3 - 15y^2 \\
\hline
3y^4 + 5y^3 - 15y^2 + 9y - 10
\end{array}
$$

| |
|---|
| 2 times $y^2 + 2y - 5$ |
| $-y$ times $y^2 + 2y - 5$ |
| $3y^2$ times $y^2 + 2y - 5$ |
| Adding down the columns |

Example 10 suggests a second shortcut for multiplying binomials. From the second step of that example, we construct the diagram shown in Figure 29. Here, $2x^2$ is the product of the *First* terms, $3x$ is the product of the *Outer* terms, $-2x$ is the product of the *Inner* terms, and $-3$ is the product of the *Last* terms. This is sometimes called the FOIL (**F**irst **O**uter **I**nner **L**ast) method.

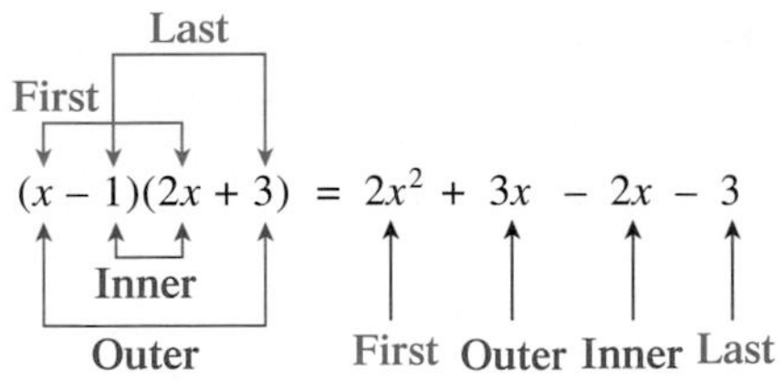

$$(x - 1)(2x + 3) = 2x^2 + 3x - 2x - 3$$

First Outer Inner Last

**FIGURE 29**

**EXAMPLE 12**   *Multiplying binomials using the FOIL method*

Perform the multiplication $(3t - 5)(t + 3)$ using the FOIL method.

**SOLUTION**

$$
\begin{aligned}
(3t - 5)(t + 3) &= 3t^2 + 9t - 5t - 15 \\
&= 3t^2 + 4t - 15
\end{aligned}
$$

Several products of binomials occur so frequently that it is helpful to recognize their forms. These are summarized as follows.

**Special binomial products**

$$(a + b)(a - b) = a^2 - b^2$$
$$(a + b)^2 = (a + b)(a + b) = a^2 + 2ab + b^2$$
$$(a - b)^2 = (a - b)(a - b) = a^2 - 2ab + b^2$$

**EXAMPLE 13**    *Using a special binomial product*

Expand $(3x - 5)^2$.

**SOLUTION**

$$(3x - 5)^2 = (3x)^2 - 2(3x)(5) + 5^2 = 9x^2 - 30x + 25$$

**EXAMPLE 14**    *A mind-reading trick*

One of your friends claims to be able to read minds. He asks you to pick a number, add 2 to it and square the result, subtract the square of your original number, multiply the result by $\frac{1}{4}$, and finally, subtract your original number. After a dramatic pause, your friend announces with great pleasure that your result is 1. How did he know?

**SOLUTION**    The mystery behind such number tricks is easily revealed using polynomial expressions. Letting $x$ represent the number you chose, we can form a polynomial expression using the verbal statements as shown below.

Add 2 and square the result:                 $(x + 2)^2$

Subtract the square of the original number:     $(x + 2)^2 - x^2$

Multiply the result by $\dfrac{1}{4}$ :           $\dfrac{1}{4}[(x + 2)^2 - x^2]$

Subtract the original number:                  $\dfrac{1}{4}[(x + 2)^2 - x^2] - x$

Simplifying this expression, we get

$$\frac{1}{4}[(x + 2)^2 - x^2] - x = \frac{1}{4}[(x^2 + 4x + 4) - x^2] - x$$

$$= \frac{1}{4}(4x + 4) - x$$

$$= (x + 1) - x$$

$$= 1$$

This shows that the polynomial $\frac{1}{4}[(x + 2)^2 - x^2] - x$ is equivalent to 1. In other words, no matter what initial number is chosen for $x$, the answer will always be 1. That's why the trick works.

## DIVIDING POLYNOMIALS

It is not too surprising that when two polynomials are added, subtracted, or multiplied, the result will always be another polynomial. A quick glance over the preceding examples will show this to be the case. However, when one polynomial is divided by another, the result will not necessarily be a polynomial. A simple example of this is the quotient $1/x^2$. Both 1 and $x^2$ are polynomials, yet the quotient $1/x^2$ simplifies to $x^{-2}$, which is not a polynomial because of the negative exponent. Occasionally, a quotient of polynomials can be expressed as a polynomial by simplification. Consider, for example, the following quotient of polynomials.

$$\frac{x^3 - 5x^2}{x} = \frac{x(x^2 - 5x)}{x} = x^2 - 5x$$

This simplification was possible because the numerator could be factored. Thus, before we look at quotients of polynomials in more detail, it will be helpful to first consider some procedures for factoring polynomials. We'll do this in Section 6, and then we'll tackle quotients of polynomials in Section 7.

## COMPLEX ARITHMETIC

Recall from Section 1 that a complex number is one of the form $a + bi$ where $i = \sqrt{-1}$. The arithmetic operations of addition, subtraction, and multiplication of complex numbers can be handled very much like the corresponding operations on polynomials. We simply treat the complex numbers as though they were binomials involving the variable $i$. The only difference is that $i^2 = -1$. As the following examples illustrate, this property is useful for simplifying any power of $i$.

$$i^3 = i^2 i = (-1)i = -i$$
$$i^4 = i^2 i^2 = (-1)(-1) = 1$$
$$i^6 = (i^2)^3 = (-1)^3 = -1$$
$$i^{13} = (i^2)^6 i = (-1)^6 i = i$$

In fact, all powers of $i$ simplify to $\pm 1$ or $\pm i$. This property, when combined with the usual rules for adding, subtracting, and multiplying binomials (e.g., the FOIL method), makes complex arithmetic a simple task.

**EXAMPLE 15**

*Subtracting complex numbers*

Write $(10 - 3i) - (7 - 6i)$ in the form $a + bi$.

**SOLUTION**

$$(10 - 3i) - (7 - 6i) = 10 - 3i - 7 + 6i = 3 + 3i$$

**EXAMPLE 16**    *Using FOIL to multiply complex numbers*

Write $(2 - i)(3 + 2i)$ in the form $a + bi$.

**SOLUTION**    Applying the FOIL method, we have

$$(2 - i)(3 + 2i) = 6 + 4i - 3i - 2i^2 = 6 + i - 2i^2$$

Since $i^2 = -1$, we can simplify further to obtain

$$6 + i - 2(-1) = 8 + i$$

---

**EXAMPLE 17**    *Simplifying powers of i*

Write $i^{1007}(4 - 2i)$ in the form $a + bi$.

**SOLUTION**    Since $i^2 = -1$, we wish to determine how many factors of $i^2$ are in $i^{1007}$. To do this, we select the largest even number less than 1007, namely 1006, and write

$$i^{1007} = i^{1006} \cdot i = (i^2)^{503} \cdot i = (-1)^{503} \cdot i = -i$$

Thus,

$$i^{1007}(4 - 2i) = -i(4 - 2i) = -4i + 2i^2 = -4i - 2 = -2 - 4i$$

---

## EXERCISES 5

**EXERCISES 1–12** □ *Determine whether or not each expression is a polynomial. For each polynomial, indicate whether it is a monomial, binomial, or trinomial and give its degree.*

1. $2a^2 - 5a + 6$
2. $-z + \sqrt{2}z^3 - 4z^5 + \sqrt{3}$
3. $\dfrac{7}{2}k^4$
4. $3x^2 - 2x + x^{1/2} - 1$
5. $2u^5 - u^{1/3} + 4u$
6. $-10$
7. $0.0005x^6 - 0.007x^3 + 0.01x^2 + 8$
8. $x^{-2}y + y^{-2}x$
9. The length of the hypotenuse of a right triangle in terms of its sides of lengths $a$ and $b$.
10. The area of a triangle in terms of its base $b$ and height $h$
11. The volume of a cylinder in terms of its base radius $r$ and height $h$
12. The surface area of a cone in terms of its base radius $r$ and height $h$

**EXERCISES 13–26** □ *Evaluate each polynomial as indicated.*

13. $4x^3 - x^2 + 2x + 1$;   $x = 2$
14. $z^4 + 3z^2 - 6z$;   $z = -1$
15. $t^2 - 2t - 5$;   $t = -\dfrac{1}{2}$
16. $9y^3 - y + 2$;   $y = \dfrac{2}{3}$
17. $\dfrac{1}{6}x^2 - x + 2$;   $x = \sqrt{2}$
18. $s^2 + 4s - 1$;   $s = \sqrt{5}$
19. $x^2 + xy + y^2$;   $x = -2, y = 3$
20. $-\dfrac{1}{2}gt^2 + v_0 t + s_0$;   $t = 2$
21. $z^2 + 4$;   $z = 2i$
22. $z^2 + 10$;   $z = 1 - 3i$
23. $p(x) = x^3 - 5x + 1$;   $p(2)$
24. $q(t) = 2t^3 + t^2 - 4$;   $q(-1)$
25. $f(x, y) = x^3 - 3x^2y + 3xy^2 - y^3$;   $f(-1, 3)$
26. $g(t, s) = t^2 - s^2 - 3t + 4s - 6$;   $g(5, 2)$

**EXERCISES 27–58** □ *Perform the indicated operation. Simplify as much as possible.*

27. $(3x^2 + 5x - 9) + (x^2 - 2x + 1)$

28. $(p^3 + 3p + 2) + (2p^3 - 3p^2 - p + 4)$

29. $(3y^3 - y + 4) - (5y^3 + 2y^2 - 6y)$

30. $(-6t^4 - t^2) - (3t^3 + 7t^2 - 1)$

31. $x\left(\frac{1}{2}x - 1\right) + 3x\left(x + \frac{5}{3}\right)$    32. $k^2\left(k - \frac{3}{2}\right) - k(k^2 + 2k)$

33. $(2a + 1)(3a - 2)$    34. $(4x - 2)(x + 6)$

35. $(4 - 3y)\left(\frac{1}{2} + 2y\right)$    36. $\left(\frac{3}{4} + m\right)(2 - 4m)$

37. $(2t - s)(2t + s)$    38. $\left(k - \frac{2}{3}\right)\left(k + \frac{2}{3}\right)$

39. $(x + \sqrt{3})(x - \sqrt{3})$    40. $(t - \sqrt{5})((t + \sqrt{5})$

41. $(2x^2 + 1)(2x^2 - 1)$    42. $(y^3 - 3x)(y^3 + 3x)$

43. $(t + 4)^2$    44. $(p - 3)^2$

45. $(3a + 2b)^2$    46. $(z^2 - 4)^2$

47. $(2x + 1)(x^2 - 6x + 2)$    48. $(t^2 + 2)(2t^3 + t^2 - 5t)$

49. $(x - y + 2)(x + y - 2)$

50. $(x - 2)(x^3 + 2x^2 + 4x + 8)$

51. $(4s^2 - s + 2)(s^2 + 3s - 1)$

52. $(2a + 3b - 1)(2a + 3b + 1)$

53. $(n + 1)(n + 2)(n + 3)$    54. $(2x + 1)(3x - 2)(1 - x)$

55. $(2a + 1)(a - 3) + (4a - 2)(a + 2)$

56. $2u(u - 5) - (u^2 - 3)(u + 1)$

57. $(a - b)^3$    58. $(a + b)^3$

**EXERCISES 59–68** □ *Perform the indicated operation. Write your answer in the form* $a + bi$.

59. $(6 + 2i) - (3 + 4i)$    60. $(9 + 7i) + (5 - 2i)$

61. $(2 + 3i)(3 + 4i)$    62. $(7 - 2i)(5 + 4i)$

63. $(\sqrt{2} - 4i)(\sqrt{2} + 4i)$    64. $(3 - i\sqrt{5})^2$

65. $i^{11}$    66. $-i^{14}$

67. $-i^{701} \cdot i^2$    68. $i^{1995}(4 - i)$

**EXERCISES 69–72** □ *The following polynomials have complex numbers as coefficients. Perform the indicated operation and simplify as much as possible.*

69. $(z + 2i)(z - 2i)$    70. $(z - 3i)(z + 3i)$

71. $(x + 2 - 3i)(x - 2 + 3i)$    72. $(x - 5 - i)(x + 5 + i)$

**EXERCISES 73–76** □ *Use Example 14 as your guide to reveal the "trick" behind each conclusion.*

73. Choose a number, double it, subtract 4, multiply by 3, divide by 6, and add 2. The result is the number you started with.

74. Choose a number, double it, subtract 1, and square the result. Now add 3 and divide by 4. Finally, subtract the product of the number you started with and one less than that number. The result is 1.

75. Choose two numbers. Subtract the smaller from the larger and multiply the result by the sum of the two numbers. Add the square of the smaller number. The result is the square of the larger number.

76. Choose two numbers. Add them together and square the result. Subtract the squares of both numbers, then divide by 2. The result is the product of the two numbers.

■ *Applications*

77. *Gas Tank Volume* The volume of a circular cylinder with base radius $r$ and height $h$ is given by the monomial $\pi r^2 h$. Find the volume of a cylindrical gas tank with a radius of 5 feet and a height of 15 feet.

78. *Gas Tank Volume* An underground gas tank is formed by attaching a hemisphere to each end of a cylinder with radius $r$ and height $h$. The volume of each hemisphere is half that of a sphere of radius $r$, or $\frac{1}{2}(\frac{4}{3}\pi r^3)$. Write out a polynomial that gives the total volume of the tank. What is the volume of a tank formed from a cylinder with a height of 15 feet and hemispherical ends of radius 5 feet?

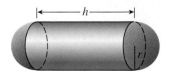

79. **Board Feet** A board foot is a unit of lumber measurement equal to 1 foot square by 1 inch thick. The number of board feet of lumber in a Ponderosa pine can be approximated by $0.0015c^3$ where $c$ is the circumference of the tree in inches at waist height. How many board feet of lumber are in a Ponderosa pine with a circumference of 60 inches? What about a Ponderosa pine with a *diameter* of 60 inches?

80. **Falling Marble** A marble is dropped from the top of the Sears Tower in Chicago. After $t$ seconds, the distance it has traveled is $16t^2$ feet, and its velocity is $32t$ feet per second. If the marble hits the ground after 9.533 seconds, how high is the Sears Tower and what is the velocity of the marble upon impact?

81. **Height of a Ball** A ball thrown in the air with a velocity of 80 feet per second has height $-16t^2 + 80t$ after $t$ seconds. What is the height of the ball after 1 second? After 2 seconds? Continue computing the height for integer values of $t$ until you find the time at which the ball hits the ground.

82. **Compound Interest** If $1500 is deposited in an account that pays interest at a rate $i$ compounded annually, the balance in the account after 3 years is given by the polynomial $1500(1 + i)^3$. Expand this polynomial to show that it has degree 3. Find the balance for interest rates of 6% and 7%.

83. **Braking Distance** Suppose you are driving on a freeway when the brake lights of the car ahead of you suddenly go on. This begins an important chain of events: your brain receives the signal to stop the car; it sends a message to your foot, which slams on the brake pedal; and the brakes dissipate energy to bring your car to a stop. From the moment the signal reaches your brain to the moment your car comes to a complete stop, your car will have traveled a certain distance. Studies have shown that this distance (measured in feet) can be approximated with the polynomial $1.1v + 0.05v^2$, where $v$ represents the speed of the car (measured in miles per hour) before the brakes are applied. How far will your car have traveled if your velocity was 65 miles per hour?

84. **Volume of a Box** A box with a square base and no top is formed by cutting squares out of the corners of a piece of cardboard and then folding up the sides (see Figure 30). If the cardboard measures $16'' \times 16''$ and the edge of each cut-out square is $x$ inches, find a polynomial that expresses the volume of the box in terms of $x$. Find the volume when $x = 1$ and again when $x = 2$.

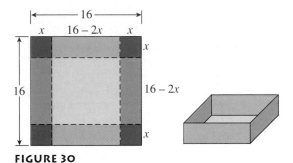

**FIGURE 30**

---

■ *Projects for Enrichment*

85. **A Mind-Reading Trick** Develop your own mind-reading trick that can be explained by using polynomial simplification.

86. **Pythagorean Triples** Three integers $a$, $b$, and $c$ are called a *Pythagorean triple* if they satisfy the Pythagorean Theorem— that is, if $a^2 + b^2 = c^2$. We denote such a triple by $(a, b, c)$. In this problem we will investigate some ways for finding Pythagorean triples.

  a. Show that the following are Pythagorean triples:

    i. (3, 4, 5)    ii. (6, 8, 10)    iii. (5, 12, 13)

  b. Find two more Pythagorean triples.

  c. Did you notice that each of the integers in the triple (6, 8, 10) is simply twice the corresponding integer in the triple (3, 4, 5)? In general, whenever $(a, b, c)$ is a Pythagorean triple, so is $(na, nb, nc)$ for any integer $n$. Show this by factoring the expression $(na)^2 + (nb)^2$ and using the fact that $a^2 + b^2 = c^2$.

The difficulty with the method in part (c) for finding new triples is that you have to know one already. One way to come up with completely new triples is to start with two arbitrary integers $m$ and $n$ and compute the quantities $m^2 - n^2$ and $2mn$. These quantities will be the first two integers of a Pythagorean triple. For example, if $m = 2$ and $n = 1$, we get $2^2 - 1^2 = 3$ and $2(2)(1) = 4$ as the first two numbers in the triple. To find the third, we can simply compute $\sqrt{3^2 + 4^2}$.

  d. Find the Pythagorean triple corresponding to

    i. $m = 3$ and $n = 1$    ii. $m = 3$ and $n = 2$

  e. Show that $(m^2 - n^2)^2 + (2mn)^2 = (m^2 + n^2)^2$ by simplifying both sides of the equation. Explain why this shows how to find the third number in the Pythagorean triple whose first two integers are $m^2 - n^2$ and $2mn$.

87. **Nested Polynomials** Not all that many years ago, a calculator with a power key such as ⌃ or $\boxed{y^x}$ would have been considered a luxury. Even today it is not uncommon to find calculators without such a key. How would one evaluate a polynomial such as $2x^5 - 3x^4 - 7x^3 + x^2 + 5x - 1$ for a given value of $x$, say $x = 1.2$, without a power key? One way would be to compute all the powers of 1.2 up to $1.2^5$ by repeated multiplication, then multiply each of the powers of 1.2 by the appropriate coefficient, and finally, add (subtract) the results to arrive at

the answer. But this would either require some pencil and paper to "store" the powers of 1.2 as you compute them—a terrible inconvenience—or you would have to recompute them as you need them. As an alternative, we will develop a method for writing the polynomial in a form that requires no exponentiation, no pencil and paper, and as few multiplications as possible. Let's begin by simplifying the expression $[(4x - 3)x + 6]x - 2$.

$$[(4x - 3)x + 6]x - 2 = [4x^2 - 3x + 6]x - 2$$
$$= 4x^3 - 3x^2 + 6x - 2$$

So the expression turns out to be a polynomial. Now compare the number of multiplications in the original expression with the number in the simplified expression. The original has only three: the first is $4x$, the second is $(4x - 3)x$, and the third is where the expression inside the brackets is multiplied by $x$. In the simplified expression, each power of $x$ requires one less multiplication than the power itself. Thus, $x^3 = x \cdot x \cdot x$ requires two multiplications, and $x^2 = x \cdot x$ requires one multiplication. We also have three multiplications for the coefficients of each power of $x$, namely $4x^3$, $3x^2$, and $6x$. That makes a total of six multiplications. So the simplified polynomial has twice as many multiplications.

**a.** Simplify the following polynomials and compare the number of multiplications.

    **i.** $[(2x - 5)x + 3]x + 7$

    **ii.** $([(3x + 4)x - 2]x + 9)x - 8$

To avoid exponents and reduce the number of multiplications, we may write polynomials in the *nested* form $(\cdots [(a_n x + a_{n-1})x + a_{n-2}]x + \cdots + a_1)x + a_0$. For example, $8x^3 + 5x^2 - 3x + 2$ can be written as follows.

$$8x^3 + 5x^2 - 3x + 2 = (8x^2 + 5x - 3)x + 2$$
$$= [(8x + 5)x - 3]x + 2$$

**b.** Write each of the following polynomials in nested form.

    **i.** $2x^5 - 3x^4 - 7x^3 + x^2 + 5x - 1$

    **ii.** $x^6 + 5x^4 - 8x^2 + 4$

**c.** Now use the nested forms to evaluate the polynomials in part (b) for $x = 1.2$.

**88.** *A Geometric Perfect Square*  The expression $(a + b)^2$ can be viewed geometrically as the area of a square with sides of length $a + b$ as illustrated below. Using only the diagram, show that $(a + b)^2 = a^2 + 2ab + b^2$.

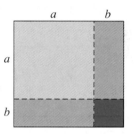

---

**■┓  *Questions for Discussion or Essay***

**89.** In Example 1 and Exercises 1–12, you identified a number of expressions as polynomials. Now explain in your own words what is *not* a polynomial. Give some examples of expressions that are not polynomials and explain why they are not. Defend or criticize the statement: One cannot understand what something is, unless one understands what something is not.

**90.** The column method was discussed in this section as a shortcut for multiplying trinomials. What advantages do you see in using such a method? Does the column method bear any resemblance to a method you have used in the past? If so, comment on the similarities and differences between the methods.

**91.** Perhaps you noticed in the examples and exercises that the sum or difference of two polynomials is again a polynomial. How does the degree of the sum or difference of two polynomials compare to the degrees of the two polynomials? Is the product of two polynomials a polynomial? If so, is there a connection between the degrees of the original polynomials and their product?

**92.** The mind-reading trick in Example 14 only works because the verbal description of the operations being performed is awkward to express in English. This awkwardness actually serves as a smoke screen that prevents the listener from "seeing" the trick. By switching to another language, algebra in this case, the smoke screen vanishes and the problem becomes almost trivial. Can you think of other examples where the language of mathematics is superior to English? Explain your examples first using words only and then using mathematics. More generally, to what extent are we limited by the language in which we think?

**93.** Explain how the distributive property is used "behind the scenes" in the FOIL method. Some mathematicians contend that such shortcuts are the essence of mathematics. Others warn that shortcuts cause important concepts to be overlooked. What do you think about using shortcuts such as the FOIL method? What other mathematical shortcuts do you know?

## FACTORING POLYNOMIALS

■ How can $6x^2 + 7x - 20$ be factored correctly with your *first* guess?

■ When is educated guessing appropriate in mathematics?

■ How can polynomial factorizations be used to find a factor of 1,000,001 without a calculator in just seconds?

■ How can $x^2 - 2$ be factored?

## COMPLETE FACTORIZATIONS

The elements in the periodic table and the prime numbers share the property that they are both basic building blocks for much more complex systems. In the case of the elements, the system is the atomic structure of our universe. The prime numbers—that is, numbers such as 2, 7, or 23 that have no factors other than 1 and themselves—are the foundation upon which the natural-number system is built. As the Greek mathematician Euclid discovered over 2000 years ago, any natural number except 1 can be written uniquely as a product of prime numbers. For example, the number 60 can be obtained by multiplying $2 \cdot 2 \cdot 3 \cdot 5$, and this is the only combination of primes that will work. The product $2 \cdot 2 \cdot 3 \cdot 5$ is called the *prime factorization* of 60.

In science as well as mathematics, we can learn a great deal about complex systems by knowing something about their basic building blocks. We have much to gain by decomposing complex structures into simpler elements, and this is certainly the case with polynomials. In fact, the process of factoring a polynomial can be described as decomposing the polynomial into a product of simpler polynomials. Some similarities between factoring polynomials and finding prime factorizations for natural numbers are shown below. The goal at each stage is to find any other factors; we stop when no more factors can be found.

| *Natural number* | *Polynomial* |
|---|---|
| 90 | $4x^2y^3 - 6xy^2$ |
| $2 \cdot 45$ | $2(2x^2y^3 - 3xy^2)$ |
| $2 \cdot 3 \cdot 15$ | $2x(2xy^3 - 3y^2)$ |
| $2 \cdot 3 \cdot 3 \cdot 5$ | $2xy^2(2xy - 3)$ |

We will consider a polynomial to be **completely factored over the integers** when it cannot be factored any further using polynomials with integer coefficients. Thus, $2xy^2(2xy - 3)$ is completely factored over the integers since no polynomial with integer coefficients can be factored out of $2xy - 3$. This complete factorization can be obtained in only one step simply by factoring out the *greatest common monomial factor* $2xy^2$. This is the most direct technique for factoring a polynomial.

| **EXAMPLE 1** | *Factoring out the greatest common monomial* |

Factor out the greatest common monomial in $12a^6 - 3a^5 + 6a^3 + 9a^2$.

**SOLUTION**    The greatest common factor of each term is $3a^2$. Factoring out $3a^2$ from each term, we get

$$12a^6 - 3a^5 + 6a^3 + 9a^2 = 3a^2(4a^4 - a^3 + 2a + 3)$$

Certain special polynomials with four or more terms can be factored by a technique known as *grouping*. Often these special polynomials will involve two variables, as is the case in the following example.

| **EXAMPLE 2** | *Factoring a polynomial by grouping* |

Factor the polynomial $xy + 2x + 3y + 6$.

**SOLUTION**    First we group the polynomial into two binomials, each of which can be factored separately. Then we apply the distributive property.

$$\begin{aligned}
xy + 2x + 3y + 6 &= (xy + 2x) + (3y + 6) \\
&= x(y + 2) + 3(y + 2) && \text{Factoring each binomial} \\
&= (x + 3)(y + 2) && \text{Factoring out the binomial } (y + 2)
\end{aligned}$$

| **EXAMPLE 3** | *Factoring a polynomial by grouping* |

Factor the polynomial $2mn^3 + 2m^2n^2 - 4mn^2 - 4m^2n$.

**SOLUTION**    Each term has a common monomial factor of $2mn$, and so we factor that out first.

$$\begin{aligned}
2mn^3 + 2m^2n^2 &- 4mn^2 - 4m^2n \\
&= 2mn[n^2 + mn - 2n - 2m] \\
&= 2mn[(n^2 + mn) - (2n + 2m)] && \text{Grouping binomials} \\
&= 2mn[n(n + m) - 2(n + m)] && \text{Factoring each binomial} \\
&= 2mn[(n - 2)(n + m)] && \text{Factoring out } (n + m) \\
&= 2mn(n - 2)(n + m)
\end{aligned}$$

> **RULE OF THUMB**    Always factor out the greatest common monomial factor before using grouping or other techniques that are discussed in this section.

## FACTORING BINOMIALS

Recall that a factorization is considered complete over the integers when no more factoring is possible using integer coefficients. Sometimes it is difficult to tell whether a polynomial is completely factored, especially if the polynomial has many terms.

However, for polynomials with three or fewer terms, there are certain clues that make the task much easier. We begin with three special binomial forms.

*Special binomial forms*

Difference of squares:  $a^2 - b^2 = (a + b)(a - b)$
Difference of cubes:   $a^3 - b^3 = (a - b)(a^2 + ab + b^2)$
Sum of cubes:      $a^3 + b^3 = (a + b)(a^2 - ab + b^2)$

**EXAMPLE 4**   *Factoring a difference of cubes*

Factor $x^3 - 27$.

**SOLUTION**

$$
\begin{aligned}
x^3 - 27 &= x^3 - 3^3 \\
&= (x - 3)(x^2 + 3x + 3^2) && \text{Applying the difference of cubes form} \\
&= (x - 3)(x^2 + 3x + 9)
\end{aligned}
$$

**EXAMPLE 5**   *Factoring a sum of cubes*

Factor $8z^3 + 125$.

**SOLUTION**

$$
\begin{aligned}
8z^3 + 125 &= (2z)^3 + 5^3 && \text{Rewriting } 8z^3 \text{ as } (2z)^3 \\
&= (2z + 5)[(2z)^2 - (2z)5 + 5^2] && \text{Applying the sum of cubes form} \\
&= (2z + 5)(4z^2 - 10z + 25) && \text{Simplifying}
\end{aligned}
$$

**EXAMPLE 6**   *Factoring a difference of squares*

Factor $100x^2 - 4y^2$.

**SOLUTION**

$$
\begin{aligned}
100x^2 - 4y^2 &= 4(25x^2 - y^2) && \text{Factoring out the common monomial} \\
&= 4[(5x)^2 - y^2] && \text{Rewriting } 25x^2 \text{ as } (5x)^2 \\
&= 4[(5x + y)(5x - y)] && \text{Applying the difference of squares form}
\end{aligned}
$$

**EXAMPLE 7**   *Factoring a binomial two different ways*

Factor $p^6 - 1$.

**SOLUTION**   The binomial $p^6 - 1$ can be viewed either as a difference of squares or a difference of cubes. We'll consider both.

Difference of squares:

$$(p^3)^2 - (1)^2 = (p^3 + 1)(p^3 - 1) \qquad \text{Factoring the difference of squares}$$
$$= [(p + 1)(p^2 - p + 1)][(p - 1)(p^2 + p + 1)] \qquad \text{Factoring the sum and difference of cubes}$$

Difference of cubes:

$$(p^2)^3 - (1)^3 = (p^2 - 1)[(p^2)^2 + p^2 + 1] \qquad \text{Factoring the difference of cubes}$$
$$= (p + 1)(p - 1)(p^4 + p^2 + 1) \qquad \text{Factoring the difference of squares}$$

In order to completely factor the difference of cubes result, you would have to recognize that $p^4 + p^2 + 1 = (p^2 - p + 1)(p^2 + p + 1)$, which is not at all obvious. So on this problem, we are better off using the difference of squares first.

## FACTORING TRINOMIALS

Some trinomials can be factored by recognizing them as *perfect-square trinomials*. A perfect-square trinomial is one that can be written as the square of a binomial, and there are two forms you should recognize. These are summarized below.

### Perfect-square trinomials

$$a^2 + 2ab + b^2 = (a + b)^2$$

$$a^2 - 2ab + b^2 = (a - b)^2$$

**EXAMPLE 8**    *Factoring a perfect-square trinomial*

Factor $x^2 - 8x + 16$.

**SOLUTION**    Notice that the outer two terms of $x^2 - 8x + 16$ are perfect squares, and the middle term (ignoring the sign) is twice the product of the square roots of the outer terms. So $x^2 - 8x + 16$ is a perfect-square trinomial.

$$x^2 - 8x + 16 = x^2 - 2(4)(x) + 4^2$$
$$= (x - 4)^2$$

**EXAMPLE 9**    *Factoring a perfect-square trinomial*

Factor $9y^2 + 30yz + 25z^2$.

**SOLUTION**    Note that $9y^2 = (3y)^2$ and $25z^2 = (5z)^2$, and the middle term $30yz$ is twice the product of $3y$ and $5z$. Thus we have

$$9y^2 + 30yz + 25z^2 = (3y)^2 + 2(3y)(5z) + (5z)^2$$
$$= (3y + 5z)^2$$

Trinomials that are not perfect squares can often be factored by "educated guessing." Here we rely on experience (and sometimes a little trial and error) to guide us to the proper choice of factors. Let's consider the trinomial $x^2 - x - 20$. We are looking for two binomial factors whose first terms multiply to give $x^2$ and whose last terms multiply to give $-20$. There are six choices:

$$(x + 1)(x - 20) \qquad (x + 2)(x - 10) \qquad (x + 4)(x - 5)$$
$$(x - 1)(x + 20) \qquad (x - 2)(x + 10) \qquad (x - 4)(x + 5)$$

If the constant terms are "too far apart," as in $(x + 1)(x - 20)$, the coefficient of the middle term of their product is much too large, $-19x$ in this case. We conclude that the constant terms of the two binomial factors should be "close together." The correct choice is

$$(x + 4)(x - 5) = x^2 - x - 20$$

Notice also that the constant terms of the two binomials are the factors of $-20$ that add up to $-1$ (the middle term). This is true in general for trinomials of the form $x^2 + bx + c$.

---

**RULE OF THUMB**    When factoring trinomials of the form $x^2 + bx + c$, look for the factors of $c$ that add up to $b$. These will be the constant terms of the binomial factors.

---

Unfortunately, the procedure is not as simple for general trinomials. However, if all else fails, we can always use trial and error to find the pair of factors that works. We'll investigate a more "scientific" method for factoring general trinomials in Exercise 51.

**EXAMPLE 10**    *Factoring a trinomial with leading coefficient* **1**

Factor $x^2 - 5x - 6$.

**SOLUTION**    Since the coefficient of $x^2$ is 1, we look for the factors of $-6$ that add up to $-5$. The correct ones are $-6$ and 1. Thus, $x^2 - 5x - 6 = (x - 6)(x + 1)$. ∎

**EXAMPLE 11**    *Factoring a trinomial with a common monomial factor*

Factor $2t^4 + 5t^3 - 3t^2$.

**SOLUTION**    First we factor out the common factor $t^2$.

$$2t^4 + 5t^3 - 3t^2 = t^2(2t^2 + 5t - 3)$$

Next we factor $2t^2 + 5t - 3$. The following are the potential factorizations of $2t^2 + 5t - 3$.

$$(2t + 1)(t - 3)$$
$$(2t - 1)(t + 3)$$
$$(2t + 3)(t - 1)$$
$$(2t - 3)(t + 1)$$

By trial and error, we discover that $2t^2 + 5t - 3 = (2t - 1)(t + 3)$. Thus, $2t^4 + 5t^3 - 3t^2 = t^2(2t - 1)(t + 3)$.

---

**EXAMPLE 12**    *Factoring a fourth-degree trinomial with leading coefficient* **1**

Factor $x^4 - 7x^2 + 12$.

**SOLUTION**    Since the leading term is $x^4$ and the middle term involves $x^2$, we are looking for a factorization of the form $(x^2 + \square)(x^2 + \square)$. The boxed terms must be factors of 12 and they must add up to $-7$; we select $-4$ and $-3$. Thus,

$$x^4 - 7x^2 + 12 = (x^2 - 4)(x^2 - 3)$$
$$= (x + 2)(x - 2)(x^2 - 3)$$

---

## EXERCISES 6

**EXERCISES 1–6** □ *Factor out the greatest common monomial.*

1. $4n^5 - 12n^3 + 24n$

2. $-16t^2 + 40t$

3. $6ab^2 - 8ab + 12a^2b$

4. $3x^2y^2 + 12x^3y - 27x^4y$

5. $w(w + 2) + 5(w + 2)$

6. $(x - 3)^2 + 4(x - 3)$

**EXERCISES 7–10** □ *Factor by grouping. You may need to regroup first.*

7. $9ab + 18b + 5a + 10$

8. $x^2 - 36y + 6xy - 6x$

9. $8p^2 - 9q^2 + 6pq - 12pq$

10. $3t^2 - 2t + 15t - 10$

**EXERCISES 11–22** □ *Factor each binomial.*

11. $x^2 - 36$

12. $t^2 - 81$

13. $4y^2 - 16z^2$

14. $-12k^2 + 27$

15. $2x^4 - 32$

16. $t^4 - 81$

17. $s^3 - 27$

18. $z^3 + 64$

19. $a^3 + 8b^3$

20. $27n^3 - m^3$

21. $2x^3 + 128$

22. $a^4b - ab^4$

**EXERCISES 23–30** □ *Factor each trinomial.*

23. $x^2 - x - 12$

24. $t^2 + 6t + 8$

25. $y^2 - 9y + 14$

26. $x^2 + 2x - 15$

27. $9m^2 - 12m + 4$

28. $4s^2 + 4s + 1$

29. $2y^2 + 5y - 3$

30. $3a^2 - 10a - 8$

**EXERCISES 31–48** □ *Factor completely.*

31. $2t^2 - 8$

32. $r^2 - rs - 2s^2$

33. $6n^2 - 19n + 10$

34. $8x^3 - 27$

35. $z^4 - 2z^2 + 1$

36. $x^2 + xy + 2x + 2y$

37. $t^6 - 64$

38. $x^2 + x - 42$

39. $p^6 - 125q^3$

40. $16a^2 + 20a + 6$

41. $81x^3 - 3$

42. $t^2 - 7t + 12$

43. $3xz + 6xw - yz - 2yw$

44. $4r^3 + 4s^3$

45. $4x^2 - 12xy + 9y^2$

46. $2ab^2 - 32ac^2$

47. $6x^3y^3 + 48$

48. $x^6 - x^3 - 20$

---

■  *Projects for Enrichment*

49. *Factoring Using Irrational and Complex Coefficients*  In all of the examples in this section, we considered a polynomial to be completely factored if it was not possible to factor it again using polynomials with integer coefficients. There are times when we may wish to factor even further, possibly using irrational numbers or even complex numbers in the factors. For example, even though the polynomial $x^2 - 2$ cannot be fac-

tored using integer coefficients, we can factor it using irrational coefficients as $(x + \sqrt{2})(x - \sqrt{2})$.

a. Factor the following polynomials into factors of degree 1, using irrational coefficients if necessary.

  i. $y^2 - 20$

  ii. $2n^2 - 7$

  iii. $4p^3 - 6p$

  iv. $a^4 - 10a^2 + 25$

Although binomials of the form $a^2 + b^2$ cannot be factored using only real coefficients, they do factor using *complex* coefficients.

   b. Use complex multiplication (see Section 3) to show that $(a - bi)(a + bi) = a^2 + b^2$.

   c. Factor the following polynomials into factors of degree 1, using complex coefficients if necessary.

      i. $x^2 + 25$         ii. $2z^4 - 32$

50. *Zeros of Polynomials* There is a strong connection between factoring a polynomial of one variable and finding its roots or zeros. A number $c$ is called a *root* or *zero* of a polynomial if the value of the polynomial is zero when $c$ is substituted for the variable. For example, 2 is a zero of the polynomial $x^3 - x^2 + x - 6$ since $2^3 - 2^2 + 2 - 6 = 0$.

   a. Show that 1 and $-3$ are zeros of $x^2 + 2x - 3$. Now factor $x^2 + 2x - 3$. Do you notice any connection between the zeros and the factors? Do you think there are any more zeros?

   b. Factor the polynomial $x^2 - 3x - 10$ to help find its zeros.

   c. The zeros of the polynomial $x^3 - 13x - 12$ are $-3$, $-1$, and 4. Use this information to factor the polynomial.

   d. Find a fourth-degree polynomial with zeros $-1$, 0, 1, and 2.

51. *A Better Educated Guess* The process of educated guessing can lead to frustration when the coefficients of the first and last terms have many factors. In such situations, the number of possible factors can be overwhelming.

   a. Three "possible" factors of $6x^2 + 7x - 20$ are listed below. How many are there all together?

$$(x + 1)(6x - 20), \quad (x - 1)(6x + 20),$$
$$(2x + 1)(3x - 20), \quad \ldots$$

Fortunately, there are other ways of obtaining factorizations for trinomials. One of these ways applies the grouping method; we'll illustrate the process with an example. Consider the trinomial $2x^2 - x - 6$. First, we multiply the coefficients of the

first and last terms, obtaining $-12$. Next, we find the pairs of factors of $-12$ that add up to $-1$, the coefficient of the middle term. The list of all possible pairs is $(-1)12$, $1(-12)$, $(-2)6$, $2(-6)$, $(-3)4$, and $3(-4)$. The pair that add up to $-1$ are $-4$ and 3. Now we write $2x^2 - x - 6$ as $2x^2 - 4x + 3x - 6$ and apply the grouping technique to obtain the following factorization.

$$
\begin{aligned}
2x^2 - x - 6 &= (2x^2 - 4x) + (3x - 6) \\
&= 2x(x - 2) + 3(x - 2) \\
&= (x - 2)(2x + 3)
\end{aligned}
$$

   b. Apply the process described above to factor the following trinomials.

      i. $3x^2 + 2x - 8$       ii. $6x^2 + 7x - 20$

      iii. $4x^2 + 4xy - 15y^2$

This process can even tell us when a trinomial cannot be factored using only integer coefficients. If none of the factors of the product of the first and last terms add up to the middle term, the trinomial is not factorable.

   c. Determine whether any of the following trinomials are not factorable using integer coefficients.

      i. $6x^2 + 14x - 9$       ii. $12x^2 - 7x + 8$

      iii. $20x^2 + 9xy - 4y^2$

You may be wondering how this mysterious factorization method works. Let's consider the product of two arbitrary binomials $(ax + b)(cx + d) = (ac)x^2 + (ad + bc)x + bd$. If we were to apply the factorization process to the trinomial $(ac)x^2 + (ad + bc)x + bd$, we would multiply the first term $ac$ by the last term $bd$ and then list all the factors, selecting the pair that add up to the middle term $ad + bc$.

   d. List all the factors of $(ac)(bd)$ and identify the two that add up to the middle term.

   e. Finish the grouping process to show that the polynomial $(ac)x^2 + (ad + bc)x + bd$ does indeed factor as $(ax + b)(cx + d)$.

**Questions for Discussion or Essay**

52. Explain how the sum of cubes factorization formula can be used to find a factor of 1,000,001 in just seconds without a calculator.

53. Compare and contrast factoring polynomials to "decomposing" chemical substances into elements. Can you think of any processes in other disciplines that involve decomposing complex structures into their basic building blocks?

54. In this section we've seen a number of special methods for factoring. Each one requires that a polynomial be in exactly the right form before the method can be applied. Discuss the importance of pattern recognition in choosing the right factoring method. Use examples to illustrate your point. Can you think of other situations in which pattern recognition is important, either in mathematics or life in general?

**SECTION 7**

# RATIONAL EXPRESSIONS

■ What possible effect could the planet Mars have on the birth of a child?

■ How can the manager of a fast-food restaurant determine the average time customers must wait in the drive-through line?

■ Why is it unlikely that we will ever live in a pollution-free world?

■ How many different ways are there to compute the average of a list of numbers?

## SIMPLIFYING RATIONAL EXPRESSIONS

A **rational expression** is simply a quotient of polynomials, such as

$$\frac{x^2 - 4}{x^2 - 3x + 2}$$

In fact, since integers are polynomials, a rational expression can be as simple as the fraction $\frac{2}{3}$. Thus, the rules for operating on rational expressions must also apply to operating on fractions. We review some of these rules below.

| Rule | Example | Comment |
|---|---|---|
| 1. $\dfrac{ac}{bc} = \dfrac{a}{b}$ | $\dfrac{16}{24} = \dfrac{2 \cdot 8}{3 \cdot 8} = \dfrac{2}{3}$ | Divide out the common factor. |
| 2. $\dfrac{a}{b} \cdot \dfrac{c}{d} = \dfrac{ac}{bd}$ | $\dfrac{4}{5} \cdot \dfrac{3}{7} = \dfrac{4 \cdot 3}{5 \cdot 7} = \dfrac{12}{35}$ | Multiply numerators and denominators. |
| 3. $\dfrac{a}{b} \div \dfrac{c}{d} = \dfrac{a}{b} \cdot \dfrac{d}{c}$ | $\dfrac{2}{3} \div \dfrac{7}{11} = \dfrac{2}{3} \cdot \dfrac{11}{7} = \dfrac{22}{21}$ | Multiply by the reciprocal of the divisor. |
| 4. $\dfrac{a}{b} + \dfrac{c}{b} = \dfrac{a + c}{b}$ | $\dfrac{4}{5} + \dfrac{7}{5} = \dfrac{4 + 7}{5} = \dfrac{11}{5}$ | Add numerators over a common denominator. |
| 5. $\dfrac{a}{b} - \dfrac{c}{b} = \dfrac{a - c}{b}$ | $\dfrac{4}{3} - \dfrac{11}{3} = \dfrac{4 - 11}{3} = \dfrac{-7}{3}$ | Subtract numerators over a common denominator. |

Note that Rule 4 may only be applied when adding fractions that have the same denominator. If the denominators are different, we must first find a *common denominator* by finding the least common multiple of the two denominators. For example, to add $\frac{5}{12}$ and $\frac{7}{20}$, we note that $12 = 3 \cdot 4$ and $20 = 4 \cdot 5$, so $3 \cdot 4 \cdot 5 = 60$ is the least common denominator. Thus,

$$\frac{5}{12} + \frac{7}{20} = \frac{5 \cdot 5}{12 \cdot 5} + \frac{7 \cdot 3}{20 \cdot 3} = \frac{25 + 21}{60} = \frac{46}{60} = \frac{23}{30}$$

The rules for operations on rational expressions are identical to those above; we simply allow the variables $a$, $b$, $c$, and $d$ to represent polynomials. Moreover, just as factoring is the key to working with fractions, so too is factoring the key to working with rational expressions.

**EXAMPLE 1**     *Simplifying a rational expression by factoring*

Simplify the rational expression $\dfrac{x^2 - 4}{x^2 - 3x + 2}$ by dividing out any common factors.

**SOLUTION**

$$\frac{x^2 - 4}{x^2 - 3x + 2} = \frac{(x + 2)(x - 2)}{(x - 1)(x - 2)} \qquad \text{Factoring both numerator and denominator}$$

$$= \frac{x + 2}{x - 1} \qquad \text{Dividing out the common factor } x - 2$$

Note that the original expression, $\dfrac{x^2 - 4}{x^2 - 3x + 2}$, is not defined for $x = 2$, whereas the simplified expression, $\dfrac{x + 2}{x + 1}$, is defined for $x = 2$. In this text, we will adopt the convention that the simplified expression is understood to be equal to the original expression only for values for which the original expression is defined.

■

**EXAMPLE 2**     *Simplifying a rational expression by factoring*

Simplify the rational expression $\dfrac{a^2 b^3 - a^2}{a^4 - a^4 b}$ by dividing out any common factors.

**SOLUTION**

$$\frac{a^2 b^3 - a^2}{a^4 - a^4 b} = \frac{a^2(b^3 - 1)}{a^4(1 - b)} \qquad \text{Factoring both numerator and denominator}$$

$$= \frac{b^3 - 1}{a^2(1 - b)} \qquad \text{Dividing out the common factor } a^2$$

$$= \frac{(b - 1)(b^2 + b + 1)}{a^2(1 - b)} \qquad \text{Factoring the numerator again}$$

$$= \frac{-(1 - b)(b^2 + b + 1)}{a^2(1 - b)} \qquad \text{Factoring } -1 \text{ out of } (b - 1)$$

$$= -\frac{b^2 + b + 1}{a^2} \qquad \text{Dividing out the common factor } (1 - b)$$

■

The final expressions in Examples 1 and 2 are said to be in **lowest terms** since no other common factors can be divided out. Whenever operations are performed on rational expressions, it is desirable to write the final answer in lowest terms. This is most easily accomplished by first factoring all terms before multiplying or dividing.

**EXAMPLE 3** *Multiplying rational expressions*

Express the product $\dfrac{y-1}{y+1} \cdot \dfrac{y^2-y-2}{3y^2-3y}$ in lowest terms.

**SOLUTION**

$$\dfrac{y-1}{y+1} \cdot \dfrac{y^2-y-2}{3y^2-3y} = \dfrac{(y-1)(y^2-y-2)}{(y+1)(3y^2-3y)}$$ 
Multiplying numerators and denominators

$$= \dfrac{(y-1)(y+1)(y-2)}{(y+1)3y(y-1)}$$ 
Factoring to check for common factors

$$= \dfrac{y-2}{3y}$$ 
Dividing out the common factors $(y+1)$ and $(y-1)$

■

**EXAMPLE 4** *Multiplying rational expressions*

Express the product $(n^2-16) \cdot \dfrac{n}{n^3-64}$ in lowest terms.

**SOLUTION**

$$(n^2-16) \cdot \dfrac{n}{n^3-64} = \dfrac{n^2-16}{1} \cdot \dfrac{n}{n^3-64}$$ 
Writing $n^2-16$ with a denominator of 1

$$= \dfrac{(n^2-16)n}{n^3-64}$$ 
Multiplying numerators and denominators

$$= \dfrac{(n-4)(n+4)n}{(n-4)(n^2+4n+16)}$$ 
Factoring to check for any common factors

$$= \dfrac{(n+4)n}{n^2+4n+16}$$ 
Dividing out the common factor $(n-4)$

$$= \dfrac{n^2+4n}{n^2+4n+16}$$ 

■

**WARNING** The final expression in Example 4 **cannot** be simplified further by dividing out $n^2+4n$. Identical expressions in the numerator and denominator of a rational expression can be divided out only if the expression is a *factor* of both numerator and denominator.

**EXAMPLE 5**     *Dividing rational expressions*

Express the quotient $\dfrac{r^2 - 36}{r^2 + r} \div \dfrac{6 - r}{r}$ in lowest terms.

**SOLUTION**

$$\dfrac{r^2 - 36}{r^2 + r} \div \dfrac{6 - r}{r} = \dfrac{r^2 - 36}{r^2 + r} \cdot \dfrac{r}{6 - r} \qquad \text{Multiplying by the reciprocal of the divisor}$$

$$= \dfrac{(r + 6)(r - 6)}{r(r + 1)} \cdot \dfrac{r}{-(r - 6)} \qquad \text{Factoring and writing } (6 - r) \text{ as } -(r - 6)$$

$$= \dfrac{r + 6}{-(r + 1)} \qquad \text{Dividing out common factors}$$

$$= -\dfrac{r + 6}{r + 1}$$

---

**EXAMPLE 6**     *Dividing rational expressions*

Express the quotient $\dfrac{\dfrac{k}{k + 4}}{k - 4}$ in lowest terms.

**SOLUTION**

$$\dfrac{\dfrac{k}{k + 4}}{k - 4} = \dfrac{\dfrac{k}{k + 4}}{\dfrac{k - 4}{1}} \qquad \text{Writing the denominator } k - 4 \text{ as a rational expression with denominator 1}$$

$$= \dfrac{k}{k + 4} \cdot \dfrac{1}{k - 4} \qquad \text{Multiplying by the reciprocal of the denominator}$$

$$= \dfrac{k}{(k + 4)(k - 4)}$$

---

## ADDING AND SUBTRACTING RATIONAL EXPRESSIONS

When adding or subtracting rational expressions such as

$$\dfrac{1}{t^2 + t} + \dfrac{2t}{t^2 - 1}$$

we must first find a common denominator. As is the case with adding fractions, we are looking for the least common multiple of the denominators. This can be found by factoring the denominators, choosing the largest power of each factor that appears, and forming the product of all the factors with the appropriate powers.

**EXAMPLE 7**   *Finding the least common denominator*

Find the least common denominator for $\dfrac{1}{x(x-2)^3}$ and $\dfrac{x+1}{x^2(x-2)^2}$.

**SOLUTION**   The largest power of $x$ is 2 and the largest power of $x-2$ is 3. Thus, the least common denominator is $x^2(x-2)^3$.

**EXAMPLE 8**   *Finding the least common denominator*

Find the least common denominator for $\dfrac{1}{u^2+2u-3}$,   $\dfrac{1}{u^2+6u+9}$,   and

$\dfrac{1}{u^2-2u+1}$.

**SOLUTION**   We first factor each denominator.

$$u^2 + 2u - 3 = (u+3)(u-1)$$
$$u^2 + 6u + 9 = (u+3)^2$$
$$u^2 - 2u + 1 = (u-1)^2$$

Now choosing the largest power of each factor, we obtain the least common denominator $(u+3)^2(u-1)^2$.

**EXAMPLE 9**   *Subtracting rational expressions*

Write the difference $\dfrac{x}{3-x} - \dfrac{x}{x-3}$ in lowest terms.

**SOLUTION**

$$\frac{x}{3-x} - \frac{x}{x-3} = \frac{x}{-(x-3)} - \frac{x}{x-3} \qquad \text{Factoring } -1 \text{ out of } 3-x$$

$$= \frac{-x}{x-3} - \frac{x}{x-3} \qquad \text{The common denominator is } x-3.$$

$$= \frac{-2x}{x-3}$$

**EXAMPLE 10**   *Adding rational expressions*

Write the sum $\dfrac{1}{t^2+t} + \dfrac{2t}{t^2-1}$ in lowest terms.

**SOLUTION**   Since $t^2 + t = t(t+1)$ and $t^2 - 1 = (t+1)(t-1)$, the least common denominator is $t(t+1)(t-1)$. Thus, we proceed as follows.

$$\frac{1}{t^2 + t} + \frac{2t}{t^2 - 1} = \frac{1}{t(t + 1)} + \frac{2t}{(t + 1)(t - 1)} \qquad \text{Factoring the denominators}$$

$$= \frac{1}{t(t + 1)} \cdot \frac{t - 1}{t - 1} + \frac{2t}{(t + 1)(t - 1)} \cdot \frac{t}{t} \qquad \text{Obtaining the common denominator}$$

$$= \frac{t - 1}{t(t + 1)(t - 1)} + \frac{2t^2}{t(t + 1)(t - 1)}$$

$$= \frac{2t^2 + t - 1}{t(t + 1)(t - 1)} \qquad \text{Adding numerators}$$

To write this expression in lowest terms, we must factor the numerator and divide out any common factors.

$$\frac{2t^2 + t - 1}{t(t + 1)(t - 1)} = \frac{(2t - 1)(t + 1)}{t(t + 1)(t - 1)}$$

$$= \frac{2t - 1}{t(t - 1)}$$

## EVALUATING RATIONAL EXPRESSIONS

In instances when we wish to evaluate a rational expression for particular values of the variable, we must be careful not to allow values that lead to a zero in the denominator. For example, in the expression

$$\frac{2}{z^2(z - 1)(z + 1)}$$

we would not want to allow $z$ to be 0, 1, or $-1$ since these values would give 0 in the denominator. Situations such as this are of particular concern in applications.

**EXAMPLE 11**    *Cleaning up pollution*

As many cities and companies have discovered in recent years, the cost of cleaning up pollution can be quite high. In fact, the cost typically increases at a greater rate as the percentage of pollutants removed gets closer to 100%. One model for the cost of removing pollutants suggests that if it costs \$1,000,000 to remove 50% of the pollutants, then to remove $x\%$ it will cost

$$\frac{1,000,000}{2 - \dfrac{x}{50}} \text{ dollars}$$

Simplify this expression and compute the cost for removing 80%, 90%, and 95% of the pollutants.

**SOLUTION**

$$\frac{1,000,000}{2 - \dfrac{x}{50}} = \frac{1,000,000}{\dfrac{100 - x}{50}}$$

$$= 1,000,000 \cdot \frac{50}{100 - x}$$

$$= \frac{50,000,000}{100 - x}$$

When $x = 80$, the cost is $\dfrac{50,000,000}{100 - 80} = \dfrac{50,000,000}{20} = 2,500,000$ dollars.

When $x = 90$, the cost is $\dfrac{50,000,000}{100 - 90} = \dfrac{50,000,000}{10} = 5,000,000$ dollars.

When $x = 95$, the cost is $\dfrac{50,000,000}{100 - 95} = \dfrac{50,000,000}{5} = 10,000,000$ dollars.

Notice that the cost increases dramatically as we get closer to 100%—it doubles going from 80 to 90%, and then doubles again going from 90 to 95%. Notice also that the cost cannot even be evaluated when $x = 100$.

## COMPLEX FRACTIONS

Often we are confronted—as we were in the original expression of the previous example and earlier in Example 6—with rational expressions whose numerators and/or denominators themselves contain rational expressions. Another example of such an expression is

$$\frac{\dfrac{1}{x} - \dfrac{1}{y}}{\dfrac{1}{x} + \dfrac{1}{y}}$$

These expressions are called *complex fractions*, and we will illustrate two methods for simplifying them. In the first we combine the fractions in the *main numerator* and the *main denominator* and then invert and multiply, much as we did in Example 6.

Main numerator:
$$\frac{1}{x} - \frac{1}{y} = \frac{y}{xy} - \frac{x}{xy}$$
$$= \frac{y - x}{xy}$$

Main denominator:
$$\frac{1}{x} + \frac{1}{y} = \frac{y}{xy} + \frac{x}{xy}$$
$$= \frac{y + x}{xy}$$

Invert and multiply:
$$\frac{\dfrac{1}{x} - \dfrac{1}{y}}{\dfrac{1}{x} + \dfrac{1}{y}} = \frac{\dfrac{y - x}{xy}}{\dfrac{y + x}{xy}}$$

$$= \frac{y - x}{xy} \cdot \frac{xy}{y + x}$$

$$= \frac{y - x}{y + x}$$

The second method is often much quicker, and it is useful when solving equations involving rational expressions. With this method, we can clear all fractions in one step simply by multiplying the main numerator and main denominator by the least common multiple of all denominators in the expression. For the complex fraction given above, we see that the least common multiple is $xy$, and so we proceed as follows.

$$\frac{\dfrac{1}{x} - \dfrac{1}{y}}{\dfrac{1}{x} + \dfrac{1}{y}} = \frac{\left(\dfrac{1}{x} - \dfrac{1}{y}\right) xy}{\left(\dfrac{1}{x} + \dfrac{1}{y}\right) xy}$$

$$= \frac{y - x}{y + x}$$

## EXERCISES 7

**EXERCISES 1–12** □ *Simplify whenever possible.*

1. $\dfrac{9ab}{15bc}$

2. $\dfrac{12xy^2z^3}{16x^2yz^2}$

3. $\dfrac{4t^6 - 10t^4}{2t^2}$

4. $\dfrac{21n^4 + 15n^3}{6n^3}$

5. $\dfrac{x^2 - 4}{x + 2}$

6. $\dfrac{4z^2 - 9}{2z + 3}$

7. $\dfrac{r^2 + 10r + 25}{2r^2 + 4r - 30}$

8. $\dfrac{3a^2 - 6a - 24}{a^2 - 6a + 8}$

9. $\dfrac{y^2 - 3y}{y - 3}$

10. $\dfrac{s^2t^3 - s^5}{st^2 - s^3}$

11. $\dfrac{x^4 - y^4}{x^3 - x^2y + xy^2 - y^3}$

12. $\dfrac{n^3 + 1}{n^2 - 1}$

**EXERCISES 13–36** □ *Perform the indicated operations and write your answer in simplest terms.*

13. $\dfrac{15x^2}{16y^3} \cdot \dfrac{12y^3}{20x^3}$

14. $\dfrac{8st^2}{25r^2t} \cdot \dfrac{10r^2t}{32s^3t^2}$

15. $\dfrac{1}{a + 1} + \dfrac{1}{a - 1}$

16. $\dfrac{u + v}{u - v} - \dfrac{u - v}{u + v}$

17. $\dfrac{x - 1}{x + 2} \cdot \dfrac{x^2 - 4}{x^2 - 1}$

18. $\dfrac{x + 3}{x + 4} \div \dfrac{x^3 + 27}{x^2 - 16}$

19. $\dfrac{1}{x^2 + 2x} - \dfrac{1}{x^2 + 4x + 4}$

20. $\dfrac{a}{a^2 - 4} + \dfrac{1}{2 - a}$

21. $(2a^2 - 8b^2) \cdot \dfrac{b - a}{a^2 - 4ab + 4b^2}$

22. $y + \dfrac{1}{y + 1}$

23. $\dfrac{1 + t}{(1 - t)^2} + \dfrac{1 - t}{t^2 - 1}$

24. $\dfrac{4x + 20}{x^2 + 8x + 16} \cdot \dfrac{x^2 + 4x}{x^2 + 7x + 10}$

25. $\dfrac{\dfrac{r - s}{r + s}}{s^2 - r^2}$

26. $\dfrac{t + 2}{t^3 + t^2} - \dfrac{1}{t + 1}$

27. $\dfrac{a+1}{a-2} \cdot \dfrac{a^2-3a+2}{a-3} \cdot \dfrac{3-a}{a^2-1}$

28. $\dfrac{x^2+2xy+y^2}{x^2-y^2} \div \dfrac{x^2-xy-2y^2}{x^2-3xy+2y^2}$

29. $\dfrac{3}{t-6} + \dfrac{3}{t+6} - \dfrac{36}{36-t^2}$

30. $\dfrac{x+5}{x^3+125} - \dfrac{1}{x+5} + \dfrac{x-5}{x^2-25}$

31. $\dfrac{\dfrac{1}{a} - b}{\dfrac{1}{b} - a}$

32. $\dfrac{\dfrac{1}{x} + \dfrac{1}{y}}{\dfrac{1}{xy}}$

33. $\dfrac{\dfrac{1}{3+h} - \dfrac{1}{3}}{h}$

34. $\dfrac{\dfrac{1}{x^2} - \dfrac{1}{4}}{x-2}$

35. $\dfrac{\dfrac{1}{x} - \dfrac{1}{c}}{x-c}$

36. $\dfrac{\dfrac{1}{x+h} - \dfrac{1}{x}}{h}$

**EXERCISES 37–38** □ *Use the operations on rational expressions to reveal the "trick" in the following number games.*

37. Subtract 10 from your age, square the result, subtract 5 times your age, divide by 20 less than your age, and add 5. The result is your age. (There is one age for which this trick fails; what is this age?)

38. Take two numbers, subtract the square of their difference from the square of their sum, and divide by 4 times either number. The result is the other number. (Under what circumstances does this trick fail?)

■ *Applications*

39. *Electrical Resistance*  If two resistors with resistance $R_1$ and $R_2$, respectively, are connected in parallel, the total resistance in the electrical circuit is given by

$$R = \dfrac{1}{\dfrac{1}{R_1} + \dfrac{1}{R_2}}$$

Show that this can be simplified to $R = \dfrac{R_1 R_2}{R_1 + R_2}$.

40. *Bouncing Ball*  A geometric series is a sum of the form $a + ar + ar^2 + ar^3 + \cdots$. One application of such a sum is in computing the total distance traveled by a bouncing ball. For example, if a ball is dropped from a height of 10 feet and it rebounds to $\frac{1}{2}$ of its previous height after each bounce, the total distance traveled would be

$$10 + 10\left(\frac{1}{2}\right)^2 + 10\left(\frac{1}{2}\right)^3 + 10\left(\frac{1}{2}\right)^4 + \cdots$$

Such a sum can be computed using a rational expression, as we'll see below.

a. Show that $\dfrac{a(1-r^2)}{1-r}$ simplifies to the sum of the first two terms of the series $a + ar + ar^2 + \cdots$.

b. Show that $\dfrac{a(1-r^3)}{1-r}$ simplifies to the sum of the first three terms of the series $a + ar + ar^2 + \cdots$.

c. What is your guess for the sum of the first four terms? Five terms? $n$ terms? How far has the ball traveled after 10 bounces?

d. It can be shown that the infinite sum

$a + ar + ar^2 + ar^3 + \cdots$ is equal to $\dfrac{a}{1-r}$ if $|r| < 1$. Use this formula to compute the total distance traveled by the ball described above.

41. *Gravitational Pull*  Astrologists claim that the alignment of celestial bodies at the time a person is born has an influence on the birth. Let's see what Newton's law of gravitation has to say about this. We will compute the gravitational pull between the planet Mars and a child born on February 12, 1995, which is the date in 1995 on which Mars was closest to Earth. According to Newton's law, the force of attraction in newtons between two bodies of masses $m_1$ and $m_2$ kilograms that are a distance $r$ meters apart is

$$\dfrac{Gm_1m_2}{r^2}$$

where $G = 6.672 \times 10^{-11}$ N·m$^2$/kg$^2$ is the universal gravitational constant.

a. Given that the mass of Mars is $6.42 \times 10^{23}$ kilograms and its distance from Earth on February 12, 1995, is $10^{11}$ meters, compute the gravitational pull between Mars and a 3.4-kilogram baby born on February 12, 1995.

b. Compute the gravitational pull between an obstetrician with mass 70 kilograms and a 3.4-kilogram baby if the obstetrician is standing 1 meter away.

c. Which of the two bodies (Mars or the obstetrician) exerts more gravitational pull on the baby?

42. *Waiting Time*  The manager of a fast-food restaurant is concerned about the amount of time a customer must wait in the drive-through line. One model suggests that, if $s$ denotes the average number of minutes for the drive-through worker to process an order and if $c$ denotes the average time between customer arrivals, then the average waiting time is given by the rational expression

$$\frac{1}{\dfrac{1}{s} - \dfrac{1}{c}}$$

(Source: *A Concrete Approach to Mathematical Modeling*, Michael Mesterton-Gibbons, Addison-Wesley, 1989.)

a. Determine the average waiting time for the following values of $c$ and $s$.

  i. $c = 10$, $s = 5$     ii. $c = 10$, $s = 6$

  iii. $c = 10$, $s = 7$   iv. $c = 10$, $s = 8$

  v. $c = 10$, $s = 9$

What happens to the waiting time as $s$ and $c$ get closer together? Does this make sense?

b. Try some examples to see what happens to the length of time if $c$ is less than $s$. Does this agree with your intuition about waiting time? Explain. What can you conclude about the waiting-time model?

## ■ Projects for Enrichment

43. *Averages*  The media and advertisers use the term "average" in many different contexts. It is not uncommon to hear or read phrases such as "the average annual income is $40,000" or "the average teenager spends 2 hours each day on the phone." What exactly is meant by the term "average," and how is it computed in each case? It may surprise you to learn that there are many different ways of measuring an average. The measure that is most commonly called an average is more properly called the *arithmetic mean*. The arithmetic mean of $n$ numbers $a_1, a_2, \ldots, a_n$ is given by

$$A = \frac{a_1 + a_2 + \cdots + a_n}{n}$$

a. During the first three basketball games of the season, Stephanie scored 12, 16, and 13 points. During the next four games, she scored 16, 22, 17, and 19 points. Compute her average for the first three games and then for the next four. How does the average of the two averages compare to the average over the seven games?

b. If $A$ is the arithmetic mean of $m$ numbers $a_1, a_2, \ldots, a_m$ and $B$ is the arithmetic mean of $n$ numbers $b_1, b_2, \ldots, b_n$, find an expression for the arithmetic mean of $A$ and $B$. Use the operations on rational expressions to show that the arithmetic mean of $A$ and $B$ is not necessarily the same as the arithmetic mean of the $m + n$ numbers $a_1, a_2, \ldots, a_m, b_1, b_2, \ldots, b_n$.

Another type of average is called the *weighted mean*. The weighted mean of $n$ numbers $a_1, a_2, \ldots, a_n$ that are weighted by the factors $w_1, w_2, \ldots, w_n$ is given by

$$W = \frac{a_1 w_1 + a_2 w_2 + \cdots + a_n w_n}{w_1 + w_2 + \cdots + w_n}$$

c. David's scores on three exams in college algebra were 76, 84, and 80, respectively. On the final, he scored 66. If each exam counts 20% and the final counts 40%, David's weighted average is

$$\frac{76 \cdot 20 + 84 \cdot 20 + 80 \cdot 20 + 66 \cdot 40}{20 + 20 + 20 + 40} = \frac{7440}{100} = 74.4$$

Find his weighted average if each exam counts 30% and the final counts 10%.

d. If each of the weighting factors is the same, $w$ say, find a simplified expression for the weighted mean of $n$ numbers $a_1, a_2, \ldots, a_n$. Does it look familiar?

A third type of average is called the *harmonic mean*. The harmonic mean of $n$ numbers $a_1, a_2, \ldots, a_n$ is given by

$$H = \frac{n}{\dfrac{1}{a_1} + \dfrac{1}{a_2} + \cdots + \dfrac{1}{a_n}}$$

e. Suppose that you and a friend are given the task of purchasing the beverages for a rather large party. You spend $48 on beverages costing $3 per 12-pack and your friend spends $48 on beverages costing $4 per 12-pack. The average price is *not* ($3 + $4)/2 = $3.50 per 12-pack. Instead, since a total of $96 is spent on $\frac{48}{3} + \frac{48}{4} = 28$ 12-packs, the actual cost is $\frac{96}{28} = $3.43$ per 12-pack. Show that the harmonic mean of $3 and $4 is equal to $3.43.

f. Show that the harmonic mean of three numbers $a$, $b$, and $c$ is $H = \dfrac{3abc}{bc + ac + ab}$.

**44. *Continued Fractions*** The term *complex fraction* was used earlier to describe rational expressions that contain fractions. A particularly interesting type of complex fraction is called a *continued fraction*. An example of a continued fraction is the expression

$$a + \cfrac{1}{b + \cfrac{1}{c}}$$

A commonly used notation for the continued fraction given above is $[a, b, c]$. For example,

$$[2, 5, 3] = 2 + \cfrac{1}{5 + \cfrac{1}{3}}$$

Any (finite) continued fraction can be written in the form $p/q$ where $p$ and $q$ are both integers. In our example from above,

$$[2, 5, 3] = 2 + \cfrac{1}{5 + \cfrac{1}{3}} = 2 + \cfrac{1}{\cfrac{16}{3}} = 2 + \cfrac{3}{16} = \frac{35}{16}$$

**a.** Simplify the expression for $[a, b, c]$ to write it in the form $p/q$ where $p$ and $q$ are both integers.

**b.** Use your simplified expression to evaluate $[6, 2, 1]$.

**c.** What is wrong with the expression $[2, 1, -1]$?

**d.** What is your guess as to the meaning of $[a, b, c, d]$? Simplify the expression you found and use it to evaluate $[1, 1, 1, 1]$.

---

■ *Questions for Discussion or Essay*

**45.** In what instances will a quotient of polynomials be a polynomial?

**46.** Because of the similarities between fractions and rational expressions, you may find that working with fractions seems easier now. Is this the case for you? It is often said about mathematics that many topics appear quite difficult at first, but by the time you see them again they seem much easier. Have you experienced this before in mathematics? Maybe you have even caught yourself making the statement ''I can't believe I couldn't figure that out before.'' Describe some mathematical topics that you found extremely challenging at first but that seemed easier to you the second time around. Have you had similar experiences in other disciplines, or is this situation unique to mathematics?

**47.** In Example 11 we looked at a mathematical model for predicting the cost of cleaning up pollution in terms of the percentage of pollutants to be removed. Do you think this model is valid? What would you say to the oil company executive who states ''Because of the *mathematically proven* unfeasibility of cleaning up 100% of the oil spill, we can only clean 85% of the spill and must leave the rest to mother nature …''? More generally, what do you think mathematical models such as this say about the likelihood of removing the existing pollution in our world?

---

**SECTION 8**

# RADICALS AND RATIONAL EXPONENTS

■ What does it mean to multiply a number by itself one-third times?

■ At what interest rate will $1 grow to $1000 in 15 years?

■ The skid marks at the scene of an accident measure 200 feet. How fast was the car traveling?

■ A woman on her way to a local restaurant to celebrate her thirtieth birthday picks up her great-great-great-great-grandson at his retirement home, and then stops off at the bank to withdraw $1,000,000 from the account that she had established by depositing $1000 on the previous day. How can this be?

## RADICALS

The volume of a cube can be obtained by cubing the side length, but how can we find the side length if the volume is given? For example, suppose that we are told that the volume of a certain cube is 125 cubic units and we are asked for the side length of the cube. The side length is a number whose cube is 125. That is, the side length is the **cube root** of 125. Since $5^3 = 125$, 5 is the cube root of 125. We define other roots in a similar fashion.

### *Definition of nth root*

> A number $y$ is an ***nth root*** of $x$ if $y^n = x$.

Since $2^2 = 4$ and also $(-2)^2 = 4$, both 2 and $-2$ are square roots of 4. The positive root, 2, is called the **principal square root** of 4. Whenever $n$ is even, there will be two $n$th roots, one positive and one negative.

### *Definition of principal root*

> For $n$ odd and $x$ any real number, the **principal *n*th root** of $x$ is the real number $y$ such that $y^n = x$. For $n$ *even* and $x > 0$, the **principal *n*th root** of $x$ is the *positive* real number $y$ such that $y^n = x$. The principal $n$th root of $x$ is denoted by $\sqrt[n]{x}$.

### *Radical notation and terminology*

> In the expression $\sqrt[n]{x}$, the symbol $\sqrt{\phantom{x}}$ is called a **radical**, the number $n$ is called the **index** of the radical, and $x$ is called the **radicand**. If no index is given, it is understood to be 2. Thus, $\sqrt{x} = \sqrt[2]{x}$.

**EXAMPLE 1**    *Evaluating radicals*

Evaluate each of the following radical expressions.

    **a.** $\sqrt[3]{27}$      **b.** $\sqrt{121}$      **c.** $\sqrt[5]{-32}$      **d.** $\sqrt[4]{\dfrac{1}{81}}$

**SOLUTION**

    **a.** $\sqrt[3]{27} = 3$, since $3^3 = 27$.

    **b.** $\sqrt{121} = 11$, since $11^2 = 121$ and $11 > 0$.

    **c.** $\sqrt[5]{-32} = -2$, since $(-2)^5 = -32$.

    **d.** $\sqrt[4]{\dfrac{1}{81}} = \dfrac{1}{3}$, since $\left(\dfrac{1}{3}\right)^4 = \dfrac{1}{81}$ and $\dfrac{1}{3} > 0$.

---

> **WARNING**    The *principal root* always refers to a single number. For example, $\sqrt{25}$ is **not** equal to $\pm 5$, but rather $\sqrt{25}$ is equal to 5.

## SIMPLIFYING RADICAL EXPRESSIONS

The rules for simplifying radical expressions can be obtained from the definition of a principal root and the properties of exponents. The most commonly used properties of radicals are listed here.

### *Properties of radicals*

Assume that $m$ and $n$ are positive integers and that $x$ and $y$ are such that all radicals are real.

| Property | Examples | Comment |
|---|---|---|
| 1. $\left(\sqrt[n]{x}\right)^n = x$ | $\left(\sqrt[3]{2}\right)^3 = 2$ | This is just the definition of *nth root*. |
| 2. For $n$ odd: <br> $\sqrt[n]{x^n} = x$ | $\sqrt[3]{7^3} = 7$ <br> $\sqrt[5]{x^{15}} = x^3$ | In general, $\sqrt[n]{x^{mn}} = \sqrt[n]{(x^m)^n} = x^m$. |
| For $n$ even: <br> $\sqrt[n]{x^n} = \lvert x \rvert$ | $\sqrt{3^2} = 3$ <br> $\sqrt[4]{(-3)^4} = 3$ <br> $\sqrt{x^2} = \lvert x \rvert$ | The absolute value arises because of the fact that the *principal root* of a positive number is positive. |
| 3. $\sqrt[n]{x}\,\sqrt[n]{y} = \sqrt[n]{xy}$ | $\sqrt{3}\,\sqrt{5} = \sqrt{15}$ | The indices of the radicals must be the same. |
| 4. $\dfrac{\sqrt[n]{x}}{\sqrt[n]{y}} = \sqrt[n]{\dfrac{x}{y}}$ | $\dfrac{\sqrt[5]{30}}{\sqrt[5]{5}} = \sqrt[5]{6}$ | The indices of the radicals must be the same. |
| 5. $\left(\sqrt[n]{x}\right)^m = \sqrt[n]{x^m}$ | $\left(\sqrt[3]{5}\right)^2 = \sqrt[3]{5^2}$ | This can de derived from Property 3. |
| 6. $\sqrt[m]{\sqrt[n]{x}} = \sqrt[mn]{x}$ | $\sqrt[4]{\sqrt[3]{x}} = \sqrt[12]{x}$ | This property reminds us of $(x^m)^n = x^{mn}$. |

Generally, a radical expression is considered to be completely simplifed whenever

- The radicand is as "small" as possible; that is, the radicand contains as few factors as possible. For example, $2\sqrt[3]{2}$ is simpler than $\sqrt[3]{16}$. In general, this means that the radicand should contain no factors that are perfect $n$th powers, where $n$ is the index.
- There are as few **nested radicals** (radicals inside of other radicals) as possible. For example, $\sqrt[15]{x}$ is simpler than $\sqrt[3]{\sqrt[5]{x}}$.
- The index of the radical is as small as possible. For example, $\sqrt{5}$ is simpler than $\sqrt[4]{25}$.
- There are no radicals in denominators. For example, $\sqrt{2}/2$ is simpler than $1/\sqrt{2}$.

| EXAMPLE 2 |

### *Simplifying radical expressions*

Simplify the following expressions.

**a.** $\sqrt{8}$       **b.** $\sqrt[3]{40}$

**SOLUTION**

**a.**
$$\sqrt{8} = \sqrt{4 \cdot 2} \qquad \text{Factoring out the largest perfect square}$$
$$= \sqrt{4}\,\sqrt{2} \qquad \text{Using Property 3}$$
$$= 2\sqrt{2} \qquad \text{Simplifying}$$

**b.**
$$\sqrt[3]{40} = \sqrt[3]{8 \cdot 5} \qquad \text{Factoring out the largest perfect cube}$$
$$= \sqrt[3]{8}\,\sqrt[3]{5} \qquad \text{Using Property 3}$$
$$= 2\sqrt[3]{5} \qquad \text{Simplifying}$$

---

**EXAMPLE 3**    *Simplifying a cube root involving variables*

Simplify $\sqrt[3]{16x^7y^9z^5}$.

**SOLUTION**

$$\sqrt[3]{16x^7y^9z^5} = \sqrt[3]{8x^6y^9z^3 \cdot 2xz^2} \qquad \text{Factoring out the largest perfect cubes}$$
$$= \sqrt[3]{8}\,\sqrt[3]{x^6}\,\sqrt[3]{y^9}\,\sqrt[3]{z^3}\,\sqrt[3]{2xz^2} \qquad \text{Using Property 3}$$
$$= \sqrt[3]{8}\,\sqrt[3]{(x^2)^3}\,\sqrt[3]{(y^3)^3}\,\sqrt[3]{z^3}\,\sqrt[3]{2xz^2}$$
$$= 2x^2y^3z\,\sqrt[3]{2xz^2} \qquad \text{Using Property 2}$$

Note that this cannot be simplified any further: the radicand contains no perfect cubes.

---

**EXAMPLE 4**    *Simplifying a square root involving variables*

Simplify $\sqrt{75a^3b^5c^6}$. Assume all variables represent positive quantities.

**SOLUTION**    We begin by factoring out the perfect squares.

$$\sqrt{75a^3b^5c^6} = \sqrt{25a^2b^4c^6 \cdot 3ab} \qquad \text{Factoring out the perfect squares}$$
$$= \sqrt{25}\,\sqrt{a^2}\,\sqrt{b^4}\,\sqrt{c^6}\,\sqrt{3ab} \qquad \text{Using Property 3}$$
$$= 5\,|a|\,|b^2|\,|c^3|\,\sqrt{3ab} \qquad \text{Using Property 2}$$
$$= 5ab^2c^3\sqrt{3ab} \qquad \begin{array}{l}\text{Assuming all variables represent}\\\text{positive quantities}\end{array}$$

---

Rational expressions with radicals in the denominator are simplified by **rationalizing the denominator**. Rationalizing the denominator consists of multiplying both numerator and denominator of an expression by a factor *chosen to eliminate any radicals in the denominator*.

**EXAMPLE 5**    *Rationalizing the denominator*

Rationalize $\dfrac{1}{\sqrt{2}}$.

**SOLUTION**

$$\frac{1}{\sqrt{2}} = \frac{1}{\sqrt{2}} \cdot \frac{\sqrt{2}}{\sqrt{2}}$$

$\sqrt{2}$ is chosen as the factor since multiplication by $\sqrt{2}$ will eliminate the radical in the denominator.

$$= \frac{\sqrt{2}}{2}$$

---

**EXAMPLE 6**    *Rationalizing the denominator*

Rationalize $\dfrac{6}{\sqrt[3]{9}}$.

**SOLUTION**

$$\frac{6}{\sqrt[3]{9}} = \frac{6}{\sqrt[3]{9}} \cdot \frac{\sqrt[3]{3}}{\sqrt[3]{3}}$$

$\sqrt[3]{3}$ is chosen as the factor to obtain a perfect cube (27) in the radicand of the denominator.

$$= \frac{6 \cdot \sqrt[3]{3}}{\sqrt[3]{27}}$$

Using Property 3

$$= \frac{6 \cdot \sqrt[3]{3}}{3}$$

Simplifying

$$= 2\sqrt[3]{3}$$

Simplifying

---

**EXAMPLE 7**    *Rationalizing the denominator by conjugate multiplication*

Simplify $\dfrac{7}{4 - \sqrt{2}}$.

**SOLUTION**    We must rationalize the denominator. If we multiply $4 - \sqrt{2}$ by its **conjugate** $4 + \sqrt{2}$, we obtain $(4 - \sqrt{2})(4 + \sqrt{2}) = 4^2 - (\sqrt{2})^2 = 16 - 2 = 14$. Thus, we multiply both numerator and denominator by $4 + \sqrt{2}$ as follows.

$$\frac{7}{4 - \sqrt{2}} = \frac{7}{4 - \sqrt{2}} \cdot \frac{4 + \sqrt{2}}{4 + \sqrt{2}}$$

Multiplying numerator and denominator by the conjugate $4 + \sqrt{2}$

$$= \frac{7 \cdot (4 + \sqrt{2})}{4^2 - (\sqrt{2})^2}$$

Using the difference of two squares formula to multiply the denominators

$$= \frac{7 \cdot (4 + \sqrt{2})}{16 - 2}$$

Simplifying

$$= \frac{7 \cdot (4 + \sqrt{2})}{14}$$

Simplifying

$$= \frac{4 + \sqrt{2}}{2}$$

Simplifying

When radical expressions are to be added or subtracted, we simplify the individual expressions first and then combine like terms.

**EXAMPLE 8**    *Addition and subtraction of radical expressions*

Simplify $\dfrac{8}{\sqrt{32}} + \dfrac{10}{\sqrt{5}} - 3\sqrt{20} - 2\sqrt{8}$.

**SOLUTION**

$$\frac{8}{\sqrt{32}} + \frac{10}{\sqrt{5}} - 3\sqrt{20} - 2\sqrt{8} = \frac{8}{\sqrt{16}\sqrt{2}} + \frac{10}{\sqrt{5}} \cdot \frac{\sqrt{5}}{\sqrt{5}} - 3\sqrt{4}\sqrt{5} - 2\sqrt{4}\sqrt{2}$$

$$= \frac{8}{4\sqrt{2}} + 2\sqrt{5} - 6\sqrt{5} - 4\sqrt{2}$$

$$= \frac{2}{\sqrt{2}} - 4\sqrt{5} - 4\sqrt{2}$$

$$= \frac{2}{\sqrt{2}} \cdot \frac{\sqrt{2}}{\sqrt{2}} - 4\sqrt{5} - 4\sqrt{2}$$

$$= \sqrt{2} - 4\sqrt{5} - 4\sqrt{2}$$

$$= -3\sqrt{2} - 4\sqrt{5}$$

---

**WARNING**   Expressions involving the roots of sums or differences don't generally simplify. For example:

$$\sqrt{x + y} \quad \text{does not equal} \quad \sqrt{x} + \sqrt{y}$$

$$\sqrt{x^2 + y^2} \quad \text{does not equal} \quad x + y$$

$$\sqrt[3]{x^3 - y^3} \quad \text{does not equal} \quad x - y$$

**SQUARE ROOTS OF NEGATIVE NUMBERS**

Recall from Section 1 that $i$ was defined to be the imaginary number such that $i^2 = -1$. Thus, we write $i = \sqrt{-1}$. This allows us to extend the definition of the principal square root to the case where the radicand is negative.

**EXAMPLE 9**    *Simplification of square roots of negative numbers*

Simplify each of the following.

**a.** $\sqrt{-16}$                                          **b.** $\sqrt{-8}$

**SOLUTION**

**a.** $\sqrt{-16} = \sqrt{16}\sqrt{-1} = 4i$                 **b.** $\sqrt{-8} = \sqrt{4}\sqrt{2}\sqrt{-1} = 2\sqrt{2}i$

## AN APPLICATION OF RADICALS

Radicals appear in many familiar mathematical applications. For example, from the Pythagorean Theorem we know that if a right triangle has sides of length $a$ and $b$ and a hypotenuse of length $c$, then $c = \sqrt{a^2 + b^2}$. Radicals also play a key role in many physics applications, such as stopping distance, periods of pendulums, the time required for an object to fall a given distance, and even Einstein's special theory of relativity.

One of the most bizarre aspects of relativity theory is *time dilation*. Although the derivation of the time-dilation formulas is well beyond the scope of this text, many time-dilation computations can be made with a knowledge of radicals. Roughly, time dilation is the tendency of physical processes to run more slowly in systems accelerated to speeds approaching that of light. Consequently, a person traveling at speeds approaching that of light would age more slowly than a counterpart on Earth. The precise relationship between Earth-time and the time of the traveler is given by

$$T = \frac{T_0}{\sqrt{1 - \dfrac{v^2}{c^2}}}$$

where $T$ is elaspsed time on Earth, $T_0$ is the time elapsed by the traveler, $v$ is the velocity of the traveler, and $c$ is the speed of light (about 186,000 miles per second).

**EXAMPLE 10**    *Time dilation*

An astronaut leaves Earth on January 1, 2001, and travels at 50% of the speed of light until the calendar on her wristwatch indicates that 2 years have elapsed. How much time has elapsed on Earth?

**SOLUTION**    Here $T_0 = 2$ years, and $v = \frac{1}{2}c$. Using the time-dilation formula, we have

$$T = \frac{2}{\sqrt{1 - \dfrac{\left(\dfrac{1}{2}c\right)^2}{c^2}}} = \frac{2}{\sqrt{1 - \dfrac{\dfrac{c^2}{4}}{c^2}}}$$

$$= \frac{2}{\sqrt{1 - \dfrac{1}{4}}} = \frac{2}{\sqrt{\dfrac{3}{4}}}$$

$$= \frac{2}{\dfrac{\sqrt{3}}{2}} = \frac{4}{\sqrt{3}} = \frac{4\sqrt{3}}{3} \approx 2.309$$

Thus, approximately 2.309 years have elapsed on Earth.

## FRACTIONAL EXPONENTS

Many of the properties of radicals bear a striking resemblance to those of exponents. For example, the property $\sqrt[m]{a}\sqrt[m]{b} = \sqrt[m]{ab}$ is very similar to the exponential property $a^m b^m = (ab)^m$. There is a good reason for this similarity—radicals can in fact be represented by expressions involving *fractional exponents*. To see this, consider the expression $2^{1/3}$. If the properties of integer exponents are to hold for fractional exponents, then we have

$$(2^{1/3})^3 = 2^{(1/3)\cdot 3}$$
$$= 2^1$$
$$= 2$$

Thus, $2^{1/3} = \sqrt[3]{2}$. More generally, we have the following definition.

**Definition of $a^{1/n}$**

For $a > 0$ and $n$ a natural number, $a^{1/n} = \sqrt[n]{a}$.

In the examples and exercises that follow, assume that all variables are positive in expressions involving fractional exponents.

**EXAMPLE 11**    *Converting from fractional exponents to radicals*

Write each of the following as a radical.
**a.** $3^{1/10}$      **b.** $2^{1/4}$      **c.** $x^{1/2}$

**SOLUTION**

**a.** $3^{1/10} = \sqrt[10]{3}$      **b.** $2^{1/4} = \sqrt[4]{2}$      **c.** $x^{1/2} = \sqrt{x}$

**EXAMPLE 12**    *Expressing radicals in exponential form*

Write the following radical expressions in exponential form.
**a.** $\sqrt{5}$      **b.** $\sqrt[3]{x}$      **c.** $\sqrt[7]{a^2}$

**SOLUTION**

**a.** $\sqrt{5} = 5^{1/2}$      **b.** $\sqrt[3]{x} = x^{1/3}$      **c.** $\sqrt[7]{a^2} = (a^2)^{1/7}$

In part (c) of the preceding example, we obtained $\sqrt[7]{a^2} = (a^2)^{1/7}$. But using the properties of exponents, this could be rewritten as $a^{2/7}$. This suggests that we *define* $a^{2/7}$ to be $\sqrt[7]{a^2}$. More generally, we have the following definition.

**Definition of $a^{m/n}$**

For $a > 0$ and $m$ and $n$ natural numbers, $a^{m/n} = \sqrt[n]{a^m}$ or $\left(\sqrt[n]{a}\right)^m$.

**EXAMPLE 13**   *Converting from fractional exponents to radicals*

Write each of the following as a radical.

**a.** $a^{3/7}$      **b.** $2^{4/3}$      **c.** $(x + y)^{16/5}$

**SOLUTION**

**a.** $a^{3/7} = \sqrt[7]{a^3}$
**b.** $2^{4/3} = \sqrt[3]{2^4}$
**c.** $(x + y)^{16/5} = \sqrt[5]{(x + y)^{16}}$

---

**EXAMPLE 14**   *Expressing radicals in exponential form*

Write each of the following radical expressions in exponential form.

**a.** $\sqrt[3]{5^2}$      **b.** $\sqrt{q^9}$      **c.** $\sqrt[4]{x^6}$

**SOLUTION**

**a.** $\sqrt[3]{5^2} = 5^{2/3}$
**b.** $\sqrt{q^9} = q^{9/2}$
**c.** $\sqrt[4]{x^6} = x^{6/4} = x^{3/2}$

---

We handle expressions with negative fractional exponents just as we handled expressions with negative integer exponents. Thus, for $a > 0$ and $m$ and $n$ natural numbers,

$$a^{-(m/n)} = \frac{1}{a^{m/n}}$$

**EXAMPLE 15**   *Simplifying an expression with a negative fractional exponent*

Evaluate $4^{-3/2}$.

**SOLUTION**

$$4^{-3/2} = \frac{1}{4^{3/2}} = \frac{1}{\sqrt{4^3}} = \frac{1}{\sqrt{64}} = \frac{1}{8}$$

---

If your calculator does not have an *n*th root key, roots can be computed by first converting the radical to a fractional exponent and then using the exponent key $\boxed{y^x}$ or $\boxed{\wedge}$. For example, $\sqrt[5]{2}$ can be computed by entering $\boxed{2}$ $\boxed{\wedge}$ $\boxed{(}$ $\boxed{1}$ $\boxed{\div}$ $\boxed{5}$ $\boxed{)}$.

We now have defined the expression $a^q$ (with $a > 0$) for all rational numbers $q$. All of the rules for integer exponents given in Section 4 hold for rational-number exponents as well. We restate them for convenience here.

**Properties of rational number exponents**

Assume $a > 0$, $b > 0$, and $p$ and $q$ are rational numbers.

| Property | Example |
|---|---|
| 1. $a^p \cdot a^q = a^{p+q}$ | $2^{1/3} \cdot 2^{4/3} = 2^{5/3}$ |
| 2. $\dfrac{a^p}{a^q} = a^{p-q}$ | $\dfrac{x^2}{x^{3/4}} = x^{2-(3/4)} = x^{5/4}$ |
| 3. $(a^p)^q = a^{pq}$ | $(x^{1/2})^{2/3} = x^{(1/2)\cdot(2/3)} = x^{1/3}$ |
| 4. $(ab)^q = a^q b^q$ | $(2x)^{5/2} = 2^{5/2} x^{5/2}$ |
| 5. $\left(\dfrac{a}{b}\right)^q = \dfrac{a^q}{b^q}$ | $\left(\dfrac{x}{3}\right)^{1/2} = \dfrac{x^{1/2}}{3^{1/2}}$ |

**EXAMPLE 16**　　*Simplifying expressions involving fractional exponents*

Simplify the following:

**a.** $x^{1/2} \cdot x^{3/4}$　　　**b.** $(a^{2/3})^{3/4}$　　　**c.** $\dfrac{p^{2/5}q^{-1/3}}{pq^{-2/3}}$

**SOLUTION**

**a.** $x^{1/2} \cdot x^{3/4} = x^{(1/2)+(3/4)}$

$\qquad\qquad = x^{5/4}$

**b.** $(a^{2/3})^{3/4} = a^{(2/3)\cdot(3/4)}$

$\qquad\qquad = a^{1/2}$

**c.** $\dfrac{p^{2/5}q^{-1/3}}{pq^{-2/3}} = p^{(2/5)-1}q^{(-1/3)-(-2/3)}$

$\qquad\qquad = p^{-3/5}q^{1/3}$

$\qquad\qquad = \dfrac{q^{1/3}}{p^{3/5}}$

---

## EXERCISES 8

**EXERCISES 1–32** □ *Simplify the given expression. Assume that all variables represent positive real numbers.*

**1.** $\sqrt{18}$

**2.** $\sqrt[3]{16}$

**3.** $\sqrt[4]{100{,}000}$

**4.** $\sqrt{5^8}$

**5.** $\sqrt[3]{3^{17}}$

**6.** $\sqrt{500}$

**7.** $\sqrt{a^2 b^5}$

**8.** $\sqrt{x^3 y^7}$

**9.** $\sqrt{28x^4 y^3 z^2}$

**10.** $\sqrt{p^9}$

**11.** $\sqrt{50a^{30}b^{20}c^{11}}$

**12.** $\sqrt{(a+b)^6}$

**13.** $\sqrt{x^7(y+z)^4}$

**14.** $\sqrt[3]{a^7}$

**15.** $\sqrt[3]{p^4 q^5 r^8}$

**16.** $\sqrt[4]{x^6 y^{13}}$

**17.** $\sqrt[3]{54s^6 t^7}$

**18.** $\sqrt[5]{32x^{11}y^{17}z^2}$

**19.** $\sqrt[3]{x^3 + 3x^2 y + 3xy^2 + y^3}$

**20.** $\sqrt{x^6 + y^6}$

**21.** $\sqrt[3]{x - y}$

**22.** $\sqrt{\dfrac{1}{5}}$

**23.** $\sqrt{\dfrac{a^4}{b^2}}$

**24.** $\dfrac{3}{\sqrt{7}}$

**25.** $\dfrac{6}{\sqrt[3]{4}}$

**26.** $\dfrac{a^2}{\sqrt[3]{a}}$

27. $\dfrac{21}{\sqrt{3}}$

28. $\dfrac{3}{\sqrt{5a}}$

29. $\dfrac{2}{3 - \sqrt{5}}$

30. $\dfrac{6}{a + \sqrt{3}}$

31. $\dfrac{a - b}{\sqrt{a} - \sqrt{b}}$

32. $\dfrac{\sqrt{x + y}}{1 - \sqrt{x + y}}$

**EXERCISES 33–36** ☐ *Rationalize the numerator. (Eliminate all radicals in the numerator.)*

33. $\dfrac{\sqrt{2}}{6}$

34. $\dfrac{\sqrt[3]{4}}{2}$

35. $\dfrac{\sqrt{x} - \sqrt{a}}{x - a}$

36. $\dfrac{1 + \sqrt{8}}{3}$

**EXERCISES 37–56** ☐ *Perform the indicated operation and simplify.*

37. $\sqrt{3} \cdot \sqrt{15}$

38. $\sqrt{5a^3b} \cdot \sqrt{10ab^3}$

39. $\sqrt[3]{2xyz} \cdot \sqrt[3]{20x^2y^4z}$

40. $(\sqrt{8})^3$

41. $\dfrac{\sqrt{10}}{\sqrt{2}}$

42. $(\sqrt[3]{a})^5$

43. $\sqrt{\dfrac{x^3y^2}{xy^3}}$

44. $\dfrac{\dfrac{\sqrt{5}}{\sqrt{5}}}{4}$

45. $(\sqrt{x} + 3)(\sqrt{x} - 3)$

46. $(\sqrt{3} + \sqrt{6})^2$

47. $(\sqrt[3]{4} - \sqrt[3]{2})^3$

48. $\sqrt{6} + \sqrt{24}$

49. $\sqrt{5} + 3\sqrt{5}$

50. $\sqrt{8} + \dfrac{1}{\sqrt{2}}$

51. $\sqrt[3]{9} - \dfrac{2}{\sqrt[3]{3}}$

52. $\sqrt{a} + \dfrac{1}{\sqrt{a}}$

53. $\sqrt{32a^3b} + \sqrt{18ab} - \sqrt{2ab}$

54. $\sqrt[3]{\sqrt{x}}$

55. $\sqrt[5]{\sqrt[3]{x^5}}$

56. $\sqrt[3]{\sqrt[4]{\sqrt[5]{x}}}$

**EXERCISES 57–64** ☐ *Convert from radical notation to exponential notation. Assume all variables are positive.*

57. $\sqrt[3]{x}$

58. $\sqrt[5]{y^2}$

59. $\sqrt{17}$

60. $\sqrt[3]{a^2}$

61. $\sqrt[4]{a^3b^6}$

62. $\sqrt[5]{(x + y)^2}$

63. $\dfrac{1}{\sqrt{x}}$

64. $\dfrac{1}{\sqrt[3]{5}}$

**EXERCISES 65–70** ☐ *Simplify the given expression.*

65. $\sqrt{-25}$

66. $\sqrt{-18}$

67. $\dfrac{2}{\sqrt{-8}}$

68. $\sqrt{-10}\sqrt{-5}$

69. $\sqrt{-18} + \sqrt{-50}$

70. $\sqrt{-12} - 2\sqrt{-27}$

**EXERCISES 71–78** ☐ *Convert the given exponential expression into radical form.*

71. $2^{3/2}$

72. $3^{-1/5}$

73. $b^{2/3}$

74. $x^{4/7}$

75. $(a - b)^{3/2}$

76. $(x^2 + y^2)^{1/2}$

77. $x^{1.3}$

78. $y^{2.5}$

**EXERCISES 79–82** ☐ *Evaluate the given exponential expression.*

79. $16^{-1/2}$

80. $27^{1/3}$

81. $32^{2/5}$

82. $8^{-2/3}$

**EXERCISES 83–94** ☐ *Simplify the given exponential expression. Express the answer in exponential form without negative exponents.*

83. $x^{1/3}x^{2/3}$

84. $a^{1/4}a^{1/3}$

85. $(x^{1/2})^{2/3}$

86. $(x^{-1/2})^{-1/3}$

87. $(81a^4b^{12})^{1/4}$

88. $\left(\dfrac{1}{8}u^6v^9\right)^{2/3}$

89. $\dfrac{z^2}{z^{3/2}}$

90. $\dfrac{x^{3/4}}{x^4}$

91. $\dfrac{x^2y^{1/3}}{x^{1/2}y}$

92. $\dfrac{(a + b)^{1/2}c}{(a + b)^{-1/2}c^{-1}}$

93. $\left(\dfrac{p^{2/3}q^{1/2}r}{p^{1/3}q^{1/4}}\right)^{-2}$

94. $(x^{1/2})^{2/3} + 2x^{-2/3}x - \dfrac{x^{4/3}}{x}$

■ **Applications**

95. *Falling Object*  If air resistance is ignored, the time in seconds required for an object to fall $s$ feet near the surface of the Earth is given by $t = \dfrac{\sqrt{s}}{4}$. How long would it take a penny dropped from an altitude of 1 mile to land?

96. *Braking Distance*  The velocity at which an automobile is traveling immediately before braking can be estimated from the braking distance using the formula $v = \sqrt{20d}$, where $v$ is the velocity of the car in miles per hour and $d$ is the distance in feet required to brake. If skid marks measure 200 feet, then how fast was the car traveling?

**97.** *Hypotenuse Length*  An isoceles right triangle with leg length $x$ is shown in Figure 31. Express the length of the hypotenuse in terms of $x$.

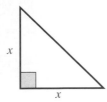

**FIGURE 31**

**98.** *Diagonal Length*  A cube with side length $x$ is shown in Figure 32. Express the length of line segment $AD$ in terms of $x$.

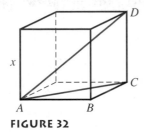

**FIGURE 32**

**99.** *Time Dilation*  Suppose that a woman deposits $1000 into a bank account earning 4% interest compounded annually and she then travels at 99.99999999% of the speed of light for 1 day.

a. Use the time-dilation formula to compute the amount of time that would have elapsed on Earth.

b. Compute the balance that she would find in her account after returning to Earth.

**100.** *Unknown Interest Rate*  If a deposit of $P$ dollars grows to a balance of $B$ dollars after $n$ years, then the annually compounded interest rate $r$ is given by

$$r = \sqrt[n]{\frac{B}{P}} - 1$$

At what interest rate will $1 grow to $1000 in 15 years?

---

### ▪️ *Projects for Enrichment*

**101.** *Heron's Formula*  Heron's formula gives the area of a triangle in terms of its side lengths. If the triangle has sides of length $a$, $b$, and $c$, and $A$ represents the area of the triangle, then

$$A = \sqrt{s(s - a)(s - b)(s - c)} \quad \text{where } s = \frac{a + b + c}{2}$$

We will derive this formula in Section 9.2.

a. Find the area of a triangle with sides of length 3, 4, and 5.

b. Find the area of a triangle with sides of length 1, 1, and 1.

c. Use Heron's formula to show that the area of an equilateral triangle of side length $x$ is given by $(\sqrt{3}/4)x^2$.

d. What is the area of a triangle with sides of length 1, 2, and 5?

e. Explain your answer to part (d).

**102.** *Technote Passing*  Suppose that a friend, located in a room 80 feet above yours, 50 feet down the hall and 35 feet across the hall, electronically transmits a note to you. Your receiver (cleverly disguised as a wristwatch) has a range of only 100 feet. Will you receive the message?

---

### ▪️ *Questions for Discussion or Essay*

**103.** Explain why $\sqrt{25} \neq \pm 5$.

**104.** Students frequently ask the question, Why is it considered simpler for the radical to be in the numerator rather than the denominator? Although we have not explicitly stated the reason for rationalizing the denominator, it can be inferred from one of the examples. Explain.

**105.** Time dilation allows us, in a sense, to travel into the future. Are there any logical difficulties with such time travel? What about traveling into the past?

**106.** It is not uncommon for algebra students to view algebra as a sort of game with arbitrary rules that the "player" must memorize. From their point of view, the statements

$\sqrt{ab} = \sqrt{a}\sqrt{b}$ and $\sqrt{a + b} = \sqrt{a} + \sqrt{b}$ seem equally valid. In fact, when told that the second equality is not in general true, they might respond, "Why not? It's just like the first one." How would you answer that question? Why isn't the second equality true? What would it take to prove that it isn't true? Who decides which algebraic statements are true and which aren't?

**107.** Some of the problems in this section look extremely difficult at first glance. Take, for example, the simplification of

$$\left(\frac{p^{2/3}q^{1/2}r}{p^{1/3}q^{1/4}}\right)^{-2}$$

Some students, however, find such problems easy (although rather long). Typically, successful algebra students remark that they solved the problem by "breaking it up into little pieces." What do they mean by this? Explain how we use this problem-solving strategy to accomplish such everyday tasks as navigating, cooking an omelette, dressing, and looking up a telephone number.

## CHAPTER REVIEW EXERCISES

**EXERCISES 1–4** □ *List all the terms that describe the given number. Choose from natural, whole, integer, rational, irrational, real, and complex.*

**1.** $-2.33$

**2.** $\dfrac{12}{3}$

**3.** A number that when squared gives $-11$

**4.** $\pi$

**EXERCISES 5–6** □ *Place the given list of numbers in ascending order. Estimate as needed.*

**5.** $1.5$, $\sqrt{2}$, $1.4$, $\sqrt{3}$, $2$

**6.** $\dfrac{2}{3}$, $0.67$, $0.66$, $\dfrac{2001}{3000}$

**EXERCISES 7–10** □ *Write the set using interval notation and sketch the graph.*

**7.** $\{x \mid -5 \le x < -1\}$

**8.** $\{x \mid x > 3\}$

**9.** $\left\{x \mid x \le \dfrac{1}{2}\right\}$

**10.** $\{x \mid \pi \le x \le 2\pi\}$

**EXERCISES 11–12** □ *Plot the given points, find the distance between them, and find the coordinates of the midpoint of the line segment connecting them.*

**11.** $(4, 5)$, $(-2, -3)$

**12.** $(-1, -2)$, $(-3, 7)$

**EXERCISES 13–14** □ *Plot the given data points and identify any trends.*

**13.**

| x | 0.7 | 1.2 | 1.9 | 2.5 | 3.2 | 3.7 |
|---|-----|-----|-----|-----|-----|-----|
| y | 2.1 | 1.5 | 1.0 | 1.2 | 2.2 | 2.6 |

**14.**

| x | 0.6 | 1.0 | 1.5 | 1.9 | 2.3 | 2.8 | 3.2 | 3.6 | 4.0 |
|---|-----|-----|-----|-----|-----|-----|-----|-----|-----|
| y | 1.9 | 2.6 | 3.0 | 3.3 | 3.5 | 3.7 | 3.8 | 3.9 | 4.0 |

**EXERCISES 15–18** □ *Answer true or false.*

**15.** The operation of subtraction is commutative.

**16.** $3 \cdot 5 + 6 \div 2 = 18$

**17.** $-x < |x|$ for all real numbers $x$

**18.** Putting on your socks and shoes is an example of a commutative operation.

**EXERCISES 19–22** □ *Perform the indicated operation and write the answer in the form $a + bi$.*

**19.** $(1 - 3i) + (4 + 2i)$

**20.** $(-2 + i) \cdot (5 - 2i)$

**21.** $\dfrac{1}{3 - i}$

**22.** $\dfrac{2 - i}{4 + 3i}$

**EXERCISES 23–26** □ *Simplify the given expression. Write your answer without using negative exponents.*

**23.** $(-3)^4$

**24.** $\left(\dfrac{4}{5}\right)^{-2}$

**25.** $\dfrac{(a^3 b)^2}{a^4 b^6}$

**26.** $\dfrac{x^{-2} y^3}{(xy^{-1})^{-3}}$

**EXERCISES 27–30** □ *Convert the given number to scientific notation.*

**27.** $31{,}400{,}000$

**28.** $0.00461$

**29.** Eleven billion

**30.** One ten-thousandth

**EXERCISES 31–32** □ *Evaluate the polynomial for the indicated value of the variable.*

**31.** $x^3 - 2x^2 - 3x$, $x = -2$

**32.** $-a^5 + 4a^2 - 4$, $a = -1$

**EXERCISES 33–38** □ *Perform the indicated operation and simplify.*

**33.** $(2t^3 + 4t) - (3t^3 + t^2 - t)$

**34.** $(8x^2 - x + 3) + (2x^2 - 4x + 6)$

**35.** $(2y + 3)(y - 4)$     **36.** $(2 - w)(4 - 3w)$

**37.** $x(x - 3) + (x + 2)(x - 1)$     **38.** $(u^2 - 3u + 7)(u - 2)$

**EXERCISES 39–46** ☐ *Factor the given polynomial.*

**39.** $9u^2 - 4$     **40.** $8x^3 + 125$

**41.** $t^2 + 4t - 32$     **42.** $2y^3x - 54x$

**43.** $9x^2 + 6x + 1$     **44.** $b^2 + 2ab + 8a + 4b$

**45.** $64s^6 - 1$     **46.** $2x^2 + 7x - 4$

**EXERCISES 47–50** ☐ *Simplify the rational expression.*

**47.** $\dfrac{x + 3}{x^2 - 9}$     **48.** $\dfrac{v^2 - v - 2}{v^2 + v - 6}$

**49.** $\dfrac{t^3 - 1}{t^3 - t}$     **50.** $\dfrac{y^2x^2 + 4y^2}{x^2y^2 + 4y^2 + 4x^2 + 16}$

**EXERCISES 51–56** ☐ *Perform the indicated operation and simplify.*

**51.** $\dfrac{3}{y + 2} - \dfrac{1}{y}$     **52.** $\dfrac{1}{x^2 + x} + \dfrac{1}{x^2 - x}$

**53.** $\dfrac{x^2}{x^2 - 9} \cdot \dfrac{3 - x}{x^2 + x}$     **54.** $\dfrac{v^3 + 8}{v^2 - 4} \div \dfrac{v^2 - 2v + 4}{v - 2}$

**55.** $\dfrac{\dfrac{1}{t} + \dfrac{1}{2}}{t + 2}$     **56.** $\dfrac{1 - \dfrac{1}{x}}{1 + \dfrac{1}{x}}$

**EXERCISES 57–62** ☐ *Simplify the given radical expression. Assume all variables represent positive numbers.*

**57.** $\sqrt[3]{a^6b^4}$     **58.** $\sqrt{32u^3v^6}$

**59.** $\dfrac{9}{\sqrt{3}}$     **60.** $\dfrac{2}{\sqrt[3]{2}}$

**61.** $\dfrac{1}{x - \sqrt{2}}$     **62.** $\dfrac{x - 4}{\sqrt{x} - 2}$

**EXERCISES 63–66** ☐ *Perform the indicated operation and simplify. Assume all variables represent positive numbers.*

**63.** $\sqrt{6s^5t} \cdot \sqrt{2st}$     **64.** $\dfrac{\sqrt[3]{54xy^2}}{\sqrt[3]{2xy}}$

**65.** $\sqrt{28} + 9\sqrt{7}$     **66.** $3\sqrt{2} - \sqrt{8}$

**EXERCISES 67–70** ☐ *Convert from radical notation to exponential notation. Assume all variables represent positive numbers. Write your answer without using negative exponents.*

**67.** $\sqrt[4]{t^2}$     **68.** $\sqrt[3]{x^2y^4}$

**69.** $\dfrac{1}{\sqrt[3]{x}}$     **70.** $\dfrac{1}{\sqrt{u^3v^4}}$

**EXERCISES 71–72** ☐ *Convert from exponential notation to radical notation.*

**71.** $x^{2/3}$     **72.** $t^{-5/2}$

**EXERCISES 73–76** ☐ *Simplify the given exponential expression. Express the answer in exponential form without negative exponents.*

**73.** $(x^{1/3})^{1/4}$     **74.** $y^{2/5}y^{3/2}$

**75.** $\dfrac{a^{2/3}b^{4/3}}{a^{5/3}b^{-2/3}}$     **76.** $\left(\dfrac{xy^{-1/2}}{x^{1/4}y}\right)^2$

**77. Simple Interest** Suppose that \$1000 is deposited in an account paying 6% simple interest. What will the balance be after 8 years?

**78. Compound Interest** If \$2500 is deposited in an account paying 6.5% interest compounded annually, what will the balance be after 10 years?

**79. Volume of Pyramid** The volume of a pyramid with height $h$ and a square base with side length $x$ is given by the monomial $\frac{1}{3}x^2h$. Find the volume of the Great Pyramid of Egypt given that the base has length 750 feet and the height is 450 feet. The dimensions of an olympic-size swimming pool are approximately $164' \times 75' \times 6\frac{1}{2}'$. How many olympic-size swimming pools could be drained into the pyramid?

The Great Pyramid at Giza near Cairo.

**80. Volume of a Box** A rectangular box with no top is to be formed by cutting squares out of the corners of a piece of cardboard and then folding up the sides (see Figure 33). If the cardboard measures $20'' \times 24''$ and the edge of each cut-out square is $x$ inches, find a polynomial that expresses the volume of the box in terms of $x$. Find the volume when $x = 2$ and again when $x = 3$.

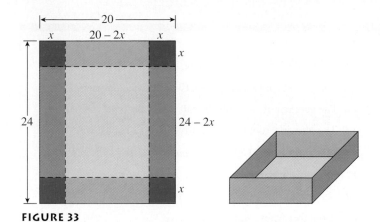

**FIGURE 33**

81. *Investment Value* Suppose an initial investment of $D$ dollars is made in some enterprise that is expected to bring in periodic cash flows of $C_1, C_2, \ldots, C_n$ at the end of each of $n$ years. If the interest rate on the investment is denoted by $i$, then the net present value of the investment at the end of the $k$th year is given by the rational expression

$$N_k = -D + \frac{C_1}{(1 + i)} + \frac{C_2}{(1 + i)^2} + \frac{C_3}{(1 + i)^3}$$
$$+ \cdots + \frac{C_k}{(1 + i)^k}$$

A negative value for $N_k$ means the enterprise is not yet profitable while a positive value for $N_k$ means it is profitable in the sense that the rate of return $i$ on the original investment has been exceeded. For an initial investment of $D = \$10,000$ and cash flows at the end of the first and second years of $C_1 = \$5000$ and $C_2 = \$7000$, find and simplify a rational expression for $N_2$ involving the variable $i$. Evaluate the expression for $i = 0.10$ (10%) and again for $i = 0.15$ (15%). For which interest rate is the enterprise profitable after 2 years?

## CHAPTER TEST

**PROBLEMS 1–6** □ *Simplify the given expression. Assume all variables represent positive numbers.*

**1.** $(-2x)^3$

**2.** $(u^{4/3}v^{-2/5})^{-1/2}$

**3.** $\left(\dfrac{a^{-3}b^4}{a^2b^{-1}}\right)^2$

**4.** $\sqrt{27x^5y^6}$

**5.** $\sqrt[4]{12a^3b^2}\;\sqrt[4]{20ab^6}$

**6.** $\dfrac{2}{\sqrt[3]{4}}$

**PROBLEMS 7–10** □ *Perform the indicated operation and simplify whenever possible.*

**7.** $(3z^4 - 2z^3) - (5z^3 - z^2)$

**8.** $(2x + 3)(3x - 2)$

**9.** $\dfrac{1}{x^2 + 2x} + \dfrac{x}{x^2 - 4}$

**10.** $\dfrac{y + 2}{y^2 - 1} \cdot \dfrac{y^2 + 2y - 3}{y^2 + 2y}$

**PROBLEMS 11–14** □ *Factor the given polynomial.*

**11.** $4x^2 - 9$

**12.** $t^2 + 2t - 15$

**13.** $u^3 + 8v^3$

**14.** $a^2 + ab - 2a - 2b$

**PROBLEMS 15–22** □ *Answer true or false.*

**15.** All rational numbers are real.

**16.** Some irrational numbers are not real.

**17.** $\pi$ is a complex number.

**18.** $-\sqrt{16}$ is a real number.

**19.** All monomials are polynomials.

**20.** The product of two polynomials is always a polynomial.

**21.** The quotient of two polynomials is always a polynomial.

**22.** The principal root of any positive number is positive.

**PROBLEMS 23–28** □ *Give an example of each.*

**23.** A rational number

**24.** An irrational number

**25.** A number whose square is negative

**26.** A polynomial in two variables with degree 5

**27.** A perfect square trinomial

**28.** A binomial

**29.** If $5000 is deposited in an account paying 5.5% interest compounded annually, what will the balance be after 20 years?

**30.** A ball thrown upward from ground level with an initial velocity of 44 feet per second has height in feet after $t$ seconds given by the polynomial $-16t^2 + 44t$. Find the height of the ball at times $t = 2$ and $t = 3$. What can you conclude about the height of the ball between 2 and 3 seconds?

**d.** Substituting 1 for $x$ in the inequality $3x + 1 < 6$, we obtain $4 < 6$, which is true. Thus, 1 is a solution of $3x + 1 < 6$.

---

The collection of all solutions of an equation or inequality is called the **solution set**. In this chapter we will describe a variety of algebraic and graphical techniques for solving—that is, finding the solution sets of—equations and inequalities.

## EQUIVALENT STATEMENTS

A pair of equations is said to be **equivalent** if the equations have the same solution sets. We will use the symbol $\Leftrightarrow$ to denote equivalence. For example, the statements $3x = 3$ and $x = 1$ are equivalent because both have $\{1\}$ as their solution set. Thus, we would write $3x = 3 \Leftrightarrow x = 1$. If the statements have different solution sets, then they are said to be **nonequivalent**. We will use the symbol $\not\Leftrightarrow$ to denote nonequivalence.

**EXAMPLE 2**

*Determining equivalence of statements*

Classify each of the following pairs of statements as being equivalent or nonequivalent.

**a.** $2x = 4$, $x = 2$    **b.** $3x + 5 = 17$, $3x = 12$

**c.** $x^2 = 9$, $x = 3$    **d.** $x + 1 < 2$, $x < 1$

**SOLUTION**

**a.** $2x = 4 \Leftrightarrow x = 2$ since the solution set of both equations is $\{2\}$.

**b.** $3x + 5 = 17 \Leftrightarrow 3x = 12$ since the solution set for both equations is $\{4\}$.

**c.** $x^2 = 9 \not\Leftrightarrow x = 3$ since the solution set of $x^2 = 9$ is $\{-3, 3\}$, whereas the solution set of $x = 3$ is $\{3\}$.

**d.** $x + 1 < 2 \Leftrightarrow x < 1$. The solution set of both inequalities is the set of all real numbers less than 1, or in interval notation, $(-\infty, 1)$.

---

Our primary technique for solving equations is to generate successively simpler, but equivalent, statements. The most common method for generating an equivalent statement from an equation is to perform the same arithmetic operation on both sides. For example, to solve the equation $x + 5 = 8$, we would subtract 5 from both sides to produce the equivalent (but simpler) equation $x = 3$. The following is a list of properties of equations that can be used to produce equivalent statements.

*Properties of equations*

| | | |
|---|---|---|
| Addition property | $x = y \iff x + c = y + c$ | You can add the same number to both sides. |
| Subtraction property | $x = y \iff x - c = y - c$ | You can subtract the same number from both sides. |
| Multiplication property | $x = y \iff cx = cy \quad (c \neq 0)$ | You can multiply both sides by the same nonzero number. |
| Division property | $x = y \iff \dfrac{x}{c} = \dfrac{y}{c} \quad (c \neq 0)$ | You can divide both sides by the same nonzero number. |

| EXAMPLE 3 | *Using properties of equality to solve equations* |

Indicate the property that is being used at each numbered step in the solution of the following equation.

$$\frac{4x + 1}{3} = 5x - 7$$

Step 1    $4x + 1 = 15x - 21$
Step 2    $4x + 22 = 15x$
Step 3         $22 = 11x$
Step 4          $2 = x$

**SOLUTION**

Step 1    Here both sides were multiplied by 3; the *multiplication property* was used.

Step 2    21 was added to both sides; the *addition property* was used.

Step 3    $4x$ was subtracted from both sides; the *subtraction property* was used.

Step 4    Both sides were divided by 11; the *division property* was used.

Special care must be taken when using the division property. For example, if both sides of the equation $x^2 = x$ are divided by $x$, the equation $x = 1$ is obtained. But $x = 1 \nLeftrightarrow x^2 = x$ since $x^2 = x$ has two solutions: 0 and 1. The problem with dividing by $x$ is that if $x$ assumes the value 0, we would be dividing by zero, and division by zero is forbidden. Thus, it is better to avoid division by variable expressions altogether.

> **WARNING**    Avoid dividing both sides of an equation by an expression involving a variable. The division property only applies when we divide by expressions that cannot be zero.

## APPROXIMATING SOLUTIONS GRAPHICALLY

There are many equations and inequalities that are extremely difficult—or even impossible—to solve by purely algebraic techniques. In such cases, we may still be able to *approximate* a solution with the aid of a graphing utility such as a graphics calculator or a computer. To do so, we must first understand the connection between graphs of equations and solutions. We begin by giving a precise definition of the graph of an equation.

### *The graph of an equation*

> The **graph of an equation** involving $x$ and $y$ is the set of all points $(x, y)$ satisfying the equation.

For example, the graph of the equation $y = x^2$ consists of all points of the form $(x, y)$ where $y = x^2$. In other words, the graph of $y = x^2$ consists of points of the form $(x, x^2)$. Thus, $(-2, 4)$, $(-1, 1)$, $(0, 0)$, $(1, 1)$, and $(2, 4)$ are all points on the graph of $y = x^2$. The graph of $y = x^2$ is shown in Figure 1.

The **$x$-intercepts of a graph** are the points where the graph intersects the $x$-axis. In other words, an $x$-intercept of a graph is a point on the graph with $y$-coordinate 0. Thus, the graph of $y = x^2$, which is shown in Figure 1, has exactly one $x$-intercept, $(0, 0)$.

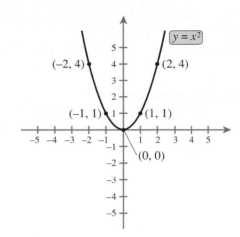

**FIGURE 1**

---

**EXAMPLE 4**   *Identifying x-intercepts*

Approximate the $x$-intercepts of the graph of the following equation, whose graph is shown in Figure 2.

$$y = \frac{2x^3 - 5x^2 - 21x + 36}{20}$$

**SOLUTION**   The graph appears to intersect the $x$-axis at the points $(-3, 0)$, $(1.5, 0)$, and $(4, 0)$. Thus, the $x$-intercepts are approximately $-3$, $1.5$, and $4$.   ∎

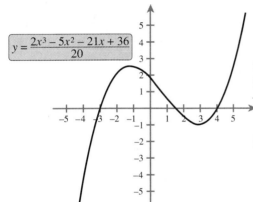

**FIGURE 2**

In the previous example we found values of $x$ such that the $y$-coordinate, $(2x^3 - 5x^2 - 21x + 36)/20$, is equal to 0. Thus, we have found solutions to the equation

$$\frac{2x^3 - 5x^2 - 21x + 36}{20} = 0$$

This suggests the following strategy for solving equations graphically.

---

■ **Steps for solving equations graphically**

1. Algebraically rearrange the equation so that it is of the form □ = 0, where □ is an expression involving the variable.
2. Graph the equation $y = $ □.
3. Find the $x$-intercepts of the graph. These are the solutions of the original equation.

---

The graphical approach to solving equations is especially convenient with the aid of a graphics calculator. However, it can still be a challenging task to adjust the view shown by the calculator so that the intercepts are clearly visible. Although a few graphics calculators have the capability of automatically finding $x$-intercepts, even they require that one select an appropriate **viewing rectangle**—that is, the portion of the graph that is shown on the calculator screen. Once a viewing rectangle is selected so that a given intercept is conspicuous, we must still determine the coordinates of the intercept to the desired degree of accuracy.

**RANGE** With all graphics calculators, the viewing rectangle can be selected by simply indicating the range of $x$-values and the range of $y$-values that are to be shown. This is done by adjusting the values of the variables Xmin, Xmax, Ymin, and Ymax, which indicate the minimum and maximum values of $x$ and $y$, respectively, as suggested by Figure 3. The first step to producing a graph is to choose appropriate initial settings for the range variables.

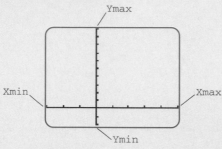

**FIGURE 3**

**ZOOM-IN** This provides us with a close-up view of a region of interest. By zooming in on an intercept repeatedly, we can approximate it with great accuracy. Figures 4 and 5 show the graph of $y = x^2 - 5$ before and after zooming in on the positive $x$-intercept.

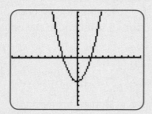

**FIGURE 4**

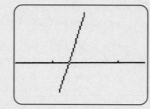

**FIGURE 5**

**ZOOM-OUT** Zooming out is the opposite of zooming in. When we zoom-out, we obtain a broader view of the graph; it is as if we are viewing the graph from a greater distance. For example, Figure 6 shows a view of the graph of $y = 0.05x^2 - 8$ in which no intercept is visible. After zooming out (Figure 7), we see two $x$-intercepts.

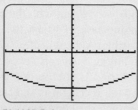

**FIGURE 6**

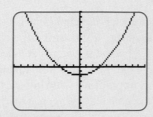

**FIGURE 7**

**ZOOM-BOX** With the zoom-box feature we can draw a box around a region of interest that will then become our viewing rectangle. Typically, the box is drawn by indicating the locations of diagonally opposite corners of the box. Figure 8 shows the graph of an equation that appears to have an $x$-intercept near $x = 4$. Figure 9 shows the zoom box that we have drawn around the region in which the $x$-intercept appears to lie. Figure 10 shows the viewing rectangle that results from the zoom box of Figure 9. Note that there are actually three $x$-intercepts!

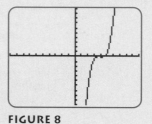

**FIGURE 8**

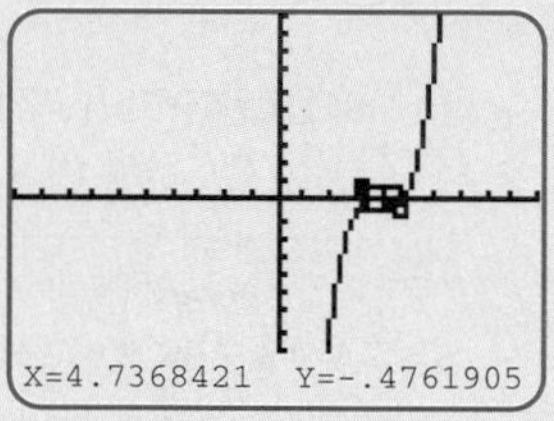

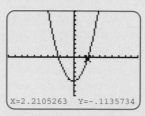

**FIGURE 9**

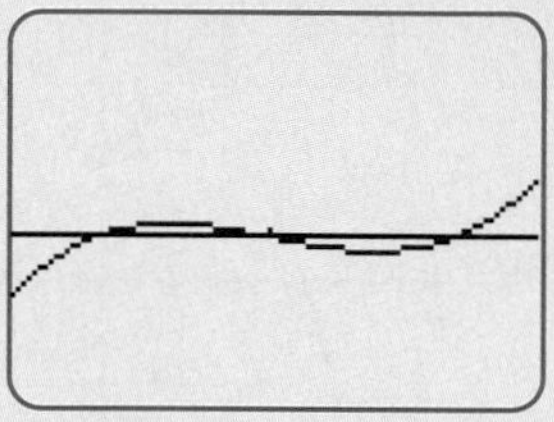

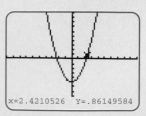

**FIGURE 10**

**TRACE** When we select the trace feature, the cursor traces points on the graph within the viewing rectangle, and the coordinates of the cursor are displayed. The trace feature is especially useful for estimating intercepts. Figure 11 shows a graph of $y = x^2 - 5$ in which the trace feature has been enabled. The cursor has been moved just to the left of the $x$-intercept, and the $x$-coordinate of the cursor is displayed as 2.2105263. In Figure 12 the cursor has been moved just to the right of the $x$-intercept, and the $x$-coordinate of the cursor is displayed as 2.4210526. Thus, the $x$-intercept lies between 2.2105263 and 2.4210526.

**FIGURE 11**

**FIGURE 12**

In the box on the preceding page, we explain some of the features of graphics calculators that are invaluable for finding intercepts. Check the graphics calculator supplement for step-by-step instructions for using these features on your calculator.

**EXAMPLE 5**    *Solving an equation graphically*

Solve $x^3 - 7x^2 = 14 - 17x$ graphically. Approximate to the nearest hundredth.

**SOLUTION**    We begin by rewriting the equation so that it is of the form $\square = 0$.

$$x^3 - 7x^2 = 14 - 17x$$

$$x^3 - 7x^2 + 17x - 14 = 0 \qquad \text{Subtracting 14 and adding } 17x \text{ to both sides}$$

**FIGURE 13**

Next we use a graphics calculator to produce a graph (such as the one shown in Figure 13) of the equation $y = x^3 - 7x^2 + 17x - 14$. Although some graphics calculators have commands like "Solve" that will automatically approximate $x$-intercepts within the viewing rectangle, we will approximate the $x$-intercepts using only the zoom and trace features.

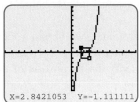

X=2.8421053   Y=-1.111111

**FIGURE 14**

There appears to be an $x$-intercept somewhere between 1 and 3. We will zoom-in on the intercept by placing a zoom box around the region where the graph appears to cross the $x$-axis, as in Figure 14. The result is the viewing rectangle shown in Figure 15.

Next we use the trace feature to determine the coordinates of the $x$-intercept. Figures 16 and 17 show the cursor just below and just above the $x$-axis, respectively. Notice that the $y$-coordinate changes from negative to positive.

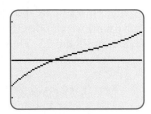

**FIGURE 15**

Thus, the intercept is between 1.9911357 and 2.0044321. The hundredths decimal place is still in question. We can gain further accuracy by zooming in a second time; this time we'll use the zoom-in feature. After zooming in near the intercept, we enable the trace feature and again move the cursor so that it is first just below the $x$-axis and then just above the $x$-axis, as shown in Figures 18 and 19.

Thus, the $x$-intercept is between 1.9997784 and 2.001108. It follows that there is an $x$-intercept at 2.00, accurate to the nearest hundredth.

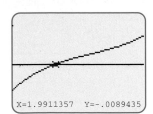

X=1.9911357   Y=-.0089435

**FIGURE 16**

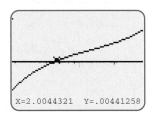

X=2.0044321   Y=.00441258

**FIGURE 17**

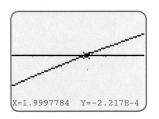

X=1.9997784   Y=-2.217E-4

**FIGURE 18**

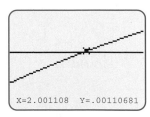

X=2.001108   Y=.00110681

**FIGURE 19**

There is a variation on the graphical technique for solving equations that can be used to solve equations like that of Example 5. If we graph $y = x^3 - 7x^2$ and $y = 14 - 17x$ on the same set of coordinate axes, then the point of intersection of the two graphs is a point $(x, y)$ such that $y = x^3 - 7x^2$ *and* $y = 14 - 17x$. Thus, $x^3 - 7x^2 = 14 - 17x$. In other words, the $x$-coordinate of the point of intersection of the two graphs is a solution to the original equation. This technique is illustrated in Figure 20.

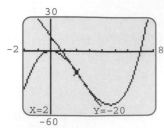

**FIGURE 20**

Although it is generally easier to rearrange the equation so that it is of the form $\square = 0$ and then find the *x*-intercepts, as demonstrated in Example 5, there are occasions when it is more convenient to graph the left and right sides of the equation and then find their point(s) of intersection.

> **RULE OF THUMB**   Initial values for the range settings are often clear from the context. For example, if *x* represents a percentage score on an exam, then set `Xmin = 0` and `Xmax = 100`. If you are not sure what the initial values for the range variables should be, select your calculator's **default** settings. For example, the calculator graphics for this text were produced using a model with default settings of `Xmin = −10`, `Xmax = 10`, `Ymin = −10`, and `Ymax = 10`.

**EXAMPLE 6**   *Solving an equation graphically*

Approximate two solutions of $y = \dfrac{x^2 - 4{,}000{,}000}{1000}$ to the nearest integer value.

**SOLUTION**   With the range variables set to their default values, we obtain a view of the graph like the one in Figure 21. The graph is completely out of view! To remedy this, we take a bird's eye view by zooming out repeatedly until a portion of the graph appears with two visible *x*-intercepts as shown in Figure 22. We now zoom-in on the negative intercept and use the trace feature to approximate its value to the nearest integer, as shown in Figure 23. Thus, one of the two solutions is approximately $-2236$.

Next we zoom-out again so that the positive intercept is again in view, as shown in Figure 24, and zoom-in on the positive intercept repeatedly until we can determine its value to the nearest integer using the trace feature, as shown in Figure 25. Thus, the other solution is approximately 2236.

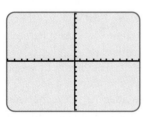

**FIGURE 21**

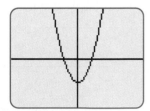

**FIGURE 22**

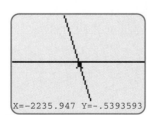

**FIGURE 23**

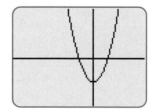

**FIGURE 24**

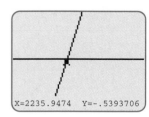

**FIGURE 25**

---

## EXERCISES 1

**EXERCISES 1–20**  $\square$  *Determine whether or not the given value is in the solution set of the equation.*

**1.** $5x + 1 = 6$; $x = 1$

**2.** $3y - 1 = 6$; $y = \dfrac{7}{3}$

**3.** $4t - 3 = 2t + 7$; $t = 4$

**4.** $\dfrac{x + 3}{2} = \dfrac{4x + 1}{3}$; $x = -1$

**5.** $0.3s + 0.5 = 0.2(s - 1)$; $s = -7$

**6.** $0.4(u + 1) = 0.3(2u - 1)$; $u = 3.5$

**7.** $(3 \times 10^7)x + (2 \times 10^8) = 5 \times 10^8$; $x = 10$

**8.** $(2.3 \times 10^{-4})r + 0.3 = (1.1 \times 10^{-4})r$; $r = -2.5 \times 10^{-4}$

**9.** $z^2 + 5z = -6$; $z = -2$      **10.** $x^2 - 9 = 0$; $x = -3$

11. $|3 - 4u| = |2u - 1|$;  $u = 1$

12. $|2 - a^2| = a^2 - 2a + 4$;  $a = 3$

13. $3t^3 + 4t^2 - t = 0$;  $t = -1$

14. $(y + 3)(y^2 + 1) = 0$;  $y = -3$

15. $x^2 = a$;  $x = \pm \sqrt{a}, a > 0$

16. $as + b = c$;  $s = \dfrac{c - b}{a}, a \neq 0$

17. $\sqrt{z^2 - 6z + 9} = \sqrt{2z} - 1$;  $z = 2$

18. $\sqrt{(x + 2)^2} = x + 2$;  $x = -4$

19. $\dfrac{6}{u - 2} = u - 1$;  $u = 4$       20. $\dfrac{3x^2}{x + 1} = \dfrac{1}{2}$;  $x = \dfrac{1}{2}$

**EXERCISES 21–30** □ *Indicate whether or not each pair of equations is equivalent. Justify your answer.*

21. $x(x + 1) = 0$
    $x + 1 = 0$

22. $2x + 3 = 0$
    $2x = -3$

23. $x^2 = 9$
    $x = 3$

24. $-x = -1$
    $x = 1$

25. $2x + 1 = 4$
    $2x = 3$

26. $(x - 5)^2 = 16$
    $x - 5 = 4$

27. $\sqrt{x} = -1$
    $(\sqrt{x})^2 = (-1)^2$

28. $x = 4$
    $\dfrac{1}{x} = \dfrac{1}{4}$

29. $7x = 3$
    $x = \dfrac{3}{7}$

30. $(x - 5)(x + 2) = 0$
    $x + 2 = 0$

**EXERCISES 31–34** □ *An equation is solved using the properties of equality. Indicate which property is being used at each lettered step.*

31.       $3x + 3 = 5$
a.          $3x = 2$
b.          $x = \dfrac{2}{3}$

32.       $-6x = 3 = -15$
a.          $-6x = -18$
b.            $x = 3$

33.          $\dfrac{2x + 3}{4} = \dfrac{3x - 1}{5}$

a.   $20\left(\dfrac{2x + 3}{4}\right) = 20\left(\dfrac{3x - 1}{5}\right)$
          $5(2x + 3) = 4(3x - 1)$
          $10x + 15 = 12x - 4$
b.       $-2x + 15 = -4$
c.          $-2x = -19$
d.             $x = \dfrac{19}{2}$

34.       $(2x - 3)(3x + 7) = (6x + 1)(x - 4)$
    $6x^2 - 9x + 14x - 21 = 6x^2 + x - 24x - 4$
       $6x^2 + 5x - 21 = 6x^2 - 23x - 4$
a.          $5x - 21 = -23x - 4$
b.          $28x - 21 = -4$
c.             $28x = 17$
d.               $x = \dfrac{17}{28}$

**EXERCISES 35–42** □ *An equation and the number of solutions is given. Use a graphics calculator to approximate the solutions to the nearest tenth.*

35. $x^3 - 6x^2 + 14x - 11 = 0$   (1 solution)

36. $x^3 - 4x^2 = 4x - 16$   (3 solutions)

37. $x^3 + 15x^2 + 88x + 100$   (3 solutions)

38. $x^3 - 2x + 1 = 0$   (3 solutions)

39. $\dfrac{7x}{x + 3} = \dfrac{4x - 1}{x - 1}$   (2 solutions)

40. $\sqrt{3x + 7} = \sqrt{x + 2} + 1$   (2 solutions)

41. $x^3 - 6x^2 = \dfrac{198}{25} - \dfrac{299}{25}x$   (3 solutions)

42. $(x - 10)^3 = \dfrac{x - 10}{25}$   (3 solutions)

■ *Applications*

43. *Hang Time*  When a team mascot jumps off of a 2-foot-high trampoline as part of a slam-dunking half-time show, her initial velocity is 20 feet per second. It can be shown that her height in feet $t$ seconds after jumping is given by $h = -16t^2 + 20t + 2$. Find the mascot's hang time—that is, the length of time she is airborne.

44. *Breaking Even*  A retail outlet specializing in the sale of trail mix estimates that when $x$ bags of the mix are sold, a profit of $x^3 - 6x^2 + 12x - 58$ dollars is earned. Estimate the number

of bags that must be sold for the retailer to break even —that is, for the profit to be zero.

45. *Unknown Interest*  An investor deposits $500 into a savings account on January 1, 1990. Her only other transaction is a deposit of $200 on January 1, 1992. The balance in the account as of January 1, 1995, was $1000. Assuming the bank pays interest compounded annually and that the interest rate has been constant, compute the interest rate. (*Hint:* Use the compound interest formula to find an expression for the accumu-

lated value of each of the deposits. Set the sum equal to 1000 and solve.)

**46.** *Snowman Dimensions* A snowman is constructed by placing three spherical snowballs on top of one another. The bottom sphere has a radius $\frac{1}{2}$ foot greater than that of the middle sphere, which, in turn, has a radius $\frac{1}{2}$ foot greater than that of the top sphere. The snowman sits in a barrel, which collects the water as the snowman melts. Based on the volume of water in the barrel, it is estimated that the snowman had a volume of 20 cubic feet. Find the radius of each of the spherical snowballs. (*Hint:* The volume of a sphere of radius $r$ is $\frac{4}{3}\pi r^3$. If $r$ represents the radius of the smallest snowball, then what are the radii of the other spheres?)

---

### ◼ Projects for Enrichment

**47.** *The Case of the Wandering Amnesiac* An amnesiac wanders into Inspector Magill's office, desperate for help in establishing his identity. The following exchange takes place.

Amnesiac:  I have absolutely no idea who I am. I woke up in a strange hotel room in this city with no idea how I got there or why. You are my last hope.

Magill:  Did you check for identification in your hotel room? Perhaps you have a driver's license, or maybe your name would be in the hotel's register.

Amnesiac:  I don't remember the name of the hotel, what room I was in, or even what street it was on. I was so panic-stricken that I simply wandered the streets randomly until I reached the police station. All I remember is that when I left my hotel room I turned right, passed five rooms and entered a stairwell. I walked down five flights of stairs, walked out of the lobby, and turned towards the setting Sun. I walked 2 miles, turned left, walked 3 miles, took another left, walked 100 yards, and turned right into the station.

Magill:  You have a pretty good memory for an amnesiac! Come with me, I'll take you to your hotel room.

**a.** How did Inspector Magill find the hotel room?

**b.** The value 30 is obtained from a given number by the following process: The given number is multiplied by 3, and 9 is then subtracted from this result. This sum is then cubed. Finally, 4 is added to the result. Find the given number, and explain how this problem can be solved in the same manner that Inspector Magill solved the mystery of the missing hotel room.

**48.** *Cardinal Numbers* Solution sets of equations vary tremendously in terms of their size, or **cardinality**. For example, the equation $x = 2$ has exactly one solution, $x = x + 1$ has no solutions, and the solution set of $x = x$ consists of all real numbers. We define the **cardinal number** of a finite set $A$ to be the number of elements in $A$, and we denote this number by $n(A)$. Thus, for example, if $A = \{0, 1, -1\}$, then $n(A) = 3$. For each of parts (a) through (d), find the cardinal number of the indicated set.

**a.** $\{0, 1, \ldots, 10\}$

**b.** {The prime numbers less than 20} (A prime number is any natural number that has no factors other than 1 and itself. The first two prime numbers are 2 and 3.)

**c.** The solution set of $|x| = 4$

**d.** The solution set of $|x| = -4$

For infinite sets, cardinal numbers are not as easy to define. We must first define the notion of equivalence. Two sets $A$ and $B$ are said to be **equivalent** if their elements can be put in a one-to-one correspondence with each other. For example, the sets $\mathbb{N} = \{1, 2, 3, \ldots\}$ (the natural numbers) and $B = \{2, 4, 6, \ldots\}$ are equivalent because of the correspondence illustrated below.

$$1, \; 2, \; 3, \ldots$$
$$\updownarrow \; \updownarrow \; \updownarrow$$
$$2, \; 4, \; 6, \ldots$$

Any set that is equivalent to $\mathbb{N}$ is said to be an **infinite countable** set. The cardinal number of an infinite countable set is denoted with the symbol $\aleph_0$ (read "aleph nought" where aleph is the first letter of the Hebrew alphabet). The set of real numbers $\mathbb{R}$ is *not* equivalent to the set of natural numbers and so it is *not* countable. In some sense, there are many more real numbers than there are natural numbers. However, and this is a seemingly odd feature of equivalence, any interval of real numbers is equivalent to $\mathbb{R}$. As a consequence, there are just as many real numbers in the interval $(0, 1)$ as there are in $\mathbb{R}$ itself! The cardinality of any set that is equivalent to $\mathbb{R}$ is denoted by $c$. In parts (e) through (h), determine the cardinality of the given set.

e. $\{1, 3, 5, \dots\}$

f. $(-1, 1)$

g. The set of Fibonacci numbers $1, 1, 2, 3, 5, 8, 13, \dots$ where each number after the second is the sum of the previous two numbers

h. The set of real numbers between $0.99998$ and $0.99999$

49. *Implications*  Statements are said to be *equivalent* if either both are true or both are false; that is, the first statement is true if and only if the second is true. For example, the statements, ''Tina works in the most populous city in California,'' and ''Tina works in Los Angeles'' are equivalent since the first statement is true if and only if the second is true. The mathematical symbol for equivalence is ''$\Leftrightarrow$.'' If statements $A$ and $B$ are equivalent, then we write $A \Leftrightarrow B$.

If the truth of statement $A$ implies the truth of statement $B$, but not necessarily vice versa, then we write $A \Rightarrow B$. Similarly, if the truth of $B$ implies the truth of $A$, then we write $A \Leftarrow B$. For example, consider the following statements.

$A$: I am a father.

$B$: I am a parent.

Clearly all fathers are parents, but not all parents are fathers. Thus $A \Rightarrow B$, but $A$ and $B$ are not equivalent.

For each pair of statements given below, indicate whether the first implies the second ($A \Rightarrow B$), the second implies the first ($A \Leftarrow B$), or both implications hold ($A \Leftrightarrow B$), in which case $A$ and $B$ are equivalent.

a. $A$: $S$ is a square.
   $B$: $S$ is a rectangle.

b. $A$: Barbara is in Congress.
   $B$: Barbara is in the Senate.

c. $A$: Bob is older than Gene.
   $B$: Gene is younger than Bob.

d. $A$: $x + 3 = 5$
   $B$: $x = 2$

e. $A$: $x = 3$
   $B$: $x^2 = 9$

f. $A$: $\dfrac{x + 1}{x} = \dfrac{1}{x}$
   $B$: $x + 1 = 1$

g. $A$: $2x = x$
   $B$: $2 = 1$

---

**⊞  *Questions for Discussion or Essay***

---

50. Two methods for solving equations graphically were described in this section. Describe each method in your own words. The authors have displayed a clear preference for one technique over the other. Which one and why?

51. Modern graphics calculators have capabilities that would seem extraordinary to algebra students of just 10 years ago. What additional features do you foresee for calculators 10 years from now? What about 25 years from now? How will these changes affect the way that algebra is taught?

52. What's wrong with the following proof that $1 = 2$?

$$x = 0$$

$$x + x = 0 + x$$

$$2x = x$$

$$\frac{2x}{x} = \frac{x}{x}$$

$$2 = 1$$

53. An **identity** is an equation or inequality that is true for all values of the variables. When mathematics students are asked to *verify an identity*, they are being asked to show that the given equation or inequality holds for all values of the variable for which the expression ''makes sense.'' It is not uncommon for

students to attempt to verify an identity by showing that it holds for some particular value of the variable. For example, some students, when asked to show that

$$(x + 1)^2 = x^2 + 2x + 1$$

would argue as follows, ''Take $x = 2$. On the left hand side, you get 9 and on the right side you get 9, so the equation is true.'' What is wrong with this reasoning? Explain how this logical error (in nonmathematical contexts!) could lead to stereotypes and prejudice.

54. A view of the graph of an equation is shown in Figure 26. Discuss the number of possible $x$-intercepts. In particular, is it possible for the graph to have two $x$-intercepts within the given viewing rectangle? Could it have three $x$-intercepts? How about four or five?

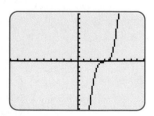

**FIGURE 26**

---

**SECTION 2**

## LINEAR EQUATIONS AND INEQUALITIES

■ How can a room's dimensions be reconstructed from its perimeter?

■ If a fly begins flying back and forth between two trains that are traveling toward each other on parallel tracks, what distance will the fly have traveled when the trains meet?

■ What property of equality is shared with human relationships?

---

### LINEAR EQUATIONS

The equation $2x - 5 = 3$ is an example of a linear equation. In general, a **linear equation** in the variable $x$ is any equation that can be put in the form $ax + b = 0$. Since the variable appears only to the first power, linear equations are usually quite easy to solve. We illustrate the process below.

$$2x - 5 = 3 \qquad \text{Original equation}$$

$$(2x - 5) + 5 = 3 + 5 \qquad \text{Adding 5 to both sides}$$

$$2x = 8$$

$$\frac{2x}{2} = \frac{8}{2} \qquad \text{Dividing both sides by 2}$$

$$x = 4$$

We have used the algebraic properties discussed in Section 1 to show that the statement $x = 4$ is equivalent to the statement $2x - 5 = 3$. Thus, 4 is the only possible solution. As a simple check, we can substitute 4 for $x$ in the original equation to get $2(4) - 5 = 8 - 5 = 3$.

---

■ *Solving linear equations*

A linear equation can be transformed to an equivalent equation of the form $x =$ (a number) by performing any combination of the following steps:

**1.** Adding or subtracting the same expression on both sides of the equation.

**2.** Multiplying or dividing by the same (nonzero) number on both sides of the equation.

---

**RULE OF THUMB**   When solving linear equations, multiply out the parentheses and collect all the variable terms on one side of the equation and all the constant terms on the other side. It doesn't matter which side you choose for the variable, although the left side is more conventional.

| EXAMPLE 1 | *Solving a linear equation* |

Solve the linear equation $2(t + 3) = 5t + 9$.

**SOLUTION**

$$2(t + 3) = 5t + 9$$

$$2t + 6 = 5t + 9 \qquad \text{Removing parentheses}$$

$$2t + 6 - 5t = 5t + 9 - 5t \qquad \text{Subtracting } 5t \text{ from both sides}$$

$$-3t + 6 - 9$$

$$-3t + 6 - 6 = 9 - 6 \qquad \text{Subtracting 6 from both sides}$$

$$-3t = 3$$

$$\frac{-3t}{-3} = \frac{3}{-3} \qquad \text{Dividing both sides by } -3$$

$$t = -1$$

**CHECK**

| $2(t + 3)$ | $5t + 9$ |
|---|---|
| $2(-1 + 3)$ | $5(-1) + 9$ |
| $2(2)$ | $-5 + 9$ |
| $4$ | $4$ ✓ |

| EXAMPLE 2 | *A contradiction* |

Solve the linear equation $\frac{1}{2}(4w + 3) = 2(w - \frac{2}{3})$.

**SOLUTION**

$$\frac{1}{2}(4w + 3) = 2\left(w - \frac{2}{3}\right)$$

$$2w + \frac{3}{2} = 2w - \frac{4}{3} \qquad \text{Removing parentheses}$$

$$\frac{3}{2} = -\frac{4}{3} \qquad \text{Subtracting } 2w \text{ from both sides}$$

Since the resulting statement is a contradiction, there is no solution. In other words, the solution set is the empty set { }.

## APPLICATIONS

An important step in using mathematics to solve real-life problems is the translation of verbal statements into mathematical statements. This step becomes much easier as you begin to recognize certain key words and phrases and their corresponding mathematical symbols. Consider, for example, the verbal statement "*the sum of* 3 *and a number is* 7." If we begin by defining $x$ to be the unknown number, then this statement translates into the algebraic statement $3 + x = 7$. Here the word "sum" becomes "+" and the word "is" becomes "=."

| EXAMPLE 3 | *Translating verbal statements into algebra* |

Translate each of the following verbal statements into algebraic statements and then solve for the variable.

**a.** 4 is 3 more than a number.
**b.** Twice a number is 4 less than the number.
**c.** The sum of three consecutive integers is 30.

**SOLUTION**

**a.** Let $y$ be the unknown number. Then

$$4 \text{ is } 3 \text{ more than } y$$

$$4 = y + 3$$

$$1 = y$$

**b.** Let $x$ be the unknown number. Then

$$\text{Twice } x \text{ is } 4 \text{ less than } x$$

$$2x = x - 4$$

$$x = -4$$

**c.** Let $x$ be the smallest of the three integers. Then the other integers are $x + 1$ and $x + 2$. Thus, we have

$$x + (x + 1) + (x + 2) = 30$$

$$3x + 3 = 30$$

$$3x = 27$$

$$x = 9$$

The three integers are therefore 9, 10, and 11.

■

Linear equations often arise in applications such as the following.

| EXAMPLE 4 | *Perimeter of a room* |

Bob is remodeling a rectangular room in his house. When he gets to the lumber yard, he realizes that he forgot to bring the room's measurements along. But he does remember that the length of the room is 1 foot more than the width and he needs 50 feet of baseboard to go around the edge of the room. What are the dimensions of the room?

**SOLUTION**    The first step, and perhaps the most important one, is to read the problem carefully and make note of the information that is given. A sketch is often helpful if the problem is geometric.

■ The room is shaped like a rectangle.
■ The length is 1 foot more than the width.
■ The perimeter of the room is 50 feet.

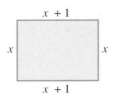

We next choose a variable to represent the quantity that is to be found, represent any other unknown quantities using that variable, and label the sketch accordingly.

$$x = \text{width of the room}$$

$$x + 1 = \text{length of the room}$$

Next we use the information given in the problem to write an equation involving the variable. It is often helpful to first verbalize the equation. Since the perimeter is the sum of the side lengths, we have

$$\text{sum of side lengths} = 50$$

$$x + (x + 1) + x + (x + 1) = 50$$

Now we can proceed to solve the equation.

$$x + (x + 1) + x + (x + 1) = 50$$

$$4x + 2 = 50$$

$$4x = 48$$

$$x = 12$$

After a solution has been found, it is important to reread the problem to see whether the solution seems reasonable and to check the solution against the verbal statement of the problem.

If the width is 12, the length must be $12 + 1 = 13$, which means the perimeter will be $12 + 13 + 12 + 13 = 50$ as required.

Thus, the dimensions are 12 feet by 13 feet.

The steps we followed to arrive at the solution to Example 4 are summarized here.

**Problem-solving strategy**

> 1. Read the problem carefully. Make note of any information. Draw a sketch if possible.
> 2. Choose a variable to represent the unknown quantity. Represent other quantities in terms of the variable. Label the sketch.
> 3. Verbalize an equation and then write it using the variable.
> 4. Solve the equation.
> 5. Check the answer.
> 6. Answer the stated problem completely.

The second step (choosing a variable) may occasionally need to be modified by choosing two or more temporary variables to represent the different quantities in the problem. Eventually, you must select one of the variables as your main variable and rewrite each of the others in terms of the main variable. We will use this strategy in the next example.

**EXAMPLE 5**    *Distance, rate, and time*

A passenger train leaves New York and travels toward Washington D.C. at a rate of 50 miles per hour. One hour later, a supertrain leaves Washington D.C. and travels toward New York on a parallel track at a rate of 200 miles per hour. If New York and Washington D.C. are 225 miles apart, at what point will the two trains pass by each other?

**SOLUTION**    A careful reading of the problem leads us to the following sketch.

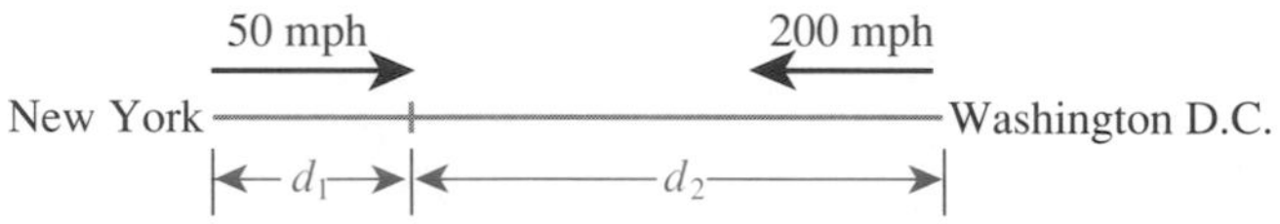

We assign the variables $t$ for the time (in hours) that the passenger train has traveled and $d_1$ for the distance traveled by the passenger train at the point where the two trains meet. Since the supertrain left 1 hour later, it has been traveling 1 hour less or $t - 1$. We represent its distance by $d_2$. Sometimes a table is a helpful way to organize all the information.

|  | *Rate* | *Time* | *Distance* |
|---|---|---|---|
| Passenger train | 50 | $t$ | $d_1$ |
| Supertrain | 200 | $t - 1$ | $d_2$ |

Since distance equals rate times time, we obtain the following two equations.

$$d_1 = 50t$$
$$d_2 = 200(t - 1)$$

But we also know that New York and Washington are 225 miles apart. This leads to the following equation.

distance traveled by passenger train + distance traveled by supertrain = 225

$$d_1 + d_2 = 225$$

By substituting for $d_1$ and $d_2$, we are able to solve for $t$.

$$50t + 200(t - 1) = 225$$
$$250t - 200 = 225$$
$$250t = 425$$
$$t = \frac{425}{250} = 1.7$$

In 1.7 hours, the passenger train will have traveled $50(1.7) = 85$ miles from New York, which is where the trains meet. As a check, the supertrain travels for 0.7 hour, which yields a distance of $200(0.7) = 140$ miles from Washington D.C. The two distances total 225 miles as required.

---

> **WARNING**   When assigning variables, we cannot use the same variable for two quantities that are different. In Example 5 it would have been incorrect to let $d$ represent the distance traveled by *both* trains since the two distances were different. Also, variables must represent numbers. Thus, in Example 5 it would have been inappropriate to write "$d_1 = $ New York."

Sometimes the only strategy needed for solving an application problem is the use of a familiar formula. In the following example we will use the formula for simple interest.

### EXAMPLE 6    *Simple interest*

Suppose a wealthy relative leaves you a sizable inheritance. You decide to save some of it for a new wardrobe for school next year. How much of the inheritance must be put into a bank paying simple interest at an annual rate of 5.25% in order to have $600 to spend on clothes after 1 year?

**SOLUTION**   Recall that the formula for working with simple interest is $B = P(1 + rt)$ where $B$ is the balance, $P$ is the principal, $r$ is the annual interest rate, and $t$ is the number of years. In this problem, we want $B = \$600$, $r = 0.0525$, and $t = 1$. Substituting these values into the formula leads to a linear equation.

$$600 = P(1 + 0.0525)$$

$$600 = 1.0525P$$

$$1.0525P = 600$$

$$P = \frac{600}{1.0525} \approx 570.07$$

So $P = \$570.07$ is the amount that must be deposited.

---

Many common formulas involving several variables are linear in more than one variable. For example, the formula for converting Celsius temperatures to Fahrenheit temperatures ($F = \frac{9}{5}C + 32$) is linear in both $C$ and $F$. Such formulas can be easily solved for any of the linear variables.

### EXAMPLE 7    *Converting to Celsius*

Disc jockey Damon reports the weather each morning on WMTH, a radio station near the border of the United States and Canada. Unfortunately, his outdoor thermometer only reports the temperature in Fahrenheit, and his listeners have demanded that he give both Fahrenheit and Celsius readings. Use the formula $F = \frac{9}{5}C + 32$ to find a formula that Damon can use to convert Fahrenheit to Celsius.

**SOLUTION**     Starting with the formula $F = \frac{9}{5}C + 32$, we wish to solve for $C$ in terms of $F$. In other words, we must isolate $C$ on one side of the equation.

$$F = \frac{9}{5}C + 32$$

$$F - 32 = \frac{9}{5}C$$

$$\frac{F - 32}{9/5} = C$$

$$\frac{5}{9}(F - 32) = C$$

The formula $C = \frac{5}{9}(F - 32)$ will allow Damon to convert any Fahrenheit temperature to Celsius.

## LINEAR INEQUALITIES

The procedure for solving linear inequalities is virtually identical to that for solving linear equations. The only difference, although a significant one, occurs when both sides of the inequality are multiplied or divided by a *negative* number. In this case, the direction of the inequality must be reversed. This is perhaps easiest to see through an example. If we begin with the true statement $1 < 2$ and multiply both sides by $-1$ without reversing the inequality, we would obtain $-1 < -2$, which is false. However, if we reverse the inequality, we obtain the true statement $-1 > -2$.

We will illustrate the procedure by solving $5 - 2x > 13$.

$$5 - 2x > 13 \qquad \text{Original inequality}$$

$$-2x > 8 \qquad \text{Subtracting 5 from both sides}$$

$$x < -4 \qquad \text{Dividing both sides by } -2 \text{ and reversing the inequality}$$

We applied algebraic properties to show that the statement $x < -4$ is equivalent to the original inequality. Thus, the solution is *the set of all real numbers x such that x is less than* $-4$. Using set-builder notation, we can write this as $\{x | x < -4\}$. In interval notation the solution set is the interval $(-\infty, -4)$. We can also graph the solution on the number line as shown in Figure 27.

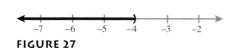

**FIGURE 27**

**Solving linear inequalities**

A linear inequality can be transformed to an equivalent inequality by performing any of the following steps:

1. Adding or subtracting the same expression on both sides of the equation.
2. Multiplying or dividing by the same **positive** number on both sides of the equation.
3. Multiplying or dividing by the same **negative** number on both sides and **reversing the inequality**.

**EXAMPLE 8**   *Solving a linear inequality*

Solve $4(t - 1) \geq t$ and graph its solution set.

**SOLUTION**

$$4(t - 1) \geq t$$

$$4t - 4 \geq t \qquad \text{Removing the parentheses}$$

$$4t \geq t + 4 \qquad \text{Adding 4 to both sides}$$

$$3t \geq 4 \qquad \text{Subtracting } t \text{ from both sides}$$

$$t \geq \frac{4}{3} \qquad \text{Dividing both sides by 3}$$

So the solution set is $[\frac{4}{3}, \infty)$ and the graph is shown at left.

---

**EXAMPLE 9**   *Solving a linear inequality*

Solve $3t - 7 < 6t + 8$ and graph its solution set.

**SOLUTION**

$$3t - 7 < 6t + 8$$

$$3t < 6t + 15 \qquad \text{Adding 7 to both sides}$$

$$-3t < 15 \qquad \text{Subtracting } 6t \text{ from both sides}$$

$$t > -5 \qquad \text{Dividing both sides by } -3 \text{ and reversing the inequality}$$

So the solution set is $(-5, \infty)$ and the graph is shown at left.

---

## EXERCISES 2

**EXERCISES 1–24** □ *Solve the linear equation and check your answer.*

**1.** $x + 8 = 10$

**2.** $y - 17 = 34$

**3.** $3x - 4 = 2$

**4.** $-2x + 7 = 9$

**5.** $\frac{1}{2}z + 13 = \frac{5}{2}$

**6.** $-\frac{2}{3}w + 8 = 5$

**7.** $3x + 4 = 6x$

**8.** $-2x - 3 = 8x + 17$

**9.** $-2(m + 3) = 9$

**10.** $-3(n - 4) = 8$

**11.** $0.71(x - 0.43) = 0.21$

**12.** $0.08(y + 0.47) = 0.6y$

**13.** $5(-2w + 9) = 3(w - 18) + 4$

**14.** $2z + 3(z - 6) = 4(z + 8)$

**15.** $(2.5 \times 10^9)q = 3.4 \times 10^{-5}$

**16.** $(-2.6 \times 10^{13})w + (7 \times 10^{12}) = 3.2 \times 10^{12}$

**17.** $-\frac{1}{4}(x + 0.35) = 0.2(x - 1) + 0.4$

**18.** $0.3\left(-2y - \frac{1}{2}\right) = \frac{1}{4}(0.3y - 7) + 9$

**19.** $-3(2p - 5) + 7(p + 1) = 13(p - 3) + 4$

**20.** $-9(w - 7) + 4(-2w + 5) + 6 = -3(3w - 8) + 40$

**21.** $(x + 2)(x - 3) = (x - 3)^2$     **22.** $(t - 1)(t - 4) = (t + 1)^2$

**23.** $(y - 1)^2 - (y - 1)^2 = 8$     **24.** $(x + 2)^2 - x^2 = x - 2$

**EXERCISES 25–44** □ *Solve the inequality. Write your answer using interval notation and graph the solution.*

**25.** $x - 8 < 5$

**26.** $t + 3 \geq -1$

**27.** $23 - 4v < 27$

**28.** $5x - 3 \leq 22$

**29.** $2z + 7 \geq 5 - 6z$

**30.** $2y - 2 > 4y - 5$

**31.** $3(x - 1) \leq 17 - (8 - x)$

**32.** $6t - 2(t + 3) < 3(t + 11)$

**33.** $\dfrac{1}{4}(x + 6) - \dfrac{3}{2}(2x - 5) > 0$

**34.** $\dfrac{1}{5}(3z - 1) + \dfrac{1}{4}(z + 2) \leq 1$

**35.** $\dfrac{1}{4}w + 1 < w$

**36.** $2 - \dfrac{2}{5}x \leq 1$

**37.** $\dfrac{3}{4}s - 2 < \dfrac{5}{8}s - 3$

**38.** $\dfrac{1}{3}y - 1 \geq -\dfrac{1}{6}y + \dfrac{1}{2}$

**39.** $2.35x - 9.16 < 16.41 - 5.72x$

**40.** $16.32 + 21.56t \geq 11.79t - 5.21$

**41.** $(z - 1)(z + 1) < (z - 3)(z + 4)$

**42.** $(x + 2)(x - 5) \leq (x - 1)(x - 3)$

**43.** $t(t + 3) - (t + 1)^2 \geq 4$

**44.** $(a - 2)^2 - a^2 < a$

**EXERCISES 45–56** ☐ *Solve the formula for the indicated variable.*

**45.** $d = rt$ for $t$

**46.** $F = ma$ for $m$

**47.** $C = 2\pi r$ for $r$

**48.** $C = \pi d$ for $d$

**49.** $I = prt$ for $t$

**50.** $A = \dfrac{1}{2}bh$ for $b$

**51.** $P = 2w + 2l$ for $l$

**52.** $A = \dfrac{1}{2}(B + b)h$ for $B$

**53.** $y = mx + b$ for $x$

**54.** $A = P(1 + r)$ for $r$

**55.** $F = G\dfrac{m_1 m_2}{r^2}$ for $m_1$

**56.** $s = -\dfrac{1}{2}gt^2 + v_0 t + s_0$ for $v_0$

**EXERCISES 57–60** ☐ *Translate each of the verbal statements into algebraic statements and then solve for the variable.*

**57.** 10 is 6 less than a number.

**58.** Twice a number is 5 more than the number.

**59.** The product of 4 and a number has the same value as the sum of 4 and that number.

**60.** A number divided by 3 has the same value as that number minus 1.

■ *Applications*

**61.** *Consecutive Integers* The sum of two consecutive integers is 431. Find the two numbers.

**62.** *Consecutive Integers* The sum of three consecutive integers is 363. Find the three numbers.

**63.** *Garden Dimensions* A rectangular garden is 20 feet longer than it is wide. If the perimeter of the garden is 180 feet, find the length and width.

**64.** *Room Dimensions* The width of a rectangular room is half its length. If the perimeter of the room is 51 feet, find the length and width.

**65.** *Exam Scores* Juan scored 85 and 88 on the first two exams of a course. He must have an average of 90 to receive an A. What must he score on the third exam to get an A?

**66.** *Bowling Scores* A world-record season bowling average of 245.63 was set by Doug Vergouven during the 1989–1990 season. (Source: *The Guinness Book of World Records*.) If Doug were to bowl 230 and 240 in the first two games of a three-game series, what must he bowl in the third game to average 246 for the three-game series?

**67.** *Job Options* Upon graduation from college, Meredith receives two job offers. One carries a combined annual salary and benefit package of $28,000 while the other has a benefit package of $6000, which represents 27% of the annual salary. Which job has the largest total salary and benefit package?

**68.** *Comparing Income* Todd and Lisa have a combined monthly gross income of $3008. If Todd's salary is 12% less than Lisa's, how much does each earn?

**69.** *Saving for a Ring* Jamal is just finishing his junior year of college and plans on proposing to his girlfriend after he graduates in 1 year. To be sure he has enough money to buy a $500 engagement ring, he wants to deposit some money in a bank account so he can pay for the ring 1 year later.

   a. What amount must he deposit in an account paying an annual interest rate of 6%?

   b. If he only has $425 to deposit, what annual interest rate must he find?

**70.** *Cost of Living Increase* Tonya's contract specifies that she is to receive a cost-of-living increase in her salary each year that is equal to the rate of inflation for the previous year. The rate of inflation in 1991 was 3.1%, and Tonya's salary *after* the cost of living increase was $21,100. What was her salary before the increase?

**71.** *Radio Range* A family is using a pickup truck and a rental van for an interstate move. In order to stay in contact during the trip, they have rented CB radios with a range of 20 miles. The two vehicles start at the same time and travel in the same direction, the pickup truck at a constant speed of 55 miles per hour and the van at a constant speed of 45 miles per hour. How long before the two vehicles are beyond the range of the radios?

72. *Highway Rendezvous*  Two groups of students from the same school are traveling in different cars to Florida during spring break. The first travels at an average speed of 55 miles per hour. The second leaves 30 minutes after the first and travels at an average speed of 65 miles per hour. How long will it take the second car to catch up to the first?

73. *Indy Racing*  During the first 300 miles of a 500-mile race, Michael averaged 180 miles per hour. If Al averaged 178 miles per hour during the same time period, how far behind Michael was Al at the instant Michael passed the 300-mile mark? If Michael continues at the same rate, what speed would Al have to average for the remainder of the race in order to win?

74. *Paving Crews*  Two paving crews are resurfacing 81 miles of freeway. The first crew starts at one end of the 81-mile stretch and completes 7 miles each week. The second crew starts 1 week later at the other end and completes 8 miles each week. How long will it take to complete the resurfacing? How much did each crew complete?

75. *Traveling Fly*  Two trains are on parallel tracks headed toward each other, one at a rate of 6 miles per hour and the other at a rate of 4 miles per hour. At the instant the trains are 2 miles apart, a fly begins flying back and forth between the two trains at a rate of 10 feet per second. What total distance has the fly traveled by the time the trains meet?

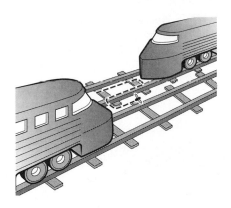

## Projects for Enrichment

76. *Relative Motion*  Suppose baseball great Roger Clemens is traveling on the back of a flatbed truck going 60 mph when he throws a fastball, which under normal circumstances would have a speed of 95 mph. How fast will the ball travel, relative to an observer at the side of the road, if it is thrown

    a. in the direction the truck is moving?

    b. in the direction opposite that of the truck?

    c. If the observer runs out into the road immediately after the truck passes and if the ball is thrown toward the observer at the instant the truck is 60 feet in front of the observer, how long will it take the ball to reach the observer?

    How fast will the ball travel relative to Roger Clemens if it is thrown

    d. in the direction the truck is moving?

    e. in the direction opposite that of the truck?

    Now suppose the truck is going 110 miles per hour when Roger throws the ball. Answer parts (a) through (e) again.

77. *Equivalence Relations*  Equality in the set of real numbers is an example of an *equivalence relation* because it has the following three properties.

   I. Reflexive property: $a = a$ for all real numbers $a$.

  II. Symmetric property: If $a = b$, then $b = a$.

 III. Transitive property: If $a = b$ and $b = c$, then $a = c$.

a. For each of the following, identify the property that is being used.

   i. If $15x = 3$, then $3 = 15x$.

  ii. If $x + 2 = 8$ and $8 = 2(x - 1)$, then $x + 2 = 2(x - 1)$.

 iii. $9y = 9y$

 iv. If $5t = s$ and $5t = r$, then $r = s$.

More generally, an **equivalence relation** is any relation that satisfies properties (I) through (III) with ''$=$'' replaced by that relation. Equality is only one of many relations that you have likely seen in the past. Other relations include $\equiv$ (congruence for triangles), $\approx$ (similarly for triangles), $<$, $\leq$, $>$, and $\geq$. However, not all of these relations are equivalence relations.

b. Identify which of the above relations are equivalence relations. For those that are not, indicate which of the three properties do not hold.

Nonalgebraic relations can also be defined. For example, if $A$ and $B$ represent people, we can define the relation $A \sim B$ to mean such things as $A$ ''looks like'' $B$, $A$ ''likes'' $B$, $A$ ''weighs the same as'' $B$, or $A$ ''is married to'' $B$.

c. Identify which of the above human relations are equivalence relations. For those that are not, indicate which of the three properties do not hold.

d. Discuss three more examples of human relations. At least one must be an equivalence relation.

## Questions for Discussion or Essay

**78.** When solving problems such as Example 5 that involve distance, rate, and time, we must make the assumption that the rate is constant. What does your experience driving (or riding in) a car tell you about such an assumption? If the rate is not constant, what meaning does "rate" have in the equation *distance = rate × time*?

**79.** Checking your solutions is a good way to see whether the solutions you found are correct. However, it does not provide a way to check whether you found all of the solutions. Why is that? How do we know for sure that the solutions we found are the only ones?

**80.** The equations in Exercises 21–24 don't appear to be linear. Do they satisfy the definition of linear equations? Explain.

**81.** A majority of algebra students find word problems difficult. What makes them so difficult? Is it the case that word problems require a deeper level of understanding than standard problems? There are currently computer programs called *computer algebra systems* that can easily solve virtually all of the standard problems in this text (and far more difficult problems as well), but there is no computer program currently available that will interpret and solve story problems. Why not?

---

## SECTION 3

# ABSOLUTE-VALUE EQUATIONS AND INEQUALITIES

- How many years would it take the world's heaviest man to reach a weight of 200 pounds on a standard weight-loss program?
- How much variation in distance will improperly inflated car tires cause?
- How can distance be estimated with a bicycle tire?

### ABSOLUTE-VALUE EQUATIONS

Equations involving absolute value, such as $|x| = 4$, can be solved by considering distances. According to the definition of absolute value, the equation $|x| = 4$ is equivalent to the statement "the distance between $x$ and 0 is 4." Since there are two values for $x$ that are a distance 4 from 0, namely 4 and $-4$, the solution set for the equation $|x| = 4$ is $\{4, -4\}$. These observations generalize to the following rule for solving absolute-value equations.

### Solving absolute-value equations

If $c > 0$, then the absolute-value equation $|u| = c$ is equivalent to $u = -c$ or $u = c$.

**EXAMPLE 1**    *Solving an absolute-value equation*

Solve $|2t| = 10$.

**SOLUTION**    Applying the rule given above, we obtain two equations to solve separately.

$$|2t| = 10$$

$$2t = -10 \quad \text{or} \quad 2t = 10$$

$$t = -5 \quad \text{or} \quad t = 5$$

**CHECK**

| $t = -5$ | | $t = 5$ | |
|---|---|---|---|
| $\lvert 2t \rvert$ | 10 | $\lvert 2t \rvert$ | 10 |
| $\lvert 2(-5) \rvert$ | | $\lvert 2(5) \rvert$ | |
| $\lvert -10 \rvert$ | | $\lvert 10 \rvert$ | |
| 10 | ✓ | 10 | ✓ |

---

**EXAMPLE 2**    *Solving an absolute-value equation*

Solve $\lvert 3y + 4 \rvert = -2$.

**SOLUTION**   We cannot apply our rule to $\lvert 3y + 4 \rvert = -2$ since the number on the right-hand side is not positive. In fact, this equation is a contradiction since there are no values of $y$ that will make $\lvert 3y + 4 \rvert$ a negative number. The solution set is the empty set.

---

**EXAMPLE 3**    *Solving an absolute-value equation*

Solve $\lvert \frac{1}{2}x - 1 \rvert = \frac{1}{3}$.

**SOLUTION**   We again apply the rule given above to obtain two equations that can be solved separately.

$$\left\lvert \frac{1}{2}x - 1 \right\rvert = \frac{1}{3}$$

$$\frac{1}{2}x - 1 = -\frac{1}{3} \quad \text{or} \quad \frac{1}{2}x - 1 = \frac{1}{3}$$

$$\frac{1}{2}x = \frac{2}{3} \quad \text{or} \quad \frac{1}{2}x = \frac{4}{3}$$

$$x = \frac{4}{3} \quad \text{or} \quad x = \frac{8}{3}$$

**CHECK**

| $x = \frac{4}{3}$ | | $x = \frac{8}{3}$ | |
|---|---|---|---|
| $\left\lvert \frac{1}{2}x - 1 \right\rvert$ | $\frac{1}{3}$ | $\left\lvert \frac{1}{2}x - 1 \right\rvert$ | $\frac{1}{3}$ |
| $\left\lvert \frac{1}{2}\left(\frac{4}{3}\right) - 1 \right\rvert$ | | $\left\lvert \frac{1}{2}\left(\frac{8}{3}\right) - 1 \right\rvert$ | |
| $\left\lvert \frac{2}{3} - 1 \right\rvert$ | | $\left\lvert \frac{4}{3} - 1 \right\rvert$ | |
| $\left\lvert -\frac{1}{3} \right\rvert$ | | $\left\lvert \frac{1}{3} \right\rvert$ | |
| $\frac{1}{3}$ | ✓ | $\frac{1}{3}$ | ✓ |

Absolute-value equations can be formed from statements involving distances. Given two numbers $a$ and $b$, the distance between $a$ and $b$ can be found by computing $|a - b|$. Notice that it doesn't matter whether or not $a$ is larger than $b$; either way the absolute value gives the proper positive distance.

**EXAMPLE 4**    *Absolute value and distance*

Find all numbers $x$ such that the distance between 5 and twice $x$ is 9.

**SOLUTION**    The distance between 5 and twice $x$ is given by $|5 - 2x|$. We set this equal to 9 and solve.

$$|5 - 2x| = 9$$

| $5 - 2x = -9$ | or | $5 - 2x = 9$ |
|---|---|---|
| $-2x = -14$ | or | $-2x = 4$ |
| $x = 7$ | or | $x = -2$ |

All of the absolute-value equations that we have encountered so far have been relatively simple and were solved easily by algebraic means alone. More complex absolute-value equations are usually easier to solve graphically, as in the following example.

**EXAMPLE 5**    *Solving an absolute-value equation graphically*

Solve $|x + 2| = 5 - |x - 1|$ graphically.

**SOLUTION**    We begin by writing the equation in the form $\square = 0$.

$$|x + 2| = 5 - |x - 1|$$
$$|x + 2| + |x - 1| - 5 = 0$$

Next we use a graphics calculator to plot the graph of $y = |x + 2| + |x - 1| - 5$ and estimate the $x$-intercepts, as shown in Figure 28. Using the trace feature, we find that the $x$-intercepts are 2 and $-3$, and it is easily verified that these satisfy the original equation. Thus, the equation has two solutions, 2 and $-3$.

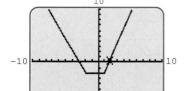

**FIGURE 28**

## COMPOUND INEQUALITIES

A **compound inequality** is formed by joining two linear inequalities with the words *and* or *or*. A simple example of a compound inequality is the statement

$$x > -3 \quad \text{and} \quad x < 2$$

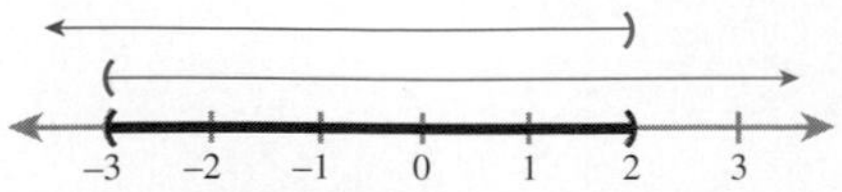

The solution set for this statement is the intersection (the portion common to both) of the intervals $(-3, \infty)$ and $(-\infty, 2)$, namely the interval $(-3, 2)$. The graph of this solution set is shown in Figure 29. To solve more complicated compound inequalities using the word *and*, we simply solve each inequality separately and take the intersection of the solution sets.

**FIGURE 29**

| EXAMPLE 6 | *Solving a compound inequality with "and"* |

Solve the compound inequality $6u - 8 \leq 16$ **and** $1 - 4u \leq -3$ and graph the solution set.

**SOLUTION**    We first solve each inequality separately.

$$6u - 8 \leq 16 \qquad \text{and} \qquad 1 - 4u \leq -3$$
$$6u \leq 24 \qquad \text{and} \qquad -4u \leq -4$$
$$u \leq 4 \qquad \text{and} \qquad u \geq 1$$

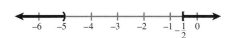

By intersecting the two sets $(-\infty, 4]$ and $[1, \infty)$, we get the solution set $[1, 4]$. The graph of this set is shown in red at left.    ■

Compound inequalities formed with the word *or* are solved in a similar way, except instead of intersecting the two sets, we combine them; that is, we form their **union**. The union of two sets $A$ and $B$ is denoted $A \cup B$.

| EXAMPLE 7 | *Solving a compound inequality with "or"* |

Solve the compound inequality $\frac{1}{5}y + 4 < 3$ **or** $-5y - 3 \leq y$ and graph the solution set.

**SOLUTION**    First we solve each inequality separately.

$$\frac{1}{5}y + 4 < 3 \qquad \text{or} \qquad -5y - 3 \leq y$$

$$\frac{1}{5}y < -1 \qquad \text{or} \qquad -5y \leq y + 3$$

$$y < -5 \qquad \text{or} \qquad -6y \leq 3$$

$$y < -5 \qquad \text{or} \qquad y \geq -\frac{1}{2}$$

The solution is formed by taking the union of these two solutions to obtain $(-\infty, -5) \cup [-\frac{1}{2}, \infty)$, as shown in the graph at left.    ■

> **RULE OF THUMB**    The solution of a compound inequality formed with the word "and" is found by solving each part separately and then taking the **intersection** of the two sets. The solution of a compound inequality formed with the word "or" is found by solving each part separately and then taking the **union** of the two sets.

Some compound inequalities are disguised as **double inequalities** such as $-4 \leq 3x - 2 < 7$. This statement is equivalent to $-4 \leq 3x - 2$ *and* $3x - 2 < 7$, although it is more convenient to leave it as a double inequality so both inequalities can be solved together, as shown in Example 8.

**EXAMPLE 8**   *Solving a double inequality*

Solve the double inequality $-4 \le 3x - 2 < 7$.

**SOLUTION**   We must isolate $x$ as the middle term.

$$-4 \le 3x - 2 < 7$$

$$-2 \le 3x < 9 \qquad \text{Adding 2 to all sides}$$

$$-\frac{2}{3} \le x < 3 \qquad \text{Dividing all sides by 3}$$

The solution set is $[-\frac{2}{3}, 3)$, as shown in the graph at left.

Compound inequalities can often be used to solve application problems involving a range of solution values. You can spot such problems by the presence of such phrases as "at least," "at most," "no more than," "no less than," and so forth.

**EXAMPLE 9**   *Weight loss*

A company is marketing a diet and exercise program that it claims will result in a weight loss of at least 1.5 pounds per week. For health reasons, they also warn that one should not lose more than 5 pounds per week. If Robert Earl Hughes (b. 1926, d. 1958) had started this program after he reached his world-record weight of 1069 pounds, over what range of weeks would his weight have dropped to 203 pounds (his weight at age 6)?

**SOLUTION**   For each week, Robert would lose at least $\frac{3}{2}$ pounds and at most 5 pounds, and so we have

$$\text{weight loss} \ge \frac{3}{2} \cdot (\text{number of weeks}) \quad \textbf{and} \quad \text{weight loss} \le 5 \cdot (\text{number of weeks})$$

Let $x$ represent the number of weeks that Robert remained on the program. In dropping from 1069 to 203, he would have a weight loss of 866 pounds in $x$ weeks. This leads to the following inequalities.

$$866 \ge \frac{3}{2}x \qquad \text{and} \qquad 866 \le 5x$$

$$\frac{2}{3} \cdot 866 \ge x \qquad \text{and} \qquad \frac{866}{5} \le x$$

$$577\frac{1}{3} \ge x \qquad \text{and} \qquad 173\frac{1}{5} \le x$$

If the company's claims are correct, it would take anywhere from $173\frac{1}{5}$ to $577\frac{1}{3}$ weeks for Robert to reach 203 pounds.

## ABSOLUTE-VALUE INEQUALITIES

Inequalities involving absolute value can be solved by rewriting them as compound inequalities, as we demonstrate in Table 1.

**TABLE 1**

*Absolute-value inequalities*

| Original inequality | Distance interpretation | Compound inequality | Interval and graph |
|---|---|---|---|
| $\lvert x \rvert < 3$ | Numbers whose distance from 0 is less than 3 | $-3 < x < 3$ | $(-3, 3)$ 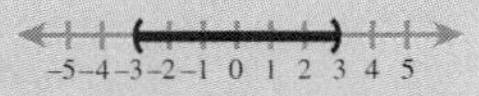 |
| $\lvert x \rvert > 3$ | Numbers whose distance from 0 is greater than 3 | $x < -3$   or   $x > 3$ | $(-\infty, -3) \cup (3, \infty)$ 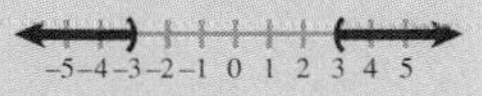 |

The inequalities in Table 1 lead us to two general properties for dealing with absolute-value inequalities.

*Solving absolute-value inequalities*

If $c > 0$, then

> **Property 1.** $\lvert u \rvert < c$   is equivalent to $-c < u < c$
>
> **Property 2.** $\lvert u \rvert > c$   is equivalent to $u < -c$ **or** $u > c$

These properties are also valid if $<$ is replaced by $\leq$ and $>$ is replaced by $\geq$.

**EXAMPLE 10**

*Solving an absolute-value inequality*

Solve $\lvert 2x + 1 \rvert < 3$.

**SOLUTION**   We first apply Property 1 and write $\lvert 2x + 1 \rvert < 3$ as the compound inequality $-3 < 2x + 1 < 3$. Next we solve for $x$.

$$-3 < 2x + 1 < 3$$
$$-4 < 2x < 2$$
$$-2 < x < 1$$

The solution is the interval $(-2, 1)$.

**EXAMPLE 11**

*Solving an absolute-value inequality*

Solve $\lvert 4 - y \rvert \leq -\frac{1}{2}$.

**SOLUTION**   The properties for solving absolute-value inequalities do not apply to $\lvert 4 - y \rvert \leq -\frac{1}{2}$ since the number on the right-hand side is negative. This inequality is a contradiction since absolute values are never negative, so the solution is the empty set.

**EXAMPLE 12**    *Solving an absolute-value inequality*

Solve $|1 - x| - 5 \geq -2$.

**SOLUTION**    The inequality must be in the form $|u| \geq c$ before we apply Property 2.

$$|1 - x| - 5 \geq -2$$

$$|1 - x| \geq 3$$

According to Property 2, this is equivalent to the compound inequality $1 - x \leq -3$ **or** $1 - x \geq 3$. Now we solve each inequality separately.

$$1 - x \leq -3 \quad \text{or} \quad 1 - x \geq 3$$

$$-x \leq -4 \quad \text{or} \quad -x \geq 2$$

$$x \geq 4 \quad \text{or} \quad x \leq -2$$

Finally, we form the union of these two intervals to obtain $(-\infty, -2] \cup [4, \infty)$.

Absolute-value inequalities are useful for expressing measurement errors. For example, if the length $l$ of a metal pin is to be 4 centimeters with an error of no more than 0.01 centimeter, then we can write $|l - 4| \leq 0.01$ to indicate that $l$ and 4 may not be more than 0.01 centimeter apart. By solving the inequality, we obtain the allowable range of $l$ values.

$$|l - 4| \leq 0.01$$

$$-0.01 \leq l - 4 \leq 0.01 \qquad \text{Property 1}$$

$$3.99 \leq l \leq 4.01 \qquad \text{Adding 4}$$

This answer could also have been expressed as $4 \pm 0.01$ centimeters, without even considering absolute values. In the following example, however, the solution is not as obvious.

**EXAMPLE 13**    *Bicycle tires and measurement error*

The circumference of a bicycle tire is to be 85 inches with an error of no more than 0.3 inch. Find the acceptable range of values for the radius of the tire.

**SOLUTION**    The circumference of a circle in terms of its radius is $C = 2\pi r$. Thus, if the circumference is to be 85 inches with a maximum error of $\pm 0.3$ inch, we write $|2\pi r - 85| \leq 0.3$ and proceed as follows.

$$|2\pi r - 85| \leq 0.3$$

$$-0.3 \leq 2\pi r - 85 \leq 0.3 \qquad \text{Property 1}$$

$$84.7 \leq 2\pi r \leq 85.3 \qquad \text{Adding 85}$$

$$\frac{84.7}{2\pi} \leq r \leq \frac{85.3}{2\pi} \qquad \text{Dividing by } 2\pi$$

$$13.48 \leq r \leq 13.58 \qquad \text{Approximating}$$

So the radius $r$ must be between 13.48 and 13.58 inches.

# EXERCISES 3

**EXERCISES 1–10** □ *Find the solution set for the absolute-value equation and check your answer.*

1. $|x| = 6$

2. $|t| = 3$

3. $|3y| = \dfrac{5}{2}$

4. $\left|\dfrac{1}{2}x\right| = 3$

5. $|2u - 5| = 1$

6. $|1 + 3v| = \dfrac{1}{5}$

7. $\left|4 - \dfrac{1}{2}z\right| = 3$

8. $\left|\dfrac{2}{3}a + 4\right| = 16$

9. $|5.3x - 1.7| = 10.5$

10. $|104.6 - 99.1t| = 1347.2$

**EXERCISES 11–16** □ *Write the verbal statement as an absolute-value equation and then solve it.*

11. The distance between 5 and $x$ is 7.

12. The distance between $-3$ and $x$ is 2.

13. The distance between $-\frac{2}{3}$ and twice $x$ is $\frac{7}{3}$.

14. The distance between 1.5 and three times $x$ is 2.3.

15. The distance between 2 and half of $x$ is 5.

16. The distance between $\frac{1}{2}$ and a third of $x$ is 6.

**EXERCISES 17–24** □ *Solve the compound inequality and graph the solution.*

17. $x + 3 > 1$ and $x - 2 < 4$

18. $\dfrac{1}{3}y \leq -1$ or $6y + 1 \geq -5$

19. $5 \leq 8t - 3 \leq 9$

20. $3u - 2 \leq 1$ or $4u + 1 > 9$

21. $-2x + 7 \leq 1$ and $\dfrac{1}{2}x \leq \dfrac{9}{4}$

22. $3w + 2 < -4$ and $-3w - 2 < 4$

23. $2z \geq 5$ or $\dfrac{1}{4}(z - 1) \geq 1$    24. $-3 < 1 - \dfrac{1}{2}x < 1$

**EXERCISES 25–44** □ *Solve the absolute-value inequality. Express your answer in interval notation.*

25. $|x| < 5$

26. $|u| \geq 2$

27. $|2w| < 6$

28. $\left|\dfrac{x}{7}\right| \leq 1$

29. $\dfrac{1}{3}|z| > 4$

30. $|-3x| < 1$

31. $|s - 5| \leq 2$

32. $|2u - 1| \geq \dfrac{5}{3}$

33. $|3x + 4| \geq \dfrac{1}{2}$

34. $|7t + 1| \leq 2$

35. $|-2r + 1| < 8$

36. $|-8p - 3| \geq 2$

37. $\left|\dfrac{1}{2}x + \dfrac{3}{4}\right| > 4$

38. $|0.3w - 0.4| \leq 0.1$

39. $|0.2y - 5| < 1.63$

40. $\left|\dfrac{4}{5}z + 9\right| > \dfrac{1}{3}$

41. $\left|\dfrac{3z - 2}{5}\right| \leq 1$

42. $-2\left|\dfrac{4x - 3}{5}\right| > -1$

43. $|6t - 4| + 3 < 8$

44. $\dfrac{1}{3}|3y - 7| + 8 \leq 10$

**EXERCISES 45–48** □ *Write the verbal statement as an absolute-value inequality and then solve it.*

45. The distance between $x$ and 1 is less than 5.

46. The distance between $t$ and $-2$ is greater than 4.

47. The distance between twice $w$ and $-3$ is greater than 6.

48. The distance between $-3$ times $x$ and 4 is less than 2.

**EXERCISES 49–52** □ *Use a graphics calculator to help you determine the answer.*

49. Plot the graphs of $y = |3x - 2|$ and $y = 5$. How is the solution set of $|3x - 2| = 5$ represented on the graph? What about the solution sets of $|3x - 2| > 5$ or $|3x - 2| < 5$?

50. Plot the graphs of $y = |3x - 2|$ and $y = -1$. How can you tell from these graphs that $|3x - 2| < -1$ has no solutions while the solution set of $|3x - 2| > -1$ is $(-\infty, \infty)$?

51. Find all solutions of $|x - 1| + |x - 2| = 3|x - 3|$.

52. Find all solutions of $|x| = |x - 2|$.

## ■ Applications

53. *Truck Rental* A U-Haul truck can be rented for $19.95 per day plus $0.39 per mile. The total cost for 1 day can be expressed as $C = 0.39x + 19.95$, where $x$ represents the number of miles traveled. What range of miles can be traveled if the total rental expense must be between $75 and $100?

54. *Rate of Return* An investment consultant guarantees that an initial investment of $10,000 will yield a balance between $15,000 and $18,000 after 5 years. Assuming simple interest, what range of interest rates are possible?

**55. *Tire Radius*** If the rear tires of a certain car average 750 revolutions per minute, the distance traveled in 1 hour over the course of many trials varies between 59.80 and 60.69 miles. Find the range of values for the radius of the tires. (*Hint:* The distance traveled is the circumference of the tire, $2\pi r$, times the number of revolutions.)

**56. *Copy Machine Cost*** A copy machine has a purchase cost of $1200 and has an estimated page cost of 2¢ per page. What is the minimum number of pages that would have to be copied before the purchase of the copy machine would be more cost-effective than renting a machine at a cost of 5¢ per page?

**57. *Scale Accuracy*** The manufacturer of a bathroom scale claims the scale is accurate to within $\frac{1}{2}$ pound. If Susan's true weight is 95 pounds, write an absolute-value inequality that expresses the possible error in the weight $W$ given by the scale.

**58. *Scale Accuracy*** A grocery store has a scale located in the produce department for customers to weigh the produce they buy. A sign next to the scale warns that the scale may be off by as much as 2 ounces and so it should only be used to estimate the weight. If you buy a bag of apples that weighs 3.5 pounds (according to the inaccurate scale) and if the apples cost $0.79 per pound, how much variation in price could there be when you take it to the cash register where there is an accurate scale?

**59. *Temperature Conversion*** On a recent trip outside of the United States, Drew begins to feel ill. The only thermometer he can find gives readings in Celsius and is accurate to within 1° Celsius. If the thermometer gives a reading of 38° Celsius, what range of Fahrenheit temperatures does this indicate? The relationship between degrees Celsius $C$ and degrees Fahrenheit $F$ is given by $C = \frac{5}{9}(F - 32)$.

**60. *Tire Spokes*** If a bicycle tire is to have a circumference of 85 inches with a maximum possible error of 0.3 inch, the hub of the tire is 1.5 inches in diameter, and the rim and tire have a combined radial thickness of 1 inch (see Figure 30), find the allowable range of values for the length of a spoke.

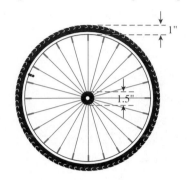

**FIGURE 30**

---

### ◼ *Projects for Enrichment*

**61. *The Triangle Inequality*** When the notion of absolute value is extended to the complex numbers, we encounter a very important inequality statement known as the **triangle inequality**. To define absolute value for a complex number $a + bi$, we rely again on the notion of distance. Recall that for a real number $a$, $|a|$ is the distance from $a$ to 0. In a like manner, if we view $a + bi$ as the point $(a, b)$ on the coordinate plane, we can define $|a + bi|$ as the distance from $(a, b)$ to $(0, 0)$, as illustrated in Figure 31. By the Pythagorean Theorem, this distance is $\sqrt{a^2 + b^2}$. Thus, we define $|a + bi| = \sqrt{a^2 + b^2}$.

**a.** Find the indicated absolute value.

    **i.** $|4 + 3i|$    **ii.** $|1 - 2i|$    **iii.** $|5 - \sqrt{2}i|$    **iv.** $|9i|$

The triangle inequality states that if $u$ and $v$ are any two complex numbers, then $|u + v| \le |u| + |v|$. To illustrate this, and for the remainder of the project, let $u = 3 + 2i$ and $v = 1 + 3i$. Then $u + v = 4 + 5i$.

**b.** Compute $|u|$, $|v|$, and $|u + v|$ to show that $|u + v| \le |u| + |v|$.

We have plotted $u$ and $u + v$ in Figure 32. The distances $|u|$ and $|u + v|$ have also been labeled. To see where the distance $|v|$ is in the sketch, we need the formula for the distance between two points. If $P(a, b)$ and $Q(c, d)$ are two points, then the distance between $P$ and $Q$ is $\sqrt{(a - c)^2 + (b - d)^2}$.

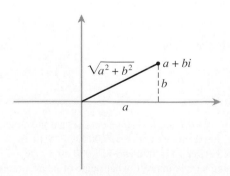

**FIGURE 31**

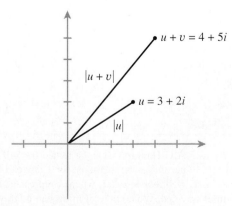

**FIGURE 32**

c. Find the distance between the point (3, 2) represented by $u$ and the point (4, 5) represented by $u + v$. Then show that this distance is the same as $|v|$.

d. Draw a sketch that includes all three of $|u|$, $|v|$, and $|u + v|$. Use the sketch to explain why $|u + v| \le |u| + |v|$. Why is this inequality called the triangle inequality?

62. ***Estimating Distance with a Bicycle*** In this project we will discover how an ordinary bicycle tire can be used as a measuring device. You will need a bicycle to perform this experiment.

   a. Measure the circumference of a tire to the closest $\frac{1}{16}$ of an inch by marking the distance that the bicycle travels in one revolution of the tire. Perform this experiment five times.

   b. Explain why there was any variance in the results of the five trials from part (a).

   c. Compute the average of the five results from part (a).

   d. Compute the difference between the largest and smallest results in part (a). We will assume that the actual circumference deviates from the average obtained in part (c) by no more than this difference.

   e. Let $C$ be the actual (unknown) circumference of the tire. Express the range of values of $C$ as the solution set to an absolute-value inequality and as an interval.

   f. Suppose that you compute the distance traveled on a lengthy trip by counting revolutions of your tire. Using the result from part (c), you estimate that the trip was 10 miles. Express the range of values of the actual trip length as the solution to an absolute-value inequality and as an interval.

   g. Take the bicycle to a standard $\frac{1}{4}$-mile track. Count the

smallest integer number of revolutions required to travel completely around the track in the inside lane, then compute the distance by which the bicycle has overshot the starting point. For example, it might happen that 180 revolutions are required, with the bicycle ending up 1 foot past the starting line at the end of the 180th revolution.

   h. Express the total distance traveled in part (g) in feet by adding the "overshoot" to $\frac{1}{4}$ mile.

   i. Use the result from part (h) to estimate the circumference of the tire.

   j. Assuming that the distance around the inside track is $\frac{1}{4}$ mile with a maximum error of 1 inch, and that the error involved in computing the overshoot is negligible, find the maximum error in computing the circumference by this method. Now express the range of values of the circumference as the solution set to an absolute-value inequality and as an interval.

   k. Now suppose that, using the value of the circumference found in part (i), you find that a certain trip was 10 miles long. Use the result of part (j) to express the range of values of the actual trip length as the solution set to an absolute-value inequality and as an interval. What's the maximum error in computing the trip length by this method?

   l. Compare and contrast the results from the two methods of computing the circumference.

   m. What physical conditions could affect the accuracy of your results?

![]  ***Questions for Discussion or Essay***

63. In Example 9 we considered a diet and exercise program that claimed a weight loss of $1\frac{1}{2}$ pounds each week. This implies that weight loss will occur in a linear way, which is based on the assumption that calories will continue to be burned at the same rate. Why isn't this valid? What modification could be made to the weight loss model?

64. When solving an absolute-value inequality of the form $|u| > c$, the first step, according to Property 2, is to write the statement

"$u > c$ or $u < -c$." It is not uncommon for beginning algebra students to ask at this juncture, Where did the absolute-value bars go? How would you respond to this question?

65. Many students, when solving an absolute-value inequality like $|x + 3| > 2$, will write down $-2 > x + 3 > 2$ as their first step and then arrive at $-5 > x > -1$ as an answer. Even if you know nothing about absolute value, what is it about this answer that tells you something is wrong?

---

**SECTION 4**

## POLYNOMIAL EQUATIONS AND INEQUALITIES

■ How many football fields could be covered with the material needed to construct a hot-air balloon capable of lifting the Statue of Liberty?

■ How can you determine a bullet's muzzle velocity with only a stopwatch?

■ If $u(u + 2) = 8$, why it is **not** true that either $u = 8$ or $u + 2 = 8$?

■ If a missile is shot straight up at 4000 feet per second, for how long will it be out of the stratosphere?

## POLYNOMIAL EQUATIONS

Recall that a polynomial is an expression involving one or more variables raised to positive integer powers. In this section we will consider equations and inequalities involving polynomials with only one variable, such as those given here.

$$x^2 - x - 6 = 0$$

$$4x^3 = 16x$$

$$243u^5 + 32 = 0$$

$$x^2 - x - 2 < 0$$

$$-x^3 - x^2 \geq -6x$$

As in the previous two sections, we are interested in determining the solution sets of such equations and inequalities. The first technique we will investigate in this section involves factoring and the following important property.

### *The zero-product property*

For real numbers $a$ and $b$, $ab = 0$ if and only if $a = 0$ or $b = 0$.

To illustrate the use of this property, consider the polynomial equation $x^2 - x - 6 = 0$. Equations such as this with zero on the right-hand side are said to be in **standard form**. If we factor the left-hand side, we obtain $(x + 2)(x - 3) = 0$. Now applying the zero-product property, we know that either $x + 2 = 0$ or $x - 3 = 0$. Solving each of these, we obtain $x = -2$ or $x = 3$. Thus, the solution set for the original equation is $\{-2, 3\}$.

### EXAMPLE 1    *Solving a polynomial equation by factoring*

Solve $u^2 + 2u = 8$.

**SOLUTION**    We first rewrite the equation in standard form (with a zero on the right-hand side) and then factor.

$$u^2 + 2u = 8$$

$$u^2 + 2u - 8 = 0 \qquad \text{Subtracting 8 from both sides}$$

$$(u + 4)(u - 2) = 0 \qquad \text{Factoring}$$

Now we apply the zero-product property.

$$u + 4 = 0 \quad \text{or} \quad u - 2 = 0$$

$$u = -4 \quad \text{or} \qquad u = 2$$

**CHECK**

| | $u = -4$ | | | $u = 2$ | |
|---|---|---|---|---|---|
| $u^2 + 2u$ | | 8 | $u^2 + 2u$ | | 8 |
| $(-4)^2 + 2(-4)$ | | | $(2)^2 + 2(2)$ | | |
| $16 - 8$ | | | $4 + 4$ | | |
| 8 | | ✓ | 8 | | ✓ |

So our solution set is $\{-4, 2\}$.

**EXAMPLE 2**   *Solving a polynomial equation*

Solve $4x^3 = 16x$.

**SOLUTION**

$$4x^3 = 16x$$

$$4x^3 - 16x = 0$$

$$4x(x^2 - 4) = 0 \qquad \text{Factoring out } 4x$$

$$4x(x + 2)(x - 2) = 0 \qquad \text{Factoring the difference of squares}$$

$$4x = 0 \quad \text{or} \quad x + 2 = 0 \quad \text{or} \quad x - 2 = 0 \qquad \text{Applying the zero-product property}$$

$$x = 0 \quad \text{or} \qquad x = -2 \quad \text{or} \qquad x = 2$$

**CHECK**

| $x = 0$ | | $x = -2$ | | $x = 2$ | |
|---|---|---|---|---|---|
| $4x^3$ | $16x$ | $4x^3$ | $16x$ | $4x^3$ | $16x$ |
| $4(0)^3$ | $16(0)$ | $4(-2)^3$ | $16(-2)$ | $4(2)^3$ | $16(2)$ |
| $0$ | $0$ ✓ | $4(-8)$ | $-32$ | $32$ | $32$ ✓ |
| | | $-32$ | ✓ | | |

The solution set is $\{-2, 0, 2\}$.

■

---

**WARNING**   In order to use the zero-product property to solve a polynomial equation, the equation must be in the standard form $\square = 0$. Thus, in Example 1 it would **not** be correct to simply factor $u^2 + 2u = 8$ as $u(u + 2) = 8$ and then claim that $u = 8$ or $u = 6$. See Exercise 85 for more discussion of this.

---

Solutions of equations of the form $p(x) = 0$ are called zeros of the polynomial $p(x)$. Thus if $p(a) = 0$, then $a$ is a zero of $p(x)$. Recall that $p(a)$ is obtained by replacing $x$ with $a$ in the polynomial $p(x)$.

**EXAMPLE 3**   *Verifying the zeros of a polynomial*

Show that $t = -1$, $t = -\frac{2}{3}$, and $t = \frac{1}{2}$ are zeros for the polynomial $p(t) = 6t^3 + 7t^2 - t - 2$.

**SOLUTION**

$$p(-1) = 6(-1)^3 + 7(-1)^2 - (-1) - 2 = -6 + 7 + 1 - 2 = 0$$

$$p\left(-\frac{2}{3}\right) = 6\left(-\frac{2}{3}\right)^3 + 7\left(-\frac{2}{3}\right)^2 - \left(-\frac{2}{3}\right) - 2$$

$$= -\frac{48}{27} + \frac{28}{9} + \frac{2}{3} - 2 = \frac{-48 + 84 + 18 - 54}{27} = 0$$

$$p\left(\frac{1}{2}\right) = 6\left(\frac{1}{2}\right)^3 + 7\left(\frac{1}{2}\right)^2 - \left(\frac{1}{2}\right) - 2$$

$$= \frac{6}{8} + \frac{7}{4} - \frac{1}{2} - 2 = \frac{3 + 7 - 2 - 8}{4} = 0$$

As we will see in Chapter 5, we can be sure that these three zeros are the only ones since $p(t)$ is a polynomial of degree 3.

---

**EXAMPLE 4**     *Finding the zeros of a polynomial*

Find the zeros of the polynomial $p(x) = (x - 2)^3 - (x - 2)$.

**SOLUTION**    We solve the polynomial equation $(x - 2)^3 - (x - 2) = 0$ by factoring out the common binomial factor $(x - 2)$.

$$(x - 2)^3 - (x - 2) = 0$$
$$(x - 2)[(x - 2)^2 - 1] = 0$$
$$(x - 2)[(x^2 - 4x + 4) - 1] = 0$$
$$(x - 2)(x^2 - 4x + 3) = 0$$
$$(x - 2)(x - 3)(x - 1) = 0$$
$$x = 2, \; x = 3, \; x = 1$$

---

**EXAMPLE 5**     *Finding a polynomial with a given zero*

Find the value for $c$ in the polynomial $p(x) = x^2 + cx - 8$ if one of the zeros is $-4$.

**SOLUTION**    Since $-4$ is a zero, it must be true that $p(-4) = 0$. That is, when we substitute $-4$ into the polynomial, we must get zero.

$$p(-4) = 0$$
$$(-4)^2 + c(-4) - 8 = 0 \qquad \text{Substituting } -4 \text{ into the polynomial}$$
$$8 - 4c = 0$$
$$-4c = -8$$
$$c = 2$$

---

**EXAMPLE 6**     *Muzzle velocity*

Ammunition is often rated according to its *muzzle velocity*, the velocity at which the bullet leaves the gun. If a bullet is shot straight up into the air from ground level with a muzzle velocity of $v_0$ feet per second, its height in feet after $t$ seconds is approximated by the polynomial $h(t) = -16t^2 + v_0 t$. Find the muzzle velocity of a bullet that hits the ground $2\frac{1}{2}$ minutes after it is fired.

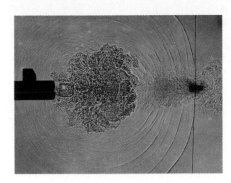

**SOLUTION**    Since the bullet hits the ground after $2\frac{1}{2}$ minutes (150 seconds), the value of the polynomial $h(t)$ must be zero when $t = 150$. Thus, $h(150) = 0$. Now we can proceed as in Example 5.

$$h(150) = 0$$
$$-16(150)^2 + v_0(150) = 0$$
$$-360{,}000 + 150v_0 = 0$$
$$150v_0 = 360{,}000$$
$$v_0 = 2400$$

So the muzzle velocity is 2400 feet per second.

---

**EXAMPLE 7**

*Approximating solutions to a polynomial equation*

Solve $x^5 - x^2 + 1 = 0$.

**SOLUTION**    Try as we might, we will not be able to factor $x^5 - x^2 + 1$ using only integer coefficients. It will thus be impossible for us to use the zero-product property. Instead, we will approximate any solutions of $x^5 - x^2 + 1 = 0$ using a graphics calculator. We begin by graphing $y = x^5 - x^2 + 1$, as shown in Figure 33. Next we zoom-in on the $x$-intercept several times and then use the trace feature to estimate the coordinates of the $x$-intercept, as shown in Figure 34. Thus, an approximate value for a solution of $x^5 - x^2 + 1 = 0$ is $-0.809$.

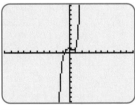

**FIGURE 33**

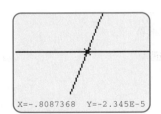

X=-.8087368    Y=-2.345E-5

**FIGURE 34**

---

It is generally more difficult to find solutions for polynomial equations when the polynomial cannot be readily factored. In the previous example we solved the equation graphically because the polynomial $x^5 - x^2 + 1$ didn't factor. Many (in some sense most) polynomial equations cannot be solved exactly by *any* algebraic technique so that the best that we can do is approximate the solutions using either a graphics calculator or some other approximation technique (see Exercise 84).

**EQUATIONS OF THE FORM** $x^n - a = 0$

For now, and again in the next section, we will consider some special types of polynomial equations that cannot be factored using integer coefficients and yet can be solved using basic algebraic techniques. The simplest type of polynomial equation that can be solved without factoring is one of the form $x^n - a = 0$. We will consider two cases:

when $n$ is even and when $n$ is odd. In both cases, we rewrite the equation in the form $x^n = a$ and then use $n$th roots to solve for $x$.

**EXAMPLE 8**    *Even nth roots*

Solve the equation $x^2 - 2 = 0$.

**SOLUTION**

$$x^2 - 2 = 0$$
$$x^2 = 2 \qquad \text{Rewriting in the form } x^n = a$$
$$\sqrt{x^2} = \sqrt{2} \qquad \text{Taking square roots on both sides}$$

We must be careful here since $\sqrt{x^2}$ does not simplify to $x$, but rather to $|x|$. Thus, we get $|x| = \sqrt{2}$ or, in other words, $x = \sqrt{2}$ or $x = -\sqrt{2}$. Our solution set is $\left\{-\sqrt{2}, \sqrt{2}\right\}$.

■

**EXAMPLE 9**    *Odd nth roots*

Solve the equation $243u^5 + 32 = 0$.

**SOLUTION**

$$243u^5 + 32 = 0$$
$$243u^5 = -32$$
$$u^5 = -\frac{32}{243}$$
$$\sqrt[5]{u^5} = \sqrt[5]{-\frac{32}{243}} \qquad \text{Taking 5th roots on both sides}$$
$$u = -\frac{2}{3}$$

■

The techniques used in the above examples generalize to the following rules for solving polynomial equations of the form $x^n = a$.

**Solving equations of the form $x^n = a$**

1. If $n$ is even and $a > 0$, then $x^n = a$ has two real solutions, $x = -\sqrt[n]{a}$ and $x = \sqrt[n]{a}$. We can condense this by writing $x = \pm\sqrt[n]{a}$.
2. If $n$ is odd, then $x^n = a$ has one real solution, $x = \sqrt[n]{a}$.

**EXAMPLE 10**    *Hot-air balloons and the Statue of Liberty*

A hot-air balloon capable of carrying two grown men has a volume of approximately 2750 cubic yards. Estimate the number of football fields that could be covered with the material needed to construct a hot-air balloon capable of lifting the Statue of Liberty.

**SOLUTION**   For approximation purposes, we will assume that 2750 cubic yards of hot air are needed for each 500 pounds of cargo. The Statue of Liberty weighs 27,000 tons or 54,000,000 pounds. The necessary volume for the hot-air balloon is thus

$$54{,}000{,}000 \text{ pounds} \cdot \frac{2750 \text{ cubic yards}}{500 \text{ pounds}} = 297{,}000{,}000 \text{ cubic yards}$$

If we assume the balloon is spherical, the volume of the sphere in terms of its radius is $V = \frac{4}{3}\pi r^3$. To find the radius of the balloon, we must solve $297{,}000{,}000 = \frac{4}{3}\pi r^3$ for $r$.

$$297{,}000{,}000 = \frac{4}{3}\pi r^3$$

$$\frac{3 \cdot 297{,}000{,}000}{4\pi} = r^3$$

$$\sqrt[3]{\frac{3 \cdot 297{,}000{,}000}{4\pi}} = r$$

Evaluating with a calculator, we obtain $r \approx 414$ yards. The amount of material can now be found by using the formula for surface area, $S = 4\pi r^2$. Using $r = 414$ yards, we obtain $S \approx 2{,}150{,}000$ square yards. A football field measures 120 yards by 40 yards (including the end zones) for a total of 4800 square yards. Thus, $2{,}150{,}000/4800 \approx 448$ football fields could be covered with the hot-air balloon material.

---

**EXAMPLE 11**   *Compound interest*

A savings account deposit of $1100 grows to $1248.74 after 3 years. If interest is compounded annually, find the interest rate.

**SOLUTION**   Recall that the formula for interest compounded annually is $B = P(1 + r)^t$. We set $B = 1248.74$, $P = 1100$, and $t = 3$.

$$1248.74 = 1100(1 + r)^3$$

$1.13522 = (1 + r)^3$   Dividing both sides by 1100

$\sqrt[3]{1.13522} = 1 + r$   Taking the cube root of both sides

$1.04318 = 1 + r$

$0.04318 = r$   Subtracting 1 from both sides

So the annual rate of interest is approximately 4.32%.

---

**POLYNOMIAL INEQUALITIES**

Consider the polynomial inequality $x^2 - x - 2 < 0$. If we let $y = x^2 - x - 2$, then we are interested in the $x$-values for which $y < 0$. Thus, the solutions of the inequality $x^2 - x - 2 < 0$ correspond to the intervals on which the graph of $y = x^2 - x - 2$ lies

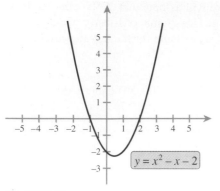

**FIGURE 35**

*below* the $x$-axis. Figure 35 shows the graph of $y = x^2 - x - 2$. Notice that the graph is below the $x$-axis for $x$ values between the two $x$-intercepts, which appear to be $-1$ and 2. Now to be precise, we should find exact values for the $x$-intercepts. We do this by solving the equation $x^2 - x - 2 = 0$.

$$x^2 - x - 2 = 0$$
$$(x - 2)(x + 1) = 0$$
$$x = 2, \ x = -1$$

Thus, the solution set of $x^2 - x - 2 < 0$ is the interval $(-1, 2)$. This suggests the following strategy for solving polynomial inequalities.

**Strategy for solving**
**polynomial inequalities**

1. Write the inequality in **standard form** with 0 on the right-hand side and a polynomial $p(x)$ on the left side [i.e., $p(x) > 0$; $p(x) < 0$; $p(x) \geq 0$; or $p(x) \leq 0$].

2. Find the $x$-intercepts of the polynomial $p(x)$. In other words, solve the equation $p(x) = 0$. If the $x$-intercepts cannot be found by algebraic means, estimate them graphically.

3. Determine where the graph of $y = p(x)$ is above or below the $x$-axis (depending on *which* inequality symbol is being used). Express the result using interval notation. If the inequality symbol is $\geq$ or $\leq$, the intervals will include their endpoints, otherwise not.

**EXAMPLE 12**   *Solving a polynomial inequality*

Solve $-x^3 - x^2 \geq -6x$.

**SOLUTION**   We begin by writing the inequality in standard form

$$-x^3 - x^2 + 6x \geq 0$$

Next we find the $x$-intercepts of $y = -x^3 - x^2 + 6x$.

$$-x^3 - x^2 + 6x = 0$$
$$-x(x^2 + x - 6) = 0$$
$$x(x + 3)(x - 2) = 0$$
$$x = 0, \ x = -3, \ x = 2$$

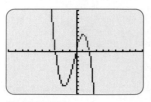

**FIGURE 36**

Then we use a graphics calculator to produce a graph of $y = -x^3 - x^2 + 6x$, as shown in Figure 36. Since the inequality symbol is $\geq$, we are interested in the $x$-values of the portion of the graph lying *on or above* the $x$-axis, shown in red in Figure 36. The graph is on or above the $x$-axis to the left of $-3$ and between 0 and 2. Thus, our solution is $(-\infty, -3] \cup [0, 2]$. Note that we include the endpoints since the inequality is "$\geq$" and not "$>$."

We recall that an alternative method for solving equations graphically is to graph the left and right sides of the equation on the same set of coordinate axes and find the

$x$-coordinates of the points of intersection. Similarly, the inequality $-x^3 - x^2 \geq -6x$ from Example 12 can be solved by graphing $y = -x^3 - x^2$ and $y = -6x$ on the same set of coordinate axes, as in Figure 37, and identifying those intervals for which the graph of $y = -x^3 - x^2$ is on or above the graph of $y = -6x$.

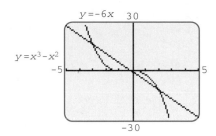

**FIGURE 37**

Now, in general, we can approximate the $x$-coordinates of the points of intersection of the graphs, but in this instance we know from our work in Example 12 that the graphs cross exactly at $x = -3$, $x = 0$, and $x = 2$. (Why?) Since the graph of $y = -x^3 - x^2$ is on or above the graph of $y = -6x$ to the left of 3 and between 0 and 2 (including $-3$, 0, and 2), the solution set is $(-\infty, -3] \cup [0, 2]$.

**EXAMPLE 13**    *Solving a polynomial inequality*

Solve $x^5 + 3x^4 + 3x^3 + 5x^2 - 7 < 0$.

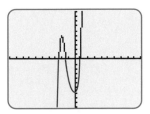

**FIGURE 38**

**SOLUTION**    We begin by finding the $x$-intercepts of $y = x^5 + 3x^4 + 3x^3 + 5x^2 - 7$. We would like to solve the equation $x^5 + 3x^4 + 3x^3 + 5x^2 - 7 = 0$, but the left side doesn't factor, and we have no algebraic technique for solving this equation. We will use the graph, shown in Figure 38, to *estimate* the $x$-intercepts. By zooming in on each of the $x$-intercepts and then using the trace feature, we produce the following approximations of the $x$-intercepts (accurate to the nearest one-hundredth): $-2.41$, $-1.33$, and $0.83$. Since the inequality symbol is $<$, we are interested in the $x$-values of the portion of the graph lying *below* the $x$-axis, which is red in Figure 38. The graph is below the $x$-axis to the left of $-2.41$ and between $-1.33$ and $0.83$. Thus, our solution is $(-\infty, -2.41) \cup (-1.33, 0.83)$.

---

**WARNING**    When finding $x$-intercepts of unfamiliar polynomials using a graphics calculator, be careful not to omit any $x$-intercepts that are out of the viewing rectangle. In Example 13 it would have been prudent to have zoomed out several times to make sure that there were no other $x$-intercepts.

## EXERCISES 4

**EXERCISES 1–30** □ *Find the real solution(s) of the given polynomial equation.*

**1.** $(x - 1)(x + 2) = 0$

**2.** $(a + 3)(a - 4)(a + 6) = 0$

**3.** $(2y + 1)(y + 3)(5y - 2) = 0$

**4.** $(3v + 2)(2v - 5) = 0$

**5.** $t^2 + 3t + 2 = 0$

**6.** $z^2 + 4z + 4 = 0$

**7.** $x^2 + 2x = 3$

**8.** $w^2 + 1 = 2w$

**9.** $(x - 2)(x^2 + 5x + 6) = 0$

**10.** $(t + 1)(t^2 - 9) = 0$

**11.** $(y + 1)(y + 4) = 4$

**12.** $x(x - 2) = 3$

**13.** $s^2 = 3s$

**14.** $5z = 2z^2$

**15.** $2x^2 + 5x - 3 = 0$

**16.** $3u^2 - 5u - 2 = 0$

**17.** $t^2 = 8$

**18.** $x^2 = 24$

**19.** $(y + 1)^2 = 3$

**20.** $(s - 2)^2 = 6$

**21.** $x^3 - 27 = 0$

**22.** $u^4 - 16 = 0$

**23.** $4t^3 - t = 0$

**24.** $8y^4 + y = 0$

**25.** $(w + 3)^3 + w^3 = 0$

**26.** $(x^2 - 1)^3 - 27 = 0$

**27.** $(x + 2)(x^2 + 5x + 9) + x^2 + x - 2 = 0$

**28.** $t^2(t + 1) + t^2(t + 3) + (t - 2)(t + 1) + (t - 2)(t + 3) = 0$

**29.** $(2w + 1)^3 + 2(2w + 1)^2 + (2w + 1) = 0$

**30.** $(x + 1)^3 - (x + 1) = 0$

**EXERCISES 31–34** □ *Show that the given numbers are zeros of the polynomial.*

**31.** $p(x) = 2x^3 - x^2 - 5x - 2;\ -1, -\dfrac{1}{2}, 2$

**32.** $p(x) = 3x^3 + 8x^2 + 3x - 2;\ -1, -2, \dfrac{1}{3}$

**33.** $p(x) = x^3 - x^2 - 2x + 2;\ -\sqrt{2}, \sqrt{2}, 1$

**34.** $p(x) = x^3 - 2x^2 - 3x + 6;\ -\sqrt{3}, \sqrt{3}, 2$

**EXERCISES 35–40** □ *Find the value(s) for c.*

**35.** One zero of $p(x) = x^2 + cx + 20$ is 5.

**36.** One zero of $p(x) = x^2 + x + c$ is $-2$.

**37.** One zero of $p(x) = cx^2 - 5x - 3$ is $-\dfrac{1}{2}$.

**38.** One zero of $p(x) = 3x^2 + cx + 6$ is $\dfrac{2}{3}$.

**39.** $p(x) = x^2 - cx + 4$ has only one zero.

**40.** $p(x) = x^2 + 8x + c$ has only one zero.

**EXERCISES 41–68** □ *Solve the polynomial inequality. Give exact answers wherever possible; when approximating, give answers accurate to the nearest hundredth.*

**41.** $(t + 3)(t - 1) \le 0$

**42.** $(x - 2)(x + 1) > 0$

**43.** $(2y - 1)(y + 2) > 0$

**44.** $(s - 3)(4s + 2) \le 0$

**45.** $x^2 + 4x \le 5$

**46.** $t^2 + 6 > -5t$

**47.** $2x^2 + 3 > 7x$

**48.** $6w \ge w^2 + 8$

**49.** $u^2 \le 8$

**50.** $x^2 \ge 25$

**51.** $x^3 < x^2 + 3$

**52.** $x^3 + 1 < 3x$

**53.** $x^4 + x > 1$

**54.** $x^4 < 1 - x^3$

**55.** $x(x + 1)(x - 3) > 0$

**56.** $(x - 1)(x + 2)(x + 4) \le 0$

**57.** $(x - 5)^2(x + 1)^3 < 0$

**58.** $(t + 8)^2(t - 2)^4 > 0$

**59.** $x(x - 2)(x + 3) < -2$

**60.** $x^4 - 2x^3 + x + 1 \le 0$

**61.** $(7x - 2)^4(2x + 3)^6 \le 0$

**62.** $(4y + 1)^3(6y - 5)^3 \ge 0$

**63.** $3t^3 + 3t^2 > 0$

**64.** $2x^4 + 6x^3 \le 0$

**65.** $2v^3 - 8v \ge 0$

**66.** $3y^4 - 27y^2 > 0$

**67.** $(x - 15)(x + 20)(x - 27) > -100$

**68.** $(x - 10)^2(x - 30) < -40$

■ *Applications*

**69.** *Throwing a Ball* A ball thrown upward with an initial velocity of 160 feet per second has height $s = -16t^2 + 160t$ feet after $t$ seconds. Solve the equation $-16t^2 + 160t = 0$ to find the time when it hits the ground.

**70.** *Falling Ball* A ball dropped from the top of the Sears Tower in Chicago has height $s = -16t^2 + 1454$ feet after $t$ seconds. How long before the ball hits the ground?

**71.** *Compound Interest* A deposit of $5000 reaches a balance of $6288.60 after 4 years. If interest is compounded annually, what is the annual percentage rate?

**72.** *Compound Interest* If you borrow $1000 from your father and pay him back $1100 after 2 years, what is the annual percentage rate, assuming interest is compounded annually?

**73.** *Circumference of a Tree* A board foot is a unit of lumber measurement equal to 1 foot square by 1 inch thick. The number of board feet in a Ponderosa pine can be approximated with the polynomial $B = 0.0015c^3$ where $c$ is the circumference of the tree in inches at waist height. What is the circumference of a Ponderosa pine with 2500 board feet?

**74.** *Radius of a Sphere* The volume of a sphere with radius $r$ is $V = \frac{4}{3}\pi r^3$. If the volume of a sphere is 200 cubic centimeters, what is its radius?

**75.** *Great Lakes Snowballs* Approximate the radius of a snowball that contains enough moisture to fill the Great Lakes. The total volume of water in the Great Lakes is approximately 5440 cubic miles and the amount of water in each cubic inch of loosely packed snow is approximately $\frac{1}{6}$ cubic inch.

**76.** *Dimensions of a Can* The volume of a cylinder with radius $r$ and height $h$ is $V = \pi r^2 h$. If a can of vegetables with a volume of 50 cubic inches is to have a height of twice the radius, find the dimensions of the can.

77. **Deer Population** A population of deer in a game preserve can be modeled by $P = 42t^2 + 36t + 1008$ where $t$ is measured in years. After how many years will the population be equal to 1620?

78. **Painting Revenue** An artist receives a revenue of $R = x(100 - x)$ dollars from the sale of $x$ paintings. For what interval of sales will the revenue exceed $1875?

79. **Height of a Projectile** A projectile is fired straight up with an initial velocity of 80 feet per second. Its height at time $t$ (in seconds) is $s = -16t^2 + 80t$. During what time period will the projectile be above 96 feet?

80. **Area of a Corral** A rancher has 200 feet of fencing with which to enclose a corral next to a stone wall (see Figure 39). If the width is $x$ feet, the resulting area is given by $A = x(200 - 2x)$. For what range of $x$ values will the resulting area be at least 3200 square feet?

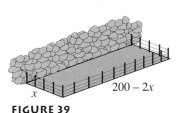

$x$     $200 - 2x$

**FIGURE 39**

81. **Volume of a Box** A rectangular package to be sent by the Postal Service can have a maximum combined length and girth (perimeter of the base) of 108 inches (see Figure 40). The volume of a box with a square base that exactly meets this restriction is $V = x^2(108 - 4x)$ cubic inches where $x$ is the width of the base in inches. Find all possible values of $x$ for which the volume is positive.

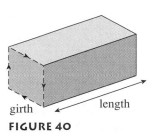

girth     length

**FIGURE 40**

82. **Missile Flight** If a missile is shot straight up at 4000 feet per second and we assume that air resistance is negligible (is it?), then the height of the missile in feet $t$ seconds after blast off is given by $h = -16t^2 + 4000t$. The stratosphere, the innermost portion of the atmosphere, extends approximately 30 miles from the Earth's surface. For what period of time will the missile be out of the stratosphere? (There are 5280 feet in 1 mile.)

## Projects for Enrichment

83. **The Effects of Shape on Surface Area**

   a. Human beings have a density approximately that of water, which weighs about 62.4 pounds per cubic foot. Estimate the volume of a 200-pound man.

   b. Estimate the radius of a ball that has the volume of a 200-pound man, and compute the surface area of such a ball.

   c. For each height given below, estimate the radius and surface area of the cylinder such that the resulting volume is that of a 200-pound man.

       i. $h = 3$ feet      ii. $h = 4$ feet

       iii. $h = 5$ feet     iv. $h = 6$ feet

       v. $h = 7$ feet

   d. Based on your results from parts (b) and (c), what type of 200-pound man would most likely have the greatest surface area: a very short obese man, a muscular 6 footer, or a slender 7 footer?

   e. Given that heat is dissipated through the skin, comment on the ideal bodily proportions for cold versus warm climates. What other factors besides surface area might affect the ability of the body to dissipate or retain heat?

84. **Locating Solutions** Even when polynomial equations cannot be solved *exactly* using factoring or some other method, it is possible to *approximate* the real solutions using a variety of computational techniques. However, the approximation techniques do require an initial guess for the solution. How does one go about coming up with an initial guess? One way is with a specialized form of the *Intermediate-Value Theorem*, abbreviated IVT. Consider a polynomial equation in the form $p(x) = 0$. If you can find two real numbers $a$ and $b$ such that $p(a)$ and $p(b)$ have opposite sign (i.e., one is positive and the other is negative), then the IVT guarantees that there is a solution to the equation somewhere between $a$ and $b$. Let's try it for the equation $x^2 - 2 = 0$. Set $p(x) = x^2 - 2$. Since $p(1) = -1$ is negative and $p(2) = 2$ is positive, the IVT says there is a solution to the equation between $x = 1$ and $x = 2$.

   a. Use the IVT to show that there is another solution to $x^2 - 2 = 0$ between $x = -2$ and $x = -1$.

Notice that the IVT doesn't tell us what the solutions are. It just shows us intervals in which they lie. We would need to use other methods to actually find or approximate the solutions.

   b. What are the actual solutions to the equation $x^2 - 2 = 0$? Show they are in the intervals we verified with the IVT.

   c. Use the IVT to show that $x^3 + 5 = 0$ has a solution in the interval $(-2, -1)$. Then find the actual solution and confirm that it indeed lies in the interval $(-2, -1)$.

**d.** Use the IVT to show that $x^2 + 2x - 2 = 0$ has a solution in the interval $(0, 1)$ and another in the interval $(-3, -2)$. In Section 5 we'll see that the quadratic formula gives us the two solutions $x = -1 - \sqrt{3}$ and $x = -1 + \sqrt{3}$. Show these are in the intervals.

**e.** Use the IVT to show that $x^3 - 3x + 1 = 0$ has solutions in the intervals $(-2, -1)$, $(0, 1)$, and $(1, 2)$.

**f.** Describe a procedure for "zeroing in" on the solution of $x^3 - 3x + 1 = 0$ that lies in the interval $(0, 1)$.

### ■▙ *Questions for Discussion or Essay*

**85.** Explain why $u(u + 2) = 0$ indicates that either $u = 0$ or $u + 2 = 0$, but $u(u + 2) = 8$ does *not* mean $u = 8$ or $u + 2 = 8$.

**86.** Earlier we mentioned the hazard of dividing equations through by variable factors and hence losing solutions. What solution would be lost if $x + 2$ was divided through on both sides of $x(x + 2) = x + 2$? Explain the correct way to solve this equation. Give another example of such an equation, other than any in the text, and discuss the proper way to solve it.

**87.** The polynomial equation $x^3 - 7x - 6 = 0$ factors as $(x - 3)(x + 1)(x + 2) = 0$ and hence has solutions $x = 3$,

$x = -1$, and $x = -2$. Do you think this works in reverse? In other words, if you are given that $x = -1$, $x = 2$, and $x = 3$ are the solutions of $x^3 - 4x^2 + x + 6 = 0$, what can you conclude about how $x^3 - 4x^2 + x + 6$ can be factored? How do you think $2x^2 - 5x + 2$ should factor given that $2x^2 - 5x + 2 = 0$ has solutions $x = \frac{1}{2}$ and $x = 2$? Explain.

**88.** Consider the statement, A polynomial equation can have only as many solutions as the degree of the polynomial. Now solve the equation $x^3 - 8 = 0$. Does your solution in any way contradict the statement? Explain.

---

## SECTION 5

# QUADRATIC EQUATIONS

■ If there is a quadratic formula for solving second-degree polynomial equations, is there a cubic formula for solving third-degree equations? What about fourth-, fifth-, and higher degree polynomial equations?

■ How can you tell within 5 seconds whether or not the polynomial $3x^2 + 10x + 6$ factors, without ever attempting to factor it?

■ A basketball player with a 48-inch vertical jump must leap 2 feet in order to dunk a basketball. What is the minimum time required for him to slam the basketball?

■ What is the radius of a 1-foot-high paint can that holds just enough paint to cover its exterior?

### ■ COMPLETING THE SQUARE

Recall from Section 4 that any equation of the form $\square^2 = d$ (with $d > 0$) is equivalent to $\square = \pm\sqrt{d}$. This fact can be used to solve some quadratic equations, as the following examples illustrate.

**EXAMPLE 1**    *Solving a quadratic equation of the form $\square^2 = d$*

Solve $(x - 13)^2 = 25$.

**SOLUTION**
$$(x - 13)^2 = 25$$
$$x - 13 = \pm\sqrt{25}$$
$$x - 13 = \pm 5$$
$$x = 13 \pm 5$$
$$x = 18, \ x = 8$$

**EXAMPLE 2**    *Solving a quadratic equation of the form □² = d*

Solve $x^2 + 6x + 9 = 10$.

**SOLUTION**

$$x^2 + 6x + 9 = 10$$
$$(x + 3)^2 = 10$$
$$x + 3 = \pm\sqrt{10}$$
$$x = -3 \pm \sqrt{10}$$
$$x = -3 - \sqrt{10}, \; x = -3 + \sqrt{10}$$

■

Note that in Example 2 we had to first express the left side as a perfect square. If the left side is not already a perfect square, then we can make it so by a process called **completing the square**. To complete the square, we add an appropriate constant to both sides so that the left side becomes a perfect square.

For example, suppose that we are given the expression $x^2 + px$ and we are asked to find a constant $c$ so that $x^2 + px + c$ will be a perfect square. If $x^2 + px + c$ is to be a perfect square, then it can be written in the form $(x + \square)^2$ for some $\square$. Multiplying out, this gives

$$x^2 + 2\square x + \square^2 = x^2 + px + c$$

From this we see that $2\square x = px$, so that $\square = p/2$, and $c = \square^2 = (p/2)^2$. Our strategy then is this: we will rearrange the terms of the quadratic equation so that the equation is of the form $x^2 + px = d$, and then we will complete the square by adding $(p/2)^2$ to both sides. We illustrate this process in the following examples.

**EXAMPLE 3**    *Solving a quadratic equation by completing the square*

Solve $x^2 + 10x + 5 = 8$ for $x$ by completing the square.

**SOLUTION**

$$x^2 + 10x + 5 = 8$$

$$x^2 + 10x = 3 \qquad \text{Subtracting 5 from both sides to eliminate the constant term.}$$

$$x^2 + 10x + \left(\frac{10}{2}\right)^2 = 3 + \left(\frac{10}{2}\right)^2 \qquad \text{Completing the square by adding the square of one-half the coefficient of } x \text{ to both sides}$$

$$x^2 + 10x + 25 = 3 + 25$$

$$(x + 5)^2 = 28 \qquad \text{Expressing the left side as a perfect square}$$

$$x + 5 = \pm\sqrt{28}$$

$$x = -5 \pm 2\sqrt{7}$$

■

If the coefficient of $x^2$ is not 1, then we simply divide both sides by this coefficient, as in the following example.

| EXAMPLE 4 | *Solving a quadratic equation by completing the square* |

Solve $2x^2 + 4x + 5 = 7 - 2x$.

**SOLUTION**

$$2x^2 + 4x + 5 = 7 - 2x$$

$$2x^2 + 6x = 2 \qquad \text{Adding } 2x \text{ and subtracting 5 on both sides}$$

$$x^2 + 3x = 1 \qquad \text{Dividing both sides by 2, the coefficient of } x^2$$

$$x^2 + 3x + \left(\frac{3}{2}\right)^2 = 1 + \left(\frac{3}{2}\right)^2 \qquad \text{Completing the square}$$

$$\left(x + \frac{3}{2}\right)^2 = 1 + \frac{9}{4} \qquad \text{Expressing the left side as a perfect square}$$

$$\left(x + \frac{3}{2}\right)^2 = \frac{13}{4}$$

$$x + \frac{3}{2} = \pm\frac{\sqrt{13}}{2}$$

$$x = -\frac{3}{2} \pm \frac{\sqrt{13}}{2}$$

$$x = \frac{-3 \pm \sqrt{13}}{2}$$

We summarize the technique of completing the square as follows.

*Solving quadratic equations by completing the square*

> 1. Write the quadratic equation so that all the variable expressions are on the left side and the constant term is on the right.
> 2. Divide both sides of the equation by the coefficient of $x^2$.
> 3. Complete the square by adding the square of one-half the coefficient of $x$ to both sides of the equation. (Divide the coefficient of $x$ by 2, square the result, and then add to both sides of the equation.)
> 4. Express the left side as the square of a binomial.
> 5. Take the square root of both sides and solve for $x$.

## THE QUADRATIC FORMULA

Since completing the square is a somewhat tedious process, we will complete the square once and for all on the general quadratic equation $ax^2 + bx + c = 0$ and derive the **quadratic formula**:

$$ax^2 + bx + c = 0$$

$$ax^2 + bx = -c \qquad \text{Subtracting } c \text{ from both sides}$$

$$x^2 + \frac{b}{a}x = -\frac{c}{a} \qquad \text{Dividing both sides by } a,\text{ the coefficient of } x^2$$

$$x^2 + \frac{b}{a}x + \left(\frac{b}{2a}\right)^2 = -\frac{c}{a} + \left(\frac{b}{2a}\right)^2 \qquad \text{Adding the square of one-half the coefficient of } x \text{ to both sides.}$$

$$\left(x + \frac{b}{2a}\right)^2 = \frac{b^2}{4a^2} - \frac{c}{a} \qquad \text{Expressing the left side as the square of a binomial}$$

$$\left(x + \frac{b}{2a}\right)^2 = \frac{b^2 - 4ac}{4a^2} \qquad \text{Simplifying}$$

$$x + \frac{b}{2a} = \pm\frac{\sqrt{b^2 - 4ac}}{2a} \qquad \text{Taking square roots}$$

$$x = \frac{-b \pm \sqrt{b^2 - 4ac}}{2a} \qquad \text{Solving for } x$$

Thus, we have obtained a solution to the general quadratic equation in terms of the coefficients $a$, $b$, and $c$.

**The quadratic formula**

The solutions to the quadratic equation $ax^2 + bx + c = 0$ are given by

$$x = \frac{-b \pm \sqrt{b^2 - 4ac}}{2a}$$

**EXAMPLE 5**

*Solving a quadratic equation with the quadratic formula*

Solve $y^2 + 3y + 1 = 0$ with the quadratic formula.

**SOLUTION**    Here $a = 1$, $b = 3$, and $c = 1$. Substituting into the quadratic formula and simplifying, we have

$$y = \frac{-b \pm \sqrt{b^2 - 4ac}}{2a}$$

$$= \frac{-3 \pm \sqrt{3^2 - 4 \cdot 1 \cdot 1}}{2 \cdot 1}$$

$$= \frac{-3 \pm \sqrt{5}}{2}$$

Thus, we obtain the two solutions $y = (-3 + \sqrt{5})/2 \approx -0.382$, and $y = (-3 - \sqrt{5})/2 \approx -2.618$.

**EXAMPLE 6**

*Solving a quadratic equation with the quadratic formula*

Solve $-10 + x^2 = 3x$ with the quadratic formula.

**SOLUTION**    We begin by writing the quadratic equation in standard form with zero on the right side and the terms in descending order.

$$-10 + x^2 = 3x$$

$$x^2 - 3x - 10 = 0$$

Now we can read off the coefficients, obtaining $a = 1$, $b = -3$, and $c = -10$. Substituting into the quadratic formula and simplifying, we obtain

$$x = \frac{-b \pm \sqrt{b^2 - 4ac}}{2a}$$

$$= \frac{-(-3) \pm \sqrt{(-3)^2 - 4 \cdot 1 \cdot (-10)}}{2 \cdot 1}$$

$$= \frac{3 \pm \sqrt{9 + 40}}{2}$$

$$= \frac{3 \pm \sqrt{49}}{2}$$

$$= \frac{3 \pm 7}{2}$$

Thus $x = \frac{10}{2} = 5$, or $x = -\frac{4}{2} = -2$

---

**EXAMPLE 7**    *An application of the quadratic equation to an equation of motion*

Suppose that a ball is thrown upward from a height of 16 feet at 16 feet per second. How long will it take for the ball to hit the ground? (Assume that the height of the ball in feet after $t$ seconds is given by $h = -16t^2 + v_0 t + h_0$, where $v_0$ is the initial velocity and $h_0$ is the initial height.)

**SOLUTION**    Using $v_0 = 16$ and $h_0 = 16$, we obtain $h = -16t^2 + 16t + 16$. The ball will hit the ground when $h = 0$, and so we must solve the quadratic equation $-16t^2 + 16t + 16 = 0$.

$$-16t^2 + 16t + 16 = 0$$

$$t^2 - t - 1 = 0 \qquad \text{Dividing through by } -16$$

$$t = \frac{-(-1) \pm \sqrt{(-1)^2 - 4(1)(-1)}}{2(1)} \qquad \text{Appling the quadratic formula}$$

$$= \frac{1 \pm \sqrt{5}}{2}$$

Thus, $t = \dfrac{1 + \sqrt{5}}{2} \approx 1.618$, or $t = \dfrac{1 - \sqrt{5}}{2} \approx -0.618$

Since the variable $t$ represents a positive quantity, the desired solution is $t \approx 1.618$ seconds.

---

**EXAMPLE 8**    *Solving a quadratic equation with the quadratic formula*

Solve $z^2 + 6z + 9 = 0$ with the quadratic formula.

**SOLUTION**   Here $a = 1$, $b = 6$, and $c = 9$. Substituting into the quadratic formula and simplifying, we obtain

$$
\begin{aligned}
z &= \frac{-b \pm \sqrt{b^2 - 4ac}}{2a} \\
&= \frac{-6 \pm \sqrt{6^2 - 4 \cdot 1 \cdot 9}}{2 \cdot 1} \\
&= \frac{-6 \pm 0}{2} \\
&= -3
\end{aligned}
$$

Note that this quadratic equation only has one solution. This is because the expression under the radical, $b^2 - 4ac$, simplified to 0.

The quadratic formula can also be used to find nonreal complex solutions of quadratic equations. Recall that we define $i = \sqrt{-1}$, so that for positive real numbers $a$,

$$\sqrt{-a} = \sqrt{a} \cdot \sqrt{-1} = \sqrt{a}\, i$$

**EXAMPLE 9**

*Using the quadratic formula when there are no real solutions*

Use the quadratic formula to find all solutions of $2t^2 + 3t + 4 = 0$.

**SOLUTION**   Here $a = 2$, $b = 3$, and $c = 4$. Substituting into the quadratic formula and simplifying we have

$$
\begin{aligned}
t &= \frac{-b \pm \sqrt{b^2 - 4ac}}{2a} \\
&= \frac{-3 \pm \sqrt{3^2 - 4 \cdot 2 \cdot 4}}{2 \cdot 2} \\
&= \frac{-3 \pm \sqrt{-23}}{4} \\
&= \frac{-3 \pm \sqrt{23}\, i}{4}
\end{aligned}
$$

Note that this quadratic equation has no real solutions, although it does have two nonreal complex solutions. The reason for this is that the expression under the radical, $b^2 - 4ac$, was negative.

## THE DISCRIMINANT

As we have seen in the previous examples, quadratic equations may have either 2, 1, or 0 *real* solutions. Since the solutions to the quadratic equation $ax^2 + bx + c = 0$ are given by

$$x = \frac{-b \pm \sqrt{b^2 - 4ac}}{2a}$$

we see that the quantity inside the radical, $b^2 - 4ac$, will determine the number of real solutions. If $b^2 - 4ac$ is positive, then $\sqrt{b^2 - 4ac}$ is a positive real number, so that

$$\frac{-b + \sqrt{b^2 - 4ac}}{2a} \quad \text{and} \quad \frac{-b - \sqrt{b^2 - 4ac}}{2a}$$

are distinct real solutions. If $b^2 - 4ac$ is zero, then we have

$$x = \frac{-b \pm \sqrt{0}}{2a} = -\frac{b}{2a}$$

so that there is only one real solution. If $b^2 - 4ac$ is negative, then $\sqrt{b^2 - 4ac}$ isn't real, so that there are no real solutions. We define the **discriminant** of the quadratic equation $ax^2 + bx + c = 0$ to be the quantity $b^2 - 4ac$. We can use the discriminant to quickly determine the number of real solutions to a quadratic equation without actually finding the solutions themselves.

*Interpretation of the discriminant*

> Let $d = b^2 - 4ac$ be the discriminant of the quadratic equation $ax^2 + bx + c = 0$.
>
> **1.** If $d > 0$, then there are two real solutions.
> **2.** If $d = 0$, then there is one real solution.
> **3.** If $d < 0$, then there are no real solutions.

**EXAMPLE 10**   *Using the discriminant to find the number of real solutions to quadratic equations*

For each quadratic equation given below, use the discriminant to find the number of real solutions.

**a.** $x^2 - 3x + 5 = 0$     **b.** $2x^2 + 3x - 1 = 0$     **c.** $9x^2 + 12x + 4 = 0$

**SOLUTION**

**a.** Here $d = (-3)^2 - 4 \cdot 1 \cdot 5 = 9 - 20 = -11$. Since $-11 < 0$, there are no real solutions.

**b.** Here $d = 3^2 - 4 \cdot 2 \cdot (-1) = 9 - (-8) = 17$. Since $17 > 0$, there are two real solutions.

**c.** Here $d = 12^2 - 4 \cdot 9 \cdot 4 = 144 - 144 = 0$. Thus, there is only one real solution.

## EQUATIONS OF QUADRATIC TYPE

Recall that a quadratic equation in the variable $x$ is of the form $ax^2 + bx + c = 0$ and can be solved by using the quadratic formula. Some equations, although not quadratic, are of **quadratic type**; that is, they can be written in the form $a\square^2 + b\square + c = 0$, where $\square$ is an expression involving $x$. Many equations of quadratic type can be solved using the quadratic formula.

Consider the equation $x^4 - 4x^2 + 1 = 0$. By rewriting the equations as $(x^2)^2 - 4(x^2) + 1 = 0$, we see that it is quadratic in the variable $x^2$. In fact, if we let $u = x^2$, then we may rewrite the equation as

$$u^2 - 4u + 1 = 0$$

# FERMAT'S LAST THEOREM

In this text we are primarily interested in real-number solutions to equations; **Diophantine equations**, named after Diophantus of Alexandria, are equations for which only integer solutions are desired. For example, the equation $a^2 + b^2 = c^2$ has integer solutions consisting of Pythagorean triples, such as $a = 3$, $b = 4$, and $c = 5$. In fact, it is easily shown that $a^2 + b^2 = c^2$ has infinitely many integer solutions.

Sometime in the early seventeenth century, the amateur mathematician Pierre Fermat was reading a copy of *Arithmetic*, written by Diophantus himself, when he made what is undoubtedly the most significant marginal note of all time: "It is impossible to separate a cube into two cubes or a fourth power into two fourth powers or, in general, any power greater than the second into powers of like degree. I have discovered a truly marvelous demonstration, which this margin is too narrow to contain." Fermat was asserting that no equation of the form $a^3 + b^3 = c^3$, or $a^4 + b^4 = c^4$, or indeed $a^n + b^n = c^n$ for any $n > 2$ has integer solutions. In subsequent years, Fermat, in correspondence with the leading mathematicians of his time, proved that the equations $a^3 + b^3 = c^3$ and $a^4 + b^4 = c^4$ never have integer solutions, but his proof of the more general case referred to in his marginal note was never found. His claim was dubbed, "Fermat's last theorem."

Over the course of the next three centuries, entire branches of mathematics arose from ill-fated attempts at proving Fermat's last theorem. In spite of the best efforts of some of the greatest intellects the world has seen, the theorem has stubbornly refused

Andrew Wiles presenting his results.

to yield a proof. The best that could be done was to establish the theorem for particular cases. For example, it has been shown that Fermat's last theorem is valid for all equations $a^n + b^n = c^n$, where $n < 4{,}000{,}000$.

In June of 1993, Andrew Wiles of Princeton University shocked the mathematical community when he announced that he had proven Fermat's last theorem. A brilliant mathematician with a distinguished record of research, Wiles had worked in relative seclusion for some 7 years on proving a conjecture, the truth of which would imply Fermat's last theorem. Upon close examination of the 200-page paper in which this conjecture is established, referees found several gaps in his proof, all but one of which was easily corrected. Wiles successfully bridged the last gap in 1994, and the complete proof has been verified by many prominent mathematicians.

Of this we can be fairly certain: Fermat did *not* have a correct proof of the theorem that bears his name. But we can forgive him this. After all, history has recorded hundreds of instances of extremely intelligent people erring in their treatment of this deceptively difficult theorem. Some respected mathematicians (such as Fermat himself in all likelihood) have produced purported proofs containing subtle flaws unnoticed by their creators. More recently, Marilyn vos Savant (who is listed in the *Guinness Book of World Records* as having the highest IQ) writing in the "*Ask Marilyn*" column appearing in Parade magazine, criticized Wiles's general approach with an invalid argument.

Solving for $u$, we obtain

$$u = \frac{-(-4) \pm \sqrt{(-4)^2 - 4 \cdot 1 \cdot 1}}{2 \cdot 1} = \frac{4 \pm \sqrt{12}}{2} = 2 \pm \sqrt{3}$$

Now using the fact that $u = x^2$, we have

$$x^2 = 2 + \sqrt{3} \quad \text{or} \quad x^2 = 2 - \sqrt{3}$$

from which it follows that

$$x = \pm\sqrt{2 + \sqrt{3}} \quad \text{or} \quad x = \pm\sqrt{2 - \sqrt{3}}$$
$$\approx \pm 1.932 \qquad\qquad \approx \pm 0.518$$

Thus, there are four solutions to $x^4 - 4x^2 + 1 = 0$.

**EXAMPLE 11**    *Solving an equation of quadratic type*

Solve the following equation of quadratic type.

$$\left(\frac{x-3}{2}\right)^2 + 5\left(\frac{x-3}{2}\right) + 5 = 0$$

**SOLUTION**    Here we let $u = \dfrac{x-3}{2}$ to obtain

$$u^2 + 5u + 5 = 0$$

Solving for $u$, we obtain

$$u = \frac{-5 \pm \sqrt{5^2 - 4 \cdot 1 \cdot 5}}{2}$$

$$= \frac{-5 \pm \sqrt{5}}{2}$$

Using the fact that $u = \dfrac{x-3}{2}$, we have

$$\frac{x-3}{2} = \frac{-5 \pm \sqrt{5}}{2}$$

$$x - 3 = -5 \pm \sqrt{5}$$

$$x = -2 \pm \sqrt{5}$$

---

## EXERCISES 5

**EXERCISES 1–6** □ *Solve the given equation.*

**1.** $x^2 = 9$               **2.** $x^2 = 16$

**3.** $(x + 3)^2 = 16$       **4.** $(y - 2)^2 = -4$

**5.** $3(t + 1)^2 = -75$     **6.** $2(x + 6)^2 = 50$

**EXERCISES 7–16** □ *Solve the given equation by completing the square.*

**7.** $x^2 + 5x + 6 = 0$      **8.** $x^2 - 4x - 21 = 0$

**9.** $x^2 - 2x - 3 = 0$      **10.** $x^2 + 4x + 1 = 0$

**11.** $y^2 + 3y + 1 = 0$      **12.** $s^2 - 5s + 2 = 0$

**13.** $x(x - 3) = 2$         **14.** $2x^2 + 6x = 8$

**15.** $4y^2 - 24y = -37$    **16.** $y(1 - y) = \dfrac{5}{4}$

**EXERCISES 17–38** □ *Solve using the quadratic formula.*

**17.** $x^2 - x - 6 = 0$       **18.** $x^2 - 4x - 21 = 0$

**19.** $6t^2 - t - 2 = 0$       **20.** $15y^2 - 4y - 4 = 0$

**21.** $x^2 + 6x = -9$        **22.** $y^2 - 14y = -49$

**23.** $u^2 - 6u + 3 = 0$      **24.** $x^2 + 4x + 2 = 0$

**25.** $2x^2 + 4x + 1 = 0$     **26.** $3t^2 - 5t - 5 = 0$

**27.** $x^2 + x = 1$           **28.** $y(2y + 7) = 1$

**29.** $x^2 - 1 = 0$            **30.** $y^2 + 1 = 0$

**31.** $w^2 + w + 1 = 0$      **32.** $2x^2 - 3x + 2 = 0$

**33.** $3y^2 - 2y + 1 = 0$

**34.** $(x + 1)^3 = (x - 1)^3 + 9x$

**35.** $2x^4 - 6x^2 + 1 = 0$

**36.** $v^4 - 2v^2 + 1 = 0$

**37.** $\left(\dfrac{w+3}{5}\right)^2 + 3\left(\dfrac{w+3}{5}\right) + 1 = 0$

**38.** $(5z + 4)^2 + 4(5z + 4) + 4 = 0$

**EXERCISES 39–48** □ *Indicate the number of real roots for each quadratic equation.*

39. $x^2 + 3x + 1 = 0$

40. $x^2 - 3x - 1 = 0$

41. $x^2 + 7 - 0$

42. $3y^2 - 5y + 3 = 0$

43. $2z^2 + 4z - 3 = 0$

44. $9w^2 + 12w + 4 = 0$

45. $4x^2 + 12x + 9 = 0$

46. $12x^2 - 11x - 36 = 0$

47. $6x^2 - 10x - 10 = 0$

48. $12s^2 + s - 6 = 0$

**EXERCISES 49–52** □ *Solve for the indicated variable.*

49. $a^2 + xa + x = 0$   (solve for $a$)

50. $w^2 + pw + q = 0$   (solve for $w$)

51. $2x^2 + tx + t^2 = 0$   (solve for $t$)

52. $x^2 + 2xy + y^2 - 3 = 0$   (solve for $x$)

■ *Applications*

53. *Dimensions of a Can*  The height of a cylindrical soft drink can is 1 inch more than three times its radius. Its lateral surface area (the area of the "side" of the can, as opposed to the top or the bottom) is $30\pi$ square inches. Find the height and the radius of the can. (The lateral surface area $L$ of a cylinder of radius $r$ and height $h$ is given by $L = 2\pi rh$.)

54. *Height of a Ball*  The height of a ball (in feet) $t$ seconds after it is thrown down from the top of the Sears Tower at 50 feet per second is given by $h = -16t^2 - 50t + 1454$. How long does it take for the ball to reach the ground?

55. *Vertical Jump*  If a basketball player has a vertical jump of $J$ inches, then it can be shown that her elevation in inches $t$ seconds after jumping is given by $h = -192t^2 + 16\sqrt{3J}\, t$.

USC and 1996 Olympic basketball star Lisa Leslie dunking.

a. If a basketball player has a vertical jump of 24 inches, how long does it take her to reach an elevation of 1 foot?

b. Find the time required for a basketball player with a vertical jump of 48 inches to attain an elevation of 1 foot.

56. *Rectangle Dimensions*  The length of a rectangle is 3 more than twice its width. Its area is 5. Find the dimensions of the rectangle.

57. *Self-painting Can*  Given that a certain paint covers approximately 400 square feet per gallon and that there are approximately 7.5 gallons per cubic foot, find the radius of a cylindrical paint can of height 1 foot that contains just enough paint to paint its own exterior. Interpret your answer. (*Hint:* The total surface area of a cylindrical can of height $h$ and radius $r$ is given by $A = 2\pi r^2 + 2\pi rh$. The volume is given by $\pi r^2 h$.)

58. *Profit*  The price $p$ (in dollars) of a product is related to the number of units sold, $x$, by the relationship $p = 11 - 0.01x$. Each unit costs \$4.00 to produce. If $x$ units are sold, the *cost* of production is the total amount of money it costs to produce $x$ units. The *revenue* generated by the product is the total amount of money obtained from sales of $x$ units. The *profit* is the revenue minus the cost.

a. Explain why the cost to produce $x$ units is $4x$, while the revenue from the sale of $x$ units is $11x - 0.01x^2$.

b. How many units must be sold to obtain a revenue of \$1000?

c. What is the profit when the revenue is \$1000?

d. How many units must be sold to generate a profit of \$1000?

59. *Multiple Deposits*
a. If \$1000 is deposited in an account at the beginning of each year for 3 years and the account pays 5% interest compounded annually, what is the balance just after the third deposit is made?

b. If \$1000 is deposited in an account at the beginning of each year for 3 years and the balance just after the third deposit is made is \$4000, what is the interest rate?

60. *Multiple Deposits*  Suppose \$5 is deposited into an account on January 1, 1996; \$15 on January 1, 1998; and \$25 on January 1, 2000; and the balance in the account just after the last deposit is \$60. Assuming that interest is compounded annually, find the interest rate.

### ◼ *Projects for Enrichment*

**61.** *Verifying the Quadratic Formula* Verify the quadratic formula by substituting

$$x = \frac{-b \pm \sqrt{b^2 - 4ac}}{2a}$$

into the general quadratic equation, $ax^2 + bx + c = 0$.

**62.** *Cubic Formulas* A general **cubic equation** is of the form $ax^3 + bx^2 + cx + d = 0$ and has three complex solutions, either one or three of which are real. Just as the quadratic formula gives the solution to a quadratic equation in terms of its coefficients, the **cubic formula** gives the solution to a cubic equation in terms of its coefficients. However, the cubic formula is considerably more difficult to use than the quadratic formula. The equation first must be placed in *normal form*, and then in *reduced form*, before the equation can be solved. Even after the equation is in reduced form, certain complications can arise (taking square roots of imaginary numbers, for example). In this project we will only deal with "nice" examples in which there are no complications. Furthermore, we will find only one of the three solutions.

I. Finding the Reduced and Normal Forms
An equation of the form $x^3 + rx^2 + sx + t = 0$ is said to be in **normal form**. Note that the coefficient of the cube term is 1. A cubic equation can be written in normal form by dividing through by the coefficient of the cube term.

**a.** Express $5x^3 + 4x^2 - 10x + 9 = 0$ in normal form.

**b.** Express $-x^3 - 3x^2 - 4x - 7 = 0$ in normal form.

A cubic equation of the form $y^3 + py + q = 0$ is said to be in **reduced form**. Note that the coefficient of the cube term is 1 and there is no square term. To transform the normal equation $x^3 + rx^2 + sx + t = 0$ into reduced form, we make the substitution $x = y - (r/3)$.

**c.** Express $x^3 - 3x^2 + x - 6 = 0$ in reduced form.

**d.** Express $2x^3 - 6x^2 + 4x + 10 = 0$ in reduced form.

II. The Cubic Formula
A real solution to the equation $y^3 + py + q = 0$ [provided that $(q^2/4) + (p^3/27) > 0$] is given by

$$y = \sqrt[3]{-\frac{q}{2} + \sqrt{\frac{q^2}{4} + \frac{p^3}{27}}} + \sqrt[3]{-\frac{q}{2} - \sqrt{\frac{q^2}{4} + \frac{p^3}{27}}}$$

**e.** Find a solution to the equation $y^3 + 2y - 1 = 0$ using the cubic formula. Evaluate your answer using a calculator.

**f.** Find a solution of the equation $3x^3 - 6x^2 + 3x - 6 = 0$ by writing it first in normal form and then in reduced form. Solve the resulting equation for $y$ and then find $x$. Evaluate your answer using a calculator and check it in the original equation.

**63.** *The Discriminant and Factoring Quadratic Equations* It can be shown that for natural numbers $n$, $\sqrt{n}$ is irrational unless $n$ is a perfect square. Thus, $\sqrt{2}$, $\sqrt{3}$, and $\sqrt{5}$ are irrational, whereas $\sqrt{4} = 2$ and $\sqrt{9} = 3$ are rational. It follows from this that if $a$, $b$, and $c$ are all integers, then $\sqrt{b^2 - 4ac}$ will be irrational unless $b^2 - 4ac$ is a perfect square.

**a.** Assume $a$, $b$, and $c$ are all integers and that $b^2 - 4ac = n^2$ for some positive integer $n$. Show that the quadratic equation $ax^2 + bx + c = 0$ has the solutions

$$x = \frac{-b + n}{2a} \quad \text{and} \quad x = \frac{-b - n}{2a}$$

**b.** If $b^2 - 4ac = n^2$, show that $ax^2 + bx + c$ factors as

$$\frac{1}{a} \cdot \left( ax + \frac{b - n}{2} \right)\left( ax + \frac{b + n}{2} \right)$$

Explain why each of the numbers $1/a$, $(b - n)/2$, and $(b + n)/2$ is rational.

**c.** It follows from part (b) that whenever the discriminant of a quadratic equation with integer coefficients is a perfect square, the quadratic can be factored using only rational coefficients. For each of the quadratics given below, indicate whether or not it can be factored using rational coefficients. If it can be factored, use the formula from part (b) to produce a factorization.

   **i.** $x^2 + 8x + 144$         **ii.** $6x^2 - 3x - 2$

   **iii.** $12x^2 + 13x - 90$         **iv.** $3x^2 + 10x + 6$

**64.** *Investigating Equations of Motion*

**a.** Drop a coin from each of the following heights and measure the time required for the coin to strike the ground. For each height, perform four trials, and then compute the average time.

   **i.** 4 feet         **ii.** 6 feet

   **iii.** 8 feet         **iv.** 10 feet

If we assume that the only force acting on a projectile is a constant gravitational force, then when the projectile is released from an initial height of $y_0$ with an initial upward velocity of $v_0$, the height $y$ after $t$ seconds is given by $y = -\frac{1}{2}gt^2 + v_0t + h$, where the constant $g$ represents the acceleration due to gravity. If we use feet for our unit of distance, seconds for our unit of time, and feet per second as our unit of velocity, then $g \approx 32$ feet/sec$^2$.

**b.** For each of the initial heights given in part (a), use the equation of motion to compute the time required for the coin to drop.

**c.** Explain any discrepancies between your answers from part (a) with your answers from part (b).

**d.** Since we are assuming that the gravitational force is constant, an object tossed into the air will slow down by

32 feet/sec². Thus, if $v$ represents the velocity of the object in feet per second, then $v = v_0 - 32t$. Compute the velocity upon impact of each of the coins from part (a).

e. Toss a coin into the air and observe how its velocity changes. At what point does the velocity appear to be zero?

f. Find a formula for the maximum height of an object in terms of its initial height and initial velocity. [*Hint:* Begin by using the result of part (e) and the formula from part (d) to find an expression for the time at which the maximum height occurs.]

g. Now use the result of part (f) to find a formula for the initial velocity in terms of the initial height and maximum height.

h. What is the initial upward velocity of an outfielder who leaps 3 feet into the air to prevent a home run?

i. How fast must a ball be thrown if it is to rise 1 mile in the air?

j. Throughout this project, we have assumed that gravity was the only force acting on the projectile and that the gravitational force was constant. What other forces might be relevant, and under what circumstances would the gravitational force vary?

## Questions for Discussion or Essay

65. We now have three techniques available for solving quadratic equations. What are they, and how do we know which technique to use? Would it ever make sense to solve a quadratic equation by using the quadratic formula if it could be solved by factoring? Why would anyone ever want to solve a quadratic equation by completing the square if the quadratic formula is available?

66. Some hand-held calculators can instantly solve any quadratic equation. Some would argue that this eliminates the need for students to learn the quadratic formula. What do you think? If inexpensive calculators can instantly solve quadratic equations, should students be forced to learn the quadratic formula? For that matter, why should any of us learn to add, subtract, multiply, or divide when $5 calculators can easily accomplish the same task?

67. A project in this exercise set is devoted to using a *cubic* formula, which gives a solution to cubic equations in terms of coefficients. There is also a quartic formula, which gives the solution to quartic equations in terms of the coefficients. Surprisingly, it was proved in the nineteenth century by the prodigies Abel and Galois that there is no quintic formula involving only the fundamental operations of arithmetic and radicals. It has also been shown that certain constructions with ruler and compass (such as trisecting an arbitrary angle) simply cannot be performed. Do you really think it's possible to *prove* that something cannot be done? Or is it the case that anything that cannot be done now will just have to wait until we become more sophisticated? Discuss some other examples, from mathematics or other disciplines, of ''proofs'' that show something *cannot* be done.

---

**SECTION 6**

# RATIONAL AND RADICAL EQUATIONS AND INEQUALITIES

■ If the life of a section of the Amazon rain forest will be extended by 20 years if the fires set by ranchers are extinguished and by 10 years if logging is banned, how long will the section of forest remain if neither cause of deforestation is eliminated?

■ If a company of 3000 employees, 35% of which are women, mandates that 60% of new hires be female, then how many employees must be added to achieve a 50–50 gender balance?

■ Why do some ''right'' techniques for solving equations produce ''wrong'' answers?

## EQUATIONS INVOLVING RATIONAL EXPRESSIONS

Equations involving rational expressions, like

$$\frac{1}{x} + \frac{2}{2 + x} = \frac{2}{x}$$

can be converted to polynomial equations by **clearing fractions**—that is, by multiplying both sides by the least common multiple of all denominators appearing in the equation. The resulting polynomial equation can then be solved using the standard techniques for solving polynomial equations, producing one or more *candidates* for solutions. The candidates must then be checked by substituting into the original equation. This is because of the fact that multiplying both sides of an equation by an expression does not necessarily give an equivalent equation. If substitution of the solution candidate into the original equation gives rise to a division by zero, then, of course, the candidate is not a solution.

**EXAMPLE 1**    *An equation giving rise to a linear equation*

Solve $\dfrac{2}{x} + \dfrac{1}{2} = \dfrac{7}{6}$.

**SOLUTION**    The least common denominator in this equation is $6x$. Multiplying both sides by $6x$ and simplifying, we have

$$6x \cdot \left( \frac{2}{x} + \frac{1}{2} \right) = 6x \cdot \frac{7}{6}$$

$$6x \cdot \frac{2}{x} + 6x \cdot \frac{1}{2} = 7x$$

$$12 + 3x = 7x$$

$$-4x = -12$$

$$x = 3$$

**CHECK**

$$
\begin{array}{c|c}
\dfrac{2}{x} + \dfrac{1}{2} & \dfrac{7}{6} \\[2ex]
\dfrac{2}{3} + \dfrac{1}{2} & \\[2ex]
\dfrac{4 + 3}{6} & \\[2ex]
\dfrac{7}{6} & \checkmark
\end{array}
$$

**EXAMPLE 2**    *An equation with no solution*

Solve $\dfrac{1}{x - 1} + \dfrac{1}{x + 1} = \dfrac{2}{x^2 - 1}$.

**SOLUTION** We begin by factoring the denominator of the expression on the right.

$$\frac{1}{x-1} + \frac{1}{x+1} = \frac{2}{(x-1)(x+1)}$$

Next we multiply both sides by the least common denominator, $(x-1)(x+1)$.

$$(x-1)(x+1)\left(\frac{1}{x-1} + \frac{1}{x+1}\right) = (x-1)(x+1)\frac{2}{(x-1)(x+1)}$$

$$(x+1) + (x-1) = 2$$

$$2x = 2$$

$$x = 1$$

**CHECK** Substituting 1 for $x$ on either the left or right side of the original equation causes a division by zero. Thus, there is no solution.

---

| **EXAMPLE 3** | *An equation giving rise to a quadratic* |

Solve $\dfrac{18}{x+2} + \dfrac{2}{x-3} = 5$.

**SOLUTION**

$$\frac{18}{x+2} + \frac{2}{x-3} = 5$$

$$(x+2)(x-3)\left(\frac{18}{x+2} + \frac{2}{x-3}\right) = (x+2)(x-3)5 \qquad \text{Multiplying by the lcd}$$

$$18(x-3) + 2(x+2) = (x^2 - x - 6)5$$

$$20x - 50 = 5x^2 - 5x - 30$$

$$-5x^2 + 25x - 20 = 0$$

$$x^2 - 5x + 4 = 0$$

$$(x-4)(x-1) = 0$$

$$x = 4, \ x = 1$$

**CHECK**

$x = 4$

$$\frac{18}{x+2} + \frac{2}{x-3} \ \bigg| \ 5$$

$$\frac{18}{4+2} + \frac{2}{4-3}$$

$$\frac{18}{6} + \frac{2}{1}$$

$$3 + 2$$

$$5 \ \bigg| \ \checkmark$$

$x = 1$

$$\frac{18}{x+2} + \frac{2}{x-3} \ \bigg| \ 5$$

$$\frac{18}{1+2} + \frac{2}{1-3}$$

$$\frac{18}{3} + \frac{2}{-2}$$

$$6 - 1$$

$$5 \ \bigg| \ \checkmark$$

---

Many problems involving distance, rate, and time (or, equivalently, work, rate of work, and time) can be solved by using equations involving rational expressions. The following rule of thumb is especially useful for solving work problems.

---

**RULE OF THUMB**   When solving work problems, find expressions for the **rate** at which the work is being accomplished. Use the assumption that when two parties work together, the rate at which work is performed is the **sum** of the rates for each party working alone.

---

**EXAMPLE 4**    *A problem involving work*

When a boy uses a toy shovel to help his mother shovel the driveway, the job takes 4 hours. If the boy were working alone, it would take him four times longer than it would his mother. How long would it take the boy's mother to shovel the driveway alone? How long would it take the boy?

**SOLUTION**   Let $x$ be the time required for the boy's mother to shovel the driveway. Then $4x$ is the time required for the boy to shovel the driveway. The rates are then as follows:

$$\text{Mother's rate: } \frac{1 \text{ job}}{x \text{ hours}} \qquad \text{Boy's rate: } \frac{1 \text{ job}}{4x \text{ hours}} \qquad \text{Rate together: } \frac{1 \text{ job}}{4 \text{ hours}}$$

Since their rate together is the sum of their rates separately, we have

$$\frac{1}{x} + \frac{1}{4x} = \frac{1}{4}$$

$$4x\left(\frac{1}{x} + \frac{1}{4x}\right) = 4x \cdot \frac{1}{4} \qquad \text{Multiplying by the lcd}$$

$$4 + 1 = x$$

$$x = 5$$

Thus, it takes the boy's mother 5 hours, whereas the boy takes $4 \cdot 5 = 20$ hours to shovel the driveway.

---

**EXAMPLE 5**    *Meeting a personnel goal—a mixture problem*

An international conglomerate currently has 3000 employees, only 30% of which are women. The company personnel director has dictated that beginning immediately, 60% of the new hires should be women until the company's target of 50% women is met. How many new employees must be hired for the company to meet its hiring goal? (Assume that no employees leave the company during the hiring period.)

**SOLUTION**   Let $x$ be the total number of new employees to be hired. Since 30% of the current employees are women and 60% of the new hires are to be women, we have

$$\text{total number of women} = \text{current number of women} \\ + \text{number of newly hired women}$$

$$= 30\% \cdot 3000 + 60\% \cdot x$$

$$= 900 + 0.6x$$

The total number of employees is just the sum of the original number of employees and the number of new employees. Thus, we have

$$\text{total number of employees} = 3000 + x$$

Now the percentage of women in the company is given by the formula

$$\text{percentage of women} = \frac{\text{total number of women}}{\text{total number of employees}} \cdot 100\%$$

or in this case

$$50\% = \frac{900 + 0.6x}{3000 + x} \cdot 100\%$$

$$0.5 = \frac{900 + 0.6x}{3000 + x}$$

Solving for $x$, we have

$$0.5(3000 + x) = 900 + 0.6x$$

$$1500 + 0.5x = 900 + 0.6x$$

$$-0.1x = -600$$

$$x = \frac{-600}{-0.1}$$

$$x = 6000$$

Thus, 6000 new employees must be hired for the company to achieve its desired gender balance. ∎

## INEQUALITIES INVOLVING RATIONAL EXPRESSIONS

Inequalities involving rational expressions can be solved in much the same way that we solved polynomial inequalities in Section 4. For example, the solution set of the inequality

$$\frac{2x^2 - 18}{x^2 - 4} > 0$$

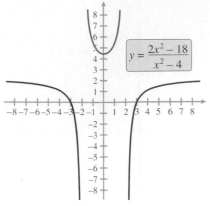

**FIGURE 41**

corresponds to the $x$-values of those portions (shown in red and blue in Figure 41) of the graph of

$$y = \frac{2x^2 - 18}{x^2 - 4}$$

that lie above the $x$-axis. It is clear that the graph is above the $x$-axis for $x < -3$ and for $x > 3$, but it's somewhat difficult to identify the endpoints of the interval corresponding to the blue portion of the graph. Let's consider this problem algebraically. The sign of

$$\frac{2x^2 - 18}{x^2 - 4}$$

is determined by the signs of the numerator and the denominator: if they are the same sign, then the quotient will be positive; if they are different signs, then the quotient will be negative. Thus, if either one changes sign, the sign of the quotient will change as well. Since the numerator changes sign at 3 and $-3$, the graph will change from one side of the $x$-axis to the other at these values. Now the denominator changes sign at $-2$ and 2, so that the graph will change sides of the $x$-axis at these values as well. The interval corresponding to the blue portion of the graph is therefore $(-2, 2)$. Thus, the solution set of the inequality is $(-\infty, -3) \cup (-2, 2) \cup (3, \infty)$. Note that when $x = 2$ or $x = -2$, the expression

$$\frac{2x^2 - 18}{x^2 - 4}$$

has a zero denominator and is therefore undefined. Thus, there is no point on the graph of

$$y = \frac{2x^2 - 18}{x^2 - 4}$$

with these $x$ values—this causes the breaks in the graph at $x = -2$ and $x = 2$. Such breaks are called **vertical asymptotes** and will be discussed in detail in Section 5.5. (Your graphics calculator may attempt to connect certain breaks; see Example 6 below.)

Our experience with the preceding example suggests the following strategy for solving rational inequalities.

*Strategy for solving rational inequalities*

1. Write the inequality in standard form with a rational expression

$$r(x) = \frac{p(x)}{q(x)}$$

on the left side and 0 on the right [i.e., $r(x) < 0$, $r(x) \leq 0$, $r(x) > 0$, or $r(x) \geq 0$].

2. Solve $p(x) = 0$. The solutions are the $x$-intercepts of $y = r(x)$. If this equation cannot be solved algebraically, approximate the $x$-intercepts with a graphics calculator.

3. Solve $q(x) = 0$. If there are any breaks in the graph, they will occur at these values. If the equation $q(x) = 0$ cannot be solved algebraically, estimate the location of the breaks graphically. Be aware that your graphics calculator may attempt to "fill-in" the breaks.

4. From the graph of $y = r(x)$, identify the portions of the graph that are either above or below the $x$-axis, depending on which inequality symbol is used. Express the solution set in interval notation using the results of steps 2 and 3 to determine the endpoints of the intervals.

**EXAMPLE 6**    *An inequality involving rational expressions*

Solve $\dfrac{5}{2} - \dfrac{9}{x + 2} < \dfrac{1}{x - 3}$.

**SOLUTION**    Subtracting $1/(x - 3)$ from both sides gives

$$\frac{5}{2} - \frac{9}{x + 2} - \frac{1}{x - 3} < 0$$

We now simplify the left side by converting to the common denominator $2(x + 2)(x - 3)$:

$$\frac{5(x + 2)(x - 3) - 9 \cdot 2 \cdot (x - 3) - 2(x + 2)}{2(x + 2)(x - 3)} < 0$$

$$\frac{5x^2 - 5x - 30 - 18x + 54 - 2x - 4}{2(x + 2)(x - 3)} < 0$$

$$\frac{5x^2 - 25x + 20}{2(x + 2)(x - 3)} < 0$$

$$\frac{5(x - 4)(x - 1)}{2(x + 2)(x - 3)} < 0$$

Setting the numerator equal to 0, we obtain $x = 4$ and $x = 1$. These are the $x$-intercepts. Setting the denominator equal to 0, we obtain $x = -2$ and $x = 3$. These are the locations of the breaks in the graph.

We now use a graphics calculator to produce a graph of

$$y = \frac{5(x - 4)(x - 1)}{2(x + 2)(x - 3)}$$

like the one shown in Figure 42. Note that the calculator didn't correctly depict the break at $x = 3$. Although the calculator supplement describes a technique for avoiding this problem, this graph is sufficient for our purposes. It is clear that the graph

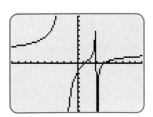

**FIGURE 42**

lies below the $x$-axis for $x$-values between $-2$ and 1, and for $x$-values between 3 and 4. Thus, the solution set is $(-2, 1) \cup (3, 4)$.

It should be noted that, in some contexts, an approximation of the solution to a rational inequality will be sufficient, and if so, much of the work shown in Example 6 can be avoided. In particular, to approximate the solution to the inequality

$$\frac{5}{2} - \frac{9}{x + 2} < \frac{1}{x - 3}$$

we need only plot the graph of

$$y = \frac{5}{2} - \frac{9}{x + 2} - \frac{1}{x - 3}$$

and estimate the intervals on which the graph appears to be below the $x$-axis.

## EQUATIONS INVOLVING RADICALS

Our primary technique for solving equations involving radicals is to eliminate the radicals to produce a polynomial equation, in much the same way as we solved equations involving rational expressions by eliminating denominators.

Radicals are eliminated by first *isolating* the radical and then raising both sides to an appropriate power. If the equation still involves a radical expression, we may need to repeat the process. When the radicals have been eliminated, the resulting equation is solved using the standard techniques for polynomial equations, producing one or more solution candidates.

It is essential that we check our candidates in the original equation. To see why, consider the equation $\sqrt{x} = -1$. If we square both sides, we obtain $x = 1$. But clearly $x = 1$ is not a solution to the original equation. Thus, the equation $\sqrt{x} = -1$ has no real-number solution. The apparent contradiction of correct algebra producing incorrect results will be explored in Exercise 69.

We summarize the technique for solving equations involving radical expressions below.

*Technique for solving equations involving radicals*

> 1. ISOLATE the radical—if there is more than one radical, isolate one of them.
> 2. EXPONENTIATE—if the radical is a square root, square both sides; if the radical is a cube root, cube both sides; and so on.
> 3. REPEAT if necessary—if a radical remains after isolating and exponentiating, then isolate the remaining radical and exponentiate again.
> 4. SOLVE the resulting equation—typically, this is a polynomial equation.
> 5. CHECK your answer by substituting into the original equation. This is important.

The following examples illustrate this technique for solving radical equations.

**EXAMPLE 7**    *Solving an equation involving a radical*

Find all real solutions of $\sqrt{3x + 1} - 4 = 0$.

**SOLUTION**

$$\sqrt{3x + 1} - 4 = 0$$

$$\sqrt{3x + 1} = 4 \qquad \text{Isolating the radical}$$

$$\left(\sqrt{3x + 1}\right)^2 = 4^2 \qquad \text{Squaring both sides to eliminate the radical}$$

$$3x + 1 = 16$$

$$3x = 15$$

$$x = 5$$

**CHECK**

$$
\begin{array}{c|c}
\sqrt{3x + 1} & 4 \\
\sqrt{3 \cdot 5 + 1} & \\
\sqrt{16} & \\
4 & \checkmark
\end{array}
$$

**EXAMPLE 8**    *A radical equation with an extraneous solution*

Find all real solutions of $\sqrt{4x + 17} - 2x = 1$.

**SOLUTION**

$$\sqrt{4x + 17} = 2x + 1 \qquad \text{Isolating the radical}$$

$$\left(\sqrt{4x + 17}\right)^2 = (2x + 1)^2 \qquad \text{Squaring both sides to eliminate the radical}$$

$$4x + 17 = 4x^2 + 4x + 1 \qquad \text{Expanding}$$

$$-4x^2 + 16 = 0 \qquad \text{Collecting all expressions on the right side}$$

$$x^2 - 4 = 0 \qquad \text{Dividing both sides by } -4$$

$$(x - 2)(x + 2) = 0$$

$$x = 2, \ x = -2$$

**CHECK**

$$
\begin{array}{c|c}
\multicolumn{2}{c}{x = 2} \\
\sqrt{4x + 17} - 2x & 1 \\
\sqrt{8 + 17} - 2 \cdot 2 & \\
\sqrt{25} - 4 & \\
5 - 4 & \\
1 & \checkmark
\end{array}
\qquad
\begin{array}{c|c}
\multicolumn{2}{c}{x = -2} \\
\sqrt{4x + 17} - 2x & 1 \\
\sqrt{-8 + 17} + 4 & \\
\sqrt{9} + 4 & \\
3 + 4 & \\
7 & \times
\end{array}
$$

Thus, the only solution is $x = 2$.

**EXAMPLE 9**     *An equation involving two radicals*

Find all real solutions of $\sqrt{x + 2} - \sqrt{x - 3} = 1$.

**SOLUTION**

$$\sqrt{x + 2} - \sqrt{x - 3} = 1$$

$$\sqrt{x + 2} = 1 + \sqrt{x - 3} \qquad \text{Isolating a radical}$$

$$x + 2 = \left(1 + \sqrt{x - 3}\right)^2 \qquad \text{Squaring both sides to eliminate the isolated radical}$$

$$x + 2 = 1 + 2\sqrt{x - 3} + (x - 3) \qquad \text{Expanding}$$

$$4 = 2\sqrt{x - 3} \qquad \text{Simplifying}$$

$$2 = \sqrt{x - 3} \qquad \text{Isolating the remaining radical}$$

$$4 = x - 3 \qquad \text{Squaring both sides to eliminate the radical}$$

$$x = 7$$

**CHECK**

$$\sqrt{x + 2} - \sqrt{x - 3} \quad \Big| \quad 1$$
$$\sqrt{7 + 2} - \sqrt{7 - 3}$$
$$\sqrt{9} - \sqrt{4}$$
$$1 \quad \Big| \quad \checkmark$$

---

**EXAMPLE 10**     *An equation involving cube roots*

Solve $\sqrt[3]{x^2 + 1} = 2$.

**SOLUTION**     Cubing both sides, we obtain

$$\left(\sqrt[3]{x^2 + 1}\right)^3 = 2^3$$

$$x^2 + 1 = 8$$

$$x^2 = 7$$

$$x = \pm\sqrt{7}$$

It is easily verified that $x = \sqrt{7}$ and $x = -\sqrt{7}$ are solutions.

---

Many problems involving distance give rise to equations involving radicals.

**EXAMPLE 11**     *A triathlon*

The first two legs of a triathlon are set up in such a way that participants must swim from checkpoint $A$ across a 4-mile wide channel and then run to checkpoint $B$, 10 miles down the coast from checkpoint $A$, as shown in Figure 43. A certain participant can swim at an average speed of 3 miles per hour and can run at an average speed of 12 miles per hour. Given that her time from checkpoint $A$ to checkpoint $B$ was 2 hours and 15 minutes, find the point where she came ashore. (Assume that she swims in a straight line.)

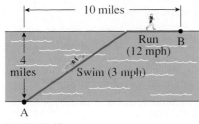

**FIGURE 43**

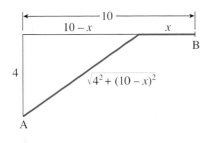

**SOLUTION**   Let $x$ be the distance in miles from the point where the athlete came ashore to checkpoint B. Then $10 - x$ is the distance from the point directly across the channel from checkpoint A to the point where she came ashore. Using the Pythagorean Theorem, the swimming distance is

$$\sqrt{4^2 + (10 - x)^2} \quad \text{or} \quad \sqrt{x^2 - 20x + 116}$$

Since the athlete swims at a rate of 3 miles per hour, the time spent swimming is

$$\frac{\text{swimming distance}}{\text{swimming rate}} = \frac{\sqrt{x^2 - 20x + 116}}{3}$$

Similarly, the time spent running is

$$\frac{\text{running distance}}{\text{running rate}} = \frac{x}{12}$$

Since the total time spent swimming and running was $2\frac{1}{4} = \frac{9}{4}$ hours, we must have

$$\text{swimming time} + \text{running time} = \frac{9}{4}$$

$$\frac{\sqrt{x^2 - 20x + 116}}{3} + \frac{x}{12} = \frac{9}{4}$$

Now we solve for $x$.

$$\frac{\sqrt{x^2 - 20x + 116}}{3} = \frac{9}{4} - \frac{x}{12} \qquad \text{Subtracting } \frac{x}{12} \text{ from both sides}$$

$$4\sqrt{x^2 - 20x + 116} = 27 - x \qquad \text{Multiplying both sides by 12}$$

$$16(x^2 - 20x + 116) = (27 - x)^2 \qquad \text{Squaring both sides}$$

$$16x^2 - 320x + 1856 = 729 - 54x + x^2$$

$$15x^2 - 266x + 1127 = 0$$

$$x = \frac{266 \pm \sqrt{266^2 - 4(15)(1127)}}{2(15)} \qquad \text{Using the quadratic formula}$$

$$x = 7, \ x = \frac{161}{15} \approx 10.7$$

Thus, the athlete landed at a point either 7 or 10.7 miles from checkpoint $B$.

When algebraic techniques fail or if an exact answer isn't necessary, we can solve equations involving rational or radical expressions graphically, as illustrated in the following example.

**EXAMPLE 12**   *Approximating the solution of a radical equation*

Use a graphics calculator to approximate the smallest positive solution of $\sqrt{x + 8} - \sqrt{x - 1} = x$ to the nearest hundredth.

**SOLUTION**   We first rewrite the equation in the standard form

$$\sqrt{x + 8} - \sqrt{x - 1} - x = 0$$

Next we graph $y = \sqrt{x + 8} - \sqrt{x - 1} - x$ with a graphics calculator, as shown in Figure 44. We zoom-in on the $x$-intercept several times until the desired level of accuracy is reached and then use the trace feature to determine the $x$-intercept to within 0.01, as shown in Figure 45. The solution is $x \approx 2.12$.

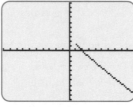

**FIGURE 44**

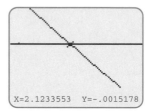

X=2.1233553   Y=−.0015178

**FIGURE 45**

---

### EXERCISES 6

**EXERCISES 1–24** □ *Solve the given equation.*

1. $2 = \dfrac{4}{x}$

2. $3 = \dfrac{2}{x - 1}$

3. $\dfrac{7}{y + 4} = \dfrac{3}{y - 4}$

4. $\dfrac{5}{y} = y$

5. $\dfrac{2}{2x + 1} - \dfrac{3}{4x + 1} = 0$

6. $\dfrac{x - 5}{3x + 2} = \dfrac{2x}{6x + 1}$

7. $\dfrac{2}{t} + \dfrac{3}{t^2 + t} = \dfrac{8}{t + 1}$

8. $\dfrac{2}{x - 1} = \dfrac{5}{x} + \dfrac{7}{x^2 - x}$

9. $\dfrac{s}{s^2 - 1} + \dfrac{2}{s + 1} = \dfrac{4}{s - 1}$

10. $\dfrac{2}{x + 3} - \dfrac{5x}{x^2 + 5x + 6} = \dfrac{3}{2x + 6}$

11. $\dfrac{3}{2x - 1} + \dfrac{x + 2}{6x^2 + x - 2} = \dfrac{2}{9x + 6}$

12. $\dfrac{x}{x + 1} = \dfrac{-6}{x - 1}$

13. $\dfrac{x}{x + 1} = \dfrac{3x - 3}{x + 5}$

14. $\dfrac{x + 3}{x - 1} = \dfrac{1}{3x} - \dfrac{1}{6}$

15. $\dfrac{x}{2x - 1} + \dfrac{3}{x} = \dfrac{2(x^2 + 1)}{2x^2 - x}$

16. $\dfrac{3x + 1}{2x - 1} + \dfrac{3x}{6x - 3} = \dfrac{13 + 8x}{x + 1}$

17. $\dfrac{1}{x + 2} + \dfrac{3}{x} = \dfrac{2x + 2}{x(x + 2)}$

18. $\dfrac{2 - 2x}{x^2 + 2x - 15} + \dfrac{4 - 3x}{x^2 - 3x} = \dfrac{5x - 1}{x^2 + 5x}$

19. $\dfrac{x + 2}{x - 1} + \dfrac{3x + 3}{x + 1} = \dfrac{2x^2 + 4x + 2}{x^2 - 1}$

20. $\dfrac{3x}{x - 3} + \dfrac{54}{x^2 - 9} = 2$

21. $\dfrac{7}{x + 5} - \dfrac{3}{x + 1} = -1$

22. $\dfrac{3x + 1}{x + 2} - \dfrac{4}{x + 1} = 3$

23. $\dfrac{x}{x + 1} + \dfrac{3}{x - 1} = \dfrac{6}{x^2 - 1}$

24. $\dfrac{1 + \dfrac{1}{x}}{1 - \dfrac{1}{x}} = 2$

**EXERCISES 25–32** □ *Solve the given inequality.*

25. $\dfrac{x + 2}{x + 3} \leq 2$

26. $\dfrac{5x}{x^2 - 9} \leq 0$

27. $\dfrac{1}{x - 2} > \dfrac{4}{x + 1}$

28. $\dfrac{x + 5}{2x - 7} \geq -1$

29. $\dfrac{x}{3x + 1} \leq \dfrac{2x - 2}{2x + 3}$

30. $x + 6 \geq -\dfrac{20}{x - 6}$

31. $1 + \dfrac{1}{x} \geq \dfrac{6}{x^2}$

32. $\dfrac{2}{x} + \dfrac{3}{x + 1} < 1$

**EXERCISES 33–52** □ *Find the exact value(s) of the real solution(s) of the given equation.*

**33.** $\sqrt{x} = 3$

**34.** $\sqrt{s} = \sqrt{7}$

**35.** $\sqrt{x} = -2$

**36.** $\sqrt{x + 5} = 2$

**37.** $\sqrt{8x - 7} = 0$

**38.** $\sqrt{3y - 4} = -3$

**39.** $\sqrt{u + 1} = \sqrt{2u - 2}$

**40.** $\sqrt{3x + 4} = \sqrt{2x - 6}$

**41.** $\sqrt{4t + 1} = -\sqrt{4t + 1}$

**42.** $\sqrt{x - 3} - \sqrt{2} - 0$

**43.** $\sqrt{\dfrac{3}{r}} - \sqrt{\dfrac{r}{3}} = 0$

**44.** $\sqrt{5x - 3} - 1 = x$

**45.** $\sqrt{x + 5} - \sqrt{x} = 1$

**46.** $\sqrt{5x + 6} + \sqrt{4x + 1} = 7$

**47.** $\sqrt{x + 13} - x = 1$

**48.** $5 = a + \sqrt{a + 1}$

**49.** $2x + 2 = \sqrt{6x + 10}$

**50.** $z = 2 + \sqrt{2z - 4}$

**51.** $3 - 2\sqrt{y + 1} = 0$

**52.** $x - 3\sqrt{4 - x} = 0$

**EXERCISES 53–58** □ *Use a graphics calculator to approximate the solution(s) to the nearest hundredth.*

**53.** $\dfrac{1}{x^2 + 1} = x$

**54.** $\dfrac{1}{x} = x - \dfrac{1}{x^2}$

**55.** $\dfrac{x + 1}{x^2 - 4} = 1 - x^2$

**56.** $\sqrt{x} = x^2 - 1$

**57.** $\sqrt{x} + \sqrt{x + 1} = x$

**58.** $\sqrt{4 - x} = x^2$

## ■ Applications

**59.** *Shadow Length* Suppose that a person $h$ feet tall is standing $d$ feet from an $l$-foot-tall lamppost. It can be shown that if the length of the person's shadow is $s$, then

$$\frac{l}{h} = \frac{s + d}{s}$$

   **a.** Solve this equation for $s$.

   **b.** How long would the shadow of a woman be if the woman is 5 feet 6 inches and is standing 30 feet from a 40-foot-tall lamppost?

**60.** *Gravitational Force* The gravitational force (in newtons) exerted by an object of mass $m_1$ kilograms on an object of mass $m_2$ kilograms with center of mass $r$ meters distant from that of the first object is given by

$$F = G\frac{m_1 m_2}{r^2}$$

where $G = 6.673 \times 10^{-11}$ N·m²/kg² is the universal gravitational constant.

   **a.** Find the force of gravitational attraction between two objects with 100-kilogram masses that are 1 meter apart.

   **b.** Solve the force equation for $m_2$.

**61.** *Harmonic Mean* The harmonic mean $h$ of three numbers $x$, $y$, and $z$ is defined by

$$h = \frac{3}{\dfrac{1}{x} + \dfrac{1}{y} + \dfrac{1}{z}}$$

   **a.** Find the harmonic mean of 3, 4, and 10.

   **b.** Solve the above equation for $z$.

   **c.** Three numbers have a harmonic mean of 10.5. One of the numbers is 7 and another is 8. What is the third number?

**62.** *Electrical Resistance* When two resistors with resistances $R_1$ and $R_2$ are in parallel, the total resistance $R$ satisfies the equation

$$\frac{1}{R} = \frac{1}{R_1} + \frac{1}{R_2}$$

   **a.** Solve this equation for $R_1$.

   **b.** Suppose that two resistors in parallel have a total resistance of 30 ohms. If the resistance of one is 75 ohms, what is the resistance of the other?

**63.** *River Current* Suppose that the record-holding racing boat *Miss Budweiser*, capable of traveling 140 miles per hour in still water, travels 40 miles upstream and back in 35 minutes.

   **a.** If $x$ denotes the speed of the current, find an expression for the time spent traveling upstream.

   **b.** Find an expression for the time traveling downstream.

   **c.** Find the speed of the current.

**64.** *Filling a Tank* Pipe A and pipe B used together require 2 hours less to fill a tank than pipe B does used alone. If pipe B requires 1 hour less than pipe A to fill the tank, find the time required for each of the pipes to fill the tank.

**65.** *AIDS Jog-a-thon* To help raise funds for AIDS research, Roberto and Danielle are participating in a jog-a-thon in which participants collect pledges based on the number of miles that they run. It takes Roberto 1 hour longer to earn $200 than it does Danielle, and together they earn $200 in 1 hour less than it takes Danielle to earn $200. Assuming that they are both running at constant speeds, how much money will they earn in a 3-hour period?

**66.** *Rain Forest Destruction* A section of the Amazon rain forest is being destroyed by two independent sources: fires set by ranchers interested in creating ranges for their cattle and logging by timber companies. It is estimated that it would take fires (without logging) 10 years longer to destroy the section of forest than both fires and logging. It would take logging (without fires) 20 years longer than both fires and logging. How long will the section of forest stand if both logging and fires continue?

67. **Running Cable**  Fiber optic cable is to be run between two office buildings that are separated by a 20-yard roadway as shown in Figure 46. The cable must start at point *A*, pass under the roadway, and then continue under the ground to point *B*. It costs $500 per yard to pass the cable under the roadway, and its costs $100 per yard to bury it under the ground the rest of the way to point *B*. If the total cost of laying the cable is

$30,000 and if the distance from point *B* to the point directly opposite *A* is 100 yards, find the point where the cable reaches the other side of the road.

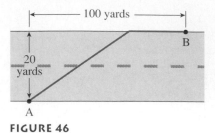

**FIGURE 46**

---

■ *Projects for Enrichment*

68. **Minimizing Triathlon Time**  In Example 11, we considered a triathlon in which an athlete swims from checkpoint *A*, across a 4-mile-wide channel, and then runs to checkpoint *B*, which is 10 miles down the coast from checkpoint *A*. Given a swimming rate of 3 miles per hour, a running rate of 12 miles per hour, and a total time of 2 hours and 15 minutes, we determined that the athlete must have landed either 7 or 10.7 miles from checkpoint *B*.

a. Does it seem reasonable that the time is the same for two different landing spots? Explain.

b. If we repeat the steps in Example 11, using a total time of 2 hours, we would arrive at the equation

$$\frac{\sqrt{x^2 - 20x + 116}}{3} + \frac{x}{12} = 2$$

where *x* denotes the distance from the point where the athlete came ashore to checkpoint *B*. Solve this equation for *x*. What does your answer indicate?

The expression

$$\frac{\sqrt{x^2 - 20x + 116}}{3} + \frac{x}{12}$$

gives the total time for the athlete to travel from *A* to *B*, assuming she comes ashore *x* miles from checkpoint *B*.

c. Evaluate the expression given above for $x = 0$, $x = 5$, and $x = 10$. Of these three landing points, which will give the athlete the best time?

d. Devise a strategy for determining where the athlete should land so that the total time is as small as possible.

69. **Equivalent Statements**  When solving equations involving either rational expressions or radicals, it is essential to check solutions. Even if no mistake is made, an answer might not be a solution of the original equation. This is because of the fact that the techniques used to solve such equations do not always yield equivalent equations.

When solving linear or quadratic equations, each step produces a statement that is equivalent to the previous statement. In this way, we can be sure that any solution of the last equation will be a solution of the first equation. This is not necessarily the case, however, with equations involving rational expressions or radicals. For example, the operation of squaring both sides of an equation will sometimes yield a nonequivalent equation: the old equation implies the new equation, but the new equation doesn't necessarily imply the old. It is for this reason that the answers to any equation involving radicals should be checked.

For each of the operations given below, indicate whether or not it *always* produces an equivalent equation. If not, give an example that demonstrates this.

a. Adding a number to both sides of an equation.

b. Multiplying both sides by zero.

c. Multiplying both sides by an expression involving a variable.

d. Dividing both sides by an expression involving a variable.

e. Cubing both sides of an equation.

f. Squaring both sides of an equation.

**70.** Consider the following statement: Equations involving radical or rational expressions are solved by first eliminating the most annoying part of the equation. What is meant by this? What would be the most "annoying" part of an equation involving radicals? Of an equation involving a rational expression? How does this approach to problem solving generalize to the solution of nonmathematical problems?

**71.** When doing algebra, it is not enough to know what can be done, one must also know what *cannot* be done. In this section we have described in some detail how to solve inequalities involving rational expressions, such as $4/x < -1$. Many students find that even after doing several such problems the cor-

rect way, the temptation to "just multiply both sides by $x$" is overwhelming. What would the final answer be if a student had erroneously multiplied both sides by $x$ and solved the resulting linear inequality? Why can't you "just multiply both sides by $x$"?

**72.** Many work problems (such as Example 4 of this section) are solved by making the assumption that the rate at which two people work together is the sum of their individual rates. Under what circumstances is this assumption valid? If a complete dental cleaning can be done in 30 minutes when *one* dentist is working, how long would it take 100,000 dentists working together to clean your teeth?

---

**SECTION 7**

# SOLVING SYSTEMS OF EQUATIONS

> ■ How can chemical equations be balanced using systems of equations?
>
> ■ How can hog lard and turkey breast be mixed to form a 90% fat-free mixture?
>
> ■ If one runner defeats another by 10 meters in the first heat of a 50-meter race and then for the second heat begins 10 meters behind the starting line to handicap himself, who will win the second heat?

## WHAT IS A SYSTEM OF EQUATIONS?

We have solved a variety of equations—linear equations, quadratic equations, equations with rational expressions, equations involving radicals, and so on. In this section we will be dealing with **systems of equations**, which consist of more than one equation and involve more than one variable. Solving a system of equations entails finding values of each of the variables so that *all* of the equations are satisfied. For example,

$$2x + 3y = 8$$
$$6x - 3y = 0$$

is a system of two equations in two unknowns. It can be shown that $x = 1$, $y = 2$ is a solution of this system. In fact, the first equation is satisfied since $2 \cdot 1 + 3 \cdot 2 = 8$, and the second equation is satisfied since $6 \cdot 1 - 3 \cdot 2 = 0$. It is customary to express the solution as $(1, 2)$, which indicates that $x = 1$, $y = 2$ is a solution of the system.

Throughout this section we will be referring to linear and nonlinear systems of equations. Linear equations involve only variables raised to the first power. For example, $2x + 3y = 5$ is linear, whereas $3x^2 + 4y^3 = 7$ is nonlinear. A **linear system of equations** is a system in which every equation is linear. A **nonlinear system of equations** includes at least one equation that is nonlinear. For example,

$$2x + 3y = 8$$
$$6x - 3y = 0$$

is a linear system of equations, whereas

$$x^2 + y^2 = 1$$
$$2x + 3y = 5$$

is a nonlinear system of equations.

**EXAMPLE 1**    *Verifying a solution of a system of nonlinear equations*

In parts (a) and (b) check to see whether the given pair of numbers satisfies the following system of equations.

$$3x^2 - 5y = 7$$
$$2x + y = 10$$

**a.** $(2, 1)$      **b.** $(3, 4)$

**SOLUTION**

**a.** Substituting $x = 2$ and $y = 1$ into the first equation, we have

$$3(2)^2 - 5(1) = 12 - 5 = 7$$

Substitution into the second equation gives

$$2(2) + 1 = 5 \neq 10$$

Since the second equation doesn't hold, $(2, 1)$ is not a solution of the system.

**b.** Substituting $x = 3$ and $y = 4$ into the first equation, we have

$$3(3)^2 - 5(4) = 27 - 20 = 7$$

Substitution into the second equation gives

$$2(3) + 4 = 6 + 4 = 10$$

Since $(3, 4)$ satisfies both equations, it is a solution of the system.

There are several methods for solving systems of equations, all of which have a common goal: to reduce the number of equations and the number of variables in a step-by-step manner until there is only one equation in one unknown. For example, three equations in three variables are reduced to two equations in two variables, which are in turn reduced to one equation in one variable. The resulting equation is then solved for the lone remaining variable, and its value is used to find the values of the previously eliminated variables.

## SOLVING SYSTEMS OF EQUATIONS BY SUBSTITUTION

The most straightforward method (although not always the shortest) for solving systems of equations is **substitution**: solving one of the equations of a system for a variable and then substituting the resulting expression in the other equations that comprise the system. This technique is illustrated in the following example.

**EXAMPLE 2**    *Solving a system of equations by substitution*

Solve the following system of equations.

(1) $$3x + y = 10$$

(2) $$2x - 3y = -8$$

**SOLUTION**    We begin by solving the first equation for $y$.

$$3x + y = 10$$

(3) $$y = 10 - 3x$$

Next we will substitute $10 - 3x$ for $y$ in equation (2).

$$2x - 3y = -8 \qquad \text{The second equation of the system}$$
$$2x - 3(10 - 3x) = -8 \qquad \text{Substituting } 10 - 3x \text{ for } y$$
$$2x - 30 + 9x = -8$$
$$11x - 30 = -8$$
$$11x = -8 + 30$$
$$11x = 22$$
$$x = 2$$

Finally, we use equation (3) and the fact that $x = 2$ to find $y$.

$$y = 10 - 3x \qquad \text{Equation (3)}$$
$$y = 10 - 3(2) \qquad \text{Substituting 2 for } x$$
$$y = 4$$

Thus, the solution is given by $(x, y) = (2, 4)$. We now verify that $(2, 4)$ is indeed a solution of the system by substitution into the original system.

**CHECK**    Equation: $3x + y = 10$          Equation: $2x - 3y = -8$
Solution: $x = 2$, $y = 4$          Solution: $x = 2$, $y = 4$

| $3x + y$ | $10$ | | $2x - 3y$ | $-8$ |
|---|---|---|---|---|
| $3(2) + 4$ | | | $2(2) - 3(4)$ | |
| $6 + 4$ | | | $4 - 12$ | |
| $10$ | ✓ | | $-8$ | ✓ |

Note that in the previous example, we chose to solve the first equation for $y$. In general, our choice of which variable to solve for (and in which equation) is determined by convenience.

In all cases, solutions to systems of equations should be checked by substitution into the original system. For the remainder of the examples in this section, we shall omit the details of the verification process.

**EXAMPLE 3**    *Solving a system of nonlinear equations by substitution*

Solve the following system of equations.

(1) $$x - y = 1$$

(2) $$6y = x^2 + 2$$

**SOLUTION**    In this case, we will solve equation (1) for $x$ and substitute into equation (2).

$$x - y = 1$$

(3) $$x = y + 1$$

We now substitute $y + 1$ for $x$ in equation (2).

| | |
|---|---|
| $6y = x^2 + 2$ | This is equation (2) |
| $6y = (y + 1)^2 + 2$ | Substituting $y + 1$ for $x$ |
| $6y = y^2 + 2y + 1 + 2$ | Expanding |
| $y^2 - 4y + 3 = 0$ | Collecting all terms on the left |
| $(y - 1)(y - 3) = 0$ | Factoring |
| $y = 1, \ y = 3$ | |

Now we use equation (3), $x = y + 1$, to find the corresponding $x$-values.

When $y = 1$:     $x = 1 + 1 = 2$

When $y = 3$:     $x = 3 + 1 = 4$

Thus, the two solutions are $(2, 1)$ and $(4, 3)$. Note that in this case *both* solutions need to be verified.

■

**EXAMPLE 4**    *Determining the weights of disks using systems of equations*

Suppose that you are at a bench press contest and have just witnessed the lightweight champion press a bar weighted with 2 large disks and 1 small disk on each side, and the weight is announced as 275 pounds. Then, the eventual middleweight champion presses a bar with 3 large disks and 1 small disk on each side, and the weight is announced as 365 pounds. The heavyweight champion lifts a bar weighted with 4 large disks and 1 small disk, but you cannot hear the announced weight. Assuming the bar weighs 45 pounds, how much did the heavyweight lift?

**SOLUTION**    Let $x$ represent the weight of the larger disk and $y$ the weight of the smaller disk. Since the lightweight champion's bar had a total of 4 large disks and 2 small disks, we have

$$4x + 2y + 45 = 275, \text{ or}$$

(1) $$4x + 2y = 230$$

Similarly, from the middleweight's lift we obtain

$$6x + 2y + 45 = 365$$

(2) $$6x + 2y = 320$$

Now solving equation (1) for $y$, we have

$$4x + 2y = 230$$
$$2y = 230 - 4x$$

(3) $$y = 115 - 2x$$

Substituting into equation (2) gives us

$$6x + 2(115 - 2x) = 320$$
$$2x + 230 = 320$$
$$2x = 90$$
$$x = 45$$

Substituting 45 for $x$ in equation (3), we obtain

$$y = 115 - 2(45)$$
$$= 25$$

Thus, the large disks weigh 45 pounds, and the small disks weigh 25 pounds. Since the heavyweight raised a total of 8 large disks and 2 small disks (plus the bar), he lifted

$$8(45) + 2(25) + 45 = 455 \text{ pounds}$$

---

## SOLVING SYSTEMS OF EQUATIONS BY ELIMINATION

A second method for solving systems of equations, **elimination**, consists of adding multiples of the equations that constitute the system in such a way that a variable is eliminated. For example, consider the following system of equations.

$$2x + y = 6$$
$$3x - y = 9$$

By the addition property of equations (if $a = b$ and $c = d$, then $a + c = b + d$), the sum of the left-hand sides of the equations will equal the sum of the right-hand sides—that is,

$$\begin{array}{r} 2x + y = 6 \\ +\ \underline{3x - y = 9} \\ 5x\quad\ = 15 \end{array}$$

This gives us $x = 3$, which can then be substituted into either of the original equations to find $y$. We substitute into the first equation:

$$2x + y = 6$$
$$2(3) + y = 6$$
$$y = 0$$

Thus, the solution is $(3, 0)$.

Note that by adding the two equations together, one of the variables (in this case, $y$) was eliminated. This gave us a simple equation involving only one variable. Of course, for an arbitrary system of two equations in two variables, there is no guarantee that simply adding the equations together will eliminate one of the variables. However, it is often the case that a variable will be eliminated upon addition if we first multiply each equation by an appropriate constant. The technique of elimination is illustrated in the following examples.

**EXAMPLE 5**    *Solving a system of linear equations by elimination*

Solve the following system of equations by elimination.

(1) $\qquad\qquad\qquad\qquad 4x + 3y = 5$

(2) $\qquad\qquad\qquad\qquad 3x + y = 4$

**SOLUTION**    Adding these equations together will eliminate neither $x$ nor $y$. In fact, addition gives

$$
\begin{array}{r}
4x + 3y = 5 \\
+ \quad 3x + \phantom{3}y = 4 \\
\hline
7x + 4y = 9
\end{array}
$$

and we're no better off than before. However, if we first multiply the second equation by $-3$, we obtain

$$(-3)(3x + y) = (-3)4$$

(3) $\qquad\qquad\qquad\qquad -9x - 3y = -12$

Since equation (3) has the term $-3y$ and equation (1) has the term $3y$, adding equations (1) and (3) will eliminate $y$.

$$
\begin{array}{rl}
4x + 3y = 5 & \text{Equation (1)} \\
+ \quad -9x - 3y = -12 & \text{Equation (3)} \\
\hline
-5x \phantom{- 3y} = -7 &
\end{array}
$$

Thus, $x = \frac{7}{5}$. We can now substitute into equation (2) to find $y$.

$$3x + y = 4$$

$$3\left(\frac{7}{5}\right) + y = 4$$

$$\frac{21}{5} + y = 4$$

$$y = 4 - \frac{21}{5}$$

$$= \frac{20}{5} - \frac{21}{5}$$

$$= -\frac{1}{5}$$

Thus, our solution is $(x, y) = (\frac{7}{5}, -\frac{1}{5})$.

In the previous example we eliminated a variable by multiplying one of the equations of a system by a constant, and then adding to the other equation. In the next example it is more convenient to multiply *both* equations by a constant before addition. Our goal is the same as before: we would like to produce a system of two equations such that when the equations are added, one of the variables will be eliminated.

**EXAMPLE 6**   *Solving a system of linear equations by elimination*

Solve the following system of equations by elimination.

$$3x - 5y = 5$$

$$4x - 7y = 6$$

**SOLUTION**   Although either variable could be eliminated, we will eliminate $x$. In order to obtain equations in which the coefficients of $x$ are opposites, we will multiply the first equation by 4 and the second equation by $-3$:

| *Original equations* | *Multiply by* | *New equations* |
|---|---|---|
| $3x - 5y = 5$ | 4 | $12x - 20y = 20$ |
| $4x - 7y = 6$ | $-3$ | $+ \quad -12x + 21y = -18$ |
| | | $y = 2$ |

Substituting 2 for $y$ in $3x - 5y = 5$ gives

$$3x - 5(2) = 5$$

$$3x - 10 = 5$$

$$3x = 15$$

$$x = 5$$

Thus, our solution is $(x, y) = (5, 2)$.

**EXAMPLE 7**     *Solving a mixture problem using elimination*

A foolish dieter, following the instructions of his nutritionist to the letter, is insistent that fat constitute exactly 10% of his diet. On a certain day, he must select from a menu consisting strictly of 92% fat-free turkey bologna and hog lard, which is virtually 100% fat. How much of each food should he consume if he wishes to eat 10 pounds of food?

**SOLUTION**     Let $x$ represent the number of pounds of turkey bologna eaten and $y$ the number of pounds of hog lard. Then we have

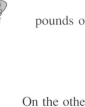

pounds of fat from turkey bologna
$$+ \text{ pounds of fat from lard } = \text{ pounds of fat in mixture}$$
$$8\% \cdot x + 100\% \cdot y = 10\% \cdot 10$$
$$0.08x + y = 1$$

On the other hand, we have

$$\text{pounds of turkey bologna } + \text{ pounds of lard } = 10 \text{ pounds}$$
$$x + y = 10$$

Thus, we must solve the system

(1)                                 $0.08x + y = 1$

(2)                                 $x + y = 10$

Multiplying equation (1) by $-1$ and adding to equation (2) gives us

$$-0.08x - y = -1$$
$$+ \quad\quad x + y = 10$$
$$\overline{\quad\quad 0.92x \quad\quad\quad = 9}$$

Solving for $x$ gives us $x = \dfrac{9}{0.92} \approx 9.8$. Substitution of this value into equation (2) gives

$$x + y = 10$$
$$9.8 + y = 10$$
$$y = 0.2$$

Thus, 9.8 pounds of turkey bolgona mixed with 0.2 pound of lard will yield 10 pounds of a mixture that is approximately 90% fat-free.

So far we have illustrated the technique of elimination for solving linear systems of equations, but it can also be an effective technique for solving nonlinear equations, as illustrated in the following examples.

**EXAMPLE 8** | *Solving a nonlinear system of equations*

Solve the following system of equations.

$$x^2 + 3y = 13$$
$$2x^2 + y^2 = 17$$

**SOLUTION** We will eliminate the variable $x$ by multiplying the first equation by $-2$ and adding it to the second equation.

| Original equations | Multiply by | New equations |
|---|---|---|
| $x^2 + 3y = 13$ | $-2$ | $-2x^2 - 6y = -26$ |
| $2x^2 + y^2 = 17$ | $1$ | $+\quad 2x^2 + y^2 = 17$ |
| | | $y^2 - 6y = -9$ |

Thus, we must solve the quadratic equation $y^2 - 6y = -9$.

$$y^2 - 6y = -9$$
$$y^2 - 6y + 9 = 0$$
$$(y - 3)^2 = 0$$
$$y = 3$$

We now substitute 3 for $y$ in the first equation, $x^2 + 3y = 13$, to find $x$.

$$x^2 + 3y = 13$$
$$x^2 + 3(3) = 13$$
$$x^2 = 4$$
$$x = 2, \ x = -2$$

We have found two solutions to our system: $(2, 3)$ and $(-2, 3)$.  ■

Elimination can be extended to systems of three equations in three unknowns as the following example illustrates.

**EXAMPLE 9** | *A system of three equations in three unknowns*

Solve the following system of equations.

(1) $$x + y + z = 4$$

(2) $$2x + y - z = 9$$

(3) $$x - 3y - 2z = -1$$

**SOLUTION** Our strategy is as follows. First we will eliminate a variable from a pair of equations. Then we will eliminate the same variable from another pair of equations. This will give us a system of two equations in two unknowns, which we will then solve. Specifically, we will eliminate $z$ by adding equation (1) to equation (2), and by adding twice equation (1) to equation (3).

$$
\begin{array}{rl}
 & x + y + z = 4 \\
+ & 2x + y - z = 9 \\
\hline
(4) \quad & 3x + 2y \phantom{+ z} = 13
\end{array}
$$

Adding equations (1) and (2) together to eliminate $z$

$$
\begin{array}{rl}
 & 2x + 2y + 2z = 8 \\
+ & x - 3y - 2z = -1 \\
\hline
(5) \quad & 3x - y \phantom{+ 2z} = 7
\end{array}
$$

Multiplying equation (1) by 2 ... and adding to equation (3) to eliminate $z$

Now we solve the system consisting of equations (4) and (5). We can eliminate $x$ from this system by multiplying equation (4) by $-1$ and adding to equation (5).

$$
\begin{array}{rl}
 & -3x - 2y = -13 \\
+ & 3x - y = 7 \\
\hline
 & -3y = -6
\end{array}
$$

It follows that $y = 2$. Substituting this value into equation (4) gives us

$$3x + 2y = 13$$
$$3x + 2 \cdot 2 = 13$$
$$3x = 9$$

Hence $x = 3$. We now substitute the values of $x$ and $y$ into equation (1) to find $z$.

$$x + y + z = 4$$
$$3 + 2 + z = 4$$
$$z = -1$$

Thus, our solution is $x = 3$, $y = 2$, and $z = -1$.

## CONSISTENT AND INCONSISTENT SYSTEMS OF EQUATIONS

Systems of equations having at least one solution are said to be **consistent**; if there is no solution, then the system is said to be **inconsistent**.

For example, suppose that you are told that the cash total of a wallet holding only fives and twenties is $120, and that the number of fives plus four times the number of twenties equals 30, and you are asked to determine the precise number of fives and twenties. If we let $f$ represent the number of fives and $t$ represent the number of twenties, we have

$$5f + 20t = 120 \qquad \text{The total value is \$120}$$
$$f + 4t = 30 \qquad \text{The number of fives plus 4 times the number of twenties is 30}$$

If we now multiply the second equation by $-5$ and add to the first equation, we obtain

$$5f + 20t = 120$$
$$\underline{+\quad -5f - 20t = -150}$$
$$0 = -30$$

This is, of course, a contradiction. Evidently we were given inconsistent data; it is impossible for both the total cash to be \$120 and for the number of fives plus 4 times the number of twenties to equal 30.

In general, when one attempts to solve an inconsistent system by either substitution or elimination, an obvious contradiction (like $0 = -30$ or $1 = 2$) will result.

# EXERCISES 7

**EXERCISES 1–42** □ *Solve the system of equations. Some systems may be inconsistent.*

**1.** $2x + y = 10$
$3x - y = 5$

**2.** $3s + t = 5$
$2s - 2t = 14$

**3.** $-3x + y = 7$
$3x + 4y = -2$

**4.** $2x + 3y = -7$
$x + 5y = -14$

**5.** $2q + 6r = 3$
$4q + 3r = 0$

**6.** $3x - 4y = 8$
$9x - 12y = 2$

**7.** $3m - n = 2$
$-4m + 2n = 2$

**8.** $3x + 5y = 2$
$5x + y = 18$

**9.** $x - 5y = 5$
$4x + 5y = 5$

**10.** $5x - y = -2$
$2x + y = -1$

**11.** $2z - 7w = -2$
$10z - 35w = -8$

**12.** $2x + 5y = 5$
$4x - 5y = 4$

**13.** $\dfrac{10}{3}a + 2b = -6$

$a + \dfrac{1}{5}b = 1$

**14.** $6a - 2b = \dfrac{7}{2}$

$3a + \dfrac{1}{2}b = \dfrac{5}{2}$

**15.** $2x + 3y = 9.8$
$3x - 5y = -8.1$

**16.** $2w - 5z = 12.9$
$3w + 4z = 4.4$

**17.** $0.5x - 0.2y = 1.36$
$1.5x + 0.4y = 5.78$

**18.** $1.2A + 4.6B = 18.5$
$-2.4A + 3.0B = 5.7$

**19.** $0.24f + 0.32g = 1$
$-0.72f - 0.96g = 2$

**20.** $(2 \times 10^{-6})x + 10^{-6}y = 33$
$3x - 5y = 3 \times 10^7$

**21.** $(4 \times 10^4)k + (1.2 \times 10^4)m = 2 \times 10^4$
$(2 \times 10^4)k + (5 \times 10^4)m = -2.1 \times 10^5$

**22.** $3x - y = 1 + 4i$   (Recall $i = \sqrt{-1}$)
$2x + y = 4 + i$

**23.** $3z_1 - z_2 = 6$
$2z_1 + 3z_2 = 4 + 11i$

**24.** $(1 + i)w + 2iz = 2 + 4i$
$3w - 2iz = 6 - 2i$

**25.** $2a - bi = 5$
$3a + b = 6 + i$

**26.** $x + y = 8$
$xy = 15$

**27.** $2x - 3y = 5$

$\dfrac{y}{x} = -1$

**28.** $2p^2 + 3q^2 = 30$
$p + q = 1$

**29.** $\dfrac{c}{d} - c = 4$

$c - 8d = 0$

**30.** $\dfrac{2}{a} + \dfrac{3}{b} = 0$

$\dfrac{4}{a} + \dfrac{12}{b} = -1$

**31.** $\dfrac{6}{p} + \dfrac{6}{q} = 33$

$-\dfrac{3}{p} + \dfrac{4}{q} = -6$

**32.** $x^2 + y^2 = 9$
$5x - y = -3$

**33.** $2(u + v)^3 - v = -11$
$(u + v)^3 - 3v = 7$

**34.** $x^4 + y^4 = 97$
$x^4 - y^4 = 65$

**35.** $\dfrac{x^2}{4} + \dfrac{y^2}{9} = 2$

$3x + y = 3$

**36.** $3a + c = 13$
$2b + c = 10$
$a + b = 1$

**37.** $3x + y + 2z = 11$
$x + 3y = 6$
$3y + z = 0$

**38.** $2x + y - z = -3$
$x + 3y = 6$
$x + 3z = 3$

**39.** $x + y + z = 2$
$2y - 3z = -7$
$3y + 2z = 9$

**40.** $a + b + c = 2$
$2a - 3b + c = 1$
$3a - 4b + 2c = 4$

**41.** $u + v + w = 1$
$2u + 2v + w = 2$
$3u + v - w = 5$

**42.** $2A - 2B + 3C = -4$
$2A + 2B - C = 8$
$3A - 4B + 5C = -9$

■ *Applications*

43. *Counting Coins*  A street musician counts the coins in his hat and discovers that he has four more nickels than dimes, and that he has a total of 80¢ in nickels and dimes. How many nickels and how many dimes does the street musician have?

44. *Counting Coins*  A budding numismatist (coin collector) has twice as many silver dollars as quarters; the total face value of the silver dollars and the quarters is $4.50. How many of each does she have?

45. *Low-fat Mixture*  A sundae is being made from 98% fat-free frozen yogurt and 80% fat-free dessert topping. How many ounces of each should be used if we wish to have a 10-ounce, 90% fat-free sundae?

46. *Achieving Racial Balance*  In a certain school district, 40% of the students are African American, but only 10% of the 1000 teachers are. Since the student/teacher ratio is also unacceptably high, the school board decides that they will correct both problems by doubling the number of teachers, and hiring in such a way that the racial makeup of the resulting faculty reflects that of the student body. The personnel director determines that all of the new hires will be from two institutions: a state university of which 60% of the students are African American and an historically black college of which 93.3% of the students are African American. How many new teachers should be hired from each of the schools?

47. *Age Comparison*  Mark is 11 times older than his son Benjamin. In 27 years, Benjamin will be $\frac{1}{2}$ of the age of his father. How old are they?

48. *Age Comparison*  Two years ago, Juanita was 2.5 times older than her younger sister LaShonda. Ten years from now, Juanita will be 1.5 times older than LaShonda. How old are they now?

49. *Ticket Sales*  For a certain Bob Dylan concert there were two types of tickets available: $30 tickets and $20 tickets. If a total of 13,000 tickets were sold and the gross receipts for the concert totaled $310,000, then how many of each kind of ticket were sold?

50. *Painting Rate*  It takes Susan 1 hour less time to paint a room than it does David. Together, they take 2 hours. How long does it take each of them working alone?

51. *Raising a Whale*  Arnold and Franco are participating in a lift-a-thon known as "Raising the Whales" to raise money to protect the blue whale, of which there are only 10–12 thousand left in existence. Their goal is to lift a total weight equal to that of the largest blue whale ever captured, approximately 209 tons. (Source: *Guinness Book of World Records*.) If Arnold were lifting alone, he could "raise a whale" in 10 fewer hours than it would take Franco working alone. Together, they can complete the task in 15 hours. How long does it take each of them working alone?

52. *Car Speed*  Car A leaves an intersection at 2:00 P.M. traveling north at a constant speed. Car B leaves the same intersection at 4:00 P.M. and travels east at a constant speed. At 6:00 P.M., the cars are 200 miles apart, and the total distance traveled by the cars is 280 miles. How fast is each car traveling?

53. *Sprint Times*  Two brothers are competing in a 50-meter race. The first time they race, the older brother wins by 10 meters. The older brother offers to start the next race 10 meters behind the starting line in order to make the race fair. Although both runners run at the same speed as before, the older brother still wins, this time by 0.5 second. What is the 50-meter time for both boys? (Assume each is running at a constant speed.)

54. *Cattle Speed*  Two ancient cows of equal speeds and well versed in the rudiments of mathematics, are standing side-by-side on a 120-foot railroad bridge that runs north–south. The cows hear the whistle of a train heading toward them from the north when the train is 1000 feet away from the bridge. The

**Get ready!**

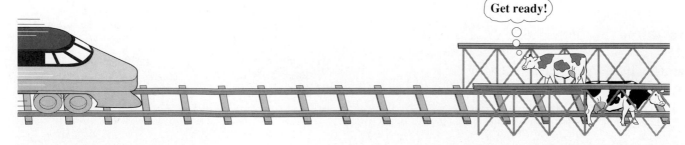

train is traveling at a speed of 53 feet per second. The impish bovines run at top speed in opposite directions to torment the engineer. Just as planned, they are each just grazed by the train as they clear the bridge. How fast can the cows run, and how far were they initially from the north end of the bridge?

55. *Unknown Constants* It is given that the variables $y$ and $x$ are related by the formula

$$y = a|x| + b|x - 1| + c|x - 2|$$

It is known that when $x = 0$, $y = 5$; when $x = 1$, $y = 6$; and when $x = 2$, $y = 1$. Find $a$, $b$, and $c$.

56. *Volume of a Box* An open box with a volume of 108 cubic inches is formed by cutting out squares with a side length of $x$ inches from the corners of an $l'' \times w''$ square sheet of cardboard and then folding along the dashed lines as shown in Figure 47. It so happens that the cut-out squares can be taped together to form a lid for the box, as shown in Figure 47. Find $l$, $w$, and $x$.

57. *Counting Coins* Your Uncle Bob pulls 15 coins (nickels, dimes, and quarters) from behind your ear. The total value of the coins is $2.10, and there are two more quarters than nickels. How many of each coin does Uncle Bob have?

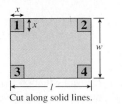

Cut along solid lines.  Fold along dashed lines . . .

. . . to make an open box.  Assemble cut-out squares to make a lid . . .  . . . that fits on the open box. The resulting closed box has a volume of 108 in³.

**FIGURE 47**

58. *Barbell Weights* A certain weight room includes a set of preweighted barbells, with the total weight printed on some of the barbells. A barbell with 2 large disks and 1 small disk on each side weighs 70 pounds; a barbell with 1 small disk on each side weighs 30 pounds; and a barbell with 3 large disks on each side weighs 75 pounds. Find the weight of the bar and each of the disks. Also, compute the weight of an (unmarked) barbell with 4 large disks and 2 small disks on each side.

---

■ *Projects for Enrichment*

---

59. *Balancing Chemical Equations* Chemical equations are a shorthand way of describing chemical changes. The left side of the chemical equation consists of the **reactants**, the chemicals that react with each other. The right side of the chemical equation consists of the **products**, the results of the chemical reaction. The left and right sides are separated by an arrow that is typically read as "yields." For example, the reaction consisting of hydrogen (in the form of hydrogen gas $H_2$) and oxygen (in the form of oxygen gas $O_2$) combining to form water would be represented as

$$H_2 + O_2 \rightarrow H_2O$$

This equation has not been balanced, however, and as such does not conform to the law of conservation of matter—there are two oxygen atoms on the left but only one on the right. The following chemical equation is a balanced version of this reaction.

$$2H_2 + O_2 \rightarrow 2H_2O$$

More generally, we must ensure that the number of atoms on each side of the yield sign are the same by affixing **integer coefficients** in front of the reactants and products. For example,

consider the unbalanced chemical equation $Al_2O_3 + F_2 \rightarrow AlF_3 + O_2F_2$. To balance the equation, we begin by placing unknown coefficients in front of each of the reactants and products.

$$wAl_2O_3 + xF_2 \rightarrow yAlF_3 + zO_2F_2$$

Now by equating the number of atoms of aluminum, fluorine, and oxygen, we obtain the following system of equations:

| | |
|---|---|
| $2w = y$ | Equating the number of aluminum atoms on each side |
| $2x = 3y + 2z$ | Equating the number of fluorine atoms on each side |
| $3w = 2z$ | Equating the number of oxygen atoms on each side |

Unfortunately, there are not enough equations to determine the values of $w$, $x$, $y$, and $z$, since there are four variables but only three equations. For the sake of convenience, we will assume that $w = 1$. This eliminates one of the variables, giving us the following system of three equations in three unknowns.

(1) $$2 = y$$

(2) $$2x = 3y + 2z$$

(3) $$3 = 2z$$

From equation (1) we have $y = 2$. From equation (3) we have $z = \frac{3}{2}$. Substituting these values into equation (2) gives us

$$2x = 3(2) + 2\left(\frac{3}{2}\right)$$

$$2x = 6 + 3$$

$$x = \frac{9}{2}$$

Thus, our balanced chemical equation now reads

$$Al_2O_3 + \tfrac{9}{2}F_2 \rightarrow 2AlF_3 + \tfrac{3}{2}O_2F_2$$

However, we would like our coefficients to be integers. Since multiplying both sides by the same constant will give an equivalent balanced equation, we simply clear fractions by multiplying both sides by the least common denominator of all fractions appearing in the equation. In this case, we multiply by 2 to obtain

$$2Al_2O_3 + 9F_2 \rightarrow 4AlF_3 + 3O_2F_2$$

Balance the following chemical equations by setting up and solving a system of equations.

a. $Na + Cl_2 \rightarrow NaCl$

b. $Ag + H_2S + O_2 \rightarrow Ag_2S + H_2O$

c. $Cu + HNO_3 \rightarrow Cu(NO_3)_2 + H_2O + NO$

d. $Ca_3(PO_4)_2 + H_3PO_4 \rightarrow Ca(H_2PO_4)_2$

**60.** *Partial Fraction Decomposition* Consider the following sum of rational expressions.

$$\frac{3}{x} - \frac{2}{x^2} + \frac{4}{x + 1}$$

We can add these terms together by finding a common denominator in the usual way. That is, since the least common denominator is $x^2(x + 1)$, we have

$$\frac{3(x)(x + 1) - 2(x + 1) + 4(x^2)}{x^2(x + 1)} =$$

$$\frac{3x^2 + 3x - 2x - 2 + 4x^2}{x^2(x + 1)} = \frac{7x^2 + x - 2}{x^2(x + 1)}$$

**Partial fraction decomposition** is the reverse of this procedure. For example, the partial fraction decomposition of

$$\frac{7x^2 + x - 2}{x^2(x + 1)} \quad \text{is} \quad \frac{3}{x} - \frac{2}{x^2} + \frac{4}{x + 1}$$

We will confine ourselves to rational expressions of the form $P(x)/Q(x)$, where $P(x)$ and $Q(x)$ are polynomials with the degree of $P(x)$ less than the degree of $Q(x)$. We will further assume that $Q(x)$ can be factored as the product of linear factors; that is, $Q(x)$ can be factored as $(x - a)^s(x - b)^t \cdots (x - c)^u$. It can be shown (it is called the Partial Fraction Decomposition Theorem) that with these restrictions, $P(x)/Q(x)$ can be written in the form

$$\frac{A_1}{x - a} + \frac{A_2}{(x - a)^2} + \cdots +$$

$$\frac{A_s}{(x - a)^s} + \frac{B_1}{x - b} + \frac{B_2}{(x - b)^2} + \cdots +$$

$$\frac{B_t}{(x - b)^t} + \cdots + \frac{C_1}{x - c} + \frac{C_2}{(x - c)^2} + \cdots + \frac{C_u}{(x - c)^u}$$

for some choice of the coefficients $A_1, A_2, \ldots, A_s, B_1, B_2, \ldots, B_t, C_1, C_2, \ldots, C_u$. This is not as confusing as it sounds. For example, according to the Partial Fraction Decomposition Theorem, the rational expression

$$\frac{3x^2 + 4x + 7}{x^3(x - 2)^2}$$

can be written in the form

$$\frac{A}{x} + \frac{B}{x^2} + \frac{C}{x^3} + \frac{D}{x - 2} + \frac{E}{(x - 2)^2}$$

with one term for each of the powers of $x$ from 1 to 3, and one term for each of the powers of $x - 2$ from 1 to 2.

We will illustrate the method of finding the partial fraction decomposition with the rational expression

$$\frac{7x^2 + x - 2}{x^2(x + 1)}$$

since we already know what the answer should be. We begin by noting that, indeed, the degree of the numerator (2) is less than the degree of the denominator (3), and that the denominator is the product of linear factors. Next we use the Partial Fraction Decomposition Theorem to conclude that

$$\frac{7x^2 + x - 2}{x^2(x + 1)} = \frac{A}{x} + \frac{B}{x^2} + \frac{C}{x + 1}$$

for some choice of constants $A$, $B$, and $C$. Clearing all denominators by multiplying both sides through by $x^2(x + 1)$ gives us

$$7x^2 + x - 2 = Ax(x + 1) + B(x + 1) + Cx^2$$

$$7x^2 + x - 2 = Ax^2 + Ax + Bx + B + Cx^2$$

Collecting like terms on the right-hand side gives us

$$7x^2 + x - 2 = (A + C)x^2 + (A + B)x + B$$

Now equating like powers of $x$ on the left- and right-hand sides gives us the following system of equations

(1)                          $7 = A + C$

(2)                          $1 = A + B$

(3)                          $-2 = B$

This system is especially easy to solve. From equation (3) we have $B = -2$. Substituting into equation (2), we get $1 = A + -2$, so that $A = 3$. Upon substituting $A = 3$ into

equation (1), we get $C = 4$. Thus, as expected,

$$\frac{7x^2 + x - 2}{x^2(x + 1)} = \frac{3}{x} + \frac{-2}{x^2} + \frac{4}{x + 1}$$

Find the partial fraction decompositions for each of the following rational expressions.

a. $\dfrac{9x^2 + 4x + 1}{x(x + 1)^2}$

b. $\dfrac{x^3 + 8x^2 + 11x + 6}{x(x + 1)^3}$

c. $\dfrac{x^3 + 10x^2 - 12x + 5}{x^2(x - 1)^2}$

## ▮▮ *Questions for Discussion or Essay*

**61.** Explain how a system of four equations in four variables could be solved.

**62.** Explain the connection between the systems of equations in the first column and the corresponding set of statements in the second column.

| | *System* | *Statements* |
|---|---|---|
| **A.** | $x + y = 2$ | Edward is the father of Mark. |
| | $2x + 2y = 4$ | Mark is the son of Edward. |
| **B.** | $x + y = 3$ | I am a crook. |
| | $x + y = 5$ | I am not a crook. |
| **C.** | $2x + 3y = 8$ | I am thinking of one of the Kennedy brothers. |
| | $3x - y = 1$ | I am thinking of an assassinated president. |

**63.** Some of the applied problems require natural-number solutions. For example, in a problem involving the number of coins, it would be unacceptable to have a final answer like "2.3 quarters, 4.7 dimes, and $-\sqrt{2}$ nickels." How do you suppose the authors *construct* systems of equations in such a way that the answers are guaranteed to be of the proper form? For example, can you construct a system of equations involving the variables $x$ and $y$, such that the solution is $x = 2$ and $y = -5$?

**64.** One of the most common mistakes made by beginning algebra students solving applied problems is that of not carefully defining variables. Consider the following problem.

There are four more than twice as many nickels as dimes. Together the nickels and dimes are worth $1.00. How many of each are there?

One key but often overlooked restriction is that any variable introduced should represent a *number*; many students make the mistake of defining variables to be inanimate objects, barnyard animals, and people. For example, when solving the above problem, it is not uncommon for students to begin this problem by writing something like, "Let $N$ = nickels, and let $D$ = dimes." The problem is that "nickels" is *not* a number. Thus, the student is forced to assume that $N$ represents a number in some way *associated* with "nickels." But there is more than one number associated with "nickels"! Explain how this difficulty leads to the following *incorrect* solution.

> Since there are four more than twice as many nickels as dimes, we have $N = 2D + 4$. Since the nickels and dimes together are worth 100¢, we have $N + D = 100$. Thus, we must solve the following system.
>
> (1)                  $N = 2D + 4$
>
> (2)                  $N + D = 100$
>
> Substituting from equation (1) into equation (2), we have
>
> $$(2D + 4) + D = 100$$
> $$3D + 4 = 100$$
> $$3D = 96$$
> $$D = 32$$
>
> Thus, $N = 2D + 4 = 2 \cdot 32 + 4 = 68$.

## CHAPTER REVIEW EXERCISES

**EXERCISES 1–4** □ *Determine whether or not the given value is in the solution set of the equation or inequality.*

1. $x^5 + 2x - 3x^2 - 1 = 23; \quad x = 2$

2. $3\sqrt{x - 7} - \dfrac{2x}{11} = 4; \quad x = 11$

3. $4|\sqrt{5x + 4} - 7| + 16 = 0; \quad x = 1$

4. $x^2 - 3x \le 4; \quad x = -1$

**EXERCISES 5–8** □ *Indicate whether or not each pair of equations or inequalities is equivalent.*

5. $x^2 + 2x = -1, \quad x + 2 = -\dfrac{1}{x}$

6. $3x + 5 = 7, \quad 3x = 2$

7. $(x + 3)^2 = 4, \quad x + 3 = 2$

8. $-x < 1, \quad x < -1$

**EXERCISES 9–12** □ *Solve the given equation with a graphics calculator. Give answers accurate to the nearest hundredth.*

9. $x^7 - x^3 + 1 = 0$

10. $4x^3 - 2x - 1 = 0$

11. $x^4 - 3x = 1$

12. $x^3 - 2x^2 = 5x - 6$

**EXERCISES 13–16** □ *Solve the given equation.*

13. $3x + 4 = 9$

14. $\frac{2}{3}(x - 4) = \frac{1}{6}(x - 1)$

15. $2(x + 1) - 3(2x - 2) = 1$

16. $(x + 1)(x - 2) = (x - 3)^2$

**EXERCISES 17–20** □ *Solve the inequality and graph the solution.*

17. $-3x - 4 < 8$

18. $2x + 1 \ge 5$

19. $\frac{1}{3}(x + 2) - \frac{1}{2}(x - 2) > 0$

20. $-2 < \dfrac{2x - 3}{4} \le 6$

**EXERCISES 21–22** □ *Solve for the indicated variable (assume all variables are nonzero).*

21. $ax + by = c, \quad$ for $x$

22. $y = mx - mx_1 + y_1, \quad$ for $m$

**EXERCISES 23–26** □ *Find the solution set for each of the following equations and inequalities.*

23. $|2x - 5| = 3$

24. $|x^2 + 12x + 16| = -16$

25. $|-3x + 4| < 2$

26. $|2x + 4| > 1$

**EXERCISES 27–30** □ *Find the real solution(s) of the given polynomial equation.*

27. $x(2x - 1)(x + 3) = 0$

28. $x^2 + 5x + 6 = 0$

29. $v^4 - 16 = 0$

30. $(x - 2)^3 - (x - 2) = 0$

**EXERCISES 31–34** □ *Solve the given polynomial inequality. Find exact values whenever possible. If you must approximate, give answers correct to the nearest hundredth.*

31. $(s + 1)(s + 5) > 0$

32. $4t^3 > 4t^2$

33. $x^4 - 2x < 1$

34. $x^3 - 2x^2 \le 5x - 6$

**EXERCISES 35–36** □ *Solve the given quadratic equation by completing the square.*

35. $x^2 + x - 12 = 0$

36. $2x^2 - 4x = 9$

**EXERCISES 37–40** □ *Solve using any method.*

37. $w^2 + 4w + 2 = 0$

38. $x^2 + x + 2 = 0$

39. $6y^2 - y - 2 = 0$

40. $x^4 - 3x^2 = 4$

**EXERCISES 41–44** □ *Indicate the number of real solutions for each quadratic equation given below.*

41. $2x^2 + 3x - 4 = 0$

42. $2x^2 - 3x + 4 = 0$

43. $x^2 + 4\sqrt{5}x + 20 = 0$

44. $2x^2 - 5 = 0$

**EXERCISES 45–50** □ *Solve the given rational equation or inequality.*

45. $\dfrac{2}{x} = \dfrac{3}{x + 1}$

46. $\dfrac{3x - 1}{2x} = \dfrac{3x + 4}{2x + 1}$

47. $\dfrac{x}{2x + 1} = \dfrac{x + 2}{5x}$

48. $\dfrac{x^2}{x + 1} = -3 + \dfrac{1}{x + 1}$

49. $\dfrac{3x - 9}{x - 1} > 0$

50. $\dfrac{2x^2 - 2}{x^2 - 16} < 0$

**EXERCISES 51–54.** □ *Solve the given equation involving radicals.*

51. $\sqrt{3x - 5} = 1$

52. $\sqrt{2x - 3} + 7 = 5$

53. $\sqrt{3a + 10} - a = 2$

54. $\sqrt{x - 1} + \sqrt{x + 4} = 1$

**EXERCISES 55–58** □ *Solve the given system of equations.*

55. $x + 5y = 13$
    $2x + 3y = 12$

56. $3x - y = -7$
    $4x + y = -7$

57. $x + 2y^2 = 6$
    $x + y^2 = 5$

58. $x + 2y + z = 11$
    $y - z = -1$
    $x + z = 5$

59. **Consecutive Integers** Find two consecutive odd integers whose sum is 236.

60. **CD Price** A compact disc (CD) costs \$12.19 including 6% sales tax. What is the price of the CD before sales tax?

61. **Interception Risk** Since many professional football players can jump and extend their hands to a height of 11 feet or more, a passed football is at some risk of interception whenever its

height is 11 feet or below. If the quarterback releases the ball from a height of 6 feet with an upward velocity of 30 feet per second, then the height of the ball in feet $t$ seconds after it is released is given by $h = -16t^2 + 30t + 6$. Find the time interval for which the football is at risk of being intercepted.

**62.** *Fortune 500* A corporate CEO would like her company's total sales to surpass 625 million dollars, which she estimates would be enough to be listed in the *Fortune 500*. If their revenue (in millions of dollars) for the sale of $x$ units of a product is given by $R = 2x(50 - x)$, then for what interval of unit sales will the company attain her goal?

**63.** *Rectangle Dimensions* The length of a rectangle is $\frac{1}{2}$ foot less than 6 times its width. Its area is 3 square feet. Find the dimensions of the rectangle.

**64.** *Car Wash Rate* It takes Daniel 1 hour longer to wax the car than it does his brother Roberto. Together it takes them 50 minutes to wax the car. How long does it take each working separately?

**65.** *Triangle Dimensions* What are the dimensions of a right triangle if the length of one side is 3 inches less than twice the length of the other and the length of the hypotenuse is 51 inches?

**66.** *Counting Coins* A woman has 11 coins in her pocket, all of which are either nickels or dimes. If the value of the coins is 75¢, how many of each type of coin does she have?

---

## CHAPTER TEST

**PROBLEMS 1–8** □ *Answer true or false.*

**1.** If $(x + 1)(x - 4) = 4$, then either $x = 3$ or $x = 6$.

**2.** If $(x + 1)(x - 4) = 0$, then either $x = -1$ or $x = 4$.

**3.** If $|x - 2| = 3$, then $x$ is a distance of 3 units from 2.

**4.** If $|x| > 1$, then $-1 > x > 1$.

**5.** The equations $x^2 = 1$ and $x = 1$ are equivalent.

**6.** If $a^2 = b^2$, then $a = b$.

**7.** All polynomial equations can be solved using the zero-product property.

**8.** A solution of a system of equations in the variables $x$ and $y$ is any pair of numbers $(x, y)$ that satisfy at least one of the equations.

**PROBLEMS 9–12** □ *Give an example of each of the following.*

**9.** A quadratic equation with 3 as its only solution

**10.** A polynomial equation of degree 3 with solutions 1, 2, and 3

**11.** An inconsistent system of equations

**12.** An absolute-value inequality with no solution

**PROBLEMS 13–22** □ *Solve the given equation or inequality. Give exact answers wherever possible. When approximating, give answers accurate to the nearest hundredth.*

**13.** $-3t + 5 = 9t - 1$

**14.** $|2x + 4| = 7$

**15.** $|3 - 2y| > 1$

**16.** $u^3 - 4u = 0$

**17.** $x^4 - 2x^2 + x - 1 = 0$

**18.** $\dfrac{x}{x + 2} < 0$

**19.** $2x^2 - 4x = 8$

**20.** $x^2 - 5 \le 4x$

**21.** $\dfrac{2}{x + 4} = \dfrac{3}{x - 2}$

**22.** $\sqrt{2x + 1} + 1 = x$

**23.** Solve the following system of equations.

$$3x - 2y = -1$$
$$2x + 3y = 21$$

**24.** The sum of three consecutive even integers is 306. Find the numbers.

**25.** A Habitat for Humanity crew can build a house in 3 days less time than it takes a local construction company. When both crews work together, a house is built every 2 days. How long does it take each crew working alone?

# RELATIONS
# AND
# THEIR
# GRAPHS

■ The Arecibo radio telescope located near Arecibo, Puerto Rico, is the largest single telescope of any kind. Electromagnetic transmissions from celestial objects are reflected from the 305-meter-diameter parabolic dish to the antenna suspended above it. Radio telescopes, although employed for such exotic purposes as identifying black holes and quasars, determining the chemical composition of distant galaxies, and even searching for extraterrestrial life, are based on the same principle as the automobile headlight: the reflective properties of parabolas.

## RELATIONS

- How can naive use of a graphics calculator be dangerous?
- How is searching with a graphics calculator like exploring the skies with a view finder and a powerful telescope?
- Does television really cause actors to look 10 pounds heavier?
- How can a graphics calculator be used to estimate the date on which the average temperature in Orlando is highest?

### RELATIONS

The ancient Greeks showed that the area of a circle and its radius are related by the equation $A = \pi r^2$, Newton found that mass and acceleration are related by the equation $F = ma$, and Einstein discovered that mass and energy are related by the equation $E = mc^2$. The Ideal Gas Law, $PV = nRT$, relates the pressure, volume, number of moles, and temperature of a gas. All of these are examples of **relations**. More generally, any equation in two or more variables defines a relation between the variables.

An equation involving only the variables $x$ and $y$ defines a **relation in $x$ and $y$**. Examples of relations in $x$ and $y$ include $y = 2x + 4$, $x^2 + y^2 = 25$, and $x^2 y - 3xy - 5y^2 = 5$. The **graph** of a relation in $x$ and $y$ is the set of points $(x, y)$ in the coordinate plane such that $x$ and $y$ satisfy the relation. For example, the graph of the relation $x^2 + y^2 = 25$ is the set $\{(x, y) | x^2 + y^2 = 25\}$, which is a circle of radius 5 centered at the origin.

**EXAMPLE 1**    *Showing points are on the graph of a relation*

Show that the points $(5, 0)$ and $(3, 4)$ are on the graph of $x^2 + y^2 = 25$.

**SOLUTION**    Since $5^2 + 0^2 = 25 + 0 = 25$, $(5, 0)$ is on the graph of $x^2 + y^2 = 25$. Similarly, $3^2 + 4^2 = 9 + 16 = 25$, and so $(3, 4)$ is on the graph also.

**EXAMPLE 2**    *Finding points on the graph of a relation*

Find all points on the graph of $x^2 y - 3xy - 5y^2 = 5$ with $y$-coordinate 1.

**SOLUTION**    We substitute $y = 1$ into the relation $x^2 y - 3xy - 5y^2 = 5$, obtaining

$$x^2 \cdot 1 - 3x \cdot 1 - 5 \cdot 1^2 = 5$$
$$x^2 - 3x - 10 = 0$$
$$(x - 5)(x + 2) = 0$$
$$x = 5, \ x = -2$$

Thus, the points on the graph of $x^2 y - 3xy - 5y^2 = 5$ with $y$-coordinate 1 are $(5, 1)$ and $(-2, 1)$.

The simplest technique for graphing a relation is **plotting points**. To do this, we choose a convenient value for one of the variables, and then use the relation to determine the value of the other variable. We then graph or "plot" the corresponding point. The process is repeated as necessary, and then the points are connected by drawing a smooth curve through the plotted points. This, by the way, is essentially what is done by a graphics calculator when it creates a graph.

**EXAMPLE 3**   *Graphing a line*

Sketch a graph of the relation $y = -\frac{1}{2}x + 1$.

**SOLUTION**   We choose a few convenient values of $x$, and compute the corresponding $y$ values.

| $x$ | $y = -\frac{1}{2}x + 1$ |
|-----|-------------------------|
| 0   | 1                       |
| 2   | 0                       |
| 4   | $-1$                    |
| $-2$ | 2                      |
| $-4$ | 3                      |

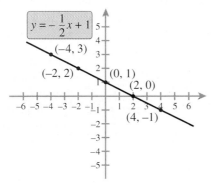

**FIGURE 1**

Thus, the points $(0, 1)$, $(2, 0)$, $(4, -1)$, $(-2, 2)$, and $(-4, 3)$ are on the graph of $y = -\frac{1}{2}x + 1$. After plotting these points and "connecting the dots," we obtain the graph shown in Figure 1.

The points at which the graph of a relation crosses an axis are called the **intercepts**. More specifically, the points where the graph crosses the $x$-axis are called the ***x*-intercepts**, and the points where the graph crosses the $y$-axis are called the **$y$-intercepts**. In the previous example, $(2, 0)$ is the only $x$-intercept and $(0, 1)$ is the only $y$-intercept. Since intercepts are a key feature of the graph, they should be included among the points that are plotted.

Intercepts can be found by using the fact that a point on the $x$-axis has $y$-coordinate 0, and a point on the $y$-axis has $x$-coordinate 0. Thus, to find any $x$-intercepts, we set $y = 0$ and solve for $x$. Similarly, to find $y$-intercepts, we set $x = 0$ and solve for $y$.

**EXAMPLE 4**   *Graphing a horizontal parabola*

Sketch a graph of $x = y^2 - 3y - 4$.

**SOLUTION**   We begin by finding the intercepts. To find any $x$-intercepts, we set $y = 0$.

$$x = (0)^2 - 3(0) - 4 = -4$$

Thus, the only $x$-intercept is $(-4, 0)$. Next we set $x = 0$ to find any $y$-intercepts.

$$0 = y^2 - 3y - 4$$
$$0 = (y + 1)(y - 4)$$
$$y = -1, \ y = 4$$

Thus, the $y$-intercepts are $(0, -1)$ and $(0, 4)$. To find other points on the graph, we choose convenient values of $y$ and then compute corresponding values for $x$.

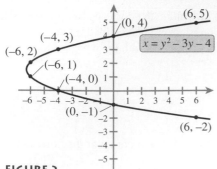

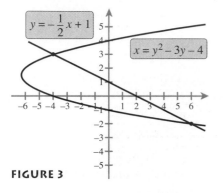

**FIGURE 2**

**FIGURE 3**

| $y$ | $x = y^2 - 3y - 4$ |
|-----|--------------------|
| −2  | 6                  |
| 1   | −6                 |
| 2   | −6                 |
| 3   | −4                 |
| 5   | 6                  |

Thus, the points $(6, -2)$, $(-6, 1)$, $(-6, 2)$, $(-4, 3)$, and $(6, 5)$ are on the graph of $x = y^2 - 3y - 4$. We plot these points and the intercepts, and then connect them with a smooth curve to obtain the graph in Figure 2.

■

Notice that if the graphs of the relations $y = -\frac{1}{2}x + 1$ and $x = y^2 - 3y - 4$ from Examples 3 and 4 are plotted on the same coordinate system, they will intersect at two points (see Figure 3). In general, a point $(a, b)$ is an **intersection point** of the graph of two relations if it lies on the graphs of both relations. In other words, $(a, b)$ is an intersection point if $x = a$ and $y = b$ satisfy the equations of both relations simultaneously. Thus, in order to find points of intersection, we solve the equations simultaneously, as demonstrated in the following example.

**EXAMPLE 5**    *Finding intersection points of relations*

Find the coordinates of the points of intersection of $y = -\frac{1}{2}x + 1$ and $x = y^2 - 3y - 4$.

**SOLUTION**    We find the points of intersection by solving the following system of equations.

$$(1) \qquad\qquad y = -\frac{1}{2}x + 1$$

$$(2) \qquad\qquad x = y^2 - 3y - 4$$

We first solve equation (1) for $x$.

$$y = -\frac{1}{2}x + 1$$

$$y - 1 = -\frac{1}{2}x$$

$$(3) \qquad\qquad -2y + 2 = x$$

Now we substitute $-2y + 2$ for $x$ in equation (2).

$$-2y + 2 = y^2 - 3y - 4$$
$$0 = y^2 - y - 6$$
$$0 = (y + 2)(y - 3)$$
$$y = -2, \; y = 3$$

Substituting $y = -2$ and $y = 3$ into equation (3) yields $x = -2(-2) + 2 = 6$ and $x = -2(3) + 2 = -4$, respectively. Thus, the points of intersection are $(6, -2)$ and $(-4, 3)$, as shown in Figure 3.

### CHOICE OF SCALE

The appearance of a graph depends greatly on the choice of **scale** for both the $x$ and the $y$ axes. Whether graphing by hand or using a graphics calculator, the scale should be chosen so that the key features of the graph stand out. Shown in Figures 4–7 are four views of the graph of $5000y = 5000x^3 - 150x^2 + x$ using different scales. Notice that if we were to look only at Figures 4–6, we would conclude that the relation had only one $x$-intercept, whereas Figure 7 clearly shows that there are three $x$-intercepts.

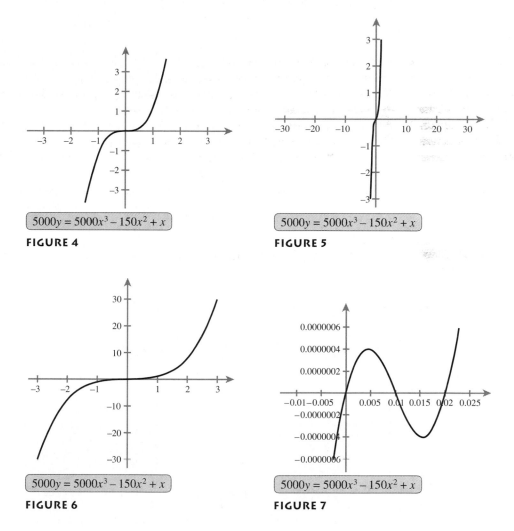

$$5000y = 5000x^3 - 150x^2 + x$$

**FIGURE 4**

$$5000y = 5000x^3 - 150x^2 + x$$

**FIGURE 5**

$$5000y = 5000x^3 - 150x^2 + x$$

**FIGURE 6**

$$5000y = 5000x^3 - 150x^2 + x$$

**FIGURE 7**

Choosing a good scale is as much art as it is science. It often takes a little trial and error to create a graph that "feels right." In general, the graph should be drawn so that the points that are plotted fill most of the graph area, as illustrated in the following example.

EXAMPLE 6    *Choosing an appropriate scale*

Graph the relation $y = 2560x^4 - 160x^2$ using the following table of values.

| $x$ | $2560x^4 - 160x^2$ |
|---|---|
| $-0.4$ | $39.9$ |
| $-0.3$ | $6.3$ |
| $-0.2$ | $-2.3$ |
| $-0.1$ | $-1.3$ |
| $0$ | $0$ |
| $0.1$ | $-1.3$ |
| $0.2$ | $-2.3$ |
| $0.3$ | $6.3$ |
| $0.4$ | $39.9$ |

**SOLUTION**   We begin by noting that the $x$-values range from $-0.4$ to $0.4$, and the $y$-values range from $-2.3$ to $39.9$. If we choose the same scale on the $x$- and $y$-axes, we would obtain something like the graph in Figure 8. Note that at the given scale, all of the plotted points are within a narrow vertical strip; the points are practically on top of each other! It is virtually impossible to depict the graph of the given relation using this scale. Instead, we should choose the scales on the $x$- and $y$-axes independently. The scale should be chosen so that the points of interest are spread out. The graph in Figure 9 shows the same points plotted on a set of coordinate axes with a different scale. Upon connecting the dots with a smooth curve, we obtain the sketch in Figure 10.

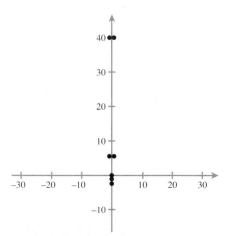

**FIGURE 8**

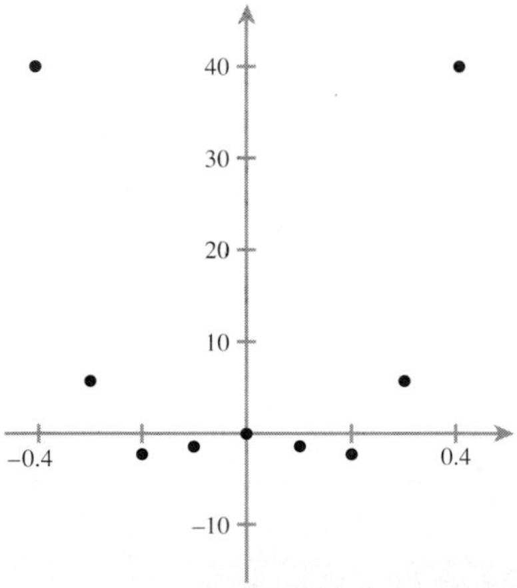

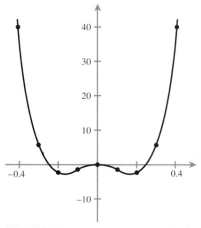

**FIGURE 9**

**FIGURE 10**

GRAPHICS
CALCULATORS

Relations that can be solved for $y$—that is, relations that can be written in the form $y = \square$, where $\square$ is an expression involving only the variable $x$—can be plotted with the aid of a graphics calculator. Such relations are called **functions** and will be explored in detail in Chapter 4. The following example illustrates the procedure for graphing a relation with a graphics calculator.

**EXAMPLE 7**

*Graphing a relation with a graphics calculator*

Graph the relation $4y + 48x = 12x^2 + 56$.

**SOLUTION**    We begin by solving the relation for $y$.

$$4y + 48x = 12x^2 + 56$$
$$4y = 12x^2 - 48x + 56$$
$$y = 3x^2 - 12x + 14$$

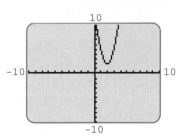

**FIGURE 11**

Next, we enter the equation $y = 3x^2 - 12x + 14$ and plot the graph. One view of the graph is shown in Figure 11.

■

Care must be taken to choose an appropriate scale when using a graphics calculator, just as when sketching graphs of relations by hand. The scale on a graphics calculator can be adjusted either by using the range variables or by zooming. (Note that some calculators also have an auto-scale feature that automatically selects a scale for the $y$-axis for a given range of values on the $x$-axis.) The range variables Xmin, Xmax, Ymin, and Ymax determine the portion of the graph that is to be viewed, whereas zooming provides several convenient shortcuts for adjusting the view. Regardless of the method you use, your goal should be to produce a plot in which the scale is large enough to show all of the features of interest and yet small enough so that these features stand out. This is often an impossible task; it is like asking for a map that clearly shows each of the seven continents, as well as the back streets and alleyways of Punxsutawney, Pennsylvania! Thus, we must either narrow our focus and show a small portion of the graph in great detail, or we must show a large portion of the graph that doesn't show fine detail. Alternatively, we may choose to produce several views of the graph, each of which shows one or more essential features. This process is illustrated in the following examples.

**EXAMPLE 8**

*Narrowing the range of the viewing rectangle*

Use a graphics calculator to graph $2y + 2x = 2x^3 - x^2 + 1$.

**SOLUTION**    First we solve the equation for $y$, to obtain

$$y = x^3 - \frac{x^2}{2} - x + \frac{1}{2}$$

Next we set the range variables to the values Xmin=-10, Xmax=10, Ymin=-10, and Ymax=10. The plot is shown in Figure 12. Although this plot does give some idea of what the graph looks like, the "humps" do not stand out. We can remedy this by selecting a smaller viewing rectangle. We adjust the range variables so that Xmin=-2, Xmax=2, Ymin=-2, and Ymax=2. This yields the plot in Figure 13.

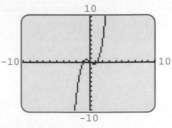

**FIGURE 12**

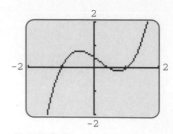

**FIGURE 13**

---

**EXAMPLE 9**     *Broadening the range of the viewing rectangle*

The average (Fahrenheit) temperature in Orlando, Florida over the course of the year can be approximated with the polynomial function

$$y = 0.018x^4 - 0.45x^3 + 2.93x^2 - 1.5x + 61.5$$

where $x$ is the number of months after January 1 (so $x = 0$ corresponds to January 1, $x = 1$ corresponds to February 1, etc.). What are the lowest and highest average temperatures, and when do they occur?

**SOLUTION**     By setting $x = 0$, we determine that the $y$-intercept is 61.5, which suggests that the range of $y$-values should include 61.5. Moreover, since we are only interested in a 12-month period, the range in $x$-values should be limited to the interval $[0, 12]$. This information, together with our intuition about Florida temperatures, suggests that we try the range values Xmin=0, Xmax=12, Ymin=55, and Ymax=90. The result is shown in Figure 14. Using the trace feature, we see in Figure 15 that the minimum temperature is approximately 61.3° when $x = 0.25$ (the second week of January). Figure 16 shows that the maximum temperature is approximately 84.2° when $x = 6.3$ (the second week of July).

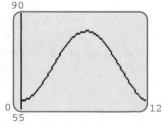

**FIGURE 14**

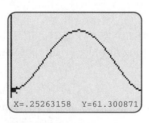

**FIGURE 15**

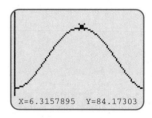

**FIGURE 16**

---

**RULE OF THUMB**     When selecting initial values for the range variables, it is often helpful to find the $y$-intercept of the relation and then choose values for Ymin and Ymax so that Ymin $\leq y$-intercept $\leq$ Ymax.

In many cases, it is difficult to select an appropriate viewing rectangle by manually setting the range variables. In the following example, we will use a zoom box to choose the viewing rectangle.

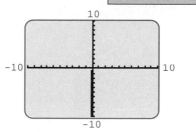

**FIGURE 17**

**EXAMPLE 10**

*Finding intercepts using zoom boxes*

Approximate the $x$-intercepts of the relation $y = 1000x^3 - 15x^2 + 0.0002$.

**SOLUTION**   The plot of this relation with a viewing rectangle defined by Xmin=−10, Xmax=10, Ymin=−10, and Ymax=10 is given in Figure 17. Although it appears that the graph crosses the $x$-axis somewhere near the origin, it is impossible to tell *exactly* where the graph crosses, or even how many times it crosses. The sequence of plots shown in Figures 18 and 19 illustrates how zoom boxes can give us a better view of the graph near the $x$-intercepts. Notice that at each step, we construct a zoom box around the region where the graph appears to cross the $x$-axis.

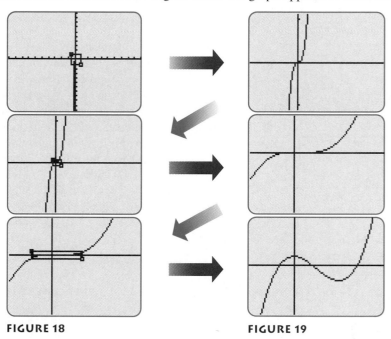

**FIGURE 18**          **FIGURE 19**

Now using the trace feature, we can obtain the approximate $x$-coordinates of the intercepts, namely $-0.0034$, $0.0042$, and $0.0141$ as shown in Figures 20–22. If more accuracy is desired, we can zoom-in near each intercept.

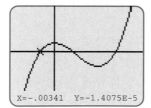

**FIGURE 20**

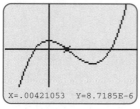

**FIGURE 21**

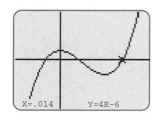

**FIGURE 22**

As we mentioned earlier, a relation must be in the form $y = \square$, where $\square$ is an expression involving only the variable $x$, before a graphics calculator will be useful. In Examples 7 and 8, it was possible to write the relations in this form by solving for $y$. In the following example, solving for $y$ yields two equations, which we refer to as the

**branches** of the relation. The graph of a relation consists of the graphs of all of its branches.

**EXAMPLE 11**    *Graphing a circle using a graphics calculator*

Graph the circle with equation $x^2 + y^2 = 9$.

**SOLUTION**    We begin by solving the relation for $y$.

$$x^2 + y^2 = 9$$
$$y^2 = 9 - x^2$$
$$y = \pm\sqrt{9 - x^2}$$

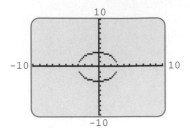

**FIGURE 23**

Thus, we have two branches, $y = \sqrt{9 - x^2}$ and $y = -\sqrt{9 - x^2}$. We enter each of the branches and plot both graphs simultaneously, as shown in Figure 23. Note that the plot appears to be more *elliptical* than it does circular. This is because the calculator screen is wider than it is high, and so whenever the range variables Xmin and Xmax are given the same values as Ymin and Ymax, the units on the $x$-axis get spaced further apart than on the $y$-axis. With most curves, we would most likely never notice this phenomenon, but it is striking with circles. To an extent, we can compensate for this effect by changing the range variables so that the values for Xmin and Xmax are further apart than Ymin and Ymax. In Figure 24 we used Xmin=−15, Xmax=15, Ymin=−10, and Ymax=10. On some calculators, these are referred to as the ''square'' range setting. The ''square'' setting give us a more nearly circular graph.

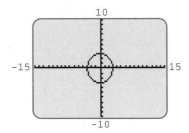

**FIGURE 24**

---

## EXERCISES 1

**EXERCISES 1–4** □ *Show that the given points are on the graph of the relation.*

1. $x^2 + y^2 = 100$;  $(8, 6), (0, -10), \left(5\sqrt{2}, 5\sqrt{2}\right)$

2. $x^2 + xy + y^2 = 7$;  $(1, 2), (-2, 3), \left(\sqrt{7}, 0\right)$

3. $y = \dfrac{x}{x + 1}$;  $(-2, 2), \left(-\dfrac{4}{3}, 4\right), \left(\sqrt{2}, 2 - \sqrt{2}\right)$

4. $x = \sqrt{y^2 - 3y}$;  $(0, 3), (2, -1), \left(\sqrt{10}, 5\right)$

**EXERCISES 5–10** □ *Find all points with the given x- or y-coordinate that lie on the graph of the relation.*

5. $3x + 4y = 5$, $x$-coordinate 3

6. $2x - 7y = 9$, $y$-coordinate 1

7. $3x^2y + 4y^2 + 7y = 38$, $y$-coordinate 1

8. $x^2y + 2xy^2 + 2x + 6y = 0$, $x$-coordinate 2

9. $\dfrac{x^2}{4} - \dfrac{y^2}{9} = 1$, $x$-coordinate 4

10. $\dfrac{x^2}{32} + \dfrac{y^2}{50} = 1$, $y$-coordinate 5

**EXERCISES 11–12** □ *Complete the table of values for the relation. Then use the coordinates of the given points to construct a graph of the relevant portion of the relation. Be sure to choose an appropriate scale.*

11. $y = 1000(x^3 + 1)$

| $x$ | $y$ |
|------|------|
| −0.1 | |
| 0.0 | |
| 0.1 | |
| 0.2 | |
| 0.3 | |

12. $x = y^3 - 3y^2$

| $x$ | $y$ |
|------|------|
| | −1 |
| | 0 |
| | 2 |
| | 3 |
| | 4 |

**EXERCISES 13–20** □ *Find the intercepts for each of the following relations and sketch the graph by plotting points.*

13. $y = 2x - 4$

14. $x = 3y + 6$

15. $2x + 3y = 24$

16. $4x - 5y = 40$

17. $x = y^2 - 9$

18. $y = x^2 - 4$

19. $y + 4 = (x - 3)^2$

20. $x + 1 = (y - 2)^2$

**EXERCISES 21–28** □ *Approximate all of the x-intercepts to the nearest hundredth using a graphics calculator. The number of x-intercepts is indicated in parentheses.*

**21.** $y = x^3 - 7x^2 - 9x + 63$   (3)

**22.** $y = x^3 - 2x^2$   (2)

**23.** $y^3 = x^4y^2 + x^3 - 4x^2 - 14x + 44$   (3)
(*Hint:* Begin by setting $y = 0$, and then solve the resulting equation for $x$ using a graphics calculator.)

**24.** $x^2y = x^4 - 8x^2 - 425x + 3y$   (2)
(*Hint:* Begin by setting $y = 0$, and then solve the resulting equation for $x$ using a graphics calculator.)

**25.** $y = x^4 - 4x^3 - \dfrac{x^2}{10,000} + \dfrac{x}{2500}$   (4)

**26.** $y = x^3 - 90x^2 - 1000x + 200$   (3)

**27.** $y = x^2 - 10x + 11 - \dfrac{22}{x + 2}$   (3)

**28.** $y = \sqrt{x^4 + 1} - \sqrt{x^2 + 1} - \sqrt{2x^2 + 3}$   (2)

**EXERCISES 29–34** □ *Match each relation with its graph.*

**29.** $y = -\dfrac{x^3}{8} + 2x^2 - 6x$

**30.** $y = 4x^2 - 16x$

**31.** $y = x^3 - \dfrac{x}{4}$

**32.** $y = \dfrac{x^3}{2} - \dfrac{x^4}{24}$

**33.** $y = 4x^4 + 12x^3 + 13x^2 + 6x + 1$

**34.** $y = 2x^2 + 3x + 1$

a. 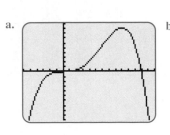   b.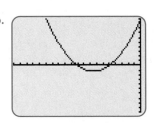

c. 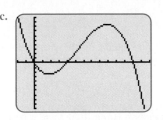   d.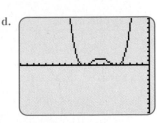

e. 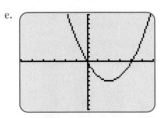   f.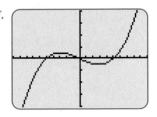

**EXERCISES 35–44** □ *Solve the relation for y and use a graphics calculator to sketch the graph.*

**35.** $y - 2x = x^2 + 2$        **36.** $y + 3x^2 = x^3 + 4$

**37.** $x = \dfrac{2}{y} - 1$        **38.** $x = \dfrac{1}{y + 1}$

**39.** $x^3 + y^3 = 8$        **40.** $x^2 - y^3 - x = 2$

**41.** $36x^2 + 9y^2 = 324$        **42.** $25x^2 - 36y^2 = 900$

**43.** $x = (y - 3)^2 - 5$        **44.** $(x + y)^2 = 2x - 3$

**EXERCISES 45–50** □ *Find the point(s) of intersection of the given pair of relations.*

**45.** $2x - y = 1$;  $3x + y = 9$

**46.** $x - 4y = 8$;  $-3x - 2y = 4$

**47.** $y + x^2 = 4$;  $y - x = 2$

**48.** $y = x^2 + 2x$;  $y = -2x - 3$

**49.** $x^2 + y^2 = 2$;  $x = y^2$

**50.** $x^2 + y^2 = 25$;  $x + y = 1$

■ *Applications*

**51.** *U.S. Population*  The U.S. population for the years 1950–1990 can be approximated with the relation

$$y = 0.0002x^3 - 0.02x^2 + 3x + 151,$$

where $y$ represents the population (in millions) and $x$ denotes the year (with $x = 0$ corresponding to 1950). Use an appropriate scale to plot the graph of the relation, and approximate the year when the population reached 190 million.

**U.S. Population 1950-1990**

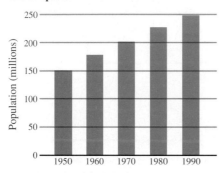

Data Source: U.S. Bureau of the Census

**52.** *$CO_2$ Concentration*  The concentration of $CO_2$ in the atmosphere during the years 1960–1980 can be approximated with the relation $y = 0.0007x^3 + 0.02x^2 + 0.8x + 316$, where $y$ represents the concentration of $CO_2$ (measured in parts per million) and $x$ denotes the year (with $x = 0$ corresponding to 1960). Use an appropriate scale to plot the graph of the relation, and approximate the year when the concentration first exceeded 330 parts per million.

**$CO_2$ Concentration 1960-1980**

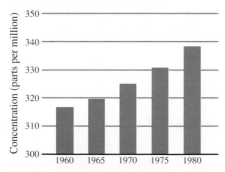

Data Source: *World Resources 1992-1993*
A Report by the World Resources Institute
Dr. Allen L. Hammond, Ed
Oxford University Press, 1992.

**53.** *AIDS Cases*  The number of new AIDS cases reported during the years 1982–1989 can be approximated with the relation $y = -45x^4 + 446x^3 - 517x^2 + 2026x + 984$, where $y$ represents the number of new cases and $x$ denotes the year (with $x = 0$ corresponding to 1982). Use an appropriate scale to plot

the graph of the relation, and approximate the year when the number of new cases was 30,000. What appears to be wrong with this model for years outside of the interval 1982–1989?

**New AIDS Cases 1982-1989**

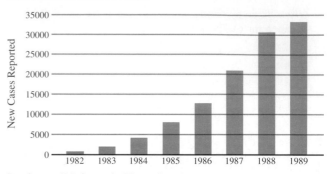

Data Source: U.S. Centers for Disease Control

**54.** *California Population Density*  The population density of California for the years 1920–1990 can be approximated with either of the two relations $y = 0.0005x^3 - 0.04x^2 + 2.8x + 22$ or $y = 0.00054x^3 + 0.072x^2 - 0.11x + 22$. In both, $y$ represents the density (people per square mile) and $x$ the year (with $x = 0$ corresponding to 1920). For which years do these relations predict the same density?

**California Population Density 1920-1990**

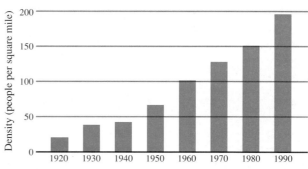

Data Source: U.S. Bureau of the Census

**55.** *Volume of a Box*  A box with a square base is to be constructed to hold a volume of 100 cubic inches. If we denote the base length by $x$ and the height by $y$, then we must have $100 = x^2y$.

**a.** Plot the graph of this relation.

**b.** Determine the base length for a box with a height of 4 inches. Illustrate this on your graph from part (a).

**56.** *Waiting Time*  A fast-food restaurant has determined that an acceptable average time for a customer to wait in line is 1 minute. A result from queuing theory suggests that the average length of time in minutes that a customer waits in line is given by $1/(x - y)$, where $x$ is the number of customers served per hour and $y$ is the number of customers who arrive per hour. Thus, the relation

$$1 = \frac{1}{x - y}$$

describes the relationship between $x$ and $y$ when the average waiting time is 1 minute.

a. Plot the graph of this relation and describe its shape.

b. During the busiest time of the day, customers arrive at a rate of 30 per hour. How many customers must be served per hour so that the time in line is still 1 minute? Illustrate this on your graph from part (a).

57. *Revenue and Cost* The manufacturers of a product estimate that if they sell $x$ units, then the total revenue (the money that

they take in) will be given by $R = x^3 + 50x^2 - 200x + 70$, and the total cost (the money they must spend) will be $C = 2x^3 + 11x^2 + 60x + 370$. Profit is defined as Revenue − Cost.

a. Give an expression for the profit in terms of $x$, the number of units sold.

b. Estimate the number of products that must be sold for the manufacturer to break even—that is, for the profit to be zero. Since $x$ represents the number of units sold, you need only consider positive values of $x$.

c. Estimate the number of units that must be sold for the profit to be as large as possible. What is the largest possible profit?

---

### ◼ *Projects for Enrichment*

---

58. *Hidden Intercepts*

a. Graph the relation $y = 10x^4 \sin(0.1/x)$ with a graphics calculator. (The expression "sin" comes from trigonometry and will be defined in Chapter 7, but you needn't have any knowledge of either the meaning of sin or trigonometry in general in order to solve this problem. Just use the $\boxed{\text{SIN}}$ key on your calculator when entering the relation.) If you use range settings of Xmin=−10, Xmax=10, Ymin=−10, and Ymax=10, your graph should look like the one shown in Figure 25.

b. Use the graph you found in part (a) to estimate the number of $x$-intercepts of the relation.

c. Adjust the viewing rectangle in order to obtain a plot similar to the one shown in Figure 26. List the values of the range variables.

d. On the basis of the graph in part (c), how many $x$-intercepts do you think there are?

e. What lesson can be learned from the disparity of your answers to parts (b) and (d)?

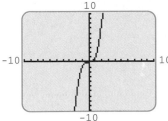

**FIGURE 25**

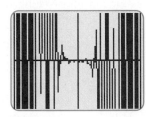

**FIGURE 26**

59. *Relations and Fractal Curves* In this section we considered relations that were defined by equations in the variables $x$ and $y$. More generally, a **relation** is any set of ordered pairs $(x, y)$. For example, the sets

$A = \{(9, -3), (4, -2), (1, -1), (0, 0), (1, 1), (4, 2), (9, 3)\}$

$B = \{(x, y)|y = x^2\}$

$C = \{(x, x^3)|x = 0, \pm1, \pm2, \ldots\}$

$D = \{(x, y)|x = 0, 1, 2, 3 \text{ and } y \geq x\}$

$E = \{(x, y)|x^2 + y^2 \leq 1\}$

are all relations. Thus, a relation can be a finite set of points, an infinite set of points defined by an equation, or a set of points arising from some other description. In any case, the graph of a relation $R$ is the plot of the points in the set $R$.

a. Plot the graphs of the relations $A$–$E$ given above.

It may not always be possible to describe a relation with concise set notation as we did with the relations $A$–$E$. Consider the set of points obtained by the following process. Start with a line segment such as that shown in Figure 27. Divide the line segment into thirds as shown in Figure 28. Take out the middle third of the segment and replace it with two line segments that form two sides of an equilateral triangle, as in Figure 29. Notice that the total length of the new figure is $\frac{4}{3}$ times the length of original line segment. Now repeat this process on each of the four line segments. That is, take out the middle third of each line segment and replace it with two line segments as shown in Figure 30. Another repetition of this process leads to the curve shown in Figure 31. If this process is repeated indefinitely, the result is a *fractal curve* similar to that shown in Figure 32. The set of points that form this curve is a relation.

**FIGURE 27**

**FIGURE 28**

**FIGURE 29**

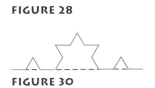

**FIGURE 30**

**FIGURE 31**

**FIGURE 32**

An interesting fact about the resulting fractal curve is that it has infinite length. This is because the total length of the curve at each stage is $\frac{4}{3}$ times the length of the preceding stage and the process is repeated infinitely many times.

b. Assuming the length of the line segment in Figure 27 is 1 unit, show that the area of the triangle in Figure 29 is $\sqrt{3}/36$.

c. Compute the total area of the regions in Figures 30 and 31 bounded above by the solid lines and below by the dashed line.

d. Even though the total area at each successive stage increases, the area of the region bounded by the fractal curve and the dashed line is finite. In fact, the area is less than $\sqrt{3}/6$. Explain how we know this.

e. On the basis of what you've seen, explain why one might make the claim that the area of a given island is finite but the coastline has infinite length.

---

■▊ *Questions for Discussion or Essay*

60. Compare the process by which an intercept of a relation is located using a graphics calculator with the method that one would use to find a particular crater on the Moon using a finder (a small wide-angled telescope) and a powerful telescope. Explain why the powerful telescope alone would be useless.

61. Why is it usually difficult for you to graph by hand a relation that *cannot* be solved for either $x$ or $y$? Why do you suppose a graphics calculator cannot graph such relations either?

62. When graphing a relation by plotting points, how do you know when you have plotted enough points? Also, how do you know what happens between the points that you have plotted?

63. A television image of a person is, in a sense, the graph of a relation. Of course, as with all graphs of relations, the appearance of the graph depends on the choice of scale. It is often said that television "adds 10 pounds" to a subject; explain this effect.

---

## SECTION 2

# GRAPHICAL TECHNIQUES

■ How can symmetry be used in data compression?

■ What does the message "ti bib dod" mean to Inspector Magill?

■ How could you construct a face with two left sides?

■ In what sense can one relation be the "mirror image" of another?

■ What types of symmetry are possessed by the human body?

---

## TRANSLATIONS

Many graphs are easily sketched by recognizing that they are merely shifts or **translations** of familiar graphs. For example, consider the graphs of $y = x^3$ and $y = (x - 5)^3$ shown in Figure 33. It is evident from the graphs and also from the tables of values given below that the $y$-values are the same whenever the $x$-values for $y = (x - 5)^3$ are 5 greater than those for $y = x^3$. The net effect is that the graph of $y = (x - 5)^3$ has the same shape as the graph of $y = x^3$, but it is translated (shifted) 5 units to the right.

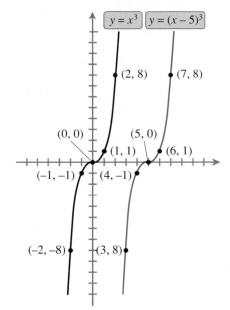

**FIGURE 33**

| $x$ | $y = x^3$ |
|---|---|
| $-2$ | $-8$ |
| $-1$ | $-1$ |
| $0$ | $0$ |
| $1$ | $1$ |
| $2$ | $8$ |

| $x$ | $y = (x - 5)^3$ |
|---|---|
| $3$ | $-8$ |
| $4$ | $-1$ |
| $5$ | $0$ |
| $6$ | $1$ |
| $7$ | $8$ |

**EXAMPLE 1**    *Translating with a graphics calculator*

Plot the graph of $y = \sqrt{x}$ together with the graphs of $y = \sqrt{x + 3}$ and $y - 2 = \sqrt{x}$, and describe how the last two graphs may be obtained from the first by translating.

**SOLUTION**    The graphs of $y = \sqrt{x}$ and $y = \sqrt{x + 3}$ are shown in Figure 34. Notice that the graph of $y = \sqrt{x + 3}$ has the same shape as that of $y = \sqrt{x}$, but it is translated 3 units to the left. To plot the graph of $y - 2 = \sqrt{x}$, we first solve for $y$ to obtain $y = \sqrt{x} + 2$. The graphs of $y = \sqrt{x}$ and $y = \sqrt{x} + 2$ are shown in Figure 35. The graph of $y = \sqrt{x} + 2$ has the same shape as that of $y = \sqrt{x}$, but it is translated 2 units up.

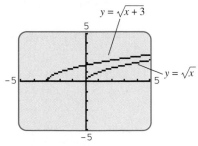

**FIGURE 34**

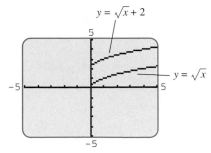

**FIGURE 35**

In general, if we substitute $x - h$ for $x$ in a relation, the graph of the resulting relation will be translated $h$ units to the right. For example, if we substitute $x - 2$ for $x$, the graph will be translated 2 units to the right; if we substitute $x + 3$ for $x$, the graph will be translated $-3$ units to the right (3 units to the left), and so on. Similarly, if we substitute $y - k$ for $y$, the graph will be translated $k$ units upward, whereas substitution of $y + k$ for $y$ will translate the graph $k$ units downward. We summarize these translations in the following box.

*Translations*

| Substitution | Resulting translation |
|---|---|
| $x - h$ in place of $x$ | $h$ units to the right |
| $x + h$ in place of $x$ | $h$ units to the left |
| $y - k$ in place of $y$ | $k$ units up |
| $y + k$ in place of $y$ | $k$ units down |

**EXAMPLE 2**    *Translating a parabola*

The graph of $y = x^2$ is given. Use it to graph $y + 4 = (x - 7)^2$.

**SOLUTION**    The relation $y + 4 = (x - 7)^2$ is the result of substituting $x - 7$ for $x$ and $y + 4$ for $y$. Thus, its graph can be obtained by translating the graph of $y = x^2$ 7 units to the right and 4 units down, as shown in Figure 36.

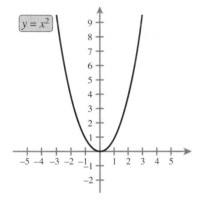

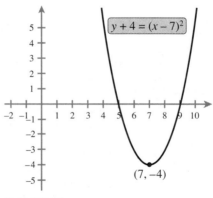

**FIGURE 36**

**EXAMPLE 3**    *Graphing the translate of an ellipse*

The graph of $\dfrac{x^2}{4} + \dfrac{y^2}{9} = 1$ is given in Figure 37. Use it to graph the following relation.

$$\frac{(x + 2)^2}{4} + \frac{(y - 1)^2}{9} = 1$$

**SOLUTION**    The relation $\dfrac{(x + 2)^2}{4} + \dfrac{(y - 1)^2}{9} = 1$ is obtained from $\dfrac{x^2}{4} + \dfrac{y^2}{9} = 1$ by substituting $x + 2$ for $x$ and $y - 1$ for $y$. Thus, its graph will be found by translat-

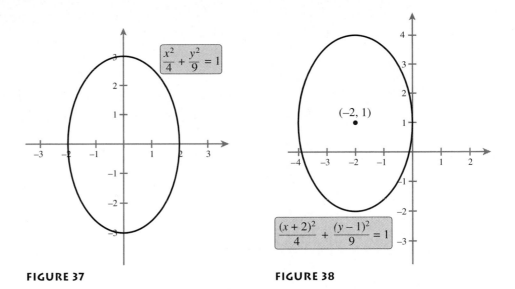

**FIGURE 37**                    **FIGURE 38**

ing the graph of $\dfrac{x^2}{4} + \dfrac{y^2}{9} = 1$ two units to the left and one unit up as shown in Figure 38.

---

## REFLECTIONS

As Figure 39 illustrates, the **reflection about the x-axis** of the point $(x, y)$ is the point $(x, -y)$. Thus, if we substitute $-y$ for $y$ in a relation, the new graph will be a reflection of the old graph about the x-axis. Similarly, the **reflection about the y-axis** of the point $(x, y)$ is the point $(-x, y)$, and so substitution of $-x$ for $x$ in a relation will cause a reflection about the y-axis. **Reflection about the origin** is defined by reflecting about both the x-axis and the y-axis. Thus, the reflection about the origin of the point $(x, y)$ is the point $(-x, -y)$, and it follows that if we substitute $-x$ for $x$ and $-y$ for $y$ in a relation, the new graph will be the reflection of the old graph about the origin.

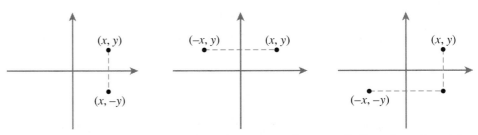

Reflect about the *x*-axis        Reflect about the *y*-axis        Reflect about the origin

**FIGURE 39**

**EXAMPLE 4**    *Using reflections to find relations*

The graph of $y - x^4 - 3x^3 - x^2 + 3x = 0$ is given in Figure 40. Parts (a), (b), and (c) each show a reflection of this graph. Use this information to find the equation for the relation corresponding to each of the reflected graphs.

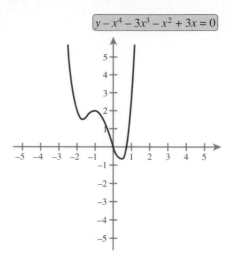

$$y - x^4 - 3x^3 - x^2 + 3x = 0$$

**FIGURE 40**

a.
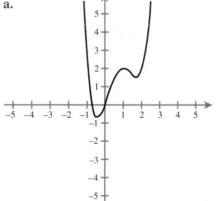

b.
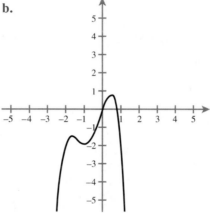

c.
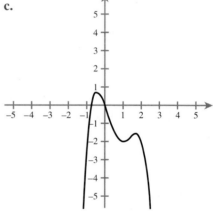

**SOLUTION**

a. This is a reflection about the $y$-axis of the original graph. Thus, its equation is obtained by replacing $x$ with $-x$.

$$y - x^4 - 3x^3 - x^2 + 3x = 0 \qquad \text{The original relation}$$

$$y - (-x)^4 - 3(-x)^3 - (-x)^2 + 3(-x) = 0 \qquad \text{Replacing } x \text{ with } -x$$

$$y - x^4 + 3x^3 - x^2 - 3x = 0 \qquad \text{Simplifying}$$

b. This is a reflection about the $x$-axis of the original graph. Thus, its equation is obtained by replacing $y$ with $-y$.

$$y - x^4 - 3x^3 - x^2 + 3x = 0 \qquad \text{The original relation}$$

$$-y - x^4 - 3x^3 - x^2 + 3x = 0 \qquad \text{Replacing } y \text{ with } -y$$

c. This graph has been obtained by reflecting the original graph about the origin. We find its equation by replacing $x$ with $-x$ and $y$ with $-y$.

$$y - x^4 - 3x^3 - x^2 + 3x = 0 \qquad \text{The original relation}$$

$$-y - (-x)^4 - 3(-x)^3 - (-x)^2 + 3(-x) = 0 \qquad \begin{array}{l}\text{Replacing } x \text{ with } -x \\ \text{and } y \text{ with } -y\end{array}$$

$$-y - x^4 + 3x^3 - x^2 - 3x = 0 \qquad \text{Simplifying}$$

Reflections can also be used as a graphing tool. In the following examples, we use a given graph of a relation to graph its reflection.

**EXAMPLE 5**   *Graphing a reflection*

The graph of $y = x^2 + 3$ is given in Figure 41. Use it to graph $y = -x^2 - 3$.

**SOLUTION**   We can rewrite $y = -x^2 - 3$ as $-y = x^2 + 3$. This relation can be obtained from $y = x^2 + 3$ by substituting $-y$ for $y$. Thus, its graph is a reflection about the $x$-axis of the graph of $y = x^2 + 3$; that is, we simply flip the graph of $y = x^2 + 3$ upside down, as shown in Figure 42.

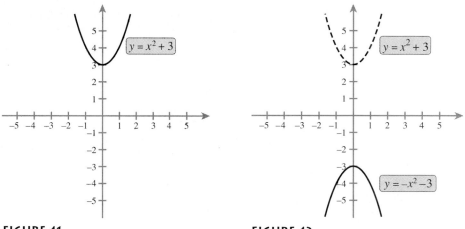

**FIGURE 41**                                        **FIGURE 42**

The graphs of some relations can be obtained by recognizing them as a combination of reflections *and* translations, as illustrated in the following example.

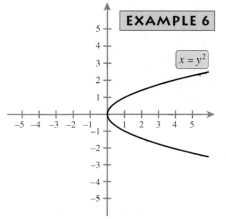

**FIGURE 43**

**EXAMPLE 6**   *Graphing a combination of reflections and translations*

The graph of $x = y^2$ is given in Figure 43. Use it to sketch the graph of $x - 3 = -(y + 2)^2$.

**SOLUTION**   The graph of $x - 3 = -(y + 2)^2$ can be obtained from the graph of $x = y^2$ by first reflecting about the $y$-axis and then translating 3 units to the right and 2 units down, as indicated by the following steps.

| Step | Relation | Substitution(s) | New relation | Effect |
|------|----------|-----------------|--------------|--------|
| 1 | $x = y^2$ | $-x$ for $x$ | $-x = y^2$ or $x = -y^2$ | Reflection about $y$-axis |
| 2 | $x = -y^2$ | $x - 3$ for $x$, and $y + 2$ for $y$ | $(x - 3) = -(y + 2)^2$ | Translation 3 units to the right and 2 units down |

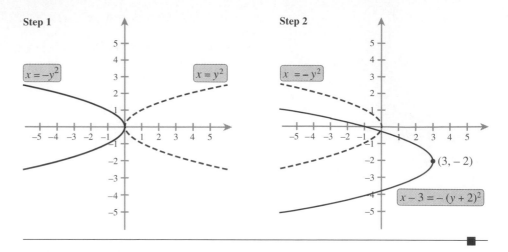

We now summarize the three principal types of reflections.

| Type of reflection | Substitution | Graphical illustration |
|---|---|---|
| About $y$-axis | $-x$ for $x$ | |
| About $x$-axis | $-y$ for $y$ | |
| About origin | $-x$ for $x$ *and* $-y$ for $y$ | |

**SYMMETRY**

As we have seen, the reflection about the $y$-axis of the graph of a relation can be obtained by substituting $-x$ for $x$ in the relation. Now suppose that we wish to reflect $y = x^2$ about the $y$-axis. If we substitute $-x$ for $x$, we obtain $y = (-x)^2 = x^2$. Thus, the reflection about the $y$-axis of the graph of $y = x^2$ is itself! Whenever the graph of a relation remains the same after undergoing some transformation (such as a reflection), we say that the graph is **symmetric** with respect to that transformation. Thus, we say that the graph of $y = x^2$ is symmetric with respect to reflection about the $y$-axis, or simply **symmetric with respect to the $y$-axis.**

The symmetry of the graph of $y = x^2$ with respect to the $y$-axis can be seen graphically as well. As Figure 44 shows, reflection of the graph of $y = x^2$ about the $y$-axis will leave it unchanged. This is because the right-hand portion of the graph is just a mirror image of the left-hand portion.

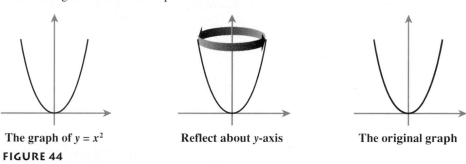

**The graph of $y = x^2$**          **Reflect about $y$-axis**          **The original graph**

**FIGURE 44**

More generally, the graph of a relation will be symmetric with respect to the $y$-axis if reflecting about the $y$-axis leaves the graph unchanged. Algebraically, this means that the relation resulting from substituting $-x$ for $x$ is equivalent to the original relation: in other words, $(-x, y)$ is on the graph of the relation whenever $(x, y)$ is on the graph. Geometrically, this means that the left- and right-hand sides of the graph are mirror images of one another. Figure 45 shows several graphs that are symmetric with respect to the $y$-axis.

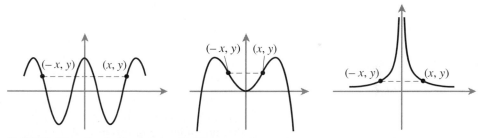

**FIGURE 45** ■ Graphs Symmetric with Respect to the $y$-axis

**EXAMPLE 7**    *Testing for symmetry with respect to the $y$-axis*

Without graphing, indicate whether or not the relation $x^2 + 2y^2 = 4y - x^4$ is symmetric with respect to the $y$-axis.

**SOLUTION**    We replace $x$ by $-x$ to test for symmetry with respect to the $y$-axis.

$$x^2 + 2y^2 = 4y - x^4 \qquad \text{The original relation}$$
$$(-x)^2 + 2y^2 = 4y - (-x)^4 \qquad \text{Substituting } -x \text{ for } x$$
$$x^2 + 2y^2 = 4y - x^4 \qquad \text{Simplifying}$$

Since the resulting relation is equivalent to the original, the graph of $x^2 + 2y^2 = 4y - x^4$ is symmetric with respect to the $y$-axis.

In a similar fashion, the graph of a relation is said to be **symmetric with respect to the x-axis** if the graph is unchanged by reflecting about the x-axis. In other words, a graph is symmetric with respect to the x-axis if the top half and the bottom half are mirror images of one another. Algebraically, this means that substituting $-y$ for $y$ in the relation yields an equivalent relation. Figure 46 shows the graphs of several relations that are symmetric with respect to the x-axis.

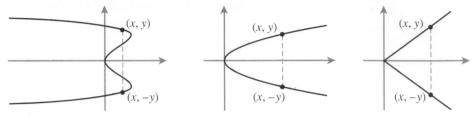

**FIGURE 46** ■ Graphs Symmetric with Respect to the x-axis

If the graph of a relation stays the same after reflection about the origin, then it is said to be **symmetric with respect to the origin**. Thus, the graph of a relation will be symmetric with respect to the origin if substitution of $-x$ for $x$ *and* $-y$ for $y$ yields an equivalent equation. Figure 47 shows the graphs of some relations that are symmetric with respect to the origin.

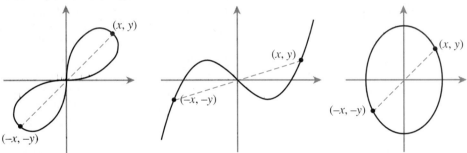

**FIGURE 47** ■ Graphs Symmetric with Respect to the Origin

Symmetry can be a powerful tool for graphing relations. If we are aware of the symmetries that the graph possesses, then we can greatly reduce the number of points that are plotted. For instance, if we know that a relation is symmetric with respect to the y-axis, then we need only plot enough points to obtain the right half of the graph; the left half of the graph will just be the reflection about the y-axis of the right half. The following examples illustrate this technique.

**EXAMPLE 8**    *Using symmetry to complete the graph of a relation*

A portion of the graph of $\dfrac{x^2}{16} + \dfrac{y^2}{9} = 1$ is shown in Figure 48. Use symmetry to complete the graph.

**SOLUTION**    We begin by noting that if either $x$ is replaced by $-x$ or $y$ is replaced by $-y$, we will obtain an equivalent relation. Thus, the graph must be symmetric with respect to the x-axis and the y-axis. Since the graph is symmetric with respect to the y-axis, the left-hand portion and the right-hand portion will be mirror images of one another about the y-axis. Reflecting the given portion of the graph about the y-axis, we obtain Figure 49. Finally, since the graph is symmetric with respect to the x-axis, the bottom portion of the graph will be a mirror image of the top portion. Reflecting about the x-axis gives us the final graph in Figure 50.

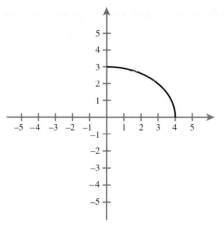

**FIGURE 48**

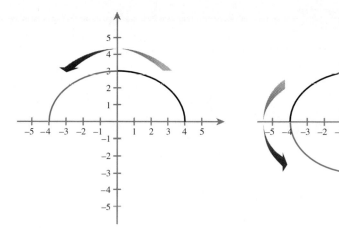

**FIGURE 49** 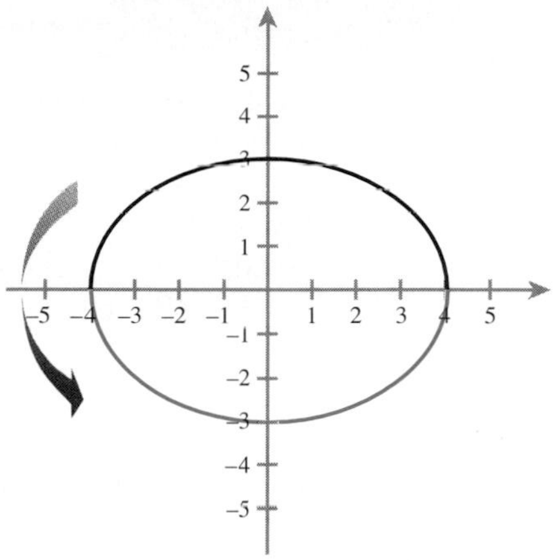                                    **FIGURE 50**

---

**EXAMPLE 9**   *Using symmetry to complete the graph of a relation*

A portion of the graph of $y = x^3 - x$ is shown in Figure 51. Use symmetry to complete the graph.

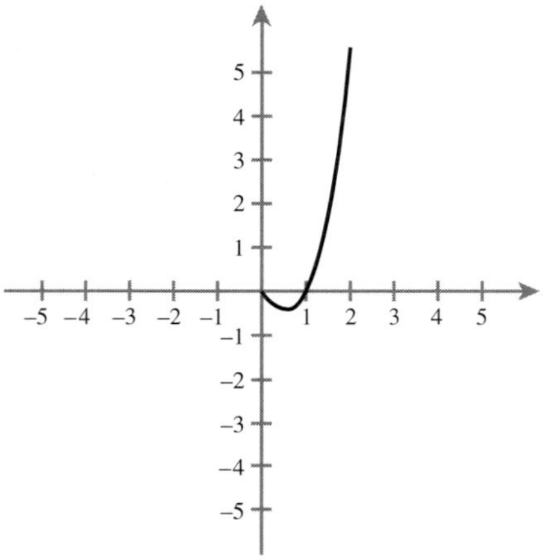

**SOLUTION**   We test for the three types of symmetry as follows.

| *y-axis* | *x-axis* | *Origin* |
|---|---|---|
| $y = (-x)^3 - (-x)$ | $(-y) = x^3 - x$ | $(-y) = (-x)^3 - (-x)$ |
| $y = -x^3 + x$ | $y = -x^3 + x$ | $-y = -x^3 + (-x)$ |
|  |  | $y = x^3 - x$ |

Since only the test for symmetry with respect to the origin yielded the same equation, the relation is symmetric with respect to the origin, but not the *y*-axis or *x*-axis. To obtain the graph, we reflect about the origin by first reflecting about the *y*-axis to obtain the red curve shown in Figure 52, and then reflecting the red curve about the *x*-axis to obtain the final graph shown in Figure 53.

**FIGURE 51**

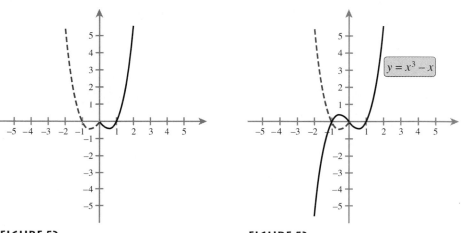

**FIGURE 52**                                    **FIGURE 53**

# EXERCISES 2

**EXERCISES 1–4** ☐ *Use the given relation and its graph to sketch the graph of the translated relations.*

1. $y = x^3$

   a. $y = (x + 2)^3$

   b. $y + 1 = (x - 3)^3$

   c. $y = (x + 4)^3 - 2$

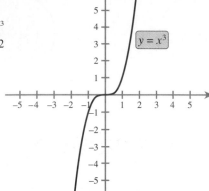

3. $\dfrac{x^2}{16} + \dfrac{y^2}{9} = 1$

   a. $\dfrac{x^2}{16} + \dfrac{(y - 1)^2}{9} = 1$

   b. $\dfrac{(x - 4)^2}{16} + \dfrac{(y + 2)^2}{9} = 1$

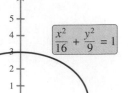

2. $x = y^2$

   a. $x = (y - 3)^2$

   b. $x - 4 = (y + 1)^2$

   c. $x = (y - 5)^2 + 2$

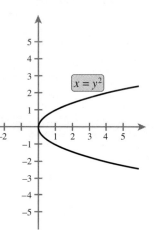

4. $\dfrac{y^2}{4} - x^2 = 1$

   a. $\dfrac{y^2}{4} - (x - 2)^2 = 1$

   b. $\dfrac{(y + 1)^2}{4} - (x + 3)^2 = 1$

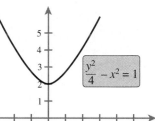

**EXERCISES 5–8** ☐ *A relation and its graph are given, together with the graph of a translation of the relation. Use the graph of the translated relation to write its equation.*

5. Original relation: $y = \sqrt{x}$　　　　　　　Translated relation:

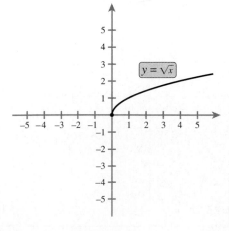

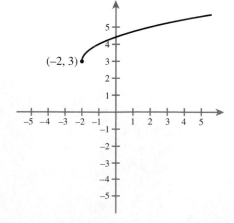

**6.** Original relation: $x = \sqrt[3]{y}$          Translated relation:

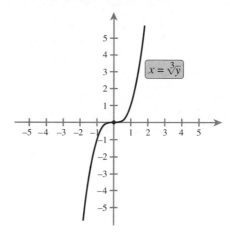

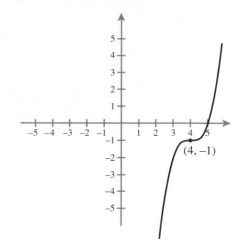

$(4, -1)$

**7.** Original relation: $xy = 4$          Translated relation:

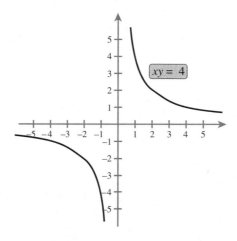

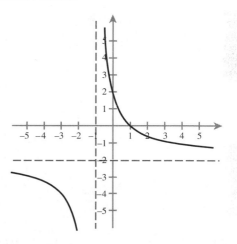

**8.** Original relation: $y^2x = 1$          Translated relation:

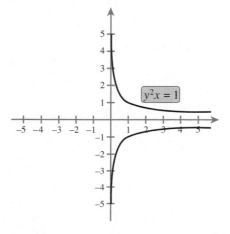

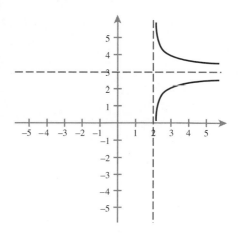

**EXERCISES 9–12** □ *Use the given relation and its graph to sketch the graph of the reflected relations.*

**9.** $x = \sqrt{y} - 2$

   **a.** $-x = \sqrt{y} - 2$

   **b.** $x = \sqrt{-y} - 2$

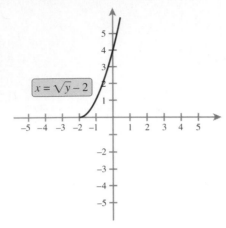

**11.** $x^3 - y^3 = 1$

   **a.** $y^3 - x^3 = 1$

   **b.** $x^3 + y^3 = 1$

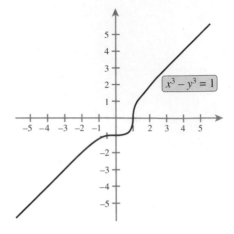

**10.** $y = x^3 + 1$

   **a.** $-y = x^3 + 1$

   **b.** $y = -x^3 + 1$

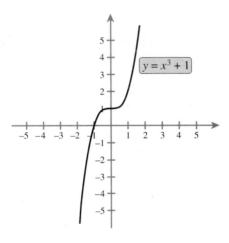

**12.** $\sqrt[3]{x} + \sqrt[3]{y} = \dfrac{1}{2}$

   **a.** $\sqrt[3]{x} - \sqrt[3]{y} = \dfrac{1}{2}$

   **b.** $\sqrt[3]{y} = \dfrac{1}{2} + \sqrt[3]{x}$

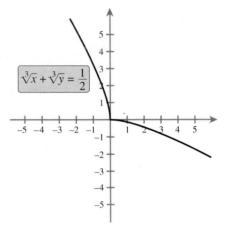

**EXERCISES 13–14** □ *A relation and its graph are given, together with the graph of a reflection of the relation. Use the graph of the reflected relation to write its equation.*

**13.** Original relation: $x^2 + 3x + 5 = y^3$    Reflected relation:

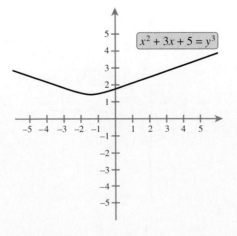

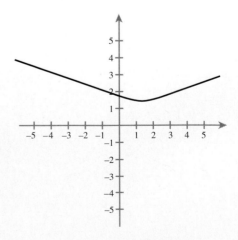

**14.** Original relation: $y + 1 = \dfrac{x + 5}{x^2 - 8x + 17}$   Reflected relation:

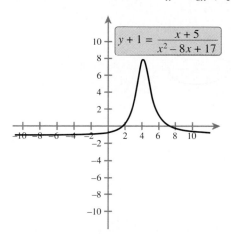

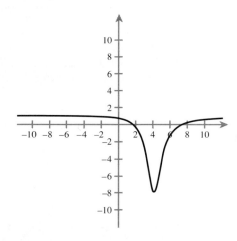

**EXERCISES 15–16** ☐ *Use the given relation and its graph to sketch the translated and reflected relation.*

**EXERCISES 17–20** ☐ *Check the given graph for symmetry with respect to the y-axis, the x-axis, and the origin.*

**15.** $x = y^3$

   **a.** $x = -(y + 2)^3$

   **b.** $x + 3 = -(y - 4)^3$

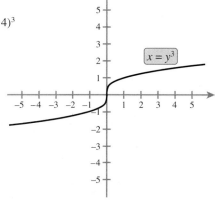

**17.**

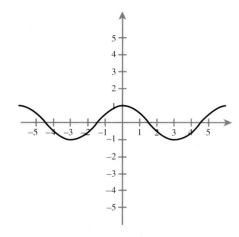

**16.** $y = x^2$

   **a.** $y = -(x - 1)^2$

   **b.** $y - 1 = -(x + 3)^2$

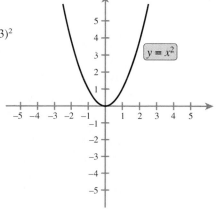

**18.**

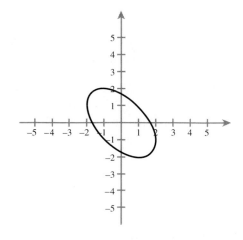

**19.**

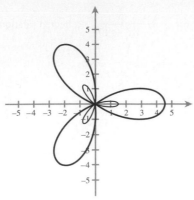

**20.**

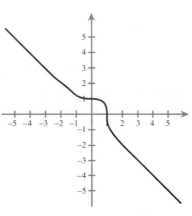

**28.** $xy = 5$

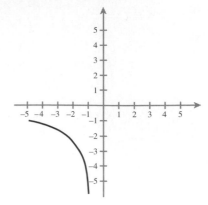

**29.** $10y = x^5$

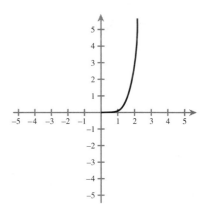

**EXERCISES 21–26** ☐ *Check the given relation for symmetry with respect to the x-axis, the y-axis, and the origin.*

**21.** $x = y^2$          **22.** $x^2 + y^3 = 10$

**23.** $y = \dfrac{1}{x}$          **24.** $y^2 = x^4 + x^2 + 2$

**25.** $x^3 - y^3 = 1$          **26.** $y = |x|$

**EXERCISES 27–34** ☐ *A portion of the graph of a relation is shown. Test the relation for symmetry with respect to the x-axis, the y-axis, and the origin, and then use symmetry to complete the graph.*

**27.** $y = x^4$

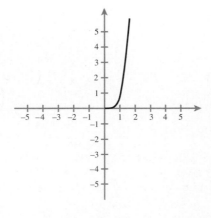

**30.** $y^2 = x - 2$

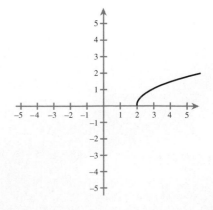

**31.** $y^2 - x^2 = 4$

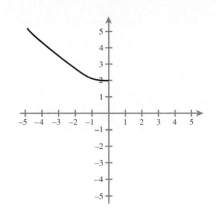

**34.** $|x| + |y| = 4$

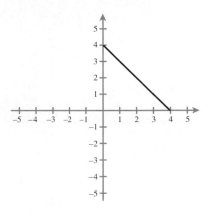

**32.** $x^6 + y^6 = 729$

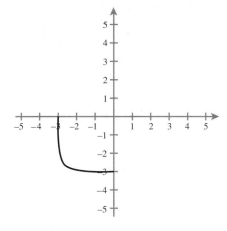

**EXERCISES 35–38** □ *Use a graphics calculator to determine how the relations in parts (a) through (c) have been obtained by translating and/or reflecting the given relation.*

**35.** $y = x^2 + x - 2$

    **a.** $y = x^2 + x - 4$

    **b.** $y = x^2 - 5x + 4$

    **c.** $y = -x^2 - x$

**36.** $y = x^3 - x$

    **a.** $y = x^3 - 9x^2 + 26x - 24$

    **b.** $y = x^3 - x + 4$

    **c.** $y = -x^3 + x + 3$

**37.** $y = \dfrac{5}{x^2 + 1}$

    **a.** $y = \dfrac{3x^2 + 8}{x^2 + 1}$

    **b.** $y = \dfrac{-5}{x^2 + 2x + 2}$

    **c.** $y = \dfrac{2x^2 + 8x + 15}{x^2 + 4x + 5}$

**38.** $y = \sqrt[3]{x^2 + 1}$

    **a.** $y = \sqrt[3]{x^2 - 4x + 5}$

    **b.** $y = \sqrt[3]{-x^2 - 1} - 2$

    **c.** $y = \sqrt[3]{x^2 + 2x + 2} + 2$

**33.** $x^2 y + 2y = 10$

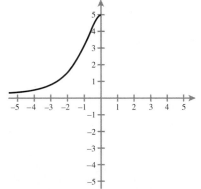

■ *Applications*

The graph shown in Figure 54 (the graph of the standard normal probability function) is extremely important in probability theory. The area under the curve between $x = a$ and $x = b$ (the shaded region in Figure 54) is the probability that the variable $x$ lies in the interval $[a, b]$. Although a detailed discussion of this graph is beyond the scope of this book, the graph is, as the figure suggests, symmetric with respect to the $y$-axis. Also, the area under the entire curve is 1. These facts may be useful in Exercises 39–42.

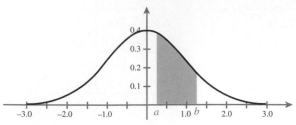

**FIGURE 54**

**39.** Find the area under the curve between 0 and ∞.

**40.** Find the area under the curve between −∞ and 0.

**41.** Given that the area under the curve between 0 and 1 is approximately 0.3413, compute the areas under the curve over the following intervals.

    **a.** $(1, \infty)$    **b.** $(-\infty, -1)$    **c.** $(-1, 1)$    **d.** $(-\infty, 1)$

**42.** Given that the area under the curve between −∞ and 2 is approximately 0.9773, compute the areas under the curve over the following intervals.

    **a.** $(2, \infty)$    **b.** $(-2, 2)$    **c.** $(-\infty, -2)$    **d.** $(0, 2)$

## ■ *Projects for Enrichment*

**43.** *Inverses of Relations* The **inverse of a relation** in $x$ and $y$ is obtained by switching the variables in the equation that defines the relation. For example, the inverse of the relation defined by $x^3 + 3y^2 = 12$ is given by $y^3 + 3x^2 = 12$.

**a.** Find the inverses of the following relations.

    **i.** $y = x^2 + 2x - 3$

    **ii.** $\dfrac{3x - 4y}{x^2 + 3y^3} = 2$

    **iii.** $3x^2 + 5xy + 3y^2 = 10$

In Figure 55 we have plotted the points $A(-4, 2)$, $B(1, 3)$, and $C(4, 5)$. We then plotted the points $A'(2, -4)$, $B'(3, 1)$, and $C'(5, 4)$ obtained from $A$, $B$, and $C$, respectively, by exchanging the $x$- and $y$-coordinates. It is easily seen that $A'$, $B'$, and $C'$ are obtained from $A$, $B$, and $C$ by reflecting about the line $y = x$. Thus, whenever we exchange the $x$- and $y$-coordinates, we reflect the point (or points) about the line $y = x$. Since the inverse of a relation is obtained by exchanging the $x$- and $y$-coordinates, the graph of a relation and its inverse are reflections about the line $y = x$. Shown in Figure 56 is the graph of a relation and its inverse.

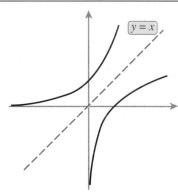

**FIGURE 56**

**b.** For each of the graphs given below, sketch the graph of the inverse.

**i.**

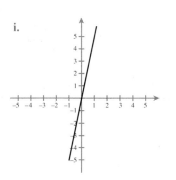

**ii.**

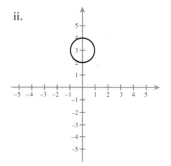

**iii.**

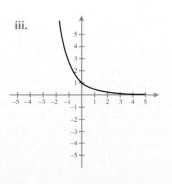

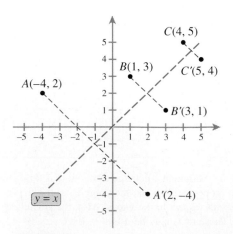

**FIGURE 55**

There are many relations that cannot be graphed directly using a graphics calculator, because they cannot easily be solved for $y$. If, however, the relation can be solved for $x$, a graphics calculator *can* produce a graph of its inverse. This graph can then be reflected about the line $y = x$ to obtain a graph of the original relation. Consider, for example, the relation $x - y^3 + 3y = 0$. Although it is not easy to solve this relation for the variable $y$, it is trivial to rewrite the equation as $x = y^3 - 3y$. Now the inverse of this relation is $y = x^3 - 3x$, which can be plotted using a graphics calculator. With a viewing rectangle defined by Xmin=-15, Xmax=15, Ymin=-10, and Ymax=10, we obtain the plot shown in Figure 57. Finally, we reflect about the line $y = x$ to obtain the graph of $x = y^3 - 3y$, as shown in Figure 58.

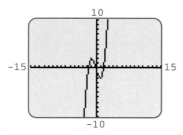

**FIGURE 57**

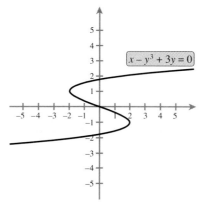

**FIGURE 58**

c. Use a graphics calculator and your understanding of inverse relations to graph the following relations.

   **i.** $x = \sqrt{36 - y^2}$

   **ii.** $x = \sqrt{y^2 - 36}$

   **iii.** $x = \dfrac{4}{y^2 + 1}$

   **iv.** $x - y^3 + 4y^2 = 0$

44. ***Rotation by 90°*** In this project we explore rotations by 90° and their connections with reflections. Pictured in Figure 59 is an arbitrary point $P$ with coordinates $(x, y)$, and a point $P'$ with coordinates $(x', y')$, the result of rotating the point $P$ by 90° counterclockwise.

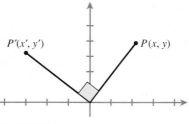

**FIGURE 59**

a. Use elementary geometry, trial and error, or any other technique to find a simple formula for $x'$ and $y'$ in terms of $x$ and $y$.

b. Use the formula obtained in part (a) to find the coordinates of the point obtained by rotating $(4, 6)$ by 90° counterclockwise.

c. Find a formula for the coordinates of the points obtained by rotating $(x, y)$ counterclockwise by 270° by applying three consecutive 90° counterclockwise rotations. Then give a formula for a 90° *clockwise* rotation.

d. Use the formula obtained in part (a) to show that a 90° counterclockwise rotation followed by a reflection about the $y$-axis is equivalent to a reflection about the line $y = x$.

e. Show that reflection about the $y$-axis followed by a 90° counterclockwise rotation is equivalent to reflecting about the line $y = x$ and then reflecting about the origin. Compare your results to those from part (d).

  ■ *Questions for Discussion or Essay*

45. Suppose that it were desired to electronically transmit a graph over a data line. Explain how the symmetry of a graph could be used to condense the transmission.

46. A face with two left sides can be constructed from the photo in Figure 60 by taking the mirror image of the left side of the face and attaching it on top of the right side. The result is shown in Figure 61. The same photo is used to create a face with two right sides in Figure 62. What do these two-faced photos suggest about symmetry in the human body? Give some additional evidence that supports your conclusion.

**FIGURE 60**

**FIGURE 61**

**FIGURE 62**

**47.** Most forms of life possess symmetries. For example, the human body is roughly *bilaterally symmetric*, starfish are *rotationally symmetric*, and sand dollars are *radically symmetric*. Perhaps the most significant and the least obvious symmetry is *symmetry of scale*. An object is said to possess a symmetry of scale if, in some sense, certain structures are repeated on various scales. For example, the branching pattern of the veins of a leaf resembles the pattern of the stems of the leaves on a branch, which resembles the pattern of the branches on the trunk. The human circulatory, respiratory, and nervous systems all possess a symmetry of scale. Discuss the meaning of symmetry of scale in the context of each of these systems.

Bilateral symmetry in a Burchell's zebra

Symmetry of scale in a tree fern

Rotational symmetry in a starfish

Radial symmetry in organ pipe coral

**48.** Inspector Magill has just been called to the scene of a homicide. The victim is lying in the middle of a carpeted floor, and the only objects near him are a pen and a hand-held mirror. A strange message is written on his forehead. It reads "ti bib dod." Almost immediately, Magill knows the first name of the murderer. Explain how the notation of reflection might have helped Magill decipher the message. Who is the murderer?

**49.** A graph is said to be **periodic** if it is symmetric with respect to a translation. That is, it remains unchanged when a certain translation take place. For example, the given graph is periodic since it remains unchanged when translated 2 units to the left (or right). The term periodic is most often used when one of the variables in a relation is time; real-world phenomena are said to be periodic if they recur at regular time intervals. The time interval required for such a phenomenon to recur is called a **period**. For each area given below, give an example of a periodic phenomenon and its corresponding period.

**a.** Biology

**b.** Astronomy

**c.** Business

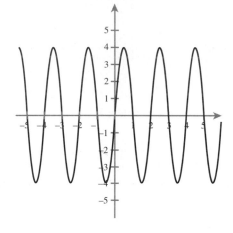

SECTION 3

# LINEAR RELATIONS

■ How can linear relations be used to compute invisible entries in mathematical tables?

■ In what sense are degrees Celsius and degrees Fahrenheit "linearly related" to one another?

■ If children grow at a steady rate, then why aren't there more 3-inch newborns?

■ How do accountants determine the value of an asset that depreciates in value?

## SLOPE

The **slope** of a nonvertical line is the number of units the line rises for every unit of horizontal change. For example, consider the line *l* shown in Figure 63. Note that in passing from (1, 3) to (2, 5), or from (2, 5) to (3, 7), the *x*-value is increased by 1 unit whereas the *y*-value is increased by 2 units. In fact, it doesn't matter which point on the line we start with, if the *x*-coordinate is increased by 1 unit, the *y*-coordinate will be increased by 2 units. Thus, the slope of *l* is 2.

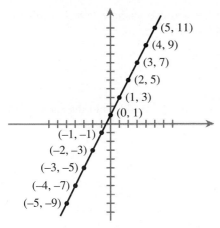

**FIGURE 63**

Since the $y$-value is increased by 2 units for every 1 unit of change in the $x$-value, the change in $y$-coordinates and the change in $x$-coordinates will always be in a 2 to 1 ratio. For example, consider the points $(-2, -3)$ and $(4, 9)$. The difference in the $x$-coordinates is 6, and the difference in the $y$-coordinates is 12, exactly twice the difference in the $x$-coordinates.

We can define the slope of an arbitrary line as the ratio of the difference in the $y$-coordinates to the difference in the $x$-coordinates for any two points on the line. If the two points are $(x_1, y_1)$ and $(x_2, y_2)$, then the difference in the $y$-coordinates is $y_2 - y_1$, and the difference in the $x$-coordinates is $x_2 - x_1$, as shown in Figure 64. Thus, we define the slope, which is usually denoted by the letter $m$, as follows.

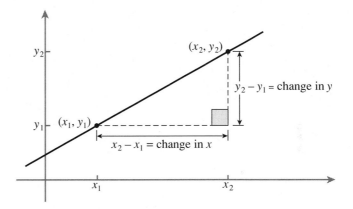

**FIGURE 64**

**Definition of the slope of a line**

The slope of the line passing through the points $(x_1, y_1)$ and $(x_2, y_2)$ (where $x_1 \neq x_2$) is defined by

$$m = \frac{\text{change in } y}{\text{change in } x} = \frac{y_2 - y_1}{x_2 - x_1}$$

The reason for the restriction $x_1 \neq x_2$ is that if $x_1 = x_2$, then $x_2 - x_1 = 0$, so that the fraction $\dfrac{y_2 - y_1}{x_2 - x_1}$ would have a denominator of zero. The restriction $x_1 \neq x_2$ implies that we cannot define slope for *vertical* lines, since all points on a vertical line have the same $x$-coordinate.

The slope of a line is a measure of its steepness; the steeper the line, the greater the slope. For example, a line with slope 1 will rise 1 unit for every unit that we move to the right, whereas a line with slope 5 will rise 5 units for every unit that we move to the right. Thus, a line with slope 5 is much steeper than a line with slope 1. If the slope of a line is negative, then moving 1 unit to the right will result in a *negative* change in the $y$-coordinate; that is, the line will descend. Figure 65 shows several lines along with their slopes.

It is an essential point that for a given line, the quantity

$$m = \frac{y_2 - y_1}{x_2 - x_1}$$

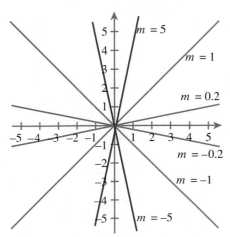

**FIGURE 65**

is always the same; it doesn't matter which two points $(x_1, y_1)$ and $(x_2, y_2)$ are chosen.

This is illustrated in Figure 66, in which it can be seen that triangle *PQR* is similar to triangle *STU*. An immediate consequence of this similarity is that

$$\frac{QR}{PR} = \frac{TU}{SU}$$

In other words, computing the slope with points *P* and *Q* would yield the same result as computing the slope with points *S* and *T*.

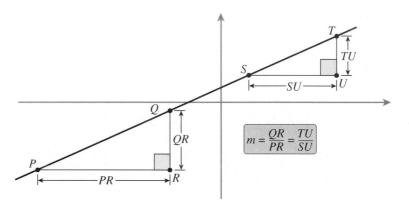

**FIGURE 66**

---

**WARNING**   When computing the slope of a line given two points on the line, it is immaterial *which* point is considered to be $(x_1, y_1)$ and which is considered to be $(x_2, y_2)$. **It is essential, however, that the order of subtraction be consistent from numerator to denominator.** In other words, we may compute slope as

$$m = \frac{y_2 - y_1}{x_2 - x_1} \text{ or } m = \frac{y_1 - y_2}{x_1 - x_2}, \text{ but } \textbf{not} \quad m = \frac{y_2 - y_1}{x_1 - x_2} \text{ or } m = \frac{y_1 - y_2}{x_2 - x_1}$$

---

**EXAMPLE 1**   *Finding the slope of a line given two points on the line*

The points $(-1, 3)$ and $(2, 4)$ are on the line *l*. Find the slope of *l*.

**SOLUTION**   We take $(x_1, y_1)$ to be the point $(-1, 3)$ and $(x_2, y_2)$ to be the point $(2, 4)$. Then

$$m = \frac{y_2 - y_1}{x_2 - x_1}$$

$$= \frac{4 - 3}{2 - (-1)}$$

$$= \frac{1}{3}$$

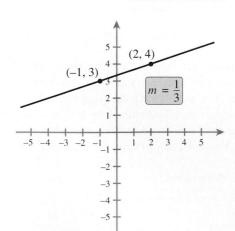

**FIGURE 67**

The graph of *l* is shown in Figure 67. Note that since *m* is positive, the line is rising. Because *m* is relatively small, the line rises gently.

■

| EXAMPLE 2 | *Finding the slope of a horizontal line* |

Find the slope of the line shown in Figure 68.

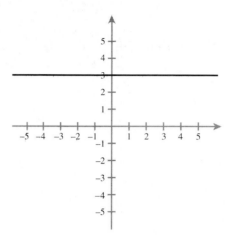

**FIGURE 68**

**SOLUTION**   We choose two points on the line, say (2, 3) and (5, 3). Then we have

$$m = \frac{3 - 3}{5 - 2} = \frac{0}{3} = 0$$

It isn't difficult to see that the slope of any horizontal line will be zero. All *y*-coordinates of points on a horizontal line are the same, so that the difference in the *y*-coordinates of two points on a horizontal line is always zero. Thus, the slope will be

$$m = \frac{\text{change in } y}{\text{change in } x} = \frac{0}{\text{change in } x} = 0$$

Similarly, since all *x*-coordinates of points on a vertical line are the same, the difference in the *x*-coordinates of two points on a vertical line is always zero. Thus, the slope of a vertical line is undefined. We highlight these facts here.

*Slope facts*

| Description of line | Slope |
|---|---|
| Rising from left to right | positive |
| Falling from left to right | negative |
| Horizontal | zero |
| Vertical | not defined |

## PARALLEL AND PERPENDICULAR LINES

Since slope is a measure of the steepness of a line, it is not surprising that parallel lines have the same slope. Figure 69 shows three parallel lines, all with slope 2.

The relationship between the slopes of perpendicular lines is slightly more complicated. In Figure 70 a line *l* with slope *m* is shown, along with a line *l'* perpendicular

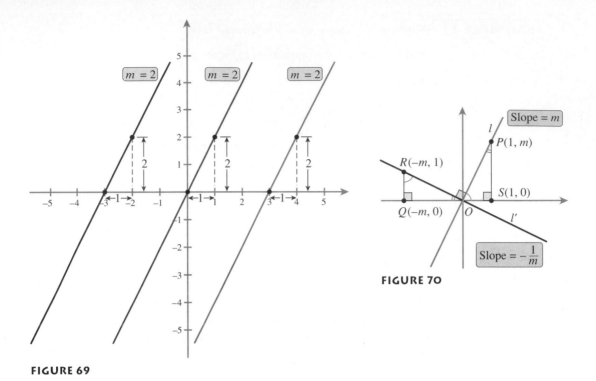

**FIGURE 69**

**FIGURE 70**

to $l$. For convenience, we have assumed that the lines cross at the origin. Now the point $(1, m)$ is on line $l$, and using elementary geometry, it can be shown that the point $(-m, 1)$ is on line $l'$. To compute the slope of $l'$, we will use the points $(0, 0)$ and $(-m, 1)$. If we let $m'$ denote the slope of $l'$, then we have

$$m' = \frac{1 - 0}{-m - 0} = -\frac{1}{m}$$

Thus, the slope of $l'$ is the negative reciprocal of the slope of $l$.

We summarize the relationships between parallel and perpendicular lines as follows.

**Slopes of parallel and perpendicular lines**

1. Parallel lines have the same slope.
2. The slopes of perpendicular lines are negative reciprocals of one another; that is, the product of the slopes of perpendicular lines is $-1$.

**EXAMPLE 3**    *Finding the slope of a line parallel to a given line*

The line $l$ passes through the points $(2, 3)$ and $(4, 7)$. The line $l'$ is parallel to $l$. Find the slope of $l'$.

**SOLUTION**    We can find the slope of $l$ directly:

$$m = \frac{7 - 3}{4 - 2} = \frac{4}{2} = 2$$

Since $l'$ is parallel to $l$, and since parallel lines have the same slope, the slope of $l'$ is also 2.

---

**EXAMPLE 4**    *Graphing a line passing through a given point and perpendicular to a given line*

The point $(3, 4)$ is on $l$. Graph $l$ given that $l$ is perpendicular to the line $l'$ passing through $(-2, 5)$ and $(3, 6)$.

**SOLUTION**    Let $m$ be the slope of $l$, and $m'$ be the slope of $l'$. Then

$$m' = \frac{6 - 5}{3 - (-2)} = \frac{1}{5}$$

Since $l$ and $l'$ are perpendicular to one another,

$$m = -\frac{1}{m'} = -\frac{1}{1/5} = -5$$

Thus, $l$ is the line passing through $(3, 4)$ with slope $-5$. Now beginning with the point $(3, 4)$, and moving 1 unit to the right and 5 units down, we obtain the point $(4, -1)$. To graph $l$, we simply connect the points $(3, 4)$ and $(4, -1)$, as shown in Figure 71.

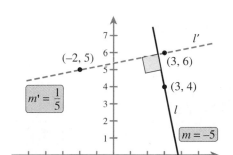

**FIGURE 71**

---

## EQUATIONS OF LINES

Let $l$ be a line with slope 2 and passing through the point $(-2, -3)$, and let $(x, y)$ be an arbitrary point on $l$, as shown in Figure 72. Now using the points $(-2, -3)$ and $(x, y)$ to compute the slope we obtain

$$m = \frac{y - (-3)}{x - (-2)} = \frac{y + 3}{x + 2}$$

Since the slope was given as 2, we must have

$$\frac{y + 3}{x + 2} = 2$$

Multiplying through by $x + 2$, we obtain

$$y + 3 = 2(x + 2)$$

More generally, if $(x_1, y_1)$ is a given point on a line with slope $m$, then for an arbitrary point $(x, y)$ on the line, we have

$$\frac{y - y_1}{x - x_1} = m$$

Multiplying through by $x - x_1$, we obtain

$$y - y_1 = m(x - x_1)$$

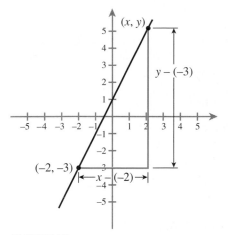

**FIGURE 72**

This is called the **point-slope equation of the line**, and it enables us to produce an equation of a line given the coordinates of a particular *point* on the line and the *slope* of the line.

**Point-slope equation of a line**

A line with slope $m$ passing through the point $(x_1, y_1)$ has equation
$y - y_1 = m(x - x_1)$.

**EXAMPLE 5** *Finding the equation of a line*

Find an equation of the line with slope $-2$ passing through the point $(1, -5)$. Graph the line with a graphics calculator.

**SOLUTION**   Using the point-slope equation with $m = -2$, $x_1 = 1$, and $y_1 = -5$, we obtain the following equation.

$$y - y_1 = m(x - x_1)$$
$$y - (-5) = -2(x - 1)$$
$$y + 5 = -2(x - 1)$$

Solving for $y$, we obtain

$$y = -2(x - 1) - 5$$
$$= -2x - 3$$

The graph of $y = -2x - 3$ is shown in Figure 73.

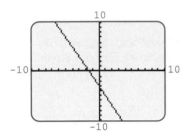

**FIGURE 73**

The point-slope equation of a line can be used to find the equation of a line when either the slope or a point on the line is given indirectly. For example, if two points on the line are given (but not the slope), we can find the equation by first computing the slope, and then using either one of the points in the point-slope equation. The following examples illustrate how the point-slope equation can be used when either the slope or a point on the line is not given.

**EXAMPLE 6** *Finding the equation of a line given two points on the line*

Find an equation of the line passing through the points $(-2, 6)$ and $(2, 7)$.

**SOLUTION**   We begin by finding the slope.

$$m = \frac{7 - 6}{2 - (-2)} = \frac{1}{4}$$

Now that we have the slope we can select either $(-2, 6)$ or $(2, 7)$ as the point $(x_1, y_1)$. We will use $(-2, 6)$. From the point-slope equation, we have

$$y - y_1 = m(x - x_1)$$
$$y - 6 = \frac{1}{4}(x + 2)$$

**EXAMPLE 7** | *Finding the equation of a line given the slope and the y-intercept*

Find an equation of the line with $y$-intercept 3 and slope $\frac{1}{2}$, the graph of which is shown in Figure 74.

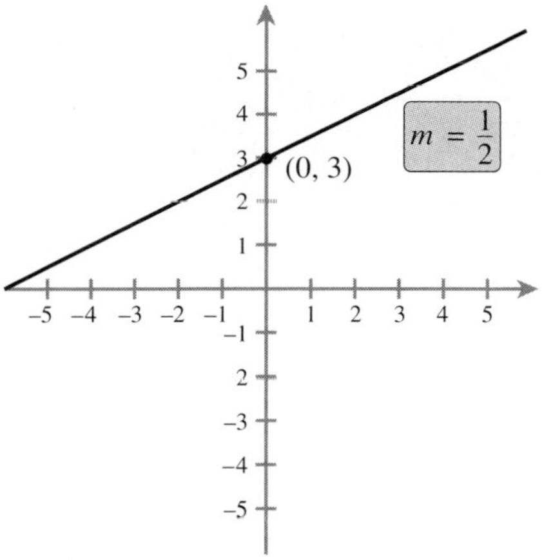

**SOLUTION** Since the $y$-intercept is 3, we know that the graph crosses the $y$-axis at the point $(0, 3)$. Thus, we have

$$y - y_1 = m(x - x_1)$$

$$y - 3 = \frac{1}{2}(x - 0)$$

$$y = \frac{1}{2}x + 3$$

**FIGURE 74**

In the previous example, we saw that the point-slope equation could be used when the slope and the $y$-intercept were given. More generally, if the line $l$ has slope $m$ and $y$-intercept $b$, then the point $(0, b)$ lies on $l$, and so the equation of $l$ is

$$y - b = m(x - 0)$$
$$y - b = mx$$
$$y = mx + b$$

This is called the **slope-intercept equation of a line**.

*Slope-intercept equation of a line*

> A line with slope $m$ and $y$-intercept $b$ has equation $y = mx + b$.

**EXAMPLE 8** | *Finding the slope-intercept equation of a line*

Find the equation of the line with slope $-4$ and $y$-intercept 6.

**SOLUTION** We apply the slope-intercept equation with $m = -4$ and $b = 6$.

$$y = mx + b$$
$$= -4x + 6$$

The slope-intercept equation of a line is often useful for extracting information about the line from its equation. For example, if an equation of a line is given, its slope can be determined by writing the equation in slope-intercept form. The slope-intercept equation is also especially useful for graphing. In particular, if an equation of a line is given and we wish to graph it using a graphics calculator, we must first place the equation in slope-intercept form. The following examples illustrate the utility of the slope-intercept form.

**EXAMPLE 9**      *Finding the slope given the equation of a line*

Find the slope and the $y$-intercept of the line with equation $5x + 4y = 8$.

**SOLUTION**   We begin by writing the equation in slope-intercept form. To do this, we solve for $y$.

$$5x + 4y = 8$$
$$4y = -5x + 8$$
$$y = -\frac{5}{4}x + 2$$

Thus, the slope is $-\frac{5}{4}$ and the $y$-intercept is 2.

■

**EXAMPLE 10**      *Finding the equation of a line through a given point and parallel to a given line*

Find the equation of the line $l$ that passes through $(-3, 5)$ and is parallel to the line with equation $2x + y = 10$.

**SOLUTION**   We begin by finding the slope of the line with equation $2x + y = 10$. To do this, we write its equation in slope-intercept form.

$$2x + y = 10$$
$$y = -2x + 10$$

Thus, we can conclude that the line $2x + y = 10$ has slope $-2$. Since $l$ is parallel to this line, it also has slope $-2$. Thus, $l$ has slope $-2$ and passes through the point $(-3, 5)$. Since we know the slope and a point on the line, we will use the point-slope equation.

$$y - y_1 = m(x - x_1)$$
$$y - 5 = -2[x - (-3)]$$
$$y - 5 = -2(x + 3)$$

If desired, we can also express the equation in slope-intercept form as follows.

$$y - 5 = -2x - 6$$
$$y = -2x - 1$$

■

**EXAMPLE 11**      *Graphing a line given its equation*

Graph the line $l$ with equation $2x - y = 4$.

**SOLUTION**   We begin by placing the equation in slope-intercept form by solving for $y$.

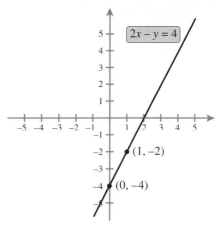

**FIGURE 75**

$$2x - y = 4$$
$$-y = -2x + 4$$
$$y = 2x - 4$$

We can now conclude that $l$ has slope 2 and $y$-intercept $-4$. To graph $l$, we begin by plotting the $y$-intercept $(0, -4)$. Next we plot one additional point by substituting a convenient value (such as 1) for $x$. With $x = 1$, we have

$$y = 2x - 4$$
$$= 2 \cdot 1 - 4$$
$$= -2$$

Thus, $(1, -2)$ is a second point on $l$. Next we simply "connect the dots," as shown in Figure 75.

The slope-intercept equation can be used to derive the equation of a horizontal line. Since the slope of a horizontal line is zero, we have

$$y = mx + b$$
$$= 0x + b$$
$$= b$$

It's not necessary, however, to use the slope-intercept equation to produce the equation of a horizontal line. Points on a given horizontal line must have the same $y$-coordinates, but there is no restriction on their $x$-coordinates ($x$ can be anything). Thus, the equation of a horizontal line is $y = c$, where $c$ is a constant.

The equation of a vertical line can be found in a similar fashion. Since points on a vertical line have no restriction on their $y$-coordinates, but their $x$-coordinates must all be the same, the equation of a vertical line is $x = c$, where $c$ is a constant.

**The equations of vertical and horizontal lines**

> Horizontal lines have equations of the form $y = c$, where $c$ is a constant.
>
> Vertical lines have equations of the form $x = c$, where $c$ is a constant.

**EXAMPLE 12**    *Finding the equation of a horizontal line*

Find the equation of the horizontal line passing through the point $(2, 5)$.

**SOLUTION**    Since the line is horizontal, we know that its equation is of the form $y = c$. Since the point $(2, 5)$ is on the line, $c$ must be 5. Thus, the equation of the line is $y = 5$.

**EXAMPLE 13**    *Finding the equation of a vertical line*

Find the equation of the line $l$ shown in Figure 76.

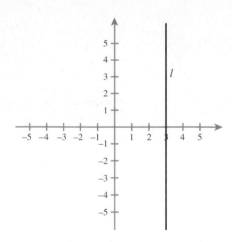

**FIGURE 76**

**SOLUTION**   Clearly every point on the line $l$ has $x$-coordinate 3. Thus, the equation of $l$ is $x = 3$.

■

One minor deficiency of the point-slope and slope-intercept equations is that neither is sufficiently general to include every possible line. Both the point-slope equation, $y - y_1 = m(x - x_1)$, and slope-intercept equation, $y = mx + b$, involve $m$, the slope of the line. The difficulty is that vertical lines do not have a slope, so that neither equation is applicable for vertical lines. Primarily for this reason, we define the **standard equation of a line** as follows.

*Standard equation of a line*

> An equation of the form $Ax + By = C$ is said to be the **standard equation of a line**. Every line has an equation of this form, and, conversely, every equation of this form is that of a line, provided $A$ and $B$ are not both 0. If the equation of a line is a standard equation, then it is said to be in **standard form**.

Note that even vertical lines can be written as equations in standard form. For example, the vertical line $x = 3$ from Example 13 is already in standard form, with $A = 1$, $B = 0$, and $C = 3$.

**EXAMPLE 14**   *Finding standard equations of lines*

Find a standard equation for the line passing through $(2, 5)$ with slope $-2$.

**SOLUTION**   Since we are given a point on the line, $(2, 5)$, and the slope of the line, $-2$, we will use the point-slope equation.

$$y - 5 = -2(x - 2)$$

Now we rearrange to obtain a standard equation.

$$y - 5 = -2x + 4$$
$$2x + y = 9$$

■

LINEAR
INTERPOLATION

It is often the case that we have no algebraic relation to express the precise relationship between two variables. Perhaps our only information is in the form of a limited number of data points. In order to use this data to make predictions, we must make certain assumptions regarding the (unknown) algebraic relation. **Linear interpolation** is the process of predicting values based on the assumption that a relation is linear "between" data points; **linear extrapolation**, on the other hand, is based on the assumption that a relation will extend in a linear fashion outside of the range of the data points. These techniques are illustrated in the following example.

**EXAMPLE 15**

*Linear interpolation*

**TABLE 1**

*Average weight for boys*

| Age (years) | Weight (pounds) |
|:-----------:|:---------------:|
| 10 | 69 |
| 11 | 77 |
| 12 | 83 |
| 13 | 92 |
| 14 | 107 |

Table 1 gives the average weight for boys ages 10–14 (Data Source: *1992 World Almanac*). Use linear interpolation or extrapolation to predict the average weight for a boy at each of the following ages.

**a.** 11 years 3 months
**b.** 14 years 6 months
**c.** 25 years

**SOLUTION**   Let $t$ represent the age in years and $w$ the weight in pounds.

**a.** We will interpolate by assuming the relation between $w$ and $t$ is linear between the data points $(11, 77)$ and $(12, 83)$. We have

$$m = \frac{83 - 77}{12 - 11} = \frac{6}{1} = 6$$

Using the point-slope equation, we have

$$w - 77 = 6(t - 11)$$
$$w = 6t + 11$$

Now 3 months = 0.25 year, so that we are interested in $t = 11.25$. Thus

$$w = 6(11.25) + 11 = 78.5$$

This is illustrated in Figure 77.

**b.** Since $t = 14.5$ (14 years 6 months) lies outside our given data, we will use linear extrapolation. We will find the equation of the line passing through the last two data points, $(13, 92)$ and $(14, 107)$, and we will assume that this relation holds for $t = 14.5$ as well. The slope of this line is given by

$$m = \frac{107 - 92}{14 - 13} = 15$$

The line thus has equation

$$w - 92 = 15(t - 13)$$
$$w = 15t - 103$$

When $t = 14.5$, we have $w = 15(14.5) - 103 = 114.5$.

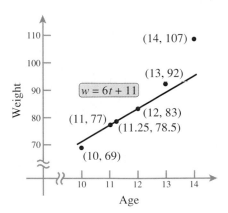

**FIGURE 77**

**c.** Again $t = 25$ is outside the range of the given data, so we will extrapolate from the last two data points. From part (b) we know the equation of the line is $w = 15t - 103$. Evaluating at $t = 25$ gives us

$$w = 15(25) - 103 = 272$$

Obviously, the average weight of 25-year-old "boys" is not 272 pounds. This illustrates the danger of extrapolating too far from the original data.

■

## EXERCISES 3

**EXERCISES 1–12** □ *Find the slope of the line passing through the given pair of points.*

**1.** $(1, 3)$ and $(2, 7)$

**2.** $(3, 4)$ and $(-2, 4)$

**3.** $(-2, 6)$ and $(3, -5)$

**4.** $(3, -4)$ and $(-7, -1)$

**5.** $(1996, 1000)$ and $(1998, 1100)$

**6.** $(1492, 1)$ and $(1997, 1000)$

**7.** $\left(\frac{1}{2}, 2\right)$ and $\left(-1, 3\frac{1}{2}\right)$

**8.** $\left(\frac{2}{3}, -3\right)$ and $\left(-\frac{3}{5}, -2\right)$

**9.** $(2, 3.5)$ and $(4.6, 7.9)$

**10.** $(1.02, 100)$ and $(0.98, 101)$

**11.** $(35, 1.5 \times 10^7)$ and $(40, 2.0 \times 10^7)$

**12.** $(2.3 \times 10^{-2}, 3.5 \times 10^{10})$ and $(4.5 \times 10^{-2}, 5.7 \times 10^{10})$

**EXERCISES 13–16** □ *The graph of a line is given. Estimate the slope of the line.*

**13.**

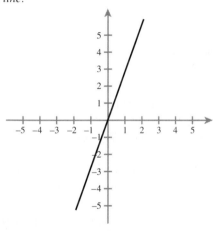

**15.**

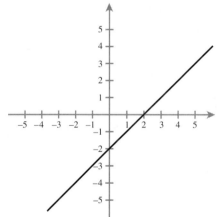

**14.**

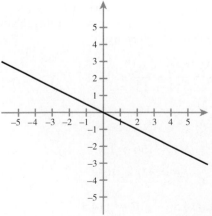

**16.**

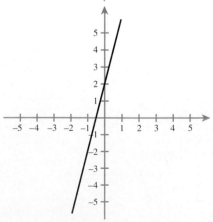

**EXERCISES 17–20** □ *Find the slope of the line l′ with the given property.*

**17.** The line $l'$ is parallel to the line $l$, where $l$ is a line with slope $-3$.

**18.** The line $l'$ is parallel to the line $l$, on which lie the points $(2, 7)$ and $(-1, 9)$.

**19.** The line $l'$ is perpendicular to the line $l$, which has a slope of $\frac{3}{5}$.

**20.** The line $l'$ is parallel to the line $l$ on which lie the points $(3, 4)$ and $(-2, 6)$.

**EXERCISES 21–22** □ *Use the given information to determine the slopes of the lines $l_1$, $l_2$, $l_3$, and $l_4$.*

**21.** The lines $l_1$, $l_2$, $l_3$, and $l_4$ have slopes $\frac{2}{3}$, $-2$, $0$, and $3$, but not necessarily in that order.

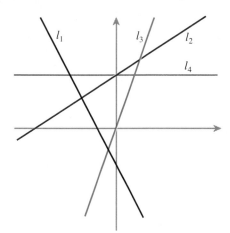

**22.** The lines $l_1$, $l_2$, $l_3$, and $l_4$ have slopes $1$, $4$, $-\frac{1}{2}$, and $-3$, but not necessarily in that order.

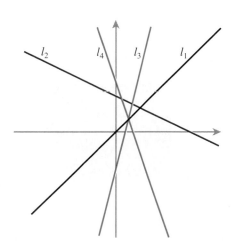

**EXERCISES 23–24** □ *Find the slope of each of the lines $l_1$, $l_2$, and $l_3$.*

**23.**

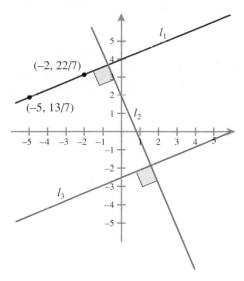

**24.**

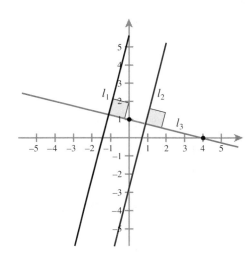

**EXERCISES 25–34** □ *Find the equation of the line with the given slope m that passes through the indicated point P. Express your final answer in slope-intercept form if possible.*

**25.** $m = 1$, $P = (2, 4)$          **26.** $m = 3$, $P = (-1, 5)$

**27.** $m = -1$, $P = (2, 8)$        **28.** $m = 0$, $P = (0, 1)$

**29.** $m$ is undefined, $P = (3, 7)$   **30.** $m = 7$, $P = (-1, -2)$

**31.** $m = 5$, $P = \left(\frac{1}{2}, \frac{2}{3}\right)$      **32.** $m = -\frac{2}{5}$, $P = (-6, 2)$

**33.** $m = 0$, $P = (2, 1)$         **34.** $m$ is undefined, $P = (3, 7)$

**EXERCISES 35–46** □ *Find the equation of the line passing through the indicated pair of points. Express your final answer in slope-intercept form if possible.*

**35.** $(1, 4)$ and $(-2, 3)$

**36.** $(2, 2)$ and $(4, 7)$

**37.** $(2, 0)$ and $(0, 4)$

**38.** $(-3, 0)$ and $(0, 2)$

**39.** $(3, -14)$ and $(3, 7)$

**40.** $(19, 5)$ and $(23, 5)$

**41.** $(1.2, 3.4)$ and $(4.6, 5.9)$

**42.** $(\frac{1}{2}, \frac{3}{5})$ and $(-\frac{1}{4}, \frac{2}{5})$

**43.** $(3, \frac{7}{8})$ and $(-2, \frac{1}{8})$

**44.** $(1, 0.000002)$ and $(1, 20{,}000)$

**45.** $(\pi, 2 \times 10^{100})$ and $(\pi, 2 \times 10^{-10})$

**46.** $(1995, 1.2 \times 10^9)$ and $(1997, 1.4 \times 10^9)$

**EXERCISES 47–52** □ *Find the slope and the y-intercept of the line with the given equation. Then graph the line.*

**47.** $2x - 3y = 8$

**48.** $3x + 6y = 9$

**49.** $x = 3y + 4$

**50.** $4 - 3y = 2x$

**51.** $y - 3 = -2(x + 3)$

**52.** $y + 2 = 3(x - 7)$

**EXERCISES 53–60** □ *Find the equation of the indicated line. Express your final answer in slope-intercept form $y = mx + b$ if possible.*

**53.** Slope: $\frac{2}{3}$
y-intercept: 2

**54.** Slope: $-4$
y-intercept: $-3$

**55.** Parallel to: $y = 4x + 2$
y-intercept: $(0, 5)$

**56.** Parallel to: The line containing $(0, 3)$ and $(4, 0)$
Point on line: $(-2, 5)$

**57.** Perpendicular to: $x - 3y = 8$
Point on line: $(2, -4)$

**58.** Perpendicular to: $4x + 8y = 7$
Point on line: $(-2, -2)$

**59.** Parallel to: $x = 6$
Point on line: $(3, 8)$

**60.** Perpendicular to: $x = 19$
y-intercept: $(0, 7)$

**EXERCISES 61–64** □ *Write the given equation in standard form*

**61.** $y = 3x + 4$

**62.** $y - 8 = -2(x - 3)$

**63.** $y - \frac{2}{3} = 2 (x - \frac{1}{2})$

**64.** $x = 2y + 8$

**EXERCISES 65–68** □ *Find the point of intersection of the given lines.*

**65.** $y = 3x + 5$ and $y = -2x - 8$

**66.** $2x - 3y = 6$ and $3x + 4y = 12$

**67.** $x = 2$ and $y = 5$

**68.** $y = m_1x + b_1$ and $y = m_2 x + b_2$   (Under what circumstances do these lines not intersect?)

**69.** If three points $P$, $Q$, and $R$ lie on a line, then the slope of the line through $P$ and $Q$ will be the same as the slope of the line through $Q$ and $R$. If these slopes are not the same, then the points do not lie on a line. Use this fact to determine whether the following triples of points lie on a line.

a. $(0, -1)$, $(2, 5)$, $(4, 1)$

b. $(3, 7)$, $(0, 1)$, $(2, 6)$

**70.** Find all points $P$ on the graph of $y = x^2 + 4$ so that the line joining $P$ and $(0, 1)$ is parallel to the line $8x - 2y = 7$.

**71.** Show that the reflection of a line about either axis is still a line. (*Hint:* Use the standard equation of a line.)

**72.** Show that the translation of a line is still a line. [*Hint:* Use the standard equation of a line.]

■ *Applications*

**73.** *Mail-order Cost*  It costs $1000 in overhead per month to run a mail-order business. In addition, each customer order costs $5.00 to produce. Express the total monthly cost $C$ in terms of $N$, the number of orders received.

**74.** *Temperature Conversion*  The relationship between degrees Fahrenheit ($F$) and Celsius ($C$) is linear. At atmospheric pressure, water freezes at 32°F and 0°C, whereas water boils at 212°F and 100°C.

a. Find a relation involving the variables $F$ and $C$.

b. Room temperature is usually taken to be 72°F. What is room temperature in degrees Celsius?

c. For decades, the normal human body temperature was assumed to be 98.6°F (this figure has since been adjusted upward). What is 98.6°F in degrees Celsius?

d. The temperature at the surface of the Sun can reach 5500°C. Express this in degrees Fahrenheit.

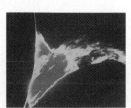

The surface of the sun during a solar flare.

e. Absolute zero is the lowest possible temperature. On the Celsius scale, absolute zero is $-273°$. What is absolute zero in degrees Fahrenheit?

**75.** *Growth Rate*  Suppose that a child were to grow at a constant rate from conception to maturity. Thus if $h$ represents the height (or length) of the child in inches and $t$ represents the number of years *since conception*, then $h$ and $t$ are related linearly. Assume that at $t = 0$, $h = 0$. Assume further that the child is fully grown at 16 years old, at which time she is $5'6''$ tall.

   **a.** Find an equation relating $h$ and $t$.

   **b.** Use the result of part (a) to estimate the length of the baby at birth.

   **c.** Estimate the height of the child at age 12.

**76.** *Sales Commission*  Each month a salesman earns a $2000 salary plus a 10% commission on all sales over $10,000. Let $P$ be the salesman's pay and $S$ be his monthly sales.

   **a.** Give a relation between $S$ and $P$.

   **b.** Graph the relation of part (a).

   **c.** Use the result of part (a) to find the salesman's total pay in a month when he had sales of $25,000.

**77.** *Machinery Depreciation*  When computing the value of a piece of machinery for tax purposes, it is common practice to use linear depreciation. With linear depreciation, we assume that the value decreases at a constant rate so that value $V$ and time $T$ are related linearly. Suppose that in 1993, a crane was purchased for $200,000. After 10 years, the crane will have only a salvage value of $20,000. Use linear depreciation to estimate the value of the crane in 1997.

**78.** *Computer Depreciation*  Suppose that a computer system purchased new in 1992 for $2500 is worth $1500 in 1995. Assuming linear depreciation—that is, assuming the value of the computer system and its age are related linearly—estimate the value of the computer in 1997.

**79.** *Population Interpolation*  The U.S. population was approximately 227 million in 1980 and 249 million in 1990.

   **a.** Use linear interpolation to estimate the population in 1987

   **b.** Use linear extrapolation to predict the population in 2000 (the next census year).

**80.** *GNP Interpolation*  The U.S. Gross National Product (GNP) was approximately $2,730 billion in 1980 and $5,470 billion in 1990.

   **a.** Use linear interpolation to estimate the GNP in 1988.

   **b.** Use linear extrapolation to predict the GNP in 1999.

### ■ *Projects for Enrichment*

**81.** *Distance Between a Line and a Point*  In this project we will derive a formula for the distance between a line $l$ and a point $P$ not on $l$. We will assume that $l$ is nonvertical. Suppose then that $l$ has equation $y = mx + b$, and that $P$ has coordinates $(c, d)$, as shown in Figure 78. Clearly the shortest path between $P$ and $l$ is along $l'$, the line perpendicular to $l$ through $P$. Then if $Q$ is the point of intersection of $l$ and $l'$, the distance between $P$ and $l$ is given by $PQ$.

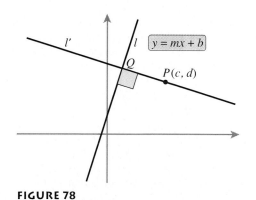

**FIGURE 78**

   **a.** What is the slope of $l'$?

   **b.** What is the equation of $l'$?

   **c.** Find the coordinates of $Q$.

   **d.** Find an expression for the distance between $l$ and $P$. Your answer should involve only the variables $m$, $b$, $c$, and $d$.

   **e.** Use the formula obtained in part (d) to find the distance between the line with equation $y = 4x + 7$ and the point $(2, 5)$.

   **f.** Use the formula obtained in part (d) to find the distance between the parallel lines $y = mx + b_1$ and $y = mx + b_2$. [*Hint:* Consider the point $(0, b_1)$ on the first line.]

**82.** *Interpolating Logarithms*  In this project we will learn how to make linear interpolations from tabular data. In particular, we will be estimating logarithms the way our parents did: tabular interpolation. Although this method has little value today as a method for computing logarithms since $7 calculators can perform the computations instantly, the technique of tabular interpolation itself is still useful in other contexts. Logarithms will be studied extensively in Chapter 6.

   Table 2 gives the logarithms of the numbers 1.1, 1.2, ..., 9.8, 9.9 to four decimal places of accuracy. For example, suppose that you wish to find the logarithm of 3.5. You simply read off the entry in the third row and the fifth column, namely 0.5441.

   Now suppose that you wish to find the logarithm of a number that falls *between* these numbers. For example, the logarithm of 2.33. According to Table 2, the logarithm of 2.3 is 0.3617 and the logarithm of 2.4 is 0.3802, so it seems reasonable to assume that the logarithm of 2.33 is somewhere between 0.3617 and 0.3802. To estimate the logarithm of 2.33 by the method of *linear interpolation*, we simply assume that the graph of the relation $y = \log x$ ("log $x$" stands for "logarithm of $x$") is linear between 2.3 and 2.4.

TABLE 2

| N | 0 | 1 | 2 | 3 | 4 | 5 | 6 | 7 | 8 | 9 |
|---|---|---|---|---|---|---|---|---|---|---|
| 1 | .0000 | .0414 | .0792 | .1139 | .1461 | .1761 | .2041 | .2304 | .2553 | .2788 |
| 2 | .3010 | .3222 | .3424 | .3617 | .3802 | .3979 | .4150 | .4314 | .4472 | .4624 |
| 3 | .4771 | .4914 | .5051 | .5185 | .5315 | .5441 | .5563 | .5682 | .5798 | .5911 |
| 4 | .6021 | .6128 | .6232 | .6335 | .6435 | .6532 | .6628 | .6721 | .6812 | .6902 |
| 5 | .6990 | .7076 | .7160 | .7243 | .7324 | .7404 | .7482 | .7559 | .7634 | .7709 |
| 6 | .7782 | .7853 | .7924 | .7993 | .8062 | .8129 | .8195 | .8261 | .8325 | .8388 |
| 7 | .8451 | .8513 | .8573 | .8633 | .8692 | .8751 | .8808 | .8865 | .8921 | .8976 |
| 8 | .9031 | .9085 | .9138 | .9191 | .9243 | .9294 | .9345 | .9395 | .9445 | .9494 |
| 9 | .9542 | .9590 | .9683 | .9685 | .9731 | .9777 | .9823 | .9868 | .9912 | .9956 |

a.  i. Find the equation of the line that passes through the points (2.3, 0.3617) and (2.4, 0.3802).

   ii. Now use this equation to estimate the value of $y$ when $x = 2.33$.

b. Estimate the logarithm of 8.86.

c.  i. Find the equation of the line that passes through $(a, b)$ and $(a + 0.1, c)$.

   ii. Suppose that $x$ is between $a$ and $a + 0.1$, and that $b$ and $c$ represent the logarithms of $a$ and $a + 0.1$, respectively. Use the result of part (i) to estimate the logarithm of $x$.

   iii. Now use the table and the formula obtained in part (ii) to estimate the logarithm of 7.74.

## Questions for Discussion or Essay

83. Explain why it doesn't matter whether one uses

$$\frac{y_2 - y_1}{x_2 - x_1} \quad \text{or} \quad \frac{y_1 - y_2}{x_1 - x_2}$$

to compute the slope.

84. A line is an idealized, abstract construction. But do lines really *exist?* To what extent does a sketch of a line (whether done by hand or with a graphics calculator) differ from the line itself? Are rays of light lines? More generally, what connections are there between the abstract spaces of mathematics (like the coordinate plane and the number line) and physical space?

85. It can easily be shown that if the lines $l_1$ and $l_2$ have equations $a_1x + b_1y = c_1$ and $a_2x + b_2y = c_2$, respectively, then either $l_1$ and $l_2$ are the same line, intersect in a single point, or are parallel. However, as you might recall from high school geometry, lines may also be **skew**—nonintersecting and yet not parallel either. How do you explain this apparent contradiction?

86. One of the most common abuses of mathematics is to assume that some particular quantity and time are related linearly with very little evidence. In essence, a variable will vary linearly with time if it changes the same amount for each unit of time—that is, if its rate of change is *constant*. Which of the following variables do you think will have a relationship with time that is approximately linear? Give reasons to support your answers.

a. The world's population

b. The national debt

c. The price of a coke

d. The world record time for the 100-meter run

e. The number of floating-point operations per second of the world's fastest computer

f. The planet Earth's total reserves of crude oil

g. The number of AIDS patients in the United States

h. The total number of solar eclipses of which there is a human record

## SECTION 4

## CIRCLES AND PARABOLAS

■ How can parabolas help a Hollywood stunt driver or a cliff diver in Acapulco?

■ What maximum height could be reached by a human cannonball?

■ How can Inspector Magill use equations of circles to foil the efforts of a kidnapper?

■ How is it possible to slice a parabola out of a cone?

### THE CONIC SECTIONS

Throughout the remainder of this chapter, we will be considering an important category of relations known as the *conic sections*. The four basic conic sections are circles, parabolas, ellipses, and hyperbolas. There are several ways to study conic sections. A purely geometric approach was developed by the ancient Greeks around 350 B.C. They described the conic sections as curves that are formed by the intersection of a plane with a double cone, as shown in Figure 79.

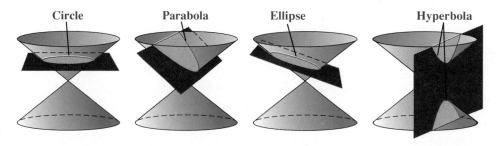

Circle          Parabola          Ellipse          Hyperbola

**FIGURE 79**

An algebraic approach to the study of conic sections was made possible after the development of analytic geometry in the seventeenth century. Using this approach, the conic sections are described as relations of the form

$$Ax^2 + Bxy + Cy^2 + Dx + Ey + F = 0$$

A third approach, one we will consider in more detail in Section 6, describes each conic as a *locus* or set of points satisfying certain geometric properties. For example, a circle is defined as the set of all points $(x, y)$ that are the same distance from a fixed point.

### CIRCLES

The standard form for the equation of a circle can be derived from the formula for the distance between two points.

### *Distance formula*

The distance between the points $(x_1, y_1)$ and $(x_2, y_2)$ is given by

$$d = \sqrt{(x_2 - x_1)^2 + (y_2 - y_1)^2}$$

Suppose that $(x, y)$ is a point on the circle of radius 3 centered at $(1, 2)$ as in Figure 80. Then the distance between $(x, y)$ and $(1, 2)$ is given by

$$\sqrt{(x - 1)^2 + (y - 2)^2}$$

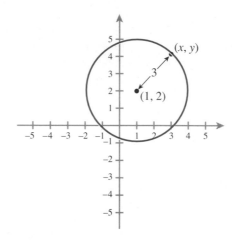

**FIGURE 80**

On the other hand, since the radius is 3, we have

$$\sqrt{(x - 1)^2 + (y - 2)^2} = 3$$

and hence

$$(x - 1)^2 + (y - 2)^2 = 3^2$$

More generally, for a circle of radius $r$ centered at $(h, k)$, we have the following equation.

**Standard equation of a circle**

> The standard form for the equation of a circle with radius $r$ and center $(h, k)$ is
>
> $$(x - h)^2 + (y - k)^2 = r^2$$

**EXAMPLE 1**   *Finding the equation of a circle*

Find the equation of the circle with radius 2 and center $(3, -5)$.

**SOLUTION**   We use the standard equation of a circle with $h = 3$, $k = -5$, and $r = 2$.

$$(x - 3)^2 + [y - (-5)]^2 = 2^2$$
$$(x - 3)^2 + (y + 5)^2 = 4$$

**EXAMPLE 2**   *Graphing a circle*

Graph the circle with equation $(x + 4)^2 + (y - 3)^2 = 16$.

**SOLUTION**   We begin by rewriting our equation in standard form.

$$[x - (-4)]^2 + (y - 3)^2 = 4^2$$

We then simply "read-off" $h = -4$, $k = 3$, and $r = 4$. Thus, our circle is centered at $(-4, 3)$ and has radius 4. The graph is shown in Figure 81.

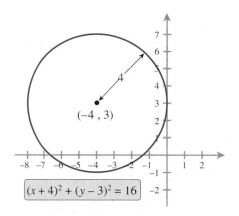

**FIGURE 81**

---

**EXAMPLE 3**    *Graphing a circle centered away from the origin*

Graph the circle with equation $x^2 + y^2 - 2x + 4y - 20 = 0$.

**SOLUTION**    At first glance, it might not even be clear that this is the equation of a circle. In order to find the center and radius, we must write the equation in standard form. To do this, we complete the square in both $x$ and $y$ as follows.

$$x^2 - 2x + y^2 + 4y = 20 \qquad \text{Grouping the } x \text{ and } y \text{ terms}$$

$$\left(x^2 - 2x + \Box\right) + \left(y^2 + 4y + \Box\right) = 20 \qquad \begin{array}{l}\text{Preparing to complete}\\ \text{the square in } x \text{ and } y\end{array}$$

$$\left(x^2 - 2x + \boxed{1}\right) + \left(y^2 + 4y + \boxed{4}\right) = 20 + 1 + 4 \qquad \text{Adding 1 and 4 to both sides}$$

$$(x - 1)^2 + (y + 2)^2 = 25 \qquad \text{Expressing in standard form}$$

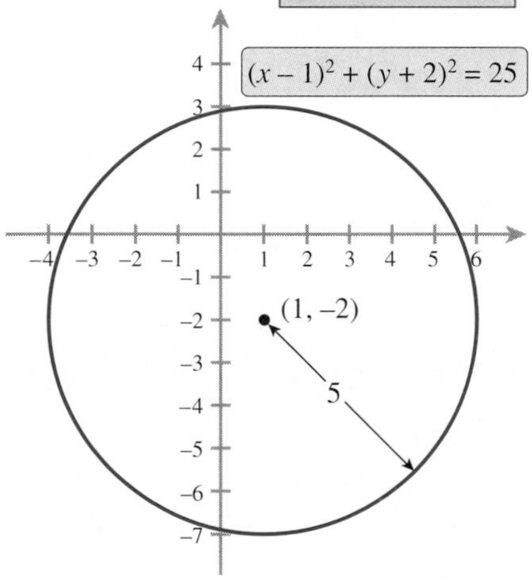

**FIGURE 82**

Thus, the circle is centered at $(1, -2)$, with radius 5. Its graph is shown in Figure 82.

$(x - 1)^2 + (y + 2)^2 = 25$

---

**EXAMPLE 4**    *Intersecting circles*

Two children are playing with gas-powered model airplanes that are restricted to circular paths by lengths of string that run from the children to their planes. If the children are 30 feet apart and the two planes have control strings that are 16 and 20 feet long, respectively, identify the point(s) where a collision could occur. We will assume that the two planes are flying at the same constant height.

**SOLUTION**    There are several techniques that could be used to solve this problem. We will illustrate how it can be solved using equations of circles. We first construct a coordinate system with one child at the origin and the other at the point $(30, 0)$, as shown in Figure 83. The paths of the planes are circles centered at the points (0, 0) and (30, 0) and with radii 16 and 20, respectively. The equations of these paths are given below.

(1) $$x^2 + y^2 = 16^2$$

(2) $$(x - 30)^2 + y^2 = 20^2$$

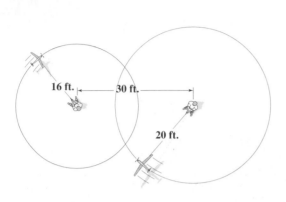

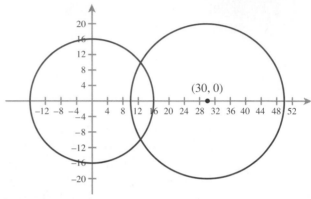

**FIGURE 83**

The intersection points of the circles can be found by solving the two equations simultaneously. We multiply equation (1) by $-1$ and add to equation (2) to eliminate $y^2$.

$$-x^2 - y^2 = -16^2$$
$$+ (x - 30)^2 + y^2 = 20^2$$
$$\overline{\phantom{xxxxxxxxxxxxxxxxxxxx}}$$
$$(x - 30)^2 - x^2 = 20^2 - 16^2$$
$$x^2 - 60x + 900 - x^2 = 400 - 256$$
$$-60x + 900 = 144$$
$$-60x = -756$$
$$x = 12.6$$

To find $y$, we substitute $x = 12.6$ into equation (1) and solve for $y$.

$$(12.6)^2 + y^2 = 16^2$$
$$y^2 = 256 - 158.76 = 97.24$$
$$y = \pm\sqrt{97.24} \approx \pm 9.86$$

So a collision could occur at the points (12.6, $-9.86$) and (12.6, 9.86).

## PARABOLAS

Geometrically, a parabola is a curve that is formed when a cone is cut with a plane parallel to the side of the cone, as shown earlier in Figure 79. If we restrict our attention to parabolas in the $xy$-coordinate system, parabolas can be described algebraically as relations in $x$ and $y$. We will consider two types of parabolas, those that open up or down, such as $y = x^2$, and those that open right or left, such as $x = y^2$, as shown in Figures 84 and 85.

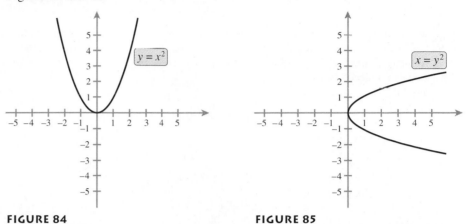

**FIGURE 84**                                                    **FIGURE 85**

The lowest point on a parabola that opens upward, or the highest point on a parabola that opens downward, is called the **vertex**. A parabola that opens upward or downward has a **vertical axis of symmetry** through its vertex, as shown in Figure 86. If a parabola opens left or right, its vertex is the point furthest to the right or left, respectively, and it has a **horizontal axis of symmetry** through its vertex, as we see in Figure 87.

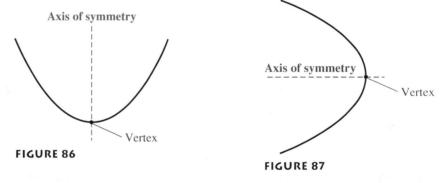

**FIGURE 86**

**FIGURE 87**

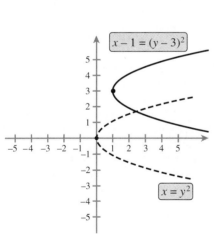

**FIGURE 88**

The standard equations of parabolas with vertical and horizontal axes can be obtained by considering translations. Recall from Section 2 that the graph of a relation is translated $h$ units horizontally and $k$ units vertically if we replace $x$ with $x - h$ and $y$ with $y - k$. Thus, the graph of the relation $x - 1 = (y - 3)^2$ is obtained from the graph of $x = y^2$ by translating 1 unit to the right and 3 units up, as shown in Figure 88. From this, we see that the graph of $x - 1 = (y - 3)^2$ is a parabola with a horizontal axis of symmetry and vertex $(1, 3)$. More generally, we have the following equations.

■ *Standard equations of a*
  *parabola*

> The standard forms for the equation of a parabola with vertex $(h, k)$ are given below.
>
> $$y - k = a(x - h)^2 \qquad \text{(vertical axis of symmetry)}$$
> $$x - h = a(y - k)^2 \qquad \text{(horizontal axis of symmetry)}$$
>
> A parabola with a vertical axis opens upward if $a > 0$ and downward if $a < 0$. A parabola with a horizontal axis opens to the right if $a > 0$ and to the left if $a < 0$.

## COMETARY COLLISIONS

From ancient times, astronomers and amateur stargazers alike have been fascinated with comets. One reason for this must certainly be the fact that many comets have been visible to the naked eye. Indeed, comet sightings have been recorded as early as the fifteenth century B.C., long before the invention of the telescope. Ironically, early comet sightings were almost always associated with catastrophe: the assassination of Julius Caesar in 44 B.C., the fall of Jerusalem to the Romans in A.D. 70, and the Norman conquest of England in A.D. 1066, to name a few. In more modern times, as comets came to be understood as astronomic phenomena, such superstitious beliefs began to wane. However, one fear has remained even to this day— that a comet may one day collide with the Earth.

Devastation at Tunguska, Siberia

How likely is such a collision? Most astronomers would likely argue that the chance of a collision with any specific comet is virtually zero. However, at least one astronomer, Brian Marsden, has suggested that there is ''a small but not negligible chance'' that the comet Swift-Tuttle will collide with the Earth on August 14, 2126. Other astronomers, whether or not they agree with Marsden, would likely admit that if the known number of comets is taken into account, a collision at some time in the future is inevitable. As evidence of this likelihood, one need only look to the spectacular collision of the comet Shoemaker-Levy 9 with Jupiter in July 1994. Further evidence can perhaps be found closer to home. Some scientists have theorized that a cometary collision caused the extinction of the dinosaur. There are even some who suggest a more recent collision. In 1908, a powerful explosion leveled thousands of acres of trees in the Tunguska

River basin in central Siberia. The accompanying aftershock was detected by seismic recording equipment over 500 miles away. Eyewitnesses described a bright blue fire-ball, trailing smoke, that some experts have concluded was a small fragment from the comet Encke. Since no conclusive evidence of a crater was found, it has been suggested that the fragment exploded several miles above the surface of the Earth.

The threat of a cometary collision has been taken seriously by a committee headed by NASA astronomer David Morrison. They have begun a project called Spaceguard Survey, the purpose of which is to search for potentially troublesome comets and asteroids. One essential ingredient of such an early detection system is a thorough understanding of the path of a comet. Here is where conic sections come into play. Over 650 comets have been observed in our solar system, and each has been identified as having a path that is either elliptical, parabolic, or hyperbolic, with the Sun as a focus. Comets with elliptical orbits will reappear at regular (though perhaps quite lengthy) time intervals. Certainly the most famous comet with an elliptical orbit is Halley's comet, which has a period of approximately 76 years. It last appeared in 1986, and it will appear again in 2061. The Swift-Tuttle comet referenced above also has an elliptical orbit. It last passed the Earth in 1992, and it will pass again in 2126. Comets with parabolic or hyperbolic paths are seen only once, and so it is likely that many have been missed, and many have yet to be seen. The threat of a collision is greatest for those comets that have parabolic or hyperbolic paths, since they are more difficult to detect and track.

**EXAMPLE 5**    *Graphing a parabola*

Graph the parabola with equation $y - 2 = (x + 1)^2$, identify the vertex, and find the equation of the axis of symmetry.

**SOLUTION**    The vertex is the point $(-1, 2)$ and the axis of symmetry is the vertical line $x = -1$. The graph is shown in Figure 89. Note that we could also write the equation in the form $y = (x + 1)^2 + 2$ and plot the graph with a graphics calculator. However, the location of the vertex would be difficult to determine precisely with the trace feature.

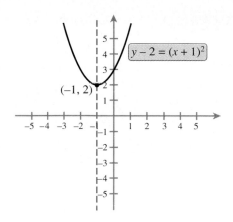

**FIGURE 89**

---

**EXAMPLE 6**   *Graphing a parabola*

Graph the parabola with equation $x = -2y^2 - 12y - 16$, identify the vertex, and find the equation of the axis of symmetry.

**SOLUTION**   Even though the equation is not in standard form, we can tell that the axis is horizontal because of the $y^2$ term. To locate the vertex, we must put the equation in standard form. This is done by completing the square.

$$x = -2y^2 - 12y - 16$$

$$x + 16 = -2y^2 - 12y$$

$$x + 16 = -2(y^2 + 6y + \Box)$$

$$x + 16 - 18 = -2(y^2 + 6y + \boxed{9}) \qquad \text{Adding } (-2)9 \text{ to both sides}$$

$$x - 2 = -2(y + 3)^2$$

From the standard form, we see that the vertex is the point $(2, -3)$. The axis of symmetry is the horizontal line $y = -3$. The graph is shown in Figure 90.

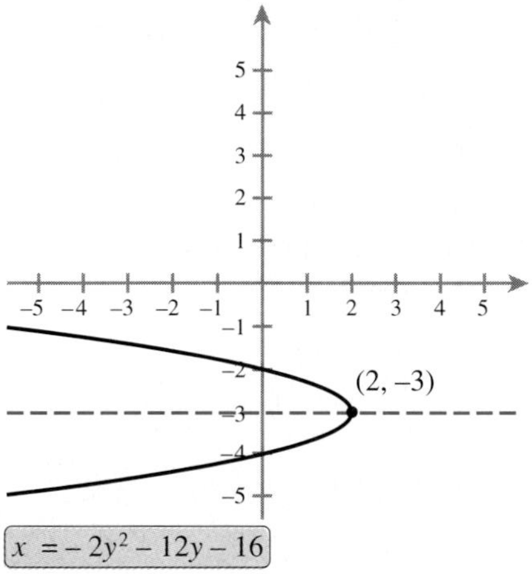

(2, –3)

$x = -2y^2 - 12y - 16$

**FIGURE 90**

---

Equations of parabolas are often given in the general form $y = ax^2 + bx + c$ or $x = ay^2 + by + c$, as was the parabola in Example 6. In such instances, it is convenient to have a simple rule for finding the vertex of the parabola without completing the square. The following rule can be verified by completing the square (see Exercise 48).

■ *Finding the vertex of a parabola*

| Parabola form | Axis of symmetry | Vertex |
|---|---|---|
| $y = ax^2 + bx + c$ | vertical | $\left(-\dfrac{b}{2a}, k\right)$, where $k$ is found by setting $x = -\dfrac{b}{2a}$ |
| $x = ay^2 + by + c$ | horizontal | $\left(h, -\dfrac{b}{2a}\right)$, where $h$ is found by setting $y = -\dfrac{b}{2a}$ |

**EXAMPLE 7**    *Finding the vertex of a parabola*

Find the coordinates of the vertex of $x = 3y^2 - 2y + 6$.

**SOLUTION**    According to the rule given above, the $y$-coordinate of the vertex is

$$y = -\frac{b}{2a} = -\frac{(-2)}{2(3)} = \frac{2}{6} = \frac{1}{3}$$

The $x$-coordinate is found by substituting $y = \frac{1}{3}$ into $x = 3y^2 - 2y + 6$.

$$x = 3\left(\frac{1}{3}\right)^2 - 2\left(\frac{1}{3}\right) + 6 = \frac{17}{3}$$

Thus, the vertex is $(\frac{17}{3}, \frac{1}{3})$.

■

**EXAMPLE 8**    *Finding the maximum height of a human cannonball*

The muzzle velocity of the renowned human cannonball Emanual Zacchini was estimated to be as much as 36 feet per second (Source: *The Guinness Book of World Records*). If we assume that this muzzle velocity was possible with the cannon pointed straight up and if we assume that Zacchini's initial height was 5 feet, then his height in feet after $t$ seconds would have been given by $s = -16t^2 + 36t + 5$. Find Zacchini's maximum height given these assumptions.

**SOLUTION**    The maximum height will occur at the vertex of the parabola $s = -16t^2 + 36t + 5$. Using the rule given above for finding the $x$-coordinate of the vertex, we can find the time at which the maximum height is reached.

$$t = -\frac{b}{2a} = -\frac{36}{-32} = \frac{9}{8} \text{ seconds}$$

We find the maximum height by setting $t = \frac{9}{8}$ in the equation $s = -16t^2 + 36t + 5$.

$$s = -16\left(\frac{9}{8}\right)^2 + 36\left(\frac{9}{8}\right) + 5 = 25.25 \text{ feet}$$

■

When a parabola has a vertical axis of symmetry, it is possible to estimate the coordinates of the vertex using a graphics calculator. In order to find the vertex of the parabola from Example 8, we would first graph the equation $y = -16x^2 + 36x + 5$ with a suitable choice of range variables (we chose `Xmin=-1`, `Xmax=4`, `Ymin=-5`, and `Ymax=30`), and then use the trace feature to obtain the approximate coordinates of the vertex, as shown in Figure 91. Note that this graph does not give the path of the human cannonball, but rather it shows how the height varies with time.

The equation of a parabola can be determined from its graph if three pieces of information are known: the orientation of the axis of symmetry, the location of the vertex, and the location of one additional point on the graph.

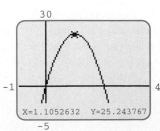

**FIGURE 91**

**EXAMPLE 9**    *Finding the equation of a parabola*

Find the standard equation of the parabola that has a vertical axis of symmetry, vertex $(1, -2)$, and passes through the point $(6, 3)$.

**SOLUTION**   Since the parabola has a vertical axis of symmetry and vertex $(1, -2)$, its equation must have the form

$$y + 2 = a(x - 1)^2$$

for some value of $a$. To find the correct value for $a$, we use the fact that the parabola passes through $(6, 3)$. Thus, we set $x = 6$ and $y = 3$ and solve for $a$.

$$3 + 2 = a(6 - 1)^2$$

$$5 = 25a$$

$$\frac{1}{5} = a$$

The equation of the parabola is

$$y + 2 = \frac{1}{5}(x - 1)^2$$

---

# EXERCISES 4

**EXERCISES 1–12** □ *Find the center and radius of the circle and sketch its graph.*

1. $x^2 + y^2 = 25$

2. $x^2 + y^2 = 10$

3. $2x^2 + 2y^2 = 16$

4. $9 - x^2 = y^2$

5. $(x + 4)^2 + (y - 1)^2 = 4$

6. $(x - 3)^2 + (y + 1)^2 = 16$

7. $x^2 - 4x + y^2 + 6y + 13 = 4$

8. $x^2 + 2x + y^2 - 4y = -4$

9. $x^2 + y^2 - 8x + 2y - 15 = 0$

10. $x^2 + y^2 + 6x + 6y + 8 = 0$

11. $4x^2 + 12x + 4y^2 + 8y = 3$

12. $9x^2 + 9y^2 + 36x - 12y + 31 = 0$

**EXERCISES 13–20** □ *Find the equation of the circle with the given properties.*

13. Center $(-1, 3)$ and radius 4

14. Center $(4, 0)$ and radius 2

15. Center $(4, -3)$ and passes through the origin

16. Center $(-3, -6)$ and passes through the point $(-3, 0)$

17. A diameter with endpoints $(-4, 2)$ and $(2, 8)$

18. A diameter with endpoints $(-3, 6)$ and $(5, -6)$

19. Passes through the points $(2, 0)$, $(0, -2)$, and $(4, -2)$   (*Hint:* Graph the points. Where is the center?)

20. Passes through the points $(-1, -2)$, $(2, 1)$, and $(-4, 1)$

**EXERCISES 21–32** □ *Find the vertex of the parabola and sketch its graph.*

21. $y = 2x^2$

22. $x = -4y^2$

23. $y^2 - 3x = 0$

24. $x^2 + 5y = 0$

25. $4y^2 + 7x = 0$

26. $2x^2 - 3y = 0$

27. $y - 3 = 3(x + 5)^2$

28. $x + 4 = -(y - 2)^2$

29. $x^2 - 4x - 4y + 8 = 0$

30. $y^2 + x - 6y + 10 = 0$

31. $x = 2y^2 + 6y + 2$

32. $y = -x^2 - x + 1$

**EXERCISES 33–38** □ *Find the equation of the parabola with the given properties.*

33. Vertical axis of symmetry, vertex $(0, 0)$, and passes through $(-1, 2)$

34. Horizontal axis of symmetry, vertex $(0, 0)$, and passes through $(4, 3)$

35. Horizontal axis of symmetry, vertex $(-3, 4)$, and passes through $(-5, 0)$

36. Vertical axis of symmetry, vertex $(-2, -4)$, and passes through $(0, -6)$

37. Vertical axis of symmetry and passes through $(0, 0)$, $(-1, 3)$, and $(4, 8)$

38. Horizontal axis of symmetry and passes through $(0, 0)$, $(-2, -10)$, and $(2, 2)$

**EXERCISES 39–42** ☐ *Solve the equation for y in terms of x and plot the resulting equation on a graphics calculator. Identify the coordinates of the vertex.*

**39.** $x^2 - 4x - y = 0$      **40.** $-x^2 + 5x + y = 0$

**41.** $x^2 + 8x - 2y + 22 = 0$    **42.** $x^2 - 6x + 3y + 12 = 0$

**EXERCISES 43–46** ☐ *Use a graphics calculator to plot the pair of equations on the same coordinate system. Identify the resulting conic and find its standard equation.*

**43.** $y = \pm\sqrt{x + 3}$      **44.** $y = \pm\sqrt{8x - x^2}$

**45.** $y = 2 \pm \sqrt{16 - 6x - x^2}$    **46.** $y = -4 \pm \sqrt{2x - 5}$

**47.** The **tangent line $l$ to the circle $C$ at the point $P$** is the line through $P$ that intersects no other point of the circle $C$, as shown in Figure 92.

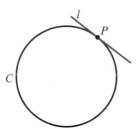

**FIGURE 92**

**a.** Graph the circle $x^2 + y^2 = 25$ with center at the origin and radius 5. Plot the point $(3, 4)$ on the circle.

**b.** A line through $(3, 4)$ will have equation $y - 4 = m(x - 3)$. Find $m$ such that the system

$$y - 4 = m(x - 3)$$
$$x^2 + y^2 = 25$$

has only the solution $(3, 4)$. (*Hint:* Solve the first equation for $y$, and then substitute into the second equation. Then choose $m$ so that the resulting quadratic equation has 3 as its only solution.)

**c.** Give the equation of the tangent line to the circle $x^2 + y^2 = 25$ at the point $(3, 4)$.

**d.** Find the equation of the tangent line through $(3, 4)$ by using the fact that the tangent line is perpendicular to the radius with endpoints $(0, 0)$ and $(3, 4)$.

**48. a.** Complete the square on the right side of the equation $y = ax^2 + bx + c$ to show that the $x$-coordinate of the vertex is $x = -b/(2a)$. What is the $y$-coordinate? Explain why it is only necessary to remember how to compute the $x$-coordinate.

**b.** Why is it not necessary to complete the square on $x = ay^2 + by + c$ to show that the $y$-coordinate of the vertex is $y = -b/(2a)$?

■ *Applications*

**49.** *Radar Search* Two radar stations are located 50 miles apart along a coastline at the points marked $A$ and $B$ in Figure 93. A ship is in distress at a point $C$, which is 20 miles from $A$ and 40 miles from $B$. Find the location of the ship relative to $A$. (*Hint:* Use point $A$ as the origin of a coordinate system and find the points of intersection of the circle centered at $A$ with radius 20 and the circle centered at $B$ with radius 40.)

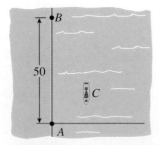

**FIGURE 93**

**50.** *Holding Patterns* Airport $A$ is 20 miles due west of airport $B$, as shown in Figure 94. A plane is circling airport $A$ with a radius of 12 miles and an altitude of 5000 feet. A second plane is circling airport $B$ with a radius of 10 miles and an altitude of 4000 feet. At which point(s) could the planes be 1000 feet apart? (*Hint:* Use airport $A$ as the origin of a coor-

dinate system and find the point(s) of intersection of the circle centered at $A$ with radius 12 and the circle centered at $B$ with radius 10.)

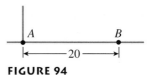

**FIGURE 94**

**51.** *Ferris Wheel* An amusement park has a ferris wheel with a radius of 30 feet. The center of the wheel is 35 feet above the ground (see Figure 95).

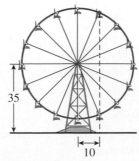

**FIGURE 95**

Olympic time for Barcelona–1992

a. Set up a coordinate system with the origin on the ground directly below the center of the wheel, and find an equation for the circular edge of the wheel.

b. If the ferris wheel becomes stuck with one passenger remaining near the top, and the distance from the point on the ground directly beneath the stranded passenger to the point on the ground directly beneath the center of the ferris wheel is 10 feet (see Figure 95), what is the length of the shortest ladder that will reach the person?

52. *Kidnap Caper* A suitcase containing $250,000 in ransom money has just been picked up by a kidnapper. Little does he realize that his fate has just been sealed; Inspector Magill is on the case and a miniature radio transmitter is in the handle of the suitcase. Using two receivers that can detect the distance to the signal, Magill and an assistant are able to monitor the movement of the signal. When the signal finally stops moving, it is 2 miles away from Magill's position and 3 miles away from his assistant's position. If Magill is 4 miles west of his assistant, use intersecting circles to determine the location(s) where police should be sent.

53. *Path of a Ball* A ball is thrown at an angle of 45° and with an initial velocity of 80 feet per second. Its path is parabolic and satisfies the equation

$$y = -\frac{x^2}{200} + x$$

where $y$ is the height in feet when the ball is a horizontal distance of $x$ feet from where it was thrown. Find the maximum height of the ball and the total horizontal distance it travels before hitting the ground.

54. *Stunt Car* The stunt coordinator for a movie is planning a stunt with a car speeding off the top floor of a parking garage. It can be shown that the path of the falling car will be parabolic with an equation of the form

$$y = h - \frac{16}{v^2}x^2$$

where $h$ is the height of the garage in feet, $v$ is the velocity of the car in feet per second, and $y$ the height of the car in

feet when its horizontal distance away from the edge of the garage is $x$ feet. If the height of the garage is 100 feet and the car must land 200 feet from the base of the garage, determine the required velocity of the car.

55. *Cliff Diving* The highest regularly performed dives are off La Quebrada in Acapulco, Mexico. (Source: *Guinness Book of World Records*). The take-off point is 87.5 feet above the water and, because of the presence of base rocks, the divers must land 27 feet out from the take-off point (see Figure 96).

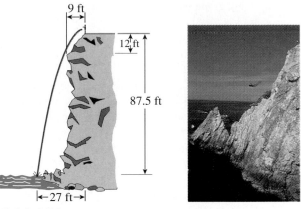

**FIGURE 96**

a. Assuming the path of a diver is a parabola with the vertex located at the take-off point, choose a coordinate system with the origin at water level directly below the take-off point and find an equation for the path.

b. Suppose that 12 feet below the take-off point, a large rock juts out 9 feet from the vertical line passing through the take-off point. By what horizontal distance will the diver clear the rock?

56. *Suspension Bridge* When a cable is suspended between two towers of a suspension bridge in such a way that it bears a uniform load, the shape of the cable will be parabolic.

a. Find the equation of the shape of the cable if the towers are 200 feet apart and 40 feet high and the cable touches the road midway between the towers (see Figure 97).

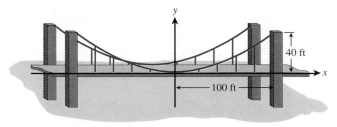

**FIGURE 97**

b. Use your equation to determine the total length of cable needed for the vertical sections of cable spaced at 20-foot intervals between the towers.

◾ *Projects for Enrichment*

57. *Similarity Among Circles and Parabolas* Geometric objects that are the same shape but possibly different sizes are said to be **similar**. For example, two triangles are said to be similar if corresponding sides are proportional. Clearly all circles are the same shape. Thus, we say that all circles are similar to one another. Surprisingly, all parabolas are similar to one another as well.

   I. Similarity of circles

   a. Using graph paper, graph the circle with equation $x^2 + y^2 = 1$.

   b. Now beginning with the equation $x^2 + y^2 = 1$, substitute $x/2$ for $x$ and $y/2$ for $y$, and simplify the resulting equation.

   c. Now graph $x^2 + y^2 = 1$ and the equation obtained in part (b) on the same set of coordinate axes. What effect did the substitution have on the size and shape of the resulting graph?

   d. How would the graphs of $x^2 + y^2 = 1$ and $(x/a)^2 + (y/a)^2 = 1$ compare in size and shape?

   e. Substituting $x/a$ for $x$ and $y/a$ for $y$ is equivalent to changing the scale of the graph by a factor of $a$. We will refer to such substitutions as **scale substitutions**. Show that the circle with equation $x^2 + y^2 = r^2$ is similar to the circle with equation $x^2 + y^2 = 1$ by showing that one can be obtained from the other by a scale substitution.

   f. In part (e) we have shown that all circles centered at the origin are similar to the circle centered at the origin with radius 1, and hence to each other. Explain why we can conclude from this that all circles, whether centered at the origin or not, are similar to one another.

   II. Similarity of parabolas

   a. Find the general equation of a parabola with a vertical axis of symmetry and vertex at the origin.

   b. Show that all such parabolas are similar to the parabola with equation $y = x^2$.

   c. Conclude that all parabolas (no matter what axis of symmetry or vertex they might have) are similar to one another.

   d. Explain the following statement: "All parabolas are the same shape; it is as if we are viewing the same parabola from different distances."

58. *The Path of a Baseball* If a baseball is hit toward center field from a height of 3 feet, at an initial angle of 45°, and with an initial velocity of 114 feet per second, will it clear a 12-foot-high center field fence 400 feet away? In order to answer this question, we must know something about the path of the baseball. The location of the ball at a given point in time can be expressed as an ordered pair $(x, y)$ where $x$ is the horizontal distance of the ball from home plate in feet and $y$ is its height in feet. Since both $x$ and $y$ depend on time, we must give equations for both. If we ignore air resistance, it can be shown that after $t$ seconds, the position of the ball will be given by

(1) $$x = 57\sqrt{2}\,t$$

(2) $$y = -16t^2 + 57\sqrt{2}\,t + 3$$

For example, when $t = 0$, $x = 0$ and $y = 3$, indicating the ball leaves home plate at a height of 3 feet. When $t = 1$, $x = 57\sqrt{2} \approx 80.6$ feet and $y = -16 + 57\sqrt{2} + 3 \approx 67.6$ feet. Thus, after 1 second, the ball is 80.6 feet from home plate (as measured from a point on the ground beneath the ball) and is 67.6 feet high.

a. Complete the following table, plot the points $(x, y)$, and connect the points with a smooth curve. What shape does the graph have?

|   | 0 | 1 | 2 | 3 | 4 | 5 |
|---|---|---|---|---|---|---|
| $t$ |   |   |   |   |   |   |
| $x$ |   |   |   |   |   |   |
| $y$ |   |   |   |   |   |   |

We wish to find a relation in $x$ and $y$ whose graph gives the path of the ball.

**b.** Solve equation (1) for $t$ and substitute into equation (2) to obtain an equation of the form $y = ax^2 + bx + c$.

**c.** Sketch the graph of the equation you found in part (b). What shape is the path of the ball?

**d.** How far is the ball from home plate when it reaches its maximum height? What is the maximum height?

**e.** What is the height of the ball when it is 400 feet from home plate? Will the ball clear the center field fence?

---

**Questions for Discussion or Essay**

**59.** In Example 4 we used intersecting circles to find the points where a collision between two model airplanes could occur. Explain how this problem could be solved using the Pythagorean Theorem and the diagram in Figure 98. Which technique is more natural? Which technique is more difficult? Explain your answers. If one technique is easier, why should the other method even be discussed?

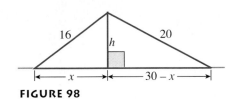

**FIGURE 98**

**60.** We have seen three methods for locating the vertex of a parabola of the form $y = ax^2 + bx + c$. One approach was to complete the square to put the equation into standard form. A

second approach was to compute $x = -b/(2a)$ and then substitute into $y = ax^2 + bx + c$ to find $y$. The third approach was to graph the equation with a graphics calculator and then approximate the coordinates with the trace feature. Discuss the pros and cons of these three methods.

**61.** Describe a procedure for plotting the graph of a parabola of the form $x = ay^2 + by + c$ with a graphics calculator. Is this procedure any easier than sketching the graph with standard techniques? Why or why not?

**62.** In Example 9 we saw that it was possible to determine the equation of a parabola by knowing only the axis of symmetry, the vertex, and one other point. What if only two of these three are known? For example, suppose the vertex and one other point are known but not the axis. How many parabolas are possible? What if the axis and the vertex are known but not a point? What does this suggest about the amount of information that is needed to determine a unique parabola?

---

 **SECTION 5**

# ELLIPSES AND HYPERBOLAS

- What is the path of a sonic boom?
- How can hyperbolas help explain shadows on a wall?
- Where is the center of Earth's orbit if it's not the Sun?

---

**ELLIPSES**

An ellipse is a curve that is formed when a cone is cut with a plane, as suggested by Figure 99. If we restrict our attention to the $xy$-coordinate system, ellipses can be described algebraically as relations in $x$ and $y$. The standard form for the equation of

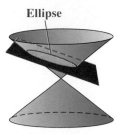

Ellipse

**FIGURE 99**

an ellipse centered at the origin, and with orientation parallel to one of the coordinate axes, is given below. It will be derived from the geometric definition of an ellipse in Section 6.

**Standard equation of an ellipse centered at the origin**

$$\frac{x^2}{a^2} + \frac{y^2}{b^2} = 1$$

The graph of an ellipse centered at the origin can be obtained by first locating the $x$- and $y$-intercepts and then connecting the intercepts with a smooth curve. To find the $x$-intercepts, we set $y = 0$ and solve for $x$. This yields $x = \pm a$. The $y$-intercepts are found by setting $x = 0$ and solving for $y$. We obtain $y = \pm b$. If $a > b$, the graph will have a horizontal orientation, such as the one shown in Figure 100. If $b > a$, the graph will have a vertical orientation, such as the one shown in Figure 101.

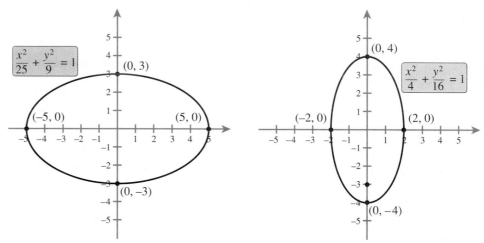

**FIGURE 100**                              **FIGURE 101**

The **major** axis is the longer of the two line segments connecting the intercepts, the **vertices** are the endpoints of the major axis, and the **minor axis** is the line segment connecting the intercepts that are not vertices. Thus, the ellipse in Figure 100 has a horizontal major axis with vertices $(5, 0)$ and $(-5, 0)$ and a minor axis connecting $(0, -3)$ and $(0, 3)$. The ellipse in Figure 101 has a vertical major axis with vertices $(0, 4)$ and $(0, -4)$ and a minor axis connecting $(-2, 0)$ and $(2, 0)$.

**EXAMPLE 1**     *Graphing an ellipse centered at the origin*

Write the equation of the ellipse $9x^2 + 4y^2 = 36$ in standard form, sketch its graph, and identify the vertices.

**SOLUTION**     To write the equation in standard form, we divide through by 36 as follows.

$$9x^2 + 4y^2 = 36$$

$$\frac{9x^2}{36} + \frac{4y^2}{36} = 1$$

$$\frac{x^2}{4} + \frac{y^2}{9} = 1$$

$$\frac{x^2}{2^2} + \frac{y^2}{3^2} = 1$$

Thus $a = 2$ and $b = 3$. The graph is shown in Figure 102. Since the major axis is vertical, the vertices have coordinates $(0, -3)$ and $(0, 3)$.

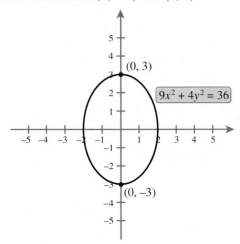

**FIGURE 102**

The standard equation of an ellipse centered at a point $(h, k)$ can be derived by applying a translation to the equation of an ellipse centered at the origin. Indeed, in order to move the center from $(0, 0)$ to $(h, k)$ we must translate $h$ units horizontally and $k$ units vertically. This is accomplished by replacing $x$ with $x - h$ and $y$ with $y - k$. The resulting equation is given here.

**Standard equation of an ellipse centered at $(h, k)$**

$$\frac{(x - h)^2}{a^2} + \frac{(y - k)^2}{b^2} = 1$$

The graph of an ellipse centered at $(h, k)$ can be obtained by viewing $(h, k)$ as the origin of an imaginary coordinate system and then applying the process used for an ellipse centered at the origin.

**EXAMPLE 2**    *Graphing an ellipse centered away from the origin*

Graph the ellipse with equation

$$\frac{(x - 2)^2}{25} + \frac{(y + 3)^2}{4} = 1$$

Identify the center, the major axis, and the vertices.

**SOLUTION**    This ellipse has center $(2, -3)$. Since $a = \sqrt{25} = 5$, two points on the ellipse are located 5 units to the left and right of the center. Similarly, since $b = \sqrt{4} = 2$, two points are located 2 units above and below the center. Connecting these points with a smooth curve yields the graph in Figure 103. The major axis is horizontal and connects the vertices $(-3, -3)$ and $(7, -3)$.

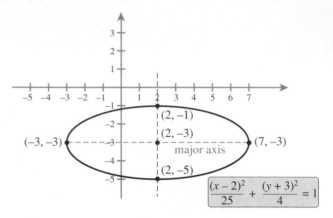

**FIGURE 103**

---

**EXAMPLE 3**    *Graphing an ellipse centered away from the origin*

Sketch the graph of the ellipse $9x^2 + y^2 + 36x + 27 = 0$ and identify the center, the major axis, and the vertices.

**SOLUTION**    We must first put the equation in standard form by completing the square.

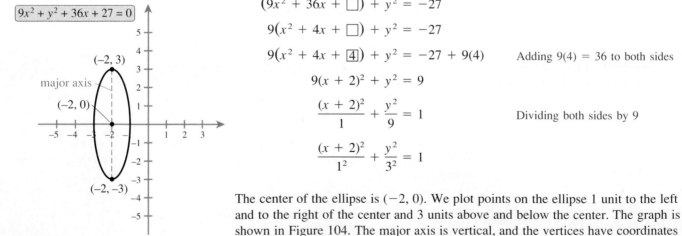

$$9x^2 + y^2 + 36x + 27 = 0$$

$$(9x^2 + 36x + \square) + y^2 = -27$$

$$9(x^2 + 4x + \square) + y^2 = -27$$

$$9(x^2 + 4x + \boxed{4}) + y^2 = -27 + 9(4) \qquad \text{Adding } 9(4) = 36 \text{ to both sides}$$

$$9(x + 2)^2 + y^2 = 9$$

$$\frac{(x + 2)^2}{1} + \frac{y^2}{9} = 1 \qquad \text{Dividing both sides by 9}$$

$$\frac{(x + 2)^2}{1^2} + \frac{y^2}{3^2} = 1$$

The center of the ellipse is $(-2, 0)$. We plot points on the ellipse 1 unit to the left and to the right of the center and 3 units above and below the center. The graph is shown in Figure 104. The major axis is vertical, and the vertices have coordinates $(-2, 3)$ and $(-2, -3)$.

**FIGURE 104**

---

**EXAMPLE 4**    *An application involving an ellipse*

A train tunnel through a mountain is semi-elliptical with height 18 feet at the center and width 24 feet at the base (see Figure 105). A train is hauling a load of flatcars

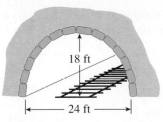

**FIGURE 105**

that are piled high with freight. If the flatcars are 10 feet wide and the load reaches a height of 16 feet above the ground, will the flatcars clear the opening of the tunnel?

**SOLUTION**  To determine the clearance, we must know the height of the tunnel at the edge of the flatcar. We first construct a coordinate system with the *x*-axis on the ground and the origin at the center of the tracks. Next we find the equation of the ellipse that forms the tunnel opening. Using the dimensions of the tunnel opening, we know that the ellipse has a vertical major axis of length 2(18) and a horizontal minor axis of length 24. Thus, $a - 12$ and $b = 18$. The equation is given as follows.

$$\frac{x^2}{12^2} + \frac{y^2}{18^2} = 1$$

The edge of a 10-foot-wide flatcar corresponds to $x = 5$, and so the height of the tunnel at the edge of the flatcar is given by the *y*-value of the point on the ellipse when $x = 5$. Thus, we set $x = 5$ and solve for *y*.

$$\frac{5^2}{12^2} + \frac{y^2}{18^2} = 1$$

$$\frac{y^2}{324} = 1 - \frac{25}{144}$$

$$y^2 = 324\left(1 - \frac{25}{144}\right)$$

$$y = \pm\sqrt{324\left(1 - \frac{25}{144}\right)} \approx \pm 16.36$$

Thus, the height of the tunnel above the edge of a flatcar is approximately 16.36 feet, which is just enough for the load to clear the tunnel opening.

◼

**HYPERBOLAS**

A hyperbola is a curve that is formed when a double cone is cut by a plane parallel to the axis of the cone, as shown in Figure 106. We will restrict our attention to hyperbolas in the *xy*-coordinate system and, in particular, to those that have horizontal or vertical axes such as the ones shown in Figures 107 and 108.

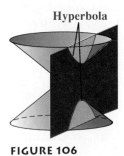

**FIGURE 106**

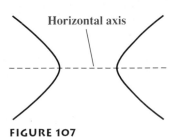

**FIGURE 107**

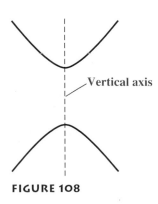

**FIGURE 108**

The standard forms for the equations of hyperbolas centered at the origin are given below. They can be derived from the geometric definition of a hyperbola given in Section 6.

**Standard equations of hyperbolas centered at the origin**

$$\frac{x^2}{a^2} - \frac{y^2}{b^2} = 1 \quad \text{(horizontal axis)}$$

$$\frac{y^2}{b^2} - \frac{x^2}{a^2} = 1 \quad \text{(vertical axis)}$$

The **vertices** of a hyperbola centered at the origin are its intercepts. For example, if a hyperbola has a horizontal axis, we locate the $x$-intercepts by setting $y = 0$ to obtain $x = \pm a$. Thus, the vertices are $(a, 0)$ and $(-a, 0)$, as shown in Figure 109. There are no $y$-intercepts since setting $x = 0$ leads to nonreal solutions. A hyperbola with a vertical axis has vertices $(0, -b)$ and $(0, b)$ as seen in Figure 110. The line segment connecting the vertices is called the **transverse axis,** shown in gold in the figures.

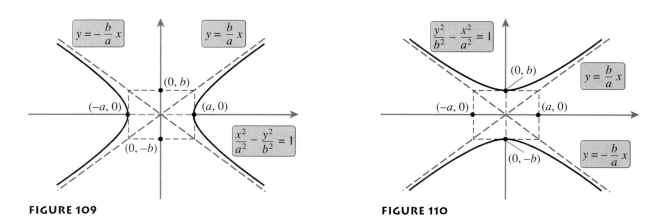

**FIGURE 109**                     **FIGURE 110**

Notice in Figures 109 and 110 that the graphs of both hyperbolas approach the lines $y = (b/a)x$ and $y = -(b/a)x$ as $x$ gets large. These lines are called **asymptotes**, and they serve as useful aids in sketching graphs of hyperbolas. The asymptotes for a hyperbola centered at the origin can be located by constructing a rectangle with sides passing through the points $(-a, 0)$, $(a, 0)$, $(0, b)$, and $(0, -b)$. The extended diagonals of this rectangle are the asymptotes; their equations are $y = (b/a)x$ and $y = -(b/a)x$.

**Asymptotes of a hyperbola centered at the origin**

The asymptotes for a hyperbola centered at the origin are

$$y = \frac{b}{a}x \quad \text{and} \quad y = -\frac{b}{a}x$$

**EXAMPLE 5** *Graphing a hyperbola centered at the origin*

Sketch the graph of the hyperbola

$$\frac{x^2}{16} - \frac{y^2}{9} = 1$$

Identify the asymptotes, vertices, and transverse axis.

**SOLUTION** From the form of the equation, we know that the hyperbola has a horizontal axis with $a = \sqrt{16} = 4$ and $b = \sqrt{9} = 3$. Thus, the vertices are the $x$-intercepts $(-4, 0)$ and $(4, 0)$ and the transverse axis is the line segment connecting these two points. To find the asymptotes, we construct the rectangle whose sides pass through the vertices $(-4, 0)$ and $(4, 0)$ on the $x$-axis and the points $(0, -3)$ and $(0, 3)$ on the $y$-axis. The asymptotes are then formed by extending the diagonals of the rectangle. The equations of the asymptotes are $y = \frac{3}{4}x$ and $y = -\frac{3}{4}x$. The graph is shown in Figure 111.

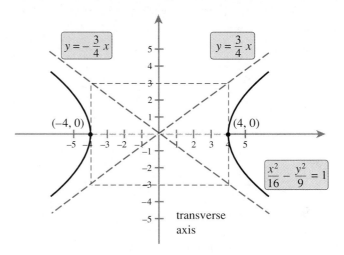

**FIGURE 111**

The standard equations of a hyperbola centered at a point $(h, k)$ are found by applying translations to the equations for hyperbolas centered at the origin.

*Standard equations of a hyperbola centered at $(h, k)$*

$$\frac{(x - h)^2}{a^2} - \frac{(y - k)^2}{b^2} = 1 \quad \text{(horizontal axis)}$$

$$\frac{(y - k)^2}{b^2} - \frac{(x - h)^2}{a^2} = 1 \quad \text{(vertical axis)}$$

As with ellipses, the graph of a hyperbola centered at $(h, k)$ can be obtained by viewing $(h, k)$ as the origin of a coordinate system and then applying the process used for a hyperbola centered at the origin.

**EXAMPLE 6**      *Graphing a hyperbola centered away from the origin*

Sketch the graph of the hyperbola

$$\frac{(x - 2)^2}{4} - \frac{(y - 1)^2}{16} = 1$$

Identify the asymptotes, vertices, and transverse axis.

**SOLUTION**    The center of the hyperbola is the point (2, 1). The axis is horizontal with $a = 2$ and $b = 4$. Thus, the vertices are 2 units to the left and to the right of the center, at (0, 1) and (4, 1). The transverse axis is the line segment connecting the two vertices. To locate the asymptotes, we construct a rectangle with horizontal sides 4 units above and below the center and vertical sides 2 units to the left and to the right of the center. The asymptotes are the extended diagonals of this rectangle. They have slope $m = \pm(b/a) = \pm 2$, and they pass through the center (2, 1). Their equations are thus $y - 1 = \pm 2(x - 2)$. The graph is shown in Figure 112.

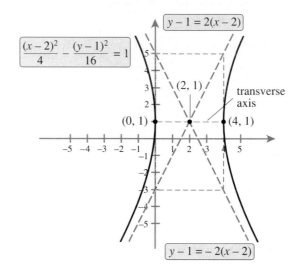

**FIGURE 112**

---

**EXAMPLE 7**      *Graphing a hyperbola centered away from the origin*

Write the equation of the hyperbola $16x^2 - y^2 + 64x - 2y + 67 = 0$ in standard form and sketch its graph.

**SOLUTION**    In order to put the equation in standard form, we must complete the square in $x$ and $y$.

$$16x^2 - y^2 + 64x - 2y + 67 = 0$$

$$\left(16x^2 + 64x + \Box\right) + \left(-y^2 - 2y + \Box\right) = -67$$

$$16\left(x^2 + 4x + \Box\right) - \left(y^2 + 2y + \Box\right) = -67$$

$$16\left(x^2 + 4x + \boxed{4}\right) - \left(y^2 + 2y + \boxed{1}\right) = -67 + 64 - 1 \qquad \begin{array}{l}\text{Adding 16(4) and} \\ -(1)\text{ to both sides}\end{array}$$

$$16(x + 2)^2 - (y + 1)^2 = -4$$

$$-4(x + 2)^2 + \frac{(y + 1)^2}{4} = 1 \qquad \text{Dividing both sides by } -4$$

$$\frac{(y + 1)^2}{4} - 4(x + 2)^2 = 1$$

$$\frac{(y + 1)^2}{4} - \frac{(x + 2)^2}{1/4} = 1$$

$$\frac{(y + 1)^2}{2^2} - \frac{(x + 2)^2}{(1/2)^2} = 1$$

The center of the hyperbola is at $(-2, -1)$. The axis is vertical with $a = \frac{1}{2}$ and $b = 2$. The vertices are 2 units above and below the center, at $(-2, -3)$ and $(-2, 1)$. To sketch the hyperbola, we construct a rectangle with center $(-2, -1)$, horizontal sides 2 units above and below the center, and vertical sides $\frac{1}{2}$ unit to the left and to the right of the center. The asymptotes are formed by extending the diagonals, as shown in Figure 113.

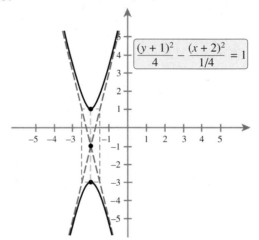

**FIGURE 113**

## THE GENERAL EQUATION FOR A CONIC SECTION

The equations of the ellipse and the hyperbola in Examples 3 and 7 both have the general form $Ax^2 + Cy^2 + Dx + Ey + F = 0$. In fact, the equations of all four types of conic sections we have seen—circles, parabolas, ellipses, and hyperbolas—can be written in this general form.

### *The general equation of a conic section*

The general form for the equation of a conic section with a vertical or horizontal axis is

$$Ax^2 + Cy^2 + Dx + Ey + F = 0$$

Note that not all equations of the form $Ax^2 + Cy^2 + Dx + Ey + F = 0$ correspond to the familiar conic sections. Consider the equations

$$x^2 - y^2 = 0 \quad \text{and} \quad x^2 + y^2 + 1 = 0$$

The first equation can be rewritten as $x = \pm y$, and so it yields the equations of two lines. The second can be rewritten as $x^2 + y^2 = -1$, which has no solutions and hence no graph. These two equations are examples of **degenerate** conic sections. Throughout the remainder of the section, we will consider only the general equations of **nondegenerate** conic sections, namely circles, parabolas, ellipses, and hyperbolas. In these cases, it is possible to recognize the type of conic by inspecting the coefficients of $x^2$ and $y^2$. The following observations can be verified by rewriting the standard equations of the conics in general form.

*Identifying conic sections*

A nondegenerate conic section of the form $Ax^2 + Cy^2 + Dx + Ey + F = 0$ can be identified as follows.

| Condition | Verbal description | Resulting conic |
|-----------|-------------------|-----------------|
| $A = C$ | The squared terms have equal coefficients. | circle |
| $AC = 0$ | There is only one squared term. | parabola |
| $AC > 0$ | The coefficients of the squared terms have the same sign. | ellipse |
| $AC < 0$ | The coefficients of the squared terms have opposite sign. | hyperbola |

**EXAMPLE 8**    *Identifying conic sections*

Identify each of the following nondegenerate conics as a circle, a parabola, an ellipse, or a hyperbola.
a. $4x^2 - 9y^2 - 8x - 36y - 68 = 0$
b. $y^2 + 8x + 6y + 25 = 0$
c. $9x^2 - 36x + 31 = -4y^2 - 8y$

**SOLUTION**

a. Since both squared terms are present and the coefficients have opposite sign, this is a hyperbola.

b. Since only one squared term is present, this is a parabola.

c. We first rewrite the equation in general form as

$$9x^2 + 4y^2 - 36x + 8y + 31 = 0$$

Since the coefficients of the squared terms have the same sign but are not equal, this is an ellipse.

If the equation of a conic section is given in general form, its graph can be obtained either by writing the equation in standard form and using the techniques described earlier, or by solving the equation for $y$ and using a graphics calculator.

**EXAMPLE 9**    *Graphing a conic section using a graphics calculator*

Identify the conic $-x^2 + y^2 + 3x + 4y - 5 = 0$ and obtain its plot with a graphics calculator. In the case of a parabola, ellipse, or hyperbola, approximate the coordinates of any vertices.

**SOLUTION**    Since the coefficients of $x^2$ and $y^2$ have opposite signs, this is a hyperbola. We solve for $y$ by completing the square in $y$.

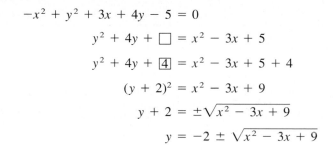

$$-x^2 + y^2 + 3x + 4y - 5 = 0$$

$$y^2 + 4y + \square = x^2 - 3x + 5$$

$$y^2 + 4y + \boxed{4} = x^2 - 3x + 5 + 4$$

$$(y + 2)^2 = x^2 - 3x + 9$$

$$y + 2 = \pm\sqrt{x^2 - 3x + 9}$$

$$y = -2 \pm \sqrt{x^2 - 3x + 9}$$

Thus, the two branches of the hyperbola are

$$y = -2 + \sqrt{x^2 - 3x + 9} \quad \text{and} \quad y = -2 - \sqrt{x^2 - 3x + 9}$$

After entering both equations and graphing, we obtain the graph shown in Figure 114. The coordinates of the vertices, found using the trace feature, are approximately $(1.5, 0.6)$ and $(1.5, -4.6)$.

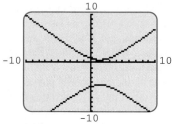

**FIGURE 114**

---

## EXERCISES 5

**EXERCISES 1–14** $\square$ *Sketch the graph of the ellipse and identify its center, vertices, and major and minor axes.*

**1.** $\dfrac{x^2}{9} + \dfrac{y^2}{25} = 1$          **2.** $\dfrac{x^2}{16} + \dfrac{y^2}{9} = 1$

**3.** $4x^2 + 9y^2 = 36$          **4.** $16x^2 + 9y^2 = 144$

**5.** $9x^2 + 4y^2 = 1$          **6.** $16x^2 + 25y^2 = 1$

**7.** $3x^2 + 2y^2 = 6$          **8.** $5x^2 + 8y^2 = 40$

**9.** $\dfrac{(x + 3)^2}{4} + \dfrac{(y - 1)^2}{16} = 1$   **10.** $\dfrac{(x - 2)^2}{25} + \dfrac{(y + 2)^2}{9} = 1$

**11.** $9x^2 + 4y^2 - 18x - 24y + 9 = 0$

**12.** $25x^2 + 16y^2 - 200x - 32y + 16 = 0$

**13.** $x^2 + 36y^2 + 4x - 72y + 4 = 0$

**14.** $16x^2 + y^2 + 4y - 12 = 0$

**EXERCISES 15–26** $\square$ *Sketch the graph of the hyperbola and identify its center, vertices, transverse axis, and asymptotes.*

**15.** $\dfrac{x^2}{9} - \dfrac{y^2}{16} = 1$          **16.** $\dfrac{y^2}{25} - \dfrac{x^2}{16} = 1$

**17.** $4y^2 - 9x^2 = 36$          **18.** $25x^2 - 16y^2 = 400$

**19.** $12x^2 - 3y^2 = -24$          **20.** $5x^2 - 3y^2 = 15$

**21.** $\dfrac{(x - 1)^2}{4} - \dfrac{(y - 4)^2}{9} = 1$   **22.** $\dfrac{(y + 3)^2}{16} - \dfrac{(x + 2)^2}{25} = 1$

**23.** $4x^2 - 9y^2 - 16x + 54y - 101 = 0$

**24.** $25x^2 - 16y^2 - 50x + 160y + 25 = 0$

**25.** $4x^2 - 25y^2 - 32x + 164 = 0$

**26.** $x^2 - 4y^2 + 4x - 24y - 36 = 0$

**EXERCISES 27–34** $\square$ *Identify the conic section as either a circle, parabola, ellipse, or hyperbola.*

**27.** $2x^2 - x - 3y + 4 = 0$          **28.** $3x^2 + 2y^2 = 7x - 7y - 5$

**29.** $4x^2 + 4y^2 = 2x - 6y + 10$   **30.** $4y^2 - 9x + 23y - 9 = 0$

**31.** $-5x^2 - 11x = y^2$          **32.** $x^2 - 2y^2 + 5 = 0$

**33.** $-2x^2 + 4y^2 - 15x + 8 = 0$

**34.** $-3x^2 + 2 = 3y^2 - 5y$

**EXERCISES 35–40** $\square$ *Solve the equation for y and plot the resulting equation(s) on a graphics calculator. In the case of a parabola, ellipse, or hyperbola, approximate the coordinates of any vertices.*

**35.** $5x^2 + 8y^2 = 40$          **36.** $11y^2 - 3x^2 = 33$

**37.** $y^2 - 3x^2 + 4x + 9 = 0$   **38.** $2x^2 + y^2 - 5x - 8 = 0$

**39.** $2x^2 + y^2 + 3x - 6y + 4 = 0$

**40.** $3x^2 - y^2 + 8x - 8y - 7 = 0$

**EXERCISES 41–44** $\square$ *Plot the two branches of the conic section with a graphics calculator. Identify the resulting conic section and find the general form of its equation.*

**41.** $y = \pm\sqrt{\dfrac{x^2}{3} - 2}$          **42.** $y = \pm\sqrt{9 - 3x^2}$

**43.** $y = 1 \pm \sqrt{2 - 8x - 2x^2}$   **44.** $y = -3 \pm \sqrt{x^2 - 10x}$

■ *Applications*

45. *Earth's Orbit*  The Earth's orbit is elliptical with major axis length 186 million miles and minor axis length 185.8 million miles. Write an equation for the path of the orbit, assuming the center is located at the origin. It may surprise you to find out the Sun is not at the center of the orbit. The Sun is in fact on the major axis, a distance of 4.3 million miles from the center. Sketch a graph of the orbit of the Earth and locate the Sun on the major axis.

46. *Mars' Orbit*  The orbit of Mars is elliptical with major axis length 283.5 million miles and minor axis length 278.5 million miles. Write an equation for the path of the orbit, assuming the center is located at the origin. The Sun is on the major axis, a distance of 26.5 million miles from the center. Sketch a graph of the orbit of Mars and locate the Sun on the major axis.

47. *Railroad Bridge*  The arch of a railroad bridge over a two-lane highway has the shape of a semi-ellipse (see Figure 115). The distance from the base of the arch on one side of the road to the base on the other is 50 feet, and the height of the arch at the center is 20 feet.

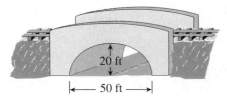

**FIGURE 115**

a. Find the equation of the ellipse that gives the shape of the arch. Assume that the *x*-axis is on the ground and runs perpendicular to the centerline of the highway, the origin is on the centerline, and the *y*-axis is vertical.

b. What is the height of the arch 5 feet from the base?

c. Could a tractor-trailer with a height of 14 feet and a width of 10 feet pass under the bridge without going over the centerline of the road?

48. *Sonic Boom*  The British-French Concorde travels at speeds exceeding the speed of sound. As a consequence, it creates a conical shock wave more commonly known as a "sonic boom." The region on the ground that is affected by the sonic boom has a boundary that is hyperbolic, as can be seen in Figure 116. When the Concorde is traveling at Mach 2 (twice the speed of sound) and at an altitude of 65,000 feet, the vertex of the hyperbolic boundary is approximately 24 miles from the point on the ground directly beneath the nose of the plane, as shown in Figure 116. If we were to set up a coordinate system with the origin on the ground directly beneath the nose of the plane, the *x*-axis lying on the ground parallel to the path of the plane, and the *y*-axis also lying on the ground, then the asymptotes for the hyperbolic boundary would have equations $y = \pm\frac{1}{2}x$. Find the equation of the hyperbolic boundary. The region affected by the sonic boom extends as far as 55 miles behind the point on the ground below the plane. What is the width of the affected region when $x = -55$?

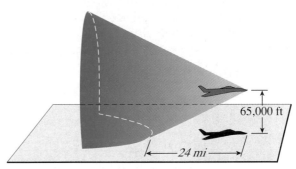

**FIGURE 116**

■ *Projects for Enrichment*

49. *Degenerate Conics*  Recall that the geometric definitions of the conic sections are based on the intersection of a plane with a double cone. The degenerate conics—a line, a pair of intersecting lines, and a point—arise geometrically when the plane passes through the vertex of the cone. The equations of the degenerate conics are all special cases of the general equation $Ax^2 + Cy^2 + Dx + Ey + F = 0$. For example, if *A* and *C* are both zero, the resulting equation $Dx + Ey + F = 0$ is a line.

a. Explain why each of the following equations has the indicated description.

   i. $2x^2 + 3y^2 = 0$;   a point

   ii. $4x^2 - 9y^2 = 0$;   a pair of intersecting lines

   iii. $x^2 + 2y^2 + 1 = 0$;   no graph

b. State some general conditions under which equations of the form $Ax^2 + Cy^2 + F = 0$ will be

   i. a point     ii. a pair of intersecting lines

c. Complete the square on the equation $Ax^2 + Cy^2 + Dx + Ey + F = 0$ to determine a test for characterizing the equation as

   i. a point     ii. a pair of intersecting lines

   [*Hint:* Your test should involve the expression $(D^2/4A) + (E^2/4C) - F$ and it should take into account the signs of *A* and *C*.]

d. Use the test you developed in part (c) to determine which, if any, of the following equations are degenerate conics.

i. $2x^2 - 3y^2 - 4x + 6y - 1 = 0$

ii. $2x^2 + y^2 - 4x + 2y + 10 = 0$

iii. $2x^2 + 3y^2 - 4x + 6y + 5 = 0$

50. *Rotated Conic Sections* Equations of the form $Ax^2 + Bxy + Cy^2 + Dx + Ey + F = 0$, for $B \neq 0$, are conic sections with **inclined** (tilted) axes. Such conics are called **rotated conics**, and it can be shown using trigonometry that the nondegenerate ones can be classified using the **discriminant** $B^2 - 4AC$ as shown here.

| Condition | Resulting conic |
|-----------|-----------------|
| $B^2 - 4AC < 0$ | ellipse |
| $B^2 - 4AC > 0$ | hyperbola |
| $B^2 - 4AC = 0$ | parabola |

For example, the equation $xy - 1 = 0$ is that of a hyperbola since $B^2 - 4AC = 1^2 - 4(0)(0) = 1$, which is greater than zero.

a. Each of the following is an example of a nondegenerate conic section. Use the discriminant to determine the type of conic.

  i. $x^2 - xy - 1 = 0$

  ii. $x^2 + xy + y^2 = 7$

  iii. $x^2 + 2xy = 4x - y^2$

In most cases, rotated conics cannot be sketched by hand without some knowledge of trigonometry. The hyperbola $xy - 1 = 0$ is one of the few exceptions since its graph is easy to sketch just by plotting a few points. First we solve for $y$ in terms of $x$ to obtain $y = 1/x$. The graph of this equation, shown in Figure 117, can be obtained by plotting a few points.

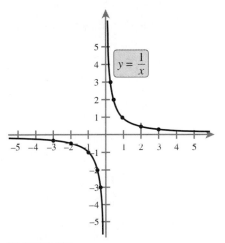

**FIGURE 117**

b. Identify the vertices and asymptotes for the hyperbola in Figure 117.

For most other rotated conics, we must either use trigonometry to obtain the angle of the rotation, or we must rely on a graphical tool such as a graphics calculator. Since the use of trigonometry is beyond the scope of this chapter, we will only consider the use of a graphics calculator. Consider the ellipse $2x^2 - xy + y^2 = 4$. To graph this ellipse, we must first solve the equation for $y$. We do this using the quadratic formula.

$$2x^2 - xy + y^2 = 4$$
$$y^2 - xy + (2x^2 - 4) = 0$$

Rewriting as a quadratic equation in $y$

$$y = \frac{-(-x) \pm \sqrt{(-x)^2 - 4(1)(2x^2 - 4)}}{2(1)}$$

Applying the quadratic formula with $b = -x$ and $c = 2x^2 - 4$

$$= \frac{x \pm \sqrt{16 - 7x^2}}{2}$$

Notice that there are two equations that must be graphed, namely $y = \frac{1}{2}(x + \sqrt{16 - 7x^2})$ and $y = \frac{1}{2}(x - \sqrt{16 - 7x^2})$. The plot of these two equations is shown in Figure 118.

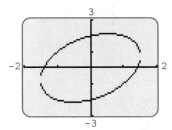

**FIGURE 118**

c. Solve each of the following equations of conics for $y$ in terms of $x$ and plot the graphs of the resulting pairs of equations. Approximate the coordinates of the vertices in each case.

  i. $x^2 - xy - 1 = 0$

  ii. $x^2 + xy + y^2 = 7$

  iii. $x^2 + 2xy = 4x - y^2$

### ◼️ *Questions for Discussion or Essay*

**51.** A lamp with a shade is placed close to a wall in a dark room. When the lamp is turned on, the outline of the light on the wall will look similar to that shown in Figure 119. Consult Figure 106 to help you explain why the outline forms a hyperbola.

**FIGURE 119**

**52.** Explain why a circle can be viewed as a special case of an ellipse.

**53.** In Example 9 we considered a method for graphing a conic section using a graphics calculator. This involved completing the square in $y$ in order to solve for $y$. For what types of conic equations is this approach quicker than finding the graph by hand? When might it actually be easier to obtain the graph entirely by hand? Give some specific examples to support your answers.

**54.** In Example 8 we concluded that the equation $9x^2 + 4y^2 - 36x + 8y + 31 = 0$ was an ellipse. However, if we change the constant term to 49 instead of 31, it is not an ellipse. Complete the square on $9x^2 + 4y^2 - 36x + 8y + 49 = 0$ to show this. What will the graph of the equation look like? Why? What does this example say about testing conic sections by looking only at the coefficients of the squared terms?

---

## SECTION 6

# GEOMETRIC PROPERTIES OF CONIC SECTIONS

◼️ How can an ellipse be accurately drawn with only a pencil, a piece of string, and two thumbtacks?

◼️ How can hyperbolas help navigate a ship?

◼️ What is the best shape for a room in which secrets must be whispered across a great distance?

◼️ Why are the shapes of conic sections often found in the mirrors of reflective telescopes?

◼️ How can hyperbolas be used to help Inspector Magill locate a stolen car?

### ◼️ ELLIPSES

We have seen that an ellipse is a curve that is formed when a cone is cut with a plane (see Figure 99 in Section 5). It can be shown that any ellipse formed in this manner can also be formed by applying the following definition.

### ◼️ *Definition of an ellipse*

An **ellipse** is the set of all points $(x, y)$ such that the sum of the distances from two fixed points (called the **foci**) is constant, as shown in Figure 120.

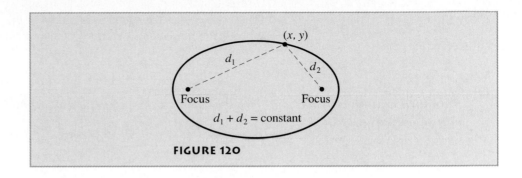

**FIGURE 120**

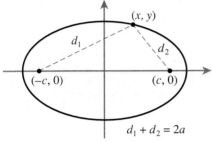

**FIGURE 121**

To derive the standard form for the equation of an ellipse centered at the origin with axes parallel to the coordinate axes, we will consider an ellipse whose foci are at $(-c, 0)$ and $(c, 0)$ as shown in Figure 121. For any point $(x, y)$ on the ellipse, the sum of the distances to the two foci must be constant. To simplify later computations, we will denote this constant sum by $2a$. Using the distance formula for each of the two distances $d_1$ and $d_2$, we have

$$d_1 = \sqrt{(x + c)^2 + y^2} \quad \text{and} \quad d_2 = \sqrt{(x - c)^2 + y^2}$$

Since $d_1 + d_2 = 2a$, it follows that

$$\sqrt{(x + c)^2 + y^2} + \sqrt{(x - c)^2 + y^2} = 2a$$
$$\sqrt{(x + c)^2 + y^2} = 2a - \sqrt{(x - c)^2 + y^2}$$

Now we square both sides and collect terms.

$$(x + c)^2 + y^2 = 4a^2 - 4a\sqrt{(x - c)^2 + y^2} + (x - c)^2 + y^2$$
$$x^2 + 2cx + c^2 + y^2 = 4a^2 - 4a\sqrt{(x - c)^2 + y^2} + x^2 - 2cx + c^2 + y^2$$
$$4a\sqrt{(x - c)^2 + y^2} = 4a^2 - 4cx$$
$$a\sqrt{(x - c)^2 + y^2} = a^2 - cx$$

Once again we square both sides and collect terms.

$$a^2[(x - c)^2 + y^2] = (a^2 - cx)^2$$
$$a^2(x^2 - 2cx + c^2 + y^2) = a^4 - 2a^2cx + c^2x^2$$
$$a^2x^2 - 2a^2cx + a^2c^2 + a^2y^2 = a^4 - 2a^2cx + c^2x^2$$
$$a^2x^2 - c^2x^2 + a^2y^2 = a^4 - a^2c^2$$
$$(a^2 - c^2)x^2 + a^2y^2 = a^2(a^2 - c^2)$$

Finally, we let $b^2 = a^2 - c^2$.

$$b^2x^2 + a^2y^2 = a^2b^2$$
$$\frac{x^2}{a^2} + \frac{y^2}{b^2} = 1$$

Thus, we see that the equation is precisely the same as that given in Section 5. By applying a translation, we arrive at the standard equation for an ellipse centered at $(h, k)$.

**Standard equation of an ellipse centered at $(h, k)$**

The standard form for the equation of an ellipse centered at $(h, k)$ is

$$\frac{(x - h)^2}{a^2} + \frac{(y - k)^2}{b^2} = 1$$

If $a > b$, the major axis is horizontal with length $2a$ (see Figure 122), while if $b > a$, the major axis is vertical with length $2b$ (see Figure 123). The vertices are the endpoints of the major axis. The foci lie on the major axis a distance of $c$ units from the center, where $c = \sqrt{|a^2 - b^2|}$.

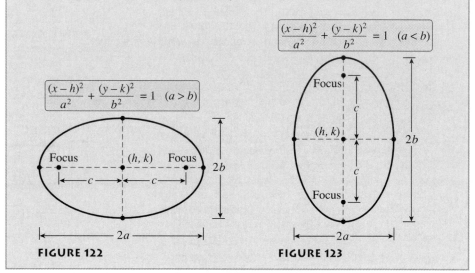

FIGURE 122    FIGURE 123

**EXAMPLE 1**    *Locating the foci of an ellipse*

Determine the coordinates of the foci of the ellipse

$$\frac{(x - 2)^2}{9} + \frac{(y + 3)^2}{25} = 1$$

**SOLUTION**    This ellipse has center $(2, -3)$, a vertical major axis, and vertices $(2, 2)$ and $(2, -8)$, as shown in Figure 124. The foci are on the major axis a distance of

$$c = \sqrt{|9 - 25|}$$
$$= \sqrt{16}$$
$$= 4$$

units from the center, at the points $(2, -7)$ and $(2, 1)$.

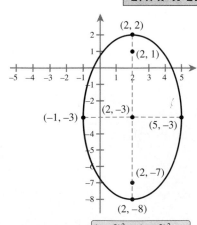

**FIGURE 124**    $\dfrac{(x-2)^2}{9} + \dfrac{(y+3)^2}{25} = 1$

**EXAMPLE 2**   *Finding the equation of an ellipse*

Find the equation of the ellipse with foci $(\pm 4, 0)$ and a major axis length of 12.

**SOLUTION**   Since the foci are on the *x*-axis at equal distances from the origin, the ellipse is centered at the origin and has a horizontal major axis. Thus, $c = 4$ and $a = 12/2 = 6$. Since $b < a$, $c = \sqrt{a^2 - b^2}$ and so

$$4 = \sqrt{6^2 - b^2}$$

$$16 = 36 - b^2$$

$$b^2 = 20$$

The equation of the ellipse is

$$\frac{x^2}{36} + \frac{y^2}{20} = 1$$

---

## HYPERBOLAS

Recall that a hyperbola is a curve that is formed when a double cone is cut by a plane parallel to the axis of the cone. It can be shown that this definition is equivalent to the one given below.

*Definition of a hyperbola*

A **hyperbola** is the set of all points $(x, y)$ such that the difference of the distances from two fixed points (called the **foci**) is constant, as shown in Figure 125.

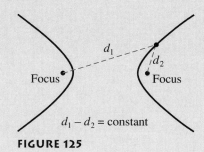

$$d_1 - d_2 = \text{constant}$$

**FIGURE 125**

The standard form for the equation of a hyperbola centered at the origin and with a vertical or horizontal axis can be derived in much the same way as we did for an ellipse. We will leave the details for Exercise 58. By applying a translation, we arrive at the following familiar equations.

**Standard equations of a hyperbola centered at (h, k)**

The standard forms for the equation of a hyperbola centered at $(h, k)$ are

$$\frac{(x - h)^2}{a^2} - \frac{(y - k)^2}{b^2} = 1 \quad \text{(horizontal axis)}$$

$$\frac{(y - k)^2}{b^2} - \frac{(x - h)^2}{a^2} = 1 \quad \text{(vertical axis)}$$

A horizontal transverse axis has length $2a$ (the brown dashed line in Figure 126), while a vertical transverse axis has length $2b$ (the brown dashed line in Figure 127). The vertices are the endpoints of the transverse axis. The foci are on the line extending from the transverse axis, a distance of $c$ units from the center, where $c = \sqrt{a^2 + b^2}$.

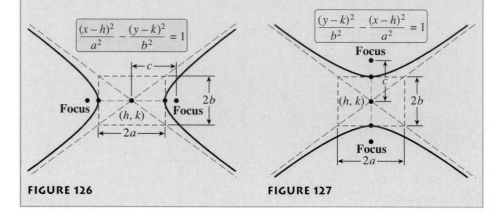

**FIGURE 126**    **FIGURE 127**

**EXAMPLE 3**   *Locating the foci of a hyperbola centered at the origin*

Determine the coordinates of the foci of the hyperbola

$$\frac{x^2}{16} - \frac{y^2}{9} = 1$$

**SOLUTION**   This hyperbola has horizontal transverse axis and vertices $(-4, 0)$ and $(4, 0)$. The distance from the center to each focus is $c = \sqrt{16 + 9} = 5$, and so the foci are located at $(-5, 0)$ and $(5, 0)$, as shown in Figure 128.

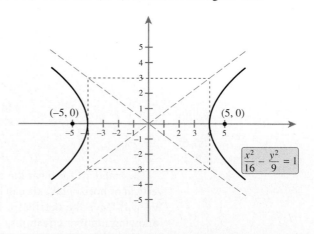

**FIGURE 128**

## PARABOLAS

*Definition of a parabola*

A parabola is a curve that is formed when a cone is cut with a plane parallel to the side of the cone. Equivalently, we have the following definition.

A **parabola** is the set of all points $(x, y)$ that are equidistant from a fixed point (called the **focus**) and a fixed line (called the **directrix**), as shown in Figure 129.

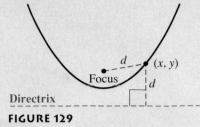

**FIGURE 129**

The standard form for the equation of a parabola with vertex at the origin and with a vertical or hoirzontal axis can be derived in much the same way as we did for an ellipse.

*Standard equations of a parabola with vertex at (h, k)*

The standard forms for the equation of a parabola with vertex at $(h, k)$ are

$$y - k = a(x - h)^2 \quad \text{(vertical axis of symmetry, Figure 130)}$$

$$x - h = a(y - k)^2 \quad \text{(horizontal axis of symmetry, Figure 131)}$$

The focus is on the axis of symmetry in the *interior* of the parabola at a distance of

$$p = \left| \frac{1}{4a} \right|$$

units from the vertex. The directrix is perpendicular to the axis of symmetry and is a distance of $p$ units from the vertex on the side of the parabola opposite the focus.

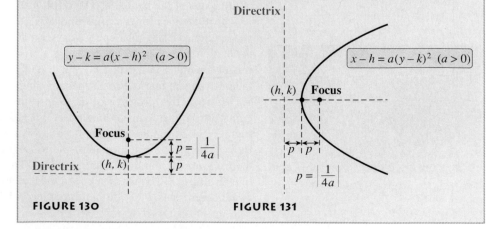

**FIGURE 130**                       **FIGURE 131**

**EXAMPLE 4**    *Locating the directrix and focus of a parabola*

Determine the directrix and the coordinates of the focus for the parabola
$x - 1 = -\frac{1}{8}(y + 2)^2$.

**SOLUTION**    The vertex of the parabola $x - 1 = -\frac{1}{8}(y + 2)^2$ is $(1, -2)$, and the axis is horizontal, as shown in Figure 132. Since $a = -\frac{1}{8}$, the parabola opens to the left and the focus is located $|1/(4a)| = |-2| = 2$ units from the vertex at the point $(-1, -2)$. The directrix is located 2 units to the right of the vertex and thus has equation $x = 3$.

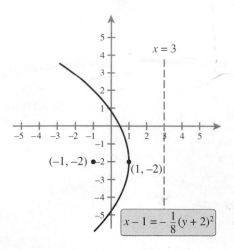

**FIGURE 132**

## THE REFLECTIVE PROPERTIES OF CONIC SECTIONS

If sound or light is emitted from the focus of a conic section, certain important reflective properties can be observed. In the case of a parabola, the reflected sound or light will follow a path parallel to the axis of the parabola, as shown in Figure 133. This property of parabolas is utilized in the construction of search lights. The reflector for the search light is formed by revolving a parabola around its axis, and the light source is placed at the focus of the parabola. The result is a light beam that does not dissipate.

The reflective property of parabolas is also used in telescopes, satellite dishes, microwave antennas, and solar energy devices. When the parabolic reflector of these devices is positioned so that the incoming light rays, sound waves, or radio waves are (approximately) parallel to the axis, they will be reflected toward the focus. This results in an image or signal that is "focused" at a single point.

The reflective property of an ellipse involves both foci. When sound or light is emitted from one focus of an ellipse, it is reflected toward the other focus, as shown in Figure 134. An interesting application of this property can be found in "whispering rooms" such as those located at the Museum of Science and Industry in Chicago or the Capitol building in Washington, D.C. These rooms have walls (or ceilings) shaped like an ellipse. If you stand at one of the foci and whisper something to a friend standing at the other focus, you will be heard clearly by your friend but nobody else.

The Very Large Array of radio antennas in New Mexico

A more useful application of the reflective property of an ellipse can be found in a medical procedure called *shockwave lithotripsy* used in the treatment of kidney stones. If a shockwave transmitter and an elliptic reflector are placed in such a way that the transmitter and kidney stone are located at the two foci of the reflector, the shock waves will be directed at the kidney stone causing it to shatter, while leaving the rest of the body unharmed.

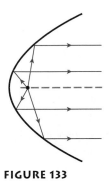

**FIGURE 133**

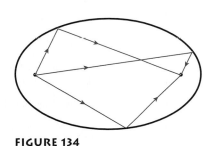

**FIGURE 134**

**FIGURE 135**

The reflective property of hyperbolas also involves both foci. If an incoming light ray or sound wave is aimed at one focus of a hyperbola, it is reflected toward the other focus, as shown in Figure 135. Hyperbolic reflectors are often used as auxiliary mirrors in telescopes to direct the image to the eyepiece of the telescope where it can be magnified. Two schematic diagrams of telescopes are given in Figures 136 and 137. Notice that in both the main reflector is parabolic. Also in both, the hyperbolic reflector is positioned so that its focus coincides with that of the parabolic mirror. In Figure 136 the eyepiece of the telescope is located at the other focus of the hyperbolic reflector. In Figure 137 an elliptic reflector is added with one of its foci at the second focus of the hyperbolic reflector and its other focus at the eyepiece.

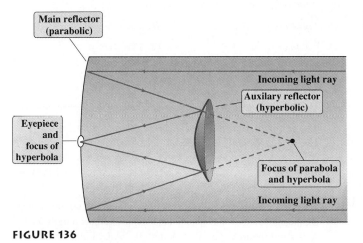

**FIGURE 136**

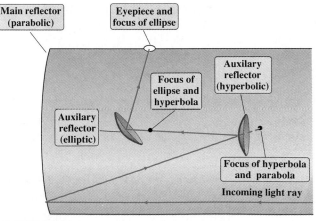

**FIGURE 137**

# EXERCISES 6

**EXERCISES 1–14** □ *Locate the coordinates of the foci. Note that these equations were considered in Exercises 1–14 of Section 5.*

1. $\dfrac{x^2}{9} + \dfrac{y^2}{25} = 1$

2. $\dfrac{x^2}{16} + \dfrac{y^2}{9} = 1$

3. $4x^2 + 9y^2 = 36$

4. $16x^2 + 9y^2 = 144$

5. $9x^2 + 4y^2 = 1$

6. $16x^2 + 25y^2 = 1$

7. $3x^2 + 2y^2 = 6$

8. $5x^2 + 8y^2 = 40$

9. $\dfrac{(x + 3)^2}{4} + \dfrac{(y - 1)^2}{16} = 1$

10. $\dfrac{(x - 2)^2}{25} + \dfrac{(y + 2)^2}{9} = 1$

11. $9x^2 + 4y^2 - 18x - 24y + 9 = 0$

12. $25x^2 + 16y^2 - 200x - 32y + 16 = 0$

13. $x^2 + 36y^2 + 4x - 72y + 4 = 0$

14. $16x^2 + y^2 + 4y - 12 = 0$

**EXERCISES 15–26** □ *Locate the coordinates of the foci. Note that these equations were considered in Exercises 15–26 of Section 5.*

15. $\dfrac{x^2}{9} - \dfrac{y^2}{16} = 1$

16. $\dfrac{y^2}{25} - \dfrac{x^2}{16} = 1$

17. $4y^2 - 9x^2 = 36$

18. $25x^2 - 16y^2 = 400$

19. $12x^2 - 3y^2 = -24$

20. $5x^2 - 3y^2 = 15$

21. $\dfrac{(x - 1)^2}{4} - \dfrac{(y - 4)^2}{9} = 1$

22. $\dfrac{(y + 3)^2}{16} - \dfrac{(x + 2)^2}{25} = 1$

23. $4x^2 - 9y^2 - 16x + 54y - 101 = 0$

24. $25x^2 - 16y^2 - 50x + 160y + 25 = 0$

25. $4x^2 - 25y^2 - 32x + 164 = 0$

26. $x^2 - 4y^2 + 4x - 24y - 36 = 0$

**EXERCISE 27–38** □ *Locate the coordinates of the focus and find the equation of the directrix. Note that these equations were considered in Exercises 21–32 of Section 4.*

27. $y = 2x^2$

28. $x = -4y^2$

29. $y^2 - 3x = 0$

30. $x^2 + 5y = 0$

31. $4y^2 + 7x = 0$

32. $2x^2 - 3y = 0$

33. $y - 3 = 3(x + 5)^2$

34. $x + 4 = -(y - 2)^2$

35. $x^2 - 4x - 4y + 8 = 0$

36. $y^2 + x - 6y + 10 = 0$

37. $x = 2y^2 + 6y + 2$

38. $y = -x^2 - x + 1$

**EXERCISES 39–44** □ *Find an equation for the ellipse centered at the origin with the given properties.*

39. Horizontal major axis of length 8 and vertical minor axis of length 6

40. Vertices $(0, \pm 4)$ and minor axis of length 2.

41. Vertices $(0, \pm \sqrt{6})$ and foci $(0, \pm \sqrt{2})$

42. Foci $(\pm \sqrt{2}, 0)$ and major axis of length 4

43. Vertices $(\pm 6, 0)$ and passes through the point $(-4, 2)$

44. Foci $(0, \pm 3)$ and passes through the point $(4, \frac{12}{5})$

**EXERCISES 45–48** □ *Find an equation for the hyperbola centered at the origin with the given properties.*

45. Vertices $(0, \pm 4)$ and foci $(0, \pm 5)$

46. Vertices $(\pm 6, 0)$ and asymptotes $y = \pm \frac{4}{3}x$

47. Foci $(0, \pm 2\sqrt{5})$ and asymptotes $y = \pm \frac{1}{2}x$

48. Vertices $(\pm 3, 0)$ and passes through the point $(5, 4)$

**EXERCISES 49–52** □ *Find an equation for the parabola with vertex at the origin and with the given properties.*

49. Directrix $y = 3$

50. Focus $(0, 2)$

51. Focus $(-4, 0)$

52. Directrix $x = -\frac{3}{2}$

## ■ *Applications*

**EXERCISES 53–54** □ *An Astronomical Unit (A.U.) refers to the mean distance from the Earth to the Sun. When dealing with elliptical orbits of celestial bodies with the Sun located at one focus, the* **perihelion** *distance is the distance from the Sun to the closest vertex, while the* **aphelion** *distance is the distance from the Sun to the most distant vertex.*

53. *Comet Encke* Some scientists have speculated that a piece of comet Encke fell to Earth in 1911, causing massive destruction

in a remote region of Siberia. (See mathematical note on page 224 for more details.) The orbit of comet Encke is an ellipse with the Sun located at one focus. The major axis has length 4.4 A.U., and the minor axis has length 2.2 A.U. Find the perihelion and aphelion distances for comet Encke.

54. *Halley's Comet* The orbit of Halley's comet is an ellipse with the Sun located at one focus. The major axis has length 36.2 A.U., and the minor axis has length 9.1 A.U. Find the perihelion and aphelion distances for Halley's comet.

Halley's comet in space

**55. Arecibo Radio Satellite** The cross-section view of the Arecibo radio satellite, the world's largest satellite dish, is parabolic and has an equation of the form $y = ax^2$. If the dish is 1000 feet across the top and 167 feet deep at the center, where must the receiver be located?

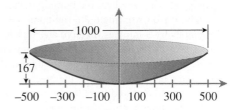

**56. Whispering Room** An elliptical whispering room is to be constructed with a maximum width of 30 feet and a maximum length of 50 feet. Locate the positions at which two people should stand if they wish to whisper to each other and be heard.

**57. Long Range Navigation** LORAN uses radio signals transmitted by two stations to track the locations of ships. When signals are sent simultaneously from the two stations, the ship can determine the difference in the time it takes to receive the two signals. By assuming the ship lies on a hyperbola with foci at the transmitting stations, this difference in time can be used to

determine the position of the ship. Suppose transmitting stations $A$ and $B$ are located 200 miles apart on a coastline as shown in Figure 138 and a ship is located at a point $P$.

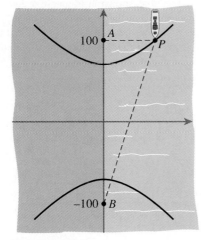

**FIGURE 138**

a. If the ship receives the signal from station $A$ 800 microseconds (1 microsecond $= 10^{-6}$ second) before the signal from $B$, find the equation of the hyperbola shown in Figure 138. Assume the signals travel at a rate of 980 feet per microsecond. [*Hint:* If the equation is

$$\frac{y^2}{b^2} - \frac{x^2}{a^2} = 1$$

then the difference of the distances from any point $(x, y)$ on the hyperbola to the foci is $2b$.]

b. If it is known that the ship is due east of station $A$, determine the location of the ship.

---

■ **Projects for Enrichment**

---

**58. Deriving the Equation for a Hyperbola** Earlier in this section, we applied the distance formula to find the standard equation of an ellipse centered at the origin with foci $(\pm c, 0)$. In this project we will derive the standard equation for a hyperbola centered at the origin with foci $(\pm c, 0)$.

a. Let $(x, y)$ represent a point on the hyperbola. Write an equation that states that the difference of the distances from $(x, y)$ to $(-c, 0)$ and $(c, 0)$ is equal to the constant $2a$.

b. Simplify the equation as much as possible by eliminating the radicals.

c. Use the substitution $b^2 = c^2 - a^2$ to put the equation in the form

$$\frac{x^2}{a^2} - \frac{y^2}{b^2} = 1$$

**59. Drawing Ellipses** In this project you will investigate how to draw ellipses of varying sizes. For materials you will need a large piece of cardboard, blank paper, two thumbtacks, several long pieces of string, and a pencil. Lay a piece of paper on top of the cardboard. Cut a piece of string a little longer than 4 inches and attach each end around a thumbtack so that 4 inches of string is between the tacks. Stick the tacks in the cardboard 3 inches apart so that there is some slack in the string. Place the tip of the pencil as shown in Figure 139 and, keeping the string taut and the tip of the pencil on the paper, carefully move the pencil along the string. The pencil will trace out the path of an ellipse with foci located at the thumbtacks. The length of the major axis is equal to the total length of the string between the tacks.

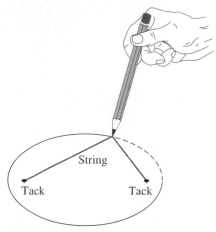

**FIGURE 139**

a. Use the fact that the string is 4 inches long and the tacks are 3 inches apart to find the equation of the ellipse you drew. Assume the $x$-axis passes through the foci and the origin is midway between them.

b. Draw an ellipse with foci a distance of 3 inches apart and major axis length of 5 inches.

c. Draw an ellipse with foci 3 inches apart and major axis length of 6 inches.

d. What general observation can you make about the shape of the ellipse if the distance between the foci is held constant but the length of the major axis (i.e., the length of the string) is increased?

e. Compute the ratio $c/a$ for each of the three ellipses you constructed. This ratio is called the *eccentricity of the ellipse*. What is the largest possible value of the eccentricity? Why? What is the connection between the size of the eccentricity and the shape of the ellipse? What is the smallest the eccentricity could be? What would the shape of the ellipse be then?

60. **The Case of the Stolen Car**  A victim of a carjacking calls Inspector Magill's office at 9:00 A.M. seeking help in recovering a stolen car. The following telephone conversation takes place.

Car theft?

Victim:  This morning on my way to work, I was forced out of my car by two masked men. I overheard one say to the other that they would take the car to "the usual place" and store it there until noon. My car was equipped with an electronic homing device; unfortunately, it isn't working quite right and only emits a signal every 5 minutes. Is there any way to locate the source of the signal before the car is moved again?

Magill:  Yes, but I will need time to round up three electronic receivers that can determine the precise time at which the signal is received from the homing device. I will also need three mobile phones and the help of two assistants.

Victim:  I have a friend who works at a high-tech electronics company and can borrow the receivers.

At 10:15, the victim and his friend arrive at Magill's office with the three receivers. After synchronizing the timing devices in the receivers, Magill takes the two friends to positions labeled $A$ and $B$ in Figure 140 and he drives to position $C$. His instructions to the two friends are simply to record the exact time at which the first signal after 11:00 is received from the homing device and then call him on the mobile phones. By 11:30, the car has been recovered and the criminals have been apprehended. We will reconstruct the method used by Magill to locate the car.

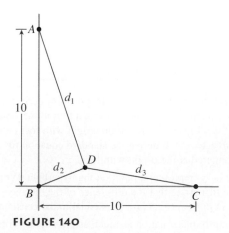

**FIGURE 140**

The times recorded by the receivers gave the following information.

- ■ It took $3.22 \times 10^{-5}$ second longer to receive the signal at point $A$ than at point $B$.
- ■ It took $2.15 \times 10^{-5}$ second longer to receive the signal at point $C$ than at point $B$.

a. If $D$ denotes the position of the car and $d_1$, $d_2$, and $d_3$ denote the distances to $A$, $B$, and $C$, respectively, determine the differences $d_1 - d_2$ and $d_3 - d_2$ (round to the nearest integer).

Assume that the signal travels at the speed of light, which is 186,000 miles per second. Recall that a hyperbola is the set of all points such that the difference of the distances from the foci is constant. Point $D$ must thus be one of the points on the hyperbola for which the difference of the distances from $A$ and $B$ is $d_1 - d_2$, and it must also be a point on the hyperbola for which the difference of the distances from $C$ and $B$ is $d_3 - d_2$. In other words, if we treat $A$ and $B$ as the foci of one hyperbola and $B$ and $C$ as the foci of a second hyperbola, point $D$ will be a point of intersection. We simply need to find the equations of the two hyperbolas to determine their points of intersection. We will demonstrate how to find the one with foci at $B$ and $C$ and leave the other for you. We first assign a coordinate system

to Figure 140. It is convenient to let the origin be at point $B$. Since $B$ and $C$ are 10 miles apart, the center of the hyperbola is at $(5, 0)$. Thus, we are looking for an equation of the form

$$\frac{(x - 5)^2}{a^2} - \frac{y^2}{b^2} = 1$$

Since the foci are 5 miles from the center, $c = 5$. The distance $d_3 - d_2$ must be twice the distance from the center to the vertices, so $2a = d_3 - d_2$. Finally, $b = \sqrt{c^2 - a^2} = \sqrt{25 - a^2}$.

b. Justify each of the computations suggested above and then complete the computations to show that the equation of the hyperbola with vertices $B$ and $C$ is

$$\frac{(x - 5)^2}{4} - \frac{y^2}{21} = 1$$

c. Show that the equation of the second hyperbola is

$$\frac{(y - 5)^2}{9} - \frac{x^2}{16} = 1$$

d. Estimate the coordinates of the intersection points of these two hyperbolas and determine the location of the car.

■■ *Questions for Discussion or Essay*

61. In Figure 137 a reflective telescope is shown that uses a parabolic main mirror and both an elliptical and hyperbolic auxiliary mirror to route the incoming light to the eyepiece. Can a telescope be constructed with a parabolic main mirror and only one auxiliary mirror so that it still directs the incoming light to the eyepiece? Explain, using a sketch if necessary.

62. An elliptical "pool" table with only one pocket is being used for a trick-shooting demonstration. The pool shark carefully places a ball on the table but hits it in a seemingly arbitrary direction. Time after time, the ball bounces off the cushion directly into the pocket. Explain how this could happen.

63. Using only the information given in Exercise 54, is it possible to determine how close Halley's comet comes to the Earth? Explain.

64. A rough sketch of a hyperbola can be drawn in the following way. Attach two strings of different lengths to a ballpoint pen near the tip and attach the other ends of the strings to a piece of cardboard using thumbtacks as shown in Figure 141. Keep-

ing the string taut and the tip of the pen on the cardboard, slowly twirl the pen so the string begins to wind around the tip. As the lengths of string shorten, the pen will follow a hyperbolic path. Explain why this is so.

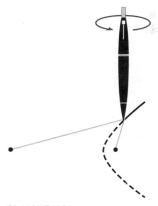

**FIGURE 141**

## CHAPTER REVIEW EXERCISES

**EXERCISES 1–4** □ *Find all points with the given x- or y-coordinate that lie on the graph of the relation.*

1. $4x - 7y = 29$, $x$-coordinate 2

2. $2x - 3y^2 = 1$, $y$-coordinate $-3$

3. $x^2 + y^2 = 13$, $y$-coordinate $-2$

4. $xy^2 + 4x^2 - 3y = 0$, $x$-coordinate 1

**EXERCISES 5–8** □ *Find the intercepts of the given relation and plot the graph.*

5. $y = \dfrac{1}{2}x + 1$      6. $3x + 6y = 12$

7. $x = y^2 - 16$       8. $y + 9 = (x + 1)^2$

**EXERCISES 9–12** □ *Approximate the x-intercepts of the graph of the given relation to the nearest hundredth. The number of x-intercepts is indicated in parentheses.*

9. $y = x^3 - 3x^2 + x$   (3)      10. $y = 2x^4 - 4x^3 + 3$   (2)

11. $x^2 + xy = 2$   (2)       12. $100x^3 - x^2 - y = 0$   (2)

**EXERCISES 13–16** □ *Find the points of intersection of the given pair of relations.*

13. $3x - y = 2$;   $x + 2y = -10$

14. $y = x^2 - 2x$;   $x + y = 2$

15. $x^2 + y^2 = 20$;   $y = x^2$

16. $x^2 - y^2 = 4$;   $x = y + 1$

**EXERCISES 17–20** □ *Use the given relation and its graph to sketch the graph of the indicated relation, formed by translating and/or reflecting the original relation.*

17. $y = x^2$

   a. $y = (x - 2)^2$

   b. $y = x^2 + 3$

   c. $y - 4 = (x - 2)^2$

   d. $y = -x^2$

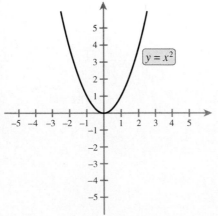

18. $x = y^3$

   a. $x = (y - 1)^3$

   b. $x = y^3 + 2$

   c. $-x = y^3$

   d. $-x = (y - 1)^3$

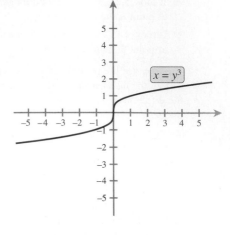

19. $xy^2 + x = 4$

   a. $x(y + 2)^2 + x = 4$

   b. $xy^2 + x = -4$

   c. $(x - 1)(y - 3)^2 + x - 1 = 4$

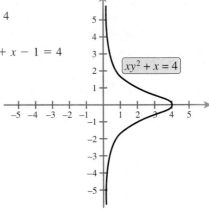

20. $x^2 + y^3 = 16$

   a. $(x - 2)^2 + y^3 = 16$

   b. $(x - 2)^2 + (y - 1)^3 = 16$

   c. $x^2 - y^3 = 16$

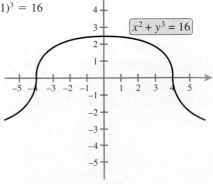

**EXERCISES 21–22** □ *A relation and its graph are given, together with the graph of a translation of the relation. Find an equation for the translated relation.*

**21.** Relation: $x^2 + y^4 = 16$      Translated relation:

**22.** Relation: $y = x^3$      Translated relation:

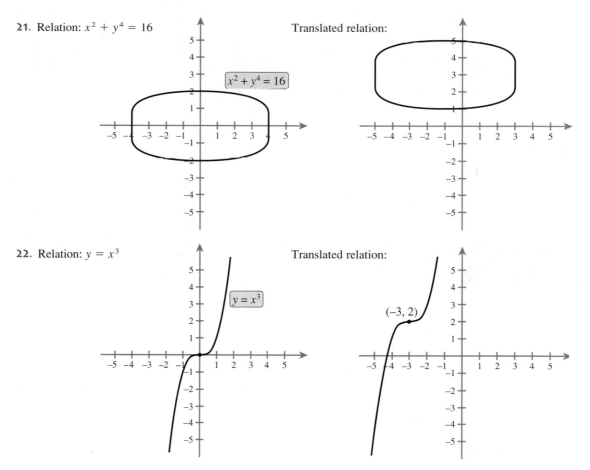

**EXERCISES 23–24** □ *A relation and its graph are given, together with the graph of a reflection of the relation. Find an equation for the reflected relation.*

**23.** Relation: $y = x^4 - 4x^2 + 2$      Reflected relation:

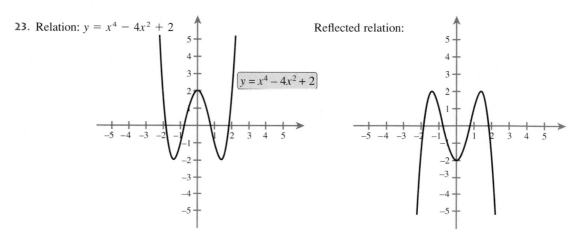

**24.** Relation: $y = x^3 - 3x^2 + 2$

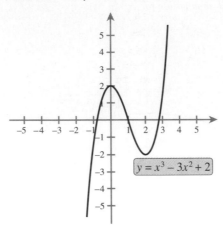

$y = x^3 - 3x^2 + 2$

Reflected relation:

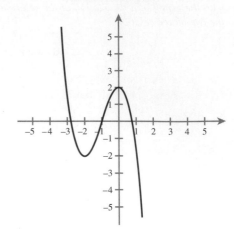

**EXERCISES 25–28** ☐ *A relation along with a portion of its graph are shown. First test the relation for symmetry, and then use the symmetry to complete the graph of the relation.*

**25.** Relation: $x = y^4 - 4y^2$

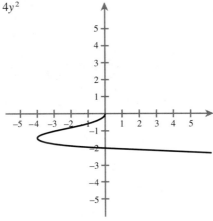

**27.** Relation: $xy + x^3y = 5$

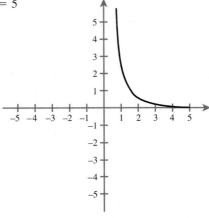

**26.** Relation: $y = 4x^2 - x^4$

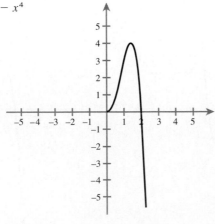

**28.** Relation: $9x^2 + y^4 = 81$

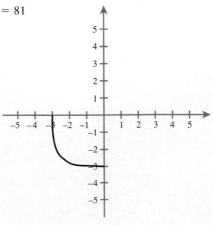

**EXERCISES 29–32** □ *Find the slope of the line satisfying the given properties.*

29. Passing through (2, 5) and (−1, 4)

30. Passing through (2, 5) with *y*-intercept 5

31. With equation $y = 4x - 7$

32. With equation $3x + 4y = 6$

**EXERCISES 33–48** □ *Find the equation of the line satisfying the indicated properties. Express your answer in slope-intercept form $y = mx + b$ if possible.*

33. Slope 3 and passing through (1, 5)

34. Slope $-\frac{1}{2}$ and passing through (−4, 8)

35. Slope undefined and passing through (2, $\pi$)

36. Slope 0 and passing through (5, −2)

37. Passing through (−1, 6) and (3, −2)

38. Passing through (2, 4) and (−2, 7)

39. Passing through ($\frac{1}{2}$, 3) and (−2, $\frac{3}{4}$)

40. Passing through (4.1, −1.2) and (2.3, 6.6)

41. With slope −2 and *y*-intercept 2

42. Slope $\frac{1}{3}$ and *x*-intercept −1

43. With *y*-intercept 5 and *x*-intercept 1

44. Slope undefined and *y*-intercept 0

45. Parallel to $y = -3x + 1$ and passing through (7, 1)

46. Passing through (−2, 4) and perpendicular to the line containing (3, −1) and (5, −1)

47. Perpendicular to $2x - 3y = 1$ and passing through (−4, 2)

48. Parallel to $5x + 2y = 10$ and passing through (0, 0)

**EXERCISES 49–54** □ *Find the center and radius of the circle and sketch its graph.*

49. $x^2 + y^2 = 16$

50. $x^2 + y^2 = \frac{1}{4}$

51. $(x - 2)^2 + (y + 1)^2 = 9$

52. $(x + 3)^2 + (y - 2)^2 = 25$

53. $x^2 + y^2 + 10x - 8y + 32 = 0$

54. $x^2 + y^2 - 2x - 6y + 6 = 0$

**EXERCISES 55–58** □ *Find the equation of the circle with the given properties.*

55. Center (2, 5) and radius 3

56. Center (−2, 1) and radius 4

57. Center (2, 4) and passes through (−4, −4)

58. The points (5, 5) and (−1, −3) are endpoints of a diameter.

**EXERCISES 59–64** □ *Find the vertex of the parabola and sketch the graph.*

59. $x = 2y^2$

60. $x^2 + 3y = 0$

61. $y + 4 = -2(x - 3)^2$

62. $2y - 4 = (x + 1)^2$

63. $x = 3y^2 + 6y + 7$

64. $y = 2x^2 + 6x + 1$

**EXERCISES 65–70** □ *Sketch a graph of the given ellipse. Identify the center, major axis, minor axis, and vertices.*

65. $\dfrac{x^2}{49} + \dfrac{y^2}{64} = 1$

66. $2x^2 + 3y^2 = 1$

67. $9(x - 3)^2 + 25(y - 2)^2 = 225$

68. $\dfrac{(x + 2)^2}{64} + \dfrac{(y - 3)^2}{100} = 1$

69. $x^2 + 4y^2 - 2x + 24y + 21 = 0$

70. $16x^2 + 9y^2 + 128x + 18y + 121 = 0$

**EXERCISES 71–76** □ *Sketch a graph of the given hyperbola. Identify the center, transverse axis, vertices, and asymptotes.*

71. $\dfrac{x^2}{25} - \dfrac{y^2}{9} = 1$

72. $8y^2 - 12x^2 = 1$

73. $4(y - 1)^2 - 9(x + 5)^2 = 36$

74. $\dfrac{(x + 1)^2}{4} - \dfrac{(y + 2)^2}{16} = 1$

75. $25x^2 - 16y^2 - 150x + 64y - 239 = 0$

76. $x^2 - 4y^2 - 6x + 32y = 51$

**EXERCISES 77–82** □ *Identify the conic section as either a circle, parabola, ellipse, or hyperbola.*

77. $4x^2 + y^2 - 1 = 0$

78. $4x^2 - 8x - y^2 + 8y - 16 = 0$

79. $y + 3x^2 + 2x = 0$

80. $-x^2 - y^2 + 2x + 4y = 0$

81. $x - 2y^2 - 2y + 10 = 0$

82. $x^2 - 8x + 4y^2 + 8y - 16 = 0$

**EXERCISES 83–86** □ *Use a graphics calculator to plot the pair of equations on the same coordinate system. Identify the resulting conic and find its standard equation.*

83. $y = \pm\sqrt{16 - x^2}$

84. $y = \pm\sqrt{x^2 - 16}$

85. $y = 2 \pm \sqrt{x - 4}$

86. $y = 4 \pm \sqrt{16 - 4x^2}$

**EXERCISES 87–90** □ *Find the coordinates of the foci of the ellipse from the given exercise.*

87. Exercise 65

88. Exercise 66

89. Exercise 67

90. Exercise 68

**EXERCISES 91–94** □ *Find the coordinates of the foci of the hyperbola from the given exercise.*

91. Exercise 71

92. Exercise 72

93. Exercise 73

94. Exercise 74

**EXERCISES 95–98** □ *Find the coordinates of the focus and the equation of the directrix for the parabola from the given exercise.*

95. Exercise 59

96. Exercise 60

97. Exercise 61

98. Exercise 62

**EXERCISES 99–104** □ *Find the equation of the conic with the indicated properties.*

99. Ellipse with horizontal major axis of length 6, vertical minor axis of length 4, and center at the origin

100. Ellipse with vertices at $(4, 0)$ and $(-4, 0)$ and foci at $(\sqrt{5}, 0)$ and $(-\sqrt{5}, 0)$

101. Hyperbola with vertices at $(5, 0)$ and $(-5, 0)$ and foci at $(\pm\sqrt{34}, 0)$

102. Hyperbola with vertices at $(0, \pm\sqrt{5})$ and passing through $(4, 5)$

103. Parabola with vertex at the origin and focus at $(0, \frac{1}{4})$

104. Parabola with horizontal axis, vertex at $(4, 3)$, and passing through $(2, 4)$

105. *Breaking Even*  Suppose that a company selling $x$ units of a product has a profit given by $x^3 - 34x^2 + 388x - 1480$ dollars. Find the number of units that must be sold to realize a profit of zero.

106. *Locating a City*  Memphis, Tennessee, lies 358 miles due north of New Orleans, Louisiana. The distance from New Orleans to Birmingham, Alabama, is 312 miles, and the distance from Memphis to Birmingham is 217 miles. Express the location of Birmingham as a point on a coordinate system with the origin located at New Orleans and the $y$-axis passing through Memphis.

107. *CD Depreciation*  A compact disc purchased in 1993 for $15.00 can be resold in 1995 for $9.50. Assuming that the value of the compact disc decreases linearly, what will be its value in 1998?

108. *Whispering Room*  A whispering room is in the shape of an ellipse that is twice as long as it is wide. If the foci are 1.86 feet from the vertices, find the distance from each focus to the center of the room.

## CHAPTER TEST

1. Find the vertex and intercepts of the parabola $x = y^2 + 4y$ and sketch its graph.

2. The graph of a relation is shown in Figure 142 with a viewing rectangle defined by `Xmin=-10`, `Xmax=10`, `Ymin=-10`, `Ymax=10`. Match the viewing rectangles labeled i–iii with the range values in parts (a) through (c).

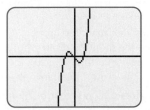

**FIGURE 142**

i.

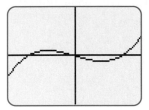

ii.

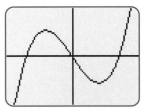

iii.

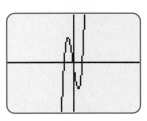

   **a.** `Xmin=-2`, `Xmax=2`, `Ymin=-2`, `Ymax=2`

   **b.** `Xmin=-2`, `Xmax=2`, `Ymin=-10`, `Ymax=10`

   **c.** `Xmin=-10`, `Xmax=10`, `Ymin=-2`, `Ymax=2`

3. The graph of $8x = y^4$ is given in Figure 143. Sketch the graphs of the relations in parts (a) and (b) that are formed by translating and/or reflecting $8x = y^4$.

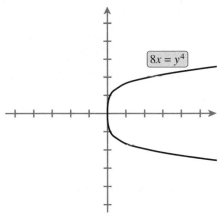

$8x = y^4$

**FIGURE 143**

   **a.** $8(x - 1) = (y + 3)^4$

   **b.** $8x = -y^4$

4. A portion of the graph of $x^3 - y^2 = 0$ is shown in Figure 144. Test the relation for symmetry and then use symmetry to complete the graph.

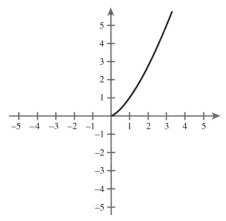

**FIGURE 144**

5. Find the equation of the circle with the points $(-3, 0)$ and $(5, 6)$ lying at opposite ends of a diameter.

**PROBLEMS 6–7** □ *Find the equation of the line with the given properties.*

6. Passes through the points $(-2, 3)$ and $(-2, 7)$

7. Passes through the point $(-1, 4)$ and is perpendicular to the line $4x + 3y = 12$

**PROBLEMS 8–9** □ *Sketch a graph of the given conic section. Identify the center, vertices, and foci. In the case of a hyperbola, identify the asymptotes and give their equations.*

8. $4x^2 - 9y^2 = 36$

9. $4x^2 + y^2 - 16x + 6y + 9 = 0$

**PROBLEMS 10–17** □ *Answer true or false.*

10. A relation can have at most one $y$-intercept.

11. A relation can have infinitely many $x$-intercepts.

12. If a relation is reflected first about the $x$-axis and then about the $y$-axis, the result is the same as when it is reflected first about the $y$-axis and then about the $x$-axis.

13. A relation can be symmetric with respect to both the $x$- and $y$-axis.

14. If a line $l$ has slope $m$ and if a line $l'$ with slope $m'$ is perpendicular to $l$, then $mm' = 1$.

15. If a line has equation $Ax + By + C = 0$, then the line must have slope $A$.

16. All conic sections with vertical or horizontal axes have equations of the form $Ax^2 + Cy^2 + Dx + Ey + F = 0$.

17. The foci of an ellipse always lie on the minor axis.

**PROBLEMS 18–23** □ *Give an example of each.*

18. A relation that has no $x$-intercept

19. A relation that is symmetric about the $x$-axis

20. A line that is perpendicular to $y = x$

21. A parabola that opens to the left

22. A conic section with two vertices on the $x$-axis

23. An application of a reflective property of a conic section

24. A company that manufactures bicycle frames has determined that the monthly profit from the sale of $x$ frames is given by $y = -0.005x^3 + 1.5x^2 + 50x - 15{,}000$, where $y$ is the profit in dollars.

   a. Approximate the number of frames that must be sold for the profit to be zero.

   b. Estimate the number of frames that must be sold to make the largest profit. What is the largest possible profit?

# FUNCTIONS AND THEIR GRAPHS

■ The Philippine volcano Mount Pinatubo erupted on
June 15, 1991, spewing millions of tons of sulfur dioxide
and ash into Earth's atmosphere. Rising, the hot gas
combined with water vapor to form a haze of sulfuric
acid droplets, blocking sunlight and cooling the Earth's
atmosphere. Such naturally occurring phenomena
complicate the task of measuring the effects of
greenhouse gasses (such as carbon dioxide) on global
warming. Atmospheric scientists believe that volcanic
eruptions (along with mankind's emissions of CFCs)
contribute to the depletion of stratospheric ozone. In this
chapter we will mathematically model both carbon
dioxide and atmospheric ozone levels, and discuss the
difficulty of modeling the Earth's atmosphere.

## SECTION 1

# FUNCTIONS

■ What will the volume of the sphere of influence of Martin Luther King's "I Have a Dream" speech be at the turn of the century?

■ How fast must a cyclist ride in order to generate enough power to replace a small power plant?

■ In what sense is $x/x \neq 1$?

■ Why can't the graph of a relation of the form $y = \square$, where $\square$ is an expression involving $x$, be a circle?

## DEFINITION OF FUNCTION

Correspondences between sets of objects are very common in mathematics as well as in the world around us. Table 1 gives some examples of such correspondences.

**TABLE 1**

*Correspondences between sets*

| To each human … | there corresponds … | a biological mother. |
| To each license plate number in a given state … | there corresponds … | a vehicle. |
| To each package weight … | there corresponds… | a first-class Postal Service delivery rate. |
| To each positive real number … | there corresponds … | the square of that number. |

Each of the examples in Table 1 has the form

To each element of a set $D$ …    there corresponds …    an element of a set $R$.

By using circles to represent the sets $D$ and $R$ and an arrow to represent the correspondence, we obtain the diagram in Figure 1. The more detailed diagram in Figure 2 shows how the elements in the set $D$ might correspond with the elements in $R$. Figure 2 suggests that to each element of $D$ there corresponds *exactly* one element of $R$. Correspondences with this property are called *functions*.

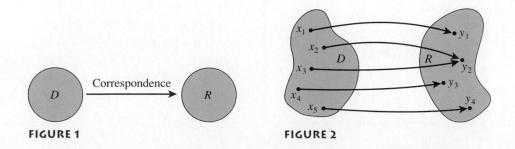

**FIGURE 1**

**FIGURE 2**

**Definition of a function**

> A **function** *f* from a set *D* to a set *R* is a correspondence or rule that assigns to each element *x* of *D* exactly one element *y* of *R*. The set *D* is called the **domain** of the function, and the elements *x* of *D* are the **input values** of *f*. The elements *y* of *R* that correspond to the input values are the **output values**. The set of all possible output values is called the **range** of the function *f*.

**EXAMPLE 1**    *Examples of functions*

Verify that each of the examples given in Table 1 are functions and identify the domain and range for each.

**SOLUTION**

a. Since each human has exactly one biological mother, this correspondence is a function. The domain is the set of all humans. The range is the set of all mothers.

b. You can imagine the resulting chaos if the same license number was assigned to two different vehicles. So it is by necessity that to each license number there corresponds exactly one vehicle, and thus this correspondence is a function. The domain is the set of all current license numbers, and the range is the set of all licensed vehicles.

c. For any given package weight, there is only one first class postal rate, and so this correspondence is a function. The domain is the set of all allowable package weights (up to 70 pounds for the Postal Service). The range is the set of all first-class package rates.

d. Each positive real number has exactly one square. The domain is the set of all real numbers. Since the square of a real number can never be negative, the range is the set of nonnegative real numbers.

**EXAMPLE 2**    *An example of a nonfunction*

Explain why the correspondence between the set *D* of all telephone numbers and the set *R* of all telephones is not a function.

**SOLUTION**    It is not uncommon for several telephones to be connected to the same line and therefore have the same number. Thus, there are elements of the set *D* that are paired with more than one element of *R*, and so the correspondence is not a function.

**RELATIONS AS FUNCTIONS**

Correspondences between sets of real numbers are often defined by an equation. For example, the correspondence given earlier between a positive real number and its square can be defined by the equation $y = x^2$. This equation specifies that to each input value *x* there corresponds the output value $y = x^2$, and so the equation defines a function. More generally, an equation of the form

$$y = \square$$

where $\square$ is an expression involving *x*, defines *y* as a function of *x*.

The domain of such a function is the set of all values that may be assigned to the input variable $x$, or the **independent variable** as it is sometimes called. The range is the set of all resulting values for the output variable $y$, otherwise known as the **dependent variable**.

Recall that a relation can be defined by an equation involving two or more variables. Thus, an equation that defines $y$ as a function of $x$ is an example of a relation in the variables $x$ and $y$. However, not all relations in $x$ and $y$ define $y$ as a function of $x$. For $y$ to be a function of $x$, there must correspond exactly one $y$ for each $x$.

**EXAMPLE 3**    *Testing a relation to see if it defines a function*

Determine whether or not the relation $y - x^3 = 0$ defines $y$ as a function of $x$.

**SOLUTION**    Solving for $y$, we have $y = x^3$. This shows that for each value of $x$ there is exactly one value for $y$. Thus, this relation does indeed define $y$ as a function of $x$.

**EXAMPLE 4**    *Testing a relation to see if it defines a function*

Determine whether or not the relation $y^2 - x = 0$ defines $y$ as a function of $x$.

**SOLUTION**    We first solve for $y$.

$$y^2 - x = 0$$
$$y^2 = x$$
$$y = \pm\sqrt{x}$$

This indicates that for each positive value of $x$ there are *two* values for $y$, namely $y = \sqrt{x}$ and $y = -\sqrt{x}$. Thus, this relation does *not* define $y$ as a function of $x$.

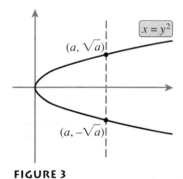

**FIGURE 3**

The graph of the relation $y^2 - x = 0$ from Example 4 shows us very clearly why it does not define $y$ as a function of $x$. By rewriting the relation as $x = y^2$ and plotting points, we obtain the graph shown in Figure 3. Notice that for any positive value of $a$, both of the points $(a, \sqrt{a})$ and $(a, -\sqrt{a})$ are on the graph: there are two distinct $y$-values corresponding to the same $x$-value. Graphically, this means that the vertical line $x = a$ intersects the graph twice. This is not allowed if the relation is to be a function, since a function pairs each $x$ with exactly one $y$. This suggests the following test.

**The vertical line test**

A relation defines $y$ as a function of $x$ if and only if no vertical line crosses the graph of the relation more than once.

**EXAMPLE 5**    *Using the vertical line test*

Use the vertical line test to identify which of the following relations define $y$ as a function of $x$.

**a.** $\dfrac{x^2}{9} + \dfrac{y^2}{4} = 1$

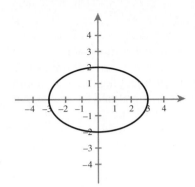

**b.** $y + 2 = (x + 1)^3$

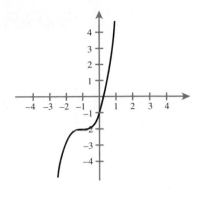

**c.** $y = \sqrt{4 - x^2}$

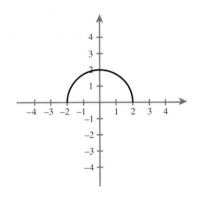

**d.** $-4x^2 + 9y^2 + 16x - 18y = 43$

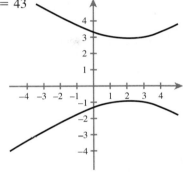

**SOLUTION**

**a.** $(x^2/9) + (y^2/4) = 1$ does not define $y$ as a function of $x$ since some vertical lines intersect the graph in two points.

**b.** $y + 2 = (x + 1)^3$ does define $y$ as a function of $x$ since no vertical line intersects the graph in more than one point.

**c.** $y = \sqrt{4 - x^2}$ does define $y$ as a function of $x$ since no vertical line intersects the graph in more than one point.

**d.** $-4x^2 + 9y^2 + 16x - 18y = 43$ does not define $y$ as a function of $x$ since vertical lines intersect the graph in two points.

---

## FUNCTION NOTATION

As we observed earlier, an equation of the form $y = \square$, where $\square$ is an expression involving $x$, defines a function. For this reason, it is convenient to use the **function notation** $f(x)$, read $f$ of $x$, to represent the output value $y$ for a given input value $x$. Thus, instead of writing $y = \sqrt{x}$, we write $f(x) = \sqrt{x}$. We view $f$ as the name of the function that associates the output $f(x)$ with the input $x$.

**EXAMPLE 6**   *Using function notation*

Write each of the following functions using function notation.
**a.** $h$ is the function that assigns to each nonzero real number its reciprocal.
**b.** $g$ is the function that assigns to each real number $t$ the number $t^3 - t$.
**c.** $P$ is the function that assigns to each real number one less than its square.

**SOLUTION**

**a.** The reciprocal of a number $x$ is the number $1/x$. Thus, $h(x) = 1/x$.

**b.** $g(t) = t^3 - t$.

**c.** One less than the square of a real number $x$ is the number $x^2 - 1$. Thus,
$P(x) = x^2 - 1$.

---

**EXAMPLE 7**   *Constructing a function from a formula*

Express the area of a circle as a function of its circumference.

**SOLUTION**   The area of a circle in terms of its radius is $A = \pi r^2$. The circumference of a circle in terms of its radius is $C = 2\pi r$. Solving the latter equation for $r$, we obtain $r = C/(2\pi)$. Substituting for $r$ in the area formula gives

$$A = \pi r^2 = \pi \left( \frac{C}{2\pi} \right)^2 = \pi \frac{C^2}{4\pi^2} = \frac{C^2}{4\pi}$$

Thus, $A(C) = \dfrac{C^2}{4\pi}$.

---

Example 7, in which we constructed a function named $A$ with $C$ as the independent variable, illustrates that not all functions are named $f$ and that the independent variable need not be $x$. In fact, the independent variable simply serves as a place holder and thus could be replaced by any symbol. For example, the function $f(x) = x^2 - 3x$ is exactly the same as the function $f(\square) = (\square)^2 - 3(\square)$: with either definition the output is obtained by subtracting three times the input from the square of the input. Thus, in order to find the value of the function $f(x) = x^2 - 3x$ when $x = -1$, we simply replace all occurrences of $x$ with $-1$ to obtain

$$f(-1) = (-1)^2 - 3(-1) = 1 + 3 = 4$$

**WARNING**   The function notation $f(-1)$ represents the output value of the function $f$ when $x = -1$. It *does not* mean we are to multiply $f$ by $-1$.

**EXAMPLE 8**   *Evaluating functions*

Let $g(t) = t^3 - t$. Find the following function values.
**a.** $g(-2)$     **b.** $g(\overset{\circ}{\wedge})$     **c.** $g(x + 1)$

**SOLUTION**

a. $g(-2) = (-2)^3 - (-2) = -8 + 2 = -6$

b. $g(\text{☺}) = (\text{☺})^3 - \text{☺}$

c. $g(x + 1) = (x + 1)^3 - (x + 1)$
$$= (x^3 + 3x^2 + 3x + 1) - (x + 1)$$
$$= x^3 + 3x^2 + 3x + 1 - x - 1$$
$$= x^3 + 3x^2 + 2x$$

Unless otherwise specified, we will assume that the domain and range of a function are sets of real numbers. Furthermore, if no restrictions are specifically placed on the input values of a function, the domain is assumed to be the set of all valid real-number input values. The range is then the set of all resulting output values.

**EXAMPLE 9** *Finding domain*

Find the domain of

$$g(t) = \frac{t}{t^2 - 1}$$

**SOLUTION** All input values of $t$ are valid except those that lead to a zero in the denominator. To determine where this occurs, we set $t^2 - 1 = 0$ and solve for $t$.

$$t^2 - 1 = 0$$
$$(t + 1)(t - 1) = 0$$
$$t = -1, \ t = 1$$

So the domain of $g$ is the set of all real numbers $t$ except $-1$ and $1$. This set can be written in set-builder notation as $\{t \mid t \neq -1, t \neq 1\}$. In interval notation, we write $(-\infty, -1) \cup (-1, 1) \cup (1, \infty)$.

**EXAMPLE 10** *Finding domain*

Find the domain of $h(x) = \sqrt{2x - 3}$.

**SOLUTION** Since we are only considering values of $x$ for which $h(x)$ is real, the valid input values of $x$ are those for which $2x - 3 \geq 0$. Solving this linear inequality for $x$ yields $x \geq \frac{3}{2}$. Thus, the domain of $h$ is $\left[\frac{3}{2}, \infty\right)$.

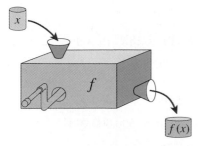

The process of evaluating a function can be illustrated using a "function machine." When an input value $x$ is fed into the machine, the machine operates on the number and produces an output value $f(x)$. If the function $f$ is defined by a formula, then the function "machinery" is completely revealed. If a formula is not known, then the function can seem like a "black box." In the following example, sample input and output values are given, and we wish to open up the black box to reveal the function machinery that produced them.

EXAMPLE 11    *Constructing a function from a table*

An overnight package delivery service charges the following rates for selected weights of packages.

| Weight (lb) | 2 | 4 | 8 | 14 |
|---|---|---|---|---|
| Rate ($) | 9.00 | 12.50 | 19.50 | 30.00 |

Plot the data and construct a function $R(x)$ giving rate as a function of weight that agrees with this data, then use this function to estimate the rate for a package weighing 12 pounds.

SOLUTION    A plot of the data, using weight as the $x$-coordinate and the rate as the $y$-coordinate, is shown in Figure 4. We wish to define a function using a formula that is consistent with these data. Since the points appear to be on a straight line, we find the equation of the line connecting the points (2, 9) and (14, 30).

$$\text{Slope} = \frac{y_2 - y_1}{x_2 - x_1} = \frac{30 - 9}{14 - 2} = \frac{21}{12} = \frac{7}{4}$$

From the point-slope form of the equation of a line, we obtain the equation

$$y - 9 = \frac{7}{4}(x - 2) \quad \text{or} \quad y = \frac{7}{4}x + \frac{11}{2}$$

It is easily verified that the remaining data points satisfy this equation also. Thus, the desired rate function is $R(x) = \frac{7}{4}x + \frac{11}{2}$ where $x$ is the weight and $R(x)$ is the corresponding rate (in dollars). For a package weighing 12 pounds, we have $R(12) = \frac{7}{4}(12) + \frac{11}{2} = 26.5$, or $26.50.

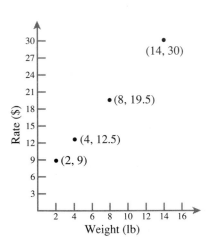

FIGURE 4

EXAMPLE 12    *Computing the power output of a cyclist*

For a 170-pound cyclist on a 30-pound bicycle pedaling at $v$ miles per hour, the power output in watts is given by $P(v) = 0.0178678v^3 + 2.01168v$.
a. Find the power output if the cyclist is traveling at 20 mph.
b. A small power plant will produce around 500,000,000 watts of power. Find the speed required for a single cyclist to generate enough power to replace a small power plant. Answer to the nearest mile per hour.

SOLUTION
a. Evaluating $P(20)$, we have

$$P(20) = 0.0178678(20)^3 + 2.01168(20) \approx 183.176 \text{ watts}$$

b. We set $P(v) = 500,000,000$ and solve for $v$.

$$P(v) = 500,000,000$$
$$0.0178678v^3 + 2.01168v = 500,000,000$$
$$0.0178678v^3 + 2.01168v - 500,000,000 = 0$$

The human-powered Gossamer Albatross in flight.

Next we plot the graph of $y = 0.0178678v^3 + 2.01168v - 500{,}000{,}000$ with a graphics calculator, and search for $x$-intercepts. After some experimentation, we settle on the view shown in Figure 5. Next we zoom-in on the $x$-intercept until we can estimate it to the nearest mile per hour, as shown in Figure 6. Thus, a cyclist capable of traveling at 3036 miles per hour would generate enough power to replace a small power plant.

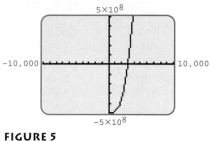

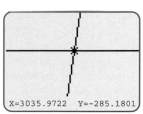

**FIGURE 5**
**FIGURE 6**

## EXERCISES 1

**EXERCISES 1–4** □ *Determine whether the most natural correspondence from the set D to the set R defines a function.*

1. $D$ is the set of fingerprints and $R$ is the set of people.

2. $D$ is the set of chemical elements and $R$ is the set of atomic weights.

3. $D$ is the set of last names in the phone book and $R$ is the set of phone numbers.

4. $D$ is the set of students completing college algebra and $R$ is the set of final grades.

**EXERCISES 5–10** □ *Determine whether the given relation defines y as a function of x.*

5. $3x + 4y = 12$
6. $y^2 - x^2 = 0$

7. $x^2 + y^2 = 1$
8. $x^2 + x - y = 0$

9. $0 = x - y^3$
10. $x = \sqrt{y}$

**EXERCISES 11–16** □ *Use the vertical line test to determine whether the given relation defines y as a function of x.*

11. $y = -x^2 + 1$

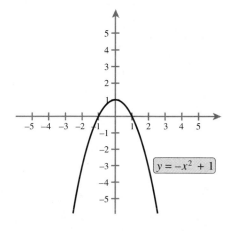

12. $\dfrac{x^2}{16} + \dfrac{y^2}{9} = 1$

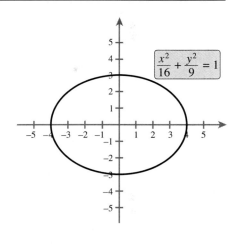

13. $x = y^3 - 4y$

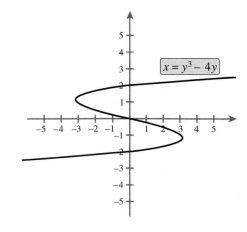

**14.** $x = y^3$

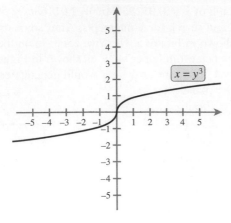

**15.** $xy = 4$

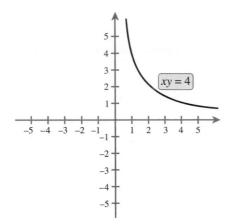

**16.** $|x| - |y| = 0$

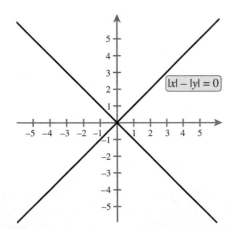

**EXERCISES 17–24** ☐ *Evaluate the given function as indicated.*

**17.** $f(x) = x^2 + 3x + 2$

   **a.** $f(2)$

   **b.** $f(-3)$

   **c.** $f(0)$

**18.** $g(t) = t^3 - 4t$

   **a.** $g(1)$

   **b.** $g(-2)$

   **c.** $g(3)$

**19.** $f(x) = \dfrac{3x + 2}{4x - 5}$

   **a.** $f(4)$

   **b.** $f(5)$

   **c.** $f(-t)$

**20.** $h(y) = \dfrac{y^2 - 3}{y + 7}$

   **a.** $h(2)$

   **b.** $h(-6)$

   **c.** $h(x)$

**21.** $x(t) = 3t^2$

   **a.** $x(t^2)$

   **b.** $x(t - 1)$

   **c.** $x\left(\dfrac{5}{t}\right)$

**22.** $f(x) = \dfrac{2x + 2}{2x - 3}$

   **a.** $f(t)$

   **b.** $f(x - 1)$

   **c.** $f\left(\dfrac{x}{2}\right)$

**23.** $g(x) = (x + 1)^2$

   **a.** $g(x - 1)$

   **b.** $g(x^2)$

   **c.** $g\left(\dfrac{1 - a}{a}\right)$

**24.** $f(x) = x^2, \ g(y) = 2y + 1$

   **a.** $f(g(t))$

   **b.** $g(f(t))$

   **c.** $f(g(c^2))$

**EXERCISES 25–28** ☐ *Express the given function using function notation.*

**25.** $h$ assigns to every real number the absolute value of the number.

**26.** $g$ assigns to every real number twice the cube of the number.

**27.** $f$ assigns to every real number in its domain the reciprocal of 3 more than the number.

**28.** $p$ assigns to each real number in its domain the reciprocal of the square root of 1 more than the number.

**EXERCISES 29–40** ☐ *Find the domain of the function.*

**29.** $f(x) = x^2$

**30.** $g(x) = \dfrac{4}{x}$

**31.** $h(x) = \dfrac{3}{x - 4}$

**32.** $r(s) = |s|$

**33.** $r(t) = \dfrac{5t^2 + 3t + 1}{t^2 - 9}$

**34.** $p(x) = \dfrac{x + 1}{x^2 - 5x + 6}$

**35.** $f(x) = \sqrt{x + 1}$

**36.** $g(t) = \sqrt{t - 3}$

**37.** $f(x) = \dfrac{\sqrt{x + 2}}{x^2 - 4x + 3}$

**38.** $g(y) = \dfrac{2}{\sqrt{2y - 6}}$

**39.** $f(x) = \dfrac{2}{|x| - 3}$

**40.** $g(x) = \dfrac{x}{x}$

**EXERCISES 41–44** ☐ *Find the range of the given function.*

**41.** $f(x) = x^2$

**42.** $h(x) = \sqrt{x}$

**43.** $g(x) = 3x$

**44.** $f(x) = 3x^2 + 1$

**EXERCISES 45–48** □ *Find a formula f(x) for a function that satisfies the given correspondences between input and output values. There may be more than one correct answer.*

45. $f(0) = -1$, $f(1) = 1$, $f(2) = 3$

46. $f(0) = 2$, $f(1) = -1$, $f(2) = -4$

47. $f(1) = 1$, $f(2) = \frac{1}{4}$, $f(3) = \frac{1}{9}$

48. $f(1) = 1$, $f(4) = 2$, $f(9) = 3$

**EXERCISES 49–52** □ *Find a function f for which the given set is the domain. There may be more than one correct answer.*

49. $(-\infty, 7]$

50. All real numbers except $-4$ and $1$

51. All real numbers except $0$ and $3$

52. $[-2, \infty)$

53. Suppose that the range of $g(x)$ is the set of all real numbers less than 4, and the domain of $g(x)$ is the set of all real numbers. Find the domain of

$$\frac{1}{g(x) - 5}$$

54. Define the function $f(x)$ by

$$f(x) = \frac{1}{q(x)}$$

where $q$ is another function. There is one number that is guaranteed *not* to be an element of the range of $f$, no matter how $q$ is defined. What number is it?

55. a. Use the quadratic formula to solve the equation $y = x^2 + 4x + 6$ for $x$.

b. For what values of $y$ does the equation of part (a) have a real-valued solution?

c. What is the range of the function $g$ defined by $g(x) = x^2 + 4x + 6$?

56. a. Find the domain of

$$f(x) = \frac{|x|}{x}$$

b. Evaluate $f(2)$, $f(-3.76)$, and $f(10^{2345})$.

c. Find the range of $f$.

■ *Applications*

57. *Volume of a Sphere* The volume of a sphere with radius $r$ is given by $V = \frac{4}{3}\pi r^3$. Express the volume of a sphere as a function of its diameter.

58. *Area of a Circle* The area of a circle with radius $r$ is given by $A = \pi r^2$. Express the area of a circle as a function of its diameter.

59. *Area of a Square* A square is inscribed in a circle. Express the area of the square as a function of the radius of the circle.

60. *Volume of a Cube* A cube is inscribed in a sphere. Express the volume of the cube as a function of the radius of the sphere.

61. *Area of a Triangle* A point $P(x, y)$ lies on the line $y = \frac{1}{2}x$, as shown. Express the area of the shaded triangle in terms of $x$.

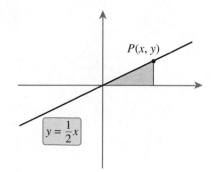

62. *Area of a Rectangle* A point $P(x, y)$ lies on the line $y = -\frac{3}{4}x + 3$, as shown. Express the area of the shaded rectangle in terms of $x$.

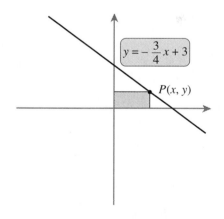

63. *Area of a Triangle* A line passing through the point $(2, 1)$ has intercepts $(a, 0)$ and $(0, b)$, which form the vertices of a right triangle as shown in the graph on the following page. Express the area of the triangle as a function of $a$. What is the domain of this function?

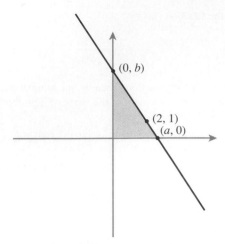

**64. Area of a Triangle** Two perpendicular lines intersect at the point (2, 3). Their *x*-intercepts form the vertices of a right triangle, as shown. Express the area of the triangle as a function of the length of its hypotenuse. What is the domain of this function?

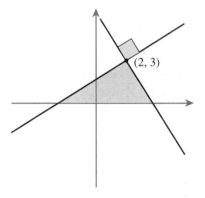

**65. Volume of a Box** A box with a square base and no top is formed by cutting squares out of the corners of a piece of cardboard and then folding up the sides. If the cardboard measures 16″ × 16″ and the length of the edge of each cut-out square is denoted by *x*, express the volume of the resulting box as a function of *x*. What is the domain of this function? (*Hint:* Include in the domain only those values that make sense in the context of this problem.)

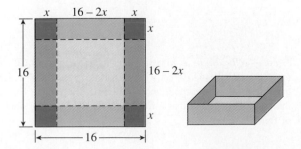

**66. Area of a Pasture** A rancher wishes to enclose a rectangular pasture at the base of a cliff: 200 linear feet of fencing are available, and no fencing is needed along the cliff side of the pasture. Express the area of the pasture as a function of *x*, the dimension perpendicular to the cliff face.

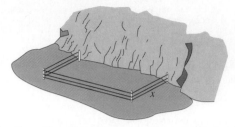

**67. Saving for a Vacation** The Reynolds estimate that in 4 years they will need $2000 for a family vacation to Disney World. They intend to shop around for the best interest rate they can find and then deposit the necessary principal so that it will grow to $2000 after 4 years. Use the formula for interest compounded annually to express the principal as a function of the interest rate. What is the domain of this function? Is it likely they could find an account that pays an interest rate that is not in the domain? Explain.

**68. Height of a Can** A research and development team for a company that produces canned foods is in the process of designing a can that must have a volume of 400 square centimeters. Use the formula $V = \pi r^2 h$ for the volume of a cylinder with radius *r* and height *h* to express the height of the can as a function of the radius. What is the domain of this function?

**69. Sphere of Influence** Electromagnetic waves, like those in radio and television transmissions for example, travel at the speed of light, often in all directions. Although our radios pick up a signal almost instantaneously, the signal continues to travel through space at the speed of light. Thus, for example, Dr. Martin Luther King's famous 1963 "I have a Dream" speech is even now reaching new portions of the galaxy. The radio waves from the broadcast began traveling through space at the speed of light in all directions from the point of origin. Since light travels at a rate of approximately $5.9 \times 10^{12}$ miles per year, the distance the waves would have traveled after *t* years would be given by $r = (5.9 \times 10^{12})t$. Moreover, when

Dr. Martin Luther King delivers his "I have a dream" speech.

the waves are a distance $r$ from the point of origin, the volume of space (ignoring the fact that some of the signal will be blocked by the Earth itself) through which they would have traveled would be given by $V = \frac{4}{3}\pi r^3$. Express the volume $V$ as a function of $t$, and compute the volume of space through which the speech will have traveled in the year 2000.

70. *Radius of an Oil Spill* An oil tanker develops a leak and begins spilling oil at a rate of 10,000,000 cubic centimeters per minute. The volume of oil that has spilled after $t$ minutes is thus given by $V = 10,000,000t$ cubic centimeters. Assuming that the resulting oil spill has a circular shape with radius $r$ and

In spite of containment efforts, oil spills continue to cause immeasurable damage to the environment.

a constant thickness $h$, the volume can also be expressed as $V = \pi r^2 h$. If $h = 1$ centimeter, express the radius as a function of time and compute the radius of the spill after 2 hours.

71. *Subsidized Housing* An apartment complex receives a government subsidy and in return must base its monthly rent on family size and monthly income. Selected income and rent amounts are given below for a family of four. Plot the points and construct a function $R(x)$ that gives the monthly rent for a family of four with an income of $x$ dollars per month.

| Income ($) | 800 | 950 | 1150 | 1400 |
|---|---|---|---|---|
| Rent ($) | 100 | 160 | 240 | 340 |

72. *Cab Fare* A cab charges the following fares for selected distances. Plot the points and construct a function $F(x)$ that gives as output values the fare that would be charged for any given distance $x$.

| Distance (mi) | 1 | 3 | 7 | 13 |
|---|---|---|---|---|
| Fare ($) | 1.50 | 2.50 | 4.50 | 7.50 |

## ◼ *Projects for Enrichment*

73. *The Formal Definition of a Function* In this section we defined a function as a *rule* that assigns to each element $x$ in a set $D$, exactly one element $y$ in a set $R$. Functions can also be defined by simply considering sets of ordered pairs. We will say that **a function $F$** is a set of ordered pairs $(x, y)$ such that if $(x_1, y)$ and $(x_2, y)$ are elements of $F$, then $x_1 = x_2$. In other words, there can be no ordered pairs in the set $F$ that have the same $x$-coordinate. For example, the set

$$F = \{(-2, 4), (-1, 1), (0, 0), (1, 1), (2, 4)\}$$

is a function (it's okay to have points with the same $y$-coordinates), but

$$G = \{(4, -2), (1, -1), (0, 0), (1, 1), (4, 2)\}$$

is not a function (there cannot be points with the same $x$-coordinate).

a. Determine which of the following sets are functions. Justify your answer using the definition given above.

   i. $C = \{(-27, -3), (-8, -2), (-1, 1), (0, 0), (1, 1), (8, 2), (27, 3)\}$

   ii. $E = \{(4, 0), (0, 5), (-4, 0), (0, -5)\}$

   iii. $H = \{(x, y)|x^2 + y^2 = 1\}$

   iv. $S = \{(t, s)|s = -16t^2\}$

The **domain** of a function $F$ is the set of $x$-coordinates of points in $F$. The **range** is the set of $y$-coordinates. For example, the domain of $F = \{(-2, 4), (-1, 1), (0, 0), (1, 1), (2, 4)\}$ is $\{-2, -1, 0, 1, 2\}$, and the range is $\{0, 1, 4\}$.

b. Find the domain and range of the following functions.

   i. $F = \{(-3, -2), (-2, 0), (-1, -2), (0, -4), (1, -6)\}$

   ii. $P = \{(0, 0), (1, 1), (2, 0), (3, 1), (4, 0), (5, 1), (6, 0), \ldots\}$

   iii. $H = \{(x, y)|y = x^2 + 1\}$

   iv. $G = \{(x, y)|y = \sqrt{4 - x}\}$

The graph of a function $F$ is simply the plot of the points in the set $F$.

c. Plot the graphs of the functions given in part (b).

Any function that can be defined by a rule can also be defined as a set of points. For example, the function $g(x) = \sqrt{x}$ can be defined as the set of points

$$G = \{(x, y)|y = \sqrt{x}\}$$

and the function

$$F = \{(-2, 4), (-1, 1), (0, 0), (1, 1), (2, 4)\}$$

can be defined as a rule by $f(x) = x^2$, $x = -2, -1, \ldots, 2$.

d. Define the following functions using a rule of the form $f(x) = \square$, where $\square$ is an expression involving $x$. You may need to restrict the domain.

   i. $\{(-2, -4), (-1, -2), (0, 0), (1, 2), (2, 4)\}$

   ii. $\{(x, y)|y = 1/x\}$

e. Define the following functions as a set of points.

    i. $f(x) = x^3$; $x = 0, 1, \ldots, 5$

    ii. $h(x) = \sqrt{1 - x^2}$

**74.** *Functions of Several Variables*  Many of the correspondences between real-world quantities require the use of more than one variable. Consider, for example, the correspondence that assigns to each rectangle of length $l$ and width $w$ an area $A$ equal to $lw$. The equation $A = lw$ defines the area as a function of two variables that we can write as $A(l, w) = lw$. In this case, each pair of positive values $(l, w)$ is associated with *exactly one* area $A$. For example, a rectangle with length $l = 8$ and width $w = 5$ yields an area of $A(8, 5) = 8 \cdot 5 = 40$, and no other area is possible.

In general, we define **a function $f$ of two variables** to be a rule or correspondence that assigns to a pair of numbers $(x, y)$ exactly one number $f(x, y)$. The **domain** of a function of two variables is the set of all pairs $(x, y)$ that can be used as input values. The **range** is the set of all corresponding output values. The domain of the function $f(x, y) = x^2 + y^2$ is the set of all pairs $(x, y)$ such that $x$ and $y$ are any pair of real numbers. The range is the set of all positive real numbers.

a. Evaluate the following functions for the indicated values.

    i. $d(r, t) = rt$; $d(55, 4)$; $d(20, \frac{1}{4})$

    ii. $f(x, y) = x^2 + 2xy + y^2$; $f(2, 3)$; $f(-1, 5)$

    iii. $h(u, v) = 1/(u - v)$; $h(10, 8)$; $h(0.3, 0.4)$

    iv. $g(x, y) = \sqrt{1 - x^2 - y^2}$; $g(0, 1)$; $g(\frac{1}{2}, \frac{1}{4})$

b. Describe the domain of each of the functions in part (a).

Functions of three or more variables are also possible. Consider the correspondence defined by the equation $B = P(1 + r)^t$, where $B$ is the balance in an account after a principal of $P$ dollars is deposited for $t$ years at an annual rate of interest $r$. So $B$ is a function of the three variables $P$, $t$, and $r$, and we write $B(P, r, t) = P(1 + r)^t$. Notice here that each triple set of values $(P, r, t)$ leads to *exactly one* balance $B$. For example, if $P = 1000$, $t = 4$, and $r = 0.05$, then

$$B(1000, 4, 0.05) = 1000(1 + 0.05)^4 \approx 1215.51$$

and this is the only balance possible for these values of $P$, $t$, and $r$.

c. Give a precise definition for a function of three variables and also for its domain and range.

d. Evaluate the following functions for the indicated values.

    i. $V(l, w, h) = lwh$; $V(4, 6, 3)$; $V(5, \frac{3}{2}, 8)$

    ii. $g(x, y, z) = \sqrt{x^2 + y^2 + z^2}$; $g(0, -4, 3)$; $g(-2, 1, 5)$

    iii. $f(u, v, w) = \dfrac{uv + vw + uw}{uvw}$; $f(1, 2, 3)$; $f(-4, 3, \frac{1}{3})$

    iv. $x(a, b, c) = \dfrac{-b + \sqrt{b^2 - 4ac}}{2a}$; $x(1, -7, -10)$; $x(2, 0, -9)$

e. Describe the domain of each of the functions in part (d).

---

### ◼️ *Questions for Discussion or Essay*

**75.** A graphics calculator is capable of graphing equations of the form $y =$ an expression involving $x$. As we have seen in this section, such equations define $y$ as a function of $x$, and their graphs pass the vertical line test. Explain how a graphics calculator can be used as an aid in graphing an equation such as $x^2 + y^2 = 1$, even though its graph does not pass the vertical line test and so does not define $y$ as a function of $x$.

**76.** Rational expressions are simplified by dividing out equal factors of the numerator and denominator. Thus, for example,

$$\frac{x^2 - 1}{x + 1} = \frac{(x + 1)(x - 1)}{x + 1} = x - 1$$

However, if we define

$$f(x) = \frac{x^2 - 1}{x + 1} \quad \text{and} \quad g(x) = x - 1$$

these two functions are not the same. Explain why this is so. In what sense is $\dfrac{x}{x} \neq 1$?

**77.** In our discussion on relations and functions, we were careful to say that an equation "defines $y$ as a function of $x$" rather than just saying that the equation was a function. What does it mean to define $x$ as a function of $y$? Give an example of an equation that defines $x$ as a function of $y$ but that does not define $y$ as a function of $x$. Now explain why it is not meaningful to say that an equation either is or is not a function.

**78.** In Example 12 we saw that a cyclist would have to travel at a rate of 3036 miles per hour to produce enough power to run a small power plant. If we agree that 20 miles per hour is a "normal" speed, the cyclist would have to travel 152 times faster than normal. If instead of one cyclist traveling 152 times faster than normal we had many cyclists traveling at normal speed, how many cyclists does your intuition tell you it should take to run the plant? Now compute how many it would take by using the facts from Example 12 that a cyclist traveling at 20 miles per hour can produce approximately 183 watts of power and a small power plant produces around 500,000,000 watts. What seems surprising about the answer? Can you explain why there is such a disparity between your intuitive answer and the actual number?

**SECTION 2**

## GRAPHS OF FUNCTIONS

- Why does a soft drink can have the shape it does?
- What are the optimum dimensions of a box to be sent by mail?
- Why are functions satisfying $f(-x) = -f(x)$ said to be *odd*?
- How can the graph of a function be used to ensure that a bicycle racer receives a smooth handoff from a support vehicle?

### GRAPHS OF FUNCTIONS

The **graph of a function** $f$ is the set of points $(x, y)$ such that $y = f(x)$. In other words, the graph of $f$ is simply the graph of the relation $y = f(x)$. Thus, the techniques developed for graphing relations can also be applied to graphing functions.

The graph of a function depicts the correspondence between input values in the domain and output values in the range. Indeed, if $(x, y)$ is a point on the graph of a function $f$, then $y$ is the value of the function at $x$, as shown in Figure 7.

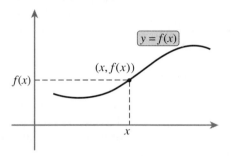

**FIGURE 7**

**EXAMPLE 1**    *Graphing a function*

Sketch the graph of the function $f(x) = \frac{1}{2}x - 3$.

**SOLUTION**    Replacing $f(x)$ with $y$, we obtain $y = \frac{1}{2}x - 3$, a straight line with $y$-intercept $-3$ and slope $\frac{1}{2}$. The graph is shown in Figure 8.

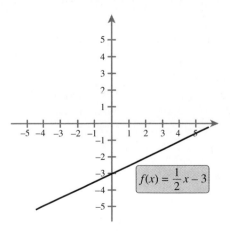

**FIGURE 8**

Graphs of functions are easily plotted using a graphics calculator. In fact, a graphics calculator is really just a superhuman point-plotter. It chooses regularly spaced input values in the interval determined by Xmin and Xmax and then computes the corresponding output values according to the given formula. As long as the output values are in the interval specified by Ymin and Ymax, the input-output pairs are plotted as points on the display.

**EXAMPLE 2**    *Determining domain and range using a graphics calculator*

Use a graphics calculator to plot the graph of the function $f(x) = \sqrt{x^2 - 16}$, and use the graph to determine the domain and range of $f$.

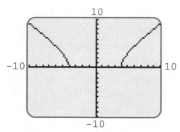

**FIGURE 9**

**SOLUTION**    We enter the expression $y = \sqrt{x^2 - 16}$ and plot it with a graphics calculator. The display should look like Figure 9. The domain consists of the set of all $x$-coordinates of points on the graph. Since no portion of the graph lies between $x = -4$ and $x = 4$, we conclude that the domain is $(-\infty, -4] \cup [4, \infty)$. The numbers $-4$ and $4$ are included because both values can be substituted into the function. The range consists of the set of all $y$-coordinates of points on the graph. Since the graph lies on or above the $x$-axis, all $y$-coordinates are positive or zero. Thus, the range is $[0, \infty)$.

## ZEROS OF A FUNCTION

Recall that the $x$-intercepts of the graph of a relation are those points where the graph crosses the $x$-axis. The $x$-intercepts of the graph of $y = f(x)$ are called the **zeros** of $f$. Thus, a zero of a function $f$ is any number $c$ for which $f(c) = 0$. In other words, a zero of a function is an input value that yields zero as the output value.

**EXAMPLE 3**    *Finding zeros algebraically*

Find the zeros of the function $f(x) = x^2 - 4$ and illustrate them graphically.

**SOLUTION**    To find the zeros, we set $f(x) = 0$ and solve for $x$.

$$f(x) = 0$$
$$x^2 - 4 = 0$$
$$(x + 2)(x - 2) = 0$$
$$x = -2, \ x = 2$$

The graph of $f$ is that of $y = x^2$ translated down 4 units. Note that it crosses the $x$-axis at the two zeros $x = -2$ and $x = 2$.

$f(x) = x^2 - 4$

**EXAMPLE 4**    *Finding zeros graphically*

Use a graphics calculator to estimate the three zeros of $f(x) = x^3 + 3x^2 - 3$ to the nearest hundredth.

**SOLUTION**   We enter and graph the expression $y = x^3 + 3x^2 - 3$, obtaining the graph shown in Figure 10. It is evident from the graph that there are three zeros ($x$-intercepts). Next we estimate each zero by zooming in on the zero until the trace feature can be used to determine its value to the nearest hundredth. Figures 11 and 12 show the graph after zooming in on the leftmost zero several times and enabling the trace feature. Note that in Figure 11 the cursor is just to the left of the zero, whereas in Figure 12, the cursor is just to the right of the zero. Thus, the zero is between $-2.533684$ and $-2.531579$, and hence we estimate the zero to be $-2.53$. We then repeat the process with the other two zeros, obtaining estimates of $x = -1.35$ and $x = 0.88$.

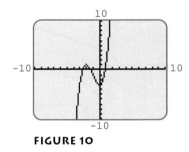

**FIGURE 10**

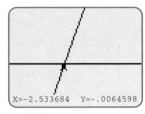

**FIGURE 11**

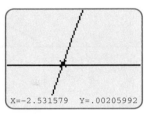

**FIGURE 12**

## TRANSLATIONS

Recall from our work with translations of relations that replacing $x$ with $x - h$ in a relation results in a horizontal shift of the graph by $h$ units (to the right if $h > 0$, to the left if $h < 0$). Thus, the graph of $y = f(x - h)$ will be the graph of $y = f(x)$, translated $h$ units horizontally. Similarly, if we replace $y$ by $y - k$ in a relation, then the graph will be shifted $k$ units vertically (up if $k > 0$, down if $k < 0$). Consequently, the graph of $y - k = f(x)$, or equivalently $y = f(x) + k$, will be a vertical shift by $k$ units of the graph of $y = f(x)$. These observations are summarized below.

**Translations of functions**

> The graph of $y = f(x - h) + k$ is a translation $h$ units horizontally and $k$ units vertically of the graph of $y = f(x)$. If $h$ is positive, then the translation is to the right; if $h$ is negative, then the translation is to the left. If $k$ is positive, then the translation is upward; if $k$ is negative, then the translation is downward.

**EXAMPLE 5**   *Graphing translated functions*

Use the graph of $f(x) = \sqrt{x}$ to sketch the graphs of the following functions.

**a.** $g(x) = \sqrt{x} + 3$          **b.** $g(x) = \sqrt{x + 2}$          **c.** $g(x) = \sqrt{x - 1} - 2$

**SOLUTION**

a. Since $g(x) = f(x - 0) + 3$, we have $h = 0$ and $k = 3$. Thus, the graph of $g$ is obtained by translating the graph of $f$ three units upward, as shown in Figure 13.

b. Since $g(x) = f(x + 2) + 0$, we have $h = -2$ and $k = 0$. Thus, the graph of $g$ is obtained by translating the graph of $f$ two units to the left, as shown in Figure 14.

c. Here $g(x) = f(x - 1) - 2$, so that $h = 1$ and $k = -2$. Thus, the graph of $g$ is obtained by translating the graph of $f$ one unit to the right and two units down, as shown in Figure 15.

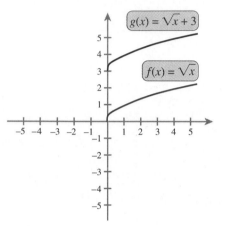

**FIGURE 13**

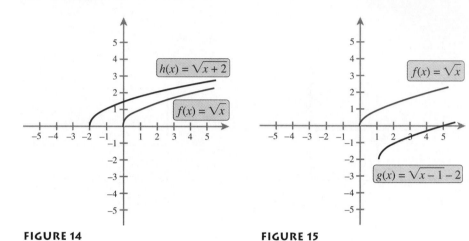

**FIGURE 14**                    **FIGURE 15**

---

**EXAMPLE 6**     *Constructing a translated function*

The graph of $f(x) = x^3$ is shown in Figure 16 together with the graph of a translated function $h(x)$. Write the equation for $h$.

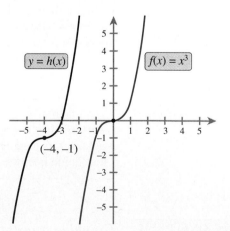

**FIGURE 16**

**SOLUTION**     The graph of $h$ is that of $f$ shifted 4 units to the left and 1 unit down. Thus, $h(x) = f(x + 4) - 1 = (x + 4)^3 - 1$.

## REFLECTIONS

Recall that if we replace $x$ with $-x$ in a relation, then the graph will be reflected about the $y$-axis. Thus, the graph of $y = f(-x)$ will be a reflection about the $y$-axis of the graph of $y = f(x)$. Similarly, since replacing $y$ with $-y$ results in a reflection about the $x$-axis, the graph of $-y = f(x)$, or equivalently $y = -f(x)$, will be a reflection of the graph of $y = f(x)$ about the $x$-axis. If we substitute $-x$ for $x$ *and* $-y$ for $y$, then we obtain a reflection about the origin. Thus, the graph of $-y = f(-x)$, or equivalently $y = -f(-x)$, will be a reflection about the origin of the graph of $y = f(x)$. These observations are summarized below.

**Reflections of functions**

| Function | Graph |
|---|---|
| $h(x) = f(-x)$ | reflection about the $y$-axis of the graph of $f$ |
| $h(x) = -f(x)$ | reflection about the $x$-axis of the graph of $f$ |
| $h(x) = -f(-x)$ | reflection about the origin of the graph of $f$ |

**EXAMPLE 7**   *Graphing reflected functions*

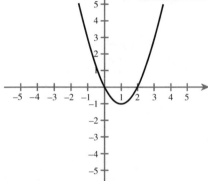

FIGURE 17

Use the graph of $f(x) = x^2 - 2x$ shown in Figure 17 to sketch the graphs of the following functions.
a.  $h(x) = (-x)^2 - 2(-x) = x^2 + 2x$
b.  $h(x) = -(x^2 - 2x) = -x^2 + 2x$

**SOLUTION**

a.  We note that $h(x) = f(-x)$, since $f(-x) = (-x)^2 - 2(-x) = x^2 + 2x$. Thus, $h$ is the reflection of $f$ about the $y$-axis, as shown in Figure 18.
b.  Here $h(x) = -f(x)$, since $-f(x) = -(x^2 - 2x) = -x^2 + 2x$. Thus, $h$ is the reflection of $f$ about the $x$-axis, as shown in Figure 19.

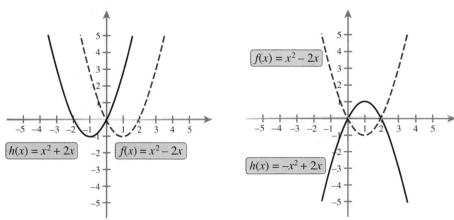

FIGURE 18                                        FIGURE 19

## SYMMETRY

We recall that if the graph of a relation remains the same after reflecting about the $y$-axis, then the graph is said to be symmetric with respect to the $y$-axis. To test for symmetry with respect to the $y$-axis, we simply replace $x$ with $-x$ and check to see whether or not the resulting equation is equivalent to the original equation. Thus, for the graph of a function $f$ to be symmetric with respect to the $y$-axis, the equations $y = f(x)$ and $y = f(-x)$ must be equivalent—that is, $f(-x) = f(x)$. Such functions are said to be **even**. Similarly, functions that are symmetric with respect to the origin can be shown to satisfy the equation $f(-x) = -f(x)$ and are said to be **odd**. We summarize these facts below.

### *Even and odd functions*

| Type of function | Graphical property | Algebraic property |
|---|---|---|
| Even | symmetry with respect to the $y$-axis | $f(-x) = f(x)$ for each $x$ in the domain of $f$ |
| Odd | symmetry with respect to the origin | $f(-x) = -f(x)$ for each $x$ in the domain of $f$ |

**EXAMPLE 8**    *Identifying even and odd functions*

Use graphical as well as algebraic tests to determine whether the following functions are even or odd.

**a.** $f(x) = |x|$        **b.** $f(x) = -x^2 + 6x$        **c.** $f(x) = x^3 - 3x$

**SOLUTION**

| *Function* | *Graph* | *Algebraic test* | *Conclusion* |
|---|---|---|---|
| $f(x) = \|x\|$ | | $f(-x) = \|-x\| = \|x\| = f(x)$ | Even |
| $f(x) = -x^2 + 6x$ | | $f(-x) = -(-x)^2 + 6(-x)$ $= -x^2 - 6x$ This is not equal to $f(x)$ or $-f(x)$. | Neither even nor odd |
| $f(x) = x^3 - 3x$ | | $f(-x) = (-x)^3 - 3(-x)$ $= -x^3 + 3x$ $= -f(x)$ | Odd |

> **RULE OF THUMB**    Notice that the function $f(x) = x^3 - 3x$ from part (c) of Example 8 includes only odd powers of $x$ and was determined to be an odd function. This was not a coincidence. It can be shown (see Exercise 50) that a polynomial function is even if all the powers of $x$ are even and that it is odd if all the powers of $x$ are odd.

**EXAMPLE 9**    *Completing the graph of a function*

Complete the graph of the function $f$ with the given property.

a. $f$ is even                                              b. $f$ is odd

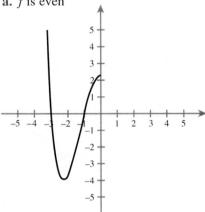
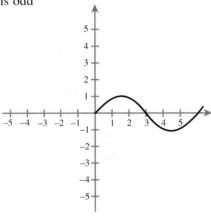

**SOLUTION**

a. Since an even function is symmetric with respect to the $y$-axis, we simply reflect the graph across the $y$-axis, as shown in Figure 20.

b. Since an odd function is symmetric with respect to the origin, we first reflect the graph across the $x$-axis to obtain the dashed portion shown in Figure 21, and then we reflect the *dashed* portion across the $y$-axis to obtain the complete graph of $f$ in Figure 22.

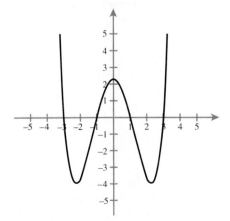

**FIGURE 20**

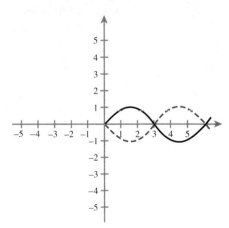

**FIGURE 21**

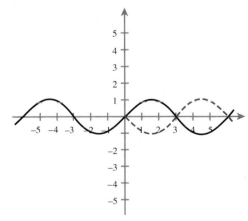

**FIGURE 22**

## INCREASING AND DECREASING FUNCTIONS

In addition to showing symmetry, graphs can be used to unveil many other important characteristics of functions. Consider, for example, the function $f(x) = x^3 - 3x^2$ graphed in Figure 23. Notice that as we move from left to right, the graph rises until we get to $x = 0$, falls in the interval from $x = 0$ to $x = 2$, and then rises to the right of $x = 2$. Thus, the function $f$ is *increasing* in the interval $(-\infty, 0)$, *decreasing* in the interval $(0, 2)$, and *increasing* in the interval $(2, \infty)$. More generally, we make the following definitions.

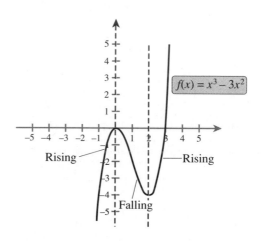

$$f(x) = x^3 - 3x^2$$

**FIGURE 23**

*Increasing and decreasing functions*

A function $f$ is said to be **increasing** or **decreasing** on an interval $I$ according to the following properties.

| Type of behavior | Graphical property | Algebraic property |
|---|---|---|
| Increasing | As we move from left to right in the interval $I$, the graph rises. | For any $x_1$ and $x_2$ in the interval $I$ with $x_1 < x_2$, we have $f(x_1) < f(x_2)$. |
| Decreasing | As we move from left to right in the interval $I$, the graph falls. | For any $x_1$ and $x_2$ in the interval $I$ with $x_1 < x_2$, we have $f(x_1) > f(x_2)$. |

The points where a function changes from increasing to decreasing, or decreasing to increasing, are called *turning points*. Other points on the graph of a function near a turning point are either below or above the turning point. For this reason, we say that a function has a *relative maximum* or *relative minimum* at a turning point.

**Relative extrema**

The value of $f$ at $x = c$ is called a **relative maximum** if it is the largest value of $f$ *near* $c$, as shown in Figure 24. The value of $f$ at $x = c$ is called a **relative minimum** if it is the smallest value of $f$ near $c$. The term **relative extremum** (plural "extrema") is used to refer to either a relative maximum or a relative minimum.

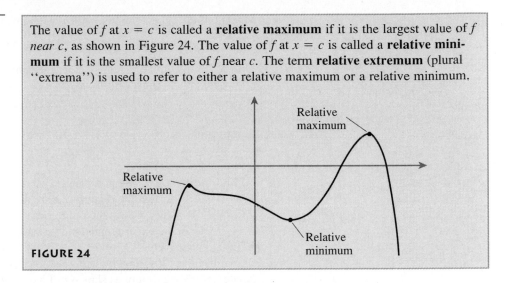

**FIGURE 24**

---

**EXAMPLE 10**   *Using a graphics calculator to find relative extrema*

Use a graphics calculator to approximate the locations of the relative extrema for the function $f(x) = \frac{1}{8}x^3 - x^2 + 2$. Also, indicate the intervals on which $f$ is increasing and the intervals on which it is decreasing.

**SOLUTION**   The graph of $y = \frac{1}{8}x^3 - x^2 + 2$, shown in Figure 25, has two turning points: a relative maximum at $x = 0$ and a relative minimum at a point in the fourth quadrant. Since $f(0) = 2$, there is a relative maximum of 2 at $x = 0$. Using the trace feature, as shown in Figure 26, we find that $f$ has a relative minimum of approximately $-7.48$ near $x = 5.37$. (If more accuracy is desired, we can zoom-in on the turning points.)

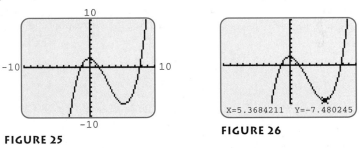

**FIGURE 25**

**FIGURE 26**

Since the graph of $f$ is rising for values of $x$ less than 0 and also for values of $x$ greater than 5.37, $f$ is increasing on $(-\infty, 0) \cup (5.37, \infty)$. Since the graph of $f$ is falling between $x = 0$ and $x = 5.37$, $f$ is decreasing on $(0, 5.37)$.

---

**EXAMPLE 11**   *Maximizing volume with a graphics calculator*

A rectangular package to be sent by the Postal Service can have a maximum combined length and girth (perimeter of the base) of 108 inches (see Figure 27). What are the dimensions of a box with a square base and with the largest possible volume that exactly meets this restriction?

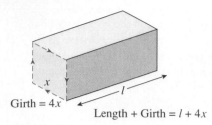

Girth = 4$x$

Length + Girth = $l + 4x$

**FIGURE 27**

**SOLUTION**    The volume of a box with a square base and length $l$ is

$$V = x^2 l$$

In order to maximize volume, we let the sum of the length and girth be the largest allowable value. Thus, $l + 4x = 108$, or

$$l = 108 - 4x$$

Substituting $l = 108 - 4x$ into $V = x^2 l$, we obtain

$$V = x^2(108 - 4x)$$

To find the correct viewing rectangle, we note that we must have $x \geq 0$ and $108 - 4x \geq 0$ so that the dimensions are nonnegative. Solving the second inequality gives $x \leq 27$. Thus, we set the range variables Xmin and Xmax to 0 and 27, respectively. After some experimentation, we settle on values for Ymin and Ymax of 0 and 12,000, respectively. A plot of $y = x^2(108 - 4x)$ with this viewing rectangle is shown in Figure 28. Finally, using the trace feature, we obtain the approximate coordinates of the maximum as shown in Figure 29. If more accuracy is desired, we can zoom-in on the maximum a few times and then use the trace feature, as shown in Figure 30, which suggests that the maximum occurs at approximately $x = 18$.

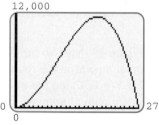

**FIGURE 28**

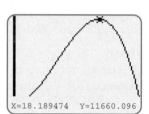

**FIGURE 29**

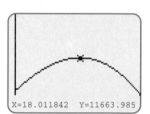

**FIGURE 30**

Computing the corresponding values of $l$ and $V$, we find that

$$l = 108 - 4x \qquad \text{and} \qquad V = x^2 l$$
$$= 108 - 4 \cdot 18 \qquad\qquad = 18^2 \cdot 36$$
$$= 36 \qquad\qquad\qquad = 11{,}664$$

Thus, the box should be $18'' \times 18'' \times 36''$, which will result in a volume of 11,664 cubic inches.

■

---

## EXERCISES 2

**EXERCISES 1–4** □ *Sketch the graph of the function and locate the points on the graph corresponding to the indicated values of x.*

**1.** $f(x) = -2x + 5$; $x = 0, x = 2, x = 4$

**2.** $f(x) = \frac{2}{3}x - 2$; $x = -3, x = 0, x = 3$

**3.** $f(x) = (x + 1)^2 - 3$; $x = -2, x = -1, x = 0$

**4.** $f(x) = -(x - 2)^2 + 4$; $x = 1, x = 2, x = 3$

**EXERCISES 5–12** □ *Use a graphics calculator to sketch the graph of the function and then use the graph to help identify the domain and range of the function.*

**5.** $f(x) = 2x^2 - 3$        **6.** $f(x) = x^4 + 2$

**7.** $f(x) = x^4 - 2x^2$      **8.** $f(x) = 3x^2 - 5x$

**9. a.** $f(x) = \sqrt{x + 4}$        **c.** $f(x) = x\sqrt{x + 4}$

   **b.** $f(x) = x + \sqrt{x + 4}$     **d.** $f(x) = \dfrac{x}{\sqrt{x + 4}}$

**10.**  a. $f(x) = \sqrt{x^2 + 1}$

    b. $f(x) = x + \sqrt{x^2 + 1}$

    c. $f(x) = x\sqrt{x^2 + 1}$

    d. $f(x) = \dfrac{x}{\sqrt{x^2 + 1}}$

**11.**  a. $f(x) = \sqrt{9 - x^2}$

    b. $f(x) = x + \sqrt{9 - x^2}$

    c. $f(x) = x\sqrt{9 - x^2}$

    d. $f(x) = \dfrac{x}{\sqrt{9 - x^2}}$

**12.**  a. $f(x) = \sqrt{x^2 + 2x - 8}$

    b. $f(x) = x + \sqrt{x^2 + 2x - 8}$

    c. $f(x) = x\sqrt{x^2 + 2x - 8}$

    d. $f(x) = \dfrac{x}{\sqrt{x^2 + 2x - 8}}$

**EXERCISES 13–22** □ *Find the exact values of all of the real zeros of the given function.*

**13.** $f(x) = 3x + 1$

**14.** $p(s) = 4s + 6$

**15.** $g(x) = x^2 - 2x - 8$

**16.** $x(t) = t^2 - 9$

**17.** $h(y) = |2y + 5|$

**18.** $f(x) = \sqrt{3x - 9}$

**19.** $f(x) = \dfrac{x^2 - 4}{x^{345} + 345x^{23} - 172{,}896}$

**20.** $g(x) = |2x - 8| - 2$

**21.** $f(a) = 2 - \dfrac{6}{a}$          **22.** $h(y) = 9 - \sqrt{y}$

**EXERCISES 23–28** □ *Use a graphics calculator to estimate the zeros of each of the following functions to the nearest hundredth.*

**23.** $f(x) = 2x^3 - 3x^2 - 20x + 2$

**24.** $h(x) - x^3 - 2x^2 - 14x + 30$

**25.** $f(x) = \dfrac{2}{x} - 3$

**26.** $g(x) = x - 5 - \dfrac{1}{x^2 + x - 2}$

    (*Hint:* There are three in all; one is between $-5$ and $0$.)

**27.** $f(x) = \sqrt{2x + 5} - \sqrt{x^2 + 1}$

**28.** $g(x) = \sqrt{4x + 6} - x - 2$

**EXERCISES 29–32** □ *A function and its graph are given. For each translated or reflected graph, determine the corresponding function.*

**29.** $f(x) = |x|$

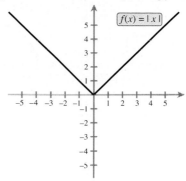

a.

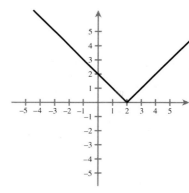

b.

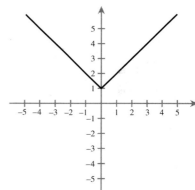

c.

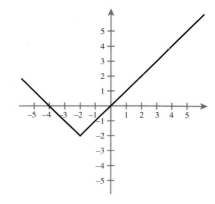

**30.** $g(x) = \sqrt{9 - x^2}$

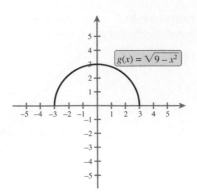

**31.** $h(x) = x^3 - 3x^2 + 2$

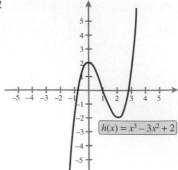

a.

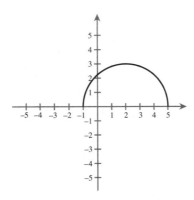

a.

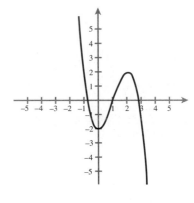

b.

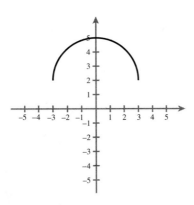

b.

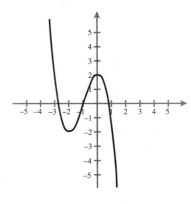

c.

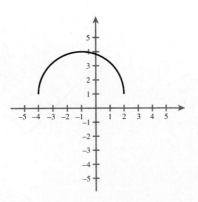

**32.** $f(x) = 2x^3 - 6x + 1$

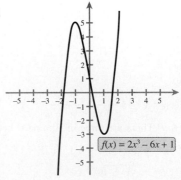

a.

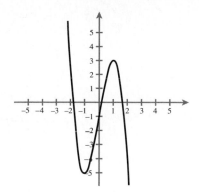

b.

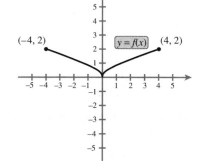

**35.** a. $h(x) = f(x - 3) + 2$
 b. $h(x) = f(-x) - 1$
 c. $h(x) = -f(x + 1)$
 d. $h(x) = f(-x + 2) - 3$

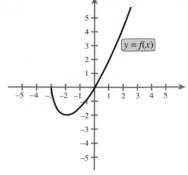

**36.** a. $h(x) = f(x + 2) - 3$
 b. $h(x) = f(-x) - 1$
 c. $h(x) = -f(x + 1)$
 d. $h(x) = -f(x - 2) + 3$

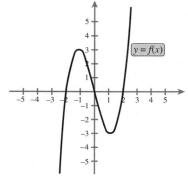

**EXERCISES 33–36** ☐ *Use the graph of the function f to sketch the graph of the function h.*

**33.** a. $h(x) = f(x - 2)$
 b. $h(x) = f(x) + 3$
 c. $h(x) = f(x + 1) - 4$
 d. $h(x) = f(-x)$
 e. $h(x) = -f(x) + 2$
 f. $h(x) = -f(x - 1)$

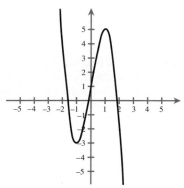

**34.** a. $h(x) = f(x) - 2$
 b. $h(x) = f(x + 3)$
 c. $h(x) = f(x + 1) - 4$
 d. $h(x) = f(-x)$
 e. $h(x) = -f(x) + 2$
 f. $h(x) = -f(x - 1)$

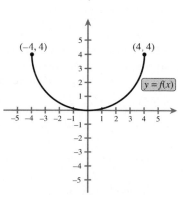

**EXERCISES 37–42** ☐ *Determine whether the given function is even, odd, or neither.*

**37.** $f(x) = x^4 - 4x^2$

**38.** $f(x) = 2x^3 - 8x + 1$

**39.** $f(x) = x\sqrt{x^2 + 1}$

**40.** $f(x) = x^{2/3}$

**41.** $f(x) = x^4 - \dfrac{x}{12}$

**42.** $f(x) = |x + 2| - |x - 2|$

**EXERCISES 43–46** ☐ *Complete the graph of the function f by assuming that the function is (a) even and (b) odd.*

**43.**

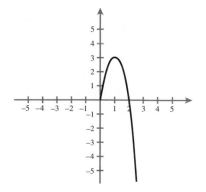

**44.**

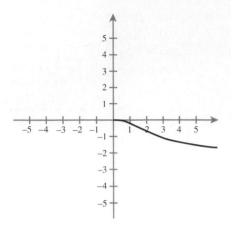

**45.**

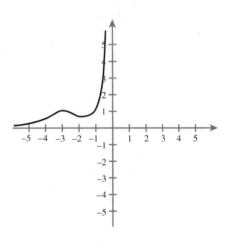

**46.**

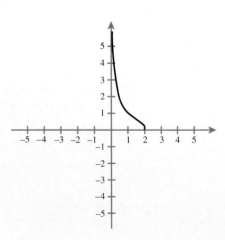

**47.** A portion of the graph of $g(x) = f(x - 3)$ is shown. Complete the graph of $g$ given that $f$ is (a) even and (b) odd.

**48.** Show that if an odd function is defined at $x = 0$, then the graph of the function must pass through the origin. Explain why this fact is not contradicted by the given graph of an odd function.

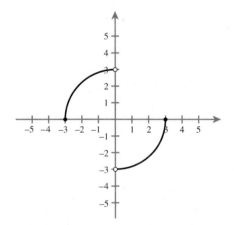

**49.** If $n$ is even, show that $f(x) = x^n$ is an even function. If $n$ is odd, show that $f$ is an odd function.

**50.** Generalize Exercise 49 to show that if $f(x)$ is a polynomial, then $f$ is even if all the powers of $x$ are even, and $f$ is odd if all the powers of $x$ are odd.

**EXERCISES 51–58** □ *Use a graphics calculator to approximate the intervals on which the given function is increasing, the intervals on which it is decreasing, and the coordinates of any relative extrema.*

**51.** $f(x) = x^2 - 4x + 5$     **52.** $g(x) = -x^2 - 6x - 5$

**53.** $h(x) = 2x^3 - 3x^2 - 6x + 5$     **54.** $p(x) = x^3 - 4x^2 + x + 1$

**55.** $q(x) = \dfrac{5}{x^2 - 4x + 5}$     **56.** $r(x) = \dfrac{-3}{x^2 - 6x + 10}$

**57.** $f(x) = x^4 - 3x^2 + x$     **58.** $g(x) = -2x^4 + 5x^2 + 7$

■ *Applications*

**59.** *Translating a Population Function* Suppose a certain state's deer population can be approximated with the function $P(x) = x^2 + 14x + 100$, where $x$ is the year ($x = 0$ corresponds to 1985) and $P(x)$ is the population (in hundreds) for that year. Use a horizontal shift to find a function $Q(x)$ that gives the same population approximations but for which $x = 0$ corresponds to 1990.

**60.** *Translating a Speed Record Function* The 1-mile automobile speed records for the years 1906–1939 can be approximated with the function $f(x) = 0.29x^2 - 2.7x + 134$, where $x$ is the year ($x = 0$ corresponds to 1906) and $f(x)$ is the speed record in miles per hour. Use a horizontal shift to find a function $g(x)$ that gives the same speed approximations but for which $x = 0$ corresponds to 1910.

**One-Mile Speed Record 1906-1939**

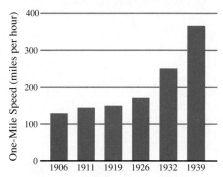

Data Source: *1993 World Almanac*

**61.** *Translating a Height Function* The height of a ball thrown from ground level with an initial velocity of 80 feet per second is given by $f(t) = -16t^2 + 80t$, where $f(t)$ is the height in feet and $t$ is the time in seconds. Find a function $g(t)$ that gives the height of the ball at time $t$ if it is thrown with the same initial velocity from the top of a 20-foot building. How do the graphs of $f$ and $g$ compare?

**62.** *Translating a Cost Function* A clothing company has daily production costs given by $C(x) = 20,000 - 12x + 0.05x^2$, where $C(x)$ is the total cost in dollars and $x$ is the number of units produced. Because of a new labor contract, the daily cost is expected to rise $2000, independent of the number of units produced. Find the new cost function. How does its graph compare to the original function?

**63.** *Maximum Profit* An electronics company has determined that the cost for producing $x$ security systems per month is $C(x) = 50x + 4050$ and the revenue from the sale of $x$ systems per month is $R(x) = 200x - 0.5x^2$. How many systems must be sold each month to break even (i.e., for the profit to be zero)? Determine the number of systems that must be sold each month for the profit to be as large as possible.

**64.** *Maximum Height* A ball is thrown up into the air from the top of a 112-foot building with an initial velocity of 96 feet per second. Its height (in feet) after $t$ seconds is given by $s(t) = -16t^2 + 96t + 112$. Find the maximum height of the ball. When does the ball hit the ground?

**65.** *Maximum Area* A point $P(x, y)$ lies on the line $y = -\frac{3}{4}x + 3$, as shown in Figure 31. Find the location of $P$ so that the area of the shaded rectangle is as large as possible. (*Hint:* First find a function that expresses the area in terms of $x$.)

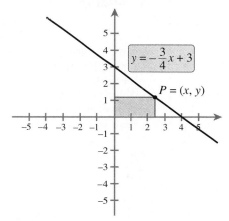

**FIGURE 31**

**66.** *Maximum Area* A line passing through the point (2, 1) has intercepts $(a, 0)$ and $(0, b)$, which are the vertices of a right triangle, as shown in Figure 32. Find the values of $a$ and $b$ so that the area of the shaded triangle is as small as possible. (*Hint:* First find a function that expresses the area in terms of $a$.)

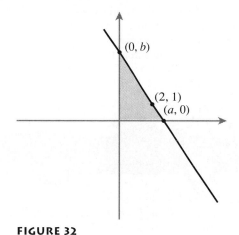

**FIGURE 32**

**67.** *Maximum Area* A rancher wishes to enclose a rectangular pasture at the base of cliff, as shown in Figure 33: 200 linear feet of fencing are available, and no fencing is needed along the cliff side of the pasture. Find the dimensions of the pasture that will have the largest possible area. (*Hint:* First find a function that expresses the area in terms of *x*.)

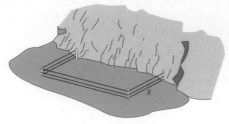

**FIGURE 33**

**68.** *Maximum Volume* A box with a square base and no top is formed by cutting squares out of the corners of a piece of cardboard and then folding up the sides (see Figure 34). If the cardboard measures 16″ × 16″, find the dimensions of the box that will have the largest possible volume. (*Hint:* First find a function that expresses the volume in terms of *x*.)

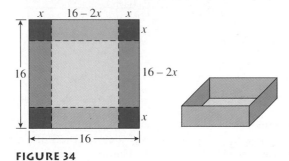

**FIGURE 34**

**69.** *Water Bottle Handoff* A bicycle racer passes a check point at a constant rate of 33 feet per second. At that instant, her support car accelerates from a dead stop in order to overtake her and hand her a water bottle. The car's distance (in feet) from the check point after *t* seconds is given by the function $f(t) = \frac{4}{33}(33t^2 - t^3)$ for *t* from 0 to 33.

**a.** Plot the graph of *f* and find the maximum distance of the car away from the check point.

**b.** Find a function $g(t)$ that gives the racer's distance in feet from the check point after *t* seconds.

**c.** Determine when the racer and support car meet. This can be done algebraically, but you may wish to plot the graphs of *f* and *g* as well. How does this time compare to the time at which the support car is furthest from the check point? Does this outcome seem reasonable? Explain.

**d.** Calculus can be used to show that the velocity of the support car as a function of *t* is given by $v(t) = \frac{4}{11}(22t - t^2)$, where *t* is again measured in seconds and the velocity is in feet per second. Find the velocity of the support car at the instant the racer and car meet. Is your answer a surprise? Why or why not?

**70.** *Modeling $CO_2$ Concentration* The atmospheric concentration of $CO_2$ has shown an overall upward trend in recent years, but it also oscillates quite predictably during the course of each year. For the years 1980–1982, for example, the concentration can be approximated by the function

$$f(x) = 36x^4 - 142x^3 + 175x^2 - 67x + 340$$

where *x* denotes the year (*x* = 0 corresponds to April 1980) and $f(x)$ is the concentration of $CO_2$ in parts per million. Find the intervals on which the function *f* is increasing and those on which it is decreasing. At approximately what time of the year is the concentration the lowest? At what time is it the highest?

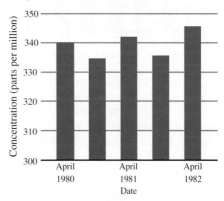

**$CO_2$ Concentration 1980-1982**

Data Source:   *The Challenge of Global Warming*
Dean Edwin Abrahamson, ed.
Island Press, Washington, D.C., 1989

**71.** *Modeling Gold Reserves* Gold reserves in the United States for the years 1973–1990 can be approximated with the function $f(x) = -0.0026x^4 + 0.099x^3 - 1.14x^2 + 2.85x + 275$, where *x* denotes the year (*x* = 0 corresponds to 1973) and $f(x)$ is the gold reserve for that year measured in millions of fine troy ounces. Find the intervals on which *f* is increasing and those on which it is decreasing. According to this model, for what year was the gold reserve highest?

**U.S. Gold Reserves 1973-1989**

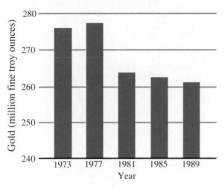

Data Source:  *1991 World Almanac*

**International Gold Reserves 1979-1989**

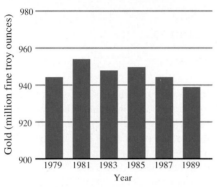

Data Source: *1991 World Almanac*

**72.** *Modeling Gold Reserves* The international gold reserves for the years 1979–1990 can be approximated with the function $f(x) = -0.01x^4 + 0.24x^3 - 2x^2 + 5.8x + 946$, where $x$ denotes the year ($x = 0$ corresponds to 1979) and $f(x)$ is the gold reserve for that year measured in millions of fine troy ounces. Find the intervals on which $f$ is increasing and those on which it is decreasing. According to this functional model, for what year was the gold reserve highest?

**73.** *Modeling Waste Production* The average number of pounds of waste produced each day by each person in the United States for the years 1960–1988 can be approximated with the polynomial function $f(x) = 0.0001x^3 - 0.0043x^2 + 0.089x + 2.66$, where $x$ denotes the year ($x = 0$ corresponds to 1960) and $f(x)$ is the number of pounds of waste. Determine the intervals where $f$ is increasing and the intervals where it is decreasing.

**Waste Production 1960-1988**

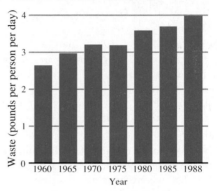

Data Source: U.S. Environmental Protection Agency

---

🔲  *Projects for Enrichment*

**74.** *Designing a Soft Drink Can* An ordinary soft drink can has a volume of 355 cubic centimeters and, for the sake of simplicity, may be assumed to be a right circular cylinder. In this project you will determine the dimensions of the cylindrical can that will provide the required volume with the least amount of aluminum.

   **a.** Write out the formula for computing the surface area of a cylindrical can in terms of its radius $r$ and height $h$.

   **b.** Write out the formula for computing the volume of a cylindrical can in terms of its radius $r$ and height $h$.

   **c.** Use the volume formula from part (b) and the fact that the volume must be 355 cubic centimeters, to obtain an expression for $h$ in terms of $r$. Substitute this expression into the surface area formula of part (a). The result should be a func-

tion $S(r)$ that gives the surface area of a can of radius $r$ with a volume of 355.

   **d.** Use a graphics calculator to approximate the value for $r$ that gives the minimum value for $S(r)$. This is the radius of the can with minimum surface area. Find the height that corresponds to the radius you found.

   **e.** Approximate the actual radius and height of an ordinary soft drink can. How do the actual dimensions compare to the ones you found in part (d)? What explanation can you give for the discrepancy?

   **f.** Now suppose the top of the can is twice as thick as the sides and bottom. Find the dimensions of the can that will minimize the amount of aluminum in the can.

**g.** Experiment with values for the thickness of the top (compared to the sides) until the dimensions of the can with a minimum amount of aluminum approximately coincide with the actual dimensions of a soft drink can. Does your value for the relative thickness of the top seem reasonable? Do you think this explains why the dimensions of a soft drink can are the way they are, or have other important factors been ignored?

**75.** *Stretching and Shrinking a Function* We have already seen how the graph of a function $y = f(x)$ is affected by the modifications $f(x + c)$, $f(x - c)$, $f(x) + c$, and $f(x) - c$. In this project you will investigate the effects of the modifications $cf(x)$ and $f(cx)$.

**a.** Sketch the graph of $f(x) = x^2 + 1$.

**b.** Sketch the graph of each of the following functions.

   **i.** $h(x) = 2(x^2 + 1)$   [i.e., $h(x) = 2f(x)$]

   **ii.** $h(x) = 4(x^2 + 1)$

   **iii.** $h(x) = \frac{1}{2}(x^2 + 1)$

   **iv.** $h(x) = \frac{1}{4}(x^2 + 1)$

**c.** Describe how each of the graphs in part (b) compare to the graph in part (a). You may find it helpful to consider what happens to several specific points on the graph of $f$. For example, the point $(2, 5)$ on the graph of $f$ is "moved" away from the $x$-axis to the point $(2, 10)$ on the graph in (i) of part (b).

**d.** Carefully write out a rule that describes how the graph of $h(x) = cf(x)$ can be obtained from the graph of a function $f$. You may find it convenient to consider two cases, one where $c > 1$ and the other where $0 < c < 1$. Explain why it is not necessary to consider $c < 0$.

**e.** Try out your rule by sketching the graphs of $y = cf(x)$ for $c = 3$ and $c = \frac{1}{3}$ and the function $f$ shown in Figure 35.

**f.** Sketch the graph of each of the following functions.

   **i.** $h(x) = (2x)^2 + 1$   [i.e., $h(x) = f(2x)$]

   **ii.** $h(x) = (4x)^2 + 1$

   **iii.** $h(x) = (\frac{1}{2}x)^2 + 1$

   **iv.** $h(x) = (\frac{1}{4}x)^2 + 1$

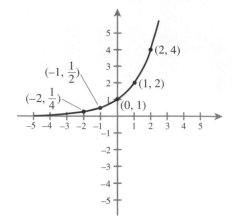

**FIGURE 35**

**g.** Describe how each of the graphs in part (f) compare to the graph in part (a). You may find it helpful to first consider what happens to several specific points on the graph of $f$. For example, the point $(2, 5)$ on the graph of $f$ is "moved" toward the $y$-axis to the point $(1, 5)$ on the graph in (i) of part (f). In other words, the same $y$-coordinate is obtained from an $x$-coordinate that is half the distance to the $y$-axis.

**h.** Carefully write out a rule that describes how the graph of $h(x) = f(cx)$ can be obtained from the graph of a function $f$. You may find it convenient to consider two cases, one where $c > 1$ and the other where $0 < c < 1$. Explain why it is not necessary to consider $c < 0$.

**i.** Try out your rule by sketching the graphs of $y = f(cx)$ for $c = 3$ and $c = \frac{1}{3}$ and the function $f$ shown in Figure 35.

---

■■  *Questions for Discussion or Essay*

---

**76.** An even function is symmetric with respect to the $y$-axis. An odd function is symmetric with respect to the origin. Why isn't there a term for a function that is symmetric with respect to the $x$-axis?

**77.** We have observed that polynomial functions with all even powers of $x$ are even and polynomial functions with all odd powers of $x$ are odd. Why can't the functional notions of *even* and *odd* be defined this way instead of using the more abstract properties $f(-x) = f(x)$ and $f(-x) = -f(x)$?

**78.** To what extent are the graphical concepts of translation, reflection, and symmetry needed as tools for sketching the graphs of

functions? If you feel that they are no longer needed as graphing tools because of the graphics calculator, what purpose do they serve? If, on the other hand, you feel that they are still needed as graphing tools, when and how should they be used in conjunction with (or instead of) a graphics calculator?

**79.** When zooming in on relative extrema, it can happen that the graph will eventually appear as a horizontal line across the entire graphics calculator display. Experiment with zoom boxes of varying dimensions to see how this can be avoided. What is your conclusion as to the best dimensions of a zoom box for zooming in on extrema?

---

**SECTION 3**

# COMBINATIONS AND INVERSES OF FUNCTIONS

■ How can the beating of a butterfly's wings in China create a tropical storm in Mexico?

■ If $f$ and $g$ are functions and not just real numbers, what is meant by $f + g$, $f - g, fg$, or $f/g$?

■ If $f(x) = 3x + 2$, why isn't $f^{-1}(x) = \dfrac{1}{3x + 2}$?

■ What do you get if you choose a number, square it, add 1, and then repeat this process (squaring and adding 1) 8 times?

---

## ARITHMETIC COMBINATIONS OF FUNCTIONS

Just as numbers can be combined by the basic arithmetic operations of addition, subtraction, multiplication, and division, so too can functions. We can define the sum of two functions $f$ and $g$ by the formula $(f + g)(x) = f(x) + g(x)$. The operations of subtraction, multiplication, and division of functions are defined in a similar fashion, as summarized below.

*Arithmetic operations on functions*

| Operation | Definition |
|---|---|
| Addition | $(f + g)(x) = f(x) + g(x)$ |
| Subtraction | $(f - g)(x) = f(x) - g(x)$ |
| Multiplication | $(fg)(x) = f(x)g(x)$ |
| Division | $\left(\dfrac{f}{g}\right)(x) = \dfrac{f(x)}{g(x)}$ |

**EXAMPLE 1**　　*Finding combinations of functions*

Let $f(x) = x^2 - 4$ and $g(x) = x + 2$. Compute each of the following.

**a.** $(f + g)(x)$　　　　　　　　　**b.** $(f - g)(x)$

**c.** $(fg)(x)$　　　　　　　　　　**d.** $\left(\dfrac{f}{g}\right)(x)$

**SOLUTION**

**a.** $(f + g)(x) = f(x) + g(x)$
$$= (x^2 - 4) + (x + 2)$$
$$= x^2 + x - 2$$

**b.** $(f - g)(x) = f(x) - g(x)$
$$= (x^2 - 4) - (x + 2)$$
$$= x^2 - x - 6$$

**c.** $(fg)(x) = f(x)g(x)$
$$= (x^2 - 4)(x + 2)$$
$$= x^3 + 2x^2 - 4x - 8$$

**d.** $\left(\dfrac{f}{g}\right)(x) = \dfrac{f(x)}{g(x)}$

$$= \dfrac{x^2 - 4}{x + 2}$$

$$= \dfrac{(x - 2)(x + 2)}{x + 2}$$

It is tempting to simplify this expression to $x - 2$ by canceling the common factor $x + 2$. This, however, would be incorrect, since $x - 2$ is defined for all real numbers $x$, but

$$\dfrac{(x - 2)(x + 2)}{x + 2}$$

is undefined for $x = -2$. We could, however, express our final answer as $(f/g)(x) = x - 2$ for $x \neq -2$.

■

In keeping with our convention of assuming that the domain (unless otherwise specified) is the set of real numbers for which the formula "makes sense," the domain of $f + g$ is the set of real numbers $x$ for which both $f(x)$ and $g(x)$ are defined. Thus, the domain of $f + g$ will consist of those real numbers that are in the domains of both $f$ **and** $g$. Similarly, the domains of $f - g$ and $fg$ consist of all real numbers that are contained in the domains of both $f$ and $g$. However, since the expression $f(x)/g(x)$ is not defined when $g(x) = 0$, the domain of $f/g$ consists of those real numbers $x$ in the domains of both $f$ and $g$ such that $g(x) \neq 0$.

---

**EXAMPLE 2**

*Finding the domains of combinations of functions*

Let $f(x) = x - 2$, and $g(x) = \sqrt{x - 1}$. Find the domains of each of the following functions.

**a.** $(f + g)(x)$                 **b.** $(f - g)(x)$

**c.** $(fg)(x)$                    **d.** $\left(\dfrac{f}{g}\right)(x)$

**e.** $\left(\dfrac{g}{f}\right)(x)$

**SOLUTION**

**a–c.** The domain of $f + g$, $f - g$, and $fg$ is the set of real numbers $x$ such that $x$ is in the domains of both $f$ and $g$. The domain of $f$ is the set of all real numbers. In order for $g(x) = \sqrt{x - 1}$ to be defined, we require that $x - 1 \geq 0$. Thus, we have $x \geq 1$. In interval notation, then, the domain of $g$ is $[1, \infty)$. Since the domain of $f$ is all real numbers and the domain of $g$ is $[1, \infty)$, the set of real numbers that are in the domains of both $f$ and $g$ is $[1, \infty)$. Thus, the domain of $f + g$, $f - g$, and $fg$ is $[1, \infty)$.

**d.** As already noted, both $f$ and $g$ are defined on the interval $[1, \infty)$. However, since $g(1) = \sqrt{0} = 0$, 1 is excluded from the domain of $f/g$. Thus, the domain of $f/g$ is $(1, \infty)$.

**e.** Since $f(2) = 2 - 2 = 0$, the number 2 is excluded from the domain of $g/f$. Thus, the domain of $g/f$ consists of all real numbers in the interval $[1, \infty)$, with the exception of 2., i.e. $[1, 2) \cup (2, \infty)$.

---

## DYNAMICAL SYSTEMS AND CHAOS THEORY

**S**uppose we enter a number in our calculator and repeatedly hit the $x^2$ key. What will happen? The result depends on the value of the original number. If the original number is less than 1, then the result becomes smaller each time we hit the $x^2$ key, and we get closer and closer to 0. On the other hand, if the original number is greater than 1, then the result becomes larger and larger as we repeatedly strike the $x^2$ key. This is an example of a **dynamical system**: a process is repeated over and over again, and the output from one step is the input for the next.

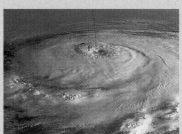
A hurricane as photographed from a satellite

Our lives are tapestries of dynamical systems. The amount that we eat influences our metabolism, which affects the amount that we eat, which influences our metabolism…. Our personalities determine our friends, who influence our personalities, which determine our friends…. The amount that we sleep, our moods, and our eating, drinking and exercise habits are all bound together in an elaborate dynamical system. Every ecosystem is, in essence, a dynamical system. On every day in every acre of tropical rain forest, an intricate dance is played out among millions of competing, cooperating and coexisting dynamical systems. The Earth's atmosphere is an enormous dynamical system, the complexity of which has thus far vexed our most brilliant atmospheric scientists. In fact, one could argue that Earth's atmosphere is the second most complex dynamical system known, second only to what is perhaps the most complex structure in the known universe, the human brain. Our very thoughts are the product of the interactions and interconnections between literally billions of neurons. Synapses are being reinforced here and weakened there, leaving in their wake the stuff of which thoughts and fears, joy and sorrow are made.

In the last few decades, computers have provided us with a lens through which dynamical systems can be viewed, the results of millions of iterations of certain processes revealed on computer screens in a matter of minutes. These powerful instruments have illuminated largely unexplored mathematical territory and have given rise to the new science of **chaos**, which has caused nothing less than a paradigm shift in our understanding of the way in which dynamical systems unfold.

For an illustration of chaos, consider a billiard ball on a frictionless pool table with bumpers that absorb no energy. One would guess that the path of the billiard ball is completely determined by the initial velocity, position, and spin of the ball—the ball, without friction to slow it, will bounce around the table in a completely determined pattern. But, in fact, small, seemingly negligible factors will, after surprisingly few caroms, give rise to enormous variation. Surely one needn't account for the gravitational pull of Jupiter or the vibrations from a tree limb falling 20 miles away, or the change in air pressure near the ball caused by the breeze from the hamster wheel at the house next door, but one must if accuracy is desired after the first few caroms. Shockingly small changes in conditions can have enormous consequences.

This important aspect of chaos, extreme sensitivity to initial conditions, is often called **the butterfly effect**. The name arises from the degree to which atmospheric conditions can be chaotic; minor perturbations can become small eddies that can give rise to storms. It is said that the beating of a butterfly's wings at the right location at the right time can cause a tropical storm a thousand miles away a few days later. There is a certain degree of unpredictability in the weather that no computer, no matter how powerful, and that no measuring instruments, no matter how sensitive, will ever overcome.

## COMPOSITION OF FUNCTIONS

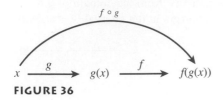

**FIGURE 36**

Functions can also be combined by *composing* or "stringing" them together—using the output of one function as the input for another. The **composition** of two functions $f$ and $g$, written $f \circ g$, is defined by the formula $(f \circ g)(x) = f(g(x))$. The function $f \circ g$ takes an input value $x$ and produces $f(g(x))$ as the corresponding output value, as shown in Figure 36.

We may also think of $f \circ g$ as the machine formed by first inputting $x$ in the machine $g$, yielding the output $g(x)$, and then inputting $g(x)$ into the machine $f$, giving $f(g(x))$ as a final product, as shown in Figure 37. The function machine helps to illustrate what must occur in order for $(f \circ g)(x) = f(g(x))$ to be defined. First, $g(x)$ must be defined, so that $x$ must be in the domain of $g$. Second, since we are using $g(x)$ as an input for $f$, $g(x)$ must be in the domain of $f$. Thus, we have the following formal definition of composition.

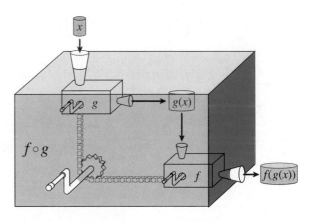

**FIGURE 37**

***Definition of composition of functions***

> The **composition** of the functions $f$ and $g$, written $f \circ g$, is defined by $(f \circ g)(x) = f(g(x))$. The domain of $f \circ g$ consists of those real numbers $x$ in the domain of $g$ such that $g(x)$ is in the domain of $f$.

**EXAMPLE 3**     *Computing the composition of two functions*

Given that $f(x) = 2x^2 + 4x + 5$, and $g(x) = 2x + 1$, compute $f(g(x))$.

**SOLUTION**

$$
\begin{aligned}
(f \circ g)(x) &= f(g(x)) \\
&= f(2x + 1) && \text{Replacing } g(x) \text{ with } 2x + 1 \\
&= 2(2x + 1)^2 + 4(2x + 1) + 5 && \text{Replacing each } x \text{ in} \\
& && 2x^2 + 4x + 5 \text{ with } 2x + 1 \\
&= 2(4x^2 + 4x + 1) + 8x + 4 + 5 \\
&= 8x^2 + 16x + 11
\end{aligned}
$$

RULE OF THUMB   When evaluating the composition of functions, begin on the inside and work your way out. That is, at each stage, evaluate the innermost function first.

**EXAMPLE 4**    *Computing compositions of functions*

Let

$$f(x) = \frac{x + 1}{x - 1} \quad \text{and} \quad g(x) = x^2$$

Compute the following compositions and find their domains.

a.  $(f \circ g)(x)$          b.  $(g \circ f)(x)$

**SOLUTION**

a.  $(f \circ g)(x) = f(g(x))$
$$= f(x^2)$$
$$= \frac{x^2 + 1}{x^2 - 1}$$

Since this expression is defined for all real numbers except for $x = \pm 1$, and $g(x)$ is defined for all real numbers $x$, the domain is the set of all real numbers except 1 and $-1$.

b.  $(g \circ f)(x) = g(f(x))$
$$= g\left(\frac{x + 1}{x - 1}\right)$$
$$= \left(\frac{x + 1}{x - 1}\right)^2$$
$$= \frac{(x + 1)^2}{(x - 1)^2}$$

Since this expression and $f$ are defined for all real numbers except $x = 1$, the domain is the set of all real numbers except for 1.

$\blacksquare$

**EXAMPLE 5**    *Computing compositions of functions*

Let $f(x) = \sqrt{x}$ and $g(x) = x^2 + 1$. Compute the following compositions and find their domains.

a.  $(f \circ g)(x)$          b.  $(g \circ f)(x)$

**SOLUTION**

a.  $(f \circ g)(x) = f(g(x))$
$$= f(x^2 + 1)$$
$$= \sqrt{x^2 + 1}$$

Since $g(x)$ is defined for all real numbers $x$ and the expression under the radical, $x^2 + 1$, is always positive, the domain of $f \circ g$ is the set of all real numbers.

**b.** $(g \circ f)(x) = g(f(x))$
$$= g(\sqrt{x})$$
$$= (\sqrt{x})^2 + 1, \text{ for } x \geqslant 0$$
$$= x + 1, \text{ for } x \geq 0$$

Note that although $x + 1$ is defined for all real numbers $x$, $(g \circ f)(x)$ is not since the inside function, $f(x) = \sqrt{x}$, is only defined for nonnegative $x$. Thus, the domain of $g \circ f$ is the set of all nonnegative real numbers.

Notice that in the previous examples, $f \circ g$ and $g \circ f$ are different functions. Thus, composition of functions is noncommutative—the order is important.

---

> **WARNING**   Don't confuse $f \circ g$ and $g \circ f$—the order matters when composing functions.

---

## INVERSE FUNCTIONS

Consider the functions

$$f(x) = 2x + 1 \quad \text{and} \quad g(x) = \frac{x - 1}{2}$$

If we form the composition $f \circ g$, we have

$$(f \circ g)(x) = f(g(x))$$
$$= f\left(\frac{x - 1}{2}\right)$$
$$= 2\left(\frac{x - 1}{2}\right) + 1$$
$$= x - 1 + 1$$
$$= x$$

Thus, if we input $x$ into the function $g$ to form $g(x)$ and then input $g(x)$ into the function $f$ to form $f(g(x))$, we obtain $x$, our original input! It is as though the function $f$ "undid" whatever $g$ did to $x$. In fact, $g$ "undoes" $f$ in precisely the same fashion. That is, if we begin by inputting $x$ into the function $f$ to form $f(x)$ and then input $f(x)$ into the function $g$ to form $g(f(x))$, we again obtain $x$, as shown below.

$$(g \circ f)(x) = g(f(x))$$
$$= g(2x + 1)$$
$$= \frac{(2x + 1) - 1}{2}$$
$$= \frac{2x}{2}$$
$$= x$$

Whenever functions $f$ and $g$ undo each other in this sense, they are said to be *inverses* of one another, as defined below.

**Definition of $f^{-1}$**

> A function $g$ satisfying $(g \circ f)(x) = x$ for all $x$ in the domain of $f$ and $(f \circ g)(x) = x$ for all $x$ in the domain of $g$, is said to be the **inverse** of the function $f$. We write $g = f^{-1}$, or equivalently, $f = g^{-1}$.

**EXAMPLE 6**    *Verification of inverse functions*

Show that the functions

$$f(x) = \frac{3x + 2}{4} \quad \text{and} \quad g(x) = \frac{4x - 2}{3}$$

are inverses of one another.

**SOLUTION**    We must show that $(f \circ g)(x) = (g \circ f)(x) = x$.

$$(f \circ g)(x) = f(g(x)) \qquad\qquad (g \circ f)(x) = g(f(x))$$

$$= f\!\left(\frac{4x - 2}{3}\right) \qquad\qquad = g\!\left(\frac{3x + 2}{4}\right)$$

$$= \frac{3\left(\dfrac{4x - 2}{3}\right) + 2}{4} \qquad\qquad = \frac{4\left(\dfrac{3x + 2}{4}\right) - 2}{3}$$

$$= \frac{4x - 2 + 2}{4} \qquad\qquad = \frac{3x + 2 - 2}{3}$$

$$= \frac{4x}{4} \qquad\qquad = \frac{3x}{3}$$

$$= x \qquad\qquad = x$$

Thus, $f$ and $g$ are indeed inverses.

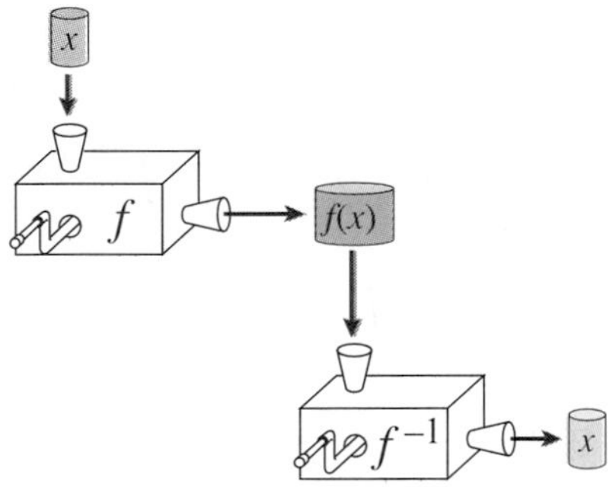

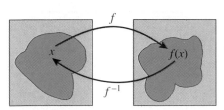

**FIGURE 38**

**FIGURE 39**

The inverse of a function is, in a sense, its opposite; it *undoes* what the function *does*, and vice versa, as illustrated in Figure 38. It is also useful to think of $f$ and $f^{-1}$ as mappings between sets, as depicted in Figure 39. Here we see that the domain of $f$ is the range of $f^{-1}$ and that the domain of $f^{-1}$ is the range of $f$. In other words, the input values for $f$ are the output values of $f^{-1}$, and vice versa.

The inverse of a function $f$ can be viewed as that function, which when given an output of the function $f$, determines the input. However, it isn't always possible to determine the input to a function from the output, and for this reason, not all functions have inverses. For example, consider the function $f(x) = x^2$. If it is given that the output of this function is 9, we cannot determine the input: it might have been either 3 or $-3$. Thus, the squaring process cannot be reversed; if the result of a squaring is known, the original number cannot be specified with certainty (unless the result of the squaring is zero, in which case the original number must have been zero). Thus, the function $f$

doesn't have an inverse. In fact, the only functions that *do* have inverses are those for which distinct inputs yield distinct outputs. We define **one-to-one functions** to be such functions. More formally, we have the following definition.

***Definition of one-to-one functions***

If the function $f$ has the property that $f(a) = f(b)$ only if $a = b$, then $f$ is said to be **one-to-one**, or 1–1.

A function can easily be tested to see whether or not it is 1–1 by looking at its graph. If the function *is* 1–1, then for each $y$-value, there will be only one $x$-value. Thus, a line with equation $y = c$ (a horizontal line) will be crossed at most once by the graph of a 1–1 function. If a 1–1 function $f$ were to cross $y = c$ two or more times, then there would be two or more distinct $x$ values $a$ and $b$ such that $f(a) = f(b) = c$, and $f$ would, by definition, fail to be 1–1. This gives us the following test for 1–1 functions.

***Horizontal line test***

A function $f$ is 1–1 if and only if no horizontal line crosses the graph of $f$ more than once.

**EXAMPLE 7**    *Testing to see whether a function is* 1–1

The graphs of four functions are shown. Determine which are the graphs of 1–1 functions.

a.

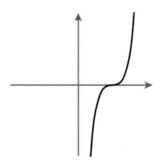

b.

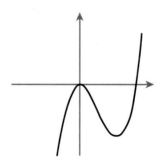

c.

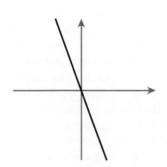

d.

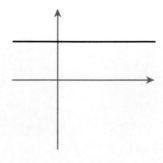

## SOLUTION

**a.** No horizontal line crosses this graph more than once; the function is thus 1–1.

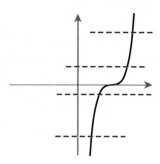

**b.** Several lines are shown that cross the graph more than once; this function is not 1–1.

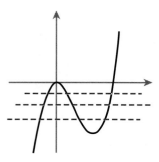

**c.** No horizontal line crosses this graph more than once; hence this function is 1–1.

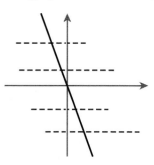

**d.** The graph of this function is itself a horizontal line. Therefore, there is one horizontal line (the graph itself) that intersects the graph more than once. In fact, it intersects itself infinitely many times. This function is not 1–1.

We have seen that if a function is 1–1, then it has an inverse. We now turn our attention to computing the inverse. Let $f$ be a 1–1 function, and let $y = f(x)$. Since the inverse of a function returns the input to $f$ when given the output, we can compute the inverse of $f$ by solving the equation $y = f(x)$ for $x$. In fact, if we apply $f^{-1}$ to both sides of the equation $y = f(x)$, we obtain

$$y = f(x)$$

$$f^{-1}(y) = f^{-1}(f(x)) \qquad \text{Applying } f^{-1} \text{ to both sides}$$

$$f^{-1}(y) = x \qquad \text{Simplifying by using the definition of } f^{-1}$$

Of course, once we know $f^{-1}(y)$, we can compute $f^{-1}(x)$ by simply substituting $x$ for $y$ in the formula that defines $f^{-1}(y)$. This suggests the following strategy for computing inverses.

**Steps for computing $f^{-1}(x)$ for a 1–1 function $f$**

> 1. Solve the equation $y = f(x)$ for $x$; this will give $x = f^{-1}(y)$.
> 2. Substitute $x$ for $y$ to find $f^{-1}(x)$.

The following examples illustrate this technique.

**EXAMPLE 8**    *Computing the inverse of a linear function*

Find $f^{-1}(x)$ if $f(x) = 3x + 2$.

**SOLUTION**    We begin by solving the equation $y = 3x + 2$ for $x$.

$$y = 3x + 2$$

$$y - 2 = 3x \qquad \text{Subtracting 2 from both sides}$$

$$x = \frac{y - 2}{3} \qquad \text{Dividing both sides by 3 and exchanging the left and right sides}$$

Thus, we have

$$f^{-1}(y) = \frac{y - 2}{3}$$

Substituting $x$ for $y$, we obtain

$$f^{-1}(x) = \frac{x - 2}{3}$$

**EXAMPLE 9**    *Finding the inverse of a function involving radicals*

Find $f^{-1}(x)$ if $f(x) = \sqrt[3]{2x - 7}$.

**SOLUTION**    We let $y = f(x)$ and solve for $x$ in terms of $y$.

$$y = \sqrt[3]{2x - 7}$$

$$y^3 = (\sqrt[3]{2x - 7})^3 \qquad \text{Cubing both sides}$$

$$y^3 = 2x - 7$$

$$y^3 + 7 = 2x$$

$$2x = y^3 + 7$$

$$x = \frac{y^3 + 7}{2}$$

Thus $f^{-1}(y) = \dfrac{y^3 + 7}{2}$, so that $f^{-1}(x) = \dfrac{x^3 + 7}{2}$.

## GRAPHS OF INVERSE FUNCTIONS

Suppose that $(a, b)$ is on the graph of $y = f(x)$. Then $b = f(a)$, from which it follows that $a = f^{-1}(b)$. Thus, $(b, a)$ is a point on the graph of $y = f^{-1}(x)$. As Figure 40 suggests, the point $(b, a)$ is the reflection of the point $(a, b)$ about the line $y = x$. Thus, the graph of $y = f^{-1}(x)$ will be the reflection of the graph of $y = f(x)$ about the line $y = x$.

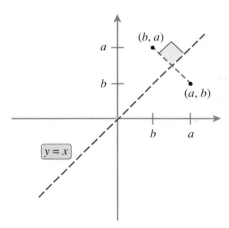

**FIGURE 40**

**EXAMPLE 10**    *Using the graph of f to graph f⁻¹*

The graph of a function $f$ is given in Figure 41. Show that $f^{-1}$ exists and graph it.

**SOLUTION**    Since the graph passes the horizontal line test, $f^{-1}$ exists. To graph $f^{-1}$, we simply reflect the graph of $f$ about the line $y = x$ as shown in Figure 42.

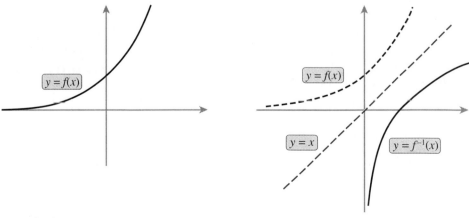

**FIGURE 41**                                      **FIGURE 42**

# EXERCISES 3

**EXERCISES 1–10** □ *Find the following combinations. Specify the domain if it is anything other than all real numbers.*

a. $(f + g)(x)$    b. $(f - g)(x)$    c. $(fg)(x)$    d. $\left(\dfrac{f}{g}\right)(x)$

1. $f(x) = x,\ g(x) = 5$

2. $f(x) = 3x + 2,\ g(x) = x - 7$

3. $f(x) = x^2,\ g(x) = 3x + 1$

4. $f(x) = 2x + 5,\ g(x) = x^2 + 1$

5. $f(x) = x,\ g(x) = x$

6. $f(x) = \dfrac{2}{x},\ g(x) = 3x$

7. $f(x) = 2x - 2,\ g(x) = \dfrac{2}{x + 5}$

8. $f(x) = \dfrac{1}{x},\ g(x) = -\dfrac{1}{x}$

9. $f(x) = \sqrt{x - 2},\ g(x) = x - 4$

10. $f(x) = 3x,\ g(x) = \sqrt{2x - 6}$

**EXERCISES 11–26** □ *Compute both $(f \circ g)(x)$ and $(g \circ f)(x)$. Specify the domain if it is anything other than all real numbers.*

11. $f(x) = 2x + 3,\ g(x) = 4x - 5$

12. $f(x) = 3 - x,\ g(x) = 5x + 1$

13. $f(x) = x^2,\ g(x) = 2x + 7$

14. $f(x) = 3x - 4,\ g(x) = x^3$

15. $f(x) = \dfrac{3}{x},\ g(x) = x^2$

16. $f(x) = \dfrac{4}{x + 1},\ g(x) = 2x + 4$

17. $f(x) = x + 1,\ g(x) = x^3 - 3x$

18. $f(x) = x^2 + 3x,\ g(x) = x - 5$

19. $f(x) = x^2 + 3x + 1,\ g(x) = x + h$

20. $f(x) = \dfrac{1}{x},\ g(x) = \dfrac{1}{x}$

21. $f(x) = \sqrt{x},\ g(x) = x^4$

22. $f(x) = |x|,\ g(x) = 4x - 1$

23. $f(x) = 3x + 5,\ g(x) = \sqrt{x - 2}$

24. $f(x) = 2x + 3,\ g(x) = \sqrt{2x - 5}$

25. $f(x) = \dfrac{3}{x^2 - 4},\ g(x) = x - 2$

26. $f(x) = \dfrac{1}{x^2 - 9},\ g(x) = 3$

**EXERCISES 27–36** □ *Determine whether the functions f and g are inverses of one another by evaluating $f(g(x))$ and $g(f(x))$.*

27. $f(x) = x + 2;\ g(x) = x - 2$

28. $f(x) = 3x;\ g(x) = \dfrac{x}{3}$

29. $f(x) = 2x + 1;\ g(x) = \dfrac{1}{2}x - 1$

30. $f(x) = \dfrac{1}{4}x - 3;\ g(x) = 4x + 12$

31. $f(x) = (x + 1)^3;\ g(x) = \sqrt[3]{x} - 1$

32. $f(x) = \sqrt[5]{x - 2};\ g(x) = (x + 2)^5$

33. $f(x) = \dfrac{2}{x - 1};\ g(x) = \dfrac{x}{2} + 1$

34. $f(x) = \dfrac{1}{x} - 4;\ g(x) = \dfrac{1}{x + 4}$

35. $f(x) = \sqrt{x - 6};\ g(x) = x^2 + 6, x \ge 0$

36. $f(x) = 4 - x^2, x \ge 0;\ g(x) = \sqrt{x + 4}$

**EXERCISES 37–46** □ *Determine whether the function is 1–1.*

37. $f(x) = -2x + 5$    38. $g(x) = \dfrac{x + 3}{4}$

39. $f(x) = x^2$    40. $f(x) = \sqrt[3]{x}$

41. $h(x) = x^3 + 2x + 1$    42. $f(x) = x^3 - x^2$

43. $g(x) = \dfrac{x^4}{12} - x^3$    44. $f(x) = x^5 + 1$

45. $h(x) = |x - 3|$    46. $f(x) = \sqrt{x + 2}$

**EXERCISES 47–56** □ *Determine whether the function f is 1–1. If it is, find $f^{-1}(x)$.*

47. $f(x) = \dfrac{1}{3}x - 1$    48. $f(x) = \dfrac{-x + 3}{4}$

49. $f(x) = x^4$    50. $f(x) = x^3$

51. $f(x) = x^2 + 1,\ x \ge 0$    52. $f(x) = (x + 4)^2$

53. $f(x) = \sqrt{2x + 5}$    54. $f(x) = \sqrt{4 - x^2}$

55. $f(x) = \dfrac{x^2}{x^2 + 1}$    56. $f(x) = \dfrac{x}{x + 4}$

**EXERCISES 57–62** ☐ *Determine whether $f^{-1}$ exists and, if so, sketch its graph.*

57.

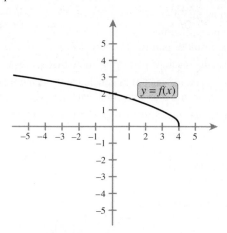

60.

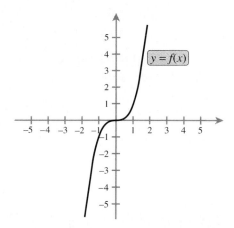

58.

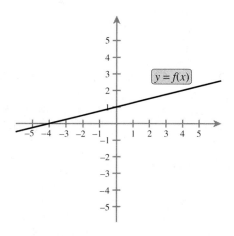

61.

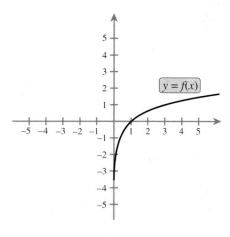

59.

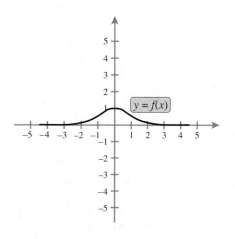

62.

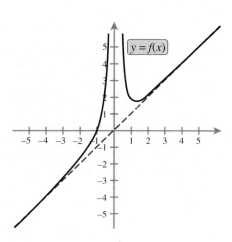

**EXERCISES 63–68** □ *These exercises deal with the* **iterates** *of a function f, which are defined in the following way.*

$$f^1(x) = f(x)$$

$$f^2(x) = f(f(x))$$

$$f^3(x) = f(f(f(x)))$$

$$\vdots$$

*Thus, the* nth *iterate of a function f, denoted* $f^n(x)$, *is found by composing f with itself n times. For example, if* $f(x) = 3x$, *then* $f^2(x) = f(f(x)) = f(3x) = 3(3x) = 9x$.

**63.** If $f(x) = x + 5$, find $f^3(x)$.

**64.** If $f(x) = 2x + 1$, find $f^2(x)$.

**65.** If $f(x) = 2x$, find $f^4(x)$.

**66.** If $f(x) = x^2$, find $f^3(x)$.

**67.** If $f(x) = x^2$, then $f^n(a)$ can be found by entering $a$ into a calculator and pressing the $x^2$ key $n$ times. Use this method to find $f^{10}(0.98)$ and $f^{10}(1.02)$. What happens to $f^n(0.98)$ and $f^n(1.02)$ as $n$ gets larger?

**68.** If $f(x) = x^2 + 1$, then $f^n(a)$ can be found by entering $a$ into a calculator, pressing the $x^2$ key, adding 1, and then repeating the procedure $n$ times. Compute $f^8(a)$ for the values $a = -1$ and $a = 0.5$. What happens to $f^n(a)$ as $n$ gets larger?

**69.** Show that composition of functions is associative; that is, show that $f \circ (g \circ h) = (f \circ g) \circ h$.

**70. a.** Simplify $f^{-1} \circ (f \circ g)$.

**b.** Suppose that $f$ and $f \circ g$ are given. Explain how the result of part (a) could be used to find $g$.

**c.** Employ the technique you described in part (b) to find $g(x)$, if it is given that $(f \circ g)(x) = 6x + 21$ and $f(x) = 3x + 9$.

**71.** Define $(f \circ g \circ h)(x) = f(g(h(x)))$.

**a.** Show that $f^{-1} \circ f \circ p = p$, for any function $p$.

**b.** Show that $p \circ f \circ f^{-1} = p$, for any function $p$.

**c.** Suppose that the function $f$ has *two* inverses, $g$ and $h$. Use the result of part (a), to show that $g \circ f \circ h = h$.

**d.** Use the result of part (b) to show that $g \circ f \circ h = g$.

**e.** Use parts (c) and (d) to conclude that $g$ and $h$ are the same function and that a function can have at most one inverse.

**72.** An arbitrary linear function can be written as $f(x) = ax + b$. Compute $f^{-1}$.

**73.** Find a function $f$ such that $f^{-1} = f$. (*Hint:* Start by looking for an operation that, when performed twice in a row, returns you to the original number.)

**74.** It can be shown that

$$1 + x + x^2 + x^3 + \cdots + x^n = \frac{1 - x^{n+1}}{1 - x} \quad \text{for } x \neq 1$$

**a.** Find a formula for
$$1 + (x + 1) + (x + 1)^2 + (x + 1)^3 + \cdots + (x + 1)^n.$$

**b.** Find a formula for $1 - x + x^2 - x^3 + \cdots + (-x)^n$.

**75.** Find a combination of rotations by 90° and reflections about the axes that yield a reflection about the line $y = x$.

**76.** Produce functions $f$ and $g$ such that $f \circ g$ is nowhere defined, even though $g$ is defined everywhere. [*Hint:* Start by finding a function $f$ that is defined only for $x > 0$. Then select a function $g$ such that $g(x) < 0$.]

**77.** Which 1–1 functions have graphs that are symmetric with respect to the $y$-axis?

■ *Applications*

**78.** *Company Sales* A company that markets office equipment has two sales people, Pete and Tina. Over a 10-week period, Pete's sales can be approximated with the function $S_1(t) = -3t^2 + 36t + 92$, and Tina's sales can be approximated with $S_2(t) = 4t^2 - 40t + 200$. In both cases, $t$ is measured in weeks ($t = 1$ is the first week), and the function value for a given $t$ is the sales for that week, in hundreds of dollars.

**a.** Graph both $S_1$ and $S_2$ and describe the sales pattern for Pete and Tina. What are the maximum and minimum sales for both for the 10-week period?

**b.** Write a function that gives the total sales for the company for the 10-week period. Plot the graph of this function and describe the total sales over this period. What are the maximum and minimum sales?

**79.** *Refinery Production* An oil company owns two refineries. Over a 6-month period, the number of barrels of crude oil refined at each can be approximated with the functions $B_1(t) = 75t^2 - 450t + 1800$ and $B_2(t) = -85t^2 + 510t + 900$, where $t$ is measured in months ($t = 1$ is the first month), and the function value for a given $t$ is the production for that month, in thousands of barrels.

**a.** Graph both $B_1$ and $B_2$ and describe the production pattern for each refinery. What are the maximum and minimum production levels for each refinery for the 6-month period?

**b.** Write a function that gives the total production for the oil company over the 6-month period. Plot the graph of this function and describe the total production over this period. What are the maximum and minimum production levels?

### Projects for Enrichment

**80.** *Computing Iterates of a Function Graphically* In Exercises 63–68 we defined the iterate $f^n(x)$ to be the function formed by composing $f$ with itself $n$ times. Suppose that for a given function $f$ and a real number $x_0$, we wish to determine the behavior of $f^n(x_0)$ as $n$ gets larger and larger. Does $f^n(x_0)$ jump around seemingly at random, does it converge to a single point, or does it bounce back and forth between two values? To answer such questions, we could simply compute $f(x_0), f^2(x_0), f^3(x_0)$, and so on until, if we are lucky, we discern a pattern. Computing $f^n(x_0)$ can be a time-consuming process, even with the aid of computers. There is, however, a way to use the graph of the function to estimate the iterates of a function. For example, suppose that we are interested in computing iterates of the function $f(x) = 2x(1 - x)$ with $x_0 = 0.2$. We begin by graphing the function $f$ and also the line $y = x$. Next we locate $x_0 = 0.2$ on the $x$-axis and move vertically to locate the point $(0.2, f(0.2))$ on the graph. We then move horizontally until we hit the line $y = x$. We should now be at the point $(f(0.2), f(0.2))$. We then move vertically until we hit the graph of $f$; we are now at $(f(0.2), f(f(0.2)))$, or more briefly, $(f(0.2), f^2(0.2))$. We again move horizontally until we hit the line $y = x$, which leaves us at $(f^2(0.2), f^2(0.2))$. We continue in this way—moving vertically until we hit the graph of the function, then moving horizontally until we hit the line $y = x$—until we have found the desired iterate. The graph in Figure 43 illustrates this procedure. It is evident from this graph that as $n$ gets larger and larger, $f^n(0.2)$ is approaching 0.5.

**a.** Let $f(x) = 2x(1 - x)$ as previously stated. Use a calculator to compute $f(0.2), f^2(0.2), f^3(0.2)$, and $f^4(0.2)$, confirming that these iterates are indeed approaching 0.5.

**b.** Use the graphical technique illustrated in Figure 43 to determine the first four iterates of $f$ with $x_0 = 0.8$.

**c.** Now let $g(x) = 3.5x(1 - x)$. Use the graphical technique to determine the first 10 iterates of $g$ with $x_0 = 0.3$. A sketch of the graph of $g$ (along with the line $y = x$) is shown in Figure 44. Describe what happens to $g^n(x)$ as $n$ gets larger and larger.

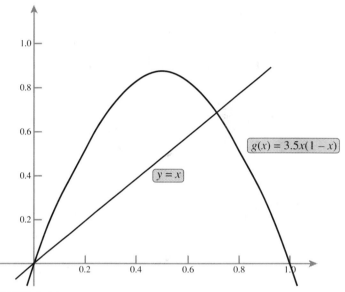

$g(x) = 3.5x(1 - x)$

$y = x$

**FIGURE 44**

**81.** *Solving Functional Equations* If $x$, $y$, and $z$ are real numbers, then the equation $xy = z$ can be solved for $y$ by multiplying both sides by $x^{-1}$:

$$xy = z$$
$$x^{-1}xy = x^{-1}z$$
$$y = x^{-1}z$$

This strategy can be modified for solving certain **functional equations**, equations in which the unknown is a function. The basic idea is that instead of multiplying by the multiplicative inverse, we compose with the functional inverse. For example, if the functions $f$ and $h$ are known and $f$ has an inverse, then the equation $f \circ g = h$ can be solved for $g$ as follows:

$$f \circ g = h$$
$$f^{-1} \circ f \circ g = f^{-1} \circ h$$
$$g = f^{-1} \circ h$$

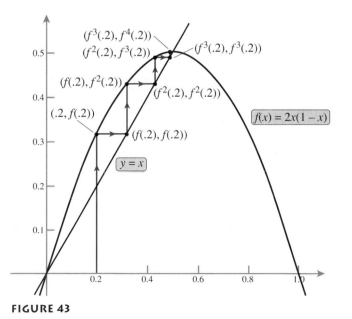

$(f^3(.2), f^4(.2))$
$(f^2(.2), f^3(.2))$
$(f^3(.2), f^3(.2))$
$(f(.2), f^2(.2))$
$(f^2(.2), f^2(.2))$
$(.2, f(.2))$
$f(x) = 2x(1 - x)$
$(f(.2), f(.2))$
$y = x$

**FIGURE 43**

a. Find a function $g$ such that $f \circ g = h$, where $f(x) = 2x - 4$ and $h(x) = \dfrac{2x - 10}{x - 2}$.

b. Find a function $f$ such that $f \circ g = h$, where $g(x) = 2x - 1$ and $h(x) = \dfrac{10x - 2}{4x - 3}$.

c. Consider the functional equation $f \circ g = h$, with $f(x) = 3x + 4$, and $h(x) = \dfrac{7x + 1}{x + 1}$. The following is an *incorrect* solution of this functional equation.

| | |
|---|---|
| $f \circ g = h$ | The original equation |
| $f^{-1} \circ f \circ g = h \circ f^{-1}$ | Composing with $f^{-1}$ on both sides |
| $g = h \circ f^{-1}$ | Simplifying |
| $g(x) = (h \circ f^{-1})(x)$ | Evaluating both sides at $x$ |
| $g(x) = h(f^{-1}(x))$ | Using the definition of composition of functions |

| | |
|---|---|
| $g(x) = h\left(\dfrac{x - 4}{3}\right)$ | Substituting $\dfrac{x - 4}{3}$ for $f^{-1}(x)$ |
| $= \dfrac{7\left(\dfrac{x - 4}{3}\right) + 1}{\left(\dfrac{x - 4}{3}\right) + 1}$ | Evaluating, using the fact that $h(x) = \dfrac{7x + 1}{x + 1}$ |
| $= \dfrac{7x - 25}{x - 1}$ | Simplifying |

If we check our solution by evaluating $(f \circ g)(x)$, we find that

$$(f \circ g)(x) = f\left(\frac{7x - 25}{x - 1}\right)$$
$$= 3\left(\frac{7x - 25}{x - 1}\right) + 4$$
$$= \frac{25x - 79}{x - 1}$$
$$\neq h(x)$$

What went wrong?

---

82. Discuss the analogy between functions and their inverses on the one hand, and numbers and their multiplicative inverses on the other.

83. What is the relationship between a function and its inverse with respect to the following properties: always positive, always negative, always increasing, always decreasing, 1–1, even, and odd? In other words, for example, if a function is always positive, then what can you say about its inverse?

84. If $f$ and $g$ are two functions of the form $h(x) = mx + b$, what can be said about $f + g$, $f - g$, $fg$, $f/g$, $f \circ g$, and $g \circ f$? Justify your answer.

---

## SECTION 4

# LINEAR, QUADRATIC, AND PIECEWISE FUNCTIONS

■ How long must one wait after drinking alcoholic beverages to ensure that the blood alcohol level falls below the legal limit?

■ Will a baseball hit at 144 feet per second (about 98 miles per hour) at an angle of 60° hit the ceiling of the Houston Astrodome?

■ Who would be willing to pay more for $10,000 worth of insurance, a wealthy person or a poor person?

■ How can a single function be used to compute federal income tax even when there are several tax brackets?

■ What are step functions and how can one be used to construct a code?

## LINEAR FUNCTIONS

Recall that the graph of an equation of the form $y = mx + b$ is a line with slope $m$ and $y$-intercept $b$. In fact, any nonvertical line can be put in this form, and it is for this reason that we define a **linear function** to be a function of the form $f(x) = mx + b$. Such functions are certainly the easiest to study, and they are also among the most useful for applications.

**EXAMPLE 1**    *Linear cost and revenue functions*

A sidewalk hot dog vendor has a fixed daily cost of $60 for the rental of a cart and a vendor permit. The variable costs (for the hot dogs, buns, and condiments) average 30¢ per hot dog. Each hot dog sells for $1.50.
a. Construct a linear cost function that gives total cost for the day as a function of the number of hot dogs sold.
b. Construct a linear revenue function that gives the total revenue for the day as a function of the number of hot dogs sold.
c. How many hot dogs would have to be sold in a given day in order to break even? Illustrate this graphically.

**SOLUTION**

a. If $x$ represents the number of hot dogs sold in 1 day, then

$$\text{Total cost} = \text{Fixed cost} + \text{Variable cost}$$
$$= 60 \qquad\quad + \text{(Cost per hot dog)(Number of hot dogs)}$$
$$= 60 \qquad\quad + (0.3)x$$

So the total cost function is $C(x) = 0.3x + 60$.

b. Again, if $x$ represents the number of hot dogs sold in 1 day, then

$$\text{Total revenue} = \text{(Revenue per hot dog)(Number of hot dogs)}$$
$$= (1.50)x$$

So the total revenue function is $R(x) = 1.5x$.

c. To break even, the total revenue must equal the total cost. Thus,

$$R(x) = C(x)$$
$$1.5x = 0.3x + 60$$
$$1.2x = 60$$
$$x = 50$$

So 50 hot dogs must be sold in 1 day to break even that day. Graphically, 50 is the $x$-coordinate of the point of intersection of the graphs of $R$ and $C$, as shown in Figure 45.

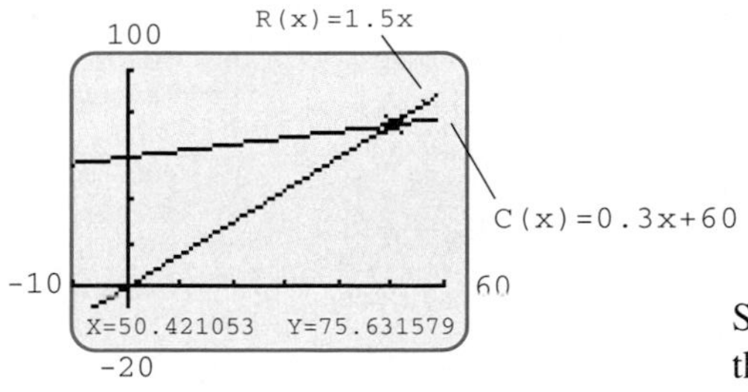

```
        100     R(x)=1.5x



                              C(x)=0.3x+60
  -10                     60
    X=50.421053   Y=75.631579
        -20
```

**FIGURE 45**

## QUADRATIC FUNCTIONS

After linear functions, the next simplest functions are the **quadratic functions**, which are functions of the form $f(x) = ax^2 + bx + c$, where $a$, $b$, and $c$ are constants and $a \neq 0$. To obtain the graph of a quadratic function, we let $y = f(x)$ so that

$y = ax^2 + bx + c$. You may recognize this equation as that of a parabola that opens either upward (if $a > 0$) or downward (if $a < 0$), as shown in Figure 46.

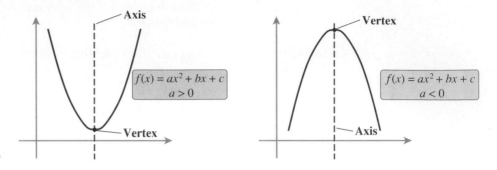

**FIGURE 46**

The **vertex** is the lowest or highest point on the parabola, and the **axis** is the vertical line passing through the vertex. The coordinates of the vertex can be found by completing the square in the equation $y = ax^2 + bx + c$. After simplifying, we obtain

$$y - \left( c - \frac{b^2}{4a} \right) = a \left( x + \frac{b}{2a} \right)^2$$

Thus, the vertex is the point

$$\left( -\frac{b}{2a}, \, c - \frac{b^2}{4a} \right)$$

Since this formula is rather difficult to memorize, we can simply remember that the vertex has $x$-coordinate $h = -b/2a$, so the $y$-coordinate of the vertex is then $f(h)$. We summarize these results below.

*Quadratic functions*

■ A function of the form $f(x) = ax^2 + bx + c \, (a \neq 0)$ is called a quadratic function, and its graph is a parabola opening upward if $a > 0$, and downward if $a < 0$.

■ The vertex of the parabola is the point $\left( -\dfrac{b}{2a}, f\left( -\dfrac{b}{2a} \right) \right)$.

**EXAMPLE 2**    *Graphing a quadratic function*

Sketch the graph of $f(x) = 2x^2 + 12x + 13$ and identify the vertex.

**SOLUTION**    Here $a = 2$ and $b = 12$. Thus, the vertex will have $x$-coordinate

$$-\frac{b}{2a} = -\frac{12}{2 \cdot 2} = -3$$

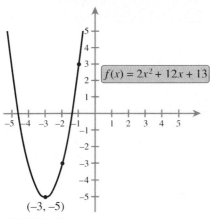

$f(x) = 2x^2 + 12x + 13$

$(-3, -5)$

**FIGURE 47**

The $y$-coordinate of the vertex is then given by

$$f(-3) = 2(-3)^2 + 12(-3) + 13$$
$$= 18 - 36 + 13$$
$$= -5$$

Thus, the vertex is $(-3, -5)$. Finally, we plot several points and then use symmetry with respect to the axis of the parabola to obtain the sketch in Figure 47.

| $x$ | $y$ |
|-----|-----|
| $-3$ | $-5$ |
| $-2$ | $-3$ |
| $-1$ | $3$ |

**EXAMPLE 3**

### *The path of a baseball*

Houston Astrodome

The Astrodome in Houston, Texas, rises to a maximum height of 208 feet above the playing field. If a baseball is hit from a height of 3 feet with an initial velocity of 144 feet per second and an initial angle of 60°, and if we ignore air resistance, the ball will follow the path given by the quadratic function $f(x) = -\frac{1}{324}x^2 + \sqrt{3}x + 3$, where $x$ is the distance in feet from home plate and $f(x)$ is the corresponding height in feet above the playing field. Assuming the ball is hit fair and the playing field extends at least 300 feet from home plate, show that the ball will hit the dome ceiling somewhere above the field.

**SOLUTION**    If we initially ignore the presence of the ceiling, the maximum height of the ball will occur at the vertex of $f(x) = -\frac{1}{324}x^2 + \sqrt{3}x + 3$. The $x$-coordinate of the vertex is given by

$$x = -\frac{b}{2a} = -\frac{\sqrt{3}}{2\left(-\dfrac{1}{324}\right)} = 162\sqrt{3} \approx 281$$

The height of the ball at this value of $x$ is

$$f(281) = -\frac{1}{324}(281)^2 + \sqrt{3}(281) + 3 \approx 246$$

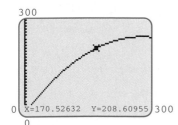

**FIGURE 48**

So disregarding the ceiling, the ball will reach a height of approximately 246 feet at a distance of approximately 281 feet from home plate. Since the ceiling has a maximum height of 208 feet, we can conclude that the ball will hit the ceiling somewhere above the playing field. A graphical view of this can be obtained by plotting $y = -\frac{1}{324}x^2 + \sqrt{3}x + 3$ and then using the trace feature as shown in Figure 48. We see that the height of the ball (given by the $y$-coordinate) reaches 208 feet by the time the distance from home plate (given by the $x$-coordinate) has exceeded 170 feet.

# PIECEWISE FUNCTIONS

Occasionally, it is desirable to use two or more formulas to define a function. Consider, for example, the function $f$ defined below.

$$f(x) = \begin{cases} -x, & x \le -1 \\ 2x + 3, & x > -1 \end{cases}$$

This *piecewise* definition indicates that for an input value less than or equal to $-1$, the output value is computed using the formula $y = -x$. For an input value greater than $-1$, the output value is computed using the formula $y = 2x + 3$. Thus, $f(-4) = -(-4) = 4$, while $f(1) = 2(1) + 3 = 5$. The graph of $f$ is obtained by plotting the line $y = -x$ for $x \le -1$, and the line $y = 2x + 3$ for $x > -1$, as shown in Figure 49.

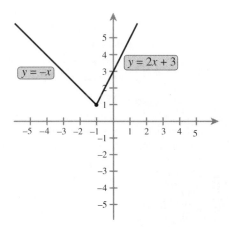

**FIGURE 49**

**EXAMPLE 4**  *Evaluating and graphing a piecewise function*

Sketch the graph of the function

$$f(x) = \begin{cases} x + 1, & x < 2 \\ x^2 - 4, & x \ge 2 \end{cases}$$

and evaluate it at $x = -3$, $x = 2$, and $x = 5$.

**SOLUTION**    The graph of $f$ is obtained by plotting the line $y = x + 1$ for $x < 2$ and the parabola $y = x^2 - 4$ for $x \ge 2$, as shown in Figure 50. Notice that an open circle is used at the point $(2, 3)$ to indicate that the point is not included on the graph while a closed circle is used at $(2, 0)$ to indicate that this point is included. To compute a function value for a given $x$, we simply note whether $x < 2$ or $x \ge 2$ and use the corresponding ''piece'' of the function definition.

$$f(-3) = (-3) + 1 = -2 \qquad -3 < 2, \text{ so we use } x + 1$$
$$f(2) = (2)^2 - 4 = 0 \qquad 2 \ge 2, \text{ so we use } x^2 - 4$$
$$f(5) = (5)^2 - 4 = 21 \qquad 5 \ge 2, \text{ so we use } x^2 - 4$$

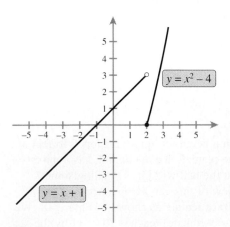

**FIGURE 50**

Piecewise functions can also be plotted with the aid of a graphics calculator. Although some graphics calculators have the ability to plot piecewise functions directly, we can simply graph all of the expressions used to define the piecewise function, while recognizing that for a given $x$-value, only one of the graphs applies.

**EXAMPLE 5**    *Graphing a piecewise function with a graphics calculator*

Graph the following function with the aid of a graphics calculator.

$$f(x) = \begin{cases} -x^3 + 6x^2 - 9x + 4, & x < 3 \\ x - 3, & x \geq 3 \end{cases}$$

**SOLUTION**    Unless our graphics calculator has the ability to plot piecewise functions, we will begin by plotting the graphs of both $y = -x^3 + 6x^2 - 9x + 4$ and $y = x - 3$ on the same screen, as shown in Figure 51. Now we recognize that only the portion of the graph of $y = -x^3 + 6x^2 - 9x + 4$ for which $x < 3$ (shown in red) is actually part of the graph of $f$; moreover, only the portion of the graph of $y = x - 3$ for which $x \geq 3$ (shown in blue) is part of the graph of $f$. Figure 52 shows the graph of $f$.

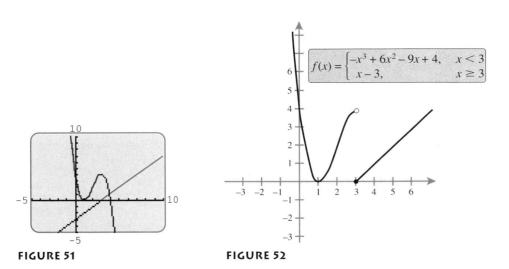

**FIGURE 51**          **FIGURE 52**

---

**EXAMPLE 6**    *Constructing and graphing a piecewise function*

A moving van rental company has two different rate schedules for a certain size of moving van. For vans used locally (a total distance of less than 100 miles), the rate is $40 plus 39¢ per mile. For long-distance rentals (a total distance of 100 miles or more), the rate is $200 plus 49¢ per mile for each mile over 400.
**a.** Construct a piecewise function that can be used to compute the rate for any distance.
**b.** Plot the graph with the aid of a graphics calculator.
**c.** Determine how far one can move with $250.

**SOLUTION**

a.  Let $x$ represent the total distance traveled. If $x < 100$, the rate is $40 + 0.39x$. If $100 \leq x \leq 400$, the rate is a fixed \$200. If $x > 400$, the rate is \$200 plus \$0.49 for each mile over 400, or $200 + 0.49(x - 400)$. The resulting function $f$ is

$$f(x) = \begin{cases} 40 + 0.39x, & 0 \leq x < 100 \\ 200, & 100 \leq x \leq 400 \\ 200 + 0.49(x - 400), & x > 400 \end{cases}$$

Values of the function $f$ give the rental cost (in dollars) for moving $x$ miles.

b.  We begin by graphing all three of the formulas that are used to define the function $f$, as shown in Figure 53. Note that only the red portion of each graph is actually part of the graph of $f$. In Figure 54 we show the graph of $f$.

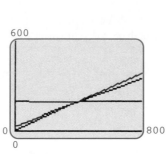

**FIGURE 53**

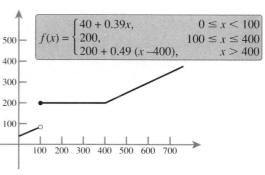

**FIGURE 54**

c.  From the graph we see that the $y$-coordinate attains the value 250 for an $x$-value larger than 400. Thus, we use the formula for $f$ corresponding to $x > 400$, set $f(x)$ equal to 250, and solve for $x$.

$$200 + 0.49(x - 400) = 250$$

$$0.49(x - 400) = 50$$

$$x - 400 = \frac{50}{0.49}$$

$$x = \frac{50}{0.49} + 400$$

Thus, the total distance is $x = \dfrac{50}{0.49} + 400 \approx 502$ miles.

■

Two special examples of piecewise functions are the absolute-value function and the so-called step functions. Although the **absolute-value function** $f(x) = |x|$ can be written as a single equation, it is equivalent to

$$f(x) = \begin{cases} -x, & x < 0 \\ x, & x \geq 0 \end{cases}$$

So the graph of $f(x) = |x|$ is found by graphing $y = -x$ for $x < 0$ and $y = x$ for $x \geq 0$, as shown in Figure 55.

A **step function** is a piecewise function that takes only constant values over each interval of its domain. An example of a step function is the function $g$ whose definition is given below and whose graph is shown in Figure 56.

$$g(x) = \begin{cases} 1, & x \leq -3 \\ -1, & x > -3 \end{cases}$$

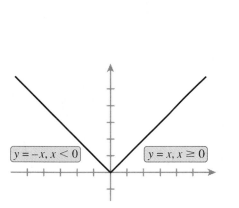

**FIGURE 55**                    **FIGURE 56**

Another example of a step function is the **greatest integer function** $f(x) = [\![x]\!]$, where $[\![x]\!]$ is defined as follows.

*The greatest integer function*

> For any real number $x$, $[\![x]\!]$ denotes the greatest integer less than or equal to $x$.

Applying this definition, we see that $[\![2.6]\!] = 2$, $[\![8]\!] = 8$, $[\![-4.1]\!] = -5$. Note in particular that the greatest integer function does not simply round or truncate the decimal places. More generally, we observe that for every real number $x$ between two consecutive integers $n$ and $n + 1$, $[\![x]\!] = n$. Thus, the graph of the greatest integer function is constant between each pair of consecutive integers, as shown in Figure 57.

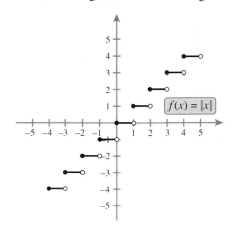

**FIGURE 57**

# EXERCISES 4

**EXERCISES 1–12** □ *Identify the type of function that is given (piecewise, quadratic, linear, or none of these).*

1. $f(x) = 3x^2 + 4$

2. $g(x) = 2x^3 + 5x + 7$

3. $h(x) = \dfrac{x^2 - 5x + 3}{x + 1}$

4. $p(x) = \dfrac{2x + 5}{x - 4}$

5. $q(x) = 2^x$

6. $r(x) = |x|$

7. $f(x) = 3x + 9$

8. $g(x) = 2$

9. $f(t) = 4^{13}t + 0.001$

10. $r(s) = 2s^2 - 1$

11. $f(x) = \begin{cases} x, & x < 0 \\ x^2, & x > 0 \end{cases}$

12. $g(p) = 3p$

**EXERCISES 13–16** □ *Plot the graph of the linear function and identify the slope and y-intercept.*

13. $f(x) = 2x - 1$

14. $g(x) = -3x + 2$

15. $h(x) = \dfrac{-3x + 4}{4}$

16. $g(x) = \dfrac{x - 6}{2}$

**EXERCISES 17–20** □ *Find the linear function that satisfies the given properties.*

17. The graph of $h$ has slope $-\frac{1}{2}$ and passes through the point $(2, 4)$.

18. The graph of $f$ passes through the points $(-2, 3)$ and $(2, -3)$.

19. The graph of $g$ has $y$-intercept $-4$ and $x$-intercept 3.

20. The graph of $f$ is the perpendicular bisector of the line segment connecting $(-2, 1)$ and $(4, 3)$.

**EXERCISES 21–28** □ *Find the coordinates of the vertex of the graph of the given quadratic function.*

21. $f(x) = 3x^2 - 4$

22. $g(x) = x^2 + 4x + 9$

23. $f(x) = 2x^2 - 6x + 10$

24. $g(x) = ax^2 + c$

25. $m(p) = p^2 + 16p + 10$

26. $f(x) = 2^5x^2 + 2^7x$

27. $P(y) = 1 - 2y + y^2$

28. $h(x) = 3x + x^2 + 5$

**EXERCISES 29–34** □ *Sketch the graph of the piecewise function and evaluate it as indicated.*

29. $f(x) = \begin{cases} -x - 4, & x \le 0 \\ 2x - 4, & x > 0 \end{cases}$

    $f(-2), f(0), f(3)$

30. $g(x) = \begin{cases} 3x + 2, & x < 0 \\ -x + 2, & x \ge 0 \end{cases}$

    $g(-4), g(0), g(1)$

31. $h(x) = \begin{cases} x^2 - 1, & x < 2 \\ \frac{1}{2}x + 1, & x \ge 2 \end{cases}$

    $h(0), h(2), h(4)$

32. $g(x) = \begin{cases} -\frac{2}{3}x + 2, & x \le -3 \\ 4 - x^2, & x > -3 \end{cases}$

    $g(-4), g(-3), g(0)$

33. $f(x) = \begin{cases} x, & x < -1 \\ 1, & -1 \le x \le 1 \\ x + 2, & x > 1 \end{cases}$

    $f(-3), f(-1), f(4)$

34. $h(x) = \begin{cases} -x - 1, & x < -2 \\ x + 3, & -2 \le x \le 0 \\ 3, & x > 0 \end{cases}$

    $h(-3), h(0), h(6)$

**EXERCISES 35–38** □ *Use the graphs of f and g to sketch the graph of h.*

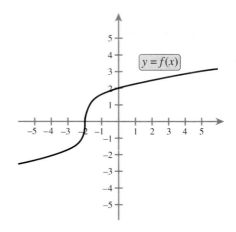

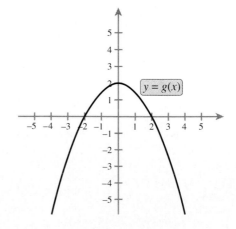

35. $h(x) = \begin{cases} f(x), & x \le -2 \\ g(x), & x > -2 \end{cases}$

36. $h(x) = \begin{cases} g(x), & x \le 0 \\ f(x), & x > 0 \end{cases}$

**37.** $h(x) = \begin{cases} g(x), & x \le -2 \\ x, & -2 < x < 0 \\ f(x), & x \ge 0 \end{cases}$

**38.** $h(x) = \begin{cases} f(x), & x < -2 \\ 0, & -2 \le x \le 2 \\ g(x), & x > 2 \end{cases}$

**EXERCISES 39–42** □ *Use a graphics calculator to help you graph the given piecewise function.*

**39.** $f(x) = \begin{cases} -x^2 + 2x, & x \le 1 \\ (x - 2)^2, & x > 1 \end{cases}$

**40.** $g(x) = \begin{cases} -x^2 - 4x - 3, & x < -2 \\ x^2 + 4x + 5, & x \ge -2 \end{cases}$

**41.** $g(x) = \begin{cases} x^3 + 2x^2, & x < -1 \\ 1, & -1 \le x \le 1 \\ x^3 - 2x^2 + 2, & x > 1 \end{cases}$

**42.** $g(x) = \begin{cases} \dfrac{2}{x^2 + 1}, & x < -1 \\ -x, & -1 \le x \le 1 \\ \dfrac{-2}{x^2 + 1}, & x > 1 \end{cases}$

**43.** Let $f(x) = [\![x + 2]\!]$. Sketch the graph of $f$.

**44.** Let $h(x) = [\![x]\!] - 2$. Sketch the graph of $h$.

**45.** For an interval $[a, b]$, define

$$f_{[a,b]}(x) = \begin{cases} 1, & a \le x \le b \\ 0, & \text{otherwise} \end{cases}$$

**a.** Graph $f_{[0,1]}$

**b.** Graph $f_{[0,2]} + f_{[1,3]}$

**c.** Graph $f_{[0,1]} + f_{[1,2]} + f_{[2,3]} + \cdots$

**46.** Let $f(x) = \dfrac{|x|}{x}$.

**a.** Find the domain of $f$.

**b.** Evaluate the following.

    **i.** $f(2)$        **ii.** $f(5)$

    **iii.** $f(-2)$     **iv.** $f(-1)$

**c.** Find the range of $f$.

**d.** Graph $f$.

**47.** Let $g(x) = ((x))$, where $((x))$ denotes the fractional part of $x$, so that, for example, $((3.456)) = 0.456$, $((2\tfrac{7}{8})) = \tfrac{7}{8}$, and so on.

**a.** Evaluate the following:

    **i.** $g(2.8)$     **ii.** $g(4.7)$     **iii.** $g(\tfrac{23}{5})$

**b.** Graph $g$ for $x \ge 0$.

**c.** Simplify $[\![x]\!] + ((x))$ for $x \ge 0$.

**48.** The function "sgn," the signum function, is defined by

$$\text{sgn}(x) = \begin{cases} 1, & x > 0 \\ -1, & x < 0 \end{cases}$$

**a.** Graph sgn$(x)$.

**b.** Let $f(x) = \text{sgn}(x - 2)$. Graph $f$.

---

■ *Applications*

---

**49.** *Rental Profit* A landlord owns 50 apartments that can be rented on a monthly basis. He has fixed overhead costs of $5000 per month, plus $80 per month for every apartment that is rented. He has determined that all of the apartments will be rented if he charges $300 per month. However, for every $10 increase in rent, two fewer apartments will be rented.

**a.** Find a linear cost function that expresses the landlord's total monthly cost as a function of the number of apartments rented.

**b.** Find a linear *demand* function that expresses the rental price as a function of the number of apartments rented. [*Hint:* Find two points $(x, p)$, where $x$ is the number of units rented and $p$ is the corresponding monthly rent, and then find the equation of the line passing through the two points.]

**c.** Use the demand function from part (b) to find a revenue function that expresses the landlord's monthly revenue as a function of the number of apartments rented.

**d.** Use the cost and revenue functions to find the profit function.

**e.** How many apartments should be rented, and at what price, for the landlord to maximize his profit?

**50.** *Ticket Profit* A university has a football stadium that can hold 80,000 people. A recent analysis has determined that the cost to staff the stadium for a game is $50,000 plus 25¢ for each person in attendance. It has also been determined that all the seats will be filled if the average ticket price is $10, but 5000 fewer people will attend for each $1 increase in ticket price.

**a.** Find a linear cost function that expresses the university's total cost to staff the stadium as a function of the number of people in attendance.

**b.** Find a linear *demand* function that expresses the ticket price as a function of the number of people in attendance. [*Hint:* Find two points $(x, p)$, where $x$ is the number of people in attendance and $p$ is the corresponding ticket price, and then find the equation of the line passing through the two points.]

c. Use the demand function from part (b) to find a revenue function that expresses the university's revenue as a function of the number of people in attendance.

d. Use the cost and revenue functions to find the profit function.

e. How many people must attend, and at what ticket price, for the university to maximize profit?

51. *Path of a Baseball* If a baseball is hit from a height of 3 feet with an initial velocity of 100 feet per second and an initial angle of 45°, and if we ignore air resistance, the ball will follow the path given by the quadratic function $f(x) = -\frac{2}{625}x^2 + x + 3$, where $x$ is the distance in feet from home plate and $f(x)$ is the corresponding height in feet above the playing field.

a. Find the maximum height of ball.

b. Will the ball clear a 12-foot-high fence located 300 feet from home plate?

52. *Path of a Baseball* If a baseball is thrown from a height of 5 feet with an initial velocity of 80 feet per second and an initial angle of 30°, and if we ignore air resistance, the ball will follow the path given by the quadratic function

$$f(x) = -\frac{1}{300}x^2 + \frac{\sqrt{3}}{3}x + 5$$

where $x$ is the distance in feet from where the ball is thrown and $f(x)$ is the corresponding height in feet.

a. Find the maximum height of ball.

b. How far will the ball travel before it hits the ground?

53. *Farm Acreage* The number of acres of U.S. farmland during the years 1930–1990 can be approximated with the quadratic function $f(x) = -0.21x^2 + 11.9x + 987$, where $x$ is the year ($x = 0$ corresponds to 1930) and $f(x)$ is the total acreage in millions of acres. At what time was the number of acres highest? How many acres were there at that time?

**U.S. Farm Acreage 1930-1990**

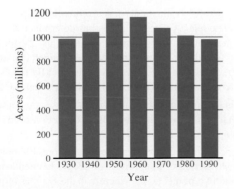

Data Source:  U.S. Department of Agriculture

54. *Alcohol Consumption* The average number of gallons of alcohol consumed by each adult in the United States for the years 1970–1990 can be approximated with the quadratic function $f(x) = -0.053x^2 + 1.17x + 35.6$, where $x$ is the year ($x = 0$ corresponds to 1970) and $f(x)$ is the average number of gallons consumed by each adult in that year. For what year was the alcohol consumption the greatest, and what was the average rate of consumption at that time?

**Alcohol Consumption 1970-1990**

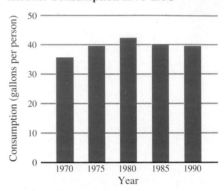

Data Source:  U.S. Department of Agriculture

55. *Telephone Rate* Suppose the cost of a phone call from New York to Los Angeles is 95¢ for the first minute and 65¢ for each additional minute or fraction thereof. Then the cost of a phone call lasting $x$ minutes can be computed using the function $C(x) = 0.95 - 0.65[\![1 - x]\!]$. Plot the graph of this function. What is the longest phone call that can be made if the cost cannot exceed $5?

56. *Postal Rate* In 1991, the U.S. postal rate for first-class mail was raised to 29¢ for the first ounce and 23¢ for each additional ounce or fraction thereof. The rate for an item weighing $x$ ounces can be computed with the function $C(x) = 0.29 - 0.23[\![1 - x]\!]$. Plot the graph of this function. What is the heaviest first-class item that can be sent if the cost cannot exceed $3.00?

57. *Tax Rates* Fred and Irma are trying to decide if they should file a joint tax return or if they should file separately. If they file separately, Fred has a taxable income of $40,000 and Irma has a taxable income of $50,000. If they file jointly, their combined taxable income is $85,000. The function $S$ below computes taxes for separate returns and the function $J$ computes taxes for joint returns. In both cases, $x$ denotes the taxable income. Compute the total taxes they will pay if they file separately. How does this amount compare to the taxes they will pay if they file jointly?

$$S(x) = \begin{cases} 0.15x, & 0 \leq x < 17,900 \\ 2685 + 0.28(x - 17,900), & 17,900 \leq x < 43,250 \\ 9783 + 0.31(x - 43,250), & x \geq 43,250 \end{cases}$$

$$J(x) = \begin{cases} 0.15x, & 0 \le x < 35{,}800 \\ 5370 + 0.28(x - 35{,}800), & 35{,}800 \le x < 86{,}500 \\ 19{,}566 + 0.31(x - 86{,}500), & x \ge 86{,}500 \end{cases}$$

**58.** *Tax Rates* The tax rates for a single person filing a U.S. tax return in 1992 are given in Table 2. Construct a piecewise function that expresses the total tax that must be paid as a function of taxable income.

**59.** *Secret Code* A coded message from a terrorist organization is intercepted by an intelligence satellite. The message reads EFPSNKXEFL. As a leading authority on terrorist activity, Inspector Magill is called in to help break the code. He imme-

diately recalls an incident from the previous day when a member of the same organization was arrested for the possession of an illegal assault rifle. Among other items found in his possession was a piece of paper upon which was written

$$f(x) = (3x + 5) - 26 \left[\!\!\left[ \frac{3x + 5}{26} \right]\!\!\right]$$

Within minutes, Magill has broken the code and foiled a hijacking plan. What was the message? (*Hint:* First assign to each letter of the alphabet the number corresponding to its position. For example, A↔1, B↔2, ....)

**TABLE 2**

*1992 Tax rates (single)*

| If taxable income is over ... | but not over ... | the tax is ... | of the amount over ... |
|---|---|---|---|
| $0 | $21,450 | 15% | $0 |
| $21,450 | $51,900 | $3,217.50 + 28% | $21,450 |
| $51,900 | | $11,743.50 + 31% | $51,900 |

## Projects for Enrichment

**60.** *Driving Under the Influence* Concern over the incidence of alcohol-related motor-vehicle accidents has prompted lower limits on the legal level of alcohol in the bloodstream. Unfortunately, most people are unaware of the relationship between alcohol consumption and the level of alcohol in the bloodstream. Indeed, many are unaware that even small quantities of alcohol can raise the level beyond legal limits. In this project we will investigate the relationship between alcohol consumption and the alcohol level in the bloodstream, and we will also see what bearing body weight has on the relationship.

Ethanol, or grain alcohol, is the form of alcohol that is found in beer, wine, and other liquors. It is eliminated from the body by the liver via an enzymatic process at the constant rate of 12 grams per hour. We will assume that each serving of beer, wine, or other liquors contains 18 grams of ethanol. We will further assume that the ethanol enters the bloodstream immediately after the drink is consumed.

**a.** Suppose one drink is consumed at time $t = 0$. Find a linear function $A_1(t)$ that gives the number of grams of ethanol remaining in the bloodstream after $t$ hours. How long will it take for all the ethanol to be eliminated?

**b.** Now suppose that one drink is consumed at time $t = 0$ and a second is consumed 1 hour later at time $t = 1$. Find a piecewise function $A_2(t)$ that gives the number of grams of ethanol remaining in the bloodstream after $t$ hours. Note that one piece will correspond to the interval $0 \le t < 1$ and the

second will correspond to $t > 1$. Plot the graph of $A_2$. How long will it take for all the ethanol to be eliminated?

**c.** Suppose that drinks are consumed at a regular rate of 2 per hour for 2 hours (i.e., at times $t = 0$, $t = \frac{1}{2}$, $t = 1$, $t = \frac{3}{2}$, and $t = 2$). Find a piecewise function $A_3(t)$ that gives the number of grams of ethanol remaining after $t$ hours, and plot the graph of $A_3$.

**d.** Repeat part (c) to find a function $A_4$ for consumption at a rate of 3 drinks per hour.

If $A(t)$ denotes the number of grams of ethanol in the bloodstream at time $t$, then the percentage levels $L(t)$ of blood serum ethanol at time $t$ are approximated by

$$L(t) = \frac{A(t)}{310 \times (\text{body weight in pounds})} \quad \text{for adult males}$$

and

$$L(t) = \frac{A(t)}{250 \times (\text{body weight in pounds})} \quad \text{for adult females}$$

**e.** For each of the functions $A_1$–$A_4$ in parts (a) through (d), find the functions $L_1(t)$ through $L_4(t)$ for an adult male weighing 160 pounds and an adult female weighing 130 pounds.

f. In many states, the legal limit is 0.1%. In other words, one is legally under the influence if $L(t) \geq 0.001$. For each of the functions $L_1(t)$ for $L_4(t)$ in part (e), find the time interval during which the blood serum ethanol level is at or above 0.001. It may be helpful to sketch the graphs of the functions.

g. Discuss some of the limitations of this model for computing blood serum ethanol level.

61. *Utility Functions* Financial decisions that involve uncertainty are the province of **utility theory**. In this project you will construct a utility function by making a series of simple decisions.

The value of the utility function $u(x)$ is an indicator of the relative value the individual in question places on having wealth $x$. We will begin by supposing that you have a wealth of $50,000. Since utility functions are only used for determining relative values, we will arbitrarily set $u(0) = 0$ and $u(50,000) = 50,000$, so that two points on the graph of $u$ are determined.

a. Suppose that a neighborhood tough offered you the following deal. If you pay him ''protection,'' then he will guarantee that nothing happens to you. If you do not pay, then there is a 50–50 chance that you will lose all your money. What is the *maximum amount $G$* that you would be willing to pay for such insurance? What factors would influence this amount?

b. We will now use the points $(0, 0)$ and $(50,000, 50,000)$ on the graph of $u$ and your response to part (a), to find a third point on the graph.

After paying $G$, your wealth will be $50,000 - G$, and the utility that you associate with this level of wealth will be $u(50,000 - G)$. We can compute $u(50,000 - G)$ as follows. Since $G$ is the maximum that you would be willing to pay for insurance, you evidently view the following situations as equally desirable.

(1) Paying $\$G$ in insurance so that your wealth is $50,000 - G$, with associated utility $u(50,000 - G)$.

(2) Facing even odds of losing everything that you own by not buying any insurance. Since the utility of losing everything is $u(0)$, and the utility of keeping your current level of wealth is $u(50,000)$, the utility of this situation is the average of the two, namely

$$\frac{u(0) + u(50,000)}{2}$$

Use the fact that situations (1) and (2) are *equally* desirable to compute $u(50,000 - G)$.

c. We can now repeat the method used in part (b) to find additional points on the graph of $u$. In fact, if $(x, u(x))$ and $(y, u(y))$ are points on the graph of $u$ with $x < y$, then we can find a third point on the graph of $u$ as follows. Decide on the maximum amount of insurance $H$ that you would be willing to pay to protect yourself from a 50–50 chance of dropping from a wealth of $y$ to a wealth of $x$. Then $u(y - H)$ can be computed by an argument similar to that used in part (b) to compute $u(50,000 - G)$. Use this technique repeatedly until you have determined a total of six points on the graph of $u$, and plot them on graph paper. Connect consecutive points with line segments.

d. Answer the following questions using only the graph of $u$ constructed in part (c).

i. How much insurance would you be willing to pay to protect against the possibility of dropping from a wealth of $32,000 to a wealth of $16,000? [*Hint:* If we let this amount be $x$, then

$$u(32,000 - x) = \frac{u(32,000) + u(16,000)}{2}$$

Use the graph of $u$ to find $32,000 - x$, and then determine $x$.]

ii. Suppose that your current level of wealth is $13,000. How much would you be willing to pay for a 50–50 chance of winning $30,000?

iii. How much would you be willing to pay to insure against a 50–50 chance of losing $10,000 if you currently have a wealth of $50,000?

iv. How much would you be willing to pay to insure against a 50–50 chance of losing $10,000 if you currently have a wealth of $15,000?

e. Based on your experience with part (d), who would be willing to pay more to insure a $10,000 automobile: an extremely wealthy person or a relatively poor person? Explain your answer.

f. How much do you suppose a multibillionaire like Bill Gates (founder and CEO of Microsoft) would be willing to pay for a 50–50 chance of winning $100,000? How about someone with a personal wealth of $55,000?

### Questions for Discussion or Essay

62. The function $f(x) = -\frac{1}{324}x^2 + \sqrt{3}x + 3$ in Example 3 takes into account the initial velocity, angle, and height of the ball, and also the force of gravity. What other factors might also affect the path of the ball? How do these factors compare in significance to the ones that were taken into account? How will these factors influence such things as the maximum height of

the ball and the maximum distance the ball travels before it hits the ground (assuming it doesn't hit the ceiling first)?

63. How can you tell whether a graphical image is the plot of the graph of one piecewise defined function, or of several functions?

**64.** Some quadratic functions have a maximum, whereas others have a minimum. How can you determine whether a given quadratic function has a maximum or a minimum without graphing the function?

**65.** Some people view taxes as payment for services rendered by the government: defending our borders, educating our children, protecting our environment, insuring our elderly and indigent, and so forth. Certainly the cost of a quart of milk is not dependent on one's income, nor is the cost of virtually any commodity available in the marketplace. Why then are taxes dependent on income? If we let $i$ represent income and $T(i)$ represent the corresponding income tax, then what kind of function would $T$ be if all individuals were taxed the same amount? What if all individuals were taxed at a flat rate? Some have suggested that anyone earning under a certain cut-off amount should not pay any tax, but that anyone earning above the cut-off should be taxed at a flat rate on the money they earn above the cut-off level. What kind of function would $T(i)$ be under these circumstances? What method of tax computation do you feel would be most fair?

---

**SECTION 5**

# MODELING WITH FUNCTIONS AND VARIATION

■ What is the gravitational pull exerted on the Moon by 450-pound wrestler Big Van Vader?

■ If a man is 48 times as tall as a grasshopper is long and a grasshopper can jump 30 inches, how far could a man-sized grasshopper jump?

■ What will the total health-care expenditures be in the year 2000?

■ How much is a graduate degree worth?

■ What concentration of the ozone-depleting chemical CFC-11 will be present in the atmosphere in the year 2000?

## MATHEMATICAL MODELS

A **mathematical model** is a mathematical description of the behavior of some aspect of the real world. Mathematical models have been formulated for such diverse phenomena as signal transmission in the human nervous system, the aerodynamics of a hummingbird, the spread of AIDS, deforestation of the Amazon rain forest, the jet stream, the path of Halley's comet, and the entire global economy. All of the major theories of the physical sciences are, in essence, mathematical models. Examples include Einstein's general theory of relativity, which models gravity, and superstring theory, which is a model of the very fabric of space itself.

Real-world phenomena tend to be so extraordinarily complex and subject to so many random influences that any description, mathematical or otherwise, is necessarily incomplete. Mathematical models attempt to capture the salient features of a real-world phenomenon, but just as a model train is an oversimplified version of an actual train, so too is a mathematical model an oversimplification. Nonetheless, the accuracy and utility of certain mathematical models is startling. For example, our current model of the motion of the Earth, the Sun, and the Moon is so good that it correctly predicts the time of eclipses to within seconds, decades in advance. On the other hand, in spite of an astronomical amount of data collection, the use of the most powerful supercomputers in existence, and the efforts of some of the most brilliant minds on the planet, existing models of the Earth's atmosphere are not good enough to reliably predict the weather even 3 or 4 days in advance!

You have already seen many examples of mathematical models in the text—from falling objects and stopping distance in previous chapters to $CO_2$ levels and alcohol consumption in this chapter. However, in these earlier examples, we were concerned primarily with how a model could be used to solve a real-world problem. In this section we investigate not just how models are used to solve problems, but also how models are constructed.

## VARIATION

Many of the most important mathematical models arising in the natural sciences can be conveniently described using the language of **variation**. For example, the force acting on an object *varies directly as* the acceleration that the object is undergoing; the strength of the gravitational pull between two bodies *varies inversely as* the square of the distance between them and *varies jointly as* the masses of the bodies. The terminology of variation is summarized here.

*Variation terminology*

| Type of variation | Equation | Terminology |
|---|---|---|
| Direct variation | $y = kx$ | "y varies directly as x," "y is proportional to x," or "y is directly proportional to x" |
| Inverse variation | $y = \dfrac{k}{x}$ | "y varies inversely as x," or "y is inversely proportional to x" |
| Joint variation | $z = kxy$ | "z varies jointly as x and y" |

In all cases, the constant $k$ is called the **constant of proportionality**.

If $y$ varies directly as $x$, then we can conclude that $y = kx$ for some constant $k$. Note, however, that this information alone is not enough to completely specify the relationship between the two variables: additional data is required to determine $k$, the constant of proportionality, as illustrated in the following examples.

**EXAMPLE 1**   *Computing the constant of proportionality*

It is given that $y$ varies directly as $x$ and that $y = 3$ when $x = 2$. Find the constant of proportionality.

**SOLUTION**   Since $y$ varies directly as $x$, we may write $y = kx$, for some constant $k$. Since $y = 3$ when $x = 2$, we have $3 = k \cdot 2$, from which it follows that $k = \frac{3}{2}$. Therefore, $y = \frac{3}{2}x$.

**EXAMPLE 2**   *Computing the constant of proportionality*

Suppose that $z$ varies jointly with $x$ and $y$ and that $z = 40$ when $x = 2$ and $y = 5$. Find the constant of proportionality.

**SOLUTION**   Since $z$ varies jointly with $x$ and $y$, we know that $z = kxy$ for some constant $k$. Substituting 2 for $x$, 5 for $y$, and 40 for $z$ gives us

$$40 = k \cdot 2 \cdot 5$$
$$40 = 10k$$
$$k = 4$$

Therefore, $z = 4xy$.

Once the constant of proportionality is determined, the model is complete and can then be used to make estimates, as in the following examples.

**EXAMPLE 3** *Estimating the time of a trip*

The time that it takes to travel a certain distance varies inversely with the average speed. If the trip takes 30 minutes at an average speed of 25 miles per hour, how long will it take at 100 miles per hour?

**SOLUTION** Let $t$ represent the time for the trip in minutes and $r$ represent the average rate of speed in miles per hour. Then since the time varies inversely with the rate, we have

$$t = \frac{k}{r}$$

Using the fact that $t = 30$ when $r = 25$ gives us

$$30 = \frac{k}{25}$$

and so

$$k = 25 \cdot 30 = 750$$

Thus, we have $t = 750/r$. When $r = 100$, this gives us $t = 750/100 = 7.5$ minutes.

**EXAMPLE 4** *Estimating body weight as a function of height*

Robert Wadlow, who reached a world record height of 8′11″ shortly before his death at the age of 22, towers above actresses Maureen O'Sullivan and Ann Morris.

If all men were shaped similarly, then the weight of a man would be proportional to the cube of his height. Suppose that the weight of a man 5 feet 10 inches tall is 170 pounds. What would the weight of a similarly shaped 8-foot-tall man be?

**SOLUTION** Let $w$ represent the weight of a man in pounds and $h$ his height in inches. Since weight is proportional to the cube of height, we have $w = kh^3$. Now 5 feet 10 inches is 70 inches, so that we have

$$170 = k \cdot 70^3$$

$$k = \frac{170}{70^3} \approx 0.0004956$$

Thus, $w = 0.0004956h^3$. Since 8 feet is 96 inches, the weight is given by

$$w = 0.0004956 \cdot 96^3 \approx 438.5 \text{ lb}$$

Often it is convenient to express the relation between several variables as a combination of two or more of the standard types of variation, as in the following examples.

**EXAMPLE 5**    *Expressing the relation between several variables*

It is given that $w$ varies jointly as the square of $x$ and the cube of $y$ and inversely as $z$. When $x = 2$, $y = 3$, and $z = 36$, $w = 9$. Find $w$ when $x = 1$, $y = 2$, and $z = 10$.

**SOLUTION**    Combining the variation statements, we have

$$w = k\frac{x^2 y^3}{z}$$

Now substituting $x = 2$, $y = 3$, $z = 36$, and $w = 9$, we can solve for $k$.

$$9 = k\frac{2^2 3^3}{36}$$

$$9 = 3k$$

$$k = 3$$

Thus,

$$w = 3\frac{x^2 y^3}{z}$$

Finally, substituting $x = 1$, $y = 2$, and $z = 10$, we can determine the value of $w$.

$$w = 3 \cdot \frac{1^2 2^3}{10}$$

$$= 3 \cdot \frac{4}{5}$$

$$= \frac{12}{5}$$

■

**EXAMPLE 6**    *Finding the gravitational pull of an object*

The gravitational pull between two objects varies jointly as their masses and inversely as the square of the distance between them. The weight of an object on Earth is, by definition, the gravitational pull between the object and Earth. Find the gravitational pull that 450-pound professional wrestler Big Van Vader exerts on the Moon, which weighs $1.62064 \times 10^{23}$ pounds and is 234,912 miles from Earth.

**SOLUTION**    We begin by finding the general equation of variation. Let $d$ be the distance between two objects, let $m_1$ and $m_2$ be their masses (or weights), and let $F$ be the force due to gravity. Since $F$ varies jointly with $m_1$ and $m_2$ and inversely with the square of $d$, we have

$$F = k\frac{m_1 m_2}{d^2}$$

In order to find the constant $k$, we will choose a scenario in which all of the quantities $F$, $m_1$, $m_2$, and $d$ are known. Since the gravitational force between a 1-pound object on the surface of the Earth is 1 pound, if we knew the weight of the Earth and the distance from the Earth's surface to its center, we would have enough information to evaluate $k$. After consulting an almanac, we discover the Earth has a weight of approximately $1.3176 \times 10^{25}$ pounds and that the radius of the Earth is approximately 3963 miles. Thus, we have $F = 1$ pound, $d = 3963$ miles, $m_1 = 1$ pound, and $m_2 = 1.3176 \times 10^{25}$ pounds. Substituting into the equation of variation, we find

$$1 \text{ lb} = k\frac{1 \text{ lb} \cdot 1.3176 \times 10^{25} \text{ lb}}{(3963 \text{ mi})^2}$$

and solving for $k$, we obtain

$$k = \frac{(3963)^2 \text{ mi}^2}{1.3176 \times 10^{25} \text{ lb}} \approx 1.19197 \times 10^{-18} \frac{\text{mi}^2}{\text{lb}}$$

Now to find the gravitational pull between Big Van Vader and the Moon, we use the value of $k$ found above and let $m_1 = 450$ pounds, $m_2 =$ the weight of the Moon $\approx 1.62064 \times 10^{23}$ pounds, and $d =$ the distance from the Earth to the Moon $\approx 234{,}912$ miles. Thus, the force is given by

$$F = 1.19197 \times 10^{-18} \frac{\text{mi}^2}{\text{lb}} \cdot \frac{450 \text{ lb} \cdot 1.62064 \times 10^{23} \text{ lb}}{(234{,}912 \text{ mi})^2}$$

$$= \frac{1.19197 \times 10^{-18} \cdot 450 \cdot 1.62064 \times 10^{23}}{(234{,}912)^2} \text{ lb}$$

$$= 0.00158 \text{ lb}$$

## DATA FITTING

Although some sophisticated mathematical models are constructed from underlying principles, many are formed simply by finding a function that fits a given set of data. Often, the *form* of the function (e.g., linear or quadratic) is either known or assumed, and the given data are to be used to determine the function precisely. Suppose that we wish to find a linear function that is consistent with the following data.

| $x$ | $y$ |
|-----|-----|
| 2   | 5   |
| 3   | 7   |

One approach would be to find the slope of the line connecting $(2, 5)$ and $(3, 7)$ and then use the point-slope form for the equation of a line. However, we will use a method that is more general and can be applied in many other situations. We first note that since $y$ is to be a linear function of $x$, we must have $y = a + bx$ for some choice of constants $a$ and $b$. (In the context of data fitting, it is conventional to represent linear functions in the form $y = a + bx$ rather than $y = mx + b$.) Substituting 2 for $x$ and 5 for $y$, we have

$$5 = a + b(2)$$

Now substituting 3 for $x$ and 7 for $y$ gives us

$$7 = a + b(3)$$

Thus, we must solve the following system.

(1)                    $a + 2b = 5$

(2)                    $a + 3b = 7$

Subtracting the first equation from the second gives us $b = 2$. Substituting 2 for $b$ in equation (1) gives

$$a + 2(2) = 5$$
$$a + 4 = 5$$
$$a = 1$$

Thus, we have $y = 1 + 2x$.

**EXAMPLE 7**    *Finding a parabola passing through two points*

Suppose that the variable $y$ is a function of the variable $x$ of the form $y = x^2 + bx + c$ and that we are provided with the following data.

| $x$ | $y$ |
|-----|-----|
| 2   | 4   |
| 4   | 8   |

Find values for $b$ and $c$ so that $y = x^2 + bx + c$ is consistent with the data.

**SOLUTION** We will use the two data points to form two equations in the two unknowns $b$ and $c$. Using the point $(2, 4)$, we set $x = 2$ and $y = 4$ to obtain

$$4 = 2^2 + b \cdot 2 + c$$

$$4 = 4 + 2b + c$$

$$2b + c = 0$$

Similarly, the point $(4, 8)$ gives us

$$8 = 4^2 + b \cdot 4 + c$$

$$8 = 16 + 4b + c$$

$$4b + c = -8$$

Thus, we must solve the following system of equations.

(1) $\qquad\qquad\qquad\qquad\qquad\qquad 2b + c = 0$

(2) $\qquad\qquad\qquad\qquad\qquad\qquad 4b + c = -8$

Subtracting the first equation from the second gives us $2b = -8$, or $b = -4$. Substituting $-4$ for $b$ in equation (1) gives us $2(-4) + c = 0$, so that $c = 8$. Thus, we have $y = x^2 - 4x + 8$.

---

**EXAMPLE 8**

*Finding a parabola passing through three points*

It is known that $s$ can be expressed as a quadratic function $f(t)$. Find $f(t)$ and sketch its graph given the following data.

**a.**

| $t$ | $s$ |
|-----|-----|
| 0 | 13 |
| 1 | 7 |
| 5 | 23 |

**b.** Find the value of $t$ that minimizes $s$.

**SOLUTION**

**a.** An arbitrary quadratic function of $t$ is of the form $f(t) = at^2 + bt + c$. We will use the three given data points to form a system of three equations in the three unknowns $a$, $b$, and $c$:

Using $(0, 13)$: $\qquad\qquad\qquad 13 = a \cdot 0^2 + b \cdot 0 + c$

$$c = 13$$

Using $(1, 7)$: $\qquad\qquad\qquad 7 = a \cdot 1^2 + b \cdot 1 + c$

$$a + b + c = 7$$

Using $(5, 23)$: $\qquad\qquad\qquad 23 = a \cdot 5^2 + b \cdot 5 + c$

$$25a + 5b + c = 23$$

Thus, we must solve the following system of equations.

(1) $$c = 13$$

(2) $$a + b + c = 7$$

(3) $$25a + 5b + c = 23$$

Substituting $c = 13$ in equations (2) and (3) gives us

(4) $$a + b + 13 = 7 \quad \Rightarrow \quad a + b = -6$$

(5) $$25a + 5b + 13 = 23 \quad \Rightarrow \quad 25a + 5b = 10$$

Now multiplying equation (4) by $-5$ and adding to equation (5), we have

$$
\begin{aligned}
-5a - 5b &= 30 \\
+ \quad 25a + 5b &= 10 \\
\hline
20a &= 40 \\
a &= 2
\end{aligned}
$$

Substituting 2 for $a$ in equation (4) gives $2 + b = -6$; thus $b = -8$. Hence $a = 2$, $b = -8$, and $c = 13$, so that $f(t) = 2t^2 - 8t + 13$. The graph of any quadratic function, $f(t)$ in particular, is a parabola. We can find the vertex using the techniques of Section 4. The $x$-coordinate of the vertex is given by

$$t = -\frac{b}{2a} = -\frac{-8}{2 \cdot 2} = 2$$

The $y$-coordinate is

$$f(2) = 2(2)^2 - 8(2) + 13 = 5$$

Thus, the vertex is (2, 5). After plotting the three given data points, we obtain the graph shown in Figure 58.

**b.** The minimum value clearly occurs at the vertex. Thus, the minimum occurs at $t = 2$, and the minimum value is 5.

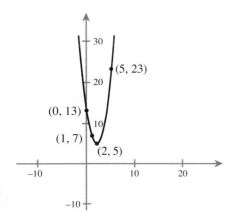

**FIGURE 58**

In the previous examples, the form of the model was given, and just enough data was given to determine the model precisely. In practice, the data will not fit the model precisely; the model can only approximate the data. Our task then is to choose the model that most nearly approximates the given data. For example, suppose that we are given several data points that we would like to model with a linear function. Unless we are extremely fortunate, the given points will not lie on any one line. Therefore, we must choose the line that, in some sense, best approximates the given data.

The data in Table 3 gives the total of all health-care expenditures in the United States in billions of dollars for the years from 1985 to 1990 (Data Source: *1992 World Almanac*). If we let $t$ be the number of years since 1985 (so that 1985 corresponds to $t = 0$, 1986 corresponds to $t = 1$, etc.), and $E$ be the total U.S. health-care expenditures in

**TABLE 3**

| Year | 1985 | 1986 | 1987 | 1988 | 1989 | 1990 |
|---|---|---|---|---|---|---|
| *U.S. health-care expenditure (in billions of dollars)* | 422.6 | 454.8 | 494.1 | 546.0 | 602.8 | 666.2 |

billions of dollars, then we obtain the graph of the data points $(t, E)$ shown in Figure 59. Now it appears that no line passes through all of these points, so the best that we can do is to find a line that "almost" passes through all of these points. Shown in Figure 60 is the line that appears to the authors to best fit the data; your eyes may tell you something different.

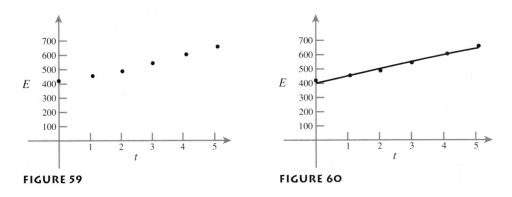

**FIGURE 59**                                      **FIGURE 60**

The primary advantage of this method is its simplicity; we simply drew the line that "seemed" to come the closest to passing through the given points. The disadvantage is obvious; there is no guarantee that the line we have chosen is the "best" line. What we need is a mathematical method for selecting a line that best fits given data. One such method is called the **least-squares best fit**.

## LEAST-SQUARES BEST FIT

Suppose that we are given three data points $(x_1, y_1)$, $(x_2, y_2)$ and $(x_3, y_3)$ as shown in Figure 61, and we would like to find the equation $y = a + bx$ of the line that in some sense most nearly fits the data. If the line were to pass through each of the three points, then the vertical distances $d_1$, $d_2$, and $d_3$ would all be zero. In fact, the distances $d_1$, $d_2$, and $d_3$ are a measure of the error involved in approximating the data points

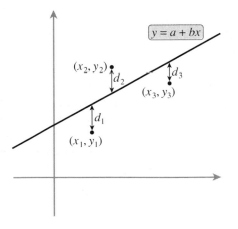

**FIGURE 61**

with the line. The line with the least-squares best fit is obtained by selecting $a$ and $b$ in such a way as to minimize the sum of the squares of the errors—that is, by minimizing the quantity $d_1^2 + d_2^2 + d_3^2$. Of course, we could repeat the above description with any number of data points. In general, we have the following definition.

---

**Least-squares best fit**

If data points $(x_1, y_1)$, $(x_2, y_2)$, ..., $(x_n, y_n)$ are given, then the **line with the least-squares best fit**, or **regression line**, is that line for which the sum of the squares of the errors, $d_1^2 + d_2^2 + \cdots + d_n^2$, is as small as possible. The process of finding the coefficients $a$ and $b$ for the line $y = a + bx$ with the least-squares best fit is called **linear regression**.

Fortunately, graphics calculators will perform linear regressions for us. The general steps to follow are outlined here. For a detailed description of how these steps are done on your calculator, refer to the graphics calculator supplement.

The **linear regression** feature of a graphics calculator can be used to find a line of the form $y = a + bx$ that has the best least-squares fit to a collection of data points $(x, y)$. There are several steps to the process.

**Step 1.** Clear any existing statistical data.

**Step 2.** Enter the points $(x, y)$ as statistical data.

**Step 3.** Select the linear regression option from the list of available regression options.

After step 3 has been completed, the calculator will compute values for $a$ and $b$, along with a third number $r$, the *coefficient of correlation*. Although a full discussion of the significance of $r$ is beyond the scope of this text, when $r$ is close to either 1 or $-1$, the regression line is a good fit to the data.

---

**EXAMPLE 9**    *Performing a linear regression with a graphics calculator*

Find the equation of the line of the form $y = a + bx$ with the least-squares best fit to the points $(1, 5)$, $(2, 6)$, and $(6, 13)$.

**SOLUTION**    We begin by clearing any existing statistical data and entering the points $(1, 5)$, $(2, 6)$, and $(6, 13)$ as statistical data. Next we select the linear regression feature. The calculator should return a screen much like the one in Figure 62, which indicates that $a \approx 3.07$, $b \approx 1.64$, and $r \approx 0.997$. Thus, the regression line has equation $y = 3.07 + 1.64x$. Since the value of $r$ is very close to 1, the line is a good fit.

```
LinReg
 a=3.071428571
 b=1.642857143
 c=.997176465
■
```

**FIGURE 62**

| EXAMPLE 10 | *Finding the least-squares best fit with real data* |

**a.** Use the data in Table 3 to find a linear model with the best least-squares fit.

**b.** Use the model developed in part (a) to predict the total cost of health care in the year 2000.

**SOLUTION**

**a.** Let $t$ represent the number of years after 1985, and $E(t)$ the total U.S. health-care expenditure (in billions of dollars) for year $t$. Then we wish to obtain a model of the form $E(t) = a + bt$. We begin by clearing any existing statistical data from the calculator and then entering the following data points.

$$(0, 422.6), \ (1, 454.8), \ (2, 494.1), \ (3, 546.0), \ (4, 602.8), \ (5, 666.2)$$

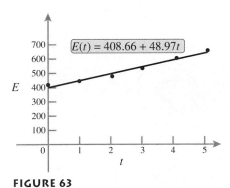

Next we employ the linear regression feature to obtain $a \approx 408.66$, $b \approx 48.97$, and $r \approx 0.99$. We conclude that the linear model with the least-squares best fit to the given data has equation $E(t) = 408.66 + 48.97t$, and since the coefficient of correlation $r$ is nearly 1, the fit is a good one. Figure 63 shows a graph of the least-squares best fit line along with a plot of the data points themselves.

**b.** Recall that $t$ represents the number of years after 1985. Thus, in the year 2000, $t = 15$. Substituting 15 for $t$ in the linear model gives

$$E = 408.66 + 48.97t = 408.66 + 48.97 \cdot 15 \approx 1143.2$$

**FIGURE 63**

Thus, according to this model, the total of all health-care expenditures in the year 2000 will be approximately 1143.2 billion dollars. ∎

## EXERCISES 5

**EXERCISES 1–6** ☐ *Find the constant of proportionality.*

**1.** $y$ varies directly as $x$, and $y = 8$ when $x = 2$.

**2.** $z$ varies directly as $y$, and $z = 10$ when $y = 5$.

**3.** $z$ varies inversely as $x$, and $z = 12$ when $x = 4$.

**4.** $y$ varies inversely as $t$, and $y = 8$ when $t = \frac{1}{2}$.

**5.** $w$ varies jointly as $x$ and $y$, and $w = 15$ when $x = \frac{1}{2}$ and $y = 3$.

**6.** $z$ varies jointly as $s$ and $t$, and $z = 2$ when $s = 10$ and $t = 20$.

**EXERCISES 7–12** ☐ *Solve the variation problem.*

**7.** $w$ varies directly as $x$. Find $w$ when $x = 4$, given that $w = 9$ when $x = 3$.

**8.** $y$ varies directly as $x$. Find $y$ when $x = 6$, given that $y = 10$ when $x = 3$.

**9.** $q$ varies jointly as $x$ and $y$. Find $q$ when $x = 2$ and $y = 5$, given that $q = 90$ when $x = 3$ and $y = 5$.

**10.** $A$ varies jointly as $x$ and $y$. Find $A$ when $x = 12$ and $y = 5$, given that $A = 18$ when $x = 3$ and $y = 10$.

**11.** $V$ varies directly as the square of $x$ and inversely as the cube of $y$. When $x = 4$ and $y = 3$, $V = 2$. Find $V$ when $x = 3$ and $y = 2$.

**12.** $W$ varies jointly as $x$ and $y$ and inversely as $z$. When $x = 7$, $y = 2$, and $z = 4$, $W = 2$. Find $W$ when $x = 5$, $y = 3$, and $z = 1$.

**EXERCISES 13–16** ☐ *Find the equation of the parabola that passes through the given points.*

**13.** $(0, 5)$, $(3, 3)$, $(6, 5)$       **14.** $(-1, -4)$, $(0, 3)$, $(2, 11)$

**15.** $(-1, -12)$, $(2, 12)$, $(3, 12)$       **16.** $(1, -9)$ $(2, 0)$, $(4, 36)$

**EXERCISES 17–20** ☐ *Find the equation of the line with the least-squares best fit.*

**17.** $(1, 3)$, $(4, 4)$, $(7, 6)$

**18.** $(-1, -4)$, $(-2, -2)$, $(-4, 0)$

**19.** $(-1, -6)$, $(0, -5)$, $(1, -4)$, $(2, -1)$

**20.** $(1, 2)$, $(2, 5)$, $(3, 4)$, $(4, 7)$

**EXERCISES 21–24** □ *Find the line with the least-squares best fit to the given data points, and then use the equation of the line to predict the unknown y value.*

**21.**

| x | 0 | 5 | 10 | 15 | 20 | 25 |
|---|---|---|----|----|----|----|
| y | 90 | 101 | 116 | 130 | 138 | ? |

**22.**

| x | 0 | 4 | 8 | 12 | 16 | 20 |
|---|---|---|---|----|----|----|
| y | 35 | 45 | 56 | 69 | 78 | ? |

**23.**

| x | 0 | 1 | 2 | 3 | 4 | 5 |
|---|---|---|---|---|-----|-----|
| y | 80 | 45 | 12 | ? | −50 | −78 |

**24.**

| x | 0 | 2 | 4 | 6 | 8 | 10 |
|---|---|---|---|-----|-----|------|
| y | 200 | 130 | ? | −10 | −60 | −120 |

■ *Applications*

**25.** *Board Feet* Suppose that the number of board feet of lumber in a Ponderosa pine varies directly as the cube of its circumference at waist height. If a Ponderosa pine with a circumference of 100 inches yields 1500 board feet of lumber, how much can be obtained from one with a circumference of 120 inches?

**26.** *Falling Ball* The distance traveled by a ball dropped from the top of a tall building varies directly as the square of the time since its release. If a ball has fallen 64 feet after 2 seconds, how far will it have fallen after 3 seconds?

**27.** *Kinetic Energy* The kinetic energy of a body in motion varies jointly as the mass of the object and the square of its velocity. If the kinetic energy of a 70-kilogram man running at a rate of 10 meters per second is 3500 joules, find the kinetic energy of the 50-kilogram woman chasing him at a rate of 11 meters per second.

**28.** *Potential Energy* The potential energy of an object at rest near the surface of the Earth varies jointly as its mass and height above the ground. If an object with a mass of 10 kilograms is 3 meters above the ground and has a potential energy of 294 joules, find the potential energy of an object with a mass of 15 kilograms that is 12 meters above the ground.

**29.** *Coulomb's Law* According to Coulomb's law, the electric force exerted between two charged particles at rest varies jointly as the charges of the particles and inversely as the square of the distance between the particles. Two electrons, each with a charge of $1.6 \times 10^{-19}$ coulomb, are 1 centimeter apart and exert a force of $2.3 \times 10^{-24}$ newton. How much force is exerted if the electrons are 2 centimeters apart?

**30.** *Grasshopper Jump* If a $1\frac{1}{2}$-inch-long grasshopper can jump 30 inches, and if jumping ability varies directly as length, how far could a 6-foot-tall human jump? Do you think jumping ability varies directly as length? Explain.

**31.** *Ant Strength* If an ant weighing 0.01 gram can lift an object weighing 0.2 gram, and if strength varies directly as body weight, how much could a 150-pound human lift? Do you think strength varies directly as body weight? Explain.

Leaf cutter ants

**32.** *Skidding Vehicles* The force needed to prevent a vehicle from skidding on a circular curve varies jointly as the weight of the vehicle and the square of the speed and inversely as the radius of the curve. If it takes 4840 pounds of force to keep a certain car traveling at a certain speed from skidding on a curve, how much force will it take to prevent a truck that is twice as heavy but is traveling half as fast from skidding on the same curve?

33. *Powerlifting Records*  The following table shows several weight classes of the men's world powerlifting bench press records. Find the line with the least-squares best fit to the data and estimate the bench press record for the 100-kilogram weight class. The actual record is 261.5 kilograms, held by Ed Coan.

| Weight class (kg) | 52 | 60 | 75 | 90 |
|---|---|---|---|---|
| Bench press record (kg) | 155 | 180 | 217.5 | 255 |

Data Source: *1993 Guinness Book of World Records.*

34. *Blood Serum Ethanol Level*  Ethanol is a form of alcohol found in beer, wine, and other liquors. It is eliminated from the body by the liver, via an enzymatic process, at a constant rate. The following table shows the blood serum ethanol level for a 150-pound male taken at half-hour intervals following four 12-ounce beers. Find the line with the least-squares best fit to the data and predict when the level is below 0.001 (the legal limit for operating a motor vehicle in many states).

| Time (minutes) | 0 | 30 | 60 | 90 |
|---|---|---|---|---|
| Blood serum ethanol ($\times 10^{-3}$) | 1.56 | 1.42 | 1.29 | 1.15 |

35. *Ozone Depletion*  Chlorofluorocarbons (CFCs) are a family of chemicals used as refrigerants and in the manufacture of numerous consumer products. The increasing concentration of CFCs in the environment has been shown to be a cause in the depletion of the ozone. The following table shows the rise in atmospheric concentration of one member of this chemical family, CFC-11, from 1980–1990. Find the line with the least-squares best fit to the data and predict the concentration of CFC-11 in 2000.

| Year | 1980 | 1982 | 1984 | 1986 | 1988 | 1990 |
|---|---|---|---|---|---|---|
| Concentration of CFC-11 (parts per trillion) | 180 | 195 | 215 | 230 | 255 | 275 |

Data Source: *1993 Environmental Almanac.*

36. *Water Consumption*  The use of bottled water in the United States has shown a steady increase in recent years. The following table shows the per capita (i.e., per person) consumption for the years 1985–1990. Find the line with the least-squares best fit to the data and predict the per capita consumption in 2000.

| Year | 1985 | 1986 | 1987 | 1988 | 1989 | 1990 |
|---|---|---|---|---|---|---|
| Bottled water consumption (gallons per person per year) | 4.4 | 5.1 | 5.7 | 6.4 | 7.3 | 8.0 |

Data Source: *1993 Environmental Almanac.*

37. *Prison Population*  The following table shows the number of inmates in federal and state prisons in 1980, 1985, and 1990. Find a quadratic model that expresses the number of prisoners as a function of the year, with 1980 corresponding to year 0. Predict the number of inmates in the year 2000.

| Year | 1980 | 1985 | 1990 |
|---|---|---|---|
| Number of inmates (in thousands) | 320 | 480 | 740 |

Data Source: U.S. Bureau of Justice.

38. *Average Income*  The average yearly earnings for a year-round full-time worker are related to the amount of education attained by an individual. Use the following data from 1990 to find a quadratic model that expresses the average earnings as a function of the number of years of education. According to the model you found, what average earnings could be expected for someone with a master's degree (usually 17 years of education)? What educational level corresponds to the smallest average earnings? Does this seem reasonable?

| Years of education | 8 | 12 | 16 |
|---|---|---|---|
| Average earnings | $16,000 | $24,000 | $36,000 |

Data Source: U.S. Bureau of the Census.

## Projects for Enrichment

39. *Transformed Least-Squares Fit*  We have seen how the least-squares method can be used to find the line of the form $y = a + bx$ that has the least-squares best fit to a set of data points. Here we consider a slight extension that will enable us to fit a function of the form $y = a + b \cdot u(x)$, where $u$ is some expression involving $x$. Consider, for example, the set of data points shown in Table 4 and graphed in Figure 64.

From the graph we suspect that the data points lie on the graph of a function of the form $y = a + bx^2$. Notice that if we define $u = x^2$, this function has the form $y = a + bu$, which is linear

**TABLE 4**

| x | 0 | 1 | 2 | 3 |
|---|---|---|---|---|
| y | −1.3 | −0.8 | 0.7 | 3.2 |

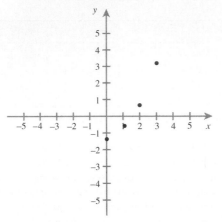

**FIGURE 64**

in $u$ (or $x^2$) and $y$. Thus, we may be successful in looking for a linear fit for the data points that are *transformed* in Table 5 by pairing each $y$ with $x^2$. Indeed, the plot of $x^2$ versus $y$ in Figure 65 does look linear. So we find the line with the least-squares best fit to the data in Table 5. We find $a = -1.3$ and $b = 0.5$. Our model is thus $y = -1.3 + 0.5x^2$, and it is not hard to check that this function fits the original data points very well.

**TABLE 5**

| $x^2$ | 0 | 1 | 4 | 9 |
|---|---|---|---|---|
| y | −1.3 | −0.8 | 0.7 | 3.2 |

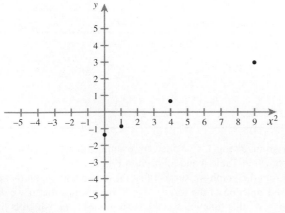

**FIGURE 65**

a. Use the linear regression feature of your calculator to verify that $a = -1.3$ and $b = 0.5$ for the data in Table 5.

For each of the sets of data in parts (b) through (e), find a function of the given form that has the best transformed least-squares fit. Plot each function along with the data points to show that the fit is good.

b. $y = a + bx^3$

| x | 0 | 1 | 2 | 3 |
|---|---|---|---|---|
| y | 2.2 | 1.8 | 0.8 | −2.4 |

c. $y = a + bx^4$

| x | 1 | $\sqrt{2}$ | 2 | $\sqrt{5}$ |
|---|---|---|---|---|
| y | −4.1 | −3.9 | −2.8 | −1.8 |

d. $y = a + b\sqrt{x}$

| x | 1 | 2 | 3 | 4 |
|---|---|---|---|---|
| y | 11 | 13 | 14 | 16 |

e. $y = a + b\dfrac{1}{x}$

| x | −3 | −1 | 2 | 4 |
|---|---|---|---|---|
| y | 6 | 7 | 4 | 5 |

40. *Tape Counters* Many audio and video cassette players are equipped with a counter that is incremented as the tape advances, thus providing a numerical reference for indexing a tape. On some models, the counter is *digital* and gives the actual elapsed time (or perhaps the length of tape that has been played, which can easily be related to the elapsed time). If the counter is *mechanical*, however, it is not as easy to relate the counter reading to the elapsed time. For example, suppose that during the playing of one side of a 90-minute cassette (45 minutes per side), you know that the counter goes from 0 to 900. You also know that your favorite song on a particular 90-minute cassette begins at the 30-minute mark ($\frac{2}{3}$ of the way through the side). However, after you fast-forward the tape until the counter reads 600, the beginning of the song is not anywhere close by. What happened? Two observations will help us explain this phenomenon: the tape must move at a constant speed across the read head, and a mechanical counter is operated by either the supply drive or the take-up drive (see Figure 66).

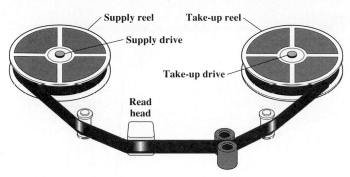

**FIGURE 66**

a. Why must the tape move at a constant speed across the read head?

b. Suppose that the take-up drive operates the counter. Use the fact that the tape must move at a constant speed to explain why the take-up reel, and hence the counter, will begin to move more slowly as the tape advances. (Hint: Think about the amount of tape on the take-up reel as the tape advances.) What happens if the supply drive operates the counter?

For the remainder of this project, we will investigate models for relating the counter reading to the elapsed time. Our goal is to find a function $y = f(x)$ that gives the elapsed time $y$ that corresponds to a counter reading $x$. Since this function will likely vary from one cassette player to another, we must restrict our attention to a specific cassette player.

c. Find a cassette player with a mechanical counter and identify whether the take-up drive or the supply drive operates the counter. Explain how you determined this.

d. Start the counter at 0 and play a tape for 5 minutes, recording the counter reading at the end of each minute. Write your data in table form, with the first column labeled $x$ (counter reading) and the second column labeled $y$ (time in minutes). Plot the data points on an $xy$-coordinate system. Do the points appear to have a linear or parabolic pattern, or neither?

e. Use the linear regression feature of your graphics calculator to find the least-squares best fit model of the form $y = a + bx$ for the data you collected in part (d). Use the model to estimate the counter readings that correspond to 6 and 10 minutes, and then test the estimate by playing the tape for 10 minutes and recording the actual counter readings. How well does the model work?

f. Find a quadratic model of the form $y = ax^2 + bx + c$ that fits the data you found in part (d) as closely as possible. If your graphics calculator has a quadratic fit feature, you may use that. Otherwise, pick three data points and find the parabola that passes through those three points. Use the model to estimate the counter readings that correspond to 6 and 10 minutes, and test the estimates by comparing them to the actual values. Does the quadratic model work any better than the linear model from part (e)? Was this what you expected? Explain.

## Questions for Discussion or Essay

41. Explain what is meant by the statement ''three distinct nonlinear points completely determine a parabola.'' Is the statement true? Why or why not? What would happen if you tried to fit a function of the form $f(x) = ax^2 + bx + c$ to three points that lie on a straight line?

42. Discuss the primary hazards in the use of mathematical models for predicting real-world phenomena. What safeguards must be taken to minimize the risks?

43. The area of a circle varies directly as the square of the radius. If the radius of a circle is doubled, does this mean that the area will double? Explain, using an example to illustrate. In general, if a quantity $y$ varies directly as the $n$th power of $x$, what happens to $y$ if $x$ is doubled? Justify your answer.

44. If the surface area of an object of a given shape varies directly as the square of its length and the volume varies directly as the cube of its length, then what can you say about the ratio of volume to surface area? How does this discussion relate to the rate at which small hamburgers grill compared to large ones? What about the rate at which large ice cubes melt compared to small ones?

45. If all men were shaped similarly, then body weight $w$ would vary directly with the cube of height $h$, so that $w = kh^3$, as mentioned in Example 4. Clearly, men are not all of the same shape. But what about the shape of the ''typical'' 5-foot-tall man? Is it the same as that of the ''typical'' 6-foot-tall man, or ''typical'' 7-foot-tall man? Using the units of pounds for weight and feet for height, estimate appropriate constants of proportionality for 5 footers, 6 footers, and 7 footers. Explain your results.

# CHAPTER REVIEW EXERCISES

**EXERCISES 1–4** ☐ *Determine whether the given relation defines y as a function of x.*

1. $y - x^2 + 1 = 0$

2. $x - y^3 = 1$

3. $x^2 + y^2 = 5$

4. $x = y^4 - 2$

**EXERCISES 5–10** ☐ *Evaluate as indicated.*

5. $g(x) = 2x^2 - 3x + 1$
   a. $g(4)$
   b. $g(-3)$

6. $h(x) = \dfrac{x}{2x + 5}$
   a. $h(0)$
   b. $h(-5)$

7. $f(x) = \dfrac{x^2}{x^2 + 1}$
   a. $f(-x)$
   b. $f(x + 1)$

8. $g(x) = (x + 2)^2$
   a. $g(x - 2)$
   b. $g\left(\dfrac{1}{x}\right)$

9. $h(x) = \begin{cases} 4 - x^2, & x \le 2 \\ -x + 4, & x > 2 \end{cases}$
   a. $h(0)$
   b. $h(5)$
   c. $h(a)$ if $a > 2$

10. $f(x) = \dfrac{|x - 3|}{x - 3}$
    a. $f(-2)$
    b. $f(10)$
    c. $f(c)$ if $c < 3$

**EXERCISES 11–14** ☐ *Find the domain of the function.*

11. $f(x) = x^3 + 1$

12. $g(x) = \sqrt{3 - x}$

13. $h(t) = \dfrac{t}{\sqrt{2t + 1}}$

14. $f(x) = \dfrac{x^2}{4 - x^2}$

**EXERCISES 15–18** ☐ *Use a graphics calculator to sketch the graph of the function and then use the graph to help you identify the domain and range.*

15. $h(x) = \sqrt{x^2 - 4}$

16. $f(x) = \dfrac{2x^2}{x^2 + 1}$

17. $f(x) = \dfrac{|1 - x|}{1 - x}$

18. $g(x) = \dfrac{\sqrt{x + 1}}{(\sqrt{x}) + 1}$

**EXERCISES 19–20** ☐ *A function and its graph are given. For each translated or reflected graph, give a formula for the function.*

19. $f(x) = 2x^3 - 6x + 1$

a.

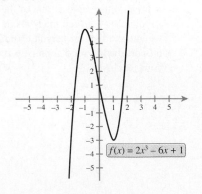

$f(x) = 2x^3 - 6x + 1$

20. $g(x) = x^3 - 3x^2 + 1$

a.

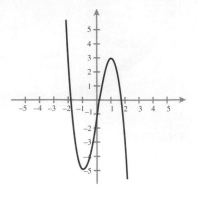

b.

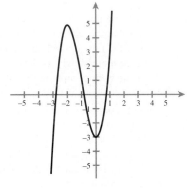

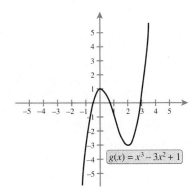

$g(x) = x^3 - 3x^2 + 1$

a.

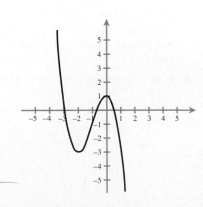

b.

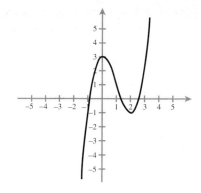

**EXERCISES 21–22** ☐ *Use the graph of f to sketch the graph of h.*

21.

a. $h(x) = f(x + 3)$

b. $h(x) = f(x - 2) + 1$

c. $h(x) = -f(x) - 2$

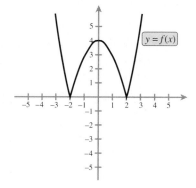

22.

a. $h(x) = f(x) - 3$

b. $h(x) = f(x + 1) - 4$

c. $h(x) = f(-x) + 2$

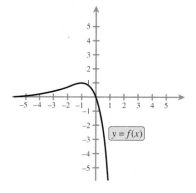

**EXERCISES 23–26** ☐ *Determine whether the given function is even, odd, or neither.*

23. $f(x) = x(4x^2 - 1)$

24. $g(x) = x^4 - x$

25. $f(x) = |x + 1|$

26. $h(x) = \sqrt{x^2 - 1}$

**EXERCISES 27–30** ☐ *Find the exact values of all the real zeros of the given function.*

27. $h(x) = 4x - 9$

28. $f(x) = x^2 - x - \dfrac{5}{16}$

29. $g(x) = 3 - \dfrac{8}{x}$

30. $g(x) = |3x - 1| - 5$

**EXERCISES 31–34** ☐ *Use a graphics calculator to estimate the following features of the given function to the nearest hundredth.*

a. *Zeros*

b. *Coordinates of the relative extrema*

c. *Intervals on which the function is increasing or decreasing*

31. $f(x) = x^3 - 3x + 1$

32. $g(x) = 3x^4 + 4x^3 - 12x^2 + 2$

33. $h(x) = \dfrac{5}{x^2 - 4x + 6}$

34. $g(x) = |x^3 - x^2 + 1|$

**EXERCISES 35–40** ☐ *Find the following combinations of the functions f and g. Specify the domain if it is anything other than the set of all real numbers.*

a. $(f + g)(x)$

b. $(fg)(x)$

c. $(f/g)(x)$

d. $(f \circ g)(x)$

e. $(g \circ f)(x)$

35. $f(x) = x^2;\ g(x) = 2x - 1$

36. $f(x) = 4x + 3;\ g(x) = \dfrac{2}{x}$

37. $f(x) = \dfrac{1}{x - 4};\ g(x) = x^3$

38. $f(x) = \sqrt{x + 2};\ g(x) = x^2 - 1$

39. $f(x) = 2x^2;\ g(x) = \sqrt{2x + 4}$

40. $f(x) = \dfrac{x}{x - 1};\ g(x) = \dfrac{2}{x}$

**EXERCISES 41–44** ☐ *Show that the functions f and g are inverses of one another by evaluating f(g(x)) and g(f(x)).*

41. $f(x) = x - 5;\ g(x) = x + 5$

42. $f(x) = 3x - 1;\ g(x) = \dfrac{x + 1}{3}$

43. $f(x) = x^3 - 4;\ g(x) = \sqrt[3]{x + 4}$

44. $f(x) = (x - 2)^2, x \geq 2;\ g(x) = \sqrt{x} + 2$

**EXERCISES 45–50** ☐ *Determine whether the function is 1–1. If so, compute its inverse.*

45. $f(x) = 3x - 2$

46. $g(x) = x^2 + 5$

47. $h(x) = x^3 - x + 1$

48. $g(x) = (x + 2)^3 + 3$

49. $f(x) = \sqrt{x - 4}$

50. $h(x) = \dfrac{1}{x - 1}$

**EXERCISES 51–52** ☐ *Use the graph of f to sketch the graph of f⁻¹.*

**51.**

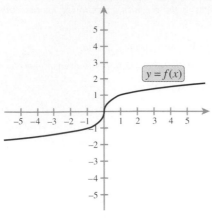

**52.**

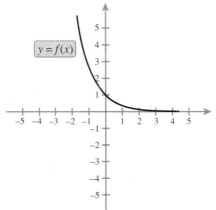

**EXERCISES 53–58** ☐ *Identify the type of function that is given (linear, quadratic, piecewise, or none of these), sketch its graph, and locate its intercepts.*

**53.** $h(x) = -(x - 3)^2 + 1$     **54.** $g(x) = \dfrac{1}{2}x - 4$

**55.** $f(x) = \dfrac{x + 2}{x - 2}$     **56.** $h(x) = \dfrac{1}{4}x^4 - 2x^2 + 2$

**57.** $g(x) = \begin{cases} 2x + 3, & x \le 1 \\ 5 - x^2, & x > 1 \end{cases}$

**58.** $f(x) = \begin{cases} x^2 + 4x, & x \le -2 \\ x, & x > -2 \end{cases}$

**EXERCISES 59–62** ☐ *Sketch the graph of the given quadratic function and locate the exact coordinates of its vertex and intercepts.*

**59.** $f(x) = (x + 1)^2 - 4$     **60.** $f(x) = -2(x - 1)^2$

**61.** $g(x) = x^2 - 8x + 17$     **62.** $h(x) = -3x^2 + 12x - 16$

**EXERCISES 63–66** ☐ *Solve the variation problem.*

**63.** $z$ varies directly as $a$. Find $z$ when $a = 3$, given that $z = 20$ when $a = 4$.

**64.** $y$ varies inversely as the square of $x$. Find $y$ when $x = \frac{1}{4}$, given that $y = 3$ when $x = 3$.

**65.** $T$ varies jointly as $x$ and $y$. If $T = 1$ when $x = \frac{1}{2}$ and $y = \frac{1}{3}$, find $T$ when $x = 2$ and $y = 3$.

**66.** $P$ varies jointly as $x$ and the square of $y$ and inversely as the cube of $z$. If $P = 5$ when $x = 4$, $y = 5$, and $z = 2$, find $P$ when $x = 2$, $y = 3$, and $z = 3$.

**EXERCISES 67–68** ☐ *Find the equation of the parabola that passes through the given points.*

**67.** $(0, 8)$, $(1, 5)$, $(2, 0)$     **68.** $(-4, -1)$, $(2, 5)$, $(4, 15)$

**EXERCISES 69–70** ☐ *Find the equation of the line with the least-squares best fit to the given points.*

**69.** $(1, 2)$, $(3, 3)$, $(6, 5)$

**70.** $(-3, 5)$, $(-1, 3)$, $(2, 1)$, $(5, -2)$

**EXERCISES 71–72** ☐ *Find the line with the least-squares best fit to the given data points, and then use the equation of the line to predict the unknown y value.*

**71.**

| $x$ | 0 | 10 | 20 | 30 | 40 | 50 |
|---|---|---|---|---|---|---|
| $y$ | 42 | 67 | 88 | 118 | 135 | ? |

**72.**

| $x$ | 0 | 3 | 6 | 9 | 12 | 15 |
|---|---|---|---|---|---|---|
| $y$ | 205 | 175 | 150 | ? | 90 | 65 |

**73.** *Area of a Circle* Express the area of a circle as a function of its circumference.

**74.** *Area of a Rectangle* A point $P(x, y)$ lies on the parabola $y = 4 - x^2$, as shown in Figure 67. Express the area of the shaded rectangle as a function of $x$. What is the domain of the function? Plot the graph of the function and estimate the value of $x$ that will yield the largest possible area for the rectangle.

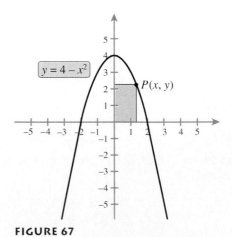

**FIGURE 67**

**75.** *Height of a Balloon* A hot-air balloon is rising vertically from a point 200 feet from an observer on the ground (Figure 68). Express the height $h$ of the balloon as a function of its distance $d$ from the observer. What domain of $d$ values makes sense for the function $h$?

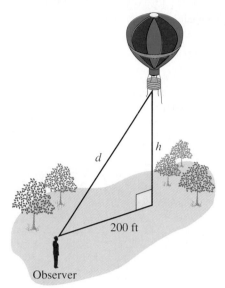

200 ft

Observer

**FIGURE 68**

**76.** *Minimum Cost* A company has determined that the total production cost for manufacturing $x$ units is
$C(x) = 0.01x^3 - 4.5x^2 + 475x + 36,000$. Estimate the intervals where this function is increasing or decreasing. Estimate the number of units that must be manufactured to minimize the cost.

**77.** *Path of a Football* If a football is kicked from ground level with an initial velocity of 64 feet per second and an initial angle of 45°, and if we ignore air resistance, the ball will follow the path given by the quadratic function $f(x) = -\frac{1}{128}x^2 + x$, where $x$ is the distance in feet from where the ball was kicked and $f(x)$ is the corresponding height of the ball in feet.

   **a.** Find the maximum height of the ball.

   **b.** Assuming the ball is kicked straight, will it clear a 10-foot-high goal post 30 yards away? What is the furthest distance from which the ball could be kicked in order to clear the goal post?

**78.** *Overnight Rate* Suppose an overnight package delivery service charges $8.50 for the first pound and $2.00 for each additional pound or fraction thereof. Then the rate for a package weighing $x$ pounds can be determined using the function $C(x) = 8.5 - 2[\![1 - x]\!]$. Plot the graph of this function. What is the heaviest package that can be sent if the cost cannot exceed $31?

**79.** *Volunteer Model* The percentage of the adult population doing volunteer work is related to educational level, as suggested by the following table. Find the line with the least-squares best fit to the data and predict the percentage of the population with 18 years of education who are involved in volunteer work.

| Years of education | 10 | 12 | 14 | 16 |
|---|---|---|---|---|
| Percentage doing volunteer work | 8.3% | 18.8% | 28.1% | 38.4% |

Data Source: U.S. Bureau of Labor.

**80.** *Accidental Death Rate* The death rate for work-related non-manufacturing accidental deaths has been on the decline in recent years. The following table shows the death rates for the years 1970, 1975, 1980, 1985, and 1990. Find the line with the least-squares best fit to the data and predict the death rate for the years 1995 and 2020. Are your answers reasonable? Explain.

| Year | 1970 | 1975 | 1980 | 1985 | 1990 |
|---|---|---|---|---|---|
| Death rate (per 100,000) | 21 | 17 | 15 | 12 | 10 |

Data Source: *1991 Accident Facts*, National Safety Council.

**81.** *Compact Disc Sales* The following table shows the number of CDs sold during each of the years 1985, 1988, and 1989. Find a quadratic model that expresses the number of CDs sold as a function of the year, with 1985 corresponding to year 0. Predict the number of CDs sold in 1992.

| Year | 1985 | 1988 | 1989 |
|---|---|---|---|
| CDs sold (millions) | 23 | 150 | 207 |

Data Source: *Inside the Recording Industry: A statistical overview*, Recording Industry Association of America.

## CHAPTER TEST

**PROBLEMS 1–8** □ *Answer true or false.*

1. If a horizontal line intersects the graph of a relation in more than one point, the relation does not define $y$ as a function of $x$.

2. Unless otherwise specified, we assume that the domain of a function $f$ is the set of all input values such that the expression $f(x)$ is defined.

3. If $h(x)$ is defined by $h(x) = f(x - 2) + 3$, then the graph of $h$ can be obtained by translating the graph of $f$ two units to the right and three units upward.

4. An even function cannot be 1–1.

5. Only functions satisfying the horizontal line test are 1–1, and only 1–1 functions have inverse functions.

6. The graph of a piecewise function is really just the graph of several different functions plotted on the same set of coordinate axes.

7. If $y$ varies directly as $x$, then $y$ is also a linear function of $x$.

8. Linear regression is the process of finding the equation of the line that passes through a given set of data points.

**PROBLEMS 9–14** □ *Give an example of each.*

9. A linear function and its inverse

10. A quadratic function and the coordinates of the vertex of its graph

11. A piecewise function that is linear over each of its three pieces

12. A set of three data points such that the regression line passes through all of the points

13. A function, the graph of which is symmetric with respect to the origin

14. A polynomial function that is neither even nor odd

15. Determine whether the relation $(1/y) + x = 2$ defines $y$ as a function of $x$. If it does, find the domain of the function.

16. Verify or disprove that

$$f(x) = \frac{3x + 1}{2} \quad \text{and} \quad g(x) = \frac{2x - 1}{3}$$

are inverses of one another.

17. Given that

$$f(x) = \begin{cases} 3x + 1, & x \le 2 \\ 4x - 5, & x > 2 \end{cases} \quad \text{and} \quad g(x) = x^2$$

compute each of the following quantities.

    **a.** $f(3)$    **b.** $(f \circ g)(\sqrt{2})$    **c.** $(g \circ f)(3)$

18. Find the domain of the function

$$f(x) = \frac{\sqrt{x - 2}}{x - 3}$$

19. Use the graph of $f$ given in Figure 69 below to produce the graph of $h(x)$, where $h(x) = f(x - 1) - 3$.

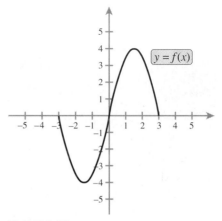

**FIGURE 69**

20. Find the coordinates of all zeros and turning points of $f(x) = x^3 - 6x^2$ to the nearest hundredth, and classify each turning point as a relative maximum or a relative minimum. In addition, indicate the intervals on which $f$ is increasing and the intervals on which it is decreasing.

21. Find the coordinates of the vertex of $f(x) = 3x^2 - 6x + 10$ and sketch the graph.

22. It is given that $z$ varies inversely as $x$ and directly as $y$. When $x = 2$ and $y = 3$, $z = 5$. Find $z$ when $x = 3$ and $y = 4$.

23. It is given that $f(x) = ax^2 + bx$, and that $(2, -4)$ and $(-1, 5)$ are points on the graph of $f$. What is $f(3)$?

24. The *1994 Guinness Book of World Records* gives the following world records for women's track and field events:

■ 100 meters:  Delorez Florence Griffin Joyner (USA)  10.49 seconds

■ 200 meters:  Delorez Florence Griffin Joyner (USA)  21.34 seconds

■ 400 meters:  Marita Koch (East Germany)  47.60 seconds

Use linear regression to estimate the world-record time for 800 meters.

# POLYNOMIAL AND RATIONAL FUNCTIONS

■ This mountain range exists only in cyberspace; it is a computer-generated *fractal landscape*. Fractals are self-similar, that is, a small portion of a fractal appears to be a scaled down version of the whole. Compare the appearance of the mountain peak with the mountain itself, and in turn the mountain with the entire range of mountains. Fractal landscapes can be made to so closely resemble reality because so much of nature is intrinsically fractal. Coastlines, clouds, ocean waves, and ferns are all essentially self-similar. In this chapter we will investigate polynomial functions, an important tool for generating such strikingly beautiful fractal images.

## POLYNOMIAL FUNCTIONS

■ Why do polynomials fail to accurately model real-world phenomena, such as the spread of AIDS, for extended periods of time?

■ How can polynomials be used to generate stunning fractal images?

■ How can we ever be certain that a graphics calculator is showing us all the important graphical features of a polynomial?

Many of the functions we have seen, including

$$f(x) = \frac{1}{2}x - 3$$

$$g(x) = 2x^2 + 4x + 5$$

$$h(x) = -\frac{1}{4}x^4 - \frac{2}{3}x^3 + \frac{5}{2}x^2 + 6x$$

were examples of polynomials. In general, a **polynomial function** has the form

$$f(x) = a_n x^n + a_{n-1}x^{n-1} + \cdots + a_1 x + a_0$$

where $n$ is a nonnegative integer, called the **degree** of the polynomial, $a_0, a_1, \ldots, a_n$ are real numbers, and $a_n \neq 0$.

Polynomial functions of degree 1, such as $f(x) = \frac{1}{2}x - 3$, and polynomial functions of degree 2, such as $g(x) = 2x^2 + 4x + 5$, can be graphed by hand using the techniques for lines and parabolas. On the other hand, it is often quite tedious to sketch the graphs of polynomial functions of degree 3 or higher. For this reason, we will often rely on a graphics calculator to help us with polynomial graphs. However, as we have seen many times in the past, an inappropriate viewing rectangle can have a disastrous effect on a graph, and so it is still necessary to have a basic understanding of the graphical features of a polynomial. We will investigate some of these features here.

### END BEHAVIOR OF POLYNOMIALS

The behavior of the graph of a function to the far right and to the far left (i.e., for large positive and negative values of the independent variable) is called the **end behavior** of the function. Consider the polynomial function $f(x) = x^3$. For large positive values of $x$, $f(x)$ will be very large. For example, $f(1000) = 1000^3 = 1,000,000,000$. In fact, as $x$ gets larger and larger, $f(x)$ becomes arbitrarily large. We say that $f(x)$ approaches infinity as $x$ approaches infinity, and we write $f(x) \to \infty$ as $x \to \infty$. Similarly, for large negative values of $x$, $f(x)$ will be a very large negative number. For example, $f(-1000) = -1,000,000,000$. In this case, $f(x) \to -\infty$ as $x \to -\infty$. We could convince ourselves of these facts by graphing $f(x)$. Notice that the graph "shoots up" on the far right and "shoots down" on the far left, as shown in Figure 1.

Let us now determine the end behavior of the polynomial function $f(x) = x^2$. When a large positive number is squared, the result is a very large positive number. For example, $1000^2 = 1,000,000$. If a large negative number is squared, the result is also a large positive number. For example, $(-1000)^2 = 1,000,000$. Thus, $f(x) \to \infty$ as $x \to \infty$ and $f(x) \to \infty$ as $x \to -\infty$. These facts can also be determined by noting that

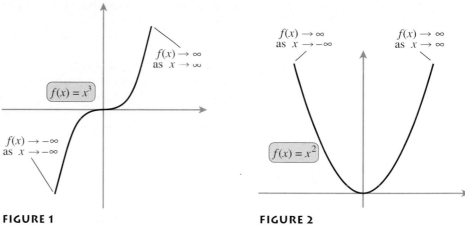

**FIGURE 1**                                              **FIGURE 2**

since the graph of $x^2$ is a parabola opening upward, the function values are approaching infinity on the far right and the far left, as shown in Figure 2.

By generalizing the observations made above, it can be shown that the end behavior of any polynomial of the form $f(x) = x^n$ depends only on whether $n$ is even or odd. When $n$ is odd, $x^n$ will behave like $x^3$, and when $n$ is even, $x^n$ will behave like $x^2$. We summarize these facts as follows.

**■ End behavior of $f(x) = x^n$**

> If $n$ is even, then $f(x) \to \infty$ as $x \to \infty$, and $f(x) \to \infty$ as $x \to -\infty$.
> If $n$ is odd, then $f(x) \to \infty$ as $x \to \infty$, and $f(x) \to -\infty$ as $x \to -\infty$.

**EXAMPLE 1**    *Determining the end behavior of monomials*

Determine the end behavior of each of the following functions and match each to its graph.

**a.** $f(x) = x^4$                          **b.** $g(x) = x^7$
**c.** $h(x) = -2x^3$                        **d.** $p(x) = -3x^2$

**i.** 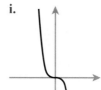    **ii.** 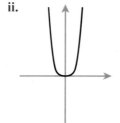    **iii.** 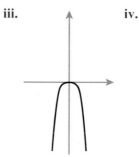    **iv.**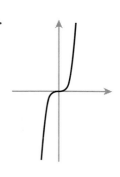

**SOLUTION**

**a.** Since the exponent is even, $f(x) \to \infty$ as $x \to \infty$ and also as $x \to -\infty$. The graph is (ii).

**b.** Since the exponent is odd, $g(x) \to \infty$ as $x \to \infty$ and $g(x) \to -\infty$ as $x \to -\infty$. The graph is (iv).

**c.** The exponent is odd, but multiplication by $-2$ changes the sign. Thus, $h(x) \to -\infty$ as $x \to \infty$, and $h(x) \to \infty$ and $x \to -\infty$. The graph is (i).

**d.** The exponent is even, but multiplication by $-3$ changes the sign. Thus, $p(x) \to -\infty$ as $x \to \infty$ and also as $x \to -\infty$. The graph is (iii).

---

Thus far, we have dealt only with *monomials*—that is, polynomials having only one term. We now turn our attention to finding the end behavior of polynomials in general. It is tempting to simply graph the function using a graphics calculator and then to note whether the graph is rising or falling on the far right and on the far left of the viewing rectangle. The difficulty is in knowing how *wide* the viewing rectangle should be, as we see in the following example.

**EXAMPLE 2**    *Investigating the end behavior of a polynomial using a graphics calculator*

Investigate the end behavior of $f(x) = x^3 - 1{,}000{,}000x^2$ using a graphics calculator.

**SOLUTION**    We have chosen a very naive approach; we begin with a standard viewing rectangle and make adjustments until we are confident that we are, in fact,

| Plot | Range | Problem | Solution |
|------|-------|---------|----------|
| | Xmin=-10 <br> Xmax=10 <br> Ymin=-10 <br> Ymax=10 | The graph is invisible because of the choice of scale. In fact, the graph is "on top of" the $y$-axis. | We must increase the range of $y$-values. |
| | Xmin=-10 <br> Xmax=10 <br> Ymin=-1,000,000 <br> Ymax=1,000,000 | The range of $y$-values is still too small; the graph is cut off at the bottom. Also, the view is too narrow. | We will take a wider view by increasing the range of $x$-values. We will also make the range of $y$-values *huge*. |
| | Xmin=-1,000,000 <br> Xmax=1,000,000 <br> Ymin=-1E18 <br> Ymax=1E18 | We see the graph beginning to turn up at the far right, but our view is too narrow to determine if this continues. | We will widen our view. Again we must increase our range of $y$-values to produce a usable plot. |
| | Xmin=-1E10 <br> Xmax=1E10 <br> Ymin=-1E30 <br> Ymax=1E30 | The graph would seem to indicate that $f(x) \to \infty$ as $x \to \infty$, and $f(x) \to -\infty$ as $x \to -\infty$. | |

witnessing the "true" end behavior of the function. The main steps that we took are shown in the table on the previous page; considerations of space prevent us from showing all of the missteps that we took along the way.

Even after going through this extraordinary effort, we are still not sure that we have answered the question correctly! For all we know, an even wider viewing rectangle might show yet another turn in the graph. Again, we see the limitations of the graphics calculator. In this case, there is an easy way to determine the end behavior of a polynomial without using a graphics calculator at all.

The end behavior of $f(x) = x^3 - 1,000,000x^2$ can be viewed as a contest between the monomials $x^3$ and $1,000,000x^2$. As $x$ approaches infinity, $x^3$ approaches infinity, whereas $-1,000,000x^2$ approaches negative infinity. The end behavior of the sum of these two monomials, namely $f(x)$, depends upon which of them is dominant. As we will see in Exercise 37, the monomial of highest degree, which in this case is $x^3$, "wins." Thus, $x^3$ approaches infinity faster than $-1,000,000x^2$ approaches negative infinity. In fact, the end behavior of $f(x)$ is identical to that of $x^3$. More generally, we have the following.

**End behavior of polynomials**

> The end behavior of a polynomial is identical to that of its term of highest degree.

**EXAMPLE 3**    *Finding the end behavior of a polynomial*

Find the end behavior of each of the polynomials given below.
**a.** $f(x) = x^4 - 2x^3 + 4x - 5$        **b.** $g(x) = 7 + 4x - 3x^2 - 2x^3$
**c.** $h(x) = -7x^6 - 4x^5 + 3x + 1$

**SOLUTION**

**a.** Since the term of highest degree is $x^4$, the end behavior of $f(x)$ will be identical to that of $x^4$. Thus, $f(x) \to \infty$ as $x \to \infty$ and $f(x) \to \infty$ as $x \to -\infty$.

**b.** The term of highest degree is $-2x^3$, so the end behavior of $g(x)$ will be identical to that of $-2x^3$. Thus, $g(x) \to -\infty$ as $x \to \infty$, and $g(x) \to \infty$ as $x \to -\infty$.

**c.** The term of highest degree is $-7x^6$, so the end behavior of $h(x)$ is identical to that of $-7x^6$. Since $-7x^6 \to -\infty$ as $x \to \infty$ and $-7x^6 \to -\infty$ as $x \to -\infty$, the same is true of $h(x)$. Thus, $h(x) \to -\infty$ as $x \to \infty$ and $h(x) \to -\infty$ as $x \to -\infty$.

## ZEROS OF POLYNOMIALS

Recall that a zero of a function $f$ is a number $c$ such that $f(c) = 0$. In other words, the zeros of a function are simply those input values that produce zero as an output value. We have also seen that the real zeros of a function coincide with its $x$-intercepts. Indeed, if $c$ is a real number for which $f(c) = 0$, then $(c, 0)$ is a point on both the graph of $f$ and also the $x$-axis, and is thus an $x$-intercept. The zeros, and hence the $x$-intercepts, of a polynomial function can often be found by factoring, as we see in the following example.

**EXAMPLE 4**    *Finding the zeros of a polynomial function*

Find the real zeros of $f(x) = 2x^3 + 6x^2 - 20x$ and locate the $x$-intercepts of $f$ on its graph.

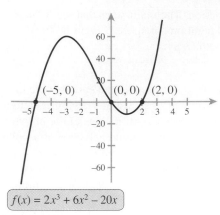

$$f(x) = 2x^3 + 6x^2 - 20x$$

**FIGURE 3**

**SOLUTION**   We first factor $f(x)$ as completely as possible.

$$
\begin{aligned}
f(x) &= 2x^3 + 6x^2 - 20x \\
&= 2x(x^2 + 3x - 10) && \text{Factoring out the monomial } 2x \\
&= 2x(x + 5)(x - 2) && \text{Factoring}
\end{aligned}
$$

Now $2x(x + 5)(x - 2) = 0$ if and only if $x = 0$, $x = -5$, or $x = 2$. Thus, these are the only zeros of $f$, and they correspond to the $x$-intercepts $(0, 0)$, $(-5, 0)$, and $(2, 0)$. The graph of $f$ is shown in Figure 3.

Notice that the zeros of the polynomial function $f$ in Example 4 were found by setting the factors of $f(x)$ equal to zero. This suggests an important connection between the zeros and factors of a polynomial. We will investigate this connection more thoroughly in upcoming sections—for now we simply note the following.

**The number of zeros of a polynomial**

> If $f$ is a polynomial function with real coefficients and degree $n$ ($n \geq 1$), then $f$ has at most $n$ real zeros. Moreover, if $n$ is odd, then $f$ must have at least one real zero.

**EXAMPLE 5**   *Finding the zeros of a polynomial with a graphics calculator*

Estimate the zeros of $f(x) = 0.05x^4 + 0.75x^3 - 0.2x^2 - 3x + 2$ to the nearest tenth using a graphics calculator.

**SOLUTION**   The graph shown in Figure 4 suggests that $f$ has three zeros near the origin. By zooming in and using the trace feature, we estimate these in Figures 5–7 to be $x \approx -2.3$, $x \approx 0.7$, and $x \approx 1.6$. Since the term of highest degree is $0.05x^4$, we know $f(x) \to \infty$ as $x \to -\infty$, and so $f$ must turn upward to the left of the viewing rectangle. After zooming out, we see in Figure 8 that a fourth zero is to the left of the origin. We zoom-in near this zero and obtain $x \approx -15.0$, as in Figure 9. We can be

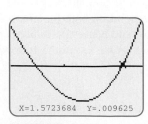

**FIGURE 4**

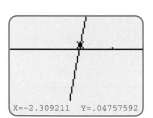

**FIGURE 5**

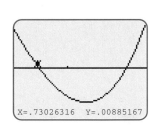

**FIGURE 6**

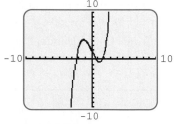

**FIGURE 7**

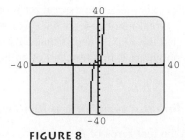

**FIGURE 8**

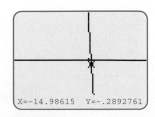

**FIGURE 9**

confident that we have not missed any zeros since a fourth-degree polynomial can have at most four zeros.

The main reason that zeros of polynomials are so important is that they arise as solutions of polynomial equations. For example, the solutions of the polynomial equation $2x^3 + 6x^2 = 20x$ are the same as the solutions of $2x^3 + 6x^2 - 20x = 0$. Thus, the solutions of $2x^3 + 6x^2 = 20x$ coincide with the zeros of the polynomial $f(x) = 2x^3 + 6x^2 - 20x$. We saw in Example 4 that $f$ has exactly three zeros ($x = -5$, $x = 0$, and $x = 2$), and so these are the only solutions of the polynomial equation. In general, if $g(x)$ and $h(x)$ are two polynomials, then the solutions of the polynomial equation $g(x) = h(x)$ can be found by finding the zeros of the polynomial $f(x) = g(x) - h(x)$.

**EXAMPLE 6**    *An application involving zeros*

Yearly income for the U.S. mining industry for the years 1975–1990 can be approximated with the polynomial

$$I(x) = 0.069x^3 - 1.7x^2 + 11.3x + 21.2$$

where $x$ is measured in years after 1975 and the income $I(x)$ is measured in billions of dollars. (Data Source: U.S. Department of Commerce.) The graph of $I(x)$ is shown in Figure 10. According to this polynomial model, when was the income 35 billion dollars?

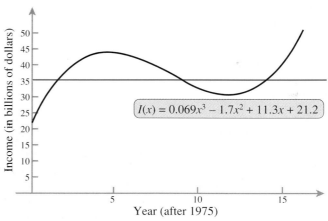

**FIGURE 10**

**SOLUTION**    It is clear from Figure 10 that there were three times between 1975 and 1990 when the income was 35 billion dollars. We could try to estimate these times from the graph in Figure 10, but instead we will use a graphics calculator and the notion of zeros. We first set $0.069x^3 - 1.7x^2 + 11.3x + 21.2 = 35$. By rewriting, this becomes $0.069x^3 - 1.7x^2 + 11.3x - 13.8 = 0$. Thus, we wish to find the zeros of the polynomial $f(x) = 0.069x^3 - 1.7x^2 + 11.3x - 13.8$. We plot the graph of $f$ as shown in Figure 11 and use the trace feature to obtain the approximations $x \approx 1.6$, $x \approx 9.2$, and $x \approx 13.9$. By rounding to the nearest integer, we see that these values roughly correspond to the years 1977, 1984, and 1989.

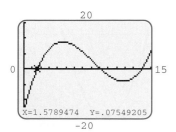

**FIGURE 11**

## MORPHING

**W**hen you see faces from men of different races blend from one to another in a commercial for Schick razors, Michael Jackson transform into a panther in a music video, or the T-2000 robot metamorphose from liquid metal to human in Terminator II, you are witnessing morphing, a spectacular graphic image manipulation technique. The most striking aspect of a morphing sequence is its fluidity; one image melts into another so seamlessly that our eyes tell us that the transformation, no matter how logic-defying, is actually taking place.

Although morphing has only existed since the 1980s, it is now commonplace in several visual media. Films, television commercials, television series, and music videos have employed morphing techniques to create startling special effects, and now software is available that enables those with personal computers to create their own morphing sequences.

Let us consider the steps required to create a morph sequence from an original to a final image. First the original and final images are digitized, that is the images are encoded as numerical data, in much the same way as images and music are stored on compact discs. Then the animator defines a correspondence between points on the initial and final images: for example, a point on the nose of a tiger might correspond to a point on the

A visual effect created with the aid of morphing in the 1994 film The Mask.

nose of an eagle; a certain point on the hindquarters of the tiger might correspond to a point on the wing of the eagle; and so forth. Next the images are blended from one to the other, step by step, in such a way that the intermediate steps seem plausible.

For example, when morphing a tiger into an eagle, we don't want to see a cavity form in the middle of the tiger's body which is then gradually filled in by feathers, nor do we want to see a paw *abruptly* change into a talon, or see a creature with whiskers on one side of its face and feathers on the other. The production of a smooth, plausible morph requires a combination of sophisticated mathematics, computational power, and often the artistic touch and intuition of the animator.

Mathematically, morphing is essentially an interpolation problem: a set of data points is given and intermediate values are to be determined. There are many ways in which data points can be "connected," and so there are many ways in which the morphing sequence can be generated. One of the most popular methods of interpolating is to use polynomial functions. Because of the fact that polynomial functions are "smooth" (their graphs have no breaks or sharp changes of direction), the resulting morphing sequence will likewise appear smooth.

## TURNING POINTS OF POLYNOMIALS

There is a close connection between the zeros of a polynomial and its turning points. To see this, suppose $c_1$ and $c_2$ are two zeros of a polynomial function $f$, and suppose further that $f$ has no other zeros between $c_1$ and $c_2$. Then the graph of $f$ must cross the $x$-axis at $x = c_1$ and $x = c_2$ but nowhere between them. Although the precise shape of $f$ between $c_1$ and $c_2$ is impossible to determine from this limited information, we show two possibilities in Figures 12 and 13. Notice that in either case, $f$ has a turning point somewhere between $c_1$ and $c_2$. In fact, this is always the case: a polynomial will always

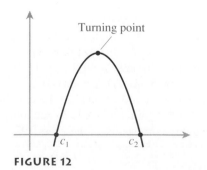

**FIGURE 12**

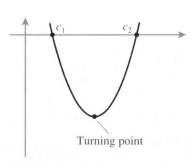

**FIGURE 13**

have at least one turning point between consecutive zeros. So if $f$ has $m$ distinct real zeros, then $f$ must have at least $m - 1$ turning points. It can also be shown, using techniques from calculus, that a polynomial of degree $n$ has at most $n - 1$ turning points. These facts are summarized below.

**The number of turning points of a polynomial**

> Suppose $f$ is a polynomial of degree $n$ with $m$ distinct real zeros, where $1 \le m \le n$. Then $f$ has at least $m - 1$ and at most $n - 1$ turning points. If the degree $n$ is even, then $f$ has at least one turning point.

**EXAMPLE 7**    *Finding the turning points of a polynomial with a graphics calculator*

Use a graphics calculator to estimate the coordinates of *all* of the turning points of $f(x) = \frac{1}{4}x^4 - 7x^3 + 10x^2$.

**SOLUTION**    Two turning points are visible in the graph of $f$ shown in Figure 14. After zooming in and using the trace feature, we estimate these points to be $(0, 0)$ and $(1, 3.25)$. See Figure 15 for the latter.

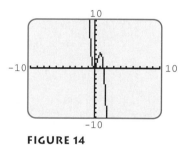

**FIGURE 14**

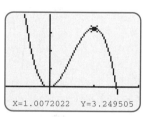

X=1.0072022    Y=3.249505

**FIGURE 15**

Now the graph of $f$ is decreasing on the far right of the viewing rectangle. On the other hand, since the degree of $f$ is even, we know that $f(x) \to \infty$ as $x \to \infty$. Thus, the graph must turn upward somewhere to the right of the viewing rectangle. After a process of trial and error, we obtain a view of the graph in Figure 16 that shows the third turning point. Again, we zoom-in on the turning point and estimate its coordinates in Figure 17 to be approximately $(20, -12,000)$.

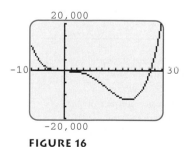

**FIGURE 16**

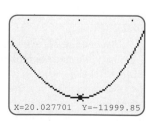

X=20.027701    Y=-11999.85

**FIGURE 17**

Since $f$ has degree 4, there can be at most $4 - 1 = 3$ turning points. Thus, we have found *all* of the turning points of $f$, namely $(0, 0)$, $(1, 3.25)$, and $(20, -12,000)$.

## EXERCISES 1

**EXERCISES 1–4** ☐ *Determine the end behavior of the given polynomial and match it with its graph.*

**1.** $f(x) = x^6$

**i.**

**2.** $g(x) = -x^5$

**3.** $p(x) = -2x^4$

**4.** $h(x) = 4x^3$

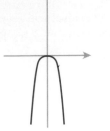

**ii.**

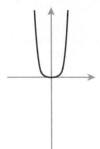

**iii.**

**iv.**

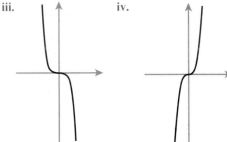

**EXERCISES 5–12** ☐ *Determine the end behavior of the given polynomial.*

**5.** $f(x) = x^2 - 3x + 7$

**6.** $h(x) = -2x^3 + 6x^2 + 3x - 1$

**7.** $p(x) = -4x^4 + 7x - 1$     **8.** $g(x) = -3x^2 + 4x^4$

**9.** $h(x) = 10{,}000x^3 - 0.001x^5$    **10.** $f(x) = 2x^5 + 3x^2 - 100$

**11.** $g(x) = 10^7 + x^3$

**12.** $h(x) = -0.000001x^2 + 1{,}000{,}000x$

**EXERCISES 13–16** ☐ *Indicate*

a. *the maximum number of real zeros of the given polynomial, and*

b. *the maximum number of turning points*

**13.** $P(x) = mx + b$        **14.** $P(x) = ax^2 + bx + c$

**15.** $P(x) = ax^5 + bx^4 + cx^3 + dx^2 + ex + f$

**16.** $P(x) = ax^6 + bx^5 + cx^4 + dx^3 + ex^2 + fx + g$

**EXERCISES 17–22** ☐ *Factor the polynomial to find its zeros.*

**17.** $f(x) = x^2 - 2x - 8$     **18.** $f(x) = x^2 + 8x + 15$

**19.** $f(x) = 3x^3 - 3x$       **20.** $f(x) = 4x^3 - 16x^2 + 16x$

**21.** $f(x) = -x^4 - 12x^3 - 36x^2$    **22.** $f(x) = x^4 - 25x^2$

**EXERCISES 23–32** ☐ *Use a graphics calculator to estimate (to the nearest hundredth) the zeros and the coordinates of the turning points of the polynomial.*

**23.** $f(x) = x^3 - 6x^2 + 5x + 13$

**24.** $g(x) = x^3 + 9x^2 + 20x - 1$

**25.** $h(x) = \dfrac{x^3}{8} + 2x^2 + 2x + 3$

**26.** $g(x) = \dfrac{x^3}{4} - 5x^2 + \dfrac{x}{4} - 2$

**27.** $p(x) = \dfrac{x^4}{4} - 3x^3 - \dfrac{x^2}{2} + 9x$

**28.** $f(x) = \dfrac{x^4}{4} + 4x^3 + 10x^2 + 1$

**29.** $h(x) = 0.1x^3 - 0.7x^2 - 5.6x - 4$

**30.** $f(x) = 0.05x^3 + 0.55x^2 - 5.8x + 8$

**31.** $g(x) = 0.1x^5 - 0.01x^6 - 1$

**32.** $q(x) = 4x - 0.001x^4 + 9$

■ **Applications**

**33.** *AIDS Cases* The number of AIDS cases reported for each of the years 1983–1990 can be approximated with the polynomial function $f(x) = -143x^3 + 1810x^2 - 187x + 2331$, where $x$ is the year ($x = 0$ corresponds to 1983) and $f(x)$ is the number of cases reported in that year. According to the function $f$, in approximately what year will 40,000 cases be reported? When will the number of new cases peak? Do you think this estimate is valid? What graphical feature of polynomials can help explain why this polynomial model works only for a limited time period? What interpretation can be given to $f(x)$ when $x$ is not an integer?

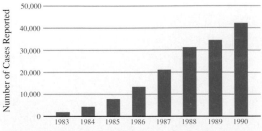

**AIDS Cases 1983–1990**

Data Source: U.S. Centers for Disease Control

**34.** *Family Size*  The average number of people per family unit in the U.S. for the years 1940–1980 can be approximated by the polynomial function $f(x) = 0.000002x^4 - 0.0002x^3 + 0.006x^2 - 0.06x + 3.76$, where $x$ is the year ($x = 0$ corresponds to 1940) and $f(x)$ is the average number of people per family unit during that year. Does $f$ have any turning points? If so, approximate the coordinates. Can you give any explanations for the trends described by the function? Does it appear as though the model is valid after 1990? What graphical feature of polynomials can help explain why this polynomial model works only for a limited time period?

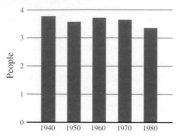

**People Per Family Unit 1940–1980**

Data Source:  U.S. Bureau of the Census

**35.** *Paper Waste*  The percentage of paper waste remaining after recycling during the years 1960–1985 can be approximated by the polynomial function $f(x) = 0.0042x^3 - 0.13x^2 + x + 32$, where $x$ is the year ($x = 0$ corresponds to 1960) and $f(x)$ is the percentage of waste remaining in that year. Approximate the coordinates of any turning points.

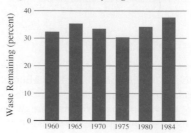

**Paper Waste after Recycling 1960–1984**

Data Source:  1990 Universal Almanac

**36.** *Fertility Rates*  The fertility rate in the U.S. (the number of live births per 1000 women of childbearing age) for the years 1930–1990 can be approximated with the polynomial function $f(x) = 0.000114x^4 - 0.013x^3 + 0.44x^2 - 3.6x + 87$, where $x$ is the year ($x = 0$ corresponds to 1930) and $f(x)$ is the fertility rate for that year. Approximate the coordinates of any turning points. Why does this model appear to be of limited utility for years after 1990?

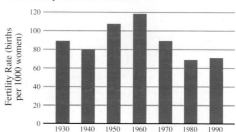

**U.S. Fertility Rate 1930–1990**

Data Source:  U.S. Bureau of the Census

---

### ◼ *Projects for Enrichment*

**37.** *Relating the Magnitudes of Monomials with Their Degrees*  In this project we will explore the relationship between the degree of a monomial and its magnitude. Let $f(x) = 0.0001x^3$ and $g(x) = 1000x^2$.

  **a.** Compute $f(10)$ and $g(10)$; and $f(100)$ and $g(100)$. Explain why $g(x)$ is larger than $f(x)$ for these relatively small values of $x$.

  **b.** Find a value of $x$ (other than zero) for which $f(x) = g(x)$.

  **c.** For what values of $x$ will $f(x)$ be greater than $g(x)$?

  **d.** Explain why $(f(x) - g(x)) \to \infty$ as $x \to \infty$.

  **e.** Suppose that $P(x) = ax^n$ and $Q(x) = bx^{n+1}$, where both $a$ and $b$ are positive. For what values of $x$ is $Q(x) > P(x)$?

  **f.** Explain why the end behavior of a polynomial is determined by the term of highest degree.

**38.** *Complex-valued Polynomials and the Mandelbrot Set*  Up until now we have considered only *real-valued* functions with domains in the set of real numbers. In other words, every func-

tion was assumed to have only real numbers as input and output values. But we can also consider functions that allow nonreal input values and yield complex output values that may or may not be real. For example, suppose $f$ is defined as $f(x) = x^2$, and we allow both real and nonreal input values. Using $x = i$ and $x = 1 + i$, we obtain $f(i) = i^2 = -1$ and $f(1 + i) = (1 + i)^2 = 1 + 2i + i^2 = 2i$. We say that $f$ is a *complex-valued* function or, since $f$ is a polynomial, a complex-valued polynomial. We can also consider functions that have nonreal numbers in their definitions. An example would be $f(x) = x^2 + ix$. Here we compute $f(i) = i^2 + i^2 = -2$ and $f(1 + i) = (1 + i)^2 + i(1 + i) = -1 + 3i$.

  **a.** Evaluate each of the following functions at $x = i$.

  i. $f(x) = x^2 - x$

  ii. $f(x) = x^3 + 2x^2 - 3i$

  iii. $f(x) = 3x^4 - 2x^2 + x$

  iv. $f(x) = x^5 - 2ix^4 + 3x^2 - 5ix + 1$

**b.** Evaluate each of the following functions at $x = 1 + i$ and $x = a + bi$.

   **i.** $f(x) = x^2 + 1$        **ii.** $f(x) = x^2 + i$

   **iii.** $f(x) = x^2 + (1 + i)$    **iv.** $f(x) = x^2 + x$

The complex-valued polynomial $f(x) = x^2 + c$ generates a surprisingly intricate *fractal* image called the **Mandelbrot set**. To see how this happens, we must first consider the geometric representation of a complex number in the **complex plane**. The complex plane looks very much like the rectangular coordinate plane, but instead of the $x$-axis we have the real axis, and instead of the $y$-axis we have the imaginary axis. A complex number $a + bi$ is represented on the complex plane as the point with coordinates $(a, b)$. The complex plane is shown in Figure 18 with several complex numbers plotted. See Section 9.3 for a more detailed treatment of the complex plane.

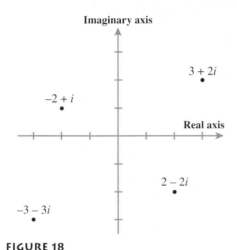

**FIGURE 18**

The Mandelbrot set is defined to be the set of numbers $c$ in the complex plane for which the sequence $c, f(c), f(f(c))$, $f(f(f(c))), \ldots$ is *bounded* in the sense that if we were to plot the numbers in the sequence in the complex plane, they would all be within a fixed distance of the origin. Notice that each number in the sequence (starting with the second) is found by using the previous number in the sequence as the input value for $f$. We show below how to compute a few terms of the sequence for two different values of $c$.

$c = -2, f(x) = x^2 - 2$    $c = 1 + i, f(x) = x^2 + 1 + i$

$f(-2) = 2$              $f(1 + i) = 1 + 3i$

$f(2) = 2$                $f(1 + 3i) = -7 + 7i$

$f(2) = 2$                $f(-7 + 7i) = 1 - 97i$

$\vdots$                      $\vdots$

Now the sequence $-2, 2, 2, \ldots$ is bounded, and so $-2$ is a member of the Mandelbrot set. However, the sequence $1 + i$, $1 + 3i, -7 + 7i, 1 - 97i, \ldots$ is not bounded because each successive point is more than twice as far from the origin as the previous point. Thus, $1 + i$ is not in the Mandelbrot set.

**c.** Determine whether the following numbers are in the Mandelbrot set.

   **i.** $-1$      **ii.** $2$

   **iii.** $i$      **iv.** $\frac{1}{4}i$

If it were possible to plot all of the points in the complex plane that belong to the Mandelbrot set, we would obtain a graph similar to the one shown in Figure 19. A more interesting "graph" can be obtained if we assign colors to the points that are not in the Mandelbrot set [this is done on the basis of how quickly the sequence $c, f(c), f(f(c)), \ldots$ moves away from the origin]. One such image is shown in Figure 20. One of the fascinating properties of the Mandelbrot set (the black region) is that it contains an infinite number of miniature copies of itself. These can be found by zooming in near the boundary of the set. Also near the boundary is an endless collection of stunning images, two of which are shown in Figures 21 and 22.

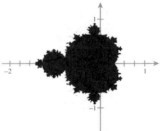

**FIGURE 19**

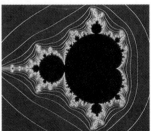

**FIGURE 20**

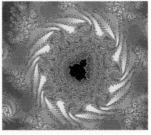

**FIGURE 21**

**FIGURE 22**

■ **Questions for Discussion or Essay**

39. Use your knowledge of the end behavior of polynomials to explain why a fourth-degree polynomial can have one or three turning points, but never two. More generally, what can you say about the possible number of turning points for polynomials of degree $n$, where $n$ is even? What about polynomials of degree $n$, where $n$ is odd?

40. When zooming in to find a turning point, it often happens that the view shown by the calculator is "too flat": the entire graph appears horizontal, and thus it is impossible to pin down the $x$-coordinate of the turning point. Explain how this problem can be remedied using either a zoom box or by adjusting the range variables.

41. In the real world there are many quantities that tend to level off over time. For example, the level of radioactivity present in a person exposed to radiation approaches zero as time goes on. Give three examples of real-world phenomena that tend to level off, and explain why polynomials don't make good models for such phenomena.

42. In Example 6, we considered a polynomial function $I(x) = 0.069x^3 - 1.7x^2 + 11.3x + 21.2$ that approximated the yearly income of the U.S. mining industry for the years 1975–1990. For example, $I(5) \approx 43.8$ billion dollars is the approximate yearly income for 1980. What interpretation can be given to the values of $I(x)$ if $x$ is not an integer? For example, what does it mean if $I(5.5) \approx 43.4$?

---

## SECTION 2

# DIVISION OF POLYNOMIALS

■ How can you divide two polynomials without writing down a single variable?

■ How could you evaluate a 17th-degree polynomial function for any given value of $x$ using only a $\boxed{\times}$ key and a $\boxed{+}$ key 17 times each?

■ If every person with a flu virus passes it on to three other people, how many people will have had the flu 20 days after the first person became infected?

---

### DIVIDING POLYNOMIALS

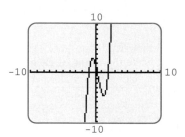

**FIGURE 23**

We have seen that the exact values of the zeros of a polynomial can be found by factoring. But what if a polynomial cannot be easily factored? Consider, for example, the polynomial $f(x) = 3x^3 - 4x^2 - 5x + 2$. It is not at all clear how to factor this polynomial, if indeed it can be factored. However, using a graphics calculator, we see in Figure 23 that there are three zeros. Two of these zeros, $x = -1$ and $x = 2$, can be determined from the graph and verified by substitution into $f$. The third zero can at least be approximated using the trace feature. However, if we wish to find it exactly, we must take a different approach. Intuition suggests that since $x = -1$ and $x = 2$ are zeros, $f(x)$ should have factors $(x + 1)$ and $(x - 2)$ (we will see why this is the case in Section 3). In other words, there is a polynomial $q(x)$ such that $f(x) = (x + 1)(x - 2)q(x)$. As we see in the following example, this information allows us to find the remaining zero.

---

**EXAMPLE 1**    *Factoring a polynomial using long division*

Given that $(x + 1)$ and $(x - 2)$ are factors of $f(x) = 3x^3 - 4x^2 - 5x + 2$, factor $f(x)$ completely and determine the zeros.

**SOLUTION**    Since $(x + 1)$ and $(x - 2)$ are both factors of $f(x)$, the product of these two factors, $(x + 1)(x - 2) = x^2 - x - 2$, is also a factor. Thus we have

$$3x^3 - 4x^2 - 5x + 2 = (x^2 - x - 2)q(x)$$

for some polynomial $q(x)$. Now to find $q(x)$ we divide both sides of the equation by $x^2 - x - 2$ to obtain

$$\frac{3x^3 - 4x^2 - 5x + 2}{x^2 - x - 2} = q(x)$$

We can simplify this expression for $q(x)$ by performing **long division**, a process for dividing two polynomials that is similar to ordinary long division of natural numbers, as illustrated here.

$$
\begin{array}{r}
3x - 1 \\
x^2 - x - 2 \overline{)\, 3x^3 - 4x^2 - 5x + 2} \\
\underline{3x^3 - 3x^2 - 6x} \\
-x^2 + x + 2 \\
\underline{-x^2 + x + 2} \\
0
\end{array}
$$

Multiplying $x^2 - x - 2$ by $3x$ to obtain $3x^3 - 3x^2 - 6x$
Subtracting the second line from the first
Multiplying $x^2 - x - 2$ by $-1$ to obtain $-x^2 + x + 2$

This shows that

$$\frac{3x^3 - 4x^2 - 5x + 2}{x^2 - x - 2} = 3x - 1$$

and so

$$3x^3 - 4x^2 - 5x + 2 = (x^2 - x - 2)(3x - 1)$$
$$= (x + 1)(x - 2)(3x - 1)$$

Setting the factors of $f(x)$ equal to 0, we find that the three zeros of $f$ are $x = -1$, $x = 2$, and $x = \frac{1}{3}$.    ∎

---

**RULE OF THUMB**    When performing long division, write the terms of each polynomial in descending order. For example, if computing $\dfrac{3x - 4x^2 + 1 - 2x^3}{1 + 3x}$, begin by rewriting as $\dfrac{-2x^3 - 4x^2 + 3x + 1}{3x + 1}$.

In Example 1, $x^2 - x - 2$ divided evenly into $3x^3 - 4x^2 - 5x + 2$ since $x^2 - x - 2$ was a factor of $3x^3 - 4x^2 - 5x + 2$. In general, it is always possible to divide a given polynomial by another of lesser degree, even if the second is not a factor of the first. This observation is formalized in the **division algorithm** given here.

**Division algorithm**

Suppose $f(x)$ and $d(x)$ are polynomials, $d(x) \neq 0$, and the degree of $d(x)$ is less than or equal to the degree of $f(x)$. Then there are unique polynomials $q(x)$ and $r(x)$ such that

$$f(x) = d(x) \cdot q(x) + r(x) \quad \text{or} \quad \frac{f(x)}{d(x)} = q(x) + \frac{r(x)}{f(x)}$$

where either $r(x) = 0$ or the degree of $r(x)$ is less than the degree of $d(x)$. The polynomials $f(x)$, $d(x)$, $q(x)$, and $r(x)$ are referred to, respectively, as the **dividend**, **divisor**, **quotient**, and **remainder**.

**EXAMPLE 2** *Using long division to find a quotient and remainder*

Use long division to find the quotient and remainder when dividing $f(x) = x^3 - 5x^2 + 20$ by $d(x) = x - 3$. In other words, find $q(x)$ and $r(x)$ so that $x^3 - 5x^2 + 20 = (x - 3)q(x) + r(x)$.

**SOLUTION** Since there is no $x$ term in the polynomial $x^3 - 5x^2 + 20$, we insert $0x$ to ensure that terms with like powers line up. We divide as follows:

$$
\begin{array}{r}
x^2 - 2x - 6 \phantom{xxxx} \\
x - 3 \overline{)x^3 - 5x^2 + 0x + 20} \\
\underline{x^3 - 3x^2} \phantom{xxxxxxxxx} \\
-2x^2 + 0x \phantom{xxxx} \\
\underline{-2x^2 + 6x} \phantom{xxxx} \\
-6x + 20 \\
\underline{-6x + 18} \\
2
\end{array}
$$

Thus, $q(x) = x^2 - 2x - 6$ and $r(x) = 2$, and so

$$x^3 - 5x^2 + 20 = (x - 3)(x^2 - 2x - 6) + 2$$

Equivalently, we can write

$$\underbrace{\frac{\overbrace{x^3 - 5x^2 + 20}^{\text{Dividend}}}{\underbrace{x - 3}_{\text{Divisor}}}} = \overbrace{x^2 - 2x - 6}^{\text{Quotient}} + \overbrace{\frac{2}{x - 3}}^{\text{Remainder}}$$

## SYNTHETIC DIVISION

When the divisor is of the form $x - c$, as it was in Example 2, the long division process can be simplified by eliminating unnecessary symbols and space. The following table shows several successive simplifications for Example 2.

| Remove all x's | Remove unnecesssary numbers and "+" signs | Condense by bringing numbers up | Combine top line with bottom line |
|---|---|---|---|
| $\begin{array}{r} 1\;-2\;-6 \\ -3)\overline{1\;-5\;+0\;+20} \\ 1\;-3 \\ \overline{\quad -2\;+0} \\ \overline{\quad\; -2\;+6} \\ \overline{\quad\quad -6\;+20} \\ \overline{\quad\quad\; -6\;+18} \\ \overline{\quad\quad\quad\; 2} \end{array}$ | $\begin{array}{r} 1\;-2\;-6 \\ -3)\overline{1\;-5\quad 0\quad 20} \\ \underline{-3} \\ -2 \\ \underline{6} \\ -6 \\ \underline{18} \\ 2 \end{array}$ | $\begin{array}{r} 1\;-2\;-6 \\ -3)\overline{1\;-5\quad 0\quad 20} \\ \underline{-3\quad 6\quad 18} \\ -2\;-6\quad 2 \end{array}$ | $\begin{array}{r} -3 \rfloor 1\;-5\quad 0\quad 20 \\ \underline{-3\quad 6\quad 18} \\ 1\;-2\;-6\quad 2 \end{array}$ |

$$\begin{array}{c|cccc} 3 & 1 & -5 & 0 & 20 \\ & & 3 & -6 & -18 \\ \hline & 1 & -2 & -6 & 2 \end{array}$$

**FIGURE 24**

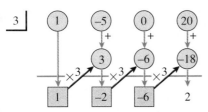

**FIGURE 25**

We can further simplify the procedure if we omit the "$-$" from the divisor $-3$. This changes the signs of the numbers in the second row and, as a consequence, allows us to add down the columns instead of subtracting. The final version of this **synthetic division** procedure is shown in Figure 24. To summarize the procedure, the numbers circled in the top row of Figure 25 are the coefficients of the dividend $x^3 - 5x^2 + 20$. The number 3 on the far left is from the divisor $x - 3$. The number 1 in the bottom row is carried down from the top row. Each boxed number is used to obtain the number to its right by following the steps below.

1. Multiply the boxed number by 3 and place the product in the circle above and to the right.
2. Add the two circled numbers to obtain the boxed number in the bottom row.

The resulting boxed numbers in the bottom row are the coefficients of the quotient $x^2 - 2x - 6$, and the unboxed 2 is the remainder. Note that the degree of the quotient $x^2 - 2x - 6$ is 1 less than the degree of the dividend $x^3 - 5x^2 + 20$.

**EXAMPLE 3**    *Using synthetic division to find a quotient and remainder*

Use synthetic division to find the quotient and remainder when $3x^4 - 12x^2 + 8x + 4$ is divided by $x + 2$.

**SOLUTION**    Since the divisor is $x + 2$, we place $-2$ to the far left. The top row is formed using the coefficients of the dividend $3x^4 - 12x^2 + 8x + 4$, including a 0 for the coefficient of the missing $x^3$ term. After bringing the leading coefficient 3 down to the bottom row, we proceed as in Figure 25, multiplying by $-2$ and then adding at each stage.

$$\begin{array}{c|ccccc} -2 & 3 & 0 & -12 & 8 & 4 \\ & & -6 & 12 & 0 & -16 \\ \hline & 3 & -6 & 0 & 8 & -12 \end{array}$$

Since the divisor $x + 2$ has degree 1 and the dividend $3x^4 - 12x^2 + 8x + 4$ has degree 4, the quotient will have degree 3. Using the first four numbers in the bottom row as the coefficients, we obtain the quotient $3x^3 - 6x^2 + 0x + 8$. The remainder is $-12$. Thus,

$$3x^4 - 12x^2 + 8x + 4 = (x + 2)(3x^2 - 6x^2 + 8) - 12$$

■

**EXAMPLE 4**   *Using synthetic division to factor a polynomial*

Given that $x + 4$ and $x - 1$ are factors of the polynomial
$f(x) = 4x^4 + 16x^3 - 3x^2 - 13x - 4$, use synthetic division to factor $f(x)$
and then find the zeros.

**SOLUTION**   We first apply synthetic division to divide
$4x^4 + 16x^3 - 3x^2 - 13x - 4$ by $x + 4$.

$$
\begin{array}{r|rrrrr}
-4 & 4 & 16 & -3 & -13 & -4 \\
   &   & -16 & 0 & 12 & 4 \\
\hline
   & 4 & 0 & -3 & -1 & 0
\end{array}
$$

Since the remainder is 0, we have confirmed that $x + 4$ is indeed a factor of $f(x)$.
The first four numbers in the bottom row are the coefficients of the quotient
$4x^3 - 3x - 1$, and so $f(x) = (x + 4)(4x^3 - 3x - 1)$. Now if $x - 1$ is a factor of
$f(x)$, it must also be a factor of $4x^3 - 3x - 1$. Thus, we use synthetic division to
divide $x - 1$ into $4x^3 - 3x - 1$.

$$
\begin{array}{r|rrrr}
1 & 4 & 0 & -3 & -1 \\
  &   & 4 & 4 & 1 \\
\hline
  & 4 & 4 & 1 & 0
\end{array}
$$

Since the remainder is 0, $x - 1$ is indeed a factor. The first three numbers in the
bottom row are the coefficients of $4x^2 + 4x + 1$. Since $4x^2 + 4x + 1 = (2x + 1)^2$,
it follows that

$$4x^4 + 16x^3 - 3x^2 - 13x - 4 = (x + 4)(x - 1)(2x + 1)^2$$

Thus, the zeros of $f$ are $x = -4$, $x = 1$, and $x = -\frac{1}{2}$.

■

Notice that in all of the synthetic division examples, the remainder turned out to be
zero or some other constant. The division algorithm guarantees that this will always be
the case. Indeed, since the divisor $x - c$ is a polynomial of degree 1, and since the
remainder's degree must be less than that of the divisor, the remainder must have degree
zero and so must be a constant. In light of this, the division algorithm for a polynomial
$f(x)$ and divisor $x - c$ can be written as

$$f(x) = (x - c)q(x) + k$$

Now if we set $x = c$, we obtain $f(c) = k$. In other words, the value of $f$ at $x = c$ is
nothing more than the remainder obtained by dividing $f(x)$ by $x - c$. This provides
justification for the Remainder Theorem.

**Remainder Theorem**

| If a polynomial $f(x)$ is divided by $x - c$, the remainder is $f(c)$. |
|---|

**EXAMPLE 5**   *Evaluating a polynomial with the Remainder Theorem*

Use the Remainder Theorem to evaluate $f(x) = 2x^3 - 5x^2 + x - 4$ at $x = -3$.

**SOLUTION**     We use synthetic division to find the remainder when $2x^3 - 5x^2 + x - 4$ is divided by $x + 3 = x - (-3)$.

$$
\begin{array}{r|rrrr}
-3 & 2 & -5 & 1 & -4 \\
   &   & -6 & 33 & -102 \\
\hline
   & 2 & -11 & 34 & -106
\end{array}
$$

Thus, $f(-3) = -106$.     ∎

## EXERCISES 2

**EXERCISES 1–12**  □  *Use long division to find the quotient* $q(x)$ *and remainder* $r(x)$ *when* $f(x)$ *is divided by* $d(x)$.

**1.** $f(x) = 2x^2 - 5x - 12$;  $d(x) = x - 4$

**2.** $f(x) = 3x^2 + x - 10$;  $d(x) = x + 2$

**3.** $f(x) = 4x^3 - 3x^2 + 20x - 15$;  $d(x) = x^2 + 5$

**4.** $f(x) = 2x^3 + 4x^2 - 1$;  $d(x) = x^2 + 2x - 1$

**5.** $f(x) = 2x^4 - 7x^3 - 6x + 9$;  $d(x) = 2x^2 - x + 3$

**6.** $f(x) = 9x^4 - 3x^3 + 2x - 4$;  $d(x) = 3x^2 - 2$

**7.** $f(x) = x^2 + 1$;  $d(x) = x + 1$

**8.** $f(x) = 2x^2 - x + 4$;  $d(x) = x - 2$

**9.** $f(x) = 3x^3 - 4x + 2$;  $d(x) = x^2 - 3$

**10.** $f(x) = x^3$;  $d(x) = x^2 + 1$

**11.** $f(x) = x^4$;  $d(x) = x^2 + x + 1$

**12.** $f(x) = 3x^4 + x^2 - 1$;  $d(x) = x^2 - x + 3$

**EXERCISES 13–24**  □  *Use synthetic division to find the quotient* $q(x)$ *and remainder* $r(x)$ *when* $f(x)$ *is divided by* $d(x)$.

**13.** $f(x) = 2x^2 + 9x - 18$;  $d(x) = x + 6$

**14.** $f(x) = 5x^2 - 13x - 6$;  $d(x) = x - 3$

**15.** $f(x) = x^3 - 2x^2 - 2x + 4$;  $d(x) = x - 2$

**16.** $f(x) = x^3 + 3x^2 - 9x + 5$;  $d(x) = x + 5$

**17.** $f(x) = 3x^4 - 8x^3 - 10x + 3$;  $d(x) = x - 3$

**18.** $f(x) = 2x^4 + x^3 - 4x^2 + 1$;  $d(x) = x + \dfrac{1}{2}$

**19.** $f(x) = x^2 - 2$;  $d(x) = x + 1$

**20.** $f(x) = x^2 + 2x$;  $d(x) = x - 2$

**21.** $f(x) = x^3 + 6x^2 - 16x + 2$;  $d(x) = x + 8$

**22.** $f(x) = 1 - 2x^3$;  $d(x) = x - 4$

**23.** $f(x) = 1 + x^3 - 4x^4$;  $d(x) = x - \dfrac{1}{3}$

**24.** $f(x) = 2x^4 - 14x^3 + 5$;  $d(x) = x - 7$

**EXERCISES 25–32**  □  *Use the given factor(s) of the polynomial to find the remaining factors, and then identify the zeros.*

**25.** $x + 2$;  $f(x) = x^3 + 6x^2 + 3x - 10$

**26.** $x - 4$;  $f(x) = x^3 - 3x^2 - 10x + 24$

**27.** $x - 3$;  $p(x) = 2x^3 - 9x^2 + 7x + 6$

**28.** $x + 1$;  $h(x) = 3x^3 + 10x^2 + x - 6$

**29.** $x + 4$ and $x - 1$;  $g(x) = x^4 + 9x^3 + 22x^2 - 32$

**30.** $x - 6$ and $x + 3$;  $f(x) = x^4 - 2x^3 - 27x^2 + 108$

**31.** $2x + 1$ and $x - 8$;  $h(x) = 6x^4 - 47x^3 - 13x^2 + 38x + 16$

**32.** $5x - 1$ and $x - 1$;  $p(x) = 10x^4 + 13x^3 - 43x^2 + 23x - 3$

**EXERCISES 33–38**  □  *Use synthetic division and the Remainder Theorem to find the indicated function value.*

**33.** $f(x) = 2x^3 - x^2 - 4x + 6$

   **a.** $f(3)$     **b.** $f(-2)$     **c.** $f\left(\dfrac{1}{2}\right)$

**34.** $g(x) = 3x^3 + 7x^2 - 18x + 4$

   **a.** $g(-4)$     **b.** $g(2)$     **c.** $g\left(-\dfrac{1}{3}\right)$

**35.** $h(x) = 3x^4 - 2x^3 - 4x + 3$

   **a.** $h(-1)$     **b.** $h(5)$     **c.** $h(-2.1)$

**36.** $h(x) = -x^4 + 3x^2 + 5x - 10$

   **a.** $h(3)$     **b.** $h(-1)$     **c.** $h(1.4)$

**37.** $g(x) = \dfrac{1}{2}x^4 - \dfrac{5}{3}x^3 + \dfrac{2}{3}x^2 - \dfrac{9}{2}x$

   **a.** $g(4)$     **b.** $g(-2)$

**38.** $f(x) = \dfrac{2}{5}x^4 - \dfrac{21}{2}x^2 + \dfrac{5}{2}x + \dfrac{3}{4}$

   **a.** $f(5)$     **b.** $f(-5)$

■ *Projects for Enrichment*

39. *Factors of $f(x) = x^n - a^n$ and Sums of the Form*
$1 + a + a^2 + \cdots + a^n$ In this project we will use synthetic
division to investigate the factors of $x^n - a^n$, and then see how
this helps us find sums of the form $1 + a + a^2 + \cdots + a^n$. To
begin, we consider the polynomial $f(x) = x^2 - a^2$. Using the
difference of squares formula, we can factor this as
$f(x) = (x - a)(x + a)$. Next we consider $f(x) = x^3 - a^3$. Using
the difference of cubes formula, we know this factors as
$f(x) = (x - a)(x^2 + ax + a^2)$. Notice that both polynomials
have $(x - a)$ as a factor.

a. Complete the synthetic division started below to divide
   $f(x) = x^4 - a^4$ by $x - a$, and then write out the factored
   form of $f(x)$.

$$\underline{a\mid}\ \ \begin{array}{ccccc} 1 & 0 & 0 & 0 & -a^4 \\ & a & a^2 & & \end{array}$$
$$\overline{\qquad 1 \quad a \qquad\qquad\qquad}$$

b. Use synthetic division to divide $f(x) = x^5 - a^5$ by $x - a$,
   and write out the factored form of $f(x)$.

c. Complete the synthetic division started below to divide
   $f(x) = x^n - a^n$ by $x - a$. Write out the factored form of
   $f(x)$.

$$\underline{a\mid}\ \ \begin{array}{cccccc} 1 & 0 & 0 & \cdots & 0 & a^n \\ & a & a^2 & & & \end{array}$$
$$\overline{\qquad 1 \quad a \qquad\qquad\qquad\qquad}$$

$$\overbrace{\qquad\qquad}^{n-1 \text{ zeros}}$$

d. By setting $x = 1$ in the factored form of $f(x)$ from part (c),
   show that

$$1 - a^n = (1 - a)(1 + a + a^2 + \cdots + a^{n-1})$$

   and thus

(1)    $$\frac{1 - a^n}{1 - a} = 1 + a + a^2 + \cdots + a^{n-1}$$

Equation (1) enables us to find sums of the form
$1 + a + a^2 + \cdots + a^n$. Indeed, if we replace $n$ with $n + 1$ in
equation (1), we obtain

(2)    $$1 + a + a^2 + \cdots + a^n = \frac{1 - a^{n+1}}{1 - a}$$

Now suppose we wish to find the sum
$1 + 3 + 3^2 + 3^3 + \cdots + 3^6$. In equation (2) we set $a = 3$ and
$n = 6$ to obtain

$$1 + 3 + 3^2 + 3^3 + \cdots + 3^6 = \frac{1 - 3^7}{1 - 3}$$

$$= \frac{1 - 2187}{-2}$$

$$= 1093$$

e. Use equation (2) to find the following sums.

   i. $1 + 4 + 4^2 + 4^3 + \cdots + 4^8$

   ii. $1 + \dfrac{1}{2} + \left(\dfrac{1}{2}\right)^2 + \cdots + \left(\dfrac{1}{2}\right)^{10}$

   iii. $1 + 6 + 36 + 216 + \cdots + 10{,}077{,}696$

   iv. $1 + \dfrac{2}{3} + \dfrac{4}{9} + \cdots + \dfrac{4096}{531{,}441}$

f. A job pays \$1 in the first month, \$2 in the second month, \$4
   in the third month, and so on. How much will the job pay
   over the course of 5 years?

g. Suppose a highly contagious flu virus is brought into the
   United States by a returning tourist. In order to model the
   spread of this virus, we will assume that the flu is passed to
   three other people on the first day, and it continues to spread
   in such a way that each person who gets the flu on a given
   day passes it to three other people the next day. If this pattern continues, how many people will have had the flu after
   10 days? After 20 days? What's wrong with this model?

40. *Synthetic Division with Nonreal Coefficients* Synthetic divi-
    sion can be applied even when the dividend or divisor involves
    nonreal coefficients. Suppose, for example, that we wish to
    divide $x^2 - 5x + 1$ by $x - i$. Using synthetic division, we pro-
    ceed as follows.

$$\underline{i\mid}\ \ \begin{array}{ccc} 1 & -5 & 1 \\ & i & -1 - 5i \end{array}$$
$$\overline{\quad 1 \quad -5 + i \quad -5i \quad}$$

So $\dfrac{x^2 - 5x + 1}{x - i} = x + (-5 + i) + \dfrac{-5i}{x - i}$.

a. Use synthetic division to find the quotient and remainder
   when the polynomial $f(x)$ is divided by $d(x)$.

   i. $f(x) = x^2 + 3x + 5$; $d(x) = x - 2i$

   ii. $f(x) = 2x^3 - 4x^2 + 1$; $d(x) = x + i$

   iii. $f(x) = x^3 - x^2 + x + 1$; $d(x) = x - (1 + i)$

   iv. $f(x) = x^4 - ix^3 + 4x^2 - 4ix + 2$; $d(x) = x - i$

b. Use synthetic division to show that the polynomial $f(x)$ has
   the given factor(s).

   i. $f(x) = x^3 - 2x^2 + x - 2$; $(x - i)$

   ii. $f(x) = x^3 - 4x^2 + 4x - 16$; $(x + 2i)$

   iii. $f(x) = x^4 + 2x^3 + 3x^2 + 2x + 2$; $(x + 1 - i)$ and
        $(x + i)$

   iv. $f(x) = x^4 - 6ix^3 - 7x^2 + 18$; $(x - 3i)$

c. Use synthetic division and the Remainder Theorem to evaluate the polynomial $f(x)$ for the given values of $x$.

   i. $f(x) = x^3 - 5x^2 + 3x - 1$; $x = 2i$

   ii. $f(x) = x^3 - x^2 + 4$; $x = 1 + i$

   iii. $f(x) = 2x^3 + 7x + 16$; $x = 1 - 2i$

   iv. $f(x) = x^4 - ix^3 - 4x^2 - 3ix + 8$; $x = 2 + i$

■ᴵᵇ *Questions for Discussion or Essay*

41. Explain how the trace feature on your graphics calculator could be used to evaluate a polynomial function $f(x)$ at specific values for $x$. Discuss the advantages and disadvantages of this technique, as compared to the synthetic division process described in this section. Does your calculator have any built-in feature for evaluating functions? If so, how does it compare to using the trace feature or synthetic division?

42. Compare and contrast long division of polynomials with long division of real numbers.

43. In part (g) of Exercise 39, we considered a model of a spreading flu virus. We assumed that each person who gets the flu on a given day passes it to three other people the next day. Is this a realistic assumption? Why or why not? What factors affect the rate at which a contagious disease spreads? Is it likely that an entire population will contract a contagious disease? Explain.

---

## SECTION 3

# ZEROS AND FACTORS OF POLYNOMIALS

■ How can a polynomial whose graph never touches the $x$-axis have six zeros?

■ How can a graphics calculator be used as an aid in factoring polynomials?

■ If a company knows how many units of a product to sell to break even, how can its maximum profit be determined?

■ How can the zeros of a polynomial be used to model bungee jumping?

---

## THE FACTOR THEOREM

We have already seen evidence of a close connection between the factors and zeros of polynomials. In Section 1, for example, we used the factors of a polynomial to find the exact values of its zeros. Now we would like to complete the connection by showing that the zeros of a polynomial can be used to find its factors. Suppose that $x = c$ is a zero of a polynomial $f$. Then $f(c) = 0$. But according to the Remainder Theorem, $f(c)$ is the value of the remainder when $f(x)$ is divided by $x - c$. Thus, $f(x) = (x - c)q(x) + 0$, which shows that $x - c$ is a factor of $f(x)$. We have just proven the Factor Theorem.

### Factor Theorem

A polynomial function $f$ has a zero $c$ if and only if $x - c$ is a factor of $f(x)$. In other words, $f(c) = 0$ if and only if $f(x) = (x - c)q(x)$ for some nonzero polynomial $q(x)$.

### EXAMPLE 1

*Using known zeros of a polynomial to find the remaining zeros*

The polynomial function $f(x) = 5x^4 + 8x^3 - 29x^2 - 20x + 12$ has among its zeros $x = -3$ and $x = 2$. Find the remaining zeros.

**SOLUTION**    Since $f$ has zeros $x = -3$ and $x = 2$, the Factor Theorem tells us that $f(x)$ has factors $x + 3$ and $x - 2$. Thus, we use synthetic division to divide $f(x)$ by $x + 3$, and then apply synthetic division again to divide the quotient by $x - 2$.

$$
\begin{array}{r|rrrrr}
-3 & 5 & 8 & -29 & -20 & 12 \\
   &   & -15 & 21 & 24 & -12 \\
\hline
   & 5 & -7 & -8 & 4 & 0
\end{array}
$$

$$
\begin{array}{r|rrrr}
2 & 5 & -7 & -8 & 4 \\
  &   & 10 & 6 & -4 \\
\hline
  & 5 & 3 & -2 & 0
\end{array}
$$

From the last line of the division by $x - 2$, we see that $5x^2 + 3x - 2$ is a factor of $f(x)$. But $5x^2 + 3x - 2 = (x + 1)(5x - 2)$, and so we have

$$
\begin{aligned}
5x^4 + 8x^3 - 29x^2 - 20x + 12 &= (x + 3)(5x^3 - 7x^2 - 8x + 4) \\
&= (x + 3)(x - 2)\,(5x^2 + 3x - 2) \\
&= (x + 3)(x - 2)\,(x + 1)\,(5x - 2)
\end{aligned}
$$

Thus, the zeros of $f$ are $-3, 2, -1,$ and $\frac{2}{5}$.

---

**EXAMPLE 2**    *Finding a polynomial with given zeros*

Find the third-degree polynomial function $f$ with zeros $x = -3$, $x = -1$, and $x = \frac{1}{2}$ that satisfies $f(0) = 3$.

**SOLUTION**    According to the Factor Theorem, each zero $c$ corresponds to a factor of the form $x - c$. So $f(x)$ must have factors $x + 3$, $x + 1$, and $x - \frac{1}{2}$. Since the product of these three factors is a third-degree polynomial, we know that $f(x)$ cannot have any other factors involving $x$. But $f(x)$ could also have a constant factor, which we will denote by $a$. So must have the form

$$
f(x) = a(x + 3)(x + 1)\left(x - \frac{1}{2}\right)
$$

To find the value of $a$, we use the fact that $f(0) = 3$.

$$
f(0) = 3
$$

$$
a(0 + 3)(0 + 1)\left(0 - \frac{1}{2}\right) = 3 \qquad \text{Setting } x = 0 \text{ in } f(x)
$$

$$
a\left(-\frac{3}{2}\right) = 3 \qquad \text{Multiplying out}
$$

$$
a = 3\left(-\frac{2}{3}\right) = -2
$$

Finally, we multiply out the factors of $f$ to obtain the standard polynomial form.

$$f(x) = -2(x + 3)(x + 1)\left(x - \frac{1}{2}\right)$$

$$= (x + 3)(x + 1)(-2x + 1) \qquad \text{Multiplying } -2 \text{ through } \left(x - \frac{1}{2}\right)$$

$$= (x^2 + 4x + 3)(-2x + 1) \qquad \text{Multiplying } (x + 3)(x + 1)$$

$$= (-2x^3 - 8x^2 - 6x) + (x^2 + 4x + 3) \quad \text{Distributing } (-2x + 1)$$

$$= -2x^3 - 7x^2 - 2x + 3$$

In the preceding example we used the fact that three factors of the form $x - c$ will yield a third-degree polynomial. More generally, $n$ factors of this form will yield an $n$th-degree polynomial. Conversely, an $n$th-degree polynomial can have no more than $n$ such factors and so can have no more than $n$ real zeros. However, it can happen that an $n$th-degree polynomial will have fewer than $n$ real zeros. For example the fifth-degree polynomial $f(x) = x^5 - 2x^4 + x^3$, which factors as $f(x) = x^3(x - 1)^2$, has as its only zeros $x = 0$ and $x = 1$. Since the zero $x = 0$ arises from the factor $x^3$, it is said to be a zero of multiplicity 3. Likewise, since $x = 1$ arises from the factor $(x - 1)^2$, it is called a zero of multiplicity 2.

**Multiple zeros**

> If $(x - c)^k$ is a factor of a polynomial $f(x)$, but $(x - c)^{k+1}$ is not a factor, then $c$ is a **zero of multiplicity $k$**.

The multiplicity of a real zero of a polynomial affects the shape of the graph of the polynomial near the zero. Consider the polynomials $f(x) = (x - 1)^2$ and $g(x) = (x - 1)^3$ shown in Figures 26 and 27. Both have $x = 1$ as a zero, but the multiplicity is 2 for $f$ and 3 for $g$. When we compare the graphs of $f$ and $g$, we see that the graph of $f$ touches the $x$-axis and then turns around, while the graph of $g$ passes through the $x$-axis. Generally speaking, if the multiplicity of a real zero is even, the graph touches the $x$-axis at the zero and then turns around. If the multiplicity of a real zero is odd, the graph passes through the $x$-axis at the zero.

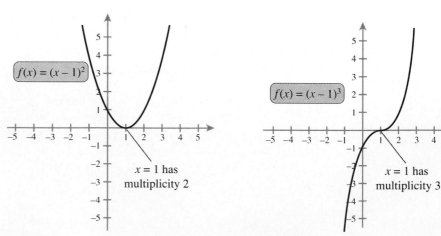

**FIGURE 26**          **FIGURE 27**

**EXAMPLE 3**   *Finding a polynomial with multiple zeros*

Find the third-degree polynomial with the given graph.

**SOLUTION**   The zero at $x = -2$ must have even multiplicity while the zero at $x = 1$ must have odd multiplicity. Since we want a third-degree polynomial, we can conclude that

$$f(x) = a(x + 2)^2(x - 1)$$

for some constant $a$. To find the value for $a$, we note that $f$ has $y$-intercept $(0, 4)$, and so $f(0) = 4$.

$$f(0) = 4$$
$$a(0 + 2)^2(0 - 1) = 4 \qquad \text{Setting } x = 0 \text{ in } f(x)$$
$$a(-4) = 4 \qquad \text{Simplifying}$$
$$a = -1$$

Thus, $f(x) = -1(x + 2)^2(x - 1) = -x^3 - 3x^2 + 4$.

∎

**EXAMPLE 4**   *Modeling bungee jumping with a cubic polynomial*

Suppose you are standing on a building 25 feet above the ground. A bungee jumper leaps from a platform 64 feet above you. She passes your level after 2 seconds, again after 6 seconds on the way back up, and after 9 seconds on the way back down. Assuming the height of the jumper can be roughly approximated with a cubic polynomial, how close to the ground does she get and how high does she rebound? Also, how well does the cubic polynomial model the motion of the bungee jumper?

**SOLUTION**   We first construct a cubic polynomial $h(t)$ that gives the height of the jumper with respect to your level. The times at which the jumper passes your level, namely $t = 2$, $t = 6$, and $t = 9$, are the zeros of the function $h$ since the height (with respect to you) at those times is 0. Thus, we know that $h(t)$ must have factors $t - 2$, $t - 6$, and $t - 9$. We must also allow for the fact that $h(t)$ may have a constant factor, and so we settle on the form $h(t) = a(t - 2)(t - 6)(t - 9)$. Since the initial height of the jump is 64 feet, we have $h(0) = 64$. This enables us to solve for $a$.

$$h(0) = 64$$
$$a(0 - 2)(0 - 6)(0 - 9) = 64$$
$$-108a = 64$$
$$a = -\frac{64}{108} = -\frac{16}{27}$$

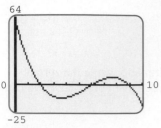

**FIGURE 28**

Thus, $h(t) = -\frac{16}{27}(t - 2)(t - 6)(t - 9)$. The graph of $h$ is shown in Figure 28. Using the trace feature, we determine that the minimum height is $-12.3$ feet below your level, or 12.7 feet above the ground, and the maximum height after the rebound is 7.5 feet above your level, or 32.5 feet above ground level.

However accurate this model might be for the first 9 seconds, it is invalid soon after. The cubic polynomial $h(t)$ approaches $-\infty$ as $t$ approaches $\infty$, but it seems unlikely that the bungee jumper will descend through the earth's crust, pass through its molten core, and then be shot out the other side.

## COMPLEX ZEROS

We have see many examples of polynomials with one or more zeros. Are there examples of polynomials with no zeros? If we consider only real-valued zeros—that is, those that correspond to $x$-intercepts—the answer is yes. A simple example is the polynomial $f(x) = x^2 + 1$. It is clear from the graph in Figure 29 that $f$ has no $x$-intercepts and hence no real zeros. However, if we set $x^2 + 1 = 0$ and solve for $x$ within the complex-number system, we see that $x = \pm i$. Thus, the imaginary numbers $i$ and $-i$ are both zeros of $f$. As it turns out, if we allow complex-valued zeros, every polynomial of degree 1 or higher will have at least one zero. This was first proven in 1799 by the famous German mathematician Carl Friedrich Gauss (he was 22 at the time). Because of its importance to algebra, his result has come to be called the **Fundamental Theorem of Algebra**.

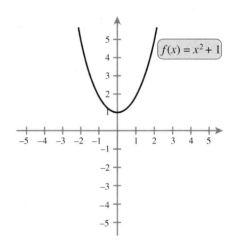

$f(x) = x^2 + 1$

**FIGURE 29**

### Fundamental Theorem of Algebra

If $f(x)$ is a polynomial of degree 1 or higher, then $f$ has at least one zero in the complex-number system.

We can extend the Fundamental Theorem of Algebra to show that a polynomial of degree $n$ must have exactly $n$ zeros (counting multiplicities). To do this, we first show that a polynomial of degree $n$ can be factored into $n$ factors of the form $x - c$. Suppose $f(x)$ is a polynomial of degree $n$, with $n \geq 1$. Then by the Fundamental Theorem of

Algebra, $f$ has a zero, call it $c_1$. By the Factor Theorem, this means $f(x)$ can be factored as

$$f(x) = (x - c_1)q_1(x)$$

where $q_1(x)$ is a polynomial of degree $n - 1$. Applying the Fundamental Theorem of Algebra to the polynomial $q_1(x)$, we know that it too must have a zero, call it $c_2$. By the Factor Theorem, $q_1(x) = (x - c_2)q_2(x)$ for some polynomial $q_2(x)$ of degree $n - 2$. Combined with the earlier factorization, we now have

$$f(x) = (x - c_1)(x - c_2)q_2(x)$$

If we continue this process, we will eventually arrive at a polynomial factor $q_n(x)$ of degree 0 (a constant), and here the process stops. The result is a *complete* factorization of $f$ into linear factors.

**■ *Linear Factorization Theorem***

> If $f(x)$ is a polynomial of degree $n$, with $n \geq 1$, then $f(x)$ can be expressed as a product of linear factors in the following way:
>
> $$f(x) = a(x - c_1)(x - c_2) \cdots (x - c_n)$$
>
> where $c_1, c_2, \ldots, c_n$ are complex numbers and $a$ is the leading coefficient of $f(x)$.

An immediate consequence of the Linear Factorization Theorem is that an $n$th-degree polynomial has $n$ zeros (counting multiplicities), each of which corresponds to a factor of the form $(x - c_i)$. However, neither the Fundamental Theorem of Algebra nor the Linear Factorization Theorem says anything about how to find the zeros or factors. For this we must still depend on techniques described earlier or on some that we will see in the next section.

**EXAMPLE 5**    ***Finding the complex zeros of a polynomial***

Given that 4 is a zero of $f(x) = x^3 - 4x^2 + 4x - 16$, find all the complex zeros of $f$ and write $f(x)$ as a product of linear factors.

**SOLUTION**    Since 4 is a zero of $f$, $x - 4$ must be a factor of $f(x)$. We apply synthetic division to find a second factor.

$$\begin{array}{r|rrrr} 4 & 1 & -4 & 4 & -16 \\ & & 4 & 0 & 16 \\ \hline & 1 & 0 & 4 & 0 \end{array}$$

So $f(x) = (x - 4)(x^2 + 4)$. Now the zeros that correspond to $x^2 + 4$ can be found as follows.

$$x^2 + 4 = 0$$

$$x^2 = -4$$

$$x = \pm\sqrt{-4} = \pm 2i$$

Thus, the zeros of $f$ are $x = 4$, $x = 2i$, and $x = -2i$, and we have

$$f(x) = (x - 4)(x - 2i)(x + 2i)$$

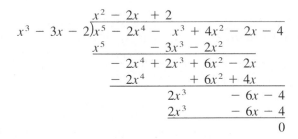

**FIGURE 30**

**EXAMPLE 6**

*Finding the complex zeros of a polynomial*

Find all the complex zeros of the polynomial $f(x) = x^5 - 2x^4 - x^3 + 4x^2 - 2x - 4$ and write it as a product of linear factors.

**SOLUTION**   We first plot the graph of $f$ to see whether any integer zeros can be easily obtained. It appears from the graph in Figure 30 that $x = -1$ and $x = 2$ are zeros of $f$, and this can be verified by substitution into $f$. It also appears that $x = -1$ has even multiplicity and so we deduce that $f(x)$ has factors $(x + 1)^2(x - 2)$. Multiplying these factors out yields $x^3 - 3x - 2$. To verify that this is a factor of $f(x)$, and also to find the remaining factors, we divide $f(x)$ by $x^3 - 3x - 2$.

$$
\begin{array}{r}
x^2 - 2x \phantom{} + 2 \\
x^3 - 3x - 2 \overline{)x^5 - 2x^4 - \phantom{0}x^3 + 4x^2 - 2x - 4} \\
\underline{x^5 \phantom{00000} - 3x^3 - 2x^2 \phantom{0000000}} \\
-2x^4 + 2x^3 + 6x^2 - 2x \\
\underline{-2x^4 \phantom{0000} + 6x^2 + 4x} \\
2x^3 \phantom{0000} - 6x - 4 \\
\underline{2x^3 \phantom{0000} - 6x - 4} \\
0
\end{array}
$$

This shows that $f(x) = (x + 1)^2(x - 2)(x^2 - 2x + 2)$. Since $x^2 - 2x + 2$ cannot be factored using integer coefficients, we find its zeros using the quadratic formula.

$$x^2 - 2x + 2 = 0$$

$$x = \frac{-(-2) \pm \sqrt{(-2)^2 - 4(1)(2)}}{2(1)} = \frac{2 \pm \sqrt{-4}}{2} = \frac{2 \pm 2i}{2} = 1 \pm i$$

Thus, $x^2 - 2x + 2 = [x - (1 + i)][x - (1 - i)]$. Putting all the factors together, we have

$$f(x) = (x + 1)^2(x - 2)[x - (1 + i)][x - (1 - i)]$$

The five zeros of $f$ are $x = -1$ (multiplicity 2), $x = 2$, $x = 1 + i$, and $x = 1 - i$.

It is no coincidence that the two nonreal zeros in Example 6, namely $x = 1 + i$ and $x = 1 - i$, differ only in the operators "+" and "−." Pairs of complex numbers of the form $a + bi$ and $a - bi$ are called **complex conjugates**. If a polynomial has real coefficients, all nonreal zeros will appear as complex conjugate pairs. In other words, if a polynomial $f(x)$ with real coefficients has a complex zero $a + bi$, then it must also have a zero $a - bi$. (See Exercise 51 for an outline of the proof.)

**EXAMPLE 7**

*Finding a polynomial with known complex zeros*

Find a polynomial of least degree with real coefficients that has zeros $x = -3$ and $x = 2 - 4i$.

**SOLUTION**   The polynomial must have a second nonreal zero $x = 2 + 4i$. The two factors corresponding to the nonreal zeros $2 - 4i$ and $2 + 4i$ are $x - (2 - 4i)$ and $x - (2 + 4i)$. Multiplying these out yields a quadratic factor with real coefficients, as shown below.

$$[x - (2 - 4i)][x - (2 + 4i)] = x^2 - (2 - 4i)x - (2 + 4i)x + (2 - 4i)(2 + 4i)$$
$$= x^2 - 4x + (2^2 - (4i)^2)$$
$$= x^2 - 4x + 4 + 16$$
$$= x^2 - 4x + 20$$

The factor corresponding to the real zero $-3$ is $x + 3$, and so we compute $(x^2 - 4x + 20)(x + 3)$.

$$
\begin{array}{r}
x^2 - \phantom{1}4x + 20 \\
\times \phantom{xxxxx} x + \phantom{1}3 \\
\hline
3x^2 - 12x + 60 \\
x^3 - 4x^2 + 20x \phantom{xxxx} \\
\hline
x^3 - \phantom{1}x^2 + \phantom{1}8x + 60
\end{array}
$$

Thus, $f(x) = x^3 - x^2 + 8x + 60$ is a polynomial of least degree with real coefficients and zeros $-3$ and $2 - 4i$.   ■

Note that if we did not require the polynomial in Example 7 to have real coefficients, a polynomial with least degree and zeros $x = -3$ and $x = 2 - 4i$ would be the second-degree polynomial $f(x) = (x + 3)[x - (2 - 4i)] = x^2 + (1 + 4i)x + (-6 + 12i)$. Complex zeros need not come in conjugate pairs if the coefficients of the polynomial are allowed to be nonreal.

## EXERCISES 3

**EXERCISES 1–6** □ *Use the given zero(s) of $f$ to factor $f(x)$ and then find the remaining zeros.*

1. $-3$; $f(x) = x^3 + 3x^2 - 4x - 12$

2. $1$; $f(x) = x^3 + 5x^2 + 3x - 9$

3. $5$; $f(x) = 2x^3 - 7x^2 - 14x - 5$

4. $-6$; $f(x) = 3x^3 + 10x^2 - 44x + 24$

5. $-\dfrac{1}{3}$, $-4$; $f(x) = 3x^4 + 7x^3 - 19x^2 + 5x + 4$

6. $\dfrac{3}{2}$, $-1$; $f(x) = 2x^4 - x^3 - 5x^2 + x + 3$

**EXERCISES 7–12** □ *Use a graphics calculator to identify the integer zeros of $f$ and then factor $f(x)$ completely to find the remaining zeros.*

7. $f(x) = 5x^3 + 3x^2 - 12x + 4$

8. $f(x) = 3x^3 - 10x^2 - x + 12$

9. $f(x) = 6x^3 - 13x^2 + x + 2$

10. $f(x) = 6x^3 + 13x^2 + 4x - 3$

11. $f(x) = x^4 - x^3 - 5x^2 + 3x + 6$

12. $f(x) = x^4 + x^3 - 7x^2 - 5x + 10$

**EXERCISES 13–20** □ *Find a polynomial function $f$ that satisfies the given properties.*

13. Degree 2; zeros $-3$ and 4

14. Degree 3; zeros $-1$, 2, and 4

15. Degree 3; zeros $-2$, 1, and 5; $f(0) = 5$

16. Degree 2; zeros $-4$ and $-1$; $f(0) = 8$

17. Degree 4; zeros 0 and 2 (both with multiplicity 2); $f(-1) = 3$

18. Degree 4; zeros $-1$ and 1 (both with multiplicity 2); $f(2) = 1$

19. Degree 5; zeros $-1$ (multiplicity 3) and $\frac{1}{2}$ (multiplicity 2)

20. Degree 5; zeros 0 (multiplicity 1), $-\frac{2}{3}$ (multiplicity 2), and 1 (multiplicity 2)

**EXERCISES 21–26** □ *Find the polynomial with the given degree and graph. Check your answer with a graphics calculator.*

**21.** Degree 2

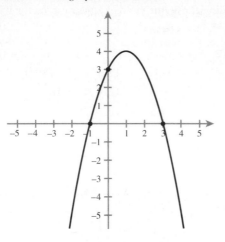

**24.** Degree 3

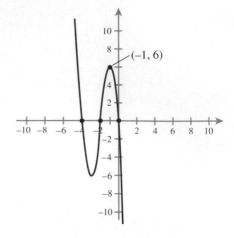

**22.** Degree 2

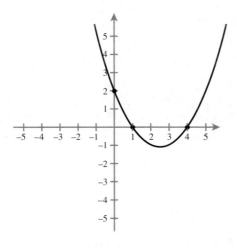

**25.** Degree 4

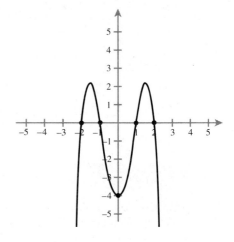

**23.** Degree 3

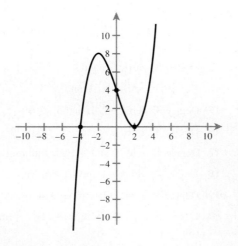

**26.** Degree 4

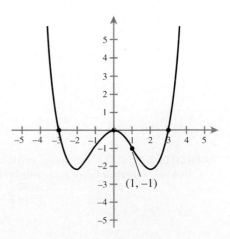

**EXERCISES 27–30** □ *Use the given zeros of f to factor f(x) over the complex numbers and then find the remaining zeros.*

**27.** $-2i$; $f(x) = x^3 - 3x^2 + 4x - 12$

**28.** $i$; $f(x) = x^3 + 2x^2 + x + 2$

**29.** $1 + i$; $f(x) = x^4 - 5x^3 + 4x^2 + 2x - 8$

**30.** $2 - i$; $f(x) = x^4 - 3x^3 - 5x^2 + 29x - 30$

**EXERCISES 31–38** □ *Find all the complex zeros of f and write f(x) as a product of linear factors.*

**31.** $f(x) = x^2 + 1$

**32.** $f(x) = x^2 - 4x + 13$

**33.** $f(x) = x^3 - 2x^2 + 5x$

**34.** $f(x) = x^3 + 2x^2 + 4x + 8$

**35.** $f(x) = x^4 + 3x^2 - 4$

**36.** $f(x) = x^4 - 4x^3 + 5x^2$

**37.** $f(x) = x^5 - 5x^4 + 9x^3 - 5x^2$

**38.** $f(x) = x^5 - x^4 - x + 1$

**EXERCISES 39–44** □ *Find a polynomial of least degree with real coefficients that has the given zeros.*

**39.** $-2$ and $2i$

**40.** $0$ and $2 + i$

**41.** $i$ and $1 - i$

**42.** $-1$, $3$, and $-4i$

**43.** $0$, $\pm\sqrt{2}$, and $-2 + 2i$

**44.** $1$, $-1 + 3i$, $1 - 3i$

■ *Applications*

**45.** *Dimensions of a Box*  A box with no top is formed by cutting squares out of the corners of a rectangular piece of cardboard and then folding up the sides (see Figure 31). The volume of the box is given by $V(x) = 4x^3 - 72x^2 + 320x$, where $x$ denotes the length of the side of a cut-out square. Find the zeros of $V$ and use them to determine the domain for $V$ (i.e., the values for $x$ that make sense in this problem) and also the dimensions of the original piece of cardboard.

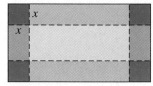

**FIGURE 31**

**46.** *Dimensions of a Box*  A box with no top is formed by cutting squares out of the corners of a rectangular piece of cardboard and then folding up the sides (see Figure 31). The volume of the box is given by $V(x) = 4x^3 - 48x^2 + 135x$, where $x$ denotes the length of the side of a cut-out square. Find the zeros of $V$ and use them to determine the domain for $V$ (i.e., the values for $x$ that make sense in this problem) and also the dimensions of the original piece of cardboard.

**47.** *Breaking Even*  A company that manufactures bicycle frames has determined that the monthly profit (in dollars) from the sale of $x$ frames is given by the polynomial function $P(x) = -0.005x^3 + 1.5x^2 + 50x - 15,000$. Find the number of frames that must be sold per month for the profit to be zero. Over what interval of $x$-values will the profit be positive?

**48.** *Breaking Even*  A company that produces copy machines has a monthly profit (in dollars) from the sale of $x$ copy machines given by $P(x) = -0.1x^3 + 20.5x^2 - 75x - 67,500$. Find the number of copy machines that must be sold per month for the profit to be zero. Over what interval of $x$-values will the profit be positive?

**49.** *Height of a Ball*  An object is projected upward from 64 feet below ground level. It passes ground level after 1 second and again after 4 seconds on its way back down. It is known that the height of the object with respect to ground level is given by a quadratic function $s(t)$, where $t$ denotes the number of seconds after the object has been projected upward. Find $s(t)$. [*Hint:* $s(t)$ has zeros $t = 1$ and $t = 4$, and $s(0) = -64$.] What is the maximum height of the ball?

**50.** *Maximum Profit*  A roofing company has determined that because of fixed operating costs, they will lose $10,000 in a given month if no roofs are completed. Moreover, because of their variable cost structure, they will break even (i.e., zero profit) if they complete either 10 roofs or 30 roofs each month. Find a quadratic function $P(x)$ that expresses the company's monthly profit in terms of the number of roofs completed. [*Hint:* $P(x)$ has zeros $x = 10$ and $x = 30$, and $P(0) = -10,000$.] How many roofs should be completed to maximize profit?

## ◢ *Projects for Enrichment*

**51.** *Complex Conjugates*  The complex conjugate of a number $a + bi$ is $a - bi$. In general, the conjugate of a complex number $z$ is denoted by $\bar{z}$. Thus, if $z = a + bi$, then $\bar{z} = a - bi$. In this project we will investigate some of the properties of complex conjugates. Throughout, we will let $z = a + bi$ and $w = c + di$.

  **a.** Show that $z + \bar{z}$ is a real number.

  **b.** Show that $z\bar{z}$ is a real number.

  **c.** Show that $\overline{z + w} = \bar{z} + \bar{w}$.

  **d.** Show that $\overline{z \cdot w} = \bar{z} \cdot \bar{w}$.

  **e.** Use part (d) to argue that $\overline{z^n} = (\bar{z})^n$.

  **f.** If $a$ is a real number, show that $\bar{a} = a$.

  **g.** If $f(x) = a_n x^n + a_{n-1} x^{n-1} + \cdots + a_1 x + a_0$ is a polynomial with real coefficients, show that $f(\bar{z}) = \overline{f(z)}$.

  **h.** If $f(x) = a_n x^n + a_{n-1} x^{n-1} + \cdots + a_1 x + a_0$ is a polynomial with real coefficients and $z$ is a zero of $f$, show that $\bar{z}$ is also a zero of $f$.

**52.** *Cardan's Formula*  The zeros of a general quadratic polynomial $f(x) = ax^2 + bx + c$ are easily found with the quadratic formula. To find the zeros of the general cubic polynomial $f(x) = ax^3 + bx^2 + cx + d$, we require **Cardan's formula**, a formula that gives the solution to cubic equations in terms of their coefficients.

  **a.** A cubic equation is said to be in **reduced form** if the squared term has coefficient 0 and the cubed term has coefficient 1. Show that the equation $ax^3 + bx^2 + cx + d = 0$ can be transformed into a cubic equation in reduced form by making the substitution $x = z - \dfrac{b}{3a}$; that is, show that an equation of the form $z^3 + pz + q = 0$ results.

  **b.** Write the cubic equation $x^3 - 9x^2 + 9x + 62 = 0$ in reduced form.

Cardan's formula can be applied only to cubic equations in reduced form. In order to express Cardan's formula concisely, we define the following three variables.

$$s = \sqrt{\frac{q^2}{4} + \frac{p^3}{27}}, \quad u = \sqrt[3]{\frac{-q}{2} + s}, \quad \text{and} \quad v = \sqrt[3]{\frac{-q}{2} - s}$$

According to Cardan's formula, the three solutions to the equation $z^3 + pz + q = 0$ are given by

$$z_1 = u + v, \quad z_2 = \frac{-(u + v) + (u - v)\sqrt{3}i}{2},$$

and

$$z_3 = \frac{-(u + v) - (u - v)\sqrt{3}i}{2}$$

  **c.** Use Cardan's formula to find the zeros of each of the following polynomials.

   **i.** $f(x) = x^3 + 63x - 316$

   **ii.** $g(x) = x^3 - 27x - 54$

   **iii.** $h(x) = x^3 - 9x^2 + 9x + 62$ [*Hint:* First write in the reduced form $z^3 + pz + q = 0$ and solve the resulting equation for $z$. Then use the relationship $x = z - \dfrac{b}{3a}$ to find the corresponding values of $x$.]

  **d.** Use Cardan's formula in conjunction with the Factor Theorem to find factorizations of each of the polynomials $f$, $g$, and $h$ defined in part (c).

## ◢ *Questions for Discussion or Essay*

**53.** The Factor Theorem states that there is a one-to-one correspondence between the zeros of a polynomial and its linear factors. In other words, each zero $c$ of a polynomial $f(x)$ corresponds to a factor $x - c$, and vice versa. This does *not* mean, however, that there is only one polynomial of a given degree that corresponds to a given set of zeros. Explain why this is so. It may be helpful to think of how two different polynomials of the same degree can be constructed so that they have the same zeros. What additional piece of information could be given so that a polynomial that corresponds to a given set of zeros is unique?

**54.** Throughout this section, we have been dealing with polynomials having real coefficients. However, there is nothing to prevent us from considering polynomials with nonreal complex coefficients. The polynomial $f(x) = x^2 - ix$ is one example. How many zeros does $f$ have? Are any of the zeros real? Do you think the Fundamental Theorem of Algebra and the Linear Factorization Theorem still hold if the polynomial has nonreal coefficients? Write out a theorem that says as much as possible about the number and type of zeros of a polynomial with complex coefficients.

55. Use the Linear Factorization Theorem to help you explain why the product of all the zeros of a polynomial with leading coefficient 1 must be equal to the constant term (or its opposite).

56. Explain why a polynomial with odd degree and real coefficients must have at least one real zero. Can anything be said along this line about polynomials of even degree with real coefficients? Explain.

57. Many physical applications involve cyclical or oscillating behavior. For example the motion of a pendulum, the path of a weight attached to a spring, alternating current, and average monthly temperatures all exhibit cyclical behavior. Discuss the limitations in using polynomials to model cyclic phenomena.

---

## SECTION 4

# REAL ZEROS OF POLYNOMIALS

■ Why is it that the polynomial $P(x) = x^{20} -$ (your age in seconds)$x^{19} +$ (your weight in kilograms)$x^5 + 2x + 2.7$ cannot possibly cross the positive $x$-axis exactly once?

■ You are told that a certain polynomial has an $x$-intercept somewhere between 1 and 1000. How could you evaluate the function at 20 points, and based on this information, determine the $x$-intercept to within three decimal places of accuracy?

■ Why is it that any integer coefficient polynomial $P(x)$ with degree $10^{1000}$ (greater than the number of atoms in the observable universe) and leading coefficient 1 satisfying $P(0) = 149$ has at most two rational zeros?

We have already seen that a graphics calculator can be a useful aid for investigating the real zeros of a polynomial. However, we have also seen that a graphics calculator has its limitations. If the zeros are not integers, we may be limited to finding approximations with the trace feature. Moreover, if the calculator is not used carefully, it can give misleading information about the number of zeros or the multiplicity of a zero. Two examples of these limitations are given in Table 1 on the following page.

### THE RATIONAL ZERO THEOREM

We first consider a test for determining the rational zeros of a polynomial. Suppose $f(x) = a_n x^n + a_{n-1}x^{n-1} + \cdots + a_1 x + a_0$ is a polynomial that has integer coefficients and a rational zero of the form $p/q$, where $p$ and $q$ have no common factor other than 1. Then $f(p/q) = 0$ and we can perform the following steps.

$$a_n\left(\frac{p}{q}\right)^n + a_{n-1}\left(\frac{p}{q}\right)^{n-1} + \cdots + a_1\left(\frac{p}{q}\right) + a_0 = 0 \qquad \text{Definition of a zero}$$

$$a_n p^n + a_{n-1}p^{n-1}q + \cdots + a_1 pq^{n-1} + a_0 q^n = 0 \qquad \text{Multiplying through by } q^n$$

$$a_n p^n + a_{n-1}p^{n-1}q + \cdots + a_1 pq^{n-1} = -a_0 q^n \qquad \text{Subtracting } a_0 q^n \text{ on both sides}$$

$$p(a_n p^{n-1} + a_{n-2}p^{n-2}q + \cdots + a_1 q^{n-1}) = -a_0 q^n \qquad \text{Factoring out } p \text{ on the left}$$

The last step shows that $p$ is a factor of the left-hand side, and so $p$ must be a factor of the right-hand side as well. But $p$ cannot be a factor of $q^n$ since $p$ and $q$ have no common

**TABLE 1**

*Limitations of Finding Zeros of Polynomials Graphically*

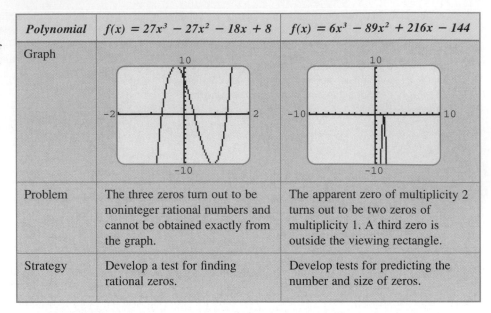

| *Polynomial* | $f(x) = 27x^3 - 27x^2 - 18x + 8$ | $f(x) = 6x^3 - 89x^2 + 216x - 144$ |
|---|---|---|
| Problem | The three zeros turn out to be noninteger rational numbers and cannot be obtained exactly from the graph. | The apparent zero of multiplicity 2 turns out to be two zeros of multiplicity 1. A third zero is outside the viewing rectangle. |
| Strategy | Develop a test for finding rational zeros. | Develop tests for predicting the number and size of zeros. |

factors other than 1. Thus, $p$ must be a factor of $a_0$. This proves the first part of the **Rational Zero Theorem**. The second part is similar.

■ *Rational Zero Theorem*

> If the polynomial $f(x) = a_n x^n + a_{n-1} x^{n-1} + \cdots + a_1 x + a_0$ has integer coefficients, then every rational zero of $f$ has the form $p/q$, where $p$ and $q$ have no common factors other than 1, $p$ is a factor of $a_0$, and $q$ is a factor of $a_n$.

According to the Rational Zero Theorem, if we find all of the factors of the constant term $a_0$ and the leading coefficient $a_n$, and form a list of the rational numbers of the form

$$\frac{\text{factor of } a_0}{\text{factor of } a_n}$$

then all the rational zeros of the polynomial will be in the list. Trial and error (perhaps with the assistance of a graphics calculator) will lead us to the actual rational zeros.

**EXAMPLE 1**    *Finding rational zeros*

Use the Rational Zero Theorem to find the rational zeros of the polynomial function $f(x) = 27x^3 - 27x^2 - 18x + 8$.

**SOLUTION**    First we list all the factors of the constant term and leading coefficient.

Factors of the constant term 8:       $\pm 1, \pm 2, \pm 4, \pm 8$
Factors of the leading coefficient 27:   $\pm 1, \pm 3, \pm 9, \pm 27$

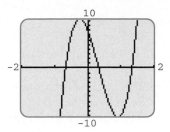

**FIGURE 32**

Next we form all quotients of the factors of 8 divided by the factors of 27.

$$\text{Possible rational zeros:} \quad \pm 1, \ \pm\tfrac{1}{3}, \ \pm\tfrac{1}{9}, \ \pm\tfrac{1}{27}$$
$$\pm 2, \ \pm\tfrac{2}{3}, \ \pm\tfrac{2}{9}, \ \pm\tfrac{2}{27}$$
$$\pm 4, \ \pm\tfrac{4}{3}, \ \pm\tfrac{4}{9}, \ \pm\tfrac{4}{27}$$
$$\pm 8, \ \pm\tfrac{8}{3}, \ \pm\tfrac{8}{9}, \ \pm\tfrac{8}{27}$$

Since there are so many possible zeros, naive trial and error will likely be quite inefficient. Instead, we use information from the graph of $f$ to help narrow down the list. From the graph in Figure 32, we know that there are no integer zeros. Moreoever, the middle zero appears to be approximately $\tfrac{1}{3}$. Thus, we begin by testing $\tfrac{1}{3}$ with synthetic division.

$$
\begin{array}{r|rrrr}
\tfrac{1}{3} & 27 & -27 & -18 & 8 \\
  &    & 9 & -6 & -8 \\
\hline
  & 27 & -18 & -24 & 0
\end{array}
$$

The remainder 0 tells us that $x = \tfrac{1}{3}$ is indeed a zero. This fact, together with the first three numbers in the bottom row, indicates that two of the factors of $f(x)$ are $x - \tfrac{1}{3}$ and $27x^2 - 18x - 24$. The quadratic factor can be factored further as

$$27x^2 - 18x - 24 = 3(9x^2 - 6x - 8) = 3(3x + 2)(3x - 4)$$

Thus, the other two zeros occur when $3(3x + 2)(3x - 4) = 0$, namely when $x = -\tfrac{2}{3}$ and $x = \tfrac{4}{3}$. So the three zeros of $f$ are $-\tfrac{2}{3}, \tfrac{1}{3}$, and $\tfrac{4}{3}$.

---

**WARNING** The Rational Zero Theorem can only be applied to polynomials with integer coefficients, and it does not guarantee that there are rational zeros.

---

**EXAMPLE 2** *Finding the rational zeros of a polynomial*

Find all the zeros of $f(x) = x^4 + \tfrac{3}{5}x^3 - \tfrac{27}{5}x^3 - 3x + 2$.

**SOLUTION** Notice that $f$ has noninteger coefficients, and so we cannot apply the Rational Zero Theorem to $f$. However, the zeros of $f$ coincide with the solutions of $x^4 + \tfrac{3}{5}x^3 - \tfrac{27}{5}x^2 - 3x + 2 = 0$, and we can clear the fractions in this equation by multiplying through by 5 to obtain $5x^4 + 3x^3 - 27x^2 - 15x + 10 = 0$. Thus, we will apply the Rational Zero Theorem to the polynomial function $g(x) = 5x^4 + 3x^3 - 27x^2 - 15x + 10$ (note that $f$ and $g$ are *different* polynomials having the same zeros).

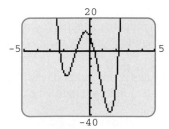

**FIGURE 33**

| | |
|---|---|
| Factors of the constant term 10: | $\pm 1, \ \pm 2, \ \pm 5, \ \pm 10$ |
| Factors of the leading coefficient 5: | $\pm 1, \ \pm 5$ |
| Possible rational zeros: | $\pm 1, \ \pm\tfrac{1}{5}, \ \pm 2, \ \pm\tfrac{2}{5}, \ \pm 5, \ \pm 10$ |

These are the only possible rational zeros of $g$ and so also of $f$. To narrow down the list, we plot the graph of $g$. (Note that we could have chosen to use $f$, but $g$ is somewhat easier to work with since it has integer coefficients.) It appears in Figure 33 that $g$ has a zero at $x = -1$, and we can easily verify this using synthetic division.

$$\begin{array}{r|rrrrr}
-1 & 5 & 3 & -27 & -15 & 10 \\
 &  & -5 & 2 & 25 & -10 \\
\hline
 & 5 & -2 & -25 & 10 & 0
\end{array}$$

So $x = -1$ is a zero and $g$ factors as $(x + 1)(5x^3 - 2x^2 - 25x + 10)$. Another zero is between 0 and 1. From our list of possible rational zeros, we see that the two choices are $\frac{1}{5}$ and $\frac{2}{5}$. Now if either of these is a zero of $g$, it must also be a zero of the factor $5x^3 - 2x^2 - 25x + 10$. Thus, we can simplify our work by applying synthetic division to this factor.

$$\begin{array}{r|rrrr}
\frac{1}{5} & 5 & -2 & -25 & 10 \\
 &  & 1 & -\frac{1}{5} & -\frac{126}{25} \\
\hline
 & 5 & -1 & -\frac{126}{5} & \frac{124}{25}
\end{array}
\qquad
\begin{array}{r|rrrr}
\frac{2}{5} & 5 & -2 & -25 & 10 \\
 &  & 2 & 0 & -10 \\
\hline
 & 5 & 0 & -25 & 0
\end{array}$$

So $\frac{1}{5}$ is not a zero, but $\frac{2}{5}$ is. The synthetic division for $\frac{2}{5}$ also tells us that a factor of $g(x)$ is $5x^2 - 25$. This factor has two irrational zeros, namely $x = \pm\sqrt{5}$. Thus, the four zeros of $g$, and $f$ also, are $x = -1$, $x = \frac{2}{5}$, $x = -\sqrt{5}$, and $x = \sqrt{5}$.

## DESCARTES' RULE OF SIGNS

Our next test gives us information about the number of real zeros of a polynomial. In Section 3, we saw that a polynomial of degree $n$ will have $n$ zeros. However, because of possible multiplicities and nonreal complex zeros, an $n$th-degree polynomial may have fewer than $n$ distinct real zeros. **Descartes' Rule of Signs** gives us more information about the number of real zeros by considering the number of times that successive coefficients of the polynomial change from positive to negative or negative to positive. These changes are referred to as **variations in sign**. The following table gives several examples.

| Polynomial | Variations in sign |
|---|---|
| $x^3 + 2x + 5$ | 0 |
| $2x^3 - x^2 - 1$ | 1 |
| $5x^3 - 3x^2 + x + 4$ | 2 |

### Descartes' Rule of Signs

Let $f(x) = a_n x^n + a_{n-1} x^{n-1} + \cdots + a_1 x + a_0$ be a polynomial with real coefficients.

1. The number of positive real zeros (counting multiplicities) is either equal to the number of variations in sign of $f(x)$ or is less than that number by an even integer.

2. The number of negative real zeros (counting multiplicities) is either equal to the number of variations in sign of $f(-x)$ or is less than that number by an even integer.

| **EXAMPLE 3** | *Using Descartes' Rule of Signs* |

Apply Descartes' Rule of Signs to the polynomial $f(x) = 2x^3 - x^2 - 1$.

**SOLUTION**    Since $f(x)$ has only one variation in sign, $f$ must have exactly one positive real zero. To test for negative zeros, we first simplify $f(-x)$.

$$f(-x) = 2(-x)^3 - (-x)^2 - 1 = -2x^3 - x^2 - 1$$

Since $f(-x)$ has no variations in sign, we can conclude that $f$ has no negative zeros. The graph of $f$ in Figure 34 shows one zero near 1. Because of Descartes' Rule of Signs we can be certain that there are no other zeros. Note that if we had just used the graph and not Descartes' Rule of Signs, we could not rule out the presence of other zeros outside the viewing rectangle.

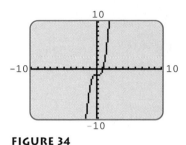

**FIGURE 34**

Since Descartes' Rule of Signs does not "pin-down" the number of zeros unless there is only one variation in sign or none at all, it is usually best to complement the information given by Descartes' Rule of Signs by graphing the polynomial with a graphics calculator.

| **EXAMPLE 4** | *Using Descartes' Rule of Signs* |

Apply Descartes' Rule of Signs to $f(x) = x^3 - 2x^2 + x + 2$.

**SOLUTION**    Since $f(x)$ has two variations in sign, $f$ has either two or no positive real zeros. Moreover, since

$$f(-x) = -x^3 - 2x^2 - x + 2$$

has only one variation in sign, $f$ has one negative zero. From the graph of $f$ in Figure 35, we see one negative zero, no positive zeros, and two turning points. Since $f$ is a cubic polynomial, and since cubic polynomials have at most two turning points, $f$ has no additional turning points. Thus the graph of $f$ will continue to increase as $x$ gets larger, and $f$ has no additional zeros.

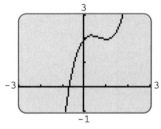

**FIGURE 35**

In the previous example, Descartes' Rule of Signs was superfluous. All the pertinent information could be obtained directly from the graph. This is not always the case, as the following example illustrates.

| **EXAMPLE 5** | *Finding zeros* |

Find the zeros of $f(x) = 6x^3 - 89x^2 + 216x - 144$.

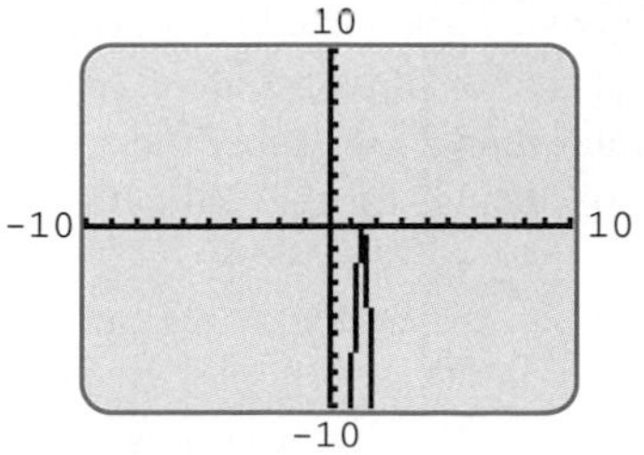

**SOLUTION**    The graph of $f$ is shown in Figure 36. At first glance, it appears that $f$ has only one zero of multiplicity 2 somewhere in the interval $[1, 2]$. However, since $f(x)$ has three variations in sign, Descartes' Rule of Signs tells us that $f$ has either one or three positive zeros (counting multiplicities), and so a single positive zero of multiplicity 2 is impossible. Thus, there must be a zero to the right of the viewing rectangle. After some experimentation, we settle on the view shown in Figure 37.

**FIGURE 36**

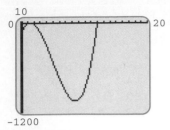

**FIGURE 37**

This suggests another zero at $x = 12$, which can be confirmed by synthetic division.

$$
\begin{array}{r|rrrr}
12 & 6 & -89 & 216 & -144 \\
   &   & 72 & -204 & 144 \\
\hline
   & 6 & -17 & 12 & 0
\end{array}
$$

Since the remainder is 0, we know that $f(x) = (x - 12)(6x^2 - 17x + 12)$. Moreover, $(6x^2 - 17x + 12) = (2x - 3)(3x - 4)$, and so $f$ has *two* more zeros at $x = \frac{3}{2}$ and $x = \frac{4}{3}$.

## UPPER AND LOWER BOUNDS TEST

Our final test will help us determine upper and lower bounds for the real zeros of a polynomial. We say that a real number $b$ is an **upper bound** for the real zeros of a polynomial if none of the zeros are greater than $b$. Similarly, a number $b$ is a **lower bound** if none of the zeros are less than $b$. Suppose we suspect that a number $b > 0$ is an upper bound for the real zeros of a polynomial $f$. How can we determine this for sure? The graph of $f$ can be misleading because there is always the possibility that a zero lies to the right of the viewing rectangle. Instead, we can apply the following test.

### *Upper and Lower Bounds Test*

Let $f(x) = a_n x^n + a_{n-1} x^{n-1} + \cdots + a_1 x + a_0$ be a polynomial with real coefficients and positive leading coefficient $a_n$.

1. If $b > 0$ and each number in the last row of the synthetic division of $f(x)$ by $x - b$ is either positive or zero, then $b$ is an upper bound for the real zeros of $f$.

2. If $b < 0$ and the numbers in the last row of the synthetic division of $f(x)$ by $x - b$ alternate in sign (with 0 counting either as positive or negative), then $b$ is a lower bound for the real zeros of $f$.

**EXAMPLE 6**   *Determining the number of zeros*

Use the Upper and Lower Bounds Test to help you determine the number of zeros of the function $f(x) = x^6 - 4x^4 + 6x^2 - 4$.

**SOLUTION**   The graph of $f$ in Figure 38 shows two zeros. However, since $f$ has degree 6, there could be as many as six zeros. Moreover, all we can tell from Descartes' Rule of Signs is that, since $f(x)$ and $f(-x)$ both have three variations in sign, $f$ could have either one or three positive zeros and either one or three negative zeros.

Since the graph suggests that there are no zeros greater than 2, we apply the Upper and Lower Bounds Test to see if $x = 2$ is an upper bound for the real zeros of $f$.

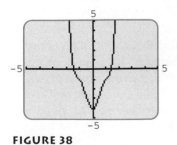

**FIGURE 38**

$$
\begin{array}{r|rrrrrrr}
2 & 1 & 0 & -4 & 0 & 6 & 0 & -4 \\
  &   & 2 & 4 & 0 & 0 & 12 & 24 \\
\hline
  & 1 & 2 & 0 & 0 & 6 & 12 & 20
\end{array}
$$

Since all the numbers in the last row are positive or zero, we know that $x = 2$ is indeed an upper bound on the zeros of $f$—there are no zeros to the right of the viewing rectangle. Next we test to see if $x = -2$ is a lower bound for the real zeros of $f$.

$$
\begin{array}{r|rrrrrrr}
-2 & 1 & 0 & -4 & 0 & 6 & 0 & -4 \\
   &   & -2 & 4 & 0 & 0 & -12 & 24 \\
\hline
   & 1 & -2 & 0 & 0 & 6 & -12 & 20 \\
   & + & - & + & - & + & - & + \\
\end{array}
$$

Since the numbers in the last row alternate in sign (note how 0 is first counted as positive and then as negative), $x = -2$ is indeed a lower bound. Thus $f$ has only the two zeros shown in Figure 38.

## APPROXIMATING REAL ZEROS

It is not uncommon for a polynomial to have no rational zeros, and in this case it is very difficult—perhaps impossible—to obtain exact values for the zeros. Fortunately, we can always resort to an approximation obtained with the trace feature of a graphics calculator.

### EXAMPLE 7    *Approximating a zero with a graphics calculator*

Show that $f(x) = x^3 + x - 1$ has exactly one real zero, which is irrational, and use a graphics calculator to approximate it to the nearest hundredth.

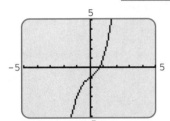

**FIGURE 39**

**SOLUTION**    We first apply Descartes' Rule of Signs. Since $f(x)$ has only one variation in sign, $f$ must have one positive zero. Since $f(-x) = -x^3 - x - 1$ has no variations in sign, $f$ has no negative zeros. Thus, $f$ has exactly one real zero. Next we apply the Rational Zero Theorem.

| | |
|---|---|
| Factors of the constant term $-1$: | $\pm 1$ |
| Factors of the leading coefficient 1: | $\pm 1$ |
| Possible rational zeros: | $\pm 1$ |

But $f(-1) = -3$ and $f(1) = 1$ so neither 1 nor $-1$ is a zero. Thus, $f$ has no rational zeros. To approximate the irrational zero, we plot the graph of $f$ as shown in Figure 39, zoom-in an appropriate number of times, and use the trace feature to find $x \approx 0.68$ as shown in Figure 40.

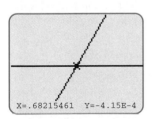

X=.68215461    Y=-4.15E-4

**FIGURE 40**

The process of repeatedly zooming in to approximate a zero is an example of an **iterative process**. Generally speaking, an iterative process is one that is applied repeatedly to obtain closer and closer approximations of some numerical value. Iterative processes that rely solely on numerical calculations are called **numerical methods**. Several such numerical methods have been developed for finding zeros of functions. We will investigate one of these methods, the **Bisection Method**, in Exercise 56.

## EXERCISES 4

**EXERCISES 1–10** □ *Use the Rational Zero Theorem to list all the possible rational zeros of the function. Use synthetic division and/or a graphics calculator to help you determine which are actually zeros.*

1. $g(x) = x^3 - 3x^2 - 4x + 12$

2. $h(x) = x^4 - 17x^2 + 16$

3. $f(x) = 4x^4 - 4x^3 - 9x^2 + x + 2$

4. $h(x) = 9x^3 + 9x^2 - 16x + 4$

5. $f(x) = x^4 - 12x^2 + 27$

6. $f(x) = x^3 + 3x^2 - 2x - 6$

7. $g(x) = 2x^5 + 3x^4 - 2x - 3$

8. $h(x) = 25x^5 - 4x^3 + 25x^2 - 4$

9. $f(x) = x^3 - \dfrac{9}{2}x^2 + \dfrac{1}{2}x + 6$

10. $g(x) = x^4 - \dfrac{10}{3}x^3 - 12x^2 + \dfrac{58}{3}x - 5$

**EXERCISES 11–18** □ *Use Descartes' Rule of Signs to determine the possible number of positive and negative zeros of the function. Check your findings with a graphics calculator.*

11. $f(x) = x^3 + 4$　　　　12. $h(x) = x^4 - 3x^2 - 1$

13. $g(x) = 2x^4 + x^2 + 3$　　14. $g(x) = 5x^3 - x^2 - 1$

15. $f(x) = x^3 - 4x^2 + 7x - 2$　16. $h(x) = 3x^3 + x^2 + 2x + 6$

17. $f(x) = x^5 - 10x^4 - 11x^3 - 5$

18. $g(x) = x^4 + 11x^3 - 12x^2 + 3$

**EXERCISES 19–24** □ *Use the Upper and Lower Bounds Test to confirm the given bounds for the zeros of the function.*

19. $f(x) = x^4 - 5x^3 - 11x^2 + 33x - 18$
　　Lower: $-4$; Upper: 7

20. $g(x) = x^4 + 3x^3 - 27x^2 + 3x - 28$
　　Lower: $-8$; Upper: 5

21. $h(x) = x^4 - 10x^3 - 5$
　　Lower: $-1$; Upper: 11

22. $g(x) = x^4 + 11x^3 - 12x^2 + 6$
　　Lower: $-13$; Upper: 1

23. $f(x) = x^4 - 62x^3 + 962x^2 - 62x + 960$
　　Lower: $-1$; Upper: 62

24. $h(x) = x^4 - 3024x^2 + x - 3026$
　　Lower: $-56$; Upper: 55

**EXERCISES 25–34** □ *Find the exact values of the real zeros of the function. Use any of the tools described in this chapter.*

25. $h(x) = x^3 + 6x^2 - x - 6$　　26. $g(x) = x^3 - 13x - 12$

27. $f(x) = x^3 - 14x^2 + 25x - 12$

28. $g(x) = x^3 + 11x^2 + 2x + 22$

29. $h(x) = x^3 + \dfrac{7}{3}x^2 - \dfrac{23}{12}x + \dfrac{1}{4}$

30. $g(x) = x^3 - \dfrac{7}{12}x^2 - \dfrac{7}{8}x - \dfrac{1}{6}$

31. $h(x) = 8x^3 + x^2 - 16x - 2$

32. $f(x) = 4x^3 - 9x^2 - 6x + 2$

33. $g(x) = x^4 + x^3 - 120x^2 - 121x - 121$

34. $h(x) = x^4 + 26x^3 + 170x^2 + 26x + 169$

**EXERCISES 35–40** □ *Find the real solutions of the polynomial equation.*

35. $x^3 + x^2 - 4x - 4 = 0$　　36. $x^3 + 3x^2 + 2x + 6 = 0$

37. $x^4 - 27 = 6x^2$

38. $2x^4 - 21x^2 - 5 = 5x^3 + 19x$

39. $8x^5 - 8x^3 = 1 - x^2$　　40. $x^5 + 18x = 11x^3$

**EXERCISES 41–46** □ *Use the Rational Zero Test to show that the function has no rational zeros. Use a graphics calculator to approximate any irrational zeros to the nearest hundredth.*

41. $f(x) = x^3 + 3x + 1$　　42. $g(x) = x^3 - x^2 - 2$

43. $h(x) = x^4 + 2x - 2$　　44. $g(x) = x^4 - x^3 - 1$

45. $h(x) = x^5 - 10x^4 - 11x^3 - 5$

46. $f(x) = x^4 + 11x^3 - 12x^2 + 3$

■ *Applications*

47. *Dimensions of a Box* A box with a square base is to be constructed so that its height is 1 inch more than twice its base length. Find the dimensions of the box if the volume must be 9 cubic inches.

48. *Dimensions of a Box* A box with a square base is to be constructed so that its height is 1 inch less than three times its base length. Find the dimensions of the box if the volume must be 100 cubic inches.

49. *Dimensions of a Box* A box with no top is formed by cutting squares out of the corners of a $10'' \times 5''$ rectangular piece of cardboard and then folding up the sides (see Figure 41). What size square must be cut out of each corner if the resulting box is to have a volume of 18 cubic inches?

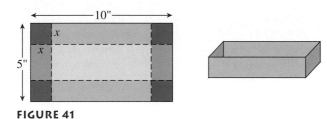

**FIGURE 41**

50. *Dimensions of a Box* A rectangular package to be sent by the U.S. Postal Service can have a maximum combined length and girth (perimeter of the base) of 108 inches (see Figure 42). Find the dimensions of the box if the volume is to be 10,800 cubic inches and if the length plus girth is exactly 108 inches.

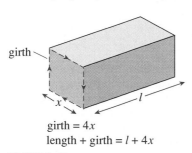

girth = $4x$
length + girth = $l + 4x$

**FIGURE 42**

51. *Cassette Cost* A company that produces blank 8 millimeter video cassettes has daily production costs that can be approximated with the function $C(x) = -4x^3 + 24x^2 - 39x + 25$ where $x$ is the number of cassettes produced (in thousands) and $C(x)$ is the total cost (in thousands of dollars). How many units can be produced for a daily cost of $11,000?

52. *Ping Pong Profit* A sporting goods manufacturer has determined that the yearly profit from the sale of $x$ ping-pong tables can be approximated by $P(x) = -8x^3 + 40x^2 + 46x - 32$

where $x$ is the number of ping-pong tables sold (in thousands) and $P(x)$ is the profit (in thousands of dollars). How many ping-pong tables should be sold to obtain a profit of $100,000?

53. *Poverty Percentage* The percentage of U.S. families below the poverty level for the years 1960–1990 can be approximated by $p(x) = -0.0013x^3 + 0.078x^2 - 1.43x + 18.1$ where $x$ is the year (with $x = 0$ corresponding to 1960) and $p(x)$ is the percentage of families below poverty level in that year. According to the function $p$, for what year(s) were 10.3% of U.S. families below the poverty level?

**U.S. Families Below Poverty Level 1960–1990**

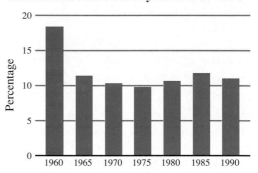

Data Source: U.S. Bureau of the Census

54. *Sunday Accidents* Of the total number of auto accidents that will occur on a given Sunday, the percentage that will have occurred by hour $x$ (with $x = 0$ corresponding to midnight) can be approximated by the polynomial function $p(x) = -0.00182x^4 + 0.0884x^3 - 1.274x^2 + 8.951x$. According to the function $p$, by what time(s) had half of the Sunday accidents occurred? During what Sunday hour do the greatest number of accidents occur?

**Sunday Accidents**

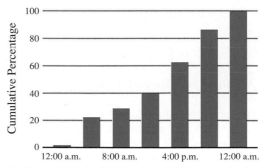

Data Source: *1988 Accident Facts*, National Safety Council

### Projects for Enrichment

**55. *Proof of the Rational Zero Theorem*** Prior to stating the Rational Zero Theorem, we showed that if $p/q$ is a zero of $f(x) = a_nx^n + a_{n-1}x^{n-1} + \cdots + a_1x + a_0$, then $p$ must be a factor of $a_0$. Complete the proof by showing that $q$ must be a factor of $a_n$.

**56. *The Bisection Method*** There often is an easy way to show that a polynomial $f$ has a zero between two numbers $a$ and $b$. We simply compute $f(a)$ and $f(b)$ and check to see if one output value is negative and the other positive. If so, then there is a zero between $a$ and $b$. For example, we can be certain that $f(x) = x^3 + x^2 - 4$ has a zero between 1 and 2 since $f(1) = -2 < 0$ and $f(2) = 8 > 0$.

**a.** Show that the given polynomial has a zero between $a$ and $b$.

    **i.** $g(x) = x^3 + x - 3$; $a = 1$, $b = 2$

    **ii.** $f(x) = -x^3 + x^2 - 2x + 9$; $a = 2$, $b = 3$

    **iii.** $h(x) = x^4 - 9x^2 + x + 4$; $a = 2$, $b = 3$

    **iv.** $g(x) = x^4 - 8x + 2$; $a = 0$, $b = 1$

    **v.** $h(x) = x^5 + 4x^2 + 3$; $a = -2$, $b = -1$

    **vi.** $f(x) = x^5 + 2x^3 + 1$; $a = -1$, $b = 0$

The Bisection Method gets its name from the fact that we repeatedly *bisect* (cut in half) intervals $[a, b]$ in which we know there is a zero. More specifically, we perform the following steps:

**1.** Find two numbers $a$ and $b$ so that $f(a)$ and $f(b)$ have opposite sign.

**2.** Compute $c = (a + b)/2$, the number halfway between $a$ and $b$.

**3.** Test to see if a zero lies between $a$ and $c$ or between $c$ and $b$. That is, compute $f(c)$ and compare its sign to that of $f(a)$ and $f(b)$. If $f(a)$ and $f(c)$ have opposite signs, then a zero is between $a$ and $c$. If $f(c)$ and $f(b)$ have opposite signs, then a zero is between $c$ and $b$.

**4.** Redefine the numbers $a$ and $b$ so they denote the two numbers (either the old $a$ and $c$ or the old $c$ and $b$) between which the zero lies.

**5.** Repeat steps 2, 3, and 4 until the zero is approximated to the desired accuracy.

Consider the function $f(x) = x^3 + x^2 - 4$. As we saw earlier, $f$ has a zero between $a = 1$ and $b = 2$. Thus, we set $c = 1.5$. Since $f(1) = -2, f(1.5) = 1.625$, and $f(2) = 8$, we know there is a zero between 1 and 1.5. So we set $a = 1$ and $b = 1.5$ and repeat the procedure.

**b.** Continue the process started above to estimate the zero of $f(x) = x^3 + x^2 - 4$ which lies between 1 and 2. Stop when $a$ and $b$ agree to two decimal places.

The repetitive procedure described above is ideally suited for a computer or programmable calculator. The pseudo code given below suggests how the program might be written. Note that the program requests values for $a$ and $b$, and also for the desired accuracy $e$. When the desired accuracy is reached, the program displays the approximate zero.

| | |
|---|---|
| INPUT $a$ | Input the left endpoint. |
| INPUT $b$ | Input the right endpoint. |
| If $f(a)*f(b) > 0$ THEN | If $f(a)$ and $f(b)$ don't have opposite |
|    OUTPUT "$f(a)$ and $f(b)$ must | signs, there may not be a zero between |
|    have opposite sign" | $a$ and $b$, so we should stop. |
|    STOP | |
| ENDIF | |
| INPUT $e$ | Input the desired accuracy. |
| WHILE $b - a > e$ DO | The program stops when $b - a \le e$ |
|    $(a + b)/2 \to c$ | Let $c$ be the point halfway between $a$ and |
|    IF $f(a)*f(c) < 0$ THEN | If $f(a)$ and $f(c)$ are opposite in sign, |
|      $c \to b$ | let $c$ be the new right endpoint; |
|    ELSE | otherwise |
|      $c \to a$ | let $c$ be the new left endpoint. |
|    ENDIF | |
| ENDWHILE | |
| OUTPUT $(a + b)/2$ | The approximate zero is $(a + b)/2$. |

**c.** Write a program for your graphics calculator that implements the Bisection Method. Test the program on the function $f(x) = x^3 + x^2 - 4$ using $a = 1$, $b = 2$, and $e = 0.0001$. You should get 1.3146 as an approximate answer. Once the program is working correctly, use it to approximate the zeros of the functions in part (a) to within $e = 0.000001$.

### Questions for Discussion or Essay

**57.** According to the Rational Zero Theorem, if $p/q$ is a zero of a polynomial $f(x)$, then $p$ must be a factor of the constant term and $q$ must be a factor of the leading coefficient. What can be said if the constant term is zero? What if the leading coefficient is 1? Explain how the Rational Zero Theorem can be used to show that $\sqrt{2}$ is an irrational number.

**58.** For what types of polynomials is Descartes' Rule of Signs most useful? Give some examples to help support your claim.

**59.** In light of all the other procedures we've seen for obtaining information about the zeros of a polynomial, discuss the usefulness of the Upper and Lower Bounds Test.

## SECTION 5

# RATIONAL FUNCTIONS

- Which rational functions have graphs that resemble lines?
- How much more might a barrel of oil cost if refineries were required to reduce their discharge by 100%?
- How could we approximate the percentage of the males between the ages of 18 and 24 who are taller than a given height?
- What is the relationship between education and unemployment?
- How can the inventory costs of a small business be minimized?

## RATIONAL FUNCTIONS AND POLYNOMIALS

A rational function is a function of the form

$$f(x) = \frac{p(x)}{q(x)}$$

where $p(x)$ and $q(x)$ are polynomials, and $q(x)$ is not identically zero. Since rational functions are quotients of polynomials, it is perhaps no surprise that they have features that are closely tied to polynomials. For example, the real zeros of a rational function, and hence its $x$-intercepts, are found by locating the zeros of the polynomial in the numerator. This is due to the simple but important fact that *a quotient is equal to zero only if its numerator is zero.*

The domain of a rational function is closely tied to the polynomial in the denominator. Since the domain of a polynomial is the set of all real numbers, the domain of a quotient of polynomials is the set of all real numbers except those for which the denominator is zero. Thus, the domain of a rational function is found by excluding the zeros of the polynomial in the denominator.

### EXAMPLE 1    *Finding the domain and zeros of a rational function*

Find the domain and zeros of the rational function $f(x) = \dfrac{x^2 - 4}{x^2 - x - 2}.$

**SOLUTION**    The values that must be excluded from the domain can be found by locating the zeros of the denominator.

$$x^2 - x - 2 = 0$$
$$(x + 1)(x - 2) = 0$$
$$x = -1, \; x = 2$$

Thus, the zeros of the denominator are $x = -1$ and $x = 2$, and so the domain of $f$ is the set of all real numbers *except $x = -1$ and $x = 2$*. To find the zeros of $f$, we simply find the zeros of the numerator.

$$x^2 - 4 = 0$$
$$(x + 2)(x - 2) = 0$$
$$x = -2, \; x = 2$$

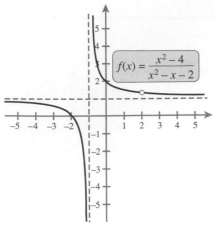

$$f(x) = \frac{x^2 - 4}{x^2 - x - 2}$$

**FIGURE 43**

Note that the solution $x = 2$ is omitted since it is not in the domain of $f$. Thus, the only zero of $f$ is $x = -2$.

**WARNING**   When finding the zeros of a rational function, be sure to omit any zeros of the numerator that are also zeros of the denominator.

Many other features of a rational function, including its graph, are not easily revealed just from our knowledge of polynomials. Consider the graph of the function $f$ from Example 1, as shown in Figure 43. There are several features of the graph that make it unlike any polynomial graph we have seen. In particular, notice the break in the graph across the line $x = -1$, and also the hole at $x = 2$. In contrast, graphs of polynomials have no breaks or holes. Notice also that the graph of $f$ levels off along the horizontal line $y = 1$ as $x$ approaches $\pm\infty$. Graphs of polynomials do not level off, but instead rise or fall without bound as $x$ approaches $\pm\infty$.

**WARNING**   Your graphics calculator may occasionally connect pieces of the graph of a rational function that should not be connected. An example of this phenomenon is shown in Figure 44, where the graph of the function $f$ from Example 1 is shown. As we can see in Figure 43, the graph of $f$ should approach the line $x = -1$, but it should not cross it. Unfortunately, your calculator does not realize this—it simply plots points and connects them with line segments. If two adjacent points happen to be on opposite sides of the line $x = -1$, they will be connected. This is what occurred in Figure 44. One way to avoid this problem is to change your calculator to **dot mode**. In this mode, your calculator will only plot points, but it will not connect them. The graphs of $f$ in Figures 45 and 46 were done in dot mode.

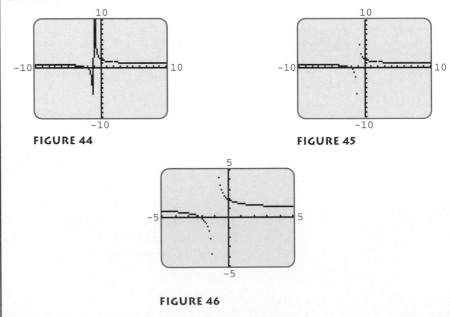

**FIGURE 44**                **FIGURE 45**

**FIGURE 46**

In the following example, we investigate the unusual behavior of the graph of a rational function in more detail.

**EXAMPLE 2**    *Investigating the graph of a rational function*

Investigate the behavior of $f(x) = \dfrac{x}{x - 2}$ near $x = 2$, and also as $x$ approaches $\pm\infty$.

**SOLUTION**    We begin by forming tables of function values for values of $x$ close to 2 on the left and the right.

$x$ approaches 2 from the right:

| $x$ | 2.1 | 2.01 | 2.001 |
|---|---|---|---|
| $f(x)$ | 21 | 201 | 2001 |

$x$ approaches 2 from the left:

| $x$ | 1.9 | 1.99 | 1.999 |
|---|---|---|---|
| $f(x)$ | $-19$ | $-199$ | $-1999$ |

Thus, as $x$ approaches 2 from the right, the function values $f(x)$ get large *without bound*. For this reason, we say that $f(x) \to \infty$ as $x \to 2$ from the right. Similarly, as $x \to 2$ from the left, $f(x) \to -\infty$. Next, we form tables of function values for values of $x$ that get further away from the origin.

$x$ approaches $\infty$:

| $x$ | 10 | 100 | 1000 |
|---|---|---|---|
| $f(x)$ | 1.25 | 1.02 | 1.002 |

$x$ approaches $-\infty$:

| $x$ | $-10$ | $-100$ | $-1000$ |
|---|---|---|---|
| $f(x)$ | 0.83 | 0.98 | 0.998 |

Since the function values $f(x)$ get closer and closer to 1, we say that $f(x) \to 1$ as $x \to \pm\infty$. Note that the behavior of $f$ near $x = 2$ and as $x \to \pm\infty$ can also be investigated graphically. In Figure 47 we show the graph of $f$ as depicted by a graphics calculator. The graph in Figure 48 identifies the behavior we have observed from the tables of function values.

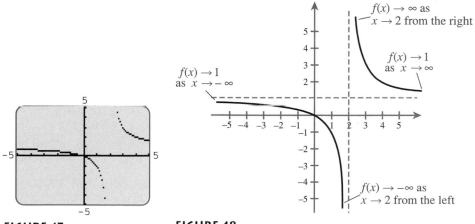

**FIGURE 47**                    **FIGURE 48**

Notice in Figure 48 that the graph of $f$ gets closer and closer to the dashed vertical line $x = 2$. We say that the line $x = 2$ is a *vertical asymptote* for the function $f$. Similarly, because the graph of $f$ approaches the horizontal line $y = 1$ as $x$ becomes large, we call the line $y = 1$ a *horizontal asymptote* for $f$. The location of such asymptotes provides important information about the graph of a rational function.

### VERTICAL AND HORIZONTAL ASYMPTOTES

Loosely speaking, a *linear asymptote* is a line that a function approaches *arbitrarily closely*. We will consider three types of linear asymptotes: vertical, horizontal, and inclined. The definitions of vertical and horizontal asymptotes follow immediately from our observations in Example 2.

### *Definition of vertical and horizontal asymptotes*

> 1. The line $x = a$ is a **vertical asymptote** for the graph of a rational function $f$ if $f(x)$ grows without bound (approaches $\infty$ or $-\infty$) as $x$ approaches $a$ from the right and from the left.
> 2. The line $y = b$ is a **horizontal asymptote** for the graph of a rational function $f$ if $f(x)$ approaches $b$ as $x$ grows without bound (approaches $\infty$ or $-\infty$).

In the following example, we provide information about the location of the asymptotes and then use the asymptotes to construct the graph of an unknown function.

### EXAMPLE 3

### *Graphing an unknown rational function*

Suppose $f$ is a rational function with a vertical asymptote $x = -3$ and a horizontal asymptote $y = 0$. Moreover, suppose $f(x) > 0$ for $x \neq -3$. Sketch a possible graph of $f$.

**SOLUTION**    Since $x = -3$ is a vertical asymptote and since $f(x)$ is always positive, we know that $f(x)$ approaches $\infty$ as $x$ approaches $-3$ from the left and the right. Since $y = 0$ is a horizontal asymptote, $f(x)$ must approach 0 (the $x$-axis) as $x$ approaches $\infty$ or $-\infty$. Moreover, since $f(x)$ is always positive, we know that the graph must approach the $x$-axis from above. A *possible* sketch for $f$ is shown in Figure 49.

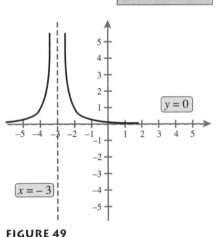

**FIGURE 49**

The task of locating asymptotes for a rational function is greatly simplified by several simple rules. To find vertical asymptotes, we need only look where the denominator is zero.

### *Locating vertical asymptotes*

> The line $x = a$ is a vertical asymptote for the rational function $f(x) = \dfrac{p(x)}{q(x)}$ if $q(a) = 0$ and $p(a) \neq 0$.

The restriction $p(a) \neq 0$ is necessary since if both $p$ and $q$ are zero at some number $a$, then $x = a$ is not necessarily an asymptote; rather, the function $f$ may have a "hole" at $x = a$. One example of this phenomenon appears in Figure 43. We will investigate such holes in detail in Exercise 58.

Horizontal asymptotes for rational functions are found by comparing the degree of the numerator to the degree of the denominator.

**Locating horizontal asymptotes**

Suppose $f$ is a rational function given by

$$f(x) = \frac{a_n x^n + a_{n-1}x^{n-1} + \cdots + a_1 x + a_0}{b_m x^m + b_{m-1}x^{m-1} + \cdots + b_1 x + b_0}$$

1. If $n < m$, then the $x$-axis is a horizontal asymptote for $f$.

2. If $n = m$, then $y = \dfrac{a_n}{b_m}$ is a horizontal asymptote for $f$.

3. If $n > m$, then $f$ has no horizontal asymptote.

**EXAMPLE 4**    *Locating vertical and horizontal asymptotes*

Locate any vertical and horizontal asymptotes for $f(x) = \dfrac{-2x}{x-4}$ and sketch the graph.

**SOLUTION**    The vertical asymptotes occur where the denominator of $f(x)$ is zero. The only such value is $x = 4$, and so this is the only vertical asymptote. Since the degree of the numerator and denominator are the same, the horizontal asymptote is found by computing the quotient of the leading coefficients of the numerator and denominator. Thus, $y = -\frac{2}{1} = -2$ is the horizontal asymptote. One view of the graph of $f$ is shown in Figure 50. The graph in Figure 51 includes the asymptotes.

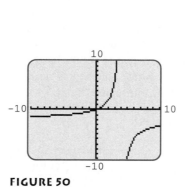

**FIGURE 50**

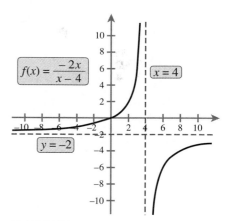

**FIGURE 51**

**EXAMPLE 5**    *Locating vertical and horizontal asymptotes*

Locate any vertical and horizontal asymptotes for $f(x) = \dfrac{x^2}{x^2 - 4}$ and sketch the graph.

**SOLUTION**    The vertical asymptotes of $f$ occur where $x^2 - 4 = 0$. By factoring, we have $(x + 2)(x - 2) = 0$, and so the vertical asymptotes are $x = -2$ and $x = 2$. The horizontal asymptote is $y = \frac{1}{1} = 1$. One view of the graph of $f$ is shown in Figure 52. The graph in Figure 53 includes the asymptotes.

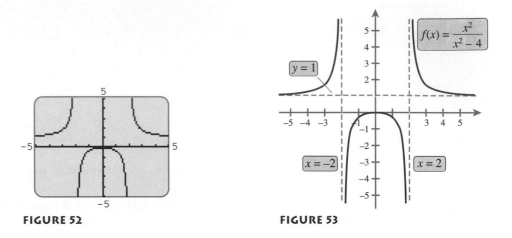

FIGURE 52                                    FIGURE 53

**EXAMPLE 6**    *Locating vertical and horizontal asymptotes*

Locate any vertical and horizontal asymptotes for $f(x) = \dfrac{80x}{16x^2 + 1}$ and sketch the graph.

**SOLUTION**    This function $f$ has no vertical asymptotes since the denominator $16x^2 + 1$ has no real zeros. The horizontal asymptote is the $x$-axis since the degree of the numerator is less than the degree of the denominator. One view of the graph is shown in Figure 54. At first glance, it appears that the $y$-axis is a vertical asymptote. However, we know that is not the case since $f$ has no vertical asymptotes. In order to see more clearly what is happening near the $y$-axis, we adjust the scale on the $x$-axis; the result is shown in Figure 55.

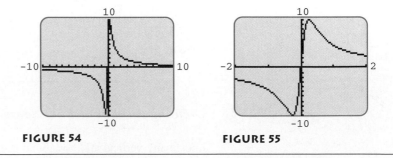

FIGURE 54                    FIGURE 55

**EXAMPLE 7**    *A rational function with no vertical or horizontal asymptotes*

Show that $f(x) = \dfrac{x^3}{x^2 - 2x + 2}$ has no vertical or horizontal asymptotes.

**SOLUTION**    The function $f$ has no horizontal asymptote since the degree of the numerator is larger than the degree of the denominator. To check for vertical asymptotes, we find the zeros of the denominator by setting $x^2 - 2x + 2 = 0$ and using the quadratic formula.

$$x = \frac{-(-2) \pm \sqrt{(-2)^2 - 4(1)(2)}}{2(1)} = \frac{2 \pm \sqrt{-4}}{2}$$

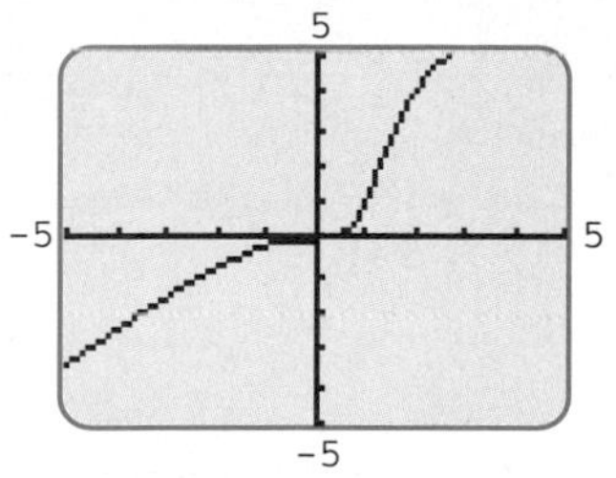

Since the only zeros of the denominator are nonreal, $f$ has no vertical asymptotes. The graph of $f$ is shown in Figure 56.

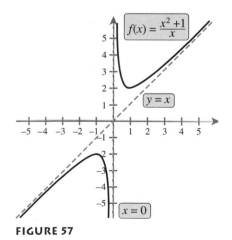

**FIGURE 56**

Several observations can be made on the basis of the examples we have seen. These are summarized below.

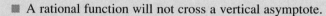

- A rational function will not cross a vertical asymptote.
- A rational function may have many vertical asymptotes, or it may not have any.
- A rational function may cross its horizontal asymptote.
- A rational function can have at most one horizontal asymptote.

## INCLINED ASYMPTOTES

It is possible for a rational function to have a linear asymptote that is neither vertical nor horizontal. Consider the function

$$f(x) = \frac{x^2 + 1}{x}$$

First note that if we divide the numerator by $x$, we can rewrite the function as

$$f(x) = x + \frac{1}{x}$$

Now for very large values of $x$, $1/x$ is a very small number. Thus, for very large values of $x$, $f(x) = x + $ (a very small number) and so $f(x) \approx x$ for very large values of $x$. Similarly, $f(x) \approx x$ for very large negative values of $x$. This shows that the graph of $f$ approaches the line $y = x$ as $x$ approaches infinity or negative infinity, as we see in Figure 57. The line $y = x$ is called an **inclined asymptote** for the function $f$. More generally, we find inclined asymptotes as follows.

**FIGURE 57**

■ *Locating inclined*
*asymptotes*

Suppose $f(x) = p(x)/q(x)$ is a rational function for which the degree of $p(x)$ is *exactly* 1 *more* than the degree of $q(x)$. Then we can use division to rewrite $f$ in the form

$$f(x) = ax + b + \frac{r(x)}{q(x)}$$

where the degree of $r(x)$ is less than the degree of $q(x)$. The line $y = ax + b$ is an inclined asymptote for $f$.

**EXAMPLE 8**     *Locating a vertical and inclined asymptote*

Find the vertical and inclined asymptotes for $f(x) = \dfrac{x^2 + 5x + 7}{x + 2}$ and sketch the graph.

**SOLUTION**     The vertical asymptote is $x = -2$. To find the inclined asymptote, we must first divide $x^2 + 5x + 7$ by $x + 2$. We will use synthetic division.

$$
\begin{array}{r|rrr}
-2 & 1 & 5 & 7 \\
   &   & -2 & -6 \\
\hline
   & 1 & 3 & 1
\end{array}
$$

From the last row we see that the quotient is $x + 3$ and the remainder is 1. Thus,

$$\frac{x^2 + 5x + 7}{x + 2} = x + 3 + \frac{1}{x + 2} \quad \text{and so} \quad f(x) = x + 3 + \frac{1}{x + 2}$$

This shows that $y = x + 3$ is an inclined asymptote. Using a graphics calculator, we obtain the view shown in Figure 58. The graph in Figure 59 includes the asymptotes.

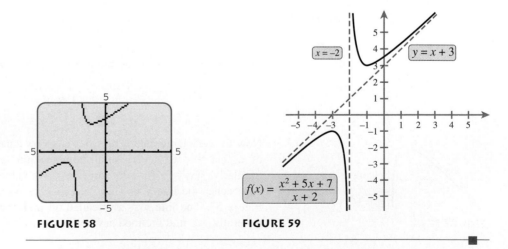

**FIGURE 58**          **FIGURE 59**

EXAMPLE 9    *Minimizing inventory cost*

An appliance retailer sells 500 microwave ovens each year. By averaging ordering costs and storage costs, the retailer has determined that if $x$ microwaves are ordered at a time, the yearly inventory cost will be

$$C(x) = 5x + 40{,}000 + \frac{4500}{x}$$

Computerized automation controls inventory costs.

Plot the graph of $C$ and approximate the number of microwaves that should be ordered each time so that the cost is as small as possible. How often should each order be placed?

**SOLUTION**    By rewriting the cost function as

$$C(x) = \frac{5x^2 + 40{,}000x + 4500}{x}$$

we see that $x = 0$ is a vertical asymptote. From the original form of the function, namely $C(x) = 5x + 40{,}000 + (4500/x)$, we see that $y = 5x + 40{,}000$ is an inclined asymptote. We wish to choose a viewing rectangle so that a relative minimum value is shown, and also so that the inclined asymptote is suggested. Since the $y$-intercept of the asymptote is 40,000, we want Ymin $< 40{,}000 <$ Ymax. After some experimentation, we settle on the view shown in Figure 60. Because we are only concerned with positive values of $x$, we look for the lowest point on the graph of $C$ to the right of the $y$-axis. Using the trace feature, we estimate the minimum cost to be \$40,300 per year when 30 microwaves are ordered each time (see Figure 61). If 500 microwaves are sold each year and 30 must be ordered each time to minimize cost, an order must be placed $\frac{500}{30} \approx 17$ times each year or every 22 days.

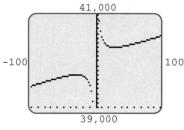

**FIGURE 60**

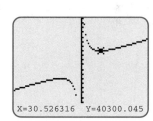

**FIGURE 61**

**EXERCISES 5**

**EXERCISES 1–8** □ *Find the domain and zeros of the rational function and match it with its graph.*

**1.** $f(x) = \dfrac{1}{x - 2}$

**3.** $f(x) = \dfrac{x + 1}{x}$

**5.** $f(x) = \dfrac{x - 2}{x^2 - 4}$

**7.** $f(x) = \dfrac{x^2 - 4}{x^2 - 1}$

**2.** $f(x) = \dfrac{x}{x - 1}$

**4.** $f(x) = \dfrac{x - 2}{x - 1}$

**6.** $f(x) = \dfrac{1}{x^2 + x - 2}$

**8.** $f(x) = \dfrac{x^2 - 1}{x^2 + x - 2}$

a.
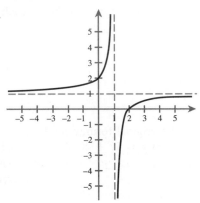

b.
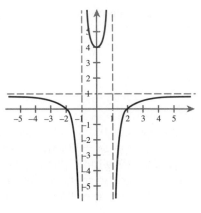

c.
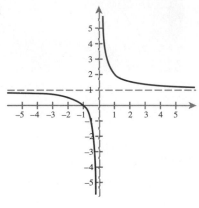

d.
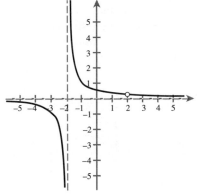

e.
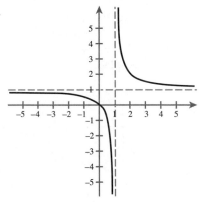

f.
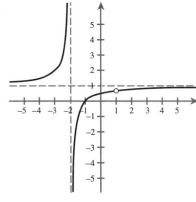

g.
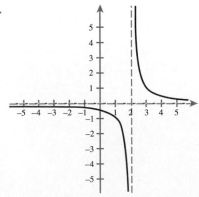

h.
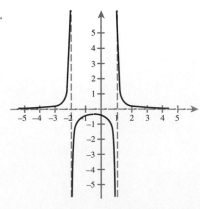

**EXERCISES 9–12** □ *Use the graph of the function f to sketch the graph of the function g.*

**9.**

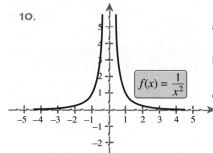

a. $g(x) = \dfrac{1}{x - 3}$

b. $g(x) = \dfrac{1}{x} + 2$

$f(x) = \dfrac{1}{x}$    c. $g(x) = -\dfrac{1}{x}$

**10.**

a. $g(x) = \dfrac{1}{x^2} - 3$

b. $g(x) = \dfrac{1}{(x + 2)^2}$

$f(x) = \dfrac{1}{x^2}$    c. $g(x) = -\dfrac{1}{x^2}$

**11.**

$y = f(x)$

a. $g(x) = f(x - 1)$

b. $g(x) = f(x) - 2$

c. $g(x) = -f(x)$

**12.**

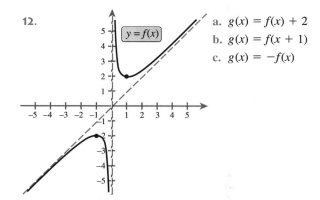

$y = f(x)$

a. $g(x) = f(x) + 2$

b. $g(x) = f(x + 1)$

c. $g(x) = -f(x)$

**EXERCISES 13–16** ☐ *Construct a possible graph for the rational function with the given properties. There may be more than one correct answer.*

**13.** $f$ has a vertical asymptote at $x = 4$, a horizontal asymptote at $y = 0$, and $f(x) < 0$ for all $x \neq 4$.

**14.** $h$ has a vertical asymptote at $x = -2$, a horizontal asymptote at $y = 0$, $h(x) > 0$ if $x < -2$, and $h(x) < 0$ if $x > -2$.

**15.** $g$ has a vertical asymptote at $x = -1$, a horizontal asymptote at $y = 2$, an $x$-intercept at $(-\frac{3}{2}, 0)$, and a $y$-intercept at $(0, 3)$.

**16.** $h$ has a vertical asymptote at $x = 2$, a horizontal asymptote at $y = -3$, and zeros at $x = 0$ and $x = 6$.

**EXERCISES 17–40** ☐ *Find any vertical and horizontal asymptotes and sketch the graph.*

**17.** $f(x) = \dfrac{1}{x + 3}$

**18.** $f(x) = \dfrac{1}{x - 2}$

**19.** $g(x) = \dfrac{2}{1 - x}$

**20.** $h(x) = \dfrac{-2}{x - 3}$

**21.** $h(x) = \dfrac{x}{x + 2}$

**22.** $g(x) = \dfrac{x - 3}{x + 4}$

**23.** $f(x) = \dfrac{3x}{x - 2}$

**24.** $g(x) = \dfrac{x + 1}{3 - 2x}$

**25.** $h(x) = \dfrac{12x + 24}{x - 15}$

**26.** $f(x) = \dfrac{1 - 16x}{x + 12}$

**27.** $g(x) = \dfrac{x}{x^2 - 9}$

**28.** $f(x) = \dfrac{x^2}{x^2 - 9}$

**29.** $h(x) = \dfrac{-1}{x^2 - 9}$

**30.** $f(x) = \dfrac{1}{x^2 + 9}$

**31.** $g(x) = \dfrac{45x^2}{9x^2 + 1}$

**32.** $h(x) = \dfrac{18}{9x^2 - 1}$

**33.** $h(x) = \dfrac{1}{(x + 1)^2}$

**34.** $f(x) = \dfrac{-x}{(x - 3)^2}$

**35.** $f(x) = \dfrac{x^2}{(x - 2)^2}$

**36.** $g(x) = \dfrac{(x - 2)^2}{x^2}$

**37.** $f(x) = \dfrac{x + 3}{x^2 - 9}$

**38.** $g(x) = \dfrac{x - 1}{1 - x^2}$

**39.** $h(x) = \dfrac{3 - x}{x^2 - 4x + 3}$

**40.** $f(x) = \dfrac{x + 1}{x^2 - 2x - 3}$

**EXERCISES 41–48** ☐ *Find the vertical and inclined asymptotes and sketch the graph.*

**41.** $g(x) = 2x + \dfrac{1}{x}$

**42.** $h(x) = -x + \dfrac{1}{x}$

**43.** $f(x) = \dfrac{x^2}{x - 1}$

**44.** $h(x) = \dfrac{x^2 + 1}{x + 1}$

**45.** $g(x) = \dfrac{-3x^2 - 6x + 1}{x + 2}$

**46.** $f(x) = \dfrac{4x^2 - 4x + 2}{x - 1}$

**47.** $g(x) = \dfrac{x^2 - 4x + 7}{x - 3}$

**48.** $h(x) = \dfrac{2x^2 + 9x + 5}{x + 4}$

■ *Applications*

**49.** *Pollution Control* Based on data from a 1973 study on the cost of reducing oil refinery discharge, the increase in refinery costs can be approximated with the function

$$f(x) = \frac{x}{1000(100 - x)}$$

where $x$ is the desired percent reduction in discharge and $f(x)$ is the corresponding additional cost in dollars per barrel crude. (Data Source: *Environment, Natural Systems, and Development*, Maynard Hufschmidt et al., The Johns Hopkins University Press, Baltimore, 1983.) Identify any vertical or horizontal asymptotes and sketch the graph of $f$. What percent reduction is possible with an increase of $0.003 per barrel crude? According to this model, is a 100% reduction attainable? Explain.

**50.** *Nerve Excitation* The minimum voltage needed to excite a nerve fiber is related to the time during which the current flows. According to data from one source, this relationship can be approximated by the function

$$f(x) = \frac{17}{x + 0.113} + 23.2$$

where $x$ is the time during which current flows (in milliseconds) and $f(x)$ is the minimum current needed to excite the nerve fiber (in millivolts). (Data Source: *The Mathematical Approach to Physiological Problems*, Douglas Riggs, M.I.T. Press, Cambridge, 1970.) Identify any vertical or horizontal asymptotes and sketch the graph of $f$. How long must the current flow to excite a nerve fiber with a voltage of 40 millivolts?

**51.** *Unemployment Rate* The percentage rate of unemployment in the United States in 1992 for people with $x$ years of education can be approximated with the function

$$f(x) = \frac{72,900}{100x^2 + 729}$$

(Data Source: U.S. Bureau of Labor Statistics.) Identify any vertical or horizontal asymptotes and sketch the graph of $f$.

What level of education corresponds to an unemployment rate of 4.9%? What happens to the unemployment rate as the level of education increases? According to this model, is there an education level that corresponds to an unemployment rate of zero? Explain.

52. *Height Distribution*  The percentage of U.S. males between the ages of 18–24 years who are within a half inch of a given height can be approximated with the function

$$f(x) = \frac{256}{(2x - 139)^2 + 16}$$

where $x$ is the height (in inches). (Data Source: U.S. National Center for Health Statistics.) Identify any vertical or horizontal asymptotes and sketch the graph of $f$. What height corresponds to the highest percentage? What interpretation could be given to this height?

7′6″ NBA center Shawn Bradley

53. *Bicycle Average Cost*  A company that manufactures bicycles has fixed costs of $100,000 and variable costs of $100 per bicycle, so that the total cost for producing $x$ bicycles is given by $C(x) = 100,000 + 100x$. The average cost per bicycle is found by dividing the total cost by the number of bicycles produced. Thus, the average cost function is

$$\overline{C}(x) = \frac{100,000 + 100x}{x}$$

Sketch the graph of the average cost function. What is the horizontal asymptote and what information can be obtained from it?

54. *Skate Average Cost*  A company that manufacturers inline skates has fixed costs of $80,000 and variable costs of $50 per pair of skates, so that the total cost for producing $x$ pairs of skates is $C(x) = 80,000 + 50x$. The average cost per pair of skates is found by dividing the total cost by the number of pairs of skates produced. Thus, the average cost function is

$$\overline{C}(x) = \frac{80,000 + 50x}{x}$$

Sketch the graph of the average cost function. What is the horizontal asymptote and what information can be obtained from it?

55. *Basketball Inventory Cost*  A sporting goods retailer sells 580 basketballs each year. By averaging ordering costs and storage costs, the retailer has determined that if $x$ balls are ordered at a time, the yearly inventory cost will be

$$C(x) = \frac{3}{2}x + 60,000 + \frac{5046}{x}$$

Plot the graph of $C$ and approximate the number of balls that should be ordered each time so that the cost is as small as possible. How often should each order be placed?

56. *Television Inventory Cost*  A retail appliance store sells 1500 television sets per year. By averaging ordering costs and storage costs, the retailer has determined that if $x$ televisions are ordered at a time, the yearly inventory cost will be

$$C(x) = 5x + 300,000 + \frac{28,200}{x}$$

Plot the graph of $C$ and approximate the number of televisions that should be ordered each time so that the cost is as small as possible. How often should each order be placed?

■  *Projects for Enrichment*

57. *Minimizing Inventory Cost*  A shoe retailer wishes to minimize the costs incurred in handling a certain style of shoe that must be ordered periodically and kept in stock as they are sold to customers. Since the expenses involved in ordering and storing the shoes depends on how often orders are made and how long items are stored, the retailer wishes to minimize the *total cost per year*. We will consider three separate costs that are involved in the total cost to the retailer.

  i. A *fixed cost* of $30 *per order* that is independent of the amount ordered. This cost includes such things as record

  keeping and other paper work, employees' time, and so on.

  ii. A *purchase cost* of $15 *per pair* (throughout, "pair" will mean "pair of shoes").

  iii. An *inventory holding cost* of $2 *per pair per year*. This cost covers the expenses of keeping the shoes in the store or warehouse.

Let $x$ denote the number of pairs ordered at a time and assume that the shoes are sold at a constant rate of 1200 per year.

Because shortages are not tolerable, the time between two consecutive orders depends only on the number ordered and the rate at which they are sold.

a. Find an expression for the number of reorders per year.

b. Find an expression for the total cost per order (not including inventory cost).

c. Use the expressions from parts (a) and (b) to find an expression for the yearly ordering costs.

d. Find an expression for the yearly inventory cost. Since $x$ pairs are ordered at a time, assume that the average number of pairs of shoes in inventory at any given time is $x/2$.

e. Use the expressions from parts (c) and (d) to obtain a function for the total cost per year. Denote this total cost function by $T(x)$.

f. Use a graphics calculator to plot the graph of $T(x)$ and approximate the value for $x$ that yields the minimum cost. This represents the number of shoes that should be ordered at a time. How often must this quantity of shoes be ordered?

g. What would change in this scenario if the retailer can only order shoes in quantities of 18?

58. *Removable and Nonremovable Discontinuities* Although a formal definition of continuity is beyond the scope of this text, we will informally say that a function $f$ is continuous over an interval $I$ if it has no breaks over $I$. In other words, a function $f$ is continuous over $I$ if you can trace its graph over the entire interval $I$ without lifting your pencil. For example, the function $f(x) = x^3 - 2x^2 - x + 2$ graphed in Figure 62 is continuous over $(-\infty, \infty)$ while the function $f(x) = 1/x$ graphed in Figure 63 is not. The break in Figure 63 occurs at $x = 0$. Notice that this is also where the function is undefined.

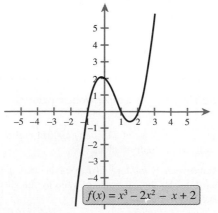

**FIGURE 62**

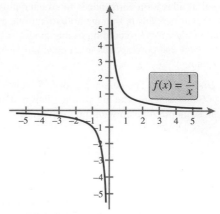

**FIGURE 63**

In general, all polynomial functions are continuous over $(-\infty, \infty)$ because they have no breaks. Rational functions, on the other hand, may or may not be continuous over all of $(-\infty, \infty)$.

a. Sketch the graphs of the following rational functions to help you determine whether they are continuous over $(-\infty, \infty)$. For those that are not, identify the location(s) of the break(s).

i. $f(x) = \dfrac{1}{x + 1}$

ii. $f(x) = \dfrac{1}{x^2 + 4}$

iii. $f(x) = \dfrac{x}{x^2 - 4}$

iv. $f(x) = \dfrac{x}{x^2 + 1}$

b. Based on your observations in part (a), state a rule for determining whether or not a rational function is continuous over $(-\infty, \infty)$, and if it is not, where the breaks occur.

The breaks in a discontinuous function are called *discontinuities*. Some discontinuities are very easy to identify from the graph while others are hard to spot. Consider the function

$$f(x) = \frac{x - 2}{x^2 - 4}$$

Using a graphics calculator, we obtain the graph in Figure 64. The discontinuity at $x = -2$ is easy to see. However, as your rule from part (b) should show, there is also a discontinuity at $x = 2$ that does not show up on the graph. A more detailed graph is given in Figure 65. Notice that the discontinuity at $x = 2$ appears as a "hole" in the graph. In some sense, the discontinuity at $x = 2$ is not as severe as the one at $x = -2$.

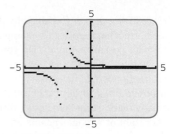

**FIGURE 64**

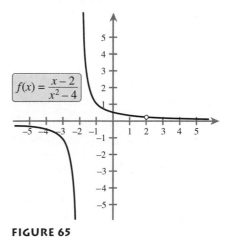

$f(x) = \dfrac{x - 2}{x^2 - 4}$

**FIGURE 65**

The discontinuity at $x = 2$ can be corrected by adding one point to the graph. For this reason, it is called a *removable discontinuity*. The discontinuity at $x = -2$ cannot be removed by simply adding one point, and so it is called a *nonremovable discontinuity*.

c. Each of the following rational functions has one removable and one nonremovable discontinuity. Locate and identify each. The graphs may be helpful for locating the nonremovable discontinuities.

i. $f(x) = \dfrac{x + 1}{x^2 - 1}$

ii. $f(x) = \dfrac{x - 3}{x^2 - x - 6}$

iii. $f(x) = \dfrac{x^2 + x - 2}{x^2 + 2x - 3}$

If a rational function has a removable discontinuity, it is usually possible to determine the coordinates of the "hole." Consider again the function $f(x) = \dfrac{x - 2}{x^2 - 4}$ and the removable discontinuity at $x = 2$. Notice that as long as $x \neq 2$, we can simplify the quotient as shown here.

$$\frac{x - 2}{x^2 - 4} = \frac{x - 2}{(x - 2)(x + 2)} = \frac{1}{(x + 2)}$$

This shows that the function $g(x) = \dfrac{1}{x + 2}$ is the same as $f(x) = \dfrac{x - 2}{x^2 - 4}$ for all values of $x$ except $x = 2$. At $x = 2$, $f$ is still undefined and has a discontinuity. However, at $x = 2$, $g(2) = 1/(2 + 2) = \frac{1}{4}$. Thus, the "hole" in $f$ occurs at $(2, \frac{1}{4})$.

d. Locate the coordinates of the "hole" for each of the functions in part (c) and then draw detailed graphs that show the removable and nonremovable discontinuities.

e. Sketch the graphs of the following functions. Label any discontinuities as either removable or nonremovable.

i. $f(x) = \dfrac{x^2 - 2}{x + 2}$

ii. $f(x) = \dfrac{x^3}{x^3 - x}$

iii. $f(x) = \dfrac{x^2 - 2x - 3}{x^3 - 2x^2 - 3x}$

---

## Questions for Discussion or Essay

59. In Exercises 53 and 54 we considered average cost functions of the form $\bar{C} = (a + bx)/x$. Could a function of this form have a minimum value? In other words, is it possible to find a production level for $x$ for which the average cost is as small as possible? Explain. Is this situation consistent with what you would expect for a company's average cost? Why or why not?

60. The line $x = a$ is a vertical asymptote for the rational function $f(x) = p(x)/q(x)$ if $q(a) = 0$ and $p(a) \neq 0$. Explain, with the help of examples, why the restriction $p(a) \neq 0$ is necessary. What can be said about the factored forms of $p(x)$ and $q(x)$ if

both are zero at $x = a$? Write a rule for finding vertical asymptotes that replaces the restriction $p(a) \neq 0$ with a statement concerning the factors of $p(x)$ and $q(x)$.

61. Rational functions of the form $f(x) = \dfrac{ax + b}{cx + d}$, $c \neq 0$, can be graphed quite easily without the use of a graphics calculator. Consider the function $f(x) = \dfrac{2x + 3}{x + 1}$. The vertical and horizontal asymptotes divide the plane into four regions, which we

have numbered in Figure 66. Experience suggests that the graph must lie either in regions 1 and 3 or 2 and 4. Why do you think this is? Once the regions have been determined, a sketch of the graph is easily completed, as shown in Figure 67. Describe a procedure for locating the horizontal and vertical asymptotes of a function $f(x) = \dfrac{ax + b}{cx + d}$, $c \neq 0$, and for deciding in which of the four regions to place the graph. Illustrate your procedure with an example.

62. In Example 3 we sketched the graph of a function $f$ with a vertical asymptote at $x = -3$, a horizontal asymptote at $y = 0$,

and with $f(x) > 0$ for $x \neq -3$. Explain why there could be more than one correct answer for this example, and give an example of another possible solution. Do you think one answer could be "more right" than another? Why or why not? What additional information would have to be given to narrow down the choice of possibilities? It has been said that one of the strengths of mathematics over some other disciplines is that an answer is either right or wrong. There is no "gray area." In your experience with mathematics, how often have you found this to be the case? Do you think this "all or nothing" phenomenon is more or less likely to be the case when mathematics is applied to the real world? Explain.

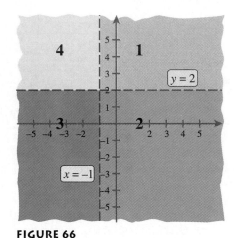

**FIGURE 66**

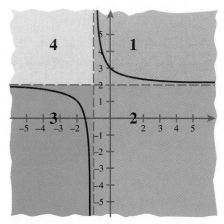

**FIGURE 67**

## CHAPTER REVIEW EXERCISES

**EXERCISES 1–6** □ *Determine the end behavior of the given polynomial.*

1. $g(x) = -3x^6$

2. $h(x) = 2x^5$

3. $f(x) = 4x^3 - x^2 + 9x + 3$

4. $g(x) = -x^4 + 5x^2 - x + 8$

5. $h(x) = x^3 + 100 - 0.00005x^4$

6. $f(x) = (2 \times 10^4)x^2 + (2 \times 10^{-4})x^3$

**EXERCISES 7–10** □ *Use a graphics calculator to estimate (to the nearest hundredth) the zeros and the coordinates of any turning points of the given polynomial.*

7. $f(x) = \dfrac{x^3}{3} - 3x^2 - 3$

8. $g(x) = \dfrac{x^4}{9} + x^2 - 4x + 2$

9. $g(x) = 0.1x^4 + x^3 - 4x^2 + x - 4$

10. $h(x) = 0.01x^5 - 0.15x^4 - 0.4x^3 + 2x^2$

**EXERCISES 11–14** □ *Use long division to find the quotient and remainder when $f(x)$ is divided by $d(x)$.*

11. $f(x) = 2x^2 + 3x - 2$;  $d(x) = x + 2$

12. $f(x) = 9x^3 - 3x^2 + 7x + 1$;  $d(x) = 3x - 2$

13. $f(x) = x^4 - 5x^2 + 2$;  $d(x) = x^2 + 4x$

14. $f(x) = x^5 + 2x^3 - x^2 - 2$;  $d(x) = x^3 - 1$

**EXERCISES 15–18** □ *Use synthetic division (and the Remainder Theorem) to find the quotient and remainder when $f(x)$ is divided by $d(x)$.*

15. $f(x) = 3x^3 - 17x^2 + 22x - 8$;  $d(x) = x - 4$

16. $f(x) = 4x^3 + 8x^2 - 9x - 13$;  $d(x) = x + 2$

17. $f(x) = -2x^4 + 4x^2 + x - 3$;  $d(x) = x + 3$

18. $f(x) = 32x^5 - 1$;  $d(x) = x - \dfrac{1}{2}$

**EXERCISES 19–22** □ *Use synthetic division to find the indicated function value.*

**19.** $f(x) = 3x^3 + 11x^2 + 2x - 6$

   a. $f\left(\dfrac{3}{2}\right)$    b. $f(-2)$

**20.** $g(x) = 5x^4 - 6x^3 - 32x^2 + 7$

   a. $g(-3)$    b. $g(3.2)$

**21.** $h(x) = -2x^4 + 5x^3 + 3x$

   a. $h(5)$    b. $h\left(\dfrac{1}{2}\right)$

**22.** $f(x) = \dfrac{1}{3}x^5 + 2x^4 + \dfrac{1}{2}x + 3$

   a. $f(-6)$    b. $f(2)$

**EXERCISES 23–28** □ *Factor the polynomial function and identify all real zeros.*

**23.** $f(x) = 2x^2 + 5x - 3$    **24.** $g(x) = 4x^3 - 4x^2 - 24x$

**25.** $h(x) = x^4 - 11x^2 + 18$

**26.** $g(x) = 4x^3 + 5x^2 - 18x + 9$

**27.** $h(x) = 6x^3 + 7x^2 - 1$

**28.** $f(x) = x^4 - 2x^3 - 5x^2 + 8x + 4$

**EXERCISES 29–36** □ *Find a polynomial function f of least degree, with real coefficients, satisfying the given properties.*

**29.** Zeros $-2$, 3, and 5

**30.** Zeros $-3$, 0, and 4; $f(1) = 10$

**31.** Zeros 1 and $-1$ (both with multiplicity 2); $f(0) = 4$

**32.** Zeros $\frac{1}{2}$ (multiplicity 2) and 0 (multiplicity 3)

**33.** Graph shown below

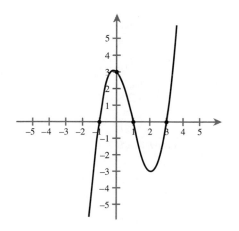

**34.** Graph shown below

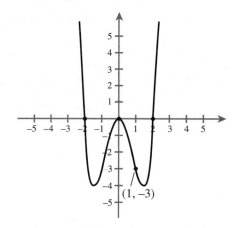

(1, –3)

**35.** Zeros $-3$ and $4i$    **36.** Zeros 0, 2, and $1 - 2i$

**EXERCISES 37–40** □ *Write the polynomial function as a product of linear factors and identify all the complex zeros.*

**37.** $f(x) = x^2 - 4x + 5$

**38.** $g(x) = x^3 + 3x^2 + 4x + 12$

**39.** $h(x) = x^4 + 7x^2 - 144$

**40.** $f(x) = x^5 - 4x^4 + 10x^3 - 12x^2 + 5x$

**EXERCISES 41–44** □ *Use the Rational Zero Theorem to list all the possible rational zeros of the polynomial. Then use synthetic division and/or a graphics calculator to help you determine which are actually zeros.*

**41.** $f(x) = x^3 - 2x^2 - 5x + 6$

**42.** $h(x) = 3x^3 - 10x^2 + 11x - 4$

**43.** $g(x) = 4x^4 - 8x^3 - x^2 + 8x - 3$

**44.** $h(x) = 9x^5 - x^3 - 9x^2 + 1$

**EXERCISES 45–48** □ *Use Descartes' Rule of Signs to determine the possible number of positive and negative zeros of the function. Check your findings with a graphics calculator.*

**45.** $h(x) = x^3 + 2x + 1$

**46.** $f(x) = x^4 - 4x^3 - x + 5$

**47.** $g(x) = -x^4 - 5x^3 + 2x^2 + 8x$

**48.** $h(x) = 2x^5 - 6x^3 + 3x^2 - 1$

**EXERCISES 49–50** □ *Use the Upper and Lower Bounds Test to confirm the given bounds for the zeros of the function.*

**49.** $f(x) = x^4 - 2x^3 - 49x^2 - 2x + 48$
   Lower: $-7$; Upper: 9

**50.** $g(x) = x^4 + 5x^3 - 299x^2 + 5x - 300$
   Lower: $-21$; Upper: 16

**EXERCISES 51–54** ☐ *Find the exact real solutions of the given polynomial equation.*

**51.** $x^3 + 4x + 12 = 7x^2$

**52.** $4x^3 + 56x^2 = x + 14$

**53.** $x^3 - \dfrac{37}{12}x^2 - \dfrac{7}{2}x - \dfrac{2}{3} = 0$

**54.** $3x^4 - 125x^2 + 19x = -57x^3 + 42$

**EXERCISES 55–58** ☐ *Show that the function has no rational zeros. Use a graphics calculator to approximate all irrational zeros to the nearest hundredth.*

**55.** $f(x) = x^3 - 5x - 1$

**56.** $g(x) = x^4 + 2x^3 - x^2 - 7$

**57.** $g(x) = 0.05x^4 + x^3 - x^2 + x - 3$

**58.** $h(x) = x^5 - 15x^4 - x + 5$

**EXERCISES 59–62** ☐ *Find the domain and zeros of the rational function and match it with its graph.*

**59.** $f(x) = \dfrac{1}{x - 3}$

**60.** $f(x) = \dfrac{x + 2}{x - 3}$

**61.** $f(x) = \dfrac{x + 2}{x^2 - 9}$

**62.** $f(x) = \dfrac{x - 3}{x^2 - 9}$

**EXERCISES 63–70** ☐ *Find the vertical and horizontal asymptotes*

**a.**

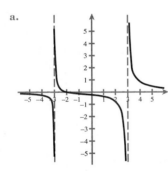

**b.**

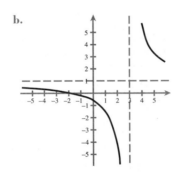

**c.**

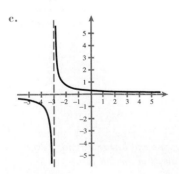

**d.**

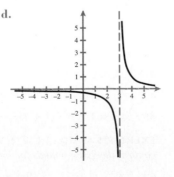

**63.** $f(x) = \dfrac{1}{x + 2}$

**64.** $h(x) = \dfrac{x + 3}{x - 1}$

**65.** $g(x) = \dfrac{2x + 5}{3x - 1}$

**66.** $h(x) = \dfrac{-1}{x^2 - 4}$

**67.** $f(x) = \dfrac{3 - x}{x^2}$

**68.** $g(x) = \dfrac{x}{x^2 + 5}$

**69.** $f(x) = \dfrac{2}{(1 - x)^2}$

**70.** $h(x) = \dfrac{x + 1}{(x - 3)^2}$

**EXERCISES 71–74** ☐ *Find the vertical and inclined asymptotes and sketch the graph.*

**71.** $h(x) = -2x - \dfrac{1}{x}$

**72.** $g(x) = \dfrac{x^2 + 2x}{x - 2}$

**73.** $h(x) = \dfrac{3x^2 + 8x - 9}{x + 3}$

**74.** $f(x) = \dfrac{x^2 - 6x + 14}{2x - 8}$

**75.** *Maximum Profit* A company has determined that its profits can be modeled with a function of the form
$f(x) = ax^2 + bx + c$, where $x$ is the number of units sold and $f(x)$ is the corresponding profit. Find $f$ given that the profit is zero for $x = 1000$ and $x = 5000$ and, because of fixed costs, the company loses \$15,000 if no units are sold. For what value of $x$ will the profit be as large as possible? What is the largest possible profit?

**76.** *Oscillating Spring* A spring is hung from a fixed point and an object is attached to its free end, as shown in Figure 68. If no external force is applied to the object, it will remain at rest in its *equilibrium* position. If the object is pulled down and then released, its position relative to equilibrium can be modeled for a brief period of time with a polynomial function of the form
$f(t) = at^4 + bt^2 + c$, where $t$ is the time after the object is

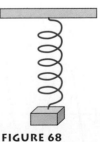

**FIGURE 68**

released and $f(t)$ is the distance away from equilibrium at that time (a negative distance indicates that the object is above equilibrium). Find $f$ if the object is pulled 6 centimeters below equilibrium and then passes equilibrium at $t = 0.11$ and again at $t = 0.27$. (*Hint:* $f$ is symmetric about the $y$-axis.) How far is the object from equilibrium when $t = 0.18$ second? At what point does the model fail to provide reasonable results?

77. *Alaskan Temperature*  The daily mean temperature in Juneau, Alaska, over a 12-month period can be approximated with the function $f(x) = 0.0134x^4 - 0.33x^3 + 2.07x^2 - 0.456x - 5.04$, where $x$ denotes the month ($x = 0$ corresponds to January) and $f(x)$ is the average Celsius temperature for that month. Approximate the zeros of $f$ in the interval $[0, 12)$. What do these zeros represent?

78. *Mathematics Degrees*  The number of bachelor's degrees conferred in mathematics during the years 1970–1990 can be approximated with the polynomial function $f(x) = -7.2x^3 + 286x^2 - 3339x + 25{,}130$, where $x$ denotes the year after 1970 and $f(x)$ is the number of degrees for that year. (Data Source: U.S. Department of Education.) Write out a function $g(x)$ whose zeros correspond to the years in which 14,500 bachelor's degrees in mathematics were conferred. Approximate the zeros. What interpretation can be given to $f(x)$ when $x$ is not an integer?

**Mathematics Degrees 1970–1990**

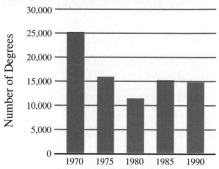

Data Source:  U.S. Department of Education

79. *Calculator Cost*  A company that manufactures calculators has determined that the total cost for producing $x$ calculators is given by $C(x) = 15{,}000 + 20x$. The average cost per calculator is thus given by

$$\overline{C}(x) = \frac{15{,}000 + 20x}{x}$$

Sketch the graph of $\overline{C}$. As the number of calculators increases, what value is the average cost approaching?

## CHAPTER TEST

**PROBLEMS 1–8** ☐ *Answer true or false.*

1. A polynomial of degree 3 with real coefficients must have at least one real zero.

2. A polynomial function of degree 4 can have at most three turning points.

3. If a polynomial function has three variations in sign, then it must have three positive zeros.

4. There is exactly one polymomial of degree three with zeros 1, −2, and 4.

5. If a polynomial of degree 5 is divided by one of degree 2, the quotient will have degree 3.

6. If $2 + 3i$ is a zero of a polynomial with real coefficients, then so is $2 − 3i$.

7. A rational function must have a vertical asymptote.

8. A rational function can have at most one horizontal asymptote.

**PROBLEMS 9–14** ☐ *Give an example of each.*

9. A polynomial function $f$ for which $f(x) \to -\infty$ as $x \to \infty$.

10. A polynomial function $f$ of least degree and with zeros −1, 3, and 4.

11. A polynomial function $f$ of least degree, with real coefficients, and with zeros 2 and −3$i$.

12. A polynomial function $f$ of degree 4, with real coefficients, and with no real zeros.

13. A rational function $f$ with vertical asymptote $x = -1$ and horizontal asymptote $y = 1$.

14. A rational function $f$ with vertical asymptote $x = 2$ and inclined asymptote $y = x - 3$.

**PROBLEMS 15–16** ☐ *Factor the given polynomial and find the exact values of all the zeros.*

15. $f(x) = 2x^4 - 7x^3 - 4x^2$

16. $f(x) = 6x^3 + x^2 - 19x + 6$

17. Find the quotient and remainder when $f(x) = 3x^4 - 12x^2 + 5x + 14$ is divided by $d(x) = x + 2$.

18. Given that $f(x) = 5x^4 - 2x^3 + 45x^2 - 18x$ has $3i$ as one of its zeros, factor $f(x)$ and find the remaining zeros.

19. Use the Rational Zero Theorem to list all the possible rational zeros of $f(x) = 2x^3 + x^2 - 12x + 9$. Use synthetic division and/or a graphics calculator to help you determine which are actually zeros.

20. Use Descartes' Rule of Signs to determine the possible number of positive and negative zeros of the function $f(x) = 2x^4 - 6x^3 + x^2 - 8x - 7$.

21. Show that $f(x) = x^3 - x - 2$ has no rational zeros, and use a graphics calculator to estimate any irrational zeros to the nearest hundredth.

22. Find any vertical or horizontal asymptotes for the function

$$f(x) = \frac{x + 1}{2x - 3}$$

and sketch its graph.

23. A company that manufactures office chairs has determined that its daily costs (in dollars) for the production of $x$ chairs is approximated by $C(x) = \frac{1}{3}x^3 - 15x^2 + 8000$.

    a. Estimate the number of chairs that can be produced for a daily cost of $6000.

    b. Estimate the number of chairs that should be produced so that the daily cost is as small as possible. What is the minimum daily cost?

# EXPONENTIAL AND LOGARITHMIC FUNCTIONS

■ As the world population has exploded, so too has humankind's influence on Earth. This satellite photograph of North America at night shows the extent of our influence; where there is light, there is humanity. And there is light from New York to Los Angeles, from Montreal to Mexico City. As the recognition that population growth profoundly affects the consumption of natural resources, environmental quality, and the entire global economy, predicting future population has become increasingly important. In this chapter we will develop mathematical models for predicting the growth of human and other populations.

# EXPONENTIAL FUNCTIONS

■ If a certain population of bacteria grows exponentially, how long would it take for it to fill a space the size of our solar system?

■ How can a doctor estimate the concentration of a medication in the bloodstream?

■ Can a bank compound interest every instant of every day without going bankrupt?

■ How quickly do college algebra students forget the material after the course ends?

## THE EXPONENTIAL FUNCTION WITH BASE A

Among the many real-world phenomena that can be described using exponential functions are population growth, the spread of disease and information, radioactive decay, concentrations of drugs in the body, sales patterns, and compounded interest. The first of these—population growth—will provide the motivation for our definition of an exponential function.

Suppose a biologist has been conducting a carefully controlled experiment on the growth of a bacteria culture. Unfortunately, a lab assistant misplaced all of the data except that which appears in Table 1. To avoid the expense of starting the experiment over, the biologist decides to construct a function that estimates the number of bacteria at any given time.

**TABLE 1**

| Time | Number of bacteria (in thousands) |
|------|-----------------------------------|
| 12:00 noon | 1 |
| 1:00 P.M. | 2 |
| 2:00 P.M. | 4 |
| 3:00 P.M. | 8 |
| 4:00 P.M. | 16 |

The pattern in Table 1 suggests that the bacteria population is doubling every hour. In fact, the values in the second column are successive powers of 2. After 1 hour, the population (in thousands) is $2^1 = 2$; after 2 hours, it is $2^2 = 4$; after 3 hours, it is $2^3 = 8$; and so on. In general, if we let $t$ represent the number of hours after 12:00 noon, then the number of bacteria after $t$ hours is $2^t$ thousand. Thus, the function that the biologist is seeking would appear to be the *exponential function*

$$f(t) = 2^t$$

where $t$ denotes the number of hours after 12:00 noon, and the population $f(t)$ is measured in thousands. Note that the function $f$ can be used to estimate the bacteria population at any time. Indeed, since 12:00 noon represents $t = 0$, times before or after 12:00 noon can be treated as negative or positive values of $t$, respectively. For example, 9:00 A.M. is 3 hours before 12:00 noon, and so we let $t = -3$ to obtain

$$f(-3) = 2^{-3} = \frac{1}{2^3} = \frac{1}{8} = 0.125$$

which indicates 125 bacteria. At 6:00 P.M., we can let $t = 6$ to see that there are

$$f(6) = 2^6 = 64$$

or 64,000 bacteria. For fractions of an hour, we can use fractional exponents. Thus, if we wish to know the size of the bacteria population at 2:30 P.M., then we should let $t$ be $2\frac{1}{2}$ or $\frac{5}{2}$. This gives

$$f\left(\frac{5}{2}\right) = 2^{5/2} = (\sqrt{2})^5 \approx 5.66$$

or 5660 bacteria. Irrational input values can be justified using methods from calculus. We will simply assume that expressions such as $2^{\sqrt{2}}$ are valid and, should the need arise, can be approximated using the power key on a calculator.

**EXAMPLE 1** *Graphing an exponential function*

Sketch the graph of the function $f(x) = 2^x$, and determine its domain and range.

**SOLUTION** The graph of $f$ can be obtained by plotting a few points and connecting these points with a smooth curve as shown in Figure 1.

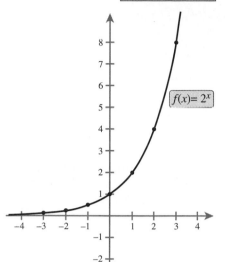

$f(x)= 2^x$

**FIGURE 1**

| $x$ | $f(x)$ |
|:---:|:---:|
| $-3$ | $\frac{1}{8}$ |
| $-2$ | $\frac{1}{4}$ |
| $-1$ | $\frac{1}{2}$ |
| $0$ | $1$ |
| $1$ | $2$ |
| $2$ | $4$ |
| $3$ | $8$ |

Alternatively, we could use a graphics calculator to obtain a graph similar to that shown in Figure 2. Recall that the domain of a function is the set of all possible input values ($x$-coordinates on the graph) and the range is the set of all possible output values ($y$-coordinates on the graph). Thus, the domain of $f$ is the set of all real numbers, and the range of $f$ is the set of positive real numbers, or $(0, \infty)$.

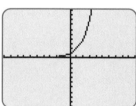

**FIGURE 2**

The function $f(x) = 2^x$ is more properly called the *exponential function with base* 2. For arbitrary bases, we give the following definition.

**Definition of the exponential function with base a**

The function $f(x) = a^x$, for $a > 0$ and $a \neq 1$, is the **exponential function with base $a$**.

Whenever $a > 1$, the graph of the exponential function $f(x) = a^x$ rises sharply as $x$ gets larger, as is shown in Figure 1, for example. However, when $a < 1$ the graph falls as $x$ gets larger, as shown by the following example.

**EXAMPLE 2** *Graphing an exponential function with a base less than 1*

Sketch the graph of the exponential function $g(x) = (\frac{1}{2})^x$ and determine its domain and range.

**SOLUTION** Once again, we can obtain the graph of $g$ by plotting a few points and connecting these points with a smooth curve, as shown in Figure 3. For example, $g(-3) = (\frac{1}{2})^{-3} = 2^3 = 8$. Other values are given as follows.

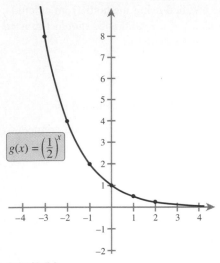

$$g(x) = \left(\frac{1}{2}\right)^x$$

**FIGURE 3**

| $x$ | $g(x)$ |
|------|--------|
| $-3$ | 8 |
| $-2$ | 4 |
| $-1$ | 2 |
| 0 | 1 |
| 1 | $\frac{1}{2}$ |
| 2 | $\frac{1}{4}$ |
| 3 | $\frac{1}{8}$ |

On the basis of the graph, we again conclude that the domain is the set of all real numbers and that the range is the set $(0, \infty)$.

Several important properties of exponential functions are illustrated in Examples 1 and 2. In both examples, we see that the domain is the set of all real numbers and that the range is the set of all *positive* real numbers. Note also that the graphs of both functions cross the $y$-axis at 1, which means that both have $y$-intercept $(0, 1)$. When the base is greater than 1, as in Example 1, the graph is rising as we move from left to right. In other words, the function is increasing. When the base is less than 1, as in Example 2, the graph is falling as we move from left to right. This means the function is decreasing. Finally, both functions are *one-to-one* (1–1) since their graphs pass the horizontal line test. These properties of the exponential function are summarized as follows.

**Properties of exponential functions**

Let $f(x) = a^x$.

1. The domain of $f$ is the set of all real numbers, and the range of $f$ is the set of all positive real numbers.

2. $f$ has $y$-intercept $(0, 1)$.

3. $f$ is increasing if $a > 1$ and decreasing if $0 < a < 1$.

4. $f$ is 1–1 (i.e., $f$ passes the horizontal line test).

Exponential functions arise frequently in applications. In the next example, we see how drug dosages can be modeled with exponential functions.

**EXAMPLE 3**    *Medication in the body*

A 4-milligram dose of a certain medication is eliminated from the body in such a way that the quantity remaining at a given time is $\frac{2}{3}$ of the amount that was there 1 hour earlier. Thus after $t$ hours, the amount remaining in the body is given by the function

$$g(t) = 4\left(\frac{2}{3}\right)^t$$

How much of the medication is remaining after 3 hours? After $5\frac{1}{2}$ hours?

**SOLUTION**   After 3 hours, the amount of medication remaining is given by

$$g(3) = 4\left(\frac{2}{3}\right)^3 = 4 \cdot \frac{8}{27} \approx 1.185 \text{ mg}$$

After $5\frac{1}{2}$ hours, the amount is

$$g\left(\frac{11}{2}\right) = 4\left(\frac{2}{3}\right)^{11/2} = 4\left(\frac{2}{3}\right)^{5.5} \approx 0.43 \text{ mg}$$

The one-to-one property of exponential functions is often useful for solving simple equations involving exponents. It can be stated in the following way.

***The one-to-one property of exponential functions***

For all real numbers $r$ and $s$, if $a^r = a^s$, then $r = s$.

**EXAMPLE 4**   *Solving an exponential equation*

Use the one-to-one property of the exponential function to solve the exponential equation $\frac{1}{4^x} = 64$.

**SOLUTION**   First we must rewrite both sides of the equation as a power of 4—that is, in the form $4^\square$. This is possible since $\frac{1}{4^x}$ can be written as $4^{-x}$ and 64 can be written as $4^3$. We proceed as follows.

$$\frac{1}{4^x} = 64$$

$$4^{-x} = 4^3 \qquad \text{Using the identity } \frac{1}{a^n} = a^{-n} \text{ and rewriting 64 as } 4^3$$

$$-x = 3 \qquad \text{Applying the one-to-one property}$$

$$x = -3$$

Notice that an exponential identity was needed in the preceding example when we claimed that $1/4^x = 4^{-x}$. Using techniques from calculus, the exponential identities that we have already seen for integer and rational exponents can be shown to be valid for any real-number exponent. For convenience, we summarize these identities again.

***Exponential identities***

Let $a$ and $b$ be positive real numbers and let $r$ and $s$ be real numbers.

1. $a^r a^s = a^{r+s}$

2. $\dfrac{a^r}{a^s} = a^{r-s}$

3. $(a^r)^s = a^{rs}$

4. $(ab)^r = a^r b^r$

5. $\left(\dfrac{a}{b}\right)^r = \dfrac{a^r}{b^r}$

6. $a^0 = 1$

7. $a^{-r} = \dfrac{1}{a^r}$

**EXAMPLE 5**    *Solving an exponential equation*

Solve the exponential equation $3^x \cdot 3^{x+1} = \dfrac{1}{27}$.

**SOLUTION**

$$3^x \cdot 3^{x+1} = \frac{1}{27}$$

$$3^{x+x+1} = \frac{1}{27} \qquad \text{Applying the exponential identity } a^r a^s = a^{r+s}$$

$$3^{2x+1} = \frac{1}{3^3} \qquad \text{Rewriting 27 as a power of 3}$$

$$3^{2x+1} = 3^{-3} \qquad \text{Using the exponential identity } \frac{1}{a^n} = a^{-n}$$

$$2x + 1 = -3 \qquad \text{Applying the one-to-one property}$$

$$2x = -4$$

$$x = -2$$

It should be noted that the technique shown in Examples 4 and 5 is practical only for exponential equations that can easily be written in the form

$$a^{\square} = a^{\bigcirc}$$

The method is not useful for such seemingly simple equations as $2^x = 25$. In this case it is not possible to rewrite 25 as an integer power of 2. We will develop an alternative approach using the inverse of the exponential function in Section 3.

## THE NATURAL EXPONENTIAL FUNCTION

The frequency with which the number $e$ appears in mathematical formulas from diverse branches of mathematics (number theory, probability, statistics, and calculus, for example) is rivaled only by that of $\pi$. Like $\pi$, $e$ is an irrational number and so cannot

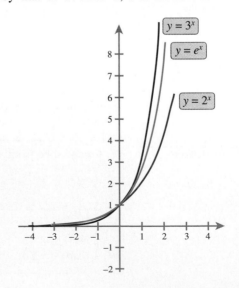

**FIGURE 4**

be written with a finite number of decimal places. Its value to five decimal places is as follows.

$$e = 2.71828\ldots$$

Perhaps the most important function in all of mathematics is the exponential function with base $e$, $f(x) = e^x$, also known as the **natural exponential function**. Since $e$ is a number between 2 and 3, the graph of the natural exponential function $f(x) = e^x$ lies between the graphs of $y = 2^x$ and $y = 3^x$, as shown in Figure 4.

| **EXAMPLE 6** | *Using the graph of $f(x) = e^x$ to graph an exponential function* |

Use the graph of $f(x) = e^x$ to sketch the graph of $g(x) = e^{x+3}$.

**SOLUTION**   Note that $g(x)$ is obtained by replacing $x$ with $x + 3$ in the function $f(x) = e^x$. Thus, the graph of $g$ can be found by translating the graph of $f$ 3 units to the left, as shown in Figure 5.

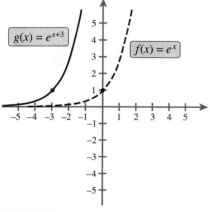

**FIGURE 5**

The number $e$ arises in a natural way from the study of **compound interest**. Earlier we computed *annually compounded interest* using the formula $B = P(1 + r)^t$, where $B$ is the balance in an account after $t$ years if a principal of $P$ dollars is deposited at an annual rate of interest $r$ compounded annually. A more general approach (see Exercise 37) would lead to the following formula.

■ ***Compound interest formula***

> If a principal of $P$ dollars is deposited in an account paying an annual rate of interest $r$ compounded $n$ times per year, then the balance $B$ in the account after $t$ years is given by
>
> $$B = P\left(1 + \frac{r}{n}\right)^{nt}$$

| **EXAMPLE 7** | *Computing compound interest* |

Suppose that instead of throwing a silver dollar across the Potomac River, George Washington had deposited it in 1776 in a bank that paid 6% interest. How much would that dollar be worth in the year 2000 if interest were compounded

**a.** annually?                     **b.** quarterly?
**c.** daily?                           **d.** every minute?

**SOLUTION**

**a.** Letting $P = 1$, $r = 0.06$, $t = 224$, and $n = 1$ in the compound interest formula, we obtain

$$B = P\left(1 + \frac{r}{n}\right)^{nt} = 1\left(1 + \frac{0.06}{1}\right)^{1(224)} = (1.06)^{224} \approx 466{,}137.26$$

**b.** Using $n = 4$, we obtain

$$B = P\left(1 + \frac{r}{n}\right)^{nt} = 1\left(1 + \frac{0.06}{4}\right)^{4(224)} = (1.015)^{896} \approx 621,689.96$$

**c.** With $n = 365$, we obtain

$$B = 1\left(1 + \frac{0.06}{365}\right)^{365(224)} \approx 686,180.14$$

**d.** Letting $n = 365 \cdot 24 \cdot 60 = 525,600$, the number of minutes in a year, we obtain

$$B = 1\left(1 + \frac{0.06}{525,600}\right)^{525,600(224)} \approx 686,937.94$$

From the preceding example it is evident that the greater the number of compounding periods, the greater the balance that accumulates. One might expect that there is no limit to the balance that would accumulate. Surprisingly, there is a limit; no matter how many compounding periods are used, George Washington's dollar could never accumulate to more than \$686,938.47 in 224 years. To see where this amount comes from, we will first examine how the natural exponential function is connected to compound interest.

Suppose you were lucky enough to find a bank that paid 100% interest and you deposited a dollar for 1 year. By setting $P = 1$, $r = 1.00$, and $t = 1$ in the compound interest formula, and by considering larger and larger values of $n$, the number of compounding periods, we obtain the following table of values (rounded to five decimal places).

| $n$ | $B = \left(1 + \dfrac{1}{n}\right)^n$ |
|---:|---|
| 1 | 2 |
| 10 | 2.59374 |
| 100 | 2.70481 |
| 1000 | 2.71692 |
| 10,000 | 2.71815 |
| 100,000 | 2.71827 |
| 1,000,000 | 2.71828 |

Notice that as the number of compounding periods increases, the balance gets closer to the number $e$. Actually, this is no coincidence since the number $e$ is *defined* to be the limiting value of $[1 + (1/n)]^n$. So if the bank compounded interest *continuously*, your dollar would increase in value to exactly $e$ dollars after 1 year. Of course, the bank would likely round this value up to \$2.72. For arbitrary values of $P, r$, and $t$, we have the following general formula for **continuously compounded interest**.

| **Continuously compounded interest** | If a principal of $P$ dollars is deposited in an account paying an annual rate of interest $r$ compounded continuously, then the balance $B$ in the account after $t$ years is given by $$B = Pe^{rt}$$ |
|---|---|

**EXAMPLE 8**    *Computing continuously compounded interest*

Suppose that George Washington deposited his dollar in 1776 in a bank that compounded interest continuously at a rate of 6%. How much would it be worth in the year 2000?

**SOLUTION**    By applying the formula for continuously compounded interest with $P = 1$, $r = 0.06$, and $t = 224$, the balance would be

$$B = 1e^{(0.06)224} = e^{13.44} \approx 686{,}938.47$$

Thus, the dollar would be worth \$686,938.47 in the year 2000.

## EXERCISES 1

**EXERCISES 1–2** □ *Sketch the graph of f and use translations or reflections to sketch the graph of g.*

**1.** $f(x) = 3^x$

   **a.** $g(x) = -3^x$          **b.** $g(x) = 3^{-x}$

   **c.** $g(x) = 3^x + 2$       **d.** $g(x) = 3^{x+2}$

**2.** $f(x) = \left(\dfrac{1}{4}\right)^x$

   **a.** $g(x) = -\left(\dfrac{1}{4}\right)^x$    **b.** $g(x) = \left(\dfrac{1}{4}\right)^{-x}$

   **c.** $g(x) = \left(\dfrac{1}{4}\right)^x - 3$   **d.** $g(x) = \left(\dfrac{1}{4}\right)^{x-2}$

**EXERCISES 3–4** □ *Determine the value for a for which $f(x) = a^x$ has the indicated graph.*

**3.**

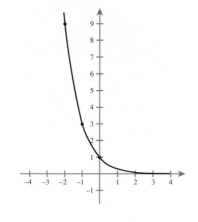

**4.**

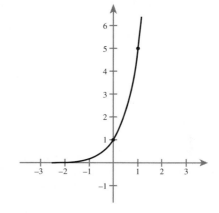

**EXERCISES 5–8** □ *Use the given graph of $f(x) = e^x$ to sketch the graph of g(x).*

**5.** $g(x) = e^{-x}$

**6.** $g(x) = -e^x$

**7.** $g(x) = e^{x-2}$

**8.** $g(x) = e^x + 3$

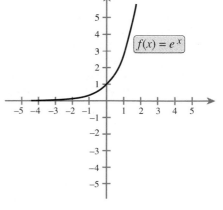

**EXERCISES 9–12** □ *Use a graphics calculator to help you sketch the graph of the given function.*

9. $P(t) = 4e^{0.1t}$

10. $N(t) = 2e^{-0.2t}$

11. $Q(t) = 1200e^{-0.05t}$

12. $P(t) = 8000e^{0.07t}$

**EXERCISES 13–22** □ *Use the one-to-one property of the exponential function to solve the given equation.*

13. $3^x = 81$

14. $4^{-x} = 64$

15. $2^{3x-4} = 32$

16. $5^{1-x} = 125$

17. $\dfrac{1}{3^x} = 27$

18. $2^{x-1} = \dfrac{1}{64}$

19. $4^{x-3} = 16 \cdot 4^{2x}$

20. $3^{x+2} = \dfrac{81}{3^x}$

21. $2^{x^2+x} = 4$

22. $5^{x^2-5x} = \dfrac{1}{625}$

**EXERCISES 23–24** □ *Determine the balance B if a principal of P dollars is deposited in an account for t years at an annual rate of interest r compounded n times per year.*

23. $P = \$1000$, $t = 10$ years, $r = 6\%$
   a. $n = 1$
   b. $n = 4$
   c. $n = 12$
   d. $n = 365$
   e. Compounded continuously

24. $P = \$2500$, $t = 5$, $r = 8\%$
   a. $n = 1$
   b. $n = 4$
   c. $n = 12$
   d. $n = 365$
   e. Compounded continuously

---

■ *Applications*

---

25. *Population Growth* A breeder who supplies pet stores with gerbils has determined that a certain gerbil population can be approximated over a 4-month period with the function

$$f(t) = 20\left(\frac{3}{2}\right)^t$$

where $t$ is measured in months. How many gerbils were initially present? Complete a table showing the gerbil population at the end of each of the first 4 months. Assuming the pattern continues beyond 4 months, predict how many gerbils will be present after 6 months and again after $10\frac{1}{2}$ months.

26. *Population Growth* An exterminating company has developed a new chemical-free method for exterminating cockroaches. They claim that a cockroach population with initial size 10,000 will number $f(t) = (10,000)2^{-t}$ after $t$ days. Plot this function. Predict how many cockroaches will remain after 1 week. Estimate the day on which the last cockroach will die.

27. *Compound Interest* On the occasion of the birth of their first grandchild, two proud grandparents deposit $5000 in a Certificate of Deposit (CD) in the child's name. The CD pays an annual interest rate of 8%, and it matures in 18 years, just in time for the child to begin college. Determine the amount that will have accumulated after 18 years if interest is compounded quarterly. What if interest is compounded continuously?

28. *Compound Interest* The grandparents in Exercise 27 wish to determine whether the CD they purchased will pay for the first year of their grandchild's college education. Assuming that 1 year at their alma mater currently costs $15,000 and the amount is likely to increase an average of 5% per year, use the formula for interest compounded annually to estimate what 1 year of college will cost in 18 years.

29. *Energy Consumption* Some experts have claimed that energy consumption during the twentieth century has grown at an exponential rate according to the formula $Q = Q_0 e^{0.07t}$ where $t$ is measured in years after 1900 and $Q_0$ is the energy consumed in 1900. Compute $Q$ in the years 1940 and 2000, and use these values to determine the percent increase in consumption from 1940 to 2000.

*Top:* A field of windmills generates electricity near Altamont, California. *Bottom:* A heliostat field focuses solar energy on a central receiving tower.

30. *Computer Transistors* Moore's law, formulated in 1965, states that the number of transistors that can be put on a computer processor chip will approximately double every 2 years.

In 1971, Intel introduced the 4004 processor with 2300 transistors. According to Moore's law, the number of transistors $t$ years after 1971 can be approximated by $N = 2300(2)^{t/2}$. Use this function to estimate the number of transistors on a processor in 1989. The actual number of transistors on the 486 processor introduced by Intel in 1989 was 1.2 million. How well does your estimate compare to this value? What does Moore's law predict for the number of transistors in the year 2001?

31. *Population Equation* Suppose we were interested in determining when the bacteria culture discussed at the beginning of this section reaches a population of 25,000. To do this, we would need to know the value of $t$ for which $2^t = 25$. Solutions to this equation can also be viewed as zeros of the function $g(x) = 2^x - 25$. Use a graphics calculator to estimate the zero(s) of $g$ to the nearest hundredth.

32. *Intersection Points* Use a graphics calculator to estimate the point(s) of intersection for the graphs of $y = 2^x$ and $y = x^2$. [*Hint:* Approximate the *three* zeros of $h(x) = 2^x - x^2$.]

33. *Density Function* The function $f(x) = (1/\sqrt{2\pi})e^{-x^2/2}$ is known as the *standard normal probability density function*, and it is important in the study of probability and statistics. Use a graphics calculator to plot the graph of $f$ and determine its maximum value. What happens to the value of $f(x)$ as $x$ approaches infinity or negative infinity?

34. *Medication Concentration* A certain medication is ingested and begins spreading throughout the bloodstream. The concentration (in nanograms per milliliter) of the medication in the bloodstream after $t$ hours is approximated by $Q(t) = 4te^{-0.8t}$.

Use a graphics calculator to plot this function. Estimate the maximum concentration of the medication and the time at which it occurs. What happens to the concentration as time increases beyond this point?

35. *Contagious Virus* A group of 100 students returns from a trip to Europe where they were exposed to a highly contagious virus. The virus begins spreading among the other 900 students in their school in such a way that $t$ days after their return, the number of students with the virus is approximated by

$$N(t) = \frac{100,000}{100 + 900e^{-0.15t}}$$

a. How many students have the virus after 2 days? After 5 days?

b. Use a graphics calculator to help you plot the graph of $N$.

c. Predict how many days it will take for 800 of the school's 1000 students to be infected.

d. What happens to $N(t)$ as $t$ approaches infinity?

36. *Knowledge Retention* Suppose a group of college algebra students finishes the semester with near 100% mastery of the material. After tracking these students for many weeks following the end of the course, the instructor discovers that the percentage of the knowledge that is retained after $t$ weeks is approximated by $P(t) = 40 + 60e^{-0.75t}$.

a. What percentage is retained after 1 week? After 5 weeks?

b. Use a graphics calculator to help you plot the graph of $P$.

c. Predict the number of weeks until only 50% is retained.

d. What happens to $P(t)$ as $t$ approaches infinity?

■ *Projects for Enrichment*

37. *Deriving Compound Interest Formulas* The formula for compound interest can be derived from the formula for simple interest. We'll consider two specific cases before tackling the general case. First, suppose that a principal of $P$ dollars is deposited in an account paying an annual rate of interest $r$ compounded annually. Then from the simple interest formula, the interest earned after the first year is given by $I = Pr(1)$. Thus, the balance at the end of the first year is $B = P + Pr = P(1 + r)$.

a. Apply the simple interest formula to the new balance $P(1 + r)$ to obtain an expression for the interest earned during the second year. Then add this interest to the balance $P(1 + r)$ and simplify to obtain the expression $P(1 + r)^2$ for the balance after 2 years.

b. Repeat part (a) for the third year to show that the balance after 3 years is $P(1 + r)^3$.

Generalizing the procedure above, we obtain a balance of $P(1 + r)^t$ after $t$ years if interest is compounded annually. Now suppose interest is compounded quarterly, or 4 times per year.

This means that the annual rate of interest $r$ must be divided equally among the 4 quarters. In other words, the quarterly rate of interest is $r/4$.

c. Use the formula for simple interest to write an expression for the interest earned on a principal $P$ during the first quarter.

d. Add the interest from part (c) to the original principal $P$ and simplify to get an expression for the balance after the first quarter. You should get $P[1 + (r/4)]$.

e. Repeat parts (c) and (d) several times to show that the balance is $P[1 + (r/4)]^4$ at the end of the first year, and $P[1 + (r/4)]^8$ at the end of the second year.

Thus, with interest compounded 4 times per year, the principal after $t$ years is $P[1 + (r/4)]^{4t}$. Finally, suppose interest is compounded $n$ times per year. From the pattern that has been established, it seems clear that the principal after $t$ years is $P[1 + (r/n)]^{nt}$.

f. What does the quantity $r/n$ represent? What about $nt$?

38. *Continuously Compounded Interest* Using the fact that $[1 + (1/m)]^m$ approaches $e$ as $m$ approaches infinity, we can derive the formula for continuously compounded interest from the compound interest formula, $B = P[1 + (r/n)]^{nt}$.

   a. Show that the formula for compound interest can be rewritten as

   $$B = P\left[\left(1 + \frac{1}{n/r}\right)^{n/r}\right]^{rt}$$

   b. Let $m = n/r$. What happens to the value of $m$ if $r$ remains fixed but $n$ approaches infinity?

   c. Substituting $m = n/r$ into the formula in part (a), we obtain

   $$B = P\left[\left(1 + \frac{1}{m}\right)^{m}\right]^{rt}$$

   Using your observation from part (b) argue that as the number of compounding periods approaches infinity, the balance gets closer in value to $Pe^{rt}$.

   d. Surprisingly, continuous compounding does not pay that much more interest than annual compounding. To see this, suppose you were to find a bank that pays 5% interest compounded continuously. If you deposit $P$ dollars, the balance after $t$ years would be $Pe^{0.05t}$. Now suppose a different bank pays 5.13% interest compounded annually. If you deposit $P$ dollars there, the balance after $t$ years would be $P(1 + 0.0513)^t$. Show that the annually compounded interest will yield a slightly higher balance.

   e. Find a value for $r$ for which quarterly compounding at an annual rate $r$ will be equivalent to continuous compounding at an annual rate of 5%.

---

### ▟▙ *Questions for Discussion or Essay*

39. Discuss the restrictions placed on the value for $a$ in the definition of the exponential function $f(x) = a^x$. In particular, consider the following questions: Why does the definition exclude the value $a = 1$? Isn't $f(x) = 1^x$ a valid function? Why isn't it considered an exponential function? Why must we have $a > 0$? What difficulties do we encounter in defining the function $f(x) = (-2)^x$?

40. Discuss the limitations of the technique used in Example 4 for solving simple exponential equations. Describe a procedure for using a graphics calculator to find approximate solutions to exponential equations.

41. As useful as exponential functions are for applications, they must be used carefully. Consider, for example, the exponential function $f(t) = 2^t$ that was used to model the bacteria population at the beginning of this section. If the growth were to continue at this exponential pace, show that the number of bacteria at the end of 1 week would be $3.74 \times 10^{50}$. Assuming an "average" size bacteria, this would be enough bacteria to fill a sphere with a diameter of that of our solar system. What is wrong with the model? What would really happen to the bacteria population over time?

42. A careful comparison of banking practices would reveal that there are many methods used by banks to compound interest. Even among banks that claim to have the same compounding periods, there are differences in when the interest is actually paid into an account. Contact several local banks and inquire about their interest rates and compounding schemes for a specific type of account. Discuss some ways for comparing banks so that you can choose the one that pays "the best" interest.

---

### SECTION 2

## LOGARITHMIC FUNCTIONS

■ At 6% interest compounded annually, how long would it take the silver dollar that Washington threw across the Potomac to accumulate to $1,000,000?

■ What makes acid rain acidic?

■ How could you cross a river with an unlimited supply of pizza boxes, a parachute, and a mathematician?

■ How can Inspector Magill estimate time of death from body temperature?

## THE LOGARITHM FUNCTION WITH BASE A

In the last section we asked the question, If George Washington had invested the silver dollar that he threw across the Potomac river at 6% interest compounded annually, how much would it be worth in the year 2000? Using the compound interest formula

$$B = P\left(1 + \frac{r}{n}\right)^{tn}$$

we found that the dollar would be worth $1.06^t$ dollars $t$ years after 1776, and hence $\$1.06^{(2000-1776)} \approx \$466,137.26$ in the year 2000. But suppose that we wanted to know the length of time required for the dollar to grow into an account worth $\$1,000,000$. We would have to solve the equation $1.06^t = 1,000,000$. Alternatively, we could find the intersection of the graphs of $y = 1.06^t$ and $y = 1,000,000$, as shown in Figure 6. Note that the graphs cross only once, so that there is a single solution to the equation $1.06^t = 1,000,000$. This solution could be described as "the power to which 1.06 must be raised in order to obtain 1,000,000." By tracing the graph of $f(t) = 1.06^t$ with a graphics calculator, as shown in Figures 7 and 8, we see that $t \approx 237$ when $y \approx 1,000,000$. Thus, the solution is approximately 237 years. So Washington's dollar would be worth $\$1,000,000$ in about the year $1776 + 237 = 2013$.

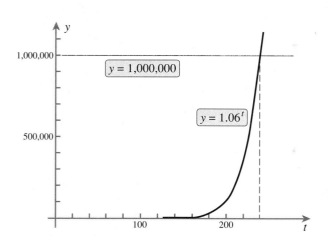

**FIGURE 6**

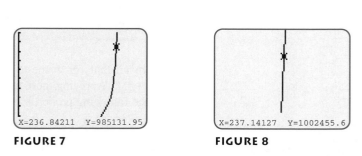

**FIGURE 7**          **FIGURE 8**

From Figure 6 we see that *any* horizontal line will intersect the graph of $1.06^t$ exactly once. Thus, for every output of this function, there is exactly one corresponding input. Recall that functions with this property are called one-to-one functions, and one-to-one functions have inverses. Inverses of exponential functions are called *logarithmic func-*

*tions*, and if we define $f(t) = 1.06^t$, then we would write "$f^{-1}(t) = \log_{1.06} t$," which is read "$f$ inverse of $t$ is the logarithm to the base 1.06 of $t$."

Of course, there is nothing special about the base 1.06. In fact, we recall from Section 1 that all exponential functions of the form $f(x) = a^x$ are one-to-one functions and hence have inverses. We make the following definition.

***Definition of logarithm base a***

> The function $g(x) = \log_a x$ for $a > 0$ and $a \neq 1$, the **logarithm function with base a**, is the inverse of the exponential function $f(x) = a^x$. That is, $y = \log_a x$ if and only if $a^y = x$.

Since logarithmic functions are inverses of exponential functions, their graphs are reflections of one another about the line $y = x$. The graphs of a few exponential functions and the logarithmic functions that are their inverses are shown in Figure 9.

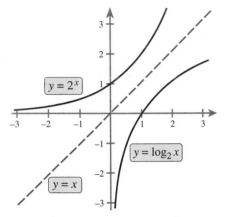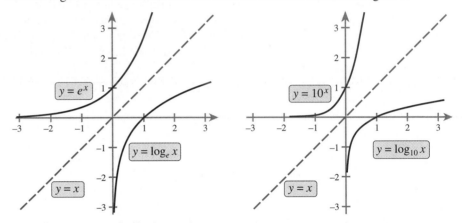

**FIGURE 9**

Since exponential and logarithmic functions are inverses of one another, the domain of one is the range of the other. Thus, since the exponential function $f(x) = a^x$ has domain $(-\infty, \infty)$ and range $(0, \infty)$, the domain and range of logarithmic functions are as follows.

> The logarithmic function $g(x) = \log_a x$ has domain $(0, \infty)$ and range $(-\infty, \infty)$.

We recall that for a one-to-one function $f$ with inverse $f^{-1}$, $f(f^{-1}(x)) = x$, and $f^{-1}(f(x)) = x$. Applying these facts to the functions $f(x) = a^x$, and $f^{-1}(x) = \log_a x$, we obtain the following properties.

***Inverse properties of logarithms and exponents***

> Let $a > 0$ and $a \neq 1$.
>
> | Property | Example |
> |---|---|
> | 1. $a^{\log_a x} = x$   for all $x > 0$ | $2^{\log_2 5} = 5$ |
> | 2. $\log_a (a^x) = x$   for all $x$ | $\log_3(3^{-8}) = -8$ |

Note that the first inverse property tells us that $a$ raised to the $\log_a x$ power equals $x$, or in other words "$\log_a x$ is the power to which $a$ must be raised in order to get $x$." This is useful for evaluating some logarithms, as illustrated by the following examples.

**EXAMPLE 1**    *Computing logarithms*

Compute each of the following quantities.
**a.** $\log_2 8$        **b.** $\log_{1/3} 9$        **c.** $\log_{123.456789} 1$

**SOLUTION**

**a.** We need to find the power to which 2 must be raised in order to get 8. Since $2^3 = 8$, $\log_2 8 = 3$.

**b.** We are interested in finding the power to which $\frac{1}{3}$ must be raised in order to get 9. Since $(\frac{1}{3})^{-2} = 9$, $\log_{1/3} 9 = -2$.

**c.** To what power must 123.456789 be raised in order to get 1? Since $123.456789^0 = 1$, $\log_{123.456789} 1 = 0$.

The result in part (c) of the preceding example holds for any base $a$. Indeed, since any number raised to the zero power is 1, we have the following general property.

> For any $a > 0$, $a \neq 1$, $\log_a 1 = 0$.

There is another way of expressing the relationship between logarithms and exponents that is useful both for solving equations involving logarithms and for deriving logarithmic identities. Suppose that we let $y = \log_a x$. Then by the definition of the logarithm function, $a^y = x$. Thus, for every logarithmic equation of the form $y = \log_a x$, there corresponds the equivalent exponential equation $a^y = x$. This correspondence is illustrated in the following examples.

**EXAMPLE 2**    *Writing logarithmic equations as exponential equations*

Write an equivalent exponential equation for each of the following logarithmic equations.
**a.** $\log_{10} 1000 = 3$        **b.** $\log_x 100 = 2.1$        **c.** $\log_3(x^2) = 12$

**SOLUTION**

**a.** $10^3 = 1000$        **b.** $x^{2.1} = 100$        **c.** $3^{12} = x^2$

> **RULE OF THUMB**    When converting exponential equations to logarithmic equations, it is helpful to keep the following rules of thumb in mind.
>
> ■ The base of the logarithm is the base of the exponent (i.e., the quantity being raised to the power).
> ■ The value of the logarithm is the exponent.

**EXAMPLE 3**    *Writing exponential equations as logarithmic equations*

For each exponential equation, write an equivalent logarithmic equation.
**a.** $0.5^{-2} = 4$      **b.** $e^0 = 1$      **c.** $1.015^{12t} = 1,000,000$

**SOLUTION**

**a.** $\log_{0.5} 4 = -2$      **b.** $\log_e 1 = 0$      **c.** $\log_{1.015} 1,000,000 = 12t$

## NATURAL AND COMMON LOGARITHMS

Special notation has developed for the two most frequently used logarithmic bases, 10 and $e$. Since most calculators can directly compute only logarithms with these bases, it is especially important to become familiar with logarithms base 10 and base $e$.

**Definitions of natural and common logarithms**

> $\ln x = \log_e x$. Base $e$ logarithms are called **natural logarithms**.
> $\log x = \log_{10} x$. Base 10 logarithms are called **common logarithms**.

**EXAMPLE 4**    *Graphing a function involving the natural logarithm*

Sketch the graph of $f(x) = \ln x$ and use it to sketch the graph of $g(x) = \ln x - 2$

**SOLUTION**    First we note that $f(x) = \ln x$ is the inverse of $y = e^x$. Thus, the graph of $f$ is the reflection of the graph of $y = e^x$ about the line $y = x$, as shown in Figure 10. Now $g(x) = \ln x - 2 = f(x) - 2$. So the graph of $g$ can be obtained by translating the graph of $f$ down 2 units. The result is shown in Figure 10.

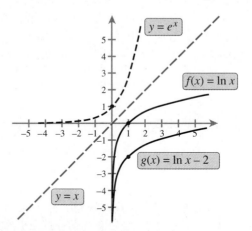

**FIGURE 10**

**EXAMPLE 5**    *An application involving natural logarithms*

According to Newton's Law of Cooling, the time it takes for an object to cool from an initial temperature $T_0$ to a temperature $T$, given that the temperature of the surrounding air is a constant $C$, is given by

$$t = k \ln\left(\frac{T_0 - C}{T - C}\right)$$

where $k$ is a constant. Suppose a turkey is taken out of the oven and is placed in air having a constant temperature of 70°F. If the turkey initially has an internal temperature of 185°F and $k = 12.5$ (for time measured in minutes), find the time required for the turkey to cool to the following temperatures.

**a.** $T = 160°F$       **b.** $T = 100°F$       **c.** $T = 80°F$

What happens to the time as $T$ gets closer to 70°F? According to this model, will the turkey ever cool to 70°F? Use a graphics calculator to sketch the graph of $t$ as a function of $T$, and then identify the domain.

**SOLUTION**    We first set $C = 70$, $T_0 = 185$, and $k = 12.5$ to obtain

$$(1) \qquad t = 12.5 \ln\left(\frac{185 - 70}{T - 70}\right) = 12.5 \ln\left(\frac{115}{T - 70}\right)$$

Now we substitute the given values for $T$.

**a.** $T = 160$:   $t = 12.5 \ln\left(\dfrac{115}{160 - 70}\right) \approx 3.1$ minutes

**b.** $T = 100$:   $t = 12.5 \ln\left(\dfrac{115}{100 - 70}\right) \approx 16.8$ minutes

**c.** $T = 80$:    $t = 12.5 \ln\left(\dfrac{115}{80 - 70}\right) \approx 30.5$ minutes

As $T$ gets closer to 70, $t$ increases without bound. The value $T = 70$ would yield an undefined expression inside the logarithm. Thus, according to the model, the turkey will never reach 70°. In practice, of course, we know that it will. We conclude that the model fails to give realistic results for values of $T$ near 70. This is illustrated by the graph in Figure 11 of

$$t = 12.5 \ln\left(\frac{115}{T - 70}\right)$$

Here we see that $t$ (the $y$-coordinate) approaches infinity as $T$ (the $x$-coordinate) approaches 70 from the right. The domain of the function is $(70, \infty)$.

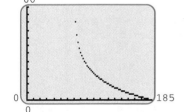

**FIGURE 11**

---

Natural logarithms are indeed natural, in that they arise naturally from problems in both applied and pure mathematics. Common logarithms, in contrast, are important primarily because of the fact that we use a base 10 number system. Prior to the advent of hand-held calculators, common logarithms were used as a computational aid. They are still used today in defining decibel ratings, the Richter scale, and the acidity level of a solution.

**EXAMPLE 6**    *Measuring the loudness of sound*

The loudness of sound is measured using a unit known as the **decibel**. The faintest sound perceptible to most individuals is assigned an intensity of $I_0$, and a sound of intensity $I$ has a decibel rating of

$$D = 10 \cdot \log\left(\frac{I}{I_0}\right)$$

A blue whale flirts with onlookers.

The sound of a blue whale (which can be heard up to 530 miles away) can reach an intensity of $6.3 \times 10^{18}$ times that of $I_0$. Find the number of decibels of this sound.

**SOLUTION**    We set $I = (6.3 \times 10^{18})I_0$ and compute $D$.

$$D = 10 \log\left(\frac{I}{I_0}\right)$$
$$= 10 \log\left(\frac{6.3 \times 10^{18}I_0}{I_0}\right)$$
$$= 10 \log(6.3 \times 10^{18})$$
$$\approx 10 \cdot 18.8$$
$$= 188$$

Thus the sound of a blue whale can reach 188 decibels.

## CHANGING BASES

It was mentioned earlier that most calculators can directly compute only natural and common logarithms. To compute a logarithm to a base other than 10 or $e$, it is necessary to *change bases* using the following formula, the proof of which is left to Exercise 63.

*The change-of-base formula*

> Let $a$ and $b$ be positive real numbers with $a \neq 1$ and $b \neq 1$. Then for $u > 0$,
>
> $$\log_a u = \frac{\log_b u}{\log_b a}$$

In practice, the change-of-base formula is used almost exclusively to change from a base $a$ to either base 10 or base $e$, as shown in the following examples.

**EXAMPLE 7**    *Changing to base* $10$

Use common logarithms to compute $\log_3 16$.

**SOLUTION**    We apply the change-of-base formula with $a = 3$, $b = 10$, and $u = 16$.

$$\log_3 16 = \frac{\log_{10} 16}{\log_{10} 3} \approx \frac{1.2041}{0.4771} \approx 2.524$$

**EXAMPLE 8**    *Changing to base* $e$

Use natural logarithms to compute $\log_3 16$.

**SOLUTION**    We apply the change-of-base formula with $a = 3$, $b = e$, and $u = 16$.

$$\log_3 16 = \frac{\ln 16}{\ln 3} \approx \frac{2.7726}{1.0986} \approx 2.524$$

Note that this result agrees with that of Example 7; it doesn't matter whether we use common logarithms or natural logarithms when we change bases.

## EXERCISES 2

**EXERCISES 1–8** ☐ *Evaluate the expression using a calculator.*

1. $\ln 15.2$

2. $\log 110$

3. $\log\left(\dfrac{2}{3}\right)$

4. $\ln \sqrt{21}$

5. $\log(2\sqrt{3})$

6. $\ln\left(\dfrac{17}{5}\right)$

7. $(\ln 0.41)^3$

8. $\log(3.52 \times 10^{47})$

**EXERCISES 9–20** ☐ *Evaluate the expression without the use of a calculator.*

9. $\log_2 16$

10. $\log_3 \dfrac{1}{9}$

11. $\ln \dfrac{1}{e^2}$

12. $\log 10{,}000$

13. $\log_6 \dfrac{1}{216}$

14. $\log_7 343$

15. $\log_{1/5} 25$

16. $\log_{3/4}\left(\dfrac{4}{3}\right)^{100}$

17. $\log_{0.1} 1000$

18. $\log_{\sqrt{2}} 16$

19. $\log(10^{100})$

20. $\ln \sqrt{e}$

**EXERCISES 21–28** ☐ *Use the change-of-base formula to rewrite the expression, first with common logarithms and then with natural logarithms. Evaluate to find a decimal approximation.*

21. $\log_2 12$

22. $\log_5 37$

23. $\log_{1/2} 108$

24. $\log_{1/3} 0.04$

25. $\log_{12} 0.341$

26. $\log_{23} 1250$

27. $\log_\pi 10$

28. $\log_\pi e$

**EXERCISES 29–34** ☐ *Use the given graph of $f(x) = \ln x$ to graph the indicated function.*

29. $g(x) = -\ln x$

30. $g(x) = \ln(-x)$

31. $g(x) = \ln(x - 3)$

32. $g(x) = 2 + \ln x$

33. $g(x) = \ln(-x) - 1$

34. $g(x) = -\ln(x + 2)$

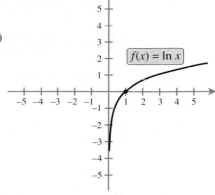

**EXERCISES 35–40** ☐ *Compute the inverse of each function, and graph both the function and its inverse on the same set of coordinate axes.*

35. $f(x) = \log_3 x$

36. $f(x) = 2 \ln x$

37. $f(x) = \ln(x + 5)$

38. $f(x) = \log(x - 4)$

39. $f(x) = \log(1000x)$

40. $f(x) = \log_5(125x)$

**EXERCISES 41–44** ☐ *Use a graphics calculator to help you sketch the graph of the given function, and find its domain.*

41. $f(x) = x \ln x$

42. $f(x) = (x^2 - 1)\ln x$

43. $f(x) = \dfrac{x}{\ln(x + 1)}$

44. $f(x) = \dfrac{\ln x}{x - 1}$

45. Graph the function

$$f(x) = \frac{\ln x}{x}$$

What value does $f(x)$ approach as $x$ becomes larger and larger?

46. Graph the function

$$f(x) = \log\left(\frac{x + 1}{2x - 8}\right)$$

Find the domain of $f$ and any vertical or horizontal asymptotes.

■ *Applications*

**EXERCISES 47–50** □ *These exercises deal with Newton's Law of Cooling, which can be expressed using the formula*

$$t = k \ln\left(\frac{T_0 - C}{T - C}\right)$$

*Here t is the time it will take for an object to cool to a temperature T given that $T_0$ is the initial temperature, C is the temperature of the surrounding air, and k is a constant.*

47. **Coffee Temperature** Find the time required for a hot cup of coffee with an initial temperature of 190°F to cool to a temperature of 80°F if the temperature of the surrounding air is 70°F and $k = 20$ (for time measured in minutes).

48. **Ice Tea Temperature** A freshly brewed pitcher of tea, currently at a temperature of 175°F, is placed in a refrigerator that has a constant temperature of 40°F. Find the time required for the tea to reach a temperature of 45°F given that $k = 50$ (for time measured in minutes).

49. **Body Temperature** Inspector Magill is called to the scene of a murder. The victim is lying in a room that has a constant temperature of 75°F. The victim's body temperature is currently 81.2°F, down from the normal body temperature of 98.6°F. After considering the height and weight of the victim, Magill arrives at a value of $k = 10$ (for time measured in hours). How long has the victim been dead?

50. **Body Temperature Revisited** After further consideration (see previous problem), Inspector Magill decides that he is not sufficiently confident in his $k$ value to fix the time of death. To determine $k$, he notes that the victim's body temperature at 3:00 p.m. Wednesday is 81.2°F and that by 4:00 a.m. Thursday it has dropped to 77.0°F. Assuming that the body was kept in a 75°F room, what value does Magill find for $k$, and what was the time of death?

**EXERCISES 51–55** □ *The following exercises deal with the loudness of sound, which is measured using* decibels. *The faintest sound perceptible to most individuals is assigned an intensity of $I_0$, and a sound of intensity I has a decibel rating of*

$$D = 10 \cdot \log \frac{I}{I_0}$$

*Find the decibel rating for the given sound.*

51. **Whisper** A whisper with $I = 110 I_0$

52. **Conversation** Quiet conversation with $I = 9000 I_0$

53. **Circular Saw** A circular saw with intensity 10 billion times that of $I_0$

54. **Auto Horn** An auto horn with intensity 100 billion times that of $I_0$

**EXERCISES 55–58** □ *The following exercises deal with the acidity of an aqueous solution, which is dependent upon the solution's hydrogen ion concentration. Since these concentrations may be very small, it is convenient to measure acidity using the formula*

$$pH = -\log[H^+]$$

*where $[H^+]$ is the hydrogen ion concentration in moles per liter. In general, the smaller the pH is, the more acidic the solution is. Find the pH for the given solution.*

55. Orange juice with $[H^+] = 2.8 \times 10^{-4}$

56. Milk with $[H^+] = 3.97 \times 10^{-7}$

57. Beer with $[H^+] = 3.16 \times 10^{-5}$

58. Lemon juice with $[H^+] = 6.3 \times 10^{-3}$

59. **Mile-thick Paper** How many times must a 0.003 inch thick piece of paper be folded to obtain a thickness of at least one mile?

60. **Doubling Your Money** Suppose that you have invested $1 into First Fantasy Bank's Double-Your-Money Account, in which your balance doubles every year. Use the graph of $y = 2^x$ to estimate the number of years required to obtain a balance of $1,000,000,000 (1 billion dollars). Then express your answer in terms of logarithms.

61. **Typing Speed** A learning model suggests that the time in weeks that it will take to learn to type $w$ words per minute (wpm) is given by

$$t = 5 \ln\left(\frac{100}{100 - w}\right)$$

a. How many weeks will it take to learn to type 60 wpm? 80 wpm?

b. What happens to the time as $w$ approaches 100? According to this model, can 100 wpm be reached? Explain.

c. Use a graphics calculator to help you sketch the graph of $t$ as a function of $w$. What is the domain of this function?

62. *Sales Growth*  A model for the sales of a new product suggests that the time in months needed for the sales to reach $S$ units is given by

$$t = 4 \ln\left(\frac{50{,}000}{50{,}000 - S}\right)$$

a. How many years will it take to reach sales of 15,000 units? 40,000 units?

b. What happens to $t$ as $S$ approaches 50,000? According to this model, can 50,000 units be sold? Explain.

c. Use a graphics calculator to help you sketch the graph of $t$ as a function of $S$. What is the domain of this function?

---

### ■ *Projects for Enrichment*

63. *Deriving the Change-of-Base Formula*  We have seen that logarithms to any base can be computed using common or natural logarithms and the change-of-base formula. In this project we will show how to derive the change-of-base formula. We first consider changing $\log_2 25$ from base 2 to base 10. We set $x = \log_2 25$ and proceed as follows.

$$x = \log_2 25$$

$$2^x = 25 \qquad \text{Converting to exponents}$$

$$(10^{\log 2})^x = 25 \qquad \text{Writing 2 as } 10^{\log 2}$$

$$10^{x \log 2} = 25 \qquad \text{Using the exponential identity } (a^r)^s = a^{rs}$$

$$x \log 2 = \log 25 \qquad \text{Converting to logarithms}$$

$$x = \frac{\log 25}{\log 2}$$

a. Follow the steps given above to show more generally that if $x = \log_a u$, then $x = \dfrac{\log u}{\log a}$.

b. Show that if $x = \log_2 25$, then $x = \dfrac{\ln 25}{\ln 2}$.

c. Show that if $x = \log_a u$, then $x = \dfrac{\ln u}{\ln a}$.

d. Show that if $x = \log_a u$, and $b$ is any other base with $b > 0$ and $b \neq 1$, then $x = \dfrac{\log_b u}{\log_b a}$.

64. *The Leaning Tower of Pizza*  It can be shown that Domino's Pizza boxes can be stacked on a table with each box overhanging the box below it so that the top box extends as far as you please past the edge of the table—without toppling! Unfortunately, producing a large overhang requires a tremendous num-

ber of boxes. It can be shown that the greatest possible overhang with $n$ $(n > 1)$ 12-inch boxes is given by

$$6 \cdot \left(1 + \frac{1}{2} + \frac{1}{3} + \frac{1}{4} + \frac{1}{5} + \cdots + \frac{1}{n-1}\right) \text{ inches}$$

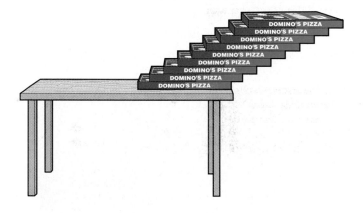

a. Compute the greatest overhang possible with 6 pizza boxes.

b. It can be shown that

$$1 + \frac{1}{2} + \frac{1}{3} + \frac{1}{4} + \frac{1}{5} + \cdots + \frac{1}{n-1} - \ln(n-1) \approx \epsilon$$

where $\epsilon$ is a constant known as **Euler's number**. Estimate Euler's number by computing

$$1 + \frac{1}{2} + \frac{1}{3} + \frac{1}{4} + \frac{1}{5} + \cdots + \frac{1}{n-1} - \ln(n-1)$$

for $n = 10$.

c. Find an approximation for the maximum overhang possible with $n$ boxes that involves Euler's number and natural logarithms. (Hint: If

$$S = 1 + \frac{1}{2} + \frac{1}{3} + \frac{1}{4} + \frac{1}{5} + \cdots + \frac{1}{n-1}$$

and $S - \ln(n-1) \approx \epsilon$, then what can be said of $S$?)

d. Using the approximation for the maximum possible overhang derived in part (c), compute the number of pizza boxes required to obtain an overhang of 10 feet.

e. Suppose that a pizza box is 2 inches thick. How high would the stack of pizza boxes in part (d) be? (Express your answer in light-years.)

f. Explain how a very narrow river could be crossed using only a parachute and Domino's Pizza boxes.

## ◄⊩ *Questions for Discussion or Essay*

65. John Napier of Scotland invented logarithms in the sixteenth century as a labor-saving device for computations. What sorts of activities in sixteenth-century Europe would have been facilitated by logarithms? In other words, who in sixteenth-century Europe would have had a need for logarithms?

66. Although Napier is generally credited with the invention of logarithms, the Swiss watchmaker Burgi discovered logarithms independently at about the same time. The mathematical literature is filled with such instances. For example, Newton and Leibniz invented calculus independently at about the same time. Are such occurrences simply coincidences? What explains the tendency for mathematicians, working independently, to make major discoveries simultaneously?

67. In Exercises 55–58 we considered the acidity of a solution as measured by the pH level. If normal rain is considered to have

High levels of sulfur in industrial emissions are a primary cause of acid rain.

a pH of approximately 5.6 and a recent rainfall is determined to have a hydrogen ion concentration of $6.3 \times 10^{-4}$, is the recent rain more or less acidic than normal rain? Which has the larger hydrogen ion concentration? If the primary causes of acid rain are sulfur dioxide and nitrogen oxides, do these compounds increase or decrease the hydrogen ion concentration of rain? In general, describe the effect on the pH level if the hydrogen ion concentration increases or decreases.

68. In this section we saw how a calculator could be used to compute the logarithm to any base of any number. Prior to the advent of the hand-held calculator, such tasks were typically accomplished using a process that required extensive logarithm tables (see Exercise 60 of Section 3 for an example of such a table). This is just one example of how calculators have changed the way in which mathematics is taught. What are some other changes that have occurred? (A discussion with your parents or an older student may help with this question.) What are some changes that you think have yet to occur, but are inevitable? Give some examples of basic skills that you think should never be replaced by a calculator.

69. In Exercises 49 and 50, Inspector Magill applied Newton's Law of Cooling to estimate the time of death. What characteristics of a body (in addition to height and weight) would influence the cooling rate? Does Newton's Law of Cooling take into account all means by which heat is transferred? If so, of what relevance is the wind chill factor, and how do you explain microwave ovens?

# SECTION 3

## LOGARITHMIC IDENTITIES AND EQUATIONS

■ How powerful would the ''bombquake'' be that would result from simultaneously detonating all of the world's nuclear warheads?

■ How can multiplication and division be performed by sliding pieces of paper?

■ How many jamboxes would it take to cause a fatal musical overdose?

■ If the entire population of China simultaneously jumped a foot off the ground, how strong would the resulting earthquake be?

## LOGARITHMIC IDENTITIES

Because of the correspondence between exponential functions and logarithmic functions, every identity involving exponential functions corresponds to an identity involving logarithmic functions. For example, suppose we begin with the exponential identity

$$a^b a^c = a^{b+c}$$

If we let $u = a^b$ and $v = a^c$, so that $b = \log_a u$ and $c = \log_a v$, then we have

$$uv = a^{b+c}$$

The corresponding logarithmic equation is

$$\log_a (uv) = b + c$$

Thus, we have the logarithmic identity

$$\log_a (uv) = \log_a u + \log_a v$$

Similarly, the exponential identities $a^b/a^c = a^{b-c}$ and $(a^b)^n = a^{bn}$ yield the logarithmic identities $\log_a(u/v) = \log_a u - \log_a v$ and $\log_a(u^n) = n \log_a u$, respectively.

The following is a summary of the logarithmic identities. Note that common and natural logarithms satisfy these identities as well.

### *Logarithmic identities*

Let $a$, $u$, and $v$ be positive real numbers with $a \neq 1$, and let $n$ be any real number.

| Arbitrary logarithms | Common logarithms | Natural logarithms |
|---|---|---|
| 1. $\log_a(uv) = \log_a u + \log_a v$ | $\log(uv) = \log u + \log v$ | $\ln(uv) = \ln u + \ln v$ |
| 2. $\log_a\left(\dfrac{u}{v}\right) = \log_a u - \log_a v$ | $\log\left(\dfrac{u}{v}\right) = \log u - \log v$ | $\ln\left(\dfrac{u}{v}\right) = \ln u - \ln v$ |
| 3. $\log_a(u^n) = n \log_a u$ | $\log(u^n) = n \log u$ | $\ln(u^n) = n \ln u$ |

Logarithmic identities can be useful for simplifying complex expressions involving logarithms. As we will see shortly, the simplification of complicated expressions involving logarithms is often the first step in solving logarithmic equations.

**EXAMPLE 1** | *Simplifying with logarithmic identities*

Express $3 \log_a(x^2) + 4 \log_a x$ as a single logarithm.

**SOLUTION**

$$3 \log_a(x^2) + 4 \log_a x = \log_a(x^2)^3 + \log_a x^4 \qquad \text{Identity 3}$$
$$= \log_a(x^6 \cdot x^4) \qquad \text{Identity 1}$$
$$= \log_a(x^{10})$$

■

**EXAMPLE 2** | *Simplifying with logarithmic identities*

Express $2 \ln x - \ln y + 6 \ln z$ as a single logarithm.

**SOLUTION**

$$2 \ln x - \ln y + 6 \ln z = (\ln x^2 - \ln y) + \ln z^6 \qquad \text{Identity 3}$$
$$= \ln \frac{x^2}{y} + \ln z^6 \qquad \text{Identity 2}$$
$$= \ln \frac{x^2 z^6}{y} \qquad \text{Identity 1}$$

■

**EXAMPLE 3** | *Expanding with logarithmic identities*

Expand $\log\left(\dfrac{x^3\sqrt{y}}{z^4}\right)$ as a sum, difference, or multiple of logarithms.

**SOLUTION**

$$\log\left(\frac{x^3\sqrt{y}}{z^4}\right) = \log(x^3\sqrt{y}) - \log z^4 \qquad \text{Identity 2}$$
$$= \log x^3 + \log y^{1/2} - \log z^4 \qquad \text{Identity 1}$$
$$= 3 \log x + \frac{1}{2} \log y - 4 \log z \qquad \text{Identity 3}$$

■

**EXAMPLE 4** | *Expanding with logarithmic identities*

Expand $\ln\left(\dfrac{3y^2}{x^5}\right)^3$ as a sum, difference, or multiple of logarithms.

**SOLUTION**

$$\ln\left(\frac{3y^2}{x^5}\right)^3 = 3 \ln\left(\frac{3y^2}{x^5}\right) \qquad \text{Identity 3}$$
$$= 3(\ln 3y^2 - \ln x^5) \qquad \text{Identity 2}$$
$$= 3(\ln 3 + \ln y^2 - \ln x^5) \qquad \text{Identity 1}$$
$$= 3(\ln 3 + 2 \ln y - 5 \ln x) \qquad \text{Identity 3}$$
$$= 3 \ln 3 + 6 \ln y - 15 \ln x$$

■

## LOGARITHMIC EQUATIONS

Equations involving one or more logarithmic expressions are called **logarithmic equations**. Many logarithmic equations can be solved by converting to exponential equations. For example, the equation

$$\ln x = 2$$

is equivalent to

$$x = e^2$$

This solution could also be found by exponentiating both sides of the original equation, as follows.

$$\ln x = 2$$

$$e^{\ln x} = e^2 \qquad \text{Exponentiating both sides}$$

$$x = e^2 \qquad \text{Using the property } a^{\log_a x} = x$$

In general, any equation of the form $\log_a \square = c$ can be solved for $\square$, either by converting to exponents or by exponentiating both sides.

**EXAMPLE 5**    *Solving a logarithmic equation*

Solve $\log_2(x + 1) = 5$.

**SOLUTION**    We convert from the base 2 logarithmic form to the base 2 exponential form.

$$\log_2(x + 1) = 5$$

$$x + 1 = 2^5$$

$$x = 32 - 1 = 31$$

Another useful technique for solving logarithmic equations is to apply the one-to-one property of logarithms stated below.

**The one-to-one property of logarithms**

For all positive real numbers $u$ and $v$, if $\log_a u = \log_a v$, then $u = v$.

**EXAMPLE 6**    *Using the one-to-one property of logarithms to solve an equation*

Solve $\log(x^2 + 2x - 5) = \log(x + 1)$.

**SOLUTION**

$$\log(x^2 + 2x - 5) = \log(x + 1)$$

$$x^2 + 2x - 5 = x + 1 \qquad \text{Using the one-to-one property of logarithms}$$

$$x^2 + x - 6 = 0$$

$$(x - 2)(x + 3) = 0 \qquad \text{Factoring}$$

$$x = 2, \ x = -3$$

**CHECK**                 $x = 2$                                            $x = -3$

| $\log(x^2 + 2x - 5)$ | $\log(x + 1)$ | $\log(x^2 + 2x - 5)$ | $\log(x + 1)$ |
| $\log[(2)^2 + 2(2) - 5]$ | $\log(2 + 1)$ | $\log[(-3)^2 + 2(-3) - 5]$ | $\log[(-3) + 1]$ |
| $\log(4 + 4 - 5)$ | $\log 3$ | $\log(9 - 6 - 5)$ | $\log(-2)$ |
| $\log 3$ | ✓ | | Not defined |

Thus, the only solution is $x = 2$.

---

> **WARNING**   Since logarithms are only defined for positive real numbers, it is important to check all solutions to logarithmic equations by substituting them into the original equation.

Some logarithmic equations must be algebraically rearranged using logarithmic identities before they can be solved, as illustrated in the following example.

**EXAMPLE 7**   *Solving an equation involving natural logarithms*

Solve $\ln 2 + 3 \ln(x - 1) = 2 \ln 4$.

**SOLUTION**   We begin by expressing each side in the form $\ln \square$.

$$\ln 2 + 3 \ln(x - 1) = 2 \ln 4$$
$$\ln 2 + \ln(x - 1)^3 = \ln 4^2 \qquad \text{Using } \ln u^n = n \ln u$$
$$\ln[2(x - 1)^3] = \ln 16 \qquad \text{Using } \ln u + \ln v = \ln uv$$
$$2(x - 1)^3 = 16 \qquad \text{Using the one-to-one property of logarithms}$$
$$(x - 1)^3 = 8$$
$$x - 1 = 2 \qquad \text{Taking the cube root of both sides}$$
$$x = 3$$

**CHECK**                 $x = 3$

| $\ln 2 + 3 \ln(x - 1)$ | $2 \ln 4$ |
| $\ln 2 + 3 \ln(3 - 1)$ | $\ln 4^2$ |
| $\ln 2 + 3 \ln 2$ | $\ln 16$ |
| $4 \ln 2$ | |
| $\ln 2^4$ | |
| $\ln 16$ | ✓ |

**EXAMPLE 8**    *Comparing sound intensity*

The decibel rating $D$ of a sound with intensity $I$ is given by

$$D = 10 \cdot \log \frac{I}{I_0}$$

where $I_0$ is the intensity of the faintest sound perceptible to the human ear. How much more intense is the 140-decibel sound of a jet taking off than a 120-decibel jack hammer sound?

**SOLUTION**    For comparison purposes, we set $I_0 = 1$. The relative intensity of the jet at take-off can be found as follows.

$$140 = 10 \log I$$

$$14 = \log I$$

$$I = 10^{14} \qquad \text{Converting to exponents}$$

For the jack hammer, we have

$$120 = 10 \log I$$

$$12 = \log I$$

$$I = 10^{12} \qquad \text{Converting to exponents}$$

Now $10^{14} = 100 \cdot 10^{12}$, and so the jet is 100 times more intense than the jack hammer.

**EXAMPLE 9**    *The nuclear earthquake*

According to the mathematician John Paulos, the TNT equivalent of all of the nuclear warheads on Earth is 25,000 megatons. The energy $E$ (in ergs) of an earthquake is related to the Richter scale reading $R$ by the equation $\log E = 11.4 + 1.5R$.

**a.** If all of the warheads were simultaneously detonated, how powerful would the resulting "bombquake" be? (A megaton is 1 million tons, and a ton of TNT carries about $3.4 \times 10^{16}$ ergs of energy.)

**b.** How many megatons of TNT would it take to create an earthquake of magnitude 7 on the Richter scale?

**SOLUTION**

**a.** We first compute the energy involved in such an explosion.

$$E = (25{,}000 \text{ megatons of TNT}) \times \frac{10^6 \text{ tons of TNT}}{1 \text{ megaton of TNT}} \times \frac{3.4 \times 10^{16} \text{ ergs}}{1 \text{ ton of TNT}}$$

$$= 8.5 \times 10^{26} \text{ ergs}$$

Using the equation $\log E = 11.4 + 1.5R$, we have

$$\log(8.5 \times 10^{26}) = 11.4 + 1.5R$$

$$26.929 = 11.4 + 1.5R$$

$$1.5R = 15.529$$

$$R = 10.353$$

A loaded ICBM missile is an ominous reminder of the nuclear threat.

Thus, the world's supply of nuclear warheads contains an energy equal to that of an earthquake of magnitude 10.353 on the Richter scale! This is greater than that of any recorded earthquake.

**b.** We set $R = 7$ and solve for $E$.

$$\log E = 11.4 + 1.5R$$

$$\log E = 11.4 + 1.5 \cdot 7$$

$$\log E = 21.9$$

$$E = 10^{21.9} \approx 7.943 \times 10^{21} \text{ ergs}$$

Converting to tons of TNT, we obtain

$$7.943 \times 10^{21} \text{ ergs} \times \frac{1 \text{ ton of TNT}}{3.4 \times 10^{16} \text{ ergs}} \times \frac{1 \text{ megaton of TNT}}{10^6 \text{ tons of TNT}} \approx$$

$$0.234 \text{ megaton of TNT}$$

## APPROXIMATING SOLUTIONS OF LOGARITHMIC EQUATIONS

We have seen many times in the past that approximate solutions of equations can be found by first writing the equation in the form $f(x) = 0$, then graphing the function $f$ on a graphics calculator, and finally estimating the zeros with the trace feature. However, if an equation involves nonpolynomial terms (such as $\ln x$ or $\log x$), it can be difficult to tell how many solutions the equation will have. Because of this, and because of our knowledge of polynomials, it is often helpful to first put the equation in the form $g(x) = p(x)$, where $p(x)$ is a polynomial, and then plot the graphs of $g$ and $p$ to see how many times they intersect. Solution values can be approximated either by zooming in on the intersection points of $g$ and $p$, or by finding the zeros of $f$.

**EXAMPLE 10**     *Determining the number of solutions of a logarithmic equation*

Determine the number of solutions of the equation $\frac{1}{18}x^3 + 4x + 1 = x^2 + \ln x$.

**SOLUTION**     We first regroup terms so that all the polynomial terms are on one side of the equation.

$$\frac{1}{18}x^3 + 4x + 1 = x^2 + \ln x$$

$$\frac{1}{18}x^3 - x^2 + 4x + 1 = \ln x$$

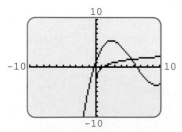

**FIGURE 12**

Next we plot the functions $p(x) = \frac{1}{18}x^3 - x^2 + 4x + 1$ and $g(x) = \ln x$ as shown in Figure 12. One intersection point can be seen near the point $(6, 2)$. Since $p(x)$ is a third-degree polynomial with a positive leading coefficient, we know that its graph must turn around and start increasing to the right of the viewing rectangle. Thus, a second intersection point is certain. With a slightly modified viewing rectangle, we see in Figure 13 that there is indeed a second intersection point. Since $p$ can have no other turning points, there can be no other intersection points. Thus, there are exactly two solutions. We could approximate these solutions either by zooming in on the points of intersection, or by graphing $y = \frac{1}{18}x^3 - x^2 + 4x + 1 - \ln x$ and estimating its $x$-intercepts.

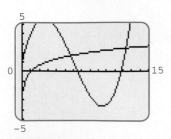

**FIGURE 13**

## EXERCISES 3

**EXERCISES 1–8** □ *Use the logarithmic identities to expand each expression, rewriting it as a sum, difference or multiple of logarithms.*

1. $\ln(x^2 y)$

2. $\log_2(xy\sqrt{z})$

3. $\log \dfrac{x^3 y^4}{z^{-3}}$

4. $\log_4\left(\dfrac{15\sqrt{q}}{\sqrt{r}}\right)$

5. $\log_\pi\left(\dfrac{e}{7x^3}\right)$

6. $\ln\left(\dfrac{xy^2}{\sqrt{z}}\right)^3$

7. $\log \sqrt{\dfrac{\sqrt{x}\,y}{z^2}}$

8. $\log_4 \sqrt{x^3 + y^3}$

**EXERCISES 9–18** □ *Use the logarithmic identities to write each expression as a single logarithm.*

9. $\log x - 2 \log y$

10. $2 \ln x + \dfrac{1}{2} \ln y$

11. $3 \ln x + 4 \ln y - \ln z$

12. $\dfrac{1}{2} \log(2x - y) + \dfrac{1}{2} \log(x + 2y)$

13. $\dfrac{1}{3} \ln(x - z) - \dfrac{2}{3} \ln(x + z)$

14. $4[\log_5(x^2 - y^2) - \log_5(x + y)]$

15. $2 \log 4 - \log 6 + 3 \log 2$

16. $\log_3 8 - 4 \log_3(x^2)$

17. $2 \ln(x + 3) - \ln(x^2 + 5x + 6)$

18. $\log(x + y) - \dfrac{1}{2} \log(x^2 + 2xy + y^2)$

**EXERCISES 19–44** □ *Solve the logarithmic equation.*

19. $\dfrac{1}{3} \log_3 x = 2$

20. $2 \log_4 x = 32$

21. $\log_x 4 = 2$

22. $\log_x \dfrac{1}{9} = 2$

23. $3 \log_5 x = 2$

24. $2 \log_2 x = 3$

25. $\log_{1/3}(2x + 1)^{1/2} = -2$

26. $\log_{1/2}(1 - 3x)^{1/3} = -1$

27. $\log_3(27^x) = -1$

28. $\log_5(5^x) = 2$

29. $2 \log_3(x - 1) + \log_3 9 = 2$

30. $\log(2x - 5) = \log x$

31. $\ln(x + 7) - \ln(x + 2) = \ln(x + 1)$

32. $\log_2(3 - x) + \log_2(2 - x) = \log_2(1 - x)$

33. $\log(2x^2 + 11x + 5) - \log(2x + 1) = 2$

34. $\log_a(x + 1) + \log_a(x - 2) = \log_a(x + 6)$

35. $\log_2(27x^3 - 1) - \log_2(9x^2 + 3x + 1) = 1$
    (Hint: Factor $27x^3 - 1$)

36. $\log_{1/4} x + \log_{1/4}(3x - 5) = -1$

37. $\log_a(x + 7) = \log_a(x + 1)$

38. $\log_5(x^7) = 7 \log_5 x$

39. $\log \sqrt{x^2 + 1999} = 3$

40. $\log \sqrt[3]{x^2 - 15x} = \dfrac{2}{3}$

41. $\ln x^2 = 2 \ln x$

42. $(\ln x)^2 = 2 \ln x$

43. $\ln(\ln x) = 1$

44. $\log[\log(x + 1)] = \log 2$

**EXERCISES 45–48** □ *Approximate the solutions of the given equation to the nearest hundredth.*

45. $x - 2 = \ln x$

46. $\ln(x + 1) = x^2 - 1$

47. $0.288x^2 = \ln(1 - x) + 0.002x^4 + 1$

48. $1.5x^2 = 0.0625x^3 + 5x + \ln x$

49. Graph the following three functions on the same coordinate system.

$$f(x) = \log x$$
$$g(x) = \log(x \cdot 10)$$
$$h(x) = \log(x \cdot 0.1)$$

What relationship do you see between the graphs of these functions? What property of logarithms explains this relationship?

50. Graph $y = \ln(e^3 x)$. Now find a number $k$ so that the graph of $y = k + \ln x$ is the same. What properties of logarithms explain this?

■ *Applications*

**EXERCISES 51–54** □ *These exercises deal with the decibel rating D of a sound with intensity I as given by $D = 10 \cdot \log(I/I_0)$, where $I_0$ is the faintest sound perceptible to the human ear.*

51. *Hearing Loss* The threshold of pain for most individuals is 120 decibels, while permanent hearing damage can be caused by even brief exposure to a sound rated at 160 decibels. How

many times more intense is a sound of 160 decibels than one of 120 decibels?

52. *Rock Concert* According to the *1990 Guinness Book of World Records*, the sound level at the mixing tower during Iron Maiden's set at the Monsters of Rock Festival reached a record 124 decibels. The loudest shout recorded by the Guinness Book was

one of 113 decibels. How many times more intense was the sound produced by Iron Maiden than that of the champion shouter?

Iron Maiden in concert.

53. **Mass Choir**  Suppose that 1000 people sing a note at precisely the same volume and pitch at exactly the same moment. Assuming that the collective sound is 1000 times more intense than that of an individual singer, how much louder is the sound of the group than that of an individual?

54. **Fatal Jambox**  It is said that a 200-decibel sound can be deadly. Suppose that a certain jambox is capable of producing 100 decibels. How many such jamboxes would it take to sustain a fatal musical overdose?

**EXERCISES 55–56** □ *These exercises deal with the Richter measure R for an earthquake of intensity I as given by $R = \log(I/I_0)$, where $I_0$ is the energy level of a certain very small earthquake.*

55. **Comparing Earthquakes**  The earthquake that shook Armenia on December 7, 1988, registered 6.8 on the Richter scale. The San Francisco earthquake of April 18, 1906, registered 8.3 on the Richter scale. How many times more powerful was the San Francisco earthquake than that of Armenia?

56. **Jumping Earthquake**  It has been hypothesized that if each Chinese man, woman, and child were to jump at precisely the same instant, an earthquake of magnitude 4.3 on the Richter scale would result. Assuming that the ground movement is proportional to the number of people jumping and that there are 1.1 billion people in China and 5.5 billion people in the world, of what magnitude would the earthquake be that would result if all the people on Earth were to jump simultaneously?

**EXERCISES 57–58** □ *These exercises deal with the equation $\log E = 11.4 + 1.5R$ which relates the energy E (in ergs) of an earthquake to the Richter scale reading R.*

57. **TNT Earthquake**  How many tons of TNT would it take to equal the energy output of an earthquake registering 5 on the Richter scale?

58. **Automobile Earthquake**  The total number of miles driven in the United States in 1991 was approximately $2.1 \times 10^{12}$. If we assume that each car was being driven at 40 miles per hour and had a 200-horsepower engine, then each automobile used about $1.34 \times 10^{14}$ ergs per mile. What magnitude earthquake has the same energy output as the total energy used by automobiles in 1991?

## ■ *Projects for Enrichment*

59. **Constructing a Slide Rule**  Slide rules were widely used before inexpensive electronic calculators became available. In this project we will construct a simple slide rule and use it to perform multiplications.

Begin by copying Figure 14 and separating it into two pieces

by cutting along the dashed line. Now secure the lower piece so that it will not move.

To multiply two numbers between 1 and 10 whose product is less than 10, say 2.2 and 3.1, simply slide the top piece over the bottom so that the 1 at the left of the top piece is directly

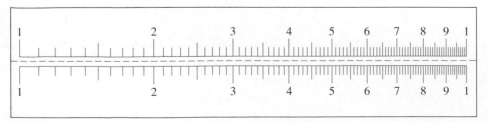

**FIGURE 14**

above 2.2 on the bottom piece. Next find 3.1 on the top piece and read off the number directly below it ($\approx 6.8$) on the bottom piece. If the numbers are such that their product exceeds 10, say 4.2 and 5.3, place the 1 at the *right-hand end* of the top piece directly above 4.2, then find 5.3 on the top piece and read off the number directly below it ($\approx 2.3$). The answer is then obtained by multiplying 2.3 by 10 to give 23.

In general, when using slide rules, we must keep track of powers of 10. For this reason, it is helpful to first convert the numbers that we are multiplying to scientific notation. For example, to compute $120 \times 4100$, we first rewrite the product as $(1.2 \times 10^2)(4.1 \times 10^3)$, then multiply 1.2 and 4.1 with the slide rule to obtain approximately 4.9, next multiply $10^2$ and $10^3$ to obtain $10^5$, and finally write our answer as $4.9 \times 10^5$.

a. Compute the following products using your makeshift slide rule.

  i. $2.5 \times 6.8$   ii. $0.0024 \times 39$   iii. $350 \times 180$

A slide rule like that depicted in Figure 14 could be formed in the following way. First, take a piece of ordinary graph paper, and mark a number line from 0 to 1. Next, compute the common logarithms of the integers from 1 to 10. Then for each integer, make a tick mark on the number line at the position corresponding to its logarithm, and label the tick mark with the corresponding integer. For example, the number 2 was written at the position 0.301, since $\log 2 \approx 0.301$. For increased accuracy, smaller tick marks could be made at locations corresponding to the logarithms of the multiples of 0.1—that is, the numbers 1.1, 1.2, ..., 9.9. Thus, the location of a number on the slide rule is determined by its logarithm.

b. Find the ratio of the distance between 1 and 2 to the distance between 2 and 4 on the slide rule.

c. What number on the slide rule is the same distance from 9 as 1 is from 2?

d. Explain the process by which numbers are multiplied with a slide rule.

60. *Logarithm Tables* John Napier developed logarithms as a labor-saving device for arithmetic calculations. Today, although logarithmic functions are extremely important in modeling real world phenomena, logarithms are rarely used as a computational aid for performing arithmetic since inexpensive electronic calculators are readily available. In this project we will investigate the methods by which our predecessors performed lengthy arithmetic computations.

**I.** *Tables*

Until fairly recently, logarithms were calculated by hand (using complicated formulas from calculus), and then tabulated in lengthy logarithm tables. Table 2 is an abbreviated version of a common (base 10) logarithm table.

To find the logarithm of a number between 1 and 10, say 3.9, we read off the entry in row 3 and column 9, namely 0.5911. Thus, $\log 3.9 \approx 0.5911$. Logarithms of numbers larger than 9.9 or smaller than 1.0 can be found by converting to scientific notation, and using the properties of logarithms. For example, we compute $\log 230$ as follows:

$$\begin{aligned} \log 230 &= \log(2.3 \times 10^2) \\ &= \log 2.3 + \log 10^2 \\ &= 0.3617 + 2 \\ &= 2.3617 \end{aligned}$$

a. Approximate each the following logarithms using Table 2.

  i. $\log 6700$   ii. $\log 0.0023$   iii. $\log 15,000,000$

To do calculations using logarithms, we must be able to compute **antilogarithms**. The antilogarithm of $x$ is simply the number $10^x$. If the number $x$ is between 0 and 1, then the

| TABLE 2 | N | 0 | 1 | 2 | 3 | 4 | 5 | 6 | 7 | 8 | 9 |
|---|---|---|---|---|---|---|---|---|---|---|---|
| *Base 10 logarithms* | **1** | .0000 | .0414 | .0792 | .1139 | .1461 | .1761 | .2041 | .2304 | .2553 | .2788 |
| | **2** | .3010 | .3222 | .3424 | .3617 | .3802 | .3979 | .4150 | .4314 | .4472 | .4624 |
| | **3** | .4771 | .4914 | .5051 | .5185 | .5315 | .5441 | .5563 | .5682 | .5798 | .5911 |
| | **4** | .6021 | .6128 | .6232 | .6335 | .6435 | .6532 | .6628 | .6721 | .6812 | .6902 |
| | **5** | .6990 | .7076 | .7160 | .7243 | .7324 | .7404 | .7482 | .7559 | .7634 | .7709 |
| | **6** | .7782 | .7853 | .7924 | .7993 | .8062 | .8129 | .8195 | .8261 | .8325 | .8388 |
| | **7** | .8451 | .8513 | .8573 | .8633 | .8692 | .8751 | .8808 | .8865 | .8921 | .8976 |
| | **8** | .9031 | .9085 | .9138 | .9191 | .9243 | .9294 | .9345 | .9395 | .9445 | .9494 |
| | **9** | .9542 | .9590 | .9683 | .9685 | .9731 | .9777 | .9823 | .9868 | .9912 | .9956 |

antilogarithm of $x$ ($10^x$) can be found by locating $x$ in Table 2, and then reading off the number whose logarithm is $x$. For example, to compute $10^{0.1139}$, we locate 0.1139 in the table, finding that 1.3 is the number whose logarithm is 0.1139. Thus, $10^{0.1139} = 1.3$. If the number $x$ does not appear in the table, then as an estimate, simply choose the nearest number in the table. For example, to compute $10^{0.52}$, we locate the entry in the table nearest to 0.52, namely 0.5185, and find that $10^{0.52} \approx 10^{0.5185} \approx 3.3$. Antilogarithms of numbers larger than 1 or smaller than 0 can be found using the properties of exponents. For example, we compute $10^{4.75}$ and $10^{-2.8}$ as follows:

$$10^{4.75} = 10^4 \cdot 10^{0.75} \qquad 10^{-2.8} = 10^{-3} \cdot 10^{0.2}$$
$$\approx 10^4 \cdot 10^{0.7482} \qquad \approx 10^{-3} \cdot 10^{0.2041}$$
$$\approx 10^4 \cdot 5.6 \qquad\qquad \approx 10^{-3} \cdot 1.6$$
$$= 56{,}000 \qquad\qquad\quad \approx 0.0016$$

b. Approximate each of the following antilogarithms using the logarithms in Table 2.

   i. $10^{3.89}$     ii. $10^{-4.05}$

**II.** *Multiplication and division*

The product or quotient of any two numbers can be estimated using the properties of logarithms and a logarithm table. For example, the identity $x \cdot y = 10^{\log(x \cdot y)} = 10^{\log x + \log y}$ can be used to compute $348 \times 5700$ as shown.

$$348 \times 5700 = 10^{\log(348 \times 5700)}$$
$$= 10^{\log 348 + \log 5700}$$
$$\approx 10^{2.5441 + 3.7559} \quad \text{Using the table to}$$
$$\qquad\qquad\qquad\qquad \text{estimate log 348 and log 5700}$$
$$= 10^{6.3}$$
$$= 10^{0.3} 10^6$$
$$\approx 2 \cdot 10^6 \quad \text{Using the table to estimate}$$
$$\qquad\qquad\qquad \text{the antilogarithm of 0.3}$$
$$= 2{,}000{,}000$$

c. Perform each of the following operations using only addition, subtraction and Table 2.

   i. $79{,}000 \times 62.5$     ii. $0.0003 \times 426.7$

d. Show that for any two positive numbers $x$ and $y$, the quotient $x/y$ is equal to the antilogarithm of the difference in their logarithms. Use this fact to compute $426/0.032$.

e. Explain how one could evaluate $3200^{2.3}$ using only a logarithm table, addition, and subtraction. (*Hint:* Start with the fact that $3200^{2.3} = 10^{\log(3200^{2.3})}$, and then use a logarithmic identity.)

---

### ▟ *Questions for Discussion or Essay*

61. When studying logarithms, many students struggle because there are so many seemingly plausible identities that are simply not true. For example, it is not true that for all $x$ and $y$, $\log_a x \cdot \log_a y = \log_a(x + y)$. What makes this nonrule so tempting to accept as true? What other nonrules involving logarithms can you develop? How would one go about showing that a nonrule (such as the one given above) is false?

62. In Exercise 56 we made the assumption that the ground movement arising from a number of people jumping at the same moment would be proportional to the number of people jumping. Is this a reasonable assumption? What other factors besides the number of people jumping would affect the resulting seismic disturbance?

63. As we have seen, the decibel rating of a sound with intensity $I$ is computed using the formula $D = 10 \log(I/I_0)$, where $I_0$ is the intensity of the faintest sound that is perceptible to the human ear. Sound intensity itself is measured in watts per square centimeter. For example, $I_0$ is by common agreement assigned the value $10^{-16}$ watts per square centimeter, and the intensity of a whisper is approximately $10^{-13}$ watts per square centimeter. Why do you think the decibel is preferred as a measure of the loudness of a sound, rather than the intensity? More generally, under what circumstances do you think logarithms would be useful in formulas that define measurement scales?

## SECTION 4

# EXPONENTIAL EQUATIONS AND APPLICATIONS

- Did Jesus wear the shroud of Turin?
- Which would you rather have, $1000 in an account paying 10% interest, or $2000 in an account paying 5% interest?
- If eliminating all sources of contamination drops the pollution level of a lake by 15% after 5 years, then how long will it take for the pollution level to drop by 50%?
- How could a coroner estimate the initial dose of a heart medication taken 24 hours before a fatal heart attack, even if later doses were given?

## EXPONENTIAL EQUATIONS

Equations such as $2^x = 2^3$, $3^x = 50$, and $2^{2x-1} = 3^x$ in which the variable appears in an exponent are called **exponential equations**. Some exponential equations can be solved using techniques we have already seen. For example, to solve $2^x = 2^3$, we use the fact that exponential functions are one-to-one to obtain $x = 3$. We can solve $3^x = 50$ by rewriting this exponential equation in its equivalent logarithmic form, $x = \log_3 50$.

Alternatively, the equation $3^x = 50$ could be solved by taking the logarithm (with any base) of both sides. Using common logarithms, we have:

$$3^x = 50$$

$$\log 3^x = \log 50$$

$$x \log 3 = \log 50$$

$$x = \frac{\log 50}{\log 3} \approx 3.56$$

Note that taking the logarithm of both sides has the effect of "bringing the variable out of the exponent." Since exponential equations are precisely those equations that have a variable in the exponent, taking the logarithm of both sides is a good way to begin solving many exponential equations. Moreover, by using either common or natural logarithms, we can obtain a decimal approximation with a calculator.

### EXAMPLE 1    *Solving an exponential equation*

Solve $7^{3x} = 15$.

**SOLUTION**

$$7^{3x} = 15$$

$$\log 7^{3x} = \log 15 \qquad \text{Taking the common logarithm of both sides}$$

$$3x \log 7 = \log 15 \qquad \text{Using the logarithmic identity } \log u^n = n \log u$$

$$(3 \log 7)x = \log 15 \qquad \text{Rearranging on the left side}$$

(continued)

$$x = \frac{\log 15}{3 \log 7} \qquad \text{Solving for } x$$

$$x \approx 0.464 \qquad \text{Using a calculator}$$

---

**EXAMPLE 2**    *Solving an exponential equation*

Solve $2^{2x-1} = 3^x$.

**SOLUTION**

$$2^{2x-1} = 3^x$$

$$\ln(2^{2x-1}) = \ln 3^x \qquad \text{Taking the natural logarithm of both sides}$$

$$(2x - 1)\ln 2 = x \ln 3 \qquad \text{Using the logarithmic identity } \ln u^n = n \ln u$$

$$(2 \ln 2)x - \ln 2 = (\ln 3)x \qquad \text{Multiplying out on the left side}$$

$$(2 \ln 2)x - (\ln 3)x = \ln 2 \qquad \begin{array}{l}\text{Rearranging so that all variable terms are on} \\ \text{one side}\end{array}$$

$$(2 \ln 2 - \ln 3)x = \ln 2 \qquad \text{Factoring out } x$$

$$x = \frac{\ln 2}{2 \ln 2 - \ln 3} \qquad \text{Solving for } x$$

$$\approx 2.409 \qquad \text{Approximating using a calculator}$$

---

When solving exponential equations such as that in Example 2, linear equations like $(2x - 1)\ln 2 = x \ln 3$ often arise. Such equations are easily solved if it is kept in mind that "ln 2" and "ln 3" are nothing more than real numbers like 5 or 7. In fact, the equation $(2x - 1)\ln 2 = x \ln 3$ is solved in exactly the same way as the equation $(2x - 1) \cdot 5 = x \cdot 7$ is solved: the parentheses are eliminated using the distributive property, all terms containing an $x$ are collected on one side, and then we solve for $x$. The following parallel solutions illustrate these similarities.

$$(2x - 1) \cdot 5 = x \cdot 7 \qquad\qquad (2x - 1) \cdot \ln 2 = x \cdot \ln 3$$

$$10x - 5 = 7x \qquad\qquad (2 \ln 2)x - \ln 2 = (\ln 3)x$$

$$10x - 7x = 5 \qquad\qquad (2 \ln 2)x - (\ln 3)x = \ln 2$$

$$3x = 5 \qquad\qquad (2 \ln 2 - \ln 3)x = \ln 2$$

$$x = \frac{5}{3} \qquad\qquad x = \frac{\ln 2}{2 \ln 2 - \ln 3}$$

**EXAMPLE 3**    *Solving an exponential equation*

Solve $e^{2x} - 3e^x - 4 = 0$.

**SOLUTION**    We first note that $e^{2x} = (e^x)^2$, and so the expression $e^{2x} - 3e^x - 4$ is quadratic in $e^x$.

$$e^{2x} - 3e^x - 4 = 0$$

$$(e^x)^2 - 3e^x - 4 = 0 \qquad \text{Writing in quadratic form}$$

$$(e^x - 4)(e^x + 1) = 0 \qquad \text{Factoring}$$

$$e^x - 4 = 0, \qquad e^x + 1 = 0 \qquad \text{Setting each factor to zero}$$
$$e^x = 4, \qquad e^x = -1$$
$$x = \ln 4, \qquad x = \ln(-1)$$

Since $\ln(-1)$ is undefined, the equation has one solution, namely $x = \ln 4$.

---

**EXAMPLE 4**  *Compound interest*

Suppose that \$1000 is deposited into an account paying 10% interest, and on the same day \$2000 is deposited into an account paying 5% interest. How long will it take for the balance in the 10% account to exceed the balance in the 5% account? Assume that interest is compounded annually.

**SOLUTION**  Let $t$ be the number of years required for the balances to be equal. Using the formula for compound interest, $B = P(1 + r)^t$, the balance in the 5% account is given by $B = 2000 \cdot 1.05^t$, whereas the balance in the 10% account is given by $B = 1000 \cdot 1.1^t$. Equating the balances, we have

$$\text{Balance on \$2000 at 5\%} = \text{balance on \$1000 at 10\%}$$

$$2000 \cdot 1.05^t = 1000 \cdot 1.1^t$$

$$2 \cdot 1.05^t = 1.1^t \qquad\qquad \text{Dividing both sides by 1000}$$

$$\log(2 \cdot 1.05^t) = \log(1.1^t) \qquad \begin{array}{l}\text{Taking the common logarithm} \\ \text{of both sides}\end{array}$$

$$\log 2 + \log 1.05^t = \log(1.1^t) \qquad \text{Using a logarithmic identity}$$

$$\log 2 + t \log 1.05 = t \log 1.1 \qquad \text{Using a logarithmic identity}$$

$$t \log 1.05 - t \log 1.1 = -\log 2 \qquad \text{Rearranging}$$

$$t(\log 1.05 - \log 1.1) = -\log 2 \qquad \text{Factoring out the variable}$$

$$t = \frac{-\log 2}{\log 1.05 - \log 1.1} \qquad \text{Solving for the variable}$$

$$\approx 14.9 \text{ years} \qquad \begin{array}{l}\text{Approximating using a} \\ \text{calculator}\end{array}$$

---

## EXPONENTIAL GROWTH AND DECAY

Exponential functions show up in many surprising places. In Section 1 we saw examples involving population growth, concentrations of drugs in the bloodstream, and compound interest, but those are only a few of the many applications where exponential functions are useful. In general, it can be shown that an exponential function of the form $Q(t) = Q_0 e^{kt}$ or $Q(t) = Q_0 e^{-kt}$ is involved whenever a quantity $Q$ increases or decreases at a rate proportional to the size of $Q$. Here $Q_0$ denotes the initial quantity (when $t = 0$) and $k$ is a positive constant. If the quantity $Q$ satisfies $Q(t) = Q_0 e^{kt}$, it is said to **grow exponentially**, whereas if it satisfies $Q(t) = Q_0 e^{-kt}$, it is said to **decay exponentially**.

**EXAMPLE 5**  *Population growth*

The population of Florida was 9.7 million in 1980 and 12.9 million in 1990. Assuming the population is growing exponentially, estimate the population of Florida in the year 2010.

Rapid population growth and urban expansion threaten the fragile ecology of the Florida everglades.

**SOLUTION:** Assuming exponential growth with an initial population of 9.7 million, we conclude that the population of Florida is modeled by a function of the form

$$P(t) = 9.7e^{kt}$$

where $t$ is the number of years after 1980 and $P(t)$ is the population in millions at time $t$. Since the population was 12.9 million in 1990, we can find $k$ as follows.

Population after 10 years $= 12.9$

$$P(10) = 12.9$$

$$9.7e^{k(10)} = 12.9$$

$$e^{10k} = \frac{12.9}{9.7}$$

$$\ln e^{10k} = \ln\left(\frac{12.9}{9.7}\right) \qquad \text{Taking the natural logarithm of both sides}$$

$$10k = \ln\left(\frac{12.9}{9.7}\right) \qquad \text{Using the logarithmic identity } \ln e^x = x$$

$$k = \frac{\ln\left(\dfrac{12.9}{9.7}\right)}{10} \approx 0.0285$$

Thus, we have $P(t) = 9.7e^{0.0285t}$. Since the year 2010 corresponds to $t = 30$, we obtain a population of

$$P(30) = 9.7e^{0.0285(30)} \approx 22.8 \text{ million}$$

■

Many radioactive substances decay at a rate that is proportional to the amount of the substance, and thus decay exponentially according to $Q(t) = Q_0 e^{-kt}$. Here $Q(t)$ denotes the amount of the substance at time $t$, $Q_0$ denotes the initial amount of the substance, and $k$ is a constant determined by the rate of decay. The **half-life** of a substance that decays exponentially in this way is defined to be the time required for half of the substance to decay.

**EXAMPLE 6**

### Carbon-14 dating

Carbon-14 (C-14) decays exponentially with a half-life of approximately 5715 years. If an artifact originally had 12 grams of C-14 and now has only 4 grams remaining, how old is the artifact?

**SOLUTION**  Let $Q(t)$ represent the quantity of C-14 remaining at time $t$. Since there were initially 12 grams of C-14 and the quantity is decaying exponentially, we may write $Q(t) = 12e^{-kt}$. We next find the constant $k$ by using the half-life of 5715 years. Since there were originally 12 grams, there would be only 6 grams remaining after 5715 years.

Quantity of C-14 remaining after 5715 years $= 6$

$$Q(5715) = 6$$

$$12e^{-5715k} = 6 \qquad \text{Setting } t = 5715 \text{ in } Q(t)$$

$$e^{-5715k} = \frac{1}{2} \qquad \text{Dividing by 12}$$

The Dead Sea Scrolls, consisting mostly of thousands of fragments such as this, have been carbon dated to around the time of Christ.

$$\ln(e^{-5715k}) = \ln \frac{1}{2}$$    Taking the natural logarithm of both sides

$$-5715k = \ln \frac{1}{2}$$    Using the logarithmic identity $\ln e^a = a$

$$k = \frac{\ln \frac{1}{2}}{-5715} \approx 0.000121$$    Solving for $k$

Thus, $Q(t) = 12e^{-0.000121t}$. Since there are currently 4 grams remaining, we have

Quantity of C-14 remaining after $t$ years $= 4$

$$Q(t) = 4$$

$$12e^{-0.000121t} = 4$$

$$e^{-0.000121t} = \frac{1}{3}$$    Dividing both sides by 12

$$\ln(e^{-0.000121t}) = \ln \frac{1}{3}$$    Taking the natural logarithm of both sides

$$-0.000121t = \ln \frac{1}{3}$$    Using the logarithmic identity $\ln e^a = a$

$$t = \frac{\ln \frac{1}{3}}{-0.000121} \approx 9079.44 \text{ years}$$    Solving for $t$

## CARBON-14 DATING AND THE SHROUD OF TURIN

The Shroud of Turin, a linen relic bearing the image of a man, first appeared in fourteenth-century France. Although the Roman Catholic Church took no official position on its authenticity, it was widely believed to be the burial shroud of Jesus. The origins of the shroud had been questioned from the very beginning, but there had been no way to disprove its authenticity. In fact, the Catholic Church allowed a team of investigators to conduct a scientific investigation on the shroud in 1978, and no conclusive evidence of fraud was produced, although one member of the team concluded that the image on the shroud was, in fact, a very thin watercolor.

In order to protect the possibly sacred shroud, the Church had not allowed **carbon-14 dating** to be performed. This test, if conducted using the technology available in 1978, would have caused the destruction of an unacceptably large portion of the shroud. By 1988, technological developments made it possible to conduct carbon-14 dating on the shroud with only minimal destruction. The results demonstrated that the shroud dated from the fourteenth century and hence could not have been the burial shroud of Jesus.

Carbon-14 is a radioactive isotope of carbon that is present in all living organisms. After death, the carbon-14 gradually decays to an isotope of nitrogen. As a consequence, the quantity of carbon-14 slowly decreases. By measuring the remaining carbon-14, scientists can determine how much time has passed since the organism died and can thus date pieces of wood, cloth, hides, bones, and other organic matter.

Since carbon-14 decays exponentially, we know that the amount remaining $t$ years after the death of the organism is given by $A(t) = A_0e^{-kt}$, where $A_0$ is the amount of carbon-14 prior to death. It can be shown (see Example 6) that since the half-life of carbon-14 is 5715 years, $k = 0.000121$. Thus, $A(t) = A_0e^{-0.000121t}$.

EXAMPLE 7     *Pollution in Lake Michigan*

Each year America's waterways are inundated with millions of pounds of toxic waste.

Under suitable conditions, and assuming all incoming pollution is stopped, the concentration of pollution in a lake decays exponentially over time. If it takes 5 years for the pollution concentration in Lake Michigan to decrease to 85% of its current level, how many years would it take for the pollution concentration to decrease to 50% of its current level?

**SOLUTION**     Let $Q(t) = Q_0e^{-kt}$ denote the concentration of pollution in Lake Michigan at time $t$. We must first determine the value of $k$. Since the concentration in 5 years will be 85% of the initial concentration $Q_0$, we obtain the following equation.

$$\text{Concentration after 5 years} = 85\% \text{ of } Q_0$$

$$Q(5) = 0.85Q_0$$

$$Q_0e^{-k(5)} = 0.85Q_0$$

Now we solve for $k$.

$$Q_0e^{-k(5)} = 0.85Q_0$$

$$e^{-5k} = 0.85 \qquad \text{Dividing both sides by } Q_0$$

$$\ln e^{-5k} = \ln 0.85 \qquad \text{Taking the natural logarithm of both sides}$$

$$-5k = \ln 0.85$$

$$k = \frac{\ln 0.85}{-5} \approx 0.0325$$

Thus, we have $Q(t) = Q_0e^{-0.0325t}$. To find how long it takes for the concentration to reach 50% of its present level, we proceed as follows.

$$\text{Concentration after } t \text{ years} = 50\% \text{ of } Q_0$$

$$Q_0e^{-0.0325t} = 0.50Q_0$$

$$e^{-0.0325t} = 0.5 \qquad \text{Dividing both sides by } Q_0$$

$$\ln e^{-0.0325t} = \ln 0.5 \qquad \text{Taking the natural logarithm of both sides}$$

$$-0.0325t = \ln 0.5$$

$$t = \frac{\ln 0.5}{-0.0325} \approx 21.3 \text{ years}$$

◼

APPROXIMATING
SOLUTIONS OF
EXPONENTIAL
EQUATIONS

There are many exponential equations that cannot be solved exactly using the methods discussed in this section. Consider, for example, the exponential equation $e^x = x + 2$. If we were to attempt to solve this equation by taking the natural logarithm of both sides, we would obtain the equation $x = \ln(x + 2)$. Unfortunately, this second equation is no easier to solve than the first. In fact, neither equation can be solved algebraically. However, we can at least approximate the solution(s) using a graphics calculator.

**EXAMPLE 8**    *Approximating the solutions of an exponential equation*

Use a graphics calculator to approximate the solutions of $e^x = x + 2$ to the nearest hundredth.

**SOLUTION**    We rewrite the equation $e^x = x + 2$ in the form $e^x - x - 2 = 0$, plot the graph of $f(x) = e^x - x - 2$, and approximate the zeros of $f$ using the trace feature. The graph of $f$ and the approximate zeros $x = -1.84$ and $x = 1.15$ are shown in Figures 15–17. We can be confident that there are no other solutions since the graphs of $y = e^x$ and $y = x + 2$ intersect only twice, as shown in Figure 18. Alternatively, we could have *begun* by graphing $y = x + 2$ and $y = e^x$, and then approximated the solutions by zooming in on the points of intersection and using the trace feature.

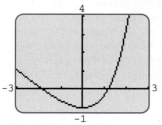

**FIGURE 15**

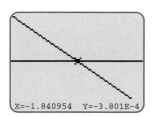

**FIGURE 16**

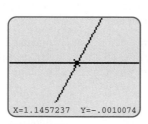

**FIGURE 17**

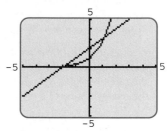

**FIGURE 18**

---

**EXERCISES 4**

**EXERCISES 1–20** □ *Solve the exponential equation.*

**1.** $2^x = 16$

**2.** $3^{-x} = \dfrac{1}{9}$

**3.** $25^{3x} = 125^{x+1}$

**4.** $3^{x^2+1} = 27$

**5.** $7^x = 10$

**6.** $5^x = 10$

**7.** $3^{-x} = 11$

**8.** $3^{x-1} = 2^x$

**9.** $\left(\dfrac{3}{5}\right)^x = 6^{2-x}$

**10.** $0.4^{2+x} = 1.5^{3x-1}$

**11.** $e^x = 2^{x+1}$

**12.** $5e^{-2.3x} = 4$

**13.** $e^{2x-1} = \pi^{2x}$

**14.** $e^{x/2} = 3^{x-4}$

**15.** $\dfrac{1}{2^{x+2}} = \dfrac{1}{8}$

**16.** $\left(\dfrac{2}{3}\right)^{-x+1} = \dfrac{81}{16}$

**17.** $e^{2x} - 5e^x = -6$

**18.** $4^x - 3 \cdot 2^x + 2 = 0$

**19.** $1.06^t = 1000$

**20.** $1000e^{0.05t} = 2000$

**EXERCISES 21–24** □ *Approximate the solution(s) of the given equation to the nearest hundredth.*

**21.** $e^{-x} = x$

**22.** $4 - x^2 = e^{x-2}$

**23.** $\dfrac{1}{25}x^4 + 4 = x^2 + e^{x+1}$

**24.** $\dfrac{1}{8}x^3 + \dfrac{7}{8}x^2 = x + 2 - e^x$

■ *Applications*

**25.** *Compound Interest* An account is established with a principal of $200 paying 4% interest compounded semi-annually. How long will it take for the balance to reach $300?

**26.** *Tripling Money* How long does it take for money to triple in an account paying 5% interest compounded annually?

**27.** *Doubling Money* An account paying interest compounded continuously at the rate $r$ doubles in size every 12 years. Find $r$.

**28.** *Competing Accounts* Suppose that Bob deposits $1000 in an account paying 6% interest compounded quarterly at the same time that Janet deposits $1200 in an account paying 6% interest compounded annually. How long will it take for Bob's balance to exceed Janet's?

29. *Competing Accounts*  Suppose that $200 was deposited on January 1, 1991, into an account that earned 5% interest compounded annually. Suppose further that $200 was deposited on January 1, 1992, into a different account that earned 6% interest compounded annually. In what month of what year will the balance in the account earning 6% overtake the balance in the account earning 5%?

30. *Competing Accounts*  On January 1, 2000, $1000 is deposited in an account paying 15% simple interest and $100 is deposited in an account paying 5% interest compounded annually. In what month of what year will the balance in the simple interest account be overtaken by that in the compound interest account?

31. *Compact Disc Inflation*  Suppose that the average price of a compact disc was $12.00 on January 1, 1992, and increases at the rate of 3% per year. Give the month and the year in which the average price of a compact disc will be $100.

32. *Annual Inflation*  Suppose that inflation is 4.5% annually. Then goods or services that cost $P$ dollars today will, on the average, cost $P(1.045)^t$ dollars in $t$ years. Assume that tuition at a certain prestigious university is $15,000 per year in 1994. In what year will the tuition reach $20,000?

33. *Consumer Price Index*  The Consumer Price Index (CPI) was 99.6 in 1983 and 140.3 in 1992. In other words, certain goods and services that cost $99.60 in 1983 cost $140.30 in 1992. Assuming exponential growth, what will the CPI be in 2001? When will it reach 230?

34. *Dollar Devaluation*  One dollar in 1992 was worth approximately 21¢ in 1960 dollars. Assuming this is an exponential decrease, when will a dollar be worth 1¢ in 1960 dollars?

35. *Population Growth*  At 8:00 A.M. on July 4, 2001, a colony of 1000 of the smallest free-living entity, *Mycoplasma laidlawii* (average weight ≈ $10^{-16}$ gram), is established. If the population were to double every hour, at what time would the collective mass of the colony exceed that of the largest organism on Earth, the 190-ton blue whale? (*Hint:* If $P$ represents the population of the colony and $t$ the number of hours since the establishment of the colony, then $P = 1000 \cdot 2^t$. A ton is 2000 pounds, and 1 pound is 454 grams.)

36. *Population Decline*  The population of a certain endangered species of owl is declining exponentially. There are currently 500 living specimens, while just 10 years ago there were 10,000. If the population continues to decline exponentially, how long will it be until there is only a single owl left?

37. *Population Decline*  In 1990, Gary, Indiana topped the list of cities in the United States with the largest percentage loss in population between 1980 and 1990. (Data Source: U.S. Bureau of the Census.) The population of Gary was 152,968 in 1980 and 116,646 in 1990. Assuming the population is decreasing exponentially, what will the population be in the year 2000?

38. *Population Growth*  The fastest growing city in the United States between the years 1980 and 1990 was Moreno Valley, California. (Data Source: U.S. Bureau of the Census.) The population of Moreno Valley was 28,309 in 1980 and 118,779 in 1990. Assuming exponential growth, what will the population be in the year 2000?

39. *Carbon Dating*  Physicists found that a certain Egyptian mummy contained only 25% of its original carbon-14. Given that the half-life of carbon-14 is 5715 years, how old is the mummy?

40. *Carbon Dating*  The well-preserved corpse (see photo) of a bronze age man was found frozen in ice in the Austrian Alps in September 1991. If only 53.3% of the carbon-14 is found, and the half-life of carbon-14 is 5715 years, how old is the body?

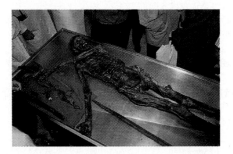

41. *Pollution Turnover*  The rate of pollution turnover in Lake Erie is quite rapid. Suppose that if all pollution inflow into Lake Erie were stopped, the concentration of pollution would decrease to 80% of its present level in 6 months. Assuming the pollution concentration decays exponentially, estimate how many years it would take for the concentration to decrease to 10% of its present level.

42. *Styrofoam Decay*  Suppose that Styrofoam cups are biodegraded at an exponential rate and that after 10 years, 99.5% of the cup remains. Compute the half-life of a Styrofoam cup.

## ■ *Projects for Enrichment*

**43.** *A Medical Mystery* A patient was admitted to the hospital early in the morning complaining of chest pains. An initial dose of the heart medicine digoxin was administered at 9 a.m. although, due to an oversight, the amount was not recorded. Subsequent doses in a quantity sufficient to immediately raise the digoxin level in the blood stream an additional 0.5 ng/ml were administered at 9 p.m. that evening and again at 9 a.m. the next morning. Unfortunately, the patient's condition did not improve, and at 9:05 a.m. he had a fatal heart attack. A test run at that time showed a level of 1.5 ng/ml in his blood stream. In order to avoid a large malpractice settlement, the hospital must prove that the level of the drug was always within the therapeutic range, which is between 0.8 and 1.6 ng/ml. It is known from many clinical observations that a given dose of digoxin decays exponentially over time and is reduced to half its original level in the body in 36 hours.

   **a.** Determine the level of digoxin in the blood stream after the initial dose.

   **b.** Illustrate graphically whether or not the level was always within the therapeutic range.

**44.** *Comparing Interest Rates and Compounding Periods* Suppose you are trying to decide whether to invest in an account that pays 5% compounded quarterly or 4.9% compounded monthly. Naturally, one way would be to determine which would yield the highest balance at the end of a given period of time. Another way would be to convert both rates to their equivalent rates compounded annually. In other words, we would like to find the annually compounded interest rates that would yield the same balances as 5% compounded quarterly and 4.9% compounded monthly. For this task, it would be helpful to have a formula that "converts" an interest rate $r$ compounded $n$ times a year to its equivalent rate $s$ compounded annually. To that end, suppose a principal $P$ is deposited at an interest rate $r$ compounded $n$ times a year. Suppose that at the same time, a principal $P$ is deposited at an interest rate $s$ compounded annually. For the rates $r$ and $s$ to be equivalent, they must yield equal balances at the end of any time $t$. Thus, we must have

$$(1) \qquad P\left(1 + \frac{r}{n}\right)^{nt} = P(1 + s)^t$$

   **a.** Solve equation (1) for $s$ to show that

$$(2) \qquad s = \left(1 + \frac{r}{n}\right)^n - 1$$

   The quantity $s$ in equation (2) is called the **effective annual interest rate**.

   **b.** Compute the effective annual interest rates that correspond to 5% compounded quarterly and 4.9% compounded monthly. Which of the two rates would you choose? Why?

   **c.** Solve equation (2) for $r$ and use the resulting equation to find the interest rate $r$ that, when compounded daily, would be equivalent to an effective annual rate of 6%.

   **d.** Use equation (2) and your graphics calculator to estimate the number of compounding periods that would be needed for an interest rate of 5.85% to have an effective rate of 6%. (*Hint:* Set $s = 0.06$ and $r = 0.058$, then write the equation in the form $\square = 0$, and finally estimate the solution of this equation with your graphics calculator. You will need to use a very small scale for $y$.)

Continuously compounded interest rates can also be converted to an effective annual interest rate. If a principal $P$ is deposited at an interest rate $r$ compounded continuously, the effective annual interest rate is the value $s$ that satisfies the equation

$$(3) \qquad Pe^{rt} = P(1 + s)^t$$

   **e.** Solve equation (3) for $s$ to show that

$$(4) \qquad s = e^r - 1$$

   **f.** Find the effective annual rate corresponding to 4.8% compounded continuously.

   **g.** Solve equation (4) for $r$ and use the resulting formula to find the continuously compounded interest rate $r$ that would be equivalent to an effective annual rate of 6%.

   **h.** What does your solution to part (f) say about the number of compounding periods that would be necessary for an interest rate of 5.8% to have an effective rate of 6%?

## ■■ *Questions for Discussion or Essay*

**45.** When a substance is decaying exponentially, the rate of decay is proportional to the amount of the substance. One consequence of this is that for a radioactive substance, such as radium, a 1-gram sample will take exactly the same amount of time to decay as two 0.5-gram samples or ten 0.1-gram samples. Does ice melt exponentially? Support your answer. What does the rate of melting of ice depend upon?

**46.** Consider the equation $2^x = 10$. If we take the base 2 logarithm of both sides, we obtain

$$\log_2 2^x = \log_2 10$$

$$x = \log_2 10$$

Although this method yields a correct answer, in practice, it is not the method of choice. Why? What is the method of choice?

**47.** In Exercise 33 we made an assumption about the exponential growth of the Consumer Price Index and observed that the costs for goods and services increase dramatically over time. What additional information would be necessary to test this assumption? What would be the inevitable effect of exponential CPI growth on our monetary system? The yearly rise in the CPI is known more commonly as inflation. Do you think inflation is one of the certainties in life, along with death and taxes? Why or why not?

**48.** Suppose that the function $f(t) = 60{,}000e^{0.2t}$ gives the population of a city at time $t$ and the function $h(t) = 3t + 85{,}000$ measures the number of "people units" of housing available in the city at time $t$. Thus, the population is growing exponentially, whereas the available housing is growing at a constant rate. Plot the two functions and approximate the time at which their graphs intersect. What does this point represent? What do these functions imply about the housing situation in the city over time? Is this realistic? Explain.

---

**SECTION 5**

# MODELING WITH EXPONENTIAL AND LOGARITHMIC FUNCTIONS

■ Is hunting necessary as a means for limiting the exponential growth of a deer population?

■ How much faster do pedestrians walk in Mexico City than in New York City?

■ When will the number of transistors on a microprocessor exceed the number of neurons in the human brain?

■ How much slower are the reflexes of a 60 year old than those of a 19 year old?

## CONSTRUCTING MATHEMATICAL MODELS

Throughout the text we have seen examples of how mathematics can be used to model real-world phenomena. However, in most of these instances, we were concerned primarily with using a given model to solve a real-world problem. With few exceptions, we allowed ourselves the luxury of assuming that a given mathematical model would provide accurate results. We now wish to consider in more detail the work that goes on "behind the scenes" in the construction and validation of a mathematical model.

The construction of a mathematical model is often preceded by detailed observations of some naturally occurring event, as was the case with the bacteria population example in Section 1. The actual construction may then be as simple as drawing a graph based on those observations or as complex as finding a formula that provides results similar to the observations. The goal is to provide a model that will accurately describe trends, predict future outcomes, and perhaps even provide insight into how future outcomes can be controlled or at least influenced. Unfortunately, because of the compromise that must often be made between accuracy and simplicity, even the best mathematical models have limitations. Thus, it is important to test and possibly modify models as the need arises. We will discuss these issues as we study the development and testing of mathematical models.

**EXAMPLE 1**   *Modeling a deer population graphically*

Consider the following scenario: In 1992, a certain state's Department of Wildlife recommended a ban on deer hunting because of a dangerously low deer population.

Now, 5 years later, the department wishes to reevaluate its recommendation. Table 3 gives estimates of the deer population since 1992 ($t = 0$). Construct a graphical model to illustrate any trends in the data and to predict what might happen in future years. Use the graph to find a rough approximation for the population in 1998.

**TABLE 3**

| Year       | 0      | 1      | 2      | 3      | 4      | 5      |
|------------|--------|--------|--------|--------|--------|--------|
| Population | 10,000 | 11,500 | 13,200 | 15,100 | 17,400 | 20,100 |

Increasing deer populations and human encroachment on natural habitat forces deer into urban and suburban settings.

**SOLUTION**   The graph in Figure 19 shows that the population growth is occurring in a "smooth" way. Unless a dramatic change occurs, it is likely that the deer population will continue to grow—at least for a few years—according to the pattern that has been established. By fitting a smooth curve to the data as shown in Figure 20, we obtain a rough estimate of 24,000 deer in 1998.

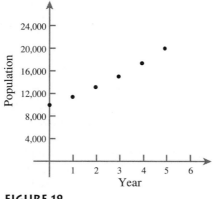

**FIGURE 19**

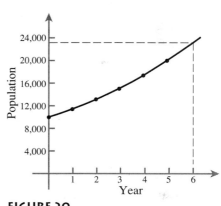

**FIGURE 20**

As useful as graphs are for displaying data, describing trends, and suggesting future behavior, they do have limitations when it comes to making precise predictions. The curve-fitting approach used in the preceding example involves some uncertainty in deciding what the graph should look like beyond the known data points. Is it linear, parabolic, exponential, or something altogether different? It would be useful to have an explicit mathematical formula that captures the relationship between the quantities involved.

**EXAMPLE 2**   *Constructing an exponential model*

Suppose that on the basis of the graph of the data in Table 3, the Department of Wildlife has determined that the growth of the deer population is exponential. Use the first two data values to construct a population model of the form $P(t) = ae^{bt}$, and then estimate the population after 6 years and again after 10. Here $P(t)$ denotes the population after $t$ years.

**SOLUTION**    We first use the fact that the population was 10,000 when $t = 0$.

$$\text{Population at time } 0 = 10,000$$
$$P(0) = 10,000$$
$$ae^{b(0)} = 10,000$$
$$a = 10,000$$

Thus, $P(t) = 10,000e^{bt}$. Next we use the population value 11,500 for $t = 1$.

$$\text{Population at year } 1 = 11,500$$
$$P(1) = 11,500$$
$$10,000e^{b(1)} = 11,500$$

$$e^b = \frac{11,500}{10,000} = 1.15 \qquad \text{Dividing both sides by 10,000}$$

$$b = \ln 1.15 \approx 0.1398 \qquad \text{Taking the natural logarithm of both sides}$$

Thus, $P(t) = 10,000e^{0.1398t}$. At 6 years, we have $P(6) = 10,000e^{0.1398(6)} \approx 23,136$. At 10 years, $P(10) \approx 40,471$.

    Notice that the model in Example 2 was constructed using only two of the six available data points. Slightly different models would result if we were to use different combinations of data points. For example, if we were to use the points (0, 10,000) and (3, 15,100), we would obtain the exponential model $P(t) = 10,000e^{0.1374t}$. This raises an important question. Is there a way to construct an exponential model that takes all available data points into consideration? Fortunately, the answer is yes. The **exponential regression** feature of your graphics calculator is capable of finding a model of the form $P(t) = ae^{bt}$ that "fits" all available data points as closely as possible.

**Exponential regression** can be used to find an exponential function of the form $y = ae^{bx}$ (on some calculators, $y = ab^x$) that best fits a collection of data points $(x, y)$. There are several steps to the process.

**1.** Clear any existing statistical data.

**2.** Enter the points $(x, y)$ as statistical data.

**3.** Select exponential regression from the list of available regression options.

After step 3 has been completed, the calculator will compute values for $a$ and $b$. Note that if your calculator finds $a$ and $b$ for a model of the form $y = ab^x$, you can obtain a model of the form $y = ae^{cx}$ by letting $c = \ln b$. For a detailed description of how these steps are done on your calculator, refer to the graphics calculator supplement.

**EXAMPLE 3**   *Using exponential regression to fit a model*

Use exponential regression to find a model of the form $P(t) = ae^{bt}$ that best fits the data in Table 3.

**SOLUTION**   Following the steps outlined above, we clear any existing statistical data in the calculator, enter the population data from Table 3 as statistical data (using time as the $x$-coordinate and population as the $y$-coordinate), and then select the exponential regression option. The computed values for $a$ and $b$ are $a \approx 9992$ and $b \approx 0.1391$. Thus, the best-fit model is $P(t) = 9992e^{0.1391t}$.

## TESTING MATHEMATICAL MODELS

Before a model can be used with confidence, it must be tested for accuracy. When actual data is available, we can test a mathematical model for accuracy by comparing its predictions to the actual data. With models such as the ones in Examples 2 and 3, this can be done either by computing tables of actual and predicted data values or by graphing the model along with the actual data points.

**EXAMPLE 4**   *Testing models*

Compare the population data in Table 3 with the populations predicted by the following models.

**a.** $P(t) = 10,000e^{0.1398t}$          **b.** $P(t) = 9992e^{0.1391t}$

**SOLUTION**

**a.** Table 4 shows how the actual data values compare with those predicted by $P(t) = 10,000e^{0.1398t}$ for $t = 0$ through $t = 5$. Figure 21 shows the actual population data plotted as points along with the graph of $P(t) = 10,000e^{0.1398t}$. The data and the graph show very close agreement. In fact, because of the scale that was chosen, the graph appears to pass through all the data points.

**TABLE 4**

| $t$ | *From Table 3* | $P(t) = 10,000e^{0.1398t}$ |
|---|---|---|
| 0 | 10,000 | 10,000 |
| 1 | 11,500 | 11,500 |
| 2 | 13,200 | 13,226 |
| 3 | 15,100 | 15,210 |
| 4 | 17,400 | 17,493 |
| 5 | 20,100 | 20,117 |

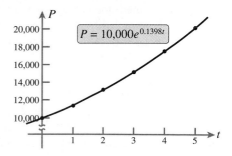

**FIGURE 21**

b. Table 5 shows how the actual data values compare with the ones predicted by $P(t) = 9992e^{0.1391t}$ for $t = 0$ through $t = 5$. In Figure 22 we have used a graphics calculator to plot the function $P(t) = 9992e^{0.1391t}$ together with a scatter plot of the actual population data. The data and graph show very close agreement.

**TABLE 5**

| $t$ | *From Table 3* | $P(t) = 9992e^{0.1391t}$ |
|-----|----------------|--------------------------|
| 0 | 10,000 | 9,992 |
| 1 | 11,500 | 11,483 |
| 2 | 13,200 | 13,197 |
| 3 | 15,100 | 15,166 |
| 4 | 17,400 | 17,430 |
| 5 | 20,100 | 20,031 |

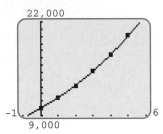

**FIGURE 22**

If a more precise measure of accuracy is desired, one can compute the *absolute error* or *deviation* between the actual data and the predicted values. For example, using the model $P(t) = 10,000e^{0.1398t}$ for the population data in Table 3, the absolute error at $t = 2$ is

$$|(\text{Actual population at } t = 2) - (\text{predicted population at } t = 2)| =$$
$$|13,200 - 13,226| = 26$$

If we compute the absolute error for all known data values, then we can use either the sum of the absolute errors or the largest of the absolute errors as a single measure of the accuracy of the model.

**EXAMPLE 5**    *Computing absolute error*

Determine the absolute errors between the data values given in Table 3 and those predicted by the following two models. Find the maximum absolute error and also the sum of the absolute errors. Which of the two models is more accurate?

a. $P(t) = 10,000e^{0.1398t}$        b. $P(t) = 9992e^{0.1391t}$

**SOLUTION**

a. Using Table 4, we obtain the following absolute errors for the model $P(t) = 10,000e^{0.1398t}$. The maximum absolute error is 110 when $t = 5$. The sum of the absolute errors is 246.

| $t$ | *From Table 3* | $P(t) = 10,000e^{0.1398t}$ | Absolute error |
|---|---|---|---|
| 0 | 10,000 | 10,000 | $\|10,000 - 10,000\| = 0$ |
| 1 | 11,500 | 11,500 | $\|11,500 - 11,500\| = 0$ |
| 2 | 13,200 | 13,226 | $\|13,200 - 13,226\| = 26$ |
| 3 | 15,100 | 15,210 | $\|15,100 - 15,210\| = 110$ |
| 4 | 17,400 | 17,493 | $\|17,400 - 17,493\| = 93$ |
| 5 | 20,100 | 20,117 | $\|20,100 - 20,117\| = 17$ |

**b.** Using Table 5, we obtain the following absolute errors for $P(t) = 9992e^{0.1391t}$. The maximum absolute error is 69 when $t = 5$. The sum of the absolute errors is 193.

| $t$ | *From Table 3* | $P(t) = 9992e^{0.1391t}$ | Absolute error |
|---|---|---|---|
| 0 | 10,000 | 9,992 | 8 |
| 1 | 11,500 | 11,483 | 17 |
| 2 | 13,200 | 13,197 | 3 |
| 3 | 15,100 | 15,166 | 66 |
| 4 | 17,400 | 17,430 | 30 |
| 5 | 20,100 | 20,031 | 69 |

Since the maximum absolute error and the sum of the absolute errors are smaller for the model $P(t) = 9992e^{0.1391t}$, we conclude that it is the more accurate of the two.

■

The absolute errors found in the preceding example are quite small, especially when the size of the population is taken into consideration. However, the acceptable error must also be weighed against the simplicity of a model. A model that is extremely accurate but difficult to use may not serve the purpose for which it was designed. In the case of our deer population example, we were fortunate in finding a simple exponential model that was also extremely accurate. This is not completely unexpected. Numerous statistical studies have shown that uninhibited population growth tends to be exponential in nature. In fact, the British economist and sociologist Thomas Malthus (1766–1834) had proposed exponential population models as early as 1798. The Malthusian model, as it has come to be called, has the form $P(t) = P_0e^{kt}$ where $t$ denotes time, $P_0$ represents the **initial population**, and $k$ is the **growth rate constant**.

Many mathematical models are not as easily verified as our Malthusian deer population model, especially when observed data is not readily available for testing purposes. For example, it would be difficult—if not impossible—to rigorously verify mathematical models that are intended to predict the effect of ozone depletion. In such cases, one must be careful about depending too heavily on the model's predictions. Even if a model has been thoroughly verified, one should be wary about placing too much confidence in its predictions.

PREDICTING WITH
MATHEMATICAL
MODELS

Once we have verified that a model is accurate and have judged it to be sufficiently simple, the next step is to use the model to make inferences about future behavior. After all, the purpose for constructing and verifying a model is not just to showcase its accuracy and simplicity, but to use it to gain insight into what might happen in the future.

**EXAMPLE 6**     *Predicting with exponential models*

Suppose that on the basis of historical data, the Department of Wildlife (see Example 1) estimates that the natural resources in the state can support at most 240,000 deer. Proponents of deer hunting argue that the ban on deer hunting should be lifted when the population reaches half of that limiting value to prevent deaths by starvation and also auto accidents that would likely occur as the population nears the limiting value. Use the population model $P(t) = 9992e^{0.1391t}$ to predict when the deer population will reach 120,000.

**SOLUTION**     We want to determine the value of $t$ at which the population predicted by $P(t) = 9992e^{0.1391t}$ will be 120,000. Thus, we must set $P(t)$ equal to 120,000 and solve for $t$.

$$9992e^{0.1391t} = 120,000$$

$$e^{0.1391t} = \frac{120,000}{9992} \qquad \text{Dividing both sides by 9992}$$

$$\ln e^{0.1391t} = \ln\left(\frac{120,000}{9992}\right) \qquad \text{Taking the natural logarithm of both sides}$$

$$0.1391t = \ln\left(\frac{120,000}{9992}\right) \qquad \text{Using the logarithmic identity } \ln(e^x) = x$$

$$t = \frac{\ln\left(\dfrac{120,000}{9992}\right)}{0.1391} \approx 17.9$$

Thus, our model predicts that it will take approximately 17.9 years for the deer population to reach 120,000.

Notice that although the Department of Wildlife has estimated that the natural resources can support a maximum of 240,000 deer, our model actually suggests that the population growth has no bound. For example, the model predicts that in 25 years the deer population will be $9992e^{(0.1391)25}$, or approximately 324,000, and still rising. Does this mean that the estimated limiting population of 240,000 is wrong? Or is our model wrong? Perhaps our estimate of the limiting population is too low, but clearly there *is* a limiting population: if the deer population grew without bound then eventually the entire state (and the airspace above it) would be nothing but a gigantic mound of deer. Thus, the problem with the model is the assumption that the deer population will continue to grow as it has over the first 5 year period. Due to limited natural resources, a slower population growth would be a certain though gradual outcome. It is not that our model is ''wrong''; it worked quite well for short time periods. But in order to extend its range of applicability, we must modify our Malthusian exponential model to reflect this new assumption of limited growth.

## MODIFYING MATHEMATICAL MODELS

Although there has been human settlement on what is now Istanbul for thousands of years, the city has not reached its limiting population.

The inability of a model to make long-term predictions is a problem that is common to exponential models and many other mathematical models. Thus, we must be wary of predictions that are far from the known data. We must also be on the lookout for ways in which a model can be modified to make it more accurate for a larger domain. One such modification for population models is possible using the logistic model developed by the Dutch mathematical biologist Pierre-François Verhulst (1804–1849). It has the general form

$$P(t) = \frac{a}{b + ce^{-kt}}$$

The logistic model improves on our earlier Malthusian model in that it imposes a limit on the size of a population. Such a limit is a natural consequence of competition for food, living space, and other natural resources.

**EXAMPLE 7**

### A logistic model for population growth

Using techniques from calculus, the following logistic model can be constructed for modeling a deer population with a limiting population of 240,000.

$$P(t) = \frac{240,000}{1 + 23e^{-0.1398t}}$$

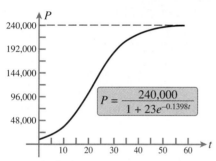

**FIGURE 23**

The graph of $P$ versus $t$ is shown in Figure 23. Plot the graph with a graphics calculator and predict when the population is growing the fastest and when the population will reach 90% of the limiting population of 240,000.

**SOLUTION**   One view of the graph of $P(t)$ is shown in Figure 24. The population is growing the fastest where the graph is rising most rapidly. In other words, the growth rate is largest where the graph is the steepest. Using the trace feature, we determine that this occurs when $t$ (the $x$-coordinate) is approximately 23, as shown in Figure 25. To find when the population will reach 90% of the limiting population, or 216,000 deer, we locate the point on the graph where the $y$-coordinate is approximately 216,000. After zooming in several times, we see in Figure 26 that $x \approx 38$.

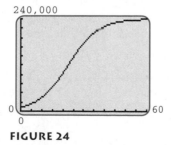

**FIGURE 24**

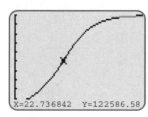

**FIGURE 25**

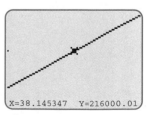

**FIGURE 26**

Although the logistic model for population growth appears to be more realistic than our earlier exponential model, it too does not account for all of the many factors that may affect population growth. For example, the logistic model suggests continued (though slower) growth and so it would not accurately model the deer population if hunting were again allowed. We will ask you to supply some other overlooked factors in Exercise 24.

---

## LOGARITHMIC MODELS

Logarithmic models, though not as common as exponential models, are useful for such tasks as measuring sound and shock intensity and modeling human response systems. In the following example, we consider a human memory model.

**EXAMPLE 8**   *A logarithmic memory model*

In a study on human memory, subjects were given an initial spelling exam on obscure words in the English language and then retested at 1-month intervals for the next year. The average scores for the initial exam and the next 3 months, out of a possible score of 20, are given in the following table.

Dominic O'Brien is listed in the Guinness Book of Records for memorizing 35 decks of playing cards (1820 cards) with only 2 errors.

| Month | 0 | 1 | 2 | 3 |
|-------|---|---|---|---|
| Score | 17 | 14 | 12 | 11 |

Construct a model of the form $y = a + b \log(t + 1)$ that approximates the average score $y$ as a function of time $t$ measured in months. Test the model for accuracy and use it to predict the average score after 6 months.

**SOLUTION**   We wish to select constants $a$ and $b$ so that the resulting model $y = a + b \log(t + 1)$ most nearly fits the data. Close inspection of the data, or even plotting a graph of the data, probably will not give us much insight into an appropriate choice of constants: we just don't have enough familiarity with functions of the form $f(t) = a + b \log(t + 1)$ to predict the effect on the graph of adjusting the constants $a$ and $b$. We can simplify things a great deal, however, by making the substitution $x = \log(t + 1)$ to obtain the linear equation $y = a + bx$. It is a much easier task to find the equation of a *line* that best fits a collection of data. To find $a$ and $b$, we can compute $x = \log(t + 1)$ for $t = 0$ to $t = 3$, plot the resulting points $(x, y)$, and finally approximate the $y$-intercept and slope of the straight line that best fits these points. (The slope of the line $y = a + bx$ is $b$, and the $y$-intercept is $a$.)

| Month ($t$) | 0 | 1 | 2 | 3 |
|-------------|---|-------|-------|-------|
| $x = \log(t + 1)$ | 0 | 0.301 | 0.477 | 0.602 |
| Score ($y$) | 17 | 14 | 12 | 11 |

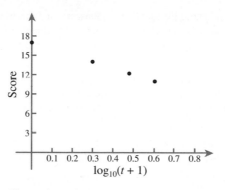
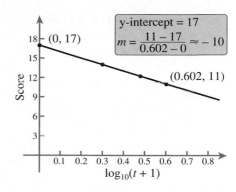

Since the $y$-intercept is approximately 17 and the slope is approximately $-10$, the model is $y = 17 - 10x = 17 - 10 \log(t + 1)$. (Note that we could also use the linear regression feature of a graphics calculator to obtain the model $y = 16.99 - 10.13 \log(t + 1)$, which is very close to the one we found.) The following table compares the actual test scores for the first 3 months with those predicted by the model.

| Month | 0 | 1 | 2 | 3 |
|---|---|---|---|---|
| Actual score | 17 | 14 | 12 | 11 |
| Predicted | 17 | 13.99 | 12.23 | 10.98 |
| Absolute error | 0 | 0.01 | 0.23 | 0.01 |

With a maximum absolute error of 0.23, this model is quite accurate. For $t = 6$, the model predicts an average score of $y = 17 - 10 \log(6 + 1) \approx 8.5$.

## EXERCISES 5

1. *Rabbit Population* A population of rabbits has been observed over a 5-month period, and the following data have been collected.

| Month | 0 | 1 | 2 | 3 | 4 | 5 |
|---|---|---|---|---|---|---|
| Rabbits | 20 | 24 | 30 | 36 | 45 | 54 |

   a. Plot the data in the table on a coordinate system with an appropriate scale.

   b. Visually fit a smooth curve to the data points you plotted in part (a).

   c. Use the curve in part (b) to estimate the population after 6 months.

2. *City Population* The population of a city has been observed over a 5-year period, and the following data have been collected.

| Year | 0 | 1 | 2 | 3 | 4 | 5 |
|---|---|---|---|---|---|---|
| Population | 125,000 | 127,500 | 130,100 | 132,700 | 135,400 | 138,200 |

   a. Plot the data on a coordinate system with an appropriate scale.

   b. Visually fit a smooth curve to the data points you plotted in part (a).

   c. Use the curve in part (b) to estimate the population after 6 years.

3. *Rabbit Population* The population of rabbits in Exercise 1 has been determined to be growing exponentially. Using the fact that there are 20 rabbits initially and 24 rabbits after 1 month, find an exponential function of the form $P(t) = ae^{bt}$ that estimates the number of rabbits after $t$ months. Compare the values given by your function for the first 5 months with the values given in the table in Exercise 1. What is the largest absolute error? What is the sum of the absolute errors?

4. *City Population* The population of the city in Exercise 2 has been determined to be growing exponentially. Using the fact that there are 125,000 people initially and 127,500 people after 1 year, find an exponential function of the form $P(t) = ae^{bt}$ that estimates the population of the city after $t$ years. Compare the values given by your function for the first 5 years with the values given in the table in Exercise 2. What is the largest absolute error? What is the sum of the absolute errors?

5. *Rabbit Population*  Repeat Exercise 3, this time using the fact that there are 20 rabbits initially and 54 rabbits after 5 months. How does your model differ from that of Exercise 3? Is it more or less accurate?

6. *City Population*  Repeat Exercise 4, this time using the fact that there are 125,000 people initially and 138,200 people after 5 years. How does your model differ from that of Exercise 4? Is it more or less accurate?

7. *Rabbit Population*  Use the exponential regression feature of your graphics calculator to find a model of the form $P(t) = ae^{bt}$ that best fits the rabbit population data given in Exercise 1. How does the accuracy of this model compare to that in Exercise 3?

8. *City Population*  Use the exponential regression feature of your graphics calculator to find a model of the form $P(t) = ae^{bt}$ that best fits the population data given in Exercise 2. How does the accuracy of this model compare to that in Exercise 4?

9. *Alaska Population*  Population data for Alaska for the census years 1960, 1970, 1980, and 1990 are given below.

| Year | 1960 | 1970 | 1980 | 1990 |
|---|---|---|---|---|
| Population | 226,000 | 303,000 | 402,000 | 550,000 |

Data Source: U.S. Bureau of the Census

Construct and test an exponential model of the form $P(t) = ae^{bt}$ that estimates the population $t$ years after 1960. According to the model, what will the population be in the year 2010? During what year will the population first exceed 1,500,000?

10. *New Hampshire Population*  Population data for New Hampshire for the census years 1960, 1970, 1980, and 1990 are given below.

| Year | 1960 | 1970 | 1980 | 1990 |
|---|---|---|---|---|
| Population | 607,000 | 728,000 | 921,000 | 1,109,000 |

Data Source: U.S. Bureau of the Census

Construct and test an exponential model of the form $P(t) = ae^{bt}$ that estimates the population $t$ years after 1960. According to the model, what will the population be in the year 2010? During what year will the population first exceed 2,000,000?

11. *Sales Decline*  In an effort to cut spending, a company decides to stop all advertising and promotions for one of its products. Over the next 4 months, the following monthly sales of this product are observed.

| Month ($t$) | 0 | 1 | 2 | 3 | 4 |
|---|---|---|---|---|---|
| Sales (in thousands) ($S$) | 80 | 72 | 66 | 61 | 58 |

Construct and test a model of the form $S(t) = ae^{bt}$ that estimates the sales for month $t$. What does your model predict for the sales for month 6? During what month will the sales reach 20,000?

12. *Sales Increase*  Due to unacceptable sales figures, a company decides to start a new promotion for one of its products. The following increases in monthly sales are observed over a 4-month period.

| Month ($t$) | 0 | 1 | 2 | 3 | 4 |
|---|---|---|---|---|---|
| Sales (in thousands) ($S$) | 58 | 61 | 65 | 68 | 73 |

Construct and test a model of the form $S(t) = ae^{bt}$ that estimates the sales for month 6. What does your model predict for the sales for month 6? During what month will the sales reach 100,000?

13. *Computer Viruses*  In a study on computer viruses conducted in 1991, the following data was obtained concerning the percentage of American companies that encountered one or more computer viruses during 1990 and 1991.

| | 0 | 1 | 2 | 3 |
|---|---|---|---|---|
| Quarter | (Oct–Dec 1990) | (Jan–Mar 1991) | (Apr–Jun 1991) | (Jul–Sept 1991) |
| Percent | 8 | 19 | 26 | 40 |

Data Source: Dataquest/National Computer Security Association 11/91

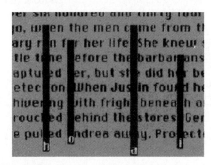

The Falling Letters virus causes characters to fall to the bottom of the screen.

Construct and test a model of the form $p(t) = ae^{bt}$ that estimates the percentage after $t$ quarters. When does the model predict that 90% of American companies will have encountered a computer virus? Why is this model not useful for very long? What modification would you suggest?

14. *Transistor Growth*  The following table shows the approximate number of transistors in various Intel microprocessors introduced since 1971.

| Year | 1971 | 1974 | 1978 | 1982 | 1985 | 1989 | 1993 |
|---|---|---|---|---|---|---|---|
| Transistors (in thousands) | 2 | 6 | 25 | 122 | 208 | 1200 | 3100 |

Data Source: Intel Technology Briefing

Construct and test a model of the form $N(t) = ae^{bt}$ that estimates the number of transistors $t$ years after 1971. How many transistors are predicted for the year 2001? The human brain is estimated to have between 10 and 100 billion neurons. According to the model, when will the number of transistors exceed 100 billion?

15. *Spreading a Rumor* A rumor begins to spread around a college campus. The number of people $N$ who have heard the rumor after $t$ hours can be approximated using the logistic model

$$N = \frac{2000}{1 + 499e^{-0.3t}}$$

   a. How many students started the rumor?

   b. How many students have heard the rumor after 10 hours?

   c. How long before 500 students have heard the rumor?

   d. Plot the graph of $N$ and estimate the limiting number of students who will hear the rumor.

   e. During which hour did the greatest number of students hear the rumor?

16. *Contagious Virus* A group of tourists returns home after a 2-week tour, and all have unknowingly been exposed to a contagious flu virus. Suppose that if left untreated, the spread of the flu virus in their city can be approximated using the logistic model

$$Q = \frac{150,000}{1 + 2999e^{-0.05t}}$$

where $N$ is the number of people with the flu virus after $t$ days.

   a. How large was the group of tourists?

   b. How many people will be infected with the virus after 14 days?

   c. How long before 10,000 people are infected?

   d. Plot the graph of $Q$ and estimate the limiting number of people who will be infected.

   e. During which day did the greatest number of people become infected?

17. *Automobile Speed Record* The 1-mile automobile speed records for the years 1906–1983 can be approximated with the logistic model

$$S = \frac{700}{1 + 6.8e^{-0.057t}}$$

where $S$ is the one-mile speed record in miles per hour for year $t$ ($t = 0$ corresponds to 1906).

   a. During what year did the speed record first exceed 400 miles per hour?

   b. According to this model, what is the upper bound on the speed record?

18. *Memory Loss* A high school algebra class is given a test on new material and then retested monthly on the same material. The average scores for the first test and those of the next 3 months are given below.

| Month | 0 | 1 | 2 | 3 |
|---|---|---|---|---|
| Average score | 75.0 | 70.5 | 67.8 | 66.0 |

Using $t$ to denote the month and $y$ to denote the score, construct a logarithmic model of the form $y = m \log(t + 1) + b$ that best fits the data in the table. Test the model against the data in the table and use it to predict the average score after 7 months.

19. *Walking Speed* Various studies have found a correlation between the size of a city and the average walking speed of pedestrians. One such study obtained the following data. (Source: Marc and Helen Bornstein, ''The Pace of Life,'' *Nature* 259 (19 February 1976) 557–559.)

| Population | 5500 | 14,000 | 71,000 | 138,000 | 342,000 |
|---|---|---|---|---|---|
| Velocity (ft/sec) | 3.3 | 3.7 | 4.3 | 4.4 | 4.8 |

Using $P$ to denote population and $v$ to denote velocity, construct a logarithmic model of the form $v = m \log P + b$ that best fits the data in the table. Test the model against the data in the table and use the model to predict the walking speed in Little Rock, Arkansas (population 176,000), New York (population 7,300,000), and Mexico City (population 20,000,000).

20. *Reflex Speed* Numerous studies have shown a relationship between age and reflex speed. One such study obtained the following data. (Source: Hodgkins, Jean, ''Reaction time and speed of movement in males and females of various ages'', Res. Quart. Amer. Assoc. Hlth. Phys. Educ. Recr. 1963 34(3) 335–343.)

| Median age | 19 | 27 | 45 | 60 |
|---|---|---|---|---|
| Reflex time (sec) | 0.18 | 0.20 | 0.23 | 0.25 |

Using $a$ to denote age and $t$ to denote reflex time, construct a logarithmic model of the form $t = m \log a + b$ that best fits the data in the table. Test the model against the data and use it to predict your reflex time.

## Projects for Enrichment

**21.** *Comparing Models* In this section we considered methods for constructing and testing exponential models of the form $y = ae^{bx}$. Similar methods can be used to construct and test models of many different forms. Two specific examples are **linear models** of the form $y = ax + b$ and **power models** of the form $y = ax^b$. In some cases, where it is not known which model form is needed, it may be necessary to construct and test several different models for the same set of data. Then it is possible to select the model that ''works'' the best.

a. As of 1990, the three largest cities in the United States were New York, Los Angeles, and Chicago. Census data for each of these cities for the years 1880–1930 are given in the following table.

| Year | New York | Los Angeles | Chicago |
|------|----------|-------------|---------|
| 1880 | 1,912,000 | 11,000 | 503,000 |
| 1890 | 2,507,000 | 50,000 | 1,100,000 |
| 1900 | 3,437,000 | 102,000 | 1,699,000 |
| 1910 | 4,767,000 | 319,000 | 2,185,000 |
| 1920 | 5,620,000 | 577,000 | 2,702,000 |
| 1930 | 6,930,000 | 1,238,000 | 3,376,000 |

For each city, use your graphics calculator to construct and test models of the form $y = ax + b$ and $y = ae^{bx}$ to see which fits the data best. Use the models that fit the best to predict the population of each city in 1940, and compare your prediction with the actual values (New York: 7,455,000; Los Angeles: 1,504,000; Chicago: 3,397,000). What is your conclusion as to the accuracy of the models?

b. Early in the seventeenth century, the German astronomer Johannes Kepler discovered three laws that govern the motion of the planets around the Sun. The first states that the shape of each planet's orbit around the Sun is an ellipse with the Sun at one focus. The second states that a line from a planet to the Sun sweeps out equal areas in equal times. Thus, referring to Figure 27, if it takes the same amount of time for a planet to travel from $A$ to $B$ as it does from $A'$ to $B'$, then the areas of $ABF$ and $A'B'F$ are equal. Kepler's

third law relates a planet's average distance from the Sun to its orbital period. The data in the following table give the average distance and the orbital period for each of the nine planets.

| Planet | Average distance (miles) | Period (days) |
|--------|--------------------------|---------------|
| Mercury | 36,000,000 | 88 |
| Venus | 67,000,000 | 225 |
| Earth | 93,000,000 | 365 |
| Mars | 142,000,000 | 687 |
| Jupiter | 484,000,000 | 4,329 |
| Saturn | 887,000,000 | 10,753 |
| Uranus | 1,784,000,000 | 30,660 |
| Neptune | 2,795,000,000 | 60,150 |
| Pluto | 3,671,000,000 | 90,670 |

Construct and test models of the form $y = ae^{bx}$ and $y = ax^b$ where $y$ is the period and $x$ is the distance. Which model is more accurate? State Kepler's third law.

**22.** *Rates of Change* Recall the deer population function $f(t) = 10,000e^{0.1398t}$ from Example 2. It would be of interest to know the rate at which the deer population is growing at any given instant in time. This is referred to as the *instantaneous rate of change* of the function, and it can be determined using calculus. But we can also use the *average rate of change* to approximate the instantaneous rate by measuring how much it has changed over small intervals of time.

**The average rate of change of $f$ from**
$$t = a \text{ to } t = b \text{ is given by } \frac{f(b) - f(a)}{b - a}.$$

For example, from $t = 1$ to $t = 5$, the average rate is

$$\frac{f(5) - f(1)}{5 - 1} \approx \frac{20,117 - 11,500}{4} = \frac{8617}{4} \approx 2154$$

So the deer population is changing at an average rate of 2154 deer per year from years 1 to 5.

a. Complete the following table of average rates of change.

| Time interval | Average rate of change |
|---------------|------------------------|
| $t = 1$ to $t = 4$ | |
| $t = 1$ to $t = 2$ | |
| $t = 1$ to $t = 1.1$ | |
| $t = 1$ to $t = 1.01$ | |
| $t = 1$ to $t = 1.001$ | |

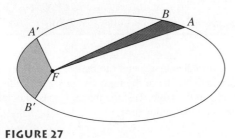

**FIGURE 27**

b. Notice that the average rate of change is approaching a certain number as the length of the time interval decreases. What number is it approaching? This number is the **instantaneous rate of change at $t = 1$**.

c. Repeat parts (a) and (b) to find the instantaneous rate of change at $t = 2$.

d. Repeat parts (a) and (b) to find the instantaneous rate of change at $t = 3$.

e. Complete the table at right. $r(t)$ denotes the instantaneous rate of change at $t$.

f. What does the final column imply about the relationship between $f(t)$ and the instantaneous rate of change at $t$? This

| $t$ | $f(t)$ | $r(t)$ | $\dfrac{r(t)}{f(t)}$ |
|---|---|---|---|
| 1 | | | |
| 2 | | | |
| 3 | | | |

relationship is characteristic of exponential functions. How does this relationship help explain why exponential functions are useful for modeling population growth?

---

### ◼️ *Questions for Discussion or Essay*

23. Discuss the pros and cons of graphical models such as the one we constructed in Example 1.

24. What are some factors that have not been considered in the logistic model for the deer population discussed in this section? What modifications would have to be made in the model to account for these factors? What are the dangers in using mathematical models such as the ones considered here to justify or condemn deer hunting?

25. Describe in your own words the various steps involved in the construction, testing, and use of an exponential model.

26. Exponential models for population growth have been validated using both theoretical and experimental observations. The theoretical explanation is based on the fact that, left unchecked, populations tend to increase at a rate proportional to the size of the population. Experimental observations have confirmed that exponential population models often work quite well for limited periods of time, but that eventually the growth rate levels off. Describe several factors that prevent unlimited growth of human populations.

---

## CHAPTER REVIEW EXERCISES

**EXERCISES 1–2** ☐ *Sketch the graph of f and use it to sketch the graph of g.*

1. $f(x) = 2^x$

   a. $g(x) = 2^{-x}$

   b. $g(x) = 2^x + 2$

   c. $g(x) = 2^{x+1}$

2. $f(x) = \left(\dfrac{1}{3}\right)^x$

   a. $g(x) = -\left(\dfrac{1}{3}\right)^x$

   b. $g(x) = \left(\dfrac{1}{3}\right)^x - 4$

   c. $g(x) = \left(\dfrac{1}{3}\right)^{x-2}$

**EXERCISES 3–6** ☐ *Solve the given equation without using logarithms.*

3. $3^x = 27$

4. $\dfrac{1}{4^x} = 16$

5. $5^{1-2x} = 25$

6. $3^{x^2+4x} = \dfrac{1}{27}$

**EXERCISES 7–12** ☐ *Evaluate each logarithmic expression without using a calculator.*

7. $\log_2 \dfrac{1}{4}$

8. $\log_3 81$

9. $\log 0.01$

10. $\ln e^3$

11. $\log_5 \sqrt{5}$

12. $\log_\pi 1$

**EXERCISES 13–16** ☐ *Use the change-of-base formula to rewrite each expression, first with common logarithms and then with natural logarithms. Evaluate to find a decimal approximation.*

13. $\log_4 10$

14. $\log_{2/5} 130$

15. $\log_{\sqrt{2}} e$

16. $\log_{100} 25$

**EXERCISES 17–20** ☐ *Sketch the graph of the given function and find its domain.*

17. $f(x) = \log_5 x$

18. $f(x) = 3 + \log x$

19. $f(x) = \ln(x - 2)$

20. $f(x) = -\ln(5x)$

**EXERCISES 21–24** ☐ *Use logarithmic identities to expand each expression, rewriting it as a sum, difference or multiple of logarithms.*

**21.** $\log_2(xy^2)$

**22.** $\log_3\left(\dfrac{x^2}{z}\right)$

**23.** $\log \sqrt[3]{x^2 y}$

**24.** $\ln\left(\dfrac{x^3 y^2}{\sqrt{z}}\right)$

**EXERCISES 25–28** ☐ *Use logarithmic identities to rewrite each expression as a single logarithm.*

**25.** $2 \log x + 3 \log y$

**26.** $4 \log_2 p - 3 \log_2 q$

**27.** $\dfrac{1}{3} \ln x + \dfrac{2}{3} \ln y - \ln z$

**28.** $3 \log 4 - \log 6 + 2 \log 2$

**EXERCISES 29–36** ☐ *Solve the logarithmic equation.*

**29.** $\log_2 x = -1$

**30.** $\log_8 x = \dfrac{1}{3}$

**31.** $\log_5(2x + 4) = \log_5 10$

**32.** $\ln(3x + 5) = \ln x$

**33.** $\log(5y - 2) = 1$

**34.** $2 \log_4 x - \log_4(x + 1) = 1 - \log_4 3$

**35.** $\ln(x - 1) + \ln(x + 2) = \ln 4$

**36.** $\log \sqrt{x + 1000} = 2$

**EXERCISES 37–44** ☐ *Solve the exponential equation.*

**37.** $5^x = 10$

**38.** $e^x = 2$

**39.** $1.05^t = 10$

**40.** $3^{-2x} = 9$

**41.** $4e^{0.1x} = 20$

**42.** $350 - 280e^{-0.05t} = 180$

**43.** $3^{x+2} = 2^{2x}$

**44.** $2^{2x} - 5 \cdot 2^x + 6 = 0$

**EXERCISES 45–48** ☐ *Estimate the solutions of the given equation to the nearest hundredth.*

**45.** $e^x = 8 - x^2$

**46.** $\ln x = x^3 - 3x^2$

**47.** $\ln(3x) + 0.02x^3 + 4x = x^2$

**48.** $-\dfrac{x^3}{9} + e^{x+2} = 2x^2 + 5x$

**49.** *Compound Interest* Suppose that $50 is deposited into an account paying 4% interest compounded quarterly. Find the balance in the account after 3 years.

**50.** *Compound Interest* If $42 is deposited into an account paying 3% compounded continuously, then what is the balance after 4 years?

**51.** *Doubling Money* How long does it take money that is deposited into an account paying 3% interest compounded monthly to double?

**52.** *Sound Intensity* The decibel rating of a sound of intensity $I$ is given by $D = 10 \log (I/I_0)$, where $I_0$ is the intensity of a sound that is just perceptible. How many times more intense is a sound of 100 decibels than one of 50 decibels?

**53.** *Population Growth* A certain bacteria colony is growing exponentially. Initially there are 10 bacteria. After 1 hour, there are 25. How long will it take until the population of the colony surpasses the 1 million mark?

**54.** *Radioactive Decay* Suppose that a sample of a radioactive substance having a half-life of 12,000 years was placed in a time capsule 2000 years ago. What percentage of the original sample remains today?

**55.** *Infectious Disease* A community of astronauts living on a base on Mars is exposed to an infectious disease. The number of people who have contracted the disease $t$ days from the initial exposure is given by

$$Q = \frac{200{,}000}{1 + 1999e^{-0.08t}}$$

a. How many astronauts have the disease after 5 days?

b. How many days will it take before 1000 astronauts have become infected?

c. Plot the graph of $Q$ and estimate the limiting number of astronauts that will become infected.

d. During what day did the greatest number of astronauts become infected?

**56.** *Memory Model* It is hypothesized that if a subject were introduced to 100 guests at a cocktail party, the number of people whose names the subject would remember after $t$ months would be given by $y = m \log(t + 1) + b$. A certain subject remembered 50 people immediately after the party, and 36 people 2 years after the party. How many guests will the subject remember 10 years after the party? How many guests did the subject remember 1 year after the party?

**57.** *Car Depreciation* A car purchased new for $14,000 in 1990 depreciates in value each year as shown in the following table.

| Year | 1990 | 1991 | 1992 | 1993 | 1994 |
|------|------|------|------|------|------|
| Value | $14,000 | $12,000 | $10,400 | $8900 | $7700 |

Construct and test an exponential model of the form $V(t) = ae^{bt}$ that estimates the value of the car $t$ years after 1990. When will the car be worth $2000?

**58.** *U.S. Population Growth* U.S. census data for the years 1790–1840 are given below. Construct and test an exponential model of the form $P(t) = ae^{bt}$ that estimates the population $t$ years after 1790. Use the model to predict the population in 1850. How does the predicted value compare to the actual population of 23.2 million in 1850?

| Year | 1790 | 1800 | 1810 | 1820 | 1830 | 1840 |
|------|------|------|------|------|------|------|
| Population (in millions) | 3.9 | 5.3 | 7.2 | 9.6 | 12.9 | 17.1 |

Data Source: U.S. Bureau of the Census

## CHAPTER TEST

**PROBLEMS 1–8** □ *Answer true or false.*

**1.** $\log_a x \cdot \log_a y = \log_a x + \log_a y$ for all $x > 0$ and $y > 0$.

**2.** $\log_a(x^y) = y \log_a x$ for all $x > 0$ and $y > 0$.

**3.** The domain of $\log_a(x + 3)$ is $\{x | x \geq -3\}$.

**4.** The logarithm of a number cannot be negative.

**5.** The logarithm of a negative number is not defined.

**6.** Natural logarithms are simply logarithms with base $e$.

**7.** It is possible that a sample of a substance undergoing exponential decay will take longer to decay from 100 grams to 50 grams than from 50 grams to 25 grams.

**8.** Both $f(x) = a^x$ and $g(x) = \log_a x$ are one-to-one functions.

**PROBLEMS 9–12** □ *Give an example of each of the following.*

**9.** An exponential function that is always **decreasing**

**10.** A number not in the domain of $\ln(x - 5)$, but in the domain of $\ln x$

**11.** An inverse of an exponential function

**12.** A reason why exponential growth of populations of bacteria colonies cannot continue indefinitely

**PROBLEMS 13–16** □ *Solve the given equation.*

**13.** $2^{x^2+x} = 4$

**14.** $\log(x + 1) = 12$

**15.** $3^{x-1} = 2^x$

**16.** $5e^{0.3x} = 10$

**PROBLEMS 17–18** □ *Evaluate the given logarithm.*

**17.** $\log_2 32$

**18.** $\log_3 \dfrac{1}{27}$

**PROBLEMS 19–21** □ *Graph the given function.*

**19.** $f(x) = e^{x-2}$

**20.** $g(x) = \left(\dfrac{2}{3}\right)^x$

**21.** $f(x) = \ln(x + 3)$

**22.** Express $\log\left(\dfrac{x^4 y^2}{z^{-2}}\right)$ in terms of logarithms of $x$, $y$, and $z$.

**23.** Write $3 \ln x - 4 \ln y + 5 \ln z^2$ as a single logarithm.

**24.** Suppose that $300 is deposited into an account paying 6% interest compounded annually. How long does it take the balance to grow to $500?

**25.** Suppose that a radioactive sample is decaying exponentially and that there were 10 grams of the sample on January 1, 1992 and 7 grams on January 1, 1995. In what year will the mass of the sample first drop below 1 gram?

**26.** The approximate numbers of new AIDS cases reported in each of the years 1983–1986 are given in the following table.

| Year | 1983 | 1984 | 1985 | 1986 |
|------|------|------|------|------|
| New AIDS cases | 2100 | 4400 | 8200 | 13,100 |

Data Source: U.S. Centers for Disease Control

Construct and test a model of the form $N(t) = ae^{bt}$ that estimates the number of new cases reported $t$ years after 1983. Use the model to predict the number of new cases reported in 1987. How does the prediction compare to 21,100, the approximate number of new cases reported in 1987?

# TRIGONOMETRIC FUNCTIONS

■ The science of surveying is a marriage of mathematics and measurement. Trigonometric techniques are used to complete a positional database, the framework of which is formed from measurements of angles and distances. Our ability to accurately determine position is only limited by the precision of our rulers, and modern rulers are capable of astonishing feats. The horizontal laser in the accompanying photograph is used to measure the distance from the earth to the moon (the bright spot at the left) to within 3 centimeters, while the vertical beam is used to determine satellite positions to within even narrower margins.

## ANGLES AND THEIR MEASUREMENTS

■ Why do figure skaters tuck in their arms while spinning?
■ How fast does the tip of Big Ben's minute hand move?
■ How can the linear speed of a bike be determined from the rate at which the pedals are moving?
■ Could the bizarre occurrences in the Bermuda Triangle be explained by the fact that the sum of its angles is greater than 180°?
■ Why doesn't light travel in straight lines?

### ANGLES

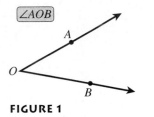

**∠AOB**

*A*

*O*

*B*

**FIGURE 1**

In geometry, an **angle** is defined as the set of points determined by two rays or half-lines, called **sides**, that have a common endpoint, called the **vertex**. If *A* is a point on one side and *B* is a point on the other, and if *O* is the vertex, as shown in Figure 1, then we refer to the angle as ∠*AOB*.

In trigonometry, it is convenient to think of angles in terms of rotated rays. To form an angle, we start with two rays that are initially in the same position. One of the rays—the **initial side**—is fixed, while the other—the **terminal side**—is rotated to its final position, as illustrated in Figure 2. It is customary to denote angles with lowercase Greek letters, such as $\alpha$, $\beta$, and $\theta$, and to indicate the direction of rotation with curved arrows. Several examples are given in Figure 3. Note in the case of the angles $\alpha$ and $\beta$ in Figure 3 that it is possible for two or more angles to have the same initial and terminal sides. Such angles are said to be **coterminal**.

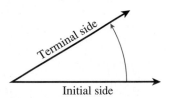

Terminal side

Initial side

**FIGURE 2**

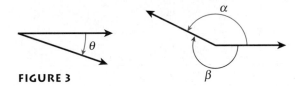

$\theta$

$\alpha$

$\beta$

**FIGURE 3**

An angle in a rectangular coordinate system is said to be in **standard position** if its vertex is at the origin and if its initial side coincides with the positive *x*-axis. Angles formed by a counterclockwise rotation are considered **positive**, while those formed by a clockwise rotation are considered **negative**. See Figure 4 for an example of each.

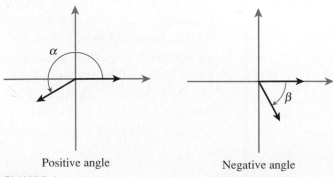

Positive angle

Negative angle

**FIGURE 4**

# DEGREE MEASURE

The measure of an angle indicates the amount of rotation from the initial to the terminal side. The most common unit of angle measure is the **degree**, and it is denoted by the symbol °.

### *Definition of degree measure*

An angle measure of **1 degree** (1°) is equivalent to $\frac{1}{360}$ of a complete revolution about the vertex. In other words, a measure of 360 degrees (360°) is equivalent to a full revolution.

Figure 5 shows some common angles and their corresponding degree measures.

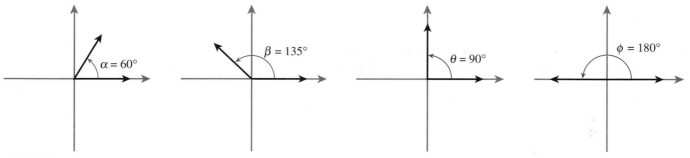

**FIGURE 5**

The angles $\alpha$, $\beta$, $\theta$, and $\phi$ given in Figure 5 can be placed into four categories: acute, obtuse, right, and straight, respectively. These categories are summarized as follows.

### *Angle categories*

An angle $\theta$ is said to be **acute**, **obtuse**, **right**, or **straight** according to the following conditions:

| Type of angle | Condition |
|---|---|
| Acute | $0 < \theta < 90°$ |
| Obtuse | $90° < \theta < 180°$ |
| Right | $\theta = 90°$ |
| Straight | $\theta = 180°$ |

Angles can also be classified according to the quadrant in which the terminal side lies. Thus, angles between 0° and 90° are said to lie in quadrant I, angles between 90° and 180° are said to lie in quadrant II, and so on. This is summarized in Figure 6. The angles 0°, 90°, 180°, and 270° are called **quadrant angles**.

Recall that two angles are said to be coterminal if they have the same initial and terminal sides. In fact, if a given angle is in standard position, infinitely many coterminal angles can be found by adding or subtracting multiples of 360° to the degree measure of the given angle.

| Quadrant II | Quadrant I |
|---|---|
| $90° < \theta < 180°$ | $0° < \theta < 90°$ |
| Quadrant III | Quadrant IV |
| $180° < \theta < 270°$ | $270° < \theta < 360°$ |

**FIGURE 6**

**EXAMPLE 1**     *Finding and sketching coterminal angles*

Find and sketch an angle that is coterminal with the given angle.

**a.** $\alpha = 45°$          **b.** $\theta = 210°$

**SOLUTION**   In each case we add or subtract a multiple of 360° to obtain a coterminal angle.

**a.** We choose to add 360° to obtain an angle $\beta$ given by

$$\beta = 45° + 360° = 405°$$

The angles $\alpha$ and $\beta$ are shown in Figure 7.

**b.** This time we subtract 360° to obtain an angle $\phi$ given by

$$\phi = 210° - 360° = -150°$$

The angles $\theta$ and $\phi$ are shown in Figure 8.

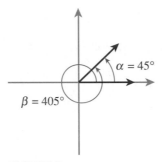

**FIGURE 7**                    **FIGURE 8**

Two positive angles are said to be **complementary** (and are called complements of each other) if the sum of their measures is 90°. For example, angles with measures 35° and 55° are complementary since $35° + 55° = 90°$. Two positive angles are said to be **supplementary** (and are called supplements of each other) if the sum of their measures is 180°. Thus, for example, angles with measures 112° and 68° are supplementary since $112° + 68° = 180°$.

**EXAMPLE 2**   *Finding complementary and supplementary angles*

Find the complementary and supplementary angles for the angle $\theta$ having measure 27°.

**SOLUTION**   To find the complement of $\theta$, we subtract 27° from 90° to obtain

$$90° - 27° = 63°$$

To find the supplement of $\theta$, we subtract 27° from 180° to obtain

$$180° - 27° = 153°$$

We will consider two ways of dividing a degree into fractional parts. When using a calculator, it is often convenient to use ordinary decimal notation. Alternatively, a degree can be divided into 60 equal parts, called **minutes**, and each minute can be divided into 60 equal parts, called **seconds**. Minutes are denoted by the symbol ′ and seconds by the symbol ″. Thus, for example, the notation 43°12′54″ denotes an angle that measures 43 degrees, 12 minutes, and 54 seconds.

**EXAMPLE 3**    *Converting to decimal degree notation*

Convert 43°12′54″ to decimal degree form.

**SOLUTION**    We first note that

$$1' = \left(\frac{1}{60}\right)^{\circ} \quad \text{and} \quad 1'' = \left(\frac{1}{3600}\right)^{\circ}$$

Thus

$$43°12'54'' = 43° + \left(\frac{12}{60}\right)^{\circ} + \left(\frac{54}{3600}\right)^{\circ}$$
$$= 43° + 0.2° + 0.015°$$
$$= 43.215°$$

**EXAMPLE 4**    *Converting to degree-minute-second notation*

Convert 110.355° to degree-minute-second notation.

**SOLUTION**    Since $1 = \dfrac{60'}{1°}$ and $1 = \dfrac{60''}{1'}$, we have

$$110.355° = 110° + 0.355°$$
$$= 110° + 0.355° \cdot \frac{60'}{1°}$$
$$= 110° + 21.3'$$
$$= 110° + 21' + 0.3'$$
$$= 110° + 21' + 0.3' \cdot \frac{60''}{1'}$$
$$= 110° + 21' + 18''$$

Thus, 110.355° = 110°21′18″.

## RADIAN MEASURE

Degree measure is used extensively in physics, engineering, navigation, surveying, and other areas. However, as we will see in later sections, degree measure is not always convenient for working with trigonometric functions. Thus, we introduce a second unit of angle measure known as the *radian*. The radian measure of an angle is defined in terms of central angles and subtended arcs. The **central angle** of a circle is an angle whose vertex is at the center of the circle. We say that the arc of a circle swept out by a central angle is the arc **subtended** by the central angle. The ratio of the length of the subtended arc to the radius of the circle is the radian measure of the central angle.

**Definition of radian measure**

Let $s$ denote the length of the arc subtended by a central angle $\theta$ in a circle of radius $r$. The **radian measure** of $\theta$ is the ratio $s/r$, as shown in Figure 9. Note that for a circle of radius $r$, 1 radian is the measure of a central angle $\theta$ that subtends an arc of length $r$.

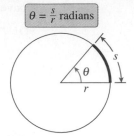

$$\theta = \frac{s}{r} \text{ radians}$$

**FIGURE 9**

It is important to note that the radian measure of a central angle $\theta$ is actually independent of the size of the circle. One way to see this is to consider the effect of increasing the radius of the circle by a certain factor. Because of similarity, the arc length subtended by angle $\theta$ would increase by the same factor. Thus, the ratio of arc length to radius would remain unchanged.

**EXAMPLE 5**    *Finding the radian measure of an angle*

Find the radian measure of a central angle $\theta$ that subtends an arc of length $2\pi$ inches on a circle of radius 2 inches.

**SOLUTION**    According to the definition of radian measure

$$\theta = \frac{s}{r} = \frac{2\pi \text{ inches}}{2 \text{ inches}} = \pi$$

Notice that radian measure is *dimensionless* in the sense that it has no unit. When it is not clear from the context, it is common practice to attach the label ''radians'' to identify the value as an angle measure.

In the previous example, we saw that a central angle of $\pi$ radians subtends an arc of length $2\pi$ on a circle of radius 2. Now the circumference of a circle of radius 2 is $2\pi(2) = 4\pi$, and so an arc of length $2\pi$ is exactly half of the circle. In other words, the central angle is 180°. Thus, we see that $180° = \pi$ radians. This leads us to the following conversion rules.

**Conversion rules**

Since $180° = \pi$ radians, it follows that $1 = \dfrac{180°}{\pi} = \dfrac{\pi}{180°}$. Thus we have the following conversion rules.

| To convert from degrees to radians | multiply by $\dfrac{\pi}{180°}$ |
| To convert from radians to degrees | multiply by $\dfrac{180°}{\pi}$ |

**EXAMPLE 6**    *Converting from degrees to radians*

Convert the following degree measures to radians.

**a.** 60°        **b.** 220.4°

**SOLUTION**

**a.** $60° = 60° \left( \dfrac{\pi}{180°} \right) = \dfrac{\pi}{3}$

**b.** $220.4° = 220.4° \left( \dfrac{\pi}{180°} \right) = \dfrac{220.4\pi}{180} \approx 3.8467$

---

**EXAMPLE 7**    *Converting from radians to degrees*

Convert the following radian measures to degrees.

**a.** $\dfrac{\pi}{6}$        **b.** 5

**SOLUTION**

**a.** $\dfrac{\pi}{6} = \dfrac{\pi}{6} \left( \dfrac{180°}{\pi} \right) = 30°$

**b.** $5 = 5 \left( \dfrac{180°}{\pi} \right) = \left( \dfrac{900}{\pi} \right)° \approx 286.48°$

---

**EXAMPLE 8**    *Finding an angle complement*

Find the radian measure of the complement of an angle of $\dfrac{\pi}{12}$ radians.

**SOLUTION**    Since $90° = \pi/2$ radians, the complement of an angle of $\pi/12$ radians is

$$\frac{\pi}{2} - \frac{\pi}{12} = \frac{6\pi}{12} - \frac{\pi}{12} = \frac{5\pi}{12}$$

---

**APPLICATIONS**

Locations of points on the surface of the Earth are often expressed in terms of *meridians of longitude* and *parallels of latitude*. A **meridian** is a circle on the Earth's surface passing through the north and south geographic poles. The **prime meridian** is the meridian passing through Greenwich, England. The **longitude** of a point is the angular "distance," measured in degrees east or west from the prime meridian to the meridian passing through the point. The **latitude** of a point is its angular "distance" north or south of the Earth's equator, measured in degrees along a meridian. In Figure 10, we illustrate the point on the surface of the Earth at 45° N latitude and 90° W longitude. In Figure 11, we show how the latitude and longitude of this point can be viewed as central angles of circles centered at the center of the Earth.

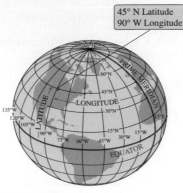

**FIGURE 10**    **FIGURE 11**

**EXAMPLE 9**    *Locating a city*

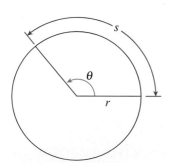

The city of Angle, Utah (really!), has latitude 38°14′57″ N and longitude 111°58′33″ W. Express its latitude and longitude in terms of decimal degrees, and describe its location relative to Salt Lake City, which has latitude 41.76° N and longitude 111.89° W.

**SOLUTION**    We first convert from degree-minute-second notation to decimal degrees.

$$\text{Latitude:} \quad 38°14′57″ = \left( 38 + \frac{14}{60} + \frac{57}{3600} \right)^{°} \approx 38.249°$$

$$\text{Longitude:} \quad 111°58′33″ = \left( 111 + \frac{58}{60} + \frac{33}{3600} \right)^{°} \approx 111.976°$$

Since both latitudes are measured north of the equator, 38.249° N is not as far north as is 41.76° N. Thus, Angle is south of Salt Lake City. Similarly, since both longitudes are measured west from the prime meridian, Angle is slightly west of Salt Lake City.

Radian measure is useful in applications involving arc length, area of circular sectors, and angular speed. For arc length, we consider a circle of radius $r$ and a central angle of $\theta$ radians. Denote by $s$ the length of the arc subtended by $\theta$, as shown in Figure 12. From the definition of radian measure, we have

$$\theta = \frac{s}{r}$$

**FIGURE 12**

By solving for $s$, we obtain the following formula for arc length.

**Arc length**

The length $s$ of an arc on a circle of radius $r$ subtended by a central angle of $\theta$ radians is given by

$$s = \theta r$$

**EXAMPLE 10**    *Finding arc length*

Find the length of the arc subtended by a central angle measuring 150° in a circle of radius 3 centimeters.

**SOLUTION**    We first convert from degrees to radians.

$$150° = 150°\left(\frac{\pi}{180°}\right) = \frac{5\pi}{6}$$

Applying the arc length formula, we have

$$s = \left(\frac{5\pi}{6}\right)3 = \frac{5\pi}{2} \approx 7.85$$

Thus, the length of the arc is approximately 7.85 centimeters.

The formula for arc length can be used to relate *linear* and *angular* speeds. Just as linear speed is a measure of how much an object's position changes in a given time period, the angular speed of an object rotating at a constant rate is a measure of how much the object's angle changes in a given time period. Thus, the angular speed of an object tells us how fast the object is rotating. For example, a second hand sweeps out 360° or $2\pi$ radians every minute, so that its angular speed is given by

$$\text{angular speed} = \frac{2\pi \text{ radians}}{1 \text{ minute}} = 2\pi \text{ radians per minute}$$

More generally, the **angular speed** $\omega$ of an object rotating at a constant speed is given by

$$\omega = \frac{\text{change in angle}}{\text{change in time}} = \frac{\theta}{t}$$

For an object rotating at a constant rate, the **linear speed** is given by

$$v = \frac{\text{distance traveled}}{\text{elapsed time}} = \frac{s}{t}$$

From the arc length formula $s = \theta r$, we thus have

$$v = \frac{s}{t}$$
$$= \frac{\theta r}{t}$$
$$= \frac{\theta}{t} \cdot r$$
$$= \omega r$$

In other words, the linear speed is found by multiplying the angular speed by the radius.

**EXAMPLE 11**   *Finding the linear speed of the tip of a helicopter blade*

A helicopter has a 20-foot-diameter main rotor that rotates at a rate of 420 revolutions per minute. Find the linear speed of the tip of each blade.

**SOLUTION**   Each revolution of the rotor corresponds to a change in angle of $2\pi$ radians, and so the angular speed of the rotor is given by

$$\omega = \frac{\text{change in angle}}{\text{change in time}} = \frac{420(2\pi) \text{ radians}}{1 \text{ minute}} = 840\pi \text{ radians per minute}$$

The tip of each blade is 10 feet from the center. Thus, the linear speed of the tip of each blade is

$$v = \omega r = (840\pi)(10) \approx 26{,}390 \text{ feet per minute}$$

   ■

Occasionally, it is convenient to express angular speed in revolutions per minute. Since there are $2\pi$ radians in one revolution, angular speed expressed in radians per minute can be converted to revolutions per minute using the relation

$$1 \text{ revolution} = 2\pi \text{ radians}$$

**EXAMPLE 12**   *Finding the angular speed of a bicycle tire*

A bicycle with 27-inch-diameter tires is moving at a constant rate of 15 miles per hour. Find the angular speed of the tires in revolutions per minute.

**SOLUTION**   Since the bicycle is moving at a constant rate, the linear speed of a point on the edge of a tire relative to the center of the wheel is the same as the rate at which the bicycle is moving. Thus, the linear speed (in inches per minute) is given by

$$v = \left(15\,\frac{\text{miles}}{\text{hour}}\right)\left(\frac{1 \text{ hour}}{60 \text{ minutes}}\right)\left(\frac{5280 \text{ feet}}{1 \text{ mile}}\right)\left(\frac{12 \text{ inches}}{1 \text{ foot}}\right) \approx 15{,}840 \text{ inches per minute}$$

It follows that the angular speed of a point on the edge of a tire is

$$\omega = \frac{v}{r} \approx \frac{15{,}840}{27} \approx 586.7 \text{ radians per minute}$$

Since there are $2\pi$ radians in each revolution, this angular speed is equivalent to

$$586.7 \frac{\text{radians}}{\text{minute}} \cdot \frac{1 \text{ revolution}}{2\pi \text{ radians}} \approx 93.4 \text{ revolutions per minute}$$

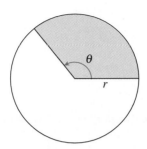

**FIGURE 13**

The area of a circular sector is proportional to its central angle. To see this, let $A$ denote the area of the shaded sector in Figure 13 formed by a central angle of $\theta$ radians in a circle of radius $r$. If we view the interior of the circle as a sector with central angle $2\pi$, then the ratio of $A$ to the area of the circle, namely $\pi r^2$, must be the same as the ratio of $\theta$ to $2\pi$. Thus,

$$\frac{A}{\pi r^2} = \frac{\theta}{2\pi}$$

and, by solving for $A$, we obtain the following area formula.

*Area of a circular sector*

> The area of a circular sector formed by a central angle of $\theta$ radians in a circle of radius $r$ is given by
>
> $$A = \frac{1}{2} r^2 \theta$$

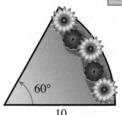

**FIGURE 14**

**EXAMPLE 13**    *Finding the area and perimeter of a circular sector*

Anton wishes to construct a decorative flower garden in the shape of a circular sector with central angle 60° and radius 10 feet (see Figure 14). In order to purchase the proper amount of topsoil and border material, he wishes to determine the area and perimeter of the garden. Find both values.

**SOLUTION**    Converting to radians, we have $60° = \pi/3$ radians. Thus, the area of the sector is

$$A = \frac{1}{2} 10^2 \left( \frac{\pi}{3} \right) \approx 52.36 \text{ square feet}$$

The perimeter $P$ is the sum of the lengths of the two radii that form the sides of the angle $\theta$, together with the length $10(\pi/3)$ of the arc subtended by $\theta = \pi/3$. Thus,

$$P = 10 + 10 + 10 \left( \frac{\pi}{3} \right) \approx 30.47 \text{ feet}$$

# EXERCISES 1

**EXERCISES 1–8** □ *Match the given angle measure with the corresponding sketch.*

1. 30°

2. −60°

3. −135°

4. 270°

5. $\dfrac{\pi}{3}$

6. $-\dfrac{\pi}{4}$

7. $-\dfrac{3\pi}{2}$

8. $\dfrac{5\pi}{6}$

g.
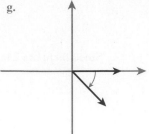

h.
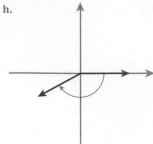

a.
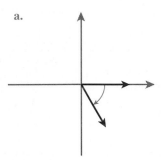

b.
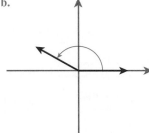

**EXERCISES 9–16** □ *Find two coterminal angles (one positive and one negative) for the given angle.*

9. 30°

10. −60°

11. −135°

12. 270°

13. $\dfrac{\pi}{3}$

14. $-\dfrac{\pi}{4}$

15. $-\dfrac{3\pi}{2}$

16. $\dfrac{5\pi}{6}$

**EXERCISES 17–20** □ *Find the complementary and supplementary angles for the given angle.*

17. 20°

18. 37°

19. $\dfrac{\pi}{6}$

20. $\dfrac{\pi}{4}$

c.
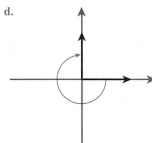

d.

**EXERCISES 21–24** □ *Convert the given angle measure to decimal form.*

21. 25°16′

22. 142°50′

23. 173°20′35″

24. 30°28′42″

**EXERCISES 25–28** □ *Convert the given angle measure to degrees-minutes-seconds form.*

25. 132.4°

26. 47.15°

27. 15.625°

28. 163.32°

e.
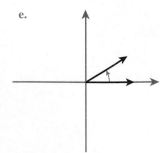

f.
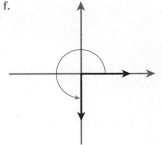

**EXERCISES 29–34** □ *Find the exact radian measure of the given angle.*

29. 30°

30. 120°

31. 225°

32. −45°

33. −150°

34. 300°

**EXERCISES 35–38** ☐ *Find the radian measure of the given angle. Round to three decimal places.*

**35.** 15°

**36.** −20°

**37.** −117.4°

**38.** 326.7°

**EXERCISES 39–44** ☐ *Find the exact degree measure of the given angle.*

**39.** $\dfrac{3\pi}{2}$

**40.** $\dfrac{11\pi}{6}$

**41.** $-\dfrac{4\pi}{3}$

**42.** $\dfrac{\pi}{9}$

**43.** $\dfrac{11\pi}{12}$

**44.** $-\dfrac{\pi}{4}$

**EXERCISES 45–48** ☐ *Find the degree measure of the given angle. Round to three decimal places.*

**45.** $-\dfrac{2\pi}{7}$ radians

**46.** $\dfrac{7\pi}{11}$ radians

**47.** 2 radians

**48.** −3 radians

**EXERCISES 49–52** ☐ *For the given central angle θ in a circle of radius r, find*

a. *the length of the arc subtended by θ*

b. *the area of the sector determined by θ*

**49.** $\theta = \dfrac{5\pi}{3}, r = 20$ meters

**50.** $\theta = \dfrac{3\pi}{4}, r = 16$ inches

**51.** $\theta = 75°, r = 4$ inches

**52.** $\theta = 300°, r = 2$ feet

■ *Applications*

**53.** *Pulley Angles*  A bucket in a well is raised by turning a crank attached to an 8-inch pulley (see Figure 15).

　　a. How far is the bucket raised if the crank is turned through an angle of $10\pi$ radians?

　　b. Through what angle must the crank be turned in order to raise the bucket 9 feet? Express your answer in degrees.

**FIGURE 15**

**54.** *Clock Angles*  Find the exact radian measure of the smaller angle made by the hands of a clock when the time is

　　a. 2:00　　b. 1:30

**55.** *Wind Turbine Speed*  A wind turbine rotates at a rate of 40 revolutions per minute. Find the linear speed of the tip of a 100-foot blade.

**56.** *Mower Blade Speed*  A 20-inch-diameter lawn mower blade rotates at a rate of 3200 revolutions per minute. Suppose that a rock is struck by the tip of the blade and is propelled away from the lawn mower with an initial speed equal to that of the linear speed of the tip of the blade. What is the initial speed of the rock?

**57.** *Big Ben's Linear Speed*  The tip of each minute hand on the four faces of Big Ben is approximately 11 feet from the center of the face. Find the linear speed of the tip of a minute hand in miles per hour.

Big Ben

**58.** *Bicycle Speed*  In a certain gear, a bicycle tire with a diameter of 27 inches makes 3 complete revolutions for each revolution of the pedals. If a cyclist is pedaling at a rate of 100 revolutions per minute, find the speed of the bicycle in miles per hour.

**59. Tire Revolutions** The tires on a certain car have a 14-inch radius. Find the number of revolutions per minute for a tire when the car is traveling at a rate of 90 feet per second (approximately 60 miles per hour).

**60. Earth's Rotational Speed** The Earth rotates about its axis approximately once every 24 hours. Find the linear speed, relative to the axis of rotation, of a point

a. on the equator.

b. with latitude 45° N. (*Hint:* Find the radius of the circle $r$ formed by all points on the surface of the Earth with latitude 45° N, as shown in Figure 16.)

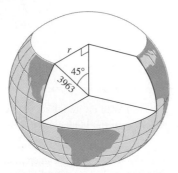

**FIGURE 16**

The apparent circular motion of celestial bodies is due to the rotation of the Earth.

**61. Stained Glass** A piece of stained glass is to be cut in the shape of a circular sector with a radius of 12 inches and a central angle of 120°.

a. Find the area of the piece of glass.

b. Find the length of a lead strip that is to be placed around the edges of the piece of glass.

**62. Estimating the Moon's Diameter** For a small angle $\theta$ and a large radius $r$, the arc length $s = \theta r$ is approximately equal to the length of the line segment connecting the endpoints of the arc. Suppose an observer estimates a 1° angle between the lines of sight to the opposite ends of the diameter of the Moon, as shown in Figure 17. Given that the distance to the Moon is

approximately 230,000 miles, estimate the diameter of the Moon.

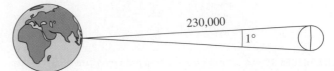

**FIGURE 17**

**63. Deflection of Light** As predicted by Einstein's theory of relativity and also observed experimentally, even light is subject to the force of gravity. Thus, when a ray of light passes by a massive heavenly body such as our Sun, the ray is *deflected* by an angle $\alpha$ that is dependent upon the mass $m$ of the body and the distance $r$ from the center of the body to the passing light ray (see Figure 18). In fact, the angle $\alpha$ satisfies

$$\alpha \approx 4.05 \frac{Gm}{rc^2}$$

where $m$ is in kilograms, $r$ is in meters, $G \approx 6.67 \times 10^{-11}$ m³/(kg · sec²) is Newton's gravitational constant, and $c \approx 3 \times 10^8$ m/sec is the speed of light. Find the angle of deflection for a light ray passing close to the surface of the Sun, in which case $m \approx 1.989 \times 10^{30}$ kilograms and $r \approx 6.95 \times 10^8$ meters. Express your answer both in radians and degrees.

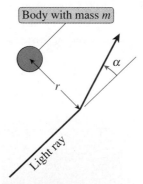

**FIGURE 18**

A solar eclipse on May 12, 1919 provides the first evidence of the gravitational bending of light.

64. *Angular Momentum*  A figure skater begins spinning with her arms extended horizontally and with 2-kilogram weights in each hand. As she brings her hands down to her sides, her angular velocity increases. This phenomenon is due to *conservation of angular momentum*. For a certain skater, the angular velocity $\omega$ satisfies

$$\omega = \left( \frac{8.4}{5.4 + 4r^2} \right)\omega_0$$

where $\omega_0$ is the initial angular velocity of the skater, and $r$ meters is the horizontal distance from the weights to the axis of revolution. If the skater is initially spinning at a rate of 3 revolutions per second, find the angular velocity after the weights are at a final distance of 0.15 meter from the axis of revolution.

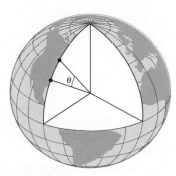

1994 Olympian Nancy Kerrigan

**EXERCISES 65–68**  □  *If two locations on the surface of the Earth have the same longitude (i.e., one lies due north of the other), the distance between them can be approximated using their latitudes. We assume that all points on the surface of the Earth with the same longitude form a circle with a radius of 3963 miles, and then find the length of the arc subtended by the central angle between the two latitudes, as suggested by Figure 19. Use this technique to approximate the distance between the following pairs of cities, which have roughly the same longitude. Answer to the nearest mile.*

65. *Distance between Phoenix and Salt Lake City*
       Phoenix:              33°30′ N latitude
       Salt Lake City:   40°45′ N latitude

66. *Distance between San Francisco and Seattle*
       San Francisco:   37°45′ N latitude
       Seattle:               47°35′ N latitude

67. *Distance between Montreal and Bogota*
       Montreal, Canada:      45°30′ N latitude
       Bogota, Colombia:      4°38′ N latitude

68. *Distance between St. Petersburg and Alexandria*
       St. Petersburg, Russia:   59°55′ N latitude
       Alexandria, Egypt:         31°13′ N latitude

**EXERCISES 69–70**  □  *It is well-known that the sum of the angles of a triangle in the plane is 180°. But for a spherical triangle, that is, a triangle on the surface of a sphere, such as the Earth, the sum of the angles is actually greater than 180°. In fact, it can be shown that if A, B, and C are the angular measures of a triangle on the surface of a sphere, then*

$$A + B + C = 180° + \left( \frac{T}{S} \right) 720°$$

*where T is the area of the triangle and S is the surface area of the sphere. In the case of triangles on the surface of the Earth, S denotes the surface area of the Earth. Note that the surface area of a sphere of radius r is given by* $4\pi r^2$.

69. *The Bermuda Triangle*  The Bermuda Triangle is a region in the Atlantic Ocean bounded by imaginary lines connecting

**FIGURE 19**

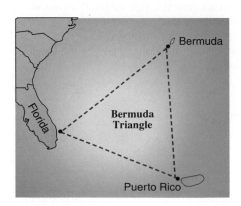

Bermuda, Puerto Rico, and Melbourne, Florida. Find the sum of the angles of the Bermuda Triangle given that its area is approximately 490,000 square miles. Express your answer in both degrees and radians. (Numerous disappearances of ships and airplanes have led to speculation about unexplainable turbulences and other atmospheric disturbances, although investigations to date have not produced any scientific evidence of unusual phenomena.)

70. *Sales Region*  A salesman covers a region roughly bounded by imaginary lines connecting Chicago, Minneapolis, and Des Moines. The spherical triangle with these three cities as vertices has angles measuring 40.64°, 58.82°, and 80.67°. Estimate the area of the region.

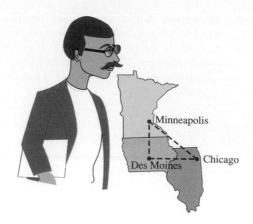

---

■ *Project for Enrichment*

71. *Bicycle Gearing*  The modern bicycle is a direct descendent of, and bears a striking resemblance to, the so-called *safety bicycle* designed by John Kemp Starley in the 1880s. Starley's 1885 Rover (see photo) was the first chain-drive bicycle with changeable front and rear sprockets. The eventual development of derailleur systems further improved upon the chain-drive concept by allowing the chain to be shifted between sprockets of different sizes. In this project, we will investigate the connection between the angular speeds of the sprockets and the resulting speed of the bicycle.

Starley's 1885 Rover

GT Bicycles' 1996 Team USA Superbike

When two sprockets of different sizes are connected by a chain, their linear speeds are forced to be the same. As a consequence, their angular speeds can be shown to be proportional. Consider the front and rear sprockets shown in Figure 20 with radii $R$ and $r$, respectively. Denote the angular speed of the front sprocket (usually called a **chain ring**) by $\omega_R$ and the angular speed of the rear sprocket by $\omega_r$. Finally, define the **gear ratio** to be the ratio $R/r$.

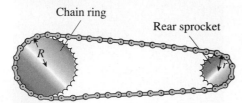

Chain ring

Rear sprocket

**FIGURE 20**

a. Show that the angular speed of the rear sprocket can be found by multiplying the angular speed of the chain ring by the gear ratio.

b. If an 8-inch-diameter chain ring has an angular speed of 80 revolutions per minute, find the angular speed of a 3-inch-diameter rear sprocket. Given that the rear tire has a 27-inch diameter, find the linear speed of the bicycle and the distance it has traveled during one revolution of the pedals.

c. Complete the following table. As before, $R$ and $r$ denote the radii of the chain ring and rear sprocket, while $v$ is the linear speed of the bike and $d$ is the distance traveled during one revolution of the pedals. Assume the angular speed of the chain ring is 80 revolutions per minute and the diameter of the rear tire is 27 inches.

| $R$ (in.) | $r$ (in.) | Gear Ratio | $v$ (mph) | $d$ (in.) |
|---|---|---|---|---|
| 8 | 4 | | | |
| 8 | 3 | | | |
| 8 | 2 | | | |
| 6 | 4 | | | |
| 6 | 3 | | | |
| 6 | 2 | | | |

d. Based on your results from part (c), describe how the linear speed of the bicycle depends upon the sizes of the chain ring and rear sprocket. How does the gear ratio make this dependency easier to describe?

Since the spacing between each link of the chain is fixed, the diameter of a sprocket is completely determined by the number of teeth it has. Thus, the gear ratio can be found by computing the ratio of the number of teeth in the chain ring to the number of teeth in the rear sprocket. For example, a 52-tooth chain ring and a 14-tooth rear sprocket would yield a gear ratio of $\frac{52}{14} \approx 3.71$.

e. Find a multispeed bicycle with at least two chain rings and at least five rear sprockets. Determine the diameter of the rear tire and the number of teeth in each chain ring and sprocket. Complete a chart that shows all of the possible chain ring/sprocket combinations, the corresponding gear ratios, and the linear distance that would be traveled for each gear ratio for a single revolution of the pedals. What does your chart suggest about the order in which you should shift gears if you wish to progress from lowest to highest in sequential order?

Many cycling experts recommend a pedaling rate of 90 revolutions per minute. This allows for a good balance between leg speed and power.

f. Measure the length of the pedal crankarm and use that length to determine the linear speed of your feet when pedaling at a rate of 90 revolutions per minute.

g. Add a column to your table in part (e) that gives the speed you would be traveling for each gear ratio if you were pedaling at a rate of 90 revolutions per minute.

### Questions for Discussion or Essay

72. In Exercise 60, we considered the linear speed of a point on the surface of the Earth at the equator. If it were possible to construct a tower of sufficient height, how high would the tower have to be in order for the tip of the tower to have a linear speed of 186,000 miles per second relative to the Earth's axis? According to Einstein's Theory of Relativity, speeds beyond the speed of light are not possible. Explain this apparent contradiction.

73. If two points on the surface of the Earth have latitude 45° N and longitudes differing by 15°, the distance between them is approximately 734 miles. On the other hand, if two points have latitude 30° N and longitudes differing by 15°, they are approximately 899 miles apart. Explain this discrepancy.

74. Suppose $\alpha$ and $\beta$ are coterminal central angles of a circle with radius $r$. Explain why the arc lengths subtended by $\alpha$ and $\beta$ need not be the same. Give an example that illustrates this.

75. The unit of length commonly referred to as a mile is more precisely called a **statute mile**. A **nautical mile**, on the other hand, is $1'$ of arc length as measured on a great circle of the Earth such as the equator. (A great circle is a circle described by the intersection of the surface of a sphere with a plane passing through the center of the sphere.) Why do you suppose nautical miles are so named? What is the radian measure corresponding to 1 nautical mile? Which is traveling faster, a boat with a rate of 40 knots (40 nautical miles per hour) or one with a rate of 40 statute miles per hour?

## SECTION 2

# TRIGONOMETRIC FUNCTIONS OF ACUTE ANGLES

- How can the length of a shadow and knowledge of trigonometry destroy a suspect's alibi?
- How can a piece of wood, a protractor, and a weight be used to estimate the height of a tall building?
- According to OHSA standards, how high should the tip of a 20-foot ladder be?
- How can the angles of elevation to known landmarks be used to determine one's location?

### RIGHT TRIANGLE DEFINITIONS

The ancient Greeks developed the notion of trigonometric functions by considering acute angles of right triangles. In fact, the word *trigonometry* was derived from the Greek words for ''triangle measurement.'' We will follow their lead by defining the trigonometric functions first in terms of the acute angles of a right triangle. In the next section we will extend our definitions to angles of any measure.

The right triangle definitions of the trigonometric functions are based on an important theorem from geometry, generally attributed to the Greek mathematician Euclid. Euclid showed that the ratio of any two sides of a given triangle must be the same as the corresponding ratio of sides of any similar triangle. Thus, for a right triangle with an acute angle $\theta$, the ratios of any two sides are dependent only upon the angle $\theta$ and not upon the size of the right triangle. To see more clearly what this statement means, let us consider the right triangle shown in Figure 21, where we have labeled the angle $\theta$, the length of the side adjacent to $\theta$ as "adj," the length of the side opposite $\theta$ as "opp," and the length of the hypotenuse as "hyp." We have also identified the six possible ratios of the three sides.

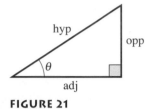

**FIGURE 21**

$$\text{Six possible ratios: } \frac{\text{opp}}{\text{hyp}}, \frac{\text{adj}}{\text{hyp}}, \frac{\text{opp}}{\text{adj}}, \frac{\text{hyp}}{\text{opp}}, \frac{\text{hyp}}{\text{adj}}, \frac{\text{adj}}{\text{opp}}$$

According to the theorem of Euclid, each of the six ratios would remain unchanged if we increase or decrease the size of the triangle, as long as we keep the measure of $\theta$ the same. Thus, each ratio is dependent only upon the angle $\theta$ and hence defines a function of $\theta$. The six functions that result, known collectively as the **trigonometric functions**, are the **sine**, **cosine**, **tangent**, **cosecant**, **secant**, and **cotangent** functions. They are usually abbreviated **sin**, **cos**, **tan**, **csc**, **sec**, and **cot**, respectively, and we define them as follows.

---

*Right triangle definitions of trigonometric functions*

Let $\theta$ be an acute angle of a right triangle, and let "opp" denote the length of the side opposite $\theta$, "adj" the length of the side adjacent to $\theta$, and "hyp" the length of the hypotenuse (Figure 22). We define the six trigonometric functions of $\theta$ as follows.

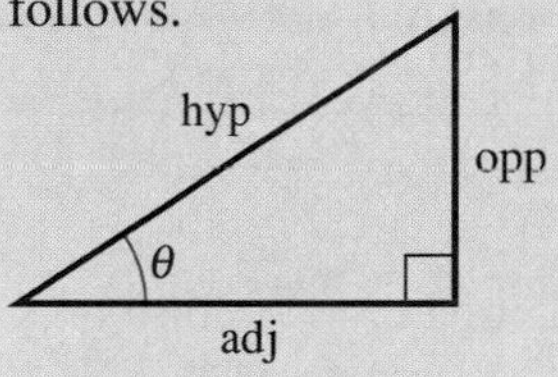

**FIGURE 22**

$$\sin \theta = \frac{\text{opp}}{\text{hyp}} \qquad \cos \theta = \frac{\text{adj}}{\text{hyp}} \qquad \tan \theta = \frac{\text{opp}}{\text{adj}}$$

$$\csc \theta = \frac{\text{hyp}}{\text{opp}} \qquad \sec \theta = \frac{\text{hyp}}{\text{adj}} \qquad \cot \theta = \frac{\text{adj}}{\text{opp}}$$

Note that since each of the lengths opp, adj, and hyp is positive, the values of the trigonometric functions for an acute angle $\theta$ are all positive.

**EXAMPLE 1**    *Evaluating trigonometric functions*

Find the values of the six trigonometric functions for the angle $\theta$ shown in Figure 23.

**SOLUTION**    The lengths of the hypotenuse and the side opposite $\theta$ are given in Figure 23. Thus, hyp = 13 and opp = 5. The length of the side adjacent to $\theta$ can be found by applying the Pythagorean Theorem.

$$\text{adj} = \sqrt{13^2 - 5^2} = \sqrt{144} = 12$$

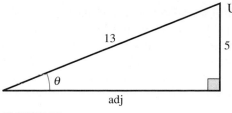

**FIGURE 23**

Using the definitions of the trigonometric functions, we obtain the following values:

$$\sin \theta = \frac{\text{opp}}{\text{hyp}} = \frac{5}{13} \qquad \cos \theta = \frac{\text{adj}}{\text{hyp}} = \frac{12}{13} \qquad \tan \theta = \frac{\text{opp}}{\text{adj}} = \frac{5}{12}$$

$$\csc \theta = \frac{\text{hyp}}{\text{opp}} = \frac{13}{5} \qquad \sec \theta = \frac{\text{hyp}}{\text{adj}} = \frac{13}{12} \qquad \cot \theta = \frac{\text{adj}}{\text{opp}} = \frac{12}{5}$$

The trigonometric function values for the angles 30°, 45°, and 60° can be found using special right triangles and some facts from geometry. We illustrate this in the following example.

**EXAMPLE 2**    *Finding trigonometric function values for 30°, 45°, and 60°*

Find the values of the sine, cosine, and tangent functions for 30°, 45°, and 60°.

**SOLUTION**    For a 45° angle, we consider the isosceles right triangle shown in Figure 24 with leg length 1. By the Pythagorean Theorem, the hypotenuse must have length $\sqrt{2}$. Now applying the definitions of the trigonometric functions, we obtain the following values:

$$\sin 45° = \frac{\text{opp}}{\text{hyp}} = \frac{1}{\sqrt{2}} = \frac{\sqrt{2}}{2} \qquad \cos 45° = \frac{\text{adj}}{\text{hyp}} = \frac{1}{\sqrt{2}} = \frac{\sqrt{2}}{2} \qquad \tan 45° = \frac{\text{opp}}{\text{adj}} = \frac{1}{1} = 1$$

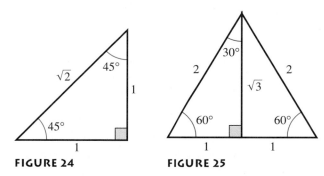

**FIGURE 24**                     **FIGURE 25**

For 30° and 60° angles, we consider the equilateral triangle shown in Figure 25 with side length 2. If we construct a line segment from the top vertex to the midpoint of the opposite side, we obtain a right triangle with acute angles of 30° and 60°. By the Pythagorean Theorem, the length of the leg opposite 60° is $\sqrt{3}$. Once again, we apply the definitions of the trigonometric functions to obtain the following values:

$$\sin 30° = \frac{\text{opp}}{\text{hyp}} = \frac{1}{2} \qquad \cos 30° = \frac{\text{adj}}{\text{hyp}} = \frac{\sqrt{3}}{2} \qquad \tan 30° = \frac{\text{opp}}{\text{adj}} = \frac{1}{\sqrt{3}} = \frac{\sqrt{3}}{3}$$

$$\sin 60° = \frac{\text{opp}}{\text{hyp}} = \frac{\sqrt{3}}{2} \qquad \cos 60° = \frac{\text{adj}}{\text{hyp}} = \frac{1}{2} \qquad \tan 60° = \frac{\text{opp}}{\text{adj}} = \frac{\sqrt{3}}{1} = \sqrt{3}$$

**TABLE 1**

*Special Angles*

| $\theta$ | $\theta$ (radians) | $\sin \theta$ | $\cos \theta$ | $\tan \theta$ |
|---|---|---|---|---|
| 30° | $\frac{\pi}{6}$ | $\frac{1}{2}$ | $\frac{\sqrt{3}}{2}$ | $\frac{\sqrt{3}}{3}$ |
| 45° | $\frac{\pi}{4}$ | $\frac{\sqrt{2}}{2}$ | $\frac{\sqrt{2}}{2}$ | 1 |
| 60° | $\frac{\pi}{3}$ | $\frac{\sqrt{3}}{2}$ | $\frac{1}{2}$ | $\sqrt{3}$ |

Table 1 summarizes the sine, cosine, and tangent values for 30°, 45°, and 60°. Because these values occur so frequently in trigonometry, we strongly recommend that they be memorized.

## FUNDAMENTAL IDENTITIES

Many important relationships can be derived from the definitions of the trigonometric functions. For example, since the sine of an angle is the ratio opp/hyp and the cosecant is the ratio opp/hyp, we see that the sine and cosecant functions are reciprocals of each other. Similar relationships hold for the cosine and secant functions and for the tangent and cotangent functions. These relationships are referred to as **ratio identities**. Two additional ratio identities relate the tangent and cotangent functions to the sine and cosine functions. For example,

$$\tan \theta = \frac{\text{opp}}{\text{adj}} = \frac{\dfrac{\text{opp}}{\text{hyp}}}{\dfrac{\text{adj}}{\text{hyp}}} = \frac{\sin \theta}{\cos \theta}$$

The eight ratio identities are summarized as follows.

### Ratio identities

$$\sin \theta = \frac{1}{\csc \theta} \qquad \cos \theta = \frac{1}{\sec \theta} \qquad \tan \theta = \frac{1}{\cot \theta}$$

$$\csc \theta = \frac{1}{\sin \theta} \qquad \sec \theta = \frac{1}{\cos \theta} \qquad \cot \theta = \frac{1}{\tan \theta}$$

$$\tan \theta = \frac{\sin \theta}{\cos \theta} \qquad \cot \theta = \frac{\cos \theta}{\sin \theta}$$

The next three identities are consequences of the Pythagorean Theorem, and so are called the **Pythagorean identities**. For an acute angle $\theta$ in a right triangle with side lengths opp, adj, and hyp as given in Figure 26, we have

$$(\text{opp})^2 + (\text{adj})^2 = (\text{hyp})^2$$

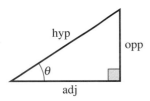

**FIGURE 26**

If we divide both sides by $(\text{hyp})^2$, we obtain

$$\left(\frac{\text{opp}}{\text{hyp}}\right)^2 + \left(\frac{\text{adj}}{\text{hyp}}\right)^2 = 1$$

or

$$(\sin \theta)^2 + (\cos \theta)^2 = 1$$

Similarly, by dividing both sides of $(\text{opp})^2 + (\text{adj})^2 = (\text{hyp})^2$ by either $(\text{opp})^2$ or $(\text{adj})^2$, we can obtain the other two Pythagorean identities stated here.

### Pythagorean identities

$$\sin^2 \theta + \cos^2 \theta = 1 \qquad 1 + \tan^2 \theta = \sec^2 \theta \qquad 1 + \cot^2 \theta = \csc^2 \theta$$

Notice that we denote $(\sin \theta)^2$ by $\sin^2 \theta$, and we use the same convention for the squares of the other trigonometric functions. In general, positive integer powers of the

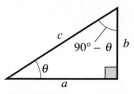

**FIGURE 27**

trigonometric functions are written using this convention. Thus, for example, we write $\tan^5 \theta$ instead of $(\tan \theta)^5$.

The final group of identities we consider here are known as the **cofunction identities**. Consider the acute angle $\theta$ shown in Figure 27. From the definition of the sine function, we have $\sin \theta = b/c$. The angle complementary to $\theta$ is $90° - \theta$, and $\cos(90° - \theta) = b/c$. Thus, $\sin \theta = \cos(90° - \theta)$. Similar identities hold for the other cofunction pairs—tangent/cotangent and secant/cosecant—as summarized here.

■ *Cofunction identities*

$$\sin \theta = \cos(90° - \theta) \qquad \tan \theta = \cot(90° - \theta) \qquad \sec \theta = \csc(90° - \theta)$$
$$\cos \theta = \sin(90° - \theta) \qquad \cot \theta = \tan(90° - \theta) \qquad \csc \theta = \sec(90° - \theta)$$

**EXAMPLE 3**    *Using identities to find trigonometric function values*

If $\theta$ is an acute angle for which $\tan \theta = \sqrt{15}$, find the exact values of the other five trigonometric functions of $\theta$.

**SOLUTION**    We will illustrate two methods. The first involves constructing a right triangle, and the second uses the trigonometric identities.

**Method 1:** We begin by sketching a right triangle with an acute angle $\theta$ and with opp $= \sqrt{15}$ and adj $= 1$, as shown in Figure 28. This ensures that $\tan \theta = \sqrt{15}$. Next, we apply the Pythagorean Theorem to find the length of the hypotenuse:

$$(\text{hyp})^2 = (\sqrt{15})^2 + 1^2$$
$$(\text{hyp})^2 = 16$$
$$\text{hyp} = \sqrt{16} = 4$$

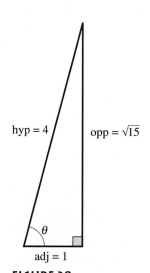

**FIGURE 28**

Finally, we apply the definitions of the trigonometric functions to obtain

$$\sin \theta = \frac{\sqrt{15}}{4} \qquad\qquad \cos \theta = \frac{1}{4}$$

$$\csc \theta = \frac{4}{\sqrt{15}} = \frac{4\sqrt{15}}{15} \qquad \sec \theta = \frac{4}{1} = 4 \qquad \cot \theta = \frac{1}{\sqrt{15}} = \frac{\sqrt{15}}{15}$$

**Method 2:** We can find $\sec \theta$ by applying the Pythagorean identity $1 + \tan^2 \theta = \sec^2 \theta$.

$$\sec^2 \theta = 1 + \tan^2 \theta$$
$$\sec^2 \theta = 1 + (\sqrt{15})^2$$
$$\sec^2 \theta = 16$$
$$\sec \theta = \sqrt{16} = 4$$

Notice that we have chosen only the positive root since all trigonometric function values are positive for acute angles. Now using the ratio identity $\cos \theta = 1/\sec \theta$, we have

$$\cos \theta = \frac{1}{\sec \theta} = \frac{1}{4}$$

Since $\tan\theta = \sin\theta/\cos\theta$, it follows that $\tan\theta\cos\theta = \sin\theta$ and so

$$\sin\theta = \tan\theta\cos\theta = \sqrt{15}\cdot\frac{1}{4} = \frac{\sqrt{15}}{4}$$

The remaining two function values, $\csc\theta$ and $\cot\theta$, can be found using the ratio identities:

$$\csc\theta = \frac{1}{\sin\theta} = \frac{4}{\sqrt{15}} = \frac{4\sqrt{15}}{15} \qquad \text{and} \qquad \cot\theta = \frac{1}{\tan\theta} = \frac{1}{\sqrt{15}} = \frac{\sqrt{15}}{15}$$

&#9632;

Notice in the preceding example that it was not necessary to find the angle $\theta$ in order to determine the remaining trigonometric function values for $\theta$. However, if exact values are not needed, an alternative approach would be to use a calculator to first find a decimal approximation for the angle $\theta$ and then find decimal approximations for the other trigonometric function values for $\theta$. The following box describes the general procedure for performing these two tasks. Refer to the calculator supplement for details on how this is done with your calculator.

**SETTING THE MODE** When using a calculator to perform trigonometric computations, it is important to be sure that the calculator is set for the proper mode of measurement. Most scientific and graphics calculators have either a [MODE] key or keys labeled [DEG] and [RAD] for setting the measurement mode. Set the mode to DEG for working in degrees or to RAD for working in radians.

**EVALUATING TRIG FUNCTIONS** Use the [SIN], [COS], or [TAN] key to find a decimal approximation for the sine, cosine, or tangent of an angle $\theta$. The cosecant, secant, or cotangent of $\theta$ can be found by taking the reciprocal of the sine, cosine, or tangent, respectively. For example, with a calculator set in degree mode, a decimal approximation for $\sin 50°$ can be found by entering [SIN] [5] [0] (or [5] [0] [SIN] on some scientific calculators). The answer, to six decimal places, is 0.766044. To find $\csc 50°$, you could then press the reciprocal key, usually labeled [$x^{-1}$] or [$1/x$], to obtain 1.305407. If an angle measure is given in radians, the process is the same except that you must be sure your calculator is set in radian mode.

**FINDING ANGLE MEASURES** Use the [$SIN^{-1}$], [$COS^{-1}$], or [$TAN^{-1}$] key (usually in conjunction with the [2nd] or [INV] key) to approximate an angle whose sine, cosine, or tangent is a given value. For example, to find an angle whose sine is 0.4, that is, to find $\theta$ so that $\sin\theta = 0.4$, you would enter [$SIN^{-1}$] [.] [4] (or [.] [4] [$SIN^{-1}$] on some scientific calculators). The answer will be given in degrees or radians, depending on the mode your calculator is in at the time the operation is performed. In degrees, the answer would be 23.578178, while in radians it would be 0.411517.

**EXAMPLE 4**    *Using a calculator to find trigonometric function values*

If $\theta$ is an acute angle for which $\sec \theta = 8$, use a calculator to find decimal approximations for $\theta$ and the other trigonometric function values for $\theta$.

**SOLUTION**    Since most calculators do not have a $\boxed{\text{SEC}^{-1}}$ key, we first apply a reciprocal identity to obtain

$$\cos \theta = \frac{1}{\sec \theta} = \frac{1}{8} = 0.125$$

Now using the $\boxed{\text{COS}^{-1}}$ key with a calculator set in degree mode, we see that $\theta \approx 82.819244°$. Finally, using the $\boxed{\text{SIN}}$, $\boxed{\text{TAN}}$, and $\boxed{\text{x}^{-1}}$ keys, we obtain

$$\sin \theta \approx 0.992157 \qquad\qquad \tan \theta \approx 7.937254$$

$$\csc \theta \approx \frac{1}{0.992157} \approx 1.007905 \qquad \cot \theta \approx \frac{1}{7.937254} \approx 0.125988$$

**APPLICATIONS**

The early development of trigonometry was motivated by problems in astronomy, navigation, and surveying. Consider, for example, the problem of a surveyor who wishes to determine the distance between two points on opposite sides of a river, labeled $A$ and $B$ in Figure 29. One approach would be to mark a third point, labeled $C$ in Figure 29, a certain distance from point $B$ and along a line perpendicular to $AB$. The surveyor could then measure the angle $\theta$ formed by the line segments $BC$ and $AC$ and use trigonometry to determine the distance from $A$ to $B$. We illustrate this process in the following example.

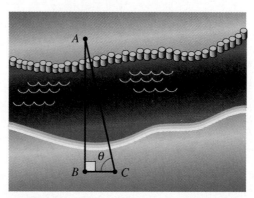

**FIGURE 29**

**EXAMPLE 5**    *Solving a right triangle*

Find all unknown side lengths and angle measures for the triangle shown in Figure 29 if it is known that the angle at vertex $C$ is 78° and the distance from $B$ to $C$ is 50 feet.

**SOLUTION**    Figure 30 shows the right triangle from Figure 29 with the known side length and angle measure labeled. The distance $c$ from $A$ to $B$ is the length of the side opposite angle $C$. The distance from $B$ to $C$, namely 50 feet, is the length of the

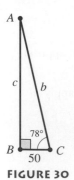

**FIGURE 30**

side adjacent to angle $C$. Thus, we need a trigonometric function that relates the angle $C$ to its opposite and adjacent sides. We choose tangent instead of cotangent since its values can be calculated directly with a calculator. Since $\tan \theta = \text{opp/adj}$, we have

$$\tan 78° = \frac{c}{50} \qquad \text{Using the definition of tangent}$$

$$50 \tan 78° = c \qquad \text{Multiplying both sides by 50}$$

$$c = 50 \tan 78°$$

$$\approx 50(4.7046) = 235.23 \qquad \text{Using a calculator to compute tan 78°}$$

Thus, the distance $c$ from $A$ to $B$ is approximately 235.23 feet. The distance $b$ is the length of the hypotenuse and so we apply the Pythagorean Theorem:

$$b = \sqrt{50^2 + c^2} \approx \sqrt{50^2 + 235.23^2} \approx 240.49$$

Finally, angle $A$ is the complement of angle $C$, and so

$$A = 90° - C = 90° - 78° = 12°$$

In many applications involving trigonometry, we consider angles formed by a horizontal line and the line of sight from a reference point on the horizontal line to an object above or below it. If the object is above the horizontal line, we use the term **angle of elevation**, and if it is below, we use the term **angle of depression**, as illustrated in Figures 31 and 32.

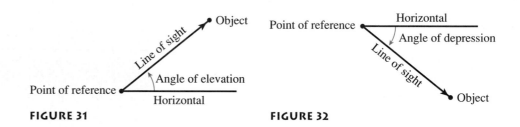

**FIGURE 31**          **FIGURE 32**

**EXAMPLE 6**    *The world's steepest railroad tracks*

According to the *Guinness Book of World Records*, the world's steepest standard-gauge railroad tracks are between Chedde and Servoz, France. On average, a train will rise 100 feet in elevation for every 1105 feet of track length traveled. Find the average angle of elevation of the tracks.

**SOLUTION**    Since the tracks rise an average of 100 feet for every 1105 of track length, we consider the right triangle shown in Figure 33. To find the average angle

**FIGURE 33**

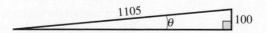

of elevation $\theta$, we note that the sine function establishes a relationship between $\theta$ and the two given distances. Thus,

$$\sin \theta = \frac{\text{opp}}{\text{hyp}} = \frac{100}{1105} \approx 0.0905$$

Now using the $\boxed{\text{SIN}^{-1}}$ key on a calculator in degree mode, we obtain

$$\theta \approx 5.19°$$

In surveying and navigation, directions are often specified in terms of bearings or courses. Generally speaking, a **bearing** gives a direction of observation while a **course** gives a direction of movement. Historically, a bearing (or course) was given as the acute angle made with a north-south line. In Figure 34, for example, ship A has a course of N 40° W, whereas ship B has a course of S 15° W. The modern approach is to express bearing (or course) as the angle measured clockwise from the north about a north-south line. Thus, ships A and B have courses of 320° and 195°, respectively, as shown in Figure 35.

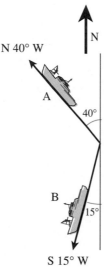

**FIGURE 34**

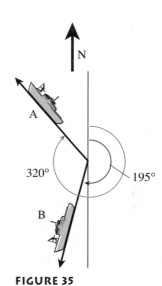

**FIGURE 35**

**EXAMPLE 7**    *Using course to compute distance*

A ship leaves port and travels at a rate of 15 miles per hour on a course of N 40° W. If the shoreline is roughly north-south, how far is the ship from shore after $3\frac{1}{2}$ hours?

**SOLUTION**    After $3\frac{1}{2}$ hours, the ship has traveled $(3.5)(15) = 52.5$ miles from port. From Figure 36, we have

$$\sin 40° = \frac{\text{opp}}{\text{hyp}} = \frac{d}{52.5}$$

and it follows that

$$d = 52.5 \sin 40° \approx 33.7 \text{ miles}$$

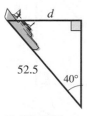

**FIGURE 36**

## EXERCISES 2

**EXERCISES 1–6** ☐ *Find the values of the six trigonometric functions for the given angle θ.*

**1.**

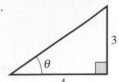

**2.**

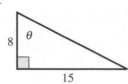

**3.**

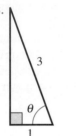

**4.**

**5.**

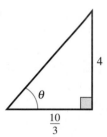

**6.**

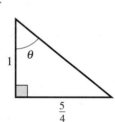

**EXERCISES 7–12** ☐ *Use a right triangle to find the other five trigonometric functions of the acute angle θ.*

**7.** $\sin \theta = \dfrac{5}{13}$     **8.** $\cos \theta = \dfrac{3}{5}$

**9.** $\sec \theta = 4$     **10.** $\cot \theta = 3$

**11.** $\tan \theta = \dfrac{2}{3}$     **12.** $\csc \theta = \dfrac{7}{4}$

**EXERCISES 13–20** ☐ *Use trigonometric identities to find the indicated trigonometric function value for the acute angle θ with the given function value.*

**13.** $\cos \theta$; $\sec \theta = 2$     **14.** $\tan \theta$; $\cot \theta = 5$

**15.** $\sin \theta$; $\cos \theta = \dfrac{1}{3}$     **16.** $\sec \theta$; $\tan \theta = \dfrac{3}{2}$

**17.** $\cot \theta$; $\sec \theta = \dfrac{25}{24}$     **18.** $\csc \theta$; $\cos \theta = \dfrac{8}{17}$

**19.** $\tan \theta$; $\cot(90° - \theta) = 8$     **20.** $\cos(90° - \theta)$; $\sin \theta = 0.25$

**EXERCISES 21–28** ☐ *Find the exact value of the given trigonometric function.*

**21.** $\sin 30°$     **22.** $\tan 45°$

**23.** $\tan \dfrac{\pi}{3}$     **24.** $\cos \dfrac{\pi}{6}$

**25.** $\sec 30°$     **26.** $\csc 60°$

**27.** $\cot \dfrac{\pi}{6}$     **28.** $\sec \dfrac{\pi}{4}$

**EXERCISES 29–38** ☐ *Use a calculator to find a decimal approximation for the given function, correct to four decimal places.*

**29.** $\sin 20°$     **30.** $\cos 75°$

**31.** $\cot 88.4°$     **32.** $\csc 15.8°$

**33.** $\sec 20°42'$     **34.** $\tan 51°11'$

**35.** $\cos \dfrac{\pi}{8}$     **36.** $\sin \dfrac{\pi}{12}$

**37.** $\csc 1$     **38.** $\cos 2$

**EXERCISES 39–44** ☐ *Use a calculator to find a decimal approximation, correct to four decimal places, for the acute angle θ measured in (a) degrees and (b) radians.*

**39.** $\cos \theta = 0.2$     **40.** $\sin \theta = 0.75$

**41.** $\tan \theta = 4.36$     **42.** $\cos \theta = 0.94$

**43.** $\csc \theta = 5.6$     **44.** $\sec \theta = 1.3$

**EXERCISES 45–54** ☐ *Find all unknown side lengths and angle measures; that is, solve the given triangle.*

**45.**

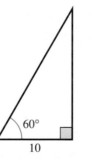

**46.**

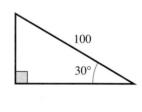

**47.**

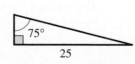

**48.**

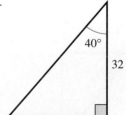

**49.**

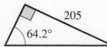

**50.**

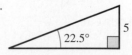

**51.**

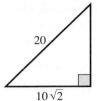

20

$10\sqrt{2}$

**52.**

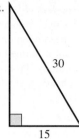

30

15

**53.**

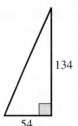

134

54

**54.**

102

48

■ *Applications*

**55.** *Height of the World's Tallest Tree* According to the *Guinness Book of World Records*, the tallest living tree is the "National Geographic Society" coast redwood in Humboldt Redwoods State Park, California. At a distance of 200 feet from the base of the tree, the angle of elevation to the top of the tree is approximately 61.3°. How tall is the tree?

Climbing champion Isabelle Patissier in action on a California redwood.

**56.** *Leaning Ladder* Based on OSHA (Occupational Safety and Health Administration) standards, an extension ladder should make a 75° angle with the ground. Using this rule of thumb, find the height of the tip of a 20-foot ladder.

**57.** *Pinpointing a Fire* A U.S. Forest Service helicopter is flying at a height of 500 feet. The pilot spots a fire in the distance with an angle of depression of 12°. Find the horizontal distance to the fire.

**58.** *Wheelchair Ramp* According to the ADA (Americans with Disabilities Act) Accessibility Guidelines, the maximum angle of a wheelchair ramp in new construction may not exceed 4.764°. If the door to a building is 5 feet above sidewalk level, how long must the ramp be in order to extend from sidewalk to door?

**59.** *Computing Area by Measuring One Distance* The length of the diagonal of a rectangular plot of land is 225 feet, and the angle made between the diagonal and one of the sides is 20°. Find the area of the plot.

**60.** *Estimating a Long Jump* Mike Powell, world record holder (as of 1996) in the long jump at approximately 29 feet 4 inches, comes upon a ravine on a hike. From a point directly opposite a rock on the other side of the ravine, he steps off a distance of 20 feet. From this vantage point, his line of sight to the rock forms an angle of 57° with the edge of the ravine (see Figure 37). Assuming the two sides of the ravine are parallel, can he make the jump?

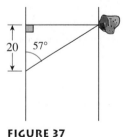

20  57°

**FIGURE 37**

Mike Powell breaks long jump record.

**61.** *Estimating Golf Distances*  On approximately level ground, the angle of elevation from a point 5 feet off the ground to the top of a 7-foot golf hole flag is 1.2°. How far away is the hole?

**62.** *Skiing Distance*  From experience, a skier knows that he cannot maintain control on slopes steeper than 30°. If he is at the top of a 5000-foot slope, skis at a constant speed of 40 feet per second, and by zigzagging is always able to keep his skis at an angle of 30° with the horizontal, how long will it take him to get to the bottom of the slope?

**63.** *Volume of a Grain Pile*  The base of a conical pile of grain has a circumference of 160 feet, and the angle made between the ground and the side of the pile is 35°. Find the volume of the pile.

**64.** *Locating a Car*  From the 102$^{nd}$ floor observation deck of the Empire State Building, an observer spots his car at an angle of depression of 5°. If the observation deck is 1250 feet high, how far is his car from the building?

**65.** *Hiking Course*  A hiker leaves a north-south highway, setting out on a constant course of 153° clockwise from north. After 3 miles, he trips over a fallen branch and sprains his ankle. How far is he from the highway, using the shortest possible route?

**66.** *Windsurfer Course*  A windsurfer starts from shore and travels 5 miles on a course of N 42° W. If the shoreline is roughly north-south, how far from shore is she?

**67.** *Supersonic Travel*  Sound travels at approximately 742 miles per hour. A fighter plane is traveling at Mach 2, that is, at twice the speed of sound, with a course of N 35° E. At 7:00 P.M. the pilot is warned that the enemy border, which forms an east-west line, is 100 miles due north. If the plane continues on its current bearing, at what time will it cross into enemy airspace?

**68.** *Calculating a Course*  A ship leaves port and travels on a constant course and with a constant speed. After 2 hours, the ship is 55 miles south and 20 miles west of the port. What is its course and speed?

**69.** *Shadow Length Mystery*  Inspector Magill is investigating a bank robbery that occurred in New York on June 20, 1997. The prime suspect claims to have been in London on that date, and makes a plea on national television for someone to corroborate his story. Several days later, a witness steps forward with a videotape showing the suspect in front of Big Ben in London. The date on the video shows June 20, 1997, and the time, 3:00 P.M., coincides with that shown on Big Ben. After a visit to London, Inspector Magill determines that the suspect's shadow in the video was 8.7 feet long. The suspect is 6 feet tall. If the angle of elevation of the Sun in London, England, at 3:00 P.M. on June 20, 1997, was 45°59′, how does Magill determine that the suspect and the witness are lying?

## Projects for Enrichment

**70.** *Measuring Angle of Elevation*  In this project, you will construct a device for measuring angles of elevation. You will need the following items:

- A long straight stick (a yardstick would be ideal)
- A protractor

- A piece of string approximately 1 foot long
- A weight that can be attached to one end of the string
- A tape measure

**a.** Attach the protractor to the stick so that its base is parallel to the side of the stick. Then attach one end of the string to the "center point" of the protractor, and attach the weight to the other end of the string. See Figure 38.

To estimate an object's angle of elevation, line the stick up with the object, allow the string to hang freely as shown in Figure 38, and record the angle indicated by the string on the protractor. The angle of elevation will be the complement of the angle indicated on the protractor.

b. Use your device to estimate the heights of five campus (or community) landmarks. Describe the steps you took, and include all relevant measurements.

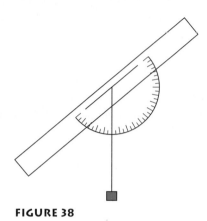

**FIGURE 38**

71. *Trigonometry to the Rescue*  A U.S. Congressman is being held hostage in Washington, D.C., by international terrorists who hope to exchange him for an imprisoned comrade. While the lone guard has left to use the restroom, the congressman is able to place a short call to his good friend Inspector Magill. Thinking quickly and realizing that time may be too short to trace the call, Magill asks the congressman if any landmarks are visible outside the groundfloor window near the phone. The Washington Monument is to the right, and the Capitol Building is to the left. Magill then asks the congressman to estimate the

angles of elevation to the top of each landmark. The congressman responds with "7° to the top of the Washington Monument" and "5° to the tip of the Capitol Building dome." Hearing the guard returning, the congressman quickly hangs up the phone. Several hours later, Magill and the FBI arrive and rescue the congressman. In the following steps, we will reconstruct the method used by Magill to estimate the location of the terrorist hideout.

a. The Washington Monument is approximately 550 feet high. Approximately how far is the terrorist hideout from the Washington Monument?

b. The tip of the dome of the Capitol Building is approximately 300 feet high. Approximately how far is the terrorist hideout from the Capitol Building?

c. The Washington Monument and the Capitol Building are approximately 7400 feet apart. By setting up a coordinate system with the Capitol Building at the origin, find the approximate location of the terrorist hideout.

d. If the congressman's estimates of the angle of elevation were off by as much as 1°, sketch the region of Washington, D.C., that would have to be searched by Magill and the FBI. (*Hint:* A compass may be useful.)

---

### ■ *Questions for Discussion or Essay*

72. Explain why the values of sin θ and cos θ must be less than 1 for any acute angle θ. On the other hand, why can tan θ be greater than 1?

73. Explain why enormous errors could result from using trigonometry to compute great distances on the surface of the Earth.

74. In Exercise 62 you were asked to compute the time it would take for a skier to get to the bottom of a slope. Explain why the answer doesn't depend upon the specific path the skier follows.

75. What can be said about the relative measures of α and β if cos α > cos β? Explain.

76. If the distance to a building is known, what additional information is required in order to obtain the height?

77. An acute angle θ in a right triangle is such that sin θ = a/b. Explain why this does not necessarily imply that the side opposite θ has length a and the hypotenuse has length b.

SECTION 3

# TRIGONOMETRIC FUNCTIONS OF REAL NUMBERS

■ What initial angle and velocity are needed to break the world's record in the javelin throw?

■ How can the average temperature in Orlando, Florida, and the blood pressure of a dog be modeled with trigonometric functions?

■ How can trigonometric functions be approximated using polynomials?

■ How can trigonometry be used to estimate the area of the state of Wyoming?

■ How can one compute the trigonometric function values of angles that are not part of any right triangle?

## TRIGONOMETRIC FUNCTIONS OF ANY ANGLE

In the previous section, we defined the trigonometric functions for acute angles. Now we would like to extend our definitions to any angle. To do so, we consider a **unit circle**, that is, a circle with radius 1, centered at the origin. Now each angle $\theta$ corresponds to a unique point $(x, y)$ on the unit circle, as suggested by Figure 39. For an acute angle $\theta$, the trigonometric definitions from the previous section give us

$$\sin \theta = \frac{\text{opp}}{\text{hyp}} = \frac{y}{1} = y \qquad \cos \theta = \frac{\text{adj}}{\text{hyp}} = \frac{x}{1} = x \qquad \tan \theta = \frac{\text{opp}}{\text{adj}} = \frac{y}{x}$$

$$\csc \theta = \frac{\text{hyp}}{\text{opp}} = \frac{1}{y} \qquad \sec \theta = \frac{\text{hyp}}{\text{adj}} = \frac{1}{x} \qquad \cot \theta = \frac{\text{adj}}{\text{opp}} = \frac{x}{y}$$

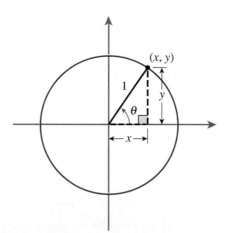

**FIGURE 39**

We extend these definitions to any angle $\theta$ simply by referring only to the coordinates of the associated point on the unit circle, avoiding any reference to a right triangle.

**Definition of trigonometric functions of any angle**

Let $\theta$ be an angle in standard position with $(x, y)$ the associated point on the unit circle on the terminal side of $\theta$. We define the six trigonometric functions of $\theta$ as follows:

$$\sin \theta = y \qquad\qquad \cos \theta = x \qquad\qquad \tan \theta = \frac{y}{x}, \; x \neq 0$$

$$\csc \theta = \frac{1}{y}, \; y \neq 0 \qquad \sec \theta = \frac{1}{x}, \; x \neq 0 \qquad \cot \theta = \frac{x}{y}, \; y \neq 0$$

Note that the fundamental trigonometric identities discussed in the previous section continue to hold. Thus, for example, $\tan \theta = \sin \theta / \cos \theta$.

---

**RULE OF THUMB**   When finding the trigonometric function values for an angle $\theta$ with associated point $P(x, y)$ on the unit circle, it is only necessary to remember that

$$\cos \theta = \text{the } x\text{-coordinate of } P \qquad \text{and} \qquad \sin \theta = \text{the } y\text{-coordinate of } P$$

The remaining function values can be found using the ratio identities.

---

**EXAMPLE 1**    *Finding angle measure and trigonometric function values given a point on the unit circle*

For the given point on the unit circle, find an angle $\theta$ in standard position whose terminal side passes through the point, and then find the trigonometric function values for that angle.

**a.** $(0, 1)$     **b.** $\left( -\dfrac{\sqrt{2}}{2}, \dfrac{\sqrt{2}}{2} \right)$

**SOLUTION**

**a.** The point $(0, 1)$ is on the $y$-axis, and so $\theta = 90°$ (see Figure 40). By the trigonometric function definitions given above, we have

$$\sin 90° = y = 1 \qquad \cos 90° = x = 0 \qquad \tan 90° = \frac{y}{x} \text{ is undefined} \atop \text{since } x = 0$$

$$\csc 90° = \frac{1}{y} = 1 \qquad \sec 90° = \frac{1}{x} \text{ is undefined} \atop \text{since } x = 0 \qquad \cot 90° = \frac{x}{y} = 0$$

**b.** The point $(-\sqrt{2}/2, \sqrt{2}/2)$ lies in the second quadrant on the line $y = -x$ which makes a 45° angle with the negative $x$-axis. Thus $\theta = 180° - 45° = 135°$ (see Figure 41). Applying the definitions for sine, cosine, and tangent, and the ratio identities for cosecant, secant, and cotangent, we obtain

$$\sin 135° = y = \frac{\sqrt{2}}{2} \qquad \cos 135° = x = -\frac{\sqrt{2}}{2} \qquad \tan 135° = \frac{y}{x}$$

$$= \frac{\sqrt{2}/2}{-\sqrt{2}/2} = -1$$

$$\csc 135° = \frac{1}{\sin 135°} \qquad \sec 135° = \frac{1}{\cos 135°} \qquad \cot 135° = \frac{1}{\tan 135°}$$

$$= \frac{1}{\sqrt{2}/2} = \sqrt{2} \qquad = \frac{1}{-\sqrt{2}/2} = -\sqrt{2} \qquad = \frac{1}{-1} = -1$$

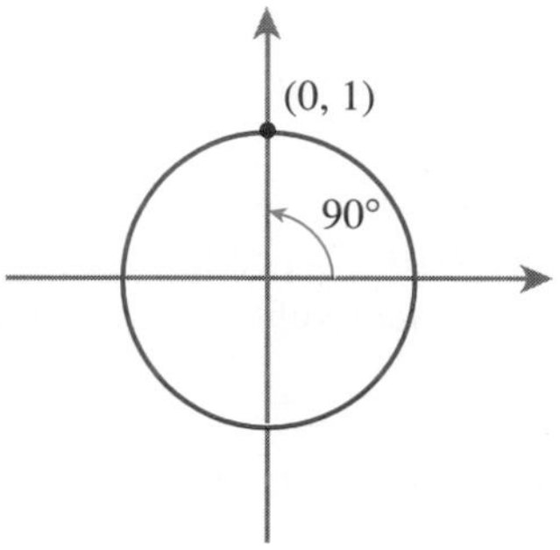

FIGURE 40 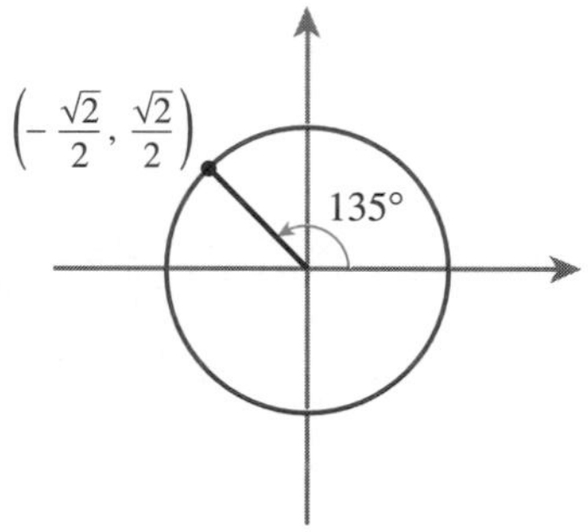                     FIGURE 41

The trigonometric function values for the other quadrant angles can be found in the same way as was done for 90° in the previous example. These values are summarized in Table 2.

**TABLE 2**
*Quadrant Angles*

| $\theta$ | $\theta$ (radians) | sin $\theta$ | cos $\theta$ | tan $\theta$ |
|---|---|---|---|---|
| 0° | 0 | 0 | 1 | 0 |
| 90° | $\frac{\pi}{2}$ | 1 | 0 | undefined |
| 180° | $\pi$ | 0 | −1 | 0 |
| 270° | $\frac{3\pi}{2}$ | −1 | 0 | undefined |

**EXAMPLE 2**    *Finding trigonometric function values for a given angle*

Find the coordinates of the point $P$ on the unit circle associated with 300°, and determine the sine, cosine, and tangent of 300°.

**SOLUTION**    The point $P$ is in the fourth quadrant, as shown in Figure 42. The line passing through $P$, perpendicular to the $x$-axis, intersects the $x$-axis at a point we denote by $Q$. Since 300° is 60° less than a full revolution, $\angle POQ = 60°$. Moreover, since $P$ lies on the unit circle, the length of the hypotenuse is 1. Thus, since $\sin 60° = \sqrt{3}/2$ and $\cos 60° = \frac{1}{2}$, the legs of $\triangle OQR$ have length $\frac{1}{2}$ and $\sqrt{3}/2$, as shown in Figure 43. Now since $P$ is in the fourth quadrant, its $x$-coordinate must be positive and its $y$-coordinate must be negative. Thus, the point $P$ has coordinates $(\frac{1}{2}, -\sqrt{3}/2)$. We can now apply the definitions of sine, cosine, and tangent to obtain

$$\sin 300° = y = -\frac{\sqrt{3}}{2}, \quad \cos 300° = x = \frac{1}{2}, \quad \tan 300° = \frac{y}{x} = \frac{-\sqrt{3}/2}{1/2} = -\sqrt{3}$$

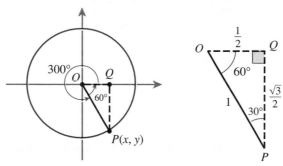

**FIGURE 42**                          **FIGURE 43**

An alternative approach to finding trigonometric function values for angles larger than 90° is to use information about the *signs* of the trigonometric function values, together with the notion of a *reference angle*. The following comparison between the values for the sine, cosine, and tangent of 60° and 300° suggests how this alternative approach works.

$$\sin 60° = \frac{\sqrt{3}}{2} \qquad \cos 60° = \frac{1}{2} \qquad \tan 60° = \sqrt{3}$$

$$\sin 300° = -\frac{\sqrt{3}}{2} \qquad \cos 300° = \frac{1}{2} \qquad \tan 300° = -\sqrt{3}$$

Notice that the only differences are the signs of sine and tangent. Thus, the trigonometric function values for 300° can be found simply by attaching the appropriate signs to the corresponding values for 60°. We now describe these ideas more fully.

## SIGNS AND REFERENCE ANGLES

The signs of the trigonometric function values for a given angle $\theta$ are determined by the quadrant in which $\theta$ lies. For example, if $\theta$ is in the second quadrant with associated point $P(x, y)$ on the unit circle, then $x < 0$ and $y > 0$. Thus,

$$\sin \theta = y > 0 \qquad \cos \theta = x < 0$$

In a similar way we obtain the signs shown in Figure 44 for the sine and cosine functions. Notice that in both quadrants II and III, exactly one of the two is positive. This fact leads us to Figure 45, where we indicate which of the sine or cosine functions are positive in the given quadrant. The signs of tangent, cotangent, cosecant, and secant can be determined from the signs of sine and cosine, using the appropriate identity.

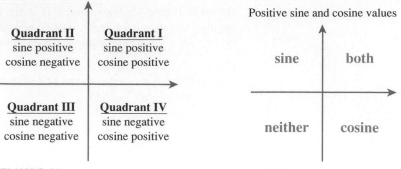

FIGURE 44                                    FIGURE 45

**EXAMPLE 3**    *Finding the sine of an angle*

Given that $\cos \theta = -\frac{1}{4}$ and $\tan \theta > 0$, find $\sin \theta$.

**SOLUTION**    We apply the Pythagorean identity $\sin^2 \theta + \cos^2 \theta = 1$ to obtain

$$\sin^2 \theta = 1 - \cos^2 \theta = 1 - \left(-\frac{1}{4}\right)^2 = \frac{15}{16}$$

$$\sin \theta = \pm \sqrt{\frac{15}{16}} = \pm \frac{\sqrt{15}}{4}$$

Since $\tan \theta = \sin \theta / \cos \theta$ is positive and $\cos \theta$ is negative, it follows that $\sin \theta$ is negative. Thus, we choose the negative square root to obtain

$$\sin \theta = -\frac{\sqrt{15}}{4}$$

■

The **reference angle** for an angle $\theta$ in standard position is the acute angle $\tilde{\theta}$ formed between the terminal side of $\theta$ and the horizontal axis. Figure 46 shows several illustrations of reference angles.

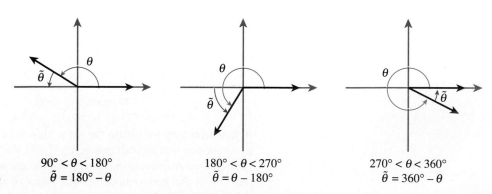

FIGURE 46

$90° < \theta < 180°$        $180° < \theta < 270°$        $270° < \theta < 360°$
$\tilde{\theta} = 180° - \theta$        $\tilde{\theta} = \theta - 180°$        $\tilde{\theta} = 360° - \theta$

**EXAMPLE 4**   *Finding reference angles*

Find the reference angle $\tilde{\theta}$ for the given angle $\theta$.

a. $\theta = 240°$          b. $\theta = \dfrac{5\pi}{6}$          c. $\theta = -225°$          d. $\theta = 5$

**SOLUTION**

a. We first sketch the angle $\theta = 240°$ in the third quadrant as shown in Figure 47. Its reference angle is the acute angle made with the negative $x$-axis. Thus,

$$\tilde{\theta} = 240° - 180° = 60°$$

b. The angle $\theta = 5\pi/6$ is in the second quadrant (see Figure 48). Its reference angle is

$$\tilde{\theta} = \pi - \frac{5\pi}{6} = \frac{\pi}{6}$$

c. As we see in Figure 49, the angle $\theta = -225°$ is in the second quadrant and is coterminal with $135°$. Thus, its reference angle is

$$\tilde{\theta} = 180° - 135° = 45°$$

d. The angle $\theta = 5$ is between $3\pi/2 \approx 4.71$ and $2\pi \approx 6.28$, and so is in the fourth quadrant (Figure 50). Thus, its reference angle is

$$\tilde{\theta} = 2\pi - 5 \approx 1.2832$$

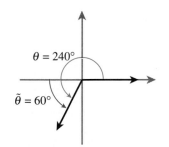

**FIGURE 47**

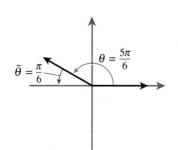

**FIGURE 48**

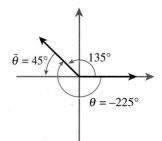

**FIGURE 49**

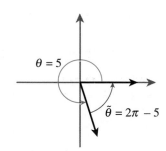

**FIGURE 50**

Reference angles can be used to evaluate trigonometric functions for angles larger than $90°$. To see how this is done, consider the nonacute angle $\theta$ with associated point $P(x, y)$ on the unit circle shown in Figure 51. From the definitions of sine, cosine, and tangent, we have

$$\sin\theta = y, \qquad \cos\theta = x, \qquad \text{and} \qquad \tan\theta = \frac{y}{x}$$

Now consider the right triangle with acute angle $\bar{\theta}$, the reference angle for $\theta$ as shown in Figure 51. The side opposite $\bar{\theta}$ has length $|y|$, the side adjacent to $\bar{\theta}$ has length $|x|$, and the hypotenuse has length 1. Thus,

$$\sin \bar{\theta} = |y|, \qquad \cos \bar{\theta} = |x| \qquad \text{and} \qquad \tan \bar{\theta} = \left| \frac{y}{x} \right|$$

So the sine, cosine, and tangent of $\theta$ and $\bar{\theta}$ differ by at most a sign, which can easily be obtained using the quadrant information on signs described earlier.

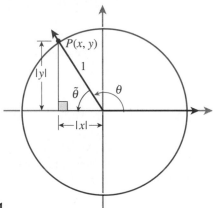

**FIGURE 51**

---

**RULE OF THUMB**    For a nonquadrant angle $\theta$ with reference angle $\bar{\theta}$, we have

$$\text{trigfunction}(\theta) = \pm\text{trigfunction}(\bar{\theta})$$

where "trigfunction" denotes any of the six trigonometric functions. In fact, we can say even more. If $\alpha$ and $\beta$ are two angles for which

$$\text{trigfunction}(\alpha) = \pm\text{trigfunction}(\beta)$$

then $\alpha$ and $\beta$ must have the same reference angle.

---

We give the following procedure for evaluating trigonometric functions using reference angles.

**Evaluating trigonometric functions using reference angles**

The value of a trigonometric function of a nonquadrant angle $\theta$ in standard position can be found as follows:

1. Sketch the angle $\theta$ to help determine its quadrant and reference angle $\bar{\theta}$.
2. Find the value of the trigonometric function for the associated reference angle $\bar{\theta}$.
3. Prefix the appropriate sign, according to the quadrant in which $\theta$ lies.

We can apply the reference angle technique to find exact trigonometric function values for any angle that has a reference angle of 30°, 45°, or 60°. Recall that these are the angles whose exact trigonometric function values were found in the previous section using 45-45-90 and 30-60-90 triangles. For reference, Table 3 summarizes the information that is useful for finding exact values of trigonometric functions.

**TABLE 3**

*Special Angles*

| $\theta$ | $\theta$ (radians) | sin $\theta$ | cos $\theta$ | tan $\theta$ |
|---|---|---|---|---|
| 0° | 0 | 0 | 1 | 0 |
| 30° | $\dfrac{\pi}{6}$ | $\dfrac{1}{2}$ | $\dfrac{\sqrt{3}}{2}$ | $\dfrac{\sqrt{3}}{3}$ |
| 45° | $\dfrac{\pi}{4}$ | $\dfrac{\sqrt{2}}{2}$ | $\dfrac{\sqrt{2}}{2}$ | 1 |
| 60° | $\dfrac{\pi}{3}$ | $\dfrac{\sqrt{3}}{2}$ | $\dfrac{1}{2}$ | $\sqrt{3}$ |
| 90° | $\dfrac{\pi}{2}$ | 1 | 0 | undefined |
| 180° | $\pi$ | 0 | $-1$ | 0 |
| 270° | $\dfrac{3\pi}{2}$ | $-1$ | 0 | undefined |
| 360° | $2\pi$ | 0 | 1 | 0 |

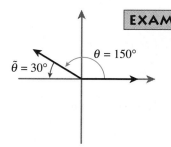

**FIGURE 52**

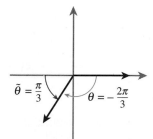

**FIGURE 53**

**EXAMPLE 5**    *Using reference angles to find trigonometric function values*

Use reference angles to find the exact values for each of the following.

**a.** $\tan 150°$        **b.** $\sin\left(-\dfrac{2\pi}{3}\right)$        **c.** $\sec\dfrac{7\pi}{4}$

**SOLUTION**

**a.** The angle $\theta = 150°$ is sketched in Figure 52 along with its reference angle $\tilde{\theta} = 30°$. Now $\tan 30° = \sqrt{3}/3$ (see Table 3), and since $\theta$ is in the second quadrant, $\tan \theta = \sin \theta/\cos \theta$ is negative ($\sin \theta > 0$ and $\cos \theta < 0$). Thus,

$$\tan 150° = -\tan 30° = -\frac{\sqrt{3}}{3}$$

**b.** The angle $\theta = -2\pi/3$ is sketched in Figure 53 along with its reference angle $\tilde{\theta} = \pi/3$. Since $\theta$ is in the third quadrant, the sine of $\theta$ is negative. Thus,

$$\sin\left(-\frac{2\pi}{3}\right) = -\sin\left(\frac{\pi}{3}\right) = -\frac{\sqrt{3}}{2}$$

**c.** The angle $\theta = 7\pi/4$ is sketched in Figure 54 along with its reference angle $\tilde{\theta} = \pi/4$. Since $\theta$ is in the fourth quadrant, the secant of $\theta$, which has the same sign as the cosine of $\theta$, is positive. Thus

$$\sec \frac{7\pi}{4} = \sec \frac{\pi}{4} = \frac{1}{\cos(\pi/4)} = \frac{1}{\sqrt{2}/2} = \sqrt{2}$$

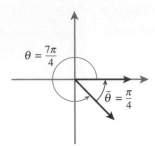

**FIGURE 54**

When exact values cannot be found, a calculator may be used to approximate trigonometric function values. As discussed in the previous section, it is important to be sure that the calculator is set in the appropriate mode for angle measures (degrees or radians).

**EXAMPLE 6**    *Approximating trigonometric function values with a calculator*

Use a calculator to approximate the given trigonometric function value.
**a.** $\cos 200°$         **b.** $\cot(-3)$

**SOLUTION**

**a.** With a calculator set in degree mode, we use the $\boxed{\text{COS}}$ key to obtain

$$\cos 200° \approx -0.939693$$

**b.** With a calculator set in radian mode, we first use the $\boxed{\text{TAN}}$ key to obtain

$$\tan(-3) \approx 0.142547$$

Next we use the reciprocal key $\boxed{x^{-1}}$ to compute

$$\cot(-3) = \frac{1}{0.142547} \approx 7.015253$$

## TRIGONOMETRIC FUNCTIONS OF REAL NUMBERS

In many real-world applications of trigonometry, as well as many areas of mathematics where trigonometry is used, the notion of an angle is not involved. For example, trigonometric functions can be used to model sound waves, electric current, and many other natural phenomena without making any references to angles. Thus, it is necessary to define trigonometric functions with real-number (rather than angle) domains. This can be done by first identifying each real number $t$ with a central angle $\theta_t$ of the unit circle having radian measure $t$. Then we may define $\sin t = \sin(\theta_t)$, $\cos t = \cos(\theta_t)$, and $\tan t = \tan(\theta_t)$. In other words, the value of a trigonometric function for a real number $t$ is simply the value of that function for an angle measuring $t$ radians.

**EXAMPLE 7**     *Evaluating a trigonometric function*

Given that $f(t) = \sin t$, find the indicated function value.

  **a.** $f(0)$        **b.** $f\!\left(\dfrac{\pi}{6}\right)$         **c.** $f(7.1)$

**SOLUTION**

**a.** Using Table 3, we obtain $f(0) = \sin 0 = 0$.

**b.** Using Table 3, we have $f\!\left(\dfrac{\pi}{6}\right) = \sin \dfrac{\pi}{6} = \dfrac{1}{2}$.

**c.** Using a calculator set in radian mode, we obtain $f(7.1) = \sin 7.1 \approx 0.728969$.

       ■

Just as with any other real-valued function, the graph of a trigonometric function $f$ is the set of points $(x, y)$ that satisfy the equation $y = f(x)$. We will study the graphs of the trigonometric functions in detail in the following sections. Here we will simply use a graphics calculator to view the graphs and to help us obtain information about the domain and range.

When plotting trigonometric functions with a graphics calculator, keep the following recommendations in mind.

  ■ Set your calculator in radian mode.

  ■ Start with your calculator's built-in TRIG range settings (if it has them), and then zoom in or out as necessary.

  ■ Use the $\boxed{\text{SIN}}$, $\boxed{\text{COS}}$, and $\boxed{\text{TAN}}$ keys for the sine, cosine, and tangent functions, respectively. For cosecant, secant, and cotangent, use the ratios $1/\sin x$, $1/\cos x$, and $1/\tan x$, respectively.

  ■ It is sometimes helpful to set your calculator in dot mode when plotting the tangent, cotangent, secant, or cosecant functions to avoid possible confusion over vertical asymptotes.

**EXAMPLE 8**     *Graphing the sine function with a graphics calculator*

Use a graphics calculator to graph $f(x) = \sin x$.

**SOLUTION**    With a graphics calculator set in radian mode, and using the built-in TRIG range settings, we enter and plot the function $y = \sin x$ to obtain the graph shown in Figure 55. Figure 56 shows the graph of the same function after changing the scale.

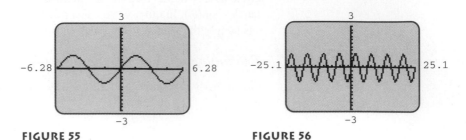

FIGURE 55                    FIGURE 56

■

It is a consequence of the fact that sin $x$ = sin(*an angle with radian measure x*) that $f(x)$ = sin $x$ is defined for all real numbers $x$. Thus, the domain of the sine function is the set of all real numbers. We can also obtain information about the range of the sine function from the graph in Figure 56. It appears that no portion of the graph lies above $y = 1$ or below $y = -1$. Thus, we suspect that the range is $[-1, 1]$. This suspicion can be confirmed by considering the definition of the sine function. Since sin $x$ is defined to be the $y$-coordinate of the point on the unit circle corresponding to a central angle with radian measure $x$, and since the $y$-coordinates of points on the unit circle vary between $-1$ and 1, the range is indeed $-1 \leq y \leq 1$.

Another important property of the sine function is suggested by Figure 56. Notice that the graph appears to be symmetric with respect to the origin. Recall that functions with this property are said to be *odd*. Algebraically, a function $f$ is odd if $f(-x) = -f(x)$ for all $x$ in the domain of $f$. This suggests that $\sin(-x) = -\sin x$. In a similar way, the graph of $g(x)$ = cos $x$ suggests that the cosine function is even, and so $\cos(-x) = \cos x$. To verify these observations, consider the angle $\theta$ shown in Figure 57 with associated point $(a, b)$ on the unit circle. The angle $-\theta$ has associated point $(a, -b)$. Thus, by the definitions of sine and cosine, we have

$$\sin(-\theta) = -b = -\sin \theta \qquad \text{and} \qquad \cos(-\theta) = a = \cos \theta$$

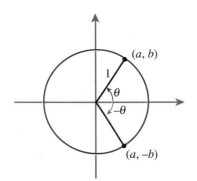

FIGURE 57

The even and odd properties of the sine and cosine functions are summarized as follows.

■ *Even/odd identities*

$$\sin(-\theta) = -\sin \theta \qquad \cos(-\theta) = \cos \theta$$

**EXERCISES 3**

**EXERCISES 1–4** □ *For the given point on the unit circle, find an angle $\theta$, $0 \leq \theta \leq 2\pi$, in standard position whose terminal side passes through the point, and then find the trigonometric function values for that angle.*

1. $(-1, 0)$

2. $\left(-\dfrac{\sqrt{2}}{2}, -\dfrac{\sqrt{2}}{2}\right)$

3. $\left(\dfrac{1}{2}, -\dfrac{\sqrt{3}}{2}\right)$

4. $\left(\dfrac{\sqrt{3}}{2}, \dfrac{1}{2}\right)$

**EXERCISES 5–8** ☐ *Find the coordinates of the point on the unit circle associated with the given angle, and use the definitions of the trigonometric functions to determine their values.*

**5.** $\theta = -90°$

**6.** $\theta = 315°$

**7.** $\theta = \dfrac{5\pi}{4}$

**8.** $\theta = 5\pi$

**EXERCISES 9–12** ☐ *Determine the quadrant in which $\theta$ lies.*

**9.** $\sin \theta < 0$, $\cos \theta > 0$

**10.** $\tan \theta > 0$, $\sin \theta < 0$

**11.** $\sec \theta < 0$, $\tan \theta > 0$

**12.** $\cot \theta < 0$, $\sin \theta > 0$

**EXERCISES 13–16** ☐ *Use the given information about the angle $\theta$ to find the indicated function value.*

**13.** $\sin \theta = \dfrac{1}{3}$, $\tan \theta < 0$; $\cos \theta = $ _____

**14.** $\tan \theta = -\dfrac{1}{2}$, $\cos \theta > 0$; $\sec \theta = $ _____

**15.** $\sec \theta = -3$, $\sin \theta < 0$; $\cot \theta = $ _____

**16.** $\cos \theta = 0.4$, $\cot \theta > 0$; $\csc \theta = $ _____

**EXERCISES 17–24** ☐ *Find the reference angle for the given angle.*

**17.** $\theta = 120°$

**18.** $\theta = -330°$

**19.** $\theta = -50°$

**20.** $\theta = 410°$

**21.** $\theta = \dfrac{5\pi}{4}$

**22.** $\theta = \dfrac{11\pi}{6}$

**23.** $\theta = -2$

**24.** $\theta = 8$

**EXERCISES 25–32** ☐ *Use reference angles to find the exact values of the given trigonometric functions.*

**25.** $\sin 210°$

**26.** $\tan(-120°)$

**27.** $\cos(-30°)$

**28.** $\csc 135°$

**29.** $\tan \dfrac{5\pi}{6}$

**30.** $\sec \dfrac{10\pi}{3}$

**31.** $\cot\left(-\dfrac{3\pi}{4}\right)$

**32.** $\sin\left(-\dfrac{\pi}{6}\right)$

**EXERCISES 33–38** ☐ *Use a calculator to find a decimal approximation, correct to four decimal places.*

**33.** $\sin 110°$

**34.** $\tan 305°$

**35.** $\sec 221.4°$

**36.** $\csc(-143.8°)$

**37.** $\cot(-2.3)$

**38.** $\cos 15.2$

**EXERCISES 39–42** ☐ *Find the indicated function values for the given function. Give exact values whenever possible.*

**39.** $f(t) = \cos t$

   **a.** $f(0)$    **b.** $f\left(-\dfrac{\pi}{4}\right)$    **c.** $f(2)$

**40.** $f(t) = \tan t$

   **a.** $f\left(\dfrac{\pi}{4}\right)$    **b.** $f\left(\dfrac{\pi}{2}\right)$    **c.** $f(-1)$

**41.** $f(t) = \csc t$

   **a.** $f\left(-\dfrac{2\pi}{3}\right)$    **b.** $f(\pi)$    **c.** $f(5.3)$

**42.** $f(t) = \sec t$

   **a.** $f\left(\dfrac{7\pi}{6}\right)$    **b.** $f(-5\pi)$    **c.** $f(21.2)$

**EXERCISES 43–46** ☐ *Use a graphics calculator to graph the given trigonometric function and then determine the domain and range.*

**43.** $f(x) = \cos x$

**44.** $f(x) = \tan x$

**45.** $f(x) = \csc x$

**46.** $f(x) = \sec x$

■ *Applications*

**47.** ***Javelin Throw*** If we ignore air resistance, the range in feet of a projectile thrown with an initial speed of $v$ feet per second, an initial angle $\theta$, and from an initial height of $h$ feet is given by

$$R = \frac{v \cos \theta}{32}\left(\sqrt{v^2 \sin^2 \theta + 64h} + v \sin \theta\right)$$

Suppose a javelin thrower releases a javelin with an initial speed of 100 feet per second from an initial height of 6 feet.

  **a.** Find the range of the javelin using initial angles of 0°, 30°, and 90°.

  **b.** Use a graphics calculator to estimate the angle that will give the maximum range. A world record javelin throw of 313.4

feet was set by Jan Zelezny in 1993. Is this range possible using the given initial velocity?

Jan Zelezny prepares to throw the javelin at the 1995 world championship games in Gothenburg, Sweden.

**48.** *Rolling Wheel* If a bicycle is traveling at a constant speed, a fixed point on the tire will travel forward at a rate that depends on the position of the point according to the function $f(\theta) = v + v \cos \theta$, where $v$ is the speed of the bicycle and $\theta$ is the angle shown in Figure 58.

  **a.** Find the forward speed of a point on the tire of a bicycle traveling at a rate of 20 miles per hour for $\theta = \pi/4$, $\theta = \pi/3$, and $\theta = \pi/2$.

  **b.** What is the maximum forward speed of the point and when is it attained? What about the minimum speed? Assume a rate of 20 miles per hour.

          **FIGURE 58**

**49.** *Average Temperature* The average monthly temperature (in degrees Fahrenheit) for Orlando, Florida can be approximated using the function

$$f(t) = 73 - 5.5 \cos\left(\frac{\pi t}{6}\right) - 13 \sin\left(\frac{\pi t}{6}\right)$$

where $t = 1$ corresponds to January 1996, $t = 2$ corresponds to February 1996, and so on.

  **a.** Compute $f(3)$ and $f(15)$ and interpret your result.

  **b.** Use a graphics calculator to estimate the month when the average temperature is highest and the month when it is lowest.

**50.** *Canine Blood Pressure* The aortic blood pressure (measured in mm Hg) of a dog can be approximated using the function

$$f(t) = -6.3 \cos(42t) + 5 \sin(42t) - 7.8 \cos(21t)$$
$$+ 16.2 \sin(21t) + 115$$

where $t$ is measured in seconds. (Data Source: *Circulation*, Bjorn Folkow and Eric Neil, Oxford: New York, 1971, p. 428.)

  **a.** Compute $f(0)$ and $f\left(\dfrac{\pi}{21}\right)$ and interpret your result.

  **b.** Use a graphics calculator to estimate the highest and lowest blood pressures.

---

### ◼ *Projects for Enrichment*

**51.** *Taylor Polynomials* The sine and cosine functions can be approximated with surprising accuracy using **Taylor polynomials**. In order to define these polynomials, we must first consider the notion of a **factorial**. We define $0! = 1$ and, for natural numbers $n \geq 1$, $n! = n(n-1)(n-2) \cdots (2)(1)$. Thus, for example, $4! = 4 \cdot 3 \cdot 2 \cdot 1 = 24$.

  **a.** Compute the following factorials.
     **i.** 3!    **ii.** 5!    **iii.** 6!

The Taylor polynomials for the sine function are defined as follows:

$$S_0(x) = x$$

$$S_1(x) = x - \frac{x^3}{3!}$$

$$S_2(x) = x - \frac{x^3}{3!} + \frac{x^5}{5!}$$

$$\vdots$$

$$S_n(x) = x - \frac{x^3}{3!} + \frac{x^5}{5!} - \cdots + (-1)^n \frac{x^{2n+1}}{(2n+1)!}$$

  **b.** Write out the Taylor polynomials $S_3$ and $S_4$. Describe in words any patterns you see in the polynomials.

  **c.** With your calculator set in radians mode, compute sin 1, $S_0(1)$, $S_1(1)$, $S_2(1)$, $S_3(1)$, and $S_4(1)$. Note that you can save yourself a lot of work by using the fact that

$$S_1(1) = S_0(1) - \frac{1}{3!}, \quad S_2(1) = S_1(1) + \frac{1}{5!}$$

and so on. In other words, each successive polynomial value can be found by adding a positive or negative term to the previous polynomial value. What do you notice about the values of $S_n(1)$ as $n$ gets larger?

  **d.** Complete the following table.

| $x$ | $\sin x$ | $S_0(x)$ | $S_1(x)$ | $S_2(x)$ | $S_3(x)$ | $S_4(x)$ |
|-----|----------|----------|----------|----------|----------|----------|
| 0.1 |          |          |          |          |          |          |
| 0.5 |          |          |          |          |          |          |
| 2   |          |          |          |          |          |          |
| 3   |          |          |          |          |          |          |

What does this table suggest about using the Taylor polynomials $S_n(x)$ to approximate values of $\sin x$? Be specific about how the sizes of $n$ and $x$ affect the approximation.

e. Use your graphics calculator to plot $y = \sin x$, $y = S_0(x)$, $y = S_1(x)$, and $y = S_2(x)$ on the same set of axes. Use an $x$-axis range of $-2 \le x \le 2$ and a $y$-axis range of $-1 \le y \le 1$. How do these graphs confirm what you found in part (d)?

The Taylor polynomials for the cosine function are defined as follows:

$$C_0(x) = 1$$

$$C_1(x) = 1 - \frac{x^2}{2!}$$

$$C_2(x) = 1 - \frac{x^2}{2!} + \frac{x^4}{4!}$$

$$\vdots$$

$$C_n(x) = 1 - \frac{x^2}{2!} + \frac{x^4}{4!} - \cdots + (-1)^n \frac{x^{2n}}{(2n)!}$$

f. Repeat parts (b) through (e) for the Taylor polynomials $C_n(x)$.

52. *Areas of Latitude/Longitude Zones* We define a **latitude zone** as a region bounded on the north and south by two parallels of latitude. Similarly, we define a **longitude zone** as a region bounded on the east and west by two meridians of longitude. A **latitude/longitude zone** is a "rectangular" region bounded on four sides by parallels of latitude and meridians of longitude. (See Figure 59.) In this project, we will develop techniques for finding the area of such zones. The area of a longitude zone bounded by longitudes with a degree difference $\theta$ is given by

$$A = \frac{\theta}{360} S$$

where $S$ is the surface area of the Earth.

a. Explain why the above equation gives the area of a longitude zone. Does the equation still hold for a zone bounded by longitudes on opposite sides of the prime meridian? Explain.

b. Find the surface area of the Earth, assuming it is a sphere with radius 3963 miles.

c. Find the area of the zone bounded by the given longitudes.
   i. 45° W, 50° W
   ii. 173° W, 150° E

d. The boundaries of time zones are often quite jagged, but they roughly coincide with meridians of longitude. Estimate the area of the Earth for which the time at a given instant is 3:00 P.M. Why do you think the boundaries of time zones don't exactly coincide with meridians of longitude?

To find the area of a latitude zone, we first convert each latitude to an angle between 0° and 180°. Given a latitude $m$, we define the **location angle** $\phi$ as follows:

$$\phi = \begin{cases} 90° - m & \text{for north latitude} \\ 90° + m & \text{for south latitude} \end{cases}$$

Now for two latitudes with corresponding location angles $\phi_1$ and $\phi_2$, the area of the latitude zone is given by

$$A = \frac{1}{2} |\cos \phi_1 - \cos \phi_2| S$$

e. Find the area of the zone bounded by the given latitudes, and determine the percentage of the Earth's surface area covered by the zone.
   i. The North Pole (90° N) and the Arctic Circle (66°30′ N).
   ii. The Tropic of Cancer (23°27′ N) and the Tropic of Capricorn (23°27′ S).

The area of a latitude/longitude zone bounded by latitudes with location angles $\phi_1$ and $\phi_2$ and longitudes having degree

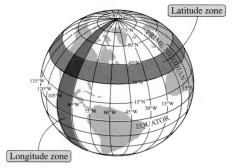

Latitude zone

Longitude zone

**FIGURE 59**

difference $\theta$ is given by

$$A = \frac{\theta}{720} \, |\cos \phi_1 - \cos \phi_1| S$$

f. Explain how the above equation is obtained.

g. Estimate the area of the state of Colorado, which is bounded by longitudes 102° W and 109° W and latitudes 37° N and 41° N.

h. Estimate the area of the state of Wyoming, which is bounded by longitudes 104° W and 111° W and latitudes 41° N and 45° N.

i. Modify the technique you used in parts (g) and (h) to estimate the area of the state of Utah. An atlas is a good reference for the boundaries.

j. Compute the areas of the following "1 degree square zones." Explain why they are not the same.
   i.  85° N, 86° N, 10° W, 11° W
   ii. 45° N, 46° N, 10° W, 11° W
  iii. 5° N, 6° N, 10° W, 11° W

## ■ *Questions for Discussion or Essay*

53. If $\alpha$ and $\beta$ are two angles for which corresponding values of the trigonometric functions agree (i.e., $\sin \alpha = \sin \beta$, $\cos \alpha = \cos \beta$, etc.), does it follow that $\alpha = \beta$? Explain.

54. If $\alpha$ and $\beta$ are two angles that have the same reference angle, is it true that $\sin \alpha = \sin \beta$? Conversely, if $\sin \alpha = \sin \beta$, does it follow that $\alpha$ and $\beta$ have the same reference angle? Explain.

55. Recall that a function $f$ is 1–1 if whenever $f(a) = f(b)$ it follows that $a = b$. Equivalently, a function $f$ is 1–1 if its graph passes the horizontal line test. Use your graphics calculator to determine whether or not any of the trigonometric functions are 1–1. Summarize your findings.

56. Use the definition of the secant function to help you explain why $\sec(\pi/2)$ is undefined.

## SECTION 4

# GRAPHS OF SINE AND COSINE

■ How can an astronaut be weighed in a weightless environment?

■ Why do soldiers march out of step whenever they cross a bridge?

■ What connection exists between ac current, music, and the ratios of sides of a right triangle?

## GRAPHS OF SINE AND COSINE

Consider the sine function $f(x) = \sin x$, where the angle $x$ is in radians. We recall from Section 3 that $\sin x$ is defined as the $y$-coordinate of the point on the unit circle corresponding to the angle $x$, as shown in Figure 60.

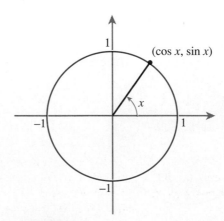

**FIGURE 60**

To graph $f(x) = \sin x$, we begin by plotting points. In Table 4, we have chosen some angles $x$ at which the sine function is known.

**TABLE 4**

| $x$ | 0 | $\dfrac{\pi}{4}$ | $\dfrac{\pi}{2}$ | $\dfrac{3\pi}{4}$ | $\pi$ | $\dfrac{5\pi}{4}$ | $\dfrac{3\pi}{2}$ | $\dfrac{7\pi}{4}$ | $2\pi$ |
|---|---|---|---|---|---|---|---|---|---|
| $y = \sin x$ | 0 | $\dfrac{\sqrt{2}}{2}$ | 1 | $\dfrac{\sqrt{2}}{2}$ | 0 | $-\dfrac{\sqrt{2}}{2}$ | $-1$ | $-\dfrac{\sqrt{2}}{2}$ | 0 |

Plotting these points and connecting with a smooth curve, we obtain the graph shown in Figure 61.

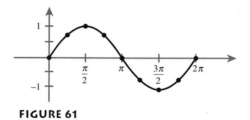

**FIGURE 61**

Now to complete the graph, we note that the angles $x$ and $x + 2\pi$ are coterminal; that is, the point on the unit circle corresponding to the angle $x$ is the same as the point on the unit circle corresponding to the angle $x + 2\pi$, as shown in Figure 62.

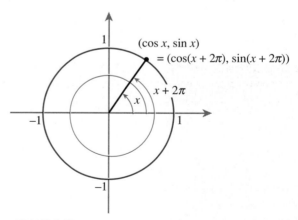

**FIGURE 62**

Thus it follows that $\sin(x + 2\pi) = \sin x$. More generally, all angles of the form $x + 2\pi k$, where $k$ is an integer, will be coterminal with the angle $x$. Consequently, the graph of the sine function will repeat itself every $2\pi$ units. So the entire graph of $\sin x$ will consist of the portion shown in Figure 61 repeated again and again, as shown in Figure 63 (using a different scale).

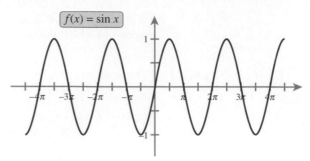

**FIGURE 63**

A similar process yields the graph of the cosine function shown in Figure 64.

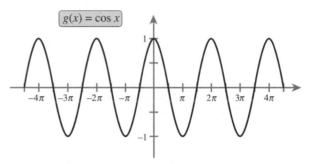

**FIGURE 64**

Functions like sine and cosine that have repeating graphs are said to be **periodic**. The following definitions are associated with periodic functions.

**Periodic functions**

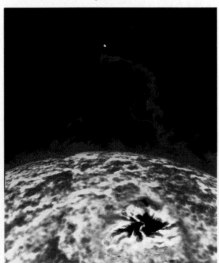

Sun spot activity varies according to an 11 year cycle.

A function $f$ is said to be **periodic** if there is a number $p$ such that $f(x + p) = f(x)$ for all $x$.

The smallest positive such number $p$ is called the **period** of $f$.

The **amplitude** of a periodic function is one-half the "height" of the graph. That is, the amplitude can be obtained by subtracting the minimum value of the function from the maximum value, and then dividing by 2:

$$\text{amplitude} = \frac{\text{max value} - \text{min value}}{2}$$

A **cycle** is the graph of a single period of a periodic function.

Thus, the functions $f(x) = \sin x$ and $g(x) = \cos x$ are periodic functions with period $2\pi$ and amplitude 1. It will prove useful to remember the appearance of one cycle of both the sine and cosine functions. Figure 65 shows the graphs of sine and cosine over the interval $[0, 2\pi]$.

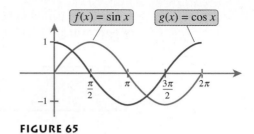

**FIGURE 65**

<table>
<tr><td>**EXAMPLE 1**</td><td>*Determining the period and amplitude*<br>*from the graph of a function*</td></tr>
</table>

Find the period and the amplitude of the function whose graph is shown in Figure 66.

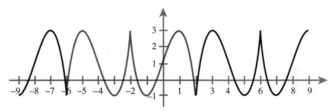

**FIGURE 66**

**SOLUTION** From Figure 66, we see that it takes 8 units for the graph to begin repeating itself. Thus, the period is 8. (A single period of the function is shown in red in Figure 66.) Since the maximum value attained by the graph is 3, and the minimum value is $-1$, the amplitude is given by

$$\frac{3 - (-1)}{2} = 2$$

<table>
<tr><td>**EXAMPLE 2**</td><td>*Estimating the period and amplitude of a periodic*<br>*function with a graphics calculator*</td></tr>
</table>

Estimate the period and amplitude of the function $f(x) = \cos x \sin x$.

**SOLUTION** We begin by plotting $y = \cos x \sin x$ with a graphics calculator, as shown in Figure 67. Next, using the trace feature, we find that the function has a maximum value of 0.5 and a minimum value of $-0.5$. Thus, the amplitude is given by

$$\frac{0.5 - (-0.5)}{2} = 0.5$$

From Figure 67, we see that the graph passes through the origin, and completes a full period more than 3 units later. Using the trace feature, as shown in Figure 68, we estimate the period to be 3.15. Upon zooming in several times, we settle on a period of approximately 3.14. (In fact, the period is $\pi$.)

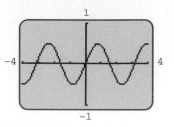

**FIGURE 67**

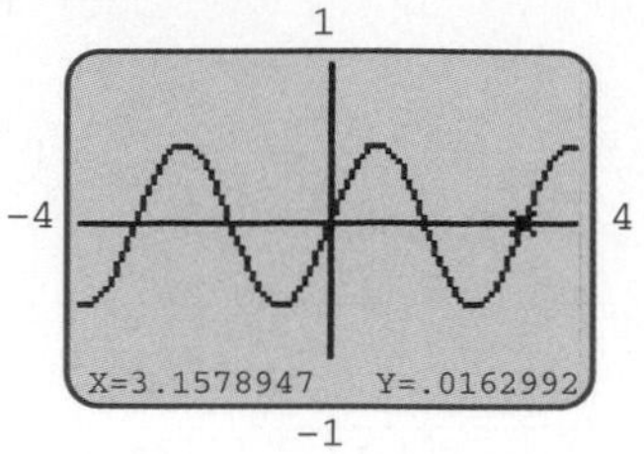

**FIGURE 68**

In the previous examples, we relied on the graphs of the trigonometric functions to provide us with information about period and amplitude. As we will see below, the period and amplitude of functions of the form $f(x) = a \sin bx$ and $g(x) = a \cos bx$ can be obtained from the values of $a$ and $b$.

## STRETCHING AND SHRINKING: FUNCTIONS OF THE FORM $f(x) = a \sin bx$ (OR $a \cos bx$)

We begin our investigation of functions of the form $f(x) = a \sin bx$ (or $a \cos bx$) by plotting the graphs of $g(x) = \sin x$ and $h(x) = \sin 2x$, as shown in Figures 69 and 70.

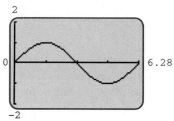

**FIGURE 69**

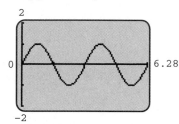

**FIGURE 70**

We have already seen that $g(x) = \sin x$ has period $2\pi$, and this is evident in the graph in Figure 69. From the graph in Figure 70, we estimate that the period of $h(x) = \sin 2x$ is $\pi$. Indeed,

$$
\begin{aligned}
h(x + \pi) &= \sin[2(x + \pi)] \\
&= \sin(2x + 2\pi) \\
&= \sin 2x \\
&= h(x)
\end{aligned}
$$

Thus, the period of $h(x) = \sin 2x$ is exactly one-half the period of $g(x) = \sin x$; in other words, the **graph of $h$ completes two cycles for every cycle of the graph of $g$.** In general, the graph of $f(x) = \sin bx$ completes a cycle $|b|$ times as fast as $g(x) = \sin x$ and so has a period of $2\pi/|b|$. A similar result holds for the cosine function. These observations generalize to the following property of periodic functions.

## *Period of f(bx)*

If $f(x)$ is a periodic function with period $p$, then $g(x) = f(bx)$ is a periodic function with period $\dfrac{p}{|b|}$.

**EXAMPLE 3**    *Finding a period*

Find the period of $f(x) = \cos \dfrac{x}{2}$.

**SOLUTION**    Since $g(x) = \cos x$ has period $2\pi$ and $f(x) = \cos bx$ with $b = \frac{1}{2}$, the period of $f$ is given by

$$\frac{2\pi}{|b|} = \frac{2\pi}{1/2} = 4\pi$$

One cycle of $f$ is shown in Figure 71.

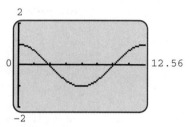

**FIGURE 71**

Next, we consider the effect of multiplying the output of the sine function by a number $a > 0$; that is, we consider the function $f(x) = a \sin x$ for $a > 0$. Essentially, since all of the $y$-values of $y = \sin x$ have been multiplied by $a$, the graph of $f(x) = a \sin x$ will be $a$ times "taller" than the graph of $g(x) = \sin x$, as suggested by Figure 72. In other words, the amplitude of $f(x) = a \sin x$ will be $a$ times that of $g(x) = \sin x$. Thus, since $g(x) = \sin x$ has amplitude 1, $f(x) = a \sin x$ has amplitude $a$.

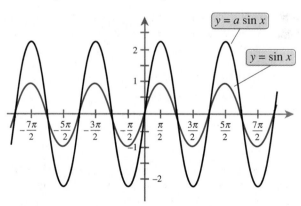

**FIGURE 72**

Finally, we consider the case that the number $a$ in the expression $f(x) = a \sin x$ is negative. Since multiplying the output of a function by $-1$ has the effect of reflecting the graph of the function about the $x$-axis, we can obtain the graph of $f(x) = a \sin x$ for $a < 0$ by graphing $y = |a|\sin bx$ and then reflecting about the $x$-axis. A similar result holds for functions of the form $g(x) = a \cos x$, where $a$ is negative. We summarize our results as follows.

**Graphs of functions of the form $f(x) = a \sin bx$ (or $f(x) = a \cos bx$)**

> If $f(x) = a \sin bx$ (or $f(x) = a \cos bx$), then the following hold:
>
> ■ The period of $f$ is given by $\dfrac{2\pi}{|b|}$.
>
> ■ The amplitude of $f$ is given by $|a|$.
>
> ■ If $a < 0$, then the graph $f$ is the reflection of the graph of $y = |a| \sin bx$ (or $y = |a| \cos bx$) about the $x$-axis.

**EXAMPLE 4**     *Finding the period and amplitude of a function of the form $f(x) = a \sin bx$*

Find the period and amplitude of the function $f(x) = -2 \sin 3x$. Then sketch two cycles of the graph.

**SOLUTION**    Here $a = -2$ and $b = 3$. Thus the amplitude is given by $|-2| = 2$, and the period is given by

$$\frac{2\pi}{|b|} = \frac{2\pi}{3}$$

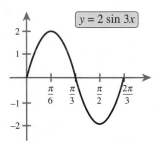

**FIGURE 73**

We begin by sketching one cycle of the sine function, and then labeling the coordinate axes to reflect the fact that the period is $2\pi/3$ and the amplitude is 2, as shown in Figure 73. Next, we repeat this basic shape, as shown dashed in red in Figure 74. Finally, since the coefficient $a$ is negative, we reflect this graph about the $x$-axis to obtain the graph shown in black in Figure 74.

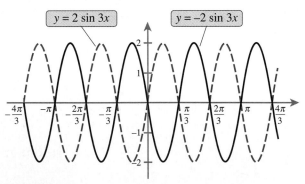

**FIGURE 74**

This graph can then be confirmed by plotting with a graphics calculator.

# RESONANCE

**W**e often hear political pundits speak of an idea (or movement) resonating with voters, by which they mean that the idea reinforces beliefs or attitudes already "in the air," so that the idea, once articulated, gathers strength as it cchos throughout the electorate. This notion of resonance as the effect created when a natural tendency is encouraged or reinforced is consistent with the scientific usage of the word.

Resonance occurs whenever an entity with a natural frequency is driven by a periodic "force" with the same frequency. For example, there is a natural frequency to the motion of a child on a swing; this is the frequency at which the child would swing if you pulled her back and released her. If you now provide her with periodic pushes, then you will, of course, affect her motion in some way. But if you push her at her natural or *resonant* frequency, then with very little effort on your part, her swings will have ever greater amplitude.

This phenomenon accounts for the shattering of crystal ware occasionally performed by accomplished singers. If you strike a crystal wine glass, it will vibrate with a certain frequency, which your ears will detect as sound of a given pitch.

This 3D computer rendering is based in part on MRI data.

This pitch is the natural frequency of the glass. Now sound of any frequency reaching the glass will impart a tendency for the glass to vibrate sympathetically, but sound vibrating at the *natural* frequency of the glass will, in effect, *encourage* the glass to vibrate at precisely the frequency at which the glass *likes* to vibrate. The effect can be extraordinary. The amplitude of the vibration of the glass becomes greater and greater until the glass shatters—no longer able to respond elastically to the strain imparted by the vibrations.

Legend has it that resonance is responsible for the 1831 collapse of the Broughton Suspension Bridge near Manchester, England. A column of soldiers marched across the bridge, and unfortunately, their cadence matched the resonant frequency of the bridge, causing it to swing wildly and finally fall. It is for this reason that soldiers break step whenever crossing a footbridge.

Although it is now viewed as unlikely, it had long been thought that the Tacoma Narrows Bridge at Puget Sound toppled due to vortices created by a gale wind that drove the bridge at a frequency matching the resonant frequency of the bridge. In 1959 (and again in 1960), an airplane resonated right out of the sky when an engine vibrated at the resonant frequency of one of the wings.

But the phenomenon of resonance is not entirely malevolent; in fact, it is responsible for one of the most potent medical diagnostic tools— Magnetic Resonance Imaging (MRI). In essence, a subject is immersed in magnetic fields in such a way as to produce resonance in the atomic nuclei of the body. When resonance occurs, the nuclei release energy in the form of radio waves. By tracing these radio waves to their source nuclei, and using the fact that the resonant frequency depends on the composition of the nuclei, a detailed three-dimensional map of the body can be formed. Since the magnetic field required to produce resonance is relatively weak, an MRI procedure results in little of the cellular damage that arises from x rays. Moreover, it provides a much clearer view of the soft tissues that appear as mere shadows on x rays.

## TRANSLATIONS

Recall from Chapters 3 and 4 that the graph of $y = f(x - h) + k$ is a translation $h$ units to the right and $k$ units up of the graph of $y = f(x)$. Applying this fact in the case of the function $y = a \sin bx$, we find that the graph of $y = a \sin[b(x - h)] + k$ is a translation of the graph of $y = a \sin bx$, $h$ units to the right and $k$ units up.

**EXAMPLE 5**   *Graphing a translate of the sine function*

Graph the function $f(x) = 2 \sin\left(x - \dfrac{\pi}{3}\right)$.

**SOLUTION**   We recognize that the function $f$ is just a translation, $\pi/3$ units to the right, of the function $g(x) = 2 \sin x$. To graph $g$, we note that it has amplitude

$|2| = 2$ and period $2\pi$. In Figure 75, we have graphed $g$ (shown in red) and then shifted the graph $\pi/3$ units to the right to obtain the graph of $f$ (shown in black).

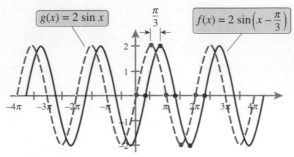

$g(x) = 2 \sin x$
$f(x) = 2 \sin\left(x - \dfrac{\pi}{3}\right)$
$\dfrac{\pi}{3}$

**FIGURE 75**

In fact, it can be shown that any function of the form $f(x) = a \sin(bx + c) + d$ is just a translate of a function of the form $g(x) = a \sin bx$, as illustrated by the following example. (As usual, corresponding results hold if the sine function is replaced by the cosine function.)

**EXAMPLE 6**

*Plotting the graph of a function of the form* $f(x) = a \sin(bx + c) + d$

Let $f(x) = 4 \sin\left(\dfrac{x}{2} + \dfrac{\pi}{2}\right) - 3$.

**a.** Find the period and the amplitude of $f$ and sketch its graph for three cycles.
**b.** Identify the intervals on which $f$ is increasing and those on which it is decreasing.
**c.** Find the maximum and minimum values of $f$.
**d.** Use a graphics calculator to estimate the first two positive zeros of $f$.

**SOLUTION**

**a.** We begin by rewriting the function so that it can be recognized as a translation of a function of the form $g(x) = a \sin bx$ as follows:

$$f(x) = 4 \sin\left(\frac{x}{2} + \frac{\pi}{2}\right) - 3$$

$$= 4 \sin\left[\frac{1}{2}(x + \pi)\right] - 3$$

In this form, it is clear that $f$ is merely a translation, $\pi$ units to the left and 3 units down, of the graph of $g(x) = 4 \sin(x/2)$, which has amplitude $|4| = 4$ and period

$$\frac{2\pi}{1/2} = 4\pi.$$

Thus, we graph $y = 4 \sin(x/2)$ (shown in Figure 76 in red) and then translate to obtain the graph of $y = f(x)$ (shown in black).

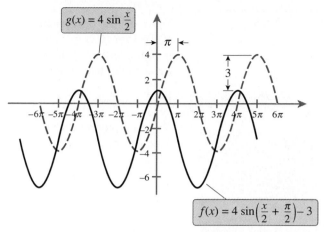

**FIGURE 76**

**b.** From the graph it is evident that $f$ is rising, and hence increasing, on intervals such as $(-6\pi, -4\pi)$, $(-2\pi, 0)$, and $(2\pi, 4\pi)$. Similarly, we see that it is falling, and hence decreasing, on intervals such as $(-4\pi, -2\pi)$ and $(0, 2\pi)$.

**c.** From the graph (or considerations of amplitude), we see that the maximum value of $f$ is 1 and the minimum is $-7$.

**d.** Using the trace feature of a graphics calculator, we estimate the first two positive zeros of $f$ to be $x \approx 1.44$ and $x \approx 11.12$. (See Figure 77.)

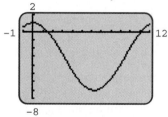
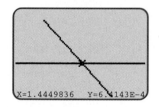
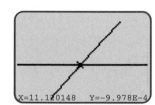

**FIGURE 77**

The **phase shift** of a function of the form $f(x) = a \sin[b(x - h)]$ is defined to be the number $h$. In other words, the phase shift is the number of units by which the graph of the function $g(x) = a \sin bx$ must be translated to the right in order to obtain the graph of $f$. A similar definition holds for functions of the form $f(x) = a \cos[b(x - h)]$. In Example 6, where $f(x) = 4 \sin[\frac{1}{2}(x + \pi)] - 3$, the phase shift of the function $f$ was $-\pi$. We summarize the notions of period, amplitude, and phase shift as follows.

*Period, amplitude, and phase shift*

Let $f(x) = a \sin[b(x - h)] + d$ or $f(x) = a \cos[b(x - h)] + d$. Then the period, amplitude, and phase shift of $f$ are

$$\text{period} \qquad \frac{2\pi}{|b|}$$

$$\text{amplitude} \qquad |a|$$

$$\text{phase shift} \qquad h$$

APPLICATIONS

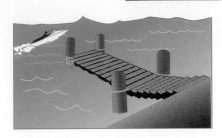

Many natural phenomena exhibit periodic motion that can be described by functions of the form $f(x) = a \sin(bx + c) + d$ or $f(x) = a \cos(bx + c) + d$. Examples include sound, alternating electric current, water waves, and light. In the following example, we consider the motion of a water wave.

**EXAMPLE 7**    *Water waves*

A certain pier piling has a scale for measuring water level. As a consequence of waves from a passing motor boat, the water level in inches relative to the zero point of the scale is approximately given over a period of time by the function

$$f(t) = 10 \cos\left(\frac{\pi}{4}t - \frac{\pi}{2}\right) + 4$$

where $t$ is in seconds. Sketch the graph of $f$ and identify the period, amplitude, and phase shift.

**SOLUTION**    We begin by expressing $f(t)$ in standard form

$$f(t) = 10 \cos\left(\frac{\pi}{4}t - \frac{\pi}{2}\right) + 4$$

$$= 10 \cos\left[\frac{\pi}{4}(t - 2)\right] + 4$$

Thus the graph of $f$ is a translation 4 units up and 2 units to the right of the graph of $g(t) = 10 \cos \frac{\pi}{4} t$, which has amplitude $|10| = 10$ and period $\frac{2\pi}{\pi/4} = 8$. To graph $g$, we sketch one cycle of the cosine function, labeling the axes to reflect the amplitude of 10 and the period of 8, and then repeat this shape, as shown in Figure 78. Finally, we obtain the graph of $f$ by shifting the graph of $g$ four units upward and two units to the right, as shown in black in Figure 79. The phase shift of $f$ is 2.

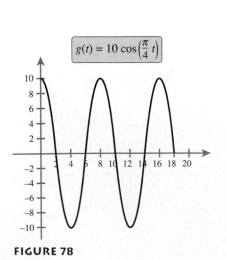

**FIGURE 78**

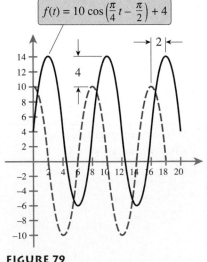

**FIGURE 79**

**EXERCISES 4**

**EXERCISES 1–8** ☐ *Find the period and the amplitude of the function whose graph is given.*

**1.**

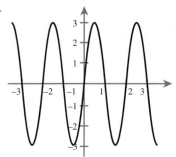

**2.**

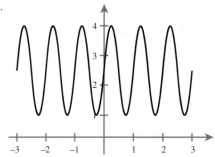

**3.**

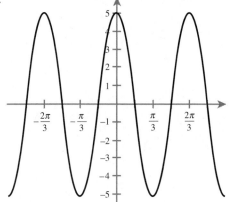

**4.**

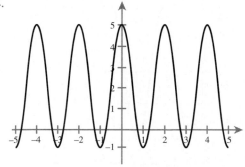

**5.**

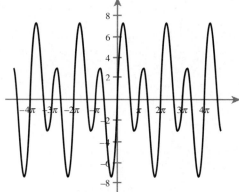

**6.**

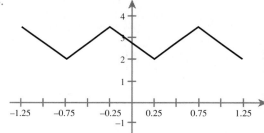

**7.**

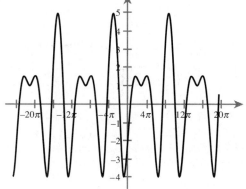

**8.**

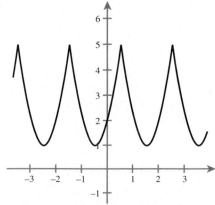

**EXERCISES 9–14** ☐ *Find the period and the amplitude of the given function. Use a graphics calculator as necessary.*

9. $f(x) = 5 \sin 4x$

10. $g(x) = 3 \cos\left(2x - \dfrac{\pi}{2}\right)$

11. $h(x) = \sin 3x - 2 \cos x$

12. $f(x) = \cos 4x + 2 \cos x + 3$

13. $g(x) = 2^{\sin 2x} - 3^{\cos x}$

14. $h(x) = 1 + 2 \sin^2 3\pi x$

**EXERCISES 15–22** ☐ *Graph at least three cycles of the given function. Indicate the amplitude, period, and phase shift.*

15. $h(x) = \cos 2x$

16. $g(t) = \sin \pi t$

17. $f(y) = 2 \sin 2\pi y$

18. $f(x) = -3 \cos 4x$

19. $p(t) = 3 \cos(2t + 1)$

20. $h(\theta) = 2 \sin\left(\pi\theta - \dfrac{\pi}{2}\right)$

21. $g(x) = -4 \sin\left(3x - \dfrac{\pi}{4}\right) + 4$

22. $f(t) = \dfrac{3}{2} \cos\left(\dfrac{1}{2}t + \pi\right) - 1$

**EXERCISES 23–30** ☐ *Find the amplitude, period, phase shift, and maximum and minimum values of the function.*

23. $f(x) = 3 \sin \pi x$

24. $g(x) = 4 \cos 2\pi x$

25. $p(t) = -4 \sin(2t + 1)$

26. $q(r) = 5 \sin(r - 2)$

27. $h(t) = 6 \sin\left(2t + \dfrac{\pi}{2}\right)$

28. $f(x) = -3 \sin(2x - \pi)$

29. $g(y) = 5 - 2 \cos(3y - 2)$

30. $r(t) = 5 \sin\left(2t + \dfrac{\pi}{4}\right) + 3$

**EXERCISES 31–36** ☐ *Indicate whether the function is increasing, decreasing, or both on the given interval.*

31. $f(x) = \sin 2x; \left(\dfrac{\pi}{4}, \dfrac{3\pi}{4}\right)$

32. $g(x) = \cos 3x; \left(\pi, \dfrac{4\pi}{3}\right)$

33. $h(x) = 2 \cos(\pi x - \pi); \left(-\dfrac{1}{10}, \dfrac{1}{10}\right)$

34. $f(x) = -3 \sin\left(x - \dfrac{\pi}{4}\right) + 1; (0, 1)$

35. $g(x) = e^{\sin x}; \left(0, \dfrac{\pi}{2}\right)$

36. $h(x) = 2 \cos^3 x - 4 \cos x; (3.1, 3.2)$

**EXERCISES 37–42** ☐ *Each of the following graphs is that of a function of the form $f(x) = a \sin(bx + c)$. Find a, b, and c.*

37.

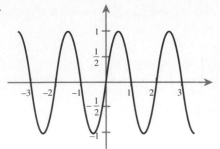

38.

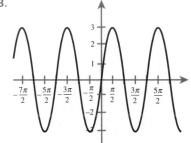

39.

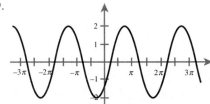

40.

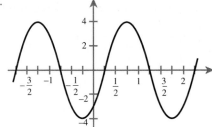

41.

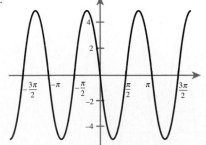

42.

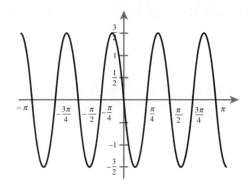

**EXERCISES 43–48** □ *Use a graphics calculator to determine which of the following are identities.*

43. $\sin(-x) = \sin x$

44. $\cos(-x) = \cos x$

45. $\sin 3x - 3 \sin x = 4 \sin^3 x$

46. $\cos 4x = 4 \cos x$

47. $\sin \dfrac{x}{2} \cos \dfrac{x}{2} - \dfrac{\sin x}{2} = 0$

48. $\sin 2x = 2 \sin x \cos x$

**EXERCISES 49–52** □ *Find a function of the form* $f(x) = a \sin(bx + c)$ *that satisfies the given properties.*

49. Amplitude 3, period $\pi$, and phase shift $\dfrac{\pi}{3}$

50. Amplitude 2, period 2, and phase shift 1

51. Amplitude $\sqrt{2}$, period $\dfrac{\pi}{4}$, and phase shift $\dfrac{\pi}{8}$

52. Amplitude 6, period 3, and phase shift $-\dfrac{\pi}{4}$

**EXERCISES 53–56** □ *Use a graphics calculator to estimate the first two positive zeros of the given function.*

53. $f(x) = 4 \sin(3x + 1) + 2$

54. $g(x) = 1 - 3 \cos(2x - 5)$

55. $h(x) = \dfrac{4}{5} \cos 20x - \dfrac{1}{2}$

56. $f(x) = 0.1 - 0.2 \sin 4\pi x$

■ *Applications*

57. *Musical Sound Wave* A tuning fork that produces a C on the musical scale creates a wave on an oscilloscope that can be modeled by the function $f(x) = 0.002 \sin 528\pi x$. Find the period and amplitude of this wave.

58. *Electromagnetic Wave* A certain electromagnetic wave can be modeled by the function $g(t) = \sin[(2 \times 10^5)\pi t]$. Find the period and amplitude of this wave.

59. *Spring Motion* When an object hung from a certain stretched spring is released, its displacement in centimeters from the equilibrium (resting) position $t$ seconds after release is approximated by

$$y(t) = 5 \sin\left(4\pi t - \frac{\pi}{2}\right)$$

a. Find the period of $y$.

b. Find the maximum displacement of the object.

60. *Sunset Times* Sunset times in Indianapolis, Indiana, can be roughly approximated using the function

$$f(t) = 89 \sin\left(\frac{2\pi}{365}t + \frac{3\pi}{2}\right) + 409$$

where $t$ is the day of the year and $f(t)$ is the corresponding sunset time, in minutes, after 12:00 P.M.

a. Find the period and amplitude of $f$.

b. Find the earliest and latest sunset times, and estimate the days on which they occur.

61. *AC Voltage* The voltage of a 110-volt outlet actually varies according to the function

$$V(t) = 110\sqrt{2} \sin 120\pi t$$

where $t$ is in seconds.

a. Find the amplitude and period.

b. What is the first positive time for which the voltage is exactly 110 volts?

62. *Bobbing Buoy* A buoy is released at the surface of still water and begins to bob up and down. For a short period of time, its motion can be approximated with the function

$$d(t) = 10 - 10 \cos(\sqrt{980t})$$

where $d(t)$ is the depth of the bottom surface of the buoy in centimeters $t$ seconds after it is released.

a. Find the period of $d(t)$.

b. Find the maximum depth of the bottom surface of the buoy.

63. *Ferris Wheel Motion* The height in feet of a certain passenger on a ferris wheel is given by

$$y(t) = 55 + 50 \sin\left(\frac{\pi t}{15} - \frac{\pi}{2}\right)$$

where $t$ is the time in seconds, and $t = 0$ coincides with the time at which the wheel was set in motion.

a. Find the initial height of the passenger.

b. Find the maximum and minimum heights of the passenger and the first times at which those heights are attained.

c. How long does it take for the wheel to make one complete revolution?

64. *Weighing an Astronaut* The mass of an astronaut in a "weightless" environment can be determined by setting the astronaut in motion on an oscillating machine, measuring the period of the motion, and calculating the mass from the period. Suppose the astronaut's displacement in centimeters is given by the function

$$x(t) = a \cos\left(\sqrt{\frac{260}{m}}\, t\right)$$

where $m$ is the mass of the astronaut in kilograms and $t$ is in seconds. If the period is measured to be 0.32 second, find the mass of the astronaut.

---

## ◾ *Project for Enrichment*

65. *Modeling Average Monthly Precipitation*

 I. Consider the function $f(t) = A + B \cos(Ct - \theta)$ where $A$, $B$, and $C$ are positive real numbers.

 a. Find the maximum value of $f$.

 b. Find the minimum value of $f$.

 c. Find the period of $f$.

 II. The average monthly precipitation for Duluth, Minnesota is given in Figure 80.

 a. Explain why a function of the form
 $f(t) = A + B \cos(Ct - \theta)$ could be considered a candidate for modeling this data.

 b. Use the results of part I to find values for $A$, $B$, and $C$.

 c. Find $\theta$ so that the precipitation predicted by the model for each month differs from the actual rainfall by no more than 0.5 inch.

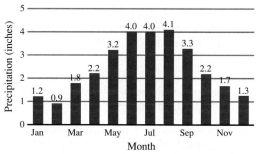

**Monthly Precipitation**
Duluth, Minnesota

**FIGURE 80**

---

## ◾ *Questions for Discussion or Essay*

66. Which of the following phenomena would you expect to be periodic? What would you expect the period to be? Explain.

a. Lunar stages

b. National debt

c. Retail sales

d. Number of people attending a house of worship on a given day

Give some additional examples, ones not considered in class or the text, of periodic phenomena.

67. What makes for a good clock? Write a short paper describing devices that have been used throughout the ages to tell time. What characteristic do all of these clocks have in common?

68. Explain why the function $\sin[2x - (\pi/3)]$ does not have a phase shift of $\pi/3$.

**SECTION 5**

# GRAPHS OF OTHER TRIGONOMETRIC FUNCTIONS

■ If a ball is thrown so that it lands 100 feet away, what is the length of its journey?

■ How can it be that the period of the tangent function is $\pi$ when $\tan x = \dfrac{\sin x}{\cos x}$

and yet both sine and cosine have periods of $2\pi$?

■ If the angle of elevation from a viewer to an object approaches $\pi/2$, what can be said about the height of the object?

■ At what angle should a 6-foot-tall outfielder release a baseball if she wishes it to travel as far as possible?

## GRAPHS OF TANGENT AND COTANGENT

Let us now consider the graph of $f(x) = \tan x$. In Table 5, we have computed values of $y = \tan x$ for selected values of $x$ between 0 and $\pi/2$. Note that for values of $x$ near $\pi/2$, $\sin x$ is nearly 1 and $\cos x$ is a very small number. Thus, for $x$ near $\pi/2$, we have

$$\tan x = \frac{\sin x}{\cos x} \approx \frac{1}{\text{very small number}} = \text{very large number}$$

**TABLE 5**

| $x$ | $\tan x = \dfrac{\sin x}{\cos x}$ |
|---|---|
| 0 | 0 |
| $\dfrac{\pi}{4}$ | 1 |
| $\dfrac{\pi}{3}$ | $\sqrt{3} \approx 1.73$ |
| 1.5 | $\approx 14.10$ |
| 1.56 | $\approx 92.62$ |
| 1.57 | $\approx 1255.77$ |
| 1.5707 | $\approx 10{,}381.33$ |
| $\dfrac{\pi}{2}\,(\approx 1.5708)$ | undefined |

In other words, as $x$ approaches $\pi/2$, $\tan x$ approaches infinity. Thus, the graph of $f(x) = \tan x$ has a vertical asymptote at $x = \pi/2$. Moreover, since the sine function is odd and the cosine function is even, we have

$$\tan(-x) = \frac{\sin(-x)}{\cos(-x)} = \frac{-\sin x}{\cos x} = -\tan x$$

Thus, the tangent function is odd, and hence its graph is symmetric with respect to the origin. In Figure 81, we have sketched a graph of $y = \tan x$ for $x$ between $0$ and $\pi/2$ (shown in red), and then reflected this portion of the graph about the origin to obtain the portion of the graph of the function between $-\pi/2$ and $0$ (shown in blue).

Next, we consider the period of the tangent function. From Figure 82 we see that $\sin(x + \pi) = -\sin x$, and $\cos(x + \pi) = -\cos x$.

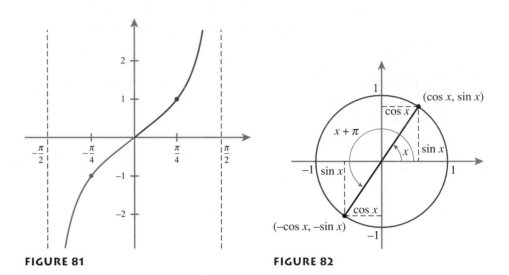

**FIGURE 81**                    **FIGURE 82**

Thus, we have

$$\tan(x + \pi) = \frac{\sin(x + \pi)}{\cos(x + \pi)}$$

$$= \frac{-\sin x}{-\cos x}$$

$$= \frac{\sin x}{\cos x}$$

$$= \tan x$$

Therefore, the tangent function is periodic with period $\pi$. Figure 83 shows several cycles of the graph of $y = \tan x$. The tangent function is undefined (and its graph has a vertical asymptote) wherever $\cos x = 0$. From Figure 83, we see that this happens whenever $x$ is of the form $\pi/2 + \pi k$, where $k$ is an integer. Similarly, since $\cot x = \cos x/\sin x$, the graph of the $\cot x$ will have vertical asymptotes wherever $\sin x = 0$, namely at integer

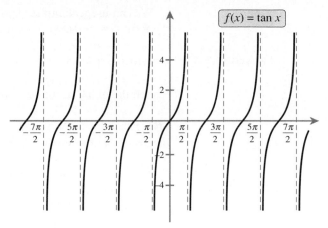

**FIGURE 83**

multiples of $\pi$. The period of the cotangent function is also $\pi$, since

$$\cot(x + \pi) = \frac{1}{\tan(x + \pi)}$$

$$= \frac{1}{\tan x}$$

$$= \cot x$$

The graph of $f(x) = \cot x$ is shown in Figure 84.

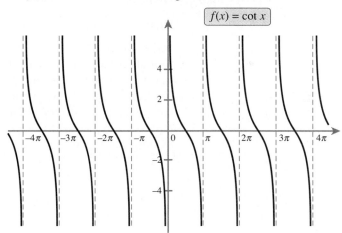

**FIGURE 84**

We summarize the most important aspects of the graphs of the tangent and cotangent functions as follows.

*Graphs of tangent and cotangent*

| Function | Vertical asymptotes | Period |
|---|---|---|
| $f(x) = \tan x$ | $x = \dfrac{\pi}{2} + \pi k$, where $k$ is an integer | $\pi$ |
| $f(x) = \cot x$ | $x = \pi k$, where $k$ is an integer | $\pi$ |

We can use techniques similar to those of the previous section to graph functions of the form $f(x) = \tan(bx + c)$ or $g(x) = \cot(bx + c)$, as the following examples illustrate.

**EXAMPLE 1**    *Graphing a translation of the tangent function*

Sketch the graph of the function $f(x) = \tan\left(x - \dfrac{\pi}{4}\right)$

**SOLUTION**    The graph of $f$ is just a translation $\pi/4$ units to the right of the graph of $y = \tan x$. The graph is shown in Figure 85.

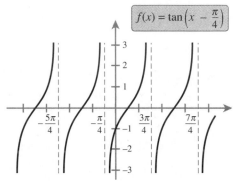

**FIGURE 85**

**EXAMPLE 2**    *Graphing a function of the form $f(x) = \tan bx$*

Graph the function $f(x) = \tan 2x$.

**SOLUTION**    Recall from Section 4 of this chapter that replacing $x$ by $bx$ in the formula for a periodic function has the effect of dividing the period by a factor of $|b|$. Since the period of the tangent function is $\pi$, the period of $f(x) = \tan 2x$ is then $\pi/2$. Thus, the graph of $f(x) = \tan 2x$ can be obtained from the graph of $y = \tan x$ by compressing by a factor of 2 in the horizontal direction. Hence the vertical asymptotes occur not at multiples of $\pi/2$ as with $y = \tan x$, but rather at multiples of $\pi/4$, as shown in Figure 86.

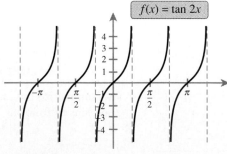

**FIGURE 86**

From the previous example we surmise that any function of the form $f(x) = \tan bx$ has period $\pi/|b|$. The graph of a function of the form $f(x) = \tan[b(x - h)]$ is merely a translation, $h$ units to the right, of the graph of $g(x) = \tan bx$. Similar results hold for functions of the form $f(x) = \cot[b(x - h)]$.

**EXAMPLE 3**    *Graphing a function of the form* $f(x) = \tan(bx + c)$

Sketch a graph of $f(x) = \tan\left(\dfrac{x}{2} - \dfrac{\pi}{4}\right)$.

**SOLUTION**    We have

$$f(x) = \tan\left(\frac{x}{2} - \frac{\pi}{4}\right)$$

$$= \tan\left[\frac{1}{2}\left(x - \frac{\pi}{2}\right)\right]$$

Thus, $f$ is a translation, $\pi/2$ units to the right, of the graph of $g(x) = \tan(x/2)$. Now the function $g$ has period

$$\frac{\pi}{1/2} = 2\pi$$

The graph of $g$ is shown in Figure 87. The graph of $f$ shown in Figure 88 is obtained by shifting the graph of $g$ to the right $\pi/2$ units.

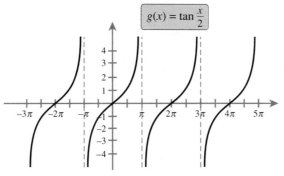

**FIGURE 87**                                                        **FIGURE 88**

Notice that in each of the previous examples, one cycle of the (translated) tangent function occured between each consecutive pair of vertical asymptotes. This suggests another method for quickly graphing functions of the form $f(x) = \tan(bx + c)$. We find the period as before, and then find an asymptote by setting

$$bx + c = \frac{\pi}{2}$$

Additional vertical asymptotes can be located using the period. Finally, the graph can be completed simply by sketching cycles of the tangent function between each consecutive pair of asymptotes.

**EXAMPLE 4**    *Finding asymptotes*

Find the asymptotes of $f(x) = \tan(\pi x + \pi)$.

**SOLUTION**    Using the technique suggested above, we locate one asymptote as follows.

$$\pi x + \pi = \frac{\pi}{2}$$

$$\pi x = -\frac{\pi}{2}$$

$$x = -\frac{1}{2}$$

Now the period of $f$ is given by

$$\frac{\pi}{\pi} = 1$$

Thus, the asymptotes are spaced 1 unit apart, at $x = -\frac{1}{2} + k$ for any integer $k$. Using a graphics calculator in dot mode, we obtain the graph shown in Figure 89.

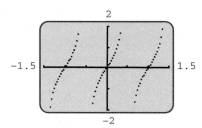

**FIGURE 89**

---

**GRAPHS OF SECANT AND COSECANT**

Since $\sec x = 1/\cos x$, the graph of $f(x) = \sec x$, like that of $\tan x$, will have vertical asymptotes wherever $\cos x = 0$—that is, at numbers of the form $\pi/2 + \pi k$, where $k$ is an integer. Moreover, since $|\cos x| \leq 1$, it follows that $|\sec x| = |1/\cos x| \geq 1$. In other words, $\sec x$ is either greater than or equal to 1, or less than or equal to $-1$. The period of the secant function is the same as that of the cosine function, $2\pi$. In fact, we have

$$\sec(x + 2\pi) = \frac{1}{\cos(x + 2\pi)}$$

$$= \frac{1}{\cos x}$$

$$= \sec x$$

Similar arguments hold for the cosecant function; the graphs of $\sec x$ and $\csc x$ are shown in Figure 90.

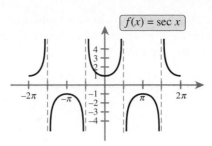

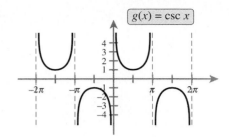

**FIGURE 90**

Since both the cosecant function and the secant function have periods of $2\pi$, any function of the form $f(x) = \sec bx$ or $g(x) = \csc bx$ will have period $2\pi/|b|$.

**EXAMPLE 5**  *Graphing a function of the form* csc *bx*

Sketch a graph of the function $f(x) = \csc \dfrac{x}{2}$.

**SOLUTION**  The period of $f$ is given by

$$\frac{2\pi}{1/2} = 4\pi$$

and so the graph of $f$ will repeat "half as fast" as the graph of $y = \csc x$. Thus, whereas the graph of $y = \csc x$ has vertical asymptotes at multiples of $\pi$, the graph of $f(x) = \csc(x/2)$ will have vertical asymptotes at multiples of $2\pi$. The graph of $f$ is shown in Figure 91.

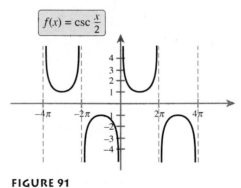

**FIGURE 91**

---

**EXAMPLE 6**  *Graphing a function of the form* sec(*bx* + *c*)

Sketch a graph of the function $f(x) = \sec\!\left(3x + \dfrac{\pi}{2}\right)$.

**SOLUTION**  We begin by rewriting as follows:

$$f(x) = \sec\!\left(3x + \frac{\pi}{2}\right)$$

$$= \sec\!\left[3\!\left(x + \frac{\pi}{6}\right)\right]$$

Thus, the graph of $f$ will be a translation, $\pi/6$ units to the left, of the graph of the function $g(x) = \sec 3x$, which has a period of $2\pi/3$. The graphs of both $g$ and $f$ are shown in Figure 92.

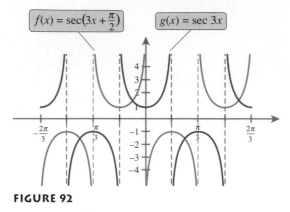

**FIGURE 92**

For convenience, we provide the following summary of the graphs of tangent, cotangent, secant, and cosecant.

*Graphs of tangent, cotangent, secant, and cosecant.*

| Function | Period | Vertical asymptotes | Graph of a single cycle |
|----------|--------|---------------------|-------------------------|
| $\tan x$ | $\pi$ | $\dfrac{\pi}{2} + \pi k$ | |
| $\cot x$ | $\pi$ | $\pi k$ | |
| $\sec x$ | $2\pi$ | $\dfrac{\pi}{2} + \pi k$ | |
| $\csc x$ | $2\pi$ | $\pi k$ | |

## EXERCISES 5

**EXERCISES 1–8**  *Match the given function with its graph.*

1. $f(x) = \tan 2x$

2. $f(x) = \cot \pi x$

3. $f(x) = \sec \dfrac{\pi x}{2}$

4. $f(x) = \tan\left(2x + \dfrac{\pi}{2}\right)$

5. $f(x) = \cot\left(\pi x - \dfrac{\pi}{2}\right)$

6. $f(x) = \sec\left(x + \dfrac{\pi}{3}\right)$

7. $f(x) = \tan(2\pi x - \pi)$

8. $f(x) = \csc 3x$

a.

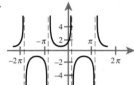

b.

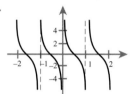

c.

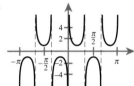

d.

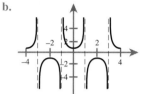

e.

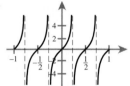

f.

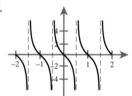

g.

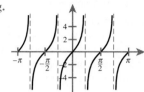

h.

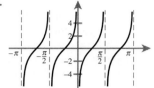

**EXERCISES 9–22**  *Graph at least three cycles of the given function. Indicate the asymptotes and determine the period.*

9. $f(x) = \tan 3x$

10. $f(x) = \tan \pi x$

11. $f(x) = \cot \dfrac{\pi x}{2}$

12. $f(x) = \cot 2x$

13. $f(x) = \tan\left(x - \dfrac{\pi}{2}\right)$

14. $f(x) = \tan\left(2x + \dfrac{\pi}{2}\right)$

15. $f(x) = \cot\left(\pi x + \dfrac{\pi}{4}\right)$

16. $f(x) = \cot\left(\dfrac{x}{2} - \dfrac{\pi}{6}\right)$

17. $f(x) = \sec 4x$

18. $f(x) = \sec\left(x - \dfrac{\pi}{2}\right)$

19. $f(x) = \sec(\pi x + \pi)$

20. $f(x) = \csc 3x$

21. $f(x) = \csc\left(2x - \dfrac{\pi}{2}\right)$

22. $f(x) = \sec\left(\dfrac{x}{4} - \dfrac{\pi}{8}\right)$

**EXERCISES 23–26**  *Use a graphics calculator to estimate the smallest positive zero of the given function.*

23. $f(x) = \tan x + \tan 2x$

24. $f(x) = \tan x - \sec x$

25. $f(x) = 1 - \tan x - \cot x$

26. $f(x) = \sec x - \sec 2x$

■ **Applications**

27. *Space Shuttle Height*  A space shuttle launch is being observed from a location 5 miles from the launch pad. The height $h$ of the shuttle can be determined using the angle of elevation $\theta$ from the observer to the shuttle (see Figure 93).

   a. Express $h$ as a function of the angle $\theta$.

   b. For what values of $\theta$ is the function $h$ valid?

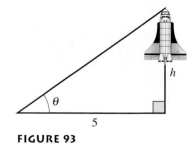

**FIGURE 93**

Space Shuttle

c. Sketch the graph of *h* over the interval you found in part (b).

d. Interpret the behavior of the graph of *h* near $\theta = \pi/2$.

28. *Tracking an Oil Tanker* An oil tanker is traveling parallel to a north-south coastline at a distance of 1000 yards from shore. The distance *d* from a Coast Guard station to a point on shore directly inland from the tanker can be determined using the angle $\theta$ between the shoreline and the line of sight to the tanker (see Figure 94).

a. Express *d* as a function of the angle $\theta$.

b. For what values of $\theta$ is the function *d* valid?

c. Sketch the graph of *d* over the interval you found in part (b).

d. Interpret the behavior of the graph of *d* near $\theta = \pi$.

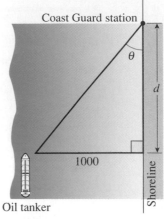

**FIGURE 94**

---

### ◼ *Project for Enrichment*

29. *Projectile Distance* If someone were to ask you the question "How far can you throw a baseball?" you would probably answer with an estimate of your **range**—that is, with an estimate of the horizontal distance from the point where the ball is thrown to the point where it hits the ground. However, there are two other interpretations of the word *distance* as it relates to projectile motion. One is the maximum height of the projectile, and the other is the length of the projectile's path. Our goal in this project is to investigate the range, maximum height, and path length of a projectile. Specifically, we wish to see how these quantities are affected by the initial angle at which an object is propelled into the air.

If an object is thrown from ground level with an initial angle $\theta$ and an initial speed of *v* feet per second, its range (ignoring air resistance) is given by $(v^2 \sin 2\theta)/32$. For the following problems, assume $v = 64$ feet per second (approximately 44 miles per hour). You may also find a graphics calculator useful.

a. Find the range for initial angles of $\pi/6$, $\pi/4$, and $\pi/3$.

b. What initial angle would be required for a range of 100 feet?

c. What initial angle will give the greatest range?

The maximum height of an object thrown from ground level with an initial angle $\theta$ and an initial speed of *v* feet per second is given by $(v^2 \sin^2 \theta)/64$. Once again, assume $v = 64$ feet per second.

d. Find the maximum height for initial angles of $\pi/6$, $\pi/4$, and $\pi/3$.

e. What initial angle would be required for a maximum height of 40 feet?

f. What initial angle will give the greatest maximum height?

Using techniques from calculus, it can be shown that the length of a projectile's path is given by

$$\frac{v^2}{32}\left[\sin\theta - \cos^2\theta\left(\ln\left[\tan\frac{\pi - 2\theta}{4}\right]\right)\right]$$

Assume as before that $v = 64$ feet per second.

g. For what values of $\theta$ is this expression meaningful?

h. Find the length of the ball's path for initial angles of $\pi/6$, $\pi/4$, and $\pi/3$.

i. What initial angle would be required for a path length of 100 feet?

j. What initial angle will give the greatest path length? How does this angle compare to the angles that give the greatest range or maximum height?

■ᵓᵇ  *Questions for Discussion or Essay*

30. Explain why the secant and cosecant functions have no *x*-intercepts.

31. The tangent, cotangent, secant, and cosecant functions are all periodic, and yet they are rarely used to model periodic phenomena. Why do you suppose it is that sine and cosine are the trigonometric functions of choice for modeling periodic phenomena?

32. Explain why the equation tan *x* = *a* has a solution *x* in the interval $(-\pi/2, \pi/2)$ no matter how large (or small) the value of *a*.

33. In Exercise 29, a project on projectile distance, a simplifying assumption was made that gravity was the only force acting on the ball. In practice, what other factors might influence the path of the ball? Under what circumstances would these other factors be significant? How accurate are these formulas for modeling the path of a golf ball? A rock? A frisbee?

---

## SECTION 6

# INVERSE TRIGONOMETRIC FUNCTIONS

■ Where is the best seat for viewing in a movie theater?

■ Why aren't road signs the most readable when we are closest to them?

■ If the sine function has no inverse, then what is meant by the inverse sine function?

■ How can the optimal position of a solar panel be determined in an urban area?

---

### THE INVERSE SINE

Consider the problem of determining the angle *x* shown in Figure 95. We may use the right triangle definition of the sine function to obtain

$$\sin x = \frac{4}{5}$$

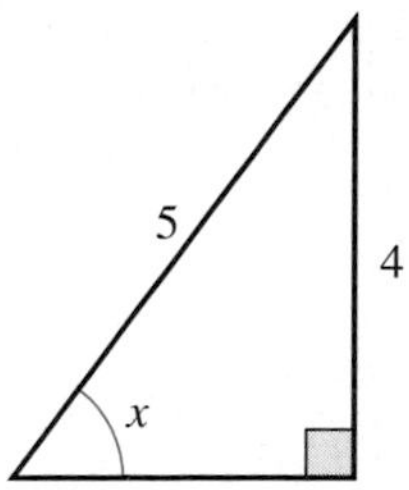

from which it follows that *x* is the angle whose sine is $\frac{4}{5}$. To find this angle, we have seen that we can use the $\boxed{\text{SIN}^{-1}}$ key on a calculator, obtaining the approximation

$$x \approx 0.9273 \text{ (or } 53.13°)$$

The calculator produces a single answer. But as Figure 96 shows, the equation $\sin x = \frac{4}{5}$ actually has infinitely many solutions. Thus, it is impossible to speak of *the* angle whose sine is $\frac{4}{5}$. So how does the calculator ''know'' which solution to give?

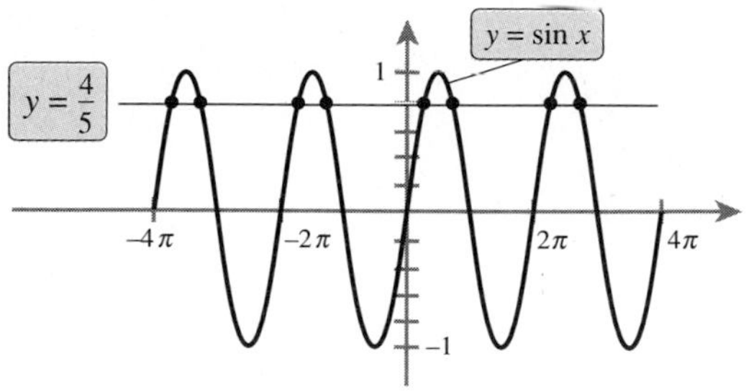

**FIGURE 95**

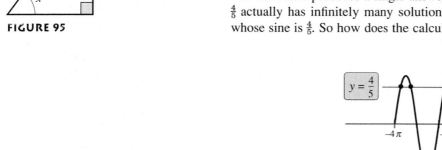

**FIGURE 96**

In the language of functions, we would say that since the function $f(x) = \sin x$ is not 1–1 (i.e., more than one input gives the same output), $f$ has no inverse. Recall that 1–1 functions satisfy the horizontal line test: any horizontal line crosses the graph of a 1–1 function at most once. Figure 96 clearly shows just how badly the graph of $f(x) = \sin x$ fails to satisfy the horizontal line test—some horizontal lines cross the graph infinitely many times! However, if the sine function is restricted to the interval $[-\pi/2, \pi/2]$, as shown in Figure 97, then the horizontal line test will indeed be satisfied. This function, obtained by restricting the sine function, *will have* an inverse, which we call $\arcsin x$ or $\sin^{-1}x$. Thus given any number $x$ between $-1$ and $1$, **arcsin $x$** will represent the *unique* angle between $-\pi/2$ and $\pi/2$ whose sine is $x$. More formally, we have the following definition.

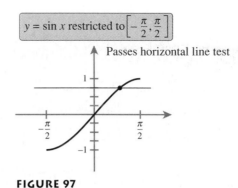

**FIGURE 97**

**Definition of the inverse sine function**

The **inverse sine function**, denoted by **arcsin** or **$\sin^{-1}$**, is defined by

$$y = \arcsin x \quad \text{if and only if} \quad \sin y = x$$

where $-1 \le x \le 1$ and $-\pi/2 \le y \le \pi/2$. The domain of $y = \arcsin x$ is $[-1, 1]$ and the range is $[-\pi/2, \pi/2]$.

**EXAMPLE 1**    *Finding exact values of the inverse sine function*

Find exact values of the following quantities.

**a.** $\sin^{-1}\dfrac{1}{2}$    **b.** $\arcsin\left(-\dfrac{\sqrt{3}}{2}\right)$    **c.** $\sin^{-1}(-3)$

**SOLUTION**

**a.** We are interested in the unique angle between $-\pi/2$ and $\pi/2$ whose sine is $\frac{1}{2}$. In other words, if we let $y = \sin^{-1}\frac{1}{2}$, then we have

$$\sin y = \frac{1}{2} \quad \text{and} \quad -\frac{\pi}{2} \le y \le \frac{\pi}{2}$$

Clearly $\pi/6$ satisfies both conditions. Thus, we conclude that

$$\sin^{-1}\frac{1}{2} = \frac{\pi}{6}$$

**b.** Let $y = \arcsin(-\sqrt{3}/2)$. Then from the definition of the inverse sine function, we have

$$\sin y = -\frac{\sqrt{3}}{2} \quad \text{and} \quad -\frac{\pi}{2} \le y \le \frac{\pi}{2}$$

Thus, we are looking for an angle $y$ between $-\pi/2$ and $\pi/2$ whose sine is $-\sqrt{3}/2$. Since $\sin(-\pi/3) = -\sqrt{3}/2$, and $-\pi/3$ is between $-\pi/2$ and $\pi/2$, it follows that

$$\arcsin\left(-\frac{\sqrt{3}}{2}\right) = -\frac{\pi}{3}$$

**c.** We are interested in an angle whose sine is $-3$. But the sine of any angle is between $-1$ and $1$. Thus, there is no angle whose sine is $-3$, and $\sin^{-1}(-3)$ is undefined. We recall that the domain of the inverse sine function is the interval $[-1, 1]$ so that $\sin^{-1} x$ is only defined for values of $x$ from $-1$ to $1$.  ■

The graph of the inverse sine function can be determined by creating a table of values for the relation $y = \sin^{-1} x$, $-\pi/2 \le y \le \pi/2$, and plotting points. Since the relation $y = \sin^{-1} x$ is equivalent to $x = \sin y$, we will select convenient $y$-values from $-\pi/2$ to $\pi/2$ and then find the corresponding $x$-values. The graph is shown in Figure 98.

| $x = \sin y$ | $y$ |
|---|---|
| $-1$ | $-\dfrac{\pi}{2}$ |
| $-\dfrac{\sqrt{3}}{2}$ | $-\dfrac{\pi}{3}$ |
| $-\dfrac{\sqrt{2}}{2}$ | $-\dfrac{\pi}{4}$ |
| $-\dfrac{1}{2}$ | $-\dfrac{\pi}{6}$ |
| $0$ | $0$ |
| $\dfrac{1}{2}$ | $\dfrac{\pi}{6}$ |
| $\dfrac{\sqrt{2}}{2}$ | $\dfrac{\pi}{4}$ |
| $\dfrac{\sqrt{3}}{2}$ | $\dfrac{\pi}{3}$ |
| $1$ | $\dfrac{\pi}{2}$ |

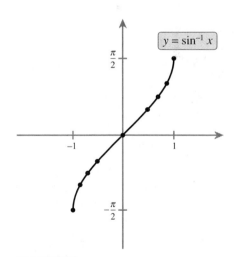

**FIGURE 98**

Although point plotting can be instructive, the graph of the inverse sine function can be obtained much more easily. Recall from Chapter 4 that the graph of the inverse of any 1–1 function $f$ is simply the reflection of the graph of $f$ about the line $y = x$. In the case of the inverse sine function, it is the graph of the sine function over the interval $[-\pi/2, \pi/2]$ that is to be reflected about the line $y = x$, as shown in Figure 99.

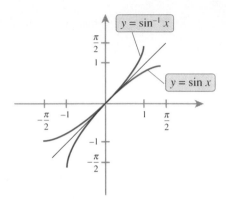

**FIGURE 99**

**EXAMPLE 2**    *Graphing a translation of the inverse sine function*

Sketch the graph of $y = \sin^{-1}(x + 1)$.

**SOLUTION**    Note that the equation $y = \sin^{-1}(x + 1)$ can be obtained from the equation $y = \sin^{-1} x$ by substituting $x + 1$ in place of $x$. Since replacing $x$ with $x + 1$ in an equation has the effect of translating the graph 1 unit to the left, the graph of $y = \sin^{-1}(x + 1)$ will be just a translation 1 unit to the left of the graph of $y = \sin^{-1} x$, as shown in Figure 100.

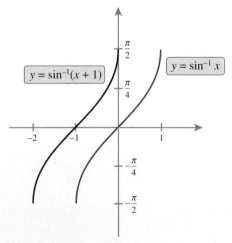

**FIGURE 100**

## THE INVERSE COSINE AND TANGENT FUNCTIONS

The inverse cosine and tangent functions may be defined in much the same way as the inverse sine function. Both the cosine and tangent functions fail to be 1–1 and thus have no inverse as Figure 101 shows. But just as with the sine function, this situation can be remedied by restricting the domains to appropriate intervals, as shown in Figure 102.

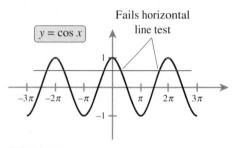

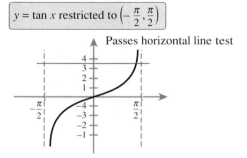

**FIGURE 101**

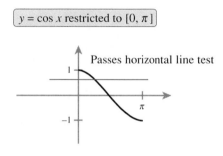

**FIGURE 102**

Note that the cosine function has been restricted to the interval $[0, \pi]$, and the tangent function to $(-\pi/2, \pi/2)$. For convenience, we offer the definitions and graphs of each of the inverse sine, cosine, and tangent functions.

### *Definition of inverse trigonometric functions*

| Function | Definition |
|---|---|
| $y = \arcsin x$ | $y = \arcsin x$ if and only if $\sin y = x$, $-\dfrac{\pi}{2} \le y \le \dfrac{\pi}{2}$ |
| $y = \arccos x$ | $y = \arccos x$ if and only if $\cos y = x$, $0 \le y \le \pi$ |
| $y = \arctan x$ | $y = \arctan x$ if and only if $\tan y = x$, $-\dfrac{\pi}{2} < y < \dfrac{\pi}{2}$ |

■ *Graphs of inverse*
*trigonometric functions*

| Function | Domain | Range | Graph |
|---|---|---|---|
| $y = \arcsin x$ | $-1 \le x \le 1$ | $-\dfrac{\pi}{2} \le y \le \dfrac{\pi}{2}$ | |
| $y = \arccos x$ | $-1 \le x \le 1$ | $0 \le y \le \pi$ | |
| $y = \arctan x$ | $-\infty < x < \infty$ | $-\dfrac{\pi}{2} < y < \dfrac{\pi}{2}$ | |

**EXAMPLE 3**   *Finding exact values of the inverse cosine function*

Find exact values of the following quantities.

**a.** $\cos^{-1} 1$      **b.** $\arccos\left(-\dfrac{\sqrt{2}}{2}\right)$

**SOLUTION**

**a.** We are interested in an angle in the interval $[0, \pi]$ whose cosine is 1. To be precise, if we let $y = \cos^{-1} 1$, then $y$ must satisfy the conditions

$$\cos y = 1 \quad \text{and} \quad 0 \le y \le \pi$$

Since 0 satisfies these conditions, we have

$$\cos^{-1} 1 = 0$$

**b.** Let $y = \arccos(-\sqrt{2}/2)$. Then the following conditions must hold:

$$\cos y = -\frac{\sqrt{2}}{2} \quad \text{and} \quad 0 \le y \le \pi$$

Since $\cos(3\pi/4) = -\sqrt{2}/2$, and $3\pi/4$ is between $0$ and $\pi$, we see that $y = 3\pi/4$. Thus,

$$\arccos\left(-\frac{\sqrt{2}}{2}\right) = \frac{3\pi}{4}$$

**EXAMPLE 4**    *Finding exact values of the inverse tangent function*

a. $\tan^{-1}\sqrt{3}$        b. $\arctan(-1)$

**SOLUTION**

a. Let $y = \tan^{-1}\sqrt{3}$. By the definition of the inverse tangent function, $y$ is the unique angle between $-\pi/2$ and $\pi/2$ for which the tangent is $\sqrt{3}$. Thus, we have

$$\tan y = \sqrt{3} \qquad \text{and} \qquad -\frac{\pi}{2} < y < \frac{\pi}{2}$$

Since $\tan(\pi/3) = \sqrt{3}$, and $\pi/3$ is between $-\pi/2$ and $\pi/2$, we have $y = \pi/3$. Thus,

$$\tan^{-1}\sqrt{3} = \frac{\pi}{3}$$

b. Since $\tan(-\pi/4) = -1$ and $-\pi/4$ is between $-\pi/2$ and $\pi/2$, we have

$$\arctan(-1) = -\frac{\pi}{4}$$

## SOLVING TRIGONOMETRIC EQUATIONS

Although $\sin^{-1} x$ only gives a single solution of the equation $\sin x = y$, other solutions can be found by considering reference angles.

**EXAMPLE 5**    *Solving a trigonometric equation with the inverse sine function*

Estimate all solutions of $3 \sin x = 2$ in the interval $[0, 2\pi)$.

**SOLUTION**    We begin by solving for $\sin x$.

$$3 \sin x = 2$$

$$\sin x = \frac{2}{3}$$

One solution can be found by taking the inverse sine of both sides

$$x = \sin^{-1}\frac{2}{3} \approx 0.7297$$

As Figure 103 shows, there is also an angle between $\pi/2$ and $\pi$ with a sine of $\frac{2}{3}$, namely $\pi - \sin^{-1}\frac{2}{3} \approx 2.4119$. Thus, our solutions are approximately 0.7297 and 2.4119.

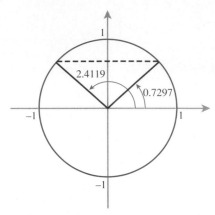

**FIGURE 103**

---

**EXAMPLE 6**     *Solving a trigonometric equation*

Use inverse trigonometric functions to find all solutions of each of the following equations on the interval $[0, 2\pi)$.

**a.** $\cos x = \frac{1}{4}$          **b.** $\tan x = -5$

**SOLUTION**

**a.** We are interested in finding all angles from 0 to $2\pi$ having a cosine of $\frac{1}{4}$. Clearly there are two such angles, one between 0 and $\pi/2$, and the other between $3\pi/2$ and $2\pi$ as shown in Figure 104. By definition, $\cos^{-1}\frac{1}{4}$ is an angle $\theta$ between 0 and $\pi$ satisfying $\cos\theta = \frac{1}{4}$. Thus, we have labeled the solution in the first quadrant as $\cos^{-1}\frac{1}{4}$. The solution between $3\pi/2$ and $\pi$ has $\cos^{-1}\frac{1}{4}$ as its reference angle and is equal to $2\pi - \cos^{-1}\frac{1}{4}$. Thus, our two solutions are $\cos^{-1}\frac{1}{4}$ and $2\pi - \cos^{-1}\frac{1}{4}$. Using a calculator, we find that our solutions are approximately 1.3181 and 4.9651, respectively.

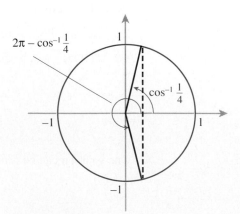

**FIGURE 104**

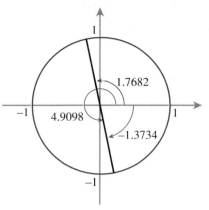

**FIGURE 105**

**b.** Since the tangent function is negative in both the second and fourth quadrants, the equation $\tan x = -5$ will have two solutions between 0 and $2\pi$: one between $\pi/2$ and $\pi$, and another between $3\pi/2$ and $2\pi$. Using the inverse tangent function, we have

$$\tan x = -5$$
$$x = \arctan(-5)$$
$$\approx -1.3734$$

Of course, this solution is not between 0 and $2\pi$, but it does have the same reference angle, namely 1.3734, as the two desired solutions. The angle in the second quadrant with reference angle 1.3734 is $\pi - 1.3734 \approx 1.7682$. The solution between $3\pi/2$ and $2\pi$ is given by $2\pi - 1.3734 \approx 4.9098$. Figure 105 shows our two solutions.

---

## COMPOSING TRIGONOMETRIC AND INVERSE TRIGONOMETRIC FUNCTIONS

Recall that a function and its inverse "undo" one another. In other words, if $f$ is a 1–1 function (so that $f^{-1}$ exists), then

$$f^{-1}(f(x)) = x \qquad \text{and} \qquad f(f^{-1}(x)) = x$$

If we let $f(x) = \sin x$ restricted to the interval $[-\pi/2, \pi/2]$, then $f^{-1}(x) = \arcsin x$, and we have

$$\arcsin(\sin x) = x, \quad \text{provided} \ -\frac{\pi}{2} \le x \le \frac{\pi}{2}$$

and

$$\sin(\arcsin x) = x, \quad \text{provided} \ -1 \le x \le 1$$

Similar results hold for the restrictions of the cosine and tangent functions and their inverses. We summarize all such results as follows.

| *Sine/Inverse Sine* | |
|---|---|
| *Property* | *Restriction* |
| $\sin(\sin^{-1} x) = x$ | $-1 \le x \le 1$ |
| $\sin^{-1}(\sin x) = x$ | $-\dfrac{\pi}{2} \le x \le \dfrac{\pi}{2}$ |

| *Cosine/Inverse Cosine* | |
|---|---|
| *Property* | *Restriction* |
| $\cos(\cos^{-1} x) = x$ | $-1 \le x \le 1$ |
| $\cos^{-1}(\cos x) = x$ | $0 \le x \le \pi$ |

| *Tangent/Inverse Tangent* | |
|---|---|
| *Property* | *Restriction* |
| $\tan(\tan^{-1} x) = x$ | None |
| $\tan^{-1}(\tan x) = x$ | $-\dfrac{\pi}{2} < x < \dfrac{\pi}{2}$ |

**EXAMPLE 7**    *Simplifying with the inverse properties*

Evaluate the following quantities.

**a.** $\cos\left(\cos^{-1}\dfrac{3}{4}\right)$          **b.** $\arctan\left(\tan\dfrac{\pi}{7}\right)$          **c.** $\sin^{-1}\left(\sin\dfrac{12\pi}{13}\right)$

## SOLUTION

**a.** By the inverse properties for cosine, we have

$$\cos\left(\cos^{-1}\frac{3}{4}\right) = \frac{3}{4}$$

**b.** From the inverse properties for tangent, we have

$$\arctan\left(\tan\frac{\pi}{7}\right) = \frac{\pi}{7}$$

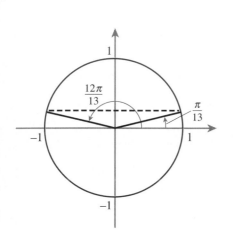

**FIGURE 106**

**c.** In this case, we *cannot* conclude that $\sin^{-1}[\sin(12\pi/13)] = 12\pi/13$, since $12\pi/13$ is not between $-\pi/2$ and $\pi/2$. Instead, let us put $y = \sin^{-1}[\sin(12\pi/13)]$. Then by definition of the inverse sine function, we have

$$\sin y = \sin\frac{12\pi}{13} \qquad \text{and} \qquad -\frac{\pi}{2} \le y \le \frac{\pi}{2}$$

Thus, we seek to find an angle $y$ between $-\pi/2$ and $\pi/2$ whose sine is the same as that of $12\pi/13$. As Figure 106 indicates, $\pi/13$ is the desired angle. We conclude that

$$\sin\left(\sin\frac{12\pi}{13}\right) = \frac{\pi}{13}$$

---

**EXAMPLE 8**

*Simplifying an expression involving the inverse sine function*

Simplify $\sin(\arccos x)$.

**SOLUTION**    Let $\theta = \arccos x$. Our task then is to simplify $\sin\theta$. From the definition of the inverse cosine function, we have

$$\cos\theta = x \qquad \text{and} \qquad 0 \le \theta \le \pi$$

Thus, the problem can be summarized as follows: *Find a simplified expression for* $\sin\theta$ *given that* $\cos\theta = x$ *and* $0 \le \theta \le \pi$.

Using the identity $\sin^2\theta + \cos^2\theta = 1$, we can solve for $\sin\theta$ as follows.

$$\sin^2\theta + \cos^2\theta = 1$$
$$\sin^2\theta + x^2 = 1$$
$$\sin^2\theta = 1 - x^2$$
$$\sin\theta = \pm\sqrt{1 - x^2}$$

Since $\theta$ is between 0 and $\pi$, $\sin\theta \ge 0$. Thus,

$$\sin\theta = \sqrt{1 - x^2}$$

There is an alternative, more intuitive approach that works well in the case where $\theta$ is an acute angle. We sketch a right triangle having an angle $\theta$ whose cosine is $x$ and

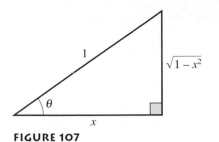

**FIGURE 107**

then use the right triangle definition of the trigonometric functions to compute sin $x$. We begin by labeling one of the acute angles of a right triangle as $\theta$. For cos $\theta$ to be equal to $x$, we label the adjacent side $x$ and the hypotenuse 1. By the Pythagorean Theorem, the side opposite $\theta$ must be $\sqrt{1 - x^2}$. The sketch is shown in Figure 107. From the triangle it is clear that

$$\sin \theta = \frac{\text{opp}}{\text{hyp}} = \frac{\sqrt{1 - x^2}}{1} = \sqrt{1 - x^2}$$

**EXAMPLE 9**    *Evaluating an expression involving the inverse tangent function*

Evaluate csc[arctan($-2$)].

**SOLUTION**    Although a calculator could be used to find an approximate value, we will show how to find the exact value using a technique similar to that of Example 8.

Let $\theta = \arctan(-2)$. Then we are interested in evaluating csc $\theta$. By definition of the inverse tangent function, we have

$$\tan \theta = -2 \qquad \text{and} \qquad -\frac{\pi}{2} < \theta < \frac{\pi}{2}$$

Since tan $\theta < 0$, we know that $\theta$ must be between $-\pi/2$ and 0. Our task can then be summarized as follows: *Find* csc $\theta$ *given that* tan $\theta = -2$ *and* $-\pi/2 < \theta < 0$. We will accomplish this task using a reference angle.

Let $\tilde{\theta}$ be the reference angle for $\theta$. Then tan $\tilde{\theta} = 2$. Since cosecant is negative in the fourth quadrant, we will find csc $\tilde{\theta}$ and then attach a negative sign. In Figure 108, we have sketched a right triangle having $\tilde{\theta}$ as one of its acute angles. In order that tan $\tilde{\theta} = 2$, we have labeled the side opposite $\tilde{\theta}$ as 2, and the side adjacent to $\tilde{\theta}$ as 1. By the Pythagorean Theorem, the hypotenuse is $\sqrt{5}$. From the definition of the cosecant function, we have csc $\tilde{\theta} = \sqrt{5}/2$. Thus, csc $\theta = -\sqrt{5}/2$ and it follows that

$$\csc[\arctan(-2)] = -\frac{\sqrt{5}}{2}$$

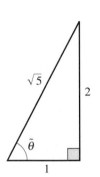

**FIGURE 108**

---

**EXERCISES 6**

**EXERCISES 1–16** □ *Find the exact value of the given quantity.*

1. arcsin 1

2. $\cos^{-1} 0$

3. $\cos^{-1}\left(-\frac{1}{2}\right)$

4. arctan($-1$)

5. $\tan^{-1} 1$

6. $\sin^{-1}(-1)$

7. arctan $\sqrt{3}$

8. $\arccos\dfrac{\sqrt{3}}{2}$

9. $\sin^{-1}\left(-\dfrac{\sqrt{2}}{2}\right)$

10. $\arctan\left(-\dfrac{\sqrt{3}}{3}\right)$

11. arccos($-1$)

12. $\sin^{-1}\dfrac{1}{2}$

13. $\tan^{-1} 0$

14. arccos 2

15. arcsin $\pi$

16. $\tan^{-1}(-\sqrt{3})$

**EXERCISES 17–24** □ *Use inverse trigonometric functions to approximate all solutions of the given equation on the interval* $[0, 2\pi)$.

17. $\sin x = \dfrac{1}{3}$

18. $\cos x = -\dfrac{1}{5}$

19. $\tan x = -20$

20. $\sin x = \dfrac{1}{4}$

21. $\cos x = 0.7$

22. $\tan x = 0.3$

23. $\sin x = \dfrac{7}{6}$

24. $\cos x = -\pi$

**EXERCISES 25–30** □ *Simplify the given quantity. Assume* $x > 0$.

25. $\tan(\arccos x)$

26. $\sin(\tan^{-1} 3x)$

27. $\sec\left(\sin^{-1}\dfrac{2}{x}\right)$

28. $\cot\left(\arctan\dfrac{1 + x}{1 - x}\right)$

29. $\cos\left(\arctan\dfrac{\sqrt{x^2 - 4}}{2}\right)$

30. $\csc(\cos^{-1}\sqrt{1 - 9x^2})$

**EXERCISES 31–40** □ *Simplify the given expression.*

31. $\sin\left(\arcsin\dfrac{1}{3}\right)$

32. $\cos^{-1}\left(\cos\dfrac{\pi}{8}\right)$

33. $\arctan\left[\tan\left(-\dfrac{2\pi}{5}\right)\right]$

34. $\cos(\arccos 0.2)$

35. $\tan[\tan^{-1}(-7)]$

36. $\arcsin\left[\sin\left(-\dfrac{2\pi}{3}\right)\right]$

37. $\arccos\left[\cos\left(-\dfrac{3\pi}{4}\right)\right]$

38. $\tan^{-1}\left(\tan\dfrac{11\pi}{6}\right)$

39. $\sin^{-1}\left(\sin\dfrac{3\pi}{2}\right)$

40. $\cos^{-1}[\cos(-\pi)]$

**EXERCISES 41–48** □ *Evaluate the given quantity.*

41. $\sin\left(\arctan\dfrac{4}{3}\right)$

42. $\tan\left(\sin^{-1}\dfrac{12}{13}\right)$

43. $\sec\left[\cos^{-1}\left(-\dfrac{1}{8}\right)\right]$

44. $\cos(\tan^{-1} 3)$

45. $\cot\left(\sin^{-1}\dfrac{2\sqrt{5}}{5}\right)$

46. $\csc\left[\arccos\left(-\dfrac{4}{5}\right)\right]$

47. $\cos[\arctan(-\sqrt{6})]$

48. $\sec\left(\arcsin\dfrac{\sqrt{3}}{3}\right)$

**EXERCISES 49–54** □ *Sketch the graph of the given function.*

49. $y = \sin^{-1}(x - 2)$

50. $y = \tan^{-1}(x) - \dfrac{\pi}{2}$

51. $y = \cos^{-1}(x) - \pi$

52. $y = \sin^{-1}(x + 2)$

53. $y = \tan^{-1}(x - 1) + \dfrac{\pi}{2}$

54. $y = \cos^{-1}(x + 1) - \pi$

**EXERCISES 55–58** □ *Use a graphics calculator to estimate the solutions of the given equation*

55. $\arcsin x = x + \dfrac{1}{2}$

56. $2\cos^{-1} x = x^2 + 1$

57. $\arctan x = 1 - x^2$

58. $\arcsin x = \arccos x$

**EXERCISES 59–62** □ *Use a graphics calculator to help determine whether or not the given equation is an identity.*

59. $\arcsin^2 x + \arccos^2 x = 1$

60. $\arcsin x + \arccos x = \dfrac{\pi}{2}$

61. $\cos(2\cos^{-1} x) = 2x^2 - 1$

62. $\arctan x = \dfrac{\arcsin x}{\arccos x}$

■ **Applications**

63. *Rising Balloon* A hot-air balloon is rising straight into the air at a constant rate of 6 feet per second.

   a. Find the height of the balloon after 8 seconds.
   b. Find the angle of inclination to the balloon after 8 seconds from an observer located a horizontal distance of 100 feet away from where the balloon was launched.
   c. Express the angle of inclination from the observer to the balloon as a function of time.

64. *Altitude of the Sun* The altitude of the Sun is defined to be the angle made between the line of sight to the Sun and the horizontal.

   a. Find the altitude of the sun if the length of the shadow of a 6-foot man is 3.5 feet.

b. If the man in part (a) is standing near the Washington Monument, and if the length of the monument's shadow is approximately 324 feet, find the height of the Washington Monument.

65. *Air-Sea Rescue* A plane flying at a constant altitude of 2000 feet is in search of a boat in distress. Based on information from a flare sighting, the pilot anticipates that she will fly directly over the boat.

   a. Find the angle of depression from the plane to the boat when the horizontal distance between the plane and the boat is 1200 feet.

   b. Express the angle of depression from the plane to the boat as a function of the horizontal distance from the plane to the boat.

   c. If the pilot is scanning ahead of the plane with an angle of depression of 60°, at what horizontal distance will she spot the boat?

66. *Filming a Stunt* A cameraman is filming a scene in which a stunt man falls from a height of 50 feet. The camera is positioned 20 feet away from the point on the ground where the stunt man will land, and the camera is 6 feet above the ground.

   a. Find the angle of inclination of the camera when the stunt man is 30 feet above the ground.

   b. Express the angle of inclination of the camera as a function of the height of the stunt man in feet.

   c. Given that the stunt man's height in feet $t$ seconds after he begins to fall is $h = 50 - 16t^2$, express the angle of inclination of the camera as a function of $t$.

67. *Sunshine Angle* A solar panel is to be placed in an empty lot between two buildings with heights 70 feet and 30 feet, respec-

tively, and which are 140 feet apart. Denote by $\theta$ the angle made between the lines running from the panel to the top of each building, as shown in Figure 109. A straightforward computation would show that the angle $\theta$ can be written as a function of the distance $x$ in feet between the panel and the shorter building as follows.

$$\theta = \pi - \tan^{-1}\left(\frac{30}{x}\right) - \tan^{-1}\left(\frac{70}{140 - x}\right)$$

   a. Complete the following table of angles for the given distances.

| x (feet) | 20 | 50 | 100 |
|---|---|---|---|
| θ (radians) | | | |

What do you notice about the angles as the distance increases?

   b. Use a graphics calculator to estimate the distance for which the angle is as large as possible. Explain why this would be the most desirable place to put the panel.

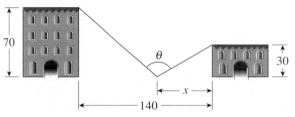

**FIGURE 109**

68. *Road Sign Viewing Angle* The bottom edge of a 5-foot-high freeway sign is 20 feet above the eye level of a motorist heading toward it, as shown in Figure 110. A straightforward computation would reveal that the vertical viewing angle $\theta$ can be

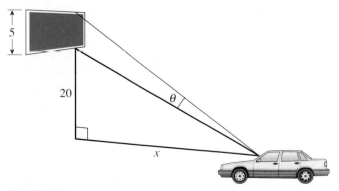

**FIGURE 110**

expressed as a function of the distance $x$ in feet from the motorist to a point directly beneath the sign as follows.

$$\theta = \tan^{-1}\left(\frac{25}{x}\right) - \tan^{-1}\left(\frac{20}{x}\right)$$

a. Complete the following table of viewing angles for the given distances.

| $x$ (feet) | 50 | 20 | 10 |
|---|---|---|---|
| $\theta$ (radians) | | | |

What do you notice about the angles as the distance decreases?

b. Use a graphics calculator to estimate the distance for which the viewing angle is as large as possible.

## Project for Enrichment

69. *The Best Seat in the House* When you walk into a movie theater, where do you look first for an available seat? If you're like most people, you probably look for an empty row somewhere in the middle of the theater, not too close to the front. But suppose there are no empty seats in the middle. In fact, suppose the only available seats are along the right wall of the theater. Which row would you choose? Would you be surprised to find out that there is a mathematical way of finding the ''best'' row? There is, provided we define the best row to be the one in which the viewing angle to the screen is as large as possible. Consider the diagram shown in Figure 111. The screen is 40 feet wide, and the available seats are in a column 30 feet away from the right edge of the screen. From a distance of $x$ feet from the front of the theater, the viewing angle $\theta$ is the angle made between the lines of sight to the left and right edges of the screen. The angle $\alpha$ is the angle made between the column of available seats and the line of sight to the right edge of the screen. The angle $\beta$ is the angle made between the column of available seats and the line of sight to the left edge of the screen.

a. Express the angle $\alpha$ as a function of $x$.

b. Express the angle $\beta$ as a function of $x$.

c. Express the angle $\theta$ as a function of $x$.

d. Use the function you found in part (c) to find $\theta$ for the following values of $x$.
   i. $x = 10$   ii. $x = 50$   iii. $x = 100$

e. Use a graphics calculator to plot the graph of $\theta$ as a function of $x$.

f. Use a graphics calculator to estimate the value of $x$ for which $\theta$ is the given value.
   i. $\theta = 15°$   ii. $\theta = 30°$   iii. $\theta = 45°$

g. Use a graphics calculator to find the distance $x$ for which $\theta$ is as large as possible.

h. Explain how the optimal value for $x$ would change if the distance from the right edge of the screen to the column of empty seats were smaller or larger than 30 feet.

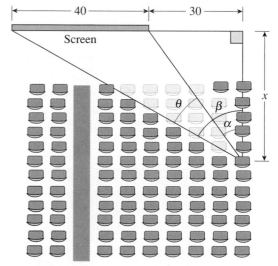

**FIGURE 111**

## Questions for Discussion or Essay

70. In this section we defined inverses for non 1–1 functions by first restricting the domain so that the resulting function was 1–1. In the case of $f(x) = \sin x$, we restricted the domain to the interval $[-\pi/2, \pi/2]$. Are there other intervals to which we could have restricted the sine function? Is the sine function 1–1 on the interval $[-\pi/4, \pi/4]$? What disadvantage would there be to defining the inverse sine function by ''$y = \sin^{-1} x$ if and only if $\sin y = x$ and $-\pi/4 \le x \le \pi/4$''?

71. The function $f(x) = x^2$ fails to be 1–1 since its graph fails the horizontal line test. How could the domain of $f$ be restricted so that the resulting function would become 1–1? What would the inverse function be?

72. We did not define any of $\cot^{-1} x$, $\sec^{-1} x$, or $\csc^{-1} x$, but each could be defined in a fashion similar to that of $\cos^{-1} x$, $\sin^{-1} x$, $\tan^{-1} x$. Define each of these functions.

**73.** Exercises 68 and 69 both relate to the concept of viewing angle. One would expect that the optimal position from which to view an object would be where the viewing angle is as large as possible. But in practice, other considerations affect the opti- mal viewing angle. What factors (besides viewing angle) deter- mine the best seat from which to view a theater screen? In the case of the road sign, how does the motion of the car affect the ideal position from which to read a road sign?

## CHAPTER REVIEW EXERCISES

**EXERCISES 1–4** □ *Match the given angle measure with the corresponding sketch.*

1. $60°$

2. $-\dfrac{\pi}{6}$

3. $-\dfrac{3\pi}{2}$

4. $315°$

a.
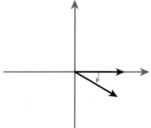

b.

c.

d.

**EXERCISES 5–8** □ *Find two coterminal angles (one positive and one negative) for the given angle.*

5. $60°$

6. $-\dfrac{\pi}{6}$

7. $-\dfrac{3\pi}{2}$

8. $225°$

**EXERCISES 9–14** □ *Convert to radian measure.*

9. $60°$

10. $315°$

11. $-210°$

12. $270°$

13. $18.3°$

14. $-72.4°$

**EXERCISES 15–20** □ *Convert to degree measure.*

15. $-\dfrac{5\pi}{6}$

16. $\dfrac{5\pi}{4}$

17. $\dfrac{\pi}{8}$

18. $-\dfrac{3\pi}{2}$

19. $\dfrac{4\pi}{13}$

20. $4$

**EXERCISES 21–24** □ *Find the values of the six trigonometric functions for the given angle θ.*

21.
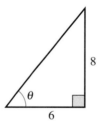

22.
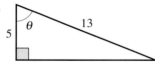

23.
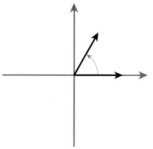

24.

**EXERCISES 25–28** □ *Use a right triangle to find the other five trigonometric functions of the acute angle θ.*

25. $\cos\theta = \dfrac{15}{17}$

26. $\tan\theta = \dfrac{3}{4}$

27. $\csc\theta = 2$

28. $\sin\theta = \dfrac{1}{\sqrt{3}}$

**EXERCISES 29–32** □ *Find all unknown side lengths and angle measures; that is, solve the given triangle.*

**29.**

**30.**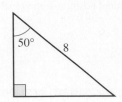

**31.**

**32.**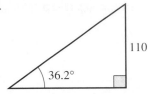

**EXERCISES 33–36** □ *Find the reference angle for the given angle.*

**33.** $\theta = 210°$

**34.** $\theta = \dfrac{7\pi}{6}$

**35.** $\theta = -\dfrac{3\pi}{4}$

**36.** $\theta = -20°$

**EXERCISES 37–42** □ *Use reference angles to find the exact values of the given trigonometric functions.*

**37.** $\sin\dfrac{2\pi}{3}$

**38.** $\cot\left(-\dfrac{\pi}{6}\right)$

**39.** $\sec(-150°)$

**40.** $\csc 210°$

**41.** $\tan\dfrac{5\pi}{4}$

**42.** $\cos 120°$

**EXERCISES 43–48** □ *Use a calculator to find a decimal approximation for the given function, correct to four decimal places.*

**43.** $\sin(-200°)$

**44.** $\tan 95°$

**45.** $\sec\dfrac{7\pi}{10}$

**46.** $\csc 4.1$

**47.** $\cot(125°10'15'')$

**48.** $\cos(-32°25'30'')$

**EXERCISES 49–52** □ *Use the given information about the angle $\theta$ to find the indicated function value.*

**49.** $\tan \theta = -2$, $\cos \theta < 0$; $\sin \theta = $ _____

**50.** $\sin \theta = \dfrac{1}{4}$, $\tan \theta < 0$; $\cos \theta = $ _____

**51.** $\cos \theta = -\dfrac{2}{3}$, $\sin \theta > 0$; $\tan \theta = $ _____

**52.** $\cot \theta = 10$, $\sec \theta > 0$; $\csc \theta = $ _____

**EXERCISES 53–56** □ *Use a calculator to find a decimal approximation, correct to four decimal places, for the acute angle $\theta$ measured in (a) degrees and (b) radians.*

**53.** $\tan \theta = 3.6$

**54.** $\cos \theta = 0.25$

**55.** $\sin \theta = 0.32$

**56.** $\cot \theta = 0.41$

**EXERCISES 57–64** □ *Match the given function with its corresponding graph.*

**57.** $f(x) = \sin \pi x$

**58.** $f(x) = \sin(\pi x + \pi)$

**59.** $f(x) = \cos 2x$

**60.** $f(x) = \cos\left(\dfrac{2x - \pi}{2}\right)$

**61.** $f(x) = \tan\left(x + \dfrac{\pi}{3}\right)$

**62.** $f(x) = \cot\left(x - \dfrac{\pi}{3}\right)$

**63.** $f(x) = \sec\dfrac{x}{2}$

**64.** $f(x) = \sec\dfrac{\pi x}{2}$

**a.**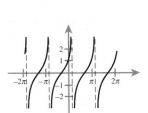

**b.**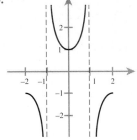

**c.**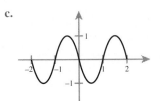

**d.**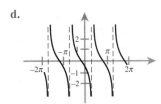

**e.**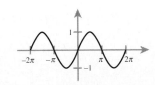

**f.**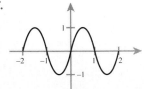

**g.**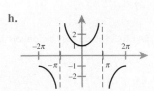

**h.**

**EXERCISES 65–68** □ *Find the period and the amplitude of the function with the given graph.*

**65.**

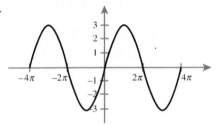

**66.**

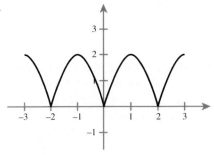

**67.**

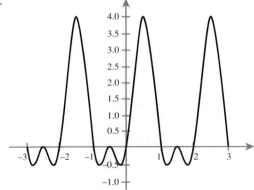

**68.**

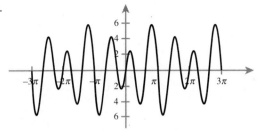

**EXERCISES 69–72** □ *Find the period and the amplitude of the given function. Use a graphics calculator as necessary.*

**69.** $g(x) = 2 \sin(3x + \pi)$

**70.** $f(x) = -3 \cos(\pi x - 1)$

**71.** $h(x) = 2 \sin x + \cos(2x)$

**72.** $g(x) = \sin(3x) - \cos(4x)$

**EXERCISES 73–86** □ *Graph three cycles of the given function. Identify any of the following features that apply: amplitude, period, phase shift, or asymptotes.*

**73.** $f(x) = \sin\dfrac{\pi x}{2}$

**74.** $g(x) = \cos 3x$

**75.** $h(t) = \cot 2t$

**76.** $g(\theta) = -\tan\dfrac{\pi\theta}{4}$

**77.** $f(t) = \sec 2\pi t$

**78.** $f(t) = \csc 4t$

**79.** $g(x) = 3 \cos\left(\dfrac{x}{2}\right) - 1$

**80.** $h(t) = 4 - 2 \sin 3t$

**81.** $y = \tan\left(x - \dfrac{\pi}{4}\right)$

**82.** $y = \cot(2x + \pi)$

**83.** $y = \csc(\pi x - \pi)$

**84.** $y = \sec\left(x + \dfrac{\pi}{6}\right)$

**85.** $y = -2 \cos\left(\pi x + \dfrac{\pi}{2}\right)$

**86.** $y = 4 \sin\left(\dfrac{x}{2} - \dfrac{\pi}{3}\right)$

**EXERCISES 87–90** □ *Each of the graphs given below is that of a function of the form $f(x) = a \sin(bx + c)$. Find a, b, and c.*

**87.**

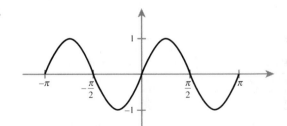

**88.**

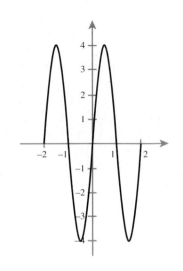

**89.**

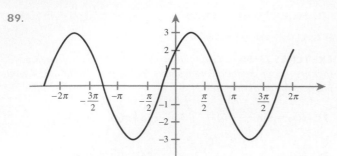

**90.**

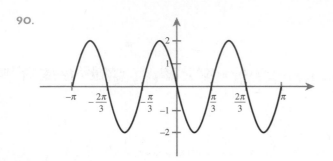

**EXERCISES 91–96** ☐ *Find the exact value of the given quantity.*

**91.** arccos 1

**92.** $\tan^{-1} 1$

**93.** $\sin^{-1} \dfrac{\sqrt{3}}{2}$

**94.** $\cos^{-1}\left(-\dfrac{1}{2}\right)$

**95.** $\arctan\left(-\dfrac{\sqrt{3}}{3}\right)$

**96.** $\arcsin(-2)$

**EXERCISES 97–100** ☐ *Use inverse trigonometric functions to approximate all solutions of the given equation on the interval* $[0, 2\pi)$.

**97.** $\cos x = -\dfrac{1}{4}$

**98.** $\sin x = \dfrac{2}{3}$

**99.** $\sin x = \dfrac{1}{\sqrt{3}}$

**100.** $\tan x = -14$

**EXERCISES 101–104** ☐ *Simplify the given expression. Assume* $x > 0$.

**101.** $\sec(\arcsin x)$

**102.** $\cot(\cos^{-1} x)$

**103.** $\cos\left(\tan^{-1} \dfrac{x}{3}\right)$

**104.** $\sin\left(\arctan \dfrac{1}{x}\right)$

**EXERCISES 105–112** ☐ *Evaluate the given quantity.*

**105.** $\cos\left(\tan^{-1} \dfrac{3}{4}\right)$

**106.** $\tan\left(\sin^{-1} \dfrac{12}{13}\right)$

**107.** $\csc\left[\arccos\left(-\dfrac{1}{3}\right)\right]$

**108.** $\sec(\arctan \sqrt{2})$

**109.** $\tan(\tan^{-1} 16)$

**110.** $\sin[\arcsin(-0.1)]$

**111.** $\sin^{-1}(\sin 3\pi)$

**112.** $\arccos\left(\cos \dfrac{5\pi}{2}\right)$

**EXERCISES 113–116** ☐ *Sketch the graph of the given function.*

**113.** $y = \arctan(x + 1)$

**114.** $y = \sin^{-1}(x - 3)$

**115.** $y = \arccos(x) + \dfrac{\pi}{2}$

**116.** $y = \tan^{-1}(x - 2) - \pi$

**117.** *Pulley Angle*  A 4-inch-diameter pulley is used to reel in 20 feet of electrical cable. Through how many radians does the pulley turn, and to how many revolutions does this correspond?

**118.** *Bicycle Speed*  A bicycle tire with a diameter of 20 inches makes $2\frac{1}{2}$ complete revolutions for each revolution of the pedals. If a child is pedaling at a rate of 40 revolutions per minute, find the speed of the bicycle in miles per hour.

**119.** *World's Largest Clock Face*  According to the *Guinness Book of World Records*, the world's largest clock face—101 feet in diameter—is that of the floral clock in Matsubara Park, Toi, Japan. Find the linear speed, in miles per hour, of the tip of the minute hand if it is 41 feet from the center of the clock.

**120.** *Yacht Distance*  The yacht *Jeffmiller* leaves port on a course of 48° and travels for 25 miles. It then travels due south for an unknown distance and returns to port by traveling due west. Find the distance traveled by the ship.

**121.** *Viewing Distance*  The Colossus of Rhodes, one of the Seven Wonders of the World, was a 120-foot statue of the Greek god Apollo at the entrance to the harbor of Rhodes. If an ancient observer was situated so that the angle of elevation to the top

of the statue was 35°, then how far was the observer from the statue?

122. *Volume of a Feeding Trough* A rectangular piece of metal with dimensions 3 feet by 20 feet is to be formed into a feeding trough by bending up two sides so that each forms an angle $\theta$ with the horizontal, as shown in Figure 112. The volume of the trough in cubic feet as a function of the angle $\theta$ is given by $f(\theta) = 20 \sin \theta(\cos \theta + 1)$.

   a. Find the volume of the trough for $\theta = \pi/4$, $\theta = \pi/3$, and $\theta = \pi/2$.

   b. Use a graphics calculator to estimate the value of $\theta$ for which the volume is as large as possible.

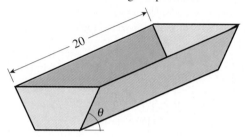

**FIGURE 112**

123. *Weight Fluctuations* A certain dieter's weight has been found to "yo-yo" according to the formula

$$w(t) = 10 \cos\left(\frac{\pi}{6} t\right) + 200$$

where $t$ represents the number of months that have elapsed since January 1.

   a. Find the period of $w(t)$. Interpret this result.

   b. Find the amplitude of $w$.

   c. What is the greatest weight that the dieter attains in the course of the year? When does the dieter reach this weight?

   d. What is the lightest weight that the dieter attains in the course of the year? When does the dieter reach this weight?

124. *Angle of Elevation* A package dropped from a helicopter descends straight down in such a way that its height in feet $t$ seconds after being dropped is given by $y(t) = 2000 - 16t^2$. From the perspective of an observer on the ground 100 feet from the point of impact, the angle of elevation of the package is $\theta$. Express $\theta$ as a function of $t$, and find $\theta$ 10 seconds after the package has been dropped.

## CHAPTER TEST

**PROBLEMS 1–10** □ *Answer true or false.*

**1.** Given any angle $\theta$, there are exactly two angles that are coterminal with $\theta$.

**2.** $\sin(\arcsin x) = x$ for $-1 \le x \le 1$.

**3.** The function $f(x) = \arcsin x$ has period $2\pi$.

**4.** If $\alpha$ and $\beta$ are coterminal, then $\alpha = \beta$.

**5.** If $\alpha$ and $\beta$ are coterminal, then $\sin \alpha = \sin \beta$.

**6.** The phase shift of the function $f(x) = a \sin(bx + c)$ is $-c$.

**7.** The period of $f(x) = a \sin(bx + c)$ depends only on the value of $b$.

**8.** The function $f(x) = e^{-x} \sin x$ is periodic.

**9.** A reference angle is always acute.

**10.** The graphs of $f(x) = a \sin(bx)$ and $g(x) = -a \sin(bx)$ have the same amplitude.

**PROBLEMS 11–14** □ *Give an example of each of the following.*

**11.** A trigonometric function with amplitude 3 and period $\dfrac{\pi}{2}$

**12.** A nonacute angle $\theta$ whose tangent is 1

**13.** A trigonometric function with an asymptote at $\pi$

**14.** The side lengths of a right triangle with an angle measuring $40°$

**15.** Find the length of the arc subtended by a central angle of $25°$ in a circle of radius 5.

**16.** Find the values of the six trigonometric functions for the angle $\theta$ shown in Figure 113.

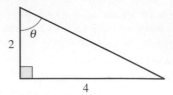

**FIGURE 113**

**17.** From a distance of 50 feet away from the base of a radio tower, the angle of elevation to the top of the tower is $75°$. Find the height of the tower.

**18.** Find the exact value of $\cos\left(\dfrac{7\pi}{6}\right)$.

**19.** Graph three periods of $f(x) = 2 \sin(3x - \pi) + 1$. Identify the amplitude, period, and phase shift.

**20.** Use a graphics calculator to estimate the period and amplitude of the function

$$f(x) = 4 \sin\left(\frac{\pi x}{2}\right) - 3 \cos(\pi x)$$

**21.** Graph three periods of the function $g(x) = \tan(\pi x - \pi)$. Identify the period and locations of asymptotes.

**22.** Find the exact value of $\arccos\left(\cos\dfrac{8\pi}{5}\right)$.

**23.** Use inverse trigonometric functions to find all solutions of $\sin x - \frac{1}{3} = 0$ on the interval $[0, 2\pi)$.

# TRIGONOMETRIC IDENTITIES AND EQUATIONS

■ This illustration depicts a double star system at the heart of the globular cluster NGC 6624, in the constellation Sagittarius. The intense gravitational pull of the very dense neutron star (upper right) distorts its less massive white dwarf counterpart (lower left), cannibalizing its gasses across a narrow accretion bridge. The stars rotate about each other every 11 minutes, locked in an intricate celestial dance punctuated by nuclear fusion explosions and accompanied by sporadic outbursts of X-rays. The script of this distant drama is encoded in these X-rays and the ultraviolet radiation that reaches the Hubble Space Telescope, but can only be decoded by mathematics that is, in essence, an extension of trigonometry.

## SECTION 1

# FUNDAMENTAL TRIGONOMETRIC IDENTITIES

■ Can an equation have infinitely many solutions and yet not be an identity?

■ What is the shape of the Gateway Arch in St. Louis?

■ How many trigonometric identities are there?

■ How do engineers determine the angle at which a turn in a highway is to be banked?

■ How many ways are there of expressing tan $x$ using only the six trigonometric functions?

## TYPES OF TRIGONOMETRIC EQUATIONS

As you may recall from Chapter 2, equations can be classified into three categories according to the nature of the solution set. Equations such as $2(x + 1) = 2x + 2$ that are always true are called **identities**; equations that are sometimes true, like $x = 2$, are called **conditional equations**; and those having an empty solution set, like $x + 1 = x + 2$, are said to be **contradictions**.

Most identities involving nontrigonometric functions are transparently true once a few basic rules are understood. For example, the equation $2(x + 1) = 2x + 2$ is clearly an identity; it is just a special case of the distributive law. Similarly, the equation $2^{x+5} = 2^x 2^5$ is a special case of the rule $a^{x+y} = a^x a^y$. But consider the trigonometric equation

$$\frac{\sin^2 x}{\cos x} + \frac{\cos^2 x}{\sin x} = \sec x + \csc x - \cos x - \sin x$$

As it turns out, this equation is also an identity. But who among us can glance at such a complex equation and immediately recognize the left and right sides as being equivalent?

The problem is that the trigonometric functions are, in a sense, redundant. Any expression involving tangent can be rewritten using cotangent; an expression involving sec $x$ can be replaced with 1/cos $x$, and so on. In fact, any expression involving trigonometric functions can be rewritten in infinitely many ways, and each time we rewrite a trigonometric expression, we are establishing an identity. Thus, there are infinitely many identities, all but a handful of which are so complex that we cannot immediately recognize them as such.

Fortunately, there is an easy way to expose identities. If the left and right sides of an equation are just different representations of the same function (as is the case with an identity), then their graphs should be identical. Thus, if an equation is an identity, when we graph both the left and right sides of the equation on the same set of coordinate axes, we should see just one graph. On the other hand, if we see two distinct graphs that intersect in one or more places, then we know that the equation is sometimes (but not always) true and is thus a conditional equation. Finally, if the graphs do not intersect, then the equation is a contradiction. In the examples that follow, this technique has been used to suggest the appropriate classification.

**EXAMPLE 1**    *Classifying a trigonometric equation with a graphics calculator*

Use a graphics calculator to decide whether the following equation is a contradiction, a conditional equation, or an identity.

$$2 \sin x = 2 - 2 \cos x$$

**SOLUTION**    Figure 1 shows the graph of $y = 2 \sin x$. Figure 2 depicts the graph of $y = 2 \sin x$ and $y = 2 - 2 \cos x$ on the same set of coordinate axes. Since the graphs intersect but are not identical, the equation is conditional.

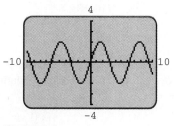

**FIGURE 1**                              **FIGURE 2**

**EXAMPLE 2**    *Classifying a trigonometric equation with a graphics calculator*

Use a graphics calculator to decide whether the following equation is a contradiction, a conditional equation, or an identity.

$$(\sin x + \cos x)^2 = 1 + \sin 2x$$

**SOLUTION**    Figure 3 shows the graph of $y = (\sin x + \cos x)^2$. In Figure 4 we have graphed $y = 1 + \sin 2x$ on the same set of coordinate axes. Since the graphs are indistinguishable (there appears to be just a single graph), we conclude that the equation is most likely an identity.

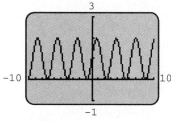
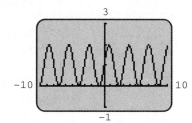

**FIGURE 3**                              **FIGURE 4**

**EXAMPLE 3**    *Classifying a trigonometric equation with a graphics calculator*

Use a graphics calculator to decide whether the following equation is a contradiction, a conditional equation, or an identity.

$$2 - \sin x = \cos x$$

**SOLUTION**    The graphs of $y = 2 - \sin x$ and $y = \cos x$ are shown in Figure 5. Since the graphs do not appear to intersect, we deduce that the equation is a contradiction.

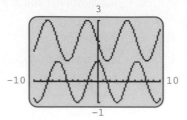

**FIGURE 5**

As usual, there are perils in relying on technology alone to settle mathematical questions. For example, consider the equation

$$1000 \sin \frac{x}{1000} = x$$

If we graph $y_1 = 1000 \sin(x/1000)$ and $y_2 = x$ on the same set of coordinate axes with the standard viewing rectangle, we obtain the graph shown in Figure 6. From this we would conclude that our equation is an identity. But if we change the viewing rectangle to $[-10{,}000, 10{,}000] \times [-10{,}000, 10{,}000]$, we obtain the graph shown in Figure 7.

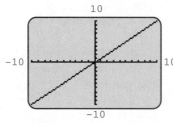

**FIGURE 6**

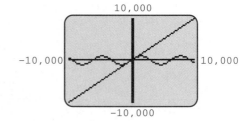

**FIGURE 7**

With this viewing rectangle, it is quite clear that the equation is *not* an identity. Evidently, for relatively small values of $x$, $1000 \sin(x/1000)$ is nearly identical to $x$, which caused us to be misled by our first graph. The point is that graphical information, although providing us with valuable insight, is not enough to verify an identity. Instead, an identity is verified by showing that the two sides of an equation are equivalent using previously established identities.

Among the infinitely many trigonometric identities, a mere handful are sufficiently simple and useful to merit memorization. These fundamental identities can then be used to verify other, more complex, identities. We have already encountered these fundamental identities in the previous chapter, but we restate them here for convenience. It is absolutely essential that these identities be memorized.

***Pythagorean identities***

$$\sin^2 \theta + \cos^2 \theta = 1 \qquad 1 + \tan^2 \theta = \sec^2 \theta \qquad 1 + \cot^2 \theta = \csc^2 \theta$$

**Ratio identities**

$$\sin\theta = \frac{1}{\csc\theta} \qquad \cos\theta = \frac{1}{\sec\theta} \qquad \tan\theta = \frac{1}{\cot\theta}$$

$$\csc\theta = \frac{1}{\sin\theta} \qquad \sec\theta = \frac{1}{\cos\theta} \qquad \cot\theta = \frac{1}{\tan\theta}$$

$$\tan\theta = \frac{\sin\theta}{\cos\theta} \qquad \cot\theta = \frac{\cos\theta}{\sin\theta}$$

## VERIFYING IDENTITIES

To verify an identity, we must show that the equation is true for all values of the variable for which both sides are defined. To do this, we begin with one side of the equation and rewrite it either by applying a basic trigonometric identity or by algebraically rearranging. We repeat this process as many times as necessary until at last we arrive at the expression on the other side of the equation. Equivalently, we can work on both sides (independently) until we are able to show that both sides are equivalent to a third expression.

Of course, there is little chance for success if each step we take is randomly chosen. Thus, it is important to have a strategy in mind. Although some trigonometric identities are quite difficult to verify and require a great deal of ingenuity, most of those we will encounter can be verified by applying a few basic rules of thumb, which we list here.

**Verifying identities**

**Rule 1** Wherever possible, simplify an expression rather than make it more complex. Thus, when given an equation in which one side is much more complex than the other, begin working on the complex side. If both sides are complex, simplify both sides independently.

**Rule 2** If all else fails, convert both sides to sines and cosines. This is not always the "slickest" way of verifying an identity, but it almost always works.

**Rule 3** Look for opportunities to exploit factoring, particularly the difference of squares formula $a^2 - b^2 = (a - b)(a + b)$. For example, the expression

$$\frac{\cos^2 x - \sin^2 x}{\cos x + \sin x}$$

can be simplified by first factoring the numerator:

$$\frac{\cos^2 x - \sin^2 x}{\cos x + \sin x} = \frac{(\cos x - \sin x)(\cos x + \sin x)}{\cos x + \sin x}$$

$$= \cos x - \sin x$$

**WARNING**   It is often tempting to perform an algebraic operation on both sides of the equation at the same time, like squaring both sides. This is not a legitimate method of verification. (See Exercise 79 for an exploration of this issue.)

In the following examples, pay special attention to the *strategy* employed. It is relatively easy to follow the individual steps but much more difficult (and more rewarding) to understand the rationale behind the chosen strategy.

### EXAMPLE 4   *Verifying an identity*

Verify that the following equation is an identity.

$$\sin x \cot x = \cos x$$

**SOLUTION**   We will begin with the left side since it is more complex than the right.

$$\sin x \cot x = \sin x \, \frac{\cos x}{\sin x} \qquad \text{Using } \cot x = \frac{\cos x}{\sin x}$$

$$= \cos x$$

---

### EXAMPLE 5   *Verifying an identity*

Verify that the following equation is an identity.

$$\frac{1 - \cos x}{\cos x} (\sec x + 1) = \tan^2 x$$

**SOLUTION**   We will work with the left side since it is much more complex than the right.

$$\frac{1 - \cos x}{\cos x} (\sec x + 1) = \left( \frac{1}{\cos x} - \frac{\cos x}{\cos x} \right)(\sec x + 1)$$

$$= (\sec x - 1)(\sec x + 1) \qquad \text{Using } \sec x = \frac{1}{\cos x}$$

$$= \sec^2 x - 1$$

$$= (\tan^2 x + 1) - 1 \qquad \text{Using } \sec^2 x = \tan^2 x + 1$$

$$= \tan^2 x$$

---

Many identities can be established in several different (but equally simple) ways. This reflects the fact that at any stage of the verification process, there is no single *right* step; there are, however, *sensible* steps for which the likelihood of success is high.

### EXAMPLE 6   *Verifying an identity*

Verify that the following equation is an identity.

$$1 = (1 - \cos^2 u)(1 + \cot^2 u)$$

**SOLUTION**    Since the right side is more complex than the left, we will begin on the right. There are several ways to attack this identity—we will show two.

**Method 1**

$$(1 - \cos^2 u)(1 + \cot^2 u) = \sin^2 u \csc^2 u \qquad \text{Using Pythagorean identities}$$

$$= \sin^2 u \, \frac{1}{\sin^2 u} \qquad \text{Using } \csc u = \frac{1}{\sin u}$$

$$= 1$$

**Method 2**

$$(1 - \cos^2 u)(1 + \cot^2 u) = \sin^2 u \left( 1 + \frac{\cos^2 u}{\sin^2 u} \right) \qquad \begin{array}{l} \text{Using } 1 - \cos^2 u = \sin^2 u \\ \text{and } \cot u = \cos u / \sin u \end{array}$$

$$= \sin^2 u + \cos^2 u \qquad \text{Multiplying through}$$

$$= 1$$

For cases in which the two sides of the equation are of comparable complexity and there is no obvious way to convert one to the other, converting all trigonometric functions to expressions involving sine and cosine is often the best strategy.

**EXAMPLE 7**    *Verifying an identity*

Verify that the following equation is an identity.

$$\frac{\tan \theta}{1 - \sec \theta} = \frac{1}{\cot \theta - \csc \theta}$$

**SOLUTION**    In this case, both sides are equally complex. We convert both sides to sines and cosines, beginning with the left side.

$$\frac{\tan \theta}{1 - \sec \theta} = \frac{\dfrac{\sin \theta}{\cos \theta}}{1 - \dfrac{1}{\cos \theta}} \qquad \text{Using ratio identities}$$

$$= \frac{\dfrac{\sin \theta}{\cos \theta}}{\dfrac{\cos \theta - 1}{\cos \theta}} \qquad \text{Simplifying the denominator}$$

$$= \left( \frac{\sin \theta}{\cos \theta} \right) \left( \frac{\cos \theta}{\cos \theta - 1} \right) \qquad \text{Inverting and multiplying}$$

$$= \frac{\sin \theta}{\cos \theta - 1}$$

On the other hand, the right side simplifies as follows.

$$\frac{1}{\cot\theta - \csc\theta} = \frac{1}{\dfrac{\cos\theta}{\sin\theta} - \dfrac{1}{\sin\theta}} \qquad \text{Using ratio identities}$$

$$= \frac{1}{\dfrac{\cos\theta - 1}{\sin\theta}}$$

$$= \frac{\sin\theta}{\cos\theta - 1}$$

Since both sides of the original equation simplify to the same expression, the equation is an identity.

$\blacksquare$

**EXAMPLE 8**     *Verifying an identity*

Verify that the following equation is an identity.

$$\frac{\cos\alpha}{1 - \sin\alpha} = \sec\alpha + \tan\alpha$$

**SOLUTION**     We begin by converting the right side to sines and cosines.

$$\sec\alpha + \tan\alpha = \frac{1}{\cos\alpha} + \frac{\sin\alpha}{\cos\alpha} \qquad \text{Using ratio identities}$$

$$= \frac{1 + \sin\alpha}{\cos\alpha}$$

Unfortunately, this is not the same as the expression on the left side. Since the denominator of the desired expression is $1 - \sin\alpha$, we will multiply the numerator and denominator of $(1 + \sin\alpha)/\cos\alpha$ by $1 - \sin\alpha$ as follows:

$$\left(\frac{1 + \sin\alpha}{\cos\alpha}\right)\left(\frac{1 - \sin\alpha}{1 - \sin\alpha}\right) = \frac{1 - \sin^2\alpha}{\cos\alpha(1 - \sin\alpha)}$$

$$= \frac{\cos^2\alpha}{\cos\alpha(1 - \sin\alpha)} \qquad \text{Using } 1 - \sin^2\alpha = \cos^2\alpha$$

$$= \frac{\cos\alpha}{1 - \sin\alpha}$$

Thus,

$$\sec\alpha + \tan\alpha = \frac{1 + \sin\alpha}{\cos\alpha} = \frac{\cos\alpha}{1 - \sin\alpha}$$

$\blacksquare$

## MATHEMATICAL APPLICATIONS OF TRIGONOMETRIC IDENTITIES

Trigonometric identities can be used to simplify complicated trigonometric expressions. In fact, we have already done this in the course of establishing trigonometric identities. There are, however, other contexts in which we might wish to simplify a trigonometric expression. For example, suppose we are given the function

$$f(x) = \frac{\cos^2 x}{1 - \sin x}$$

and we wish to determine its period, its maximum and minimum values, and its amplitude. Of course, these quantities could be *estimated* by inspecting the graph of $f$, but if we knew that this function could be rewritten as

$$f(x) = 1 + \sin x$$

then we would immediately recognize it as nothing more than a vertical translation (1 unit upward) of the function $\sin x$. As such, we would conclude that $f$ has period $2\pi$, maximum value 2, minimum value 0, and amplitude 1. The process by which the expression $\cos^2 x/(1 - \sin x)$ is simplified to obtain the expression $1 + \sin x$ is the subject of the following example.

**EXAMPLE 9**    *Simplifying a trigonometric expression*

Simplify the following expression.

$$\frac{\cos^2 x}{1 - \sin x}$$

**SOLUTION**    We notice that the numerator involves $\cos^2 x$, which can easily be rewritten in terms of $\sin x$ using the Pythagorean identity $\sin^2 x + \cos^2 x = 1$. From there, we simplify algebraically:

$$\frac{\cos^2 x}{1 - \sin x} = \frac{1 - \sin^2 x}{1 - \sin x}$$
$$= \frac{(1 - \sin x)(1 + \sin x)}{1 - \sin x} \qquad \text{Factoring a difference of squares}$$
$$= 1 + \sin x$$

Facility with the basic trigonometric identities enables us to express any trigonometric function in terms of any other, although there may be an unknown sign. For example, from the identity $\sin^2 x + \cos^2 x = 1$, we conclude that $\cos^2 x = 1 - \sin^2 x$. From this it follows that $\cos x = \pm\sqrt{1 - \sin^2 x}$. In some instances, additional information is provided that enables us to determine the sign (plus or minus) as well.

**EXAMPLE 10**     *Expressing one trigonometric function in terms of another*

Write $\tan x$ in terms of $\csc x$. Assume that $0 \le x < \dfrac{\pi}{2}$.

**SOLUTION**    It would be nice if one of the fundamental identities directly related the tangent and cosecant functions, but none do. However, there *are* fundamental identities that connect the tangent and cotangent functions and the cosecant and cotangent functions. Specifically, we have

(1)
$$\tan x = \frac{1}{\cot x}$$

(2)
$$\cot^2 x + 1 = \csc^2 x$$

In essence, equations (1) and (2) allow us to use cotangent as an intermediary between tangent and cosecant. We begin by solving equation (2) for $\cot x$.

$$\cot^2 x + 1 = \csc^2 x$$
$$\cot^2 x = \csc^2 x - 1$$
$$\cot x = \pm\sqrt{\csc^2 x - 1}$$

Since $0 \le x < \pi/2$, and the cotangent function is positive in the first quadrant, we have

$$\cot x = +\sqrt{\csc^2 x - 1}$$

Now from equation (1), we have

$$\tan x = \frac{1}{\cot x}$$

$$= \frac{1}{\sqrt{\csc^2 x - 1}}$$

■

## EXERCISES 1

**EXERCISES 1–12** □ *Use a graphics calculator to determine whether the given equation is an identity, a contradiction, or a conditional equation.*

**1.** $\sec^2 x - \csc^2 x = \tan^2 x - \cot^2 x$

**2.** $\cos x = 1$

**3.** $\sin^2 x + \cos^2 x = 1.45$

**4.** $\sec x \cot x = \csc x$

**5.** $2 \cos x = \sin x$

**6.** $\sin 2x + \sin^2 x = 2$

**7.** $\cos 2x + \cos x = 3$

**8.** $\cos 3x + \sin 3x = 3x$

**9.** $2 \cot 2x = \cot x - \tan x$

**10.** $\sin x = 1.45 - \cos x$

**11.** $\cos x = 1 - \dfrac{x^2}{2} + \dfrac{x^4}{4}$

**12.** $\sin\left(x + \dfrac{\pi}{6}\right) = \dfrac{\sqrt{3}}{2} \sin x + \dfrac{1}{2} \cos x$

**EXERCISES 13–22** □ *First classify the equation as an identity, conditional equation, or contradiction. If the equation is an iden-*

*tity, verify it. If it is a conditional equation, then estimate the smallest positive solution.*

**13.** $2 \sin^2 x - 1 = 1 - 2 \cos^2 x$    **14.** $\cos^3 x + \sin^3 x = 1$

**15.** $\sin(x - \pi) = \sin x - \pi$    **16.** $\dfrac{1}{\cos v} - \cos v = \dfrac{\sin^2 v}{\cos v}$

**17.** $\cos x \sin x = \cos x + \sin x$

**18.** $\sin\left(x - \dfrac{\pi}{2}\right) = \sin x - \sin \dfrac{\pi}{2}$

**19.** $\tan x \cot x - 1 = 0$    **20.** $\cos x = 1.1 - \dfrac{x^2}{2} + \dfrac{x^4}{24}$

**21.** $\sin 4x = 4 \sin x$    **22.** $\dfrac{\cot \theta}{\csc \theta + 1} = \dfrac{\csc \theta - 1}{\cot \theta}$

**EXERCISES 23–58** □ *Verify the given identity algebraically.*

**23.** $\dfrac{\tan x}{\sin x} = \sec x$    **24.** $\dfrac{\cot x}{\cos x} = \csc x$

**25.** $\dfrac{\sec x}{\csc x} = \tan x$    **26.** $\dfrac{\tan x}{\cot x} = \sec^2 x - 1$

**27.** $\sin^2 x = 1 - \cos^2 x$    **28.** $\sec^2 x - \tan^2 x = 1$

**29.** $(1 + \cos s)(1 - \cos s) = \sin^2 s$

**30.** $\tan^2 w + 8 = \sec^2 w + 7$

**31.** $\tan^2 \beta - \sin^2 \beta = \tan^2 \beta \sin^2 \beta$

**32.** $\cos x - \sin x = \dfrac{1 - 2 \sin^2 x}{\cos x + \sin x}$

**33.** $\cos t + \sin t = \dfrac{\tan t + 1}{\sec t}$

**34.** $\tan x + \cot x = \sec x \csc x$

**35.** $\sec^2 x + \csc^2 x = \sec^2 x \csc^2 x$

**36.** $\dfrac{\cos u}{\sec u} + \dfrac{\sin u}{\csc u} = 1$

**37.** $\sin u + \cos u \cot u = \csc u$

**38.** $\sec^2 \theta = \sec \theta \csc \theta \tan \theta$

**39.** $(\sin^2 \alpha - 1)(\cot^2 \alpha + 1) = 1 - \csc^2 \alpha$

**40.** $\dfrac{1}{\csc^2 x} + \dfrac{1}{\sec^2 x} = 1$

**41.** $\dfrac{\csc x + \cot x}{\csc x - \cot x} = (\csc x + \cot x)^2$

**42.** $\dfrac{\sin \theta \tan \theta}{1 - \cos \theta} = \sec \theta + 1$

**43.** $\dfrac{1}{1 - \sin x} = \sec^2 x + \sec x \tan x$

**44.** $\dfrac{1 - \cos x}{1 + \cos x} = 2 \csc^2 x - 2 \cot x \csc x - 1$

**45.** $\dfrac{\sin^2 x}{\cos x} + \dfrac{\cos^2 x}{\sin x} = \sec x + \csc x - \cos x - \sin x$

**46.** $\dfrac{\tan^4 x - 1}{\tan^2 x - 1} = \sec^2 x$

**47.** $\dfrac{\sin y \tan y}{\tan y - \sin y} = \dfrac{\tan y + \sin y}{\sin y \tan y}$

**48.** $\dfrac{1}{\tan x + 1} + \dfrac{1}{\cot x + 1} = 1$

**49.** $\sec^4 z - 2 \tan^2 z \sec^2 z + \tan^4 z = 1$

**50.** $\dfrac{\sin w}{\sec w + 1} + \dfrac{\sin w}{\sec w - 1} = 2 \cot w$

**51.** $(\csc \phi - \cot \phi)^2 = \dfrac{1 - \cos \phi}{1 + \cos \phi}$

**52.** $\sin^3 v + \cos^3 v = (1 - \sin v \cos v)(\sin v + \cos v)$

**53.** $|\cos x| = \sqrt{1 - \sin^2 x}$

**54.** $|\tan x| = \sqrt{\sec^2 x - 1}$

**55.** $\ln|\cot s| = -\ln|\tan s|$

**56.** $\ln|\tan x| = \ln|\sin x| - \ln|\cos x|$

**57.** $\ln|\sec w + \tan w| = -\ln|\sec w - \tan w|$

**58.** $\ln|\csc \theta + \cot \theta| = -\ln|\csc \theta - \cot \theta|$

**EXERCISES 59–64** □ *Use fundamental trigonometric identities to write the given expression as a single trigonometric function.*

**59.** $\dfrac{\sin^2 x + \cos^2 x}{\cos x}$    **60.** $\dfrac{\tan^2 x + 1}{\sec x}$

**61.** $\dfrac{\tan x \cot x}{\sin x}$    **62.** $\cos x \sec^2 x \cot x$

**63.** $\dfrac{\sin^2 x - \cos^2 x}{\sin x + \cos x} + \dfrac{1}{\sec x}$

**64.** $(\csc x - 1)(\csc x + 1)\tan x$

**EXERCISES 65–72** □ *Use fundamental trigonometric identities to rewrite the first function in terms of the second.*

**65.** $\tan x$; $\cot x$    **66.** $\sec x$; $\cos x$

**67.** $\cos x$; $\sin x$    **68.** $\sec x$; $\tan x$

**69.** $\cot x$; $\cos x$    **70.** $\tan x$; $\sin x$

**71.** $\sec x$; $\sin x$    **72.** $\cot x$; $\cos x$

■ *Application*

**73.** *Banked curves*  For vehicles to safely travel around curves at high speeds, it is often necessary to bank the curves (Figure 8). By considering frictional and centripetal forces, it can be shown

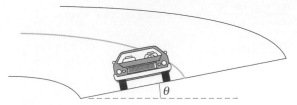

**FIGURE 8**

that the proper angle $\theta$ for banking a curve with radius $r$ feet so that a vehicle can travel safely around the curve at a velocity of $v$ feet per second is given by the equation

$$\sin \theta = \frac{v^2}{32r} \cos \theta$$

**a.** Solve this equation for $v$ and simplify as much as possible.

**b.** What is the safest speed that a vehicle can travel around a curve with a radius of 2400 feet and banked at an angle of 5°?

■ *Projects for Enrichment*

**74.** *Hyperbolic Functions*  The trigonometric functions are often called **circular functions** because of the fact that cos $\theta$ and sin $\theta$ can be defined to be the $x$- and $y$-coordinates, respectively, of the point on the unit circle corresponding to the angle $\theta$. If we begin with a hyperbola (instead of a circle), we obtain the **hyperbolic functions**: cosh $x$, sinh $x$, tanh $x$, sech $x$, csch $x$, and coth $x$. Notice that the notation for the hyperbolic functions is quite similar to that of the trigonometric functions.

The hyperbolic functions arise in many physical contexts. In particular, a cable of uniform density strung between supports (such as a power or telephone line) will hang in the shape of a **catenary**, the graph of cosh $x$. Moreover, the St. Louis Arch is in the shape of an "upside-down" catenary. In this project we will investigate the algebraic properties of hyperbolic functions and the hyperbolic analogs of the trigonometric identities. The hyperbolic functions are defined as follows.

$$\cosh x = \frac{e^x + e^{-x}}{2} \qquad \text{sech } x = \frac{2e^x}{e^{2x} + 1}$$

$$\sinh x = \frac{e^x - e^{-x}}{2} \qquad \text{csch } x = \frac{2e^x}{e^{2x} - 1}$$

$$\tanh x = \frac{e^{2x} - 1}{e^{2x} + 1} \qquad \text{coth } x = \frac{e^{2x} + 1}{e^{2x} - 1}$$

**a.** Use the definitions of the hyperbolic functions to verify the following identities.

**i.** $\tanh x = \dfrac{\sinh x}{\cosh x}$     **ii.** $\coth x = \dfrac{1}{\tanh x}$

**iii.** $\text{sech } x = \dfrac{1}{\cosh x}$     **iv.** $\text{csch } x = \dfrac{1}{\sinh x}$

**v.** $\cosh^2 x - \sinh^2 x = 1$     **vi.** $\text{sech}^2 x + \tanh^2 x = 1$

**vii.** $\coth^2 x - \text{csch}^2 x = 1$     **viii.** $\cosh(-x) = \cosh x$

**ix.** $\sinh(-x) = -\sinh x$

**b.** Use the identities from part (a) to verify the following identities.

**i.** $\dfrac{\text{csch } x}{\text{sech } x} = \coth x$

**ii.** $\cosh^2 x + \sinh^2 x = 2\cosh^2 x - 1$

**iii.** $\dfrac{\tanh x}{\cosh x - \sinh x \tanh x} = \sinh x$

**iv.** $\cosh x + \sinh x = \dfrac{1}{\cosh x - \sinh x}$

**c.** Write the corresponding trigonometric identity for each hyperbolic identity given in part (a).

**d.** Show that the point (cosh $x$, sinh $x$) is on the hyperbola $x^2 - y^2 = 1$.

St. Louis' Gateway Arch

Cable hanging in the shape of a catenary

### Questions for Discussion or Essay

**75.** Explain how any polynomial in $\sin x$ and $\cos x$, such as $\cos^5 x + \cos^4 x \sin^3 x + 3 \cos x \sin^2 x - 4 \sin^3 x$, could be rewritten so that $\sin x$ never appears to a power higher than 1.

**76.** If an equation is true for infinitely many values, is it an identity? Explain.

**77.** In the text it was stated that most *algebraic* identities are transparently true, like $x + 2 = 2 + x$, whereas most trigonometric identities require effort to verify. We have seen, however, a category of nontrigonometric function for which there were several nontrivial identities. To what type of function are we referring? List three identities involving these functions.

**78.** Consider the identity $\tan x \cos x = \sin x$. This identity is true for virtually all values of $x$. However, there are values of $x$ for which the expression on the left side of the equal sign is not equivalent to the expression on the right-hand side. For which values of $x$ does this identity not hold? Is the equation still worthy of being called an identity if it doesn't hold for certain values of $x$?

**79.** In the text it was stated that identities may be verified either by simplifying one side of an equation until it looks like the other, or simplifying both sides separately until they are equal to a common expression. What is wrong with the following "proof" of the "identity" $\sin x = -\sqrt{1 - \cos^2 x}$?

$$\sin x = -\sqrt{1 - \cos^2 x}$$
$$\sin^2 x = 1 - \cos^2 x$$
$$\sin^2 x + \cos^2 x = 1$$
$$1 = 1$$

Evaluate both sides of the equation

$$\sin x = -\sqrt{1 - \cos^2 x}$$

at $x = \pi/2$. Is this equation an identity? Explain.

**80.** Suppose that the tangent of an angle $\theta$ is known. Which of the other five trigonometric functions can be determined at $\theta$? What if you also know that $\sin \theta$ is positive?

---

## SECTION 2

## SUM AND DIFFERENCE IDENTITIES

■ What note do our ears hear when two instruments play a note of the same frequency?

■ Is the sine of the sum of two angles equal to the sum of the sines of the angles?

■ What connection is there between the ratios of sides in a triangle and the chemical composition of distant galaxies?

### SUM AND DIFFERENCE IDENTITIES FOR SINE

Suppose that the sine and cosine of the angles $x$ and $y$ are known. Is this sufficient information to compute $\sin(x + y)$? That is, can we somehow express $\sin(x + y)$ in terms of $\cos x$, $\sin x$, $\cos y$, and $\sin y$? A natural (but dead wrong!) guess would be that $\sin(x + y) = \sin x + \sin y$. This erroneous guess arises from viewing the expression "$\sin(x + y)$" as somehow representing the product of something called "sin" with $(x + y)$ and then applying the distributive law. Remember, the expression $\sin(x + y)$ denotes the sine function evaluated at $x + y$.

There is, in fact, a formula for $\sin(x + y)$, as well as for $\sin(x - y)$, $\cos(x + y)$, and $\cos(x - y)$. We begin by using right triangles to derive the formula for $\sin(x + y)$ for the special case where $x$, $y$, and $x + y$ are all acute angles. (A more traditional approach for proving these formulas, one that is somewhat less intuitive but works for all angles $x$ and $y$, is considered in Exercise 54. In Section 4 of the next chapter we will consider a third approach using complex numbers.)

In Figure 9, we have adjoined two right triangles, one with an angle of $x$ and another with an angle of $y$, in such a way as to form an angle of $x + y$. The angle measures and labels shown in blue are by assumption, whereas those in red follow from elemen-

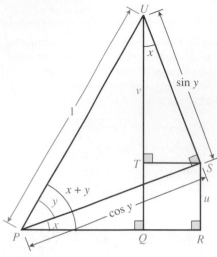

**FIGURE 9**

tary geometry and right triangle trigonometry. Thus, for example, it can be shown that $\angle TUS = x$ using elementary geometry, whereas $\angle QPS = x$ by design. The fact that $US = \sin y$ and $PS = \cos y$ follows from applying the definitions of sine and cosine, respectively, to the right triangle $PSU$.

Without confirming all of the details, let us assume that the diagram in Figure 9 is accurate. We are interested in an expression for $\sin(x + y)$, so we will focus on $\triangle PQU$, a right triangle with one angle of measure $x + y$. From the definition of sine, we have

$$\sin(x + y) = \frac{QU}{PU}$$

$$= \frac{u + v}{1}$$

$$= u + v$$

Now if we can find expressions for $u$ and $v$ in terms of sines and cosines of $x$ and $y$, we will be done. We begin with $u$. From $\triangle RPS$, we see that

$$\sin x = \frac{u}{\cos y}$$

and so it follows that $u = \sin x \cos y$. To find an expression for $v$, we consider $\triangle UTS$. We have

$$\cos x = \frac{v}{\sin y}$$

Thus, $v = \sin y \cos x$. Collecting our results, we have

$$\sin(x + y) = u + v$$

$$= \sin x \cos y + \sin y \cos x$$

We can obtain the difference formula for sine by substituting $-y$ for $y$ in the sum formula as follows.

$$\sin[x + (-y)] = \sin x \cos(-y) + \sin(-y)\cos x$$

Now using the even/odd identities for sine and cosine [i.e., $\sin(-x) = -\sin x$ and $\cos(-x) = \cos x$], we arrive at

$$\sin(x - y) = \sin x \cos y - \sin y \cos x$$

We summarize the sum and difference identities as follows.

■ *Sum and difference identities for sine*

$$\sin(x + y) = \sin x \cos y + \sin y \cos x$$

$$\sin(x - y) = \sin x \cos y - \sin y \cos x$$

Thus, if the sines and cosines of two angles are known, the sines of the sums and differences can be computed.

**EXAMPLE 1**    *Applying the sum and difference identities*

Given that $\sin u = \frac{3}{5}$ and $\cos u = \frac{4}{5}$, compute $\sin[u - (3\pi/2)]$.

**SOLUTION**    According to the difference identity for sine, we have

$$\sin\left(u - \frac{3\pi}{2}\right) = \sin u \cos \frac{3\pi}{2} - \sin \frac{3\pi}{2} \cos u$$

$$= \left(\frac{3}{5}\right)(0) - (-1)\left(\frac{4}{5}\right)$$

$$= \frac{4}{5}$$

**EXAMPLE 2**    *Computing the sine of 75°*

Find an exact value for $\sin 75°$.

**SOLUTION**    We begin by searching for two special angles, the sum or difference of which is 75°. After some trial and error, we happen upon the relation $75° = 30° + 45°$. Thus, we will use the sum identity for sine.

$$\sin 75° = \sin(30° + 45°)$$

$$= \sin 30° \cos 45° + \sin 45° \cos 30°$$

$$= \left(\frac{1}{2}\right)\left(\frac{\sqrt{2}}{2}\right) + \left(\frac{\sqrt{2}}{2}\right)\left(\frac{\sqrt{3}}{2}\right)$$

$$= \frac{\sqrt{2} + \sqrt{6}}{4}$$

In the previous examples the sines and cosines of both angles were either given or known. However, if we know the sine (or cosine) of an angle and its quadrant, we can determine the cosine (or sine) using the identity $\sin^2 x + \cos^2 x = 1$.

**EXAMPLE 3**    *Applying the sum and difference identities*

Find $\sin(u + v)$ given that $\sin u = \frac{12}{13}$ and $\cos v = \frac{1}{2}$, where both $u$ and $v$ are acute.

**SOLUTION**    The sum identity for $\sin(u + v)$ involves $\cos u$ and $\sin v$, neither of which is given. However, each can easily be computed using the identity $\sin^2 x + \cos^2 x = 1$. In the case of $\cos u$, we have

$$\sin^2 u + \cos^2 u = 1$$

$$\left(\frac{12}{13}\right)^2 + \cos^2 u = 1$$

$$\frac{144}{169} + \cos^2 u = 1$$

$$\cos^2 u = \frac{25}{169}$$

$$\cos u = \pm\frac{5}{13}$$

Since $u$ is in the first quadrant, we conclude that $\cos u = \frac{5}{13}$. We compute $\sin v$ in a similar fashion.

$$\sin^2 v + \cos^2 v = 1$$

$$\sin^2 v + \frac{1}{4} = 1$$

$$\sin^2 v = \frac{3}{4}$$

$$\sin v = \pm\frac{\sqrt{3}}{2}$$

Again, since $v$ is in the first quadrant, we conclude that $\sin v = \sqrt{3}/2$. Thus, we have

$$\sin(u + v) = \sin u \cos v + \sin v \cos u$$

$$= \left(\frac{12}{13}\right)\left(\frac{1}{2}\right) + \left(\frac{\sqrt{3}}{2}\right)\left(\frac{5}{13}\right)$$

$$= \frac{12 + 5\sqrt{3}}{26}$$

## SUM AND DIFFERENCE IDENTITIES FOR COSINE

The difference identity for sine can be used to find the sum identity for cosine using $\cos u = \sin[(\pi/2) - u]$. In particular, with $u = x + y$, we have

$$\cos(x + y) = \sin\left[\frac{\pi}{2} - (x + y)\right]$$

$$= \sin\left[\left(\frac{\pi}{2} - x\right) - y\right]$$

$$= \sin\left(\frac{\pi}{2} - x\right)\cos y - \sin y \cos\left(\frac{\pi}{2} - x\right)$$

$$= \cos x \cos y - \sin y \sin x$$

We can now find the difference identity for cosine by substituting $-y$ for $y$, and again applying the even/odd identities for sine and cosine, to obtain

$$\cos[x + (-y)] = \cos x \cos(-y) - \sin(-y)\sin x$$

$$\cos(x - y) = \cos x \cos y + \sin y \sin x$$

We summarize these results as follows.

*Sum and difference identities for cosine*

$$\cos(x + y) = \cos x \cos y - \sin x \sin y$$

$$\cos(x - y) = \cos x \cos y + \sin x \sin y$$

---

**EXAMPLE 4**    *Applying the sum and difference identities for sine and cosine*

Given that $\cos u = \frac{3}{5}$, $\sin u = -\frac{4}{5}$, $\cos v = \frac{4}{5}$, and $\sin v = \frac{3}{5}$, compute each of the following quantities.

**a.** $\cos(u + v)$        **b.** $\cos(u - v)$

**SOLUTION**

**a.** From the sum identity for cosine, we have

$$\cos(u + v) = \cos u \cos v - \sin u \sin v$$
$$= \left(\frac{3}{5}\right)\left(\frac{4}{5}\right) - \left(-\frac{4}{5}\right)\left(\frac{3}{5}\right)$$
$$= \frac{12}{25} + \frac{12}{25}$$
$$= \frac{24}{25}$$

**b.** From the difference identity for cosine, we have

$$\cos(u - v) = \cos u \cos v + \sin u \sin v$$
$$= \left(\frac{3}{5}\right)\left(\frac{4}{5}\right) + \left(-\frac{4}{5}\right)\left(\frac{3}{5}\right)$$
$$= \frac{12}{25} - \frac{12}{25}$$
$$= 0$$

---

On occasion, complex expressions can be simplified using the sum or difference identities.

**EXAMPLE 5**    *Simplifying with a sum identity*

Simplify $\cos 59° \cos 31° - \sin 59° \sin 31°$.

**SOLUTION**    This can be simplified using the sum identity for cosine, namely

$$\cos(u + v) = \cos u \cos v - \sin u \sin v$$

with $u = 59°$ and $v = 31°$ we obtain

$$\cos 59° \cos 31° - \sin 59° \sin 31° = \cos(59° + 31°)$$
$$= \cos(90°)$$
$$= 0$$

---

There are countless identities involving special angles that are easily verified using sum or difference formulas.

| EXAMPLE 6 | *Verifying a trigonometric identity* |

Verify that the following equation is an identity.

$$\cos(x - \pi) = -\cos x$$

**SOLUTION**    From the difference identity for cosine, we have

$$\cos(x - \pi) = \cos x \cos \pi + \sin x \sin \pi$$
$$= \cos x \cdot (-1) + \sin x \cdot 0$$
$$= -\cos x$$

## SUM AND DIFFERENCE IDENTITIES FOR TANGENT

Since the tangent of an angle is simply the ratio of its sine to its cosine, that is,

$$\tan x = \frac{\sin x}{\cos x}$$

the sum identity for tangent can be derived from those of sine and cosine. In particular, we have

$$\tan(x + y) = \frac{\sin(x + y)}{\cos(x + y)}$$
$$= \frac{\sin x \cos y + \sin y \cos x}{\cos x \cos y - \sin x \sin y}$$

Dividing numerator and denominator by $\cos x \cos y$, we obtain

$$\tan(x + y) = \frac{\dfrac{\sin x \cos y}{\cos x \cos y} + \dfrac{\sin y \cos x}{\cos x \cos y}}{\dfrac{\cos x \cos y}{\cos x \cos y} - \dfrac{\sin x \sin y}{\cos x \cos y}}$$
$$= \frac{\dfrac{\sin x}{\cos x} + \dfrac{\sin y}{\cos y}}{1 - \left(\dfrac{\sin x}{\cos x}\right)\left(\dfrac{\sin y}{\cos y}\right)}$$
$$= \frac{\tan x + \tan y}{1 - \tan x \tan y}$$

Using the fact that the tangent function is odd—that is, that $\tan(-x) = -\tan x$—we obtain the difference identity for tangent:

$$\tan(x - y) = \tan[x + (-y)]$$
$$= \frac{\tan x + \tan(-y)}{1 - \tan x \tan(-y)}$$
$$= \frac{\tan x - \tan y}{1 + \tan x \tan y}$$

We summarize the sum and difference identities for tangent as follows.

---

**Sum and difference identities for tangent**

$$\tan(x + y) = \frac{\tan x + \tan y}{1 - \tan x \tan y}$$

$$\tan(x - y) = \frac{\tan x - \tan y}{1 + \tan x \tan y}$$

---

**EXAMPLE 7**    *Applying the sum and difference identities for tangent*

Given that $\tan u = 2$ and $\tan v = \frac{1}{4}$, compute the following quantities.
**a.** $\tan(u + v)$      **b.** $\tan(u - v)$

**SOLUTION**

**a.**

$$\tan(u + v) = \frac{\tan u + \tan v}{1 - \tan u \tan v}$$

$$= \frac{2 + \dfrac{1}{4}}{1 - 2 \cdot \dfrac{1}{4}}$$

$$= \frac{9/4}{1/2}$$

$$= \frac{9}{4} \cdot \frac{2}{1}$$

$$= \frac{9}{2}$$

**b.**

$$\tan(u - v) = \frac{\tan u - \tan v}{1 + \tan u \tan v}$$

$$= \frac{2 - \dfrac{1}{4}}{1 + 2 \cdot \dfrac{1}{4}}$$

$$= \frac{7/4}{3/2}$$

$$= \frac{7}{4} \cdot \frac{2}{3}$$

$$= \frac{7}{6}$$

# SPECTRAL ANALYSIS

A musical score encodes information regarding several well-defined aspects of the music—including pitch, amplitude, and duration. But the score contains no information regarding one of the most interesting (and difficult to define) musical qualities, **timbre**. Roughly, timbre is that quality that allows us to distinguish between different musical instruments. Mathematically, music arises from periodic sound waves. The graph or **waveform** associated with a pure tone, for example, is simply that of a function of the form $y(t) = a \cos wt$. On the other hand, the waveform associated with the same note played on a rich-sounding instrument (like an oboe or a violin) will have the same period but will be much more complex. In essence, it is the intricacy of the waveform that determines the richness of the timbre.

Now we can produce fairly complex waveforms by taking sums of cosine functions with different periods. For example, consider the graph of the function

$$f(t) = 4 \cos t + 3 \cos 3t + 2 \cos 5t + 5 \cos 7t$$

shown in the accompanying figure. *Which* periods are selected determines the **spectrum** of a waveform.

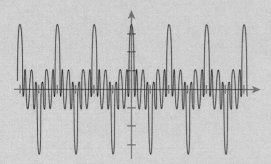

**Spectral analysis** is the process by which one identifies the constituent tones ($4 \cos t$, $3 \cos 3t$, etc.) of an underlying function given a limited amount of data. Since each musical instrument has a unique spectral "signature," knowledge of this signature enables us to electronically reconstruct or **synthesize** music that sounds like a particular instrument. But the sound being analyzed need not be music in order for spectral analysis to be fruitful. For example, spectral analysis is a powerful tool in speech recognition.

Moreover, spectral analysis is performed on many different kinds of data—the waveform needn't correspond to a sound at all. Important information can be gleaned and mathematical models constructed by examining the spectra associated with such diverse phenomena as earthquakes, tides, and sunspots. One of the most intriguing areas of application of spectral analysis is in **image processing**. By analyzing the spectrum of a digital photo-

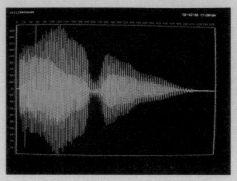

A voice print as it appears on a computer monitor.

graph, important features become apparent. For example, it is sometimes possible to sharpen a fuzzy satellite photograph to improve the level of detail visible.

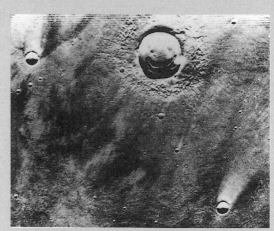

Image enhancement sharpens the detail of the "Happy face crater" on Mars, as viewed by the Viking orbiter 2 spacecraft.

Perhaps the most astounding application of spectral analysis is **spectroscopy**, in which the spectrum of electromagnetic radiation (light, infrared radiation, etc.) is determined; this enables us to determine the chemical composition of unknown substances. Spectroscopy techniques have even been used to determine the chemical composition of distant galaxies.

Spectral analysis provides us with a compelling example of the unreasonable effectiveness of mathematics. How mysterious and wonderful it is that a subject arising from the study of triangles should prove to be a powerful tool for exploring music, understanding earthquakes, and analyzing starlight.

## EXERCISES 2

**EXERCISES 1–6** ☐ *Use the given information to compute the exact values of the indicated trigonometric functions.*

1. $\cos u = \dfrac{3}{5}, 0 \le u < \dfrac{\pi}{2}; \sin v = \dfrac{12}{13}, 0 \le v < \dfrac{\pi}{2}$

    a. $\cos(u - v)$        b. $\cos(u + v)$
    c. $\sin(u - v)$        d. $\sin(u + v)$

2. $\cos u = -\dfrac{3}{5}, \dfrac{\pi}{2} \le u < \pi; \cos v = \dfrac{5}{13}, \dfrac{3\pi}{2} \le v < 2\pi$

    a. $\cos(u - v)$        b. $\cos(u + v)$
    c. $\sin(u - v)$        d. $\sin(u + v)$

3. $\sin r = -\dfrac{4}{5}, \pi \le r < \dfrac{3\pi}{2}; \sin s = \dfrac{3}{5}, 0 \le s \le \dfrac{\pi}{2}$

    a. $\cos(r - s)$        b. $\cos(r + s)$
    c. $\sin(r - s)$        d. $\sin(r + s)$

4. $\sin \alpha = \dfrac{15}{17}, \dfrac{\pi}{2} \le \alpha < \pi; \cos \beta = -\dfrac{3}{5}, \dfrac{\pi}{2} \le \beta < \pi$

    a. $\cos(\alpha - \beta)$        b. $\cos(\alpha + \beta)$
    c. $\sin(\alpha - \beta)$        d. $\sin(\alpha + \beta)$

5. $\sec v = \dfrac{5}{4}, 0 \le v < \dfrac{\pi}{2}; \csc w = -\dfrac{13}{12}, \pi \le w < \dfrac{3\pi}{2}$

    a. $\cos(v - w)$        b. $\cos(v + w)$
    c. $\sin(v - w)$        d. $\sin(v + w)$

6. $\csc \theta_1 = -\dfrac{17}{8}, \dfrac{3\pi}{2} \le \theta_1 < 2\pi; \sec \theta_2 = -\dfrac{5}{4}, \dfrac{\pi}{2} \le \theta_2 < \pi$

    a. $\cos(\theta_1 - \theta_2)$        b. $\cos(\theta_1 + \theta_2)$
    c. $\sin(\theta_1 - \theta_2)$        d. $\sin(\theta_1 + \theta_2)$

**EXERCISES 7–16** ☐ *Rewrite the given expression as a trigonometric function of a single angle.*

7. $\sin 62° \cos 32° - \sin 32° \cos 62°$

8. $\cos 100° \cos 10° + \sin 100° \sin 10°$

9. $\cos \dfrac{\pi}{8} \cos \dfrac{5\pi}{8} - \sin \dfrac{\pi}{8} \sin \dfrac{5\pi}{8}$

10. $\dfrac{\tan \dfrac{\pi}{5} + \tan \dfrac{\pi}{20}}{1 - \tan \dfrac{\pi}{5} \tan \dfrac{\pi}{20}}$

11. $\sin \dfrac{\pi}{5} \cos \dfrac{\pi}{15} + \sin \dfrac{\pi}{15} \cos \dfrac{\pi}{5}$

12. $\sin 152° \cos 12° - \sin 12° \cos 152°$

13. $\cos y \cos 2y - \sin y \sin 2y$

14. $\sin \dfrac{u}{2} \cos \dfrac{u}{2} + \sin \dfrac{u}{2} \cos \dfrac{u}{2}$

15. $\dfrac{\tan 4a - \tan 2a}{1 + \tan 4a \tan 2a}$

16. $\cos 4x \cos x + \sin 4x \sin x$

**EXERCISES 17–20** ☐ *Find the exact value of the given trigonometric function.*

17. $\sin \dfrac{\pi}{12}$ $\left( \text{Hint: } \dfrac{\pi}{12} = \dfrac{\pi}{3} - \dfrac{\pi}{4} \right)$

18. $\cos 105°$ *(Hint: $105° = 60° + 45°$)*

19. $\tan \dfrac{11\pi}{12}$              20. $\cos 75°$

**EXERCISES 21–34** ☐ *Simplify the given trigonometric expression.*

21. $\cos(x + \pi)$            22. $\sin(x + \pi)$

23. $\sin \left( \dfrac{\pi}{2} + x \right)$        24. $\cos \left( x - \dfrac{\pi}{4} \right)$

25. $\cos \left( \dfrac{\pi}{3} - x \right)$        26. $\sin \left( \dfrac{\pi}{6} - x \right)$

27. $\cos \left( x + \dfrac{2\pi}{3} \right)$        28. $\sin \left( x + \dfrac{\pi}{4} \right)$

29. $\tan \left( x - \dfrac{\pi}{4} \right)$        30. $\tan(x + \pi)$

31. $\sin \left( \arcsin \dfrac{3}{5} + \arccos \dfrac{3}{5} \right)$    32. $\cos \left( \arcsin \dfrac{4}{5} + \dfrac{\pi}{4} \right)$

33. $\tan \left( \dfrac{\pi}{4} - \arctan \dfrac{25}{24} \right)$    34. $\sin \left( \arcsin \dfrac{5}{13} - \arccos \dfrac{3}{5} \right)$

**EXERCISES 35–48** ☐ *Verify the given identity using the sum and difference identities.*

35. $\sec(u + v) = \dfrac{\sec v \csc u}{\cot u - \tan v}$

36. $\cot(u + v) = \dfrac{\cot u \cot v - 1}{\cot u + \cot v}$

37. $\cos \left( \dfrac{\pi}{2} - x \right) = \sin x$

38. $\sin \left( \dfrac{\pi}{2} - x \right) = \cos x$

39. $\cot(u - v) = \dfrac{\cot u \cot v + 1}{\cot v - \cot u}$

40. $\csc(u + v) = \dfrac{\csc u \sec v}{1 + \tan v \cot u}$

41. $\cos(-x) = \cos x$ *(Hint: $-x = 0 - x$)*

42. $\sin(-x) = -\sin x$

43. $\sin u \sin v = \dfrac{1}{2}[\cos(u - v) - \cos(u + v)]$

44. $\cos u \cos v = \dfrac{1}{2}[\cos(u + v) + \cos(u - v)]$

45. $\sin u \cos v = \dfrac{1}{2}[\sin(u + v) + \sin(u - v)]$

46. $\cos 2x = \cos^2 x - \sin^2 x$ (*Hint:* $2x = x + x$)

47. $\sin 2x = 2 \sin x \cos x$

48. $\tan 2x = \dfrac{2 \tan x}{1 - \tan^2 x}$

49. Suppose that $u$ and $v$ are angles in the first quadrant. Show that $u + v$ is also in the first quadrant if and only if $\tan u \tan v < 1$. [*Hint:* What can be said about $\tan(u + v)$ when $u + v$ is in the first quadrant?]

50. Consider the triangle with vertices $O(0, 0)$, $P(a, b)$, and $Q(c, d)$, as shown in Figure 10. Show that $\triangle POQ$ is a right

triangle if $ac + bd = 0$. [*Hint:* Begin with the fact that $\cos \theta = \cos(\alpha - \beta)$.]

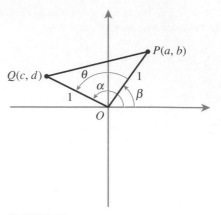

**FIGURE 10**

51. Produce formulas for $\cos(x + y + z)$ and $\sin(x + y + z)$ in terms of $\cos x$, $\cos y$, $\cos z$, $\sin x$, $\sin y$, and $\sin z$.

---

■ *Projects for Enrichment*

52. **Product-Sum Identities** In this section we have seen that the sine and cosine of the sum or difference of two angles can be written in terms of products involving the sine or cosine of the two angles. In this project, we will use these sum and difference identities to develop additional identities involving sums and products of sine and cosine. These are often referred to as the **product-sum identities**. We first consider the product-sum identities involving cosine.

a. Recall that for any two angles $x$ and $y$,

$$\cos(x - y) = \cos x \cos y + \sin x \sin y$$
$$\cos(x + y) = \cos x \cos y - \sin x \sin y$$

By adding these two equations together, show that

$$\cos x \cos y = \frac{1}{2}\cos(x - y) + \frac{1}{2}\cos(x + y)$$

b. Let $a = x - y$ and $b = x + y$. Solve this system of equations to find expressions for $x$ and $y$ in terms of $a$ and $b$. Then make appropriate substitutions into the product-sum identity in part (a) to show that

$$\cos a + \cos b = 2 \cos\left(\frac{a + b}{2}\right)\cos\left(\frac{a - b}{2}\right)$$

c. Follow steps similar to those outlined in parts (a) and (b) to show that the following product-sum identities hold. (*Hint:*

In the first step, subtract the sum and difference identities for cosine.)

$$\sin x \sin y = \frac{1}{2}\cos(x - y) - \frac{1}{2}\cos(x + y)$$

$$\cos a - \cos b = -2 \sin\left(\frac{a + b}{2}\right)\sin\left(\frac{a - b}{2}\right)$$

d. Use the sum and difference identities for sine to show that the following product-sum identities hold.

$$\sin x \cos y = \frac{1}{2}\sin(x - y) + \frac{1}{2}\sin(x + y)$$

$$\sin a + \sin b = 2 \sin\left(\frac{a + b}{2}\right)\cos\left(\frac{a - b}{2}\right)$$

$$\sin a - \sin b = 2 \cos\left(\frac{a + b}{2}\right)\sin\left(\frac{a - b}{2}\right)$$

53. **Music** Sound is created whenever a disturbance causes changes in air pressure at frequencies in the audible range (20–20,000 Hz). Musical tones are created when the oscillations of air pressure are periodic. As a consequence, musical tones can be modeled with functions of the form

$$f(t) = A \cos(\omega t + C)$$

where $A$ is the amplitude and $\omega$ is the frequency. Changes in the amplitude correspond to changes in the loudness of the

A vibrating guitar string creates a musical tone.

sound, and changes in the frequency correspond to changes in pitch.

In this project we will investigate what happens when two musical tones of the same frequency are out of phase. To that end, consider two musical tones with corresponding functions $f(t) = A_1 \cos(\omega t)$ and $g(t) = A_2 \cos(\omega t + \theta)$. The resulting sound is given by

$$h(t) = f(t) + g(t)$$
$$= A_1 \cos(\omega t) + A_2 \cos(\omega t + \theta)$$

We would like to show that this sound is again a musical tone. In other words, we would like to show that $h(t)$ can be written in the form

$$h(t) = A \cos(\omega t + C)$$

As a first step, we will consider two specific functions $f$ and $g$.

a. Let $f(t) = 3 \cos(\pi t)$ and $g(t) = 4 \cos[\pi t + (\pi/2)]$. Plot $h(t) = f(t) + g(t)$ and estimate its amplitude and phase shift. Then produce a function of the form $k(t) = A \cos(\pi t + C)$ whose graph is nearly identical to that of $h$.

Next, we consider the general case.

b. Use the sum identity for cosine to rewrite $A_1 \cos(\omega t) + A_2 \cos(\omega t + \theta)$ in terms of sines and cosines of $\omega t$ and $\theta$.

c. Use the sum identity for cosine to rewrite $A \cos(\omega t + C)$ in terms of sines and cosines of $\omega t$ and $C$.

d. Define

$$A = \sqrt{A_1^2 + A_2^2 + 2 A_1 A_2 \cos \theta}$$

and let $C$ be an angle for which

$$\cos C = \frac{A_1 + A_2 \cos \theta}{A}$$

Show that

$$\sin^2 C = \frac{A_2^2(1 - \cos^2 \theta)}{A}$$

and, assuming $A_2 > 0$ and $\theta$ is in the first quadrant,

$$\sin C = \frac{A_2 \sin \theta}{A}.$$

e. Combine your results from parts (b), (c), and (d) to show that

$$A_1 \cos(\omega t) + A_2 \cos(\omega t + \theta) = A \cos(\omega t + C)$$

Now we not only know that the sound resulting from two musical tones is again a musical tone, but we can also determine how its amplitude depends on the original tones.

f. Suppose two tones with the same frequency and amplitude are in phase. What is the amplitude of the resulting tone? What if the tones are in phase with different amplitudes? [*Hint:* Consider what happens in part (c) if $\theta = 0$.]

g. Suppose two tones with the same frequency and amplitude are out of phase by $\pi$ radians. What is the amplitude of the resulting tone? Does this seem reasonable? What potential uses do you see for this outcome? How might it lead to the notion of "anti-noise"?

54. *Deriving the Sum and Difference Formulas*  In this project we will derive the sum and difference formulas for sine and cosine using an approach that involves the distance formula. Consider two angles $u$ and $v$ as suggested by Figure 11. For convenience, we will assume that $0 < v < u < 2\pi$.

a. Find an expression for the distance between $P$ and $R$.

b. Find an expression for the distance between $Q$ and $S$.

c. From geometry we know that the arcs $RP$ and $SQ$ have central angles with the same measure, namely $u - v$, and hence the line segments $RP$ and $SQ$ have the same length. Use this

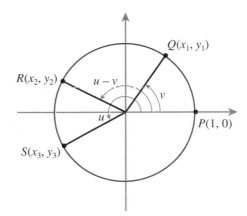

**FIGURE 11**

fact and the distances you found in parts (a) and (b) to show that

$$x_2 = x_3 x_1 + y_3 y_1$$

d. Now show that

$$\cos(u - v) = \cos u \cos v + \sin u \sin v$$

e. Write $\cos(u + v)$ as $\cos[u - (-v)]$ and use the result of part (d) to show that

$$\cos(u + v) = \cos u \cos v - \sin u \sin v$$

f. Use a cofunction identity to show that

$$\sin(u + v) = \cos\left[\left(\frac{\pi}{2} - u\right) - v\right]$$

and then use the result from part (d) to show that

$$\sin(u + v) = \sin u \cos v + \cos u \sin v$$

g. Write $\sin(u - v)$ as $\sin[u + (-v)]$ and use the result from part (f) to show that

$$\sin(u - v) = \sin u \cos v - \cos u \sin v$$

## Questions for Discussion or Essay

55. Show that $\sin(\pi + 0) = \sin \pi + \sin 0$, but $\sin[(\pi/2) + \pi] \neq \sin(\pi/2) + \sin \pi$. What can be said in general about the statement $\sin(x + y) = \sin x + \sin y$? In general, what can you say about the use of examples to prove or disprove a statement in mathematics?

56. Use the sum identities to show that $\cos(x + \pi) = -\cos x$ and $\sin(x + \pi) = \sin x$. Explain how these could be proven without

using the sum identities. (*Hint:* Sketch the angles $x$ and $x + \pi$ on the unit circle.)

57. Prior to this section, how many acute angles did you know exact trigonometric function values for? Using the techniques of this section, how many more can you find exact values for?

58. Explain how angle addition formulas could be used to compute $\sec(x \pm y)$, $\csc(x \pm y)$, and $\cot(x \pm y)$.

---

## SECTION 3

# DOUBLE-ANGLE AND HALF-ANGLE IDENTITIES

- Why does a straw in a glass of water look bent?
- How can cubic equations be solved using trigonometry?
- How can the cosine of $\dfrac{\pi}{8}$ be determined?
- If one line makes twice the angle with the horizontal as a second line, is it twice as steep?

---

### THE DOUBLE-ANGLE IDENTITIES

A simple application of the sum identities for sine, cosine, and tangent gives us the **double-angle identities**, which enable us to express any trigonometric function of twice an angle in terms of trigonometric functions of the angle itself. For example, the sum identity for sine tells us that $\sin(x + y) = \sin x \cos y + \sin y \cos x$. Replacing $y$ with $x$ gives us the double-angle identity for sine.

$$\sin(x + y) = \sin x \cos y + \sin y \cos x$$

$$\sin(x + x) = \sin x \cos x + \sin x \cos x$$

$$\sin 2x = 2 \sin x \cos x$$

A double-angle identity for the cosine function can be derived in a similar fashion. This time, we begin with the sum identity for cosine, as follows.

$$\cos(x + y) = \cos x \cos y - \sin x \sin y$$

$$\cos(x + x) = \cos x \cos x - \sin x \sin x$$

$$\cos 2x = \cos^2 x - \sin^2 x$$

$$= (1 - \sin^2 x) - \sin^2 x \quad \textbf{or} \quad = \cos^2 x - (1 - \cos^2 x)$$

$$= 1 - 2 \sin^2 x \qquad\qquad = 2 \cos^2 x - 1$$

Note that this establishes *three* formulas for cos 2x.

The double-angle identity for tangent also derives from the corresponding sum identity.

$$\tan(x + y) = \frac{\tan x + \tan y}{1 - \tan x \tan y}$$

$$\tan(x + x) = \frac{\tan x + \tan x}{1 - \tan x \tan x}$$

$$\tan 2x = \frac{2 \tan x}{1 - \tan^2 x}$$

**Double-angle identities for sine, cosine, and tangent**

$$\sin 2x = 2 \sin x \cos x$$

$$\cos 2x = \cos^2 x - \sin^2 x = 2 \cos^2 x - 1 = 1 - 2 \sin^2 x$$

$$\tan 2x = \frac{2 \tan x}{1 - \tan^2 x}$$

The double-angle identities allow us to find the trigonometric function values of twice an angle, provided we have the appropriate trigonometric values for the angle itself. We illustrate the use of these identities in the following examples.

**EXAMPLE 1**  *Using the double-angle identity for sine*

Compute sin 2x given that $\sin x = -\frac{4}{5}$ and $\cos x = \frac{3}{5}$.

**SOLUTION**

$$\sin 2x = 2 \sin x \cos x$$

$$= 2\left(-\frac{4}{5}\right)\frac{3}{5}$$

$$= -\frac{24}{25}$$

**EXAMPLE 2**  *Computing the cosine of twice an angle*

Find cos 2θ, given that $\cos \theta = \frac{2}{5}$.

**SOLUTION**    Since $\cos \theta$ is given but $\sin \theta$ is unknown, we select the form of the double-angle identity for cosine involving only the cosine function.

$$\cos 2\theta = 2 \cos^2 \theta - 1$$
$$= 2\left(\frac{2}{5}\right)^2 - 1$$
$$= \frac{8}{25} - 1 = -\frac{17}{25}$$

■

**EXAMPLE 3**    *Computing the tangent of a double angle*

Evaluate $\tan 2x$, given that $\tan x = 2$.

**SOLUTION**

$$\tan 2x = \frac{2 \tan x}{1 - \tan^2 x}$$
$$= \frac{4}{1 - 4}$$
$$= -\frac{4}{3}$$

■

Thus far, we have been given exactly the data that we needed in order to use one of the double-angle identities. In some cases, it will be necessary to perform some preliminary computations to produce all the necessary information. In the next example, we consider the computation of $\sin 2\theta$ when $\sin \theta$ is given, but $\cos \theta$ is not.

**EXAMPLE 4**    *An application of the double-angle formula for sine*

Given that $\sin \theta = \frac{3}{5}$ and that $\frac{\pi}{2} < \theta < \pi$, compute $\sin 2\theta$.

**SOLUTION**    To use the double-angle identity for sine, we first must find $\cos \theta$.

$$\sin^2 \theta + \cos^2 \theta = 1$$
$$\left(\frac{3}{5}\right)^2 + \cos^2 \theta = 1$$
$$\frac{9}{25} + \cos^2 \theta = 1$$
$$\cos^2 \theta = \frac{16}{25}$$
$$\cos \theta = \pm\frac{4}{5}$$

Since $\theta$ is in the second quadrant and the cosine function is negative there, we have

$$\cos \theta = -\frac{4}{5}$$

Now we use the double-angle identity for sine.

$$\sin 2\theta = 2 \sin \theta \cos \theta$$

$$= 2\left(\frac{3}{5}\right)\left(-\frac{4}{5}\right)$$

$$= -\frac{24}{25}$$

Infinitely many new identities can be verified with the aid of the double-angle identities for sine, cosine, and tangent. In particular, any trigonometric function of $nx$ can be expressed in terms of trigonometric functions of $x$.

**EXAMPLE 5**   *Deriving an identity for* $\sin 3x$

Express $\sin 3x$ in terms of $\sin x$ and $\cos x$.

**SOLUTION**   We begin by recognizing that $3x = 2x + x$ and applying the sum identity for sine.

$$\sin 3x = \sin(2x + x)$$

$$= \sin 2x \cos x + \sin x \cos 2x \qquad \text{Applying the sum identity for sine}$$

$$= (2 \sin x \cos x)\cos x + \sin x(\cos^2 x - \sin^2 x) \qquad \text{Applying the double-angle identity for sine}$$

$$= 2 \sin x \cos^2 x + \sin x \cos^2 x - \sin^3 x$$

$$= 3 \sin x \cos^2 x - \sin^3 x$$

Among the most useful identities that can be established using the double-angle identities are the **half-angle identities**.

## THE HALF-ANGLE IDENTITIES

Intuitively, since each of the double-angle identities relates trigonometric functions of a given angle to trigonometric functions of *twice* that angle, we would expect to be able to produce a relationship between trigonometric functions of an angle and trigonometric functions of *half* the given angle. Such is indeed the case.

Half-angle identities for sine and cosine can be derived using forms of the double-angle identity for cosine, as follows.

$$\cos 2u = 2 \cos^2 u - 1 \qquad\qquad \cos 2u = 1 - 2 \sin^2 u$$

$$2 \cos^2 u = 1 + \cos 2u \qquad\qquad 2 \sin^2 u = 1 - \cos 2u$$

$$\cos^2 u = \frac{1 + \cos 2u}{2} \qquad\qquad \sin^2 u = \frac{1 - \cos 2u}{2}$$

$$\cos u = \pm\sqrt{\frac{1 + \cos 2u}{2}} \qquad\qquad \sin u = \pm\sqrt{\frac{1 - \cos 2u}{2}}$$

Now we replace $u$ with $x/2$ to obtain

$$\cos\frac{x}{2} = \pm\sqrt{\frac{1 + \cos x}{2}} \quad \text{and} \quad \sin\frac{x}{2} = \pm\sqrt{\frac{1 - \cos x}{2}}$$

Note that, in practice, we must use additional information to determine whether the appropriate symbol should be "+" or "−."

A formula for $\tan(x/2)$ can be obtained by dividing the expressions for $\sin(x/2)$ and $\cos(x/2)$.

$$\tan\frac{x}{2} = \frac{\sin\dfrac{x}{2}}{\cos\dfrac{x}{2}}$$

$$= \pm\frac{\sqrt{\dfrac{1 - \cos x}{2}}}{\sqrt{\dfrac{1 + \cos x}{2}}}$$

$$= \pm\sqrt{\frac{1 - \cos x}{1 + \cos x}}$$

Fortunately, two alternative expressions for $\tan(x/2)$ can be derived (see Exercise 40), neither one of which involves the awkward "±" symbol.

$$\tan\frac{x}{2} = \frac{1 - \cos x}{\sin x} = \frac{\sin x}{1 + \cos x}$$

**EXAMPLE 6**    *Verifying an identity using a half-angle identity*

Verify the identity $\sec^2\dfrac{x}{2} = \dfrac{2}{1 + \cos x}$.

**SOLUTION**    Since $\sec^2 u = 1/\cos^2 u$, we are able to apply the half-angle identity for cosine as follows.

$$\sec^2\frac{x}{2} = \frac{1}{\left(\cos\dfrac{x}{2}\right)^2}$$

$$= \frac{1}{\left(\pm\sqrt{\dfrac{1 + \cos x}{2}}\right)^2}$$

$$= \frac{1}{\dfrac{1 + \cos x}{2}}$$

$$= \frac{2}{1 + \cos x}$$

For convenience, we summarize the half-angle identities here.

**Half-angle identities for sine, cosine, and tangent**

$$\cos\frac{x}{2} = \pm\sqrt{\frac{1 + \cos x}{2}}$$

$$\sin\frac{x}{2} = \pm\sqrt{\frac{1 - \cos x}{2}}$$

$$\tan\frac{x}{2} = \frac{1 - \cos x}{\sin x} = \frac{\sin x}{1 + \cos x}$$

**EXAMPLE 7**    *Using the half-angle identity for sine*

Compute $\sin\dfrac{\theta}{2}$ given that $\cos\theta = \frac{1}{4}$, and that $\dfrac{3\pi}{2} < \theta < 2\pi$.

**SOLUTION**    One approach would be to use the $\boxed{\cos^{-1}}$ key on a calculator to estimate the reference angle for $\theta$, then determine the angle $\theta$ from the given quadrant information, and finally estimate $\sin(\theta/2)$ using the $\boxed{\text{SIN}}$ key. Instead, we will illustrate how the half-angle identity for sine can be used, thus eliminating the need to actually find $\theta$.

From the half-angle identity for sine, we have

$$\sin\frac{\theta}{2} = \pm\sqrt{\frac{1 - \cos\theta}{2}}$$

$$= \pm\sqrt{\frac{1 - \frac{1}{4}}{2}}$$

$$= \pm\sqrt{\frac{\frac{3}{4}}{2}}$$

$$= \pm\sqrt{\frac{3}{8}}$$

$$= \pm\frac{\sqrt{3}}{\sqrt{8}} \cdot \frac{\sqrt{2}}{\sqrt{2}}$$

$$= \pm\frac{\sqrt{6}}{4}$$

To determine the appropriate sign, we proceed as follows:

$$\frac{3\pi}{2} < \theta < 2\pi$$

$$\frac{3\pi}{4} < \frac{\theta}{2} < \pi$$

Thus $\theta/2$ is in the second quadrant, and since the sine is positive there, we have

$$\sin\frac{\theta}{2} = \frac{\sqrt{6}}{4}$$

**EXAMPLE 8**     *Applying the half-angle identity for tangent*

Find the exact value of $\tan\left(\dfrac{\pi}{8}\right)$.

**SOLUTION**    Using the half-angle identity for tangent $\left(\text{with } x = \dfrac{\pi}{4}\right)$, we have

$$
\begin{aligned}
\tan\frac{\pi}{8} &= \tan\left(\frac{1}{2}\cdot\frac{\pi}{4}\right) \\[2mm]
&= \frac{1 - \cos\dfrac{\pi}{4}}{\sin\dfrac{\pi}{4}} \\[2mm]
&= \frac{1 - \dfrac{\sqrt{2}}{2}}{\dfrac{\sqrt{2}}{2}} \\[2mm]
&= \frac{2}{\sqrt{2}} - 1 \\[2mm]
&= \sqrt{2} - 1
\end{aligned}
$$

## EXERCISES 3

**EXERCISES 1–8**  □  *Use the given information to compute*

a. $\sin 2\theta$    b. $\cos 2\theta$    c. $\tan 2\theta$

**1.** $\sin\theta = \dfrac{3}{5}$, $\cos\theta = -\dfrac{4}{5}$

**2.** $\sin\theta = -\dfrac{24}{25}$, $\cos\theta = \dfrac{7}{25}$

**3.** $\sin\theta = -\dfrac{3}{5}$, $\pi \le \theta < \dfrac{3\pi}{2}$

**4.** $\cos\theta = \dfrac{5}{13}$, $0 \le \theta < \dfrac{\pi}{2}$

**5.** $\tan\theta = \dfrac{8}{15}$, $0 \le \theta < \dfrac{\pi}{2}$

**6.** $\csc\theta = -\dfrac{5}{3}$, $\dfrac{3\pi}{2} \le \theta < 2\pi$

**7.** $\cot\theta = -\dfrac{12}{5}$, $\dfrac{\pi}{2} \le \theta < \pi$

**8.** $\sin\theta = \dfrac{1}{2}$, $0 \le \theta < \dfrac{\pi}{2}$

**EXERCISES 9–16**  □  *Use the given information to compute*

a. $\sin\dfrac{x}{2}$    b. $\cos\dfrac{x}{2}$    c. $\tan\dfrac{x}{2}$

**9.** $\cos x = \dfrac{7}{25}$, $0 \le x < \dfrac{\pi}{2}$

**10.** $\cos x = -\dfrac{7}{25}$, $\pi \le x < \dfrac{3\pi}{2}$

**11.** $\cos x = -\dfrac{1}{9}$, $\dfrac{\pi}{2} \le x < \pi$

**12.** $\cos x = \dfrac{119}{169}$, $0 \le x < \dfrac{\pi}{2}$

**13.** $\sec x = \dfrac{9}{7}$, $\dfrac{3\pi}{2} \le x < 2\pi$

**14.** $\sec x = -8$, $\dfrac{\pi}{2} \le x < \pi$

**15.** $\sin x = -\dfrac{\sqrt{15}}{8}$, $\dfrac{3\pi}{2} \le x < 2\pi$

**16.** $\sin x = -\dfrac{4\sqrt{5}}{9}$, $\pi \le x < \dfrac{3\pi}{2}$

**EXERCISES 17–20**  □  *Find the exact value of the given quantity.*

**17.** $\sin 15°$

**18.** $\cos\dfrac{\pi}{8}$

**19.** $\sec\dfrac{\pi}{8}$

**20.** $\cot 165°$

**EXERCISES 21–36** □ *Verify that the given equation is an identity.*

**21.** $\tan 2x = \dfrac{2\cot x}{\cot^2 x - 1}$

**22.** $\sec 2x = \dfrac{1}{\cos^4 x - \sin^4 x}$

**23.** $\sin x \cos x \csc 2x = \dfrac{1}{2}$

**24.** $2\sin x \sin 2x = \cos x - \cos 3x$

**25.** $\tan 2x = \dfrac{2\sin x}{2\cos x - \sec x}$

**26.** $2\sin^2 x + \cos 2x = 1$

**27.** $2\tan x \cot 2x = 2 - \sec^2 x$

**28.** $\dfrac{\sin 4x}{4} = \sin x \cos x - 2\sin^3 x \cos x$

**29.** $2\csc 30x = \csc 15x \sec 15x$

**30.** $\dfrac{\cos\left(2x + \dfrac{\pi}{2}\right)}{\cos x} = -2\sin x$

**31.** $\sin\left(2x - \dfrac{\pi}{3}\right) + \dfrac{\sqrt{3}}{2} = \sin x(\cos x + \sqrt{3}\sin x)$

**32.** $\tan\left(-\dfrac{x}{2}\right) = \cot x - \csc x$

**33.** $\cot\dfrac{x}{2} = \csc x + \cot x$

**34.** $\sin x \sec\dfrac{x}{2} = \pm\sqrt{2 - 2\cos x}$

**35.** $\tan\dfrac{3x}{2} = \dfrac{3\sin x - 4\sin^3 x}{4\cos^3 x - 3\cos x + 1}$

**36.** $4\cos^4\dfrac{x}{2} - 2\cos^2 x = \sin^2 x + 2\cos x$

■ *Applications*

**37.** *TV Area*  The area of the TV screen shown in Figure 12, in terms of its diagonal $d$ and the angle $\theta$ made between the bottom of the screen and diagonal, is given by $A = d^2 \sin\theta\cos\theta$.

  **a.** Express the area in terms of a single trigonometric function.

  **b.** Find the area of a 27-inch TV given that $\theta = 40°$.

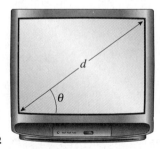

**FIGURE 12**

**38.** *Projectile Range*  If an object is thrown from ground level with an initial angle $\theta$ and an initial speed of $v$ feet per second, its range in feet is given by $(v^2 \sin\theta\cos\theta)/16$.

  **a.** Express the range in terms of a single trigonometric function.

  **b.** Find the range of an object thrown with an initial angle of 30° and an initial speed of 50 feet per second.

**39.** *Refraction of Light*  When light waves pass from one medium to another of different density, they bend. This phenomenon is called *refraction*. Figure 13 illustrates light passing from air to a liquid of greater density. The result is that the *angle of incidence* $\alpha$ is greater than the *angle of refraction* $\beta$. In general,

these angles are related by the formula

$$\frac{c_1}{c_2} = \frac{\sin\alpha}{\sin\beta}$$

where $c_1$ and $c_2$ denote the speed of light in air and the liquid, respectively. If the angle of incidence is twice the angle of refraction, that is, if $\alpha = 2\beta$, show that $c_2 = \frac{1}{2}c_1 \sec\beta$.

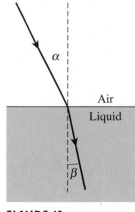

**FIGURE 13**

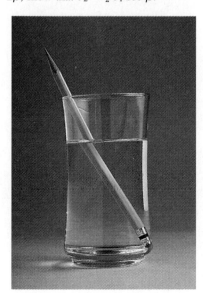

Refraction of light causes the apparent break in the pencil.

■ *Projects for Enrichment*

**40.** *Half-angle Formula for Tangent* In this project, we will complete the derivation of the half-angle formula for tangent.

   **a.** Verify the identity

$$\tan u = \frac{2(1 - \cos^2 u)}{\sin 2u}$$

   **b.** Use your result of part (a) to show that

$$\tan \frac{x}{2} = \frac{1 - \cos x}{\sin x}$$

   **c.** Verify the identity

$$\frac{1 - \cos x}{\sin x} = \frac{\sin x}{1 + \cos x}$$

   which shows that

$$\tan \frac{x}{2} = \frac{\sin x}{1 + \cos x}$$

**41.** *Cubics and Cosines* A cubic equation is one that can be written in the form $x^3 + rx^2 + sx + t = 0$. By making the substitution $x = y - (r/3)$, it is possible to write such an equation in the form $y^3 - py + q = 0$. Thus, if we can develop a technique for solving equations of the form $y^3 - py + q = 0$, we will be able to solve any cubic equation. We will concentrate our efforts on the case where $(q^2/4) - (p^3/27) < 0$, since a technique has already been considered in Exercise 62 of Section 2.5 for the case where $(q^2/4) - (p^3/27) > 0$.

   To begin, we note that if $(q^2/4) - (p^3/27) < 0$, then

$$\left| q\sqrt{\frac{27}{4p^3}} \right| < 1$$

Let $\phi$ be an angle for which

$$\cos \phi = -q\sqrt{\frac{27}{4p^3}}$$

and let

$$y = \frac{2\sqrt{3p}}{3} \cos \frac{\phi}{3}$$

The following steps will show that $y$ is a solution to the equation $y^3 - py + q = 0$.

   **a.** Find an expression for $\cos 3\theta$ that only involves $\cos \theta$. [*Hint:* $\cos 3\theta = \cos(2\theta + \theta)$.]

   **b.** Use your result from part (a) to show that

$$\cos^3 \frac{\phi}{3} = \frac{1}{4}\left( \cos \phi + 3 \cos \frac{\phi}{3} \right)$$

   **c.** Use the fact that

$$y = \frac{2\sqrt{3p}}{3} \cos \frac{\phi}{3}$$

   together with your result from part (b), to show that

$$y^3 = \frac{8\sqrt{3p^3}}{9} \left[ \frac{1}{4}\left( \cos \phi + 3 \cos \frac{\phi}{3} \right) \right]$$

   **d.** Use the fact that

$$\cos \phi = -q\sqrt{\frac{27}{4p^3}}$$

   and your result from part (c) to show that $y^3 - py + q = 0$.

Using similar steps, it can be shown that the other two solutions to the cubic are given by

$$y = \frac{2}{3}\sqrt{3p} \cos\left( \frac{\phi}{3} + 120° \right)$$

$$y = \frac{2}{3}\sqrt{3p} \cos\left( \frac{\phi}{3} + 240° \right)$$

   **e.** Solve the given cubic equation.
      **i.** $y^3 - 9y + 9 = 0$
      **ii.** $x^3 - 3x - 1 = 0$
      **iii.** $\frac{1}{8} t^3 - 6t + 4 = 0$
      **iv.** $x^3 + 3x^2 - 9x - 3 = 0$ (*Hint:* Substitute $x = y - 1$ and simplify.)

**42.** If $x$ and $y$ are coterminal, does it follow that $\sin \dfrac{x}{2} = \sin \dfrac{y}{2}$? Explain.

**43.** If $0 < \theta < 2\pi$ and $\cos \dfrac{\theta}{2} > 0$, what can be said about $\theta$? Explain.

**44.** Explain how the formula for $\sin 2x$ could be obtained from the formula for $\cos 2x$.

**45.** Explain why there are infinitely many angles whose trigonometric function values can be found exactly.

**46.** Consider a line with equation $y = mx + b$ and let $\theta_m$ denote the angle made between the line and the positive $x$-axis. Describe the connection between $m$ and $\theta_m$. (*Hint:* Use the tangent function.) When is it the case that doubling $\theta_m$ results in doubling $m$?

**47.** Figure 14 suggests a proof of the identity $\sin 2x = 2 \sin x \cos x$. Can you find it? (*Hint:* Compute the area of the triangle in Figure 14 two different ways.)

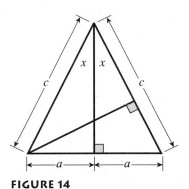

**FIGURE 14**

---

# CONDITIONAL TRIGONOMETRIC EQUATIONS

- How can the distance between any two points on Earth be determined?
- How can trigonometry be used to predict the flight of a golf ball?
- At what angle must a ball be thrown so that its range equals its maximum height?
- What does the sound of a violin look like?

In the previous three sections we have been verifying trigonometric identities; that is, we have been showing that certain equations are *always* satisfied, no matter what the value of the variable. In this section we will discuss techniques for solving **conditional trigonometric equations**, equations involving trigonometric functions that are true only for certain values of the variable. Here the task will be to find all solutions of the equation, or perhaps all solutions lying in a certain interval.

By and large, trigonometric equations are solved in much the same way that algebraic equations are solved. We can solve them analytically (by hand) using the properties of equality, or we can solve them graphically. There are, however, two aspects of solving trigonometric equations that distinguish them from ordinary algebraic equations. First of all, since the trigonometric functions are periodic, most trigonometric equations have infinitely many solutions. For example, solutions to the equation $\sin x = \frac{1}{3}$ correspond to points of intersection of the graphs of $y = \sin x$, and $y = \frac{1}{3}$, as shown in Figure 15. Clearly there are infinitely many solutions.

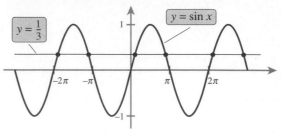

**FIGURE 15**

Secondly, if a trigonometric equation involves more than one trigonometric function, then we may need to employ one or more trigonometric identities in order to obtain exact solutions. (When approximating solutions graphically, it isn't necessary to employ trigonometric identities.)

Solving trigonometric equations will require us to synthesize much of what we have already learned about trigonometric functions. For convenience, the facts that are used most extensively in this section are summarized as follows.

**Tools for solving trigonometric equations**

**Properties of Equality**
- The addition, subtraction, and multiplication principles
- The zero-product property; if $ab = 0$, then $a = 0$ or $b = 0$

**Signs of Trigonometric Functions**
- The sign of both $\cos \theta$ and $\sin \theta$ can be determined by considering the sign of the $x$- and $y$-coordinates of the point on the unit circle corresponding to the angle $\theta$.
- The sign of any other trigonometric function can be determined by expressing the function in terms of sine and cosine.

**Reference Angles**
- $\sin x = \pm \sin y$ implies that $x$ and $y$ have the same reference angle. Similar statements hold for the other trigonometric functions.

**Special Angles**
- The sines and cosines of the angle $\pi/6$, $\pi/4$, and $\pi/3$ are known. All other trigonometric function values for these angles can be determined from basic identities.

**Periodicity**
- The sine, cosine, secant, and cosecant functions have period $2\pi$; the tangent and cotangent functions have period $\pi$.

**Identities**
- All of the identities included in the first three sections of this chapter can be used to simplify or rewrite trigonometric expressions.

**Graphical Techniques**
- Solutions of an equation of the form $f(x) = g(x)$ can be approximated either by graphing both $f$ and $g$ and estimating their points of intersection or (preferably) by graphing the function $f - g$ and estimating its zeros.

In the following example we solve a simple trigonometric equation by first finding a single solution using our knowledge of the special angles and then exploiting our knowledge of reference angles and periodicity to find the other solutions.

**EXAMPLE 1**    *Solving a linear equation in* **sin** *x*

**a.** Find all solutions of $\sin x = \frac{1}{2}$ in the interval $[0, 2\pi]$.
**b.** Find all solutions of the equation $\sin x = \frac{1}{2}$.

**SOLUTION**

**a.** Since $\sin(\pi/6) = \frac{1}{2}$, we know that the reference angle of $x$ is $\pi/6$. Since $\sin x$ is positive for angles $x$ in the second quadrant, the angle in the second quadrant with reference angle $\pi/6$ will also be a solution. As Figure 16 shows, this angle is $5\pi/6$. Thus the solution set is given by $\{\pi/6, 5\pi/6\}$.

**b.** Since the sine function is periodic with period $2\pi$, the following equations are true for all integers $k$.

$$\sin\left(\frac{\pi}{6} + 2\pi k\right) = \sin\frac{\pi}{6} = \frac{1}{2}$$

$$\sin\left(\frac{5\pi}{6} + 2\pi k\right) = \sin\frac{5\pi}{6} = \frac{1}{2}$$

Thus, the solution set consists of all numbers of the form $\pi/6 + 2\pi k$ or $5\pi/6 + 2\pi k$, where $k$ is an integer.

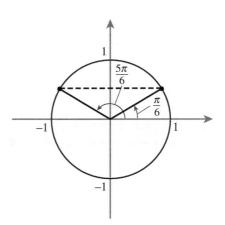

**FIGURE 16**

In some instances, the trigonometric functions appearing in an equation have arguments other than just "$x$." In the following two examples, the trigonometric functions involve double angles. In such cases, we first find the double angle and then simply divide by 2 to find the unknown angle.

**EXAMPLE 2**    *Solving a trigonometric equation involving multiple angles*

Find all solutions of $\cot 2x - 1 = 0$.

**SOLUTION**    Solving for $\cot 2x$, we obtain

$$\cot 2x = 1$$

If we let $\theta = 2x$, then our equation becomes

$$\cot \theta = 1$$

Since the period of the cotangent function is $\pi$, we confine our initial search to values of $\theta$ in the interval $[0, \pi)$. One solution is given by $\pi/4$, and since the cotangent function is negative in the second quadrant, $\pi/4$ is the only solution in the interval $[0, \pi)$. Moreover, since the cotangent function has period $\pi$, all solutions of $\cot \theta = 1$ are of the form

$$\theta = \frac{\pi}{4} + \pi k, \quad \text{for } k \text{ an integer}$$

Since $\theta = 2x$, we have

$$2x = \frac{\pi}{4} + \pi k, \quad \text{for } k \text{ an integer}$$

Dividing by 2, we obtain

$$x = \frac{\pi}{8} + \frac{\pi}{2} k, \quad \text{for } k \text{ an integer}$$

Checking, we see that

$$\cot\left[2\left(\frac{\pi}{8} + \frac{\pi}{2} k\right)\right] - 1 = \cot\left(\frac{\pi}{4} + \pi k\right) - 1 = \cot\left(\frac{\pi}{4}\right) - 1 = 0$$

**EXAMPLE 3**    *Projectile Motion*

A projectile launched from ground level at an initial speed of $v_0$ feet per second with an initial angle $\theta$ has range (ignoring air resistance)

$$R = \frac{v_0^2 \sin 2\theta}{32}$$

Find the initial angle $\theta$ if a golf ball is struck with an initial speed of 200 feet per second and first hits the ground 150 yards from the tee.

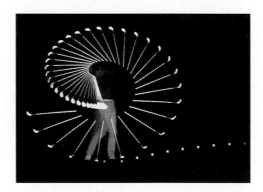

**SOLUTION**    Setting $R = 450$ (150 yards is 450 feet) and $v_0 = 200$, we obtain the equation

$$450 = \frac{200^2 \sin 2\theta}{32}$$

Solving for $\sin 2\theta$, we have

$$\sin 2\theta = \frac{32 \cdot 450}{200^2}$$

$$= 0.36$$

Now one solution of the equation $\sin 2\theta = 0.36$ can be obtained using the inverse sine function, which gives

$$2\theta = \sin^{-1}(0.36) \approx 21.1°$$

It follows that any other angle $2\theta$ must have a reference angle of $21.1°$. Since the sine function is positive in the first and second quadrants, we are interested in an angle between $90°$ and $180°$ with reference angle $21.1°$. This angle is $180° - 21.1° = 158.9°$. Thus, we have

$$2\theta \approx 21.1° \quad \text{or} \quad 2\theta \approx 158.9°$$

and so

$$\theta \approx 10.55° \quad \text{or} \quad \theta \approx 79.45°$$

Many complicated trigonometric equations can be simplified using the zero-product property, which states that $ab = 0$ if and only if either $a = 0$ or $b = 0$. Thus, the solution set of a trigonometric equation of the form $f(\theta)g(\theta) = 0$ is formed by pooling together (taking the union of) the solution sets of the equations $f(\theta) = 0$ and $g(\theta) = 0$.

**EXAMPLE 4**   *Solving a trigonometric equation in factored form*

Find all solutions of $(\sin x - 2)(\csc x - 2) = 0$ in the interval $[0, 2\pi)$.

**SOLUTION**   By the zero-product property, we have that either $\sin x - 2 = 0$ or $\csc x - 2 = 0$. The first equation gives us $\sin x = 2$, which is impossible since $-1 \le \sin x \le 1$. The second equation can be solved as follows:

$$\csc x - 2 = 0$$

$$\csc x = 2$$

$$\frac{1}{\sin x} = 2$$

$$\sin x = \frac{1}{2}$$

Since $\sin(\pi/6) = \frac{1}{2}$, all other solutions of $\sin x = \frac{1}{2}$ must have $\pi/6$ as their reference angle. Since the sine function is positive in the first and second quadrants, we are looking for the angle between $\pi/2$ and $\pi$ with reference angle $\pi/6$, namely $5\pi/6$. Thus, our two solutions are $x = \pi/6$ and $x = 5\pi/6$. Checking these in the original equation, we see that

$$\left(\sin \frac{\pi}{6} - 2\right)\left(\csc \frac{\pi}{6} - 2\right) = \left(\frac{1}{2} - 2\right)(2 - 2) = 0$$

and

$$\left(\sin \frac{5\pi}{6} - 2\right)\left(\csc \frac{5\pi}{6} - 2\right) = \left(\frac{1}{2} - 2\right)(2 - 2) = 0$$

**EXAMPLE 5**   *Solving a quadratic equation in cosine*

Find all solutions of $3 \cos^2 x = 2 \cos x + 1$ in the interval $[0, 2\pi)$.

**SOLUTION**   We begin by recognizing that this is a quadratic equation in $\cos x$. Our strategy is to solve this quadratic equation for $\cos x$ and then finally to solve for $x$.

$$3 \cos^2 x = 2 \cos x + 1$$

$$3 \cos^2 x - 2 \cos - 1 = 0$$

$$(3 \cos x + 1)(\cos x - 1) = 0$$

$$\cos x = -\frac{1}{3} \quad \text{or} \quad \cos x = 1$$

Now one solution to the equation $\cos x = -\frac{1}{3}$ can be obtained by taking the inverse cosine of both sides

$$\cos x = -\frac{1}{3}$$

$$x = \cos^{-1}\left(-\frac{1}{3}\right)$$

$$\approx 1.9106$$

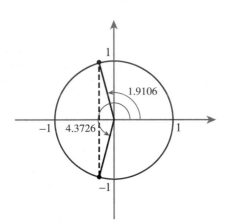

**FIGURE 17**

From Figure 17 we see that $2\pi - 1.9106 \approx 4.3726$ has the same cosine as 1.9106. Thus, the equation $\cos x = -\frac{1}{3}$ gives two solutions, $x \approx 1.9106$ and $x \approx 4.3726$. The equation $\cos x = 1$ gives the solution $x = 0$. Therefore, the original equation has a total of three solutions in $[0, 2\pi)$: $x = 0$, $x \approx 1.9106$, and $x \approx 4.3726$. Each of these can easily be checked in the original equation.

---

Of course, if we are only interested in approximate solutions to trigonometric equations, it is often much easier to employ graphical techniques. It is important to recognize that only a few of the infinitely many solutions will be visible from the graph. All other solutions can be determined by considering periodicity.

**EXAMPLE 6**   *Estimating solutions of a trigonometric equation graphically*

Estimate all solutions of $\sin 3x - 2 \cos 2x = 3 \sin x - 2$ given that
**a.** $x$ lies in $[0, 2\pi)$.
**b.** $x$ is any real number.

**SOLUTION**

**a.** We begin by subtracting $3 \sin x - 2$ from both sides in order to write the equation in the standard form $f(x) = 0$.

$$\sin 3x - 2 \cos 2x = 3 \sin x - 2$$

$$\sin 3x - 2 \cos 2x - 3 \sin x + 2 = 0$$

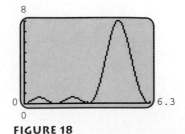

**FIGURE 18**

Next we plot the graph of $f(x) = \sin 3x - 2 \cos 2x - 3 \sin x + 2$ using a graphics calculator. Figure 18 suggests that there are four distinct zeros. However, the fourth one occurs at $2\pi$ and hence is outside of our interval. Zooming in to estimate the values of the other three zeros, we obtain the following approximations

$$x \approx 0, \quad x \approx 1.57, \quad \text{and} \quad x \approx 3.14$$

These numbers are suspiciously close to the values $0$, $\pi/2$, and $\pi$, respectively. Substitution of $0$, $\pi/2$, and $\pi$ into the equation confirms that they are, indeed, solutions.

**b.** In Figure 19, we have zoomed out to estimate the period of $f$. Close inspection reveals that the graph of $f$ repeats itself approximately every 6.28 units. Again, this value is suspiciously close to $2\pi$, and it is easily shown that $f(x + 2\pi) = f(x)$. Thus, the period is $2\pi$. Since the period is $2\pi$ and since $0$, $\pi/2$, and $\pi$ are solutions, numbers of the following forms are also solutions for any integer $k$.

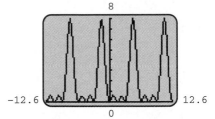

**FIGURE 19**

$$x = 0 + 2\pi k = 2\pi k$$

$$x = \frac{\pi}{2} + 2\pi k$$

$$x = \pi + 2\pi k$$

If an equation involves more than one trigonometric function but cannot be written in factored form, then it may be necessary to use a trigonometric identity to produce an equation involving just a single trigonometric function. For example, an equation involving both $\sin x$ and $\cos x$ may require the use of the identity $\sin^2 x + \cos^2 x = 1$ in order to produce an equation involving just $\sin x$ (or $\cos x$). Similarly, an equation involving both $\cos 2x$ and $\cos x$ can be rewritten as an equation involving only $\cos 2x$ by employing the double-angle identity for cosine.

**EXAMPLE 7**    *Solving an equation involving tangent and secant*

Find all solutions of $\tan x + 1 = \sec x$.

**SOLUTION**    Here we need to find an identity that relates the tangent and secant functions so that we can rewrite the equation in terms of a single trigonometric function. The identity $\tan^2 x + 1 = \sec^2 x$ will do nicely, but since it involves the squares of $\tan x$ and $\sec x$, we will begin by squaring both sides of our equation.

$$\tan x + 1 = \sec x$$

$$(\tan x + 1)^2 = \sec^2 x$$

$$\tan^2 x + 2 \tan x + 1 = \tan^2 x + 1 \qquad \text{Using } \sec^2 x = \tan^2 x + 1$$

$$2 \tan x = 0$$

$$\tan x = 0$$

Since $\tan x = \sin x/\cos x$, $\tan x = 0$ whenever $\sin x = 0$, which occurs at all multiples of $\pi$. Thus, the solution *candidates* of $\tan x + 1 = \sec x$ have the form $x = \pi k$, where $k$ is an integer. Note that in this example it is especially important to check our answer since squaring both sides of the equation could have introduced extraneous solutions. Indeed, if $k$ is odd, $\sec \pi k = -1$, and so for odd $k$,

$$\tan \pi k + 1 = 1 \qquad \text{but} \qquad \sec \pi k = -1$$

Thus, the solutions are limited to those of the form $x = 2k\pi$ (i.e., even multiples of $\pi$).

---

**EXAMPLE 8**   *Solving an equation involving sine and cosine*

Find all solutions of $\cos x + \sin x = 1$.

**SOLUTION**   In this case, we use the Pythagorean identity $\sin^2 x + \cos^2 x = 1$ to relate cosine and sine.

$$\cos x + \sin x = 1$$
$$(\cos x + \sin x)^2 = 1^2 \qquad \text{Squaring both sides}$$
$$\cos^2 x + 2 \sin x \cos x + \sin^2 x = 1$$
$$1 + 2 \sin x \cos x = 1 \qquad \text{Using } \sin^2 x + \cos^2 x = 1$$
$$2 \sin x \cos x = 0$$
$$\sin x = 0, \cos x = 0$$

The equation $\sin x = 0$ is satisfied when $x$ is an integer multiple of $\pi$—that is, $x = \pi k$ for an integer $k$; and $\cos x = 0$ is satisfied for $x = \pi/2 + \pi k$. Thus, our solution candidates have the forms $x = \pi k$ and $x = \pi/2 + \pi k$. By checking these solutions in the original equation, we see that $x = \pi k$ is not a solution if $k$ is odd, and $x = \pi/2 + \pi k$ is not a solution for odd $k$. Thus, the solutions of $\cos x + \sin x = 1$ have either of the two forms $x = 2\pi k$ or $x = \pi/2 + 2\pi k$, where it is understood that $k$ represents an arbitrary integer.

---

**EXAMPLE 9**   *Solving a trigonometric equation involving multiple angles*

Find the solutions of $\cos 2x = 2 \cos x$, where $x$ is in the interval $[0, 2\pi)$.

**SOLUTION**   This problem is made more difficult by the presence of the double angle $2x$. Thus, we will eliminate it by using one of the double-angle identities for cosine.

$$\cos 2x = 2 \cos x$$
$$2 \cos^2 x - 1 = 2 \cos x \qquad \text{Using } \cos^2 x = 2\cos^2 x - 1$$
$$2 \cos^2 x - 2 \cos x - 1 = 0$$

This last equation is quadratic in $\cos x$, but it does not factor with integer coefficients. Thus, we first make the substitution $u = \cos x$.

$$2 \cos^2 x - 2 \cos x - 1 = 0$$
$$2u^2 - 2u - 1 = 0$$

Next we apply the quadratic formula to obtain

$$u = \frac{2 \pm \sqrt{(-2)^2 - 4(2)(-1)}}{2(2)}$$

$$= \frac{1 \pm \sqrt{3}}{2}$$

$$\approx 1.366 \text{ or } -0.366$$

It follows that $\cos x \approx 1.366$ or $\cos x \approx -0.366$. But there is no angle $x$ for which $\cos x \approx 1.366$. Thus, our only solutions are those $x$ in the interval $[0, 2\pi)$ for which $\cos x = -0.366$. Using the $\boxed{\cos^{-1}}$ key on a calculator, we obtain a solution in the second quadrant of $x \approx 1.9455$. A second solution, in the third quadrant with reference angle approximately $\pi - 1.9455$, is given by $x \approx \pi + (\pi - 1.9455) \approx 4.3377$. These two solutions can easily be checked in the original equation.

---

> **RULE OF THUMB**   When solving trigonometric equations, keep the following suggestions in mind:
>
> **1.** Convert to a single trigonometric function if possible.
> **2.** Look for opportunities to factor or apply the quadratic formula.
> **3.** Check your answers in the original equation.

As we have seen throughout this and the previous chapter, trigonometric functions are an essential tool in modeling periodic and even "quasi-periodic" (almost periodic) functions. In the following example, trigonometric functions are used to model the path of a weight hanging from a spring. The periodic nature of the trigonometric functions mirrors the fact that the weight "bobs up and down." The motion of the spring is not truly periodic, however, since the spring is gradually coming to rest.

**EXAMPLE 10**    *Harmonic Motion*

When a weight hung from a certain stretched spring is released (see Figure 20), its displacement (in inches) from the equilibrium (or resting) position $t$ seconds after the spring is released is approximated by

$$y(t) = -e^{-t}(6 \cos 6t + \sin 6t) \qquad (t > 0)$$

**a.** Find the initial position of the weight.
**b.** Estimate the time at which the weight first crosses the equilibrium position.
**c.** Estimate the highest point to which the weight rises.

**SOLUTION**

**a.** At $t = 0$, we have

$$y(0) = -e^0(6 \cos 0 + 6 \sin 0)$$
$$= -1(6 + 0)$$
$$= -6$$

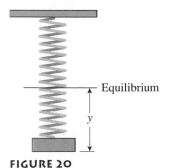

**FIGURE 20**

Evidently, the spring is released 6 inches *below* its equilibrium point.

**b.** The equilibrium point is defined to be where $y(t) = 0$. Thus, we are interested in the smallest positive value of $t$ such that $y(t) = 0$. Although this equation could be solved by analytic methods, we will solve it graphically. In Figure 21 we have shown the graph of $y(t)$ with the cursor near the first positive intercept. By zooming in repeatedly, we find that the graph of $y(t)$ first crosses the $x$-axis near 0.289.

**c.** Figure 22 shows the graph of $y(t)$ near its highest point. After zooming in on this point, we find that the maximum value is about 3.554 inches, attained after approximately 0.524 second.

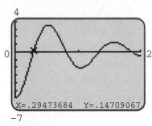

**FIGURE 21**

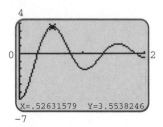

**FIGURE 22**

# EXERCISES 4

**EXERCISES 1–6** □ *Answer the following questions true or false.*

1. The equation $\cos x = \dfrac{\sqrt{3}}{2}$ has $x = -\left(\dfrac{\pi}{6}\right)$ as a solution.

2. The equation $\tan x = -1$ has $x = \dfrac{9\pi}{4}$ as a solution.

3. The equation $\csc x + 1 = 4 \sin x$ has $x = -\left(\dfrac{5\pi}{3}\right)$ as a solution.

4. The equation $\sin x\left(\cos x + \dfrac{1}{2}\right) = 0$ has $x = \dfrac{2\pi}{3}$ as a solution.

5. The equation $\sin x = \dfrac{1}{3}$ has exactly one solution in the interval $[0, \pi)$.

6. If $c \geq 1$, the equation $\sec x = c$ has a solution.

**EXERCISES 7–16** □ *Find all solutions to the given equation in the interval $[0, 2\pi)$. Give exact answers where possible.*

7. $\cos x = 0$

8. $\sin x = \dfrac{1}{2}$

9. $\tan x - 2 = 0$

10. $\cot x + \sqrt{3} = 0$

11. $3 \sin x + \sqrt{3} = \sin x$

12. $4 \cos x = \cos x + 1$

13. $(\csc x - 2)(\sec x - 2) = 0$

14. $(\tan x + 1)(\cot x - 3) = 0$

15. $\sin x \cos x = 0$

16. $\tan x(\sec x - \sqrt{2}) = 0$

**EXERCISES 17–30** □ *Find all solutions to the given equation.*

17. $\sin x = 0$

18. $\sqrt{3} \tan x = -1$

19. $2 \cos x - 1 = 0$

20. $\sqrt{2} \csc x - 2 = 0$

21. $(\sin x + 1)(\tan x - 1) = 0$

22. $(2 \sec x - 2)(2 \cos x + \sqrt{3}) = 0$

23. $\cot x \csc x - 2 \cot x = 0$

24. $(\sin x - 1)\cos x + \dfrac{\sqrt{2}}{2} \cos x = 0$

25. $\cos^2 x = 1$

26. $2 \cos^2 x - \cos x = 1$

27. $\tan^3 x - \tan x = 0$

28. $2 \sec x \sin^2 x - \sec x = 0$

29. $2 \sin^2 x = \cos x + 1$

30. $\sec x - \cos x = 0$

**EXERCISES 31–56** □ *Find all solutions to the given equation in the interval $[0, 2\pi)$. Give exact answers where possible.*

31. $\sec x = -\sqrt{2} \tan x$

32. $\csc x \tan x + 2 = 0$

33. $\sin x - \cos x = 0$

34. $\csc x(\cos x + 1) = 2 \cos x + \csc x$

35. $\tan^2 x - \sin x \sec^2 x = 0$

36. $\sqrt{3}(\cot x - \tan x) = 2$

37. $\cos 2x = \cos^2 x + 1$

38. $\csc x \sin 2x + \cos x = \dfrac{3}{2}$

39. $\cot x = \tan 2x$

40. $\cos 2x = \sin^2 x$

41. $10 \sin^2 x + \sin x - 6 = 0$

42. $\tan^2 x + \tan x - 1 = 0$

43. $\cos^2 x - 5 \cos x + 6 = 0$

44. $2 \csc^2 x + \csc x - 2 = 0$

45. $\sec^2 x - \tan x = 3$

46. $\cos^2 x = -\sin x$

47. $\cos^2 x + \sin^2 x = \dfrac{1}{2}$

48. $\tan^2 x = \sec x - 1$

49. $\cos x \sin x = \dfrac{1}{4}$

50. $\tan x \cos^2 x = -\dfrac{\sqrt{3}}{4}$

51. $\cos 2x = 1$

52. $\tan 3x = \sqrt{3}$

53. $\csc(2x + \pi) = 2$

54. $\sin\left(\dfrac{x}{2} - \dfrac{\pi}{2}\right) = \dfrac{\sqrt{2}}{2}$

55. $\cot \dfrac{x}{2} = -1$

56. $\sin \dfrac{x}{3} = 0$

**EXERCISES 57–62** □ *Find all solutions to the given equation in the interval* $[0, 2\pi)$. *Give exact answers where possible. Use a graphics calculator as necessary.*

57. $\sin^3 x - 3 \sin x + 1 = 0$

58. $\cos^3 x = \sin^2 x$

59. $x = \cos x$

60. $x^2 = \sin x$

61. $\cos^2 x = \sin^2 x$

62. $\tan^2 x - \sec^2 x = 0$

## ■ Applications

63. *Ferris Wheel Height*  The height in feet of a certain passenger on a ferris wheel is given by

$$y(t) = 55 + 50 \sin\left(\dfrac{\pi t}{15} - \dfrac{\pi}{2}\right)$$

where $t$ is the time in seconds and $t = 0$ coincides with the time at which the wheel was set in motion.

   a. Find the first two times for which the height of the passenger is 80 feet.

   b. How long does it take for the ferris wheel to complete one rotation?

64. *Spring Displacement*  A weight is attached to an elastic spring that is suspended from a ceiling. If the weight is pulled 1 inch below its rest position and released, its displacement in inches after $t$ seconds is given by

$$x(t) = 2 \cos\left(5\pi t + \dfrac{\pi}{3}\right)$$

Find the first two times for which the displacement is 1.5 inches.

65. *Electric Current*  The current (in amperes) in an electrical circuit $t$ seconds after the circuit is closed is given by the function

$$E(t) = -12 \cos 4t$$

Find the first two times for which the current is 8 amperes.

66. *Violin Sound*  The waveform of a certain violin note is approximately given by the function

$$y(t) = 165 \sin(\theta + 5.86) + 60 \sin(\theta + 1.15) \\ + 27 \sin(\theta + 0.19)$$

Find the first two times for which $y(t) = 50$. (*Source:* D. C. Miller, *The Science of Musical Sounds*, New York: Macmillan, 1916)

**EXERCISES 67–70** □ *Ignoring air resistance, an object thrown from ground level with an initial angle* $\theta$ *and an initial speed of* $v$ *feet per second has range (maximum horizontal distance)* $(v^2 \sin 2\theta)/32$ *feet and maximum height* $(v^2 \sin^2 \theta)/64$ *feet.*

67. *Football Range*  A quarterback spots a receiver in the end zone, 60 yards away. If the quarterback throws the ball with an initial velocity of 80 feet per second, what initial angle is required in order for the ball to just reach the receiver? (*Hint:* Assume the receiver catches the ball at the same height as it was thrown and then apply the expression for range given above.)

San Francisco 49er's quarterback Steve Young

68. *Soccer Ball Height*  A soccer ball is kicked with an initial velocity of 50 feet per second and reaches a maximum height of 30 feet. At what initial angle was the ball kicked?

69. *Range Equals Height*  An object is projected into the air from ground level in such a way that its range equals its maximum height. With what initial angle was it projected?

**70.** *Tennis Cannon Error*  A certain tennis cannon can be adjusted to project tennis balls at any angle between 0 and $\pi/2$. However, due to design limitations, the angle indicated on the cannon may be in error by a small amount each time it is adjusted. As a consequence, for any given angle setting, the range of a tennis ball may vary.

a. For an angle setting of $\pi/4$, denote the *angular error* by $\alpha$ and show that the difference in ranges for a tennis ball projected at angles $\pi/4 + \alpha$ and $\pi/4$ is given by

$$\frac{v^2}{32} [\cos 2\alpha - 1]$$

We will refer to this expression as the *range error*.

b. Calculate the angular error if the cannon is set at $\pi/4$, the initial velocity is 40 feet per second, and the range error is measured at 2 feet.

## ■ Projects for Enrichment

**71.** *Measuring Distance on the Earth*  Distances between points on the surface of the Earth can be computed using latitude and longitude. First we define the **location angles** $\phi$ and $\theta$ for a point with latitude and longitude $m°$ and $n°$, respectively.

$$\phi = \begin{cases} 90° - m° & \text{for north latitude} \\ 90° + m° & \text{for south latitude} \end{cases}$$

$$\theta = \begin{cases} 90° - n° & \text{for west longitude} \\ 90° + n° & \text{for east longitude} \end{cases}$$

See Figure 23 for a geometric interpretation of $\phi$ and $\theta$.

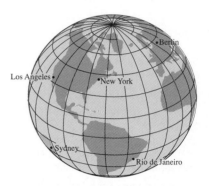

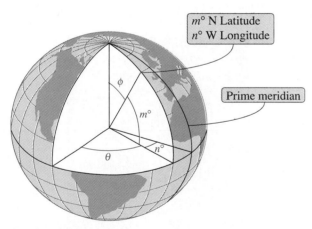

**FIGURE 23**

a. Find the location angles $\phi$ and $\theta$ for the following cities.
   i. New York: 40°42′51″ N, 74°00′23″ W
   ii. Los Angeles: 34°03′08″ N, 118°14′34″ W
   iii. Berlin: 52°32′00″ N, 13°25′00″ E
   iv. Sydney: 33°52′00″ S, 151°12′00″ E
   v. Rio de Janeiro: 22°53′43″ S, 43°13′22″ W

Now consider two points $P_1$ and $P_2$ with location angles $\phi_1$, $\theta_1$ and $\phi_2$, $\theta_2$, respectively. Let $\Phi$ denote the angle made between the line segments from the center of the Earth to each of $P_1$ and $P_2$, as shown in Figure 24.

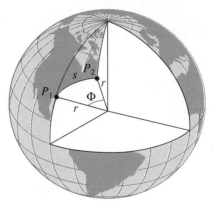

**FIGURE 24**

Note that the distance $s$ between $P_1$ and $P_2$ is the length of the arc subtended by the angle $\Phi$ in a circle of radius $r$, where $r$ is the radius of the Earth. Thus, when $\Phi$ is measured in radians, $s = \Phi r = \Phi(3963)$ miles. Moreover, it can be shown that $\Phi$ satisfies

$$\cos \Phi = \sin \phi_1 \sin \phi_2 \cos \theta_1 \cos \theta_2$$
$$+ \sin \phi_1 \sin \phi_2 \sin \theta_1 \sin \theta_2 + \cos \phi_1 \cos \phi_2$$

b. Show that the above equation can be rewritten as

$$\cos \Phi = \sin \phi_1 \sin \phi_2[\cos(\theta_1 - \theta_2)] + \cos \phi_1 \cos \phi_2$$

To find the distance $s$, we simply find $\cos \Phi$, then find $\Phi$ using a calculator set in radian mode, and finally multiply by the radius of the Earth.

c. Estimate the distances between New York and each of the other cities in part (a).

d. A city in North America with latitude $42°14'32''$ N is approximately 1600 miles from New York. Find its longitude and locate the city in an atlas. (*Hint:* In the first equation given above, substitute the location angles for New York for $\phi_1$ and $\theta_1$, the location angle for a point with latitude $42°14'32''$ N for $\phi_2$, and the angle subtended by an arc

with length 1600 miles in a circle of radius 3963 miles for $\Phi$. Solve the resulting equation for $\theta_2$ and then find the longitude.)

e. Tangent, Oregon, has latitude $44°32'24''$ N and longitude $123°06'24''$ W. A city approximately 170 miles away has longitude $123°15'48''$ W. Find its latitude and locate the city in an atlas. [*Hint:* Make substitutions similar to those suggested in the hint in part (d). You will need to use a graphics calculator to estimate the value of $\phi_2$ that satisfies the resulting equation.]

f. Discuss the effect of elevation on this technique for estimating distances between points on the Earth.

## Questions for Discussion or Essay

72. Describe how the period of a complex trigonometric function can be determined from its graph.

73. In Example 3 we considered a formula that gave the range of a golf ball in terms of the initial angle and initial velocity with which the ball was struck. Why do you think it is that dimpled golf balls fly further than undimpled balls, even when struck with the same initial angle and velocity?

74. A periodic function either has no zeros or infinitely many. Explain why this is so.

75. If a function $f$ is periodic with period $\pi/3$ and $f(x) = \frac{1}{2}$ has two solutions on the interval $[0, \pi/3]$, how many solutions are on the interval $[4\pi/3, 7\pi/3]$?

## CHAPTER REVIEW EXERCISES

**EXERCISES 1–6** □ *Classify the equation as an identity, conditional equation, or contradiction. If the equation is an identity, verify it. If it is a conditional equation, then estimate the smallest positive solution. Use a graphics calculator as necessary.*

1. $2 \sin x + \cos x = 2$

2. $\sec x \cot x \csc x = \cot^2 x + 1$

3. $\cos\left(\dfrac{\pi}{2} - x\right) = \sin x + 1$

4. $(\cos x - 2)(\sin x - 3) = 0$

5. $\dfrac{1}{1 - \cos x} = \csc^2 x + \csc x \cot x$

6. $\sin^2 x + 1 = 2 \cos^2 x$

**EXERCISES 7–16** □ *Verify the given identity.*

7. $\dfrac{\cot x}{\tan x} = \csc^2 x - 1$

8. $\dfrac{\csc A}{\sec A} = \cot A$

9. $\cot^2 y + 5 = \csc^2 y + 4$

10. $\sin x - \cos x = \dfrac{1 - 2\cos^2 x}{\cos x + \sin x}$

11. $\cos \theta + \sin \theta \tan \theta = \sec \theta$

12. $\dfrac{1}{1 - \sin u} = \sec^2 u + \sec u \tan u$

13. $\dfrac{\sec \alpha + \tan \alpha}{\sec \alpha - \tan \alpha} = (\sec \alpha + \tan \alpha)^2$

14. $\dfrac{\cot^4 y - 1}{\cot^2 y - 1} = \csc^2 y$

15. $|\sin x| = \sqrt{1 - \cos^2 x}$

16. $\ln|\sec \theta| = -\ln|\cos \theta|$

**EXERCISES 17–18** □ *Use fundamental trigonometric identities to rewrite the first function in terms of the second.*

17. $\cos x;\ \sin x$

18. $\tan x;\ \sin x$

**EXERCISES 19–20** □ *Use the given information to compute exact values of the indicated trigonometric functions.*

19. $\cos u = \dfrac{4}{5},\ 0 \le u < \dfrac{\pi}{2};\ \cos v = \dfrac{8}{17},\ 0 \le v < \dfrac{\pi}{2}$

    a. $\cos(u - v)$      b. $\cos(u + v)$
    c. $\sin(u - v)$      d. $\sin(u + v)$

20. $\sin \alpha = -\dfrac{24}{25}, \pi \le \alpha < \dfrac{3\pi}{2}; \cos \beta = \dfrac{3}{5}, \dfrac{3\pi}{2} \le \beta < 2\pi$

   a.  $\cos(\alpha - \beta)$          b.  $\cos(\alpha + \beta)$
   c.  $\sin(\alpha - \beta)$         d.  $\sin(\alpha + \beta)$

**EXERCISES 21–24** □ *Rewrite the given expression as a trigonometric function of a single angle.*

21. $\cos 10° \cos 50° - \sin 10° \sin 50°$

22. $\sin \dfrac{3\pi}{8} \cos \dfrac{\pi}{8} - \cos \dfrac{3\pi}{8} \sin \dfrac{\pi}{8}$

23. $\sin x \cos 2x + \cos x \sin 2x$

24. $\dfrac{\tan \dfrac{y}{2} + \tan \dfrac{3y}{2}}{1 - \tan \dfrac{y}{2} \tan \dfrac{3y}{2}}$

**EXERCISES 25–28** □ *Simplify the given trigonometric expression.*

25. $\sin\left(x - \dfrac{3\pi}{2}\right)$        26. $\tan(\pi - x)$

27. $\cos\left(\dfrac{\pi}{4} - x\right)$        28. $\sin\left(\arcsin \dfrac{4}{5} - \arccos \dfrac{12}{13}\right)$

**EXERCISES 29–30** □ *Verify the given identity using the sum and difference identities.*

29. $\dfrac{1}{\cot x - \cot y} = \dfrac{\sin x \sin y}{\sin(y - x)}$

30. $\cos(\alpha + \beta) \cos(\alpha - \beta) = \cos^2 \alpha - \sin^2 \beta$

**EXERCISES 31–32** □ *Use the given information to compute*

a. $\sin 2\theta$    b. $\cos 2\theta$    c. $\tan 2\theta$

31. $\tan \theta = \dfrac{3}{4}, \pi \le \theta < \dfrac{3\pi}{2}$

32. $\cos \theta = -\dfrac{5}{12}, 0 \le \theta < \dfrac{\pi}{2}$

**EXERCISES 33–34** □ *Use the given information to compute*

a. $\sin \dfrac{x}{2}$    b. $\cos \dfrac{x}{2}$    c. $\tan \dfrac{x}{2}$

33. $\sin x = -\dfrac{3}{5}, \dfrac{3\pi}{2} \le x < 2\pi$

34. $\tan x = 2, 0 \le x < \dfrac{\pi}{2}$

**EXERCISES 35–36** □ *Find the exact value of the given quantity.*

35. $\cos \dfrac{\pi}{12}$           36. $\sin 165°$

**EXERCISES 37–42** □ *Verify that the given equation is an identity.*

37. $2 \sin x = 4 \sin \dfrac{x}{2} \cos \dfrac{x}{2}$

38. $\cos 2x = \cos^4 x - \sin^4 x$

39. $\cos 4x = 8 \cos^4 x - 8 \cos^2 x + 1$

40. $\sin 3x = \sin x(3 - 4 \sin^2 x)$

41. $\tan x + \cot x = 2 \csc 2x$

42. $\cot x = \dfrac{1 + \sin 2x + \cos 2x}{1 + \sin 2x - \cos 2x}$

**EXERCISES 43–56** □ *Find all solutions to the given equation in the interval* $[0, 2\pi)$. *Give exact answers where possible.*

43. $\cos x = -1$        44. $\csc x = \sqrt{2}$

45. $(2 \sin x - 1)(\tan x - \sqrt{3}) = 0$

46. $\sin x(\cos x - \pi) = 0$

47. $2 \sin x \cos x - \sin x = 0$

48. $2 \tan^2 x \cos x - \tan^2 x = 0$

49. $2 \cos x + \sec x - 3 = 0$    50. $2 \cos^2 x - 3 \sin x = 3$

51. $\sin 2x + \sin x = 0$      52. $\tan 2x - 2 \cos x = 0$

53. $6 \sin^2 x - 2 \sin x - 1 = 0$    54. $\sec^2 x - 2 \sec x - 4 = 0$

55. $\cos 3x = -1$        56. $2 \sin 2x = 1$

**EXERCISES 57–64** □ *Find all solutions to the given equation.*

57. $\cos x = \dfrac{1}{2}$       58. $\cot x = -\sqrt{3}$

59. $(2 \sin x - \sqrt{3}) \cos x = 0$    60. $\sin^2 x \cos x - \cos x = 0$

61. $\cot^3 x + \cot x = 0$    62. $\cot^2 x + \csc^2 x = 0$

63. $\sin 2x = 1$        64. $\tan 2x = -\sqrt{3}$

**EXERCISES 65–66** □ *Find all solutions to the given equation in the interval* $[0, 2\pi)$. *Give exact answers where possible. Use a graphics calculator as necessary.*

65. $\sin 2x + \cos 3x = 1$    66. $\sin^3 x = 3 \sin x + 1$

**EXERCISES 67–68** □ *The range R in feet of a projectile launched from ground level with an initial speed v feet per second and angle* $\theta$ *is given by*

$$R = \dfrac{v^2 \sin 2\theta}{32}$$

67. ***Baseball Range*** Suppose that a baseball is thrown with an initial speed of $v = 60$ feet per second and that $R = 100$ feet. Find the initial angle $\theta$.

68. ***Projectile Range*** Suppose that a projectile fired at 32 feet per second with an initial angle $\theta$ has a range of $R$, and that if the

angle $\theta$ is increased by $\pi/3$, then the resulting range will be $R + 16$. Find the initial angle $\theta$, and then find $R$.

69. *Electric Current* The current in amperes in a certain electrical circuit at time $t$ seconds is given by

$$i(t) = 20 \cos\left(60t - \frac{\pi}{3}\right)$$

Find all times at which the current is equal to 10 amperes.

70. *Telephone Line Distance* A temporary telephone line runs from a point $P$ on the ground to a point $Q$ at the base of a building and then up the side of the building to a point $R$, as shown in Figure 25. To run the line directly from $P$ to $R$ would require 100 feet, 20 feet less than with the present arrangement. Find $\angle QPR$, and the distances $PQ$ and $QR$.

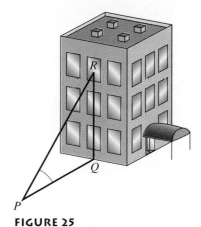

**FIGURE 25**

---

## CHAPTER TEST

**PROBLEMS 1–8** □ *Answer true or false.*

1. The equation $\sin(x + y) = \sin x + \sin y$ is true for all values of $x$ and $y$.

2. The equation $\sin(2x) = 2 \sin x$ is true for all $x$.

3. The equation $\sin(2x) = 2 \sin x$ has infinitely many solutions.

4. Any equation that has infinitely many solutions is an identity.

5. The equation $\tan^2\left(\dfrac{e^x}{2}\right) + 1 = \sec^2\left(\dfrac{e^x}{2}\right)$ is true for all $x$.

6. There is at least one value of $a$ for which the equation $\sin x = a$ has exactly two solutions.

7. The equation $\sin x = x$ has infinitely many solutions.

8. The equation

$$\frac{\tan 3x + \tan 5x}{1 - \tan 3x \tan 5x} = \frac{\tan 2x + \tan 6x}{1 - \tan 2x \tan 6x}$$

is true for all $x$.

**PROBLEMS 9–11** □ *Give an example of numbers a and b such that the equation $\cos^a x + \sin^a x = b$ is*

9. an identity.

10. a conditional equation.

11. a contradiction.

**PROBLEMS 12–14** □ *Verify the given identity.*

12. $\cos x \csc x \tan x = 1$

13. $\cos x = \sec x (1 - \sin^2 x)$

14. $\dfrac{\cos 2x}{\cos x - \sin x} = \dfrac{\csc x + \sec x}{\sec x \csc x}$

15. Simplify $\tan\left(x - \dfrac{\pi}{4}\right)$.

16. Find the exact value of $\sin\left(\dfrac{\pi}{12}\right)$.

17. Find all solutions of the equation $2 \tan x - 4 = 2$ on the interval $[0, 2\pi)$.

18. Find all solutions of the equation $9 \sin x + 2 \cos^2 x = 6$.

19. Use a graphics calculator to determine whether the equation $2 \sin^2 x \cos x = 3$ is an identity, a contradiction, or a conditional equation. If the latter, estimate the solutions on the interval $[0, 2\pi)$.

20. A child is swinging in such a way that the height of the swing $t$ seconds after being pushed is approximated by

$$h = 6 - 3 \cos\frac{\pi t}{2}$$

Find the first time at which the swing is 4.5 feet off the ground.

# APPLICATIONS OF TRIGONOMETRY

■ Stonehenge, an arrangement of enormous stone blocks on the Salisbury Plain in central southern England, is the ruin of a single stone structure, parts of which are over 4000 years old. Although the civilization that erected it and their purpose for doing so may be permanently lost in the mists of time, the alignment and orientation of the stones suggests a familiarity with the seasonal changes in position of sun and moon. In fact, some have even suggested that Stonehenge served as an astronomical computer, capable of predicting eclipses and other astronomical events. Little is known about the mathematics employed by the builders of Stonehenge, but clearly the notions of angle and distance, cornerstones of trigonometry, were familiar to its designers.

**SECTION 1**

## THE LAW OF SINES

■ How can you estimate the height of a mountain from inside a car?

■ What is the least amount of information necessary for finding all the side lengths and angle measures of a triangle?

■ How can the height of the Eiffel Tower be estimated from a plane flying overhead?

■ How can a device constructed from ordinary household objects be used to estimate distances to remote objects?

### SOLVING TRIANGLES WITH THE LAW OF SINES

We have already seen how the definitions of the trigonometric functions can be used to solve a right triangle. More specifically, we have seen that if a side length and either a second side length or an acute angle measure of a right triangle are given, trigonometric functions can be used to find all the unknown side lengths and angle measures. In this section and the next, we will develop techniques for solving triangles that do not have a right angle. Such triangles are called **oblique triangles**.

There are four different cases we must consider for solving oblique triangles. Each requires that a side length and two other parts of the triangle be given. We summarize the four cases as follows:

1. Two angles and a side are given. We refer to this case as ASA or AAS depending upon whether or not the given side is between the angles.

2. Two sides and an angle opposite one of the sides are given. We refer to this case as SSA.

3. Three sides are given. This case is referred to as SSS.

4. Two sides and the angle between are given. This case is known as SAS.

The first two cases can be solved using the **Law of Sines**, which we will consider in this section. The last two cases are most easily solved with the **Law of Cosines** and will be discussed in the next section. Throughout both sections we will follow the convention of denoting the angles of an oblique triangle by $A$, $B$, and $C$ and the lengths of the corresponding opposite sides by $a$, $b$, and $c$, as shown in Figure 1.

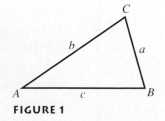

**FIGURE 1**

**Law of Sines**

> If a triangle with vertices $A$, $B$, and $C$ has side lengths $a$, $b$, and $c$ as shown in Figure 1, then
>
> $$\frac{a}{\sin A} = \frac{b}{\sin B} = \frac{c}{\sin C}$$

The first equality of the Law of Sines can be verified by considering either of the two oblique triangles shown in Figure 2. Here we let $h$ denote the length of the altitude from vertex $C$ to side $AB$. From the definition of the sine function, we have

$$\sin A = \frac{h}{b} \qquad \text{or} \qquad h = b \sin A$$

and

$$\sin B = \frac{h}{a} \qquad \text{or} \qquad h = a \sin B$$

Equating the two expressions for $h$, we obtain

$$b \sin A = a \sin B \qquad \text{or} \qquad \frac{a}{\sin A} = \frac{b}{\sin B}$$

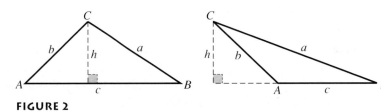

**FIGURE 2**

A similar argument, developed by letting $h$ denote the length of the altitude from vertex $A$ to side $BC$, would yield

$$\frac{b}{\sin B} = \frac{c}{\sin C}$$

Thus

$$\frac{a}{\sin A} = \frac{b}{\sin B} = \frac{c}{\sin C}$$

**EXAMPLE 1**    *Solving an AAS triangle*

Find all unknown side lengths and angle measures for the triangle shown in Figure 3.

**SOLUTION**    Since the angles must add up to 180°,

$$A = 180° - B - C = 180° - 107° - 32° = 41°$$

To find $a$, we use the fact that $c$, $A$, and $C$ are known and apply the Law of Sines.

$$\frac{a}{\sin A} = \frac{c}{\sin C} \qquad \text{Applying the Law of Sines}$$

$$\frac{a}{\sin 41°} = \frac{17}{\sin 32°} \qquad \text{Substituting } A, c, \text{ and } C$$

$$a = \frac{17}{\sin 32°}(\sin 41°) \approx 21.05 \qquad \text{Solving for } a$$

Similarly, we find $b$ using the known values for $c$, $B$, and $C$.

$$\frac{b}{\sin B} = \frac{c}{\sin C}$$

$$\frac{b}{\sin 107°} = \frac{17}{\sin 32°}$$

$$b = \frac{17}{\sin 32°}(\sin 107°) \approx 30.68$$

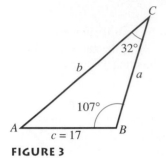

**FIGURE 3**

---

**RULE OF THUMB**    As a simple check of your work, be sure that the longest side is opposite the largest angle and the shortest side is opposite the smallest angle.

**EXAMPLE 2**    *Using an ASA triangle to find the height of a mountain*

A motorist is traveling on a straight and level highway at a constant speed of 60 miles per hour. A mountain top with an angle of elevation of 10° is visible straight ahead. Five minutes later, the angle of elevation to the top of the mountain is 15°. Find the height of the mountain top (in feet) relative to the highway.

**SOLUTION**    At a rate of 60 miles per hour, the motorist will travel 5 miles in 5 minutes. Thus, if the first elevation reading of 10° is recorded at point $A$ and the second reading of 15° is recorded at point $B$, the distance between $A$ and $B$ will be 5 miles (see Figure 4). We wish to find the height $h$ of the mountain relative to the road.

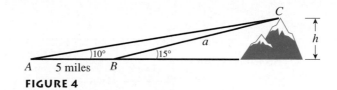

**FIGURE 4**

As a first step, we apply the Law of Sines to the ASA triangle $ABC$ shown in Figure 4 to find the side length $a$. Notice that

$$\angle ABC = 180° - 15° = 165°$$

and so

$$\angle ACB = 180° - (10° + 165°) = 5°$$

Thus, from the Law of Sines, we have

$$\frac{a}{\sin 10°} = \frac{5}{\sin 5°}$$

$$a = \frac{5}{\sin 5°}(\sin 10°) \approx 9.96 \text{ miles}$$

Now by applying the definition of sine to the right triangle having $a$ as the length of the hypotenuse and $h$ as the length of a side, we obtain

$$\sin 15° = \frac{h}{a}$$

or

$$h = a \sin 15° \approx (9.96)\sin 15° \approx 2.58 \text{ miles}$$

Converting from miles to feet, we have

$$2.58 \text{ miles} \times \frac{5280 \text{ feet}}{1 \text{ mile}} \approx 13,600 \text{ feet}$$

When two sides of a triangle and an angle opposite one of the sides are known (i.e., the SSA case), finding the remaining parts of the triangle may not be as straightforward as it was in the previous examples. This is due to the fact that three different outcomes are possible, depending on the given side lengths and angle measure: (1) There may be no triangle satisfying the given conditions; (2) there may be a unique triangle satisfying the given conditions; or (3) there may be two distinct triangles satisfying the given conditions. Table 1 illustrates the possibilities, given angle $A$ and sides $a$ and $b$.

**TABLE 1**

*SSA possibilities*

| Angle A | Condition $(h = b \sin A)$ | Triangles possible | Example |
|---------|---------------------------|--------------------|---------|
| Acute | $a < h$ | 0 | 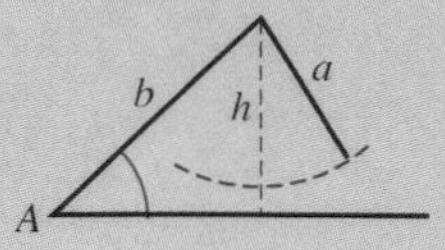 |
| Acute | $a = h$ | 1 | 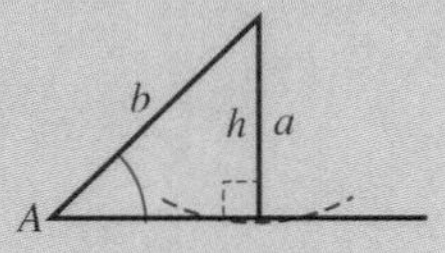 |
| Acute | $h < a < b$ | 2 | 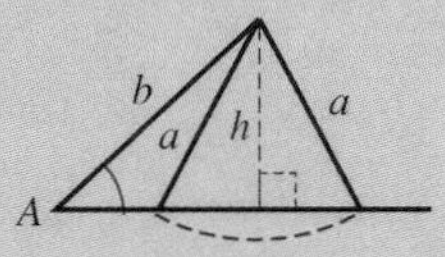 |
| Acute | $a \geq b$ | 1 | 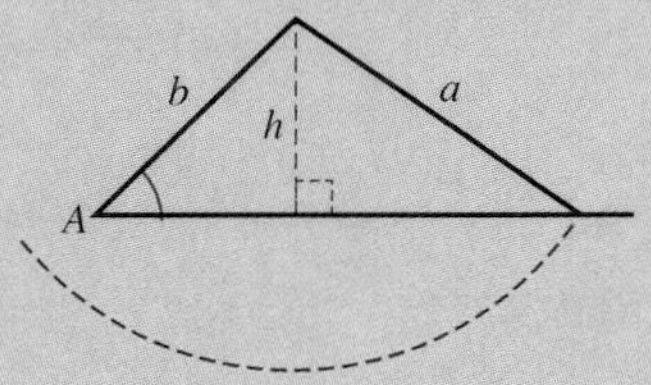 |
| Obtuse | $a \leq b$ | 0 | 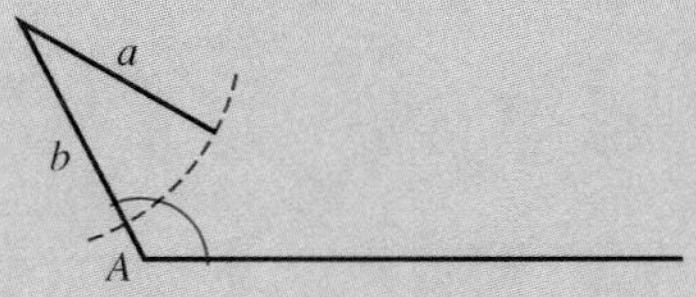 |
| Obtuse | $a > b$ | 1 | 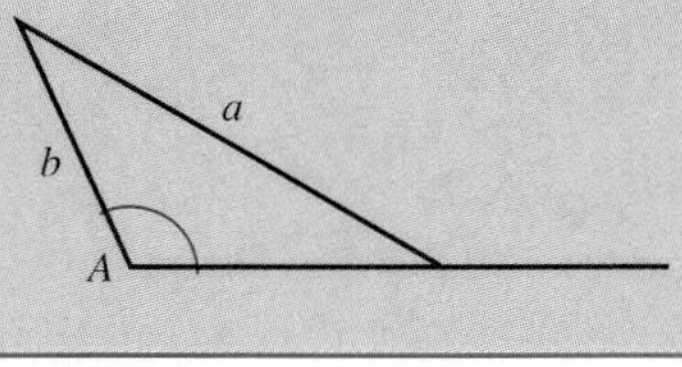 |

Note that it is not necessary to memorize the conditions given in Table 1. As we will see in the following examples, these situations become apparent as the solutions develop.

**EXAMPLE 3**    *Solving an SSA triangle with a unique solution*

Find the unknown side length and angle measures for the triangle *ABC*, given that $a = 90$, $c = 100$, and $C = 65°$, as shown in Figure 5.

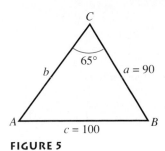

**FIGURE 5**

**SOLUTION**   Using the Law of Sines and the known values for $a$, $c$, and $C$, we obtain

$$\frac{a}{\sin A} = \frac{c}{\sin C}$$

$$\frac{90}{\sin A} = \frac{100}{\sin 65°}$$

$$\sin A = \frac{90 \sin 65°}{100} \approx 0.8157 \qquad \text{Solving for } \sin A$$

Since the sine function is positive in the first and second quadrants, there are two angles between 0° and 180° whose sine is 0.8157. Using the $\boxed{\text{SIN}^{-1}}$ key on a calculator set in degree mode, we find that the angle in the first quadrant whose sine is 0.8157 is approximately 54.66°. The angle in the second quadrant whose sine is 0.8157 is approximately $180° - 54.66° = 125.34°$. However, this could not be the angle $A$ in the triangle $ABC$ since the sum of 125.34° and $C = 65°$ exceeds 180°. Thus, $A \approx 54.66°$ and

$$B = 180° - A - C \approx 180° - 54.66° - 65° = 60.34°$$

To find $b$, we apply the Law of Sines once more, using the known values for $c$, $B$, and $C$.

$$\frac{b}{\sin B} = \frac{c}{\sin C}$$

$$b = \frac{c \sin B}{\sin C} \approx \frac{100 \sin 60.34°}{\sin 65°} \approx 95.88$$

---

**EXAMPLE 4**   *Solving an SSA triangle with no solutions*

Show that there is no triangle $ABC$ with $a = 0.3$, $b = 0.6$, and $A = 42°$.

**SOLUTION**   From the Law of Sines, we know that if a triangle exists with the given side lengths and angle measure, then

$$\frac{a}{\sin A} = \frac{b}{\sin B}$$

$$\frac{0.3}{\sin 42°} = \frac{0.6}{\sin B}$$

$$\sin B = \frac{0.6 \sin 42°}{0.3} \approx 1.338$$

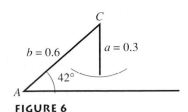

**FIGURE 6**

However, there is no angle $B$ whose sine is greater than 1. Thus, there is no triangle with the given side lengths and angle measure. In Figure 6, we see that side $a$ is too short to reach side $c$, no matter what value is chosen for angle $C$.

| EXAMPLE 5 | *Solving an SSA triangle with two solutions* |

Find two triangles $ABC$ with $a = 5$, $b = 3$, and $B = 31°$.

**SOLUTION**   We begin by applying the Law of Sines to find angle $A$.

$$\frac{a}{\sin A} = \frac{b}{\sin B}$$

$$\frac{5}{\sin A} = \frac{3}{\sin 31°}$$

$$\sin A = \frac{5 \sin 31°}{3} \approx 0.8584$$

Now there are two angles between $0°$ and $180°$ whose sine is $0.8584$. Using the
$\boxed{\text{SIN}^{-1}}$ key on a calculator set in degree mode, we find the angle in the first quadrant
to be approximately $59.1°$. The angle in the second quadrant is approximately
$180° - 59.1° = 120.9°$. In this case, unlike Example 3, the second angle yields a
second triangle since the sum of $120.9°$ and $B = 31°$ is less than $180°$. Thus, we
consider two cases, the first with $A \approx 59.1°$, and the second with $A \approx 120.9°$.

**Case 1:** For $A \approx 59.1°$, we have

$$C = 180° - A - B \approx 180° - 59.1° - 31° = 89.9°$$

The remaining side, $c$, can now be found using the Law of Sines.

$$c = \frac{b \sin C}{\sin B} \approx \frac{3 \sin 89.9°}{\sin 31°} \approx 5.82$$

This triangle is shown in Figure 7.

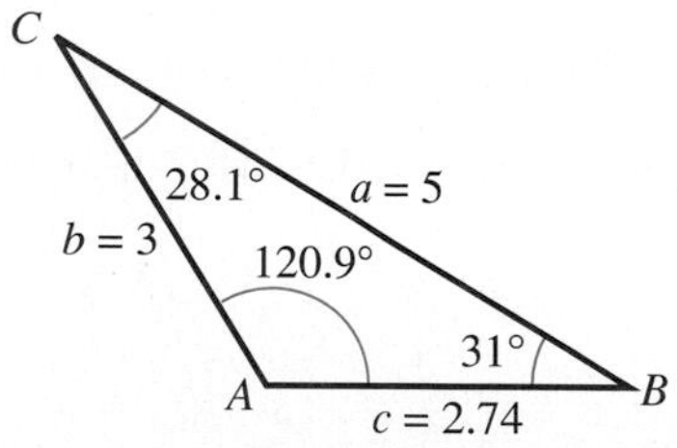

**Case 2:** For $A \approx 120.9°$, we have

$$C = 180° - A - B \approx 180° - 120.9° - 31° = 28.1°$$

and, from the Law of Sines,

$$c = \frac{b \sin C}{\sin B} \approx \frac{3 \sin 28.1°}{\sin 31°} \approx 2.74$$

This triangle is shown in Figure 8.

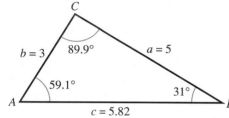

**FIGURE 7**

**FIGURE 8**

■

> **RULE OF THUMB**   Always check for a possible second triangle when solving
> an SSA triangle.

## AREA OF TRIANGLES

In proving the Law of Sines, we used the fact that the altitude $h$ of either triangle shown in Figure 9 is given by

$$h = b \sin A$$

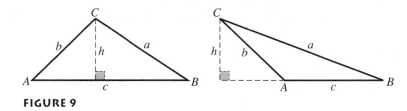

**FIGURE 9**

Thus, the area of both triangles is given by

$$\text{Area} = \frac{1}{2} \text{(base)(height)} = \frac{1}{2} c(b \sin A) = \frac{1}{2} bc \sin A$$

Similarly, we can show that

$$\text{Area} = \frac{1}{2} ab \sin C = \frac{1}{2} ac \sin B$$

Note that in each case, the area involves two sides of the triangle and the angle between them. This leads us to the following general formula for the area of a triangle.

### Area of a triangle

Let $s_1$ and $s_2$ denote the lengths of two sides of a triangle, and let $\theta$ denote the angle between the two sides. Then the area of the triangle is given by

$$\text{Area} = \frac{1}{2} s_1 s_2 \sin \theta$$

**EXAMPLE 6**

*Finding the area of an oblique triangle*

Find the area of the triangle shown in Figure 10.

**SOLUTION**    Since we are given two sides and the angle between, we apply the area formula with $s_1 = 75$, $s_2 = 50$, and $\theta = 130°$.

$$\text{Area} = \frac{1}{2} s_1 s_2 \sin \theta$$

$$= \frac{1}{2} (75)(50) \sin 130°$$

$$\approx 1436.3 \text{ square feet}$$

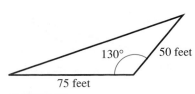

**FIGURE 10**

## EXERCISES 1

**EXERCISES 1–8** □ *Find all unknown side lengths and angle measures; that is, solve the given triangle.*

**1.**
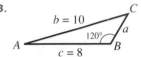

**2.**
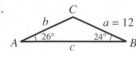

**3.**
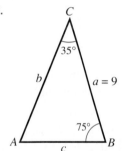

**4.**
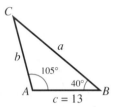

**5.**
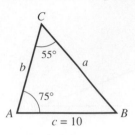

**6.**
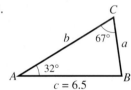

**7.**
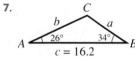

**8.**
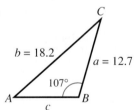

**EXERCISES 9–28** □ *Solve the given triangle. Assume angles A, B, and C, and sides a, b, and c are labeled as shown in Figure 11.*

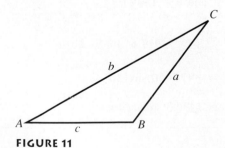

**FIGURE 11**

**9.** $c = 18, A = 17°, C = 86°$

**10.** $b = 7.5, c = 3.2, B = 140°$

**11.** $b = 20, c = 10, C = 30°$

**12.** $b = 80, A = 28°, C = 74°$

**13.** $a = 42, B = 7°, C = 51°$

**14.** $a = 15, A = 68°, B = 43°$

**15.** $a = 23, c = 16, A = 122°$

**16.** $a = 14, b = 18, A = 115°$

**17.** $b = 2, c = 9, B = 38°$

**18.** $c = 2.4, A = 116°, B = 32°$

**19.** $a = 14, c = 12, C = 40°$

**20.** $a = 6.3, b = 5.8, B = 55°$

**21.** $a = 12.5, c = 14, A = 140°$

**22.** $b = 150, B = 23°, C = 17°$

**23.** $b = 70, A = 62°, C = 53°$

**24.** $b = 9, c = 4, B = 34°$

**25.** $a = 300, A = 43°, B = 117°$

**26.** $a = 5, b = 40, A = 26°$

**27.** $b = 0.13, c = 0.9, C = 52°$

**28.** $a = 1, b = \sqrt{2}, A = 45°$

**EXERCISES 29–34** □ *Find the area of the given triangle.*

**29.**
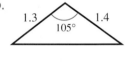

**30.**
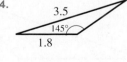

**31.**

**32.**
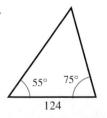

**33.**
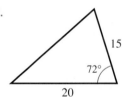

**34.**
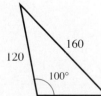

■ *Applications*

**35.** *Length of a Power Line* A power company would like to run a high-voltage power line across a canyon. It must extend from point *A* on the south side of the canyon to point *C* on the north side, as shown in Figure 12. A surveyor has marked point *B* on the south side of the canyon 100 feet from point *A* and has determined that $\angle CAB = 42°$ and $\angle ABC = 110°$. What is the minimum length of power line that will be needed?

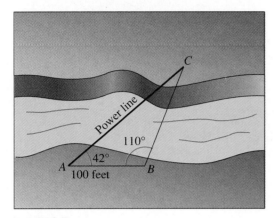

**FIGURE 12**

**36.** *Height of a Radio Tower* Two guy wires for a radio tower make angles with the horizontal of 58° and 49°, as shown in Figure 13. If the ground anchors for each wire are 150 feet apart, find

a. the length of each wire, assuming the wires are taut.

b. the height of the tower.

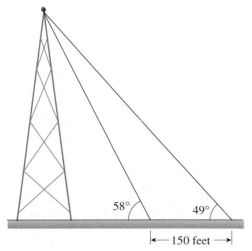

**FIGURE 13**

Eiffel Tower

**37.** *Height of the Eiffel Tower* An airplane is flying at a constant altitude of 3650 feet and at a constant speed of 660 feet per second (approximately 450 miles per hour) on a path that will take it directly over the Eiffel Tower, as shown in Figure 14. At a certain point, the angle of depression to the top of the tower is 22°. Four seconds later, the angle of depression to the top of the tower is 34°. Estimate the height of the Eiffel Tower.

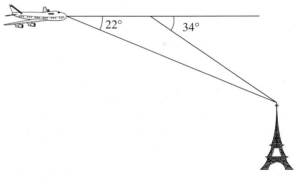

**FIGURE 14**

**38.** *Fire Lookout* Two fire lookout towers are located 1 mile apart on a line running east-west. A campfire is spotted at a bearing of N 38° E from the western tower and N 29° W from the other. A forest ranger is dispatched immediately from the tower nearest the fire to put it out. If the ranger travels at a rate of 3 miles per hour, how long will it take her to get to the fire?

**39.** *Ship in Distress* Two Coast Guard stations are located 10 miles apart on a coastline that runs north-south. A distress signal is received from a ship with a bearing of N 43° E from the southern station and a bearing of S 56° E from the northern station. How far is the ship from each station? How far is the ship from shore?

**40.** *Great Pyramid Ramp*  It has been speculated that the Egyptian pyramids were built with the aid of long ramps. Given that the sides of the Great Pyramid at Giza make an angle of 51.9° with the ground, what would be the length of a ramp with a 7° angle of incline that reaches 100 feet along the face of one side?

The pyramids at Giza

**41.** *Triangular Plot*  A plot of land is bounded on two sides by rivers, as shown in Figure 15. If the rivers meet at an angle of 35° and the lengths of the sides bounded by the rivers are 1300 feet and 650 feet, find the area of the plot. The owner of the plot would like to sell it for $2000 per acre. How much is the asking price, given that 1 acre is 43,560 square feet?

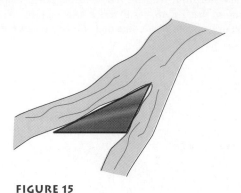

**FIGURE 15**

---

### Project for Enrichment

**42.** *Constructing a Device to Measure Angles*  In this project, you will construct a device to measure line-of-sight angles and then use it to estimate various distances and heights. An example of such a device is shown in Figure 16. Depending on the materials and tools you have available, yours may not look or operate exactly the same as this one, but it should be based on the same principles. Our apparatus is made from three sticks. The two shaded sticks are both 12 inches from pivot point to pivot point. The ruled stick is somewhat longer than 24 inches. The leftmost shaded stick, which we will refer to as the *pivot stick*, is anchored to the ruled stick in such a way that it is able to pivot. The second shaded stick, the *slider stick*, is anchored to one end of the pivot stick so that it can also pivot, but the other end of the slider stick must be allowed to slide along the ruled stick. The angle made between the pivot stick and the ruled stick can be read off below the right end of the slider stick.

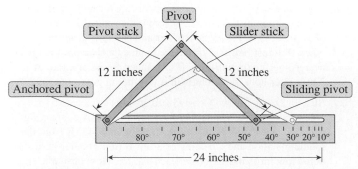

**FIGURE 16**

To determine the placement of the degree measure marks on the ruled stick, consider the isosceles triangle shown in Figure 17.

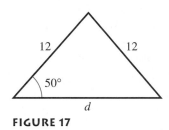

**FIGURE 17**

**a.** Find the length *d* shown in Figure 17. Explain how this value would help you locate the grid mark for 50° on the ruled stick.

**b.** Find the distances that correspond to each of the degree measure grid marks on the ruled stick.

**c.** Construct a device similar to the one shown in Figure 16. Mark the ruled stick with degree measures in increments of 5° from 85° down to 5°.

**d.** Choose a well-known tall landmark on your campus. Measure an appropriate distance away from the base of the landmark. From that point, use your device to estimate the angle of elevation to the top of the landmark. This can be done by keeping the ruled stick horizontal and ''lining up'' the pivot stick with the top of the landmark. Now estimate the height of the landmark. Explain your steps in detail.

e. Choose a location on or near your campus where the distance between two points would be difficult or impossible to measure using standard tools (e.g., a tape measure). Identify these points as $A$ and $B$. Mark a point $C$ a reasonable dis-

tance from point $A$ and measure that distance. From point $A$, use your device to estimate $\angle BAC$. From point $C$, estimate $\angle BCA$. Now use the Law of Sines to estimate the distance between $A$ and $B$. Explain your steps in detail.

 *Questions for Discussion or Essay*

43. A Rule of Thumb in this section states that the largest side of a triangle is opposite the largest angle. Use the Law of Sines to explain why this is so.

44. Exercise 35 asks for the *minimum length* of power line needed to span a canyon. Why might more actually be needed?

45. At the beginning of this section, we listed four categories of triangles that can be solved using the Law of Sines and the Law of Cosines. These were labeled AAS (or ASA), SSA, SSS, and

SAS. What do you suppose is meant by an AAA triangle, and why is it not listed as a fifth case?

46. Explain why the Law of Sines is not effective for computing distances between cities, even if all necessary angles are known.

47. The Law of Sines was presented in this section as a tool for solving oblique triangles, that is, nonright triangles. What happens in the case of right triangles?

---

**SECTION 2**

# THE LAW OF COSINES

- How can one find the area of a triangle if the three side lengths are known?
- If the hypotenuse and two legs of a right triangle are related by the equation $c^2 = a^2 + b^2$, what can be said about the relationship between sides of an oblique triangle?
- How can two people talking on the telephone pinpoint the location of a lightning flash?
- How can the area of a pentagon be useful for finding the area of a circle?

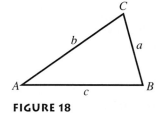

**FIGURE 18**

In the previous section we used the Law of Sines to solve oblique triangles of the form AAS or SSA. The remaining two cases—SAS and SSS—are more easily solved with the **Law of Cosines**. As in the previous section, we denote the angles of an oblique triangle by $A$, $B$, and $C$ and the lengths of the corresponding opposite sides by $a$, $b$, and $c$, as shown in Figure 18.

■ *Law of Cosines*

If a triangle with vertices $A$, $B$, and $C$ has side lengths $a$, $b$, and $c$, as shown in Figure 18, then the following equations hold:

$$a^2 = b^2 + c^2 - 2bc \cos A$$
$$b^2 = a^2 + c^2 - 2ac \cos B$$
$$c^2 = a^2 + b^2 - 2ab \cos C$$

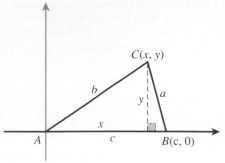

**FIGURE 19**

The Law of Cosines can be derived using the distance formula. We begin by placing the angle $A$ in standard position, as indicated in Figure 19, with vertex $B$ at the point $(c, 0)$. If we denote the coordinates of vertex $C$ by $(x, y)$ and use the definitions of sine and cosine, we see that

$$\cos A = \frac{x}{b} \quad \text{and} \quad \sin A = \frac{y}{b}$$

Thus,

$$x = b \cos A \quad \text{and} \quad y = b \sin A$$

Note that these expressions for $x$ and $y$ are valid even if $A$ is obtuse. Now by applying the formula for the distance between vertices $B$ and $C$, we obtain

$$a = \sqrt{(x - c)^2 + (y - 0)^2}$$

$$a^2 = (b \cos A - c)^2 + (b \sin A)^2 \qquad \text{Squaring both sides and substituting for } x \text{ and } y$$

$$= b^2 \cos^2 A - 2bc \cos A + c^2 + b^2 \sin^2 A$$

$$= b^2(\cos^2 A + \sin^2 A) + c^2 - 2bc \cos A$$

$$= b^2 + c^2 - 2bc \cos A \qquad \text{Using the identity } \sin^2 A + \cos^2 A = 1$$

Thus, $a^2 = b^2 + c^2 - 2bc \cos A$, which is our first version of the Law of Cosines. The other two versions can be derived in a similar fashion by successively placing angles $B$ and $C$ in standard position.

---

**RULE OF THUMB**   It is not necessary to memorize each of the three equations in the Law of Cosines. You simply need to remember that the expression appearing to the right of the equal sign involves two sides and the included angle. The side opposite the angle appears to the left of the equal sign. In equation form we have

$$(\text{the side opposite the angle})^2 = (\text{side 1})^2 + (\text{side 2})^2$$
$$- 2(\text{side 1})(\text{side 2}) \cos(\text{the included angle})$$

---

Note that the equation given above involves four quantities. If any three of the four quantities are given, we can solve for the fourth. In particular, if two sides and the included angle are known, we can solve for the third side, or if the three side lengths are known, we can solve for any angle. We illustrate these possibilities in the following examples.

**EXAMPLE 1**   *Solving an SAS triangle*

Find the unknown side length and angle measures for the triangle in Figure 20.

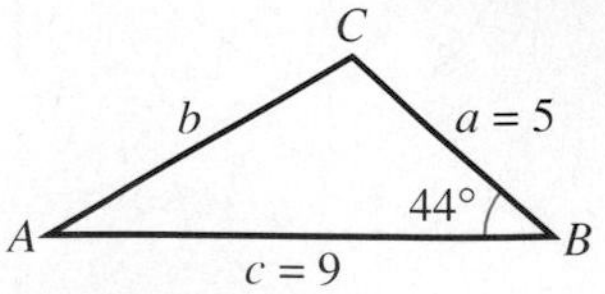

**SOLUTION**   We first apply the Law of Cosines to find $b$.

$$b^2 = a^2 + c^2 - 2ac \cos B = 5^2 + 9^2 - 2(5)(9) \cos 44° \approx 41.26$$

$$b \approx \sqrt{41.26} \approx 6.42$$

**FIGURE 20**

Next, we use the Law of Sines to find $A$.

$$\frac{a}{\sin A} = \frac{b}{\sin B}$$

$$\frac{5}{\sin A} = \frac{6.42}{\sin 44°}$$

$$\sin A = \frac{5 \sin 44°}{6.42} \approx 0.5410 \qquad \text{Solving for } \sin A$$

Thus, $A \approx \sin^{-1} 0.5410 \approx 32.8°$ and it follows that

$$C \approx 180 - 32.8° - 44° = 103.2°$$

---

**EXAMPLE 2**   *Solving an SSS triangle*

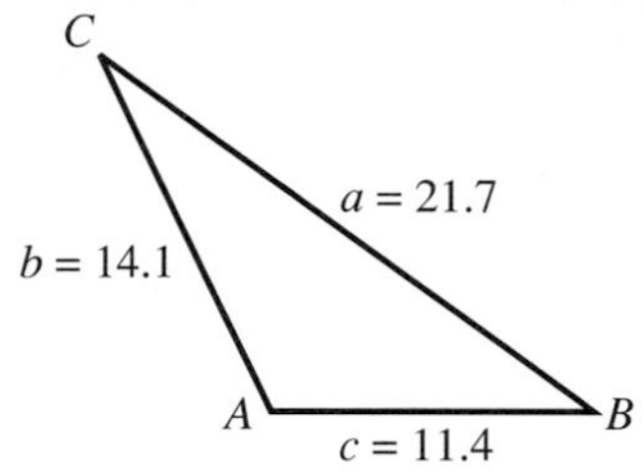

Find the angles for the triangle shown in Figure 21.

**SOLUTION**   We first find $A$ using the Law of Cosines.

$$a^2 = b^2 + c^2 - 2bc \cos A$$

$$\cos A = \frac{b^2 + c^2 - a^2}{2bc} \qquad \text{Solving for } \cos A$$

$$= \frac{14.1^2 + 11.4^2 - 21.7^2}{2(14.1)(11.4)}$$

$$\approx -0.4421$$

$C$

$a = 21.7$

$b = 14.1$

$A$

$c = 11.4$

$B$

**FIGURE 21**

Thus, $\cos A \approx -0.4421$ and it follows that $A \approx \cos^{-1}(-0.4421) \approx 116.2°$. Next we find $B$ using the Law of Sines.

$$\frac{a}{\sin A} = \frac{b}{\sin B}$$

$$\frac{21.7}{\sin 116.2°} = \frac{14.1}{\sin B}$$

$$\sin B \approx \frac{14.1 \sin 116.2°}{21.7} \approx 0.5830 \qquad \text{Solving for } \sin B$$

So $B \approx \sin^{-1} 0.5830 \approx 35.7°$. Finally we have

$$C \approx 180° - 116.2° - 35.7° = 28.1°$$

---

In general, the Law of Cosines can be applied whenever two sides and an angle are known. Thus, as we see in the following example, the Law of Cosines can be used for SSA triangles. (Note, however, that the Law of Sines is often more efficient for solving SSA triangles.)

**EXAMPLE 3**    *Solving an SSA triangle using the Law of Cosines*

Use the Law of Cosines to find the unknown side length and angle measures for the triangle *ABC* with $b = 78$, $c = 152$, and $B = 26°$.

**SOLUTION**    Since $b$ is the length of the side opposite angle $B$, we have

$$b^2 = a^2 + c^2 - 2ac \cos B$$

$$78^2 = a^2 + 152^2 - 2a(152) \cos 26°$$

$$6084 = a^2 + 23{,}104 - 273.23a$$

$$0 = a^2 - 273.23a + 17{,}020$$

Now the last equation is quadratic in $a$, so we apply the quadratic formula to obtain

$$a = \frac{273.23 \pm \sqrt{273.23^2 - 4(1)17{,}020}}{2} \approx \frac{273.23 \pm 81.08}{2}$$

or

$$a \approx 177.2, \quad a \approx 96.1$$

Thus, there are two triangles satisfying the given conditions, one with $a \approx 177.2$, and the other with $a \approx 96.1$. We consider both cases.

**Case 1:** Using $a \approx 177.2$ and the Law of Cosines, we have

$$a^2 = b^2 + c^2 - 2bc \cos A$$

$$\cos A = \frac{b^2 + c^2 - a^2}{2bc}$$

$$\approx \frac{78^2 + 152^2 - 177.2^2}{2(78)(152)}$$

$$\approx -0.0933$$

$$A \approx \cos^{-1}(-0.0933) \approx 95.4°$$

Thus, $C = 180° - A - B \approx 180° - 95.4° - 26° = 58.6°$. The resulting triangle is shown in Figure 22.

**Case 2:** For $a \approx 96.1$, we have

$$a^2 = b^2 + c^2 - 2bc \cos A$$

$$\cos A = \frac{b^2 + c^2 - a^2}{2bc}$$

$$\approx \frac{78^2 + 152^2 - 96.1^2}{2(78)(152)}$$

$$\approx 0.8415$$

$$A \approx \cos^{-1}(0.8415) \approx 32.7°$$

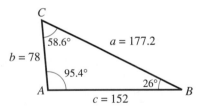

**FIGURE 22**

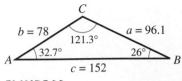

**FIGURE 23**

Thus, $C = 180° - A - B \approx 180° - 32.7° - 26° = 121.3°$. The resulting triangle is shown in Figure 23.

**EXAMPLE 4** | *Computing distance*

Two ships start from the same point and sail in different directions, one on a course of 40° clockwise from north at a rate of 6 miles per hour and the other on a course of 150° clockwise from north at a rate of 8 miles per hour. How far apart are the ships after $2\frac{1}{2}$ hours?

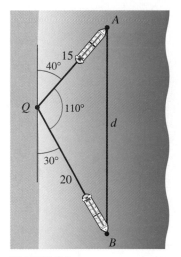

**FIGURE 24**

**SOLUTION**   We first draw a sketch showing the courses and relative positions of the ships (see Figure 24). The position of the ship with a course of 40° is labeled $A$ in Figure 24, while the position of the ship with a course of 150° is labeled $B$. Thus, $\angle AQB = 150° - 40° = 110°$. After $2\frac{1}{2}$ hours, ship $A$ is $(2.5)(6) = 15$ miles from the starting point, and ship $B$ is $(2.5)(8) = 20$ miles from the starting point. The distance $d$ between the two ships can be computed using the Law of Cosines. We have

$$d^2 = 15^2 + 20^2 - 2(15)(20)\cos 110° \approx 830.212$$

and so

$$d \approx \sqrt{830.2} \approx 28.8 \text{ miles}$$

Thus, the ships are approximately 28.8 miles apart after $2\frac{1}{2}$ hours.

■ HERON'S FORMULA

The Law of Cosines can be used to derive a formula for the area of a triangle in terms of its side lengths. The formula is named after the Greek mathematician Heron of Alexandria.

■ *Heron's formula*

> The area of a triangle with side lengths $a$, $b$, and $c$ is given by the formula
>
> $$\text{Area} = \sqrt{s(s - a)(s - b)(s - c)}$$
>
> where $s$ is the **semiperimeter**
>
> $$s = \frac{a + b + c}{2}$$

To verify this formula, we begin with the formula from Section 9.1 for the area of a triangle in terms of two sides and the included angle. Here we follow the usual convention of denoting the angle opposite side $a$ by $A$.

$$\text{Area} = \frac{1}{2}\, bc \sin A$$

$$(\text{Area})^2 = \frac{1}{4}\, b^2 c^2 \sin^2 A \qquad\qquad \text{Squaring both sides}$$

$$= \frac{1}{4}\, b^2 c^2 (1 - \cos^2 A) \qquad\qquad \text{Using the identity } \sin^2\theta = 1 - \cos^2\theta$$

$$= \frac{1}{4}\, b^2 c^2 (1 + \cos A)(1 - \cos A) \qquad \text{Factoring the difference of squares}$$

From the Law of Cosines, we know that

$$\cos A = \frac{b^2 + c^2 - a^2}{2bc}$$

Thus,

$$(\text{Area})^2 = \frac{1}{4}\, b^2 c^2 \left(1 + \frac{b^2 + c^2 - a^2}{2bc}\right)\left(1 - \frac{b^2 + c^2 - a^2}{2bc}\right)$$

$$= \frac{1}{4}\, b^2 c^2 \left[\frac{(b^2 + 2bc + c^2) - a^2}{2bc}\right]\left[\frac{a^2 - (b^2 - 2bc + c^2)}{2bc}\right]$$

$$= \frac{1}{16}\, [(b + c)^2 - a^2][a^2 - (b - c)^2]$$

$$= \frac{1}{16}\, [(b + c) + a][(b + c) - a][a - (b - c)][a + (b - c)]$$

$$= \left(\frac{a + b + c}{2}\right)\left(\frac{a + b + c}{2} - a\right)\left(\frac{a + b + c}{2} - b\right)\left(\frac{a + b + c}{2} - c\right)$$

By substituting $s = \frac{1}{2}(a + b + c)$ and taking square roots on both sides, we obtain

$$\text{Area} = \sqrt{s(s - a)(s - b)(s - c)}$$

**EXAMPLE 5**  *Using Heron's formula to compute area*

Find the area of the triangle shown in Figure 25.

**SOLUTION**    The semiperimeter is

$$s = \frac{12 + 26 + 18}{2} = \frac{56}{2} = 28$$

Thus, using Heron's formula, we have

$$\text{Area} = \sqrt{s(s - 12)(s - 26)(s - 18)}$$
$$= \sqrt{28(28 - 12)(28 - 26)(28 - 18)}$$
$$= \sqrt{8960}$$
$$\approx 94.7 \text{ square inches}$$

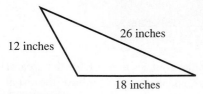

12 inches

26 inches

18 inches

**FIGURE 25**

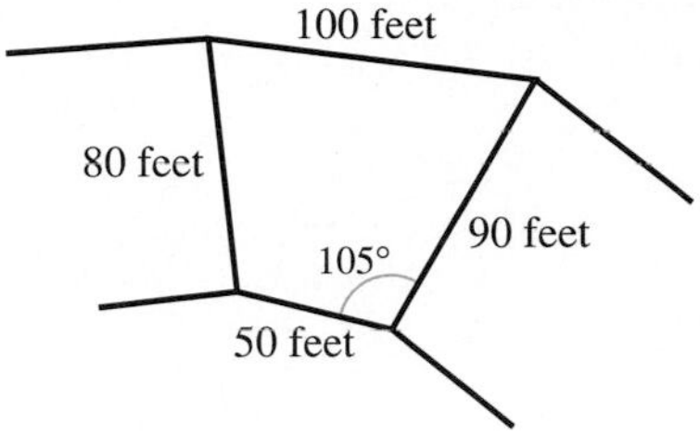

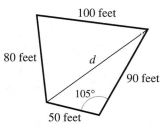

100 feet

80 feet

105°    90 feet

50 feet

**FIGURE 26**

**EXAMPLE 6**   *Finding the area of a quadrilateral*

A plot of land at the end of a cul-de-sac in a new subdivision has the shape of the quadrilateral shown in Figure 26. Find its area.

**SOLUTION**   By constructing a diagonal between two opposite vertices, we can compute the area as the sum of the areas of two triangles. Let $d$ denote the length of the diagonal from the lower left vertex to the upper right vertex, as shown in Figure 27. Using the Law of Cosines, we have

$$d^2 = 50^2 + 90^2 - 2(50)(90)\cos 105°$$

and so

$$d = \sqrt{50^2 + 90^2 - 2(50)(90)\cos 105°} \approx 113.7 \text{ feet}$$

100 feet

80 feet    $d$

105°    90 feet

50 feet

**FIGURE 27**

Now we apply Heron's formula to the triangle with side lengths 50, 90, and 113.7 feet to obtain

$$s = \frac{50 + 90 + 113.7}{2} = 126.85$$

and

$$\text{Area} = \sqrt{s(s - 50)(s - 90)(s - 113.7)}$$
$$= \sqrt{126.85(76.85)(36.85)(13.15)}$$
$$\approx 2173.4 \text{ square feet}$$

Similarly, for the triangle with side lengths 80, 100, and 113.7 feet, we have

$$s = \frac{80 + 100 + 113.7}{2} = 146.85$$

and

$$\text{Area} = \sqrt{s(s - 80)(s - 100)(s - 113.7)}$$
$$= \sqrt{146.85(66.85)(46.85)(33.15)}$$
$$\approx 3904.7 \text{ square feet}$$

Finally, adding the two areas together, we obtain the total area of

$$2173.4 + 3904.7 = 6078.1 \text{ square feet}$$

### EXERCISES 2

**EXERCISES 1–8** □ *Use the Law of Cosines to solve the given tri-angle. Assume angles A, B, and C, and sides a, b and c are labeled as shown in Figure 28.*

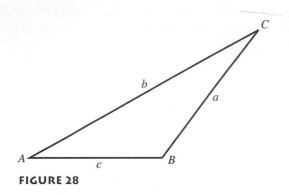

**FIGURE 28**

1. $a = 12$, $b = 9$, $C = 59°$

2. $a = 70$, $c = 85$, $B = 125°$

3. $a = 5$, $b = 8$, $c = 12$

4. $a = 12.5$, $b = 13.4$, $c = 14.7$

5. $b = 26$, $c = 4$, $A = 35°$

6. $a = 23$, $b = 17$, $A = 118°$

7. $b = 15$, $c = 20$, $B = 63°$

8. $a = 23$, $b = 19$, $C = 70°$

**EXERCISES 9–28** □ *Solve the given triangle by any means. Assume angles A, B, and C, and sides a, b, and c are labeled as shown in Figure 28.*

9. $b = 130$, $c = 180$, $A = 120°$

10. $a = 4.2$, $c = 3.8$, $B = 38°$

11. $a = 30$, $b = 35$, $c = 50$

12. $b = 28$, $c = 32$, $C = 165°$

13. $a = 23$, $B = 65°$, $C = 44°$

14. $a = 13$, $b = 15$, $c = 23$

15. $a = 26$, $b = 14$, $A = 115°$

16. $a = 0.12$, $c = 0.25$, $B = 130°$

17. $a = 1000$, $b = 2000$, $C = 27°$

18. $b = 36$, $A = 25°$, $C = 42°$

19. $a = 240$, $b = 175$, $c = 300$

20. $a = 56$, $c = 8$, $C = 155°$

21. $b = 14$, $c = 9$, $B = 28°$

22. $a = 1.7$, $b = 2.8$, $c = 3.5$

23. $a = 7$, $b = 25$, $A = 28°$

24. $b = 600$, $c = 500$, $B = 75°$

25. $a = 35$, $b = 70$, $c = 14$

26. $a = 80$, $A = 37°$, $B = 79°$

27. $b = 1.8$, $B = 104°$, $C = 39°$

28. $a = 140$, $b = 65$, $c = 73$

**EXERCISES 29–40** □ *Solve the given triangle by any means.*

29.
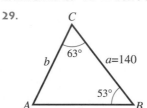

30.
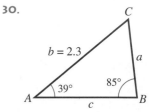

31.
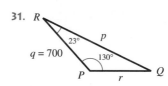

32.
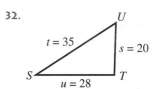

33.
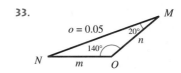

34.
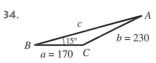

35.
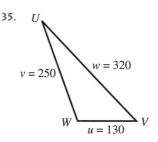

36.
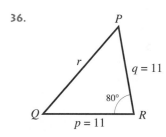

37.
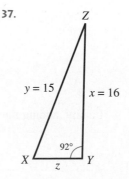

38.
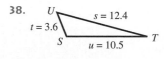

**39.**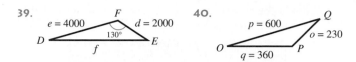

$e = 4000$   $F$   $d = 2000$
$D$   $130°$   $E$
$f$

**40.**

$p = 600$   $Q$
$o = 230$
$O$   $P$
$q = 360$

**41.** $a = 12$, $b = 14$, $c = 20$

**42.** $a = 10$, $b = 8$, $c = 5$

**43.** $b = 18$, $c = 15$, $B = 75°$

**44.** $a = 0.8$, $c = 1.6$, $C = 150°$

**EXERCISES 41–46** ☐ *Use Heron's formula to find the area of the given triangle. Assume angles A, B, and C, and sides a, b, and c are labeled as shown in Figure 28.*

**45.** $a = 1.4$, $b = 2.8$, $C = 16°$

**46.** $b = 120$, $c = 200$, $A = 138°$

■ *Applications*

**47.** *Ship Distance* Two ships leave port at the same time. One travels in the direction N 35° E at a rate of 25 miles per hour. The other travels in the direction S 70° E at a rate of 30 miles per hour. How far apart are the ships after 90 minutes?

**48.** *Hiking Distance* A hiker leaves a certain point and walks in the direction 60° (clockwise from north) for 7 miles and then in the direction 320° for 5 miles. If the hiker wishes to head directly back toward the starting point, in what direction should she head, and how far will she have to walk?

**49.** *Locating Lightning* Margaret and Elizabeth live 1 mile apart. While talking to each other on the phone during an electrical storm, they both notice a bolt of lightning in the sky between their two homes. Margaret hears the thunder 5 seconds after the flash, while Elizabeth hears the thunder 4 seconds after the flash. If the speed of sound is 1088 feet per second, describe the location of the lightning bolt relative to Margaret's position.

**50.** *Baseball Diamond* In a major-league baseball diamond, the bases form a square with side length 90 feet. The pitcher's mound is 60.5 feet from home plate. Find the distance from the pitcher's mound to each of the other three bases.

**51.** *Room Partition* A 10′ × 12′ × 8′ rectangular room is to be partitioned using a triangular piece of canvas. Each of the vertices of the triangle is to be anchored at three corners of the room, as shown in Figure 29. Find the area of the piece of canvas and the measure of each vertex angle.

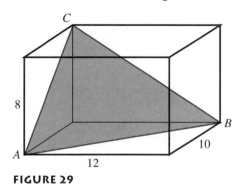

**FIGURE 29**

**52.** *Isosceles Triangle Area* An isosceles triangle is to be constructed with a perimeter of 24 inches. Let $x$ denote the lengths of the two equal legs.

a. Use Heron's formula to express the area of the triangle as a function of $x$. For what values of $x$ is this function defined?

b. Use a graphics calculator to plot the graph of the function you found in part (a). For what value of $x$ is the area largest? What do you notice about the resulting triangle?

**53.** *Minimizing Train Distances*  A passenger train traveling due west at a rate of 80 miles per hour passes a junction at 3:00 P.M. At the same instant, a freight train traveling 90 miles per hour is 100 miles away from the junction on a set of tracks that head straight toward the junction in the direction 140° (clockwise from north).

a. Find the distance between the trains at 3:30 P.M.

b. Find expressions for $p(t)$ and $f(t)$, the distances from the junction of the passenger train and freight train, respectively, $t$ hours after 3:00 P.M.

c. Use $p(t)$, $f(t)$, and the Law of Cosines to express the distance $d$ between the trains as a function of $t$.

d. Use a graphics calculator to estimate the time at which the trains are closest to each other.

---

### ■ *Project for Enrichment*

---

**54.** *Regular n-gons*  A **regular $n$-gon** is a polygon with $n$ sides of equal length. Some familiar examples are the regular 3-gon, also known as an equilateral triangle, the regular 4-gon, also known as a square, and the regular 5-gon, also known as a regular pentagon. In this project we will consider the area and perimeter of a regular $n$-gon that is inscribed in a circle of radius $r$. Our goal is to "discover" the area and circumference formulas for a circle, in much the same way as it was discovered by the ancient Greeks. We begin with a regular pentagon inscribed in a circle of radius 1, as shown in Figure 30.

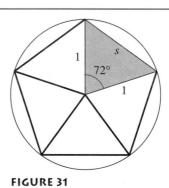

**FIGURE 31**

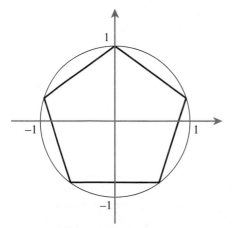

**FIGURE 30**

By constructing line segments from the center of the circle to each of the five vertices of the pentagon, we can divide the pentagon into five triangular regions of equal area (see Figure 31). The *central angle* of each of these five triangles is 360°/5 = 72°, and each of the sides adjacent to the central angles has length 1, the length of the radius.

a. Use the area formula from the previous section to find the area of the shaded triangle in Figure 31. Then find the area of the pentagon.

b. Use the Law of Cosines to find $s$, the length of one side of the pentagon. Then find the perimeter of the pentagon.

Now suppose a pentagon is inscribed in a circle of radius $r$. If we divide the pentagon into five triangles as we did above, then the central angles will still be 72°, but the sides adjacent to these angles will have length $r$.

c. Repeat parts (a) and (b) to find the area and perimeter of a pentagon inscribed in a circle of radius $r$.

Finally, we consider the more general case of a regular $n$-gon inscribed in a circle of radius $r$. Proceeding as above, we divide the $n$-gon into $n$ triangles of equal area. The central angle of each triangle will be 360°/$n$, and the sides adjacent to these angles will have length $r$ (see Figure 32).

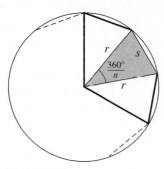

**FIGURE 32**

d. Find a formula for the area of the shaded triangle in Figure 32. Then find a formula for the area of the regular *n*-gon.

e. Find a formula for *s* in Figure 32. Then find a formula for the perimeter of the regular *n*-gon.

As the ancient Greeks observed many centuries ago, for very large values of *n*, the area and perimeter of a regular *n*-gon

closely approximate the area and circumference of a circumscribed circle. Thus, the area and circumference of a circle of radius *r* can be approximated using the formulas developed in parts (d) and (e).

f. Complete the following table for a regular *n*-gon inscribed in a circle of radius *r*.

| *n* | *10* | *100* | *1000* |
|---|---|---|---|
| Area | | | |
| Perimeter | | | |

What connection do you see between the values in this table and the area and circumference of a circle of radius *r*?

g. Describe how the area and perimeter formulas for a regular *n*-gon could be used to approximate $\pi$.

▪▟  *Questions for Discussion or Essay*

55. We have seen that an SSA triangle can be solved using either the Law of Sines or the Law of Cosines. Discuss the advantages and disadvantages of the two methods for solving SSA triangles. Which method do you prefer? Why?

56. The Law of Cosines can be described as a generalization of the Pythagorean Theorem. Explain why this is so.

57. When finding an unknown angle using the Law of Sines, one must eventually solve an equation of the form $\sin \theta = k$. On the other hand, when finding an unknown angle using the Law

of Cosines, one must eventually solve an equation of the form $\cos \theta = k$. Use these facts to explain why the Law of Sines can involve ambiguity while the Law of Cosines never does.

58. By now you should be familiar with three formulas for finding the area of a triangle, one involving the base and height, another involving two sides and the angle between, and the third involving three sides. Describe these three formulas in detail, discuss the advantages and disadvantages of each, and give real-world examples of circumstances in which each might be used.

**SECTION 3**

# TRIGONOMETRIC FORM OF COMPLEX NUMBERS

■ If a real number can be represented graphically as a point on a number line, what is the graphical representation of a complex number?

■ How can angles be useful for finding the product or quotient of complex numbers?

■ What famous mathematical hypothesis has eluded proof for over 130 years, in spite of overwhelming computer evidence?

■ How can complex numbers be used to find the equation of a rotated parabola?

## THE COMPLEX PLANE

Recall that the complex number system consists of all numbers that can be written in the form $a + bi$ where $a$ and $b$ are real numbers and $i = \sqrt{-1}$. Thus, each complex number can be associated with an ordered pair of real numbers $(a, b)$, which in turn can be represented graphically on a rectangular coordinate system known as the **complex plane**. The complex plane looks very much like the rectangular coordinate system we are already familiar with. However, instead of an $x$-axis and $y$-axis, the complex plane has a **real axis** and an **imaginary axis**. A complex number $a + bi$ is represented graphically as a point on the complex plane located above (or below) the number $a$ on the real axis and to the right (or left) of the number $b$ on the imaginary axis. For example, the number $3 + 2i$ is located above 3 on the real axis and to the right of 2 on the imaginary axis. Figure 33 shows $3 + 2i$ and several other points in the complex plane.

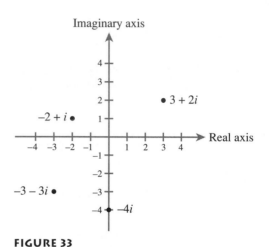

**FIGURE 33**

Just as with real numbers, the absolute value of a complex number can be viewed as its distance from zero. More precisely, the absolute value of $a + bi$ is its distance from the origin in the complex plane. The algebraic definition is as follows.

**Definition of absolute value**

> The **absolute value** of the complex number $a + bi$ is given by
> $$|a + bi| = \sqrt{a^2 + b^2}$$

**EXAMPLE 1**   *Finding the absolute value of a complex number*

Represent the given complex number graphically, and find its absolute value.
**a.** $3 - i$     **b.** $2i$

**SOLUTION**

**a.** The graphical representation of $3 - i = 3 + (-1)i$ is shown in Figure 34. Its absolute value is

$$|3 - i| = \sqrt{3^2 + (-1)^2} = \sqrt{9 + 1} = \sqrt{10}$$

**b.** The number $2i$ is shown on the imaginary axis in Figure 34. Its absolute value is

$$|2i| = \sqrt{2^2} = \sqrt{4} = 2$$

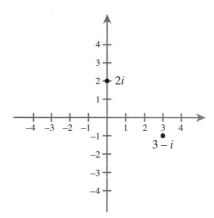

**FIGURE 34**

## TRIGONOMETRIC FORM OF COMPLEX NUMBERS

It is often convenient to write complex numbers in terms of sine and cosine rather than in the standard form $a + bi$. Consider the geometric representation of $a + bi$ shown in Figure 35. Here $\theta$ denotes the angle (measured counterclockwise) between the positive real axis and the line segment connecting $a + bi$ and the origin. Note also that $r = \sqrt{a^2 + b^2}$ is the length of the line segment from $a + bi$ to the origin. From the definitions of sine and cosine, we have

$$\cos \theta = \frac{a}{r} \qquad \text{and} \qquad \sin \theta = \frac{b}{r}$$

or

$$a = r \cos \theta \qquad \text{and} \qquad b = r \sin \theta$$

Thus,

$$a + bi = r \cos \theta + (r \sin \theta)i = r(\cos \theta + i \sin \theta)$$

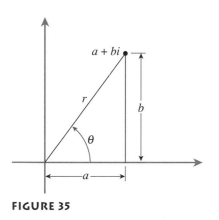

**FIGURE 35**

The expression $r(\cos \theta + i \sin \theta)$ is known as the trigonometric form of the complex number $a + bi$.

*Trigonometric form of a complex number*

The **trigonometric form** of a complex number $z = a + bi$ is

$$z = r(\cos \theta + i \sin \theta)$$

where $a = r \cos \theta$ and $b = r \sin \theta$. It follows that $r = \sqrt{a^2 + b^2}$ and $\tan \theta = b/a$. The number $r$ is called the **modulus** of $z$, and $\theta$ is called the **argument**.

Note that because there are infinitely many angles coterminal with a given argument $\theta$, the trigonometric form of a complex number is not unique. Typically, $\theta$ is taken to be an angle in the interval $[0, 2\pi)$.

**EXAMPLE 2**    *Writing complex numbers in trigonometric form*

Write the given complex number in trigonometric form.
a. $-5 + 5\sqrt{3}i$     b. $4 - i$

**SOLUTION**

a. The modulus of $-5 + 5\sqrt{3}i$ is

$$r = \sqrt{(-5)^2 + (5\sqrt{3})^2} = \sqrt{100} = 10$$

The argument can be found by noting that

$$\tan \theta = \frac{b}{a} = \frac{5\sqrt{3}}{-5} = -\sqrt{3}$$

Since $-5 + 5\sqrt{3}i$ is in the second quadrant and $\tan \theta = -\sqrt{3}$, we take $\theta = 120°$ or, in radians,

$$\theta = \frac{2\pi}{3}$$

Thus, the trigonometric form of $-5 + 5\sqrt{3}i$ is

$$10\left( \cos \frac{2\pi}{3} + i \sin \frac{2\pi}{3} \right)$$

The graphical representation is given in Figure 36.

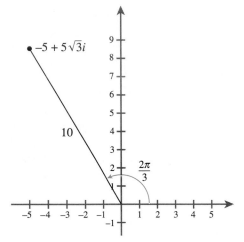

**FIGURE 36**

b. The modulus of $4 - i$ is

$$r = \sqrt{4^2 + (-1)^2} = \sqrt{17}$$

The argument $\theta$ is an angle in the fourth quadrant with

$$\tan \theta = \frac{b}{a} = -\frac{1}{4}$$

and so we take

$$\theta = \tan^{-1}\left( -\frac{1}{4} \right) \approx -14°$$

or, equivalently, $\theta = 346°$. Thus, the trigonometric form of $4 - i$ is approximately

$$\sqrt{17}[\cos 346° + i \sin 346°]$$

The graphical representation is given in Figure 37.

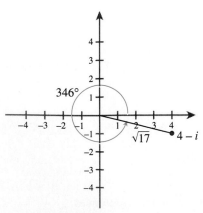

**FIGURE 37**

| **EXAMPLE 3** | *Writing a complex number in standard form* |

Write the number $2\left(\cos \dfrac{3\pi}{2} + i \sin \dfrac{3\pi}{2}\right)$ in standard form.

**SOLUTION**   Evaluating, we have

$$2\left(\cos \frac{3\pi}{2} + i \sin \frac{3\pi}{2}\right) = 2[0 + (-1)i] = -2i$$

The graphical representation is given in Figure 38.

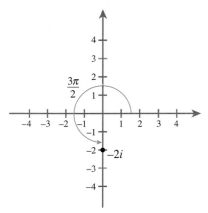

**FIGURE 38**

---

## MULTIPLICATION AND DIVISION OF COMPLEX NUMBERS

The trigonometric form of a complex number is particularly useful when computing products, quotients, powers, and roots of complex numbers. We will consider products and quotients in this section, and powers and roots in the next. Let $z_1$ and $z_2$ be complex numbers with trigonometric forms

$$z_1 = r_1(\cos \theta_1 + i \sin \theta_1) \qquad \text{and} \qquad z_2 = r_2(\cos \theta_2 + i \sin \theta_2)$$

Then the product of $z_1$ and $z_2$ is

$$
\begin{aligned}
z_1 z_2 &= r_1 r_2 (\cos \theta_1 + i \sin \theta_1)(\cos \theta_2 + i \sin \theta_2) \\
&= r_1 r_2 (\cos \theta_1 \cos \theta_2 + i \sin \theta_1 \cos \theta_2 + i \cos \theta_1 \sin \theta_2 + i^2 \sin \theta_1 \sin \theta_2) \\
&= r_1 r_2 [(\cos \theta_1 \cos \theta_2 - \sin \theta_1 \sin \theta_2) + i(\sin \theta_1 \cos \theta_2 + \cos \theta_1 \sin \theta_2)]
\end{aligned}
$$

Finally, applying the sum identities for sine and cosine, we have

$$z_1 z_2 = r_1 r_2 [\cos(\theta_1 + \theta_2) + i \sin(\theta_1 + \theta_2)]$$

In words, the product of $z_1$ and $z_2$ is the complex number whose modulus is the product of the moduli of $z_1$ and $z_2$ and whose argument is the sum of the arguments of $z_1$ and $z_2$. Using a similar procedure, we can show that the quotient of $z_1$ and $z_2$ is the complex number whose modulus is the quotient of the moduli of $z_1$ and $z_2$ and whose argument is the difference of the arguments of $z_1$ and $z_2$ (see Exercise 57). These facts are summarized as follows.

---

**Products and quotients in trigonometric form**

If $z_1 = r_1(\cos\theta_1 + i\sin\theta_1)$ and $z_2 = r_2(\cos\theta_2 + i\sin\theta_2)$, then

$$z_1 z_2 = r_1 r_2[\cos(\theta_1 + \theta_2) + i\,\sin(\theta_1 + \theta_2)]$$

$$\frac{z_1}{z_2} = \frac{r_1}{r_2}[\cos(\theta_1 - \theta_2) + i\,\sin(\theta_1 - \theta_2)], \qquad z_2 \neq 0$$

---

**EXAMPLE 4**     *Computing products and quotients of complex numbers*

Perform the indicated operation.

a. $4(\cos 18° + i\sin 18°) \cdot 3(\cos 47° + i\sin 47°)$     b. $\dfrac{3i}{2 - 2i}$

**SOLUTION**

a. According to the rule for computing products, we multiply the moduli and add the arguments to obtain

$$4(\cos 18° + i\,\sin 18°) \cdot 3(\cos 47° + i\,\sin 47°) = 4 \cdot 3[\cos(18° + 47°) + i\,\sin(18° + 47°)]$$
$$= 12(\cos 65° + i\,\sin 65°)$$

b. We will compute the quotient in two different ways, first using the algebraic method discussed in Section 1.3, and then by converting to trigonometric form.

**Method 1:** $\quad \dfrac{3i}{2 - 2i} = \dfrac{3i}{2 - 2i} \cdot \dfrac{2 + 2i}{2 + 2i} \qquad$ Multiplying by 1

$$= \frac{6i + 6i^2}{2^2 - (2i)^2} \qquad \text{Using the difference of squares formula}$$

$$= \frac{6i - 6}{4 - (-4)} \qquad \text{Using the fact that } i^2 = -1$$

$$= \frac{-6 + 6i}{8}$$

$$= -\frac{3}{4} + \frac{3}{4}i$$

**Method 2:** We first convert to trigonometric form. The numerator $3i$ is on the positive imaginary axis, so its modulus is 3 and its argument is 90° or $\pi/2$. Thus,

$$3i = 3\left(\cos\frac{\pi}{2} + i\,\sin\frac{\pi}{2}\right)$$

The denominator $2 - 2i$ is in the fourth quadrant with modulus $\sqrt{2^2 + (-2)^2} = 2\sqrt{2}$ and argument 315° or $7\pi/4$. So

$$2 - 2i = 2\sqrt{2}\left(\cos\frac{7\pi}{4} + i\,\sin\frac{7\pi}{4}\right)$$

Thus,

$$\frac{3i}{2-2i} = \frac{3\left(\cos\dfrac{\pi}{2} + i\,\sin\dfrac{\pi}{2}\right)}{2\sqrt{2}\left(\cos\dfrac{7\pi}{4} + i\,\sin\dfrac{7\pi}{4}\right)}$$

$$= \frac{3}{2\sqrt{2}}\left[\cos\left(\frac{\pi}{2} - \frac{7\pi}{4}\right) + i\,\sin\left(\frac{\pi}{2} - \frac{7\pi}{4}\right)\right]$$

$$= \frac{3\sqrt{2}}{4}\left[\cos\left(-\frac{5\pi}{4}\right) + i\,\sin\left(-\frac{5\pi}{4}\right)\right]$$

$$= \frac{3\sqrt{2}}{4}\left(-\frac{\sqrt{2}}{2} + \frac{\sqrt{2}}{2}\,i\right)$$

$$= -\frac{3}{4} + \frac{3}{4}\,i$$

## THE RIEMANN HYPOTHESIS

In 1859, the German mathematician Bernhard Riemann published an eight-page paper that is considered to be one of the most important mathematical papers ever published. In this work, Riemann stated six conjectures, or unproven statements, and went on to show how these could be used to prove the famous Prime Number Theorem, which roughly states that for large integers $N$ there are approximately $N/\log N$ prime numbers less than $N$. Since then, five of the six conjectures have been proven. The sixth has become known as the Riemann Hypothesis, and it is widely regarded as the major unsolved problem in pure mathematics. Ironically, the Prime Number Theorem itself was proven in 1896 by the French mathematician Jacques Hadamard using an alternative approach.

Bernhard Reimann

The Riemann Hypothesis involves the complex zeros of the **zeta function**

$$\zeta(z) = 1 + \frac{1}{2^z} + \frac{1}{3^z} + \frac{1}{4^z} + \cdots$$

Simply put, the Riemann Hypothesis asserts that every zero of the zeta function (with certain trivial exceptions) lies on the verti-

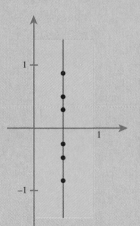

cal line in the complex plane passing through $\frac{1}{2}$ on the real axis as shown.

It has been proven that the zeta function has an infinite number of zeros and that all of the important ones lie in the shaded strip shown in the figure. Moreover, it has been shown using a computer that over 70 million solutions lie on the line $z = \frac{1}{2}$. However, in spite of this evidence, no one has proven that *all* solutions lie on this line.

Why is a proof of this hypothesis so desirable that it "*would instantaneously catapult its fortunate creator to the summit of the mathematical world*"? * In large part, the significance of this hypothesis derives from the stature of the mathematical giants who have attempted but failed to prove it. Moreover, a proof would automatically lead to more precise information about the distribution of prime numbers, a problem that has fascinated and frustrated mathematicians for centuries.

*Douglas M. Campbell and John C. Higgins, *Mathematics: People, Problems, Results* (Wadsworth: Belmont, CA, 1984), p. 150.

**EXERCISES 3**

**EXERCISES 1–10** ☐ *Find the absolute value of the given complex number.*

1. $3 + 4i$

2. $-8 - 6i$

3. $1 + 3i$

4. $-1 - i$

5. $1 - \sqrt{3}\, i$

6. $\sqrt{7} + 3i$

7. $-15$

8. $\pi i$

9. $-\dfrac{3}{5} + \dfrac{4}{5} i$

10. $\dfrac{5}{13} + \dfrac{2}{13} i$

**EXERCISES 11–20** ☐ *Represent the complex number graphically and write it in trigonometric form.*

11. $1 + i$

12. $2\sqrt{3} + 2i$

13. $-4i$

14. $3.5$

15. $-6$

16. $3i$

17. $1 - \sqrt{3}i$

18. $8 - 8i$

19. $-2 - 3i$

20. $3 + i$

**EXERCISES 21–34** ☐ *Represent the complex number graphically and write it in the form $a + bi$.*

21. $4(\cos 90° + i \sin 90°)$

22. $3\left(\cos \dfrac{\pi}{6} + i \sin \dfrac{\pi}{6}\right)$

23. $2(\cos 120° + i \sin 120°)$

24. $0.5\left(\cos \dfrac{3\pi}{2} + i \sin \dfrac{3\pi}{2}\right)$

25. $9\left(\cos \dfrac{3\pi}{4} + i \sin \dfrac{3\pi}{4}\right)$

26. $5[\cos(-270°) + i \sin(-270°)]$

27. $1[\cos(-\pi) + i \sin(-\pi)]$

28. $\pi(\cos 240° + i \sin 240°)$

29. $6(\cos 380° + i \sin 380°)$

30. $10(\cos 4 + i \sin 4)$

31. $\sqrt{2}(\cos 100° + i \sin 100°)$

32. $\sqrt{3}\left(\cos \dfrac{11\pi}{3} + i \sin \dfrac{11\pi}{3}\right)$

33. $\dfrac{3}{2}(\cos 2 + i \sin 2)$

34. $7(\cos 340° + i \sin 340°)$

**EXERCISES 35–48** ☐ *Perform the indicated operation.*

35. $2(\cos 30° + i \sin 30°) \cdot 3(\cos 40° + i \sin 40°)$

36. $(3 - i)(4 + 2i)$

37. $(7 - i)/(1 + 2i)$

38. $\dfrac{16(\cos 135° + i \sin 135°)}{2(\cos 15° + i \sin 15°)}$

39. $\dfrac{\sqrt{5}\left(\cos \dfrac{\pi}{4} + i \sin \dfrac{\pi}{4}\right)}{2\sqrt{5}\left(\cos \dfrac{\pi}{3} + i \sin \dfrac{\pi}{3}\right)}$

40. $\sqrt{2}(\cos 83° + i \sin 83°) \cdot 2\sqrt{2}(\cos 97° + i \sin 97°)$

41. $(-5 + 3i)(1 - i)$

42. $\dfrac{13}{2 + 3i}$

43. $\dfrac{3 + 5i}{2 - 4i}$

44. $(2 + 4i)(2 - 4i)$

45. $5\left(\cos \dfrac{3\pi}{4} + i \sin \dfrac{3\pi}{4}\right) \cdot 3\left(\cos \dfrac{\pi}{2} + i \sin \dfrac{\pi}{2}\right)$

46. $\dfrac{27(\cos 20° + i \sin 20°)}{3(\cos 110° + i \sin 110°)}$

47. $\dfrac{4.2(\cos 3\pi + i \sin 3\pi)}{0.6\left(\cos \dfrac{\pi}{4} + i \sin \dfrac{\pi}{4}\right)}$

48. $0.8(\cos 2 + i \sin 2) \cdot 1.2(\cos 4 + i \sin 4)$

**EXERCISES 49–56** ☐ *Perform the indicated operation by first converting to trigonometric form. Check your answer by performing the operation without converting.*

49. $(2 - 2i)(-3 + 3i)$

50. $\dfrac{6}{1 - \sqrt{3}i}$

51. $\dfrac{-8 - 8i}{2 + 2i}$

52. $3i(4 + 4i)$

53. $(\sqrt{3} + i)(2 - 2\sqrt{3}i)$

54. $\dfrac{-4 + 4\sqrt{3}i}{-1 + i}$

55. $\dfrac{2i}{4 - 3i}$

56. $(2 - i)(2 + i)$

57. Show that if $z_1 = r_1(\cos \theta_1 + i \sin \theta_1)$ and $z_2 = r_2(\cos \theta_2 + i \sin \theta_2)$, then

$$\frac{z_1}{z_2} = \frac{r_1}{r_2}[\cos(\theta_1 - \theta_2) + i \sin(\theta_1 - \theta_2)], \; z_2 \neq 0$$

(*Hint:* Multiply the numerator and denominator of $z_1/z_2$ by $\cos \theta - i \sin \theta$.)

**58.** Show that if $z = r(\cos \theta + i \sin \theta)$, then

$$\frac{1}{z} = \frac{1}{r} (\cos \theta - i \sin \theta)$$

(*Hint:* Write the numerator 1 in trigonometric form.)

**59.** Show that the complex conjugate of

$$z = r(\cos \theta + i \sin \theta) \quad \text{is} \quad \bar{z} = r[\cos(-\theta) + i \sin(-\theta)]$$

**60.** Show that if $z = r(\cos \theta + i \sin \theta)$, then

$$-z = r[\cos(\theta + \pi) + i \sin(\theta + \pi)]$$

---

## ◢ *Projects for Enrichment*

**61.** *Quaternions* The quaternions were discovered in 1843 by the Irish mathematician William Hamilton (1805–1865). He initially envisioned them as a tool for the exploration of physical space. This was not to be; the eventual development of vectors (see Section 5) all but eliminated the need for quaternions as an application tool. However, the quaternions are still of interest as an extension of the complex number system.

A quaternion is an expression of the form $a + bi + cj + dk$ where $a$, $b$, $c$, and $d$ are real numbers. We will see that $i$, $j$, and $k$ have properties very much like the imaginary number $i$. Equality, addition, and multiplication for quaternions are defined as follows:

(1)   $(a_1 + a_2i + a_3j + a_4k) = (b_1 + b_2i + b_3j + b_4k)$
   if and only if $a_n = b_n$ for $n = 1, 2, 3, 4$

(2)   $(a_1 + a_2i + a_3j + a_4k) + (b_1 + b_2i + b_3j + b_4k)$
       $= (a_1 + b_1) + (a_2 + b_2)i$
         $+ (a_3 + b_3)j + (a_4 + b_4)k$

(3)   $(a_1 + a_2i + a_3j + a_4k)(b_1 + b_2i + b_3j + b_4k)$
       $= (a_1b_1 - a_2b_2 - a_3b_3 - a_4b_4)$
         $+ (a_1b_2 + a_2b_1 + a_3b_4 - a_4b_3)i$
         $+ (a_1b_3 - a_2b_4 + a_3b_1 + a_4b_2)j$
         $+ (a_1b_4 + a_2b_3 - a_3b_2 + a_4b_1)k$

Don't be too concerned with the complexity of these statements. The first two suggest that quaternions can be treated much like polynomials in the variables $i$, $j$, and $k$. To see that this carries through to multiplication also, we consider certain multiplicative properties of $i$, $j$, and $k$.

**a.** Show that $i^2 = j^2 = k^2 = -1$. (*Hint:* $i$ can be written as $0 + 1i + 0j + 0k$. Now use the rule for multiplying quaternions to show that $i^2 = i \cdot i = -1$. Repeat for $j$ and $k$.)

**b.** Show that the following hold.

$$ij = k, \qquad jk = i, \qquad ki = j$$
$$ji = -k \qquad kj = -i, \qquad ik = -j$$

Is quaternion multiplication commutative? Why or why not?

The products in part (b) are easy to remember if you use the sequence

$$i, j, k, i, j, k$$

The product from left to right of two adjacent elements is the next one to the right. The product from right to left of two adjacent elements is the negative of the next one to the left.

**c.** Show that $ijk = jki = kij = -1$ while $kji = ikj = jik = 1$.

**d.** Derive formula (3) using the multiplicative properties of $i$, $j$, and $k$, as shown in part (b).

**62.** *Rotated Conic Sections* In Chapter 3 we studied the equations of the conic sections: circles, ellipses, parabolas, and hyperbolas. In each case, the conic sections we dealt with were oriented with the coordinate axes. In this project, we will consider more general conic sections for which the axes are "tilted."

The following development rests heavily on the identification between the point $z = x + iy$ in the complex plane with the point $(x, y)$ in the Cartesian plane.

**a.** Figure 39 shows a complex number $z = r(\cos \theta + i \sin \theta)$ and the complex number $z'$ obtained by rotating $z$ by a clockwise angle of measure $\alpha$. Explain why

$$z' = r[\cos(\theta + \alpha) + i \sin(\theta + \alpha)]$$

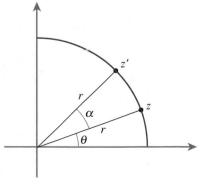

**FIGURE 39**

b. Use the multiplication formula to show that
$z' = (\cos \alpha + i \sin \alpha) \cdot z$.

c. Let $z = 1 + i$. Use the results of part (b) to find the complex numbers $z'$ obtained by rotating $z$ counterclockwise by the indicated angle $\alpha$. Express $z'$ in the form $a + bi$.

   i. $\alpha = \dfrac{\pi}{2}$        ii. $\alpha = \dfrac{\pi}{4}$

   iii. $\alpha = \pi$        iv. $\alpha = -\dfrac{\pi}{2}$

d. Starting with the formula $z' = (\cos \alpha + i \sin \alpha) \cdot z$, show that $z = (\cos \alpha - i \sin \alpha)z'$.

e. Suppose that $z = x + iy$ and $z' = x' + iy'$, where $z$ and $z'$ are as in (d). Use the results of part (d) to find formulas for $x$ and $y$ in terms of $x'$ and $y'$.

f. Figure 40 shows the graph of the parabola $P: y = x^2$ and also the graph of the parabola $P'$ obtained by rotating $P$ clockwise by 45°. Note that each point $z'$ on the graph of $P'$ can be obtained by rotating a point $z$ on the parabola $P$ through an angle of 45°. Use the formula obtained in part (e) to express $x$ and $y$ in terms of $x'$ and $y'$. Next, substitute these expressions into the equation for $P$ to obtain an equation for $P'$.

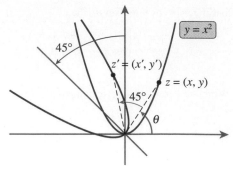

**FIGURE 40**

g. Repeat the process outlined in part (f) to find the equation of the ellipse $E'$ obtained by rotating an ellipse $E$ with equation

$$\frac{x^2}{4} + \frac{y^2}{9} = 1$$

counterclockwise by an angle of $\pi/3$. Sketch a graph of both $E$ and $E'$ on the same set of coordinate axes.

h. Let $C$ be a circle of radius $r$ centered at the origin, and $C'$ the circle obtained by rotating $C$ by a counterclockwise angle of $\theta$. Show that $C$ and $C'$ have the same equation. Explain.

---

◼◼   *Questions for Discussion or Essay*

63. Write the complex numbers

$$2\left(\cos \frac{2\pi}{3} + i \sin \frac{2\pi}{3}\right) \quad \text{and} \quad 2\left(\cos \frac{8\pi}{3} + i \sin \frac{8\pi}{3}\right)$$

in standard form. What does this say about the trigonometric form of a complex number? What if we restricted the argument $\theta$ to be an angle between 0 and $2\pi$?

64. Describe the set of complex numbers $z$ for which $|z| = 2$. In general, what can be said about the set $\{z \mid z$ is complex and $|z| = r\}$ for any $r > 0$?

65. If $z = r(\cos \theta + i \sin \theta)$, find $|z|$.

66. Describe the set of complex numbers $z$ for which the argument of $z$ is 45°. In general, what can be said about the set $\{z \mid z$ is complex and the argument of $z$ is $\theta\}$ for any $0 \leq \theta \leq 2\pi$?

---

**SECTION 4**

# POWERS AND ROOTS OF COMPLEX NUMBERS

◼ How can $(1 + i)^{16}$ be computed using only two multiplications and a power of 2?

◼ How can you draw a pentagon, a hexagon, or even a dodecagon on a graphics calculator?

◼ If $i$ is the square root of $-1$, then what is the square root of $i$?

◼ How can complex numbers be used to prove trigonometric identities?

## POWERS OF COMPLEX NUMBERS

We saw in the previous section that it is easy to find products and quotients of complex numbers given in trigonometric form. The same is true for finding powers. To see this, let $z = r(\cos\theta + i\sin\theta)$ and consider the following powers of $z$:

$$z^2 = r(\cos\theta + i\sin\theta) \cdot r(\cos\theta + i\sin\theta)$$
$$= r^2[\cos(\theta + \theta) + i\sin(\theta + \theta)]$$
$$= r^2(\cos 2\theta + i\sin 2\theta)$$
$$z^3 = z \cdot z^2$$
$$= r(\cos\theta + i\sin\theta) \cdot r^2(\cos 2\theta + i\sin 2\theta)$$
$$= r^3(\cos 3\theta + i\sin 3\theta)$$

Continuing in this manner, we obtain

$$z^4 = r^4(\cos 4\theta + i\sin 4\theta)$$
$$z^5 = r^5(\cos 5\theta + i\sin 5\theta)$$
$$\vdots$$

This pattern generalizes to the following theorem, named after the French mathematician Abraham de Moivre.

**De Moivre's Theorem**

If $z = r(\cos\theta + i\sin\theta)$ is a complex number and $n$ is a positive integer, then

$$z^n = [r(\cos\theta + i\sin\theta)]^n = r^n(\cos n\theta + i\sin n\theta)$$

**EXAMPLE 1**    *Finding powers of complex numbers*

Use de Moivre's Theorem to expand the given power.

**a.** $[3(\cos 15° + i\sin 15°)]^5$        **b.** $(\sqrt{3} - i)^6$

**SOLUTION**

**a.** Applying de Moivre's Theorem, we obtain

$$[3(\cos 15° + i\sin 15°)]^5 = 3^5[\cos(5 \cdot 15°) + i\sin(5 \cdot 15°)]$$
$$= 243(\cos 75° + i\sin 75°)$$

**b.** We first convert $\sqrt{3} - i$ to trigonometric form. The modulus and argument are computed as follows:

$$r = \sqrt{(\sqrt{3})^2 + (-1)^2} = 2$$

$$\tan\theta = \frac{-1}{\sqrt{3}} \text{ and so } \theta = \frac{11\pi}{6} \qquad \text{Using the fact that } \sqrt{3} - i \text{ is in the fourth quadrant}$$

Thus,

$$\sqrt{3} - i = 2\left(\cos\frac{11\pi}{6} + i\sin\frac{11\pi}{6}\right)$$

Now we apply de Moivre's Theorem:

$$(\sqrt{3} - i)^6 = \left[2\left(\cos\frac{11\pi}{6} + i\sin\frac{11\pi}{6}\right)\right]^6$$

$$= 2^6\left[\cos\left(6 \cdot \frac{11\pi}{6}\right) + i\sin\left(6 \cdot \frac{11\pi}{6}\right)\right]$$

$$= 64[\cos 11\pi + i\sin 11\pi)$$

$$= 64(-1 + 0i)$$

$$= -64$$

## ROOTS OF COMPLEX NUMBERS

The notion of an $n$th root of a real number is a familiar one. For example, 2 is a fourth root of 16 since $2^4 = 16$. Likewise, $-2$ is a fourth root of 16 since $(-2)^4 = 16$. Imaginary numbers also have $n$th roots. For example, $4i$ is a cube root of $-64i$ since $(4i)^3 = 64i^3 = -64i$. In general, an $n$th root of a complex number is defined as follows.

***Definition of nth root of a complex number***

A complex number $w$ is an **$n$th root** of a complex number $z$ if $w^n = z$.

It is a consequence of the Fundamental Theorem of Algebra that every complex number has $n$ $n$th roots. Some examples are given in Table 2.

TABLE 2

| Complex number | $n$ | $n$th roots |
|:---:|:---:|:---|
| 25 | 2 | $5, -5$ |
| $-64i$ | 3 | $4i, 2\sqrt{3} - 2i, -2\sqrt{3} - 2i$ |
| 16 | 4 | $2, -2, 2i, -2i$ |

Note that each $n$th root in Table 2 can be verified by computing its $n$th power, using de Moivre's Theorem if necessary. For example, after converting $2\sqrt{3} - 2i$ to trigonometric form, we can apply de Moivre's Theorem to verify that it is a third root of $-64i$ as follows:

$$(2\sqrt{3} - 2i)^3 = \left[ 4\left( \cos \frac{11\pi}{6} + i \sin \frac{11\pi}{6} \right) \right]^3$$

$$= 4^3 \left[ \cos\left( 3 \cdot \frac{11\pi}{6} \right) + i \sin\left( 3 \cdot \frac{11\pi}{6} \right) \right]$$

$$= 64\left( \cos \frac{11\pi}{2} + i \sin \frac{11\pi}{2} \right)$$

$$= 64(0 - i)$$

$$= -64i$$

More generally, if

$$z = r(\cos \theta + i \sin \theta) \qquad \text{and} \qquad w = \sqrt[n]{r}\left( \cos \frac{\theta}{n} + i \sin \frac{\theta}{n} \right)$$

then de Moivre's Theorem gives us

$$w^n = \left[ \sqrt[n]{r}\left( \cos \frac{\theta}{n} + i \sin \frac{\theta}{n} \right) \right]^n = r(\cos \theta + i \sin \theta) = z$$

So $w = \sqrt[n]{r}[\cos(\theta/n) + i \sin(\theta/n)]$ is an $n$th root of $z$. But how do we find the other $n - 1$ roots? The key is to recognize that the trigonometric form of a complex number is not unique. Indeed, adding any integer multiple of $2\pi$ to $\theta$ will yield another form for $z$. Thus, for any integer $k$,

$$z = r[\cos(\theta + 2\pi k) + i \sin(\theta + 2\pi k)]$$

and so

$$w_k = \sqrt[n]{r}\left( \cos \frac{\theta + 2\pi k}{n} + i \sin \frac{\theta + 2\pi k}{n} \right)$$

is an $n$th root of $z$. For $k = 0, 1, 2, \ldots, n - 1$, we obtain distinct roots; other integer values for $k$ will lead to $n$th roots that have already been obtained. These observations lead us to the following statement.

■ ***Formula for nth roots of a complex number***

> For any positive integer $n$, the $n$ distinct $n$th roots of the complex number $z = r(\cos \theta + i \sin \theta)$ are given by
>
> $$\sqrt[n]{r}\left( \cos \frac{\theta + 2\pi k}{n} + i \sin \frac{\theta + 2\pi k}{n} \right)$$
>
> where $k = 0, 1, 2, \ldots, n - 1$.

The $n$th roots of a complex number have a nice geometric interpretation. For example, consider the three cube roots of $-64i$ from Table 2, namely $4i$, $2\sqrt{3} - 2i$, and $-2\sqrt{3} - 2i$. The graphical representations of these roots are shown in Figure 41. Notice that the points are equally spaced around a circle of radius 4 centered at the origin. This is a consequence of the fact that the $n$th roots of $z = r(\cos\theta + i\sin\theta)$ all have modulus $\sqrt[n]{r}$ and their arguments differ by $2\pi/n$. Thus, the $n$ points will be equally spaced around a circle of radius $\sqrt[n]{r}$ centered at the origin.

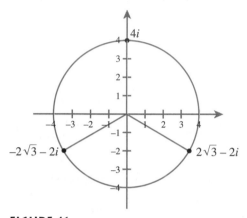

**FIGURE 41**

**EXAMPLE 2** | *Finding the nth roots of a real number*

Find the three cube roots of 1 and represent them graphically.

**SOLUTION** We first write 1 in trigonometric form as

$$1(\cos 0 + i\sin 0)$$

Next, we apply the $n$th root formula with $r = 1$, $\theta = 0$, and $n = 3$, to obtain roots of the form

$$\sqrt[3]{1}\left(\cos\frac{0 + 2\pi k}{3} + i\sin\frac{0 + 2\pi k}{3}\right) = \cos\frac{2\pi k}{3} + i\sin\frac{2\pi k}{3}$$

Finally, setting $k = 0, 1, 2$, we compute the three cube roots as follows:

$$\text{For } k = 0: \quad \cos 0 + i\sin 0 = 1$$

$$\text{For } k = 1: \quad \cos\frac{2\pi}{3} + i\sin\frac{2\pi}{3} = -\frac{1}{2} + \frac{\sqrt{3}}{2}i$$

$$\text{For } k = 2: \quad \cos\frac{4\pi}{3} + i\sin\frac{4\pi}{3} = -\frac{1}{2} - \frac{\sqrt{3}}{2}i$$

The graphical representations of these roots are shown in Figure 42.

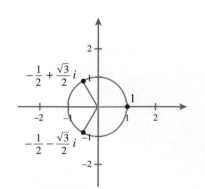

**FIGURE 42**

When the *n*th roots of 1 arise in advanced areas of mathematics, they are usually referred to as the ***n*th roots of unity**. They have the useful property that they are equally spaced around the unit circle, starting at 1 on the real axis.

**EXAMPLE 3**    *Finding the nth roots of a complex number*

Find the five fifth roots of $16\sqrt{2}(1 - i)$ and represent them graphically.

**SOLUTION**    Since $16\sqrt{2}(1 - i)$ has modulus $r = 32$ and argument $\theta$ in the fourth quadrant satisfying $\tan \theta = -1$, its trigonometric form is

$$32(\cos 315° + i \sin 315°)$$

From the *n*th root formula with $k = 0$, we see that one of the fifth roots is

$$\sqrt[5]{32}\left(\cos \frac{315°}{5} + i \sin \frac{315°}{5}\right) = 2(\cos 63° + i \sin 63°)$$

Now the five roots are equally spaced around a circle centered at the origin, so the arguments of ''consecutive'' roots must differ by $360°/5 = 72°$. Thus, we simply add $72°$ to the argument of the first root to find the second, add $72°$ to the argument of the second to find the third, and so on. The five roots are as follows, and their graphical representations are shown in Figure 43.

$$2(\cos 63° + i \sin 63°) \approx 0.9080 + 1.7820i$$

$$2(\cos 135° + i \sin 135°) \approx -1.4142 + 1.4142i$$

$$2(\cos 207° + i \sin 207°) \approx -1.7820 - 0.9080i$$

$$2(\cos 279° + i \sin 279°) \approx 0.3129 - 1.9754i$$

$$2(\cos 351° + i \sin 351°) \approx 1.9754 - 0.3129i$$

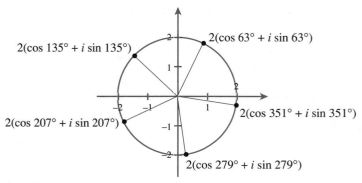

**FIGURE 43**

In Chapter 5, we saw that polynomials of degree $n$ have $n$ complex zeros (counting multiplicities). In other words, polynomial equations of degree $n$ have $n$ complex solutions. As we demonstrate in the following example, the $n$th root formula is useful for finding solutions of polynomial equations of the form

$$x^n = c$$

where $c$ is a complex number.

**EXAMPLE 4**     *Finding solutions of polynomial equations*

Find all solutions of the equation $16x^4 + 1 = 0$ and represent them graphically.

**SOLUTION**     The equation $16x^4 + 1 = 0$ is equivalent to

$$x^4 = -\frac{1}{16}$$

and so we must find the fourth roots of $-\frac{1}{16}$. We write $-\frac{1}{16}$ in trigonometric form as

$$\frac{1}{16} (\cos \pi + i \sin \pi)$$

and then apply the $n$th root formula with $r = \frac{1}{16}$, $\theta = \pi$, $n = 4$, and $k = 0$ to obtain the root

$$\sqrt[4]{\frac{1}{16}} \left( \cos \frac{\pi}{4} + i \sin \frac{\pi}{4} \right) = \frac{1}{2} \left( \frac{\sqrt{2}}{2} + \frac{\sqrt{2}}{2} i \right) = \frac{\sqrt{2}}{4} (1 + i)$$

Since the four roots must be equally spaced around a circle centered at the origin, their arguments must be $2\pi/4 = \pi/2$ radians apart. Thus, the other three roots are

$$\frac{1}{2} \left( \cos \frac{3\pi}{4} + i \sin \frac{3\pi}{4} \right) = \frac{1}{2} \left( -\frac{\sqrt{2}}{2} + \frac{\sqrt{2}}{2} i \right) = \frac{\sqrt{2}}{4} (-1 + i)$$

$$\frac{1}{2} \left( \cos \frac{5\pi}{4} + i \sin \frac{5\pi}{4} \right) = \frac{1}{2} \left( -\frac{\sqrt{2}}{2} - \frac{\sqrt{2}}{2} i \right) = \frac{\sqrt{2}}{4} (-1 - i)$$

$$\frac{1}{2} \left( \cos \frac{7\pi}{4} + i \sin \frac{7\pi}{4} \right) = \frac{1}{2} \left( \frac{\sqrt{2}}{2} - \frac{\sqrt{2}}{2} i \right) = \frac{\sqrt{2}}{4} (1 - i)$$

The graphical representations of the roots are shown in Figure 44.

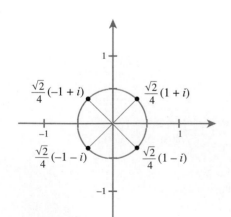

**FIGURE 44**

# EXERCISES 4

**EXERCISES 1–10** □ *Use de Moivre's Theorem to expand the given power. Express the answer in the form used in the statement of the problem.*

1. $(2 \cos 30° + i \sin 30°)^5$

2. $\left( \cos \dfrac{\pi}{6} + i \sin \dfrac{\pi}{6} \right)^{12}$

3. $[\sqrt{2}(\cos 10° + i \sin 10°)]^8$

4. $\left[ \sqrt[5]{7} \left( \cos \dfrac{\pi}{5} + i \sin \dfrac{\pi}{5} \right) \right]^5$

5. $(1 + i)^{16}$

6. $\left( \dfrac{1}{2} - \dfrac{\sqrt{3}}{2} i \right)^{15}$

7. $(-\sqrt{3} + i)^7$

8. $(\sqrt{2} - \sqrt{2} i)^5$

9. $(1 + 2i)^6$

10. $(3 - i)^{12}$

**EXERCISES 11–22** □ *Find the indicated roots, express them in the form a + bi, and represent them graphically.*

11. Square roots of $25(\cos 60° + i \sin 60°)$

12. Square roots of $\dfrac{1}{16} (\cos \pi + i \sin \pi)$

13. Fourth roots of $\dfrac{1}{16} \left( \cos \dfrac{2\pi}{3} + i \sin \dfrac{2\pi}{3} \right)$

14. Fifth roots of $100{,}000(\cos 225° + i \sin 225°)$

15. Square roots of $36i$

16. Square roots of $-49$

17. Cube roots of $27$

18. Cube roots of $-16i$

19. Sixth roots of $1$

20. Sixth roots of $-64$

21. Cube roots of $-\dfrac{\sqrt{2}}{16} + \dfrac{\sqrt{2}}{16} i$

22. Fourth roots of $-\dfrac{1}{2} - \dfrac{\sqrt{3}}{2} i$

**EXERCISES 23–28** □ *Find all solutions of the equation and represent them graphically.*

23. $x^4 + 1 = 0$

24. $8x^3 - 1 = 0$

25. $x^3 + i = 0$

26. $x^4 - 16i = 0$

27. $x^2 - (2 + 2\sqrt{3}i) = 0$

28. $x^3 + 27 - 27i = 0$

29. Write $f(x) = x^6 + 64$ as a product of linear factors.

30. Write $f(x) = x^6 - 1$ as a product of linear factors.

31. Estimate the fourth roots of $i$.

32. Estimate the fourth roots of $1 + \sqrt{3}i$.

33. Show that the cube roots of unity add up to zero.

34. Show that the fourth roots of unity add up to zero.

## ▪ *Projects for Enrichment*

35. ***Regular n-gons on a Graphics Calculator*** In this project we will devise a technique for drawing a regular *n*-gon (i.e., a polygon with *n* sides of equal length) on a graphics calculator. Let's first consider a regular 6-gon, also known as a hexagon. Our strategy will be to find the sixth roots of unity and then use these six points as the vertices of a hexagon.

    a. Show that the sixth roots of unity correspond to the six points in Figure 45.

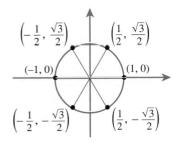

**FIGURE 45**

If we connect consecutive pairs of points as we move around the unit circle, we obtain the polygon shown in Figure 46.

   b. How do we know that the polygon shown in Figure 46 is a hexagon?

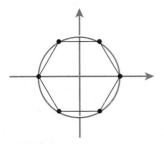

**FIGURE 46**

Next we use the line drawing feature of a graphics calculator to draw line segments between consecutive pairs of points. With one popular brand of graphics calculator, a line segment can be drawn between two points $(x_1, y_1)$ and $(x_2, y_2)$ by entering `Line(x₁,y₁,x₂,y₂)`. Thus, `Line(1,0,.5,√3/2)` would draw a line segment between points $(1, 0)$ and $(1/2, \sqrt{3}/2)$. Entering similar `Line` commands for the other five pairs of points would produce the hexagon shown in Figure 47. Note that it is necessary to use square range settings to avoid distorting the hexagon.

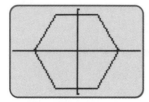

**FIGURE 47**

c. Find the eighth roots of unity and use them to draw an octagon. Write out the commands you used with your calculator to plot the eight line segments.

d. Repeat part (c) to draw a pentagon. Note that you will need to approximate most of the coordinates of the points.

e. If you have access to a programmable calculator, write a program similar to the one given here that requests a value for $n$ and then proceeds to draw a regular $n$-gon. Be sure your calculator is set in radians mode and select square range settings.

| | |
|---|---|
| `Prgm1:NGON` | |
| `:Disp "NUMBER` | Input number of sides |
| `OF SIDES"` | |
| `:Input N` | |
| `:ClrDraw` | |
| `:1 → A` | First vertex is (1, 0) |
| `:0 → B` | |
| `:1 → K` | Start with $K = 1$ |
| `:Lbl 1` | Begin loop |
| `:2πK/N → T` | Compute angle $T$ for next root |
| `:cos T → C` | Use $T$ to find the |
| `:sin T → D` | next vertex $(C, D)$ |
| `:Line(A,B,C,D)` | Draw a line between |
| | vertices $(A, B)$ and $(C, D)$ |
| `:C → A` | Rename vertex $(C, D)$ to $(A, B)$ |
| `:D → B` | |
| `:IS>(K,N)` | Increment $K$ and stop if $K > N$ |
| `:Goto 1` | End loop |

**36.** *Euler's Formula and Applications* In this project we will give Euler's formula for complex numbers, one of the most useful and profound formulas in all of mathematics. Then we will show that de Moivre's Theorem and the sum and difference identities for sine and cosine follow from Euler's formula.

Euler's formula (named in honor of the great Swiss mathematician Leonhard Euler), states that for real numbers $\theta$,

$$e^{i\theta} = \cos\theta + i\sin\theta$$

a. Apply Euler's formula to compute $e^{i\theta}$ for each of the following values of $\theta$.

   i. $\dfrac{\pi}{2}$    ii. $\dfrac{\pi}{4}$    iii. $\pi$    iv. $-\dfrac{\pi}{2}$

Suppose that a complex number $z$ has trigonometric form

$$z = r(\cos\theta + i\sin\theta)$$

Then according to Euler's formula, $z$ has the following exponential form:

$$z = re^{i\theta}$$

b. Write each of the following complex numbers in the form $z = re^{i\theta}$.

   i. $1 + i$    ii. $3i$    iii. $1 - i$    iv. $-4$

c. De Moivre's Theorem states that

$$[r(\cos\theta + i\sin\theta)]^n = r^n(\cos n\theta + i\sin n\theta)$$

   i. Express $\cos\theta + i\sin\theta$ in exponential form using Euler's formula.
   ii. Express $\cos n\theta + i\sin n\theta$ in exponential form using Euler's formula.
   iii. Prove de Moivre's Theorem using your results from parts (i) and (ii) and exponential properties.

d. Let $z_1 = e^{i\theta_1}$ and $z_2 = e^{i\theta_2}$.
   i. Explain why $z_1 z_2 = e^{i(\theta_1 + \theta_2)}$. Then use Euler's formula to write $e^{i(\theta_1 + \theta_2)}$ in terms of the sine and cosine of $\theta_1 + \theta_2$.
   ii. Use Euler's formula to write both $z_1$ and $z_2$ in the form $a + bi$. Then multiply these expressions to obtain the product $z_1 z_2$ in terms of sines and cosines of $\theta_1$ and $\theta_2$.
   iii. Use the results of (i) and (ii) to obtain the sum identities for sine and cosine.
   iv. Repeat the process described above to the quotient $z_1/z_2$ to derive the difference identities for sine and cosine.

### ■ *Questions for Discussion or Essay*

**37.** In this section we considered positive integer powers of complex numbers. How should we define $z^{-n}$ if $z$ is a complex number and $n > 0$? Select a particular nonreal complex number $z$ and a positive integer $n$ and show that your definition of $z^{-n}$ can be written as a complex number in standard form. Do you think this can always be done? Why or why not?

**38.** Describe a process by which the sixth roots of unity can be represented geometrically without applying the $n$th root for-

mula. Can your process be extended to locate the $n$th roots of unity geometrically for any $n$? Explain.

**39.** Euler's formula (see Exercise 36) states that $e^{i\theta} = \cos\theta + i\sin\theta$. For $\theta = \pi$, we obtain the formula $e^{i\pi} = -1$. This has been described as one of the most beautiful formulas in mathematics. Why do you suppose mathematicians would find this little formula so compelling?

---

## SECTION 5

# POLAR COORDINATES

> ■ In what sense is a spiral the graph of a function?
>
> ■ What mathematical function has a graph shaped like flower petals?
>
> ■ How can a single point in the plane have infinitely many coordinates?
>
> ■ What kind of symmetry do roses have?

We have seen that any point in the plane is completely determined by specifying two numbers: its $x$- and $y$-coordinates. In particular, a point $P$ with coordinates $(x, y)$ is located $x$ units to the left or right of the vertical axis and $y$ units above or below the horizontal axis. As we will see in this section, the location of the same point $P$ could also be given in terms of *polar coordinates* $(r, \theta)$, in much the same way as we described complex numbers in trigonometric form.

### ■ POLAR COORDINATE SYSTEM

O •————————————————▶
Pole                                Polar axis

**FIGURE 48**

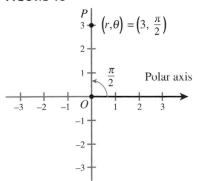

**FIGURE 49**

The **polar coordinate system** is defined in terms of a fixed point $O$, called the **origin** or **pole**, and a half-line or ray emanating from $O$, called the **polar axis**. The polar axis is usually horizontal and directed toward the right, as shown in Figure 48. To each point $P$ in the plane we assign **polar coordinates** $(r, \theta)$, where $r$ is the directed distance from $O$ to $P$, and $\theta$ is the directed angle between the polar axis and the segment $\overline{OP}$. Thus, for example, the point $P$ with rectangular coordinates $(0, 3)$ can be expressed as the ordered pair $(3, \pi/2)$ in polar coordinates since $P$ lies 3 units from the origin and $\pi/2$ is the angle made between the polar axis $\overline{OP}$, as shown in Figure 49.

There is an important distinction between rectangular and polar coordinates. Although there is only one way of specifying a point in rectangular coordinates, there are infinitely many ways of specifying a point in polar coordinates. This ambiguity arises from two different sources. First of all, the angle $\theta$ associated with a point $P$ is a *directed* angle from the polar axis to the segment $\overline{OP}$, which essentially says that $\theta$ is an angle that, in standard position, has terminal side $\overline{OP}$. But as we have already seen, there are infinitely many angles coterminal with a given angle. In the case of the point $P(0, 3)$, not only could $\theta$ be taken to be $\pi/2$, it could also be taken to be $(\pi/2) + 2\pi$, $(\pi/2) + 4\pi$, $-3\pi/2$, and so on. By convention we agree that any point of the form $(0, \theta)$ describes the origin. In other words, at the pole, $\theta$ can be anything!

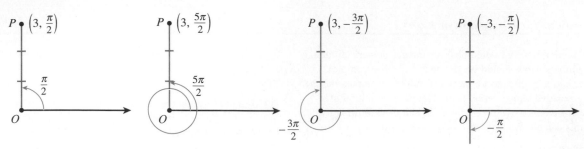

**FIGURE 50**

Secondly, since *r* is a *directed* distance (as opposed to a "distance"), it may be negative. For negative values of *r*, the distance is measured along the line segment formed by extending the terminal side of $\theta$ to the opposite side of *O*. (By this logic, a point 2 miles to the north could be described as "−2 miles to the south.") Figure 50 shows the point *P* described in several different ways.

When working with the polar coordinate system, it is convenient to use a grid of concentric circles with the pole at the common center, the polar axis directed to the right, and line segments radiating from the pole at regular angular intervals, as shown in Figure 51.

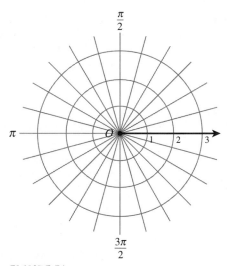

**FIGURE 51**

**EXAMPLE 1**     *Plotting points in polar coordinates*

Plot the points whose polar coordinates are given.

a. $\left(2, \dfrac{\pi}{6}\right)$     b. $\left(3, -\dfrac{2\pi}{3}\right)$     c. $\left(-\dfrac{5}{2}, \dfrac{3\pi}{4}\right)$

**SOLUTION**     The points are shown in Figures 52–54.

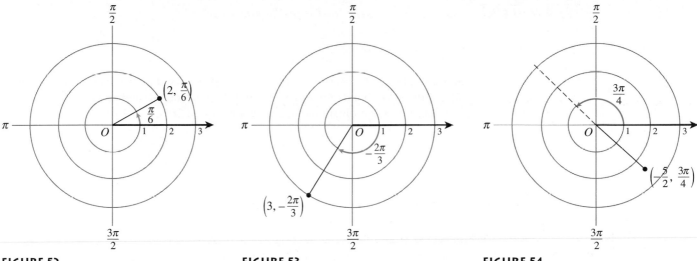

**FIGURE 52**                    **FIGURE 53**                    **FIGURE 54**

A **polar equation** is an equation relating $r$ and $\theta$. The graph of a polar equation consists of the collection of all points $(r, \theta)$ such that $r$ and $\theta$ satisfy the equation. If an equation is of the form $r = r(\theta)$, then we say that the equation defines a **polar function**. Just as the simplest and most useful categories of rectangular functions have been given special names and studied in great detail (polynomial, rational, exponential, logarithmic, etc.), so too have many polar functions been classified and explored in great detail. There are polar functions bearing names such as lemniscate, cardioid, limaçon, spiral, and rose. We will avoid such taxonomy in this section; instead our focus will be on general techniques that apply to all polar functions.

**EXAMPLE 2**    *Sketching the graph of a polar equation*

Sketch the graph of $r = 4$.

**SOLUTION**    Here $\theta$ is unrestricted, and so we are interested in the collection of points for which $r$, the distance to the origin, is 4. This is just a circle centered at the origin with radius 4, as shown in Figure 55.

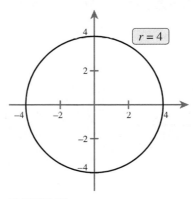

**FIGURE 55**

**EXAMPLE 3**    *Sketching the graph of a polar equation*

Sketch the graph of $\theta = \dfrac{\pi}{3}$.

**SOLUTION**    In this case $r$ is unrestricted and $\theta$ is fixed at $\pi/3$. Thus, we are looking for the set of points for which $\theta$ is $\pi/3$. By considering both positive and negative values of $r$, we obtain the line shown in Figure 56.

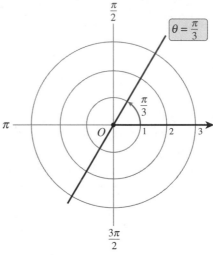

**FIGURE 56**

In each of the previous two examples, we were able to sketch the graph by using a straightforward analysis of the polar equation. However, in cases where the equation is complex and unfamiliar, this may not be possible. In such cases, we may be able to shed some light on the graph by plotting a few points, as we do in the following examples.

**EXAMPLE 4**    *Sketching the graph of a spiral*

Sketch the graph of $r = \theta$.

**SOLUTION**    We begin by making a table of values.

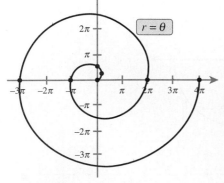

**FIGURE 57**

| $\theta$ | 0 | $\dfrac{\pi}{4}$ | $\dfrac{\pi}{2}$ | $\pi$ | $2\pi$ | $3\pi$ | $4\pi$ |
|---|---|---|---|---|---|---|---|
| $r$ | 0 | $\dfrac{\pi}{4}$ $\approx 0.79$ | $\dfrac{\pi}{2}$ $\approx 1.57$ | $\pi$ $\approx 3.14$ | $2\pi$ $\approx 6.28$ | $3\pi$ $\approx 9.42$ | $4\pi$ $\approx 12.57$ |

Plotting these points and connecting with a smooth curve, we obtain the spiral shown in Figure 57.

### EXAMPLE 5    *Sketching a polar graph*

Sketch the graph of $r = 2 \sin \theta$.

**SOLUTION**    Again, we will make a table of values and plot points. After plotting these points and connecting them with a smooth curve, we obtain the curve shown in Figure 58. In Exercise 47 we will confirm that this graph is, in fact, a circle.

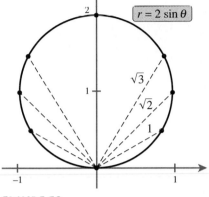

$r = 2 \sin \theta$

**FIGURE 58**

| $\theta$ | 0 | $\dfrac{\pi}{6}$ | $\dfrac{\pi}{4}$ | $\dfrac{\pi}{3}$ | $\dfrac{\pi}{2}$ | $\dfrac{\pi}{3}$ | $\dfrac{\pi}{4}$ | $\dfrac{\pi}{6}$ | $\pi$ |
|---|---|---|---|---|---|---|---|---|---|
| $r$ | 0 | 1 | $\sqrt{2}$ | $\sqrt{3}$ | 2 | $\sqrt{3}$ | $\sqrt{2}$ | 1 | 0 |

Some graphs can be fairly difficult to plot, in part because of the complications arising from negative $r$ values. The next example illustrates the process by which a complicated polar graph can be plotted.

### EXAMPLE 6    *Sketching a polar graph*

Sketch the graph of $r = 3 \cos 2\theta$.

**SOLUTION**    Because of the complexity of producing a plot of this graph, we will form it in several stages. We begin by plotting points for values of $\theta$ between 0 and $\pi/4$, and connecting them with a smooth curve to obtain the graph shown in Figure 59.

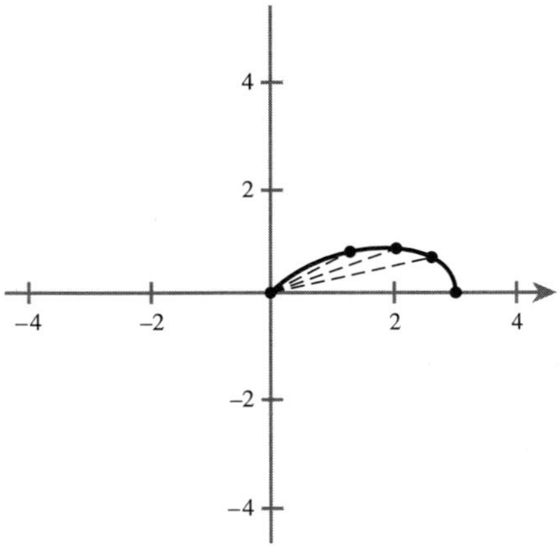

**FIGURE 59**

| $\theta$ | 0 | $\dfrac{\pi}{12}$ | $\dfrac{\pi}{8}$ | $\dfrac{\pi}{6}$ | $\dfrac{\pi}{4}$ |
|---|---|---|---|---|---|
| $r = 3 \cos 2\theta$ | 3 | $\dfrac{3\sqrt{3}}{2} \approx 2.60$ | $\dfrac{3\sqrt{2}}{2} \approx 2.12$ | $\dfrac{3}{2} = 1.5$ | 0 |

Next, we consider points for which $\theta$ lies between $\pi/4$ and $\pi/2$.

| $\theta$ | $\dfrac{\pi}{4}$ | $\dfrac{\pi}{3}$ | $\dfrac{3\pi}{8}$ | $\dfrac{5\pi}{12}$ | $\dfrac{\pi}{2}$ |
|---|---|---|---|---|---|
| $r = 3 \cos 2\theta$ | 0 | $-\dfrac{3}{2} = -1.5$ | $-\dfrac{3\sqrt{2}}{2} \approx -2.12$ | $-\dfrac{3\sqrt{3}}{2} \approx -2.60$ | $-3$ |

Notice that each of these points has a negative $r$ value. Thus, even though the angles $\theta$ all have terminal sides in the first quadrant, the corresponding points lie in the third

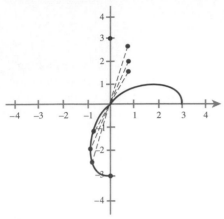

**FIGURE 60**

quadrant, as shown in Figure 60. If this process is continued to include all values of $\theta$ between 0 and $2\pi$, we obtain the graph shown in Figure 61, called a four-petaled rose.

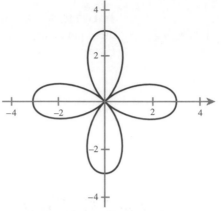

**FIGURE 61**

Some graphics calculators have the capability of plotting polar curves. In the following box, we describe the general procedure for plotting such curves. See the graphics calculator supplement for details on how to do this with your calculator.

Polar curves of the form $r = f(\theta)$ can be plotted on some graphics calculators by setting the graph mode to **polar**, entering the expression $f(\theta)$ for $r$, selecting an appropriate range for $\theta$ (the interval $0 \le \theta \le 2\pi$ works well for many polar graphs), selecting an appropriate viewing window (your calculator's "square" setting would be a good one to try), and then plotting the graph.

**EXAMPLE 7**    *Plotting a polar curve with a graphics calculator*

Use a graphics calculator to plot the graphs of $r = 2\cos\theta$ and $r = 1 + 2\sin\theta$. Estimate the points of intersection.

**SOLUTION**    We first set the graph mode to polar and then enter $r = 2\cos\theta$ and $r = 1 + 2\sin\theta$. Next, we select a range for $\theta$ by setting $\theta$min=0 and $\theta$max=$2\pi$. Finally, we select our calculator's "square" viewing rectangle and zoom-in to obtain the graph shown in Figure 62. The graph of $r = 2\cos\theta$ is a circle of radius 1 centered at $(1, 0)$. The graph of $r = 1 + 2\sin\theta$ is known as a limaçon with an inner loop. One of the intersection points is clearly at the pole. The other two can be estimated using the trace feature, as shown in Figures 63 and 64. They are approximately $(1.8, 0.4)$ and $(0.8, 1.1)$ in polar coordinates.

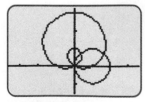

**FIGURE 62**

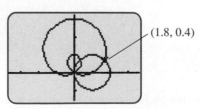

(1.8, 0.4)

**FIGURE 63**

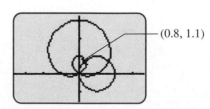

(0.8, 1.1)

**FIGURE 64**

## CONVERSION

At times it is desirable to convert from one coordinate system to another. The following relationships between rectangular and polar coordinates are easily obtained by considering Figure 65 for the case that $\theta$ is acute and $r$ is positive. With little effort, it can be shown that these relationships hold for all $\theta$.

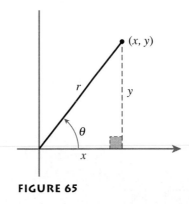

**FIGURE 65**

**Coordinate relationships**

The rectangular coordinates $(x, y)$ and polar coordinates $(r, \theta)$ of a point $P$ are related as follows:

(1)
$$x = r \cos \theta, \quad y = r \sin \theta$$

(2)
$$r^2 = x^2 + y^2, \quad \tan \theta = \frac{y}{x}, \, x \neq 0$$

The equations (1) enable us to find the rectangular coordinates of a point if the polar coordinates are known.

**EXAMPLE 8**    *Converting from polar to rectangular coordinates*

Convert the given point from polar to rectangular coordinates.

a. $\left(2, \dfrac{\pi}{6}\right)$      b. $\left(3, -\dfrac{2\pi}{3}\right)$

**SOLUTION**

a. We have $r = 2$ and $\theta = \pi/6$. Thus,

$$x = 2 \cos \frac{\pi}{6} \qquad\qquad y = 2 \sin \frac{\pi}{6}$$

$$= 2 \cdot \frac{\sqrt{3}}{2} \qquad\qquad = 2 \cdot \frac{1}{2}$$

$$= \sqrt{3} \qquad\qquad\qquad = 1$$

and so the rectangular coordinates are $\left(\sqrt{3}, 1\right)$.

**b.** With $r = 3$ and $\theta = -2\pi/3$, we have

$$x = 3 \cos\left(-\frac{2\pi}{3}\right) \qquad\qquad y = 3 \sin\left(-\frac{2\pi}{3}\right)$$

$$= 3 \cdot \left(-\frac{1}{2}\right) \qquad\qquad\quad = 3 \cdot \left(-\frac{\sqrt{3}}{2}\right)$$

$$= -\frac{3}{2} \qquad\qquad\qquad\quad = -\frac{3\sqrt{3}}{2}$$

Thus, the rectangular coordinates are $(-3/2, -3\sqrt{3}/2)$.

It is slightly more difficult to convert from rectangular to polar coordinates because of the non-uniqueness of polar coordinates. The value for $r$ can be determined using the equation $r^2 = x^2 + y^2$. An angle $\theta$ can be found by using the equation $\tan \theta = y/x$ and knowledge of the quadrant in which $\theta$ lies.

**EXAMPLE 9**    *Converting from rectangular to polar coordinates*

Convert the given point from rectangular to polar coordinates.

**a.** $\left(\dfrac{1}{3}, \dfrac{1}{3}\right)$      **b.** $(-2, 2\sqrt{3})$      **c.** $(-1, -3)$

**SOLUTION**

**a.** We begin by finding $r$.

$$r = \sqrt{x^2 + y^2}$$

$$= \sqrt{\left(\frac{1}{3}\right)^2 + \left(\frac{1}{3}\right)^2}$$

$$= \sqrt{\frac{2}{9}} = \frac{\sqrt{2}}{3}$$

From the fact that $\tan \theta = y/x$, we have

$$\tan \theta = \frac{1/3}{1/3} = 1$$

Thus, $\theta$ is an angle in the first quadrant with tangent equal to 1, and it follows that $\theta = \pi/4$. The point in polar coordinates is $(\sqrt{2}/3, \pi/4)$.

**b.** We find $r$ and $\tan \theta$ as in part (a).

$$r = \sqrt{(-2)^2 + (2\sqrt{3})^2}$$

$$= \sqrt{4 + 12}$$

$$= 4$$

$$\tan \theta = \frac{2\sqrt{3}}{-2}$$

$$= -\sqrt{3}$$

Since $\tan(\pi/3) = \sqrt{3}$, we conclude that $\theta$ has reference angle $\pi/3$. Since the point is in the second quadrant, we have $\theta = 2\pi/3$. Thus, the point is $(4, 2\pi/3)$ in polar coordinates.

c. Once again, $r$ and $\tan \theta$ are easily found.

$$r = \sqrt{(-1)^2 + (-3)^2}$$
$$= \sqrt{1 + 9}$$
$$\approx 3.1623$$

$$\tan \theta = \frac{-3}{-1}$$
$$= 3$$

Thus, $\theta$ is an angle in the third quadrant with reference angle $\arctan 3 \approx 1.2490$. It follows that

$$\theta = \pi + \arctan 3 \approx 4.3906$$

and so the point $(-1, -3)$ has approximate polar coordinates $(3.1623, 4.3906)$.

It is often the case that an equation expressed in terms of one coordinate system is much more complex than when expressed in terms of the other system. In the following examples, we see how it is possible to convert equations from one system to the other.

**EXAMPLE 10**   *Converting an equation from rectangular to polar form*

Find a polar equation for the line $y = mx + b$.

**SOLUTION**   Replacing $x$ with $r \cos \theta$ and $y$ with $r \sin \theta$, we have

$$r \sin \theta = m(r \cos \theta) + b$$

Solving for $r$ gives us

$$r \sin \theta - mr \cos \theta = b$$
$$r(\sin \theta - m \cos \theta) = b$$
$$r = \frac{b}{\sin \theta - m \cos \theta}$$

**EXAMPLE 11**   *Converting an equation from polar to rectangular form*

Convert the polar equation $r = 2 \cos \theta$ to an equivalent equation in $x$ and $y$.

**SOLUTION**   We will exploit the relation $x = r \cos \theta$ by multiplying both sides by $r$:

$$r = 2 \cos \theta$$
$$r^2 = 2r \cos \theta$$

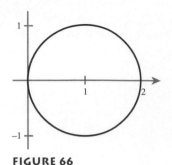

**FIGURE 66**

Thus,

$$x^2 + y^2 = 2x$$

This, by the way, is the equation of a circle of radius 1 centered at (1, 0). To see this, we complete the square as follows:

$$x^2 + y^2 = 2x$$
$$x^2 - 2x + y^2 = 0$$
$$(x^2 - 2x + 1) + y^2 = 1$$
$$(x - 1)^2 + y^2 = 1^2$$

The graph is shown in Figure 66.

■

## EXERCISES 5

**EXERCISES 1–10** □ *Plot the point given in polar coordinates and find the corresponding rectangular coordinates.*

1. $\left(2, -\dfrac{\pi}{2}\right)$     2. $\left(3, \dfrac{3\pi}{2}\right)$

3. $\left(1, \dfrac{3\pi}{4}\right)$     4. $\left(-4, \dfrac{\pi}{4}\right)$

5. $\left(-2, \dfrac{\pi}{6}\right)$     6. $\left(3, -\dfrac{\pi}{3}\right)$

7. $\left(\dfrac{3}{2}, -\dfrac{5\pi}{6}\right)$     8. $\left(2\sqrt{2}, \dfrac{7\pi}{4}\right)$

9. $(-3, -\pi)$     10. $\left(-\dfrac{5}{2}, -\dfrac{7\pi}{2}\right)$

**EXERCISES 11–20** □ *Plot the point given in rectangular coordinates and find two sets of polar coordinates with $0 \le \theta < 2\pi$.*

11. $(4, 0)$     12. $(0, -3)$

13. $(2, 2)$     14. $(-3, 3)$

15. $(1, -\sqrt{3})$     16. $(-2\sqrt{3}, 2)$

17. $(-4, -3)$     18. $(12, -5)$

19. $(3, 1)$     20. $(-2, 4)$

**EXERCISES 21–34** □ *Sketch the graph of the given polar equation.*

21. $r = 3$     22. $r = -2$

23. $\theta = -\dfrac{\pi}{3}$     24. $\theta = \dfrac{5\pi}{6}$

25. $r = 2\theta$     26. $r = 1 + \theta$

27. $r = 3 \sin \theta$     28. $r = -2 \cos \theta$

29. $r = 2 - 4 \cos \theta$     30. $r = 2 \sin \theta + 2$

31. $r = \sin 2\theta$     32. $r = \cos 4\theta$

33. $r = 2 \cos 3\theta$     34. $r = 3 \sin \dfrac{\theta}{2}$

**EXERCISES 35–42** □ *Convert the given rectangular equation to polar form.*

35. $y = 3$     36. $x = -1$

37. $y + x = 0$     38. $y = \sqrt{3}\, x$

39. $x^2 + y^2 = 9$     40. $x^2 - y^2 = 1$

41. $\ln y - \ln x = 1$     42. $e^{x^2} e^{y^2} = 2$

**EXERCISES 43–50** □ *Convert the given polar equation to rectangular form.*

43. $r = 7$     44. $\theta = \dfrac{\pi}{4}$

45. $r \cos \theta = 1$     46. $r \sin \theta = 4r \cos \theta + 2$

47. $r = 2 \sin \theta$     48. $r = 2 \cos \theta + 3 \sin \theta$

49. $r = \cos 2\theta$     50. $r = \sin 2\theta$

**EXERCISES 51–56** □ *Use a graphics calculator to plot the graphs of the given polar equations. Estimate the coordinates of the point(s) of intersection.*

51. $r = 2$
    $r = 4 \sin \theta$

52. $r = 2 \sin \theta$
    $r = 4 \cos \theta$

53. $r = 2 \cos \theta$
    $r = 1 - \sin \theta$

54. $r = 3$
    $r = 2 + 3 \cos \theta$

55. $r = \sin 2\theta$
    $r = \sin \theta$

56. $r = \cos 3\theta$
    $r = \sin \theta$

### ■ *Project for Enrichment*

**57.** *Rose Curves* In this project we will explore **rose curves**, the graphs of polar equations of the form $r = a \sin n\theta$ or $r = a \cos n\theta$, with $n \geq 2$. Because of the great number of polar graphs involved in this project, a graphics calculator with polar capabilities would be especially useful.

**a.** Graph each of the following rose curves.

   **i.** $r = \cos 2\theta$        **ii.** $r = 3 \cos 5\theta$

   **iii.** $r = 2 \sin 3\theta$        **iv.** $r = 2 \sin 4\theta$

**b.** Indicate the number of petals for each of the rose curves graphed in part (a). In general, how many petals does the graph of $a \cos n\theta$ (or $a \sin n\theta$) have?

**c.** For each of the rose curves graphed in part (a), find the angle between the line segments joining the pole to the tips of adjacent petals. How does this value depend on $n$?

**d.** For each of the polar functions $r = f(\theta)$ given in part (a), sketch the graph of

$$r = f\left(\theta - \frac{\pi}{6}\right) \quad \text{and} \quad r = f\left(\theta + \frac{\pi}{4}\right)$$

In general, describe the relationship between the graphs of $r = f(\theta)$ and $r = f(\theta - \alpha)$.

**e.** It is clear that rose curves enjoy **rotational symmetry**—that is, their graphs remain the same after rotation through certain angles. For each of the graphs produced in part (a), estimate the smallest angle $\alpha$ for which the graph would appear the same after rotating by $\alpha$.

**f.** For each of the following graphs, find a corresponding polar equation $r = f(\theta)$.

   **i.**

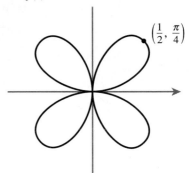

$$\left(\frac{1}{2}, \frac{\pi}{4}\right)$$

   **ii.**

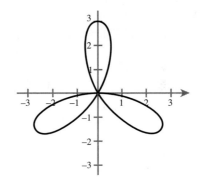

### ■■ *Questions for Discussion or Essay*

**58.** Compare and contrast the trigonometric form of complex numbers discussed in Section 9.3 with the polar coordinate representation of points in the plane.

**59.** The graph of $r = 4 \sin 5\theta$ is given in Figure 67. Explain how a graphics calculator could be used to find the point farthest from the origin, without graphing in polar coordinates.

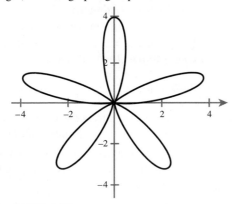

**FIGURE 67**

**60.** The graph of $r = -\sin^2 \theta$ is shown in Figure 68. Does the point with polar coordinates $(1, \pi/2)$ appear to be on the graph? Compute $r$ for $\theta = \pi/2$. Is the point $(1, \pi/2)$, in fact, on the graph of $r = -\sin^2 \theta$? Explain.

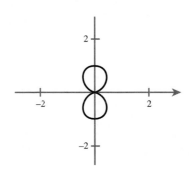

**FIGURE 68**

# SECTION 6

## VECTORS

■ What is the difference between speed and velocity?

■ How does constant wind velocity affect the path of the British-French Concorde?

■ Is there any benefit to a tug-of-war team splitting into two groups that pull in slightly different directions?

The physical quantities that we have encountered so far—time, distance, mass, and volume, to name just a few—can all be characterized using single numerical values. Thus, we might refer to a distance of 250 feet or to an object with a mass of 5 kilograms. Such quantities are referred to as **scalar** quantities; the measure of a scalar quantity is called its **magnitude**.

However, there are other physical quantities that cannot be represented with a single numerical value. Two important examples are velocity and force, both of which involve not only a magnitude, but also a **direction**. In the case of velocity, we might refer to a plane traveling with a velocity of 100 miles per hour in a northeast direction. Or we might speak of a force of 50 pounds being exerted on an object at angle of 30° with the horizontal. Quantities such as velocity and force that involve both a magnitude and a direction are called **vector** quantities.

A vector quantity is normally represented by a **directed line segment**, such as the one shown in Figure 69. The directed line segment $\overrightarrow{PQ}$ in Figure 69 has an **initial point** $P$ and a **terminal point** $Q$. The arrow indicates that the vector quantity acts in the direction from $P$ to $Q$, and the length of the segment gives the magnitude of the vector quantity. Two directed line segments are said to be **equivalent** if they have the same magnitude and direction. In Figure 70, we see a collection of directed line segments that are equivalent to $\overrightarrow{PQ}$. We use the term **vector** to refer to the common magnitude and direction shared by a set of equivalent directed line segments. In other words, a vector with a given magnitude and direction may be represented by infinitely many directed line segments, each with a different initial point. We will denote vectors by lowercase, boldface letters such as **u**, **v**, and **w**. For convenience, we will often draw vectors in **standard position** with their initial point at the origin. In Figure 71, we illustrate the standard position of the vector **v** corresponding to $\overrightarrow{PQ}$.

When a vector is represented in standard position, its magnitude and direction are completely determined by the terminal point. Thus, if a vector **v** has initial point $(0, 0)$ and terminal point $(v_1, v_2)$, we define the **component form** of the vector **v** by

$$\mathbf{v} = \langle v_1, v_2 \rangle$$

The $x$- and $y$-coordinates of the terminal point are referred to as the $x$- and $y$-**components** of the vector **v**. A vector **w** with initial point $P(x_1, y_1)$ and terminal point $Q(x_2, y_2)$ can be placed in standard position by translating $\overrightarrow{PQ}$ so that its initial point is at the origin. In so doing, the terminal point must be translated $x_1$ units horizontally and $y_1$ units vertically. Thus, we obtain the component form

$$\mathbf{w} = \langle x_2 - x_1, y_2 - y_1 \rangle$$

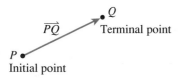

$\overrightarrow{PQ}$    $Q$  Terminal point

$P$
Initial point

**FIGURE 69**

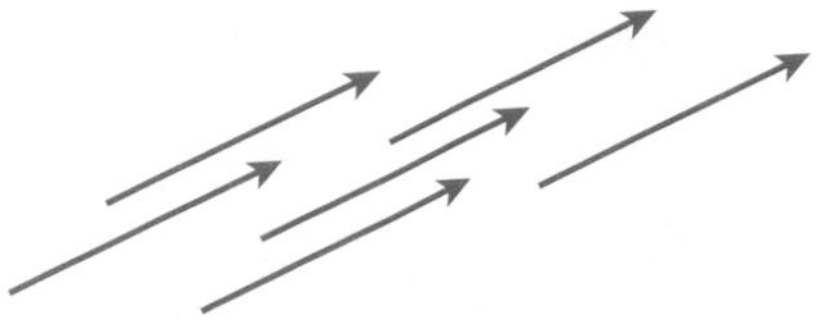

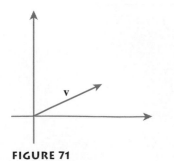

**FIGURE 70**

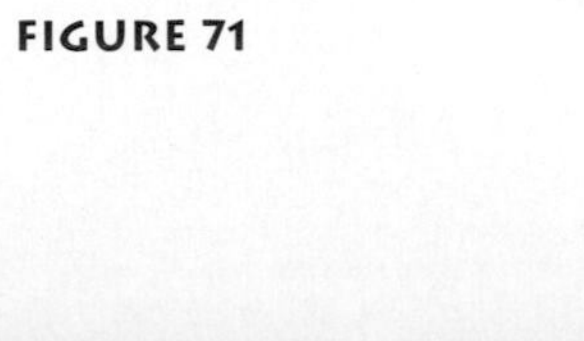

**v**

**FIGURE 71**

The component form of a vector is unique in the sense that if $\mathbf{u} = \langle u_1, u_2 \rangle$ and $\mathbf{v} = \langle v_1, v_2 \rangle$, then $\mathbf{u} = \mathbf{v}$ if and only if $u_1 = v_1$ and $u_2 = v_2$.

**EXAMPLE 1**  *Sketching vectors*

Given the points $P(-1, -2)$ and $Q(-3, 1)$,
**a.** sketch the directed line segment $\overrightarrow{PQ}$,
**b.** find the component form of the vector $\mathbf{v}$ represented by $\overrightarrow{PQ}$, and
**c.** sketch the vector $\mathbf{v}$ in standard position.

**SOLUTION**

**a.** We first locate the points $P(-1, -2)$ and $Q(-3, 1)$, and then draw a directed line segment from the initial point $P$ to the terminal point $Q$. The resulting vector $\overrightarrow{PQ}$ is shown in Figure 72.

**b.** To find the component form of $\mathbf{v}$, we subtract the *x*- and *y*-coordinates of $P$ from the *x*- and *y*-coordinates of $Q$.

$$\mathbf{v} = \langle -3 - (-1), 1 - (-2) \rangle = \langle -2, 3 \rangle$$

**c.** The vector $\mathbf{v} = \langle -2, 3 \rangle$ is shown in Figure 73 with initial point $(0, 0)$ and terminal point $(-2, 3)$.

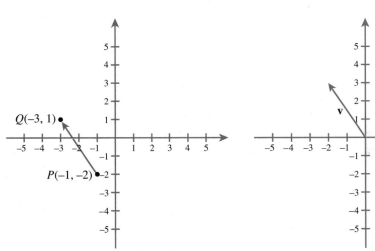

**FIGURE 72**                                    **FIGURE 73**

Since the magnitude of a vector corresponds to its length, the magnitude is simply the distance between the initial and terminal points. If the vector is given in component form, the distance formula leads us to the following statement.

**Magnitude of a vector**

> The magnitude of a vector $\mathbf{v} = \langle v_1, v_2 \rangle$, denoted by $\|\mathbf{v}\|$, is given by
>
> $$\|\mathbf{v}\| = \sqrt{v_1^2 + v_2^2}$$

**EXAMPLE 2**  *Finding the magnitude of a vector*

Sketch the vector $\mathbf{v} = \langle 4, -5 \rangle$ and find its magnitude.

**SOLUTION**  The vector $\mathbf{v} = \langle 4, -5 \rangle$ is shown in Figure 74 with initial point $(0, 0)$ and terminal point $(4, -5)$. Its magnitude is

$$\|\mathbf{v}\| = \sqrt{4^2 + (-5)^2} = \sqrt{16 + 25} = \sqrt{41}$$

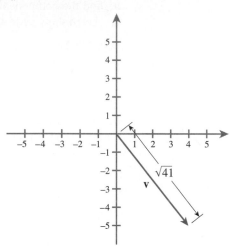

**FIGURE 74**

Notice that $\|\mathbf{v}\| > 0$ for any vector $\mathbf{v} = \langle v_1, v_2 \rangle$ as long as either $v_1$ or $v_2$ is nonzero. In the case where $\mathbf{v} = \langle 0, 0 \rangle$, $\|\mathbf{v}\| = 0$. We refer to the vector $\langle 0, 0 \rangle$ as the **zero vector** and, in keeping with our convention of writing vectors in boldface, we write

$$\mathbf{0} = \langle 0, 0 \rangle$$

## VECTOR OPERATIONS

The two most common vector operations are addition and scalar multiplication. Geometrically, the sum of two vectors $\mathbf{u}$ and $\mathbf{v}$ can be obtained by translating $\mathbf{v}$ so that its initial point lies on the terminal point of $\mathbf{u}$. The sum $\mathbf{u} + \mathbf{v}$ is a vector whose initial point is the initial point of $\mathbf{u}$ and whose terminal point is the terminal point of $\mathbf{v}$. See Figure 75 for an illustration of this.

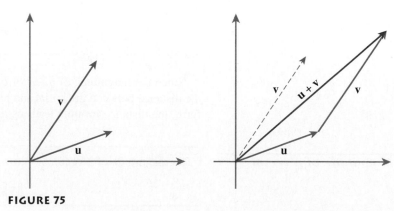

**FIGURE 75**

Scalar multiplication refers to the product of a scalar (a real number) and a vector. Geometrically, the product of a scalar $k$ and a vector $\mathbf{v}$ is a vector whose length is $|k|$ times that of $\mathbf{v}$. If $k$ is positive, $k\mathbf{v}$ has the same direction as $\mathbf{v}$, while if $k$ is negative, $k\mathbf{v}$ has the opposite direction of $\mathbf{v}$. Figure 76 gives some examples of scalar products.

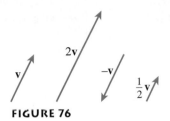

**FIGURE 76**

**EXAMPLE 3**    *Performing vector operations graphically*

Use the vectors shown in Figure 77 to sketch the indicated vector.
a. $-2\mathbf{u}$        b. $\mathbf{u} + \mathbf{v}$        c. $\mathbf{u} - 2\mathbf{v}$

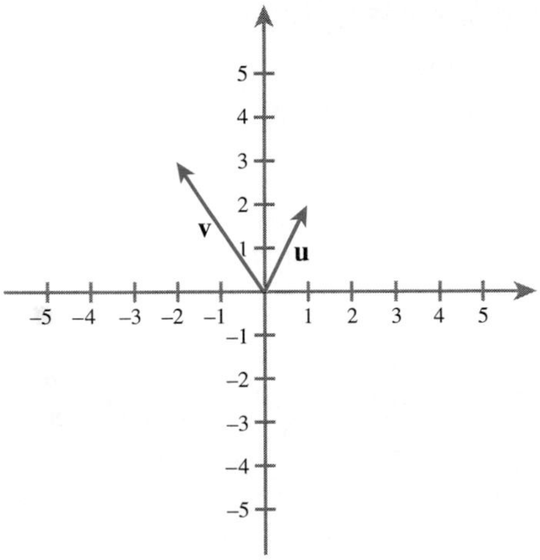

**SOLUTION**

a. The vector $-2\mathbf{u}$ is twice the length of $\mathbf{u}$ but has the opposite direction, as shown in Figure 78.

b. We first translate $\mathbf{v}$ so that its initial point lies on the terminal point of $\mathbf{u}$. The vector $\mathbf{u} + \mathbf{v}$ has the same initial point as $\mathbf{u}$ and the same terminal point as $\mathbf{v}$, as shown in red in Figure 79.

c. Notice that $\mathbf{u} - 2\mathbf{v} = \mathbf{u} + (-2\mathbf{v})$. Thus, we first sketch the vector $-2\mathbf{v}$ with its initial point at the terminal point of $\mathbf{u}$. The vector $\mathbf{u} + (-2\mathbf{v})$ has the same initial point as $\mathbf{u}$, and the same terminal point as $-2\mathbf{v}$, as shown in green in Figure 80.

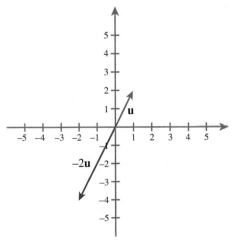

**FIGURE 77**

**FIGURE 78**

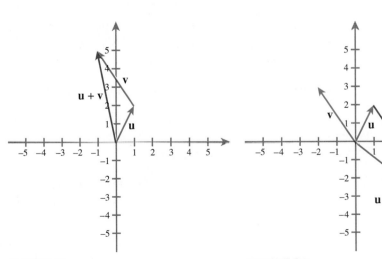

**FIGURE 79**                              **FIGURE 80**

For vectors expressed in component form, addition and scalar multiplication can be defined algebraically as follows. It's not hard to show that these definitions are consistent with the geometric definitions given earlier.

**Definitions of vector addition and scalar multiplication**

> If $\mathbf{u} = \langle u_1, u_2 \rangle$ and $\mathbf{v} = \langle v_1, v_2 \rangle$ then **vector addition** and **scalar multiplication** are defined as follows:
>
> $$\mathbf{u} + \mathbf{v} = \langle u_1 + v_1, u_2 + v_2 \rangle$$
>
> $$k\mathbf{u} = \langle ku_1, ku_2 \rangle$$

**EXAMPLE 4**     *Performing vector operations algebraically*

Perform the indicated operation on the vectors $\mathbf{u} = \langle 3, 5 \rangle$ and $\mathbf{v} = \langle 2, -4 \rangle$.

a. $\mathbf{u} + \mathbf{v}$     b. $-3\mathbf{u}$     c. $2\mathbf{u} - \frac{1}{2}\mathbf{v}$

**SOLUTION**

a. $\mathbf{u} + \mathbf{v} = \langle 3, 5 \rangle + \langle 2, -4 \rangle = \langle 3 + 2, 5 + (-4) \rangle = \langle 5, 1 \rangle$

b. $-3\mathbf{u} = -3\langle 3, 5 \rangle = \langle (-3)3, (-3)5 \rangle = \langle -9, -15 \rangle$

c. $2\mathbf{u} - \frac{1}{2}\mathbf{v} = 2\langle 3, 5 \rangle - \frac{1}{2}\langle 2, -4 \rangle = \langle 6, 10 \rangle - \langle 1, -2 \rangle = \langle 5, 12 \rangle$

Vector addition and scalar multiplication have many of the same properties as the operations of addition and multiplication for real numbers. Some of these are given here.

**Properties of vector addition and scalar multiplication**

> Let $\mathbf{u}$, $\mathbf{v}$, and $\mathbf{w}$ be vectors, and let $r$ and $s$ be scalars. Then the following properties hold:
>
> 1. $\mathbf{u} + \mathbf{v} = \mathbf{v} + \mathbf{u}$
> 2. $(\mathbf{u} + \mathbf{v}) + \mathbf{w} = \mathbf{u} + (\mathbf{v} + \mathbf{w})$
> 3. $\mathbf{u} + \mathbf{0} = \mathbf{u}$
> 4. $\mathbf{u} + (-\mathbf{u}) = \mathbf{0}$
> 5. $r(\mathbf{u} + \mathbf{v}) = r\mathbf{u} + r\mathbf{v}$
> 6. $(r + s)\mathbf{u} = r\mathbf{u} + s\mathbf{u}$
> 7. $r(s\mathbf{u}) = (rs)\mathbf{u}$
> 8. $1\mathbf{u} = \mathbf{u}$
> 9. $\|r\mathbf{u}\| = |r| \, \|\mathbf{u}\|$

**EXAMPLE 5**     *Verifying a property of vector addition*

Verify property 1; that is, given $\mathbf{u} = \langle u_1, u_2 \rangle$ and $\mathbf{v} = \langle v_1, v_2 \rangle$, show that $\mathbf{u} + \mathbf{v} = \mathbf{v} + \mathbf{u}$.

**SOLUTION**

$$
\begin{aligned}
\mathbf{u} + \mathbf{v} &= \langle u_1, u_2 \rangle + \langle v_1, v_2 \rangle \\
&= \langle u_1 + v_1, u_2 + v_2 \rangle && \text{Using the definition of vector addition} \\
&= \langle v_1 + u_1, v_2 + u_2 \rangle && \text{Applying the commutative property of addition} \\
&= \langle v_1, v_2 \rangle + \langle u_1, u_2 \rangle && \text{Using the definition of vector addition} \\
&= \mathbf{v} + \mathbf{u}
\end{aligned}
$$

# UNIT VECTORS

In cases where one is only interested in the direction of a vector and not its magnitude, it is usually more convenient to work with vectors of length 1. Such vectors are called unit vectors. More formally, a **unit vector** is any vector **u** for which $\|\mathbf{u}\| = 1$. In cases where a nonzero vector is given that is not a unit vector, it is possible to form a unit vector that has the same direction as the given vector. This can be done by "dividing" the given vector by its length, in the following sense.

■ *Forming a unit vector*

If **v** is a nonzero vector, then the vector **u** given by

$$\mathbf{u} = \frac{1}{\|\mathbf{v}\|}\,\mathbf{v}$$

is a unit vector in the same direction as **v**.

The fact that **u** is a unit vector can be verified using the scalar multiplication property $\|r\mathbf{v}\| = |r|\,\|\mathbf{v}\|$ as follows:

$$\left\|\frac{1}{\|\mathbf{v}\|}\,\mathbf{v}\right\| = \left|\frac{1}{\|\mathbf{v}\|}\right|\,\|\mathbf{v}\| = \frac{1}{\|\mathbf{v}\|}\,\|\mathbf{v}\| = 1$$

**EXAMPLE 6**    *Finding a unit vector in a given direction*

Find and sketch the unit vector **u** in the direction of $\mathbf{v} = \langle -1, \sqrt{3}\rangle$.

**SOLUTION**    The magnitude of **v** is

$$\|\mathbf{v}\| = \sqrt{(-1)^2 + (\sqrt{3})^2} = \sqrt{1 + 3} = \sqrt{4} = 2$$

To find the unit vector **u** in the direction of **v**, we multiply **v** by $1/\|\mathbf{v}\|$.

$$\mathbf{u} = \frac{1}{\|\mathbf{v}\|}\,\mathbf{v} = \frac{1}{2}\langle -1, \sqrt{3}\rangle = \left\langle -\frac{1}{2}, \frac{\sqrt{3}}{2}\right\rangle$$

Since **u** has magnitude 1, its terminal point must lie on the unit circle. A sketch is shown in Figure 81.

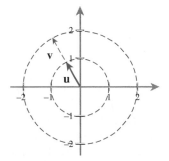

**FIGURE 81**

Two unit vectors of special significance are the so-called standard unit vectors *i* and *j*. As illustrated in Figure 82, these two vectors point in the directions of the positive *x*- and *y*-axes.

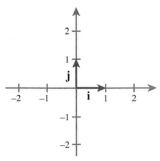

**FIGURE 82**

■ *Standard unit vectors*

The **standard unit vectors** are denoted **i** and **j**, and they are defined as

$$\mathbf{i} = \langle 1, 0 \rangle \qquad \text{and} \qquad \mathbf{j} = \langle 0, 1 \rangle$$

The standard unit vectors **i** and **j** can be viewed as the building blocks from which all other vectors (in the plane) can be formed. In particular, a vector $\langle a, b \rangle$ can be written as

$$\begin{aligned} \langle a, b \rangle &= \langle a, 0 \rangle + \langle 0, b \rangle \\ &= a\langle 1, 0 \rangle + b\langle 0, 1 \rangle \\ &= a\mathbf{i} + b\mathbf{j} \end{aligned}$$

The expression "$a\mathbf{i} + b\mathbf{j}$" is called a **linear combination** of **i** and **j**. More generally, a linear combination of two vectors **u** and **v** is an expression of the form $a\mathbf{u} + b\mathbf{v}$.

We've shown, then, that any vector can be written as a linear combination of the standard unit vectors **i** and **j**. Thus, for example, $\langle -3, 5 \rangle = -3\mathbf{i} + 5\mathbf{j}$. Notice that the coefficients of **i** and **j** in the linear combination are nothing more than the *x*- and *y*-components of the vector.

**EXAMPLE 7**     *Linear combinations of standard unit vectors*

Write the vector $\mathbf{v} = \langle 4, -2 \rangle$ as a linear combination of the standard unit vectors **i** and **j**.

**SOLUTION**     Applying the observation made above, we write

$$\mathbf{v} = \langle 4, -2 \rangle = 4\mathbf{i} - 2\mathbf{j}$$

---

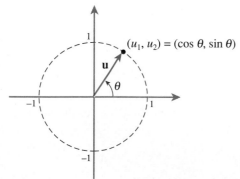

**FIGURE 83**

Vectors can also be represented in what is known as *trigonometric form*. We begin by considering a unit vector **u**. Let $\theta$ denote the angle measured counterclockwise from the positive *x*-axis to **u** as shown in Figure 83. The angle $\theta$ is referred to as the **direction angle** for the vector **u**. Since **u** is a unit vector, its terminal point, $(u_1, u_2)$ say, lies on the unit circle. From the definitions of sine and cosine, we know that $u_1 = \cos \theta$ and $u_2 = \sin \theta$. Thus,

$$\mathbf{u} = \cos \theta \mathbf{i} + \sin \theta \mathbf{j}$$

Now suppose **v** is any other nonzero vector with direction angle $\theta$. Then **v** has the same direction as **u** but is $\|\mathbf{v}\|$ times as long. That is, $\mathbf{v} = \|\mathbf{v}\| \mathbf{u} = \|\mathbf{v}\|(\cos \theta \mathbf{i} + \sin \theta \mathbf{j})$. Based on these observations, we define the trigonometric form of a vector **v** as follows.

■ *Trigonometric form of a vector*

Let **v** be a nonzero vector with direction angle $\theta$. Then the trigonometric form for **v** is

$$\mathbf{v} = \|\mathbf{v}\|(\cos \theta \mathbf{i} + \sin \theta \mathbf{j})$$

Notice how closely the trigonometric form of a vector parallels the trigonometric form of a complex number. In fact, if we let $r = \|\mathbf{v}\|$, then the $x$- and $y$-components of the vector $\mathbf{v}$ are $r \cos \theta$ and $r \sin \theta$, respectively, which are precisely the forms given in Section 9.3 for the real and imaginary parts of a complex number.

In practice, we determine the direction angle from the fact that if $\mathbf{v} = v_1\mathbf{i} + v_2\mathbf{j} = \|\mathbf{v}\|(\cos \theta \mathbf{i} + \sin \theta \mathbf{j})$, then

$$\frac{v_2}{v_1} = \frac{\|\mathbf{v}\|\sin \theta}{\|\mathbf{v}\|\cos \theta} = \frac{\sin \theta}{\cos \theta} = \tan \theta$$

Here we see yet another parallel with polar coordinates.

**EXAMPLE 8**  *Finding the trigonometric form of a vector*

Find the trigonometric form of the vector $\mathbf{v} = -3\mathbf{i} + 3\mathbf{j}$.

**SOLUTION**  The magnitude of $\mathbf{v}$ is

$$\|\mathbf{v}\| = \sqrt{(-3)^2 + 3^2} = \sqrt{18} = 3\sqrt{2}$$

The direction angle $\theta$ is in the second quadrant and satisfies

$$\tan \theta = \frac{3}{-3} = -1$$

Thus, $\theta = 3\pi/4$. The trigonometric form of $\mathbf{v}$ is

$$\mathbf{v} = \|\mathbf{v}\|(\cos \theta \mathbf{i} + \sin \theta \mathbf{j}) = 3\sqrt{2}\left(\cos \frac{3\pi}{4} \mathbf{i} + \sin \frac{3\pi}{4} \mathbf{j}\right)$$

## APPLICATIONS

Entire textbooks in physics, mathematics, and engineering have been devoted to applications of vectors. Vector techniques are even applied in social sciences such as economics and psychology. Still, the seminal notion of vector arose from considering the physical quantities velocity and force. In each of the following examples, note how vector addition provides a mechanism for accounting for the combined effects of several competing influences.

**EXAMPLE 9**  *Resultant velocity vector*

In the absence of wind, the British-French Concorde is flying at a speed of 1,320 miles per hour on a course of 120° (clockwise from north). If no adjustment is made when the plane encounters wind with a speed of 80 miles per hour in the direction 60°, what will the resultant velocity (i.e., the resultant speed and direction) of the plane be?

**SOLUTION**  If we let $\mathbf{v}_1$ denote the velocity vector of the Concorde in the absence of wind, then $\mathbf{v}_1$ has magnitude 1,320 and direction angle 330° as measured counterclockwise from the positive $x$-axis. Thus,

$$\mathbf{v}_1 = 1{,}320(\cos 330° \mathbf{i} + \sin 330° \mathbf{j}) = 660\sqrt{3}\mathbf{i} - 660\mathbf{j}$$

British-French Concorde

Similarly, if $\mathbf{v}_2$ denotes the wind velocity vector, then

$$\mathbf{v}_2 = 80(\cos 30°\mathbf{i} + \sin 30°\mathbf{j}) = 40\sqrt{3}\mathbf{i} + 40\mathbf{j}$$

The vectors $\mathbf{v}_1$ and $\mathbf{v}_2$ are shown in Figure 84. The resultant velocity vector $\mathbf{v}$ is the sum of $\mathbf{v}_1$ and $\mathbf{v}_2$, as shown in Figure 85. Algebraically, we have

$$\begin{aligned}\mathbf{v} &= \mathbf{v}_1 + \mathbf{v}_2 \\ &= (660\sqrt{3}\mathbf{i} - 660\mathbf{j}) + (40\sqrt{3}\mathbf{i} + 40\mathbf{j}) \\ &= 700\sqrt{3}\mathbf{i} - 620\mathbf{j}\end{aligned}$$

The resulting speed of the Concorde is

$$\|\mathbf{v}\| = \sqrt{(700\sqrt{3})^2 + (-620)^2} \approx 1361.8 \text{ miles per hour}$$

The resulting direction angle is in the third quadrant and satisfies

$$\tan\theta = \frac{-620}{700\sqrt{3}}$$

Thus $\theta \approx -27.1°$, which corresponds to a course of $117.1°$.

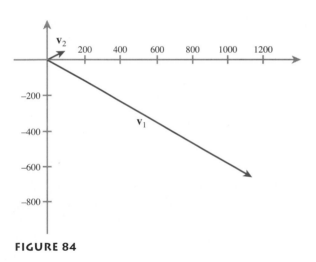

**FIGURE 84**

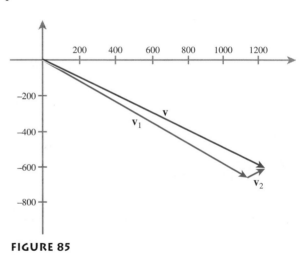

**FIGURE 85**

Newton's second law states that an object's acceleration is proportional to the net force acting on it. As a consequence, if an object is at rest, the net force acting on it must be zero. In other words, the vector sum of the external forces acting on an object at rest must be zero. We apply this principle in the following example.

**EXAMPLE 10**    *Tug-of-war*

In a certain tug-of-war competition, four pieces of rope are attached to a ring. Each team has control over two of the ropes, and each team can decide if they should split into two groups, one pulling on each rope, or if they should remove one rope and all pull on the remaining one. Suppose one team splits into two groups, one pulling with a magnitude of 400 pounds, and the other pulling with a magnitude of 500 pounds at an angle of 10° counterclockwise from the first. Suppose the second team is pulling on a single rope. Describe the force that is being applied by the second team if neither team is moving.

**SOLUTION**   Let $\mathbf{F}_1$ and $\mathbf{F}_2$ be the 400- and 500-pound forces, respectively, exerted by the divided team, and let $\mathbf{F}_3$ be the force exerted by the second team. We next set up a coordinate system with the ring at the origin. For convenience, we place the force $\mathbf{F}_1$ with magnitude 400 pounds pointing in the positive $x$ direction. The force $\mathbf{F}_2$ with magnitude 500 pounds thus has direction angle 10° (see Figure 86).

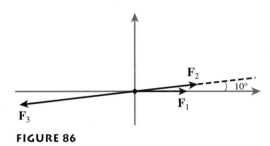

**FIGURE 86**

We next write the force vectors $\mathbf{F}_1$ and $\mathbf{F}_2$ in component form as follows:

$$\mathbf{F}_1 = 400\mathbf{i}$$

$$\mathbf{F}_2 = 500(\cos 10°\mathbf{i} + \sin 10°\mathbf{j}) \approx 492.40\mathbf{i} + 86.82\mathbf{j}$$

Since the ring is not moving, the sum of the vector forces acting on it is zero. Thus

$$\mathbf{F}_1 + \mathbf{F}_2 + \mathbf{F}_3 = \mathbf{0}$$

and so

$$\mathbf{F}_3 = -\mathbf{F}_1 - \mathbf{F}_2 \approx -400\mathbf{i} - (492.40\mathbf{i} + 86.82\mathbf{j}) = -892.40\mathbf{i} - 86.82\mathbf{j}$$

The magnitude of $\mathbf{F}_3$ is

$$\|\mathbf{F}_3\| \approx \sqrt{(-892.4)^2 + (-86.82)^2} \approx 896.61$$

and its direction angle $\theta$ is in the third quadrant and satisfies

$$\tan \theta \approx \frac{-86.82}{-892.40} \approx 0.09729$$

Thus, the second team must be pulling with a magnitude of approximately 896.61 pounds and in the direction $\theta \approx 180° + \tan^{-1} 0.09729 = 185.56°$.

## EXERCISES 6

**EXERCISES 1–6** □ (a) *Sketch the directed line segment $\overrightarrow{PQ}$,*
(b) *find the component form of the vector* **v** *represented by $\overrightarrow{PQ}$, and*
(c) *sketch the vector* **v** *in standard position.*

1. $P(2, 5), Q(3, 7)$

2. $P(-1, 0), Q(-3, -2)$

3. $P(-3, 1), Q(2, -2)$

4. $P\left(\frac{1}{4}, \frac{1}{2}\right), Q\left(\frac{1}{2}, 0\right)$

5. $P\left(0, -\frac{1}{2}\right), Q\left(-\frac{1}{4}, 1\right)$

6. $P(6, -3), Q(11, 1)$

**EXERCISES 7–14** □ *Find the magnitude of the vector*

7. $\langle 4, 0 \rangle$

8. $\langle 3, -4 \rangle$

9. $\langle -12, -5 \rangle$

10. $\langle 4, -2\sqrt{5} \rangle$

11. $\langle 3, 5 \rangle$

12. $\langle -2, 1 \rangle$

13. $\langle \sqrt{5}, -\sqrt{7} \rangle$

14. $\left\langle -\frac{4}{9}, \frac{1}{3} \right\rangle$

**EXERCISES 15–20** □ *Use Figure 87 to sketch the graph of the indicated vector.*

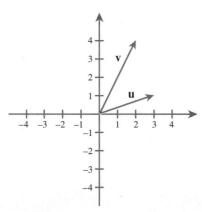

**FIGURE 87**

15. $-\mathbf{u}$

16. $2\mathbf{v}$

17. $\mathbf{u} + \mathbf{v}$

18. $2\mathbf{u} - \mathbf{v}$

19. $3\mathbf{u} - 2\mathbf{v}$

20. $-\mathbf{u} + 3\mathbf{v}$

**EXERCISES 21–26** □ *Use Figure 88 to sketch the graph of the indicated vector.*

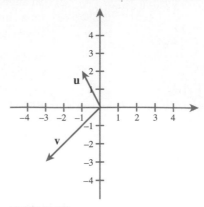

**FIGURE 88**

21. $2\mathbf{u}$

22. $-\frac{1}{3}\mathbf{v}$

23. $\mathbf{u} + \mathbf{v}$

24. $\mathbf{v} - \mathbf{u}$

25. $3\mathbf{u} - \mathbf{v}$

26. $-2\mathbf{u} + \mathbf{v}$

**EXERCISES 27–32** □ *Perform the indicated operations. Express your answer in the form in which the vectors are given.*

27. $\mathbf{u} = \langle 1, 3 \rangle, \mathbf{v} = \langle 2, 5 \rangle$
    a. $-\mathbf{u}$     b. $\mathbf{u} + \mathbf{v}$     c. $-2\mathbf{u} + \mathbf{v}$

28. $\mathbf{u} = \langle -2, 2 \rangle, \mathbf{v} = \langle 3, 0 \rangle$
    a. $-\frac{1}{2}\mathbf{v}$     b. $3\mathbf{u} + \mathbf{v}$     c. $\frac{1}{2}\mathbf{u} - \frac{1}{3}\mathbf{v}$

29. $\mathbf{u} = 2\mathbf{i} + \mathbf{j}, \mathbf{v} = 3\mathbf{i} - \mathbf{j}$
    a. $\sqrt{2}\mathbf{v}$     b. $-2\mathbf{u} - \mathbf{v}$     c. $5\mathbf{u} + 7\mathbf{v}$

30. $\mathbf{u} = -\frac{1}{2}\mathbf{i} + 2\mathbf{j}, \mathbf{v} = -\mathbf{i} - \frac{1}{2}\mathbf{j}$
    a. $-\mathbf{u}$     b. $-2\mathbf{u} + 4\mathbf{v}$     c. $\mathbf{u} - \mathbf{v}$

31. $\mathbf{u} = \frac{1}{5}\mathbf{i} - \frac{1}{5}\mathbf{j}, \mathbf{v} = \mathbf{j}$
    a. $5\mathbf{u}$     b. $10\mathbf{u} - \mathbf{v}$     c. $\mathbf{u} + \frac{1}{5}\mathbf{v}$

32. $\mathbf{u} = \langle 2, 3 \rangle, \mathbf{v} = 3\langle -1, -2 \rangle$
    a. $-\frac{1}{3}\mathbf{v}$     b. $\mathbf{u} - \mathbf{v}$     c. $\frac{1}{2}(\mathbf{u} - \mathbf{v})$

**EXERCISES 33–38** □ *Find the unit vector in the direction of the given vector.*

33. $\sqrt{3}\mathbf{i}$

34. $\langle 0, 4 \rangle$

35. $\langle -3, 4 \rangle$

36. $5\mathbf{i} - 12\mathbf{j}$

37. $-2\mathbf{i} - 3\mathbf{j}$

38. $\left\langle \frac{1}{2}, -\frac{1}{2} \right\rangle$

**EXERCISES 39–46** ☐ *Find the component form of the vector* **v**. *Assume that θ denotes the angle made with the positive x-axis.*

**39.** $\|\mathbf{v}\| = 1$, $\theta = \dfrac{\pi}{3}$    **40.** $\|\mathbf{v}\| = 2$, $\theta = 45°$

**41.** $\|\mathbf{v}\| = \dfrac{1}{2}$, $\theta = -120°$    **42.** $\|\mathbf{v}\| = 3$, $\theta = \pi$

**43.** $\|\mathbf{v}\| = 5$, $\theta = -\dfrac{\pi}{2}$    **44.** $\|\mathbf{v}\| = \dfrac{1}{3}$, $\theta = 60°$

**45.** $\|\mathbf{v}\| = 2$, $\theta = 110°$    **46.** $\|\mathbf{v}\| = 4$, $\theta = \dfrac{\pi}{5}$

**EXERCISES 47–52** ☐ *Write the given vector* **v** *in the standard trigonometric form* $r(\cos\theta\mathbf{i} + \sin\theta\mathbf{j})$, *where* $r = \|\mathbf{v}\|$ *and θ is the angle made with the positive x-axis.*

**47.** $7\mathbf{i}$    **48.** $-3\mathbf{j}$

**49.** $2\sqrt{3}\mathbf{i} - 2\mathbf{j}$    **50.** $2\mathbf{i} + 2\mathbf{j}$

**51.** $-\dfrac{1}{8}\mathbf{i} - \dfrac{1}{8}\mathbf{j}$    **52.** $-4\mathbf{i} + 4\sqrt{3}\,\mathbf{j}$

**EXERCISES 53–60** ☐ *Verify the given property of vector addition and/or scalar multiplication. Assume that* $\mathbf{u} = \langle u_1, u_2 \rangle$, $\mathbf{v} = \langle v_1, v_2 \rangle$, *and* $\mathbf{w} = \langle w_1, w_2 \rangle$, *and r and s are scalars.*

**53.** $(\mathbf{u} + \mathbf{v}) + \mathbf{w} = \mathbf{u} + (\mathbf{v} + \mathbf{w})$

**54.** $\mathbf{u} + \mathbf{0} = \mathbf{u}$

**55.** $\mathbf{u} + (-\mathbf{u}) = \mathbf{0}$    **56.** $r(\mathbf{u} + \mathbf{v}) = r\mathbf{u} + r\mathbf{v}$

**57.** $(r + s)\mathbf{u} = r\mathbf{u} + s\mathbf{u}$    **58.** $r(s\mathbf{u}) = (rs)\mathbf{u}$

**59.** $1\mathbf{u} = \mathbf{u}$    **60.** $\|r\mathbf{u}\| = |r|\,\|\mathbf{u}\|$

---

■ *Applications*

**61.** *Airplane Velocity* An airplane with a speed in still air of 250 miles per hour is attempting to fly due north but is being blown off course by a 50 mile per hour wind blowing in from the northeast (towards the direction S 45° W). What is the resulting velocity of the plane?

**62.** *Cooperative Canine* A woman is walking her three dogs, holding all three leashes in one hand. The first dog pulls north with a force of 30 pounds, and the second pulls east with a force of 25 pounds. The third dog, empathizing with the plight of its mistress, pulls with the appropriate force and direction to exactly counteract the effects of its colleagues. In what direction and with how much force did the third dog pull?

**63.** *Gulf Stream Velocity* The Gulf Stream is a warm ocean current of the northern Atlantic Ocean off North America. It has an average surface velocity of 4 miles per hour, and peak surface velocities have been known to exceed 5.5 miles per hour. Suppose an ocean liner with a velocity of 12 miles per hour on a course of 30° (clockwise from north) enters the Gulf Stream at a point where its velocity is 4 miles per hour with a bearing of 45°. Find the resultant velocity (speed and direction) of the liner.

**64.** *Jet Stream Velocity* The jet stream is a wind current that generally moves in a west-to-east direction at speeds often exceeding 250 miles per hour and at altitudes above 50,000 feet. Suppose an F-15 fighter with a velocity of 1200 miles per hour on a course of 345° (clockwise from north) encounters a jet stream current with a velocity of 200 miles per hour with a bearing of 280°. Find the resultant velocity (speed and direction) of the fighter.

F-15 Fighter

**65.** *Sled Dog Team*  Two sled dogs are pulling on the same sled, the first with 100 pounds of force, and the second with 120 pounds of force. The first dog is pulling in the correct direction, but the second is aiming 20° off course. Find the magnitude of the net force acting on the sled. How far off course will the sled be?

**66.** *Frictional Force*  A 200-pound block is at rest on a smooth ramp that makes an angle of 30° with the horizontal. The external forces acting on the block are the force due to gravity **W**, which has magnitude 200 and acts downward; the *normal* force **N** of the ramp pushing upward on the block with magnitude $100\sqrt{3}$ and perpendicular to the ramp; and the frictional force **F**, which acts parallel to the ramp, as shown in Figure 89. Find the magnitude of the frictional force. (*Hint:* Set up a coordinate system with the origin at the block.)

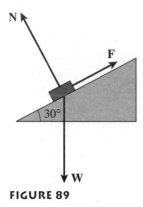

**FIGURE 89**

**67.** *Suspended Crate*  A crate is suspended by two ropes as shown in Figure 90. The external forces acting on the crate are the force due to gravity **W**, which acts downward, the pull **F**$_1$ of the rope making an angle of 45° with the horizontal, and the pull **F**$_2$ of the rope making an angle of 30° with the horizontal. If **F**$_1$ and **F**$_2$ have magnitudes 448 and 366, respectively, find the weight of the crate; that is, find the magnitude of **W**. (*Hint:* Set up a coordinate system with the origin at the crate.)

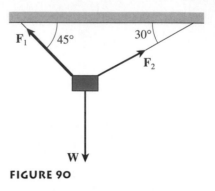

**FIGURE 90**

---

■ *Projects for Enrichment*

---

**68.** *Dot Product and Work*  In this section, we introduced two operations on vectors, namely vector addition and scalar multiplication. In this project, we consider a third operation on vectors, the dot product. The **dot product** of two vectors **u** = $\langle u_1, u_2 \rangle$ and **v** = $\langle v_1, v_2 \rangle$ is denoted by **u** · **v** and is defined by

$$\mathbf{u} \cdot \mathbf{v} = u_1 v_1 + u_2 v_2$$

Thus, for example, the dot product of **u** = $\langle -3, 2 \rangle$ and **v** = $\langle 1, -4 \rangle$ is

$$\mathbf{u} \cdot \mathbf{v} = (-3)1 + 2(-4) = -11$$

Note that the dot product of two vectors is a scalar, not a vector.

**a.** Find the dot product of the given pair of vectors.
  i. **u** = $\langle 2, -1 \rangle$, **v** = $\langle 5, 7 \rangle$
  ii. **u** = $\langle -4, 3 \rangle$, **v** = $\langle 2, -2 \rangle$
  iii. **u** = $\langle 3, -2 \rangle$, **v** = $\langle 6, 9 \rangle$

The dot product has a number of important applications, one of which is in finding the angle between vectors. The angle

between two nonzero vectors **u** and **v** is defined to be the angle $\theta$ with $0 \leq \theta \leq \pi$ that is formed when the vectors are in standard position, as shown in Figure 91. Using the Law of Cosines, we can show that $\theta$ satisfies

(1) $$\cos \theta = \frac{\mathbf{u} \cdot \mathbf{v}}{\|\mathbf{u}\| \, \|\mathbf{v}\|}$$

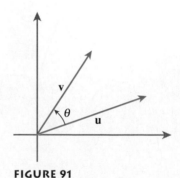

**FIGURE 91**

To see this, consider the two vectors $\mathbf{u} = \langle u_1, u_2 \rangle$ and $\mathbf{v} = \langle v_1, v_2 \rangle$ shown in Figure 91. The vector $\mathbf{u} - \mathbf{v}$ is the vector whose initial point is the terminal point of $\mathbf{v}$ and whose terminal point is the terminal point of $\mathbf{u}$. Now consider the triangle $ABC$ formed by $\mathbf{u}$, $\mathbf{v}$, and $\mathbf{u} - \mathbf{v}$, as shown in Figure 92. Note that the side lengths $a$, $b$, and $c$ are given by $\|\mathbf{u} - \mathbf{v}\|$, $\|\mathbf{v}\|$, and $\|\mathbf{u}\|$, respectively, and $\theta$ is the angle opposite the side with length $\|\mathbf{u} - \mathbf{v}\|$. Thus, we may apply the Law of Cosines to obtain

$$\|\mathbf{u} - \mathbf{v}\|^2 = \|\mathbf{u}\|^2 + \|\mathbf{v}\|^2 - 2\|\mathbf{u}\| \, \|\mathbf{v}\| \cos\theta$$

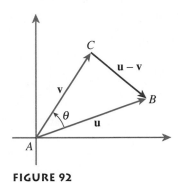

**FIGURE 92**

b. Express $\|\mathbf{u}\|^2$, $\|\mathbf{v}\|^2$, and $\|\mathbf{u} - \mathbf{v}\|^2$ in terms of $u_1$, $u_2$, $v_1$, and $v_2$, and solve for $\cos\theta$. Then simplify to obtain the equation (1).

To find the angle $\theta$ between the vectors $\mathbf{u} = \langle -3, 2 \rangle$ and $\mathbf{v} = \langle 1, -4 \rangle$, we proceed as follows:

$$\cos\theta = \frac{\mathbf{u} \cdot \mathbf{v}}{\|\mathbf{u}\| \, \|\mathbf{v}\|} = \frac{-11}{\sqrt{13}\sqrt{17}} = \frac{-11}{\sqrt{221}}$$

$$\theta = \cos^{-1}\left(\frac{-11}{\sqrt{221}}\right) \approx 137.73°$$

c. Find the angle between each pair of vectors.
   i. $\mathbf{u} = \langle 2, -1 \rangle$, $\mathbf{v} = \langle 5, 7 \rangle$
   ii. $\mathbf{u} = \langle -4, 3 \rangle$, $\mathbf{v} = \langle 2, -2 \rangle$
   iii. $\mathbf{u} = \langle 3, -2 \rangle$, $\mathbf{v} = \langle 6, 9 \rangle$

d. What geometric relationship holds for the vectors in part (c)(iii)? State a general rule that relates the dot product to this relationship.

Another important application of the dot product is in computing work done by a constant force. When an object is moved a distance of $d$ feet by a constant force of magnitude $F$ applied in the direction of motion, the work done by the force is defined

to be the product $Fd$. For example, the work done in lifting a 50-pound box of books to a height of 3 feet is

$$W = Fd = (50 \text{ lb})(3 \text{ ft}) = 150 \text{ foot-pounds}$$

More generally, the work done by a constant force $\mathbf{F}$ in moving an object from some point $P$ to some point $Q$, regardless of whether or not $\mathbf{F}$ is applied in the direction of motion, is given by

$$W = \mathbf{F} \cdot \overrightarrow{PQ}$$

Suppose you are pulling an object along flat ground by a rope attached to the front of the object. If you are pulling with a force of 80 pounds at an angle of 30° with the horizontal, then the force is given by

$$\mathbf{F} = 80(\cos 30° \mathbf{i} + \sin 30° \mathbf{j}) = 40\sqrt{3}\mathbf{i} + 40\mathbf{j}$$

If you move the object a horizontal distance of 10 feet, say from $P(0, 0)$ to $Q(10, 0)$, the *displacement vector* is

$$\overrightarrow{PQ} = 10\mathbf{i}$$

The work is found by computing the dot product of the force vector and the displacement vector

$$\begin{aligned} W &= \mathbf{F} \cdot \overrightarrow{PQ} \\ &= (40\sqrt{3}\mathbf{i} + 40\mathbf{j}) \cdot (10\mathbf{i}) \\ &= (40\sqrt{3})(10) \\ &\approx 693 \text{ foot-pounds} \end{aligned}$$

e. Find the work done by a child pulling a wagon 100 feet along flat ground with a constant force of 50 pounds if the wagon handle makes an angle of 45° with the horizontal.

f. Two lumberjacks are pulling a log along flat horizontal ground by ropes attached to the front of the log. One is pulling with a force of 80 pounds at an angle of 30° with the horizontal, while the other is pulling with a force of 100 pounds at an angle of 45°. Find the work done in moving the log a distance of 50 feet. (*Hint:* First find the resultant force.)

■   *Questions for Discussion or Essay*

**69.** We've defined vectors as being quantities having both a magnitude and direction. But we've also seen that vectors (in the plane) can be represented as an ordered pair of numbers $\langle a, b \rangle$. In fact, a loose definition of a vector is ''a quantity requiring two or more numbers for a full description.'' According to this definition, which of the following human traits are best quantified as vectors and which by scalars? Support your answers.

  **a.** SAT performance          **b.** Height

  **c.** Beauty                          **d.** Weight

  **e.** Strength                        **f.** Blood Pressure

  **g.** Age

**70.** We have seen trigonometric form in three different contexts: polar coordinates, complex numbers, and now vectors. Discuss the similarities and differences between the three.

**71.** In Example 9, we made the assumption that the velocity vector of the Concorde would be the vector sum of the prevailing wind and the velocity the plane would have in the absence of wind. What factors determine how the flight of an airplane depends upon air currents? Is it legitimate to assume that the air blows at a constant speed in a constant direction? How fast would the airplane travel if there were no air whatsoever?

**72.** In Example 10, we considered a tug-of-war in which a team was allowed to split into two groups, pulling at slightly different angles. Discuss some reasons why a group may or may not decide to split into two groups in this way.

## CHAPTER REVIEW EXERCISES

**EXERCISES 1–4** □ *Use the Law of Sines to solve the given triangle. Assume angles A, B, and C, and sides a, b, and c are labeled as shown in Figure 93.*

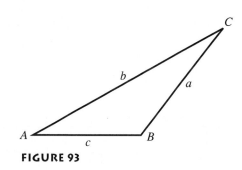

**FIGURE 93**

**1.** $c = 12, A = 47°, C = 52°$

**2.** $a = 25, B = 39°, C = 112°$

**3.** $a = 1.2, b = 1, B = 42°$

**4.** $a = 100, b = 55, A = 140°$

**EXERCISES 5–8** □ *Use the Law of Cosines to solve the given triangle. Assume angles A, B, and C, and sides a, b, and c are labeled as shown in Figure 93.*

**5.** $b = 19, c = 15, A = 10°$

**6.** $a = 5.8, b = 3.4, C = 64°$

**7.** $a = 5, b = 8, c = 9$

**8.** $a = 8, c = 12, A = 50°$

**EXERCISES 9–12** □ *Solve the given triangle by any means. Assume angles A, B, and C, and sides a, b, and c are labeled as shown in Figure 93.*

**9.** $a = 335, b = 260, c = 540$

**10.** $a = 6, A = 133°, B = 24°$

**11.** $a = 2, b = 7, A = 41°$

**12.** $b = 18, c = 16, C = 30°$

**EXERCISES 13–16** □ *Solve the given triangle by any means.*

**13.**

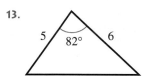

**14.**

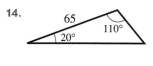

**15.**

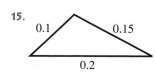

**16.**

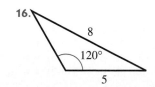

**EXERCISES 17–20** □ *Find the area of the given triangle.*

**17.**

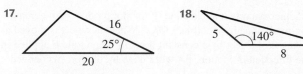

**18.**

**19.**

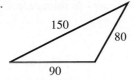

150
80
90

**20.**

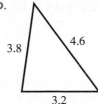

3.8     4.6
3.2

**EXERCISES 21–24** □ *Find the absolute value of the given complex number.*

**21.** $4 - 3i$

**22.** $\sqrt{2} + \sqrt{2}i$

**23.** $-9 + i$

**24.** $\dfrac{1}{2} - \dfrac{3}{2}i$

**EXERCISES 25–28** □ *Represent the complex number graphically and write it in trigonometric form.*

**25.** $-2 - 2i$

**26.** $\sqrt{3} - i$

**27.** $\dfrac{3}{2}i$

**28.** $-2 + i$

**EXERCISES 29–32** □ *Represent the complex number graphically and write it in the form a + bi.*

**29.** $2(\cos 330° + i \sin 330°)$

**30.** $3\left(\cos \dfrac{5\pi}{4} + i \sin \dfrac{5\pi}{4}\right)$

**31.** $\sqrt{3}\left(\cos \dfrac{\pi}{3} + i \sin \dfrac{\pi}{3}\right)$

**32.** $\dfrac{5}{2}(\cos 270° + i \sin 270°)$

**EXERCISES 33–36** □ *Perform the indicated operation.*

**33.** $5\left(\cos \dfrac{\pi}{3} + i \sin \dfrac{\pi}{3}\right) \cdot 2\left(\cos \dfrac{\pi}{4} + i \sin \dfrac{\pi}{4}\right)$

**34.** $\sqrt{3}(\cos 51° + i \sin 51°) \cdot \sqrt{27}(\cos 102° + i \sin 102°)$

**35.** $\dfrac{12(\cos 300° + i \sin 300°)}{4(\cos 100° + i \sin 100°)}$

**36.** $\dfrac{2.4(\cos \pi + i \sin \pi)}{0.8\left(\cos \dfrac{\pi}{6} + i \sin \dfrac{\pi}{6}\right)}$

**EXERCISES 37–40** □ *Use de Moivre's Theorem to expand the given power. Express the answer in the same form used in the statement of the problem.*

**37.** $[5(\cos 10° + i \sin 10°)]^4$

**38.** $\left[\sqrt{2}\left(\cos \dfrac{\pi}{4} + i \sin \dfrac{\pi}{4}\right)\right]^{10}$

**39.** $\left(1 - \sqrt{3}i\right)^5$

**40.** $(1 - i)^8$

**EXERCISES 41–44** □ *Find the indicated roots, express them in the form a + bi, and represent them graphically.*

**41.** Square roots of $4(\cos 120° + i \sin 120°)$

**42.** Third roots of $8i$

**43.** Fourth roots of $81$

**44.** Sixth roots of $-1$

**EXERCISES 45–46** □ *Find all solutions of the equation and represent them graphically.*

**45.** $x^3 + 1 = 0$

**46.** $x^4 - i = 0$

**EXERCISES 47–48** □ *Plot the point given in polar coordinates and find the corresponding rectangular coordinates.*

**47.** $(2, \pi)$

**48.** $\left(-4, \dfrac{5\pi}{3}\right)$

**EXERCISES 49–50** □ *Plot the point given in rectangular coordinates and find two sets of polar coordinates with $0 \le \theta < 2\pi$.*

**49.** $(-3, 3)$

**50.** $(4\sqrt{3}, 4)$

**EXERCISES 51–56** □ *Sketch the graph of the given polar equation.*

**51.** $r = 5$

**52.** $\theta = -\dfrac{\pi}{4}$

**53.** $r = 3\theta$

**54.** $r = \cos \theta - 1$

**55.** $r = 3 \sin 2\theta$

**56.** $r = 2 \cos \theta - 1$

**EXERCISES 57–60** □ *Convert the given rectangular equation to polar form.*

**57.** $y = 2x$

**58.** $x = 5$

**59.** $x^2 + y^2 = 1$

**60.** $\dfrac{x^2}{4} - \dfrac{y^2}{9} = 1$

**EXERCISES 61–64** □ *Convert the given polar equation to rectangular form.*

**61.** $r = -1$

**62.** $\theta = \dfrac{\pi}{2}$

**63.** $r = 3 \sin \theta$

**64.** $r^2 = \tan \theta$

**EXERCISES 65–66** □ *Use a graphics calculator to plot the graph of the given polar equations. Estimate the coordinates of the point(s) of intersection.*

**65.** $r = 3$
$r = 4 \cos \theta$

**66.** $r = 3 \sin \theta$
$r = 1 - 4 \cos \theta$

**EXERCISES 67–68** ☐ (a) *Sketch the directed line segment $\overrightarrow{PQ}$*, (b) *find the component form of the vector* **v** *represented by $\overrightarrow{PQ}$, and* (c) *sketch the vector* **v** *in standard position.*

**67.** $P(4, -2)$, $Q(-1, -1)$     **68.** $P\left(-\dfrac{7}{2}, 3\right)$, $Q\left(-\dfrac{3}{2}, 1\right)$

**EXERCISES 69–70** ☐ *Find the magnitude of the vector.*

**69.** $\langle -6, 8 \rangle$     **70.** $\langle 4, -1 \rangle$

**EXERCISES 71–72** ☐ *Use Figure 94 to sketch the graph of the indicated vector.*

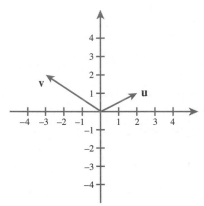

**FIGURE 94**

**71.** $\mathbf{u} - \mathbf{v}$     **72.** $3\mathbf{u} + \mathbf{v}$

**EXERCISES 73–74** ☐ *Perform the indicated operations. Express your answer in the same form in which the vectors are given.*

**73.** $\mathbf{u} = \langle -4, 2 \rangle$, $\mathbf{v} = \langle -2, 3 \rangle$

   a. $-2\mathbf{u}$    b. $\mathbf{u} - 2\mathbf{v}$    c. $\dfrac{1}{2}\mathbf{u} + \dfrac{1}{2}\mathbf{v}$

**74.** $\mathbf{u} = \mathbf{i} - 3\mathbf{j}$, $\mathbf{v} = 2\mathbf{i} + 5\mathbf{j}$

   a. $\dfrac{1}{3}\mathbf{u}$    b. $\mathbf{v} - \mathbf{u}$    c. $-2\mathbf{u} - \mathbf{v}$

**EXERCISES 75–76** ☐ *Find the unit vector in the direction of the given vector.*

**75.** $12\mathbf{i} + 5\mathbf{j}$     **76.** $\langle \sqrt{3}, -1 \rangle$

**EXERCISES 77–78** ☐ *Find the component form of the vector* **v**. *Assume that θ denotes the angle made with the positive x-axis.*

**77.** $\|\mathbf{v}\| = 2$, $\theta = \dfrac{\pi}{6}$     **78.** $\|\mathbf{v}\| = \dfrac{1}{2}$, $\theta = 150°$

**EXERCISES 79–80** ☐ *Write the given vector* **v** *in the standard trigonometric form $r(\cos \theta \mathbf{i} + \sin \theta \mathbf{j})$, where $r = \|\mathbf{v}\|$ and θ is the angle made with the positive x-axis.*

**79.** $4\mathbf{j}$     **80.** $-2\mathbf{i} - 2\sqrt{3}\mathbf{j}$

**81.** *Estimating Travel Time* A boater spots a lighthouse at a bearing of N 38° E. He then sails due east for 10 miles, at which time the lighthouse is at a bearing of N 47° W. If his boat is capable of 30 miles per hour, how long will it take him to reach the lighthouse if he heads directly for it?

**82.** *Estimating Distance* A child is looking through a telescope on the observation deck of a skyscraper. She spots her home, which is 18 miles from the skyscraper, and then swings the telescope 22.4° and sights her school, which is 14 miles from the skyscraper. How far is the girl's school from her home?

**83.** *Estimating Speed* A nature photographer traveling at a constant speed on an overhead tram spots a (virtually motionless) elk at an angle of depression of 30°. Two minutes later, she spots the same elk at an angle of depression of 34°, and estimates that she is now 2 miles from the elk. How fast does the tram move?

**84.** *Finding a Pyramid Angle* A man standing 200 feet from the base of a pyramid spots his friend, who has climbed 100 feet up the side of the pyramid. He estimates that his friend is 300 feet away. What angle does the face of the pyramid make with the ground?

## CHAPTER TEST

**PROBLEMS 1–10** □ *Answer true or false.*

1. Every real number has a real square root.

2. Every complex number has a complex square root.

3. If the three angles of a triangle are known, then the side lengths can be determined.

4. If the three sides of a triangle are known, then the angles can be determined.

5. Age is a vector quantity.

6. Trigonometric form is more convenient than standard form when adding complex numbers.

7. The length of the sum of two vectors is the sum of the lengths of the vectors.

8. For any complex number $w$, there are exactly three complex numbers $z$ for which $z^3 = w$.

9. If the point $(x, y)$ in rectangular coordinates corresponds to the point $(r, \theta)$ in polar coordinates, then $x + iy = r(\cos \theta + i \sin \theta)$.

10. The graph of $r = \theta$ crosses the graph of $\theta = \pi/4$ once.

**PROBLEMS 11–14** □ *Give an example of each of the following.*

11. Side lengths of a triangle having angles 60° and 70°

12. A cube root of $i$

13. Vectors **u** and **v** such that $\|\mathbf{u}\| = \|\mathbf{v}\| = 2$ and $\mathbf{u} + \mathbf{v} = 0$

14. A complex number with argument $\theta = 3\pi/4$.

**PROBLEMS 15–16** □ *Solve the given triangle. Assume angles A, B, and C, and sides a, b, and c are labeled as shown in Figure 95.*

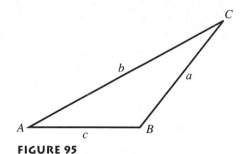

**FIGURE 95**

15. $b = 12$, $c = 8$, $A = 55°$

16. $a = 3$, $c = 4$, $C = 62°$

17. Let $z = -1 - i$.

    a. Represent $z$ graphically and write it in trigonometric form.

    b. Compute $z^6$.

    c. Approximate the square roots of $z$.

18. Let $\mathbf{u} = \langle 2, 4 \rangle$ and **v** be a vector with magnitude $\sqrt{2}$ and direction angle $\theta = \pi/4$.

    a. Express **v** in the form $a\mathbf{i} + b\mathbf{j}$.

    b. Compute $2\mathbf{u} - 3\mathbf{v}$.

19. Sketch the graph of $r = \sin 2\theta$.

20. A ship leaves a certain port at 9:00 A.M. and sails on a course of 30° (clockwise from north) at a rate of 15 miles per hour. A second ship leaves the same port at 9:30 and sails on a course of 100° at a rate of 12 miles per hour. How far apart are the two ships at 11:00?

# SYSTEMS OF EQUATIONS AND INEQUALITIES

■ Hurricane Andrew struck southern Florida and Louisiana in August 1992 causing over $20 billion in damage. With enough destructive force to topple homes, uproot trees, and change the course of rivers, hurricanes provide meteorologists with a strong incentive to accurately model the Earth's weather systems: the difference between 2 hours notice and 2 days notice of an impending hurricane can be the difference between life and death. Mathematical models of the Earth's atmosphere are incredibly complex, often involving the solution of systems consisting of thousands of linear equations. In this chapter we will explore several methods of solving such systems.

---

**SECTION 1**

## MATRICES

- When is $AB \neq BA$?
- How can $AB = 0$ if neither $A$ nor $B$ is 0?
- How can you find the coordinates of a point after it undergoes a rotation?
- If 10% of the nonsmokers take up smoking every year, while 20% of the smokers quit smoking each year, then in a town of 2000 people, how many people will smoke after 32 years, and why doesn't the answer depend on the current number of smokers?

### FUNDAMENTALS

A matrix (plural: *matrices*) is a rectangular array used to store and manipulate data. For example, the linear system of equations

$$2x + 3y + 2z = 3$$

$$4x - 7y - z = 2$$

$$2x - 4y + 3z = -4$$

can be encoded as the $3 \times 4$ matrix

$$\begin{bmatrix} 2 & 3 & 2 & 3 \\ 4 & -7 & -1 & 2 \\ 2 & -4 & 3 & -4 \end{bmatrix}$$

The performance of a company whose domestic division has quarterly earnings (in millions of dollars) of 2.3, 2.7, 3.2, and 4.1, and whose international division has quarterly earnings of 1.8, 1.2, 2.0, and 1.7, can be expressed using the $2 \times 4$ matrix

$$\begin{bmatrix} 2.3 & 2.7 & 3.2 & 4.1 \\ 1.8 & 1.2 & 2.0 & 1.7 \end{bmatrix}$$

We give a formal definition of a matrix as follows.

### Matrix definitions

- **A matrix of order $m \times n$** is a rectangular array consisting of $m$ (horizontal) **rows** and $n$ (vertical) **columns**, as shown in Figure 1.

$$\left.\begin{bmatrix} a_{11} & a_{12} & a_{13} \cdots a_{1n} \\ a_{21} & a_{22} & a_{23} \cdots a_{2n} \\ a_{31} & a_{32} & a_{33} \cdots a_{3n} \\ \vdots & \vdots & \vdots & \vdots \\ a_{m1} & a_{m2} & a_{m3} \cdots a_{mn} \end{bmatrix}\right\} m \text{ rows}$$

$\underbrace{\phantom{a_{11} \quad a_{12} \quad a_{13}}}_{n \text{ columns}}$

**FIGURE 1**

- The $m \times n$ matrix in which every entry is zero is called the **$m \times n$ zero matrix**.
- Two $m \times n$ matrices are said to be equal if corresponding entries are equal.

Note that the entries of the matrix in Figure 1 are written using a double subscript. The first subscript indicates the row of the entry, and the second subscript indicates the column of the entry. Rows are numbered from top to bottom, and columns are numbered from left to right. Thus, the entry in the $i$th row from the top and the $j$th column from the left is denoted by $a_{ij}$, as shown in Figure 2.

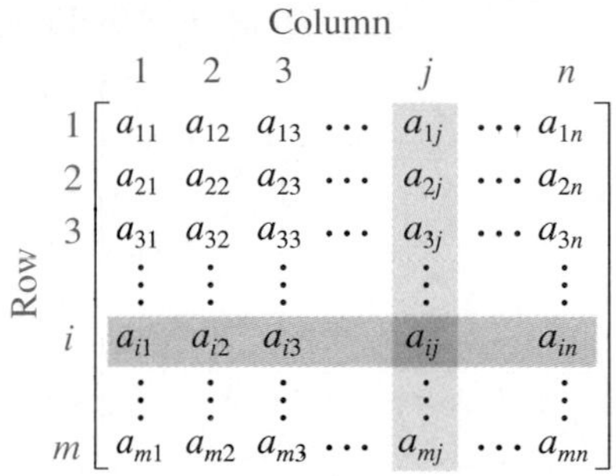

**FIGURE 2**                                    Entry $a_{ij}$ is in the $i$th row and the $j$th column.

The matrix for which the entry in row $i$ and column $j$ is $a_{ij}$ is often denoted by $(a_{ij})$. This is an especially convenient way of representing matrices with unknown or arbitrary entries. Thus, for example, we can write

$$(a_{ij}) = \begin{bmatrix} a_{11} & a_{12} & a_{13} & \cdots & a_{1n} \\ a_{21} & a_{22} & a_{23} & \cdots & a_{2n} \\ a_{31} & a_{32} & a_{33} & \cdots & a_{3n} \\ \vdots & \vdots & \vdots & & \vdots \\ a_{m1} & a_{m2} & a_{m3} & \cdots & a_{mn} \end{bmatrix} \quad \text{and} \quad (b_{ij}) = \begin{bmatrix} b_{11} & b_{12} & b_{13} & \cdots & b_{1n} \\ b_{21} & b_{22} & b_{23} & \cdots & b_{2n} \\ b_{31} & b_{32} & b_{33} & \cdots & b_{3n} \\ \vdots & \vdots & \vdots & & \vdots \\ b_{m1} & b_{m2} & b_{m3} & \cdots & b_{mn} \end{bmatrix}$$

**EXAMPLE 1**    *Identifying entries using subscript notation*

Let

$$A = \begin{bmatrix} 5 & 6 & \frac{1}{2} \\ -2 & 3 & -7 \end{bmatrix}$$

**a.** What is the order of $A$?

**b.** If $A = (a_{ij})$, identify $a_{21}$ and $a_{13}$.

**SOLUTION**

**a.** Since $A$ has 2 rows and 3 columns, it is of order $2 \times 3$.

**b.** The entry $a_{21}$ is in the second row and the first column. Thus, $a_{21} = -2$. The entry $a_{13}$ is in the first row and the third column, and so $a_{13} = \frac{1}{2}$.

If the only function of matrices were to serve as a convenient way of tabulating data, then they would be nothing more than glorified tables. It is the *algebraic* properties of matrices that set them apart from simple tables and make them especially useful as a tool for solving linear systems of equations. Under certain conditions, matrices can be added, subtracted, multiplied, and even divided.

### ADDITION AND SUBTRACTION OF MATRICES

The definitions of addition and subtraction of matrices are very intuitive; we simply add or subtract corresponding entries. Thus, if

$$A = \begin{bmatrix} 2 & -4 \\ 3 & -5 \end{bmatrix} \quad \text{and} \quad B = \begin{bmatrix} -1 & 7 \\ 2 & 3 \end{bmatrix}$$

then $A + B$ and $A - B$ are computed as shown below.

$$A + B = \begin{bmatrix} 2 & -4 \\ 3 & -5 \end{bmatrix} + \begin{bmatrix} -1 & 7 \\ 2 & 3 \end{bmatrix} \qquad A - B = \begin{bmatrix} 2 & -4 \\ 3 & -5 \end{bmatrix} - \begin{bmatrix} -1 & 7 \\ 2 & 3 \end{bmatrix}$$

$$= \begin{bmatrix} (2 + (-1)) & (-4 + 7) \\ (3 + 2) & (-5 + 3) \end{bmatrix} \qquad = \begin{bmatrix} (2 - (-1)) & (-4 - 7) \\ (3 - 2) & (-5 - 3) \end{bmatrix}$$

$$= \begin{bmatrix} 1 & 3 \\ 5 & -2 \end{bmatrix} \qquad = \begin{bmatrix} 3 & -11 \\ 1 & -8 \end{bmatrix}$$

We give the formal definition of matrix addition and subtraction here.

*Matrix addition and subtraction*

For the $m \times n$ matrices $A = (a_{ij})$ and $B = (b_{ij})$,

$A + B$ is the $m \times n$ matrix $(c_{ij})$, where $c_{ij} = a_{ij} + b_{ij}$.

$A - B$ is the $m \times n$ matrix $(d_{ij})$, where $d_{ij} = a_{ij} - b_{ij}$.

We cannot define matrix addition or subtraction unless the matrices are the same order.

**EXAMPLE 2**    *Adding and subtracting matrices*

Perform the following matrix operations.

a. $\begin{bmatrix} 0 & 4 & 5 \\ 1 & 6 & 7 \end{bmatrix} + \begin{bmatrix} -2 & 2 & 4 \\ -2 & 5 & 4 \end{bmatrix}$      b. $\begin{bmatrix} 2 & 3 \\ 4 & 5 \end{bmatrix} - \begin{bmatrix} 1 & 2 \\ 3 & 2 \end{bmatrix}$      c. $\begin{bmatrix} 1 & 2 \\ 4 & 5 \\ 3 & 6 \end{bmatrix} + \begin{bmatrix} 0 & 2 \\ 5 & 2 \end{bmatrix}$

**SOLUTION**

a. $\begin{bmatrix} 0 & 4 & 5 \\ 1 & 6 & 7 \end{bmatrix} + \begin{bmatrix} -2 & 2 & 4 \\ -2 & 5 & 4 \end{bmatrix} = \begin{bmatrix} (0 + -2) & (4 + 2) & (5 + 4) \\ (1 + -2) & (6 + 5) & (7 + 4) \end{bmatrix}$

$$= \begin{bmatrix} -2 & 6 & 9 \\ -1 & 11 & 11 \end{bmatrix}$$

b. $\begin{bmatrix} 2 & 3 \\ 4 & 5 \end{bmatrix} - \begin{bmatrix} 1 & 2 \\ 3 & 2 \end{bmatrix} = \begin{bmatrix} (2 - 1) & (3 - 2) \\ (4 - 3) & (5 - 2) \end{bmatrix}$

$$= \begin{bmatrix} 1 & 1 \\ 1 & 3 \end{bmatrix}$$

c. Since the first matrix is of order $3 \times 2$ and the second is of order $2 \times 2$, the matrices are of different orders and cannot be added together.

Matrices are often useful in applications where large quantities of data are involved. In the following example, we see how matrices can be used to organize data, and also how matrix addition can be used to "combine" the information stored in matrices.

**EXAMPLE 3**    *An application of matrix addition*

A telecommunications company has two divisions, one that deals primarily with computer networking and one that deals primarily with satellite data transmission. The costs and revenues (in millions of dollars) for the two divisions for the four quarters of 1996 are as follows. The networking division had costs of 1.5, 0.5, 1.2, and 1.1 million for the first through fourth quarters, respectively, and revenues of 1.8, 0.7, 1.4, and 1.2 million, respectively. The satellite transmission division had costs of 2.8, 1.4, 1.5, and 1.7 million for the first through fourth quarters, respectively, and revenues of 3.3, 1.8, 2.0, and 2.1 million, respectively. Organize this information in matrix form and use matrix addition to find the company's total cost and total revenue for each of the four quarters of 1996.

**SOLUTION**    We define two $4 \times 2$ matrices, one for each of the divisions. The 4 rows correspond to the 4 quarters of 1996, and the 2 columns correspond to cost and revenue. Thus, if $N$ and $S$ denote the matrices for the networking and satellite divisions, respectively, then

$$N = \begin{bmatrix} 1.5 & 1.8 \\ 0.5 & 0.7 \\ 1.2 & 1.4 \\ 1.1 & 1.2 \end{bmatrix} \quad \text{and} \quad S = \begin{bmatrix} 2.8 & 3.3 \\ 1.4 & 1.8 \\ 1.5 & 2.0 \\ 1.7 & 2.1 \end{bmatrix}$$

The sum of the two matrices is found by adding corresponding entries as shown.

$$N + S = \begin{bmatrix} (1.5 + 2.8) & (1.8 + 3.3) \\ (0.5 + 1.4) & (0.7 + 1.8) \\ (1.2 + 1.5) & (1.4 + 2.0) \\ (1.1 + 1.7) & (1.2 + 2.1) \end{bmatrix} = \begin{bmatrix} 4.3 & 5.1 \\ 1.9 & 2.5 \\ 2.7 & 3.4 \\ 2.8 & 3.3 \end{bmatrix}$$

To interpret the entries of this matrix, we express the information in tabular form.

| *Quarter* | *Cost* | *Revenue* |
| --- | --- | --- |
| 1 | 4.3 | 5.1 |
| 2 | 1.9 | 2.5 |
| 3 | 2.7 | 3.4 |
| 4 | 2.8 | 3.3 |

## SCALAR MULTIPLICATION

In the context of matrices, it is customary to refer to ordinary real or complex numbers as **scalars**. To multiply a matrix by a scalar, we simply multiply each entry of the matrix by the scalar. For example,

$$2 \cdot \begin{bmatrix} 3 & 5 \\ 7 & 8 \end{bmatrix} = \begin{bmatrix} (2 \cdot 3) & (2 \cdot 5) \\ (2 \cdot 7) & (2 \cdot 8) \end{bmatrix}$$

$$= \begin{bmatrix} 6 & 10 \\ 14 & 16 \end{bmatrix}$$

More generally, we have the following definition.

*Scalar multiplication*

> If $c$ is a real or complex number and $A = (a_{ij})$, then $cA = (b_{ij})$, where $b_{ij} = ca_{ij}$.

**EXAMPLE 4**    *Performing scalar multiplication*

Compute $-\frac{1}{2} \cdot \begin{bmatrix} 2 & -4 \\ 8 & 5 \\ -6 & 7 \end{bmatrix}$.

**SOLUTION**    We simply multiply each entry of the matrix by $-\frac{1}{2}$.

$$-\frac{1}{2} \cdot \begin{bmatrix} 2 & -4 \\ 8 & 5 \\ -6 & 7 \end{bmatrix} = \begin{bmatrix} (-\frac{1}{2} \cdot 2) & (-\frac{1}{2} \cdot (-4)) \\ (-\frac{1}{2} \cdot 8) & (-\frac{1}{2} \cdot 5) \\ (-\frac{1}{2} \cdot (-6)) & (-\frac{1}{2} \cdot 7) \end{bmatrix} = \begin{bmatrix} -1 & 2 \\ -4 & -\frac{5}{2} \\ 3 & -\frac{7}{2} \end{bmatrix}$$

**EXAMPLE 5**    *Combining scalar multiplication with subtraction*

Let

$$A = \begin{bmatrix} 1 & 4 \\ 5 & 3 \end{bmatrix} \quad \text{and} \quad B = \begin{bmatrix} 3 & 6 \\ 4 & 2 \end{bmatrix}$$

Compute $3A - 2B$.

**SOLUTION**

$$3A - 2B = 3 \cdot \begin{bmatrix} 1 & 4 \\ 5 & 3 \end{bmatrix} - 2 \cdot \begin{bmatrix} 3 & 6 \\ 4 & 2 \end{bmatrix}$$

$$= \begin{bmatrix} 3 & 12 \\ 15 & 9 \end{bmatrix} - \begin{bmatrix} 6 & 12 \\ 8 & 4 \end{bmatrix} \qquad \text{Performing the scalar multiplications}$$

$$= \begin{bmatrix} -3 & 0 \\ 7 & 5 \end{bmatrix} \qquad \text{Subtracting corresponding entries}$$

## MATRIX MULTIPLICATION

To add or subtract two matrices, we simply add or subtract corresponding entries. The product of two matrices, however, is *not* found by multiplying corresponding entries. Instead, each entry of the product matrix is the "product" of a *row* of the first matrix with a *column* of the second matrix.

### *Multiplication of a row and a column*

If

$$R = [a_1 \ a_2 \ a_3 \cdots a_n] \quad \text{and} \quad C = \begin{bmatrix} b_1 \\ b_2 \\ \vdots \\ b_n \end{bmatrix}$$

then $RC = a_1 b_1 + a_2 b_2 + \cdots + a_n b_n$. Note that the product of a row and a column is a scalar.

---

**WARNING**    A row and column must have the same number of entries in order to be multiplied.

---

**EXAMPLE 6**    *Multiplying a row by a column*

Let

$$A = \begin{bmatrix} 2 & -5 & 4 \\ -1 & 7 & 5 \end{bmatrix} \quad \text{and} \quad B = \begin{bmatrix} 1 & 2 & -3 & 5 \\ 3 & -2 & 1 & 5 \\ 5 & 4 & 0 & -7 \end{bmatrix}$$

Compute the product of the first row of $A$ with the third column of $B$.

**SOLUTION**    We must compute

$$[2 \ -5 \ 4] \cdot \begin{bmatrix} -3 \\ 1 \\ 0 \end{bmatrix}$$

According to the definition, we multiply corresponding entries and then add. Thus, we have

$$[2 \ -5 \ 4] \cdot \begin{bmatrix} -3 \\ 1 \\ 0 \end{bmatrix} = 2 \cdot (-3) + (-5) \cdot 1 + 4 \cdot 0 = -11$$

---

Note that in the preceding example, matrix $A$ has the same number of columns as matrix $B$ has rows. This ensures that the rows of $A$ have the same number of entries as the columns of $B$. In fact, whenever the number of columns of $A$ is the same as the number of rows of $B$, the matrix product $AB$ is defined. The entry in row $i$ and column $j$ of $AB$ is simply the product of the $i$th row of $A$ with the $j$th column of $B$, as stated below.

■ *Multiplication of matrices*

If $A$ has the same number of columns as $B$ has rows, then $AB = (c_{ij})$, where

$$c_{ij} = (\text{row } i \text{ of } A) \cdot (\text{column } j \text{ of } B)$$

It follows that if the order of $A$ is $m \times n$ and the order of $B$ is $n \times p$, then $AB$ has order $m \times p$.

Figure 3 depicts the relationships between the orders of the factors and the resulting product matrix.

$m \times n$
$n$ columns

$n \times p$
$p$ columns

$m \times p$
$p$ columns

**FIGURE 3**

**RULE OF THUMB** When multiplying matrices, first write their dimensions side by side in the order in which the product is to be computed. If the inner numbers are the same, the product is defined, and the order of the product is given by the outer numbers. If the inner numbers are different, then the product is not defined.

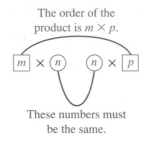

The order of the product is $m \times p$.

$\boxed{m} \times \bigcirc{n} \quad \bigcirc{n} \times \boxed{p}$

These numbers must be the same.

**EXAMPLE 7**    *Multiplication of matrices*

Multiply

$$\begin{bmatrix} 1 & 2 & 3 \\ -2 & 0 & 5 \end{bmatrix} \cdot \begin{bmatrix} 2 & 1 \\ -3 & 4 \\ 2 & 1 \end{bmatrix}$$

**SOLUTION**    Since the first factor is a $2 \times 3$ matrix and the second is a $3 \times 2$ matrix, the product will be a $2 \times 2$ matrix. Let us denote the product by $(p_{ij})$, so that we have

$$\begin{bmatrix} 1 & 2 & 3 \\ -2 & 0 & 5 \end{bmatrix} \cdot \begin{bmatrix} 2 & 1 \\ -3 & 4 \\ 2 & 1 \end{bmatrix} = \begin{bmatrix} p_{11} & p_{12} \\ p_{21} & p_{22} \end{bmatrix}$$

Table 1 illustrates the process of computing the product $(p_{ij})$. Note that the product of row $i$ from the first matrix and column $j$ from the second gives $p_{ij}$.

**TABLE 1**

| **Big picture** | **Computations** | **Update** |
|---|---|---|
| $\begin{bmatrix} \boxed{1\ 2\ 3} \\ -2\ 0\ 5 \end{bmatrix} \cdot \begin{bmatrix} \boxed{\begin{matrix}2\\-3\\2\end{matrix}}\ \begin{matrix}1\\4\\1\end{matrix} \end{bmatrix} = \begin{bmatrix} \boxed{p_{11}}\ p_{12} \\ p_{21}\ p_{22} \end{bmatrix}$<br><br>Row 1        Column 1 | $[1\ 2\ 3] \cdot \begin{bmatrix} 2 \\ -3 \\ 2 \end{bmatrix} = 1 \cdot 2 + 2 \cdot (-3) + 3 \cdot 2$<br><br>$= 2 - 6 + 6$<br>$= 2$ | $\begin{bmatrix} 2\ p_{12} \\ p_{21}\ p_{22} \end{bmatrix}$ |
| $\begin{bmatrix} \boxed{1\ 2\ 3} \\ -2\ 0\ 5 \end{bmatrix} \cdot \begin{bmatrix} \begin{matrix}2\\-3\\2\end{matrix}\ \boxed{\begin{matrix}1\\4\\1\end{matrix}} \end{bmatrix} = \begin{bmatrix} 2\ \boxed{p_{12}} \\ p_{21}\ p_{22} \end{bmatrix}$<br><br>Row 1        Column 2 | $[1\ 2\ 3] \cdot \begin{bmatrix} 1 \\ 4 \\ 1 \end{bmatrix} = 1 \cdot 1 + 2 \cdot 4 + 3 \cdot 1$<br><br>$= 1 + 8 + 3$<br>$= 12$ | $\begin{bmatrix} 2\ 12 \\ p_{21}\ p_{22} \end{bmatrix}$ |
| $\begin{bmatrix} 1\ 2\ 3 \\ \boxed{-2\ 0\ 5} \end{bmatrix} \cdot \begin{bmatrix} \boxed{\begin{matrix}2\\-3\\2\end{matrix}}\ \begin{matrix}1\\4\\1\end{matrix} \end{bmatrix} = \begin{bmatrix} 2\ 12 \\ \boxed{p_{21}}\ p_{22} \end{bmatrix}$<br><br>Row 2        Column 1 | $[-2\ 0\ 5] \cdot \begin{bmatrix} 2 \\ -3 \\ 2 \end{bmatrix} = -2 \cdot 2 + 0 \cdot (-3) + 5 \cdot 2$<br><br>$= -4 + 0 + 10$<br>$= 6$ | $\begin{bmatrix} 2\ 12 \\ 6\ p_{22} \end{bmatrix}$ |
| $\begin{bmatrix} 1\ 2\ 3 \\ \boxed{-2\ 0\ 5} \end{bmatrix} \cdot \begin{bmatrix} \begin{matrix}2\\-3\\2\end{matrix}\ \boxed{\begin{matrix}1\\4\\1\end{matrix}} \end{bmatrix} = \begin{bmatrix} 2\ 12 \\ 6\ \boxed{p_{22}} \end{bmatrix}$<br><br>Row 2        Column 2 | $[-2\ 0\ 5] \cdot \begin{bmatrix} 1 \\ 4 \\ 1 \end{bmatrix} = -2 \cdot 1 + 0 \cdot 4 + 5 \cdot 1$<br><br>$= -2 + 0 + 5$<br>$= 3$ | $\begin{bmatrix} 2\ 12 \\ 6\ 3 \end{bmatrix}$ |

Thus, the product is $\begin{bmatrix} 2 & 12 \\ 6 & 3 \end{bmatrix}$.

■

The order in which matrices are multiplied is crucial. For example, suppose that $A$ is a $2 \times 4$ matrix and $B$ is a $4 \times 3$ matrix. Then $AB$ is a $2 \times 3$ matrix, but $BA$ isn't even defined! Even if both $AB$ and $BA$ are defined, they are generally not equal.

**EXAMPLE 8**   *Multiplying $2 \times 2$ matrices in both orders*

Let

$$A = \begin{bmatrix} -1 & 2 \\ 3 & 4 \end{bmatrix} \quad \text{and} \quad B = \begin{bmatrix} 2 & -1 \\ 3 & 2 \end{bmatrix}$$

Compute both $AB$ and $BA$ to show that $AB \neq BA$.

**SOLUTION**

$$AB = \begin{bmatrix} -1 & 2 \\ 3 & 4 \end{bmatrix} \cdot \begin{bmatrix} 2 & -1 \\ 3 & 2 \end{bmatrix}$$

$$= \begin{bmatrix} (-1 \cdot 2 + 2 \cdot 3) & ((-1) \cdot (-1) + 2 \cdot 2) \\ (3 \cdot 2 + 4 \cdot 3) & (3 \cdot (-1) + 4 \cdot 2) \end{bmatrix}$$

$$= \begin{bmatrix} 4 & 5 \\ 18 & 5 \end{bmatrix}$$

$$BA = \begin{bmatrix} 2 & -1 \\ 3 & 2 \end{bmatrix} \cdot \begin{bmatrix} -1 & 2 \\ 3 & 4 \end{bmatrix}$$

$$= \begin{bmatrix} (2 \cdot (-1) + (-1) \cdot 3) & (2 \cdot 2 + (-1) \cdot 4) \\ (3 \cdot (-1) + 2 \cdot 3) & (3 \cdot 2 + 2 \cdot 4) \end{bmatrix}$$

$$= \begin{bmatrix} -5 & 0 \\ 3 & 14 \end{bmatrix}$$

Thus, $AB \neq BA$.

The matrix product, although seemingly defined in an unnatural way, has many applications. We will explore some of them in the exercise sets. The primary use of matrix multiplication in this text will be discussed in Section 4, where we solve systems of linear equations by employing matrix multiplication. In the next example we show that a system of equations is equivalent to a matrix equation.

**EXAMPLE 9**    *Expressing a system of equations in matrix form*

Show that the system of equations

$$2x + 3y + 4z = 5$$

$$3x - 5y + z = 6$$

$$4x + 7y - z = 3$$

is equivalent to the matrix equation

$$\begin{bmatrix} 2 & 3 & 4 \\ 3 & -5 & 1 \\ 4 & 7 & -1 \end{bmatrix} \cdot \begin{bmatrix} x \\ y \\ z \end{bmatrix} = \begin{bmatrix} 5 \\ 6 \\ 3 \end{bmatrix}$$

**SOLUTION**    We begin by multiplying the matrices on the left side of the equation.

$$\begin{bmatrix} 2 & 3 & 4 \\ 3 & -5 & 1 \\ 4 & 7 & -1 \end{bmatrix} \cdot \begin{bmatrix} x \\ y \\ z \end{bmatrix} = \begin{bmatrix} 5 \\ 6 \\ 3 \end{bmatrix}$$

$$\begin{bmatrix} 2x + 3y + 4z \\ 3x - 5y + z \\ 4x + 7y - z \end{bmatrix} = \begin{bmatrix} 5 \\ 6 \\ 3 \end{bmatrix}$$

Since two matrices are equal if and only if corresponding entries are equal, we have

$$2x + 3y + 4z = 5$$

$$3x - 5y + z = 6$$

$$4x + 7y - z = 3$$

The arithmetic operations of matrix addition, subtraction, and multiplication, as well as scalar multiplication, can be performed by your graphics calculator. First it is necessary to specify the dimension or order of each matrix. The next step is to enter the entries of each matrix. Finally, the matrix expression can be entered and evaluated. For details, consult the graphics calculator supplement.

Matrix multiplication arises in applications in various ways. The following example illustrates how matrix multiplication is used to compare the effect of different interest rates on the total return of an investment.

**EXAMPLE 10**

*An application of matrix multiplication*

Lachelle has $500 withheld out of each month's paycheck and placed in a 401k retirement plan. The plan offers two fund choices: a high-risk growth fund and a low-risk income fund. Lachelle has chosen to place $350 in the growth fund and $150 in the income fund. The manager of each fund determines how investments are distributed among three categories of stocks: blue chip, biotech, and international. The manager of the growth fund distributes all investments in the following way: 30% to blue chip stocks, 50% to biotech stocks, and 20% to international stocks. The manager of the income fund distributes investments in the following way: 50% to blue chip stocks, 35% to biotech stocks, and 15% to international stocks. Express the distribution of Lachelle's $500 monthly withholding as a 2 × 3 matrix, and use matrix multiplication to find the earnings of each fund for a single month given that the rate of return is

**a.** 8% for blue chip stocks, 10% for biotech stocks, and 9% for international stocks.
**b.** 9% for blue chip stocks, 13% for biotech stocks, and 7% for international stocks.

**SOLUTION**    We define a 2 × 3 matrix $A$ representing Lachelle's investment portfolio, with each row corresponding to a fund (growth and income, in that order) and each column corresponding to a stock category (blue chip, biotech, and international, in that order). Thus, we have

$$A = \begin{bmatrix} (30\% \text{ of } 350) & (50\% \text{ of } 350) & (20\% \text{ of } 350) \\ (50\% \text{ of } 150) & (35\% \text{ of } 150) & (15\% \text{ of } 150) \end{bmatrix} = \begin{bmatrix} 105 & 175 & 70 \\ 75 & 52.5 & 22.5 \end{bmatrix}$$

**a.** For a given fund, the amounts invested in blue chip, biotech, and international stocks must be multiplied by 0.08, 0.10, and 0.09, respectively, and the resulting amounts must be added together. Thus, each row of matrix $A$ must be multiplied by the $3 \times 1$ matrix

$$B = \begin{bmatrix} 0.08 \\ 0.10 \\ 0.09 \end{bmatrix}$$

Thus, the interest for both funds can be found by computing the matrix product $AB$, a task that is ideally suited to a graphics calculator. We obtain the following product.

$$AB = \begin{bmatrix} 105 & 175 & 70 \\ 75 & 52.5 & 22.5 \end{bmatrix} \begin{bmatrix} 0.08 \\ 0.10 \\ 0.09 \end{bmatrix} = \begin{bmatrix} 32.2 \\ 13.275 \end{bmatrix}$$

Thus, on Lachelle's investment of $500, the growth fund earns $32.20 and the income fund earns approximately $13.28.

**b.** We redefine the matrix $B$ as

$$B = \begin{bmatrix} 0.09 \\ 0.13 \\ 0.07 \end{bmatrix}$$

and again compute $AB$ to obtain

$$AB = \begin{bmatrix} 105 & 175 & 70 \\ 75 & 52.5 & 22.5 \end{bmatrix} \begin{bmatrix} 0.09 \\ 0.13 \\ 0.07 \end{bmatrix} = \begin{bmatrix} 37.1 \\ 15.15 \end{bmatrix}$$

Thus, the growth fund earns $37.10 and the income fund earns $15.15.

---

We now summarize some of the useful properties of matrices. Proofs of some of these properties in the case of $2 \times 2$ matrices will be developed in Exercises 55–58.

---

**Properties of matrix operations**

For scalars $x$ and $y$, and matrices $A$, $B$, and $C$:

1. $A + B = B + A$
2. $(A + B) + C = A + (B + C)$
3. $x(yA) = (xy)A$
4. $A(BC) = (AB)C$
5. $x(AB) = (xA)B = A(xB)$
6. $x(A + B) = xA + xB$
7. $A(B + C) = AB + AC$
8. $(A + B)C = AC + BC$

**EXERCISES 1**

**EXERCISES 1–12** ☐ *Perform the indicated matrix operation.*

1. $2 \cdot \begin{bmatrix} 1 & 0 \\ 3 & 5 \end{bmatrix}$

2. $-3 \cdot \begin{bmatrix} 1 & 5 & 3 \\ 4 & 0 & 2 \end{bmatrix}$

3. $\begin{bmatrix} 2 & 3 \\ 6 & 4 \end{bmatrix} + \begin{bmatrix} -4 & 8 \\ 0 & 2 \end{bmatrix}$

4. $[0 \ 1 \ 2 \ 3] + [4 \ 3 \ 1 \ 0]$

5. $\begin{bmatrix} 2 & -3 & 5 \\ -1 & 7 & 0 \end{bmatrix} - \begin{bmatrix} 3 & -5 & 2 \\ 1 & 0 & 4 \end{bmatrix}$

6. $\begin{bmatrix} -3 \\ 1 \\ 6 \end{bmatrix} + 2 \cdot \begin{bmatrix} 4 \\ -1 \\ 0 \end{bmatrix}$

7. $\begin{bmatrix} 0 & 3 \\ 6 & 7 \end{bmatrix} + \begin{bmatrix} -2 & 4 \\ 0 & 1 \end{bmatrix} - \begin{bmatrix} 2 & -5 \\ 3 & -5 \end{bmatrix}$

8. $\begin{bmatrix} a & b \\ c & d \end{bmatrix} + \begin{bmatrix} 3a & 2b \\ c & 5d \end{bmatrix}$

9. $2 \cdot \begin{bmatrix} 1 & 2 & 0 \\ 2 & 6 & 9 \\ 3 & -1 & -4 \end{bmatrix} + \begin{bmatrix} -1 & 3 & 4 \\ 3 & -3 & 0 \\ 1 & 4 & 5 \end{bmatrix}$

10. $3 \cdot \begin{bmatrix} 2 & 3 & 0 \\ 1 & -1 & 1 \\ -1 & -2 & 2 \end{bmatrix} - 2 \cdot \begin{bmatrix} 0 & -6 & -1 \\ 2 & -4 & 3 \\ -1 & 2 & -2 \end{bmatrix}$

11. $\begin{bmatrix} 2 & a & 1 \\ 0 & 1 & 2 \end{bmatrix} - a \cdot \begin{bmatrix} 1 & 1 & 1 \\ 1 & 1 & 0 \end{bmatrix}$

12. $\begin{bmatrix} 1 & x \\ x^2 & 0 \\ 2 & 1 \end{bmatrix} + \begin{bmatrix} x & 2 \\ -x^2 & 3 \\ x^2 & -2 \end{bmatrix}$

**EXERCISES 13–16** ☐ *Simplify the given quantity for*

$$A = \begin{bmatrix} 1 & 2 \\ 3 & 5 \end{bmatrix} \quad \text{and} \quad B = \begin{bmatrix} 2 & 1 \\ 4 & 9 \end{bmatrix}$$

13. $A + B$

14. $A - B$

15. $2A - 5B$

16. $3A + 2B$

**EXERCISES 17–20** ☐ *Solve the given equation for the unknown matrix X. (Hint: Solve for the variable as you would an ordinary algebraic equation, only using the corresponding matrix operations.)*

17. $2X = \begin{bmatrix} 4 & 6 \\ 8 & 2 \end{bmatrix}$

18. $3X + \begin{bmatrix} 1 & 0 \\ 0 & 1 \end{bmatrix} = \begin{bmatrix} 10 & 12 \\ 3 & 16 \end{bmatrix}$

19. $2X + \begin{bmatrix} 1 & 2 & 3 \\ 6 & 5 & 4 \end{bmatrix} = \begin{bmatrix} 0 & 0 & 0 \\ 0 & 0 & 0 \end{bmatrix}$

20. $-2 \left( X + \begin{bmatrix} 2 & 6 \\ 3 & 3 \end{bmatrix} \right) = -4X + \begin{bmatrix} 0 & 0 \\ 2 & 4 \end{bmatrix}$

**EXERCISES 21–26** ☐ *Let*

$$A = \begin{bmatrix} 1 & 2 & 4 \\ 2 & 3 & 0 \end{bmatrix}, \quad B = \begin{bmatrix} 1 & 2 \\ 3 & 4 \end{bmatrix}, \quad \text{and} \quad C = \begin{bmatrix} 3 & 7 & 9 \\ 4 & 5 & 1 \\ 2 & 6 & 4 \end{bmatrix}$$

*For each product, first indicate whether or not the product is defined. If the product is defined, give the order of the product matrix. You need not compute the product.*

21. $AB$

22. $AC$

23. $BA$

24. $BC$

25. $CA$

26. $CB$

**EXERCISES 27–44** ☐ *Compute the given products.*

27. $\begin{bmatrix} 3 & 5 \\ 4 & 2 \end{bmatrix} \begin{bmatrix} 3 \\ -3 \end{bmatrix}$

28. $\begin{bmatrix} a & b \\ c & d \end{bmatrix} \begin{bmatrix} x \\ y \end{bmatrix}$

29. $\begin{bmatrix} 2 & 3 \\ 4 & 6 \end{bmatrix} \begin{bmatrix} 1 & 2 \\ 0 & 1 \end{bmatrix}$

30. $\begin{bmatrix} 1 & 2 \\ 0 & 1 \end{bmatrix} \begin{bmatrix} 2 & 3 \\ 4 & 6 \end{bmatrix}$

31. $\begin{bmatrix} 1 & -2 \\ -5 & 7 \end{bmatrix} \begin{bmatrix} 3 & 1 \\ 0 & -4 \end{bmatrix}$

32. $\begin{bmatrix} 2 & 1 \\ 3 & 2 \end{bmatrix} \begin{bmatrix} 2 & -1 \\ -3 & 2 \end{bmatrix}$

33. $\begin{bmatrix} 2 & 0 \\ 1 & 4 \\ 2 & 1 \end{bmatrix} \begin{bmatrix} 3 & 5 \\ 1 & 7 \end{bmatrix}$

34. $\begin{bmatrix} 0 & 2 & 0 \\ 0 & 0 & 3 \\ 0 & 0 & 0 \end{bmatrix} \begin{bmatrix} 0 & 2 & 0 \\ 0 & 0 & 3 \\ 0 & 0 & 0 \end{bmatrix}$

35. $\begin{bmatrix} 1 & 0 & 0 \\ 0 & 1 & 0 \\ 0 & 0 & 1 \end{bmatrix} \cdot \begin{bmatrix} p & q \\ r & s \\ t & u \end{bmatrix}$

36. $\begin{bmatrix} a & 0 & 0 & 0 \\ 0 & b & 0 & 0 \\ 0 & 0 & c & 0 \\ 0 & 0 & 0 & d \end{bmatrix} \cdot \begin{bmatrix} p & 0 & 0 & 0 \\ 0 & q & 0 & 0 \\ 0 & 0 & r & 0 \\ 0 & 0 & 0 & s \end{bmatrix}$

37. $\begin{bmatrix} 1.8 & 3.5 & 4.6 \\ 4.8 & 1.7 & 3.2 \\ 1.7 & 2.5 & 0.7 \end{bmatrix} \begin{bmatrix} 2.6 \\ 3.4 \\ 4.9 \end{bmatrix}$

38. $\begin{bmatrix} 2.75 & 3.01 \\ 4.07 & 0.07 \end{bmatrix} \begin{bmatrix} 1.7 & 3.4 \\ 6.8 & 5.9 \end{bmatrix}$

39. $\begin{bmatrix} 1 & 0 \\ x & 1 \end{bmatrix} \begin{bmatrix} a & b \\ c & d \end{bmatrix}$

40. $\begin{bmatrix} x & 0 \\ 0 & 1 \end{bmatrix} \begin{bmatrix} a & b \\ c & d \end{bmatrix}$

41. $[a \ b \ c] \begin{bmatrix} i \\ j \\ k \end{bmatrix}$

42. $\begin{bmatrix} 0 & \dfrac{1}{b} \\ \dfrac{1}{a} & 0 \end{bmatrix} \begin{bmatrix} 0 & a \\ b & 0 \end{bmatrix}$

43. $\begin{bmatrix} 0 & 0 & 1 \\ 0 & 1 & 0 \\ 1 & 0 & 0 \end{bmatrix} \begin{bmatrix} a_{11} & a_{12} & a_{13} \\ a_{21} & a_{22} & a_{23} \\ a_{31} & a_{32} & a_{33} \end{bmatrix}$

44. $\begin{bmatrix} a_{11} & a_{12} & a_{13} \\ a_{21} & a_{22} & a_{23} \\ a_{31} & a_{32} & a_{33} \end{bmatrix} \begin{bmatrix} 0 & 0 & 1 \\ 0 & 1 & 0 \\ 1 & 0 & 0 \end{bmatrix}$

45. For real numbers $a$ and $b$, the zero product property holds: If $ab = 0$, then either $a = 0$ or $b = 0$. Show that the zero product property does *not* hold for matrices by finding a $1 \times 2$ matrix $A$ and a $2 \times 1$ matrix $B$ such that $AB - 0$, but neither $A$ nor $B$ consist entirely of zero entries.

**46.** If $A$ is an $n \times m$ matrix, then the **transpose** of $A$, denoted $A^T$, is the matrix whose rows are formed from the columns of $A$. More precisely, if $A = (a_{ij})$, then $A^T = (b_{ij})$, where $b_{ij} = a_{ji}$. For example, if

$$A = \begin{bmatrix} 2 & 4 \\ 6 & 8 \end{bmatrix}, \quad \text{then } A^T = \begin{bmatrix} 2 & 6 \\ 4 & 8 \end{bmatrix}$$

Compute the transposes of each of the following

a. $\begin{bmatrix} 1 & 2 & 3 \\ 4 & 5 & 6 \end{bmatrix}$     b. $\begin{bmatrix} a & b & c \\ d & e & f \\ g & h & i \end{bmatrix}$     c. $\begin{bmatrix} 1 & 2 & 3 \\ 2 & 5 & 6 \\ 3 & 6 & 4 \end{bmatrix}$

**EXERCISES 47–50** □ *Note that for a square matrix A, $A^2$ is defined as $A \cdot A$. Similarly $A^3 = A \cdot A \cdot A$, and so on.*

**47.** Compute $\begin{bmatrix} 1 & 2 \\ 0 & 1 \end{bmatrix}^2$

**48.** Compute $\begin{bmatrix} 0 & 1 \\ 0 & 0 \end{bmatrix}^3$

**49.** Compute $\begin{bmatrix} 1 & 0 \\ 0 & 0 \end{bmatrix}^4$

**50.** Compute $\begin{bmatrix} a & b \\ c & d \end{bmatrix}^2$

**EXERCISES 51–54** □ *Let*

$$A = \begin{bmatrix} 1 & 2 & 3 \\ 0 & 1 & 4 \\ 0 & 0 & 1 \end{bmatrix}$$

*and compute the indicated powers of A.*

**51.** $A^2$

**52.** $A^3$

**53.** $A^{32}$ (*Hint:* If you enter 3 in your calculator and then press $\boxed{x^2}$, you will obtain $3^2$. If you press $\boxed{x^2}$ again, you will obtain $(3^2)^2 = 3^4$. If you press $\boxed{x^2}$ once again, you obtain $3^8$, etc.)

**54.** $A^{33}$ (*Hint:* $A^{33} = A^{32} \cdot A$.)

**EXERCISES 55–58** □ *Confirm the given property for $2 \times 2$ matrices. Do so by letting*

$$A = \begin{bmatrix} a_1 & a_2 \\ a_3 & a_4 \end{bmatrix}, \quad B = \begin{bmatrix} b_1 & b_2 \\ b_3 & b_4 \end{bmatrix}, \quad \text{and} \quad C = \begin{bmatrix} c_1 & c_2 \\ c_3 & c_4 \end{bmatrix}$$

*and directly verifying the given identity.*

**55.** $A + B = B + A$

**56.** $(xA)B = x(AB)$, where $x$ is a scalar

**57.** $(A + B)C = AC + BC$

**58.** $x(A + B) = xA + xB$, where $x$ is a scalar

■ *Applications*

**59.** *Health-Care Claims* All employees of a computer software company, whether hourly workers or salaried personnel, select one of two health-care plans: major medical or comprehensive. Company records from 1994 show that major medical claims totaled $230,000 for hourly workers, $320,000 for dependents of hourly workers, $125,000 for salaried personnel, and $250,000 for dependents of salaried personel. The total of all claims filed for the comprehensive plan were $280,000 for hourly workers, $300,000 for dependents of hourly workers, $400,000 for salaried personnel, and $500,000 for their dependents. Express the breakdown of claims filed by hourly workers as a $2 \times 2$ matrix, and do likewise for claims filed by salaried personnel. Find and interpret the sum of these two matrices.

**60.** *Birth Distributions* A city has two major hospitals: St. Mary's and Parkview. St. Mary's recorded births of 200 boys and 210 girls for the first half of 1978 and 185 boys and 177 girls for the second half. Parkview Hospital recorded 320 girls and 307 boys for the first half and 300 girls and 295 boys for the second half. Express the births for each 6-month period as a $2 \times 2$ matrix, and compute and interpret the sum.

**61.** *Point Breakdown* The team statistician of the Charlotte Hornets recorded the following statistics from the team's last game against the Orlando Magic. The Hornets made 5 three-point shots, 40 two-point field goals, and 18 free throws. The Magic made 4 three-point shots, 37 two-point field goals, and 26 free throws. Express the shot distribution for the game as a $2 \times 3$ matrix $S$. Find a $3 \times 1$ matrix $T$ such that the final score will be given by the matrix product $ST$. What is the final score?

**62.** *Congressional Districts* Since the number of congressional districts is fixed at 435 and representation in the House of Representatives is proportional to population, it is occasionally necessary to redefine congressional districts. Population growth in a certain state has lagged behind that of the rest of the nation; thus, the state will be combining portions of two districts into one. The first district is 40% Republican and 30% Democrat, whereas the second is 25% Republican and 55% Democrat. There are 200,000 registered voters in the first district and 150,000 in the second. Express the political party percentage distribution of the two districts as a $2 \times 2$ matrix, and then find the total number of Democrats and Republicans in the new district using matrix multiplication assuming that the new district will be formed from:

a. 200,000 voters from the first district and 150,000 from the second

b. 175,000 voters from each district

■ *Projects for Enrichment*

**63.** *Linear transformation: An application of matrix multiplication* If the ray *OP* is rotated by an angle $\alpha$ about the origin, the point $P(x, y)$ is transformed to a new point $P'(x', y')$ as shown in Figure 4. We can express $x'$ and $y'$ in terms of $x$ and $y$ using matrix multiplication.

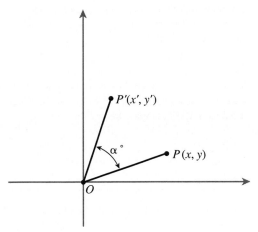

**FIGURE 4**

The point *P* in the Cartesian plane with coordinates $(x, y)$ can also be represented as the coordinate matrix

$$\begin{bmatrix} x \\ y \end{bmatrix}$$

The coordinates of $P'$ can be obtained by multiplying the coordinate matrix by a certain $2 \times 2$ matrix *A*. Thus, if the coordinates of $P'$ are $(x', y')$, then we have

$$A \begin{bmatrix} x \\ y \end{bmatrix} = \begin{bmatrix} x' \\ y' \end{bmatrix}$$

To find the appropriate matrix *A*, we use the following rules.

■ The first column of *A* represents the rotated coordinates of the point $(1, 0)$.

■ The second column of *A* represents the rotated coordinates of the point $(0, 1)$.

For example, consider a rotation by 90° counterclockwise. The point $(1, 0)$ will be rotated to the point $(0, 1)$ and so the first column of *A* is

$$\begin{bmatrix} 0 \\ 1 \end{bmatrix}$$

The point $(0, 1)$ will be rotated to the point $(-1, 0)$ so that the second column of *A* is

$$\begin{bmatrix} -1 \\ 0 \end{bmatrix}$$

Thus,

$$A = \begin{bmatrix} 0 & -1 \\ 1 & 0 \end{bmatrix}$$

**a.** Where does the point $(2, 3)$ end up if it is rotated counterclockwise by 90°?

**b.** Where does the point $(-1, 2)$ end up if it is rotated counterclockwise by 90°?

**c.** Define *A* to be the 90° counterclockwise rotation matrix as above. Compute $A^4$. Explain your answer.

**d.** Find the matrix *B* for a 45° counterclockwise rotation. [*Hint:* The point $(1, 0)$ will be rotated onto a point 1 unit from the origin on the line $y = x$; the point $(0, 1)$ will be rotated onto a point 1 unit from the origin on the line $y = -x$.]

**e.** Show directly that $B^2 = A$. Explain why this is logical.

**f.** What is the smallest whole number *n* so that $B^n = \begin{bmatrix} 1 & 0 \\ 0 & 1 \end{bmatrix}$? Explain your answer.

**64.** *Transition matrices: An application of matrix multiplication* Suppose that in a certain town, 10% of nonsmokers take up smoking each year, while 20% of smokers quit each year. If in a given year there are *x* smokers and *y* nonsmokers, then the smoking distribution of the population can be represented as

$$\begin{bmatrix} x \\ y \end{bmatrix}$$

The next year, the smokers will consist of the 80% of smokers who did *not* quit plus the 10% of former nonsmokers who take up smoking. Thus, the number of smokers in the next year will be $0.8x + 0.1y$. Similarly, the number of nonsmokers will be $0.2x + 0.9y$. Thus, after 1 year, the distribution of the population is given by

$$\begin{bmatrix} 0.8x + 0.1y \\ 0.2x + 0.9y \end{bmatrix}$$

**a.** Find a $2 \times 2$ matrix *A* such that

$$A \begin{bmatrix} x \\ y \end{bmatrix} = \begin{bmatrix} 0.8x + 0.1y \\ 0.2x + 0.9y \end{bmatrix}$$

This is a **transition matrix**.

b. Suppose that the town has a smoking distribution of

$$\begin{bmatrix} 600 \\ 1400 \end{bmatrix}$$

Compute the smoking distribution of the town after 4 years. (*Hint:* The smoking distribution after 1 year is given by

$$A \begin{bmatrix} 600 \\ 1400 \end{bmatrix}, \quad \text{after 2 years by} \quad A \cdot \left( A \begin{bmatrix} 600 \\ 1400 \end{bmatrix} \right), \quad \ldots )$$

c. Find the smoking distribution for the town after 32 years assuming its initial distribution is

$$\begin{bmatrix} 600 \\ 1400 \end{bmatrix}$$

Now find the smoking distribution after 64 years with the same initial distribution. What relationship do you find between your answers, and how can this be explained?

d. Compare the smoking distribution of two towns after 64 years, both with initial populations of 2000 people, but one having 100 smokers and the other having 1900 smokers. Explain the relationship between your results.

e. We are making numerous assumptions about the population when we employ transition matrices. For instance, we are assuming that there is no immigration either into or out of the town. What other assumptions are we making, and how reasonable are they?

f. We have assumed that the smoking transition matrix for every year is the same. Of course, this is unlikely to be true. What sort of trends would we be likely to find if we checked historical records? How would a smoking transition matrix for 1950 compare to one for 1995?

---

■▟ *Questions for Discussion or Essay*

65. How does a matrix differ from a simple table of numbers?

66. Most of the matrix products in this problem set can be determined using a graphics calculator. Why bother to learn how to multiply matrices by hand? For that matter, is there any point in learning how to multiply numbers by hand?

67. We are free to define the product of matrices in any way that we want. So why do we define them in the seemingly bizarre way that we do? Why not simply define the product of the matrices $(a_{ij})$ and $(b_{ij})$ to be $(c_{ij})$, where $c_{ij} = a_{ij}b_{ij}$?

68. It's reasonable to wonder how much gain we really get from having the graphics calculator perform matrix operations such

as matrix multiplication. To multiply a pair of $2 \times 2$ matrices by hand, a total of 12 arithmetic operations (addition or multiplication) must be performed. On the other hand, $4 + 4 = 8$ entries must be entered in the calculator. Develop a formula for the number of operations (additions and multiplications count as one each) required to multiply a pair of $n \times n$ matrices. (*Hint:* The product of two $n \times n$ matrices is an $n \times n$ matrix with $n^2$ entries. So compute the number of multiplications and additions required to determine each entry of the product matrix, and then multiply this number by $n^2$). How many entries must be entered into a calculator when multiplying two $n \times n$ matrices? In the case of the multiplication of a pair of $6 \times 6$ matrices, is the calculator worthwhile?

---

## SECTION 2

# LINEAR SYSTEMS AND MATRICES

■ How can diets be balanced by solving systems of equations?

■ How can an athlete choose an appropriate combination of activities as part of a cross-training routine?

■ Why are there never exactly two solutions to a linear system of equations?

■ How can we *quickly* find the equation of the parabola passing through three given points?

# LINEAR SYSTEMS OF EQUATIONS

A linear system of two equations in the variables $x$ and $y$ consists of two equations, each of which is linear in $x$ and $y$. Thus, a linear system of two equations in $x$ and $y$ is of the form

$$ax + by = c$$
$$dx + ey = f$$

For example, the system

$$3x - 4y = 12$$
$$2x + 9y = 1$$

is linear since both equations are linear, whereas the system

$$2x^2 - 3y = 5$$
$$x + 47y = 9$$

is nonlinear since one of the equations (the first in this case) is not linear.

We can consider linear systems consisting of any number of equations in any number of variables. For example, each of the following is a linear system of equations.

$$2x - 3y + 4z = 10 \qquad\qquad 2p - 3q + r - 5s = 3$$
$$x - y + z = 3 \qquad\qquad p + q + 2r + 3s = -4$$
$$3x + 5y - 2z = 9$$

(3 linear equations in 3 variables)     (2 linear equations in 4 variables)

More precisely, we have the following definition.

## Definition of a linear system of equations

A **linear system of $m$ equations in the $n$ variables** $x_1, x_2, x_3, \ldots, x_n$ is a system that can be expressed in the form

$$a_{11}x_1 + a_{12}x_2 + a_{13}x_3 + \cdots + a_{1n}x_n = b_1$$
$$a_{21}x_1 + a_{22}x_2 + a_{23}x_3 + \cdots + a_{2n}x_n = b_2$$
$$a_{31}x_1 + a_{32}x_2 + a_{33}x_3 + \cdots + a_{3n}x_n = b_3$$
$$\vdots \qquad\quad \vdots \qquad\quad \vdots \qquad\qquad\quad \vdots \qquad \vdots$$
$$a_{m1}x_1 + a_{m2}x_2 + a_{m3}x_3 + \cdots + a_{mn}x_n = b_m$$

where the $a_{ij}$'s and $b_i$'s are numbers. The $a_{ij}$'s are called *coefficients*.

# SOLUTION SETS OF LINEAR EQUATIONS

Consider the system of equations

$$ax + by = c$$
$$dx + ey = f$$

If a point $(x, y)$ is a solution of this system, then $(x, y)$ must satisfy each equation. Moreover, since the graph of each equation is a line, a solution point $(x, y)$ must lie on both lines and hence is a point of intersection. Now a pair of lines may intersect in a single point, but a pair of lines may also intersect in infinitely many points (if the lines are the same), or no points at all (if the lines are parallel). Thus, we see that the number of solutions to a system of two linear equations in two variables is either 0, 1, or infinity. Table 2 illustrates the range of possibilities for the solution set of a system of two equations in two variables.

**TABLE 2**

*Solution possibilities*

| *System* | *Graph* | *Points of intersection* |
|---|---|---|
| $2x - y = 1$  $x - 2y = -4$ | 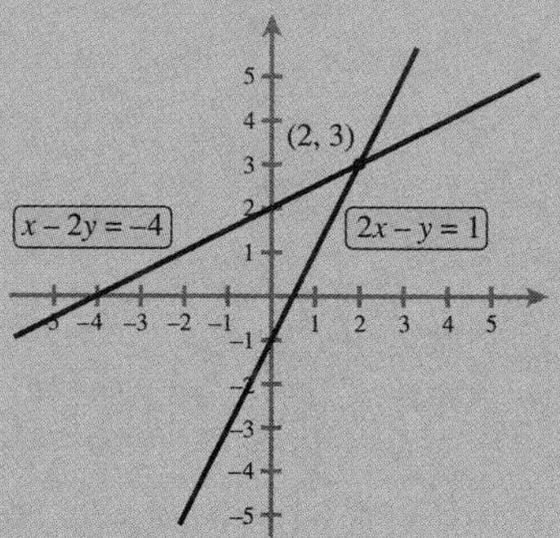 | Just one point of intersection: $(2, 3)$. |
| $2x - y = 1$  $2x - y = -3$ | 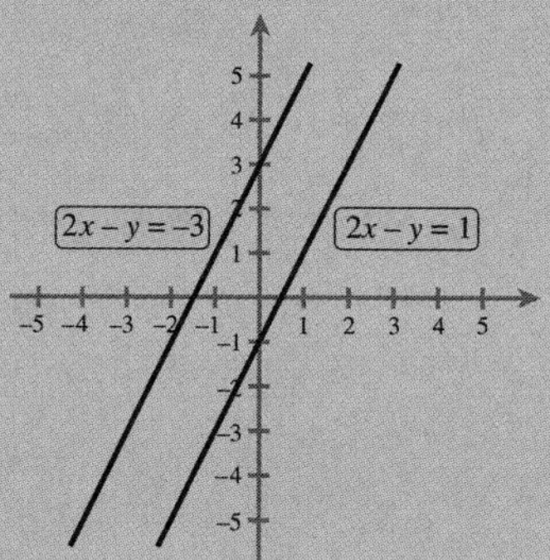 | Zero points of intersection: the lines are parallel. The equations are inconsistent. |
| $x + 2y = 3$  $-2x - 4y = -6$ | 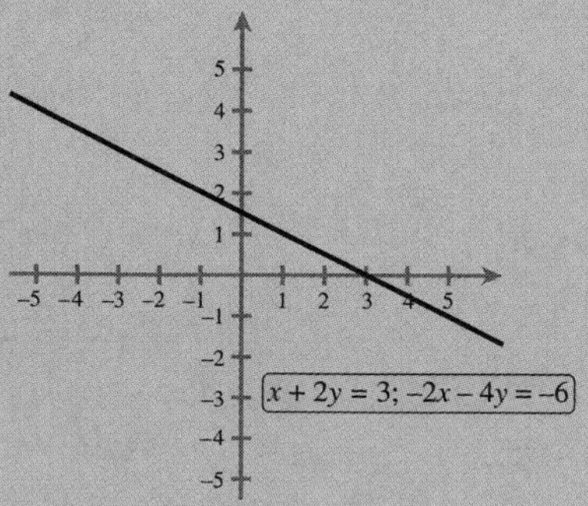 | Infinitely many points of intersection; the equations are of the same line. |

The results in Table 2 hold true for general linear systems, regardless of the number of equations or variables, as stated here.

| | |
|---|---|
| **Number of solutions to a linear system of equations** | A linear system of equations has either 0, 1, or infinitely many solutions. |

**EXAMPLE 1**    *Describing the solution set of a linear system with infinitely many solutions*

Solve the system of equations

$$x - y = 6$$
$$2x - 2y = 12$$

**SOLUTION**    Since the second equation can be obtained by multiplying both sides of the first equation by 2, there is, in essence, only one equation to be satisfied. Thus, the solution set consists of all pairs $(x, y)$ such that $x - y = 6$. If we solve this equation for $y$, we obtain $y = x - 6$, so that the solution set to the system could also be described as the set of pairs $(x, x - 6)$. Thus, for example, when $x = 1$, we obtain the point $(1, -5)$ as a solution; when $x = -3$, we obtain $(-3, -9)$, and so on.

---

## REPRESENTING SYSTEMS WITH MATRICES

The notation used in defining linear systems is suggestive of matrix notation. In fact, we have already seen (see Example 9 of Section 1) that matrices are a convenient way of expressing linear equations. As we will soon see, matrices are also an invaluable tool for *solving* linear systems of equations. The following definitions are central to our discussion.

| | |
|---|---|
| **Definition of coefficient matrix and augmented matrix** | The matrix $(a_{ij})$ is called the **coefficient matrix** of the linear system of equations |

$$a_{11}x_1 + a_{12}x_2 + a_{13}x_3 + \cdots + a_{1n}x_n = b_1$$
$$a_{21}x_1 + a_{22}x_2 + a_{23}x_3 + \cdots + a_{2n}x_n = b_2$$
$$a_{31}x_1 + a_{32}x_2 + a_{33}x_3 + \cdots + a_{3n}x_n = b_3$$
$$\vdots \qquad \vdots \qquad \vdots \qquad \qquad \vdots \qquad \vdots$$
$$a_{m1}x_1 + a_{m2}x_2 + a_{m3}x_3 + \cdots + a_{mn}x_n = b_m$$

The matrix formed by adjoining a column consisting of the $b_i$'s to the coefficient matrix $(a_{ij})$ is called the **augmented matrix** of the system. The augmented matrix is often written with a vertical bar separating the column of the $b_i$'s from the coefficient matrix, as follows.

### THE AUGMENTED MATRIX

$$\begin{bmatrix} a_{11} & a_{12} & a_{13} & \ldots & a_{1n} & b_1 \\ a_{21} & a_{22} & a_{23} & \ldots & a_{2n} & b_2 \\ a_{31} & a_{32} & a_{33} & \ldots & a_{3n} & b_3 \\ \vdots & \vdots & \vdots & & \vdots & \vdots \\ a_{m1} & a_{m2} & a_{m3} & \ldots & a_{mn} & b_m \end{bmatrix}$$

**EXAMPLE 2**   *Finding the matrix of coefficients and the augmented matrix of a linear system*

Find the coefficient and augmented matrix for the following linear system.

$$2x - 4y + 2z = 6$$
$$3y + 5x = 8$$

**SOLUTION**   We begin by writing the variables in the same order in each equation of the system.

$$2x - 4y + 2z = 6$$
$$5x + 3y \qquad = 8$$

The coefficient matrix is then

$$\begin{bmatrix} 2 & -4 & 2 \\ 5 & 3 & 0 \end{bmatrix}$$

The augmented matrix is formed by "augmenting" the coefficient matrix with a column formed from the constants to the right of the equal sign. Thus, the augmented matrix is given by

$$\left[\begin{array}{ccc|c} 2 & -4 & 2 & 6 \\ 5 & 3 & 0 & 8 \end{array}\right]$$

Conversely, given an augmented matrix (and an ordering of the variables), we can reproduce the original system of equations, as the following examples illustrate.

**EXAMPLE 3**   *Producing the system of equations corresponding to an augmented matrix*

What linear system in the variables $x$ and $y$ can be expressed using the following augmented matrix? (Assume that the entries correspond to the order $x$, $y$.)

$$\left[\begin{array}{cc|c} 2 & 5 & 1 \\ 1 & 3 & 0 \end{array}\right]$$

**SOLUTION**

$$2x + 5y = 1$$
$$x + 3y = 0$$

**EXAMPLE 4**   *Finding and solving the system of equations corresponding to an augmented matrix*

Find and solve the system of equations corresponding to the following augmented matrix. (Assume that the variables are $x$, $y$, and $z$, respectively.)

$$\begin{bmatrix} 2 & 2 & -1 & | & 10 \\ 0 & 1 & 4 & | & 13 \\ 0 & 0 & 3 & | & 6 \end{bmatrix}$$

**SOLUTION**   The augmented matrix corresponds to the following system of equations.

(1) $\qquad\qquad\qquad\qquad 2x + 2y - z = 10$

(2) $\qquad\qquad\qquad\qquad\qquad y + 4z = 13$

(3) $\qquad\qquad\qquad\qquad\qquad\qquad 3z = 6$

Solving equation (3) for $z$, we have

$$3z = 6$$
$$z = 2$$

Substituting $z = 2$ into equation (2), we find

$$y + 4z = 13$$
$$y + 4 \cdot (2) = 13$$
$$y + 8 = 13$$
$$y = 5$$

Now that $y$ and $z$ are known, we can find $x$ by substituting their values into equation (1).

$$2x + 2y - z = 10$$
$$2x + 2 \cdot 5 - 2 = 10$$
$$2x = 2$$
$$x = 1$$

Thus, the solution is given by $x = 1$, $y = 5$, and $z = 2$.

## UPPER TRIANGULAR SYSTEMS AND BACK-SUBSTITUTION

In Example 4 we saw a system of equations that was especially easy to solve. The technique we employed is called **back-substitution**. In effect, we simply solved the last equation for the last variable and then substituted this value into the second to the last equation to solve for the second to the last variable, and so on. Of course, back-substitution only works for systems of a very special form: the last equation must involve only the last variable, the second to the last equation must involve only the last two variables, and so on. In other words, the coefficient matrix must have zeros below the *main diagonal*, the diagonal running from the upper left corner of the matrix to the lower right corner. Whenever the coefficient matrix is of this form, we say that the matrix is *upper triangular*. We summarize these ideas as follows.

**Back-substitution and upper triangular matrices**

■ The **main diagonal** of a square matrix (a matrix with the same number of rows as columns) runs from the upper left corner to the lower right corner, as shown in Figure 5.

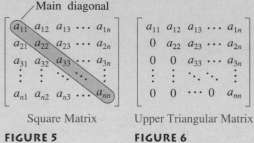

Main diagonal

Square Matrix         Upper Triangular Matrix
**FIGURE 5**          **FIGURE 6**

■ Square matrices with zeros below the main diagonal are called **upper triangular**. See Figure 6.

■ Systems of equations for which the coefficient matrix is upper triangular can be solved using **back-substitution**.

---

**EXAMPLE 5**    *Planning a meal*

The following nutritional values are given for a serving of lean ham, potatoes, and green beans.

| Nutritional component | Ham | Potatoes | Green beans |
|---|---|---|---|
| Calories | 245 | 145 | 35 |
| Carbohydrates (in grams) | 0 | 33 | 8 |
| Vitamin A (in I.U.) | 0 | 0 | 780 |

Suppose that you wish to prepare a meal consisting of ham, potatoes, and green beans with 650 calories, 80 grams of carbohydrates, and 1200 I.U. of vitamin A. How many servings of each type of food must be used?

**SOLUTION**    Let $x$ be the number of servings of ham, $y$ the number of servings of potatoes, and $z$ the number of servings of green beans. Then we have

(1)   $\begin{matrix} \text{calories from} \\ \text{ham} \end{matrix} + \begin{matrix} \text{calories from} \\ \text{potatoes} \end{matrix} + \begin{matrix} \text{calories from} \\ \text{green beans} \end{matrix} = 650$

$245x + 145y + 35z = 650$

(2)   $\begin{matrix} \text{carbohydrates from} \\ \text{ham} \end{matrix} + \begin{matrix} \text{carbohydrates from} \\ \text{potatoes} \end{matrix} + \begin{matrix} \text{carbohydrates from} \\ \text{green beans} \end{matrix} = 80$

$0x + 33y + 8z = 80$

(3)   $\begin{matrix} \text{vitamin A from} \\ \text{ham} \end{matrix} + \begin{matrix} \text{vitamin A from} \\ \text{potatoes} \end{matrix} + \begin{matrix} \text{vitamin A from} \\ \text{green beans} \end{matrix} = 1200$

$0x + 0y + 780z = 1200$

Combining the three equations, we obtain the following system of equations.

(1) $$245x + 145y + 35z = 650$$

(2) $$33y + 8z = 80$$

(3) $$780z = 1200$$

Note that this is an upper triangular system that can be solved by back-substitution. Solving equation (3) for $z$ gives

$$780z = 1200$$

$$z \approx 1.5$$

Substituting $z = 1.5$ into equation (2) and solving for $y$, we have

$$33y + 8z = 80$$

$$33y + 8(1.5) = 80$$

$$33y = 68$$

$$y \approx 2.1$$

Substituting $z = 1.5$ and $y = 2.1$ into equation (1) and solving for $x$, we obtain

$$245x + 145y + 35z = 650$$

$$245x + 145(2.1) + 35(1.5) = 650$$

$$245x = 293$$

$$x \approx 1.2$$

Thus, the meal should be made with approximately 1.2 servings of ham, 2.1 servings of potatoes, and 1.5 servings of green beans.

■

## EXERCISES 2

**EXERCISES 1–6** □ *Indicate whether the given system of equations is linear in the variables x, y, and z.*

**1.** $2x - 3y - 2z = \dfrac{1}{10}$
    $x + y + z = 2$

**2.** $x^2 - y = 0$
    $x - y = 1$

**3.** $\sqrt{x} - 3y = 7$
    $2x + 5y = 9$

**4.** $\dfrac{1}{x} + y = 0$
    $\dfrac{3}{x} - z = 2$

**5.** $x - 2y = 3$
    $x + 3yz = 1$

**6.** $\sqrt{34}y - \dfrac{1}{7}x = z$
    $x + 2y - 3z = 0$

**EXERCISES 7–12** □ *Express the given system of equations as an augmented matrix.*

**7.** $3x + 4y = 10$
    $2x - 7y = 4$

**8.** $-2x + 3y = 2$
    $x - 6y = 0$

**9.** $x = 3$
    $y = 4$

**10.** $2x - 3y + 4z = 2$
    $x + 6y + 3z = 5$
    $5x - 7y + 2z = 1$

**11.** $x - y + 2z = 1$
    $2y + 6z = 3$

**12.** $x + 2y - 3z = 4$
    $y + 3z = 0$
    $z = 6$

**EXERCISES 13–18** ☐ *Determine whether the given system of two linear equations has 0, 1, or infinitely many solutions.*

13. $2x - 3y = 4$
    $4x - 6y = 7$

14. $3x + 2y = 4$
    $2x + 3y = 1$

15. $x - y = 7$
    $2x - 2y = 14$

16. $x + \dfrac{1}{2}y = 3$

    $3x + \dfrac{3}{2}y = 7$

17. $5x - 6y = 1$
    $x - y = 3$

18. $4x + 2y = 6$
    $20x + 10y = 30$

**EXERCISES 19–24** ☐ *Solve the given system using back-substitution.*

19. $2x + 3y = 7$
    $2y = 2$

20. $3x - 6y = 4$
    $\dfrac{1}{2}y = 3$

21. $2x + y - z = 12$
    $y + z = 8$
    $3z = 3$

22. $2a + 3b + 4c = 2$
    $2b - c = 5$
    $2c = -2$

23. $q - 2r + 3s + t = 3$
    $4r + s - t = 18$
    $2s + 2t = 4$
    $3t = -6$

24. $a + b + c + d = 2$
    $b + c + d = 2$
    $c + d = 1$
    $d = 1$

**EXERCISES 25–26** ☐ *An upper triangular augmented matrix is given. Find the solution to the system using back-substitution. Assume that the variable names are x and y, in that order.*

25. $\left[\begin{array}{cc|c} 2 & -3 & -5 \\ 0 & 4 & 12 \end{array}\right]$

26. $\left[\begin{array}{cc|c} -1 & -2 & -11 \\ 0 & \frac{1}{3} & 3 \end{array}\right]$

**EXERCISES 27–30** ☐ *An upper triangular augmented matrix is given. Find the solution to the system using back-substitution. Assume that the variable names are x, y, and z, in that order.*

27. $\left[\begin{array}{ccc|c} 1 & 3 & 6 & 0 \\ 0 & 1 & 2 & 2 \\ 0 & 0 & 1 & -4 \end{array}\right]$

28. $\left[\begin{array}{ccc|c} 1 & 2 & -3 & 1 \\ 0 & 1 & 3 & 5 \\ 0 & 0 & 1 & 0 \end{array}\right]$

29. $\left[\begin{array}{ccc|c} 1 & -2 & 4 & 1 \\ 0 & 1 & -3 & 1 \\ 0 & 0 & 1 & -1 \end{array}\right]$

30. $\left[\begin{array}{ccc|c} 1 & -3 & 5 & 2 \\ 0 & 1 & -3 & \frac{1}{2} \\ 0 & 0 & 1 & 2 \end{array}\right]$

■ *Applications*

31. *Income Tax*  A family has income totaling $58,000 from three different sources: salary, payments from municipal bonds, and payments from U.S. treasury bonds. The municipal and U.S. treasury bonds are free from state income tax, while salary is subject to a 6% state income tax. The U.S. treasury bonds are free from federal tax, but both the salary and the income from the municipal bonds are subject to 20% federal tax. If the family's state tax totals $2400 and their federal tax totals $10,000, compute the income from each of the three sources.

32. *Sports Drink*  An athlete wishes to concoct a 165-calorie, 16-ounce sports beverage by mixing together orange juice (which has 13.75 calories per ounce) and water. How many ounces of each ingredient should be used?

33. *Power Breakfast*  A breakfast consisting of milk, bananas, and whole wheat toast is to have 1200 units of vitamin A, 1.5 units of vitamin $B_6$, and 1.35 units of vitamin $B_{12}$. Each serving of milk has 500 units of vitamin A, 0.1 unit of vitamin $B_6$, and 0.9 unit of vitamin $B_{12}$. Each banana has 226 units of vitamin A and 0.6 units of vitamin $B_6$, but no vitamin $B_{12}$. Each piece of whole wheat toast has 0.04 units of vitamin $B_6$, but no vitamin A or $B_{12}$. How much of each food should be consumed?

34. *Cross Training*  A 170-pound athlete cross-trains during the off season by cycling (at 10 miles per hour), playing volleyball, and weightlifting. He determines that he needs to burn-off an extra 6000 calories per week through exercise, and that he needs to ride 80 miles per week. Because of his full-time job, he will be unable to devote more than 20 hours per week to exercise. His trainer tells him that he will burn approximately 552 calories per hour while cycling at 10 miles per hour, and 234 calories per hour while playing volleyball, while the calories burned through weightlifting are negligible. Determine the number of hours that the athlete should spend in each of the three activities.

Kenyan marathon runner Cosmas N'Dete retains his title in the 1994 Boston marathom.

■ *Projects for Enrichment*

**35.** *Linear Systems with More than One Solution* In the text it was stated that the solution set to a linear system of equations will either have 0, 1, or infinitely many solutions. In this project we will prove this statement by employing matrix notation.

a. Suppose that we are given a linear system of $m$ equations in the $n$ variables $x_1, x_2, x_3, \ldots, x_n$.

$$
\begin{aligned}
a_{11}x_1 + a_{12}x_2 + a_{13}x_3 + \cdots + a_{1n}x_n &= b_1 \\
a_{21}x_1 + a_{22}x_2 + a_{23}x_3 + \cdots + a_{2n}x_n &= b_2 \\
a_{31}x_1 + a_{32}x_2 + a_{33}x_3 + \cdots + a_{3n}x_n &= b_3 \\
\vdots \qquad \vdots \qquad \vdots \qquad\qquad \vdots \qquad \vdots \\
a_{m1}x_1 + a_{m2}x_2 + a_{m3}x_3 + \cdots + a_{mn}x_n &= b_m
\end{aligned}
$$

i. Show that the system is equivalent to the matrix equation $AX = B$, where

$$
A = \begin{bmatrix}
a_{11} & a_{12} & a_{13} & \ldots & a_{1n} \\
a_{21} & a_{22} & a_{23} & \ldots & a_{2n} \\
a_{31} & a_{32} & a_{33} & \ldots & a_{3n} \\
\vdots & \vdots & \vdots & & \vdots \\
a_{m1} & a_{m2} & a_{m3} & \ldots & a_{mn}
\end{bmatrix},
$$

$$
B = \begin{bmatrix} b_1 \\ b_2 \\ b_3 \\ \vdots \\ b_m \end{bmatrix}, \quad \text{and} \quad X = \begin{bmatrix} x_1 \\ x_2 \\ x_3 \\ \vdots \\ x_n \end{bmatrix}
$$

ii. Use a property of matrices to show that if $Y$ and $Z$ are both solutions to $AX = B$, then $Y - Z$ is a solution to $AX = 0$.

iii. Now show that if $Y$ and $Z$ are solutions to $AX = B$, then so is $Y + t(Y - Z)$, where $t$ is any real number.

iv. Use the result of part (iii) to show that if a linear system of equations has two solutions, then it has infinitely many solutions.

b. For each system of equations given below, two solutions are provided. Use the results from part (a) to give three more solutions.

i.
$$
\begin{aligned}
2x + 3y - z &= 12 \\
x - y + z &= 0 \\
3x + 2y &= 12
\end{aligned}
$$

Solution 1: $x = 4$, $y = 0$, $z = -4$
Solution 2: $x = 2$, $y = 3$, $z = 1$

*Hint:* In matrix form these solutions correspond to

$$
Y = \begin{bmatrix} 4 \\ 0 \\ -4 \end{bmatrix} \quad \text{and} \quad Z = \begin{bmatrix} 2 \\ 3 \\ 1 \end{bmatrix}
$$

ii.
$$
\begin{aligned}
p + q + 3r &= 3 \\
2q + 4r &= 2
\end{aligned}
$$

Solution 1: $p = 2$, $q = 1$, $r = 0$
Solution 2: $p = 1$, $q = -1$, $r = 1$

**36.** *Interpolating with Polynomials* It can be shown that given any collection of $n$ points, there is a polynomial of degree $n - 1$ or lower that passes through the points. Suppose that we wish to find the equation of the quadratic polynomial passing through three points, say $(1, 4)$, $(2, 6)$, and $(4, 22)$. Since the polynomial is quadratic, it is of the form $p(x) = ax^2 + bx + c$, and we have $p(1) = 4$, $p(2) = 6$, and $p(4) = 22$. From the fact that $p(1) = 4$, we obtain

$$
p(1) = a \cdot 1^2 + b \cdot 1 + c = 4
$$

which gives us the equation

$$
a + b + c = 4
$$

In a similar fashion, from $p(2) = 6$ and $p(4) = 22$, we obtain the equations

$$
\begin{aligned}
4a + 2b + c &= 6 \\
16a + 4b + c &= 22
\end{aligned}
$$

thus giving us the linear system

$$
\begin{aligned}
a + b + c &= 4 \\
4a + 2b + c &= 6 \\
16a + 4b + c &= 22
\end{aligned}
$$

This system could be solved using substitution, elimination, or any of the other techniques that will be described in the next three sections. In this project we will discover an alternative way to find a polynomial passing through a given set of points that will involve solving a system of equations by back-substitution.

a. We will find the equation of the quadratic polynomial passing through the points $(1, 4)$, $(2, 6)$, and $(4, 22)$. We begin by assuming that the polynomial is of the form

$$
p(x) = a + b(x - 1) + c(x - 1)(x - 2)
$$

i. Set up a linear system involving the three variables $a$, $b$, and $c$.

ii. Solve the system obtained in part (i) by back-substitution.

iii. Now multiply out all products appearing in the expression for $p(x)$ in order to write $p(x)$ in standard form.

b. Use the approach developed in part (a) to find the cubic polynomial passing through

   i. $(-2, 25)$, $(0, 1)$, $(1, 5)$, and $(2, 19)$

   ii. $(0, 0)$, $(1, 0)$, $(4, 12)$, and $(2, -2)$

c. Find the equation of the quadratic function passing through the points $(x_1, y_1)$, $(x_2, y_2)$, and $(x_3, y_3)$ using the same technique.

d. The U.S. Census Bureau records the following data for the population of Arizona:

   1960: 1,302,161

   1970: 1,775,399

   1980: 2,716,546

   1990: 3,665,228

Let $t$ represent the year (with $t = 0$ corresponding to 1960) and $P$ the population of Arizona. Use the approach developed in part (a) to express $P$ as a cubic polynomial in the variable $t$. According to this model, when will the population of Arizona reach 40 million? What are the limitations of this model?

---

### Questions for Discussion or Essay

37. It turns out that the graph of a linear equation in three variables is a plane in three-dimensional space. Explain why three planes will meet in 0, 1, or infinitely many points. Give an example from the room that you are in that suggests three planes meeting in a single point; also describe a situation in which three planes meet in a line. Sketch both.

38. Give an example of a system of two equations in two unknowns that has exactly two solutions. (*Hint:* The graphs of the two equations should intersect exactly twice. Make one graph a line and the other an appropriate conic section.) Explain why this doesn't contradict the statement that the number of solutions to a linear system of equations is either 0, 1, or infinity.

39. Suppose that you are given a system of three linear equations and that the equations are given in no particular order. In how many different ways could the system be represented using augmented matrices? Without checking all of these different matrix representations one by one, how could you tell just by looking at the equations whether or not one of the matrix representations would be upper triangular?

40. Suppose that you were asked to write an exam for this class and you wished to produce a linear system of three equations with infinitely many solutions. If the first two equations are $2x + 3y - z = 5$ and $x + 2y + z = 3$, what might the third equation be? What if you wanted a system with no solution?

---

## SECTION 3

# GAUSSIAN AND GAUSS-JORDAN ELIMINATION

■ What is sum of the squares of the first 10,000 natural numbers, $1^2 + 2^2 + 3^2 + \cdots + 10,000^2$?

■ A baseball is hit so that it is at a height of 9 feet when it passes over the pitcher and clears the 10-foot center field wall by 4 feet. Is the ball rising or falling as it leaves the field?

■ How can a system of equations be solved without ever writing down a single variable?

## GAUSSIAN ELIMINATION

We have already seen two methods for solving systems of equations: substitution and elimination. The technique of substitution works well when the number of equations is small, or if the equations have an especially simple form. Indeed, we saw in the previous section that if the coefficient matrix of a system of equations is upper triangular (all zeros below the main diagonal), then the system can be easily solved using back-substitution. In this section we will see how systematic elimination can be used to convert augmented matrices to a form that is closely related to upper triangular, namely **row-echelon form**.

### Row-echelon form

A matrix is said to be in row-echelon form if the following conditions are satisfied.

- The first nonzero entry of each row is a 1. It is called the **leading 1**.
- Below each leading 1 are only 0's.
- Each leading 1 is to the right of the leading 1 in the row above.
- Any rows consisting entirely of 0's are at the bottom of the matrix.

### EXAMPLE 1    *Determining whether a matrix is in row-echelon form*

For each matrix given below, indicate whether it is in row-echelon form. If it is not, support your answer.

a. $\begin{bmatrix} 1 & 4 & 6 \\ 0 & 1 & 1 \\ 0 & 0 & 1 \end{bmatrix}$
b. $\begin{bmatrix} 1 & 2 & 3 \\ 0 & 1 & 5 \\ 0 & 6 & 2 \end{bmatrix}$

c. $\left[\begin{array}{ccc|c} 1 & 0 & 3 & 4 & 5 \\ 0 & 2 & 0 & 1 & 4 \end{array}\right]$
d. $\left[\begin{array}{ccc|c} 1 & 0 & 2 & 5 \\ 0 & 1 & 4 & 2 \\ 0 & 0 & 0 & 0 \end{array}\right]$

### SOLUTION

a. This matrix is in row-echelon form.

b. This matrix fails to be in row-echelon form because of the presence of a nonzero number (6) in the third row and second column directly below the leading 1 of the second row. Also, the first nonzero entry of the third row is 6 and not 1.

c. This augmented matrix is not in row-echelon form because the second row does not lead with a 1.

d. This augmented matrix is in row-echelon form. Note that there is one row consisting entirely of 0's, and it is the last row.

The following example illustrates why systems expressed as augmented matrices in row-echelon form are desirable.

**EXAMPLE 2**    *Solving a system corresponding to a matrix in row-echelon form*

Solve the system of equations corresponding to the augmented matrix. (Assume the variables are $x$, $y$, and $z$, in that order.)

$$\begin{bmatrix} 1 & 3 & 2 & | & 11 \\ 0 & 1 & 4 & | & 6 \\ 0 & 0 & 1 & | & 1 \end{bmatrix}$$

**SOLUTION**    Expressing the augmented matrix as a system of equations, we have

(1) $$x + 3y + 2z = 11$$

(2) $$y + 4z = 6$$

(3) $$z = 1$$

Using back-substitution, we substitute $z = 1$ into equation (2).

$$y + 4z = 6$$
$$y + 4 \cdot 1 = 6$$
$$y = 2$$

Now that $y$ and $z$ have been determined, we can find $x$ by substituting their values into equation (1).

$$x + 3y + 2z = 11$$
$$x + 3 \cdot 2 + 2 \cdot 1 = 11$$
$$x = 3$$

Thus, the solution is given by $x = 3$, $y = 2$, and $z = 1$.

∎

Clearly, any system of equations that is represented as an augmented matrix in row-echelon form can be solved easily. If the augmented matrix is *not* in row-echelon form, we can convert it to row-echelon form using a systematic procedure known as **Gaussian elimination**.

At each stage of Gaussian elimination, one of three **elementary row operations** is performed on the augmented matrix. These operations are interchanging two rows of the matrix, multiplying a row by a constant, and adding a multiple of one row to another. It can be shown that performing an elementary row operation on an augmented matrix results in an equivalent system of equations. The elementary row operations, their notation, and their effect on the underlying system are summarized as follows.

**■** *Elementary row operations*

| Operation | Notation | Effect on system |
|---|---|---|
| Interchanging rows $i$ and $j$ | $R_{ij}$ | Equivalent. (Two equations change places.) |
| Multiplying row $i$ by $c$, where $c \neq 0$ | $cR_i$ | Equivalent. (An equation is multiplied by a non-zero constant.) |
| Adding $c$ times row $i$ to row $j$ | $cR_i + R_j$ | Equivalent. (A multiple of an equation is added to another.) |

Note that the operation $cR_i + R_j$ changes the $j$th row, **but not the $i$th row**. In other words, the $j$th row is replaced by $cR_i + R_j$, but the $i$th row is left alone. Also, elementary row operations may be performed on any matrix, augmented or not.

**EXAMPLE 3**   *Performing elementary row operations*

For each matrix given below, perform the indicated elementary row operation.

a. $\begin{bmatrix} 1 & 3 \\ 2 & 5 \end{bmatrix}$, $-2R_1 + R_2$
b. $\left[\begin{array}{ccc|c} 3 & 6 & 5 & 3 \\ 4 & 2 & 1 & 7 \end{array}\right]$, $-\frac{1}{3}R_1$
c. $\left[\begin{array}{ccc|c} 0 & 1 & 2 & 3 \\ 3 & 6 & 1 & 5 \\ 1 & 2 & 0 & 4 \end{array}\right]$, $R_{13}$

**SOLUTION**

a. Multiplying the first row by $-2$ and adding to the second row gives us

$$\begin{bmatrix} 1 & 3 \\ (2 + (-2) \cdot 1) & (5 + (-2) \cdot 3) \end{bmatrix} = \begin{bmatrix} 1 & 3 \\ 0 & -1 \end{bmatrix}$$

b. Multiplying the first row by $-\frac{1}{3}$ gives us

$$\left[\begin{array}{ccc|c} (-\frac{1}{3})3 & (-\frac{1}{3})6 & (-\frac{1}{3})5 & (-\frac{1}{3})3 \\ 4 & 2 & 1 & 7 \end{array}\right] = \left[\begin{array}{ccc|c} -1 & -2 & -\frac{5}{3} & -1 \\ 4 & 2 & 1 & 7 \end{array}\right]$$

c. Exchanging rows 1 and 3 gives us

$$\left[\begin{array}{ccc|c} 1 & 2 & 0 & 4 \\ 3 & 6 & 1 & 5 \\ 0 & 1 & 2 & 3 \end{array}\right]$$

Gaussian elimination proceeds column by column from left to right. In each column, a leading 1 is obtained in the diagonal position, and then the entries under the leading 1 are cleared out, one by one, using elementary row operations. The following example illustrates Gaussian elimination.

| EXAMPLE 4 | *Solving a system of equations by Gaussian elimination* |

Solve the following system of equations using Gaussian elimination.

$$2y - 3z = 12$$

$$2x + 7y - 11z = 27$$

$$x + 2y - z = 0$$

**SOLUTION**    We begin by writing the system as an augmented matrix.

$$\begin{bmatrix} 0 & 2 & -3 & 12 \\ 2 & 7 & -11 & 27 \\ 1 & 2 & -1 & 0 \end{bmatrix}$$

Next we perform Gaussian elimination to convert this matrix to row-echelon form, as shown in Table 3. For emphasis, we have boxed in the entry that has been "fixed" at each step.

**TABLE 3**

| Augmented matrix | Explanation | Operation Notation |
|---|---|---|
| $\begin{bmatrix} 0 & 2 & -3 & 12 \\ 2 & 7 & -11 & 27 \\ 1 & 2 & -1 & 0 \end{bmatrix}$ | The original augmented matrix. | |
| $\begin{bmatrix} \boxed{1} & 2 & -1 & 0 \\ 2 & 7 & -11 & 27 \\ 0 & 2 & -3 & 12 \end{bmatrix}$ | Rows 1 and 3 were interchanged in order to obtain a leading 1 in row 1, column 1. | $R_{13}$ |
| $\begin{bmatrix} 1 & 2 & -1 & 0 \\ \boxed{0} & 3 & -9 & 27 \\ 0 & 2 & -3 & 12 \end{bmatrix}$ | $-2$ times the first row was added to the second row to obtain a 0 in row 2, column 1. | $-2R_1 + R_2$ |
| The first column is now completed. | | |
| $\begin{bmatrix} 1 & 2 & -1 & 0 \\ 0 & \boxed{1} & -3 & 9 \\ 0 & 2 & -3 & 12 \end{bmatrix}$ | Row 2 was multiplied by $\frac{1}{3}$ in order to obtain a leading 1 in row 2, column 2. | $\frac{1}{3}R_2$ |
| $\begin{bmatrix} 1 & 2 & -1 & 0 \\ 0 & 1 & -3 & 9 \\ 0 & \boxed{0} & 3 & -6 \end{bmatrix}$ | $-2$ times the second row was added to the third row to obtain a 0 in row 3, column 2. | $-2R_2 + R_3$ |
| The second column is now completed. | | |
| $\begin{bmatrix} 1 & 2 & -1 & 0 \\ 0 & 1 & -3 & 9 \\ 0 & 0 & \boxed{1} & -2 \end{bmatrix}$ | The third row was multiplied by $\frac{1}{3}$ in order to obtain a leading 1 in row 3, column 3. | $\frac{1}{3}R_3$ |

The matrix is now in row-echelon form. Translating back into equation form, we have

$$x + 2y - z = 0$$
$$y - 3z = 9$$
$$z = -2$$

We now use back-substitution with $z = -2$ to find $y$.

$$y - 3z = 9$$
$$y - 3(-2) = 9$$
$$y = 3$$

Finally, we use back-substitution with $z = -2$ and $y = 3$ to find $x$.

$$x + 2y - z = 0$$
$$x + 2(3) - (-2) = 0$$
$$x = -8$$

Thus, our solution is given by $x = -8$, $y = 3$, and $z = -2$.

■

Recall from Section 2 that a linear system of equations will have either 0, 1, or infinitely many solutions. Thus far, we have used Gaussian elimination to solve systems of equations with exactly one solution, but it is also effective for describing the solution sets of systems with infinitely many solutions and for revealing those systems that have no solution, as the following examples illustrate.

**EXAMPLE 5**    *Describing the solution set of a linear system with infinitely many solutions*

$$x + 3y - 5z = 2$$
$$x - y + z = 4$$
$$x + y - 2z = 3$$

**SOLUTION**   We first express the system as an augmented matrix and then use Gaussian elimination to convert the matrix to row-echelon form. In moving from one matrix to the next, we indicate the elementary row operations that are used.

$$\begin{bmatrix} 1 & 3 & -5 & | & 2 \\ 1 & -1 & 1 & | & 4 \\ 1 & 1 & -2 & | & 3 \end{bmatrix} \xrightarrow[\substack{-R_1 + R_3}]{-R_1 + R_2} \begin{bmatrix} 1 & 3 & -5 & | & 2 \\ 0 & -4 & 6 & | & 2 \\ 0 & -2 & 3 & | & 1 \end{bmatrix} \xrightarrow{-\frac{1}{4}R_2}$$

$$\begin{bmatrix} 1 & 3 & -5 & | & 2 \\ 0 & 1 & -\frac{3}{2} & | & -\frac{1}{2} \\ 0 & -2 & 3 & | & 1 \end{bmatrix} \xrightarrow{2R_2 + R_3} \begin{bmatrix} 1 & 3 & -5 & | & 2 \\ 0 & 1 & -\frac{3}{2} & | & -\frac{1}{2} \\ 0 & 0 & 0 & | & 0 \end{bmatrix}$$

Converting to equations, we have

(1) $$x + 3y - 5z = 2$$

(2) $$y - \frac{3}{2}z = -\frac{1}{2}$$

(3) $$0 = 0$$

Solving equation (2) for $y$ gives us

$$y - \frac{3}{2}z = -\frac{1}{2}$$

$$y = \frac{3}{2}z - \frac{1}{2}$$

Finally, we substitute $y = \frac{3}{2}z - \frac{1}{2}$ into equation (1) to obtain

$$x + 3\left(\frac{3}{2}z - \frac{1}{2}\right) - 5z = 2$$

$$x - \frac{1}{2}z - \frac{3}{2} = 2$$

$$x = \frac{1}{2}z + \frac{7}{2}$$

Thus, the solution set consists of triples of numbers of the form $(\frac{1}{2}z + \frac{7}{2}, \frac{3}{2}z - \frac{1}{2}, z)$.

■

**EXAMPLE 6**   *A linear system with no solution*

Solve the linear system of equations

$$x - y = 1$$
$$2x - 2y = 4$$

**SOLUTION**   Using Gaussian elimination, we have

$$\begin{bmatrix} 1 & -1 & | & 1 \\ 2 & -2 & | & 4 \end{bmatrix} \xrightarrow{-2R_1 + R_2} \begin{bmatrix} 1 & -1 & | & 1 \\ 0 & 0 & | & 2 \end{bmatrix}$$

Now the second row of this matrix corresponds to the equation $0 = 2$, which is a contradiction. Thus, the system has no solution.

■

Although the process of Gaussian elimination will always yield an equivalent matrix in row-echelon form, this matrix is not unique. In other words, the final row-echelon matrix depends on choices made during Gaussian elimination. The method described below, Gauss-Jordan elimination, yields a unique matrix; no matter what decisions are made during the course of Gauss-Jordan elimination, the final matrix will be the same.

## GAUSS-JORDAN ELIMINATION AND REDUCED ROW-ECHELON FORM

We have seen that a system of equations can be solved by first converting the augmented matrix to row-echelon form, and then back-substituting. It turns out that if the augmented matrix is processed still further, until it is in *reduced row-echelon form,* back-substitution can be avoided.

**Reduced row-echelon form**

> A matrix is said to be in **reduced row-echelon form** if the following two conditions are met.
>
> ■ The matrix is in row-echelon form.
> ■ There are only 0's directly above each leading 1.

**EXAMPLE 7**

*Identifying matrices in reduced row-echelon form*

Determine whether each of the following matrices is in reduced row-echelon form.

$$\text{a.} \begin{bmatrix} 1 & 0 & 0 & | & 3 \\ 0 & 1 & 0 & | & 2 \\ 0 & 0 & 1 & | & 5 \end{bmatrix} \qquad \text{b.} \begin{bmatrix} 1 & 0 & 0 & | & 1 \\ 0 & 1 & 2 & | & 6 \\ 0 & 0 & 1 & | & 2 \end{bmatrix} \qquad \text{c.} \begin{bmatrix} 6 & 0 & 0 & | & 3 \\ 0 & 1 & 0 & | & 5 \\ 0 & 0 & 1 & | & 1 \end{bmatrix}$$

**SOLUTION**

**a.** This matrix is in reduced row-echelon form.

**b.** This matrix is in row-echelon form but not reduced row-echelon form. The culprit is the 2 in the third column and second row: it would need to be 0 for the matrix to be in reduced row-echelon form.

**c.** This matrix is not even in row-echelon form, let alone reduced row-echelon form. The culprit is the 6 in the first row and first column. If it were a 1, then this matrix would be in reduced row-echelon form.

It is easy to see why reduced row-echelon form is desirable. For example, suppose that a system of equations is represented by the following augmented matrix, which is in reduced row-echelon form:

$$\begin{bmatrix} 1 & 0 & 0 & | & 7 \\ 0 & 1 & 0 & | & -2 \\ 0 & 0 & 1 & | & 3 \end{bmatrix}$$

Converting this augmented matrix into equation form gives

$$x = 7$$
$$y = -2$$
$$z = 3$$

The system is solved, and no back-substitution is required!

**Gauss-Jordan elimination** is a systematic procedure for converting a matrix into reduced row-echelon form using elementary row operations. First, Gaussian elimination is used to produce a matrix in row-echelon form. Next, elementary row operations are used to eliminate the nonzero entries *above* each leading 1, in much the same way that nonzero entries below leading 1's were eliminated when converting the matrix to row-echelon form. Gauss-Jordan elimination is illustrated in the following example.

**EXAMPLE 8**  *Solving a system of equations using Gauss-Jordan elimination*

Solve the following system using Gauss-Jordan elimination.

$$2x - 4y - 10z = -12$$

$$2x - 2y - 2z = 2$$

$$-3x + 4y + 11z = 8$$

**SOLUTION**

$$\begin{bmatrix} 2 & -4 & -10 & | & -12 \\ 2 & -2 & -2 & | & 2 \\ -3 & 4 & 11 & | & 8 \end{bmatrix} \xrightarrow{\frac{1}{2}R_1} \begin{bmatrix} 1 & -2 & -5 & | & -6 \\ 2 & -2 & -2 & | & 2 \\ -3 & 4 & 11 & | & 8 \end{bmatrix} \xrightarrow[3R_1 + R_3]{-2R_1 + R_2}$$

$$\begin{bmatrix} 1 & -2 & -5 & | & -6 \\ 0 & 2 & 8 & | & 14 \\ 0 & -2 & -4 & | & -10 \end{bmatrix} \xrightarrow{\frac{1}{2}R_2} \begin{bmatrix} 1 & -2 & -5 & | & -6 \\ 0 & 1 & 4 & | & 7 \\ 0 & -2 & -4 & | & -10 \end{bmatrix} \xrightarrow{2R_2 + R_3}$$

$$\begin{bmatrix} 1 & -2 & -5 & | & -6 \\ 0 & 1 & 4 & | & 7 \\ 0 & 0 & 4 & | & 4 \end{bmatrix} \xrightarrow{\frac{1}{4}R_3} \begin{bmatrix} 1 & -2 & -5 & | & -6 \\ 0 & 1 & 4 & | & 7 \\ 0 & 0 & 1 & | & 1 \end{bmatrix} \xrightarrow{2R_2 + R_1}$$

$$\begin{bmatrix} 1 & 0 & 3 & | & 8 \\ 0 & 1 & 4 & | & 7 \\ 0 & 0 & 1 & | & 1 \end{bmatrix} \xrightarrow[-4R_3 + R_2]{-3R_3 + R_1} \begin{bmatrix} 1 & 0 & 0 & | & 5 \\ 0 & 1 & 0 & | & 3 \\ 0 & 0 & 1 & | & 1 \end{bmatrix}$$

Converting to equations, we have $x = 5$, $y = 3$, and $z = 1$.    ■

## EXERCISES 3

**EXERCISES 1–8** □ *Use Gaussian elimination to write the augmented matrix in row-echelon form. (There may be more than one correct answer.)*

1. $\begin{bmatrix} 4 & 8 & | & 12 \\ 3 & 9 & | & 21 \end{bmatrix}$

2. $\begin{bmatrix} \frac{2}{3} & -\frac{2}{3} & | & \frac{4}{3} \\ -4 & 2 & | & -18 \end{bmatrix}$

3. $\begin{bmatrix} 2 & 0 & 1 & | & 0 \\ 2 & 1 & 4 & | & 1 \\ 3 & 1 & 8 & | & 1 \end{bmatrix}$

4. $\begin{bmatrix} 2 & 6 & 8 & | & -2 \\ 2 & 4 & 12 & | & 5 \\ 1 & 7 & -6 & | & -21 \end{bmatrix}$

5. $\begin{bmatrix} 2 & 4 & 2 & | & 4 \\ 3 & 4 & 1 & | & 6 \end{bmatrix}$

6. $\begin{bmatrix} 1 & 3 & 0 & | & 4 \\ \frac{1}{3} & -1 & -4 & | & \frac{16}{3} \end{bmatrix}$

7. $\begin{bmatrix} 1 & 0 & 1 & 2 & | & 1 \\ -1 & 1 & -1 & -1 & | & -1 \\ 0 & 2 & 1 & 3 & | & 0 \\ 1 & 0 & 2 & 4 & | & 3 \end{bmatrix}$

8. $\begin{bmatrix} 1 & 1 & 0 & 0 & | & 2 \\ 2 & 3 & 0 & 0 & | & 4 \\ 3 & 3 & 1 & 0 & | & 7 \\ 4 & 4 & 0 & 4 & | & 8 \end{bmatrix}$

**EXERCISES 9–18** □ *Solve the system of equations by first expressing it as an augmented matrix and then applying Gaussian elimination and back-substitution.*

9. $\begin{aligned} 2x - 6y &= -2 \\ 3x + 5y &= 11 \end{aligned}$

10. $\begin{aligned} 2u + 4v &= -6 \\ 5u - 3v &= 26 \end{aligned}$

11. $\begin{aligned} 2p + q &= \frac{5}{3} \\ 3p - 6q &= -\frac{5}{2} \end{aligned}$

12. $\begin{aligned} 4y + 6z &= -\frac{8}{5} \\ 3y - 5z &= \frac{13}{5} \end{aligned}$

13. $\begin{aligned} x + y + 2z &= 7 \\ x + 2y + 3z &= 10 \\ x - 4y + z &= 4 \end{aligned}$

14. $\begin{aligned} x - y + z &= -2 \\ x + y - z &= 0 \\ -x + y + z &= -4 \end{aligned}$

15. $\begin{aligned} 2p + r &= 2 \\ 2p + q + 4r &= 9 \\ 3p + q + 8r &= 17 \end{aligned}$

16. $\begin{aligned} 2u - 4v + 2w &= 18 \\ 3u - 5v + 2w &= 23 \\ 4u + 9v + 6w &= -13 \end{aligned}$

17. $\begin{aligned} w + x + y + 2z &= 1 \\ -w + x - y - z &= -2 \\ 2x + y + 3z &= -1 \\ w + x + 2y + 5z &= 1 \end{aligned}$

18. $\begin{aligned} 2q + 4r + 6s - 2t &= 8 \\ r - s + t &= 2 \\ r + s - t &= 3 \\ -r + s + t &= -5 \end{aligned}$

**EXERCISES 19–24** □ *Use Gauss-Jordan elimination to express each of the following matrices in reduced* row-echelon *form.*

19. $\begin{bmatrix} 1 & 2 & | & 3 \\ 0 & 1 & | & 2 \end{bmatrix}$

20. $\begin{bmatrix} 1 & 3 & | & -2 \\ 0 & 1 & | & 5 \end{bmatrix}$

21. $\begin{bmatrix} 2 & 4 & | & -8 \\ 3 & 3 & | & -3 \end{bmatrix}$

22. $\begin{bmatrix} 2 & 3 & | & 0 \\ 2 & 6 & | & -2 \end{bmatrix}$

23. $\begin{bmatrix} 1 & 2 & 2 & | & 2 \\ 0 & 1 & 3 & | & -2 \\ 0 & 0 & 1 & | & -1 \end{bmatrix}$

24. $\begin{bmatrix} 1 & 3 & 1 & | & \frac{7}{2} \\ 0 & 1 & -\frac{1}{6} & | & 0 \\ 0 & 0 & 1 & | & 2 \end{bmatrix}$

**EXERCISES 25–32** □ *Solve the system of equations by expressing it in matrix form and using Gauss-Jordan elimination to obtain a matrix in* reduced *row-echelon form.*

25. $2x - 3y = 12$
    $x + 2y = -1$

26. $3x - y = 1$
    $6x + 3y = 7$

27. $4x + 6y = 0$
    $x - y = \frac{5}{6}$

28. $2x - 4y = -\frac{13}{2}$
    $4x + 2y = 7$

29. $x - 2y + z = 9$
    $y - 3z = -10$
    $z = 3$

30. $x + 3y - 2z = -9$
    $y + 2z = 10$
    $z = 5$

31. $x + 2y + 4z = -10$
    $2x + 3y + 6z = -15$
    $x - y + z = -7$

32. $x - 3y + z = -12$
    $x + y + z = 0$
    $2x - y + z = -8$

**EXERCISES 33–36** □ *A linear system of equations is given. In each case, the system has infinitely many solutions. Use Gaussian elimination to express the system as an augmented matrix in row-echelon form, and then express the solution set in terms of one of the variables.*

33. $2x + y = 5$
    $4x + 2y = 10$

34. $3x - y = 1$
    $\frac{3}{2}x - \frac{1}{2}y = \frac{1}{2}$

35. $x + 2y + z = 5$
    $x - y + 3z = 6$
    $2x + y + 4z = 11$

36. $2x + 4y + 2z = 0$
    $3x - y + z = 1$
    $x - 5y - z = 1$

■ *Applications*

37. *Fitting a Parabola*  Find the equation of the parabola that passes through the points $(1, 5)$, $(2, 7)$, and $(3, 13)$.

38. *River Beautification*  Kyung-Bai, Antoine, and Vanessa are cleaning a river bank of trash. The time required for Vanessa to clear a 100-yard stretch is the average of the time required for Kyung-Bai and Antoine. The sum of their times is 27 hours, and the time required for Antoine to clear a 100-yard stretch is 7 less than the sum of the corresponding times for Kyung-Bai and Vanessa.

    a. Find the time required for each of them to clear a 100-yard stretch of the river bank.

    b. How long would it take them to clear a 100-yard stretch if they work together?

39. *Unknown Coins*  Suppose that 16 coins (nickels, dimes, or quarters) are used to pay a bill of $1.65 and that the number of nickels is equal to the sum of the number of dimes and quarters. Find the number of nickels, dimes, and quarters.

40. *Unknown Power*  A weightlifter is bench-pressing a total of 305 pounds using a 45-pound bar on which there are only 45-, 25-, and 5-pound weights. He uses a total of 12 weights, and the number of 5-pound weights is equal to the sum of the total number of 25- and 45-pound weights. How many weights of each type did the weightlifter use?

41. *Ancient Faiths*  When Martin Luther broke from the Roman Catholic Church in 1517, Buddhism and Hinduism were a combined 5059 years old. When Joseph Smith founded the Mormon Church in 1827, Islam and Buddhism were a combined 3557 years old. Hinduism is 975 years older than Buddhism. Find the years in which Islam, Buddhism, and Hinduism were founded.

42. *Nut Mixture*  How can 30 pounds of a mixture that is 40%

pecans, 40% almonds, and 20% cashews be formed if only the following cans of mixed nuts are available?

▨ 1-pound can: 50% pecans, 30% almonds, and 20% cashews

▨ 2-pound can: 30% pecans, 60% almonds, and 10% cashews

▨ 5-pound can: 40% pecans, 30% almonds, and 30% cashews

43. *School Integration*  In a certain community, the middle schools fall into three categories:

    (1) Schools of about 500 students in which roughly half the students are white and half the students are black.

    (2) Schools with about 450 students that are approximately half black and half Latino.

    (3) Schools with about 400 students that are 80% white and 20% black.

    A new high school with the capacity for 3750 students is being built. It is desired that the ethnic composition of the school be consistent with that of the community as a whole, which is 46.8% black, 35.2% white, and 18% Latino. How many middle schools of each type should feed into the new high school in order to achieve the desired ethnic composition?

44. *Ancestral Heritage*  A woman is $\frac{19}{32}$ Cherokee and $\frac{13}{32}$ Irish. Her mother is half Cherokee. Her maternal grandfather has the same percentage of Cherokee blood as does her paternal grandfather. Her maternal grandmother has half as much Cherokee blood as does her paternal grandmother. Compute the fractions of Cherokee and Irish blood of her father and all four of her grandparents. [Hint: Let the four variables be the fraction of Cherokee blood of each of the woman's grandparents. If one's father is $a$% Cherokee and one's mother is $b$% Cherokee, then one's percentage of Cherokee blood is given by $(a + b)/2$.]

### ■ *Projects for Enrichment*

**45.** *The Sum of the First n Squares* In this project we will develop a formula for the sum $S$ of the squares of the first $n$ natural numbers, $S = 1^2 + 2^2 + 3^2 + \cdots + n^2$. We will take advantage of the fact that in general, if $T = 1^k + 2^k + \cdots + n^k$, where $k$ and $n$ are natural numbers, then $T$ is a polynomial of degree $k + 1$ in the variable $n$ with constant term 0. In particular, $S$ can be expressed as a third-degree polynomial in $n$ with constant term 0; that is, $S = an^3 + bn^2 + cn$.

  a. Evaluate $S = 1^2 + 2^2 + \cdots + n^2$ for $n = 1, 2$, and 3.

  b. Evaluate the expression $an^3 + bn^2 + c$ for $n = 1, 2$, and 3.

  c. Use the results of parts (a) and (b) to form a system of three equations in the three variables $a$, $b$, and $c$.

  d. Use Gaussian elimination to solve the system of part (c).

  e. Express $S$ as a third-degree polynomial in $n$.

  f. Compute the sum of the squares of the first 10,000 natural numbers.

**46.** *Line Drive Home-Runs* Suppose that a batter hits a line drive that passes over the pitcher (60 feet 6 inches from home plate) at a height of 9 feet above the field, and then clears the 10-foot center field wall (330 feet from home plate) by 4 feet. A broadcaster announces that the ball "was still rising when it cleared the center field wall."

  We will investigate the possible locations at which the ball reaches its zenith, and in particular, we will determine if it is possible that the ball was still rising when it cleared the wall.

  a. Assuming that there is no wind resistance, the ball will travel in a parabolic path. Thus, if we let $y$ represent the height of the ball and $x$ represent the horizontal distance from home plate, then $y = ax^2 + bx + c$ for some constants

Deion Sanders watches to see if a long hit clears the outfield fence.

$a$, $b$, and $c$. Set up a system of equations that reflects the facts given in the introductory paragraph of this project.

  b. Solve the system from part (a). Describe the solution set in terms of the variable $c$. Note that $c$ represents the height of the ball when it was struck by the batter.

  c. The vertex of the parabola will occur when $x = -\dfrac{b}{2a}$. Find an expression for the location of the vertex in terms of the variable $c$.

  d. Compute the location of the vertex for values of $c$ ranging from 2 to 6, in increments of 1.

  e. Find the value of $c$ for which the ball's peak occurs as it crosses the fence. Is this reasonable?

### ■ *Questions for Discussion or Essay*

**47.** What similarities do you see between the techniques of Gaussian elimination and synthetic division?

**48.** If we define an arithmetic step to be either the addition, subtraction, or multiplication of two numbers, then what is the maximum number of arithmetic steps required to solve a system of three equations in three unknowns by Gaussian elimination? What about for Gauss-Jordan elimination? Support your answers.

**49.** One of the elementary row operations is multiplying a row by a nonzero constant. This operation corresponds to multiplying

both sides of an equation by a constant. Why do we insist that the constant be nonzero? After all, if we multiply both sides of an equation by 0, we obtain the equation $0 = 0$, which is certainly true.

**50.** Gaussian elimination and Gauss-Jordan elimination are examples of algorithms, step-by-step methods for solving a problem. Name two other mathematical algorithms and describe two algorithms from everyday life. When elimination was first introduced, it was done so in a nonalgorithmic way: each step was not prescribed; there was some choice involved. Discuss the merits and drawbacks to processes that are algorithmic.

**SECTION 4**

# INVERSES OF SQUARE MATRICES

■ How can linear systems of equations be solved using a graphics calculator?

■ How can you graph the letter ''M'' on a graphics calculator?

■ How can the physical laws governing the universe be discovered using matrices?

■ What kind of matrices can we divide by?

## THE IDENTITY MATRIX AND INVERSE MATRICES

A matrix for which the number of rows is the same as the number of columns is called a **square matrix**. The **main diagonal** of a square matrix runs from the upper left corner to the lower right corner, as shown in Figure 7. Thus, it consists of the entries for which the row and column are the same. The $n \times n$ matrix with 1's on the main diagonal and 0's elsewhere (see Figure 8) is called the $n$th-order **identity matrix**, and is denoted by $I_n$, or simply $I$ if the order is apparent.

An $n$th-order square matrix

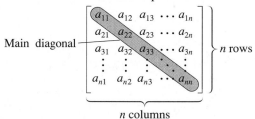

Main diagonal

$n$ rows

**FIGURE 7**

$n$ columns

The $n$th-order identity matrix $I_n$

$$I_n = \begin{bmatrix} 1 & 0 & 0 & 0 & \cdots & 0 \\ 0 & 1 & 0 & 0 & \cdots & 0 \\ 0 & 0 & 1 & 0 & \cdots & 0 \\ \vdots & \vdots & & & \ddots & \vdots \\ 0 & & & & \cdot & 0 \\ 0 & 0 & \cdots & 0 & 0 & 1 \end{bmatrix} \Big\} n \text{ rows}$$

**FIGURE 8**

$n$ columns

The following example suggests why $I$ is called the *identity* matrix.

**EXAMPLE 1**    *Multiplication of a matrix by I*

Let

$$A = \begin{bmatrix} 1 & 2 & 3 \\ 4 & 5 & 6 \\ 7 & 8 & 9 \end{bmatrix}$$

Compute $A \cdot I$ and $I \cdot A$.

**SOLUTION**

$$A \cdot I = \begin{bmatrix} 1 & 2 & 3 \\ 4 & 5 & 6 \\ 7 & 8 & 9 \end{bmatrix} \cdot \begin{bmatrix} 1 & 0 & 0 \\ 0 & 1 & 0 \\ 0 & 0 & 1 \end{bmatrix}$$

$$= \begin{bmatrix} (1 \cdot 1 + 2 \cdot 0 + 3 \cdot 0) & (1 \cdot 0 + 2 \cdot 1 + 3 \cdot 0) & (1 \cdot 0 + 2 \cdot 0 + 3 \cdot 1) \\ (4 \cdot 1 + 5 \cdot 0 + 6 \cdot 0) & (4 \cdot 0 + 5 \cdot 1 + 6 \cdot 0) & (4 \cdot 0 + 5 \cdot 0 + 6 \cdot 1) \\ (7 \cdot 1 + 8 \cdot 0 + 9 \cdot 0) & (7 \cdot 0 + 8 \cdot 1 + 9 \cdot 0) & (7 \cdot 0 + 8 \cdot 0 + 9 \cdot 1) \end{bmatrix}$$

$$= \begin{bmatrix} 1 & 2 & 3 \\ 4 & 5 & 6 \\ 7 & 8 & 9 \end{bmatrix}$$

Thus, $A \cdot I = A$.

$$I \cdot A = \begin{bmatrix} 1 & 0 & 0 \\ 0 & 1 & 0 \\ 0 & 0 & 1 \end{bmatrix} \cdot \begin{bmatrix} 1 & 2 & 3 \\ 4 & 5 & 6 \\ 7 & 8 & 9 \end{bmatrix}$$

$$= \begin{bmatrix} (1 \cdot 1 + 0 \cdot 4 + 0 \cdot 7) & (1 \cdot 2 + 0 \cdot 5 + 0 \cdot 8) & (1 \cdot 3 + 0 \cdot 6 + 0 \cdot 9) \\ (0 \cdot 1 + 1 \cdot 4 + 0 \cdot 7) & (0 \cdot 2 + 1 \cdot 5 + 0 \cdot 8) & (0 \cdot 3 + 1 \cdot 6 + 0 \cdot 9) \\ (0 \cdot 1 + 0 \cdot 4 + 1 \cdot 7) & (0 \cdot 2 + 0 \cdot 5 + 1 \cdot 8) & (0 \cdot 3 + 0 \cdot 6 + 1 \cdot 9) \end{bmatrix}$$

$$= \begin{bmatrix} 1 & 2 & 3 \\ 4 & 5 & 6 \\ 7 & 8 & 9 \end{bmatrix}$$

Thus, $I \cdot A = A$.

In the previous example we saw that the product of a third-order square matrix and $I$ was the given third-order matrix. In fact, this is true for all square matrices, as stated below. We leave the proof to the exercises.

**The identity property of $I_n$**

> If $A$ is an $n \times n$ matrix, then $A \cdot I_n = I_n \cdot A = A$.

The matrix identity $I_n$ is similar to the multiplicative identity 1 in that multiplying a matrix by $I_n$ leaves the matrix unchanged, just as multiplication by 1 leaves a number unchanged. Now the multiplicative inverse of a number $x$ is, by definition, a *number* that when multiplied by $x$ gives the multiplicative identity 1; it is commonly denoted by $x^{-1}$. What then is the multiplicative inverse of an $n \times n$ matrix $A$? Of course, it should be a *matrix* that when multiplied by $A$ (in either order) gives the multiplicative identity $I_n$. We make the following definition.

**Definition of the multiplicative inverse of a matrix**

> Let $A$ be an $n \times n$ matrix. Then the **inverse** of $A$, if it exists, is an $n \times n$ matrix $B$ satisfying $AB = I_n$ and $BA = I_n$. We say that $A$ and $B$ are inverses of one another and write $B = A^{-1}$. In this case, we say that the matrix $A$ is **invertible**. If $A^{-1}$ doesn't exist, then we say that $A$ is **noninvertible**.

| EXAMPLE 2 | *Confirming inverses* |

Let

$$A = \begin{bmatrix} 2 & 0 \\ -4 & 1 \end{bmatrix} \quad \text{and} \quad B = \begin{bmatrix} \frac{1}{2} & 0 \\ 2 & 1 \end{bmatrix}$$

Show that $A$ and $B$ are inverses of one another.

**SOLUTION**   We must show that

$$AB = BA = \begin{bmatrix} 1 & 0 \\ 0 & 1 \end{bmatrix}$$

Thus, we proceed as follows.

$$AB = \begin{bmatrix} 2 & 0 \\ -4 & 1 \end{bmatrix} \cdot \begin{bmatrix} \frac{1}{2} & 0 \\ 2 & 1 \end{bmatrix}$$

$$= \begin{bmatrix} (2 \cdot \frac{1}{2} + 0 \cdot 2) & (2 \cdot 0 + 0 \cdot 1) \\ (-4 \cdot \frac{1}{2} + 1 \cdot 2) & (-4 \cdot 0 + 1 \cdot 1) \end{bmatrix}$$

$$= \begin{bmatrix} 1 & 0 \\ 0 & 1 \end{bmatrix}$$

$$BA = \begin{bmatrix} \frac{1}{2} & 0 \\ 2 & 1 \end{bmatrix} \cdot \begin{bmatrix} 2 & 0 \\ -4 & 1 \end{bmatrix}$$

$$= \begin{bmatrix} (\frac{1}{2} \cdot 2 + 0 \cdot -4) & (\frac{1}{2} \cdot 0 + 0 \cdot 1) \\ (2 \cdot 2 + 1 \cdot -4) & (2 \cdot 0 + 1 \cdot 1) \end{bmatrix}$$

$$= \begin{bmatrix} 1 & 0 \\ 0 & 1 \end{bmatrix}$$

Note that it is not necessary to compute both $AB$ and $BA$ to show that $A$ and $B$ are inverses. It can be shown that for square matrices $A$ and $B$, if $AB = I$, then $BA = I$ also.

## COMPUTING INVERSES

We will now see how the inverse of a $2 \times 2$ matrix can be computed by solving a pair of systems of equations using Gauss-Jordan elimination. Let us consider the problem of computing the inverse of the matrix

$$A = \begin{bmatrix} 1 & 3 \\ 2 & 5 \end{bmatrix}$$

If we let

$$A^{-1} = \begin{bmatrix} a & b \\ c & d \end{bmatrix}$$

then we have

$$A \cdot A^{-1} = I$$

$$\begin{bmatrix} 1 & 3 \\ 2 & 5 \end{bmatrix} \cdot \begin{bmatrix} a & b \\ c & d \end{bmatrix} = \begin{bmatrix} 1 & 0 \\ 0 & 1 \end{bmatrix}$$

$$\begin{bmatrix} (a + 3c) & (b + 3d) \\ (2a + 5c) & (2b + 5d) \end{bmatrix} = \begin{bmatrix} 1 & 0 \\ 0 & 1 \end{bmatrix}$$

Equating corresponding entries gives us the following pair of systems of equations and their corresponding augmented matrices.

$$\text{System 1} \begin{cases} a + 3c = 1 \\ 2a + 5c = 0 \end{cases} \qquad \left[ \begin{array}{cc|c} 1 & 3 & 1 \\ 2 & 5 & 0 \end{array} \right]$$

$$\text{System 2} \begin{cases} b + 3d = 0 \\ 2b + 5d = 1 \end{cases} \qquad \left[ \begin{array}{cc|c} 1 & 3 & 0 \\ 2 & 5 & 1 \end{array} \right]$$

We solve both systems by Gauss-Jordan elimination as shown below.

### SYSTEM 1:

$$\left[ \begin{array}{cc|c} 1 & 3 & 1 \\ 2 & 5 & 0 \end{array} \right] \xrightarrow{-2R_1 + R_2} \left[ \begin{array}{cc|c} 1 & 3 & 1 \\ 0 & -1 & -2 \end{array} \right] \xrightarrow{-1 \cdot R_2} \left[ \begin{array}{cc|c} 1 & 3 & 1 \\ 0 & 1 & 2 \end{array} \right] \xrightarrow{-3R_2 + R_1} \left[ \begin{array}{cc|c} 1 & 0 & -5 \\ 0 & 1 & 2 \end{array} \right]$$

Thus, $a = -5$ and $c = 2$.

### SYSTEM 2:

$$\left[ \begin{array}{cc|c} 1 & 3 & 0 \\ 2 & 5 & 1 \end{array} \right] \xrightarrow{-2R_1 + R_2} \left[ \begin{array}{cc|c} 1 & 3 & 0 \\ 0 & -1 & 1 \end{array} \right] \xrightarrow{-1 \cdot R_2} \left[ \begin{array}{cc|c} 1 & 3 & 0 \\ 0 & 1 & -1 \end{array} \right] \xrightarrow{-3R_2 + R_1} \left[ \begin{array}{cc|c} 1 & 0 & 3 \\ 0 & 1 & -1 \end{array} \right]$$

Thus, $b = 3$ and $d = -1$. Using the values of $a$, $b$, $c$, and $d$, we find that

$$A^{-1} = \begin{bmatrix} -5 & 3 \\ 2 & -1 \end{bmatrix}$$

Notice that precisely the same elementary row operations were performed at each stage of the Gauss-Jordan elimination process for systems 1 and 2, since the coefficient matrices were the same. This suggests that we solve both systems simultaneously by applying Gauss-Jordan elimination to the following augmented matrix.

$$\left[ \begin{array}{cc|cc} 1 & 3 & 1 & 0 \\ 2 & 5 & 0 & 1 \end{array} \right]$$

When this matrix is in reduced row-echelon form, the first two columns will form the identity matrix $I_2$, the solution to system 1 will be given by the entries in the third column, and the solution to system 2 will be given by the entries in the fourth column. In other words, at the conclusion of the Gauss-Jordan elimination process, the matrix will be as follows.

$$\begin{bmatrix} 1 & 0 & a & b \\ 0 & 1 & c & d \end{bmatrix}$$

Thus, the third and fourth columns will form $A^{-1}$. We now compute $A^{-1}$ with this procedure.

$$\begin{bmatrix} 1 & 3 & 1 & 0 \\ 2 & 5 & 0 & 1 \end{bmatrix} \xrightarrow{-2R_1 + R_2} \begin{bmatrix} 1 & 3 & 1 & 0 \\ 0 & -1 & -2 & 1 \end{bmatrix} \xrightarrow{-1 \cdot R_2}$$

$$\begin{bmatrix} 1 & 3 & 1 & 0 \\ 0 & 1 & 2 & -1 \end{bmatrix} \xrightarrow{-3R_2 + R_1} \begin{bmatrix} 1 & 0 & -5 & 3 \\ 0 & 1 & 2 & -1 \end{bmatrix}$$

So

$$A^{-1} = \begin{bmatrix} -5 & 3 \\ 2 & -1 \end{bmatrix}$$

The technique described above generalizes to square matrices of arbitrary dimension, and we state this as follows.

*Computing inverses with Gauss-Jordan elimination*

1. Form the augmented matrix whose first $n$ columns constitute $A$ and whose last $n$ columns form $I_n$, symbolically $[A|I_n]$.

2. Use Gauss-Jordan elimination to write this matrix in reduced row-echelon form.

   a. If the first $n$ columns of the resulting matrix form the identity $I_n$, then the last $n$ columns will form $A^{-1}$. Symbolically, the reduced row-echelon matrix is of the form $[I_n|A^{-1}]$.

   b. If the first $n$ columns do not form the identity matrix $I_n$, then $A$ is noninvertible.

**EXAMPLE 3** *Computing inverses of matrices*

Compute the inverse of each of the following matrices. Check your answer.

a. $A = \begin{bmatrix} 1 & 0 & 1 \\ 1 & -1 & 0 \\ 0 & 2 & 1 \end{bmatrix}$      b. $B = \begin{bmatrix} 1 & 0 \\ 0 & 0 \end{bmatrix}$

## SOLUTION

**a.** We form the augmented matrix $[A|I_3]$ and perform Gauss-Jordan elimination.

$$\begin{bmatrix} 1 & 0 & 1 & | & 1 & 0 & 0 \\ 1 & -1 & 0 & | & 0 & 1 & 0 \\ 0 & 2 & 1 & | & 0 & 0 & 1 \end{bmatrix} \xrightarrow{-R_1 + R_2} \begin{bmatrix} 1 & 0 & 1 & | & 1 & 0 & 0 \\ 0 & -1 & -1 & | & -1 & 1 & 0 \\ 0 & 2 & 1 & | & 0 & 0 & 1 \end{bmatrix}$$

$$\xrightarrow{-1 \cdot R_2} \begin{bmatrix} 1 & 0 & 1 & | & 1 & 0 & 0 \\ 0 & 1 & 1 & | & 1 & -1 & 0 \\ 0 & 2 & 1 & | & 0 & 0 & 1 \end{bmatrix} \xrightarrow{-2R_2 + R_3} \begin{bmatrix} 1 & 0 & 1 & | & 1 & 0 & 0 \\ 0 & 1 & 1 & | & 1 & -1 & 0 \\ 0 & 0 & -1 & | & -2 & 2 & 1 \end{bmatrix}$$

$$\xrightarrow{-1 \cdot R_3} \begin{bmatrix} 1 & 0 & 1 & | & 1 & 0 & 0 \\ 0 & 1 & 1 & | & 1 & -1 & 0 \\ 0 & 0 & 1 & | & 2 & -2 & -1 \end{bmatrix} \xrightarrow[-R_3 + R_2]{-R_3 + R_1} \begin{bmatrix} 1 & 0 & 0 & | & -1 & 2 & 1 \\ 0 & 1 & 0 & | & -1 & 1 & 1 \\ 0 & 0 & 1 & | & 2 & -2 & -1 \end{bmatrix}$$

Thus

$$A^{-1} = \begin{bmatrix} -1 & 2 & 1 \\ -1 & 1 & 1 \\ 2 & -2 & -1 \end{bmatrix}$$

We can check this result by multiplying $A^{-1}$ by the original matrix $A$; if the product is $I$, then our result is correct.

$$A^{-1} \cdot A = \begin{bmatrix} -1 & 2 & 1 \\ -1 & 1 & 1 \\ 2 & -2 & -1 \end{bmatrix} \cdot \begin{bmatrix} 1 & 0 & 1 \\ 1 & -1 & 0 \\ 0 & 2 & 1 \end{bmatrix}$$

$$= \begin{bmatrix} (-1 \cdot 1 + 2 \cdot 1 + 1 \cdot 0) & (-1 \cdot 0 + 2 \cdot -1 + 1 \cdot 2) & (-1 \cdot 1 + 2 \cdot 0 + 1 \cdot 1) \\ (-1 \cdot 1 + 1 \cdot 1 + 1 \cdot 0) & (-1 \cdot 0 + 1 \cdot -1 + 1 \cdot 2) & (-1 \cdot 1 + 1 \cdot 0 + 1 \cdot 1) \\ (2 \cdot 1 + -2 \cdot 1 + -1 \cdot 0) & (2 \cdot 0 + -2 \cdot -1 + -1 \cdot 2) & (2 \cdot 1 + -2 \cdot 0 + -1 \cdot 1) \end{bmatrix}$$

$$= \begin{bmatrix} 1 & 0 & 0 \\ 0 & 1 & 0 \\ 0 & 0 & 1 \end{bmatrix}$$

**b.** We begin by forming the augmented matrix

$$\begin{bmatrix} 1 & 0 & | & 1 & 0 \\ 0 & 0 & | & 0 & 1 \end{bmatrix}$$

Since this matrix is already in reduced row-echelon form, and the first two columns do *not* constitute the identity matrix, $B$ is noninvertible.

## SOLVING LINEAR SYSTEMS OF EQUATIONS USING INVERSES

Although inverse matrices arise in many contexts, the most important application of the inverse for our purpose is as a tool for solving linear systems of equations. Consider the following system of equations.

$$2x + 3y + z = 6$$

$$4x - y + z = -3$$

$$x + y + \frac{1}{2}z = 1$$

This system of equations can be written in matrix form as follows.

$$\begin{bmatrix} 2 & 3 & 1 \\ 4 & -1 & 1 \\ 1 & 1 & \frac{1}{2} \end{bmatrix} \cdot \begin{bmatrix} x \\ y \\ z \end{bmatrix} = \begin{bmatrix} 6 \\ -3 \\ 1 \end{bmatrix}$$

If we define

$$A = \begin{bmatrix} 2 & 3 & 1 \\ 4 & -1 & 1 \\ 1 & 1 & \frac{1}{2} \end{bmatrix}, \quad X = \begin{bmatrix} x \\ y \\ z \end{bmatrix}, \quad \text{and} \quad B = \begin{bmatrix} 6 \\ -3 \\ 1 \end{bmatrix}$$

then we have the matrix equation $AX = B$. Now if $A$ is invertible, we can solve this equation for the matrix $X$ as follows.

| | |
|---|---|
| $AX = B$ | The original matrix equation |
| $A^{-1} \cdot (AX) = A^{-1} \cdot B$ | Multiplying both sides **on the left** by $A^{-1}$ |
| $(A^{-1} \cdot A) \cdot X = A^{-1}B$ | The associative law of matrix multiplication |
| $IX = A^{-1}B$ | The definition of the multiplicative inverse |
| $X = A^{-1}B$ | Using a property of the identity matrix. Note that the product on the right-hand side is defined since $A^{-1}$ is a $3 \times 3$ matrix and $B$ is a $3 \times 1$ matrix. |

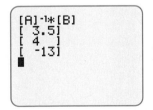

**FIGURE 9**

This technique is ideal for use with a graphics calculator since the solution involves only the operations of matrix inversion and multiplication, both of which are easily performed on a graphics calculator. After entering the matrices $A$ and $B$ in the calculator, and computing $A^{-1}B$, the calculator returns the answer shown in Figure 9.

It should be noted however that this technique will work only for systems of equations in which the coefficient matrix $A$ is both square and invertible. If either one of these conditions is not satisfied, then some other technique, such as Gaussian or Gauss-Jordan elimination, must be employed to describe the solution set.

**EXAMPLE 4**    *Finding the equation of a parabola passing through three given points*

The graph of a quadratic function $f$ includes the points $(-2, 25)$, $(1, 4)$, and $(3, 20)$. Find an expression for $f(x)$.

**SOLUTION**    Since $f$ is a quadratic function, we know that $f(x) = ax^2 + bx + c$ for some choice of constants $a$, $b$, and $c$. Since $(-2, 25)$ is a point on the graph of $f$, we know that $f(-2) = 25$; thus, we have

$$f(-2) = a(-2)^2 + b(-2) + c = 25$$
$$4a - 2b + c = 25$$

Similarly, from the facts that $(1, 4)$ and $(3, 20)$ are points on the graph of $f$, we obtain the equations

$$f(1) = a + b + c = 4$$
$$f(3) = 9a + 3b + c = 20$$

Thus, we must solve the following system of equations.

$$4a - 2b + c = 25$$
$$a + b + c = 4$$
$$9a + 3b + c = 20$$

In matrix form, we have

$$\begin{bmatrix} 4 & -2 & 1 \\ 1 & 1 & 1 \\ 9 & 3 & 1 \end{bmatrix} \cdot \begin{bmatrix} a \\ b \\ c \end{bmatrix} = \begin{bmatrix} 25 \\ 4 \\ 20 \end{bmatrix}$$

Solving for the unknown matrix

$$\begin{bmatrix} a \\ b \\ c \end{bmatrix}$$

we have

$$\begin{bmatrix} a \\ b \\ c \end{bmatrix} = \begin{bmatrix} 4 & -2 & 1 \\ 1 & 1 & 1 \\ 9 & 3 & 1 \end{bmatrix}^{-1} \cdot \begin{bmatrix} 25 \\ 4 \\ 20 \end{bmatrix}$$

Using a graphics calculator to evaluate this expression, we have

$$\begin{bmatrix} a \\ b \\ c \end{bmatrix} = \begin{bmatrix} 3 \\ -4 \\ 5 \end{bmatrix}$$

Thus $a = 3$, $b = -4$, and $c = 5$. Thus, the quadratic function $f$ is defined by

$$f(x) = 3x^2 - 4x + 5$$

The graph of $f$ along with the points $(-2, 25)$, $(1, 4)$, and $(3, 20)$ is shown in Figure 10.

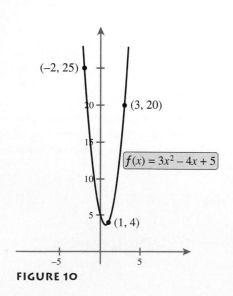

$(-2, 25)$

$(3, 20)$

$f(x) = 3x^2 - 4x + 5$

$(1, 4)$

$-5$        $5$

**FIGURE 10**

**EXERCISES 4**

**EXERCISES 1–22** □ *A matrix is given. Find the inverse if it exists. Use a graphics calculator as needed.*

1. $\begin{bmatrix} 1 & 0 \\ 2 & 1 \end{bmatrix}$

2. $\begin{bmatrix} 0.2 & 0.4 \\ 0.5 & 0.5 \end{bmatrix}$

3. $\begin{bmatrix} \frac{1}{2} & \frac{1}{3} \\ 2 & 0 \end{bmatrix}$

4. $\begin{bmatrix} 0.125 & 0.578 \\ 3.24 & 1.67 \end{bmatrix}$

5. $\begin{bmatrix} 1 & 2 \\ 2 & 4 \end{bmatrix}$

6. $\begin{bmatrix} a & 0 \\ 0 & b \end{bmatrix}$ $(a \neq 0, b \neq 0)$

7. $\begin{bmatrix} 0 & a \\ b & 0 \end{bmatrix}$ $(a \neq 0, b \neq 0)$

8. $\begin{bmatrix} x & y \\ 4x & 4y \end{bmatrix}$

9. $\begin{bmatrix} x & x+1 \\ x & x-1 \end{bmatrix}$ $(x \neq 0)$

10. $\begin{bmatrix} 1 & x \\ x^2 & x^3 \end{bmatrix}$

11. $\begin{bmatrix} 1 & 2 & 0 \\ 0 & 1 & 4 \\ 0 & 0 & 1 \end{bmatrix}$

12. $\begin{bmatrix} -\frac{1}{5} & 0 & 1 \\ \frac{2}{5} & -1 & -1 \\ -\frac{1}{5} & 1 & 0 \end{bmatrix}$

13. $\begin{bmatrix} 1 & 4 & -2 \\ 0 & 2 & -1 \\ -3 & 0 & 2 \end{bmatrix}$

14. $\begin{bmatrix} 1 & 2 & 3 \\ 4 & 5 & 6 \\ 7 & 8 & 9 \end{bmatrix}$

15. $\begin{bmatrix} -\frac{1}{12} & \frac{7}{12} & -\frac{1}{3} \\ -\frac{1}{12} & -\frac{5}{12} & \frac{2}{3} \\ \frac{5}{12} & \frac{1}{12} & -\frac{1}{3} \end{bmatrix}$

16. $\begin{bmatrix} 1.2 & -2.3 & 4.7 \\ 0 & 1.9 & -8.1 \\ 3.1 & 0 & 0 \end{bmatrix}$

17. $\begin{bmatrix} 1 & x & 0 \\ 0 & 1 & x \\ 0 & 0 & 1 \end{bmatrix}$

18. $\begin{bmatrix} 0 & x & x \\ x & 0 & x \\ x & x & 0 \end{bmatrix}$

19. $\begin{bmatrix} 1 & 2 & 3 & 4 \\ 2 & 3 & 4 & 1 \\ 3 & 4 & 1 & 2 \\ 4 & 1 & 2 & 3 \end{bmatrix}$

20. $\begin{bmatrix} 1 & 0 & 0 & 1 \\ 0 & 1 & 0 & 1 \\ 0 & 0 & 1 & 1 \\ 1 & 1 & 1 & 3 \end{bmatrix}$

21. $\begin{bmatrix} 1 & 0 & 0 & 0 & 0 & 0 & 0 \\ -1 & 1 & 0 & 0 & 0 & 0 & 0 \\ 0 & -1 & 1 & 0 & 0 & 0 & 0 \\ 0 & 0 & -1 & 1 & 0 & 0 & 0 \\ 0 & 0 & 0 & -1 & 1 & 0 & 0 \\ 0 & 0 & 0 & 0 & -1 & 1 & 0 \\ 0 & 0 & 0 & 0 & 0 & -1 & 1 \end{bmatrix}$

22. $\begin{bmatrix} 1 & 1 & 1 & 1 & 1 & 1 & 1 \\ 0 & 1 & 1 & 1 & 1 & 1 & 1 \\ 0 & 0 & 1 & 1 & 1 & 1 & 1 \\ 0 & 0 & 0 & 1 & 1 & 1 & 1 \\ 0 & 0 & 0 & 0 & 1 & 1 & 1 \\ 0 & 0 & 0 & 0 & 0 & 1 & 1 \\ 0 & 0 & 0 & 0 & 0 & 0 & 1 \end{bmatrix}$

**EXERCISES 23–28** □ *Three systems of equations are given. Solve the systems by first expressing them in matrix form as $AX = B$, and then evaluating $X = A^{-1}B$.*

23. a. $\begin{aligned} x - y &= -1 \\ x + 2y &= 12 \end{aligned}$

    b. $\begin{aligned} x - y &= -4 \\ x + 2y &= -1 \end{aligned}$

    c. $\begin{aligned} x - y &= 1 \\ x + 2y &= -\frac{1}{2} \end{aligned}$

24. a. $\begin{aligned} -2x + 6y &= 6 \\ 3x + 5y &= 5 \end{aligned}$

    b. $\begin{aligned} -2x + 6y &= 8.4 \\ 3x + 5y &= 8.4 \end{aligned}$

    c. $\begin{aligned} -2x + 6y &= -2.5 \\ 3x + 5y &= 7.25 \end{aligned}$

25. a. $\begin{aligned} 1.5x - 4y &= -12.5 \\ 2.3x + 6y &= 41.5 \end{aligned}$

    b. $\begin{aligned} 1.5x - 4y &= 2 \\ 2.3x + 6y &= 15.2 \end{aligned}$

    c. $\begin{aligned} 1.5x - 4y &= -15 \\ 2.3x + 6y &= 13.4 \end{aligned}$

26. a. $\begin{aligned} \frac{2}{5}x + \frac{3}{5}y &= 8 \\ x - y &= -5 \end{aligned}$

    b. $\begin{aligned} \frac{2}{5}x + \frac{3}{5}y &= 2 \\ x - y &= -5 \end{aligned}$

    c. $\begin{aligned} \frac{2}{5}x + \frac{3}{5}y &= 1.4 \\ x - y &= -1.5 \end{aligned}$

27. a. $\begin{aligned} x + 2y - 4z &= -8 \\ 3x - y + z &= 7 \\ 2x + 2y - z &= 6 \end{aligned}$

    b. $\begin{aligned} x + 2y - 4z &= -6 \\ 3x - y + z &= -5 \\ 2x + 2y - z &= -5 \end{aligned}$

    c. $\begin{aligned} x + 2y - 4z &= -5 \\ 3x - y + z &= 15 \\ 2x + 2y - z &= 15 \end{aligned}$

28. a. $\begin{aligned} x + y + z &= 2 \\ x + 2y - z &= 1 \\ y + z &= -1 \end{aligned}$

    b. $\begin{aligned} x + y + z &= 6 \\ x + 2y - z &= 2 \\ y + z &= 5 \end{aligned}$

    c. $\begin{aligned} x + y + z &= \frac{7}{20} \\ x + 2y - z &= \frac{4}{5} \\ y + z &= \frac{3}{20} \end{aligned}$

29. Let $A$ and $B$ be invertible matrices. Define $C = AB$ and $D = B^{-1}A^{-1}$. Show that $C$ and $D$ are inverses of one another. (*Hint:* Use the definitions of $C$ and $D$ to show that $CD = I$ and $DC = I$).

30. Show that a matrix with either a row or a column of all zeros is not invertible.

31. Show that if $A$ is an $n \times n$ matrix, then $I_n \cdot A = A$.

32. Show that $(aI)^{-1} = (1/a)I$ for any scalar $a$.

33. Let

$$A = \begin{bmatrix} a & 0 & 0 & 0 \\ 0 & b & 0 & 0 \\ 0 & 0 & c & 0 \\ 0 & 0 & 0 & d \end{bmatrix}$$

Under what conditions is $A$ invertible? If $A$ is invertible, what is its inverse?

34. Let $A$ be an invertible matrix such that $A^2 = A^4$. Show that $A = A^{-1}$.

35. The **trace** of a square matrix is defined to be the sum of the elements on the main diagonal. For each matrix given below, compute the trace.

    a. $\begin{bmatrix} 2 & -1 \\ 3 & -4 \end{bmatrix}$

    b. $\begin{bmatrix} 1 & 2 & 4 \\ 6 & 9 & 8 \\ 2 & 4 & 8 \end{bmatrix}$

    c. $aI_n$, where $a$ is a scalar

36. To find the coordinates of the point $(x, y)$ after a counterclockwise rotation of $45°$, we multiply the matrix $\begin{bmatrix} x \\ y \end{bmatrix}$ by

    $$\begin{bmatrix} \dfrac{\sqrt{2}}{2} & -\dfrac{\sqrt{2}}{2} \\ \dfrac{\sqrt{2}}{2} & \dfrac{\sqrt{2}}{2} \end{bmatrix}$$

    By what would we multiply $\begin{bmatrix} x \\ y \end{bmatrix}$ in order to find the coordinates of the point $(x, y)$ after undergoing a *clockwise* rotation of $45°$? (*Hint:* How many counterclockwise rotations by $45°$ make a clockwise rotation by $45°$?)

37. What is $I^{-1}$?

38. Consider the following system of equations.

    $$x^2 y^3 z^4 = 10$$
    $$x^3 y^2 z^5 = 9$$
    $$x^4 y^6 z^6 = 5$$

    a. Is the system linear?

    b. Take the natural logarithm of both sides of each equation. The resulting system is linear in the variables $\ln x$, $\ln y$, and $\ln z$. Solve the system for these variables.

    c. Find $x$, $y$, and $z$.

■ *Applications*

39. *Running Regimen* To safely increase aerobic capacity and burn fat, a man has developed a routine that consists of walking at 20 minutes per mile, jogging at 11 minutes per mile, and running at 8 minutes per mile. Given that walking burns 5.8 calories per minute, jogging burns 9.1 calories per minute, and running burns 14.1 calories per minute, how many minutes should he spend at each pace if he wants to

    a. burn 1200 calories, work out for 2 hours, and travel 11 miles?

    b. burn 750 calories, work out for 1.5 hours, and travel 7 miles?

40. *Investment Options* An investment company offers three different funds, each of which invests in highly volatile growth stocks, conservative but stable blue chip stocks, and income-generating bonds, but in differing proportions, as follows:

| Fund content | Growth fund | Income fund | Security fund |
|---|---|---|---|
| Growth stocks | 60% | 20% | 30% |
| Bonds | 20% | 50% | 20% |
| Blue chips | 20% | 30% | 50% |

Determine the amount that should be invested in each of the three funds for an individual investor who wants to invest

    a. $2000 in growth stocks, $1000 in bonds, and $1000 in blue chips.

    b. $1500 in growth stocks, $1500 in bonds, and $1500 in blue chips.

41. *Fitting a Cubic* Find the equation of the cubic function $f(x) = ax^3 + bx^2 + cx + d$ passing through the points $(1, 5)$, $(3, 43)$, $(5, 201)$, and $(7, 575)$.

42. *Unknown Interest Rates* Suppose that a bank has three different money market accounts (growth, domestic, and international) paying annual interest rates of $i$, $j$, and $k$ (compounded annually), respectively. Alberto, Brenda, and Carlos each have $1000 to invest for 5 years. Alberto put his money in growth for 4 years, then switched to domestic for the last year. Brenda invested her money in domestic for 1 year, and then international for 4 years. Carlos put his money in growth for 2 years, domestic for 2 years, and international for the last year. At the end of the 5 years, Alberto had $1240.05, Brenda had $1442.11, and Carlos had $1312.50.

    a. Set up a system of three equations in the three variables $i$, $j$, and $k$. [*Hint:* If, for example, a principal of $P$ is deposited into an account paying interest at the rate $l$ for 4 years and then is switched to an account paying $m$ for 3 years, at the end of 7 years the balance will be $P(1 + l)^4(1 + m)^3$.]

b. Take natural logarithms of both sides of each equation. Let $I = \ln(1 + i)$, $J = \ln(1 + j)$, and $K = \ln(1 + k)$. Use prop-

erties of logarithms to write the resulting system as a *linear* system of three equations in the variables $I$, $J$, and $K$.

c. Use the techniques of this section to solve the linear system from part (b) for the variables $I$, $J$, and $K$.

d. Use the results of part (c) and the definitions of $I$, $J$, and $K$ given in part (b) to find $i$, $j$, and $k$.

---

### ◼ *Projects for Enrichment*

43. *Dimensions*  Physical laws express relationships between quantities such as mass, length, time, velocity, acceleration, force, energy, work, and many others. Each of these quantities is assigned a unit of measurement, and the physical laws relating them are only valid if the units are consistent. When we are not concerned with the particular system of measurement—SI (metric) or U.S. Customary—we can instead consider *dimensions*. To each of the three basic quantities mass, length, and time, we associate the dimensions $M$, $L$, and $T$, respectively. Other quantities have dimensions that are products involving the three basic dimensions. For example, since velocity is equal to distance (dimension $L$) divided by time (dimension $T$), the dimension of velocity is $L/T$ or $LT^{-1}$.

a. Find the dimension of the following quantities. Express your answers in the form $M^aL^bT^c$. Assume that all constants have no dimension.

    i. Acceleration ($=$ velocity $\div$ time)

    ii. Force ($=$ mass $\times$ acceleration)

    iii. Kinetic energy ($= \frac{1}{2} \times$ mass $\times$ velocity$^2$)

    iv. Work ($=$ force $\times$ distance)

Equations that express physical laws must be *dimensionally compatible* in the sense that all terms must have the same dimension. Consider the equation $s = -\frac{1}{2}gt^2 + v_0t + s_0$ that gives the height of an object at time $t$ if it is thrown into the air with initial velocity $v_0$ and from an initial height $s_0$. Both $s$ and $s_0$ have dimension $L$. Since $g$ is the acceleration due to gravity and $-\frac{1}{2}$ is a dimensionless constant, the term $-\frac{1}{2}gt^2$ has dimension $(LT^{-2})T^2 = L$. Likewise, $v_0t$ has dimension $(LT^{-1})T = L$. So each term has dimension $L$, and the equation is dimensionally compatible.

b. Determine whether the following equations are dimensionally compatible. Assume $m$ denotes mass, $s$ distance, $t$ time, $v$ velocity, $a$ acceleration, $F$ force, and $W$ work.

    i. $F = mv + v^2$

    ii. $v^2 = v_0^2 + 2as$

    iii. $E = \frac{1}{2}mv^2 + mgs$   ($E$ denotes total energy, which has dimension $ML^2T^{-2}$.)

    iv. $W = msv + Fs$

Some physical laws can be derived from a basic form and dimensional considerations. Let us consider the velocity of a falling object. It is reasonable to expect that the velocity is dependent on the acceleration due to gravity and the total time that the object has been falling. If we assume (perhaps on the basis of observations) that velocity varies directly as some power of the acceleration due to gravity and some power of time, we obtain the possible model $v = kg^at^b$, where $v$ is velocity; $g$ is the acceleration due to gravity; $t$ is time; and $k$, $a$, and $b$ are constants. Now the dimensions of $v$, $g$, and $t$ are $LT^{-1}$, $LT^{-2}$, and $T$, respectively, and the constant $k$ is dimensionless. So the equation $v = kg^at^b$ is dimensionally compatible only if

$$LT^{-1} = (LT^{-2})^aT^b \quad \text{which implies} \quad LT^{-1} = L^aT^{-2a+b}$$

Equating exponents, we obtain the system of equations

$$a = 1$$
$$-2a + b = -1$$

Solving the system, we find $a = 1$ and $b = 1$. Thus, $v = kgt$ is dimensionally compatible.

c. Use the process described above to find models of the given form.

    i. $t = kr^ag^b$ where $t$ is the period of a pendulum with length $r$, $k$ is a constant, and $g$ is the acceleration due to gravity.

    ii. $a = kv^ir^j$ where $a$ is the acceleration of a particle traveling with velocity $v$ in a circular path of radius $r$, and $k$ is a constant.

    iii. $F = km^av^br^c$ where $F$ is the centrifugal force of a particle with mass $m$ traveling in a circular path with radius $r$ and velocity $v$, and $k$ is a constant.

    iv. $F = k\mu^av^br^c$ where $F$ is the force opposing a ball of radius $r$ as it is falling with velocity $v$ through air having viscosity coefficient $\mu$ (dimension $ML^{-1}T^{-1}$), and $k$ is a constant.

**v.** $V = kG^a m^b R^c$ where $V$ is the velocity required for an object to escape the gravitational field of a planet with mass $m$ and radius $R$, $G$ is the universal gravitational constant with dimension $M^{-1}L^3T^{-2}$, and $k$ is a constant.

**44.** *Drawing Letters with a Graphics Calculator* In this project we will learn how matrix techniques can be used to draw letters on the display of a graphics calculator. In particular, we will illustrate how the letter "M" can be drawn on a graphics calculator. To do so, we must first consider the notion of a parametrically defined function. Instead of defining $y$ as a function of $x$, we define both $x$ and $y$ as functions of $t$. For example, suppose we define

$$x = t^2 \quad \text{and} \quad y = t^3$$

In other words, to each real number $t$ corresponds the pair of numbers $(x, y)$ of the form $(t^2, t^3)$. Thus, if $t = -2$, the corresponding pair of numbers consists of $x = (-2)^2 = 4$ and $y = (-2)^3 = -8$, or the pair $(4, -8)$. Some additional values of this correspondence are shown in Table 4. If we plot the points given in the second column of Table 4 and connect the points with a smooth curve, we obtain the graph shown in Figure 11. The arrows indicate the *direction of motion* as $t$ increases. We will find functions $x(t)$ and $y(t)$ such that the graph of $(x(t), y(t))$ for $t$ in an appropriate interval will resemble the letter "M." Shown in Figure 12 is a depiction of the letter "M" superimposed on the coordinate axes.

**TABLE 4**

| $t$ | $(t^2, t^3)$ |
|---|---|
| $-2$ | $(4, -8)$ |
| $-1$ | $(1, -1)$ |
| $0$ | $(0, 0)$ |
| $1$ | $(1, 1)$ |
| $2$ | $(4, 8)$ |

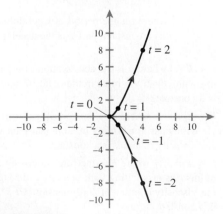

**FIGURE 11**

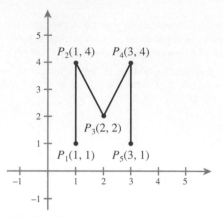

**FIGURE 12**

Let us suppose that a bug begins crawling at time 0 from point $P_1$ to point $P_2$, arriving at point $P_2$ at time $t = 1$; continuing to point $P_3$, arriving at time $t = 2$; and so on, finally arriving at point $P_5$ at time $t = 4$. Table 5 gives the $x$- and $y$-coordinates of the bug at times $t$ from 0 to 4.

**TABLE 5**

| $t$ | $x$ | $y$ |
|---|---|---|
| 0 | 1 | 1 |
| 1 | 1 | 4 |
| 2 | 2 | 2 |
| 3 | 3 | 4 |
| 4 | 3 | 1 |

**a.** With $x$ as the vertical axis and $t$ as the horizontal axis, plot the tabulated values of $x$ versus $t$. Connect the points with line segments.

**b.** It can be shown that the figure depicted in part (a) is the graph of a function of the form

$$x(t) = a|t| + b|t - 1| + c|t - 2| + d|t - 3| + e|t - 4|$$

Find the constants $a$, $b$, $c$, $d$, and $e$ by solving the system of equations arising from the fact that the graph must pass through the five indicated points.

**c.** Now using the same process as that outlined in parts (a) and (b), find constants $f$, $g$, $h$, $i$, and $j$ so that

$$y(t) = f|t| + g|t - 1| + h|t - 2| + i|t - 3| + j|t - 4|$$

and the graph of $y(t)$ passes through the points $(t, y)$ given in Table 4.

**d.** Now using the parametric function capabilities of the graphics calculator, graph $(x(t), y(t))$ for $t$ from 0 to 4.

**e.** Now repeat the procedure outlined in parts (a) through (d) to draw the letter "R."

45. It isn't hard to see that solving a system of equations by computing the inverse of the coefficient matrix takes longer than solving the system using Gauss-Jordan elimination, if all computations are done by hand. But suppose that there are six systems of equations to be solved, each with the same coefficient matrix. Assuming that a calculator is not available, explain why it would be more efficient to solve all six systems of equations by using the inverse method than it would be to use Gaussian elimination.

46. Explain why a nonsquare matrix could not have an inverse.

47. If $A$ is noninvertible but we attempt to compute $A^{-1}$ by using Gauss-Jordan elimination as outlined in the text, how will we discover that $A^{-1}$ does not exist?

48. What similarities do you find between the inverse of a matrix and the inverse of a function? A function will fail to have an inverse whenever it is not one-to-one. What would it mean for a matrix to be one-to-one? If $f$ and $g$ have inverses, then it can be shown that $(f \circ g)^{-1} = g^{-1} \circ f^{-1}$. What would the corresponding rule for matrices be?

---

**SECTION 5**

# DETERMINANTS AND CRAMER'S RULE

■ How can you quickly determine whether or not a linear system having the same number of equations as unknowns has a unique solution?

■ How can the value of a single variable be found without solving the entire system?

■ How can matrices be used to determine whether three given points lie on a line?

■ If the coordinates of the vertices of a quadrilateral are given, how can the area of the quadrilateral be determined?

---

## DETERMINANTS OF 2 × 2 MATRICES

As was mentioned in the previous section, not all square matrices have inverses. We will now describe a test for determining *which* matrices have inverses. Consider the matrix

$$A = \begin{bmatrix} 1 & 3 \\ 4 & 12 \end{bmatrix}$$

If we attempt to compute the inverse using an augmented matrix, we will be unsuccessful, as shown here.

$$\begin{bmatrix} 1 & 3 & | & 1 & 0 \\ 4 & 12 & | & 0 & 1 \end{bmatrix} \xrightarrow{-4R_1 + R_2} \begin{bmatrix} 1 & 3 & | & 1 & 0 \\ 0 & 0 & | & -4 & 1 \end{bmatrix}$$

Already we are in trouble; there is no longer any possibility that the first two columns of this augmented matrix will form the identity matrix. This problem occurred because the second row of $A$ is 4 times the first row of $A$, so that the operation $-4R_1 + R_2$

"wiped out" the second row. In fact, this problem will arise whenever one row is a multiple of another. We will show in Exercise 49 that one row of the matrix

$$\begin{bmatrix} a & b \\ c & d \end{bmatrix}$$

is a multiple of the other if and only if $ad - bc = 0$. Thus, the $2 \times 2$ matrix

$$\begin{bmatrix} a & b \\ c & d \end{bmatrix}$$

is *not* invertible whenever $ad - bc = 0$. Moreover, it can be shown that the matrix *is* invertible whenever $ad - bc \neq 0$. The quantity $ad - bc$ is called the **determinant** of the matrix $A$, and it is denoted by $\det(A)$ or $|A|$. We summarize determinants of $2 \times 2$ matrices as follows.

**Determinants of $2 \times 2$ matrices**

1. The **determinant** of the matrix

$$A = \begin{bmatrix} a & b \\ c & d \end{bmatrix}$$

is the quantity $ad - bc$, and it is denoted by

$$\det(A), \quad |A|, \quad \text{or} \quad \begin{vmatrix} a & b \\ c & d \end{vmatrix}$$

2. The determinant of a $2 \times 2$ matrix is the difference of the products of the diagonal entries, as shown below.

$$\begin{vmatrix} a & b \\ c & d \end{vmatrix} = ad - bc$$

3. A $2 \times 2$ matrix is invertible if and only if its determinant is nonzero.

In the following example, we will see how the determinant can be used to determine whether a given matrix is invertible.

**EXAMPLE 1**    *Using determinants to test for invertibility*

Use determinants to decide which of the following matrices are invertible.

a. $\begin{bmatrix} 4 & -2 \\ 3 & 5 \end{bmatrix}$    b. $\begin{bmatrix} 3 & 3 \\ 2 & 2 \end{bmatrix}$    c. $\begin{bmatrix} 1 & -1 \\ -1 & 0 \end{bmatrix}$

**SOLUTION**

a. The determinant is given by

$$\begin{vmatrix} 4 & -2 \\ 3 & 5 \end{vmatrix} = 4 \cdot 5 - (-2) \cdot 3 = 26 \neq 0$$

Thus, the matrix is invertible.

b. The determinant is given by

$$\begin{vmatrix} 3 & 3 \\ 2 & 2 \end{vmatrix} = 3 \cdot 2 - 3 \cdot 2 = 0$$

Thus, the matrix is noninvertible.

c. The determinant is given by

$$\begin{vmatrix} 1 & -1 \\ -1 & 0 \end{vmatrix} = 1 \cdot 0 - (-1)(-1) = -1 \neq 0$$

The matrix is thus invertible.

---

## EVALUATING DETERMINANTS WITH MINORS

We have seen that the determinant of a $2 \times 2$ matrix indicates whether the matrix is invertible. In a similar fashion, determinants can be used to test $3 \times 3$, $4 \times 4$, and larger matrices for invertibility. In fact, a square matrix of any order is invertible if and only if its determinant is nonzero.

We will define the determinant of a $3 \times 3$ matrix as a combination of certain $2 \times 2$ determinants; determinants of $4 \times 4$ matrices can be expressed as a combination of $3 \times 3$ determinants, and so on. We will describe the process of computing $3 \times 3$ determinants in some detail, but we will leave the computation of higher order determinants to the exercises.

The **minor of an entry** of a $3 \times 3$ matrix is defined to be the determinant of the $2 \times 2$ matrix resulting from crossing out both the row and column of the entry. For example, let the matrix $A$ be defined by

$$A = \begin{bmatrix} a_{11} & a_{12} & a_{13} \\ a_{21} & a_{22} & a_{23} \\ a_{31} & a_{32} & a_{33} \end{bmatrix}$$

Then the minor of $a_{11}$ is the determinant of the $2 \times 2$ matrix

$$\begin{bmatrix} a_{22} & a_{23} \\ a_{32} & a_{33} \end{bmatrix}$$

resulting from crossing out both the first row and the first column of $A$. Table 6 shows how the minors of all of the entries of the first row are computed.

**TABLE 6**

*Computing Minors*

| The minor of | Cross out row 1 and column 1 | ... to obtain |
|---|---|---|
| $a_{11}$ | $\begin{vmatrix} \boxed{a_{11}} & a_{12} & a_{13} \\ a_{21} & a_{22} & a_{23} \\ a_{31} & a_{32} & a_{33} \end{vmatrix}$ | $\begin{vmatrix} a_{22} & a_{23} \\ a_{32} & a_{33} \end{vmatrix} = a_{22}a_{33} - a_{23}a_{32}$ |
| The minor of $a_{12}$ | Cross out row 1 and column 2 $\begin{vmatrix} a_{11} & \boxed{a_{12}} & a_{13} \\ a_{21} & a_{22} & a_{23} \\ a_{31} & a_{32} & a_{33} \end{vmatrix}$ | ... to obtain $\begin{vmatrix} a_{21} & a_{23} \\ a_{31} & a_{33} \end{vmatrix} = a_{21}a_{33} - a_{23}a_{31}$ |
| The minor of $a_{13}$ | Cross out row 1 and column 3 $\begin{vmatrix} a_{11} & a_{12} & \boxed{a_{13}} \\ a_{21} & a_{22} & a_{23} \\ a_{31} & a_{32} & a_{33} \end{vmatrix}$ | ... to obtain $\begin{vmatrix} a_{21} & a_{22} \\ a_{31} & a_{32} \end{vmatrix} = a_{21}a_{32} - a_{22}a_{31}$ |

**EXAMPLE 2**     *Computing the minor of an entry*

Given that $(a_{ij})$ is the following matrix, find the minor of $a_{23}$.

$$\begin{bmatrix} -2 & 0 & 5 \\ 3 & 1 & -4 \\ 2 & -6 & -1 \end{bmatrix}$$

**SOLUTION**     The minor of $a_{23}$ is the determinant of the $2 \times 2$ matrix that results from crossing out row 2 and column 3 as follows.

$$\begin{bmatrix} -2 & 0 & \boxed{5} \\ \boxed{3} & \boxed{1} & \boxed{-4} \\ 2 & -6 & \boxed{-1} \end{bmatrix}$$

Thus

$$\text{Minor of } a_{23} = \begin{vmatrix} -2 & 0 \\ 2 & -6 \end{vmatrix} = (-2)(-6) - 2 \cdot 0 = 12$$

The determinant of a $3 \times 3$ matrix can be obtained by multiplying the entries of the first row by their corresponding minors and then either adding or subtracting, as described in the following box. Determinants of higher order matrices are defined in a similar fashion.

**Definition of 3 × 3 determinant**

The determinant of

$$\begin{bmatrix} a_{11} & a_{12} & a_{13} \\ a_{21} & a_{22} & a_{23} \\ a_{31} & a_{32} & a_{33} \end{bmatrix}$$

is given by

$$\begin{vmatrix} a_{11} & a_{12} & a_{13} \\ a_{21} & a_{22} & a_{23} \\ a_{31} & a_{32} & a_{33} \end{vmatrix} = a_{11} \cdot (\text{minor of } a_{11}) - a_{12} \cdot (\text{minor of } a_{12}) + a_{13} \cdot (\text{minor of } a_{13})$$

$$= a_{11} \begin{vmatrix} a_{22} & a_{23} \\ a_{32} & a_{33} \end{vmatrix} - a_{12} \begin{vmatrix} a_{21} & a_{23} \\ a_{31} & a_{33} \end{vmatrix} + a_{13} \begin{vmatrix} a_{21} & a_{22} \\ a_{31} & a_{32} \end{vmatrix}$$

This technique of computing the determinant is called **expansion about the first row**.

**EXAMPLE 3** *Computing a 3 × 3 determinant by expansion about the first row.*

Find the determinant of the matrix

$$\begin{bmatrix} 1 & 4 & 3 \\ 0 & 2 & 1 \\ -2 & 6 & 0 \end{bmatrix}$$

**SOLUTION** We multiply the entries of the first row by their corresponding minors and then add or subtract according to the definition.

$$\begin{vmatrix} 1 & 4 & 3 \\ 0 & 2 & 1 \\ -2 & 6 & 0 \end{vmatrix} = 1 \cdot \begin{vmatrix} 2 & 1 \\ 6 & 0 \end{vmatrix} - 4 \cdot \begin{vmatrix} 0 & 1 \\ -2 & 0 \end{vmatrix} + 3 \cdot \begin{vmatrix} 0 & 2 \\ -2 & 6 \end{vmatrix}$$

$$= 1(0 - 6) - 4[0 - (-2)] + 3[0 - (-4)]$$

$$= -6 - 8 + 12$$

$$= -2$$

Expansion about the first row is not the only technique for evaluating 3 × 3 determinants. In fact, expansion can be done about any row or column. To do so, we simply multiply the entries of the given row or column by their corresponding minors and then either add or subtract, depending on the position of the entry. Specifically, the sign of each term is given by the corresponding entry in the following **matrix of signs**.

$$\begin{bmatrix} + & - & + \\ - & + & - \\ + & - & + \end{bmatrix}$$

For example, to evaluate the determinant of the matrix

$$A = \begin{bmatrix} 1 & 4 & 3 \\ 0 & 2 & 1 \\ -2 & 6 & 0 \end{bmatrix}$$

from Example 3 by expanding about the second column, we would begin by forming the products of each of the entries of the second column with their corresponding minors as shown.

$$4 \cdot \begin{vmatrix} 0 & 1 \\ -2 & 0 \end{vmatrix} \qquad 2 \cdot \begin{vmatrix} 1 & 3 \\ -2 & 0 \end{vmatrix} \qquad 6 \cdot \begin{vmatrix} 1 & 3 \\ 0 & 1 \end{vmatrix}$$

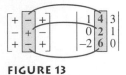

**FIGURE 13**

Next we affix the appropriate sign to each term using the matrix of signs, as suggested by Figure 13. Thus, we have

$$\begin{vmatrix} 1 & 4 & 3 \\ 0 & 2 & 1 \\ -2 & 6 & 0 \end{vmatrix} = -4 \cdot \begin{vmatrix} 0 & 1 \\ -2 & 0 \end{vmatrix} + 2 \cdot \begin{vmatrix} 1 & 3 \\ -2 & 0 \end{vmatrix} - 6 \cdot \begin{vmatrix} 1 & 3 \\ 0 & 1 \end{vmatrix}$$

$$= -4 \cdot 2 + 2 \cdot 6 - 6 \cdot 1$$

$$= -2$$

---

**RULE OF THUMB**   Since we may evaluate the determinant by expanding about *any* row or column, it is convenient to choose a row or column containing as many zeros as possible.

---

**EXAMPLE 4**   *Evaluating a 3 × 3 determinant by expansion*

Compute

$$\begin{vmatrix} 2 & 1 & 3 \\ 1 & 0 & 0 \\ 4 & 5 & 6 \end{vmatrix}$$

**SOLUTION**   Although both the second and third columns contain a zero, the second row contains two zeros. We will thus expand about the second row. Since the second row of the matrix of signs is "$- + -$," we have

$$\begin{vmatrix} 2 & 1 & 3 \\ 1 & 0 & 0 \\ 4 & 5 & 6 \end{vmatrix} = -1 \cdot \begin{vmatrix} 1 & 3 \\ 5 & 6 \end{vmatrix} + 0 \cdot \begin{vmatrix} 2 & 3 \\ 4 & 6 \end{vmatrix} - 0 \cdot \begin{vmatrix} 2 & 1 \\ 4 & 5 \end{vmatrix}$$

$$= -1(-9) + 0 - 0$$

$$= 9$$

Note that by choosing the second row, we had only one 2 × 2 determinant to evaluate.

Graphics calculators can compute determinants. Simply enter in the matrix and then select the determinant option. Check the graphics calculator supplement for the details of computing determinants with your calculator.

## A SPECIAL RULE FOR COMPUTING DETERMINANTS OF 3 × 3 MATRICES

The technique of expanding a determinant about a row or a column works for square matrices of any order. We will now describe a shortcut for computing 3 × 3 determinants. Unfortunately, this rule does not generalize to determinants of any other order.

To compute the determinant

$$\begin{vmatrix} a_1 & b_1 & c_1 \\ a_2 & b_2 & c_2 \\ a_3 & b_3 & c_3 \end{vmatrix}$$

we begin by writing down the matrix, and then forming fourth and fifth columns consisting of copies of columns 1 and 2, respectively, as shown below. Then we follow the arrows going downward to form three products, to each of which we affix a "+" sign, and follow the arrows going upward to form three products, to each of which we affix a "−" sign.

$$\begin{matrix} a_1 & b_1 & c_1 & a_1 & b_1 \\ a_2 & b_2 & c_2 & a_2 & b_2 \\ a_3 & b_3 & c_3 & a_3 & b_3 \end{matrix} \quad = a_1 b_2 c_3 + b_1 c_2 a_3 + c_1 a_2 b_3 - a_3 b_2 c_1 - b_3 c_2 a_1 - c_3 a_2 b_1$$

**EXAMPLE 5**    *Computing a 3 × 3 determinant using a special technique*

Evaluate

$$\begin{vmatrix} 1 & -2 & 3 \\ 2 & 1 & -1 \\ 1 & 0 & 4 \end{vmatrix}$$

**SOLUTION**    We recopy the first two columns to the right of the matrix and apply he shortcut rule for evaluating 3 × 3 determinants.

$$\begin{matrix} 1 & -2 & 3 & 1 & -2 \\ 2 & 1 & -1 & 2 & 1 \\ 1 & 0 & 4 & 1 & 0 \end{matrix}$$

$$\begin{aligned} &= 1 \cdot 1 \cdot 4 + (-2) \cdot (-1) \cdot 1 + 3 \cdot 2 \cdot 0 \\ &\quad - 1 \cdot 1 \cdot 3 - 0 \cdot (-1) \cdot 1 - 4 \cdot 2 \cdot (-2) \\ &= 4 + 2 + 0 - 3 - 0 - (-16) \\ &= 19 \end{aligned}$$

## CRAMER'S RULE FOR SOLVING SYSTEMS OF EQUATIONS

Earlier in this section, we learned how solutions to certain systems of linear equations could be expressed using inverse matrices. Of course, this method works only if the system is "square" (having the same number of equations as variables) and the coefficient matrix is invertible. It so happens that if precisely these same conditions are met, the solution to such a system of equations can be expressed using determinants according to **Cramer's Rule**.

### Cramer's Rule

Given a system of $n$ equations in the $n$ variables $x_1, x_2, \ldots, x_n$, expressible in matrix form as $AX = B$, let $A_i$ be the matrix obtained by replacing the $i$th column of $A$ with the $n \times 1$ matrix $B$. Then the solution to the system is given by

$$x_1 = \frac{|A_1|}{|A|}, \quad x_2 = \frac{|A_2|}{|A|}, \quad \ldots, \quad x_n = \frac{|A_n|}{|A|}$$

provided that $|A| \neq 0$.

**EXAMPLE 6**    *Solving a system of three equations in three variables with Cramer's Rule*

Solve the following system using Cramer's Rule.

$$x - y + 2z = 3$$
$$x + y - 3z = -11$$
$$2x + 3y + z = 9$$

**SOLUTION**    This system can be expressed in matrix form as $AX = B$, with

$$A = \begin{bmatrix} 1 & -1 & 2 \\ 1 & 1 & -3 \\ 2 & 3 & 1 \end{bmatrix}, \quad X = \begin{bmatrix} x \\ y \\ z \end{bmatrix}, \quad \text{and} \quad B = \begin{bmatrix} 3 \\ -11 \\ 9 \end{bmatrix}$$

Now the matrices $A_1$, $A_2$, and $A_3$ are each defined by replacing the appropriate column of $A$ with the matrix $B$. Thus,

$$A_1 = \begin{bmatrix} 3 & -1 & 2 \\ -11 & 1 & -3 \\ 9 & 3 & 1 \end{bmatrix}, \quad A_2 = \begin{bmatrix} 1 & 3 & 2 \\ 1 & -11 & -3 \\ 2 & 9 & 1 \end{bmatrix}, \quad \text{and} \quad A_3 = \begin{bmatrix} 1 & -1 & 3 \\ 1 & 1 & -11 \\ 2 & 3 & 9 \end{bmatrix}$$

According to Cramer's Rule, we have $x = \dfrac{|A_1|}{|A|}$, $y = \dfrac{|A_2|}{|A|}$, and $z = \dfrac{|A_3|}{|A|}$. Evaluation of the determinants, either by hand or with the aid of a graphics calculator gives

$$|A| = 19, \quad |A_1| = -38, \quad |A_2| = 57, \quad \text{and} \quad |A_3| = 76$$

Thus

$$x = \frac{-38}{19} = -2, \quad y = \frac{57}{19} = 3, \quad \text{and} \quad z = \frac{76}{19} = 4$$

# EXERCISES 5

**EXERCISES 1–22** □ *Evalute the given determinant.*

1. $\begin{vmatrix} 2 & 3 \\ 4 & 5 \end{vmatrix}$

2. $\begin{vmatrix} 3 & 1 \\ -2 & 0 \end{vmatrix}$

3. $\begin{vmatrix} 1 & 6 \\ 2 & 12 \end{vmatrix}$

4. $\begin{vmatrix} 2 & 3 \\ -2 & 3 \end{vmatrix}$

5. $\begin{vmatrix} 3 & 7 \\ 4 & 1 \end{vmatrix}$

6. $\begin{vmatrix} -2 & 4 \\ -4 & 8 \end{vmatrix}$

7. $\begin{vmatrix} x & (x-1) \\ x & x \end{vmatrix}$

8. $\begin{vmatrix} a & 2a \\ b & 2b \end{vmatrix}$

9. $\begin{vmatrix} 2 & 1 & 1 \\ 0 & 1 & 4 \\ 0 & 2 & 3 \end{vmatrix}$

10. $\begin{vmatrix} 1 & 0 & 4 \\ 3 & 5 & -1 \\ 2 & 0 & 6 \end{vmatrix}$

11. $\begin{vmatrix} -4 & 2 & 1 \\ -2 & 1 & 0 \\ 3 & -1 & 5 \end{vmatrix}$

12. $\begin{vmatrix} 3 & -1 & 6 \\ 2 & -4 & 1 \\ -1 & 7 & 0 \end{vmatrix}$

13. $\begin{vmatrix} 1.2 & 3.9 & 2.5 \\ 3.7 & 4.1 & 3.6 \\ 1.0 & 2.5 & 7.1 \end{vmatrix}$

14. $\begin{vmatrix} \frac{2}{3} & 4 & \frac{37}{55} \\ 1 & \frac{1}{2} & 0 \\ 0 & 0 & 3 \end{vmatrix}$

15. $\begin{vmatrix} a & b & c \\ 0 & 1 & d \\ 0 & 0 & 1 \end{vmatrix}$

16. $\begin{vmatrix} (x-1) & 2 & 1 \\ 0 & (x-2) & 2 \\ 0 & 2 & (x-3) \end{vmatrix}$

17. $\begin{vmatrix} i & j & k \\ a_1 & a_2 & a_3 \\ b_1 & b_2 & b_3 \end{vmatrix}$

18. $\begin{vmatrix} a & a & a \\ 1 & 1 & 1 \\ x & y & z \end{vmatrix}$

19. $\begin{vmatrix} 1 & 2 & 3 & 4 \\ 2 & 2 & 5 & 7 \\ 8 & 7 & 0 & 2 \\ 8 & 3 & 9 & 0 \end{vmatrix}$

20. $\begin{vmatrix} 2 & 3 & 5 & 7 \\ 11 & 13 & 17 & 19 \\ 23 & 29 & 31 & 37 \\ 41 & 43 & 47 & 53 \end{vmatrix}$

21. $\begin{vmatrix} 1 & 1 & 1 & 1 & 1 & 1 & 1 & 1 & 1 \\ 0 & 1 & 1 & 1 & 1 & 1 & 1 & 1 & 1 \\ 0 & 0 & 1 & 1 & 1 & 1 & 1 & 1 & 1 \\ 0 & 0 & 0 & 1 & 1 & 1 & 1 & 1 & 1 \\ 0 & 0 & 0 & 0 & 1 & 1 & 1 & 1 & 1 \\ 0 & 0 & 0 & 0 & 0 & 1 & 1 & 1 & 1 \\ 0 & 0 & 0 & 0 & 0 & 0 & 1 & 1 & 1 \\ 0 & 0 & 0 & 0 & 0 & 0 & 0 & 1 & 1 \\ 0 & 0 & 0 & 0 & 0 & 0 & 0 & 0 & 1 \end{vmatrix}$

22. $I_{100}$   (the $100 \times 100$ identity matrix)

**EXERCISES 23–30** □ *Use determinants to test the given matrices for invertibility.*

23. $\begin{bmatrix} -3 & 5 \\ 1 & 0 \end{bmatrix}$

24. $\begin{bmatrix} 4 & -1 \\ -8 & 2 \end{bmatrix}$

25. $\begin{bmatrix} 2 & 5 \\ 4 & 10 \end{bmatrix}$

26. $\begin{bmatrix} 11 & 2 \\ 2 & 11 \end{bmatrix}$

27. $\begin{bmatrix} 2 & 0 & 5 \\ -1 & 3 & 0 \\ -1 & 9 & 5 \end{bmatrix}$

28. $\begin{bmatrix} 4 & -3 & 1 \\ -2 & 0 & 6 \\ 1 & 0 & 2 \end{bmatrix}$

29. $\begin{bmatrix} -12 & 1 & 14 \\ 1 & -16 & 2 \\ 15 & 0 & 0 \end{bmatrix}$

30. $\begin{bmatrix} 0 & 10 & -8 \\ 11 & -1 & 3 \\ 22 & 8 & -2 \end{bmatrix}$

**EXERCISES 31–36** □ *Find the value(s) of x for which the given matrix is not invertible.*

31. $\begin{bmatrix} x & -1 \\ 1 & 2 \end{bmatrix}$

32. $\begin{bmatrix} 2 & x \\ 4 & -3 \end{bmatrix}$

33. $\begin{bmatrix} -3 & x \\ x & -3 \end{bmatrix}$

34. $\begin{bmatrix} x & 2 \\ 8 & x \end{bmatrix}$

35. $\begin{bmatrix} x & 0 & 1 \\ 1 & x & 0 \\ x & 1 & 1 \end{bmatrix}$

36. $\begin{bmatrix} 0 & x & 1 \\ x & 1 & 0 \\ 1 & 0 & x \end{bmatrix}$

**EXERCISES 37–46** □ *A system of equations is given. Solve the system using Cramer's Rule.*

37. $x + 2y = 19$
    $3x - 7y = -8$

38. $3x - 8y = -2$
    $5x + 3y = 13$

39. $2x + z = 6$
    $x - y = 4$
    $-y + z = 7$

40. $x - y + 2z = 0$
    $4x + y = 11$
    $y - 3z = 5$

41. $x + 2y - z = -3$
    $3x - 8y - 5z = -3$
    $2x + 2y + z = 10$

42. $6x - 3y + 9z = -7.65$
    $x - 2y + 11z = -4.15$
    $3y - 4z = 9.5$

43. $x + 2y + 3z + 4w = 1$
    $y - z + w = 0$
    $2z + 3w = 0$
    $z - w = 0$

44. $p + q + r + s + t = 1$
    $q + r + s + t = 0$
    $r + s + t = 1$
    $s + t = 0$
    $s = 1$

45. $ax + by = c$
    $dx + ey = f$
    (Solve for $x$ and $y$.
    Assume $ae - bd \neq 0$.)

46. $a_1x + a_2y + a_3z = 0$
    $b_1x + b_2y + b_3z = 0$
    $c_1x + c_2y + c_3z = 0$
    (Assume there is a unique solution.)

47. Show that if $A$ is a $3 \times 3$ matrix such that the matrix equation $AX = 0$ has a nonzero solution $X$, then $\det(A) = 0$.
[*Hint:* Suppose $\det(A) \neq 0$. Then $A$ is invertible. Apply $A^{-1}$ to both sides of the equation $AX = 0$ to find a contradiction.]

48. Let

$$A = \begin{bmatrix} a & b \\ c & d \end{bmatrix}$$

Compare $\det(A)$ to $\det(kA)$ for a nonzero scalar $k$. If $A$ is a $3 \times 3$ matrix, what is your guess as to how $\det(kA)$ compares to $\det(A)$?

49. Complete the following steps to show that one row of the matrix

$$\begin{bmatrix} a & b \\ c & d \end{bmatrix}$$

is a multiple of the other if and only if $ad - bc = 0$.

    a. If $c = ka$ and $d = kb$ (i.e., row 2 is a multiple of row 1), show that $ad - bc = 0$.

    b. If $ad - bc = 0$, show that $a/c = b/d$ and so $a = kc$ and $b = kd$ for some $k$. Thus, row 1 is a multiple of row 2.

50. It is a well-known fact that $\det(AB) = \det(A) \det(B)$ for square matrices $A$ and $B$. Use this fact to show that $\det(A^{-1}) = 1/\det(A)$.

## ■ Applications

**EXERCISES 51–52** □ *Fixed Costs* Assume that production cost is a quadratic function of the number of units produced. Thus, if $C$ denotes the cost and $x$ the number of units produced, then $C = ax^2 + bx + c$ for certain values of $a$, $b$, and $c$. The value of $c$ gives the fixed cost—the cost of production if no units are produced. Use the given data to set up a system of equations involving $a$, $b$, and $c$, and use Cramer's Rule to find the company's fixed costs.

51.

| x (units produced) | 10 | 50 | 100 |
|---|---|---|---|
| C (cost in dollars) | $8,000 | $5,000 | $10,000 |

52.

| x (units produced) | 150 | 200 | 300 |
|---|---|---|---|
| C (cost in dollars) | $20,000 | $25,000 | $40,000 |

53. *Hospital Breakfast* A hospital patient is served a breakfast consisting of milk, oatmeal, and English muffins. The total

mass, caloric value, and fat content were recorded as 427 grams, 311 calories, and 4.4 grams, respectively. It is later discovered that the patient has lactose intolerance, and so it becomes necessary to determine how many grams of milk were consumed. If it is known that each gram of oatmeal contains 0.54 calorie and 0.0097 gram of fat, each gram of milk contains 0.42 calorie and 0.011 gram of fat, and each gram of English muffin contains 2.37 calories and 0.01 gram of fat, use Cramer's Rule to determine the number of grams of milk consumed.

54. *Unknown Investment* A total of $10,000 is invested in three different funds. The first fund pays 5% dividends for the first year and has a 1% load fee. The second pays 10% dividends and has a 1.5% load fee. The third pays 8% dividends and has a 0.5% load fee. The total paid in dividends is $680, and the total in load fees is $110. Due to a bookkeeping error, the amount invested in the first fund is in question. Use Cramer's Rule to determine the amount.

## ■ Projects for Enrichment

55. *Performing Elementary Row Operations by Matrix Multiplication* In this project we will see that for $3 \times 3$ matrices, each of the elementary row operations, (switching rows, adding a multiple of one row to another, and multiplying a row by a constant) can be represented as multiplication by an appropriate matrix.

    a. Define

$$S_{12} = \begin{bmatrix} 0 & 1 & 0 \\ 1 & 0 & 0 \\ 0 & 0 & 1 \end{bmatrix}$$

Show that if $B$ is *any* $3 \times 3$ matrix, then $S_{12}B$ is the matrix obtained by switching the first and second rows of $B$.

    b. Find matrices $S_{13}$ and $S_{23}$ such that if $B$ is an arbitrary $3 \times 3$ matrix, then $S_{13}B$ and $S_{23}B$ are the matrices obtained by switching the first and third rows of $B$, and the second and third rows of $B$, respectively.

    c. Find $M_1(x)$, $M_2(x)$, and $M_3(x)$, such that if $B$ is an arbitrary $3 \times 3$ matrix, then $M_i(x)B$ is the matrix formed by multiplying row $i$ of $B$ by $x$.

**d.** Define $A_{13}(x)$ to be the matrix

$$\begin{bmatrix} 1 & 0 & 0 \\ 0 & 1 & 0 \\ x & 0 & 1 \end{bmatrix}$$

Show that if $B$ is an arbitrary matrix, then $A_{13}(x)B$ is the matrix obtained by adding $x$ times row 1 of $B$ to row 3 of $B$. Find expressions for $A_{12}(x)$ and $A_{23}(x)$.

**e.** Compute the determinants of all of the matrices $S_{ij}$, $M_i(x)$, and $A_{ij}(x)$.

**f.** It is a well-known fact that for square matrices $C$ and $D$, $\det(C \cdot D) = \det(C) \cdot \det(D)$. Use this fact and the results of part (e) to show that performing an elementary row operation on a matrix does not change its invertibility.

**56.** *Determinants and Plane Geometry*   In this project we will explore several geometric applications of determinants. The area of a triangle in the coordinate plane with vertices $(x_1, y_1)$, $(x_2, y_2)$, and $(x_3, y_3)$ can be shown to be the absolute value of the quantity

$$\frac{1}{2} \begin{vmatrix} x_1 & y_1 & 1 \\ x_2 & y_2 & 1 \\ x_3 & y_3 & 1 \end{vmatrix}$$

**a.** Find the area of the triangle having vertices with coordinates $(0, 0)$, $(1, 2)$, and $(3, 5)$.

**b.** Use the determinant formula for the area of a triangle to develop a formula for the area of quadrilateral $PQRS$ shown here.

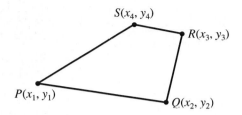

**c.** Use the formula obtained in part (b) to find the area of the quadrilateral with vertices $(2, 3)$, $(4, 7)$, $(1, 8)$, and $(0, 5)$.

**d.** If for a given triple of points $(x_1, y_1)$, $(x_2, y_2)$, $(x_3, y_3)$, the area of the resulting triangle is zero, then what does this say about the three points?

**e.** Suppose that $(x, y)$ is a point on the line determined by $(x_1, y_1)$ and $(x_2, y_2)$. Use the fact that the area of the triangle determined by $(x, y)$, $(x_1, y_1)$, and $(x_2, y_2)$ will be zero to find a determinant form of the equation of the line.

---

### ■ Questions for Discussion or Essay

**57.** Since solving linear systems of equations with inverses or Cramer's Rule is relatively easy and can be performed by most graphics calculators, is there ever a reason to use Gauss-Jordan elimination? Are there any linear systems of equations that can be solved by Gauss-Jordan elimination but not by inverses or Cramer's Rule?

**58.** Suppose that 10 people are to split a jackpot and that the shares correspond to the solution of a certain system of 10 equations in 10 unknowns. If you were given the task of computing each person's share, what method would you use? If you were one of the 10 winners, what method would you use to compute *your* share? Explain your answers.

**59.** With Cramer's Rule, the solution values to a system of linear equations are expressed as quotients of determinants. Of course, Cramer's Rule will not provide us with a solution if a determinant appearing in the denominator is zero. If that should happen, would it make sense to attempt to solve the system using the inverse of the coefficient matrix?

**60.** What is the maximum total number of arithmetic operations (additions, subtractions, and multiplications) required to compute the determinant of a $3 \times 3$ matrix by expansion? What is the minimum? What about a $4 \times 4$ matrix?

**SECTION 6**

## SYSTEMS OF INEQUALITIES

■ How can an athlete select a fast-food snack and still stay within the guidelines of a rigid training diet?

■ How is it possible to ensure that a low-fat, low-calorie diet will adequately satisfy important nutritional requirements?

■ If the graph of $ax + by = c$ is a line, what does the graph of $ax + by \leq c$ look like?

■ What can a system of inequalities tell a farmer about land utilization, or an appliance store about inventory levels, or a retiring couple about investment options?

## LINEAR INEQUALITIES

A **linear inequality** in the variables $x$ and $y$ is an inequality that can be written in one of the forms

$$ax + by \leq c, \quad ax + by < c, \quad ax + by \geq c \quad \text{or} \quad ax + by > c$$

where $a$, $b$, and $c$ are real numbers. For example, $2x - 3y > 6$ is a linear inequality in $x$ and $y$. A point $(x, y)$ is a **solution** of an inequality in $x$ and $y$ if it satisfies the inequality. Thus, the point $(4, -1)$ is a solution of $2x - 3y > 6$ since $2(4) - 3(-1) = 8 + 3 = 11$ and $11 > 6$. The **solution set** of an inequality is the set of all solutions.

**EXAMPLE 1**   *Checking solutions of linear inequalities*

Determine whether the following ordered pairs are solutions of the linear inequality $2x - 3y > 6$.
**a.** $(3, 0)$        **b.** $(-3, -5)$

**SOLUTION**

**a.** $2(3) - 3(0) = 6$ and $6 \not> 6$. Thus, $(3, 0)$ is not a solution.

**b.** $2(-3) - 3(-5) = -6 - (-15) = 9$ and $9 > 6$. Thus, $(-3, -5)$ is a solution.

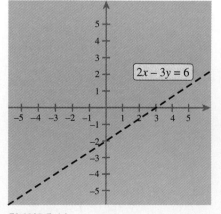

**FIGURE 14**

The **graph** of an inequality in the variables $x$ and $y$ is the set of all points $(x, y)$ in the coordinate plane that are solutions of the inequality. We can obtain the graph of any of the linear inequalities given above by first graphing the corresponding linear equation $ax + by = c$. This line divides the plane into two **half-planes**, exactly one of which contains solutions of the inequality.

Consider, for example, the inequality $2x - 3y > 6$. The graph of the corresponding linear equation $2x - 3y = 6$ is shown in Figure 14. Notice that the graph has been drawn with a dashed line. This is because the **strict inequality** ">" in the expression $2x - 3y > 6$ indicates that none of the points on the line $2x - 3y = 6$ are actually part of the solution set. If the inequality had been "≥" or "≤," then the line itself would have been part of the solution set, and so a solid line would have been used. In any

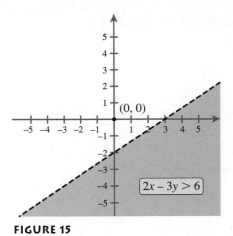

**FIGURE 15**

case, the line $2x - 3y = 6$ divides the plane into two half-planes (shaded green and blue in Figure 14), exactly one of which contains solutions to the inequality. To determine which half-plane is the correct one, we simply test a point on one side of the line to see whether the inequality is satisfied. We select $(0, 0)$ as our test point.

$$3(0) - 2(0) \overset{?}{>} 6$$

$$0 \not> 6$$

Thus, $(0, 0)$ does not satisfy the inequality, and so we know that the solutions of the inequality are on the other side of the line. This is the region we have shaded in Figure 15.

The general procedure for graphing linear inequalities is summarized as follows.

■ *Graphing a linear inequality*

> **1.** Replace the inequality with an "=" and sketch the graph of the resulting line. Use a dashed line for the strict inequalities "<" or ">" and a solid line for "≤" or "≥."
>
> **2.** Test a point on one side of the line. If the point satisfies the original inequality, then shade the half-plane containing that point. If the point does not satisfy the inequality, shade the half-plane on the other side of the line.

**EXAMPLE 2**   *Graphing a linear inequality*

Graph the linear inequality $4x + y \le 0$.

**SOLUTION**   We first replace the inequality with "=" and sketch the graph of the line $4x + y = 0$. Notice that the equation can be rewritten as $y = -4x$, which is a line passing through the origin with slope $-4$, as shown in Figure 16. Next, we test a point on one side of the line. We choose $(0, -1)$ and substitute into the original inequality.

$$4(0) + (-1) \overset{?}{\le} 0$$

$$-1 \le 0$$

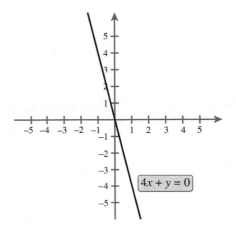

**FIGURE 16**

Since $(0, -1)$ satisfies the inequality, we shade the region containing $(0, -1)$ as shown in Figure 17.

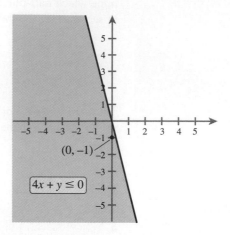

**FIGURE 17**

---

**EXAMPLE 3**     *Graphing a linear inequality*

Graph the linear inequality $y < -3$.

**SOLUTION**     The graph of $y = -3$ is a horizontal line with $y$-intercept $-3$. The points that satisfy the inequality $y < -3$ are those whose $y$-coordinate is less than $-3$—that is, those that lie below the line, as shown in Figure 18.

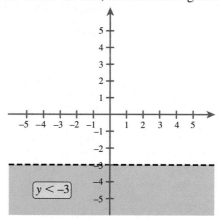

**FIGURE 18**

---

## SYSTEMS OF LINEAR INEQUALITIES

Applications involving two or more linear inequalities arise frequently in business, the natural and social sciences, and many other areas. The following is an example of a system of linear inequalities.

$$x + 2y \geq 2$$
$$-3x + 4y \leq 12$$

A **solution** of a system of inequalities in $x$ and $y$ is a point $(x, y)$ that satisfies all of the inequalities in the system. For example, the point $(3, 2)$ is a solution of the system given above since $3 + 2(2) = 7 \geq 2$ and $-3(3) + 4(2) = -1 \leq 12$. The **graph** of a system of inequalities in $x$ and $y$ is the set of points $(x, y)$ that satisfy all of the inequalities in the system. We can obtain the graph of a system of inequalities by first sketching the graph of each individual inequality on the same coordinate system. Then we determine the region that is common to each inequality. In other words, we find the intersection of all the graphs of the inequalities in the system.

| | |
|---|---|
| **EXAMPLE 4** | *Graphing a system of linear inequalities* |

Graph the system of inequalities

$$x + 2y \geq 2$$

$$-3x + 4y \leq 12$$

**SOLUTION**    We first graph the inequality $x + 2y \geq 2$. This is done by first plotting $x + 2y = 2$ as a solid line. Next we test the point $(0, 0)$ to see whether it satisfies the inequality. Since $0 + 2(0) = 0 \not\geq 2$, we shade the side of the line that does not contain $(0, 0)$. The result is shown in Figure 19. Next, on the same coordinate system but with a different color, we graph the inequality $-3x + 4y \leq 12$. To do this, we first plot the equation $-3x + 4y = 12$ as a solid line and test $(0, 0)$ to see whether it satisfies the inequality. Since $-3(0) + 4(0) = 0 \leq 12$, we shade the side of the line $-3x + 4y = 12$ that contains $(0, 0)$. The two shaded regions are shown in Figure 20. The solution set of the system of inequalities is the region that is common to both inequalities, as shown in Figure 21.

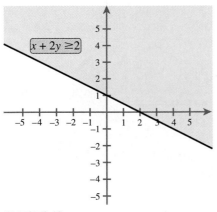

**FIGURE 19**

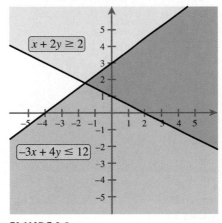

**FIGURE 20**

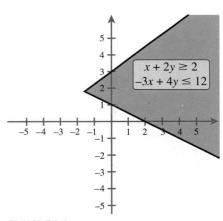

**FIGURE 21**

| | |
|---|---|
| **EXAMPLE 5** | *Graphing a system of linear inequalities* |

Graph the system of inequalities

$$x + y \leq 4$$

$$-2x + y \leq 1$$

$$y \geq -1$$

$$x \leq 2$$

**SOLUTION**    In Figures 22–25 we graph each of the four inequalities on the same coordinate system. The solution set of the system of inequalities is the region that is common to all four inequalities, as shown in Figure 26.

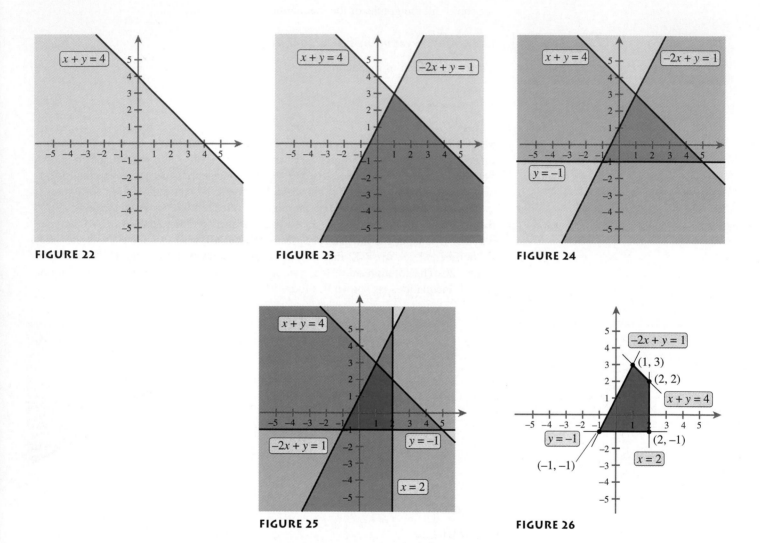

**FIGURE 22**

**FIGURE 23**

**FIGURE 24**

**FIGURE 25**

**FIGURE 26**

The four "corner points" of the region in Example 5—(1, 3), (2, 2), (2, −1), and (−1, −1)—are called **vertices**. Such points will turn out to be of considerable importance in Section 7. In general, the vertices of a system of linear inequalities are found by locating the intersection points of appropriate pairs of lines. Since the exact points of intersection may not be obvious from the graph, it is usually necessary to solve systems of linear equations. Thus, for example, the vertex (1, 3) in Example 5 could be found by solving the equations $-2x + y = 1$ and $x + y = 4$ simultaneously, as shown.

$$\begin{array}{rl} 2x - y = -1 & \quad \text{Multiplying the first equation by } -1 \text{ and} \\ +\quad x + y = 4 & \quad \text{adding to the second} \\ \hline 3x = 3 & \\ x = 1 & \end{array}$$

$$\begin{array}{rl} 1 + y = 4 & \quad \text{Substituting } x = 1 \text{ into the second equation} \\ y = 3 & \end{array}$$

**EXAMPLE 6**   *An application of systems of linear inequalities*

Andrea is training for a bike race and so wants to keep her weight down while maintaining a high intake of protein and carbohydrates. On the way home from a particularly hard workout, she decides to treat herself to a chocolate milk shake and an order of fries at her favorite fast-food restaurant. Each gram of chocolate milk shake has approximately 1.5 calories, 0.04 gram of protein, and 0.2 gram of carbohydrates. Each gram of fries has approximately 3 calories, 0.04 gram of protein, and 0.4 gram of carbohydrates. For this snack, she must not exceed 405 calories but would like to have at least 6 grams of protein and 40 grams of carbohydrates. Set up and graph a system of linear inequalities that describes how many grams of milk shake and fries she can have. Locate the vertices of the solution set.

**SOLUTION**   We assign the following variables to represent the number of grams of milk shake and fries.

$$x = \text{the number of grams of milk shake}$$

$$y = \text{the number of grams of fries}$$

To meet the requirements described above, the following inequalities must be satisfied.

$$\text{calories from milk shake} + \text{calories from fries} \leq 405$$
$$1.5x + 3y \leq 405$$

$$\text{protein from milk shake} + \text{protein from fries} \geq 6$$
$$0.04x + 0.04y \geq 6$$

$$\text{carbohydrates from milk shake} + \text{carbohydrates from fries} \geq 40$$
$$0.2x + 0.4y \geq 40$$

Since $x$ and $y$ cannot be negative, we must also have $x \geq 0$ and $y \geq 0$. Thus, the system of inequalities is

$$1.5x + 3y \leq 405$$

$$0.04x + 0.04y \geq 6$$

$$0.2x + 0.4y \geq 40$$

$$x \geq 0$$

$$y \geq 0$$

The last two inequalities tell us that we only need to be concerned with the first quadrant when we graph the first three inequalities in Figures 27–29. The resulting region is shaded in Figure 30.

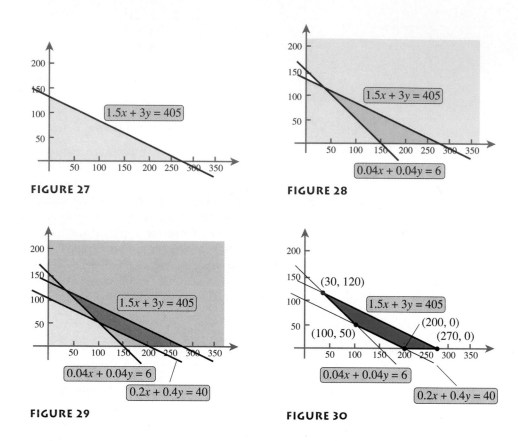

**FIGURE 27**

**FIGURE 28**

**FIGURE 29**

**FIGURE 30**

Any point $(x, y)$ in the shaded region represents a combination of milk shake and fries that Andrea can have. The vertices (200, 0) and (270, 0) are the $x$-intercepts of the lines $0.2x + 0.4y = 40$ and $1.5x + 3y = 405$. The other two vertices are found by solving the following systems of equations.

$$(100, 50): \quad 0.04x + 0.04y = 6 \qquad (30, 120): \quad 0.04x + 0.04y = 6$$
$$0.2x + 0.4y = 40 \qquad\qquad\qquad\qquad 1.5x + 3y = 405$$

From the vertex (100, 50), we conclude that one possible combination would be 100 grams of milk shake and 50 grams of fries. From the vertex (30, 120), we conclude that Andrea can have 120 grams of fries if she only drinks 30 grams of milk shake. The values of the other two vertices, (200, 0) and (270, 0), suggest that if she skips the fries altogether, she can drink anywhere from 200 to 270 grams of milk shake.

## NONLINEAR INEQUALITIES

The graph of a nonlinear inequality can be obtained in much the same way as the graph of a linear inequality. We first replace the inequality with an ''=,'' and then sketch the graph of the resulting relation. For example, to graph $4 - x^2 \geq y$, we first graph

$4 - x^2 = y$, as shown in Figure 31. As before, we use a dashed curve for the strict inequalities "$<$" or "$>$," and a solid curve for "$\leq$" or "$\geq$."

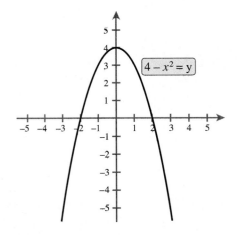

**FIGURE 31**

The graph of the relation will normally divide the plane into two or more regions. In each region, either all of the points are solutions of the inequality or none of the points are solutions. Thus, we test a point in *each* of the regions to see which are solutions.

| Test point | $4 - x^2 \overset{?}{\geq} y$ | Conclusion |
|------------|-------------------------------|------------|
| $(0, 5)$ | $4 - (0)^2 \overset{?}{\geq} 5$ <br> $4 \not\geq 5$ | The region above the parabola is not in the solution set. |
| $(0, 0)$ | $4 - (0)^2 \overset{?}{\geq} 0$ <br> $4 \geq 0$ | The region below the parabola is in the solution set. |

We conclude that the solution set consists of the region below and on the parabola $y = 4 - x^2$, as shown in Figure 32.

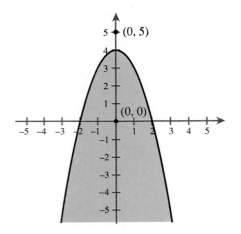

**FIGURE 32**

EXAMPLE 7    *Graphing a nonlinear inequality*

Graph the inequality $\dfrac{x^2}{9} - \dfrac{y^2}{4} > 1$.

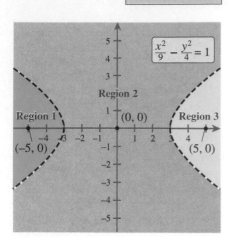

FIGURE 33

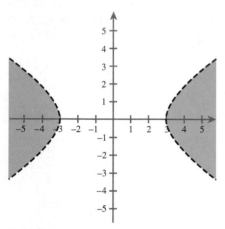

FIGURE 34

**SOLUTION**    We first plot the hyperbola $x^2/9 - y^2/4 = 1$ with dashed curves, as shown in Figure 33. The hyperbola divides the plane into three regions, and so we pick a point in each to see which satisfy the inequality.

| Test point | $\dfrac{x^2}{9} - \dfrac{y^2}{4} \overset{?}{>} 1$ | Conclusion |
|---|---|---|
| $(-5, 0)$ | $\dfrac{(-5)^2}{9} - \dfrac{(0)^2}{4} \overset{?}{>} 1$  $\dfrac{25}{9} > 1$ | The region inside the left branch is in the solution set. |
| $(0, 0)$ | $0 \not> 1$ | The region between the branches is not in the solution set. |
| $(5, 0)$ | $\dfrac{(5)^2}{9} - \dfrac{(0)^2}{4} \overset{?}{>} 1$  $\dfrac{25}{9} > 1$ | The region inside the right branch is in the solution set. |

We conclude that the solution set is inside the two branches of the hyperbola, as shaded in Figure 34.

A graphics calculator can be used to sketch the graph of an inequality that is in any of the following four forms:

$$y < f(x), \quad y \le f(x), \quad y > f(x), \quad \text{or} \quad y \ge f(x)$$

For example, the graph of $y \ge x^2 - 4x + 3$ is shown in Figure 35. Refer to the graphics calculator supplement for details on how this is done with your calculator.

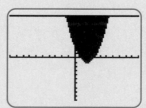

FIGURE 35

| EXAMPLE 8 | *Graphing a system of nonlinear inequalities* |

Graph the system of inequalities

$$x^2 + y^2 \leq 4$$
$$y \geq |x|$$

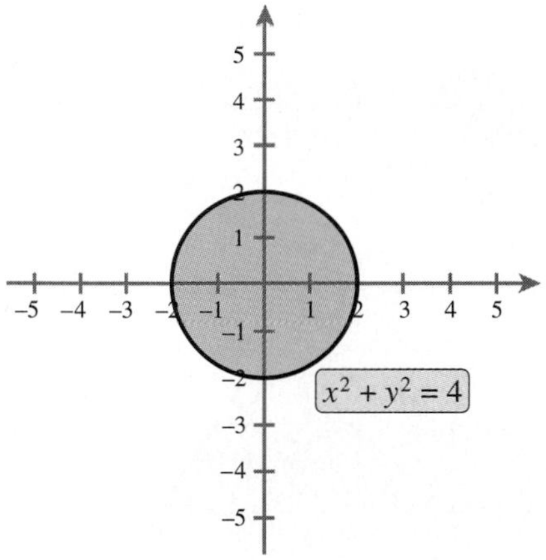

**SOLUTION** For the inequality $x^2 + y^2 \leq 4$, we plot the circle $x^2 + y^2 = 4$ and test the points $(0, 0)$ and $(4, 0)$.

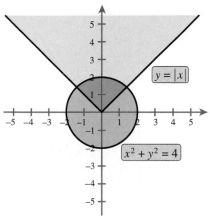

**FIGURE 36**

| Test point | $x^2 + y^2 \overset{?}{\leq} 4$ | Conclusion |
|---|---|---|
| $(0, 0)$ | $(0)^2 + (0)^2 \overset{?}{\leq} 4$ <br> $0 \leq 4$ | The region inside the circle is in the solution set. |
| $(4, 0)$ | $(4)^2 + (0)^2 \overset{?}{\leq} 4$ <br> $16 \not\leq 4$ | The region outside the circle is not in the solution set. |

The resulting region is shaded in Figure 36. Similarly, for the inequality $y \geq |x|$, we plot $y = |x|$ (on the same coordinate system) and test the points $(0, -1)$ and $(0, 1)$.

**FIGURE 37**

| Test point | $y \overset{?}{\geq} |x|$ | Conclusion |
|---|---|---|
| $(0, -1)$ | $-1 \overset{?}{\geq} |0|$ <br> $-1 \not\geq 0$ | The region below $y = |x|$ is not in the solution set. |
| $(0, 1)$ | $1 \overset{?}{\geq} |0|$ <br> $1 \geq 0$ | The region above $y = |x|$ is in the solution set. |

The intersection of the two regions is shaded in dark blue in Figure 37. ∎

## EXERCISES 6

**EXERCISES 1–6** □ *Determine which of the ordered pairs are solutions of the inequality.*

1. $x - 2y \geq 5$; $(0, 0)$, $(4, 1)$, $(-1, -3)$
2. $y < 3x + 2$; $(0, 2)$, $(-1, 1)$, $(5, 5)$
3. $4x + 3y > 12$; $(1, 4)$, $(3, 0)$, $(-2, -1)$
4. $6x \geq 3y - 2$; $(-1, -5)$, $(1, 2)$, $(0, -3)$
5. $y + x^2 \leq 4$; $(0, 2)$, $(-1, 3)$, $(5, -1)$
6. $x^2 + y^2 > 25$; $(1, -2)$, $(-4, 3)$, $(5, 5)$

**EXERCISES 7–12** □ *Match the inequality with its graph.*

**7.** $3x - 2y \geq 6$

**8.** $3x + 2y \leq 6$

**9.** $x < 3$

**10.** $y > 3$

**11.** $y - x^2 \geq 0$

**12.** $x + y^2 \leq 0$

a.

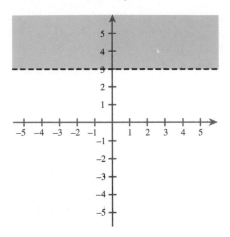

b.

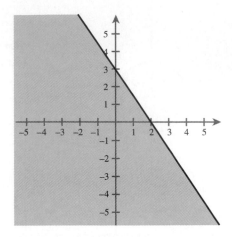

c.

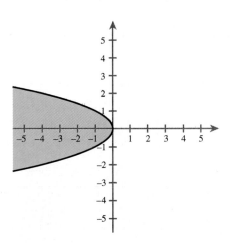

d.

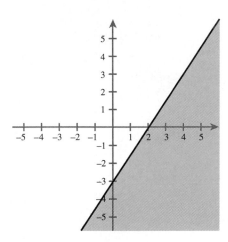

e.

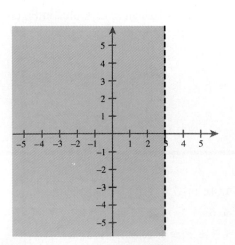

f.

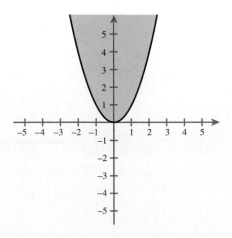

**EXERCISES 13–28** ☐ *Graph the inequality.*

**13.** $x + y < 8$

**14.** $x - y \geq 2$

**15.** $6x - 2y \leq 12$

**16.** $5x + 4y > -20$

**17.** $x \geq -5$

**18.** $y < 4$

**19.** $x > -3y$

**20.** $2x - y \leq 0$

**21.** $y < x^2 + 1$

**22.** $x^2 - y \geq 4$

**23.** $y^2 \leq 1 - x^2$

**24.** $x^2 + y^2 > 9$

**25.** $y \geq \dfrac{1}{x}$

**26.** $x - y^2 < 0$

**27.** $y > e^x$

**28.** $y - \sin x \leq 0$

**EXERCISES 29–32** ☐ *Match the graph display with the inequality that produced it.*

**a.** $y \leq -x^2 - 6x - 7$

**b.** $y \geq x^2 - 6x + 7$

**c.** $y \leq x^3 - 9x$

**d.** $y \geq -x^3 + 9x$

**29.**

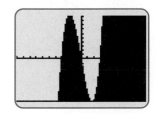

**30.**

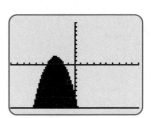

**31.**

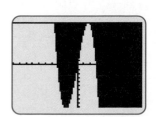

**32.**

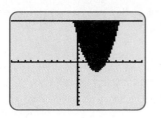

**EXERCISES 33–54** ☐ *Graph the system of inequalities.*

**33.** $x + y \leq 5$
$x - y \leq 1$

**34.** $2x + 3y > 6$
$x - 3y < 3$

**35.** $\quad y < 2x$
$x + 2y < 5$

**36.** $x - 2y \geq -6$
$x - 2y \leq 3$

**37.** $\quad 4x + y \leq 6$
$-4x - y \leq 4$

**38.** $5x - 2y < 5$
$2y > 5x - 20$

**39.** $\quad x + y \leq 2$
$-x + y \geq 2$
$x \geq -2$

**40.** $x - 2y \leq 4$
$y \geq -2x$
$y \leq 4$

**41.** $-3x + 2y < 6$
$-x + 3y > 2$
$2x + y < 3$

**42.** $\quad x - 2y > -4$
$-3x + 4y > -18$
$x - 3y > 6$

**43.** $2x + 5y \leq 10$
$x + y \leq 3$
$x \geq 0$
$y \geq 0$

**44.** $2x + 3y \leq 15$
$3x + y \leq 12$
$x \geq 0$
$y \geq 0$

**45.** $2x + y \geq 8$
$2x - y \leq 8$
$x \geq 3$
$y \leq 4$

**46.** $y \geq 3x - 5$
$y \leq 2x$
$x \leq 3$
$y \geq 1$

**47.** $-2x + y \geq 1$
$y \leq 4 - x^2$

**48.** $x - y^2 > 0$
$x + y < 2$

**49.** $xy < 1$
$y < x$

**50.** $x^2 + y^2 \leq 9$
$x - y \leq 0$

**51.** $x^2 + y^2 \leq 16$
$x^2 + y^2 \geq 4$

**52.** $y \leq \sqrt{x}$
$y \geq \dfrac{1}{4}(x^2 - 2x)$

**53.** $y > x^3 - x$
$y < 6x$
$x \geq 0$

**54.** $y \leq e^x$
$y \geq 1$
$x \geq 0$

**EXERCISES 55–64** ☐ *Write out an inequality or a system of inequalities with the indicated region as the solution set.*

**55.**

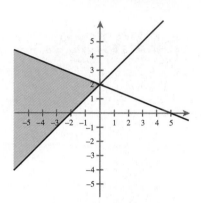

**56.**

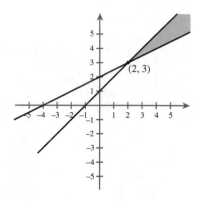

**57.**

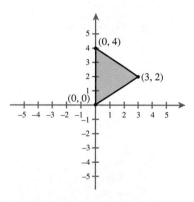

**58.**

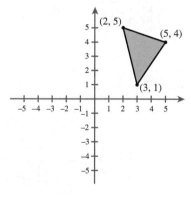

**59.**

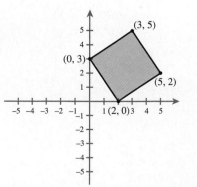

**60.**

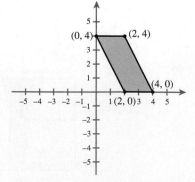

**61.**

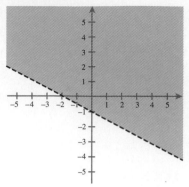

**62.**

**63.**

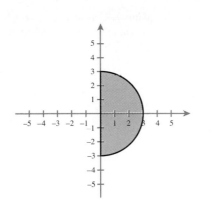

**64.**

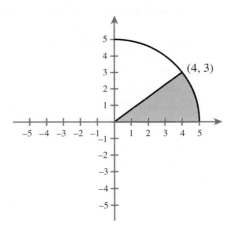

## ■ Applications

**65. *Retirement Planning*** Juan and Rosa are nearing retirement and want to invest up to $24,000 of their savings in two different mutual funds. Their financial adviser suggests they put at least $6000 in each fund, but because one of the funds has more risk, the amount they invest in it should be no more than half the amount invested in the other fund. Set up and graph a system of inequalities that describes the various amounts they can invest in each fund. Locate the vertices of the solution set.

**66. *Low-fat Diet*** Aleshia is on a low-fat, low-calorie diet, but she prefers 2% milk over skim milk and won't give up her mozzarella cheese stick snacks. Each 8-ounce serving of 2% milk has approximately 120 calories, 3 grams of saturated fats, and 300 grams of calcium. Each 1-ounce serving of mozzarella cheese has approximately 80 calories, 3 grams of saturated fats, and 200 grams of calcium. To stay within the guidelines of her diet, she must not exceed 480 calories and 15 grams of saturated fats from milk and cheese. Her daily goal for calcium from milk and cheese is at least 600 grams. Set up and graph a system of inequalities that describes how many servings of milk and cheese she can have. Locate the vertices of the solution set.

**67. *Land Allocation*** Roy has 1000 acres of land available to raise corn and soybeans, although he may leave some unplanted. Each acre of corn costs $100 and requires 2 hours of labor. Each acre of soybeans costs $80 and requires 1 hour of labor. Roy does not wish his costs for the two crops to exceed $88,000 and he has at most 1600 hours of labor available for the two crops. Also, to feed his own cattle, he must plant at least 200 acres of corn. Set up and graph a system of inequalities that describes how many acres of corn and oats Roy should plant. Locate the vertices of the solution set.

**68. *Inventory Control*** A TV and appliance store carries a large inventory of TVs and refrigerators. Each TV requires 10 cubic feet of storage space and costs the store $500. Each refrigerator requires 50 cubic feet of storage space and costs the store $1000. The store has at most 1500 cubic feet of warehouse space and $48,000 of inventory capital available for TVs and refrigerators. Finally, because of demand, it is necessary to stock at least eight refrigerators and at least twice as many TVs as refrigerators. Set up and graph a system of inequalities that describes how many TVs and refrigerators should be kept in stock. Locate the vertices of the solution set.

## ■ Projects for Enrichment

**69. *Systems of Inequalities and Flight Paths*** Inspector Magill has been called in to settle a dispute between an airline company and a group of environmentalists. The airline company had been given a court order that prohibits its planes from flying over a triangular region that is the sole breeding ground for an endangered bird species. The environmentalists claim that certain flights are still flying over this region, and so Magill must determine whether this is indeed the case. His first task is to describe the region algebraically. By setting up a coordinate system with the origin at one of the vertices of the region, Magill has determined that the other two vertices are at (20, 80) and (120, 30), as shown in Figure 38.

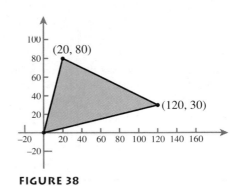

**FIGURE 38**

a. Find a system of three linear inequalities whose graph is the same as that shown in Figure 38.

There are three daily flights that are being contested by the environmentalists. The first is a direct flight from city A located at $(-22, 35)$ to city B located at $(110, -58)$.

b. Find the equation of the line connecting cities A and B, and sketch its graph on the same coordinate system as the triangular region. Does the flight violate the court order?

The second flight is a direct flight between city A and city C located at $(140, 197)$. A plot of the line connecting these two points is shown in Figure 39. Since it is difficult to determine graphically if the line passes through the region, Magill decides to check it algebraically.

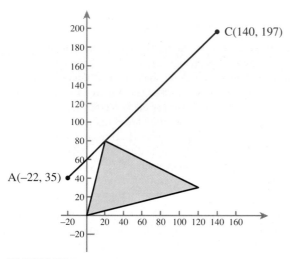

**FIGURE 39**

c. Find the equation of the line connecting cities A and C. Write the equation in the form $y = mx + b$.

d. Substitute $mx + b$ [use your values from part (c)] in place of $y$ in each of the three inequalities found in part (a). After simplifying, the following three inequalities should be obtained.

$$x \geq -76$$

$$x \geq 19$$

$$x \leq 22$$

e. In order for the flight to pass through the region, there must be a value for $x$ that satisfies each of the three inequalities given above. Does the flight pass through the region?

f. The third flight is between city B and city C. Use the technique described in parts (c) through (e) to determine whether the third flight violates the court order.

70. *Systems of Inequalities and Area* In this project, we will consider the areas of closed regions determined by systems of

inequalities. Consider, for example, the linear system given below whose solution is the triangular region shown in Figure 40.

$$x + y \leq 6$$

$$3x + y \geq 8$$

$$y \geq 2$$

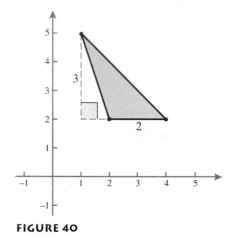

**FIGURE 40**

The area of the triangle in Figure 40 can be computed using the formula

$$A = \frac{1}{2}bh = \frac{1}{2}(2)(3) = 3$$

a. Find the area of the region bounded by the given system of inequalities.

   i.  $x + y \leq 7$          ii.  $x + 2y \leq 11$
       $x + 2y \geq 8$              $2x + 3y \leq 1$
           $x \geq 2$                 $4x + y \geq 9$
iii.  $x - 2y \geq -6$
      $x - 2y \leq 0$
     $2x - y \leq 3$
     $2x - y \geq 0$

The region in part (iv) is unusual in that one of the boundary curves is not linear and yet it is still possible to find the area using standard geometric formulas. In most cases, if one or more of the inequalities is nonlinear, it is not possible to find the area using standard formulas. Consider the following nonlinear system, which is graphed in Figure 41.

$$y \leq x^2 + 1$$

$$x \leq 2$$

$$x \geq 0$$

$$y \geq 0$$

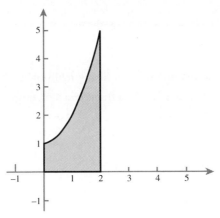

**FIGURE 41**

Since the top boundary curve is nonlinear, there is no standard formula for computing the area of the region. In fact, without the aid of calculus, we cannot compute the area exactly. Thus, we must settle for an approximation. In Figure 42 we have "inscribed" four rectangles of equal width in the region. The height of each rectangle is given by the $y$-coordinate of a point on the curve $y = x^2 + 1$. Thus, the first rectangle has height 1, the second has height $\frac{5}{4}$, and so on. Since each rectangle has width $\frac{1}{2}$, the area of each rectangle can be computed as shown below.

$$A_1 = \frac{1}{2} \cdot 1 = \frac{1}{2} \qquad A_2 = \frac{1}{2} \cdot \frac{5}{4} = \frac{5}{8}$$

$$A_3 = \frac{1}{2} \cdot 2 = 1 \qquad A_4 = \frac{1}{2} \cdot \frac{13}{4} = \frac{13}{8}$$

The sum of the areas is

$$A = \frac{1}{2} + \frac{5}{8} + 1 + \frac{13}{8} = \frac{30}{8} = \frac{15}{4}$$

and this is an approximation for the area of the shaded region in Figure 41. More accurate approximations could be found by inscribing more rectangles with smaller widths.

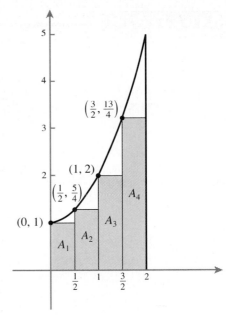

**FIGURE 42**

b. Approximate the area of the region bounded by the given system of inequalities. Use the specified number of inscribed rectangles.

　i. $y \le x^2 + 1$
　　$x \le 2$
　　$x \ge 0$
　　$y \ge 0$
　　8 rectangles

　ii. $y \le \dfrac{1}{x}$
　　$x \ge 1$
　　$x \le 4$
　　$y \ge 0$
　　6 rectangles

　iii. $y \le x^2 - 2x$
　　$y \ge 0$
　　8 rectangles (2 will have zero height)

### ■■  *Questions for Discussions or Essay*

71. When graphing systems of linear inequalities, it is helpful to be aware of the number of regions that are possible when one or more lines are drawn on the coordinate plane. For example, if one line is drawn on the coordinate plane, it divides the plane into two regions. If two lines are drawn, they divide the plane into either three or four regions, depending on whether or not the lines intersect. Discuss the connection between the number of regions formed by two lines and the solution sets of systems of two linear inequalities. What observations and connections can be made concerning the number of regions formed by three lines and the solutions of systems of three linear inequalities?

72. Describe the circumstances under which the solution set of a system of two linear inequalities is either the empty set or the entire coordinate plane. Give examples of such systems.

73. Explain in detail why a linear inequality involving "<" or ">" is graphed with a dashed line while one involving "≤" or "≥" is graphed with a solid line.

## SECTION 7

# LINEAR PROGRAMMING

■ How can a business maximize its revenue without hiring additional labor?

■ What combination of stocks and bonds will yield the highest income for a fixed investment?

■ What postwar mathematical discovery helped the U.S. Air Force allocate supplies more efficiently?

■ How can a political party attract the largest number of TV viewers?

## A CASE STUDY

A common problem in business, industry, agriculture, and many other areas, is that of optimizing the use of limited resources. Some examples are a factory that wishes to maximize its revenues within the constraints of limited manpower and/or raw materials, a hospital food service that wishes to minimize its costs within the constraints of necessary nutritional requirements, or a university that wishes to maximize the number of students it can house in a new dormitory within the constraints of limited available land and funding. Linear programming is a mathematical method that can be used to solve such problems. We will illustrate its basic ideas with an example.

Willow Woods Furniture Factory makes tables and bookcases. Each table requires 6 hours of woodworking and 3 hours of finishing and sells for $160. Each bookcase requires 4 hours of woodworking and 4 hours of finishing and sells for $200. Without hiring additional labor, there are 48 hours available each day for woodworking and 30 hours each day for finishing. Due to orders coming in, the factory must make at least 2 tables each day. The owner of the factory would like to determine how many tables and how many bookcases should be made each day to maximize revenue.

Step 1: *Read and understand the problem.*

Among the various combinations of tables and bookcases, some will naturally bring in more revenue than others. For example, if the factory sells 5 tables and 3 bookcases, the revenue is $160(5) + 200(3) = \$1400$, while if it sells 4 tables and 6 bookcases, the revenue is $160(4) + 200(6) = \$1840$. However, some combinations of tables and bookcases may exceed the constraints imposed by the available labor. The 5–3 combination would require $6(5) + 4(3) = 42$ hours of woodworking and $3(5) + 4(3) = 27$ hours of finishing, while the 4–6 combination would require $6(4) + 4(6) = 48$ hours of woodworking and $3(4) + 4(6) = 36$ hours of finishing. Thus, although the 4–6 combination brings in more revenue, it is not allowable since it requires too many hours of finishing. We are looking for the combination that will provide the highest revenue and still be within the constraints of the available labor.

Step 2: *Assign variables to the unknowns and write out the expression that is to be optimized.*

The unknowns are the number of tables and bookcases to be made. Thus, we let

$x$ = the number of tables made each day

$y$ = the number of bookcases made each day

We wish to maximize the revenue from the sale of $x$ tables and $y$ bookcases. Since each table sells for \$160 and each bookcase for \$200, the revenue $R$ can be computed as

$$R = 160x + 200y$$

This equation defines the **objective function**, and it is what we wish to maximize.

Step 3: *Write out the inequalities that correspond to the constraints.*

Woodworking constraint: Each table requires 6 hours and each bookcase requires 4 hours. The total hours may not exceed 48.

Hours of woodworking for tables + hours of woodworking for bookcases $\le$ 48

$$6x \qquad\qquad + \qquad\qquad 4y \qquad\qquad\quad \le 48$$

Finishing constraint: Each table requires 3 hours and each bookcase requires 4 hours. The total hours may not exceed 30.

Hours of finishing for tables + hours of finishing for bookcases $\le$ 30

$$3x \qquad\qquad + \qquad\qquad 4y \qquad\qquad\quad \le 30$$

Demand constraint: At least 2 tables must be made each day.

$$x \ge 2$$

Implied constraints: The number of tables and bookcases may not be negative.

$$x \ge 0$$

$$y \ge 0$$

Step 4: *Graph the system of inequalities.*

The system of inequalities obtained in step 3 is

$$6x + 4y \le 48$$

$$3x + 4y \le 30$$

$$x \ge 2$$

$$y \ge 0$$

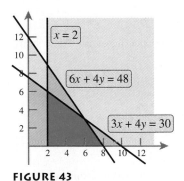

**FIGURE 43**

Notice that we have not included the constraint $x \ge 0$ since it is accounted for in the constraint $x \ge 2$. Moreover, the constraint $y \ge 0$ indicates that our region must be above the $x$-axis. Thus, we can restrict the shading in the first three inequalities to the region above the $x$-axis. The region that is common to all four inequalities is shaded dark blue in Figure 43.

The darkly shaded region in Figure 43 is called the **feasible set**, and it constitutes the set of points that satisfy all of the constraints. In the context of our problem, the feasible set consists of all the table-bookcase combinations that are within the labor constraints. For example, the point (5, 3) is in the feasible set. It corresponds to 5 tables and 3 bookcases. As we observed in step 1, this combination is within the constraints imposed by the available labor. But there are also many other combinations that are within the constraints. What we need is a way to determine which of the table-bookcase combinations in the feasible set is the one that will give the maximum revenue. For this task we note that, if there is an optimum solution, *it must occur at one of the vertices of the feasible set*. This observation will be discussed more thoroughly later in the section. For now, we will simply apply it by locating and testing each of the vertices to see which yields the maximum revenue.

Step 5: *Locate the vertices of the feasible set.*

The four vertices are shown in Figure 44. The two on the $x$-axis are simply the $x$-intercepts of the lines $x = 2$ and $6x + 4y = 48$ or (2, 0) and (8, 0), respectively. A third vertex is the intersection of the lines $x = 2$ and $3x + 4y = 30$. Setting $x = 2$ in the equation $3x + 4y = 30$ and solving for $y$ gives $y = 6$. Thus, the vertex is (2, 6). The fourth vertex is the intersection of the lines $6x + 4y = 48$ and $3x + 4y = 30$. Solving these simultaneously yields the solution (6, 3) as the vertex.

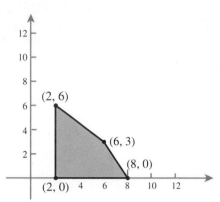

**FIGURE 44**

Step 6: *Evaluate the objective function at each of the vertices and identify the optimum solution.*

We compute the revenue at each vertex using the objective function $R = 160x + 200y$.

$$\text{At } (2, 0): \quad R = 160(2) + 200(0) = 320$$

$$\text{At } (8, 0): \quad R = 160(8) + 200(0) = 1280$$

$$\text{At } (6, 3): \quad R = 160(6) + 200(3) = 1560 \qquad \text{Maximum revenue}$$

$$\text{At } (2, 6): \quad R = 160(2) + 200(6) = 1520$$

So a maximum daily revenue of $1560 can be obtained by making 6 tables and 3 bookcases each day.

## THE GENERAL LINEAR PROGRAMMING PROBLEM

The six steps we followed to determine the maximum revenue for the Willow Woods Furniture Factory can be used as guidelines for solving a variety of optimization problems. We summarize the steps as follows.

### Solving an optimization problem

> 1. Read the problem carefully. If necessary, try some specific examples to help you understand the problem.
> 2. Assign variables to the unknowns and write out the objective function that is to be optimized.
> 3. Write out the inequalities that correspond to the constraints in the problem.
> 4. Graph the system of inequalities to obtain the feasible set.
> 5. Locate the vertices of the feasible set.
> 6. Evaluate the objective function at each vertex, and identify the optimum solution.

**EXAMPLE 1**    *A nutrition problem*

Al is on a diet. The daily fruit portion of his diet must have as few calories as possible and yet provide at least 750 units of vitamin A, 0.72 unit of vitamin $B_6$, and 60 units of vitamin C. His fruit bowl contains a few small bananas and oranges. Each banana has approximately 100 calories, 250 units of vitamin A, 0.6 unit of vitamin $B_6$, and 12 units of vitamin C. Each orange has approximately 60 calories, 250 units of vitamin A, 0.06 unit of vitamin $B_6$, and 60 units of vitamin C. What combination of bananas and oranges will yield the fewest calories?

**SOLUTION**    The unknowns are the number of bananas and oranges that Al should eat. Thus, we let

$$x = \text{the number of bananas}$$

$$y = \text{the number of oranges}$$

We wish to minimize the number of calories. Since each banana has 100 calories and each orange has 60 calories, the objective function to be minimized is

$$K = 100x + 60y$$

The nutritional requirements lead to the following inequalities.

$$\text{Vitamin A:  units from bananas + units from oranges} \geq 750$$
$$250x + 250y \geq 750$$

$$\text{Vitamin } B_6\text{: units from bananas + units from oranges} \geq 0.72$$
$$0.6x + 0.06y \geq 0.72$$

$$\text{Vitamin C:  units from bananas + units from oranges} \geq 60$$
$$12x + 60y \geq 60$$

Since $x$ and $y$ must be nonnegative, we also have $x \geq 0$ and $y \geq 0$. Thus, we have the following constraints.

$$\text{Constraints:}\quad 250x + 250y \geq 750$$
$$0.6x + 0.06y \geq 0.72$$
$$12x + 60y \geq 60$$
$$x \geq 0$$
$$y \geq 0$$

The graph of this system of inequalities is shown in Figure 45. The portion shown in dark blue is the feasible set, and the vertices shown in Figure 46 are found by locating intercepts and the intersection points of appropriate pairs of lines.

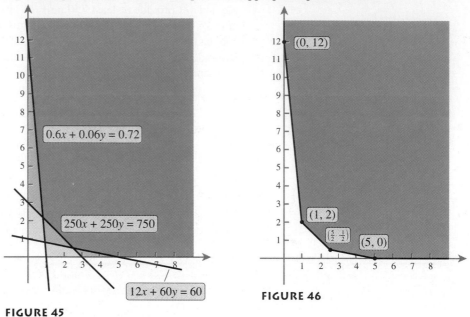

**FIGURE 45**

**FIGURE 46**

To determine the minimum number of calories, we evaluate the objective function at each vertex.

$$\text{At } (0, 12): \quad K = 100(0) + 60(12) = 720$$

$$\text{At } (1, 2): \quad K = 100(1) + 60(2) = 220 \qquad \text{Minimum}$$

$$\text{At } \left(\frac{5}{2}, \frac{1}{2}\right): \quad K = 100\left(\frac{5}{2}\right) + 60\left(\frac{1}{2}\right) = 280$$

$$\text{At } (5, 0): \quad K = 100(5) + 60(0) = 500$$

Thus, to keep the fruit calories as low as possible and still obtain the required nutrients, Al should eat one banana and two oranges during the course of the day.

The two optimization problems we have considered so far have had several features in common. Each involved two variables, and each required us to maximize or minimize a linear objective function subject to two or more linear constraints. We refer to such problems as linear programming problems in two variables. In general, they are of the following form.

*Linear programming problem in two variables*

> Maximize or minimize an objective function $P = Ax + By + C$ subject to constraints of the form $ax + by \leq c$ or $ax + by \geq c$.

As we have seen, the solution of a linear programming problem can be found by testing the objective function at each of the vertices of the feasible set. This fact is known as the Fundamental Theorem of Linear Programming, and we state it formally below.

### The Fundamental Theorem of Linear Programming

An optimal solution to a linear programming problem, if it exists, will occur at a vertex of the feasible set.

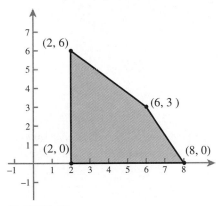

**FIGURE 47**

To see why the optimal solution occurs at a vertex, let us return briefly to the furniture factory problem. Recall that we wished to maximize the objective function $R = 160x + 200y$ on the feasible set shown in Figure 47. If we approach the problem naively, without any knowledge of the Fundamental Theorem of Linear Programming, we might choose to consider a certain specific revenue value and the corresponding ordered pairs $(x, y)$ that would yield that revenue. For example, to obtain a revenue of $1000, we must have $1000 = 160x + 200y$. This is the equation of the line graphed in Figure 48.

Since the line $1000 = 160x + 200y$ passes through the feasible set, we can see that there are points $(x, y)$ that satisfy the constraints of the problem and also yield a revenue of $1000. To determine whether there are points that yield a higher revenue and still satisfy the constraints, we next consider revenues of $1200 and $1500. The points that satisfy the constraints and yield revenues of $1200 and $1500, respectively, are the points of intersection of the lines $1200 = 160x + 200y$ and $1500 = 160x + 200y$ with the feasible set as shown in Figure 49.

Notice that the lines are all parallel and they progress further away from the origin as the revenue values increase. This suggests that we look for the line that is parallel to the other three, intersects the feasible set, and is as far from the origin as possible. This line is shown in Figure 50, and it turns out to have equation $1560 = 160x + 200y$ and it intersects the feasible set at the vertex $(6, 3)$. Thus, we see that the largest revenue of $1560 is attained at the vertex $(6, 3)$.

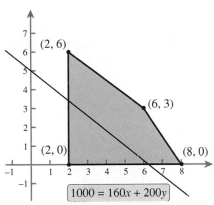

**FIGURE 48**

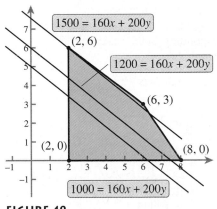

**FIGURE 49**

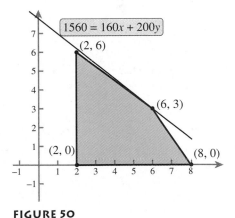

**FIGURE 50**

In general, specific values of the objective function lead to families of parallel lines. If the objective function has a maximum value, it will correspond to the line that intersects the feasible set and is furthest from, or nearest to, the origin, depending on the particular objective function $P = Ax + By + C$. This line must necessarily pass through a vertex. Likewise, if there is a minimum, it corresponds to the line that is closest to, or furthest from, the origin, again depending on the particular objective function $P = Ax + By + C$, and it must also pass through a vertex.

**EXAMPLE 2**    *Solving a linear programming problem*

Find the maximum and minimum values of $P = 3x + 6y$ subject to the following constraints.

$$x + y \leq 8$$

$$2x - y \leq 7$$

$$x + 2y \geq 6$$

$$x \geq 2$$

$$y \leq 5$$

Sketch several of the lines that correspond to specific values of the objective function, and show that the lines corresponding to the maximum and minimum values pass through vertices of the feasible set.

**SOLUTION**    The feasible set is shown in Figure 51. Testing the objective function at each of the vertices yields the following results.

At $(2, 5)$:    $P = 3(2) + 6(5) = 36$

At $(3, 5)$:    $P = 3(3) + 6(5) = 39$        Maximum

At $(5, 3)$:    $P = 3(5) + 6(3) + 33$

At $(4, 1)$:    $P = 3(4) + 6(1) = 18$        Minimum

At $(2, 2)$:    $P = 3(2) + 6(2) = 18$        Minimum

Thus, the maximum value of $P$ is 39 and it occurs at $(3, 5)$, and the minimum value of $P$ is 18, which occurs at both $(4, 1)$ and $(2, 2)$. Figure 52 shows the lines corresponding to the objective function values 18, 24, 32, and 39. Note that the lines corresponding to the maximum $P = 39$ and minimum $P = 18$ pass through the vertices found above.

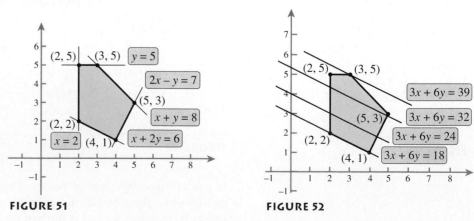

**FIGURE 51**                **FIGURE 52**

As can be seen in Example 2, the objective function can attain its maximum or minimum value at more than one vertex of a feasible set. Note also that the objective function need not have both a maximum and a minimum on a given feasible set. An example of this situation can be seen in Example 1 where the objective function has a minimum but not a maximum.

## ADVANCED TECHNIQUES OF LINEAR PROGRAMMING

In this text we investigate extremely simple linear programming problems involving several constraints and only two variables. We are using the geometric method: we graph the constraint inequalities and then evaluate the objective function at each of the vertices to determine where the maximum or minimum value is obtained. But what would we do if there were thousands of variables and tens of thousands of constraints? We are certainly not going to be able to graph the region in a thousand-dimensional space; indeed it would be difficult to solve even a three-dimensional problem in this fashion. Clearly we need an **algorithm**, or step-by-step method, that doesn't require geometric visualization.

Motivated by problems arising in the context of military planning and support for the U.S. Department of the Air Force, George Dantzig in 1947 developed the **simplex method**, a systematic method for solving linear programming problems. The simplex method has found wide application in communications, airline scheduling, and inventory control, and has hence become an integral part of management, industrial engineering, and operations research training programs throughout the world.

In the two-dimensional problems that we consider in this section, the feasible set is a region in the plane bounded by line segments, and optimal solutions exist at vertices. In the sort of multidimensional problems to which the simplex method is applied, the feasible set is a ''region'' in a multidimensional space bounded by portions of **hyperplanes**. If we visualize the feasible set as a multifaceted gemstone, then the hyperplanes that form the boundary of the feasible set are the facets of the gemstone. Just as in the two-dimensional case, optimal solutions occur at the vertices (the corners of the gemstone). The simplex method begins with a vertex in the feasible set and then at the next step selects an adjacent vertex for which the objective function is greater (in the case of a maximization problem). In this manner, step by step, the simplex method propels us ever closer to an optimal solution.

In spite of the widespread success that the simplex method has enjoyed for half a century, in recent years problems have arisen, particularly in the communications industries, for which the simplex method has proved to be inadequate. For this reason, there is an ongoing search for more efficient methods for solving high-dimensional linear programming problems. Perhaps the most significant breakthrough in this area was made in the 1980s by Narendra K. Karmarkar of AT&T Laboratories (pictured here). Karmarkar's technique involves cutting a swath through the interior of the region rather than caroming from vertex to vertex on the surface. At each stage, another interior point, closer to the optimal solution than the last, is reached. Although the improvement in computation time is often modest for low-dimensional linear programming problems, the improvement can be quite dramatic when the dimension is large. The demand for more efficient techniques for solving linear programming problems, fueled in part by the ongoing creation of the information superhighway, will likely lead to the development of ever faster algorithms.

**EXERCISES 1–6** □ *Find the maximum and minimum value of the objective function subject over the given feasible set.*

1. Objective function: $P = 8x + 10y$
   Feasible set:

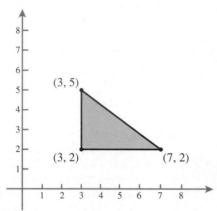

2. Objective function: $P = 10x + 4y$
   Feasible set:

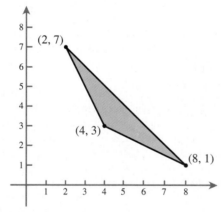

3. Objective function: $P = 4x + 5y$
   Feasible set:

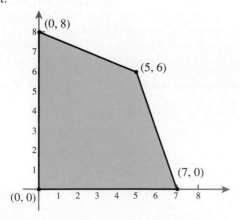

4. Objective function: $P = 6x - 2y$
   Feasible set:

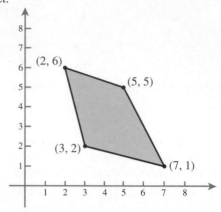

5. Objective function: $P = 4x - 4y$
   Feasible set: See Exercise 3

6. Objective function: $P = 2x + 8y$
   Feasible set: See Exercise 4

**EXERCISES 7–16** □ *Solve the linear programming problem.*

7. Maximize $P = 10x + 15y$
   subject to   $3x + 4y \le 24$
   $x + 2y \le 10$
   $x \ge 0$
   $y \ge 0$

8. Maximize $P = 14x + 30y$
   subject to   $3x + 5y \le 30$
   $2x + 4y \le 22$
   $x \ge 0$
   $y \ge 0$

9. Minimize $P = 15x + 40y$
   subject to   $2x + 5y \ge 20$
   $3x + 2y \ge 19$
   $x \ge 0$
   $y \ge 0$

10. Minimize $P = 30x + 10y$
    subject to   $2x + y \ge 8$
    $4x + y \ge 12$
    $x \ge 0$
    $y \ge 0$

11. Maximize $P = 10x + 12y$
    subject to   $x + 2y \le 17$
    $2x + 3y \le 30$
    $x \ge 3$
    $y \ge 2$

12. Maximize $P = 12x + 8y$
    subject to   $2x + y \le 18$
    $x + 2y \le 21$
    $x \ge 1$
    $y \ge 4$

13. Minimize $P = 20x + 24y$
    subject to   $2x + 3y \le 12$
    $x + y \ge 5$
    $x + 4y \ge 8$
    $x \ge 0$
    $y \ge 0$

14. Minimize $P = 10x + 4y$
    subject to   $x + 3y \ge 16$
    $x + y \le 10$
    $2x + y \ge 12$
    $x \ge 0$
    $y \ge 0$

15. Maximize $P = 2x + 10y$
    subject to   $x + y \ge 9$
    $3x - 2y \le 24$
    $x + 4y \le 36$
    $x \ge 4$
    $y \ge 3$

16. Minimize $P = 10x + 8y$
    subject to   $x + 2y \ge 12$
    $x - y \le 4$
    $3x + y \ge 16$
    $x \le 8$
    $y \le 10$

## ■ *Applications*

17. *Clothing Profits* A clothing company makes two styles of tailored suits. Style A requires 2 hours of cutting and 3 hours of sewing. Style B requires 4 hours of cutting and 2 hours of sewing. There are 56 work hours available per day for cutting and 72 hours per day for sewing. The profit on style A is $35 and the profit on style B is $40. How many of each style should be made to maximize profit?

18. *Livestock Allocation* Maple Grove Farms raises cattle and pigs. Each cow requires an investment of $480 and 10 hours of labor. Each pig requires an investment of $1400 and 9 hours of labor. The profit from each cow is $120 and the profit from each pig is $400. The resources available for raising cows and pigs are 680 hours of labor and $52,000. How many cows and pigs should Maple Grove Farms raise to maximize profit?

19. *Refinery Production* An oil company owns two refineries. The daily production for refinery 1 is fixed at 100 barrels of high-grade oil, 200 barrels of medium-grade oil, and 300 barrels of low-grade oil. The daily production for refinery 2 is fixed at 200 barrels of high-grade oil, 100 barrels of medium-grade oil, and 200 barrels of low-grade oil. An order is received for 1000 barrels of high-grade oil, 1000 barrels of medium-grade oil, and 1800 barrels of low-grade oil. How many days should each refinery be operated so that the order can be filled at the least cost, assuming refinery 1 costs $10,000 per day to operate and refinery 2 costs $9000 per day to operate?

20. *Furniture Profit* A furniture company makes chairs and sofas. Each chair requires 6 hours of carpentry, 1 hour of finishing, and 2 hours of upholstery. Each sofa requires 3 hours of carpentry, 1 hour of finishing, and 6 hours of upholstery. The total labor available each day for carpentry, finishing, and upholstery is 96 hours, 18 hours, and 72 hours, respectively. How many chairs and sofas should be made each day to maximize profit, assuming that the profit is $80 for each chair and $70 for each sofa?

21. *Vehicle Allocation* Steve can borrow either his mother's car or his father's truck to commute to school, as long as he replaces the gas he uses. After averaging his costs for a period of time, he has determined that the car costs him 4¢ per mile while the truck averages 6¢ per mile. Depending on the route he takes, he travels at least 300 miles each month but no more than 350. For insurance purposes, he must drive his father's truck at least twice as far as his mother's car. However, his father does not want him use the truck for more than 250 miles each month. How many miles should Steve drive each vehicle

in order to minimize his monthly cost? What is the least Steve will have to spend each month?

22. *Dormitory Construction* A university is making plans to build a new dormitory. Early estimates show that each single room will cost $20,000 and will require approximately 1200 cubic feet of space. Each double room will cost $25,000 and will require 1800 cubic feet of space. The university can spend at most $2,200,000. Available space and zoning restrictions limit the total room space to 150,000 cubic feet. Finally, there must be at least as many double rooms as single rooms. How many of each type of room should the dormitory have in order to maximize the number of students it can hold?

23. *Campaign Strategy* A political party is planning a half-hour television show for their incumbent candidates for state governor and U.S. Senate. Based on a pre-show survey, it is believed that 40,000 viewers will watch the show for each minute the senator is on and 60,000 viewers will watch for each minute the governor is on. The senator is a party "elder statesman" and so demands to be on the air at least one and a half times as long as the governor. The governor will not participate if her time is less than 10 minutes. And of course, the sum of their speaking times may not exceed 30 minutes. Determine the time that should be allotted to each of the two candidates to maximize the number of viewers.

24. *Power Lunch* Sam's lunches consist primarily of peanut butter sandwiches and hamburgers from his favorite fast-food restaurant. Each sandwich has 1.5 grams of saturated fats, 40 grams of carbohydrates, 9 grams of protein, and 2 grams of iron. Each hamburger has 4 grams of saturated fats, 30 grams of carbohydrates, 12 grams of protein, and 4 grams of iron. Determine how many sandwiches and hamburgers Sam must eat to minimize the amount of saturated fats but still obtain at least 110 grams of carbohydrates, 30 grams of protein, and 7 grams of iron.

25. *Hospital Meal* A hospital food service is serving a meal that includes Swiss steak and peas. Each ounce of Swiss steak costs 9¢ and has 150 units of vitamin A, 0.06 unit of vitamin $B_6$, and 3 units of vitamin C. Each ounce of peas costs 4¢ and has 120 units of vitamin A, 0.02 unit of vitamin $B_6$, and 9 units of vitamin C. From the meat and vegetable portion of the meal, each patient must receive at least 1260 units of vitamin A, 0.35 unit of vitamin $B_6$, and 45 units of vitamin C. How many ounces of Swiss steak and peas should be served to minimize the cost?

■  *Projects for Enrichment*

**26.** *Linear Programming with Three Variables* Linear programming problems can also involve more than two variables. In general, a linear programming problem in $n$ variables would involve maximizing or minimizing an objective function $P = A_1x_1 + A_2x_2 + \cdots + A_nx_n$ subject to constraints of the form $a_1x_1 + a_2x_2 + \cdots + a_nx_n \leq b$ or $a_1x_1 + a_2x_2 + \cdots + a_nx_n \geq b$. Unfortunately, such problems usually cannot be solved geometrically unless it is possible to express $n - 2$ of the variables in terms of the remaining two. In the following example, we start with three variables but are able to replace one of the variables with an expression involving the other two.

*Example* Suppose that $18,000 is to be invested in some combination of government bonds paying 6%, municipal bonds paying 8%, and stocks paying 10%. After analyzing the risks, it is decided that at most $15,000 is to be invested in government and municipal bonds, at least $10,000 in government bonds, and at most $6000 in stocks. What combination of investments will maximize the yearly income?

*Solution* We initially assign the following variables.

$x$ = amount invested in government bonds

$y$ = amount invested in municipal bonds

$z$ = amount invested in stocks

The yearly income from investing these amounts at the given rates is

$$I = 0.06x + 0.08y + 0.10z$$

We wish to maximize this objective function subject to the following constraints.

| | |
|---|---|
| $x + y \leq 15,000$ | At most $15,000 in government and municipal bonds |
| $x \geq 10,000$ | At least $10,000 in government bonds |
| $z \leq 6000$ | At most $6000 in stocks |
| $x \geq 0, \; y \geq 0, \; z \geq 0$ | |

In order to solve this problem geometrically, we must reduce it to only two variables. Since the total amount invested is $18,000, we know that

$$x + y + z = 18,000$$

so that

$$z = 18,000 - x - y$$

Substituting this expression into the objective function and the constraints gives

$$I = 0.06x + 0.08y + 0.10(18,000 - x - y)$$
$$= 1800 - 0.04x - 0.02y$$

and

$$x + y \leq 15,000$$
$$x \geq 10,000$$
$$18,000 - x - y \leq 6000 \qquad \text{Or equivalently, } x + y \geq 12,000$$
$$x \geq 0$$
$$y \geq 0$$
$$18,000 - x - y \geq 0 \qquad \text{Or equivalently, } x + y \leq 18,000$$

The feasible set corresponding to these constraints is shown in Figure 53.

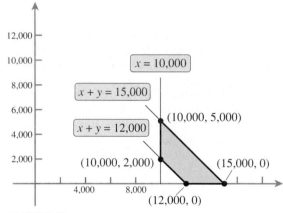

**FIGURE 53**

We now test the objective function at each of the vertices.
At (12,000, 0):
$$I = 1800 - 0.04(12,000) - 0.02(0) = 1320$$
At (15,000, 0):
$$I = 1800 - 0.04(15,000) - 0.02(0) = 1200$$
At (10,000, 2000):
$$I = 1800 - 0.04(10,000) - 0.02(2000) = 1360 \qquad \text{Maximum}$$
At (10,000, 5000):
$$I = 1800 - 0.04(10,000) - 0.02(5000) = 1300$$

So the maximum income is $1360 when $x = 10,000$ and $y = 2000$. For these values of $x$ and $y$,

$$z = 18,000 - 10,000 - 2000 = 6000$$

Thus, $10,000 should be invested in government bonds, $2000 in municipal bonds, and $6000 in stocks to achieve a maximum income of $1360.

Use the method described above to solve the following linear programming problems involving three variables.

**a.** Michael has $20,000 to invest in some combination of mutual funds, stocks, and bonds. The projected annual rates of return are 14% for mutual funds, 16% for stocks, and 10% for bonds. He does not wish to invest more than $12,000 in mutual funds and stocks, but would like to invest at least $4000 in stocks. Moreover, he does not wish to invest more than $10,000 in bonds. What combination of investments will maximize Michael's income?

b. Eliza wishes to have $2000 from her monthly paycheck deposited automatically in some combination of three different accounts at her credit union. She would like to maximize the amount of earnings from each new deposit after one month. The savings account pays $\frac{1}{3}$% per month, the checking account pays $\frac{1}{2}$% per month on the average balance but has a fee of $5 per month, and the money market account pays $\frac{2}{3}$% per month. Due to bills that must be paid during the course of the month, she must put at least $1000 in the checking account (assume that the average balance is

half of what she deposits). She would like to deposit at least $800 in savings and the money market account, although the money market account has restrictions on withdrawals, and so she does not wish to deposit more than $500 there. How should the deposit be distributed among the three accounts to maximize earnings?

c. A farmer has 1000 acres to plant in some combination of corn, soybeans, and wheat. Each acre of corn costs $100 and requires 2 hours of labor. Each acre of soybeans costs $80 and requires 1 hour of labor. Each acre of wheat costs $70 and requires 1.5 hours of labor. The total cost for the three crops may not exceed $85,000 and the labor may not exceed 1600 hours. The profits from each acre of corn, soybeans, and wheat are $170, $110, and $90, respectively. How many acres of each should be planted to maximize profit?

 **Questions for Discussion or Essay**

27. A friend relates the following story: "... so we're taking this test on Chapter 10, and there are only a couple minutes left in the hour, when 'Ivan the Terrible' announces that there is a typo in the linear programming problem I've just spent the last 10 minutes finishing. Since there isn't enough time to redo the problem, I erase everything and write a note explaining the steps I would follow if given enough time. Can you believe I only got a couple of points for the problem?" The friend shows you the test and the location of the typo. You immediately realize that, although your friend has the steps memorized, he doesn't really understand the linear programming process. Only a few simple calculations would have been necessary to correct the problem. Was the typo in the objective function or one of the constraints? Explain.

28. In examples such as the furniture factory problem, it is tempting to assume that the maximum value of the objective function will occur at the "obvious" vertex—that is, the one where the available resources have been completely utilized. Review the furniture factory problem, and explain why the solution (6, 3) is the "obvious" one. Then show that it is possible, by changing only one of the coefficients in the objective function, to make the maximum occur at the vertex (2, 6). At this vertex, which labor resource has not been fully utilized? In practice, do you think it's possible for a company to maximize revenue without using all available resources? Explain.

29. In Example 1 we minimized the objective function $K = 100x + 60y$ over the feasible set shown in Figure 54. The values of the objective function at the vertices are given in Table 7. Thus, we found the minimum to be 220 at the vertex (1, 2). Why can we not assume that the maximum is 720 at the vertex (0, 12)? What can be said about the maximum of the objective function over this feasible set? Does this contradict the Fundamental Theorem of Linear Programming? Explain.

**TABLE 7**

| Vertex | $K = 100x + 60y$ |
|--------|------------------|
| (0, 12) | 720 |
| (1, 2) | 220 |
| $\left(\frac{5}{2}, \frac{1}{2}\right)$ | 280 |
| (5, 0) | 500 |

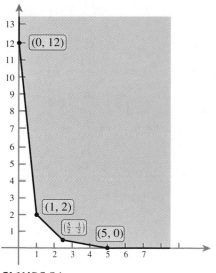

**FIGURE 54**

## CHAPTER REVIEW EXERCISES

**EXERCISES 1–4** □ *Perform the indicated matrix operation.*

1. $3 \cdot \begin{bmatrix} 2 & -1 \\ -5 & 3 \end{bmatrix}$

2. $-2 \cdot \begin{bmatrix} -3 & 0 \\ 4 & -1 \\ 0 & -2 \end{bmatrix}$

3. $\begin{bmatrix} 3 & 0 & -2 \\ 5 & -1 & 3 \end{bmatrix} + \begin{bmatrix} 0 & -1 & 4 \\ -2 & 6 & 2 \end{bmatrix}$

4. $\begin{bmatrix} 1 & 0 \\ 3 & -8 \end{bmatrix} - \begin{bmatrix} 0 & 5 \\ -1 & -2 \end{bmatrix}$

**EXERCISES 5–12** □ *Let*

$$A = \begin{bmatrix} 1 & 0 \\ -2 & 3 \end{bmatrix}, \quad B = \begin{bmatrix} 2 & -1 \\ 4 & 3 \end{bmatrix},$$

$$C = \begin{bmatrix} 4 & 0 \\ -3 & 1 \\ 0 & 2 \end{bmatrix}, \quad \text{and} \quad D = \begin{bmatrix} 3 & 1 & 0 \\ -2 & 0 & 4 \\ 1 & -2 & 1 \end{bmatrix}.$$

*Determine whether the given expression is defined. If so, evaluate it. If not, explain why not.*

5. $A + 2B$

6. $2C - D$

7. $CD$

8. $BA$

9. $DC$

10. $C(A - B)$

11. $C^2$

12. $A^2$

**EXERCISES 13–20** □ *Compute the given product.*

13. $\begin{bmatrix} -2 & 3 \\ 1 & 4 \end{bmatrix} \begin{bmatrix} 3 \\ -1 \end{bmatrix}$

14. $\begin{bmatrix} 3 & 0 \\ 2 & -4 \end{bmatrix} \begin{bmatrix} 2 & -1 \\ -4 & 6 \end{bmatrix}$

15. $\begin{bmatrix} 3 & 4 \end{bmatrix} \begin{bmatrix} 2 \\ -1 \end{bmatrix}$

16. $\begin{bmatrix} 5 & -3 \\ 0 & 4 \end{bmatrix} \begin{bmatrix} 2 & 1 & -1 \\ -1 & 3 & 0 \end{bmatrix}$

17. $\begin{bmatrix} 0 & 4 \\ -1 & 2 \\ 3 & 0 \end{bmatrix} \begin{bmatrix} 2 & 3 \\ -3 & 1 \end{bmatrix}$

18. $\begin{bmatrix} 1 & 1 & 0 \\ 0 & 1 & 1 \end{bmatrix} \begin{bmatrix} 1 & 2 & 3 \\ 0 & 1 & 2 \\ 0 & 0 & 1 \end{bmatrix}$

19. $\begin{bmatrix} 1 & -1 & 2 \\ 3 & 0 & -1 \\ 4 & 2 & 0 \end{bmatrix} \begin{bmatrix} b \\ c \\ a \end{bmatrix}$

20. $\begin{bmatrix} 1 & 2 & 4 \\ 2 & 4 & 1 \\ 4 & 1 & 2 \end{bmatrix} \begin{bmatrix} -1 & 0 & 2 \\ 0 & 2 & -1 \\ 2 & -1 & 0 \end{bmatrix}$

**EXERCISES 21–22** □ *Write the given system as an augmented matrix.*

21. $5x + y = 3$
    $x - 3y = 4$

22. $x + 2y - 2z = 3$
         $2x + y = -1$
          $3y + 2z = 6$

**EXERCISES 23–24** □ *Write out a system of equations in x, y, and z, in that order, that corresponds to the given augmented matrix, and then solve the system using back-substitution.*

23. $\left[\begin{array}{ccc|c} 1 & -4 & 1 & 3 \\ 0 & 1 & -2 & 5 \\ 0 & 0 & 1 & -2 \end{array}\right]$

24. $\left[\begin{array}{ccc|c} 1 & -2 & -1 & 4 \\ 0 & 1 & \frac{1}{2} & 2 \\ 0 & 0 & 1 & 3 \end{array}\right]$

**EXERCISES 25–28** □ *Use Gaussian elimination to write the augmented matrix in row-echelon form. There may be more than one correct answer.*

25. $\left[\begin{array}{cc|c} -1 & 3 & -2 \\ 4 & -9 & 1 \end{array}\right]$

26. $\left[\begin{array}{ccc|c} 2 & 4 & -2 & 6 \\ 3 & 5 & -1 & 4 \\ -2 & -4 & 9 & 6 \end{array}\right]$

27. $\left[\begin{array}{ccc|c} 2 & 4 & -3 & 1 \\ 3 & 0 & \frac{9}{2} & -\frac{3}{2} \end{array}\right]$

28. $\left[\begin{array}{ccc|c} 3 & 10 & 4 & 0 \\ 1 & 3 & 1 & 1 \\ -1 & -6 & -3 & -2 \end{array}\right]$

**EXERCISES 29–34** □ *Solve the system of equations by first writing it in matrix form and then applying Gaussian elimination with back-substitution.*

29. $-x + 2y = 7$
    $2x - 3y = -9$

30. $3x + 6y = 1$
    $2x + 2y = 1$

31. $x - y + z = -4$
   $3x - 2y - z = -4$
    $x - 2z = 1$

32. $2x + 4y - 2z = 9$
         $y - z = 1$
   $-x - 3y + z = -6$

33. $x - y = 4$
   $2x - z = 5$
    $y - z = 1$

34. $x + y = 0$
   $-2x - y + z = -9$
    $3x + y - z = 8$

**EXERCISES 35–36** □ *Use Gauss-Jordan elimination to write the augmented matrix in reduced row-echelon form.*

35. $\left[\begin{array}{cc|c} 1 & 2 & 4 \\ 2 & 3 & 5 \end{array}\right]$

36. $\left[\begin{array}{ccc|c} 1 & -1 & 2 & 8 \\ 0 & 1 & -4 & -9 \\ 0 & 0 & 1 & 3 \end{array}\right]$

**EXERCISES 37–40** □ *Solve the system of equations by first writing it in matrix form and then using Gauss-Jordan elimination.*

37. $3x + 5y = -5$
   $-x - 2y = 3$

38. $x - 4y = -0.5$
   $-2x + 9y = 1.25$

39. $x - y + 3z = 9$
      $y - 2z = -3$
         $z = 4$

40. $x - y + z = 0$
   $-4x + 5y - 6z = 7$
    $2x - y + z = 2$

**EXERCISES 41–48** □ *Find the inverse of the given matrix, if it exists.*

41. $\begin{bmatrix} 2 & 1 \\ -3 & -1 \end{bmatrix}$

42. $\begin{bmatrix} 0 & \frac{1}{2} \\ 8 & -\frac{1}{4} \end{bmatrix}$

43. $\begin{bmatrix} a & b \\ b & a \end{bmatrix}$

44. $\begin{bmatrix} 1 & x \\ x & x^2 \end{bmatrix}$

45. $\begin{bmatrix} 1 & -2 & 3 \\ 0 & 1 & -2 \\ 0 & 0 & 1 \end{bmatrix}$

46. $\begin{bmatrix} 0 & 4 & 8 \\ 4 & 0 & 4 \\ 8 & 4 & 0 \end{bmatrix}$

47. $\begin{bmatrix} 0 & 0 & a \\ 0 & 0 & 0 \\ a & 0 & 0 \end{bmatrix}$

48. $\begin{bmatrix} 1 & -1 & 0 & 0 \\ -1 & 1 & -1 & 0 \\ 0 & -1 & 1 & -1 \\ 0 & 0 & -1 & 1 \end{bmatrix}$

**EXERCISES 49–52** □ *Solve the system of equations by first writing them in matrix form as $AX = B$ and then evaluating $X = A^{-1}B$.*

49. a. $\begin{aligned} x + y &= 3 \\ 2x + y &= -4 \end{aligned}$

   b. $\begin{aligned} x + y &= 2 \\ 2x + y &= 0 \end{aligned}$

50. a. $\begin{aligned} 3x - 2y &= 5 \\ 4x - y &= -10 \end{aligned}$

   b. $\begin{aligned} 3x - 2y &= -2 \\ 4x - y &= 3 \end{aligned}$

51. a. $\begin{aligned} \frac{1}{2}x + \frac{3}{2}y &= 3 \\ -\frac{1}{4}x + \frac{5}{4}y &= -4 \end{aligned}$

   b. $\begin{aligned} \frac{1}{2}x + \frac{3}{2}y &= 0 \\ -\frac{1}{4}x + \frac{5}{4}y &= -3 \end{aligned}$

52. a. $\begin{aligned} x + y - z &= 3 \\ -4x - 3y + 6z &= -3 \\ -x - 2y &= 9 \end{aligned}$

   b. $\begin{aligned} x + y - z &= 1 \\ -4x - 3y + 6z &= 0 \\ -x - 2y &= 2 \end{aligned}$

**EXERCISES 53–60** □ *Evaluate the given determinant.*

53. $\begin{vmatrix} 4 & 6 \\ 3 & 2 \end{vmatrix}$

54. $\begin{vmatrix} -2 & 1 \\ 3 & 5 \end{vmatrix}$

55. $\begin{vmatrix} \sqrt{5} & 2 \\ 2 & \sqrt{5} \end{vmatrix}$

56. $\begin{vmatrix} 3 & -1 & 0 \\ -1 & 3 & 0 \\ -2 & -2 & 3 \end{vmatrix}$

57. $\begin{vmatrix} x & 0 & 1 \\ 0 & x & 0 \\ 1 & 0 & x \end{vmatrix}$

58. $\begin{vmatrix} 1 & 2 & 3 \\ 1 & 1 & 1 \\ 3 & 2 & 1 \end{vmatrix}$

59. $\begin{vmatrix} 1 & -1 & 0 & 0 \\ 3 & 4 & 0 & 0 \\ 0 & 0 & 2 & -3 \\ 0 & 0 & 1 & 1 \end{vmatrix}$

60. $\begin{vmatrix} a & b & 0 & 0 & 0 \\ b & a & 0 & 0 & 0 \\ 0 & 0 & 1 & 0 & 0 \\ 0 & 0 & 0 & 1 & 0 \\ 0 & 0 & 0 & 0 & 1 \end{vmatrix}$

**EXERCISES 61–64** □ *Solve the system of equations using Cramer's Rule.*

61. $\begin{aligned} 3x + y &= 0 \\ 5x + 3y &= -1 \end{aligned}$

62. $\begin{aligned} 2x - 3y &= 16 \\ 5x + 2y &= 2 \end{aligned}$

63. $\begin{aligned} x + y &= 0 \\ x + 2y - z &= -6 \\ -4x - 3y &= -1 \end{aligned}$

64. $\begin{aligned} x + y + z - 2w &= -6 \\ -2x - y + 3w &= 5 \\ 2x - z - 3w &= -5 \\ x - w &= -2 \end{aligned}$

**EXERCISES 65–72** □ *Graph the inequality.*

65. $x + y \geq 4$

66. $2x - y < -3$

67. $x < 2y$

68. $y \leq 3x - 4$

69. $x \geq -2$

70. $y \geq x^2 - 1$

71. $x^2 + y^2 < 4$

72. $x - y^2 \leq 0$

**EXERCISES 73–82** □ *Graph the system of inequalities.*

73. $\begin{aligned} x - y &\leq 6 \\ x + y &\leq 4 \end{aligned}$

74. $\begin{aligned} 3x + y &> 3 \\ x &< 3y \end{aligned}$

75. $\begin{aligned} 4x - 7y &\geq 14 \\ 3y &\leq x + 6 \end{aligned}$

76. $\begin{aligned} 2x + y &\leq 4 \\ y &\geq 2x - 1 \\ y &\geq -2 \end{aligned}$

77. $\begin{aligned} -x + 4y &> 8 \\ -x + y &> 0 \\ -x + 2y &> 4 \end{aligned}$

78. $\begin{aligned} -4x + 3y &\leq 24 \\ -3x + 5y &\leq 30 \\ x &\geq 2 \\ y &\geq 1 \end{aligned}$

79. $\begin{aligned} y &\geq x^2 - 2 \\ y &\leq x \end{aligned}$

80. $\begin{aligned} y + x^2 &\leq 4 \\ y &\geq x^2 \end{aligned}$

81. $\begin{aligned} x^2 + y^2 &\leq 25 \\ x &\geq 3 \end{aligned}$

82. $\begin{aligned} y &\geq 2^x \\ y &\leq 4 \\ x &\geq 0 \end{aligned}$

**EXERCISES 83–84** □ *Find the maximum and minimum value of the objective function over the given feasible set.*

83. $P = 5x + 8y$

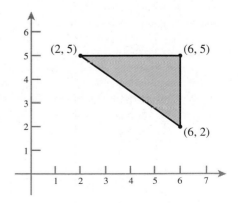

**84.** $P = 4x + 20y$

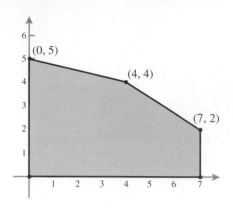

**EXERCISES 85–88** □ *Solve the linear programming problem.*

**85.** Maximize $P = 12x + 10y$
subject to   $x + y \leq 6$
$2x + y \leq 10$
$x \geq 0$
$y \geq 0$

**86.** Minimize $P = 6x + 12y$
subject to $5x + 4y \geq 35$
$5x + 3y \geq 30$
$x \geq 0$
$y \geq 0$

**87.** Minimize $P = 8x + 10y$
subject to   $2x + y \geq 10$
$x + y \geq 7$
$2x + 3y \geq 16$
$x \geq 2$
$y \geq 0$

**88.** Maximize $P = 10x + 15y$
subject to $3x + 2y \leq 21$
$x + y \leq 8$
$y \leq 2x$
$x \geq 0$
$y \geq 2$

**89.** *Estate Management* An estate manager has $100,000 to invest in three funds, a money market fund paying 4% interest per year and charging a maintenance fee of 0.2%, a mutual fund (carrying substantial risk) averaging 12% per year with an annual load of 0.5%, and a savings account with no fees paying 3% interest per year. The manager wishes to earn 6.55% on the $100,000, and he will pay only $255 in fees. How much should he invest in each account?

**90.** *Balanced Breakfast* An 800-calorie breakfast is to consist of link sausages, pancakes, and orange juice, and should contain 25 grams of protein and 90 grams of carbohydrates. Each pancake contains 23.7 grams of carbohydrates, 5.3 grams of protein, and 164 calories; each 4-ounce serving of orange juice contains 13.2 grams of carbohydrates, 0.9 gram of protein, and 55 calories; each serving of sausage links contains a negligible amount of carbohydrates, 4.7 grams of protein, and 124 calories. How many servings of each should be consumed?

**91.** *Fitting a Cubic* Find the $y$-intercept of the graph of the cubic function $f(x)$ passing through the points $(-1, 5)$, $(1, 5)$, $(2, 10)$, and $(3, 31)$.

**92.** *Unknown Coins* A coin purse contains 10 coins (quarters, dimes, and nickels) with a total value of $1.60. There are as many quarters as there are nickels and dimes together. How many of each kind of coin are there?

**93.** *Land Allocation* A farmer intends to grow corn and hay. Each acre of corn requires an investment of $125 and 2 hours of labor. Each acre of hay requires an investment of $100 and 6 hours of labor. Between available cash and a bank line of credit, the farmer has $50,000 available to invest in the two crops. He and his son have a total of 1200 hours of labor available. If each acre of corn yields revenues of $245 and each acre of hay yields revenues of $210, how many acres of each should he plant to maximize revenue? How many acres should be planted to maximize profit?

**94.** *Food Service Planning* A university food service is planning a meal that includes broiled chicken breast and a baked potato. Each ounce of chicken costs 20¢ and has 9 units of protein, 18 units of vitamin $B_6$, and 3 units of iron. Each ounce of potato costs 3¢ and has 1 unit of protein, 6 units of vitamin $B_6$, and 2 units of iron. Each serving of chicken and potato must provide at least 31 units of protein, 80 units of vitamin $B_6$, and 16 units of iron. How many ounces of chicken and baked potato should be served to minimize the cost?

**95.** *Investment Options* Damon has $10,000 to invest in selected mutual funds and municipal bonds. Because of tax reasons, his accountant has advised that he invest at least twice as much in municipal bonds as in mutual funds. However, his financial advisor suggests that he invest at least $2500 in mutual funds. If the mutual funds average a 12% return and the municipal bonds average a 10% return, how much should he invest in each to maximize his earnings?

**96.** *Studying for Finals* Gina's final exams in college algebra and sociology are scheduled for the same day. She has 12 hours available to study for the two exams. Past experience suggests that each hour of studying for college algebra requires $\frac{1}{2}$ cup of coffee while each hour of studying for sociology requires $\frac{1}{4}$ cup of coffee. Based on previous exam scores, each hour of studying for college algebra results in 15 points, and each hour of studying for sociology results in 10 points. If she must score at least 60 points on the sociology exam, and she only has 4 cups of coffee left, how many hours should she study for each exam to maximize the total score?

## CHAPTER TEST

**PROBLEMS 1–8** □ *Answer the question as true or false.*

1. Matrices must be of the same order if they are to be added.

2. Matrices must be of the same order if they are to be multiplied.

3. All matrices have inverses.

4. If $\det(A) = 0$, then the system $AX = B$ can be solved by Cramer's Rule.

5. All upper triangular square matrices with 1's on the main diagonal are in row-echelon form.

6. Some linear systems of equations have exactly two solutions.

7. The solution set of a linear inequality is a half-plane.

8. According to the Fundamental Theorem of Linear Programming, the objective function will attain its maximum value at the vertex of the feasible region that is furthest from the origin.

**PROBLEMS 9–13** □ *Give an example of each.*

9. A matrix of order $2 \times 3$

10. A noninvertible $2 \times 2$ matrix

11. An augmented matrix in row-echelon form but not reduced row-echelon form

12. A nonlinear system of equations

13. An elementary row operation

14. Let

$$A = \begin{bmatrix} 2 & -1 & 0 \\ 3 & 4 & -2 \\ 0 & -3 & 1 \end{bmatrix} \quad \text{and} \quad B = \begin{bmatrix} 0 & -2 & 5 \\ 6 & 3 & 1 \\ 4 & 0 & -1 \end{bmatrix}.$$

Compute each of the following.
   a. $A \cdot B$    b. $2A + B$    c. $A - B$

15. Express the following system as an equation involving matrices, and then solve it using an inverse matrix.

$$3x - 5y + z = 2$$
$$2x - 8y = -2$$
$$3x - z = 0$$

16. Consider the augmented matrix

$$\begin{bmatrix} 2 & -2 & 2 & | & 0 \\ 0 & 1 & -3 & | & -2 \\ 0 & 0 & 4 & | & 8 \end{bmatrix}$$

   a. Give an equivalent system of equations and solve by back-substitution.

   b. Write the matrix in reduced row-echelon form.

17. Solve the following system of equations by Gaussian elimination.

$$x - y - z = 4$$
$$3x + y + z = 4$$
$$x + 2y - z = 13$$

18. Use Cramer's Rule to solve the following system for $b$.

$$3a + b - c = 1$$
$$2a = -4$$
$$a + c = 2$$

19. In 1988, municipalities generated a total of 179.6 million tons of solid waste. Of this total, 58% was paper and yard waste; we'll classify the other 42% as miscellaneous waste. The total amount of paper waste was only 4.4 million tons less than the amount of miscellaneous waste. Find the amount of paper, yard, and miscellaneous waste. (Data Source: Environmental Protection Agency.)

20. Find the maximum value of $P = 3x - 5y$, subject to the constraints $x \leq 0$, $y \geq 0$, $y \leq x + 6$, and $y \leq \frac{1}{4}x + 3$. Graph the feasible set and clearly indicate all relevant vertices.

21. Find a system of inequalities such that the solution set is the interior of the triangle with vertices $(0, 0)$, $(2, 3)$, and $(-1, 6)$.

# INTEGER FUNCTIONS AND PROBABILITY

■ Through personal experience we develop a sense of the frequency of many everyday events and hence of the probability of their occurence. But personal experience doesn't provide us with intuition regarding events that rarely occur, such as becoming the victim of a terrorist when traveling abroad, winning a state lottery, or being struck by lightning. Surprisingly, our chances of being struck by lightning are greater than winning certain state lotteries and much greater than becoming terrorist victims. In Section 4 of this chapter we will investigate such probabilities in detail.

# SEQUENCES

■ What do flowers, pine cones, and sunflowers have in common with the family tree of a male honey bee?

■ How is it possible to predict your salary 15 years into the future?

■ If one checker is placed on the corner square of a checker board and the number of checkers in each successive square is doubled, how high will the pile of checkers in the 64th square reach?

■ Why is 10 not necessarily the next number in the sequence 1, 4, 7, …?

## NUMBER SEQUENCES

In mathematics, the term *sequence* is used to refer to an ordered list of objects. The following lists are all examples of sequences of real numbers.

$$\text{Sequence } a: -2, 0, 5, 8$$

$$\text{Sequence } b: 1, 4, 9, 16, 25, \ldots$$

$$\text{Sequence } c: \frac{1}{2}, \frac{1}{4}, \frac{1}{8}, \frac{1}{16}, \ldots$$

$$\text{Sequence } d: -1, 0, 1, 0, -1, 0, 1, \ldots$$

Sequence $a$ is an example of a **finite** sequence, while sequences $b$, $c$, and $d$ are all **infinite**. Each of the numbers in a sequence is called a **term**. Sequences are "ordered" in the sense that they have a first term, a second term, a third term, and so on. This allows us to refer to specific terms of a sequence simply by attaching subscripts to the name of the sequence. For example, the fourth term of the sequence $b$ can be denoted by $b_4$, and so $b_4 = 16$. More generally, the $n$th term of the sequence $b$ can be denoted by $b_n$. By observing that each term of the sequence $b$ is the square of its position, we can completely define the sequence $b$ by describing its $n$th term as $b_n = n^2$.

**EXAMPLE 1**      *Finding terms of a sequence*

Write out the first four terms of the sequences whose $n$th terms are defined below.

**a.** $a_n = 2n + 3$          **b.** $z_n = \dfrac{n}{n + 1}$          **c.** $r_n = \dfrac{(-1)^n}{2^n - 1}$

**SOLUTION**

**a.** $a_1 = 2(1) + 3 = 5$          $a_2 = 2(2) + 3 = 7$
    $a_3 = 2(3) + 3 = 9$          $a_4 = 2(4) + 3 = 11$

**b.** $z_1 = \dfrac{1}{1 + 1} = \dfrac{1}{2}$          $z_2 = \dfrac{2}{2 + 1} = \dfrac{2}{3}$

    $z_3 = \dfrac{3}{3 + 1} = \dfrac{3}{4}$          $z_4 = \dfrac{4}{4 + 1} = \dfrac{4}{5}$

**c.** $r_1 = \dfrac{(-1)^1}{2^1 - 1} = \dfrac{-1}{1} = -1$            $r_2 = \dfrac{(-1)^2}{2^2 - 1} = \dfrac{1}{3}$

$r_3 = \dfrac{(-1)^3}{2^3 - 1} = \dfrac{-1}{7}$            $r_4 = \dfrac{(-1)^4}{2^4 - 1} = \dfrac{1}{15}$

---

## THE INFORMATION REVOLUTION

**W**e are witnessing nothing less than the most revolutionary change in the way in which information is stored and transmitted since Gutenberg invented the printing press. Compact disc players play back music free of hiss and distortion even after thousands of plays. Single CD-ROM discs contain entire encyclopedias. Visual images are transmitted across oceans via the world wide web. Entire libraries are accessible from personal computers, and interactive games are played with thousands of players living thousands of miles apart.

Although this information revolution is dependent upon technological developments such as computer chips, fiber optic cable and compact discs, it would have been impossible if not for mathematical innovations in the ways in which information can be encoded and decoded. By **digitizing** data—that is,

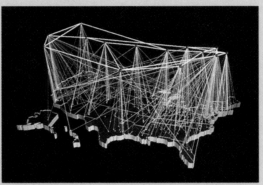

A visualization of data traffic over the National Science Foundation computer network.

converting information to a sequence of numbers, usually 1's and 0's—we are able to convert information into a language that computers can understand and manipulate, and we are able to bring to bear powerful mathematical techniques.

The accuracy of data transmission and storage is limited by the presence of **noise**—random errors with any number of possible causes: tiny flaws in the structure of a cable or a disc, electromagnetic spikes from sunspots, or even human error. Techniques for combating noise are the province of the branch of mathematics known as **error-correcting codes**, the basic principle of which is to introduce a redundancy in the way in which a message is encoded so that even with a small amount of noise, the correct message can still be determined. A simple (and too inefficient to be useful) example of an error-correcting code would be to simply repeat a digit 5 times. Thus, to transmit the digit "1", we would send "1 1 1 1 1"; to transmit "0," we would send

"0 0 0 0 0." If the sequence received was "1 1 1 1 0," we would be fairly certain that the intended message was "1" and not "0."

An important key to sustaining progress in the information revolution is to continually increase the efficiency with which we store and send data. To this end, mathematicians have developed powerful **data compression** techniques that in essence allow us to store long sequences as shorter ones. As a simple example of a data compression technique, suppose that we are dealing with sequences of 1's and 0's consisting of long runs in which a digit is repeated. Rather than send the original message, we could compress it by indicating the length of each run. Thus, the sequence consisting of 20 1's followed by 15 0's might be encoded as 20 15 (or in base 2, 10100 1111).

The exponential increase in access to information does not come without cost: the same automatic teller machine that allows us to perform financial transactions can be used by thieves to steal from our accounts; the cellular telephone system that enables us to call home from remote locations enables others to overhear our conversations; and the same medical records that can be downloaded by an emergency room physician also can be used by an unscrupulous insurance company to deny us coverage. Controlling who has access to what information poses one of the greatest challenges of the information revolution. **Cryptography** is a branch of mathematics that addresses how to transmit messages and have them remain secret, except to the intended receiver. Surprisingly, many encryption schemes involve factoring enormous numbers, and for this reason number theory, long viewed as pure mathematics with little direct applicability to the real world, is now an essential component of the information revolution.

It is no coincidence that the procedure for computing terms of a sequence is very much like that for evaluating a function. Indeed, a sequence can be viewed as a function whose domain is a set of integers. To emphasize this fact, we could define the sequence in part (a) of Example 1 by $a(n) = 2n + 3$ for $n = 1, 2, 3, \ldots$. However, it is customary to use the shorthand notation $a_n$ instead of the function notation $a(n)$. The formal definition of a sequence is given below.

**Definition of sequence**

> A **sequence** $a$ is a function whose domain is a set of integers. The function values $a(n)$ are called the **terms** of the sequence, and they are denoted $a_n$.

**EXAMPLE 2**    *Finding the nth term of a sequence*

Write an expression for the $n$th term of a sequence whose first five terms are given.

**a.** $3, 6, 9, 12, 15, \ldots$        **b.** $1, \dfrac{1}{2}, \dfrac{1}{4}, \dfrac{1}{8}, \dfrac{1}{16}, \ldots$

**SOLUTION**

**a.** Each term of the sequence is a multiple of 3. Specifically, $a_1 = 3(1)$, $a_2 = 3(2)$, $a_3 = 3(3)$, and so on. Thus, we define $a_n = 3n$.

**b.** Here each denominator is a power of 2. Thus, we first rewrite the sequence as shown here.

$$\frac{1}{2^0}, \frac{1}{2^1}, \frac{1}{2^2}, \frac{1}{2^3}, \frac{1}{2^4}, \ldots$$

Since the power of 2 in the denominator is one less than the position of the term,

$$a_n = \frac{1}{2^{n-1}}$$

The sequences given in the previous example are representatives of two important types of sequences. The sequence $3, 6, 9, 12, 15, \ldots$ has the property that each term after the first is 3 more than the one preceding it. It is an example of an *arithmetic sequence*. The sequence $1, \frac{1}{2}, \frac{1}{4}, \frac{1}{8}, \frac{1}{16}, \ldots$ has the property that each term after the first is half of the one preceding it. It is an example of a *geometric sequence*. We now investigate arithmetic and geometric sequences in more detail.

## ARITHMETIC SEQUENCES

The distinctive feature of an arithmetic sequence is that each term after the first is obtained from the preceding term by adding a fixed number. In other words, each pair of consecutive terms has a common difference.

**Definition of an arithmetic sequence**

> A sequence $a_1, a_2, a_3, \ldots, a_n, \ldots$ is **arithmetic** if there is a number $d$, called the **common difference**, such that
>
> $$a_2 - a_1 = d, \quad a_3 - a_2 = d, \quad \ldots, \quad a_n - a_{n-1} = d, \quad \ldots$$

| **EXAMPLE 3** | *Examples of arithmetic sequences* |

Each of the following are arithmetic sequences. Identify the common difference $d$ for each.

**a.** $-10, -3, 4, 11, 18, \ldots$      **b.** $2, \dfrac{7}{4}, \dfrac{3}{2}, \dfrac{5}{4}, 1, \ldots$

**SOLUTION**   Since we are given that the sequences are arithmetic, in each case we can compute the common difference using any pair of consecutive terms. We will use the first two terms.

**a.** $d = -3 - (-10) = 7$

**b.** $d = \dfrac{7}{4} - 2 = -\dfrac{1}{4}$

An arithmetic sequence is completely determined by the first term and the common difference. To see exactly what this statement means, suppose the sequence $a_1, a_2, a_3, \ldots$ is an arithmetic sequence with common difference $d$. Then each term after $a_1$ can be found by adding $d$ to the one before, as follows.

$$a_1$$
$$a_2 = a_1 + d$$
$$a_3 = a_2 + d = (a_1 + d) + d = a_1 + 2d$$
$$a_4 = a_3 + d = (a_1 + 2d) + d = a_1 + 3d$$
$$\vdots$$

Thus, each term can be found by adding a multiple of $d$ to $a_1$. Moreover, the multiple of $d$ is one less than the subscript of the term. This leads us to the following formula for the $n$th term of the sequence.

**The nth term of an arithmetic sequence**

The $n$th term of an arithmetic sequence with first term $a_1$ and common difference $d$ is

$$a_n = a_1 + (n - 1)d$$

| **EXAMPLE 4** | *Finding the nth term of an arithmetic sequence* |

Write an expression for the $n$th term of the arithmetic sequence $8, -1, -10, \ldots$.

**SOLUTION**   The common difference is $d = -1 - 8 = -9$. Using $a_1 = 8$, we apply the $n$th-term formula to find $a_n$.

$$a_n = a_1 + (n - 1)d$$
$$= 8 + (n - 1)(-9)$$
$$= -9n + 17$$

**EXAMPLE 5**    *Using given terms of an arithmetic sequence to find an unknown term*

The fifth and eleventh terms of an arithmetic sequence are 28 and 64, respectively. Find the twentieth term.

**SOLUTION**    By substituting $n = 5$ and $a_5 = 28$ into the $n$th-term formula, we obtain the equation

$$(1) \qquad\qquad 28 = a_1 + 4d$$

Similarly, substituting $n = 11$ and $a_{11} = 64$ gives

$$(2) \qquad\qquad 64 = a_1 + 10d$$

Subtracting equation (1) from equation (2) yields

$$36 = 6d$$
$$d = 6$$

Substituting $d = 6$ into equation (1) gives

$$28 = a_1 + 4(6)$$
$$a_1 = 4$$

Now that $a_1$ and $d$ are known, we apply the $n$th-term formula once more with $n = 20$ to see that

$$a_{20} = 4 + (20 - 1)(6) = 118$$

## GEOMETRIC SEQUENCES

We have defined an arithmetic sequence to be a sequence whose consecutive terms have a common difference. A *geometric sequence* is a sequence whose consecutive terms have a common ratio.

**Definition of a geometric sequence**

A sequence $a_1, a_2, a_3, \ldots, a_n, \ldots$ is **geometric** if there is a number $r \neq 0$, called the **common ratio**, such that

$$\frac{a_2}{a_1} = r, \quad \frac{a_3}{a_2} = r, \quad \ldots, \quad \frac{a_n}{a_{n-1}} = r, \quad \ldots$$

**EXAMPLE 6**    *Examples of geometric sequences*

Each of the following is a geometric sequence. Find the common ratio $r$.

**a.** 6, 18, 54, 162, 486, ...

**b.** $-\dfrac{4}{3}, \dfrac{8}{9}, -\dfrac{16}{27}, \dfrac{32}{81}, -\dfrac{64}{243}, \ldots$

**SOLUTION**    Since we are given that the sequences are geometric, in each case we can compute the common ratio using any pair of consecutive terms. We will use the first two terms.

a.  $r = \dfrac{18}{6} = 3$

b.  $r = \dfrac{\frac{8}{9}}{-\frac{4}{3}} = \dfrac{8}{9} \cdot \left(-\dfrac{3}{4}\right) = -\dfrac{2}{3}$

A formula for the *n*th term of a geometric sequence can be found using a method similar to that used for arithmetic sequences. We begin with the first term $a_1$ and use the fact that each successive term is found by multiplying the preceding one by the common ratio *r*.

$$a_1$$
$$a_2 = a_1 r$$
$$a_3 = a_2 r = (a_1 r)r = a_1 r^2$$
$$a_4 = a_3 r = (a_1 r^2)r = a_1 r^3$$
$$\vdots$$

So each term can be found by multiplying $a_1$ by an appropriate power of *r*, In fact, we see that the power is one less than the subscript of the term. Thus, we have the following formula.

**The *n*th term of a geometric sequence**

The *n*th term of a geometric sequence with first term $a_1$ and common ratio *r* is

$$a_n = a_1 r^{n-1}$$

**EXAMPLE 7**

*Using the nth-term formula for geometric sequences*

Find the 25th term of the geometric sequence having common ratio $r = \frac{1}{4}$ and first term $a_1 = 3$.

**SOLUTION**    Applying the *n*th-term formula for geometric sequences, we obtain the following.

$$a_n = a_1 r^{n-1}$$
$$a_{25} = 3 \left(\dfrac{1}{4}\right)^{25-1}$$
$$= \dfrac{3}{4^{24}}$$

**EXAMPLE 8**    *A sequence of checker stacks*

A checker is placed on a corner square of a checker board. In each successive square, the number of checkers is doubled, as shown in Figure 1. How many checkers will be on the 64th square? If a checker is approximately 5 millimeters thick, how high will the stack of checkers reach? (Note: Figure 2 shows approximately how high the stack in the 32nd square would reach.)

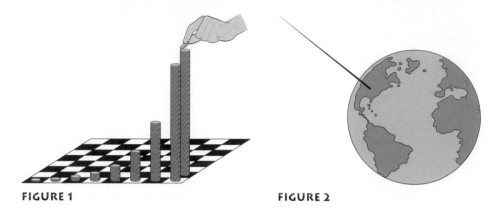

**FIGURE 1**                                             **FIGURE 2**

**SOLUTION**    The number of checkers on each successive square is given by a term of the sequence 1, 2, 4, 8, 16, .... This is a geometric sequence with first term 1 and common ratio $r = 2$. Thus, by applying the $n$th-term formula for geometric sequences, we see that $a_n = (1)2^{n-1}$. The 64th term of this sequence is

$$a_{64} = 2^{63} \approx 9.22 \times 10^{18}$$

So there are approximately 9,220,000,000,000,000,000 checkers on the 64th square. If each checker is 5 millimeters thick, the pile of checkers would reach

$$(9.22 \times 10^{18})(5 \text{ mm})(1 \times 10^{-6} \text{ km/mm}) = 4.61 \times 10^{13} \text{ km}$$

which is approximately the distance to Proxima Centauri, the closest star (other than the Sun) to the Earth.

■

**RECURSIVELY DEFINED SEQUENCES**

Arithmetic and geometric sequences have the property that each term after the first can be computed from the preceding term. In the case of arithmetic sequences, we simply add the common difference $d$, so that $a_n = a_{n-1} + d$. For geometric sequences, we multiply by the common ratio $r$, obtaining $a_n = a_{n-1}r$. Such formulas for the $n$th term of a sequence are said to be **recursive**. Generally speaking, a **recursively defined sequence** is one for which the first few terms are given and every successive term can be computed using a formula involving one or more of the preceding terms.

**EXAMPLE 9**    *Finding terms of a recursively defined sequence*

Find the first four terms of the sequence defined recursively by $a_1 = 3$ and $a_n = 2a_{n-1} + 5$ for $n > 1$.

**SOLUTION**

$$a_1 = 3$$
$$a_2 = 2a_1 + 5 = 2(3) + 5 = 11$$
$$a_3 = 2a_2 + 5 = 2(11) + 5 = 27$$
$$a_4 = 2a_3 + 5 = 2(27) + 5 = 59$$

**EXAMPLE 10**    *Finding terms of a recursively defined sequence*

Find the first six terms of the sequence defined recursively by $a_1 = 1$, $a_2 = 1$, and $a_n = a_{n-1} + a_{n-2}$ for $n > 2$.

**SOLUTION**

$$a_1 = 1$$
$$a_2 = 1$$
$$a_3 = a_2 + a_1 = 1 + 1 = 2$$
$$a_4 = a_3 + a_2 = 2 + 1 = 3$$
$$a_5 = a_4 + a_3 = 3 + 2 = 5$$
$$a_6 = a_5 + a_4 = 5 + 3 = 8$$

The recursively defined sequence 1, 1, 2, 3, 5, 8, 13, 21, 34, 55, 89, ... of the previous example is known as the **Fibonacci sequence**. It is named after the Italian mathematician Leonardo of Pisa, whose nickname was Fibonacci. The sequence, which first appeared in Fibonacci's book *Liber Abaci* (''The Book of the Abacus'') in 1202, can be described as the sequence for which the first two terms are 1 and each successive term is found by adding together the previous two terms. Despite its simplicity, the Fibonacci sequence has some fascinating mathematical properties. We will investigate some of these properties in Section 6.

The Fibonacci numbers occur with remarkable frequency in nature. Many species of flowers have petal arrangements consisting of 3, 5, 8, 13, 21, 34, 55, or 89 petals. The bracts of a pine cone spiral in sets of 8 and 13 rows. A sunflower has 34 petals, and its florets spiral out from the center in sets of 55 rows and 89 rows. The scales of a pineapple spiral in sets of 8, 13, and 21 rows.

The Fibonacci sequence is even present in the family tree of a male honey bee. As first observed by Johann Dzierzon in 1845, the male honey bee hatches from an unfertilized egg, and so has a mother but no father. Female honey bees, on the other hand, hatch from fertilized eggs, and so have a mother and a father. Thus, if we trace the family tree of the male honey bee, we see that each male "node" of the tree has only one parent branch (a mother) while each female "node" has two branches (a mother and a father). The number of nodes at each level of the family tree correspond to the Fibonacci numbers, as shown in Figure 3.

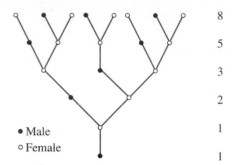

• Male
○ Female

**FIGURE 3**

8

5

3

2

1

1

---

## EXERCISES 1

**EXERCISES 1–16** ☐ *Write out the first four terms of the sequence with the given nth term.*

1. $a_n = 3n + 1$

2. $u_n = 4 - n$

3. $z_n = (-4)^n$

4. $c_n = -\dfrac{1}{2^n}$

5. $r_n = \dfrac{1}{n^2}$

6. $b_n = \dfrac{n - 1}{n + 1}$

7. $c_n = 1 + (-1)^n$

8. $x_n = (-1)^n n^2$

9. $v_n = \dfrac{(-1)^{n-1}}{n}$

10. $r_n = \dfrac{(-1)^{n+1}}{3^n}$

11. $a_n = \sqrt{n(n + 1)}$

12. $y_n = \sqrt{n + 1} - \sqrt{n}$

13. $b_n = \left(1 + \dfrac{1}{n}\right)^n$

14. $r_n = \dfrac{n^2}{2^n}$

15. $s_n = 1 + 2 + \cdots + n$

16. $a_n = 1 \cdot 2 \cdot \cdots \cdot n$

**EXERCISES 17–24** ☐ *Write an expression for the nth term of a sequence whose first few terms are given.*

17. $2, 4, 6, 8, \ldots$

18. $3, 9, 27, 81, \ldots$

19. $-\dfrac{1}{2}, \dfrac{1}{4}, -\dfrac{1}{8}, \dfrac{1}{16}, \ldots$

20. $3, 7, 11, 15, \ldots$

21. $\dfrac{1}{2}, \dfrac{2}{3}, \dfrac{3}{4}, \dfrac{4}{5}, \ldots$

22. $2 \cdot 3, 3 \cdot 4, 4 \cdot 5, 5 \cdot 6, \ldots$

23. $1 - \dfrac{1}{1}, 1 - \dfrac{1}{2}, 1 - \dfrac{1}{3}, 1 - \dfrac{1}{4}, \ldots$

24. $1 + \dfrac{1}{2}, 1 - \dfrac{1}{4}, 1 + \dfrac{1}{8}, 1 - \dfrac{1}{16}, \ldots$

**EXERCISES 25–34** ☐ *Determine whether the sequence appears to be arithmetic, geometric, or neither. If you claim it is arithmetic, find the common difference. If you think it is geometric, find the common ratio.*

25. $5, 8, 11, 14, \ldots$

26. $1, 3, 6, 10, \ldots$

27. $3, 12, 60, 360, \ldots$

28. $3, -6, 12, -24, \ldots$

29. $5, 1, -3, -7, \ldots$

30. $6, 2, \dfrac{4}{3}, \dfrac{8}{9}, \ldots$

31. $\dfrac{1}{2}, \dfrac{3}{4}, \dfrac{7}{8}, \dfrac{15}{16}, \ldots$

32. $\dfrac{4}{3}, \dfrac{8}{3}, 4, \dfrac{16}{3}, \ldots$

33. $0.2, 0.02, 0.002, 0.0002, \ldots$

34. $1, 4, 9, 16, \ldots$

**EXERCISES 35–40** ☐ *Write an expression for the nth term of the given arithmetic sequence.*

35. $7, 13, 19, 25, \ldots$

36. $8, 4, 0, -4, \ldots$

37. $100, 85, 70, 55, \ldots$

38. $2, 14, 26, 38, \ldots$

39. $-2, -\dfrac{3}{4}, \dfrac{1}{2}, \dfrac{7}{4}, \ldots$

40. $\dfrac{1}{2}, -\dfrac{1}{6}, -\dfrac{5}{6}, -\dfrac{3}{2}, \ldots$

**EXERCISES 41–46** ☐ *Use the given information about the arithmetic sequence to find the indicated term.*

41. $a_1 = 3, d = 8; a_{10}$

42. $a_1 = 22, d = -4; a_{18}$

43. $a_1 = 10, a_5 = -10; a_{15}$

44. $a_1 = 9, a_4 = 30; a_{21}$

45. $a_8 = 38, a_{17} = 92; a_{31}$

46. $a_{12} = -16, a_{20} = -40; a_{26}$

**EXERCISES 47–52** □ *Write an expression for the nth term of the given geometric sequence.*

**47.** 1, 3, 9, 27, …

**48.** 5, −10, 20, −40, …

**49.** 32, −16, 8, −4, …

**50.** $\dfrac{4}{3}, \dfrac{2}{3}, \dfrac{1}{3}, \dfrac{1}{6}, \dots$

**51.** $4, 5, \dfrac{25}{4}, \dfrac{125}{16}, \dots$

**52.** $-4, \dfrac{8}{3}, -\dfrac{16}{9}, \dfrac{32}{27}, \dots$

**EXERCISES 53–56** □ *Use the given information about the geometric sequence to find the indicated term.*

**53.** $a_1 = 2, r = 3;\ a_8$

**54.** $a_1 = 6, r = -\dfrac{1}{2};\ a_{10}$

**55.** $a_1 = 9, a_3 = 1;\ a_7$

**56.** $a_1 = \dfrac{1}{8}, a_3 = 2;\ a_6$

**EXERCISES 57–64** □ *Find the first five terms of the given recursively defined sequence.*

**57.** $a_1 = 2$ and $a_n = 1 - 2a_{n-1}$ for $n \geq 2$

**58.** $a_1 = 1$ and $a_n = 3a_{n-1} + 1$ for $n \geq 2$

**59.** $a_1 = 1$ and $a_n = na_{n-1}$ for $n \geq 2$

**60.** $a_1 = 1$ and $a_n = n + a_{n-1}$ for $n \geq 2$

**61.** $a_1 = 2$ and $a_n = \sqrt{a_{n-1}}$ for $n \geq 2$

**62.** $a_1 = 2$ and $a_n = \sqrt{(a_{n-1})^2 + 1}$ for $n \geq 2$

**63.** $a_1 = 1, a_2 = 2$, and $a_n = a_{n-1}a_{n-2}$ for $n \geq 3$

**64.** $a_1 = 1, a_2 = 2$, and $a_n = \dfrac{a_{n-1}}{a_{n-2}}$ for $n \geq 3$

■ *Applications*

**65.** *Projecting Income* Juanita is considering a new job that offers a starting base salary of $28,500, a guaranteed $1200 raise each year, and an annual bonus equal to 3% of the base salary. Determine her potential income (salary plus bonus) for each of the first three years and show that these values form an arithmetic sequence. What would her income be in the seventh year?

**66.** *Projecting Income* Vincent is considering a new job that offers a starting salary of $27,000 and a guaranteed $1500 raise each year. The benefit package (health, retirement, etc.) is 11% of the annual salary. Determine the total value (salary plus benefits) for each of the first three years and show that these totals form an arithmetic sequence. What would the total be in the eighth year?

**67.** *Membership Increase* An organization claims that its membership has increased 20% each year since it was formed with 10 charter members. How many members are there at the beginning of the twelfth year?

**68.** *Salary Increase* If you start a job with an annual salary of $30,000 and a guaranteed 5% raise each year, what will your salary be for the fifteenth year?

**69.** *Accumulated Value* Suppose $1000 is deposited in an account in which interest is compounded annually. The balances at the beginning of each of the first 4 years are $1000.00, $1055.00, $1113.03, and $1174.25. Assuming this pattern continues, write out a formula for the *n*th term of this sequence. What will the balance be at the beginning of the tenth year? What is the annual interest rate?

**70.** *Amortization Schedule* The first four entries on the amortization schedule for a $120,000 mortgage indicate monthly interest charges of $800.00, $799.46, $798.92, and $798.38. Write out a formula for the *n*th term of this sequence. How much interest will be charged in the 24th month?

**71.** *Breeding Rabbits* A pair of baby rabbits (one male and one female) is placed inside an enclosed area for the purposes of breeding. During the first month, the pair does not produce any new rabbits, but each month thereafter it produces a pair (one male and one female) of rabbits. Assuming each new pair follows the same reproductive pattern, complete the first six rows of the following table and find the number of rabbits at the end of 12 months. (*Hint:* Because of the 1-month maturation period, the number of pairs of rabbits born in a given month equals the number of pairs of rabbits at the beginning of the previous month.)

| Month | Pairs at the beginning of the month | Pairs born during the month | Pairs at the end of the month |
|---|---|---|---|
| 1 | 1 | 0 | 1 |
| 2 | 1 | 1 | 2 |
| 3 | 2 | 1 | 3 |
| 4 | 3 | | |

■ *Projects for Enrichment*

**72.** *Limiting Behavior of Sequences* Although a formal definition of the limit of a sequence is beyond the scope of this text, we will say that a sequence $a_1, a_2, a_3, \ldots$ has a limit $L$ if the terms $a_n$ approach $L$ (and only $L$) as $n$ increases without bound. In this case, we write $a_n \to L$ as $n \to \infty$. As an example, consider the sequence $a_n = 1/n$ whose first few terms are $1, \frac{1}{2}, \frac{1}{3}, \frac{1}{4}, \ldots$. Clearly, the terms of this sequence are getting closer and closer to zero. Thus, we conclude that $a_n \to 0$ as $n \to \infty$. It is sometimes possible to predict the limit of a sequence in this way, by inspecting the first "few" terms of the sequence.

**a.** Write out a sufficient number of terms for the following sequences and predict the limit.

    **i.** $a_n = \dfrac{1}{n^2}$      **ii.** $a_n = \left(\dfrac{1}{2}\right)^n$

    **iii.** $a_n = \dfrac{n}{n+1}$      **iv.** $a_n = \dfrac{2^n}{2^n - 1}$

Note that the method described above can also be very unreliable. A sequence may converge to a limit so slowly that it is difficult to tell what the limit is.

**b.** Let

$$a_n = \frac{\ln n}{\sqrt{n}}$$

    Find $a_n$ for $n = 10$, $n = 20$, $n = 50$, $n = 100$, and $n = 1000$. What do you think the limit is?

Not all sequences have a limit. For example, the terms of the arithmetic sequence $5, 8, 11, 14, \ldots$ increase without bound and so do not approach any number. The terms of the sequence $1, -1, 1, -1, 1, \ldots$ simply alternate between $1$ and $-1$ and so do not approach a *single* number. In order for a sequence to have a limit, its terms must approach a unique finite number.

**c.** Decide whether or not the following sequences have a limit. For those that do, predict the limit. For those that do not, explain why not.

    **i.** $a_n = \dfrac{1}{2}n$      **ii.** $a_n = \dfrac{1}{n} + 5$

    **iii.** $a_n = (-1)^n + 1$      **iv.** $a_n = \dfrac{(-1)^n}{n}$

    **v.** $a_n = \left(\dfrac{3}{2}\right)^n$      **vi.** $a_n = e^{-n}$

Sometimes a graphics calculator can be a useful tool for studying limits of sequences. Consider the sequence

$$a_n = \frac{n}{\sqrt{2n^2 + 1}}$$

If we define the function

$$f(x) = \frac{x}{\sqrt{2x^2 + 1}}$$

then $f(n) = a_n$ for all $n$. To investigate the limit of $a_n$, we plot the graph of $f$ in Figure 4 and use the trace feature to see what value $f(x)$ approaches as $x$ gets very large. The cursor shows that $f(16) \approx 0.706$ and so $a_{16} \approx 0.706$. Since the graph does not appear to be rising further, we conclude that the limit is approximately $0.706$. Using calculus, we could show that the exact limit is $\sqrt{2}/2$.

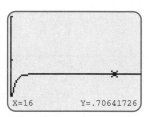

    X=16        Y=.70641726

**FIGURE 4**

**d.** Use a graphics calculator to estimate the limits of the following sequences.

    **i.** $a_n = \dfrac{2n}{3n + 1}$

    **ii.** $a_n = \dfrac{4n^2 - 2n}{7n^2 + 3}$

    **iii.** $a_n = \left(1 + \dfrac{1}{n}\right)^n$

**e.** Experiment with different viewing screens on a graphics calculator to investigate the limit of the sequence

$$a_n = \frac{0.001n^2 + n}{n + 1000}$$

What can you conclude? What potential hazards can you see in using a graphics calculator for estimating limits?

**73.** *Population Growth* Suppose a certain city has population $P_0$ and a constant rate of growth of $k$ people per thousand per year. In other words, for each group of 1000 people in the population, there will be an increase of $k$ people after 1 year. We will see in this project that constant growth rates lead to geometric sequences.

**a.** In 1990, Mesa, Arizona, had a population of approximately 288,000 and a growth rate of approximately 66 people per thousand per year. Estimate the population in 1991 and 1992.

b. Argue more generally that if the growth rate is $k$ and there are $P_n$ people at the beginning of year $n$, then there will be $P_n + k(P_n/1000)$ people at the end of year $n + 1$.

c. Rewrite the expression in part (b) to show that if $P_n$ denotes the number of people at the beginning of year $n$, then $P_{n+1} = (1 + 0.001k)P_n$.

d. Use $P_0 = 288{,}000$ (the population in 1990) and $k = 66$ to estimate the population of Mesa, Arizona, in each of the

years 1991–1995. Explain why this sequence is a geometric sequence. Estimate the population of Mesa in the year 2010.

e. In 1990, Gary, Indiana, had a population of approximately 117,000 and a growth rate of $-26$ people per thousand. Estimate the population in each of the years 1991–1995. Are the values forming a geometric sequence? Why or why not? What do you notice about the population of Gary? How should we interpret a negative growth rate?

---

### ■ *Questions for Discussion or Essay*

**74.** Suppose the instructions for a set of algebra problems reads "Find an expression for the $n$th term of *the* sequence whose first few terms are given." Explain why these are not valid instructions. To help you see the difficulty, compute the first four terms of the following sequences, and answer the question, "Why is 10 not necessarily the next number in the sequence 1, 4, 7, …?"

$$a_n = 3n - 2$$
$$a_n = n^3 - 6n^2 + 14n - 8$$

**75.** The $n$th term of an arithmetic or geometric sequence can be defined either *explicitly* or *recursively*. For example, the $n$th term of the arithmetic sequence 4, 10, 16, 22, … can be defined explicitly by $a_n = 6n - 2$ for $n = 1, 2, 3, \ldots$ or recursively by

$a_1 = 4$ and $a_n = a_{n-1} + 6$ for $n = 2, 3, \ldots$. Find both explicit and recursive definitions for the geometric sequence 5, 15, 45, …. Discuss the advantages and disadvantages of each definition.

**76.** In many of the application problems involving geometric sequences, we assumed that certain percentage rates of increase or decrease (e.g., membership increase, salary increase, interest rate, population growth or decline) were constant from one year to the next, and this allowed us to make projections well into the future. Which, if any, of these assumptions are valid? Justify your answer. If a rate cannot be assumed to be constant, are the projections of any use? Explain. Give some examples from the "real world" that involve rates of increase or decrease that could be assumed to be constant.

---

## SECTION 2

## SERIES

■ How would a 10-year-old child prodigy add up the first 100 positive integers?

■ Would you buy a $9600 car on the following terms: no money down and 48 monthly payments of $200 plus 2% per month interest on the remaining balance?

■ If you had a choice between a job with a starting salary of $32,000, a yearly raise of $3000, and a 4% year-end bonus, and one with a starting salary of $32,000 and a 6% yearly raise, which would you choose?

■ How is it that an early Greek philosopher could make the ridiculous claim that motion was impossible, and yet a simple refutation of his argument took centuries to develop?

■ If you were to win the grand prize in a sweepstakes, should you take the $1,000,000 lump sum payment or the annuity option of $100,000 each year for the next 20 years?

## SERIES AND SUMMATION NOTATION

The story is told that when the German mathematician Karl Frederich Gauss was only 10 years old, he amazed his teacher by adding together the first 100 positive integers in just seconds. Historians speculate that Gauss accomplished this feat by observing that the numbers in the sum $1 + 2 + 3 + \cdots + 98 + 99 + 100$ can be grouped in pairs that add up to 101, as follows.

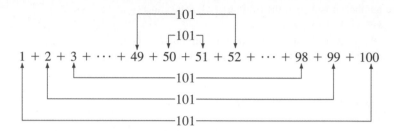

Since there are 50 such pairs, the total is simply $50(101) = 5050$. The sum $1 + 2 + 3 + \cdots + 100$ is an example of a **series**. In general, a series is a sum of a sequence of numbers, as is indicated by the following definition.

**Definition of series**

> Given an infinite sequence $a_1, a_2, a_3, \ldots, a_n, \ldots$, the sum
>
> $$a_1 + a_2 + a_3 + \cdots + a_n$$
>
> is called a **finite series**, while the sum
>
> $$a_1 + a_2 + a_3 + \cdots + a_n + \cdots$$
>
> is called an **infinite series**. Each $a_j$ is called a **term** of the series.

**EXAMPLE 1**    *Evaluating a series*

Evaluate the series $a_1 + a_2 + \cdots + a_6$ where $a_n = \dfrac{1}{n}$.

**SOLUTION**

$$\begin{aligned}
a_1 + a_2 + \cdots + a_6 &= \frac{1}{1} + \frac{1}{2} + \frac{1}{3} + \frac{1}{4} + \frac{1}{5} + \frac{1}{6} \\
&= \frac{60 + 30 + 20 + 15 + 12 + 10}{60} \\
&= \frac{147}{60}
\end{aligned}$$

Since series involving many terms can be awkward to write, we introduce a convenient shorthand notation called **summation notation**.

**■** *Summation notation*

$$\sum_{k=1}^{n} a_k = a_1 + a_2 + \cdots + a_n$$

$$\sum_{k=1}^{\infty} a_k = a_1 + a_2 + a_3 + \cdots$$

The symbol $\Sigma$ is the capital Greek letter **sigma**. The variable $k$ is called the **index variable**. The expression $k = 1$ below the sigma defines the starting value of the index variable, otherwise known as the **lower limit**. The ending value of the index variable—the **upper limit**—is given above the sigma.

**EXAMPLE 2**   *Using summation notation*

Evaluate the series $\displaystyle\sum_{k=1}^{5} a_k$ where $a_k = 2^k$.

**SOLUTION**

$$\sum_{k=1}^{5} a_k = a_1 + a_2 + a_3 + a_4 + a_5$$

$$= 2^1 + 2^2 + 2^3 + 2^4 + 2^5$$

$$= 2 + 4 + 8 + 16 + 32$$

$$= 62$$

The series in the previous example can be defined much more concisely as $\sum_{k=1}^{5} 2^k$. Here it is understood that we are adding the first five terms of the sequence defined by $a_k = 2^k$.

**EXAMPLE 3**   *Evaluating a series*

Evaluate the series $\displaystyle\sum_{k=1}^{6} k^2$.

**SOLUTION**

$$\sum_{k=1}^{6} k^2 = 1^2 + 2^2 + 3^2 + 4^2 + 5^2 + 6^2$$

$$= 1 + 4 + 9 + 16 + 25 + 36$$

$$= 91$$

**EXAMPLE 4**     *Evaluating a series*

Evaluate the series $\sum\limits_{i=2}^{4} \left(-\frac{1}{10}\right)^i$.

**SOLUTION**     Note that here we have a starting index of 2.

$$\sum_{i=2}^{4} \left(-\frac{1}{10}\right)^i = \left(-\frac{1}{10}\right)^2 + \left(-\frac{1}{10}\right)^3 + \left(-\frac{1}{10}\right)^4$$

$$= \frac{1}{100} - \frac{1}{1000} + \frac{1}{10,000}$$

$$= \frac{100 - 10 + 1}{10,000}$$

$$= \frac{91}{10,000}$$

**EXAMPLE 5**     *Writing a series with summation notation*

Write the series $1 + 4 + 9 + 16 + 25 + 36 + 49 + 64 + 81$ using summation notation.

**SOLUTION**     The terms of this series are the squares of the numbers 1–9. Thus, the general $k$th term has the form $a_k = k^2$. The first term corresponds to $k = 1$, and the last term corresponds to $k = 9$. Thus, the lower and upper limits are 1 and 9, respectively. In summation notation we have

$$\sum_{k=1}^{9} k^2$$

**EXAMPLE 6**     *Writing a series with summation notation*

Write the series $\frac{1}{3} + \frac{1}{9} + \frac{1}{27} + \frac{1}{81} + \cdots$ using summation notation.

**SOLUTION**     The denominators are successive powers of 3, starting with $3^1$. Thus, the general $k$th term has the form $a_k = 1/3^k$. The first term corresponds to $k = 1$, and so the lower limit is 1. Since the series is infinite, the upper limit is $\infty$. In summation notation we have

$$\sum_{k=1}^{\infty} \frac{1}{3^k}$$

Note that Examples 5 and 6 were not concerned with actually evaluating the given series. We now turn our attention to techniques for evaluating certain finite series. Infinite series, such as that in Example 6, will be considered in Exercise 67.

## ARITHMETIC SERIES

Recall from Section 1 that an arithmetic sequence is one whose successive terms have a common difference. The sum of the terms of an arithmetic sequence is called an

**arithmetic series.** We would like to develop a formula for evaluating *finite* arithmetic series. To that end, suppose $a_1, a_2, a_3, \ldots, a_n$ is an arithmetic sequence with common difference $d$, and let the sum $S$ be given by

$$S = a_1 + a_2 + \cdots + a_{n-1} + a_n$$

Since $a_k = a_1 + (k - 1)d$, it follows that

(1) $\qquad S = a_1 + [a_1 + d] + \cdots + [a_1 + (n - 2)d] + [a_1 + (n - 1)d]$

By writing the terms of equation (1) in reverse order, we also have

(2) $\qquad S = [a_1 + (n - 1)d] + [a_1 + (n - 2)d] + \cdots + [a_1 + d] + a_1$

Adding equations (1) and (2) leads to the following.

$$
\begin{aligned}
S &= a_1 & &+ [a_1 + d] & &+ \cdots + [a_1 + (n - 2)d] & &+ [a_1 + (n - 1)d] \\
+ \quad S &= [a_1 + (n - 1)d] & &+ [a_1 + (n - 2)d] & &+ \cdots + [a_1 + d] & &+ \quad a_1 \\
\hline
2S &= [2a_1 + (n - 1)d] & &+ [2a_1 + (n - 1)d] & &+ \cdots + [2a_1 + (n - 1)d] & &+ (2a_1 + (n - 1)d)
\end{aligned}
$$

$$\underbrace{\hphantom{2S = [2a_1 + (n - 1)d] + [2a_1 + (n - 1)d] + \cdots + [2a_1 + (n - 1)d] + (2a_1 + (n - 1)d)}}_{n \text{ terms}}$$

$$
\begin{aligned}
2S &= n[2a_1 + (n - 1)d] \\
2S &= n[a_1 + a_1 + (n - 1)d] \\
2S &= n(a_1 + a_n) & &\text{Substituting } a_n \text{ for } a_1 + (n - 1)d \\
S &= \frac{n}{2}(a_1 + a_n) & &\text{Dividing both sides by 2}
\end{aligned}
$$

Thus we have the following formula.

---

**Formula for finite arithmetic series**

> If $a_1, a_2, \ldots, a_n$ is an arithmetic sequence, then
>
> $$a_1 + a_2 + \cdots + a_n = \frac{n}{2}(a_1 + a_n)$$

The formula for finite arithmetic series suggests another approach to the problem solved by the 10-year-old Gauss.

**EXAMPLE 7**    *Using the formula for finite arithmetic series*

Find the sum of the first 100 positive integers.

**SOLUTION**    The series $1 + 2 + 3 + \cdots + 100$ is a finite arithmetic series. Thus, applying the formula for finite arithmetic series with $n = 100$, $a_1 = 1$, and $a_{100} = 100$, we obtain

$$1 + 2 + 3 + \cdots + 100 = \frac{100}{2}(1 + 100) = 5050$$

**EXAMPLE 8**     *Evaluating an arithmetic series*

Evalute $\sum\limits_{k=1}^{40} (3k + 5)$.

**SOLUTION**     The first term of the series is $3(1) + 5 = 8$. The 40th term is $3(40) + 5 = 125$. Thus,

$$\sum_{k=1}^{40} (3k + 5) = 8 + 11 + 14 + \cdots + 125$$

$$= \frac{40}{2} (8 + 125)$$

$$= 2660$$

**EXAMPLE 9**     *An unusual financing arrangement*

A1 Used Cars is offering special financing. With no money down and approved credit, you can drive away in a $9600 car and pay the balance in 48 monthly payments of only $200, plus 2% interest per month on the remaining balance. How much interest will you pay over 4 years?

**SOLUTION**     The interest payments can be viewed as the terms of an arithmetic sequence, as shown in the following table.

| *Payment number* | *Balance before payment* | *Interest charged* |
|:---:|:---:|:---:|
| 1 | $9600 | $(0.02)9600 = 192$ |
| 2 | $9400 | $(0.02)9400 = 188$ |
| 3 | $9200 | $(0.02)9200 = 184$ |
| ⋮ | ⋮ | ⋮ |
| 48 | $200 | $(0.02)200 = \phantom{00}4$ |

Applying the formula for finite arithmetic series, we obtain

$$192 + 188 + 184 + \cdots + 4 = \frac{48}{2} (192 + 4) = 4704$$

Thus, the total interest paid over 4 years would be $4704. An unusual feature of this financing scheme is that the monthly payments decrease over the life of the loan since the amount of interest decreases each month.

**GEOMETRIC SERIES**

In Section 1 we defined a geometric sequence as one whose successive terms have a common ratio. The sum of the terms of a geometric sequence is called a **geometric series**. Just as we did with arithmetic series, we would like to develop a formula for evaluating geometric series. Thus, let $a_1, a_2, \ldots, a_n$ be a geometric sequence with common ratio $r$, and let the sum $S$ be given by

$$S = a_1 + a_2 + \cdots + a_{n-1} + a_n$$

Since $a_k = a_1 r^{k-1}$, we have

(3)             $$S = a_1 + a_1 r + a_1 r^2 + \cdots + a_1 r^{n-2} + a_1 r^{n-1}$$

If we multiply equation (3) through by $r$, we obtain

(4)             $$rS = a_1 r + a_1 r^2 + a_1 r^3 + \cdots + a_1 r^{n-1} + a_1 r^n$$

Subtracting equation (4) from equation (3) simplifies as follows.

$$S - rS = a_1 - a_1 r^n$$

$$S(1 - r) = a_1(1 - r^n)$$

$$S = \frac{a_1(1 - r^n)}{1 - r} \qquad \text{Assuming } r \neq 1$$

Thus, we have the following formula.

**Formula for finite geometric series**

> If $a_1, a_2, a_3, \ldots, a_n$ is a geometric sequence with common ratio $r$ (i.e., $a_k = a_1 r^{k-1}$), and $r \neq 1$, then
>
> $$a_1 + a_2 + \cdots + a_n = \frac{a_1(1 - r^n)}{1 - r}$$

**EXAMPLE 10**

*Finding the sum of a geometric sequence*

Find the sum of the first 12 terms of the geometric sequence 5, 2, 0.8, 0.32, ....

**SOLUTION**    The formula for finite geometric series requires the first term, the common ratio, and the number of terms. The first term is $a_1 = 5$. The common ratio is

$$r = \frac{a_2}{a_1} = \frac{2}{5} = 0.4$$

The number of terms is $n = 12$. Thus,

$$a_1 + a_2 + \cdots + a_{12} = \frac{a_1(1 - r^n)}{1 - r}$$

$$= \frac{5[1 - (0.4)^{12}]}{1 - 0.4}$$

$$\approx 8.33319$$

---

**RULE OF THUMB**     As suggested by Example 10, it is helpful to think of $a_1$ and $n$ in the expression

$$\frac{a_1(1 - r^n)}{1 - r}$$

as simply the first term and the number of terms, respectively.

---

**EXAMPLE 11**     *Evaluating a geometric series*

Evalute $\sum\limits_{k=3}^{10} 4(-\frac{1}{3})^k$.

**SOLUTION**     The first term of the series is $4(-\frac{1}{3})^3 = -\frac{4}{27}$. Since the starting index is 3 and the ending index is 10, there are 8 terms in the sum. Finally, the common ratio is $-\frac{1}{3}$. Thus,

$$\sum_{k=3}^{10} 4\left(-\frac{1}{3}\right)^k = \frac{-\dfrac{4}{27}\left[1 - \left(-\dfrac{1}{3}\right)^8\right]}{1 - \left(-\dfrac{1}{3}\right)} \approx -0.111094$$

---

**EXAMPLE 12**     *Comparing contracts*

Suppose you are starting a new job that offers a choice between two contracts. The first has a starting base salary of $32,000, yearly increases of $3000 in the base salary, and a year-end bonus equal to 4% of the base salary for that year. The second has a starting salary of $32,000 and a 6% raise each year. Which job would pay the highest total salary over a 26-year period?

**SOLUTION**     The following table shows the total income for both contract options for the first 4 years.

| Year | Option 1 Bonus and fixed raise | | | Option 2 Percent raise |
|---|---|---|---|---|
| | *Base salary* | *Bonus of 4%* | *Total* | *Salary (including 6% raise)* |
| 1 | 32,000 | (0.04)32,000 = 1,280 | 33,280 | 32,000 |
| 2 | 35,000 | (0.04)35,000 = 1,400 | 36,400 | 32,000 + (0.06)32,000 = 33,920 |
| 3 | 38,000 | (0.04)38,000 = 1,520 | 39,520 | 33,920 + (0.06)33,920 = 35,955.20 |
| 4 | 41,000 | (0.04)41,000 = 1,640 | 42,640 | 35,955.20 + (0.06)35,955.20 = 38,112.51 |

The salary totals for option 1 form an arithmetic sequence with common difference $d = 36{,}400 - 33{,}280 = 3120$. Using the formula for the $n$th term of an arithmetic sequence, the salary for year 26 will be $a_{26} = 33{,}280 + (26 - 1)3120 = 111{,}280$.

Now we apply the formula for finite arithmetic series to determine the sum of the salaries for 26 years.

$$33{,}280 + 36{,}400 + \cdots + 111{,}280 = \frac{26}{2}(33{,}280 + 111{,}280)$$

$$= 1{,}879{,}280$$

The salaries for option 2 form a geometric sequence with common ratio $r = 33{,}920/32{,}000 = 1.06$. Thus, using the formula for finite geometric series, we obtain the following salary total.

$$32{,}000 + 33{,}920 + \cdots + (\text{salary for year } 26) = \frac{32{,}000(1 - 1.06^{26})}{1 - 1.06}$$

$$\approx 1{,}893{,}004.25$$

So even though option 1 starts out with larger salaries, option 2 has a slightly higher total.

## EXERCISES 2

**EXERCISES 1–4** □ *Find the sum of the first five terms of the sequence.*

**1.** $a_n = 4n + 3$

**2.** $a_n = n^2$

**3.** $a_n = \dfrac{n}{n + 1}$

**4.** $a_n = \dfrac{4}{10^n}$

**EXERCISES 5–14** □ *Find the indicated sum.*

**5.** $\displaystyle\sum_{k=1}^{6} (3k - 2)$

**6.** $\displaystyle\sum_{k=1}^{4} (1 - 5k)$

**7.** $\displaystyle\sum_{k=1}^{5} -3$

**8.** $\displaystyle\sum_{k=1}^{4} 8$

**9.** $\displaystyle\sum_{k=1}^{4} 2k^2$

**10.** $\displaystyle\sum_{k=2}^{5} k^3$

**11.** $\displaystyle\sum_{k=3}^{6} \dfrac{1}{k}$

**12.** $\displaystyle\sum_{k=0}^{4} \dfrac{1}{k + 1}$

**13.** $\displaystyle\sum_{k=0}^{4} 5\left(\dfrac{1}{2}\right)^k$

**14.** $\displaystyle\sum_{k=1}^{5} 3(-2)^k$

**EXERCISES 15–24** □ *Write the series using summation notation.*

**15.** $\dfrac{1}{1^3} + \dfrac{1}{2^3} + \dfrac{1}{3^3} + \dfrac{1}{4^3} + \dfrac{1}{5^3} + \dfrac{1}{6^3}$

**16.** $\dfrac{1}{3^1} + \dfrac{1}{3^2} + \dfrac{1}{3^3} + \dfrac{1}{3^4} + \dfrac{1}{3^5}$

**17.** $[5(1) + 2] + [5(2) + 2] + [5(3) + 2] + \cdots + [5(12) + 2]$

**18.** $[7(1)^2 - 1] + [7(2)^2 - 1] + [7(3)^2 - 1] + \cdots + [7(15)^2 - 1]$

**19.** $3 + 9 + 27 + \cdots + 2187$

**20.** $2 - 4 + 8 - 16 + \cdots - 128$

**21.** $\dfrac{1}{1} - \dfrac{1}{4} + \dfrac{1}{9} - \dfrac{1}{16} + \cdots - \dfrac{1}{256}$

**22.** $\dfrac{1}{2} + \dfrac{2}{3} + \dfrac{3}{4} + \cdots + \dfrac{20}{21}$

**23.** $\dfrac{1}{2} + \dfrac{3}{4} + \dfrac{7}{8} + \cdots + \dfrac{63}{64}$

**24.** $\dfrac{1}{1 \cdot 2} + \dfrac{1}{2 \cdot 3} + \dfrac{1}{3 \cdot 4} + \cdots + \dfrac{1}{15 \cdot 16}$

**EXERCISES 25–36** □ *Evaluate the given arithmetic series.*

**25.** $6 + 14 + 22 + \cdots + 158$  (20 terms)

**26.** $10 + 16 + 22 + \cdots + 154$  (25 terms)

**27.** $2 + 11 + 20 + \cdots + 128$

**28.** $20 + 19.4 + 18.8 + \cdots + 11$

**29.** $a_1 + a_2 + \cdots + a_{15}$, where $a_1 = 12$ and $a_2 = 8$

**30.** $a_1 + a_2 + \cdots + a_{18}$, where $a_1 = -3$ and $a_2 = 4$

**31.** $\displaystyle\sum_{k=1}^{50} (3k - 2)$

**32.** $\displaystyle\sum_{k=1}^{60} (8k + 2)$

**33.** $\sum_{k=0}^{30} (200 - 6k)$

**34.** $\sum_{k=0}^{80} (10 - 0.01k)$

**43.** $\sum_{k=1}^{8} 3(4^k)$

**44.** $\sum_{k=1}^{11} 10(0.6)^k$

**35.** $\sum_{k=10}^{25} \dfrac{5k + 3}{2}$

**36.** $\sum_{k=5}^{35} \left(\dfrac{1}{2}k + 3\right)$

**45.** $\sum_{k=0}^{9} 2\left(-\dfrac{1}{6}\right)^k$

**46.** $\sum_{k=0}^{8} \dfrac{1}{2}(-2)^k$

**EXERCISES 37–48** □ *Evaluate the given geometric series.*

**37.** The sum of the first 10 terms of the sequence 5, 15, 45, …

**38.** The sum of the first 12 terms of the sequence 20, 4, 0.8, …

**39.** $a_1 + a_2 + \cdots + a_{15}$, where $a_1 = 3$ and $a_2 = 1$

**40.** $a_1 + a_2 + \cdots + a_{10}$, where $a_1 = -2$ and $a_2 = 4$

**41.** $4 - 0.4 + 0.04 - 0.004 + \cdots - 0.0000004$

**42.** $\dfrac{1}{25} + \dfrac{1}{5} + 1 + \cdots + 78{,}125$

**47.** $\sum_{k=5}^{16} -3(0.2)^k$

**48.** $\sum_{k=3}^{12} 9\left(\dfrac{2}{3}\right)^k$

**EXERCISES 49–52** □ *Find the sum of the indicated sequence of integers.*

**49.** The first 1000 positive integers

**50.** The first 200 positive even integers

**51.** The first 100 positive odd integers.

**52.** The first 100 positive multiples of 3

■ *Applications*

**53.** *Pipe Pile*  A pile of sewer pipes has 12 layers. The first layer has 30 pipes, the second has 29, the third has 28, and so on, as shown in Figure 5. How many pipes are in the pile?

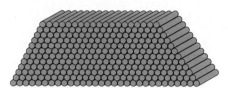

**FIGURE 5**

**54.** *Brick Laying*  A bricklayer estimates that in an 8-hour shift she can complete 15 rows of bricks in a portion of a wall near the peak of a roof, as shown in Figure 6. The first row has 24 bricks, the second has 23 bricks, and so on. How many bricks will she need that day?

**FIGURE 6**

**55.** *Auditorium Seats*  An auditorium has 25 rows of seats. The first row has 12 seats, the second row has 14 seats, the third row has 16 seats, and so on, as shown in Figure 7. How many seats are in the auditorium?

**FIGURE 7**

**56.** *Building Height*  An object is dropped from the top of the First Interstate World Center in Los Angeles. During the first second, it falls 16 feet; during the second second, it falls 48 feet; during the third second, it falls 80 feet; and so on. The object hits the ground after approximately 8 seconds. How high is the World Center?

**57.** *Advertising Expenditures*  The annual advertising expenditures by U.S. companies for the years 1980–1990 can be approximated with the model $a_n = 7.7n + 55$, where $n$ denotes the year (with $n = 0$ corresponding to 1980), and $a_n$ is the expenditure for that year (in billions of dollars). (Data Source: *Advertising Age*.) Find the total of all advertising expenditures for 1980 through 1990.

**58.** *Government Expenditures*  The annual expenditures by the federal government for the years 1980–1992 can be approximated with the model $a_n = 69.3n + 580$, where $n$ is the year (with $n = 0$ corresponding to 1980) and $a_n$ is the outlay for that year (in billions of dollars). (Data Source: U.S. Office of Management and Budget.) Find the total of all expenditures for 1980–1992.

**59.** *Accumulated Value*  Jamal deposits $100 at the beginning of each month in an account that pays 5% interest compounded monthly. Evaluate the following geometric series to find the balance in his account at the end of 4 years.

(continued)

$$B = 100 \left(1 + \frac{0.05}{12}\right)^1 + 100 \left(1 + \frac{0.05}{12}\right)^2$$
$$+ 100 \left(1 + \frac{0.05}{12}\right)^3 + \cdots + 100 \left(1 + \frac{0.05}{12}\right)^{48}$$

**60.** *Credit Card Debt*  Todd has run up a $2000 debt on a credit card that charges interest on the unpaid balance at a rate of 1.5% per month. He resolves to stop using the card and, 1 month later, begins making monthly payments of $50. His balance after the sixtieth payment can be found using the following formula. What is his balance?

$$B = 2000(1.015)^{60} - [50 + 50(1.015)^1$$
$$+ 50(1.015)^2 + \cdots + 50(1.015)^{59}]$$

**61.** *Automobile Expenses*  The fixed costs (insurance, license, depreciation, etc.) for owning and operating a car for the years 1980–1990 can be approximated with the model $a_n = 1960(1.06)^n$, where $n$ denotes the year (with $n = 0$ corresponding to 1980) and $a_n$ is the cost in dollars. (Data Source: *Motor Vehicle Facts and Figures*, Motor Vehicle Manufacturers Association of the United States, Detroit.) Find the total cost for owning and operating a car from 1980–1990.

**62.** *Mail Delivery*  The number of pieces of mail delivered by the U.S. Postal Service from 1970 to 1990 can be approximated with the model $a_n = 82(1.036)^n$, where $n$ denotes the year (with $n = 0$ corresponding to 1970) and $a_n$ is the number of

pieces of mail delivered (in billions). (Data Source: U.S. Postal Service.) Find the total number of pieces of mail delivered from 1970 through 1990.

**63.** *Loan Repayment*  Molly has just borrowed $2400 from her Uncle Dave and agrees to pay it back in 24 monthly payments of $100, plus 0.5% interest per month on the remaining balance. How much interest will Molly have to pay?

**64.** *Rumor Mill*  Suppose a disgruntled accountant in a large corporation begins his work day by telling four co-workers that the president of the corporation has been embezzling. After 1 hour, each of these four has spread the rumor to four of their friends. After the second hour, each of those has told four more, and so on. How many people have heard the rumor after 8 hours? If this pattern continues through the second day, how many will have heard the rumor after 16 hours?

**65.** *Contract Value*  Suppose a college basketball star signs a 15-year contract to play professional basketball. If the first year's salary is $1.6 million and the salary for each of the following 14 years is 10% more than the preceding year's, how much is the total contract worth?

**66.** *Job Options*  If you plan to stay with a job for 30 years, would you choose a job with a starting salary of $28,000, a yearly bonus of 3%, and a guaranteed annual raise of $4000, or one with a starting salary of $36,000 and a guaranteed raise of 4% per year? Explain.

---

◼ **Projects for Enrichment**

---

**67.** *Zeno's Paradox and Infinite Geometric Series*  Around 460 B.C., the Greek philosopher Zeno of Elea posed a paradox that has intrigued philosophers and mathematicians for centuries. He argued that motion is paradoxical in that it is seemingly not possible to travel a finite distance in a finite amount of time. In order to do so, one would first have to travel half of the distance, then half of the remaining distance, then half of the remaining distance, and so on, as suggested in Figure 8.

$A$                                              $B$

**FIGURE 8**

Thus, one would have to cover an infinite number of distances, which is clearly not possible in a finite amount of time. Or is it? Suppose the distance between $A$ and $B$ is 1 mile and you start walking from $A$ to $B$ at a rate of 1 mile per hour. At that rate, you will cover the first half mile in $\frac{1}{2}$ hour, the next fourth mile in $\frac{1}{4}$ hour, the next eighth mile in $\frac{1}{8}$ hour, and so on. Thus, your total time in traveling from $A$ to $B$ is given by the infinite sum

$$\frac{1}{2} + \frac{1}{4} + \frac{1}{8} + \cdots$$

Let us define $S_n$ to be the sum of the first $n$ terms of the infinite series given above. Then

$$S_n = \frac{1}{2} + \frac{1}{4} + \cdots + \left(\frac{1}{2}\right)^n = \sum_{k=1}^{n} \left(\frac{1}{2}\right)^k$$

**a.** Find $S_1$, $S_2$, $S_3$, and $S_4$ and explain what each represents.

**b.** Use the formula for finite geometric series to show that $S_n = 1 - (1/2^n)$.

**c.** Compute $S_5$, $S_{10}$, $S_{20}$, and $S_{40}$. What are these values approaching? Since each of these values is the sum of more and more terms of the infinite sum $\frac{1}{2} + \frac{1}{4} + \frac{1}{8} + \cdots$, what would be a reasonable guess for the value of the infinite sum? According to this value, how much time will it take to travel from $A$ to $B$? Does this time agree with the amount of time suggested by the standard formula $d = rt$?

Using calculus, a tool developed centuries after Zeno, it is possible to show that if $|r| < 1$, then the sum $S$ of the infinite geometric series $a + ar + ar^2 + ar^3 + \cdots$ is given by

$$S = \frac{a}{1 - r}$$

**d.** Use this formula to verify the value you found in part (c).

**e.** Find the sum of the following infinite geometric series.

    **i.** $\dfrac{1}{4} + \dfrac{1}{16} + \dfrac{1}{64} + \cdots$

    **ii.** $\dfrac{2}{3} + \dfrac{4}{9} + \dfrac{8}{27} + \cdots$

    **iii.** $\dfrac{1}{10} - \dfrac{1}{100} + \dfrac{1}{1000} - \dfrac{1}{10{,}000} + \cdots$

**68.** *Telescoping Series* The term *telescoping series* is used to describe a series for which much cancellation occurs in the summation. Consider, for example, the series

$$\sum_{k=1}^{5} \left( \frac{1}{k} - \frac{1}{k+1} \right)$$

If we write out the terms of this series, we see that everything cancels except part of the first term and part of the last.

$$\sum_{k=1}^{5} \left( \frac{1}{k} - \frac{1}{k+1} \right) = \left( \frac{1}{1} - \frac{1}{2} \right) + \left( \frac{1}{2} - \frac{1}{3} \right) + \left( \frac{1}{3} - \frac{1}{4} \right)$$

$$+ \left( \frac{1}{4} - \frac{1}{5} \right) + \left( \frac{1}{5} - \frac{1}{6} \right)$$

$$= 1 - \frac{1}{6} = \frac{5}{6}$$

**a.** Evaluate

$$\sum_{k=1}^{10} \left( \frac{1}{k} - \frac{1}{k+1} \right)$$

by writing out a few terms at the beginning and end of the sum and then canceling.

More generally, we have the following formula.

$$\sum_{k=1}^{n} \left( \frac{1}{k} - \frac{1}{k+1} \right) = \left( \frac{1}{1} - \frac{1}{2} \right) + \left( \frac{1}{2} - \frac{1}{3} \right) + \left( \frac{1}{3} - \frac{1}{4} \right)$$

$$+ \cdots + \left( \frac{1}{n-1} - \frac{1}{n} \right) + \left( \frac{1}{n} - \frac{1}{n+1} \right)$$

$$= 1 - \frac{1}{n+1}$$

**b.** Use the formula given above to evaluate

$$\sum_{k=1}^{20} \left( \frac{1}{k} - \frac{1}{k+1} \right)$$

**c.** How would the formula change if the sum started at $k = 2$? What about $k = 5$?

**d.** Find formulas for the following telescoping sums and test them for $n = 5$, 10, and 20.

    **i.** $\displaystyle\sum_{k=1}^{n} \frac{1}{k^2} - \frac{1}{(k+1)^2}$

    **ii.** $\displaystyle\sum_{k=1}^{n} \left( \frac{1}{k} - \frac{1}{k+2} \right)$

    **iii.** $\displaystyle\sum_{k=1}^{n} \left( \frac{1}{2k-1} - \frac{1}{2k+1} \right)$

**69.** *Geometric Series and Annuities* An **annuity** is a sequence of regular payments or receipts, all subject to some underlying interest rate. Some examples of annuities include mortgage payments, installment purchases, premium payments on insurance policies, payroll deductions for pension plans, and lease payments. We will limit our discussion to annuities for which payment is made at the end of the payment period. As a simple example, suppose you deposit $100 at the *end* of each year for the next 10 years at an interest rate of 5% compounded annually. The **future value** of a given payment is simply its value at the end of the last year, and this can be found using the standard formula for compound interest. Thus, the first payment, which earns 5% interest for 9 years, would have a future value of $100(1 + 0.05)^9$. The second would have a future value of $100(1 + 0.05)^8$, the third would be $100(1 + 0.05)^7$, and so on up to the tenth, which would have a future value of 100. The future value of the annuity is the sum of the future values of all the payments. If we denote this sum by $S$ and write the terms of the sum is reverse order, we have

$$S = 100 + 100(1 + 0.05)^1$$
$$+ 100(1 + 0.05)^2 + \cdots + 100(1 + 0.05)^9$$

**a.** Use the formula for finite geometric series to find the future value of the annuity described above.

More generally, the future value of an annuity with periodic payments $R$ made at the end of each of $n$ periods with interest rate $i$ is given by

$$S = R + R(1 + i)^1 + R(1 + i)^2 + \cdots + R(1 + i)^{n-1}$$

**b.** Use the formula for finite geometric series to show that

$$S = R \left[ \frac{(1 + i)^n - 1}{i} \right]$$

**c.** Find the future value for the following annuities.

    **i.** Eliza deposits $3000 at the end of each year for 20 years into a pension fund that earns 6% compounded annually.

    **ii.** Charlie deposits $500 at the end of each quarter for 5 years into a bank account that earns interest at a rate of 4% compounded quarterly. Note that the periodic interest rate is $i = 0.04/4 = 0.01$ and the number of periods is $n = 4(5) = 20$.

We next consider the notion of **present value**. As an example, suppose a parent wishes to invest a certain amount of money now, at 8% compounded annually, in order to provide a child with a yearly allowance of $1000 at the end of each year for the next 10 years. We can determine this amount by considering the present value of a single payment—that is, the amount that must be invested now to make a given payment. Using the formula for compound interest, the amount $A_{10}$ that must be invested now to make the payment at the end of the tenth year is found by solving $1000 = A_{10}(1 + 0.08)^{10}$. Thus, $A_{10} = 1000/(1 + 0.08)^{10}$. Similarly, $A_9 = 1000/(1 + 0.08)^9$, $A_8 = 1000/(1 + 0.08)^8$, and so on up to $A_1 = 1000/(1 + 0.08)$. The present value of the annuity—that is, the entire amount that must be invested—is simply the sum of the present values of each payment. Thus,

$$A = A_1 + A_2 + \cdots + A_{10}$$

$$= 1000 \left( \frac{1}{1 + 0.08} \right)^1 + 1000 \left( \frac{1}{1 + 0.08} \right)^2$$

$$+ \cdots + 1000 \left( \frac{1}{1 + 0.08} \right)^{10}$$

This is also a geometric series.

d. Use the formula for finite geometric series to find the present value of the annuity described above.

More generally, if an annuity has $n$ periodic payments of $R$ dollars and a periodic interest rate $i$, then its present value is given by

$$A = R \left( \frac{1}{1 + i} \right)^1 + R \left( \frac{1}{1 + i} \right)^2 + \cdots + R \left( \frac{1}{1 + i} \right)^n$$

e. Use the formula for finite geometric series to show that

$$A = R \left[ \frac{1 - (1 + i)^{-n}}{i} \right]$$

f. Use the concept of present value to solve the following.

   i. An investment contract promises to pay $5000 at the end of each year for each of the next 10 years. If you wish to earn 9% compounded annually, how much should you pay for the contract now?

   ii. Suppose you have just won the grand prize in a sweepstakes and must choose between receiving $1,000,000 now or $100,000 at the end of each year for the next 20 years. Assuming an annual yield of 10% on the annuity option, which option should you choose? Explain. What additional factors might influence your decision?

---

![icon] **Questions for Discussion or Essay**

**70.** Write out and compare the terms of the following series.

  a. $\displaystyle\sum_{k=1}^{5} \frac{1}{k}$

  b. $\displaystyle\sum_{k=3}^{7} \frac{1}{k - 2}$

Find a third way of writing the series, this time using $k = 0$ as the starting index. What can you conclude about the way in which a given series can be represented using summation notation? Explain.

**71.** In Example 12, we saw that a percentage raise eventually beat a fixed raise with a bonus. Do you think this will always be the case? How much effect does the length of time have on the situation? What are some factors that might lead you to choose the fixed-raise option even if the percentage option ends a little better? How would these factors be affected by the length of time you plan on keeping the job?

**72.** Financing terms such as those described in Example 9 are uncommon because federal law requires a lender to provide the loan's A.P.R. (annual percentage rate) as well as the total interest charge. This allows a borrower to make intelligent choices between loans with different terms. Consult a local newspaper or inquire with a local lending institution to see if you find the A.P.R. and interest charge information to be readily offered. Discuss some other factors that you would consider when choosing between different loan offers.

**73.** In Exercise 67 we considered a paradox posed by the Greek philosopher Zeno of Elea. In logic, a paradox is an assertion that is self-contradictory. An example is the statement, "This statement is false." If we assume the statement is true, we get a contradiction, and if we assume it is false, we get a contradiction. The following famous paradox was posed by the twentieth-century philosopher and mathematician Bertrand Russell: *A barber in a certain village shaves all those, and only those, who do not shave themselves. Does the barber shave himself?* What makes this a paradox? Form a similar paradox of your own and explain why it is a paradox.

# PERMUTATIONS AND COMBINATIONS

- Why should a combination lock really be called a permutation lock?
- Why is it not a good idea for a TV program manager to select a program schedule by trial and error?
- How many different five-member subcommittees can be formed in the U.S. Senate?
- How many winning combinations are possible in a state lottery?

## THE FUNDAMENTAL COUNTING PRINCIPLE

| Meat | Potato | Vegetable | Complete meal |
|------|--------|-----------|---------------|
|      |        | S         | C, B, S       |
|      | B      | P         | C, B, P       |
|      |        | G         | C, B, G       |
| C    |        | S         | C, M, S       |
|      | M      | P         | C, M, P       |
|      |        | G         | C, M, G       |
|      |        | S         | R, B, S       |
|      | B      | P         | R, B, P       |
|      |        | G         | R, B, G       |
| R    |        | S         | R, M, S       |
|      | M      | P         | R, M, P       |
|      |        | G         | R, M, G       |

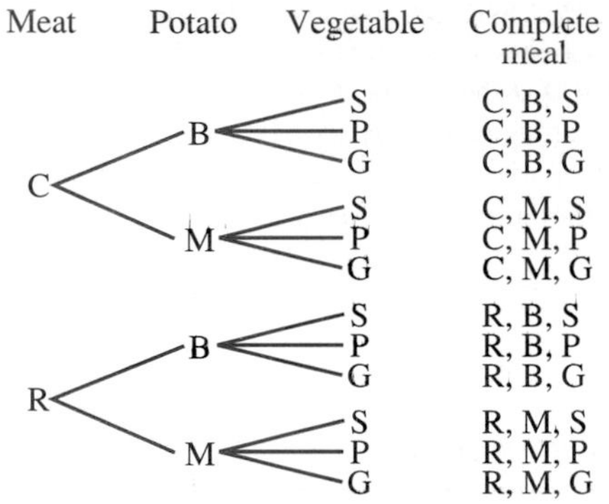

**FIGURE 9**

Suppose you are on a committee that is planning a banquet. The committee has selected a caterer that offers a choice of two meats (chicken breast or roast beef), two potatoes (baked or mashed), and three vegetables (spinach, peas, or green beans). Your task is to prepare a list of all the meals consisting of one meat, one potato, and one vegetable. In the process of creating this list, you would likely discover the need for a systematic method for finding all the possibilities without any duplication. One such method involves the use of a **tree diagram**.

We first adopt the convention that each food item is denoted by the first letter of its name (i.e., C for chicken breast, R for roast beef, and so on). We next list the two choices for meat in a column labeled ''Meat,'' as shown in Figure 9. Since each meat choice has two choices for potato, we draw two lines, called **decisions branches**, from each meat to the two possible choices for potato in the column labeled ''Potato.'' Similarly, since each potato has three possible choices for vegetable, we draw three branches from each potato to the three vegetables in the column labeled ''Vegetable.'' All of the complete meals can be found by following the different paths from meat to vegetable. There are a total of 12 possible meals.

If it is only necessary to determine the *number* of possible meals, without actually listing them, there is a much easier way. Since each of the two meats can be paired with one of the two potatoes, there are $2 \cdot 2 = 4$ distinct meat and potato pairs. For each of these 4 meat and potato pairs, there are 3 possibilities for a vegetable, for a total of $4 \cdot 3 = 12$ meals. Thus, given a meal selection task involving 2 choices for meat, 2 choices for potato, and 3 choices of vegetable, there are $2 \cdot 2 \cdot 3 = 12$ possible meals. This observation generalizes into the Fundamental Counting Principle.

**Fundamental Counting Principle**

If a task consists of $k$ parts, the first of which can be completed in $n_1$ ways, the second of which can be completed in $n_2$ ways, and so on up to the $k$th, which can be completed in $n_k$ ways, then the total number of ways in which the task can be completed is given by the product

$$n_1 n_2 \cdots n_k$$

| EXAMPLE 1 | *Counting license plates* |

The state of Illinois has license plates of the form LLL DDD, where L denotes a letter and D denotes a digit. How many different license plates of this form are possible?

**SOLUTION**    It is convenient to first draw six blank lines to represent the letters and digits that must be selected.

_____  _____  _____  _____  _____  _____

Each of the first three blanks must be filled with a letter, and so each has 26 possibilities. Each of the last three blanks must be filled with one of the digits 0–9, and so each has 10 possibilities. We represent these choices as shown below.

26    26    26    10    10    10

Thus, according to the Fundamental Counting Principle, there are a total of

$$26 \cdot 26 \cdot 26 \cdot 10 \cdot 10 \cdot 10 = 17,576,000$$

possible license plates.

| EXAMPLE 2 | *Arranging TV shows* |

A TV program manager must choose between 6 shows for 4 available time slots. He decides to try each possible arrangement for 1 week to see which draws the highest ratings. How many weeks will this take?

**SOLUTION**    We first draw four blank lines to represent the 4 time slots that must be filled. Initially, there are 6 choices for filling the first time slot, and so we place a 6 in the first blank. If we assume that the same show may not be used twice, there are only 5 choices remaining for the second time slot. Continuing in this fashion, there are 4 choices for the third slot, and 3 for the fourth.

6    5    4    3

By viewing the selection process as a four-part task, each part of which can be completed in 6, 5, 4, and 3 ways, respectively, the Fundamental Counting Principle tells us that the total number of possibilities is

$$6 \cdot 5 \cdot 4 \cdot 3 = 360$$

Thus, it would take 360 weeks, or almost 7 years, to try all the possible arrangements.

## PERMUTATIONS

In Example 2, the *ordering* of the TV shows was an essential part of the problem. That is, if the same four shows were arranged in a different order, we would consider it to be a completely different arrangement. We say that each of the ordered arrangements is a *permutation* of 6 shows taken 4 at a time. In general, we define a permutation as follows.

**Definition of permutation**

> A **permutation** of $n$ elements taken $r$ at a time is an ordered arrangement, without repetitions, of $r$ of the $n$ elements. The number of permutations of $n$ elements taken $r$ at a time is denoted by $_nP_r$.

To compute $_nP_r$, we consider an ordered arrangement of $r$ out of $n$ elements without repetitions. Following the lead of Examples 1 and 2, we draw $r$ blank lines to represent the $r$ elements that must be chosen.

$$\underbrace{\underline{\hspace{2cm}}\ \ \underline{\hspace{2cm}}\ \ \underline{\hspace{2cm}}\ \cdots\ \underline{\hspace{2cm}}}_{r \text{ blank lines}}$$

Since there are initially $n$ elements to chose from, the first blank can be filled in $n$ ways. Since repetitions are not allowed, the second blank can be filled in $n - 1$ ways. Continuing in this fashion, the third blank can be filled in $n - 2$ ways, and so on up to the $r$th blank, which can be filled in $n - (r - 1)$ or $n - r + 1$ ways.

$$\underbrace{\underline{\ \ n\ \ }\quad \underline{\ \ n-1\ \ }\quad \underline{\ \ n-2\ \ }\quad \cdots\ \underline{\ \ n-r+1\ \ }}_{r \text{ blank lines}}$$

Now by applying the Fundamental Counting Principle, we obtain the following formula.

**Permutation formula**

> For natural numbers $n$ and $r$ with $1 \le r \le n$,
>
> $$_nP_r = n(n - 1)(n - 2) \cdots (n - r + 1).$$

> **RULE OF THUMB**   You do not need to memorize the permutation formula as it appears above. Instead, just remember that the product starts with $n$ and has $r$ factors, which decrease by one.

**EXAMPLE 3**    *Evaluating permutations*

Evaluate the following permutations.
a. $_7P_3$        b. $_5P_4$

**SOLUTION**
a. $_7P_3 = \underbrace{7 \cdot 6 \cdot 5}_{3 \text{ factors}} = 210$

b. $_5P_4 = \underbrace{5 \cdot 4 \cdot 3 \cdot 2}_{4 \text{ factors}} = 120$

**EXAMPLE 4**    *Using permutations to count batting orders*

The manager of a coed softball team must assign a batting order by selecting 10 players from a team consisting of 14 players, 8 men and 6 women. Find the number of batting orders that are possible if
a. there are no restrictions on the number of men and women.
b. league rules require that 3 women must bat first.

**SOLUTION**

a. If there are no restrictions, then the manager may simply select and arrange 10 out of the 14 team members. Thus, there are

$$_{14}P_{10} = 3,632,428,800$$

different batting orders.

b. The number of ways to arrange 3 out of 6 women in the first 3 positions is $_6P_3$. Since no restrictions have been specified for the remaining 7 positions, the manager may select them out of the 11 people that remain. This may be done in $_{11}P_7$ ways. By the Fundamental Counting Principle, the total number of ways to arrange the batting order is

$$_6P_3 \cdot {}_{11}P_7 = 120 \cdot 1,663,200 = 199,584,000$$

Notice that if we let $r = n$ in the permutation formula, we obtain

$$_nP_n = n(n-1)(n-2) \cdots (1)$$

which is the product of the first $n$ positive integers. This product occurs so frequently that it is convenient to define a special notation for it—the **factorial** notation.

*Factorial notation*

> For any natural number $n$, $n! = n \cdot (n-1) \cdots 3 \cdot 2 \cdot 1$. In the case where $n = 0$, we define $0! = 1$.

**EXAMPLE 5**    *Evaluating factorials*

Evaluate the following factorial expressions.

a. $5!$      b. $\dfrac{20!}{18!}$

**SOLUTION**

a. $5! = 5 \cdot 4 \cdot 3 \cdot 2 \cdot 1 = 120$

b. $\dfrac{20!}{18!} = \dfrac{20 \cdot 19 \cdot 18 \cdot 17 \cdot 16 \cdots 2 \cdot 1}{18 \cdot 17 \cdots 2 \cdot 1} = 20 \cdot 19 = 380$

Since $_nP_r = n(n-1)(n-2) \cdots (n-r+1)$, we see that $_nP_r$ contains the first $r$ factors of $n!$. Thus, it is perhaps no surprise that the permutation formula can be rewritten using factorials. We simply multiply $_nP_r$ by a quotient whose numerator and denominator contain the factors of $n!$ that are missing from $_nP_r$.

$$\begin{aligned} _nP_r &= \frac{n(n-1)(n-2)\cdots(n-r+1)}{1} \cdot \frac{(n-r)!}{(n-r)!} \\ &= \frac{n(n-1)(n-2)\cdots(n-r+1)(n-r)(n-r-1)\cdots(3)(2)(1)}{(n-r)!} \\ &= \frac{n!}{(n-r)!} \end{aligned}$$

This verifies the following factorial formula for $_nP_r$.

*Factorial formula for $_nP_r$*

For natural numbers $n$ and $r$ with $1 \leq r \leq n$, $_nP_r = \dfrac{n!}{(n-r)!}$.

**WARNING**   The factorial symbol cannot be distributed through parentheses. Thus, the expression $(n-r)!$ in the factorial formula for $_nP_r$ *cannot* be rewritten as $n! - r!$.

**EXAMPLE 6**   *Applying the factorial formula for permutations*

Use the factorial formula for $_nP_r$ to evaluate $_7P_3$.

**SOLUTION**

$$_7P_3 = \frac{7!}{(7-3)!} = \frac{7!}{4!} = \frac{7\cdot6\cdot5\cdot\cancel{4}\cdot\cancel{3}\cdot\cancel{2}\cdot\cancel{1}}{\cancel{4}\cdot\cancel{3}\cdot\cancel{2}\cdot\cancel{1}} = 7\cdot6\cdot5 = 210$$

## COMBINATIONS

We have seen that a permutation takes into account the order in which elements are selected. In many cases, however, the order of selection is not important. Suppose, for example, you are taking a course that requires you to read and report on three out of four unrelated journal articles. In this case, you would probably not be concerned about the order in which the articles are read. Thus, the list of possible reading combinations would not include all the different orderings of the same three articles. If we denote the four articles by the letters A, B, C, and D, the possible reading combinations are

$$\{A, B, C\}, \quad \{A, B, D\}, \quad \{A, C, D\}, \quad \text{and} \quad \{B, C, D\}$$

Notice that these four combinations are simply the three-element subsets of $\{A, B, C, D\}$. More generally, a combination is defined in the following way.

*Definition of combination*

A **combination** of $n$ elements taken $r$ at a time is an $r$-element subset of a set of $n$ elements. The number of combinations of $n$ elements taken $r$ at a time is denoted by $_nC_r$ or $\dbinom{n}{r}$.

Although the definitions of permutation and combination use similar wording, it is important to remember that permutations and combinations are not the same. The difference can be summed up in one word—**order**. A permutation takes order into account, whereas a combination does not. However, $_nC_r$, the *number* of combinations of $n$ elements taken $r$ at a time, is closely related to $_nP_r$, the *number* of permutations of $n$ elements taken $r$ at a time. We first note that each combination of $r$ elements can be rearranged in $r!$ distinct ways. Thus, by the Fundamental Counting Principle, there are $_nC_r \cdot r!$ ordered arrangements of $r$ elements chosen from $n$. But $_nP_r$ also represents the number of ordered arrangements of $r$ elements chosen from $n$. Thus,

$$_nC_r \cdot r! = {_nP_r}$$

and so

$$_nC_r = \frac{_nP_r}{r!}$$

By applying the two earlier formulas for computing $_nP_r$, we obtain the following formulas for computing $_nC_r$.

**Combination formulas**

> For natural numbers $n$ and $r$ with $1 \le r \le n$,
>
> $$_nC_r = \frac{_nP_r}{r!} = \frac{n(n-1)(n-2) \cdots (n-r+1)}{r!}$$
>
> or
>
> $$_nC_r = \frac{n!}{r!(n-r)!}$$

**EXAMPLE 7**   *Evaluating combinations*

Evaluate the following combinations.

**a.** $_5C_3$      **b.** $\binom{8}{6}$

**SOLUTION**

**a.** $_5C_3 = \dfrac{_5P_3}{3!} = \dfrac{5 \cdot 4 \cdot 3}{3 \cdot 2 \cdot 1} = 10$

**b.** Recall that $\binom{8}{6}$ is an alternate notation for $_8C_6$. Thus,

$$\binom{8}{6} = \frac{8!}{6!(8-6)!} = \frac{8 \cdot 7 \cdot \cancel{6} \cdot \cancel{5} \cdot \cancel{4} \cdot \cancel{3} \cdot \cancel{2} \cdot \cancel{1}}{(\cancel{6} \cdot \cancel{5} \cdot \cancel{4} \cdot \cancel{3} \cdot \cancel{2} \cdot \cancel{1}) \cdot (2 \cdot 1)} = \frac{8 \cdot 7}{2 \cdot 1} = 28$$

**EXAMPLE 8**   *Using combinations to count Senate subcommittees*

The U.S. Senate, consisting of 100 senators, wishes to form a five-member subcommittee to study the selection process for Senate subcommittees.

a. How many different subcommittees are possible?

b. How many different subcommittees are possible if the Senate consists of 56 Democrats and 44 Republicans and each committee must have 3 Democrats and 2 Republicans?

**SOLUTION**

a. The order in which members are selected is unimportant. Thus, we wish to find the number of combinations of 100 senators taken 5 at a time.

$$_{100}C_5 = \frac{_{100}P_5}{5!} = \frac{100 \cdot 99 \cdot 98 \cdot 97 \cdot 96}{5 \cdot 4 \cdot 3 \cdot 2 \cdot 1} = 75{,}287{,}520$$

So there are a total of 75,287,520 possible subcommittees.

b. The number of ways of selecting 3 Democrats out of 56 without regard to order is $_{56}C_3$. The number of ways of selecting 2 Republicans out of 44 without regard to order is $_{44}C_2$. By the Fundamental Counting Principle, the total number of subcommittees is

$$_{56}C_3 \cdot _{44}C_2 = \frac{56 \cdot 55 \cdot 54}{3 \cdot 2 \cdot 1} \cdot \frac{44 \cdot 43}{2 \cdot 1} = 26{,}223{,}120$$

---

**EXAMPLE 9**   *Counting card hands*

A standard 52-card deck of playing cards consists of two black suits, clubs and spades, and two red suits, hearts and diamonds. Each suit has 13 cards of different face values—A (ace), 2, 3, 4, 5, 6, 7, 8, 9, 10, J (jack), Q (queen), and K (king). In certain varieties of poker, each player is dealt a hand of 5 cards.

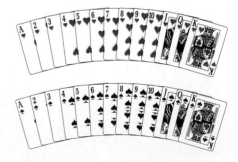

a. Find the number of 5-card poker hands that are possible.

b. A *full house* consists of a pair (two cards of the same face value from different suits) and three of a kind (three cards of the same face value from different suits), such as K-K-6-6-6. How many different full houses are possible?

**SOLUTION**

a. Since the order in which the cards are dealt is not important, the number of ways of being dealt 5 cards out of 52 is

$$_{52}C_5 = \frac{52 \cdot 51 \cdot 50 \cdot 49 \cdot 48}{5 \cdot 4 \cdot 3 \cdot 2 \cdot 1} = 2{,}598{,}960$$

b. A pair can be formed by being dealt 2 out of 4 cards of the same face value, and there are $_4C_2$ ways in which this can be done. Since there are 13 different face values in the deck, we have

the number of possible pairs $= 13 \cdot {}_4C_2$

A three of a kind must consist of 3 out of 4 cards of the same face value, but a different face value than that of the pair. Since there are 12 other face values remaining, we see that

the number of possible three of a kinds $= 12 \cdot {}_4C_3$

Thus, the number of possible full houses is

$$(13 \cdot {}_4C_2) \cdot (12 \cdot {}_4C_3) = 3744$$

---

## EXERCISES 3

**EXERCISES 1–4** □ *Draw a tree diagram to illustrate the number of choices that are possible.*

1. The math and science requirement at a certain university includes a math course (College Algebra or Calculus I), a physics course (Elementary Physics or University Physics I), and a biology course (General Biology or Zoology). How many different course combinations will satisfy the math and science requirement?

2. The option packages for a popular pickup truck include a choice of cab style (regular or extended), bed size (short or long), and drive train (two-wheel or four-wheel). How many different option packages are possible?

3. A computer store offers three sizes of monitors (14″, 15″, or 17″), two styles of computer case (desktop or tower), and two types of keyboards (programmable or with a built-in calculator). How many different computer configurations are possible?

4. A restaurant offers a single-topping pizza special with a choice of appetizer (bread sticks or garlic bread), type of crust (thin, hand-tossed, or pan), and meat topping (pepperoni or sausage). How many different meals are possible?

**EXERCISES 5–8** □ *Use the Fundamental Counting Principle to determine the number of possibilities.*

5. How many 3-digit numbers can be formed using the digits {1, 2, 3, 4, 5}
   a. if repetitions are allowed?
   b. if repetitions are not allowed?

6. How many four-letter ''words'' can be formed using the letters {a, e, i, o, u, y}
   a. if repetitions are allowed?
   b. if repetitions are not allowed?

7. The state of California has license plates of the form LLLDDDD where each L denotes a letter and each D denotes one of the digits 0–9. How many license plates of this form are possible?

8. The state of Indiana has license plates of the form NN L DDDD where NN is a number from 1 to 99 (depending upon the county in which the plate was issued), L is one of the letters A–Z (corresponding to the license branch that issued the plate), and each D is one of the digits 0–9. How many license plates of this form are possible?

**EXERCISES 9–16** □ *Evaluate the given expression.*

9. $6!$

10. $(9 - 5)!$

11. $\dfrac{10!}{7!}$

12. $\dfrac{8!}{2! \cdot 6!}$

13. $_6P_2$

14. $_9P_1$

15. $_5P_5$

16. $_8P_4$

**EXERCISES 17–20** □ *Express your answer using permutation notation and evaluate.*

17. In a survey of TV viewer preferences, respondents are asked to rank their favorite 3 out of 8 sitcoms. How many different survey responses are possible?

18. A club consisting of 12 members wishes to elect a president, vice president, secretary, and treasurer. In how many ways can this be done?

19. The manager of a Little League baseball team intends to select the starting 9 players by assigning a batting order. In how many ways can this be done if there are 13 players to choose from and

   a. there are no restrictions on the order of selection?

   b. the two children of the team's sponsor must bat first and second?

20. Four boys and six girls are attending a birthday party. In how many ways can they be lined up for ice cream

   a. if there are no restrictions on the order?

   b. if the girls are served before the boys?

**EXERCISES 21–26** □ *Evaluate the given combination.*

21. $_6C_4$

22. $\dbinom{7}{2}$

23. $\dbinom{8}{1}$

24. $_4C_4$

25. $_{10}C_6$

26. $\dbinom{9}{8}$

**EXERCISES 27–30** □ *Express your answer using combination notation and evaluate.*

27. A local election ballot allows each voter to select 3 out of 8 school board candidates. In how many ways can this be done?

28. A club consisting of 12 members wishes to appoint a 4-person committee. In how many ways can this be done?

29. The manager of a high school volleyball team must select 6 first string players from a team of 12 girls. In how many ways can this be done if

   a. there are no restrictions?

   b. exactly 2 of the 6 first string players must be chosen from the 5 seniors on the team?

30. A 5-card poker hand is dealt from a standard 52-card deck. How many hands are possible if

   a. at least three cards have the same face value?

   b. exactly three cards have the same face value? (Such a hand is called *three of a kind*.)

**EXERCISES 31–42** □ *Express your answer using permutation or combination notation if possible and evaluate.*

31. A student has 4 remaining time slots in her schedule. Ten different classes are available, each of which has a section open at those times. In how many ways can the time slots be filled?

32. An employment agency is asked to fill secretarial openings in 6 different companies. A total of 20 people have applied. In how many ways can the positions be filled?

33. A certain state lottery is played by selecting 5 different numbers from 1 through 40. In how many ways can this be done, assuming the order of selection is not important?

34. A certain state lottery is played by selecting 6 different numbers from 1 through 50. In how many ways can this be done, assuming the order of selection is not important?

35. The dial on a combination lock is marked with the numbers 1 through 40. The lock may be opened by selecting the correct sequence of 3 numbers. How many different combinations are possible

   a. if repetitions of numbers are not allowed?

   b. if repetitions of numbers are allowed?

36. The combination lock on a briefcase consists of three dials that are each marked with the digits 0–9. The lock may be opened by selecting the correct digit on each of the three dials. How many different combinations are possible

   a. if repetitions of digits are not allowed?

   b. if repetitions of digits are allowed?

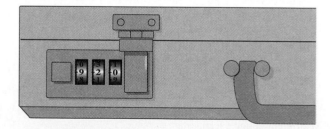

37. On a history exam, a student is asked to select 4 out of 6 essay questions. In how many ways can this be done? What if a student is also asked to select 10 out of 15 short-answer questions?

38. On a mathematics placement exam, a student is asked to select 3 out of 5 application problems. In how many ways can this be done? What if a student is also asked to select 6 out of 10 computational problems?

**39.** In a classroom of 30 students, how long would it take for every student to shake hands with each of the other students, assuming it takes 2 seconds for each shake and only two students can shake hands at a time? What about a class of 60 students?

**40.** If each of the 8 teams in a softball league must play every team twice, what is the total number of games that must be played?

**41.** In how many ways can a 5-card poker hand be dealt from a standard 52-card deck so that all 5 cards are of the same suit? (Such a hand is called a *flush*.)

**42.** In how many ways can a 5-card poker hand be dealt from a standard 52-card deck so that 4 of the 5 cards have the same face value? (Such a hand is called *four of a kind*.)

---

### ◼ Projects for Enrichment

---

**43.** *Combination Identities*  The combination formula

$$\binom{n}{r} = \frac{n!}{r!(n-r)!}$$

leads to many useful and sometimes surprising identities. We will consider a few in this project, most of which can be verified by direct simplification. One example is the identity $\binom{n}{n} = 1$. This can be verified by observing that

$$\binom{n}{n} = \frac{n!}{n!(n-n)!} = \frac{n!}{n!0!} = 1$$

The interpretation of this identity is that there is only one way to select $n$ elements (without regard to order) out of a set of $n$ elements.

**a.** Simplify the following combinations and give a verbal interpretation of the result.

**i.** $\binom{n}{1}$    **ii.** $\binom{n}{0}$    **iii.** $\binom{n}{n-1}$

**b.** Simplify the following combinations.

**i.** $\binom{n}{2}$    **ii.** $\binom{n}{n-2}$

You should have found that the two combinations in part (b) simplify to the same expression. In general,

$$\binom{n}{r} = \binom{n}{n-r} \quad \text{for any } r \le n$$

In other words, a set of $n$ elements has the same number of $r$-element subsets as it has $(n-r)$-element subsets.

**c.** Simplify $\binom{n}{n-r}$ to show that

$$\binom{n}{r} = \binom{n}{n-r}$$

Interpret this identity.

**d.** The following identities are often useful when performing complicated calculations involving combinations. Verify each.

**i.** $\binom{n}{m}\binom{m}{r} = \binom{n}{r}\binom{n-r}{m-r}$

(*Hint:* Simplify both sides of the equation.)

**ii.** $\binom{n}{r-1} + \binom{n}{r} = \binom{n+1}{r}$

(*Hint:* Find a common denominator for the expression to the left of the equal sign.)

**iii.** $\binom{n-1}{r} = \frac{r+1}{n}\binom{n}{r+1}$

**44.** *Counting Poker Hands*  In Example 9, we computed the number of possible 5-card poker hands that could be dealt from a standard 52-card deck, and we also computed the number of possible full houses. The following table lists all of the 5-card poker hands, together with a partial list of the number of possible such hands.

| Hand | Number possible |
|---|---|
| Royal flush | 4 |
| Other straight flush | 36 |
| Four of a kind | |
| Full house | 3,744 |
| Flush | |
| Straight | 10,200 |
| Three of a kind | 54,912 |
| Two pairs | |
| One pair | |
| Nothing | 1,302,540 |
| Total | 2,598,960 |

**a.** Look up and write out a description of each type of hand.

**b.** Show how combinations can be used to arrive at the number of possible hands for each type.

c. Although there are many varieties of 5-card poker, the general idea is that 5 cards are dealt to all players and there are one or more rounds of betting. After all betting is finished, the player with the hand that appears highest on the list wins all the money. Why does it make sense that the hand appearing highest on the list should win?

d. Do you think it's necessary for a poker player to know how many hands of each type are possible? Why or why not?

45. *Seating Arrangements* Three couples—Andy and Becky, Calvin and Dawn, and Eddie and Fran—are all good friends and decide to go to a concert together. They have reserved six adjoining seats next to an aisle. Determine the number of seating arrangements that are possible given the following conditions.

a. There are no restrictions.

b. The men must sit in the three seats closest to the aisle.

c. Each couple must sit together. (*Hint:* First find the number of pairs of seats where each couple can sit, then the number of arrangements for each of the three couples.)

d. The women must sit together. (*Hint:* First find the number of groups of 3 seats where the women can sit, then find the number of arrangements of the 3 women in those seats, and finally, find the number of arrangements of the men in the remaining 3 seats.)

e. Andy and Becky had a fight and refuse to sit next to each other. (*Hint:* First find the number of ways in which Andy and Becky can be separated, and then find the number of ways of arranging the remaining four people.)

After the concert, the three couples decided to go out for a late dinner. Determine the number of ways in which they can be seated around a circular table, given the following conditions. Keep in mind that we are only concerned with their positions relative to each other, not with the particular seat each person occupies.

f. There are no restrictions.

g. Each couple must sit together.

h. The women must sit together.

i. Andy and Becky had a fight and refuse to sit next to each other.

## Questions for Discussion or Essay

46. Explain why a combination lock should really be called a permutation lock. What are some other instances in which the word "combination" should perhaps be replaced with the word "permutation"? Explain your examples. To get you started, think about a TV cooking show referring to "the right combination of ingredients" or a coach referring to "a winning combination of players."

47. In Exercise 33 you were asked to determine the number of ways of selecting 5 numbers out of 40 in a state lottery. The answer to that exercise represents the number of possible winning combinations. What does this say about your chances of winning a lottery? How many different lottery combinations would you have to play in order to give yourself a 50/50 chance of winning? Explain.

48. Suppose 10 sprinters are competing in the 100-meter dash at a track meet. If the race is a qualifying round, the top three finishers advance to the next round. If the race is the final round, the top three finishers are awarded first, second, and third place. For each of these two scenarios, determine whether a permutation or combination would be used to count the number of ways of selecting 3 out of 10 sprinters. Explain in detail why you chose the method you did.

49. A basketball coach, notorious for changing his starting line-up, states that he will try every combination until he finds one that works. Assuming the team has 12 players, and that a season consists of 30 games, how many seasons will it take to try every combination? In practice, would a coach really try every possible starting line-up?

**SECTION 4**

# PROBABILITY

- How can it be that approximately 60% of the people testing positive for drug usage are actually not drug users, even though the test is 97% accurate?
- What is the probability that, in a class of 30 students, at least 2 will have the same birthday?
- How likely is it that a fatal accident would involve a driver between the ages of 18 and 21 who had been drinking?
- Which is more likely, getting struck by lightning or winning a state lottery?

## INTRODUCTION

Our language is replete with terms used to indicate the degree of certainty of uncertain events. We refer to outcomes as being "improbable," "unlikely," "a sure thing," "more likely than not," or a "long shot." **Probability** is the branch of mathematics in which degrees of certainty are quantified. Virtually every day of our lives we encounter uncertainties that can be expressed as probabilities. We are told that there is a 30% chance of thunderstorm activity, that there is a one in a million chance of winning a lottery, that one in four women will develop breast cancer, and that three out of four dentists recommend a certain sugarless gum. In addition to these everyday occurrences, probabilities arise in, and indeed are central to, such diverse areas as medicine, insurance, sociology, psychology, physics, and engineering.

In mathematical probability, a real number between 0 and 1 is associated with a possible outcome of an experiment: the greater the probability, the more likely it is that the outcome occurs. Events that have a probability of 1 are virtually certain to occur, whereas events with probability 0 virtually never occur. An event with a probability of 0.5 will occur about half the time. We give the following definitions.

## *Probability terminology and notation*

An **experiment** is any occurrence for which the outcome is uncertain. The **sample space** of an experiment is the set of all possible outcomes. The number of possible outcomes in a sample space $S$ is denoted by $n(S)$. An **event** is any subset of the sample space. The number of outcomes in an event $E$ is denoted by $n(E)$.

**EXAMPLE 1**    *An experiment involving a die*

An experiment consists of rolling an ordinary six-sided die.
**a.** Find the sample space $S$ and determine $n(S)$.
**b.** Describe the event $E$ that an even number is rolled and determine $n(E)$.

### SOLUTION

**a.** Since an outcome of the experiment consists of rolling one of the numbers 1–6, the sample space $S$ can be written as $\{1, 2, 3, 4, 5, 6\}$; $n(S)$ is the number of outcomes in $S$ and so $n(S) = 6$.

**b.** The event $E$ that an even number is rolled is the set $\{2, 4, 6\}$. Since $n(E)$ is the number of outcomes in $E$, $n(E) = 3$.

EXAMPLE 2    *A coin tossing experiment*

An experiment consists of tossing a coin two times.
a. Find the sample space $S$ and determine $n(S)$.
b. Describe the event $E$ that a tail is tossed both times and determine $n(E)$.

**SOLUTION**

a. Let $H$ denote the outcome that a head is tossed, and let $T$ denote the outcome that a tail is tossed. Thus, for example, the outcome consisting of tossing a head and then a tail can be written as $HT$. The sample space $S$ is the set $\{HH, HT, TH, TT\}$, and $n(S) = 4$.

b. The event $E$ is the set of outcomes for which a tail is tossed both times, or $\{TT\}$. Thus, $n(E) = 1$.

EXAMPLE 3    *A sample space and an event associated with the gender of children*

A couple has two children.
a. Find the sample space $S$ consisting of the possible outcomes with respect to the sexes of the children.
b. Describe the event $E$ that the children are of different sexes.

**SOLUTION**

a. Let $M$ represent a male birth and let $F$ represent a female birth. Thus, for example, the outcome in which the first child is male and the second child is female can be written $MF$. The sample space $S$ can then be expressed as the set $\{MM, MF, FM, FF\}$.

b. The two outcomes in which the children are of different sexes are $MF$ and $FM$. Thus, $E$ can be expressed as the set $\{MF, FM\}$.

When all outcomes are equally likely, we can find the probability of an event simply by dividing the number of outcomes that make up the event by the total number of outcomes in the sample space. For example, the probability of rolling a 2 on a fair die is $\frac{1}{6}$, since one of six outcomes corresponds to rolling a 2. Similarly, the probability of obtaining a head with one toss of a fair coin is $\frac{1}{2}$. More generally, we have the following definition of the probability of an event.

**Probability of an event**

If $E$ is an event in the sample space $S$ and if all of the outcomes of $S$ are equally likely, then the **probability of $E$** is denoted by $P(E)$ and is defined by

$$P(E) = \frac{n(E)}{n(S)}$$

**EXAMPLE 4**     *Computing a probability involving a die*

If an ordinary six-sided die is rolled, what is the probability that an even number is rolled?

**SOLUTION**    From Example 1 we have $S = \{1, 2, 3, 4, 5, 6\}$ and $E = \{2, 4, 6\}$. Thus,

$$P(E) = \frac{n(E)}{n(S)} = \frac{3}{6} = \frac{1}{2}$$

**EXAMPLE 5**     *Computing a probability involving two coins*

If a coin is tossed two times, what is the probability that a tail is tossed both times?

**SOLUTION**    As we saw in Example 2, $S = \{HH, HT, TH, TT\}$ and $E = \{TT\}$. Thus,

$$P(E) = \frac{n(E)}{n(S)} = \frac{1}{4}$$

**EXAMPLE 6**     *Computing a probability involving two dice*

Two ordinary six-sided dice are rolled. What is the probability that the total of the dice is 7?

**SOLUTION**    There are six possible outcomes for the first die and six possible outcomes for the second die. Thus, by the Fundamental Counting Principle, there are 36 possible outcomes in the sample space $S$. These outcomes are shown as ordered pairs in Table 1. Now for each possible outcome of the first die, there is exactly one outcome for the second die that will result in a total of 7. Thus, there are six ways of rolling a total of 7. These appear in the circled diagonal of Table 1.

**TABLE 1**

*Possible outcomes*

|  |  | 1 | 2 | 3 | 4 | 5 | 6 |
|---|---|---|---|---|---|---|---|
|  |  | | | *Second Die* | | | |
| | **1** | (1, 1) | (1, 2) | (1, 3) | (1, 4) | (1, 5) | (1, 6) |
| | **2** | (2, 1) | (2, 2) | (2, 3) | (2, 4) | (2, 5) | (2, 6) |
| *First Die* | **3** | (3, 1) | (3, 2) | (3, 3) | (3, 4) | (3, 5) | (3, 6) |
| | **4** | (4, 1) | (4, 2) | (4, 3) | (4, 4) | (4, 5) | (4, 6) |
| | **5** | (5, 1) | (5, 2) | (5, 3) | (5, 4) | (5, 5) | (5, 6) |
| | **6** | (6, 1) | (6, 2) | (6, 3) | (6, 4) | (6, 5) | (6, 6) |

So if $E$ is the event that a 7 is rolled, then $n(E) = 6$. Since $n(S) = 36$, we have

$$P(E) = \frac{n(E)}{n(S)} = \frac{6}{36} = \frac{1}{6}$$

**EXAMPLE 7**   *Computing probabilities involving three coins*

Three coins are tossed. Find the probabilities of the following events.
a. All three coins come up the same.
b. Exactly two heads are tossed.
c. At least one tail is tossed.

**SOLUTION**   We first find $n(S)$. There are two possible outcomes for each of the three coins, namely heads or tails. Thus, by the Fundamental Counting Principle, there are $2 \cdot 2 \cdot 2 = 8$ possible outcomes in the sample space $S$. Thus, $n(S) = 8$.

a. Let $E$ be the event that all three coins come up the same. Since the event $E$ consists of two outcomes, $HHH$ and $TTT$, $n(E) = 2$. Thus,

$$P(E) = \frac{n(E)}{n(S)} = \frac{2}{8} = \frac{1}{4}$$

b. Let $F$ be the event that exactly two heads are tossed. Then $F$ consists of the outcomes $HHT$, $HTH$, and $THH$. So $n(F) = 3$ and we have

$$P(F) = \frac{n(F)}{n(S)} = \frac{3}{8}$$

c. Let $G$ be the event that at least one tail is tossed. Of the 8 outcomes in $S$, the only one that doesn't have at least one tail is $HHH$. Thus, $n(G) = 7$ and we have

$$P(G) = \frac{n(G)}{n(S)} = \frac{7}{8}$$

---

**EXAMPLE 8**   *Playing the lottery*

A certain state lottery is designed so that a player selects 6 numbers from 1 to 40 (inclusive). If those 6 numbers match the 6 that are randomly drawn by the lottery (without regard to order), the player wins. Find the probability that a player wins the lottery after having selected 10 different combinations of 6 numbers.

**SOLUTION**   The sample space $S$ consists of all possible combinations of 6 numbers chosen from the numbers 1 to 40. Thus, $n(S) = {}_{40}C_6 = 3,838,380$. Let $E$ be the event that 1 of the player's 10 combinations is a winner. Then $n(E) = 10$. Thus,

$$P(E) = \frac{n(E)}{n(S)} = \frac{10}{3,838,380} \approx 0.0000026$$

---

**EXAMPLE 9**   *A 5-card poker hand*

A *straight flush* is a poker hand consisting of 5 cards in sequence from the same suit, such as A-2-3-4-5 or 8-9-10-J-Q. Find the probability of being dealt a straight flush from a standard 52-card deck.

**SOLUTION**   The sample space $S$ consists of all possible 5-card hands from a deck of 52 cards. Thus,

$$n(S) = {}_{52}C_5 = 2{,}598{,}960$$

Let $E$ be the event that a straight flush is dealt. Then $n(E)$ is the number of straight flushes that are possible in a 52-card deck. If we assume that the ace can be either low or high, then the possible straights in a given suit are A-2-3-4-5, 2-3-4-5-6, ..., 10-J-Q-K-A. Thus, there are 10 possible straights in each of the four suits, for a total of 40 possible straights. So $n(E) = 40$ and we have

$$P(E) = \frac{n(E)}{n(S)} = \frac{40}{2{,}598{,}960} \approx 0.000015$$

---

**EXAMPLE 10**  |  *Reliability of drug testing*

The president of a university with an enrollment of 30,000 is contemplating mandatory cocaine testing for all students. The test is 97% accurate in the sense that 97% of all students who are cocaine users will test positive and 97% of students who are not cocaine users will test negative. If 2% of the student body are in fact cocaine users, what is the probability that someone who tests positive for cocaine use is a cocaine user?

**SOLUTION**   We must first identify the sample space. Since we are only concerned with those students who test positive, we will let $S$ be the set of students who test positive, whether they are cocaine users or not. Next we let $E$ be the event that a person tests positive and is indeed a cocaine user. To find $P(E)$, we must determine $n(S)$ and $n(E)$. Since 2% of the 30,000 university students are cocaine users, there are $(0.02)30{,}000 = 600$ cocaine users and $30{,}000 - 600 = 29{,}400$ nonusers. Now the test will come out positive for approximately 97% of the users, but it will also come out positive for about 3% of the nonusers. Thus,

$$
\begin{aligned}
n(S) &= \text{the number of students who test positive} \\
&= \text{the number of } \textit{users} \text{ who test positive} \\
&\quad + \text{ the number of } \textit{nonusers} \text{ who test positive} \\
&\approx 97\% \text{ of } 600 + 3\% \text{ of } 29{,}400 \\
&= (0.97)600 + (0.03)29{,}400 \\
&= 1464
\end{aligned}
$$

and

$$
\begin{aligned}
n(E) &= \text{the number of students who test positive and are cocaine users} \\
&\approx (0.97)600 \\
&= 582
\end{aligned}
$$

Thus, we have

$$P(E) = \frac{n(E)}{n(S)} = \frac{582}{1464} \approx 0.4$$

So the probability that someone who tests positive for cocaine use is actually a user is less than $\frac{1}{2}$! We will ask you to discuss this surprising result in Exercise 58.

---

## INTERSECTIONS, UNIONS, AND COMPLEMENTS OF EVENTS

In many cases, we are interested in finding probabilities that involve two or more events from the same sample space, or that indirectly involve a given event. For these tasks, we make the following definitions.

*Definitions of intersection, union, and complement*

> Let $E$ and $F$ be events from the same sample space $S$. Then
>
> 1. The **intersection** of $E$ and $F$, denoted $E \cap F$ and read "*E and F*," is the event that both $E$ and $F$ occur.
> 2. The **union** of $E$ and $F$, denoted $E \cup F$ and read "*E or F*," is the event that either $E$ or $F$ (or both) occur.
> 3. The **complement** of $E$, denoted $E'$ and read "*E complement*," is the event that $E$ does *not* occur.

**EXAMPLE 11**

*Finding intersections, unions, and complements*

Suppose that a number from 1 to 10 (inclusive) is selected at random. Let $E$ be the event that the number chosen is a multiple of 3, and let $F$ be the event that the number chosen is even. Find expressions for each of the following events, and then compute the probability of the event.

a. $E \cap F$          b. $E \cup F$

c. $E'$          d. $F'$

**SOLUTION**

a. $E \cap F$ consists of all numbers from 1 to 10 that are both even *and* multiples of 3. Thus, $E \cap F = \{6\}$, and $n(E \cap F) = 1$. Since $n(S) = 10$, we have

$$P(E \cap F) = \frac{n(E \cap F)}{n(S)} = \frac{1}{10}$$

b. $E \cup F$ consists of all numbers from 1 to 10 that are either even *or* multiples of 3. Thus, $E \cup F = \{2, 3, 4, 6, 8, 9, 10\}$. Hence $n(E \cup F) = 7$, and $P(E \cup F) = \frac{7}{10}$.

c. $E'$ is the set of numbers from 1 to 10 that are *not* multiples of 3. Thus, $E' = \{1, 2, 4, 5, 7, 8, 10\}$, and $n(E') = 7$. Hence $P(E') = \frac{7}{10}$.

d. $F'$ is the set of numbers from 1 to 10 that are not in $F$—that is, those numbers that are not even. Hence $F'$ consists of the odd numbers between 1 and 10, and so $F' = \{1, 3, 5, 7, 9\}$. Thus, $n(F') = 5$, and $P(F') = \frac{5}{10} = \frac{1}{2}$.

---

If $E$ and $F$ are events from the same sample space and $E$ and $F$ have no outcomes in common, then $E \cap F = \{\}$, the empty set, and we say that the events are **mutually exclusive**. Consider, for example, an experiment in which a die is rolled. If $E$ is the event that an odd number is rolled and $F$ is the event that an even number is rolled, then $E$ and $F$ are mutually exclusive since there is no outcome that belongs to both events.

When two events $E$ and $F$ are mutually exclusive, the probability of $E \cup F$ is nothing more than the sum of the probabilites of $E$ and $F$. That is,

$$P(E \cup F) = P(E) + P(F)$$

This is because the number of outcomes in the event $E \cup F$ will be the sum of the number of outcomes in each of the events $E$ and $F$.

As an example, consider the event $E$ that a 3 is obtained by a single throw of a die, and the event $F$ that a 5 is thrown. Since these events are mutually exclusive, (we can't throw both a 3 and a 5 on the same toss), the probability of $E \cup F$ is simply the sum of the probabilities of $E$ and $F$. Since $P(E) = \frac{1}{6}$ and $P(F) = \frac{1}{6}$, we have

$$P(E \cup F) = P(E) + P(F) = \frac{1}{6} + \frac{1}{6} = \frac{1}{3}$$

**EXAMPLE 12**   *Winning the World Series*

Suppose that a preseason poll of sports writers suggests that the probability of the Chicago Cubs winning the World Series is 0.06 and the probability of the New York Mets winning is 0.13. Find the probability that one of the two teams will win.

**SOLUTION**   Let $E$ be the event that the Cubs win and let $F$ be the event that the Mets win. We would like to determine $P(E \cup F)$, the probability that $E$ or $F$ occurs. Since both teams cannot win, the events $E$ and $F$ are mutually exclusive, and so

$$P(E \cup F) = P(E) + P(F) = 0.06 + 0.13 = 0.19$$

---

Now let us consider two events that are *not* mutually exclusive. Suppose that 2 coins are tossed. Then the sample space $S$ can be represented as the set $\{HH, HT, TH, TT\}$. Let $E$ be the event that the first coin comes up heads, and let $F$ be the event that the second coin comes up heads. Thus, $E = \{HH, HT\}$ and $F = \{TH, HH\}$, so that $n(E) = n(F) = 2$. The event $E \cup F$ consists of those outcomes for which either the first coin or the second coin is a head. That is, $E \cup F = \{HH, HT, TH\}$, and thus $n(E \cup F) = 3$. But $n(E) + n(F) = 2 + 2 = 4$, so that $n(E \cup F) \neq n(E) + n(F)$. The reason for the discrepancy is that the outcome $HH$ belongs to both $E$ and $F$ so that when we add $n(E)$ and $n(F)$, we have counted this outcome twice.

In general, the expression $n(E) + n(F)$ will count all outcomes belonging to $E \cap F$ twice. To correct for this double counting, we must subtract the number of elements in the set $E \cap F$. Thus, we have

$$n(E \cup F) = n(E) + n(F) - n(E \cap F)$$

Figure 10 illustrates this formula with a **Venn diagram**.

Not surprisingly, a similar formula holds for finding the probability of the union of two events that may not be mutually exclusive. It is stated as follows.

$E \cup F$

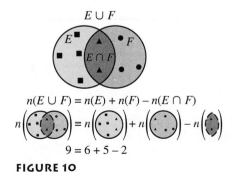

$n(E \cup F) = n(E) + n(F) - n(E \cap F)$

$9 = 6 + 5 - 2$

**FIGURE 10**

*The probability of the union of two events*

Let $E$ and $F$ be events in the same sample space. Then the probability of $E$ or $F$ occurring is given by

$$P(E \cup F) = P(E) + P(F) - P(E \cap F)$$

Note that this formula applies even if $E$ and $F$ are mutually exclusive since, in that case, $E \cap F = \{\}$, and so $P(E \cap F) = 0$.

EXAMPLE 13

*Computing the probability of the union of two events*

Suppose that a single card is selected at random from a standard 52-card deck. Find the probability that the card selected is either a face card (jack, queen, or king) or a heart.

**SOLUTION**    Let $E$ be the event that the card selected is a face card, and let $F$ be the event that the card selected is a heart. Then $E \cup F$ represents the event that the card selected is a face card or a heart. Thus, we are interested in $P(E \cup F)$. Since there are 3 face cards in each of the 4 suits, there are 12 face cards in all. Thus,

$$P(E) = \frac{n(E)}{n(S)} = \frac{12}{52} = \frac{3}{13}$$

On the other hand $\frac{1}{4}$ of the cards in the deck are hearts, so that

$$P(F) = \frac{1}{4}$$

The event $E \cap F$ consists of those outcomes in which a face card of hearts is drawn. Since there are 3 hearts that are face cards, we have

$$P(E \cap F) = \frac{n(E \cap F)}{n(S)} = \frac{3}{52}$$

We can now compute $P(E \cup F)$ as follows.

$$P(E \cup F) = P(E) + P(F) - P(E \cap F)$$
$$= \frac{3}{13} + \frac{1}{4} - \frac{3}{52}$$
$$= \frac{12 + 13 - 3}{52}$$
$$= \frac{22}{52} = \frac{11}{26}$$

EXAMPLE 14

*Probabilities involving fatal car accidents*

The following data from the U.S. Federal Highway Administration gives certain percentages for drivers involved in fatal motor-vehicle accidents in 1989:

■ 14.1% were 18–21 years old.

■ 31.9% had been drinking.

■ 5.3% were 18–21 years old and had been drinking.

If a driver involved in a fatal motor-vehicle accident is selected at random, find the probability of each of the following events.

**a.** The driver was 18–21 and had been drinking.

**b.** The driver had not been drinking.

**c.** The driver had been drinking but was not 18–21.

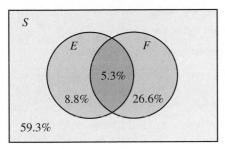

**FIGURE 11**

**SOLUTION**    Let $S$ be the sample space consisting of drivers involved in fatal accidents. Let $E$ be the event that the selected driver was 18–21, and let $F$ be the event that the driver had been drinking. According to the given data, 5.3% of the drivers in $S$ are also in $E \cap F$. Moreover, since 14.1% of the drivers are in $E$, there must be 14.1% $-$ 5.3% $=$ 8.8% of the drivers in $E$ that aren't also in $E \cap F$. Similarly, there are 31.9% $-$ 5.3% $=$ 26.6% of the drivers in $F$ that aren't also in $E \cap F$. It is convenient to summarize these percentages in a Venn diagram such as the one in Figure 11. Note that 59.3% of drivers involved in fatal accidents had not been drinking and were not 18–21 years old. Note also that the percentages themselves represent probabilities.

**a.** The event that the driver was 18–21 and had been drinking is $E \cap F$ and $P(E \cap F) = 5.3\% = 0.053$.

**b.** The event that the driver had not been drinking is $F'$ and $P(F') = 59.3\% + 8.8\% = 0.593 + 0.088 = 0.681$.

**c.** The event that the driver had been drinking but was not 18–21 is $F \cap E'$. In the Venn diagram, this is the region inside $F$ but outside $E$. Thus, $P(F \cap E') = 26.6\% = 0.266$.

---

## EXERCISES 4

**EXERCISES 1–4** □ *Find the sample space S for the given experiment and determine n(S).*

1. A tetrahedral (four-sided) die is rolled.

2. A name is drawn from a hat containing five names.

3. A coin is tossed three times.

4. A six-sided die is rolled and a coin is tossed.

**EXERCISES 5–8** □ *A standard six-sided die is rolled. Find the probability of each of the following events.*

5. Rolling a 4

6. Rolling a number larger than 4

7. Rolling a number less than 7

8. Rolling an even number

**EXERCISES 9–12** □ *A ball is selected at random from an urn containing 3 red balls, 4 blue balls, and 5 white balls. Find the probability of each of the following events.*

9. Selecting a red ball

10. Selecting a red or white ball

11. Selecting a ball that is not white

12. Selecting a yellow ball

**EXERCISES 13–18** □ *A single card is drawn from a standard 52-card deck. Compute the probability of each of the following events.*

13. Drawing the ace of spades

14. Drawing a spade

15. Drawing an ace

16. Drawing an ace or a spade

17. Drawing a red face card

18. Drawing a 3 or a 4

**EXERCISES 19–22** □ *Assume that the probability of a male birth is 0.5.*

19. A family has three children. Compute the probability that the children are of the same sex.

20. Compute the probability that in a family with three children, exactly one of the children is female.

21. Compute the probability that in a family with three children, exactly two of the children are male.

22. Compute the probability that in a family of three children, there is no girl with an older brother.

**EXERCISES 23–24** □ *Balls are selected (without replacement) from an urn containing 5 red balls and 4 blue balls.*

23. If two balls are drawn, find the probability that both are red.

24. If three balls are drawn, find the probability that two of the balls are blue and the other is red.

**EXERCISES 25–28** □ *A pair of standard six-sided dice are rolled. Find the probability of each of the following events.*

25. Rolling the same number on each die

26. Rolling ''snake-eyes'' (a 1 on each die)

**27.** Rolling a total of 10

**28.** Rolling a 6 on at least one of the dice

**EXERCISES 29–32** □ *Compute the probability of being dealt the given 5-card poker hand from a standard 52-card deck.*

**29.** Four aces

**30.** Four of a kind

**31.** A flush (all cards of the same suit)

**32.** Three of a kind

**EXERCISES 33–38** □ *A fair coin is being tossed. Compute the probability of the given event.*

**33.** Tossing 4 heads in a row

**34.** Having a streak of four heads in a row if the coin is tossed 5 times

**35.** Obtaining your first head on the third toss

**36.** Obtaining your second head on the fourth toss

**37.** Tossing exactly 3 heads out of the first 5

**38.** The first 5 tosses alternate head/tails or tails/heads

**EXERCISES 39–42** □ *An experiment is described and two events E and F are given. Find P(E) and P(F), describe the events E ∩ F, E ∪ F, E′, and F′, and find the probability of each.*

**39.** A single card is drawn from a standard 52-card deck; *E* is the event that a red card is drawn, and *F* is the event that a face card is drawn.

**40.** A six-sided die is rolled; *E* is the event that an odd number is rolled, and *F* is the event that a number less than 5 is rolled.

**41.** Two six-sided dice are rolled; *E* is the event that the sum is even, and *F* is the event that the sum is less than 8.

**42.** Two integers from 1 to 10 are chosen at random; *E* is the event that the sum is odd, and *F* is the event that the sum is greater than 10.

■ *Applications*

**43.** *NCAA Basketball Championship* Suppose that at the beginning of the NCAA basketball tournament, it has been determined that the probability that Duke will win the championship is 0.2 and the probability that Michigan will win is 0.15. Find the probability that one of these two teams will win the championship.

**44.** *Math Grades* Suppose you have determined that the probability of getting an A in this class is 0.25 and the probability of getting a B is 0.6. Find the probability that you will get an A or a B.

**45.** *Lottery Chances* A state lottery is designed so that a player selects 5 numbers from 1 to 40. What is the probability that a single choice of 5 numbers will win the lottery? What is the probability of winning if 20 combinations of numbers are chosen? How many combinations must be chosen so that the probability of winning is $\frac{1}{2}$?

**46.** *Lottery Chances* A state lottery is designed so that a player selects 6 numbers from 1 to 50. What is the probability that a single choice of 6 numbers will win? What is the probability of winning if 100 combinations of numbers are chosen? How many combinations must be chosen so that the probability of winning is $\frac{1}{2}$?

**47.** *Grade Averages* The distribution of high school grade averages for college freshman in the United States in 1990 is given in Figure 12. (Data Source: *The American Freshman: National Norms*, University of California, Los Angeles.) If a college freshman were selected at random, find the probability that the student

  **a.** had a C− to C+ average.

  **b.** did not have an A− to A+ average.

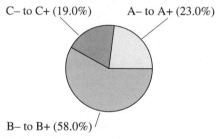

C− to C+ (19.0%)          A− to A+ (23.0%)

B− to B+ (58.0%)
**FIGURE 12**

**48.** *College Enrollment* The distribution of college enrollment in the United States in 1990 is given in Figure 13. (Data Source: *Digest of Education Statistics*, U.S. National Center for Education Statistics.) If a college student were selected at random, find the probability that the student

  **a.** was a minority.

  **b.** was not white.

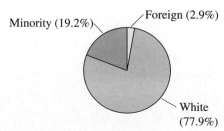

Minority (19.2%)          Foreign (2.9%)

White (77.9%)
**FIGURE 13**

**49.** *Polygraph Testing* A large fast-food chain with 10,000 employees has been experiencing a rash of employee theft and so begins requiring employees to take a polygraph test (i.e., a lie detector test). The test is 90% accurate in the sense that 90% of all employees who are stealing will fail the test and 90% of employees who are not stealing will pass the test. If 5% of the employees are in fact stealing, what is the probability that someone who fails the test is stealing?

50. *Smoking Distribution* In 1989, 12.4% of the population of the United States were black, 28.8% of the entire population were smokers, and 3.8% of the population were black smokers. If an American is selected at random, find the probability of each of the following events.

   a. The person selected is black or smokes.

   b. The person selected is nonblack.

   c. The person selected does not smoke.

   d. The person selected is black but does not smoke.

   e. The person selected is a nonblack smoker.

51. *Hispanic Californians* According to the United States census of 1990, approximately 12% of the population lives in California. Also, 25.8% of all Californians are Hispanic. Nationwide, 9% of the population is Hispanic. If an American is selected at random, find the probability of each of the following events.

   a. The person is a Californian or Hispanic.

   b. The person is a Californian Hispanic.

   c. The person is a non-Californian.

   d. The person is a Californian but not Hispanic.

   e. The person is Hispanic but not a Californian.

52. *Extracurricular Activities* At a certain university, extracurricular activities consist entirely of athletics, Greek organizations (fraternities and sororities), and clubs. You are given the following information regarding the makeup of the student body.

   ▨ 10% participate in athletics.

   ▨ 40% belong to Greek organizations.

   ▨ 45% are club members.

   ▨ 25% are involved in both Greek organizations and clubs, but not athletics.

   ▨ 2% are involved in athletics, Greek organizations, and clubs.

   ▨ 15% are non-Greek, nonathlete, club members.

   ▨ 4% are Greek athletes who do not belong to any clubs.

Assume that a student is selected at random from the student body of this university. Compute the probabilities of each of the following events.

   a. The student is involved in an extracurricular activity.

   b. The students is involved in exactly one extracurricular activity.

   c. The student is involved in a club and is an athlete, but is not a member of a Greek organization.

   d. The student is involved in exactly two extracurricular activities.

## Projects for Enrichment

53. *Assessing Psychic Phenomena* A group of 10,000 people, all of whom claim to be psychic, are assembled on New Year's Eve, 1999. Each is asked to make four predictions regarding the new year:

Nostradamus, a French astrologer and physician of the 16th century, became famous for his prediction of the death of King Henry II.

(1) The day in January of 2000 on which the highest temperature is achieved in Tempe, Arizona.

(2) The party (Democrat or Republican) that wins the presidential election of 2000.

(3) The state in which the percentage change in population is the greatest.

(4) Whether the economy will strengthen or weaken over the course of the year 2000.

   a. Find the probability of correctly guessing the outcome to the above predictions, assuming all outcomes are equally likely.

   b. Find the probability of a given individual being correct on all four predictions, if he or she is guessing at random.

   c. Find the probability of an individual making at least one incorrect prediction.

   d. Find the probability of none of the 10,000 alleged psychics being correct on all four correct predictions.

   e. Find the probability that at least one of the 10,000 alleged psychics will make all correct predictions.

   f. Is it fair to say that any individual who makes four correct predictions is psychic?

g. What is your opinion as to the credibility of psychic predictions? Can mathematics be used to prove or disprove the possibility of psychic phenomena? Explain.

54. *Coincidental Birthdays*  In this project we will investigate the probability that two people in a group have the same birthday. For convenience, we will ignore leap years.

   a. In how many ways can two people have *different* birthdays? (*Hint:* Let us call the two people Alfonso and Belinda. If we assign Alfonso any of the 365 days of the year, then that leaves only 364 possible days for Belinda's birthday. Use the Fundamental Counting Principle.)

   b. Find the number of ways in which two people can have birthdays.

   c. Find the probability of two people having different birthdays.

   d. Find an expression for the number of ways in which $n$ people can have different birthdays. Use the permutation symbol $_nP_r$.

   e. Find an expression for the probability that among a group of $n$ people, no two have the same birthday.

   f. Use the result of part (e) to find the probability that among a group of 30 people, there are at least 2 people with a common birthday.

   g. By trial and error, determine the minimum number of people required to ensure that the probability of at least one common birthday is greater than 0.5.

## Questions for Discussion or Essay

55. Estimates for the probability of getting struck by lightning in a given year range as high as 0.000007. How do you think such estimates are obtained? Do you think such estimates are valid? Why or why not? In Example 8 we saw that the probability of winning a certain state lottery after having played 10 combinations of numbers was 0.0000026, which is lower than the estimate for getting struck by lightning. Can we conclude that one is more likely to get struck by lightning than to win a state lottery? Explain.

56. A weathercaster announces that there is a 70% chance of rain on Saturday and a 30% chance of rain on Sunday, and concludes that there is a 100% chance of rain for the weekend. Is his reasoning correct? If not, what would a more reasonable estimate for the chance of rain over the weekend be?

57. In the column "Ask Marilyn" in Parade Magazine, Marilyn Vos Savant posed the following problem. Suppose that you are a contestant on a game show. You are asked to select one of three doors. Behind one of the doors is a new car and behind each of the other two doors are goats. After choosing your door, the host opens one of the remaining doors to reveal a goat. The host then gives you the choice of staying with your original selection or switching to the other door that is still closed. Ms. Vos Savant claimed that you improve your probability of winning if you switch. She was promptly flooded with letters from irate readers (many of them educators) who argued

that the probability of winning is 0.5 whether you choose to switch or not. They reasoned that there are two doors remaining—behind one is a car, behind the other is a goat—and so the probability of correctly guessing the door with the car behind it is $\frac{1}{2} = 0.5$. Surprisingly, this is not the case. You would in fact improve your probability of winning to $\frac{2}{3}$ by switching to the other door. Explain why this seems counterintuitive. How would you design an experiment to test the strategy? How relevant is the fact that the host of the game show knows which door the car is behind and so will always show you a door with a goat behind it? Suppose there were 10 doors and after your selection the host showed you what was behind all but two, the one you chose and one other. Would you switch doors? What if there were 1000 doors, and again the host showed you what was behind all but two, the one you chose and one other? What light does this shed on the original problem?

58. In Example 10 we considered a drug test that claimed to be 97% accurate and discovered that the probability that a student who tested positive was actually a user was only 0.4. In other words, it was likely that 60% of the students who tested positive were not actually users. On the other hand, the test will fail to detect only 3% of the users. What do these observations suggest about the use of such tests or the interpretation of the results?

**SECTION 5**

# THE BINOMIAL THEOREM

■ What is the coefficient of $x^2$ in the expansion of $(1 + x)^{100}$?

■ How many subsets does a set with 50 elements have?

■ How can $\sqrt{10}$ be approximated to 10 decimal places using only the four basic arithmetic operations?

■ How can you compute $11^7$ *by hand* without ever multiplying?

■ What is Pascal's triangle and how can it be used to expand $(a + b)^5$?

## EXPANDING BINOMIALS

Polynomials with just two terms, such as $a + b$, are called *binomials*. In this section we will develop the **binomial theorem**, a general formula for the expansion of $(a + b)^n$. We are already familiar with some examples of the binomial formula. Several are listed below.

$$(a + b)^0 = 1$$
$$(a + b)^1 = 1a + 1b$$
$$(a + b)^2 = 1a^2 + 2ab + 1b^2$$
$$(a + b)^3 = 1a^3 + 3a^2b + 3ab^2 + 1b^3$$

Notice that in each case the expansion of $(a + b)^n$ is a polynomial in the variables $a$ and $b$, and that each term in the polynomial is of degree $n$. For example, in the expansion of $(a + b)^3$, each term has degree 3; that is, the sum of the exponents of $a$ and $b$ is equal to 3. For convenience, we have written each expression so that the powers of $a$ are decreasing and the powers of $b$ are increasing. Thus, we would expect the expansion of $(a + b)^4$ to be of the form

$$(a + b)^4 = \Box a^4 + \Box a^3b + \Box a^2b^2 + \Box ab^3 + \Box b^4$$

where the entries in each of the boxes are constants called **binomial coefficients**.

We can determine the binomial coefficients using basic counting principles. Suppose that we were to expand $(a + b)^4$ directly by multiplication using the formula

$$(a + b)^4 = (a + b)(a + b)(a + b)(a + b)$$

The terms of the product can be formed by repeatedly selecting either an $a$ or a $b$ from each of the four factors of $(a + b)$. For example, the term $a^4$ is obtained by selecting $a$ from each of the four factors. Since this can be done in only one way, the coefficient of $a^4$ is 1. The terms of the form $a^3b$ will arise by selecting $a$ from three of the four factors and $b$ from the remaining factor. This corresponds to choosing a subset of size 3 from a set with four elements. In Section 3 we saw that this can be done in $\binom{4}{3} = 4$ ways. We illustrate the four choices as follows.

## FOUR WAYS OF OBTAINING $a^3b$

$$(\boxed{a} + b)(\boxed{a} + b)(\boxed{a} + b)(a + \boxed{b})$$
$$(\boxed{a} + b)(\boxed{a} + b)(a + \boxed{b})(\boxed{a} + b)$$
$$(\boxed{a} + b)(a + \boxed{b})(\boxed{a} + b)(\boxed{a} + b)$$
$$(a + \boxed{b})(\boxed{a} + b)(\boxed{a} + b)(\boxed{a} + b)$$

Products of the form $a^2b^2$ will arise by selecting $a$'s from two of the four factors and $b$'s from the other two factors. Thus, the coefficient of $a^2b^2$ will be $\binom{4}{2}$. Similarly, the coefficients of $ab^3$ and $b^4$ will be $\binom{4}{1}$ and $\binom{4}{0}$, respectively. Thus, we have

$$(a + b)^4 = \binom{4}{4}a^4 + \binom{4}{3}a^3b + \binom{4}{2}a^2b^2 + \binom{4}{1}ab^3 + \binom{4}{0}b^4$$

$$= \frac{4!}{4!0!}a^4 + \frac{4!}{3!1!}a^3b + \frac{4!}{2!2!}a^2b^2 + \frac{4!}{1!3!}ab^3 + \frac{4!}{0!4!}b^4$$

$$= a^4 + 4a^3b + 6a^2b^2 + 4ab^3 + b^4$$

More generally, the coefficient of $a^kb^{n-k}$ in the expansion of $(a + b)^n$ is $\binom{n}{k}$. Thus, we have the following formula.

**■ Binomial theorem**

For $n$ a nonnegative integer,

$$(a + b)^n = \binom{n}{n}a^n + \binom{n}{n-1}a^{n-1}b + \binom{n}{n-2}a^{n-2}b^2$$

$$+ \cdots + \binom{n}{1}ab^{n-1} + \binom{n}{0}b^n$$

Thus, the coefficient of $a^kb^{n-k}$ in the expansion of $(a + b)^n$ is $\binom{n}{k}$, where

$$\binom{n}{k} = \frac{n!}{k!(n-k)!}$$

**EXAMPLE 1** *Finding a given coefficient using the binomial theorem*

Find the coefficient of $x^2y^4$ in the expansion of $(x + y)^6$.

**SOLUTION**   Applying the binomial theorem with $n = 6$ and $k = 2$, the coefficient of $x^2y^4$ is given by

$$\binom{6}{2} = \frac{6!}{2!(6-2)!}$$

$$= \frac{6!}{2!4!}$$

$$= \frac{6 \cdot 5}{2} = 15$$

**EXAMPLE 2**    *Finding a given coefficient using the binomial theorem*

Find the coefficient of $x^2 y^3$ in the expansion of $(x - y)^5$.

**SOLUTION**    We begin by expressing $(x - y)^5$ as $[x + (-y)]^5$. Then according to the binomial theorem, the coefficient of $x^2(-y)^3$ will be

$$\binom{5}{2} = \frac{5!}{2!3!} = \frac{5 \cdot 4}{2} = 10$$

Thus, the term in question is $10x^2(-y)^3 = -10x^2 y^3$, and so the coefficient of $x^2 y^3$ is $-10$.

**EXAMPLE 3**    *Expanding with the binomial theorem*

Expand $(x + 2)^5$.

**SOLUTION**    According to the binomial theorem,

$$(x + 2)^5 = \binom{5}{5}x^5 + \binom{5}{4}x^4 2 + \binom{5}{3}x^3 2^2 + \binom{5}{2}x^2 2^3 + \binom{5}{1}x 2^4 + \binom{5}{0}2^5$$

$$= x^5 + \frac{5!}{4!1!}x^4 \cdot 2 + \frac{5!}{3!2!}x^3 \cdot 2^2 + \frac{5!}{2!3!}x^2 \cdot 2^3 + \frac{5!}{1!4!}x \cdot 2^4 + 2^5$$

$$= x^5 + \frac{5}{1}x^4 \cdot 2 + \frac{5 \cdot 4}{2}x^3 \cdot 4 + \frac{5 \cdot 4}{2}x^2 \cdot 8 + \frac{5}{1}x^2 \cdot 16 + 32$$

$$= x^5 + 10x^4 + 40x^3 + 80x^2 + 80x + 32$$

## TECHNIQUES FOR EVALUATING BINOMIAL COEFFICIENTS

Expanding powers of binomials using the binomial theorem can be quite tedious; fortunately, we can take advantage of several properties of binomial coefficients to reduce our work considerably. For example, if we evaluate the binomial coefficient $\binom{7}{3}$ using only factorials, we have

$$\binom{7}{3} = \frac{7!}{3!(7 - 3)!} = \frac{7!}{3!4!} = \frac{5040}{6 \cdot 24} = 35$$

which is a fairly tedious calculation. On the other hand, we can use the formula for combinations

$$\binom{n}{r} = {}_nC_r = \frac{{}_nP_r}{r!}$$

introduced in Section 3 to obtain

$$\binom{7}{3} = \frac{{}_7P_3}{3!}$$

$$= \frac{7 \cdot \cancel{6} \cdot 5}{\cancel{3} \cdot \cancel{2} \cdot 1}$$

$$= 35$$

More generally, we have the following formula.

*Alternative form for binomial coefficients*

The binomial coefficient $\binom{n}{k}$ can be computed as follows.

$$\binom{n}{k} = \frac{n(n-1)(n-2)\cdots(n-k+1)}{k(k-1)(k-2)\cdots(2)(1)}$$

Note that there are a total of $k$ factors in both the numerator and denominator.

**EXAMPLE 4**    *Using the alternative form to evaluate binomial coefficients*

Evaluate each of the following binomial coefficients using the alternative form.

a. $\binom{10}{3}$      b. $\binom{8}{4}$

**SOLUTION**

a. $\binom{10}{3} = \dfrac{10 \cdot 9 \cdot 8}{3 \cdot 2 \cdot 1}$

$= \dfrac{720}{6}$

$= 120$

b. $\binom{8}{4} = \dfrac{\cancel{8} \cdot 7 \cdot \overset{2}{\cancel{6}} \cdot 5}{\cancel{4} \cdot \cancel{3} \cdot \cancel{2} \cdot 1}$

$= 7 \cdot 2 \cdot 5$

$= 70$

As part (b) of Example 4 suggests, it is always possible to cancel the factors in the denominator. This is due to the fact that $\binom{n}{k}$ is always a natural number.

The computation of binomial coefficients can be further simplified by noting that

$$\binom{n}{n-k} = \frac{n!}{(n-k)![n-(n-k)]!}$$

$$= \frac{n!}{(n-k)!k!}$$

$$= \binom{n}{k}$$

We have thus established the following property of binomial coefficients.

**Symmetry property of binomial coefficients**

For integers $n$ and $k$ with $0 \le k \le n$,

$$\binom{n}{k} = \binom{n}{n-k}$$

Thus, the coefficient of $a^k b^{n-k}$ in the expansion of $(a+b)^n$ is the same as the coefficient of $a^{n-k} b^k$.

As a consequence of the symmetry property, one need only compute half of the binomial coefficients of $(a+b)^n$; the other half will be a mirror image of the first half.

**EXAMPLE 5**    *Using the symmetry property to simplify binomial expansions*

Find the binomial expansion of $(a+b)^6$.

**SOLUTION**    According to the binomial theorem, we have

$$(a+b)^6 = \binom{6}{6}a^6 + \binom{6}{5}a^5b + \binom{6}{4}a^4b^2 + \binom{6}{3}a^3b^3$$

$$+ \binom{6}{2}a^2b^4 + \binom{6}{1}ab^5 + \binom{6}{0}b^6$$

Evaluating the "second half" of these coefficients, we have

$$\binom{6}{0} = 1, \quad \binom{6}{1} = \frac{6}{1} = 6, \quad \binom{6}{2} = \frac{6 \cdot 5}{2 \cdot 1} = 15, \quad \text{and} \quad \binom{6}{3} = \frac{6 \cdot 5 \cdot 4}{3 \cdot 2 \cdot 1} = 20$$

By the symmetry property we have

$$\binom{6}{4} = \binom{6}{2} = 15, \quad \binom{6}{5} = \binom{6}{1} = 6, \quad \text{and} \quad \binom{6}{6} = \binom{6}{0} = 1$$

Thus,

$$(a+b)^6 = a^6 + 6a^5b + 15a^4b^2 + 20a^3b^3 + 15a^2b^4 + 6ab^5 + b^6$$

A graphics calculator can also be used to compute binomial coefficients, but note that for most calculators, the combination notation **nCr** is used instead of $\binom{n}{r}$. To compute a binomial coefficient $\binom{n}{r}$, first key in the number $n$, then select **nCr** from the menu of mathematical functions, and finally enter the number $r$.

Another convenient method for evaluating binomial coefficients arises from the observation that the binomial coefficients form a triangular array of numbers known as Pascal's triangle. Listed here are the expansions of $(a + b)^n$ for $n = 1, 2, 3,$ and 4, with the binomial coefficients boxed.

$$(a + b)^0 = \boxed{1}$$
$$(a + b)^1 = \boxed{1}a + \boxed{1}b$$
$$(a + b)^2 = \boxed{1}a^2 + \boxed{2}ab + \boxed{1}b^2$$
$$(a + b)^3 = \boxed{1}a^3 + \boxed{3}a^2b + \boxed{3}ab^2 + \boxed{1}b^3$$
$$(a + b)^4 = \boxed{1}a^4 + \boxed{4}a^3b + \boxed{6}a^2b^2 + \boxed{4}ab^3 + \boxed{1}b^4$$

Now if we write down *only* the binomial coefficients, we have the array known as Pascal's triangle shown in Figure 14.

Note that any given binomial coefficient is the sum of the two binomial coefficients immediately above it in the preceding row. In other words,

$$\binom{n + 1}{k} = \binom{n}{k - 1} + \binom{n}{k}$$

This property, which was considered in Exercise 43 of Section 3, can be used to complete the fifth and sixth rows of Pascal's triangle as shown in Figure 15. Note that the first and last entries of each row are 1's, and that all other entries are obtained by adding together the entries immediately above the given entry, as suggested by the arrows.

We refer to the top row of Pascal's triangle—the row consisting of a single 1—as row 0. The row consisting of two 1's is row 1, and so on. Since the $n$th row of Pascal's triangle consists of the binomial coefficients $\binom{n}{k}$, Pascal's triangle can be used to obtain binomial expansions, as illustrated in the following examples.

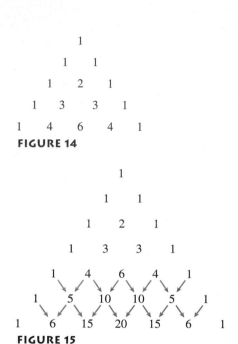

1
1    1
1    2    1
1    3    3    1
1    4    6    4    1

**FIGURE 14**

1
1    1
1    2    1
1    3    3    1
1    4    6    4    1
1    5    10    10    5    1
1    6    15    20    15    6    1

**FIGURE 15**

**EXAMPLE 6**    *Expanding with Pascal's triangle*

Expand the following using Pascal's triangle.

**a.** $(a + b)^6$        **b.** $(3x - 2y)^5$        **c.** $\left(1 + \sqrt{2}\right)^4$

**SOLUTION**

**a.** The sixth row of Pascal's triangle in Figure 15 consists of the numbers 1, 6, 15, 20, 15, 6, and 1. Thus,

$$(a + b)^6 = 1a^6 + 6a^5b + 15a^4b^2 + 20a^3b^3 + 15a^2b^4 + 6ab^5 + 1b^6$$

**b.** The fifth row of Pascal's triangle in Figure 15 consists of the numbers 1, 5, 10, 10, 5, and 1. Thus,

$$(3x - 2y)^5 = 1(3x)^5 + 5(3x)^4(-2y) + 10(3x)^3(-2y)^2 + 10(3x)^2(-2y)^3$$
$$+ 5(3x)(-2y)^4 + 1(-2y)^5$$
$$= 243x^5 - (5 \cdot 81 \cdot 2)x^4y + (10 \cdot 27 \cdot 4)x^3y^2$$
$$- (10 \cdot 9 \cdot 8)x^2y^3 + (5 \cdot 3 \cdot 16)xy^4 - 32y^5$$

$$= 243x^5 - 810x^4y + 1080x^3y^2 - 720x^2y^3$$
$$+ 240xy^4 - 32y^5$$

**c.** From Figure 15 we see that the fourth row of Pascal's triangle consists of the numbers 1, 4, 6, 4, and 1. Thus, we have

$$(1 + \sqrt{2})^4 = \mathbf{1} \cdot 1^4 + \mathbf{4} \cdot 1^3 \cdot \sqrt{2} + \mathbf{6} \cdot 1^2 \cdot (\sqrt{2})^2$$
$$+ \mathbf{4} \cdot 1 \cdot (\sqrt{2})^3 + \mathbf{1} \cdot (\sqrt{2})^4$$
$$= 1 + 4\sqrt{2} + 6 \cdot 2 + 4 \cdot 2\sqrt{2} + 4$$
$$= 17 + 12\sqrt{2}$$

■

---

## EXERCISES 5

**EXERCISES 1–12** □ *Find the coefficient of the indicated term in the expansion of the binomial.*

1. $(A + B)^4$; $AB^3$

2. $(C - D)^3$; $CD^2$

3. $(x + y)^5$; $x^3y^2$

4. $(x + y)^6$; $x^2y^4$

5. $(z + 1)^6$; $z^5$

6. $(x + 2)^3$; $x^3$

7. $(2x - 3y)^3$; $xy^2$

8. $(2a - b)^6$; $a^3b^3$

9. $\left(x + \dfrac{1}{x}\right)^4$; the constant term

10. $\left(z - \dfrac{1}{z}\right)^4$; $z^2$

11. $(1 + w^2)^5$; $w^4$

12. $(x + 2w^3)^4$; $xw^9$

**EXERCISES 13–30** □ *Expand each of the following binomial expressions.*

13. $(x + y)^5$

14. $(a - b)^5$

15. $(2x - y)^4$

16. $(3a + b)^4$

17. $(3x + 5y)^3$

18. $(u - 2w)^4$

19. $(x + 1)^5$

20. $(2s + 1)^4$

21. $(z - 2)^6$

22. $(10w - 1)^6$

23. $(1 + \sqrt{2})^3$

24. $(3 + \sqrt{3})^4$

25. $(\sqrt{2} + \sqrt{3})^4$

26. $(\sqrt{2} - \sqrt{3})^4$

27. $(x^2 - 1)^3$

28. $(y^3 + 1)^3$

29. $\left(x + \dfrac{1}{x}\right)^5$

30. $\left(y - \dfrac{1}{y}\right)^4$

**EXERCISES 31–36** □ *Simplify the given power of the indicated complex number. Recall that $i^2 = -1$.*

31. $(1 + i)^3$

32. $(2 - 3i)^3$

33. $(2 + 2i)^4$

34. $(1 - 3i)^4$

35. $\left(\dfrac{\sqrt{2}}{2} + \dfrac{\sqrt{2}}{2}i\right)^2$

36. $\left(\dfrac{3}{5} + \dfrac{4}{5}i\right)^3$

**EXERCISES 37–44** □ *Verify the given identity involving factorials or binomial coefficients.*

37. $n! \cdot n = (n + 1)! - n!$

38. $(n^2 - n)(n - 2)! = n!$

39. $\dfrac{(n + 1)!}{n!} = n + 1$

40. $\dfrac{(n - 1)!}{(n + 1)!} = \dfrac{1}{n^2 + n}$

41. $\dbinom{n - 1}{k} + \dbinom{n - 1}{k - 1} = \dbinom{n}{k}$

42. $\dbinom{n}{r} = \dfrac{n - r + 1}{r}\dbinom{n}{r - 1}$

43. $\dbinom{n}{r} = \dfrac{n}{n - r}\dbinom{n - 1}{r}$

44. $n\dbinom{n - 1}{r} = (r + 1)\dbinom{n}{r + 1}$

■ *Projects for Enrichment*

**45. The Extended Binomial Theorem** The binomial theorem tells us how to expand expressions of the form $(a + b)^n$ where $n$ is a positive integer. In this project we will consider a generalization of the binomial theorem to the case where $n$ is not an integer. We begin by extending our definition of the binomial coefficients. For a fixed natural number $k$ the expression $\binom{n}{k}$ is a polynomial in the variable $n$. For example,

$$\binom{n}{2} = \frac{n(n - 1)}{2}$$

Clearly, this polynomial is defined whether $n$ is an integer or not. Thus, for example, we could define

$$\binom{\frac{1}{2}}{2} = \frac{\frac{1}{2}\left(\frac{1}{2} - 1\right)}{2} = -\frac{1}{8}$$

Similarly we define for any real number $n$

$$\binom{n}{3} = \frac{n(n - 1)(n - 2)}{3 \cdot 2 \cdot 1},$$

$$\binom{n}{4} = \frac{n(n - 1)(n - 2)(n - 3)}{4 \cdot 3 \cdot 2 \cdot 1}, \quad \cdots$$

**a.** Compute each of the following extended binomial coefficients.

**i.** $\binom{\frac{1}{3}}{2}$     **ii.** $\binom{\frac{3}{2}}{3}$     **iii.** $\binom{\frac{6}{5}}{4}$

Using this extended notion of a binomial coefficient, it can be shown that if $|b/a| < 1$, then

$$(a + b)^n = a^n + \binom{n}{1} a^{n-1}b + \binom{n}{2} a^{n-2}b^2$$

$$+ \binom{n}{3} a^{n-3}b^3 + \binom{n}{4} a^{n-4}b^4 + \cdots$$

Notice here that if $n$ is not a natural number, the expansion has an *infinite* number of terms of the form

$$\binom{n}{k} a^{n-k}b^k$$

As an example, consider the following binomial expansion for $(a + b)^{1/2}$.

$$(a + b)^{1/2} = a^{1/2} + \binom{\frac{1}{2}}{1} a^{-1/2}b + \binom{\frac{1}{2}}{2} a^{-3/2}b^2$$

$$+ \binom{\frac{1}{2}}{3} a^{-5/2}b^3 + \cdots$$

$$= a^{1/2} + \frac{1}{2}a^{-1/2}b - \frac{1}{8}a^{-3/2}b^2 + \frac{1}{16}a^{-5/2}b^3 + \cdots$$

**b.** Find the first four terms of the expansion for the following binomial expressions.

**i.** $(a + b)^{1/3}$     **ii.** $(a + b)^{3/2}$     **iii.** $(a + b)^{6/5}$

One application of the extended binomial theorem is approximating roots, such as $\sqrt{10}$. We first write 10 as the sum of a perfect square and a "small" number, namely $9 + 1$. Now if we set $a = 9$ and $b = 1$ in the binomial expansion of $(9 + 1)^{1/2}$, we obtain

$$(9 + 1)^{1/2} = (9)^{1/2} + \frac{1}{2}(9)^{-1/2}(1) - \frac{1}{8}(9)^{-3/2}(1)^2$$

$$+ \frac{1}{16}(9)^{-5/2}(1)^3 + \cdots$$

$$= 3 + \frac{1}{2} \cdot \frac{1}{3} - \frac{1}{8} \cdot \frac{1}{27} + \frac{1}{16} \cdot \frac{1}{243} + \cdots$$

Taking only the first four terms, we thus have

$$\sqrt{10} \approx 3 + \frac{1}{6} - \frac{1}{216} + \frac{1}{3888} \approx 3.162$$

Note that $a$ and $b$ were chosen so that $a$ was a perfect square. If we were to approximate $(a + b)^{1/3}$ using this procedure, we would want $a$ to be a perfect cube. In addition, it is necessary for $a$ and $b$ to be such that $|b/a| < 1$. Finally, as we will see shortly, the accuracy of the approximation is much better if $b$ is small compared to $a$.

**c.** Use the first four terms of an appropriate binomial expansion to approximate the following roots.

**i.** $\sqrt{26}$     **ii.** $\sqrt{102}$     **iii.** $\sqrt[3]{220}$

Any square root between 9 and 16 can be approximated by writing it in the form $(9 + x)^{1/2}$ and then expanding as we did above. For example, we can approximate $\sqrt{14}$ by expanding $(9 + 5)^{1/2}$. However, if we use only the first four terms of the expansion, the accuracy may not be very good. To see this, consider the following expansion of $(9 + x)^{1/2}$.

$$(9 + x)^{1/2} = (9)^{1/2} + \frac{1}{2}(9)^{-1/2}(x) - \frac{1}{8}(9)^{-3/2}(x)^2$$

$$+ \frac{1}{16}(9)^{-5/2}(x)^3 + \cdots$$

$$= 3 + \frac{1}{6}x - \frac{1}{216}x^2 + \frac{1}{3888}x^3 + \cdots$$

The first four terms give us the polynomial

$$p(x) = 3 + \frac{1}{6}x - \frac{1}{216}x^2 + \frac{1}{3888}x^3$$

d. Use a graphics calculator to plot the graphs of

$$f(x) = \sqrt{9 + x} \quad \text{and} \quad p(x) = 3 + \frac{1}{6}x - \frac{1}{216}x^2 + \frac{1}{3888}x^3$$

on the same coordinate axes. Then use the trace feature to determine the values of $x$ for which $p(x)$ approximates $f(x)$ to at least one decimal place.

46. *Finding the Number of Subsets of a Set*  A set with 3 elements, such as $S = \{a, b, c\}$, has $2^3$ subsets. The subsets of $S$ are $\{\}$ (the empty set), $\{a\}$, $\{b\}$, $\{c\}$, $\{a, b\}$, $\{a, c\}$, $\{b, c\}$, and $\{a, b, c\}$. In this project we will demonstrate that a set with $n$ elements has $2^n$ subsets.

a. Recall from Section 3 that $\binom{n}{k}$ represents the number of subsets of size $k$ of a set with $n$ elements. Show that the *total* number of subsets of a set with $n$ elements is given by

$$\binom{n}{0} + \binom{n}{1} + \binom{n}{2} + \cdots + \binom{n}{n-1} + \binom{n}{n}$$

b. Now use the binomial formula to express the sum in part (a) as a binomial to the $n$th power.

c. Finally, simplify the expression obtained in part (b) to show that the number of subsets of a set with $n$ elements is $2^n$.

---

■ *Questions for Discussion or Essay*

47. Under what circumstances is it easier to use the binomial formula than Pascal's triangle?

48. An extremely common error is for students to simplify $(x + y)^n$ as $x^n + y^n$. Why do you think this error is so common? Are there any values of $x$ and $y$ such that $(x + y)^n = x^n + y^n$? Graph the set of points $(x, y)$ such that $(x + y)^2 = x^2 + y^2$.

49. There was a time when nearly every high school student studying algebra would learn the binomial formula. Recently, however, an increasing number of college students have never been exposed to the binomial formula. In fact, many instructors using this text will not include this section in their syllabi. How important do you feel this topic is? Why might it have been considered more important 40 years ago then it is today?

50. Explain how $(x + y)^4 - 4(x + y)^3y + 6(x + y)^2y^2 - 4(x + y)y^3 + y^4$ can be simplified within 10 seconds.

51. Write $11^7$ as $(10 + 1)^7$, and explain how the binomial theorem can be used to compute $11^7$ without using a calculator. Will this technique work for other similar computations? If so give some examples. If not, explain why not.

---

**SECTION 6**

# MATHEMATICAL INDUCTION

■ What does a cascading line of dominoes have in common with a mathematical technique for proving statements involving the natural numbers?

■ How can a statement such as $n^n \geq n!$ be shown to be true for *every* natural number $n$?

■ How can it be shown that any natural-number debt of $8 or greater could be paid using only $3 and $5 bills?

■ How can the incorrect use of a mathematical technique be used to prove that all horses are the same color?

# THE PRINCIPLE OF MATHEMATICAL INDUCTION

Many important mathematical formulas and theorems involving the natural numbers—that is, the numbers 1, 2, 3, ...—can be proved using a technique known as *mathematical induction*. Before we give a formal statement of mathematical induction, let us consider the following nonmathematical examples of induction.

■ A group of first graders are standing in a line. It is known that any first grader who gets pushed will push the first grader in front of him. It is also known that a second grade bully pushes the first grader at the back of the line. We can thus conclude that every first grader in the line gets pushed.

■ Dominoes are lined up in such a way that if a domino falls, it will knock over the domino in front of it. The first domino is pushed over. It then follows that all of the dominoes will fall.

■ Sufferers of a certain genetic disorder will pass the disorder along to all of their children. It is known that Stan Edbury has this disorder. We can thus conclude that all of Stan Edbury's descendants will have the disorder.

Notice how in each of the above examples we conclude that every member of a list (a line of first graders, a row of dominoes, the descendants of Stan Edbury) satisfies a certain property (being pushed, falling down, suffering from a genetic disorder) provided that two conditions are satisfied.

1. The property is possessed by the first member of the list.
2. If a member of the list has the property, then the next member of the list will have the property as well.

Clearly if either of the conditions is not satisfied, then our conclusion is invalid. For example, if the dominoes are spaced sufficiently far apart that a falling domino *doesn't* knock over the next domino, then even if the first domino is toppled, it doesn't follow that all of the dominoes will fall. Likewise even if our genetic disorder is 100% transmittable to one's offspring, if Stan Edbury himself doesn't have the disorder, then we cannot conclude that all of his descendants will have the disorder.

Mathematical induction is the application of this line of reasoning to statements made about the natural numbers. In this context, the list is the natural numbers, and the property is the truth of a statement about the natural numbers. If we can show that the statement is true for the first natural number (the number 1) and that the truth of the statement for a natural number implies the truth of the statement for the next natural number, then we can conclude that the statement is in fact true for all natural numbers. More formally, we have the following.

**Principle of Mathematical Induction**

> If a statement about the natural numbers is true for the number 1 and if the truth of the statement for the natural number $k$ implies the truth of the statement for $k + 1$ (the next natural number), then the statement is true for all natural numbers.

Proofs by mathematical induction consist of two steps.

*Step 1*  Show that the statement is true for $n = 1$.

*Step 2*  Show that if the statement is true for a natural number $k$, then it is true for $k + 1$.

**EXAMPLE 1**    *Establishing a formula by mathematical induction*

Use mathematical induction to prove that

$$1 + 2 + 3 + \cdots + n = \frac{n(n+1)}{2}$$

for all natural numbers $n$.

**SOLUTION**

*Step 1*  For $n = 1$, the statement reduces to

$$1 = \frac{1(1+1)}{2}$$

$$= 1$$

which is clearly true.

*Step 2*  Assume that the statement is true for the natural number $k$; that is, assume that

$$1 + 2 + 3 + \cdots + k = \frac{k(k+1)}{2} \qquad \text{Replacing } n \text{ with } k \text{ in the original statement}$$

We must show that this implies the truth of the statement for $k + 1$; that is, we must show that

$$1 + 2 + \cdots + k + (k+1) = \frac{(k+1)[(k+1)+1]}{2} \qquad \begin{array}{l}\text{Replacing } n \text{ with } k+1 \\ \text{in the original statement}\end{array}$$

We proceed as follows.

$$1 + 2 + \cdots + k + (k+1)$$

$$= (1 + 2 + \cdots + k) + (k+1)$$

$$= \frac{k(k+1)}{2} + (k+1) \qquad \begin{array}{l}\text{Using the assumption that} \\ 1 + 2 + \cdots + k = \frac{k(k+1)}{2}\end{array}$$

$$= \frac{k(k+1)}{2} + \frac{2(k+1)}{2}$$

$$= \frac{(k+1)(k+2)}{2}$$

$$= \frac{(k+1)[(k+1)+1]}{2}$$

We can thus conclude from the Principle of Mathematical Induction that

$$1 + 2 + \cdots + n = \frac{n(n+1)}{2}$$

for all natural numbers $n$.

EXAMPLE 2     *Establishing divisibility by mathematical induction*

Use mathematical induction to prove that $n^3 - n + 3$ is divisible by 3 for all natural numbers $n$.

**SOLUTION**

*Step 1*  For $n = 1$, our statement reads "$1^3 - 1 + 3$ is divisible by 3," which is true since $1^3 - 1 + 3 = 3$.

*Step 2*  Assume that the statement is true for the natural number $k$. That is, assume $k^3 - k + 3$ is divisible by 3. We must show that $(k + 1)^3 - (k + 1) + 3$ is divisible by 3 also. We start by rewriting $(k + 1)^3 - (k + 1) + 3$.

$$(k + 1)^3 - (k + 1) + 3 = (k^3 + 3k^2 + 3k + 1) - k - 1 + 3$$
$$= (k^3 - k + 3) + (3k^2 + 3k)$$
$$= (k^3 - k + 3) + 3(k^2 + k)$$

Now $3(k^2 + k)$ is divisible by 3, and we are assuming that $k^3 - k + 3$ is divisible by 3. Thus, their sum, $(k^3 - k + 3) + 3(k^2 + k)$, is divisible by 3 also.

We can thus conclude from the Principle of Mathematical Induction that $n^3 - n + 3$ is divisible by 3 for all natural numbers $n$.

◼

EXAMPLE 3     *Establishing an inequality by mathematical induction*

Use mathematical induction to prove that $n^n \geq n!$ for all natural numbers $n$.

**SOLUTION**

*Step 1*  For $n = 1$, we have $1^1 \geq 1!$ or $1 \geq 1$, which is true.

*Step 2*  Assume that the statement is true for the natural number $k$; that is, assume that $k^k \geq k!$. It must be shown that

$$(k + 1)^{k+1} \geq (k + 1)!$$

We proceed as follows.

$$(k + 1)^{k+1} = (k + 1)^k(k + 1)$$
$$> k^k(k + 1) \qquad \text{Using the fact that } (k + 1)^k > k^k$$
$$\geq k!(k + 1) \qquad \text{Using the assumption that } k^k \geq k!$$
$$= (k + 1)!$$

So $(k + 1)^{k+1} \geq (k + 1)!$.

Thus, we can conclude that $n^n \geq n!$ for all natural numbers $n$.

◼

## GENERALIZED PRINCIPLE OF MATHEMATICAL INDUCTION

Let us return to our example of the dominoes lined up in a row with the property that any domino that falls will knock over the domino in front of it. Suppose that instead of knocking over the first domino, we knock over the fifth. We cannot conclude that every domino falls, but we can conclude that every domino starting with the fifth does indeed fall. When this reasoning is applied to statements regarding the natural numbers, the following generalization of the Principle of Mathematical Induction is obtained.

### *Generalized Principle of Mathematical Induction*

If a statement about the natural numbers is true for $n_0$ and if the truth of the statement for any natural number $k \geq n_0$ implies the truth of the statement for $k + 1$, then the statement is true for all natural numbers greater than or equal to $n_0$.

The generalized principle of induction is invaluable when proving statements that do not hold true for small natural numbers $n$, but rather are true from some point on. As with the ordinary Principle of Mathematical Induction, we prove a statement using the Generalized Principle of Mathematical Induction by proceeding in two steps. In step 1 we show that the statement is true for a particular natural number $n_0$, which is usually stated in the problem. In step 2 we assume that the statement is true for some natural number $k$ greater than $n_0$, and we then show that this implies the statement is also true for $k + 1$.

**EXAMPLE 4**    *Proving an inequality using the Generalized Principle of Mathematical Induction*

Show that $n! > 20n$ for $n \geq 5$.

**SOLUTION**

*Step 1*   When $n = 5$, we have $n! = 5! = 120$ and $20n = 20 \cdot 5 = 100$. Since $120 > 100$, the statement is true for $n = 5$.

*Step 2*   Now assume that the statement is true for some natural number $k \geq 5$; that is, that $k! > 20k$ for some $k \geq 5$. We must show that $(k + 1)! > 20(k + 1)$.

$$
\begin{aligned}
(k + 1)! &= (k + 1) \cdot k! & \\
&> (k + 1) \cdot 20k & \text{Using the assumption that } k! > 20k \\
&= k \cdot 20(k + 1) & \\
&> 20(k + 1) & \text{Since } k > 1
\end{aligned}
$$

Thus, $(k + 1)! > 20(k + 1)$.

We can thus conclude from the Generalized Principle of Mathematical Induction that $n! > 20n$ for all $n \geq 5$.

## EXERCISES 6

**EXERCISES 1–24** ☐ *Use the Principle of Mathematical Induction to show that the statement is true for all natural numbers n.*

1. $1 + 3 + 5 + 7 + 9 + \cdots + (2n + 1) = (n + 1)^2$

2. $2 + 4 + 6 + 8 + \cdots + (2n) = n(n + 1)$

3. $1^2 + 2^2 + 3^2 + \cdots + n^2 = \dfrac{n(n + 1)(2n + 1)}{6}$

4. $1^3 + 2^3 + 3^3 + \cdots + n^3 = \dfrac{n^2(n + 1)^2}{4}$

5. $5 + 10 + 15 + \cdots + 5n = \dfrac{5n(n + 1)}{2}$

6. $6 + 10 + 14 + \cdots + (4n + 2) = 2n(n + 2)$

7. $1 + 2 + 2^2 + 2^3 + \cdots + 2^n = 2^{n+1} - 1$

8. $1 - 2 + 2^2 - 2^3 + \cdots - 2^{2n-1} = \dfrac{1 - 2^{2n}}{3}$

9. $1 - 2 + 2^2 - 2^3 + \cdots - 2^{2n-1} + 2^{2n} = \dfrac{1 + 2^{2n+1}}{3}$

10. $3 + 3^2 + 3^3 + \cdots + 3^n = \dfrac{3^{n+1} - 3}{2}$

11. $\dfrac{1}{1 \cdot 2} + \dfrac{1}{2 \cdot 3} + \dfrac{1}{3 \cdot 4} + \cdots + \dfrac{1}{(n - 1) \cdot n} = 1 - \dfrac{1}{n}$

12. $\dfrac{1}{1 \cdot 3} + \dfrac{1}{3 \cdot 5} + \dfrac{1}{5 \cdot 7} + \cdots + \dfrac{1}{(2n - 1) \cdot (2n + 1)} = \dfrac{n}{2n + 1}$

13. $1 \cdot 2 + 3 \cdot 4 + 5 \cdot 6 + \cdots + (2n - 1) \cdot 2n = \dfrac{n(n + 1)(4n - 1)}{3}$

14. $1 \cdot 2 + 2 \cdot 3 + 3 \cdot 4 + \cdots + n \cdot (n + 1) = \dfrac{n(n + 1)(n + 2)}{3}$

15. $1 \cdot 2^1 + 2 \cdot 2^2 + 3 \cdot 2^3 + \cdots + n \cdot 2^n = 2^{n+1}(n - 1) + 2$

16. $\dfrac{1}{2} + \dfrac{3}{2^2} + \dfrac{5}{2^3} + \cdots + \dfrac{2n - 1}{2^n} = 3 - \dfrac{2n + 3}{2^n}$

17. $n^2 + n$ is even.   18. $n^2 - n + 1$ is odd.

19. $n^3 - n$ is divisible by 3.   20. $n^3 + 2n$ is divisible by 3.

21. $n \leq 2^{n-1}$   22. $1 + 2n \leq 3^n$

23. If $a < 1$, then $a^n < 1$.   24. If $a > 1$, then $a^n > 1$.

**EXERCISES 25–30** ☐ *Use the Generalized Principle of Mathematical Induction to show that the statement is true.*

25. $2^n < n!$   for $n \geq 4$   26. $4n < 2^n$   for $n \geq 5$

27. $n + 12 \leq n^2$   for $n \geq 5$   28. $n^2 + 18 \leq n^3$   for $n \geq 4$

29. $n^2 + 4 < (n + 1)^2$   for $n \geq 2$

30. $n^3 > (n + 1)^2$   for $n \geq 3$

**EXERCISES 31–36** ☐ *These exercises deal with the Fibonacci sequence defined in Section 1 by $F_1 = 1$, $F_2 = 1$, and $F_n = F_{n-1} + F_{n-2}$ for $n > 2$. Use the Principle of Mathematical Induction (or the Generalized Principle) to prove each of the following properties of the Fibonacci sequence.*

31. $F_1 + F_2 + \cdots + F_n = F_{n+2} - 1$

32. $F_2 + F_4 + \cdots + F_{2n} = F_{2n+1} - 1$

33. $F_1 + F_3 + \cdots + F_{2n-1} = F_{2n}$

34. $F_1^2 + F_2^2 + F_3^2 + \cdots + F_n^2 = F_n F_{n+1}$

35. $F_{n+1}^2 - F_{n-1}^2 = F_{2n}$ for $n \geq 2$

36. $F_n^2 - F_{n-1}F_{n+1} = (-1)^{n+1}$   for $n \geq 2$

■ *Projects for Enrichment*

37. *Proving the Binomial Theorem* In this project we will use the technique of mathematical induction to establish the binomial theorem:

$$(a + b)^n = \binom{n}{n}a^n + \binom{n}{n - 1}a^{n-1}b + \binom{n}{n - 2}a^{n-2}b^2 + \cdots + \binom{n}{1}ab^{n-1} + \binom{n}{0}b^n$$

a. Show that the theorem is true for $n = 1$.

b. Assume that the binomial theorem is true for $n$. By writing $(a + b)^{n+1}$ as $(a + b)^n(a + b)$, show that

$$(a + b)^{n+1} = \binom{n}{n}a^{n+1} + \binom{n}{n - 1}a^n b + \binom{n}{n - 2}a^{n-1}b^2$$

$$+ \cdots + \binom{n}{1}a^2 b^{n-1} + \binom{n}{0}ab^n$$

$$+ \binom{n}{n}a^n b + \binom{n}{n - 1}a^{n-1}b^2 + \binom{n}{n - 2}a^{n-2}b^3$$

$$+ \cdots + \binom{n}{1}ab^n + \binom{n}{0}b^{n+1}$$

c. Combine like terms in the expression obtained in part (b). Then use this result to show that the coefficient of $a^{(n+1)-j}b^j$ in the expansion of $(a + b)^{n+1}$ is given by

$$\binom{n}{j} + \binom{n}{j-1}$$

d. Use the fact that

$$\binom{n}{j} + \binom{n}{j-1} = \binom{n+1}{j}$$

to complete the induction proof.

38. *Paying Debts with $3 and $5 Bills* In this project we will show that any $n$ debt, where $n$ is a natural number greater than or equal to 8, can be paid using only $3 and $5 bills.

a. Prove by induction that any natural-number debt of the form $3n + 5$, where $n \geq 1$, can be paid using only $3 and $5 bills.

b. Prove by induction that any natural-number debt of the form $3n + 6$, where $n \geq 1$, can be paid using only $3 and $5 bills.

c. Prove by induction that any natural-number debt of the form $3n + 7$, where $n \geq 1$, can be paid using only $3 and $5 bills.

d. Show that the results of parts (a) through (c) imply that any natural-number debt greater than or equal to $8 can be paid using only $3 and $5 bills.

e. There is a slightly more general version of the induction principle that states that if a statement is true for a natural number $k$ and if it can be shown that the truth of the statement for all natural numbers between $k$ and $n$ implies the truth of the statement for $n + 1$, then the statement is true for all natural numbers. Use this more general induction principle to prove that all $n$ debts can be paid using only $3 and $5 bills, provided that $n$ is a natural number greater than or equal to 8.

## ◼ Questions for Discussion or Essay

39. What is wrong with the following proof that all odd numbers are even?

*Any odd number can be written in the form $2n + 1$. Thus, we must show that all numbers of the form $2n + 1$ are even. Suppose that the statement is true for $n$. Then $2n + 1$ is even and hence divisible by 2. There is therefore a natural number $m$ such that $2n + 1 = 2 \cdot m$. Now we must show that $2 \cdot (n + 1) + 1 = 2n + 3$ is even as well. But $2n + 3 = (2n + 1) + 2 = 2 \cdot m + 2 = 2 \cdot (m + 1)$. So $2n + 3$ is divisible by 2 and is thus even. By induction we have shown that all odd numbers are even.*

40. In this question we will investigate a mysterious use of mathematical induction that seemingly proves that all horses are the same color! The theorem is that all horses are of the same color. The "proof" is as follows.

*Suppose that there is 1 horse. Obviously, it is of one color. Now suppose that it is true that any group of $n$ horses have the same color. We will demonstrate that any group of $n + 1$ horses will also be of one color as follows.*

*Consider a group of $n + 1$ horses. Remove 1 horse (whom we will call Silver) from the group so that $n$ horses are left. By our earlier assumption, all $n$ of these horses are the same color. We need only show that Silver is the same color as the rest of the horses.*

*To do this, return Silver to the group, and remove a different horse so that again $n$ horses are left. By hypothesis, these $n$ horses are all colored the same. Since Silver is one of the horses and the rest of the horses are part of the original group of $n$ horses that had the same color, Silver is the same color as*

*the other horses. Thus, by induction, all horses are the same color.*

Since the theorem is obviously false, there must be an error somewhere in the proof. Explain the error in detail.

41. It has been said that mathematical induction is not a method for discovering mathematical statements, but a technique for rigorously proving a statement that has already been discovered. Do you agree? Why or why not?

42. Inductive reasoning is a method of reasoning whereby specific examples lead to a general conclusion. An example would be the argument "Trevor studied hard and got an A on the test; LaShonda studied hard and got an A on the test; therefore, everyone who studies hard will get an A on the test." A mathematical example would be the argument "$5^2 < 2^5$, $6^2 < 2^6$, and $7^2 < 2^7$; therefore $n^2 < 2^n$ for all natural numbers $n$." What do you see as a major problem with inductive reasoning? Under what circumstances is inductive reasoning useful? What is the connection between inductive reasoning and mathematical induction?

43. It is obviously not legitimate to prove a theorem by assuming that which is to be proven. For example, consider the following "proof" that all mathematicians are males.

*Suppose that all mathematicians are men. Let Pat be a mathematician. Since all mathematicians are men, Pat is a man. The same will be true of all other mathematicians. Thus, all mathematicians are men.*

Of course, the reasoning is absurd, but isn't this precisely the same reasoning that is used when in the course of a proof by mathematical induction we assume that the statement is true for $k$? If not, how is it different?

# CHAPTER REVIEW EXERCISES

**EXERCISES 1–8** ☐ *Write out the first five terms of the sequence with the given nth term. If the sequence is arithmetic, indicate its common difference. If it is geometric, indicate its common ratio.*

**1.** $u_n = 1 - 5n$

**2.** $a_n = n + \dfrac{(-1)^n}{n}$

**3.** $c_n = ne^{-n}$

**4.** $b_n = \dfrac{n + 3}{2}$

**5.** $x_n = \dfrac{1}{2} + (-1)^n$

**6.** $a_n = \dfrac{2^n}{3^n}$

**7.** $a_1 = 2$ and $a_n = 3a_{n-1}$

**8.** $u_1 = 256$ and $u_n = \sqrt{u_{n-1}}$

**EXERCISES 9–16** ☐ *Write an expression for the nth term of a sequence whose first few terms are given. If the sequence is arithmetic, indicate its common difference. If it is geometric, indicate its common ratio.*

**9.** $2, 10, 50, 250, \ldots$

**10.** $e, 2e^2, 3e^3, 4e^4, \ldots$

**11.** $1, \dfrac{1}{2}, \dfrac{1}{3}, \dfrac{1}{4}, \ldots$

**12.** $-3, 2, 7, 12, \ldots$

**13.** $1, -\dfrac{3}{2}, -4, -\dfrac{13}{2}, \ldots$

**14.** $24, -6, \dfrac{3}{2}, -\dfrac{3}{8}, \ldots$

**15.** $1, -8, 27, -64, \ldots$

**16.** $-\dfrac{1}{2}, \dfrac{2}{3}, -\dfrac{3}{4}, \dfrac{4}{5}, \ldots$

**EXERCISES 17–20** ☐ *Find the indicated sum.*

**17.** $\displaystyle\sum_{k=1}^{4} (2k + 5)$

**18.** $\displaystyle\sum_{k=1}^{5} 3(2^k)$

**19.** $\displaystyle\sum_{k=2}^{6} \dfrac{1}{k^2}$

**20.** $\displaystyle\sum_{k=0}^{4} \dfrac{k}{k + 1}$

**EXERCISES 21–24** ☐ *Write the series using summation notation.*

**21.** $[5(1) - 2] + [5(2) - 2] + [5(3) - 2] + \cdots + [5(11) - 2]$

**22.** $1^3 + 2^3 + 3^3 + \cdots + 16^3$

**23.** $\dfrac{1}{2} - \dfrac{1}{3} + \dfrac{1}{4} - \dfrac{1}{5} + \cdots - \dfrac{1}{15}$

**24.** $4\left(\dfrac{2}{3}\right)^3 - 4\left(\dfrac{2}{3}\right)^4 + 4\left(\dfrac{2}{3}\right)^5 - 4\left(\dfrac{2}{3}\right)^6 + \cdots + 4\left(\dfrac{2}{3}\right)^{13}$

**EXERCISES 25–32** ☐ *Evaluate the given arithmetic or geometric series.*

**25.** The sum of the first 10 terms of the arithmetic sequence 2, 8, 14, 20, …

**26.** The sum of the first 11 terms of the geometric sequence 2, 8, 32, 128, …

**27.** $10 + 2 + 0.4 + 0.08 + \cdots + 0.0000256$

**28.** $10 + 2 - 6 - 14 - \cdots - 62$

**29.** $\displaystyle\sum_{k=1}^{9} 5(0.3)^k$

**30.** $\displaystyle\sum_{k=1}^{12} (8 + 0.6k)$

**31.** $\displaystyle\sum_{k=3}^{10} (1000 - 9k)$

**32.** $\displaystyle\sum_{k=8}^{20} \dfrac{1}{3}(-3)^k$

**EXERCISES 33–40** ☐ *Evaluate the given expression.*

**33.** $_9C_2$

**34.** $\dfrac{6!}{(6 - 2)!}$

**35.** $_8P_5$

**36.** $\dbinom{5}{4}$

**37.** $\dfrac{5!}{3!(5 - 3)!}$

**38.** $_{10}P_6$

**39.** $\dbinom{10}{3}$

**40.** $_7C_3$

**EXERCISES 41–46** ☐ *Express your answer first using permutation or combination notation, if possible, and then evaluate.*

**41.** A club consisting of 20 members wishes to elect a president, vice president, secretary, and treasurer. In how many ways can this be done?

**42.** A club consisting of 20 members wishes to form a four-member committee. In how many ways can this be done?

**43.** Jermaine's CD wish list has 6 rock groups and 4 rap groups. Unfortunately, he only has enough money to buy 5 of the CDs on the list. Determine the number of different purchases he can make if

 **a.** there are no other restrictions.

 **b.** he intends to buy 3 of the rock groups and 2 of the rap groups.

**44.** Alia's CD player can hold 6 CDs at a time. Her CD collection consists of 10 classical and 8 jazz CDs. Determine the number of ways in which she can program her CD player to play 6 CDs if

 **a.** there are no other restrictions.

 **b.** she would like to listen to 4 classical and 2 jazz CDs.

**45.** A true-false portion of an exam contains 10 questions. In how many different ways can this portion of the exam be completed?

**46.** A multiple-choice portion of an exam contains 10 questions, each with 4 possible answers. In how many different ways can this portion of the exam be completed?

**EXERCISES 47–50** □ *A single card is drawn from a standard 52-card deck. Compute the probability of the given event.*

**47.** Drawing a diamond

**48.** Drawing a face card

**49.** Drawing a diamond or face card

**50.** Drawing a diamond face card

**EXERCISES 51–54** □ *A pair of standard six-sided dice are rolled. Compute the probability of the given event.*

**51.** Rolling an even number on both dice

**52.** Rolling the number 6 on exactly one die

**53.** Rolling a total of 5

**54.** Rolling a total that is either odd or less than 5

**EXERCISES 55–56** □ *Find the probability that in a family of 3 children, the indicated outcome will occur. Assume that the probability of a male birth is 0.5.*

**55.** All of the children are boys.

**56.** At least 2 of the children are boys.

**EXERCISES 57–60** □ *Use Figure 16 to determine the probability that a randomly selected member of the armed services will satisfy the given criteria.*

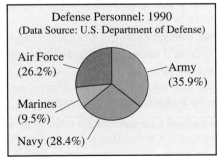

**FIGURE 16**

**57.** Is in the Army

**58.** Is in the Marines

**59.** Is in the Marines or Navy

**60.** Is not in the Air Force

**EXERCISES 61–64** □ *Find the coefficient of the indicated term in the expansion of the binomial.*

**61.** $(x + y)^8$; $x^5y^3$

**62.** $(a - b)^5$; $a^3b^2$

**63.** $(2a - 5b)^4$; $ab^3$

**64.** $(3 + v^2)^7$; $v^6$

**EXERCISES 65–70** □ *Expand each of the following binomial expressions.*

**65.** $(x - y)^4$

**66.** $(2a + 3)^5$

**67.** $\left(r + \dfrac{1}{s}\right)^3$

**68.** $(x^2 - 3y)^4$

**69.** $(\sqrt{z} + 2)^5$

**70.** $(4 + \sqrt{2})^3$

**EXERCISES 71–76** □ *Use the Principle of Mathematical Induction to show that the statement is true for all natural numbers* n.

**71.** $1 + 4 + 7 + \cdots + (3n - 2) = \dfrac{n(3n - 1)}{2}$

**72.** $3 + 9 + 15 + \cdots + (6n - 3) = 3n^2$

**73.** $1^3 + 3^3 + 5^3 + \cdots + (2n - 1)^3 = n^2(2n^2 - 1)$

**74.** $\dfrac{1}{1 \cdot 4} + \dfrac{1}{4 \cdot 7} + \dfrac{1}{7 \cdot 10} + \cdots + \dfrac{1}{(3n - 2)(3n + 1)} = \dfrac{n}{3n + 1}$

**75.** $2^{2n} - 1$ is divisible by 3.

**76.** $5^n - 1$ is divisible by 4.

**77.** *Population Projection* The population of a certain city is increasing at a rate of 5% per year. Assuming a population of 200,000 in 1995 and the growth rate remains the same, find the population for the next 3 years and show that these values form a geometric sequence. What will the population be in 2007?

**78.** *Car Price Inflation* The price of a certain model of car has increased 3% per year. If the price was initially $12,000, find the price for the next 3 years and show that these values form a geometric sequence. What will the price of the car be in the 10th year?

**79.** *Income Projection* Suppose you have been offered a job with a starting salary of $26,000 and guaranteed annual raises of 6%. Use a geometric series to determine the total income for the first 10 years.

**80.** *Sales Projection* At a year-end stockholders' meeting, a company reports that sales for 1994 totaled 1.5 million dollars. Based on a market analysis, the company predicts annual increases in sales of at least 8% through the year 2003. Use a geometric series to determine the total sales for the 10-year period from 1994 to 2003.

**81.** *Exam Possibilities* A mathematics instructor decides to write a final exam by choosing 20 of the 80 problems that appeared on exams during the semester. Use combination notation to write an expression for the number of final exams that are possible if

**a.** there are no additional restrictions.

**b.** there were four exams during the semester, each one had 20 problems, and the final exam must have 5 problems from each exam.

82. *Basketball Recruiting* A college basketball coach wishes to recruit 6 high school players to replace the current seniors on the team when they graduate. She has narrowed her list to a group of 15 prospective players, consisting of 4 guards, 6 forwards, and 5 centers. Determine the number of ways she can rank the top 6 players if

   a. there are no restrictions on positions.

   b. she would like to recruit 2 guards, 3 forwards, and 1 center.

83. *Game Show Probability* On a certain game show, you are given the 5 digits of the price of a car, but you must arrange them in the correct order to win the car. Find the probability that you will win the car if you order the digits at random.

84. *Lottery Probability* In a certain state lottery, a player selects 4 numbers from 1 through 30. Find the probability that a single choice of 4 numbers will win the lottery, assuming the order of the numbers is not important.

## CHAPTER TEST

**PROBLEMS 1–8** □ *Answer true or false.*

1. No sequence is both arithmetic and geometric.

2. Some sequences are neither arithmetic nor geometric.

3. A permutation takes order into account, whereas a combination does not.

4. If $n \geq r$, then the value of $_nC_r$ is a natural number.

5. If $E$ is an event in a sample space $S$, then $0 < P(E) < 1$.

6. If $E$ and $F$ are events in the same sample space $S$, then $P(E \cup F) = P(E) + P(F) - P(E \cap F)$.

7. Each term in the expansion of $(x + y)^n$ has degree $n + 1$.

8. Mathematical induction can be used to prove that a statement is true for all real numbers.

**PROBLEMS 9–14** □ *Give an example of each of the following.*

9. A formula for an arithmetic sequence with common difference $-4$

10. A formula for a geometric sequence with common ratio $\frac{1}{3}$

11. A series that is neither arithmetic nor geometric.

12. An experiment involving coins that has a sample space $S$ with $n(S) = 8$

13. An event that has probability $\frac{1}{3}$

14. A row from Pascal's triangle with at least 5 numbers

15. Write out the first five terms of the sequence $a_n = 3n^2 + 1$.

16. Write an expression for the $n$th term of a sequence whose first few terms are $\frac{2}{1}, \frac{3}{2}, \frac{4}{3}, \frac{5}{4}, \ldots$.

17. Evaluate $\sum_{k=1}^{7} (5k - 2)$.

18. Write the geometric series $\frac{5}{2} + \frac{5}{4} + \frac{5}{8} + \cdots + \frac{5}{64}$ using summation notation and find the sum.

19. How many different ways are there to answer the first 8 problems of this exam? What is the probability that you will get them all correct if you randomly assign answers?

20. A Little League baseball team consists of 7 second graders and 8 first graders. The coach must select 9 children to start the next game. Assuming the coach is not concerned with batting order or playing position, determine the number of ways this can be done if

   a. there are no restrictions.

   b. there must be 5 second graders and 4 first graders on the field at all times.

   How would your solution in part (a) change if the coach was arranging the 9 children in a batting order?

21. A pair of standard six-sided dice are rolled. Find the probability that the sum is a multiple of 3.

22. A single card is drawn from a standard 52-card deck. Find the probability that the card is a black card or a face card.

23. Find the coefficient of $z^4$ in the expansion of $(z + 5)^9$.

24. Expand the expression $(2x - 3y)^4$.

25. Use the Principle of Mathematical Induction to show that $1 + 3 + 5 + \cdots + (2n - 1) = n^2$ for all natural numbers $n$.

# ANSWERS

## CHAPTER 1

### Section 1 ■ page 10

**1.** Natural, whole, integer, rational, real, complex
**3.** Irrational, real, complex     **5.** Complex
**7.** Whole, integer, rational, real, complex
**9.** Rational, real, complex
**11.** Natural, whole, integer, rational, real, complex
**13.** Integer, rational, real, complex     **15.** Complex
**17.** Rational, real, complex     **19.** True     **21.** True
**23.** True     **25.** $>$     **27.** $<$     **29.** $\geq$
**31.** $-23, -22.9, 22.9, 23$     **33.** $3 + \frac{1}{10} + \frac{4}{100}, \pi, \frac{22}{7}, 3.15$
**35.** The number of miles from the earth to the sun, 1,000,000,000, the GNP of the U.S. in dollars, the number of living organisms on earth
**37.** $(-2, 4]$
**39.** $(-7, -\frac{1}{2})$
**41.** $(-\infty, -3)$
**43.** $[\sqrt{2}, \infty)$
**45.** $\{x \mid 3 < x < 5\}$     **47.** $\{x \mid 1 \leq x \leq 10^{100}\}$
**49.** $\{x \mid 1 \leq x \leq 2\}$     **51.** $\{x \mid x > 100\}$
**53.** Answers will vary.     **55.** Answers will vary.
**57.** Answers will vary.

### Section 2 ■ page 19

**1.**

**3.**
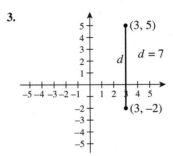
$d = 7$

**5.**
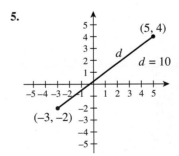
$d = 10$

**7.**
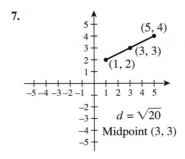
$d = \sqrt{20}$
Midpoint $(3, 3)$

A1

**9.**

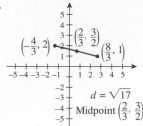

$$d = \sqrt{17}$$
$$\text{Midpoint } \left(\frac{2}{3}, \frac{3}{2}\right)$$

**11.**

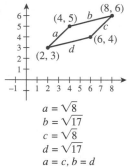

$a = \sqrt{8}$
$b = \sqrt{17}$
$c = \sqrt{8}$
$d = \sqrt{17}$
$a = c, b = d$

**13.**

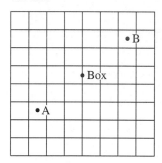

$a = \sqrt{10}$
$b = \sqrt{10}$
$c = \sqrt{20}$
$a^2 + b^2 = c^2$

**15.**

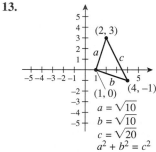

The points appear to lie on a decreasing line.

**17.**

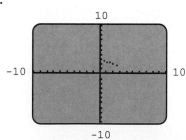

The points appear to be oscillating.

**19.** Counterclockwise; $\sqrt{89} + \sqrt{104} + \sqrt{61} \approx 27.44$

**21.**

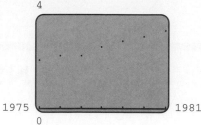

The points appear to lie roughly on a straight line; approximately \$5.25 in 1990 (answers will vary), which is \$1.45 above the actual value of \$3.80.

**23.** $\sqrt{117} + 2 \approx 12.82$ blocks traveling on Lloyd Avenue from point $A$ to the point $(3, 4)$ and then right to point $B$; 17 blocks traveling from point $A$ to the point $(5, -2)$ and then to point $B$

**25.** In the middle of the fourth row of tiles, between the 4th and 5th tiles

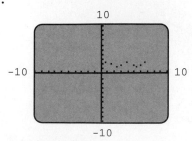

*Section 3 ▪ page 30*

**1.** $a \cdot b = b \cdot a$

**3.** The product of a number and the sum of two other numbers is the same as the sum of the products of the first number paired with each of the other two numbers

**5. a.** Not commutative; $5 \div 3 \neq 3 \div 5$
   **b.** Commutative      **c.** Commutative
   **d.** Not commutative; if Dirk loves Camille, Camille need not love Dirk.
   **e.** Commutative
   **f.** Not commutative; ''Jamal speaks to Ann'' is not the same as ''Ann speaks to Jamal''
   **g.** Not commutative; the statement ''If it is raining, then the game is cancelled'' is not equivalent to the statement ''If the game is cancelled, then it is raining''
   **h.** Commutative

**7.** No          **9.** \$9.60          **11.** 74,999,925

**13.** $a \cdot 0 = a \cdot (-1 + 1)$
$\quad\quad = a \cdot -1 + a \cdot 1$
$\quad\quad = -a + a$
$\quad\quad = 0$

**15.** $4 + 2i$          **17.** $4 - 2i$          **19.** $6 + 2i$          **21.** $-1 + i$

**23.** $5 + i$          **25.** $-\frac{1}{2}i$          **25.** $\frac{1}{2} - \frac{1}{2}i$          **29.** $2 - i$

**31.** $-4 + i$          **33.** $-\frac{3}{4} + 2i$          **35.** $-7$          **37.** $-6.5$

**39.** 16    **41.** 6    **43.** 0    **45.** 2    **47.** $-1$
**49.** $x - 1$    **51.** $-(y + 2)$    **53.** 10    **55.** $\frac{1}{15}$

## Section 4 ■ page 40

**1.** $2^6$    **3.** $x^7$    **5.** $-1$    **7.** 16    **9.** $\frac{1}{9}$    **11.** $-\frac{1}{2}$

**13.** 3    **15.** 0.00001    **17.** $\frac{1}{1.44} \approx 0.6944$    **19.** $-25$

**21.** $\frac{49}{9}$    **23.** $\frac{1}{4096}$    **25.** 4096    **27.** 1    **29.** $a^9$

**31.** $2x^5$    **33.** $\frac{1}{r^2}$    **35.** $y$    **37.** $\frac{a}{b^6}$    **39.** $x^4 y^6$

**41.** $p^6 q^4$    **43.** $\frac{ac^5}{b^4}$    **45.** $\frac{x + y}{x - y}$    **47.** $\frac{3x^6}{y^3}$    **49.** $\frac{z^3}{y^2}$

**51.** $\frac{4y^6 z^{10}}{x^6}$    **53.** 0.00304    **55.** 300,100    **57.** 7.5

**59.** $4 \times 10^6$    **61.** $9.43 \times 10^{-2}$    **63.** $5.2 \times 10^{13}$
**65.** $2.031 \times 10^4$    **67.** $1 \times 10^8$
**69.** $1.3176 \times 10^{25}$ pounds    **71.** $148    **73.** $58,558.82
**75.** The second, since the first pays $9.3 million for the whole month while the second pays $10.7 million on the last day alone

## Section 5 ■ page 50

**1.** Trinomial; degree 2    **3.** Monomial; degree 4
**5.** Not a polynomial    **7.** Polynomial; degree 6
**9.** Not a polynomial    **11.** Monomial; degree 3    **13.** 33
**15.** $-\frac{15}{4}$    **17.** $\frac{7}{3} - \sqrt{2}$    **19.** 7    **21.** 0    **23.** $-1$
**25.** $-64$    **27.** $4x^2 + 3x - 8$    **29.** $-2y^3 - 2y^2 + 5y + 4$
**31.** $\frac{7}{2}x^2 + 4x$    **33.** $6a^2 - a - 2$    **35.** $-6y^2 + \frac{13}{2}y + 2$
**37.** $4t^2 - s^2$    **39.** $x^2 - 3$    **41.** $4x^4 - 1$
**43.** $t^2 + 8t + 16$    **45.** $9a^2 + 12ab + 4b^2$
**47.** $2x^3 - 11x^2 - 2x + 2$    **49.** $x^2 - y^2 + 4y - 4$
**51.** $4s^4 + 11s^3 - 5s^2 + 7s - 2$    **53.** $n^3 + 6n^2 + 11n + 6$
**55.** $6a^2 + a - 7$    **57.** $a^3 - 3a^2 b + 3ab^2 - b^3$
**59.** $3 - 2i$    **61.** $-6 + 17i$    **63.** 18    **65.** $-i$
**67.** $i$    **69.** $z^2 + 4$    **71.** $x^2 + 5 + 12i$

**73.** $\frac{3(2x - 4)}{6} + 2$ simplifies to $x$

**75.** $(x - y)(x + y) + y^2$ simplifies to $x^2$
**77.** Approximately 1178.1 cubic feet
**79.** 324 board feet; approximately 10,050 board feet
**81.** 64 feet; 96 feet; $t = 5$ seconds
**83.** Approximately 282.75 feet

## Section 6 ■ page 59

**1.** $4n(n^4 - 3n^2 + 6)$    **3.** $2ab(6a + 3b - 4)$
**5.** $(w + 2)(w + 5)$    **7.** $(a + 2)(9b + 5)$
**9.** $(4p + 3q)(2p - 3q)$    **11.** $(x - 6)(x + 6)$
**13.** $4(y - 2z)(y + 2z)$    **15.** $2(x - 2)(x + 2)(x^2 + 4)$
**17.** $(s - 3)(s^2 + 3s + 9)$    **19.** $(a + 2b)(a^2 - 2ab + 4b^2)$
**21.** $2(x + 4)(x^2 - 4x + 16)$    **23.** $(x - 4)(x + 3)$
**25.** $(y - 7)(y - 2)$    **27.** $(3m - 2)^2$    **29.** $(y + 3)(2y - 1)$

**31.** $2(t - 2)(t + 2)$    **33.** $(3n - 2)(2n - 5)$
**35.** $(z - 1)^2(z + 1)^2$
**37.** $(t - 2)(t + 2)(t^2 - 2t + 4)(t^2 + 2t + 4)$
**39.** $(p^2 - 5q)(p^4 + 5p^2 q + 25q^2)$
**41.** $3(3x - 1)(9x^2 + 3x + 1)$    **43.** $(3x - y)(z + 2w)$
**45.** $(2x - 3y)^2$    **47.** $6(xy + 2)(x^2 y^2 - 2xy + 4)$

## Section 7 ■ page 68

**1.** $\frac{3a}{5c}$    **3.** $t^2(2t^2 - 5)$    **5.** $x - 2$    **7.** $\frac{r + 5}{2(r - 3)}$

**9.** $y$    **11.** $x + y$    **13.** $\frac{9}{16x}$    **15.** $\frac{2a}{(a + 1)(a - 1)}$

**17.** $\frac{x - 2}{x + 1}$    **19.** $\frac{2}{x(x + 2)^2}$    **21.** $\frac{2(b - a)(a + 2b)}{a - 2b}$

**23.** $\frac{4t}{(t + 1)(t - 1)^2}$    **25.** $-\frac{1}{(r + s)^2}$    **27.** $-1$

**29.** $\frac{6}{t - 6}$    **31.** $\frac{b}{a}$    **33.** $\frac{1}{3(h + 3)}$    **35.** $-\frac{1}{xc}$

**37.** $\frac{(x - 10)^2 - 5x}{x - 20} + 5$ simplifies to $x$ if $x \neq 20$

**41. a.** $1.46 \times 10^{-8}$ N
   **b.** $1.59 \times 10^{-8}$ N
   **c.** The obstetrician

## Section 8 ■ page 80

**1.** $3\sqrt{2}$    **3.** $10\sqrt[4]{10}$    **5.** $243\sqrt[3]{9}$    **7.** $ab^2\sqrt{b}$
**9.** $2x^2 yz\sqrt{7y}$    **11.** $5a^{15}b^{10}c^5\sqrt{2c}$    **13.** $x^3(y + z)^2\sqrt{x}$
**15.** $pqr^2\sqrt[3]{pq^2 r^2}$    **17.** $3s^2 t^2\sqrt[3]{2t}$    **19.** $x + y$

**21.** $\sqrt[3]{x - y}$    **23.** $\frac{a^2}{b}$    **25.** $3\sqrt[3]{2}$    **27.** $7\sqrt{3}$

**29.** $\frac{3 + \sqrt{5}}{2}$    **31.** $\sqrt{a} + \sqrt{b}$    **33.** $\frac{1}{3\sqrt{2}}$

**35.** $\frac{1}{\sqrt{x} + \sqrt{a}}$    **37.** $3\sqrt{5}$    **39.** $2xy\sqrt[3]{5y^2 z^2}$    **41.** $\sqrt{5}$

**43.** $\frac{x\sqrt{y}}{y}$    **45.** $x - 9$    **47.** $6\sqrt[3]{2} - 6\sqrt[3]{4} + 2$

**49.** $4\sqrt{5}$    **51.** $\frac{\sqrt[3]{9}}{3}$    **53.** $(4a + 2)\sqrt{2ab}$    **55.** $\sqrt[3]{x}$

**57.** $x^{1/3}$    **59.** $17^{1/2}$    **61.** $a^{3/4}b^{3/2}$    **63.** $\frac{1}{x^{1/2}}$    **65.** $5i$

**67.** $-\frac{\sqrt{2}}{2}i$    **69.** $8\sqrt{2}i$    **71.** $2\sqrt{2}$    **73.** $\sqrt[3]{b^2}$

**75.** $(a - b)\sqrt{a - b}$    **77.** $x\sqrt[10]{x^3}$    **79.** $\frac{1}{4}$    **81.** 4

**83.** $x$    **85.** $x^{1/3}$    **87.** $3ab^3$    **89.** $z^{1/2}$    **91.** $\frac{x^{3/2}}{y^{2/3}}$

**93.** $\frac{1}{p^{2/3}q^{1/2}r^2}$    **95.** $\sqrt{330} \approx 18.2$ seconds    **97.** $\sqrt{2}\,x$

**99. a.** Approximately 193.7 years    **b.** $1,990,000

*Chapter Review* ■ *page 83*

**1.** Rational, real, complex   **3.** Complex

**5.** $1.4, \sqrt{2}, 1.5, \sqrt{3}, 2$

**7.** $[-5, -1)$

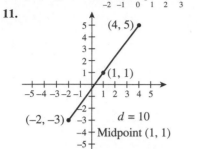

**9.** $(-\infty, \frac{1}{2}]$

**11.**

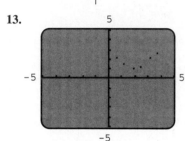

$d = 10$
Midpoint $(1, 1)$

**13.**

As the $x$-values increase, the $y$-values decrease and then increase.

**15.** False   **17.** False   **19.** $5 - i$   **21.** $\frac{3}{10} + \frac{1}{10}i$

**23.** 81   **25.** $\dfrac{a^2}{b^4}$   **27.** $3.14 \times 10^7$   **29.** $1.1 \times 10^{10}$

**31.** $-10$   **33.** $-t^3 - t^2 + 5t$   **35.** $2y^2 - 5y - 12$

**37.** $2x^2 - 2x - 2$   **39.** $(3u - 2)(3u + 2)$

**41.** $(t - 4)(t + 8)$   **43.** $(3x + 1)^2$

**45.** $(2s - 1)(2s + 1)(4s^2 - 2s + 1)(4s^2 + 2s + 1)$

**47.** $\dfrac{1}{x - 3}$   **49.** $\dfrac{t^2 + t + 1}{t(t + 1)}$   **51.** $\dfrac{2(y - 1)}{y(y + 2)}$

**53.** $-\dfrac{x}{(x + 1)(x + 3)}$   **55.** $\dfrac{1}{2t}$   **57.** $a^2 b \sqrt[3]{b}$

**59.** $3\sqrt{3}$   **61.** $\dfrac{x + \sqrt{2}}{x^2 - 2}$   **63.** $2s^3 t \sqrt{3}$   **65.** $11\sqrt{7}$

**67.** $t^{1/2}$   **69.** $x^{-1/3}$   **71.** $\sqrt[3]{x^2}$   **73.** $x^{1/12}$

**75.** $\dfrac{b^2}{a}$   **77.** \$1480

**79.** 84,375,000 cubic feet; approximately 1055.3

**81.** $N_2 = -\dfrac{1000(10i^2 + 15i - 2)}{(1 + i)^2}$; when $i = 0.10$,

$N_2 \approx 330.58$; when $i = 0.15$, $N_2 \approx -359.17$; profitable after 2 years when $i = 0.10$

*Chapter Test* ■ *page 86*

**1.** $-8x^3$   **2.** $\dfrac{v^{1/5}}{u^{2/3}}$   **3.** $\dfrac{b^{10}}{a^{10}}$   **4.** $3x^2 y^3 \sqrt{3x}$

**5.** $2ab^2 \sqrt[4]{15}$   **6.** $\sqrt[3]{2}$   **7.** $3z^4 - 7z^3 + z^2$

**8.** $6x^2 + 5x - 6$   **9.** $\dfrac{x - 1}{x(x - 2)}$   **10.** $\dfrac{y + 3}{y(y + 1)}$

**11.** $(2x - 3)(2x + 3)$   **12.** $(t + 5)(t - 3)$

**13.** $(u + 2v)(u^2 - 2uv + 4v^2)$   **14.** $(a - 2)(a + b)$

**15.** True   **16.** False   **17.** True   **18.** True

**19.** True   **20.** True   **21.** False   **22.** True

**23.** $\frac{2}{3}$, for example   **24.** $\sqrt{2}$, for example

**25.** $i$, for example   **26.** $x^3 y^2 + 2xy - y$, for example

**27.** $x^2 + 2x + 1$, for example   **28.** $x + 3$, for example

**29.** Approximately \$14,588.79

**30.** 24; $-12$; the ball hits the ground at some time between 2 and 3 seconds

---

**CHAPTER 2**

*Section 1* ■ *page 94*

**1.** Yes   **3.** No   **5.** Yes   **7.** Yes   **9.** Yes

**11.** Yes   **13.** No   **15.** Yes   **17.** Yes   **19.** Yes

**21.** Not equivalent since $x = 0$ is a solution of the first equation but not the second

**23.** Not equivalent, since $x = -3$ is a solution of the first equation but not the second

**25.** Equivalent

**27.** Not equivalent since the first equation has no real solutions, but the second equation has solution $x = 1$

**29.** Equivalent

**31. a.** Subtraction property   **b.** Division property

**33. a.** Multiplication property   **b.** Subtraction property
   **c.** Subtraction property   **d.** Division property

**35.** 1.5   **37.** $-3.1, -1.7, 19.7$   **39.** 0.2, 5.8

**41.** 1.8, 2.0, 2.2   **43.** Approximately 1.3 seconds

**45.** Approximately 8.3%

*Section 2* ■ *page 105*

**1.** $x = 2$   **3.** $x = 2$   **5.** $z = -21$   **7.** $x = \frac{4}{3}$

**9.** $m = -\frac{15}{2}$   **11.** $x = \frac{5153}{7100} \approx 0.7258$   **13.** $w = \frac{95}{13}$

**15.** $q = 1.36 \times 10^{-14}$   **17.** $x = -\frac{23}{36} \approx -0.6389$

**19.** $p = \frac{19}{4}$   **21.** $x = 3$   **23.** No solution

**25.** $(-\infty, 13)$

**27.** $(-1, \infty)$

**29.** $[-\frac{1}{4}, \infty)$

**31.** $(-\infty, 6]$

**33.** $(-\infty, \frac{36}{11})$

**35.** $\left(\frac{4}{3}, \infty\right)$

**37.** $(-\infty, -8)$

**39.** $(-\infty, 3.169)$

**41.** $(11, \infty)$

**43.** $[5, \infty)$

**45.** $t = \dfrac{d}{r}$    **47.** $r = \dfrac{c}{2\pi}$    **49.** $t = \dfrac{I}{pr}$    **51.** $l = \dfrac{P - 2w}{2}$

**53.** $x = \dfrac{y - b}{m}$    **55.** $m_1 = \dfrac{Fr^2}{Gm_2}$    **57.** $10 = x - 6; x = 16$

**59.** $4x = 4 + x; x = \frac{4}{3}$    **61.** 215 and 216

**63.** The length is 55 feet and the width is 35 feet.    **65.** 97

**67.** The second offer has a combined salary and benefit package of $28,222.22.

**69. a.** $471.70    **b.** Approximately 17.6%

**71.** 2 hours    **73.** $3\frac{1}{3}$ miles; more than 183 mph

**75.** Approximately 1.36 miles

### Section 3 ■ page 115

**1.** $x = \pm 6$    **3.** $y = \pm\frac{5}{6}$    **5.** $u = 2, u = 3$

**7.** $z = 2, z = 14$    **9.** $x \approx 2.3, x \approx -1.7$

**11.** $|x - 5| = 7; x = 12, x = -2$

**13.** $\left|2x - \left(-\frac{2}{3}\right)\right| = \frac{7}{3}; x = -\frac{3}{2}, x = \frac{5}{6}$

**15.** $\left|\dfrac{x}{2} - 2\right| = 5; x = -6, x = 14$

**17.** $(-2, 6)$

**19.** $[1, \frac{3}{2}]$

**21.** $[3, \frac{9}{2}]$

**23.** $[\frac{5}{2}, \infty)$

**25.** $(-5, 5)$    **27.** $(-3, 3)$    **29.** $(-\infty, -12) \cup (12, \infty)$

**31.** $[3, 7]$    **33.** $(-\infty, -\frac{3}{2}] \cup [-\frac{7}{6}, \infty)$    **35.** $(-\frac{7}{2}, \frac{9}{2})$

**37.** $(-\infty, -\frac{19}{2}) \cup (\frac{13}{2}, \infty)$    **39.** $(16.85, 33.15)$    **41.** $[-1, \frac{7}{3}]$

**43.** $(-\frac{1}{6}, \frac{3}{2})$    **45.** $|x - 1| < 5; (-4, 6)$

**47.** $|2w + 3| > 6; (-\infty, -\frac{9}{2}) \cup (\frac{3}{2}, \infty)$

**49.**

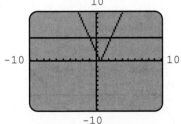

**51.** $x = 2.4, x = 6$    **53.** $141.15 < x < 205.26$

**55.** 13.4 inches $< r <$ 13.6 inches    **57.** $|W - 95| \le \frac{1}{2}$

**59.** Between 98.6 and 102.2 degrees Fahrenheit

### Section 4 ■ page 125

**1.** $x = -2, x = 1$    **3.** $y = -3, y = -\frac{1}{2}, y = \frac{2}{5}$

**5.** $t = -2, t = -1$    **7.** $x = -3, x = 1$

**9.** $x = -3, x = -2, x = 2$    **11.** $y = -5, y = 0$

**13.** $s = 0, s = 3$    **15.** $x = -3, x = \frac{1}{2}$    **17.** $t = \pm 2\sqrt{2}$

**19.** $y = -1 \pm \sqrt{3}$    **21.** $x = 3$    **23.** $t = -\frac{1}{2}, t = 0, t = \frac{1}{2}$

**25.** $w = -\frac{3}{2}$    **27.** $x = -4, x = -2$    **29.** $w = -1, w = -\frac{1}{2}$

**35.** $c = -9$    **37.** $c = 2$    **39.** $c = 4$ or $c = -4$

**41.** $[-3, 1]$    **43.** $(-\infty, -2) \cup (\frac{1}{2}, \infty)$    **45.** $[-5, 1]$

**47.** $(-\infty, \frac{1}{2}) \cup (3, \infty)$    **49.** $(-2\sqrt{2}, 2\sqrt{2})$

**51.** Approximately $(-\infty, 1.86)$

**53.** Approximately $(-\infty, -1.22) \cup (0.72, \infty)$

**55.** $(-1, 0) \cup (3, \infty)$    **57.** $(-\infty, -1)$

**59.** Approximately $(-\infty, -3.12) \cup (0.36, 1.76)$

**61.** $x = -\frac{3}{2}, x = \frac{2}{7}$    **63.** $(-1, 0) \cup (0, \infty)$

**65.** $[-2, 0] \cup [2, \infty)$

**67.** Approximately $(-20.06, 15.24) \cup (26.82, \infty)$

**69.** $t = 10$ seconds    **71.** Approximately 5.9%

**73.** Approximately 118.56 inches

**75.** Approximately 19.8 miles    **77.** Approximately 3.4 years

**79.** Between 2 and 3 seconds    **81.** Between 0 and 27 inches

### Section 5 ■ page 136

**1.** $x = -3, x = 3$    **3.** $x = -7, x = 1$    **5.** $t = -1 \pm 5i$

**7.** $x = -3, x = -2$    **9.** $x = -1, x = 3$

**11.** $x = \dfrac{-3 \pm \sqrt{5}}{2}$    **13.** $x = \dfrac{3 \pm \sqrt{17}}{2}$    **15.** $y = 3 \pm \frac{1}{2}i$

**17.** $x = -2, x = 3$    **19.** $t = -\frac{1}{2}, t = \frac{2}{3}$    **21.** $x = -3$

**23.** $u = 3 \pm \sqrt{6}$    **25.** $x = \dfrac{-2 \pm \sqrt{2}}{2}$

**27.** $x = \dfrac{-1 \pm \sqrt{5}}{2}$    **29.** $x = -1, x = 1$

**31.** $w = \dfrac{-1 \pm \sqrt{3}i}{2}$    **33.** $y = \dfrac{1 \pm \sqrt{2}i}{3}$

**35.** $x = \pm\sqrt{\dfrac{3 - \sqrt{7}}{2}}, x = \pm\sqrt{\dfrac{3 + \sqrt{7}}{2}}$

**37.** $w = \dfrac{-21 \pm 5\sqrt{5}}{2}$    **39.** 2    **41.** 0    **43.** 2

**45.** 1    **47.** 2    **49.** $a = \dfrac{-x \pm \sqrt{x^2 - 4x}}{2}$

**51.** $t = \dfrac{-x \pm |x|\sqrt{7}i}{2}$    **53.** $r = \dfrac{-1 + \sqrt{181}}{6} \approx 2.08$ inches

$h = \dfrac{1 + \sqrt{181}}{2} \approx 7.23$ inches

**55. a.** $t = \dfrac{-1 + \sqrt{2}}{4} \approx 0.1$ seconds

**b.** $t = \dfrac{2 - \sqrt{3}}{4} \approx 0.07$ seconds

**57.** $r = \frac{1}{1499} \approx 6.7 \times 10^{-4}$ feet

**59. a.** $3152.50    **b.** $r = \dfrac{-3 + \sqrt{13}}{2} \approx 0.303 = 30.3\%$

## Section 6 ■ page 150

**1.** $x = 2$    **3.** $y = 10$    **5.** $x = \frac{1}{2}$    **7.** $t = \frac{5}{6}$

**9.** $s = -6$    **11.** $x = -1$    **13.** $x = -\frac{1}{2}, x = 3$

**15.** $x = 1, x = 5$    **17.** No solution    **19.** $x = \frac{3}{2}$

**21.** $x = -5 \pm 2\sqrt{7}$    **23.** $x = -3$

**25.** $(-\infty, -4] \cup (-3, \infty)$    **27.** $(-\infty, -1) \cup (2, 3)$

**29.** $(-\infty, -\frac{3}{2}) \cup (-\frac{1}{3}, -\frac{1}{4}] \cup [2, \infty)$    **31.** $(-\infty, -3] \cup [2, \infty)$

**33.** $x = 9$    **35.** No solution    **37.** $x = \frac{7}{8}$    **39.** $u = 3$

**41.** $t = -\frac{1}{4}$    **43.** $r = 3$    **45.** $x = 4$    **47.** $x = 3$

**49.** $x = 1$    **51.** $y = \frac{5}{4}$    **53.** $x \approx 0.68$

**55.** $x = -1, x \approx -2.08$    **57.** $x \approx 4.44$

**59. a.** $s = \dfrac{dh}{l - h}$    **b.** $s = \frac{110}{23} \approx 4.78$ feet

**61. a.** $h = \frac{180}{41} \approx 4.39$    **b.** $z = \dfrac{hxy}{3xy - hx - hy}$    **c.** $z = 56$

**63. a.** $\dfrac{40}{140 - x}$    **b.** $\dfrac{40}{140 + x}$    **c.** 20 mph

**65.** \$424.26    **67.** Approximately 128 yards or 55.3 yards to the left of B

## Section 7 ■ page 163

**1.** $x = 3, y = 4$    **3.** $x = -2, y = 1$    **5.** $q = -\frac{1}{2}, r = \frac{2}{3}$

**7.** $m = 3, n = 7$    **9.** $x = 2, y = -\frac{3}{5}$    **11.** Inconsistent

**13.** $a = \frac{12}{5}, b = -7$    **15.** $x = 1.3, y = 2.4$

**17.** $x = 3.4, y = 1.7$    **19.** Inconsistent

**21.** $k = 2, m = -5$    **23.** $z_1 = 2 + i, z_2 = 3i$

**25.** $a = 2, b = i$    **27.** $x = 1, y = -1$    **29.** $c = 4, d = \frac{1}{2}$

**31.** $p = \frac{1}{4}, q = \frac{2}{3}$    **33.** $u = 3, v = -5$

**35.** $x = 2, y = -3; x = -\frac{2}{5}, y = \frac{21}{5}$

**37.** $x = \frac{9}{2}, y = \frac{1}{2}, z = -\frac{3}{2}$    **39.** $x = -2, y = 1, z = 3$

**41.** $u = 2, v = -1, w = 0$    **43.** 8 nickels and 4 dimes

**45.** $\frac{50}{9}$ ounces of yogurt and $\frac{40}{9}$ ounces of dessert topping

**47.** Mark is 33 and Benjamin is 3.

**49.** 8000 \$20 tickets and 5000 \$30 tickets

**51.** Arnold's time would be approximately 25.8 hours, and Franco's time would be approximately 35.8 hours.

**53.** The older brother's time is 10 seconds, and the younger brother's time is 12.5 seconds.

**55.** $a = 2, b = -3, c = 4$

**57.** 3 nickels, 7 dimes, and 5 quarters

## Chapter Review ■ page 168

**1.** Yes    **3.** No    **5.** Yes    **7.** No    **9.** $-1.14$

**11.** $-0.33, 1.54$    **13.** $x = \frac{5}{3}$    **15.** $x = \frac{7}{4}$

**17.** $(-4, \infty)$     (number line from $-6$ to $2$, open circle at $-4$)

**19.** $(-\infty, 10)$     (number line from $3$ to $11$, open circle at $10$)

**21.** $x = \dfrac{c - by}{a}$    **23.** $x = 1, x = 4$    **25.** $(\frac{2}{3}, 2)$

**27.** $x = -3, x = 0, x = \frac{1}{2}$    **29.** $v = 2, v = -2$

**31.** $(-\infty, -5) \cup (-1, \infty)$    **33.** $(-0.47, 1.40)$

**35.** $x = 3, x = -4$    **37.** $w = -2 \pm \sqrt{2}$

**39.** $y = -\frac{1}{2}, y = \frac{2}{3}$    **41.** 2    **43.** 1    **45.** $x = 2$

**47.** $x = -\frac{1}{3}, x = 2$    **49.** $(-\infty, 1) \cup (3, \infty)$    **51.** $x = 2$

**53.** $a = 2$    **55.** $x = 3, y = 2$

**57.** $x = 4, y = 1; x = 4, y = -1$    **59.** 117 and 119

**61.** For values of $t$ in the interval $[0, 2]$

**63.** Length 4 feet and width $\frac{3}{4}$ foot

**65.** Side lengths 24 and 45 inches

## Chapter Test ■ page 169

**1.** False    **2.** True    **3.** True    **4.** False

**5.** False    **6.** False    **7.** False    **8.** False

**9.** $x^2 - 6x + 9 = 0$, for example

**10.** $(x - 1)(x - 2)(x - 3) = 0$, for example

**11.** $\begin{matrix} x + y = 2 \\ x + y = 1 \end{matrix}$, for example

**12.** $|x + 1| < 0$, for example    **13.** $t = \frac{1}{2}$

**14.** $x = -\frac{11}{2}, x = \frac{3}{2}$    **15.** $(-\infty, 1) \cup (2, \infty)$

**16.** $u = -2, u = 0, u = 2$    **17.** $x \approx -1.71, x \approx 1.35$

**18.** $(-2, 0)$    **19.** $x = 1 \pm \sqrt{5}$    **20.** $[-1, 5]$

**21.** $x = -16$    **22.** $x = 4$    **23.** $x = 3, y = 5$

**24.** 100, 102, 104

**25.** 3 days for Habitat for Humanity and 6 days for the construction company

# CHAPTER 3

## Section 1 ■ page 180

**5.** $(3, -1)$    **7.** $(3, 1); (-3, 1)$    **9.** $(4, -3\sqrt{3}); (4, 3\sqrt{3})$

**11.**

| $x$ | $y = 1000(x^3 + 1)$ |
| --- | --- |
| $-0.1$ | 999 |
| 0.0 | 1000 |
| 0.1 | 1001 |
| 0.2 | 1008 |
| 0.3 | 1027 |

(graph showing points $(-0.1, 999)$, $(0, 1000)$, $(0.1, 1001)$, $(0.2, 1008)$, $(0.3, 1027)$ on curve $y = 1000(x^3 + 1)$)

**13.** $y$-intercept $(0, -4)$; $x$-intercept $(2, 0)$

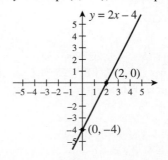

(graph of $y = 2x - 4$ with points $(2, 0)$ and $(0, -4)$)

**15.** *y*-intercept (0, 8); *x*-intercept (12, 0)

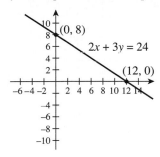

**17.** *y*-intercepts (0, −3), (0, 3); *x*-intercept (−9, 0)

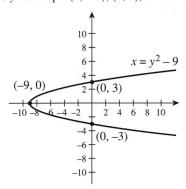

**19.** *y*-intercept (0, 5); *x*-intercepts (1, 0), (5, 0)

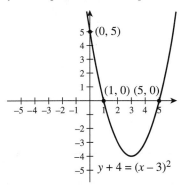

**21.** −3, 3, 7      **23.** −3.52, 2.48, 5.05
**25.** −0.01, 0, 0.01, 4      **27.** −1, 0, 9      **29.** c
**31.** f      **33.** d
**35.** $y = x^2 + 2x + 2$

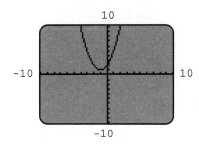

**37.** $y = \dfrac{2}{x + 1}$

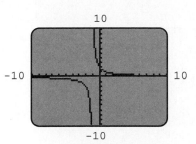

**39.** $y = \sqrt[3]{8 - x^3}$

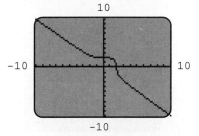

**41.** $y = \pm 2\sqrt{9 - x^2}$

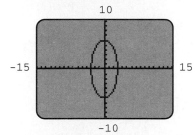

**43.** $y = 3 \pm \sqrt{x + 5}$

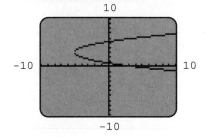

**45.** (2, 3)      **47.** (1, 3) and (−2, 0)      **49.** (1, −1) and (1, 1)
**51.**

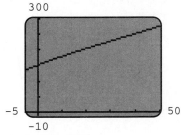

Approximately 1964

**53.**

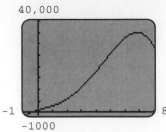

40,000

−1                    8

−1000

Approximately 1988

**55. a.**

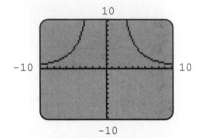

10

−10          10

−10

**b.** base length 5

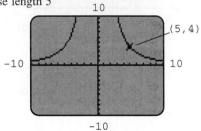

10

(5,4)

−10          10

−10

**57. a.** $P = -x^3 + 39x^3 - 260x - 300$
   **b.** 10 and 30 units
   **c.** $x \approx 22$ units; approximately \$2208

**Section 2** ■ *page 194*

**1. a.**

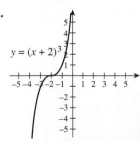

$y = (x + 2)^3$

**b.**

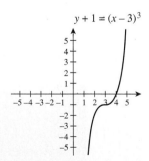

$y + 1 = (x - 3)^3$

**c.**

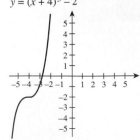

$y = (x + 4)^3 - 2$

**3. a.**

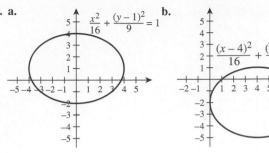

$\dfrac{x^2}{16} + \dfrac{(y-1)^2}{9} = 1$

**b.**

$\dfrac{(x-4)^2}{16} + \dfrac{(y+2)^2}{9} = 1$

**5.** $y - 3 = \sqrt{x + 2}$     **7.** $(x + 1)(y + 2) = 4$

**9. a.**

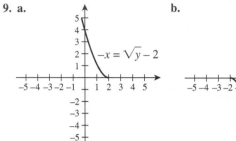

$-x = \sqrt{y} - 2$

**b.**

$x = \sqrt{-y} - 2$

**11. a.**

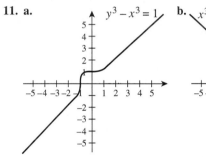

$y^3 - x^3 = 1$

**b.**

$x^3 + y^3 = 1$

**13.** $x^2 - 3x + 5 = y^3$

**15. a.**

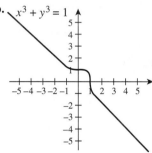

$x = -(y + 2)^3$

**b.**

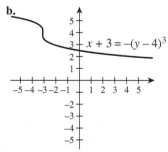

$x + 3 = -(y - 4)^3$

**17.** Symmetric with respect to the $y$-axis
**19.** Symmetric with respect to the $x$-axis
**21.** Symmetric with respect to the $x$-axis
**23.** Symmetric with respect to the origin
**25.** Not symmetric with respect to the $x$-axis, $y$-axis, or origin

**27.** Symmetric with respect to $y$-axis

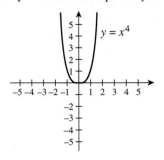

**29.** Symmetric with respect to the origin

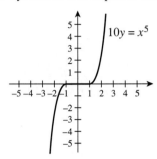

**31.** Symmetric with respect to the $x$-axis, $y$-axis and the origin

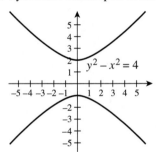

**33.** Symmetric with respect to the $y$-axis

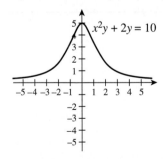

**35. a.** Translate 2 units down.
   **b.** Translate 3 units to the right.
   **c.** Reflect about the $x$-axis and translate 2 units down.
**37. a.** Translate 3 units up.
   **b.** Reflect about the $x$-axis and translate 1 unit to the left.
   **c.** Translate 2 units to the left and 2 units up.

**39.** $\frac{1}{2}$
**41. a.** 0.1587
   **b.** 0.1587
   **c.** 0.6826
   **d.** 0.8413

*Section 3* ■ *page 214*

**1.** 4    **3.** $-\frac{11}{5}$    **5.** 50    **7.** $-1$    **9.** $\frac{22}{13} \approx 1.69$
**11.** 1,000,000    **13.** 3    **15.** 1    **17.** $-3$    **19.** $-\frac{5}{3}$
**21.** Slope of $l_1 = -2$
   Slope of $l_2 = \frac{2}{3}$
   Slope of $l_3 = 3$
   Slope of $l_4 = 0$
**23.** Slope of $l_1 = \frac{3}{7}$
   Slope of $l_2 = -\frac{7}{3}$
   Slope of $l_3 = \frac{3}{7}$
**25.** $y = x + 2$    **27.** $y = -x + 10$    **29.** $x = 3$
**31.** $y = 5x - \frac{11}{6}$    **33.** $y = 1$    **35.** $y = \frac{1}{3}x + \frac{11}{3}$
**37.** $y = -2x + 4$    **39.** $x = 3$    **41.** $y \approx 0.7353x + 2.5176$
**43.** $y = \frac{3}{20}x + \frac{17}{40}$    **45.** $x = \pi$
**47.** $m = \frac{2}{3}, b = -\frac{8}{3}$

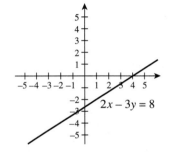

**49.** $m = \frac{1}{3}, b = -\frac{4}{3}$

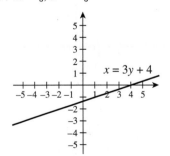

**51.** $m = -2, b = -3$

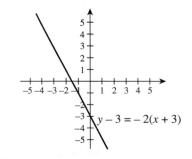

**53.** $y = \frac{2}{3}x + 2$      **55.** $y = 4x + 5$      **57.** $y = -3x + 2$

**59.** $x = 3$      **61.** $3x - y = -4$

**63.** $2x - y = \frac{1}{3}$ or $6x - 3y = 1$      **65.** $(-\frac{13}{5}, -\frac{14}{5})$

**67.** $(2, 5)$

**69. a.** Not on a line      **b.** Not on a line

**73.** $C = 5N + 1000$

**75. a.** $h = \frac{264}{67}t$

    **b.** Approximately 3 inches

    **c.** Approximately 50 inches

**77.** $128,000

**79. a.** 242.4 million      **b.** 271 million

### Section 4 ■ *page 227*

**1.** Center: $(0, 0)$, radius: 5

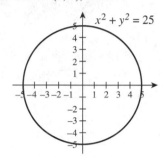

$x^2 + y^2 = 25$

**3.** Center: $(0, 0)$, radius: $2\sqrt{2}$

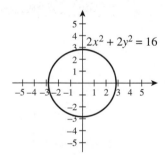

$2x^2 + 2y^2 = 16$

**5.** Center: $(-4, 1)$, radius: 2

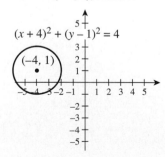

$(x + 4)^2 + (y - 1)^2 = 4$

$(-4, 1)$

**7.** Center: $(2, -3)$, radius: 2

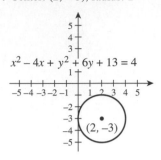

$x^2 - 4x + y^2 + 6y + 13 = 4$

$(2, -3)$

**9.** Center: $(4, -1)$, radius: $4\sqrt{2}$

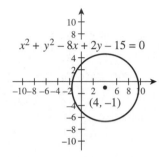

$x^2 + y^2 - 8x + 2y - 15 = 0$

$(4, -1)$

**11.** Center: $(-\frac{3}{2}, -1)$, radius: 2

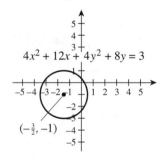

$4x^2 + 12x + 4y^2 + 8y = 3$

$(-\frac{3}{2}, -1)$

**13.** $(x + 1)^2 + (y - 3)^2 = 16$      **15.** $(x - 4)^2 + (y + 3)^2 = 25$

**17.** $(x + 1)^2 + (y - 5)^2 = 18$      **19.** $(x - 2)^2 + (y + 2)^2 = 4$

**21.** $(0, 0)$

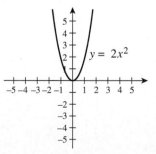

$y = 2x^2$

**23.** $(0, 0)$

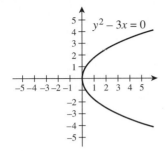

**25.** $(0, 0)$

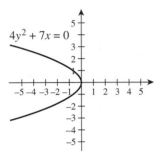

**27.** $(-5, 3)$

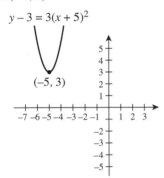

**29.** $(2, 1)$

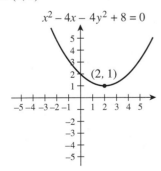

**31.** $\left(-\frac{5}{2}, -\frac{3}{2}\right)$

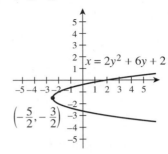

**33.** $y = 2x^2$    **35.** $x + 3 = -\frac{1}{8}(y - 4)^2$    **37.** $y = x^2 - 2x$

**39.** $y = x^2 - 4x$, vertex $(2, -4)$

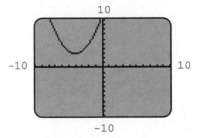

**41.** $y = \frac{1}{2}x^2 + 4x + 11$, vertex $(-4, 3)$

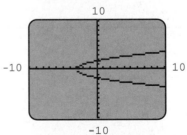

**43.** $x + 3 = y^2$, parabola

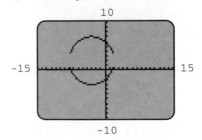

**45.** $(x + 3)^2 + (y - 2)^2 = 5^2$, circle

**47. a.**

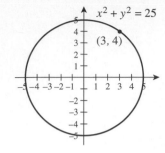

**b.** $m = -\frac{3}{4}$  **c.** $y = -\frac{3}{4}x + \frac{25}{4}$  **d.** $y = -\frac{3}{4}x + \frac{25}{4}$

**49.** Approximately 15.2 miles east and 13 miles north of point $A$

**51. a.** $x^2 + (y - 35)^2 = 30^2$  **b.** Approximately 63.3 feet

**53.** Maximum height: 50 feet
Horizontal distance traveled: 200 feet

**55. a.** $y - 87.5 = -0.12x^2$  **b.** 1 foot

## Section 5 ■ page 241

**1.** $\frac{x^2}{9} + \frac{y^2}{25} = 1$

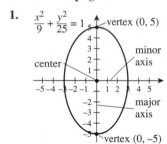

**3.**

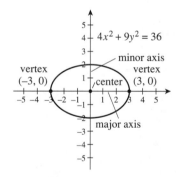

**5.**

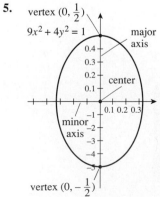

**7.**

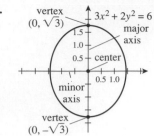

**9.**

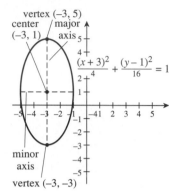

**11.**

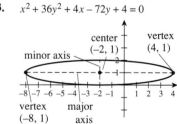

**13.** $x^2 + 36y^2 + 4x - 72y + 4 = 0$

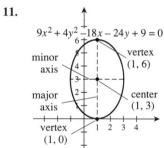

**15.**

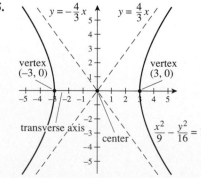

**17.**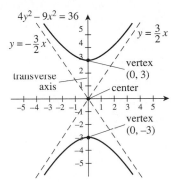

**19.**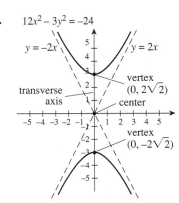

**21.**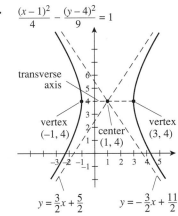

**23.**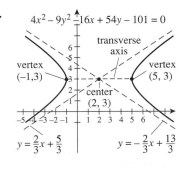

**25.**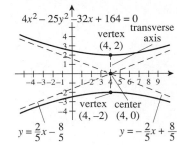

**27.** Parabola   **29.** Circle   **31.** Ellipse   **33.** Hyperbola

**35.** $y = \pm\frac{1}{4}\sqrt{80 - 10x^2}$;
vertices approximately $(\pm 2.83, 0)$

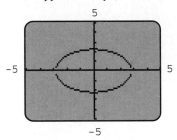

**37.** $y = \pm\sqrt{3x^2 - 4x - 9}$;
vertices approximately $(-1.19, 0)$ and $(2.52, 0)$

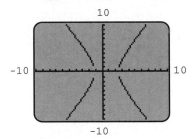

**39.** $y = 3 \pm \sqrt{-2x^2 - 3x + 5}$;
vertices approximately $(-0.75, 5.47)$ and $(-0.75, 0.53)$

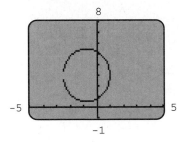

**41.** $\frac{1}{3}x^2 - y^2 - 2 = 0$, hyperbola

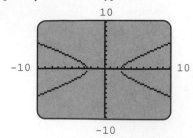

**43.** $2x^2 + y^2 + 8x - 2y - 1 = 0$, ellipse

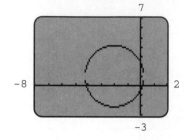

**45.** $\dfrac{x^2}{93^2} + \dfrac{y^2}{92.9^2} = 1$

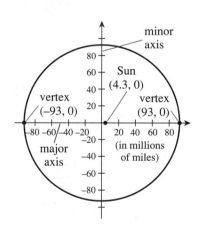

**47. a.** $\dfrac{x^2}{25^2} + \dfrac{y^2}{20^2} = 1$    **b.** 12 feet    **c.** Yes

*Section 6 ■ page 252*

**1.** $(0, 4), (0, -4)$     **3.** $(\sqrt{5}, 0), (-\sqrt{5}, 0)$

**5.** $\left(0, \dfrac{\sqrt{5}}{6}\right), \left(0, -\dfrac{\sqrt{5}}{6}\right)$     **7.** $(0, 1), (0, -1)$

**9.** $(-3, 1 + 2\sqrt{3}), (-3, 1 - 2\sqrt{3})$

**11.** $(1, 3 + \sqrt{5}), (1, 3 - \sqrt{5})$

**13.** $(-2 - \sqrt{35}, 1)(-2 + \sqrt{35}, 1)$

**15.** $(5, 0), (-5, 0)$     **17.** $(0, \sqrt{13}), (0, -\sqrt{13})$

**19.** $(0, \sqrt{10}), (0, -\sqrt{10})$

**21.** $(1 + \sqrt{13}, 4), (1 - \sqrt{13}, 4)$

**23.** $(2 - \sqrt{13}, 3), (2 + \sqrt{13}, 3)$

**25.** $\left(4, \sqrt{29}\right), \left(4, -\sqrt{29}\right)$

**27.** Focus: $(0, \frac{1}{8})$, directrix: $y = -\frac{1}{8}$

**29.** Focus: $(\frac{3}{4}, 0)$, directrix: $x = -\frac{3}{4}$

**31.** Focus: $(-\frac{7}{16}, 0)$, directrix: $x = \frac{7}{16}$

**33.** Focus: $(-5, \frac{37}{12})$, directrix: $y = \frac{35}{12}$

**35.** Focus: $(2, 2)$, directrix: $y = 0$

**37.** Focus: $(-\frac{19}{8}, -\frac{3}{2})$, directrix: $x = -\frac{21}{8}$

**39.** $\dfrac{x^2}{16} + \dfrac{y^2}{9} = 1$    **41.** $\dfrac{x^2}{4} + \dfrac{y^2}{6} = 1$    **43.** $\dfrac{x^2}{36} + \dfrac{y^2}{36/5} = 1$

**45.** $\dfrac{y^2}{16} - \dfrac{x^2}{9} = 1$    **47.** $\dfrac{y^2}{4} - \dfrac{x^2}{16} = 1$    **49.** $y = -\frac{1}{12}x^2$

**51.** $x = -\frac{1}{16}y^2$

**53.** Perihelion: approximately 0.3 A.U.
    Aphelion: approximately 4.1 A.U.

**55.** Approximately 374 feet above the vertex

**57. a.** $\dfrac{y^2}{5512} - \dfrac{x^2}{4488} = 1$

    **b.** Approximately 60.5 miles east of transmitter $A$

*Chapter Review ■ page 256*

**1.** $(2, -3)$     **3.** $(3, -2), (-3, -2)$

**5.** $x$-intercept: $(-2, 0)$, $y$-intercept: $(0, 1)$

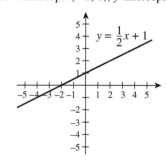

**7.** $x$-intercept $(-16, 0)$, $y$-intercepts: $(0, 4), (0, -4)$

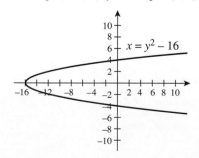

**9.** $0, 0.38, 2.62$    **11.** $1.41, -1.41$    **13.** $\left(-\frac{6}{7}, -\frac{32}{7}\right)$

**15.** $(-2, 4), (2, 4)$

**17. a.**

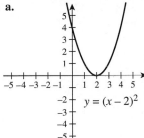

**b.**

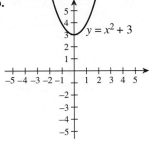

**c.**

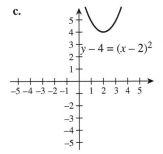

**d.**

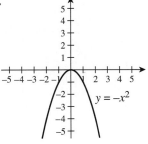

**19. a.**

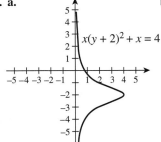

**b.**

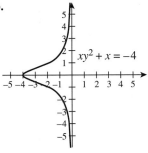

**c.**

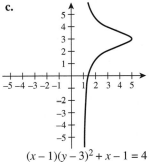

**21.** $(x + 1)^2 + (y - 3)^4 = 16$     **23.** $-y = x^4 - 4x^2 + 2$

**25.** Symmetric with respect to the $x$-axis

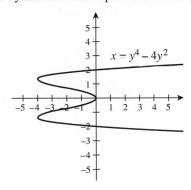

**27.** Symmetric with respect to the origin

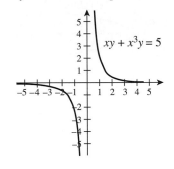

**29.** $\frac{1}{3}$     **31.** 4     **33.** $y = 3x + 2$     **35.** $x = 2$
**37.** $y = -2x + 4$     **39.** $y = \frac{9}{10}x + \frac{51}{20}$     **41.** $y = -2x + 2$
**43.** $y = -5x + 5$     **45.** $y = -3x + 22$
**47.** $y = -\frac{3}{2}x - 4$
**49.** Center: (0, 0), radius: 4

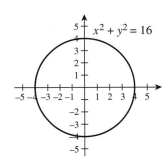

**51.** Center: (2, $-1$), radius: 3

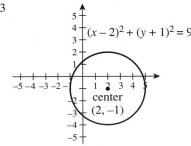

**53.** Center: $(-5, 4)$, radius: 3

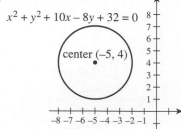

$x^2 + y^2 + 10x - 8y + 32 = 0$

center $(-5, 4)$

**55.** $(x - 2)^2 + (y - 5)^2 = 9$
**57.** $(x - 2)^2 + (y - 4)^2 = 100$
**59.** Vertex: $(0, 0)$

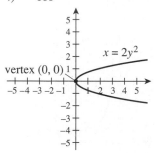

$x = 2y^2$

vertex $(0, 0)$

**61.** Vertex: $(3, -4)$

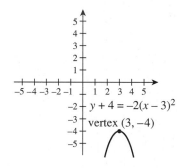

$y + 4 = -2(x - 3)^2$

vertex $(3, -4)$

**63.** Vertex: $(4, -1)$

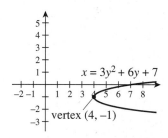

$x = 3y^2 + 6y + 7$

vertex $(4, -1)$

**65.** Center: $(0, 0)$

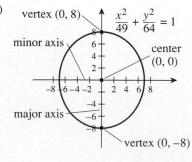

vertex $(0, 8)$

$\dfrac{x^2}{49} + \dfrac{y^2}{64} = 1$

minor axis

center $(0, 0)$

major axis

vertex $(0, -8)$

**67.**

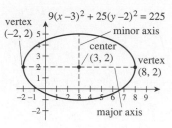

$9(x - 3)^2 + 25(y - 2)^2 = 225$

vertex $(-2, 2)$

minor axis

center $(3, 2)$

vertex $(8, 2)$

major axis

**69.**

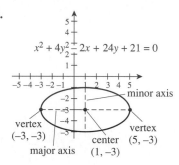

$x^2 + 4y^2 - 2x + 24y + 21 = 0$

minor axis

vertex $(-3, -3)$

center $(1, -3)$

vertex $(5, -3)$

major axis

**71.**

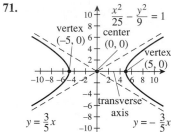

$\dfrac{x^2}{25} - \dfrac{y^2}{9} = 1$

vertex $(-5, 0)$

center $(0, 0)$

vertex $(5, 0)$

transverse axis

$y = \dfrac{3}{5}x$

$y = -\dfrac{3}{5}x$

**73.**

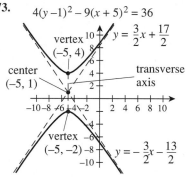

$4(y - 1)^2 - 9(x + 5)^2 = 36$

vertex $(-5, 4)$

$y = \dfrac{3}{2}x + \dfrac{17}{2}$

center $(-5, 1)$

transverse axis

vertex $(-5, -2)$

$y = -\dfrac{3}{2}x - \dfrac{13}{2}$

**75.** $25x^2 - 16y^2 - 150x + 64y - 239 = 0$

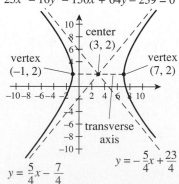

center $(3, 2)$

vertex $(-1, 2)$

vertex $(7, 2)$

transverse axis

$y = \dfrac{5}{4}x - \dfrac{7}{4}$

$y = -\dfrac{5}{4}x + \dfrac{23}{4}$

**77.** Ellipse    **79.** Parabola    **81.** Parabola
**83.** Circle, $x^2 + y^2 = 16$

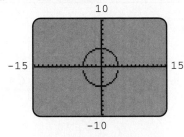

**85.** Parabola, $x - 4 = (y - 2)^2$

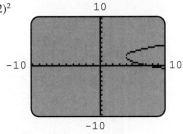

**87.** $(0, \sqrt{15})$, $(0, -\sqrt{15})$    **89.** $(7, 2)$, $(-1, 2)$
**91.** $(\sqrt{34}, 0)$, $(-\sqrt{34}, 0)$
**93.** $(-5, 1 + \sqrt{13})$, $(-5, 1 - \sqrt{13})$
**95.** Focus: $(\frac{1}{8}, 0)$; directrix: $x = -\frac{1}{8}$
**97.** Focus: $(3, -4\frac{1}{8})$; directrix: $y = -3\frac{7}{8}$
**99.** $\dfrac{x^2}{9} + \dfrac{y^2}{4} = 1$    **101.** $\dfrac{x^2}{25} - \dfrac{y^2}{9} = 1$    **103.** $y = x^2$
**105.** 10 units    **107.** \$1.25

### Chapter Test ■ *page 261*

**1.** Intercepts: $(0, 0)$, $(0, -4)$, vertex: $(-4, -2)$

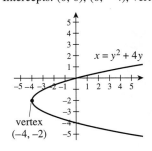

**2. a.** ii    **b.** i    **c.** iii
**3. a.**

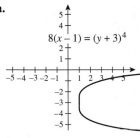

**b.**

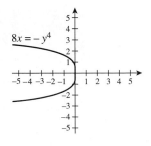

**4.** Symmetric with respect to the *x*-axis

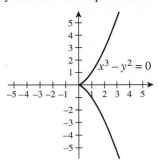

**5.** $(x - 1)^2 + (y - 3)^2 = 25$    **6.** $x = -2$
**7.** $y = \frac{3}{4}x + \frac{19}{4}$
**8.**

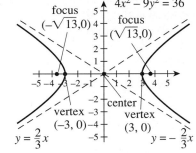

**9.**

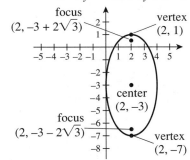

**10.** False    **11.** True    **12.** True    **13.** True
**14.** False    **15.** False    **16.** True    **17.** False
**18.** $y = 2$, for example    **19.** $x = y^2$, for example
**20.** $y = -x$, for example    **21.** $x = -y^2$, for example
**22.** $\dfrac{x^2}{16} + \dfrac{y^2}{4} = 1$, for example

**23.** A satellite dish, for example
**24. a.** Either 100 or 300 frames
     **b.** Approximately 215 frames; approximately \$15,396 profit

## CHAPTER 4

**Section 1 ■ page 271**

**1.** Function    **3.** Not a function    **5.** Function
**7.** Not a function    **9.** Function    **11.** Function
**13.** Not a function    **15.** Function
**17. a.** 12    **b.** 2    **c.** 2

**19. a.** $\frac{14}{11}$    **b.** $\frac{17}{15}$    **c.** $\frac{-3t + 2}{-4t - 5}$

**21. a.** $3t^4$    **b.** $3t^2 - 6t + 3$    **c.** $\frac{75}{t^2}$

**23. a.** $x^2$    **b.** $x^4 + 2x^2 + 1$    **c.** $\frac{1}{a^2}$

**25.** $h(x) = |x|$    **27.** $f(x) = \frac{1}{x + 3}$

**29.** All real numbers    **31.** All real numbers except 4
**33.** All real numbers except 3 and $-3$    **35.** $[-1, \infty)$
**37.** All real numbers in $[-2, \infty)$ except 1 and 3
**39.** All real numbers except 3 and $-3$
**41.** $[0, \infty)$    **43.** All real numbers

**45.** $f(x) = 2x - 1$    **47.** $f(x) = \frac{1}{x^2}$    **49.** $f(x) = \sqrt{7 - x}$

**51.** $f(x) = \frac{1}{x(x - 3)}$    **53.** All real numbers

**55. a.** $x = -2 \pm \sqrt{y - 2}$    **b.** $y \geq 2$    **c.** $[2, \infty)$

**57.** $V = \frac{\pi}{6}d^3$    **59.** $A = 2r^2$    **61.** $A = \frac{1}{4}x^2$

**63.** $A = \frac{a^2}{2(a - 2)}$; the domain is the set $(2, \infty)$

**65.** $V = x(16 - 2x)^2$; the domain is the set $(0, 8)$

**67.** $P = \frac{2000}{(1 + r)^4}$; the domain is the set of all real numbers except $-1$; no

**69.** $V \approx (8.603 \times 10^{38})t^3$; $4.358 \times 10^{43}$ cubic miles
**71.** $R(x) = \frac{2}{5}x - 220$

**Section 2 ■ page 286**

**1.**

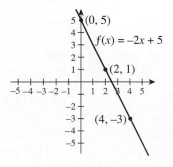

**3.**

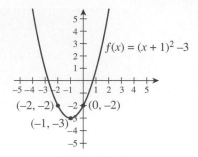

$f(x) = (x + 1)^2 - 3$

**5.**

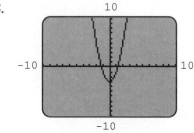

Domain: $(-\infty, \infty)$
Range: $[-3, \infty)$

**7.**

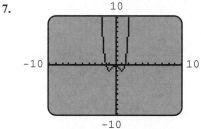

Domain: $(-\infty, \infty)$
Range: $[-1, \infty)$

**9. a.**

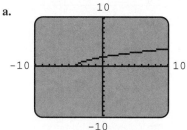

Domain: $[-4, \infty)$
Range: $[0, \infty)$

**b.**

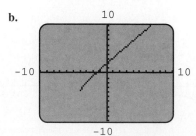

Domain: $[-4, \infty)$
Range: $[-4, \infty)$

**c.**

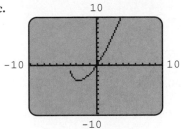

Domain: $[-4, \infty)$
Range: $[-3.08, \infty)$

**d.**

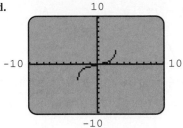

Domain: $(-3, 3)$
Range: $(-\infty, \infty)$

**d.**

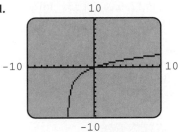

Domain: $(-4, \infty)$
Range: $(-\infty, \infty)$

**13.** $-\frac{1}{3}$ **15.** $-2, 4$ **17.** $-\frac{5}{2}$ **19.** $-2, 2$ **21.** $3$

**23.** $-2.56, 0.10, 3.96$ **25.** $0.67$ **27.** $-1.24, 3.24$

**29. a.** $h(x) = |x - 2|$
   **b.** $h(x) = |x| + 1$
   **c.** $h(x) = |x + 2| - 2$

**31. a.** $g(x) = -x^3 + 3x^2 - 2$
   **b.** $g(x) = -x^3 - 3x^2 + 2$

**33. a.**

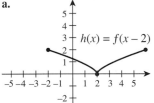

**b.**

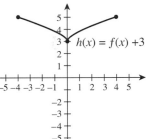

**11. a.**

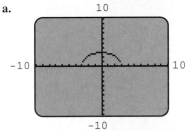

Domain: $[-3, 3]$
Range: $[0, 3]$

**c.**

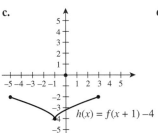

**d.**

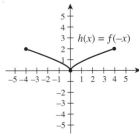

**b.**

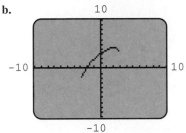

Domain: $[-3, 3]$
Range: $[-3, 4.24]$

**c.**

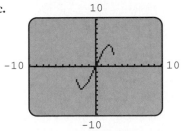

Domain: $[-3, 3]$
Range: $[-4.5, 4.5]$

**e.**

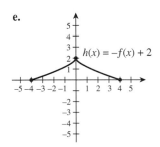

**f.**

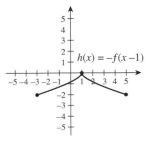

**35. a.**  $h(x) = f(x - 3) + 2$  **b.**

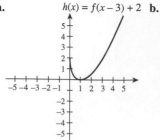

$h(x) = f(-x) - 1$

**c.**  **d.**

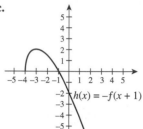

$h(x) = -f(x + 1)$

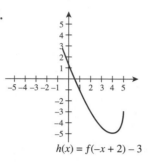

$h(x) = f(-x + 2) - 3$

**37.** Even  **39.** Odd  **41.** Neither

**43. a.**  **b.**

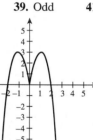
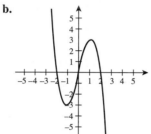

**45. a.**  **b.**

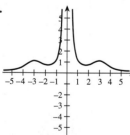
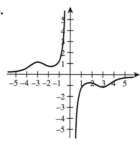

**47. a.**  **b.**

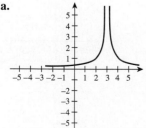
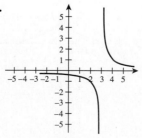

**51.** Increasing: $(2, \infty)$
Decreasing: $(-\infty, 2)$
Extrema: $(2, 1)$
**53.** Increasing: $(-\infty, -0.62)$, $(1.62, \infty)$
Decreasing: $(-0.62, 1.62)$
Extrema: $(-0.62, 7.09)$, $(1.62, -4.09)$
**55.** Increasing: $(-\infty, 2)$
Decreasing: $(2, \infty)$
Extrema: $(2, 5)$
**57.** Increasing: $(-1.3, 0.17)$, $(1.13, \infty)$
Decreasing: $(-\infty, -1.3)$, $(0.17, 1.13)$
Extrema: $(-1.3, -3.51)$, $(0.17, 0.08)$, $(1.13, -1.07)$
**59.** $Q(x) = (x + 5)^2 + 14(x + 5) + 100$
**61.** $g(t) = -16t^2 + 80t + 20$; the graph of $g$ is the graph of $f$ shifted 20 units upward
**63.** Approximately 30 or 270 units per month to break even; 150 units per month to maximize profit
**65.** $(2, \frac{3}{2})$   **67.** $50' \times 100'$
**69. a.**

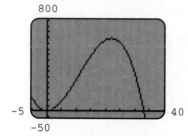

Approximately 645.33 feet
**b.** $g(t) = 33t$
**c.**

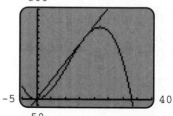

They meet at $t = 16.5$ seconds; the support car is furthest from the check point at $t = 22$ seconds.
**d.** 33 feet per second
**71.** Increasing: $[0, 1.54)$, $(11.29, 15.72)$
Decreasing: $(1.54, 11.29)$, $(15.72, 17]$
Reserves highest at $x \approx 1.52$, which corresponds to midway through 1974
**73.** Increasing: $[0, 28]$
Decreasing: none

*Section 3*  ■  *page 306*

**1. a.** $(f + g)(x) = x + 5$
  **b.** $(f - g)(x) = x - 5$
  **c.** $(fg)(x) = 5x$

**d.** $\left(\dfrac{f}{g}\right)(x) = \dfrac{x}{5}$

**3. a.** $(f + g)(x) = x^2 + 3x + 1$

  **b.** $(f - g)(x) = x^2 - 3x - 1$

  **c.** $(fg)(x) = 3x^3 + x^2$

  **d.** $\left(\dfrac{f}{g}\right)(x) = \dfrac{x^2}{3x + 1}, x \neq -\dfrac{1}{3}$

**5. a.** $(f + g)(x) = 2x$

  **b.** $(f - g)(x) = 0$

  **c.** $(fg)(x) = x^2$

  **d.** $\left(\dfrac{f}{g}\right)(x) = 1, x \neq 0$

**7. a.** $(f + g)(x) = 2x - 2 + \dfrac{2}{x + 5}, x \neq -5$

  **b.** $(f - g)(x) = 2x - 2 - \dfrac{2}{x + 5}, x \neq -5$

  **c.** $(fg)(x) = \dfrac{4x - 4}{x + 5}, x \neq -5$

  **d.** $\left(\dfrac{f}{g}\right)(x) = x^2 + 4x - 5, x \neq -5$

**9. a.** $(f + g)(x) = \sqrt{x - 2} + x - 4, x \geq 2$

  **b.** $(f - g)(x) = \sqrt{x - 2} - x + 4, x \geq 2$

  **c.** $(fg)(x) = (x - 4)\sqrt{x - 2}, x \geq 2$

  **d.** $\left(\dfrac{f}{g}\right)(x) = \dfrac{\sqrt{x - 2}}{x - 4}, x \geq 2, x \neq 4$

**11.** $(f \circ g)(x) = 8x - 7$

  $(g \circ f)(x) = 8x + 7$

**13.** $(f \circ g)(x) = 4x^2 + 28x + 49$

  $(g \circ f)(x) = 2x^2 + 7$

**15.** $(f \circ g)(x) = \dfrac{3}{x^2}, x \neq 0$

  $(g \circ f)(x) = \dfrac{9}{x^2}, x \neq 0$

**17.** $(f \circ g)(x) = x^3 - 3x + 1$

  $(g \circ f)(x) = x^3 + 3x^2 - 2$

**19.** $(f \circ g)(x) = x^2 + 2xh + h^2 + 3x + 3h + 1$

  $(g \circ f)(x) = x^2 + 3x + 1 + h$

**21.** $(f \circ g)(x) = x^2$

  $(g \circ f)(x) = x^2, x \geq 0$

**23.** $(f \circ g)(x) = 3\sqrt{x - 2} + 5, x \geq 2$

  $(g \circ f)(x) = \sqrt{3x + 3}, x \geq -1$

**25.** $(f \circ g)(x) = \dfrac{3}{x^2 - 4x}, x \neq 0, x \neq 4$

  $(g \circ f)(x) = \dfrac{-2x^2 + 11}{x^2 - 4}, x \neq -2, x \neq 2$

**27.** Inverses    **29.** Not inverses    **31.** Inverses

**33.** Not inverses    **35.** Inverses    **37.** One-to-one

**39.** Not one-to-one    **41.** One-to-one    **43.** Not one-to-one

**45.** Not one-to-one    **47.** One-to-one; $f^{-1}(x) = 3x + 3$

**49.** Not one-to-one    **51.** One-to-one; $f^{-1}(x) = \sqrt{x - 1}$

**53.** One-to-one; $f^{-1}(x) = \frac{1}{2}(x^2 - 5), x \geq 0$

**55.** Not one-to-one

**57.**

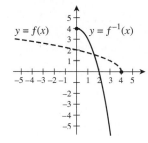

**59.** $f^{-1}$ does not exist.

**61.**

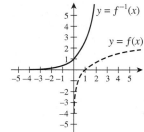

**63.** $f^3(x) = x + 15$    **65.** $f^4(x) = 16x$

**67.** $f^{10}(0.98) \approx 1.03633 \times 10^{-9}$; $f^{10}(1.02) \approx 6.40583 \times 10^8$

**73.** $f(x) = \dfrac{1}{x}$

**75.** Reflect about the $x$-axis and rotate 90° counterclockwise

**77.** There are no such 1–1 functions.

**79. a.**

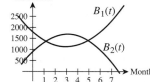

  Maximum for $B_1$: 1800 barrels
  Minimum for $B_1$: 1125 barrels
  Maximum for $B_2$: 1665 barrels
  Minimum for $B_2$: 900 barrels

  **b.** $B(t) = -10t^2 + 60t + 2700$

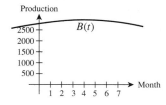

  Maximum for $B$: 2790
  Minimum for $B$: 2700

**Section 4 ■ page 318**

**1.** Quadratic     **3.** None     **5.** None     **7.** Linear

**9.** Linear     **11.** Piecewise

**13.**

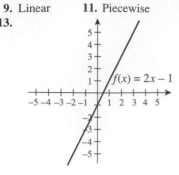

Slope 2, $y$-intercept $-1$

**15.**

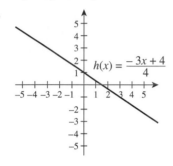

Slope $-\frac{3}{4}$, $y$-intercept 1

**17.** $h(x) = -\dfrac{1}{2}x + 5$     **19.** $g(x) = \dfrac{4}{3}x - 4$     **21.** $(0, -4)$

**23.** $(\frac{3}{2}, \frac{11}{2})$     **25.** $(-8, -54)$     **27.** $(1, 0)$

**29.**

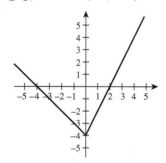

$f(-2) = -2; f(0) = -4; f(3) = 2$

**31.**

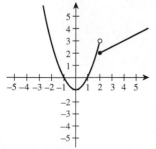

$h(0) = -1; h(2) = 2; h(4) = 3$

**33.**

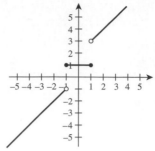

$f(-3) = -3; f(-1) = 1; f(4) = 6$

**35.**

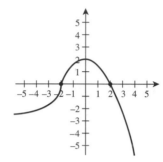

**37.**

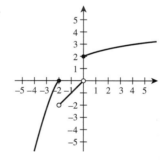

**39.**

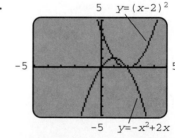     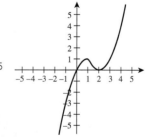

**41.**

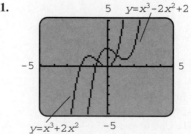     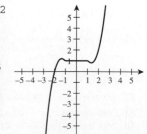

**43.**

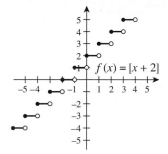

**45. a.**

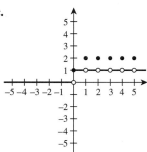

**b.**

**c.**

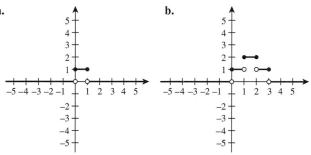

**47. a.** 0.8; 0.7; $\frac{3}{5}$

**b.**

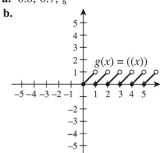

**c.** *x*

**49. a.** $C(x) = 80x + 5000$
**b.** $d(x) = -5x + 550$
**c.** $R(x) = -5x^2 + 550x$
**d.** $P(x) = -5x^2 + 470x - 5000$
**e.** 47 units; $315

**51. a.** 81.13 feet    **b.** Yes

**53.** In 1958; approximately 1155.6 million acres

**55.**

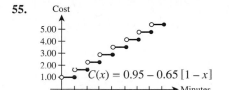

7 minutes

**57.** $20,748.50 for filing separately; $19,146.00 for filing jointly

**59.** TWA JULY TWO

*Section 5* ■ *page 333*

**1.** 4    **3.** 48    **5.** 10    **7.** 12    **9.** 60    **11.** $\frac{243}{64}$

**13.** $y = \frac{2}{9}x^2 - \frac{4}{3}x + 5$    **15.** $y = -2x^2 + 10x$

**17.** $y = 0.5x + 2.33$    **19.** $y = 1.6x - 4.8$

**21.** $y = 2.5x + 90$; $y = 152.5$ when $x = 25$

**23.** $y = -31.55x + 77.51$; $y = -17.14$ when $x = 3$

**25.** 2592 board feet    **27.** 3025 joules

**29.** $5.75 \times 10^{-25}$ N    **31.** 3000 lb

**33.** $y = 2.6x + 21.7$; 281.7 kg

**35.** $y = 9.57x + 177$ where $x = 0$ corresponds to 1980; 368.4 parts per trillion

**37.** $y = 2x^2 + 22x + 320$ where $x = 0$ corresponds to 1980; 1,560,000 inmates

*Chapter Review* ■ *page 338*

**1.** Function    **3.** Not a function

**5. a.** 21    **b.** 28

**7. a.** $\dfrac{x^2}{x^2 + 1}$    **b.** $\dfrac{x^2 + 2x + 1}{x^2 + 2x + 2}$

**9. a.** 4    **b.** $-1$    **c.** $-a + 4$

**11.** All real numbers    **13.** $[-\frac{1}{2}, \infty)$

**15.**

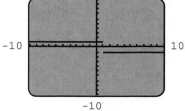

Domain: $(-\infty, -2] \cup [2, \infty)$
Range: $[0, \infty)$

**17.**

Domain: All real numbers except 1
Range: $\{-1, 1\}$

**19. a.** $h(x) = -2x^3 + 6x - 1$
  **b.** $h(x) = 2(x + 1)^3 - 6(x + 1) + 1$

**21. a.**

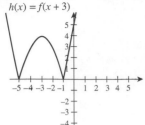

$h(x) = f(x + 3)$

  **b.**

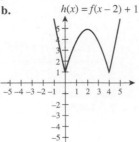

$h(x) = f(x - 2) + 1$

  **c.**

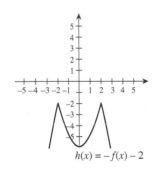

$h(x) = -f(x) - 2$

**23.** Odd     **25.** Neither     **27.** $x = \frac{9}{4}$     **29.** $x = \frac{8}{3}$

**31. a.** $-1.88, 0.35, 1.53$
  **b.** Relative maximum at $(-1, 3)$
    Relative minimum at $(1, -1)$
  **c.** Increasing on $(-\infty, -1) \cup (1, \infty)$
    Decreasing on $(-1, 1)$

**33. a.** No zeros
  **b.** Relative maximum at $(2, \frac{5}{2})$
  **c.** Increasing on $(-\infty, 2)$
    Decreasing on $(2, \infty)$

**35. a.** $(f + g)(x) = x^2 + 2x - 1$
  **b.** $(fg)(x) = 2x^3 - x^2$
  **c.** $\left(\dfrac{f}{g}\right)(x) = \dfrac{x^2}{(2x - 1)}; x \neq \frac{1}{2}$
  **d.** $(f \circ g)(x) = 4x^2 - 4x + 1$
  **e.** $(g \circ f)(x) = 2x^2 - 1$

**37. a.** $(f + g)(x) = \dfrac{1}{x - 4} + x^3; x \neq 4$
  **b.** $(fg)(x) = \dfrac{x^3}{x - 4}; x \neq 4$
  **c.** $\left(\dfrac{f}{g}\right)(x) = \dfrac{1}{x^4 - 4x^3}; x \neq 4, x \neq 0$
  **d.** $(f \circ g)(x) = \dfrac{1}{x^3 - 4}; x \neq \sqrt[3]{4}$
  **e.** $(g \circ f)(x) = \dfrac{1}{(x - 4)^3}; x \neq 4$

**39. a.** $(f + g)(x) = 2x^2 + \sqrt{2x + 4}; x \geq -2$
  **b.** $(fg)(x) = 2x^2\sqrt{2x + 4}; x \geq -2$
  **c.** $\left(\dfrac{f}{g}\right)(x) = \dfrac{2x^2}{\sqrt{2x + 4}}; x > -2$
  **d.** $(f \circ g)(x) = 4x + 8; x \geq -2$
  **e.** $(g \circ f)(x) = \sqrt{4x^2 + 4}$

**41.** $f(g(x)) = x; g(f(x)) = x$     **43.** $f(g(x)) = x; g(f(x)) = x$

**45.** One-to-one; $f^{-1}(x) = \dfrac{x + 2}{3}$     **47.** Not one-to-one

**49.** One-to-one; $f^{-1}(x) = x^2 + 4, x \geq 0$

**51.**

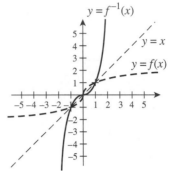

**53.** Quadratic

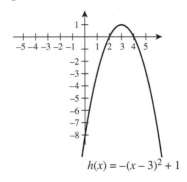

$h(x) = -(x - 3)^2 + 1$

$x$-intercepts $(2, 0)$ and $(4, 0)$; $y$-intercept $(0, -8)$

**55.** None

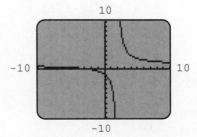

$x$-intercept $(-2, 0)$; $y$-intercept $(0, -1)$

**57.** Piecewise

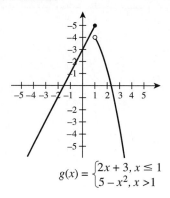

$$g(x) = \begin{cases} 2x + 3, x \le 1 \\ 5 - x^2, x > 1 \end{cases}$$

*x*-intercepts $\left(-\frac{3}{2}, 0\right)$ and $\left(\sqrt{5}, 0\right)$; *y*-intercept $(0, 3)$

**59.**

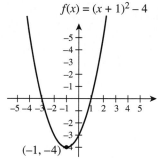

$f(x) = (x + 1)^2 - 4$

$(-1, -4)$

Vertex $(-1, -4)$; *x*-intercepts $(-3, 0)$ and $(1, 0)$; *y*-intercept $(0, -3)$

**61.**

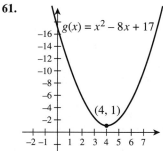

$g(x) = x^2 - 8x + 17$

$(4, 1)$

Vertex $(4, 1)$; no *x*-intercepts; *y*-intercept $(0, 17)$

**63.** 15  **65.** 36  **67.** $y = -x^2 - 2x + 8$

**69.** $y = 0.605x + 1.316$

**71.** $y = 2.37x + 42.6$; $y = 161.1$ when $x = 50$

**73.** $A = \dfrac{1}{4\pi}C^2$  **75.** $h = \sqrt{d^2 - 40{,}000}$; $[200, \infty)$

**77. a.** 32 feet  **b.** Yes; approximately 117 feet

**79.** $y = 4.98x - 41.34$; 48.3%

**81.** $y = \frac{11}{3}x^2 + \frac{94}{3}x + 23$; 422 million

***Chapter Test*** ■ *page 342*

**1.** False  **2.** True  **3.** True  **4.** True  **5.** True
**6.** False  **7.** True  **8.** False
**9.** $f(x) = 2x$; $f^{-1}(x) = \frac{1}{2}x$, for example
**10.** $f(x) = x^2$; $(0, 0)$, for example
**11.** $f(x) = \begin{cases} -x, & x < -1 \\ 0, & -1 \le x \le 1, \text{ for example} \\ x, & x > 1 \end{cases}$
**12.** $(0, 0)$, $(1, 1)$, $(2, 2)$, for example
**13.** $f(x) = x^3$, for example  **14.** $f(x) = x + 1$, for example
**15.** Function; the set of all real numbers except 2
**16.** Inverses since $f(g(x)) = g(f(x)) = x$
**17. a.** 7  **b.** 7  **c.** 49
**18.** $[2, 3) \cup (3, \infty)$

**19.**

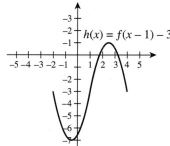

$h(x) = f(x - 1) - 3$

**20.** Zeros $x = 0$ and $x = 6$; relative maximum at $(0, 0)$; relative minimum at $(4, -32)$; increasing on $(-\infty, 0) \cup (4, \infty)$; decreasing on $(0, 4)$

**21.**

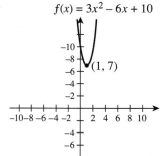

$f(x) = 3x^2 - 6x + 10$

$(1, 7)$

Vertex $(1, 7)$

**22.** $z = \frac{40}{9}$  **23.** $-3$  **24.** 1 minute 37.19 seconds

# CHAPTER 5

**Section 1 ▪ page 352**

1. $f(x) \to \infty$ as $x \to \infty$ or $-\infty$; graph: ii
3. $p(x) \to -\infty$ as $x \to \infty$ or $-\infty$; graph: i
5. $f(x) \to \infty$ as $x \to \infty$ or $-\infty$
7. $p(x) \to -\infty$ as $x \to \infty$ or $-\infty$
9. $h(x) \to \infty$ as $x \to -\infty$, $h(x) \to -\infty$ as $x \to \infty$
11. $g(x) \to -\infty$ as $x \to -\infty$, $g(x) \to \infty$ as $x \to \infty$
13. **a.** 1  **b.** 0
15. **a.** 5  **b.** 4
17. $f(x) = (x - 4)(x + 2)$
    Zeros: 4, $-2$
19. $f(x) = 3x(x - 1)(x + 1)$
    Zeros: 0, 1, $-1$
21. $f(x) = -x^2(x + 6)^2$
    Zeros: 0, $-6$
23. Zeros: 3.36, $-1.05$, 3.69
    Turning points: (0.47, 14.13), (3.53, $-0.13$)
25. Zero: $-15.04$
    Turning points: ($-10.14$, 58.04), ($-0.53$, 2.48)
27. Zeros: 0, 1.78, 11.91, $-1.70$
    Turning points: (1, 5.75), ($-1$, $-6.25$), (9, $-506.25$)
29. Zeros: $-0.80$, 11.96, $-4.16$
    Turning points: ($-2.58$, 4.07), (7.24, $-43.29$)
31. Zeros: 10.00, 1.64
    Turning point: (8.33, 668.79)
33. 1989; mid-1991; invalid
35. (5.11, 34.28), corresponding to 1965
    (15.52, 31.91), corresponding to mid-1975

**Section 2 ▪ page 360**

1. $q(x) = 2x + 3$, $r(x) = 0$  3. $q(x) = 4x - 3$, $r(x) = 0$
5. $q(x) = x^2 - 3x - 3$, $r(x) = 18$
7. $q(x) = x - 1$, $r(x) = 2$  9. $q(x) = 3x$, $r(x) = 5x + 2$
11. $q(x) = x^2 - x$, $r(x) = x$  13. $q(x) = 2x - 3$, $r(x) = 0$
15. $q(x) = x^2 - 2$, $r(x) = 0$
17. $q(x) = 3x^3 + x^2 + 3x - 1$, $r(x) = 0$
19. $q(x) = x - 1$, $r(x) = -1$  21. $q(x) = x^2 - 2x$, $r(x) = 2$
23. $q(x) = -4x^3 - \frac{1}{3}x^2 - \frac{1}{9}x - \frac{1}{27}$, $r(x) = \frac{80}{81}$
25. $f(x) = (x + 5)(x + 2)(x - 1)$
    Zeros: $-5$, $-2$, 1
27. $p(x) = (2x + 1)(x - 2)(x - 3)$
    Zeros: $-\frac{1}{2}$, 2, 3
29. $g(x) = (x + 4)^2(x + 2)(x - 1)$
    Zeros: $-4$, $-2$, 1
31. $h(x) = (2x + 1)(3x + 2)(x - 1)(x - 8)$
    Zeros: $-\frac{1}{2}$, $-\frac{2}{3}$, 1, 8
33. **a.** 39  **b.** $-6$  **c.** 4
35. **a.** 12  **b.** 1608  **c.** 88.27
37. **a.** 14  **b.** 33

**Section 3 ▪ page 369**

1. $f(x) = (x + 3)(x + 2)(x - 2)$
    Zeros: $-3$, $-2$, 2
3. $f(x) = (x - 5)(x + 1)(2x + 1)$
    Zeros: 5, $-1$, $-\frac{1}{2}$
5. $f(x) = 3(x + \frac{1}{3})(x + 4)(x - 1)^2$
    Zeros: $-\frac{1}{3}$, $-4$, 1
7. $f(x) = (x - 1)(x + 2)(5x - 2)$
    Zeros: 1, $-2$, $\frac{2}{5}$
9. $f(x) = (x - 2)(2x - 1)(3x + 1)$
    Zeros: 2, $\frac{1}{2}$, $-\frac{1}{3}$
11. $f(x) = (x - 2)(x + 1)(x^2 - 3)$
    Zeros: 2, $-1$, $\sqrt{3}$, $-\sqrt{3}$
13. $f(x) = x^2 - x - 12$, for example
15. $f(x) = \frac{1}{2}(x - 5)(x - 1)(x + 2)$
17. $f(x) = \frac{1}{3}x^2(x - 2)^2$
19. $f(x) = (x + 1)^3(x - \frac{1}{2})^2$, for example
21. $f(x) = -x^2 + 2x + 3$
23. $f(x) = \frac{1}{4}(x + 4)(x - 2)^2$
25. $f(x) = -(x + 2)(x + 1)(x - 1)(x - 2)$
27. $f(x) = (x - 3)(x - 2i)(x + 2i)$
    Zeros: 3, $2i$, $-2i$
29. $f(x) = (x - 1 - i)(x - 1 + i)(x - 4)(x + 1)$
    Zeros: $1 + i$, $1 - i$, 4, $-1$
31. Zeros: $i$, $-i$
    $f(x) = (x - i)(x + i)$
33. Zeros: 0, $1 + 2i$, $1 - 2i$
    $f(x) = x(x - 1 - 2i)(x - 1 + 2i)$
35. Zeros: $-1$, 1, $-2i$, $2i$
    $f(x) = (x + 1)(x - 1)(x + 2i)(x - 2i)$
37. Zeros: 0, 1, $2 - i$, $2 + i$
    $f(x) = x^2(x - 1)(x - 2 + i)(x - 2 - i)$
39. $f(x) = (x + 2)(x^2 + 4)$, for example
41. $f(x) = (x^2 + 1)(x^2 - 2x + 2)$, for example
43. $f(x) = x(x^2 - 2)(x^2 + 4x + 8)$, for example
45. Zeros: 0, 8, 10
    Domain: (0, 8)
    Original dimensions: 20 × 16
47. 100 or 300 frames; (100, 300)
49. $s(t) = -16t^2 + 80t - 64$; 36 feet

**Section 4 ▪ page 380**

1. Possible rational zeros: $\pm 1$, $\pm 2$, $\pm 3$, $\pm 4$, $\pm 6$, $\pm 12$
    Actual rational zeros: $-2$, 2, 3
3. Possible rational zeros: $\pm 1$, $\pm 2$, $\pm \frac{1}{2}$, $\pm \frac{1}{4}$
    Actual rational zeros: $-1$, $-\frac{1}{2}$, $\frac{1}{2}$, 2
5. Possible rational zeros: $\pm 1$, $\pm 3$, $\pm 9$, $\pm 27$
    Actual rational zeros: $-3$, 3

7. Possible rational zeros: $\pm 1, \pm 3, \pm\frac{1}{2}, \pm\frac{3}{2}$
   Actual rational zeros: $-\frac{3}{2}, -1, 1$
9. Possible rational zeros: $\pm 1, \pm 2, \pm 3, \pm 4, \pm 6, \pm\frac{1}{2}, \pm\frac{3}{2}$
   Actual rational zeros: $-1, \frac{3}{2}, 4$
11. Number of positive zeros: 0
    Number of negative zeros: 1
13. Number of positive zeros: 0
    Number of negative zeros: 0
15. Possible number of positive zeros: 1 or 3
    Actual number of positive zeros: 1
    Number of negative zeros: 0
17. Number of positive zeros: 1
    Possible number of negative zeros: 0 or 2
    Actual number of negative zeros: 0
25. $1, -1, -6$     27. $1, 12$     29. $\frac{1}{6}, \frac{1}{2}, -3$
31. $-\frac{1}{8}, -\sqrt{2}, \sqrt{2}$     33. $11, -11$
35. $x = -2, x = -1, x = 2$     37. $x = -3, x = 3$
39. $x = -1, x = -\frac{1}{2}, x = 1$     41. $-0.32$     43. $-1.49, 0.80$
45. $11.00$     47. $\frac{3}{2}$ inches $\times \frac{3}{2}$ inches $\times 4$ inches
49. $\frac{1}{2}$ inch $\times \frac{1}{2}$ inch squares should be cut out
51. $500, 2000,$ or $3500$ units
53. $1970, 1980, 1990$

### Section 5 ■ *page 392*

1. Domain: all real numbers except 2
   Zeros: none; graph: g
3. Domain: all real numbers except 0
   Zero: $-1$; graph: c
5. Domain: all real numbers except 2 and $-2$
   Zeros: none; graph: d
7. Domain: all real numbers except 1 and $-1$
   Zeros: $2, -2$; graph: b

9. a.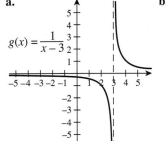
   b.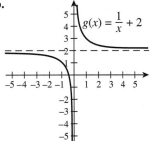
   c.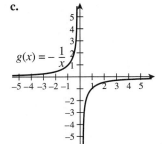

11. a.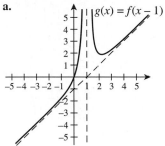
    b.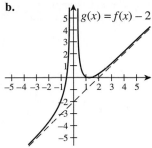
    c.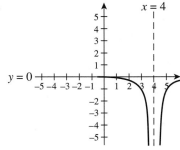

13.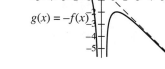

15.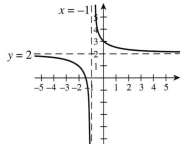

17. Vertical asymptote: $x = -3$
    Horizontal asymptote: $y = 0$

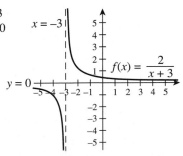

**19.** Vertical asymptote: $x = 1$
Horizontal asymptote: $y = 0$

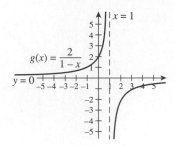

**21.** Vertical asymptote: $x = -2$
Horizontal asymptote: $y = 1$

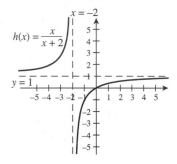

**23.** Vertical asymptote: $x = 2$
Horizontal asymptote: $y = 3$

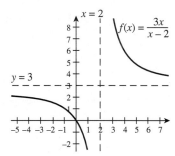

**25.** Vertical asymptote: $x = 15$
Horizontal asymptote: $y = 12$

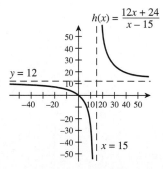

**27.** Vertical asymptotoes: $x = 3$, $x = -3$
Horizontal asymptote: $y = 0$

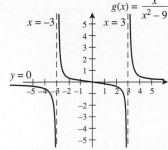

**29.** Vertical asymptotes: $x = 3$, $x = -3$
Horizontal asymptote: $y = 0$

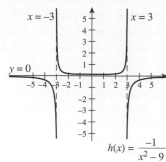

**31.** Vertical asymptote: none
Horizontal asymptote: $y = 5$

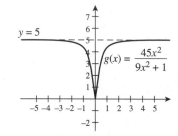

**33.** Vertical asymptote: $x = -1$
Horizontal asymptote: $y = 0$

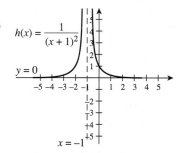

**35.** Vertical asymptote: $x = 2$
Horizontal asymptote: $y = 1$

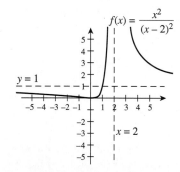

**37.** Vertical asymptote: $x = 3$
Horizontal asymptote: $y = 0$

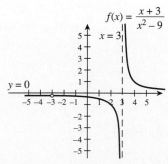

**39.** Vertical asymptote: $x = 1$
Horizontal asymptote: $y = 0$

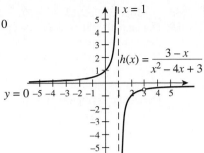

**47.** Vertical asymptote: $x = 3$
Inclined asymptote: $y = x - 1$

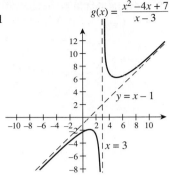

**41.** Vertical asymptote: $x = 0$
Inclined asymptote: $y = 2x$

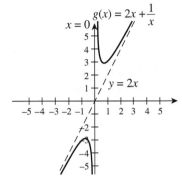

**49.** Vertical asymptote: $x = 100$
Horizontal asymptote: $y = -0.001$
A 75% reduction; 100% reduction is unattainable according to this model.

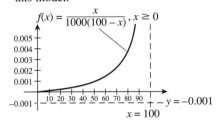

**51.** Vertical asymptote: none
Horizontal asymptote: $y = 0$
Approximately 11.9 years; unemployment rate decreases as education increases; 0% unemployment is unattainable according to this model.

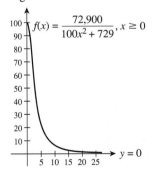

**43.** Vertical asymptote: $x = 1$
Inclined asymptote: $y = x + 1$

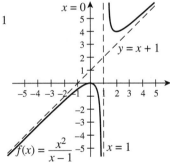

**45.** Vertical asymptote: $x = -2$
Inclined asymptote: $y = -3x$

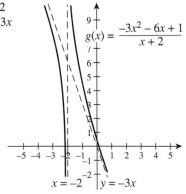

**53.** Horizontal asymptote: $y = 100$
The average cost approaches \$100 as the number of units increases.

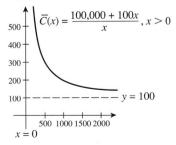

**55.** 58 balls should be ordered 10 times per year.

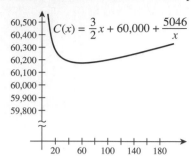

$$C(x) = \frac{3}{2}x + 60{,}000 + \frac{5046}{x}$$

*Chapter Review* ■ *page 398*

**1.** $g(x) \to -\infty$ as $x \to \infty$ or $-\infty$
**3.** $f(x) \to -\infty$ as $x \to -\infty$, $f(x) \to \infty$ as $x \to \infty$
**5.** $h(x) \to -\infty$ as $x \to -\infty$ or $\infty$
**7.** Zero: approximately 9.11
   Turning points: $(0, -3)$, $(6, -39)$
**9.** Zeros: $-13.12$, $3.12$
   Turning points: $(0.13, -3.94)$,
   $(-9.61, -417.63)$, $(1.98, -8.40)$
**11.** Quotient: $2x - 1$; remainder: 0
**13.** Quotient: $x^2 - 4x + 11$; remainder: $-44x + 2$
**15.** Quotient: $3x^2 - 5x + 2$; remainder: 0
**17.** Quotient: $-2x^3 + 6x^2 - 14x + 43$; remainder: $-132$
**19. a.** $\frac{255}{8}$   **b.** 10
**21. a.** $-610$   **b.** 2
**23.** $f(x) = (x + 3)(2x - 1)$
   Zeros: $-3, \frac{1}{2}$
**25.** $h(x) = (x - 3)(x + 3)(x^2 - 2)$
   Zeros: $3, -3, -\sqrt{2}, \sqrt{2}$
**27.** $h(x) = (x + 1)(2x + 1)(3x - 1)$
   Zeros: $-1, -\frac{1}{2}, \frac{1}{3}$
**29.** $f(x) = x^3 - 6x^2 - x + 30$, for example
**31.** $f(x) = 4(x - 1)^2(x + 1)^2$
**33.** $f(x) = x^3 - 3x^2 - x + 3$
**35.** $f(x) = x^3 + 3x^2 + 16x + 48$, for example
**37.** $f(x) = (x - 2 - i)(x - 2 + i)$
   Zeros: $2 - i, 2 + i$
**39.** $h(x) = (x - 3)(x + 3)(x - 4i)(x + 4i)$
   Zeros: $3, -3, -4i, 4i$
**41.** Possible rational zeros: $\pm 1, \pm 2, \pm 3, \pm 6$
   Actual rational zeros: $1, -2, 3$
**43.** Possible rational zeros: $\pm 1, \pm 3, \pm\frac{1}{2}, \pm\frac{1}{4}, \pm\frac{3}{2}, \pm\frac{3}{4}$
   Actual rational zeros: $1, -1, \frac{3}{2}, \frac{1}{2}$
**45.** Number of positive zeros: 0
   Number of negative zeros: 1
**47.** Number of positive zeros: 1
   Possible number of negative zeros: 2 or 0
   Actual number of negative zeros: 2
**51.** $x = -1, x = 2, x = 6$   **53.** $x = -\frac{1}{4}, x = -\frac{2}{3}, x = 4$

**55.** $-0.20, 2.33, -2.13$   **57.** $-21.00, 1.52$
**59.** Domain: all real numbers except 3
   Zeros: none; graph: d
**61.** Domain: all real numbers except 3 and $-3$
   Zero: $-2$; graph: a
**63.** Vertical asymptote: $x = -2$
   Horizontal asymptote: $y = 0$

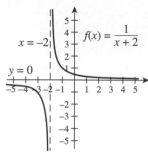

**65.** Vertical asymptote: $x = \frac{1}{3}$
   Horizontal asymptote: $y = \frac{2}{3}$

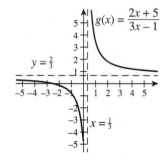

**67.** Vertical asymptote: $x = 0$
   Horizontal asymptote: $y = 0$

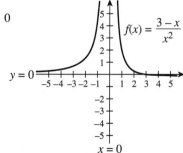

**69.** Vertical asymptote: $x = 1$
   Horizontal asymptote: $y = 0$

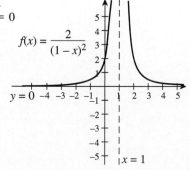

**71.** Vertical asymptote: $x = 0$
Inclined asymptote: $y = -2x$

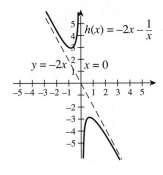

**73.** Vertical asymptote: $x = -3$
Inclined asymptote: $y = 3x - 1$

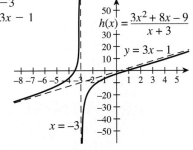

**75.** $f(x) = -\frac{3}{1000}x^2 + 18x - 15{,}000$
Largest profit of $12,000 occurs when $x = 3000$

**77.** Zeros: 2, 10.3
The months when the average temperature is zero

**79.**

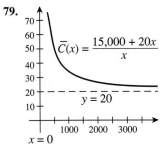

The average cost approaches $20.

*Chapter Test ■ page 402*

**1.** True    **2.** True    **3.** False    **4.** False    **5.** True
**6.** True    **7.** False   **8.** True
**9.** $f(x) = -x$, for example
**10.** $f(x) = x^3 - 6x^2 + 5x + 12$, for example
**11.** $f(x) = x^3 - 2x^2 + 9x - 18$, for example
**12.** $f(x) = x^4 + 1$, for example
**13.** $f(x) = \dfrac{x}{x + 1}$, for example
**14.** $f(x) = x - 3 + \dfrac{1}{x - 2}$, for example
**15.** $f(x) = x^2(x - 4)(2x + 1)$
    Zeros: $0, 4, -\frac{1}{2}$
**16.** $f(x) = (x + 2)(3x - 1)(2x - 3)$
    Zeros: $-2, \frac{1}{3}, \frac{3}{2}$
**17.** Quotient: $3x^3 - 6x^2 + 5$, remainder: 4
**18.** $f(x) = x(x^2 + 9)(5x - 2)$
    Zeros: $0, \frac{2}{5}, -3i, 3i$
**19.** Possible rational zeros: $\pm 1, \pm 3, \pm 9, \pm\frac{1}{2}, \pm\frac{3}{2}, \pm\frac{9}{2}$
    Actual rational zeros: $1, \frac{3}{2}, -3$
**20.** Possible number of positive zeros: 3 or 1
    Number of negative zeros: 1
**21.** Approximately 1.52
**22.** Vertical asymptote: $x = \frac{3}{2}$
    Horizontal asymptote: $y = \frac{1}{2}$

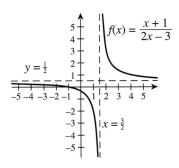

**23. a.** Approximately 14 or 42
    **b.** 30 chairs; $3500

---

**CHAPTER 6**

*Section 1 ■ page 411*

**1.**

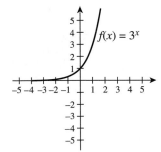

**a.** **b.**

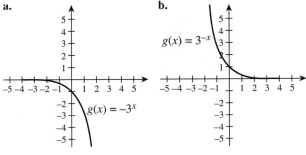

**c.**

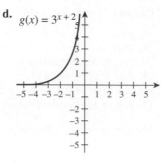

**d.** $g(x) = 3^{x+2}$

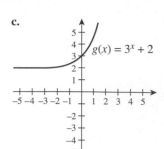

**3.** $a = \frac{1}{3}$

**5.**

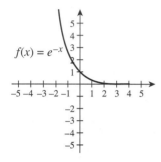

$f(x) = e^{-x}$

**7.**

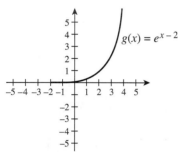

$g(x) = e^{x-2}$

**9.**

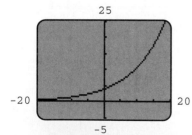

**11.**

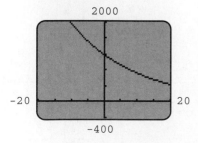

**13.** $x = 4$     **15.** $x = 3$     **17.** $x = -3$     **19.** $x = -5$
**21.** $x = -2, x = 1$
**23.** **a.** \$1790.85     **b.** \$1814.02     **c.** \$1819.40
  **d.** \$1822.03     **e.** \$1822.12
**25.** 20 gerbils initially present

| $t$ | $f(t)$ |
|---|---|
| 0 | 20 |
| 1 | 30 |
| 2 | 45 |
| 3 | 68 |
| 4 | 101 |

228 gerbils after 6 months
1412 gerbils after $10\frac{1}{2}$ months

**27.** \$20,805.70; \$21,103.48
**29.** $Q(40) \approx 16.44 Q_0$; $Q(100) \approx 1096.63 Q_0$; 6570%
**31.** Approximately 4.64
**33.**

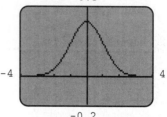

Maximum at approximately (0, 0.40); $f(x)$ approaches 0 as $x$ approaches $\pm\infty$

**35.** **a.** Approximately 130 students after 2 days;
  Approximately 190 students after 5 days
  **b.**

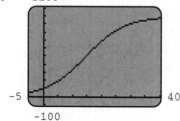

  **c.** Approximately 24 days
  **d.** $N(t)$ approaches 1000

**Section 2 ■ page 421**

**1.** 2.721295     **3.** $-0.176091$     **5.** 0.539591
**7.** $-0.708773$     **9.** 4     **11.** $-2$     **13.** $-3$
**15.** $-2$     **17.** $-3$     **19.** 100
**21.** $\dfrac{\log 12}{\log 2} = \dfrac{\ln 12}{\ln 2} \approx 3.584963$
**23.** $\dfrac{\log 108}{\log \frac{1}{2}} = \dfrac{\ln 108}{\ln \frac{1}{2}} \approx -6.754888$

**25.** $\dfrac{\log 0.341}{\log 12} = \dfrac{\ln 0.341}{\ln 12} \approx -0.432963$

**27.** $\dfrac{\log 10}{\log \pi} = \dfrac{\ln 10}{\ln \pi} \approx 2.011466$

**29.**

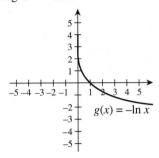

$g(x) = -\ln x$

**31.**

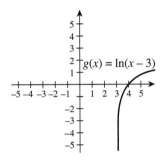

$g(x) = \ln(x - 3)$

**33.**

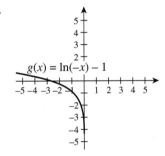

$g(x) = \ln(-x) - 1$

**35.** $f^{-1}(x) = 3^x$

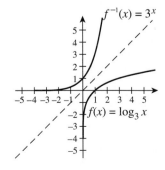

$f^{-1}(x) = 3^x$

$f(x) = \log_3 x$

**37.** $f^{-1}(x) = e^x - 5$

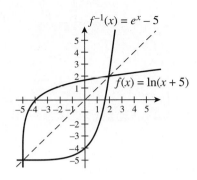

$f^{-1}(x) = e^x - 5$

$f(x) = \ln(x + 5)$

**39.** $f^{-1}(x) = \dfrac{10^x}{1000} = 10^{x-3}$

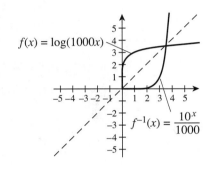

$f(x) = \log(1000x)$

$f^{-1}(x) = \dfrac{10^x}{1000}$

**41.**

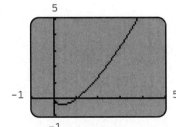

Domain: $(0, \infty)$

**43.**

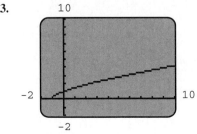

Domain: $(-1, \infty)$

**45.**

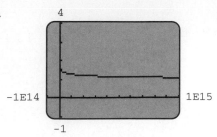

$f(x)$ approaches 0 as $x$ approaches $\infty$

**47.** Approximately 49.7 minutes
**49.** Approximately 13.4 hours
**51.** Approximately 20.4 decibles
**53.** 100 decibels      **55.** Approximately 3.55
**57.** Approximately 4.5      **59.** 25 times
**61. a.** Approximately 4.6 weeks; approximately 8 weeks
   **b.** The time gets larger without bound; no
   **c.**

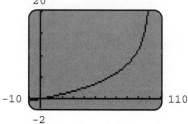

Domain: $[0, 100)$

## Section 3 ■ page 431

**1.** $2 \ln x + \ln y$      **3.** $3 \log x + 4 \log y + 3 \log z$
**5.** $\log_\pi e - \log_\pi 7 - 3 \log_\pi x$      **7.** $\frac{1}{4} \log x + \frac{1}{2} \log y - \log z$
**9.** $\log \dfrac{x}{y^2}$   **11.** $\ln \dfrac{x^3 y^4}{z}$   **13.** $\ln \sqrt[3]{\dfrac{x - z}{(x + z)^2}}$   **15.** $\log \frac{64}{3}$
**17.** $\ln \dfrac{x + 3}{x + 2}$   **19.** $x = 729$   **21.** $x = 2$
**23.** $x = \sqrt[3]{25}$   **25.** $x = 40$   **27.** $x = -\frac{1}{3}$   **29.** $x = 2$
**31.** $x = \sqrt{6} - 1$   **33.** $x = 95$   **35.** $x = 1$
**37.** No solution   **39.** $x = \pm 999$   **41.** All $x > 0$
**43.** $x = e^e$   **45.** $0.16, 3.15$   **47.** $0.59, -11.42, -2.97$
**49.**

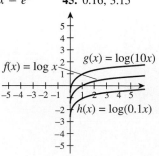

**51.** 10,000 times more intense      **53.** 30 decibels
**55.** Approximately 31.6 times more powerful
**57.** Approximately 233.6 tons

## Section 4 ■ page 441

**1.** $x = 4$      **3.** $x = 1$      **5.** $x = \dfrac{\log 10}{\log 7} \approx 1.18329$

**7.** $x = -\dfrac{\log 11}{\log 3} \approx -2.18266$      **9.** $x = \dfrac{\log 36}{\log \frac{18}{5}} \approx 2.79758$

**11.** $x = \dfrac{\ln 2}{1 - \ln 2} \approx 2.25889$

**13.** $x = \dfrac{1}{2(1 - \ln \pi)} \approx -3.45471$      **15.** $x = 1$

**17.** $x = \ln 2 \approx 0.69315$ and $x = \ln 3 \approx 1.09861$

**19.** $t = \dfrac{3}{\log 1.06} \approx 118.54959$      **21.** $0.57$

**23.** $-4.48, -2.11, 0.35$      **25.** Approximately 10.2 years
**27.** Approximately 5.78%      **29.** Approximately February, 1997
**31.** September 2063      **33.** Approximately 197.6; the year 2005
**35.** At approximately 6:33 A.M. on July 7, 2001.
**37.** Approximately 88,949      **39.** Approximately 11,400 years
**41.** Approximately 5.2 years

## Section 5 ■ page 453

**1. a.**

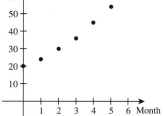

   **b.**

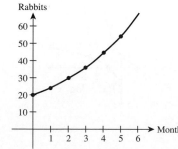

**c.**

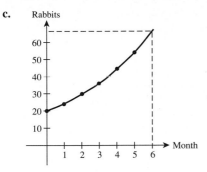

Approximately 66 rabbits

**3.** $P(t) = 20e^{0.1823t}$

| $t$ | Actual Data | $P(t) = 20e^{0.1823t}$ | Absolute Error |
|---|---|---|---|
| 0 | 20 | 20 | 0 |
| 1 | 24 | 24 | 0 |
| 2 | 30 | 28.8 | 1.2 |
| 3 | 36 | 34.6 | 1.4 |
| 4 | 45 | 41.5 | 3.5 |
| 5 | 54 | 49.8 | 4.2 |

Largest absolute error: 4.2
Sum of the absolute errors: 10.3

**5.** $P(t) = 20e^{0.1987t}$

| $t$ | Actual Data | $P(t) = 20e^{0.1987t}$ | Absolute Error |
|---|---|---|---|
| 0 | 20 | 20 | 0 |
| 1 | 24 | 24.4 | 0.4 |
| 2 | 30 | 29.8 | 0.2 |
| 3 | 36 | 36.3 | 0.3 |
| 4 | 45 | 44.3 | 0.7 |
| 5 | 54 | 54 | 0 |

Largest absolute error: 0.7
Sum of the absolute errors: 1.6
This model is more accurate.

**7.** $P(t) = 19.9e^{0.201t}$

| $t$ | Actual Data | $P(t) = 19.9e^{0.201t}$ | Absolute Error |
|---|---|---|---|
| 0 | 20 | 19.9 | 0.1 |
| 1 | 24 | 24.3 | 0.3 |
| 2 | 30 | 29.7 | 0.3 |
| 3 | 36 | 36.4 | 0.4 |
| 4 | 45 | 44.5 | 0.5 |
| 5 | 54 | 54.4 | 0.4 |

Largest absolute error: 0.5
Sum of the absolute errors: 2.0
The largest error is smaller but the sum of the errors is larger.
Both models are extremely accurate.

**9.** $P(t) = 225,300e^{0.0295t}$

| $t$ | Actual Data | $P(t) = 225,300e^{0.0295t}$ | Absolute Error |
|---|---|---|---|
| 0 | 226,000 | 225,300 | 700 |
| 10 | 303,000 | 302,600 | 400 |
| 20 | 402,000 | 406,400 | 4400 |
| 30 | 550,000 | 545,900 | 4100 |

Largest absolute error: 4400
Sum of the absolute errors: 9600
Population in 2010: approximately 985,000
First exceeds 1,500,000: 2024

**11.** $S(t) = 78.7e^{-0.0809t}$

| $t$ | Actual Data | $S(t) = 78.7e^{-0.0809t}$ | Absolute Error |
|---|---|---|---|
| 0 | 80 | 78.7 | 1.3 |
| 1 | 72 | 72.6 | 0.6 |
| 2 | 66 | 66.9 | 0.9 |
| 3 | 61 | 61.7 | 0.7 |
| 4 | 58 | 56.9 | 1.1 |

Largest absolute error: 1.3
Sum of the absolute errors: 4.6
Sales for month 6: approximately $48,400
Sales reach $20,000: month 17

**13.** $p(t) = 9.2e^{0.5142t}$

| $t$ | Actual Data | $p(t) = 9.2e^{0.5142t}$ | Absolute Error |
|---|---|---|---|
| 0 | 8 | 9.2 | 1.2 |
| 1 | 19 | 15.4 | 3.6 |
| 2 | 26 | 25.7 | 0.3 |
| 3 | 40 | 43.0 | 3.0 |

Largest absolute error: 3.6
Sum of the absolute errors: 8.1
90% encountered virus: approximately half way through the 4th quarter

**15. a.** 4 students
**b.** Approximately 77 students
**c.** Approximately 17 hours
**d.**

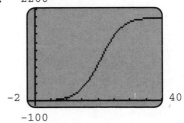

Approximately 2000 students
**e.** The 21st hour
**17. a.** 1944     **b.** 700 mph

**19.** $v = 0.81 \log P + 0.3$

| P | Actual | $v = 0.81 \log P + 0.3$ | Absolute Error |
|---|--------|------------------------|----------------|
| 5,500 | 3.3 | 3.33 | 0.03 |
| 14,000 | 3.7 | 3.66 | 0.04 |
| 71,000 | 4.3 | 4.23 | 0.07 |
| 138,000 | 4.4 | 4.46 | 0.06 |
| 342,000 | 4.8 | 4.78 | 0.02 |

Largest absolute error: 0.07
Sum of the absolute errors: 0.22
Little Rock: 4.5 ft/sec
New York: 5.9 ft/sec
Mexico City: 6.2 ft/sec

**17.**

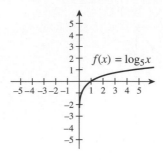

Domain: $(0, \infty)$

**19.**

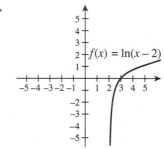

Domain: $(2, \infty)$

**21.** $\log_2 x + 2 \log_2 y$     **23.** $\frac{2}{3} \log x + \frac{1}{3} \log y$

**25.** $\log(x^2 y^3)$     **27.** $\ln\left(\dfrac{\sqrt[3]{xy^2}}{z}\right)$     **29.** $x = \frac{1}{2}$     **31.** $x = 3$

**33.** $y = \frac{12}{5}$     **35.** $x = 2$     **37.** $x = \dfrac{\ln 10}{\ln 5} \approx 1.43$

**39.** $t = \dfrac{\ln 10}{\ln 1.05} \approx 47.19$     **41.** $x = 10 \ln 5 \approx 16.09$

**43.** $x = \dfrac{2 \ln 3}{2 \ln 2 - \ln 3} \approx 7.64$     **45.** $x \approx 1.66, x \approx -2.82$

**47.** $x \approx 0.17, x \approx 5.05, x \approx 45.48$     **49.** \$56.34
**51.** Approximately 23.1 years     **53.** Approximately 12.6 hours
**55. a.** Approximately 149 astronauts
    **b.** Approximately 28.8 days
    **c.**

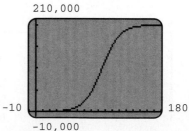

Approximately 200,000 astronauts
**d.** The 96th day

***Chapter Review*** ■ *page 457*

**1.**

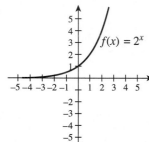

**a.**

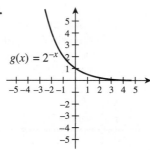

**b.**

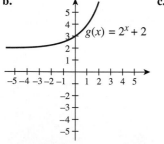

**c.**

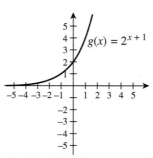

**3.** $x = 3$     **5.** $x = -\frac{1}{2}$     **7.** $-2$     **9.** $-2$     **11.** $\frac{1}{2}$

**13.** $\dfrac{\log 10}{\log 4} = \dfrac{\ln 10}{\ln 4} \approx 1.66096$

**15.** $\dfrac{\log e}{\log \sqrt{2}} = \dfrac{\ln e}{\ln \sqrt{2}} \approx 2.88539$

**57.** $V(t) = 13,978e^{-0.1495t}$

| $t$ | Actual Data | $V(t) = 13,978e^{-0.1495t}$ | Absolute Error |
|---|---|---|---|
| 0 | 14,000 | 13,978 | 22 |
| 1 | 12,000 | 12,037 | 37 |
| 2 | 10,400 | 10,366 | 34 |
| 3 | 8,900 | 8,926 | 26 |
| 4 | 7,700 | 7,687 | 13 |

Largest absolute error: 37
Sum of the absolute errors: 132
Worth $2000: approximately 13 years

*Chapter Test* ■ *page 459*

**1.** False      **2.** True      **3.** False      **4.** False      **5.** True
**6.** True      **7.** False      **8.** True
**9.** $f(t) = e^{-t}$, for example      **10.** 1, for example
**11.** log $x$, for example
**12.** The population will grow without bound.
**13.** $x = 1, x = -2$      **14.** $x = 10^{12} - 1$

**15.** $x = \dfrac{\ln 3}{\ln 3 - \ln 2} \approx 2.7095$      **16.** $x = \dfrac{\ln 2}{0.3} \approx 2.3105$

**17.** 5      **18.** −3
**19.**

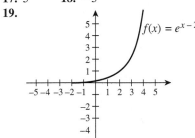

**20.**

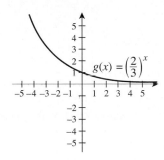

**21.**

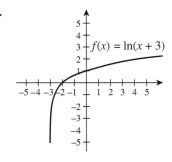

**22.** $4 \log x + 2 \log y + 2 \log z$      **23.** $\ln\left(\dfrac{x^3 z^{10}}{y^4}\right)$

**24.** Approximately 8.8 years      **25.** 2011
**26.** $N(t) = 2243e^{0.6115t}$

| $t$ | Actual Data | $N(t) = 2243e^{0.6115t}$ | Absolute Error |
|---|---|---|---|
| 0 | 2100 | 2243 | 143 |
| 1 | 4400 | 4134 | 266 |
| 2 | 8200 | 7620 | 580 |
| 3 | 13,100 | 14,046 | 946 |

Largest absolute error: 946
Sum of the absolute errors: 1935
New cases in 1987: approximately 25,889
The approximation is 4789 too large.

# CHAPTER 7

*Section 1* ■ *page 472*

**1.** e      **3.** h      **5.** c      **7.** d
**9.** 390° and −330°, for example
**11.** 225° and −495°, for example

**13.** $\dfrac{7\pi}{3}$ and $-\dfrac{5\pi}{3}$, for example

**15.** $\dfrac{\pi}{2}$ and $-\dfrac{7\pi}{2}$, for example

**17.** complement: 70°, supplement: 160°

**19.** complement: $\dfrac{\pi}{3}$, supplement: $\dfrac{5\pi}{6}$      **21.** 25.27°

**23.** 173.34°      **25.** 132°24′      **27.** 15°37′30″      **29.** $\dfrac{\pi}{6}$

**31.** $\dfrac{5\pi}{4}$      **33.** $-\dfrac{5\pi}{6}$      **35.** 0.262      **37.** −2.049 radians

**39.** 270°      **41.** −240°      **43.** 165°      **45.** −51.429°
**47.** 114.592°

**49. a.** $\dfrac{100\pi}{3}$ centimeters

   **b.** $\dfrac{1000\pi}{3}$ square centimeters

**51. a.** $\dfrac{5\pi}{3}$ inches      **b.** $\dfrac{10\pi}{3}$ square inches

**53. a.** Approximately 251.33 inches

  **b.** Approximately 773.493°

**55.** Approximately 418.88 feet per second

**57.** Approximately 0.0131 mile per hour

**59.** Approximately 736.7 revolutions per minute

**61. a.** $A \approx 150.796$ square inches

  **b.** Approximately 49.133 inches

**63.** Approximately $8.590 \times 10^{-6}$ radians or $4.922 \times 10^{-4}$ degrees

**65.** Approximately 501 miles

**67.** Approximately 2827 miles

**69.** Approximately 181.79° or 3.17 radians

*Section 2 ■ page 486*

**1.** $\sin \theta = \frac{3}{5}$    $\cos \theta = \frac{4}{5}$

  $\tan \theta = \frac{3}{4}$    $\cot \theta = \frac{4}{3}$

  $\sec \theta = \frac{5}{4}$    $\csc \theta = \frac{5}{3}$

**3.** $\sin \theta = \dfrac{2\sqrt{2}}{3}$    $\cos \theta = \dfrac{1}{3}$

  $\tan \theta = 2\sqrt{2}$    $\cot \theta = \dfrac{\sqrt{2}}{4}$

  $\sec \theta = 3$    $\csc \theta = \dfrac{3\sqrt{2}}{4}$

**5.** $\sin \theta = \dfrac{6\sqrt{61}}{61}$    $\cos \theta = \dfrac{5\sqrt{61}}{61}$

  $\tan \theta = \dfrac{6}{5}$    $\cot \theta = \dfrac{5}{6}$

  $\sec \theta = \dfrac{\sqrt{61}}{5}$    $\csc \theta = \dfrac{\sqrt{61}}{6}$

**7.** $\sin \theta = \frac{5}{13}$    $\cos \theta = \frac{12}{13}$

  $\tan \theta = \frac{5}{12}$    $\cot \theta = \frac{12}{5}$

  $\sec \theta = \frac{13}{12}$    $\csc \theta = \frac{13}{5}$

**9.** $\sin \theta = \dfrac{\sqrt{15}}{4}$    $\cos \theta = \dfrac{1}{4}$

  $\tan \theta = \sqrt{15}$    $\cot \theta = \dfrac{\sqrt{15}}{15}$

  $\sec \theta = 4$    $\csc \theta = \dfrac{4\sqrt{15}}{15}$

**11.** $\sin \theta = \dfrac{2\sqrt{13}}{13}$    $\cos \theta = \dfrac{3\sqrt{13}}{13}$

  $\tan \theta = \dfrac{2}{3}$    $\cot \theta = \dfrac{3}{2}$

  $\sec \theta = \dfrac{\sqrt{13}}{3}$    $\csc \theta = \dfrac{\sqrt{13}}{2}$

**13.** $\frac{1}{2}$    **15.** $\dfrac{2\sqrt{2}}{3}$    **17.** $\frac{24}{7}$    **19.** 8    **21.** $\frac{1}{2}$

**23.** $\sqrt{3}$    **25.** $\dfrac{2\sqrt{3}}{3}$    **27.** $\sqrt{3}$    **29.** 0.3420

**31.** 0.0279    **33.** 1.069    **35.** 0.9239    **37.** 1.1884

**39. a.** 78.4630°    **b.** 1.3694 radians

**41. a.** 77.0822°    **b.** 1.3453 radians

**43. a.** 10.2866°    **b.** 0.1795 radians

**45.**

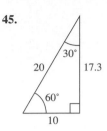

**47.**

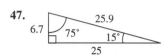

**49.**

**51.**

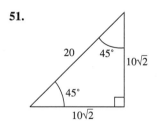

**53.**

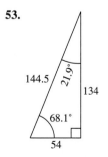

**55.** Approximately 365 feet    **57.** Approximately 2350 feet

**59.** Approximately 16,270 square feet

**61.** Approximately 95 feet or 32 yards

**63.** Approximately 12,108 cubic feet

**65.** Approximately 1.36 miles    **67.** Approximately 7:05 P.M.

**69.** The video shows the angle of elevation to be approximately 34.6°.

*Section 3* ■ *page 500*

**1.** $\theta = \pi$

   $\sin \pi = 0$          $\cos \pi = -1$

   $\tan \pi = 0$         $\cot \pi$ is undefined

   $\sec \pi = -1$      $\csc \pi$ is undefined

**3.** $\theta = \dfrac{5\pi}{3}$

   $\sin \dfrac{5\pi}{3} = -\dfrac{\sqrt{3}}{2}$     $\cos \dfrac{5\pi}{3} = \dfrac{1}{2}$

   $\tan \dfrac{5\pi}{3} = -\sqrt{3}$     $\cot \dfrac{5\pi}{3} = -\dfrac{\sqrt{3}}{3}$

   $\sec \dfrac{5\pi}{3} = 2$       $\csc \dfrac{5\pi}{3} = -\dfrac{2\sqrt{3}}{3}$

**5.** $(0, -1)$

   $\sin(-90°) = -1$     $\cos(-90°) = 0$

   $\tan(-90°)$ undefined   $\cot(-90°) = 0$

   $\sec(-90°)$ undefined   $\csc(-90°) = -1$

**7.** $\left( -\dfrac{\sqrt{2}}{2}, -\dfrac{\sqrt{2}}{2} \right)$

   $\sin \dfrac{5\pi}{4} = -\dfrac{\sqrt{2}}{2}$     $\cos \dfrac{5\pi}{4} = -\dfrac{\sqrt{2}}{2}$

   $\tan \dfrac{5\pi}{4} = 1$        $\cot \dfrac{5\pi}{4} = 1$

   $\sec \dfrac{5\pi}{4} = -\sqrt{2}$    $\csc \dfrac{5\pi}{4} = -\sqrt{2}$

**9.** Fourth quadrant     **11.** Third quadrant     **13.** $-\dfrac{2\sqrt{2}}{3}$

**15.** $\dfrac{\sqrt{2}}{4}$     **17.** $60°$     **19.** $50°$     **21.** $\dfrac{\pi}{4}$     **23.** $\pi - 2$

**25.** $-\dfrac{1}{2}$    **27.** $\dfrac{\sqrt{3}}{2}$    **29.** $-\dfrac{\sqrt{3}}{3}$    **31.** 1    **33.** 0.9397

**35.** $-1.3331$     **37.** 0.8935

**39. a.** 1    **b.** $\dfrac{\sqrt{2}}{2}$    **c.** $-0.4161$

**41. a.** $-\dfrac{2\sqrt{3}}{3}$    **b.** Undefined    **c.** $-1.2015$

**43.**

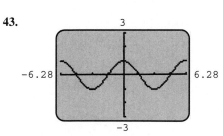

   Domain: All real numbers

   Range:  $[-1, 1]$

**45.**

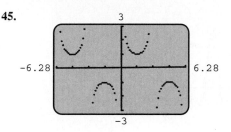

   Domain: All real numbers except multiples of $\pi$

   Range:  $(-\infty, -1] \cup [1, \infty)$

**47. a.** Approximately 61.2, 280.7, and 0 feet, respectively

   **b.** Approximately 0.7760 or 44.46°; yes

**49. a.** 60, 60

   The average temperatures for March 1996 and March 1997 are the same.

   **b.** Lowest: Approximately 59° in February

   Highest: Approximately 87° in August

*Section 4* ■ *page 504*

**1.** Period: 2

   Amplitude: 3

**3.** Period: $\dfrac{2\pi}{3}$

   Amplitude: 5

**5.** Period: $2\pi$

   Amplitude: Approximately 7.2

**7.** Period: $12\pi$

   Amplitude: Approximately 4.5

**9.** Period: $\dfrac{\pi}{2}$

   Amplitude: 5

**11.** Period: $2\pi$

   Amplitude: Approximately 2.78

**13.** Period: $2\pi$

   Amplitude: Approximately 1.87

**15.**

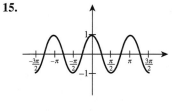

   Amplitude: 1

   Period: $\pi$

   Phase shift: 0

**17.**

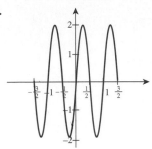

Amplitude: 2
Period: 1
Phase shift: 0

**19.**

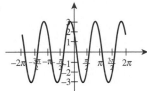

Amplitude: 3
Period: $\pi$
Phase shift: $-\frac{1}{2}$

**21.**

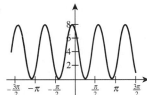

Amplitude: 4
Period: $\dfrac{2\pi}{3}$

Phase shift: $\dfrac{\pi}{12}$

**23.** Amplitude: 3
Period: 2
Phase shift: 0
Maximum: 3
Minimum: −3

**25.** Amplitude: 4
Period: $\pi$
Phase shift: $-\frac{1}{2}$
Maximum: 4
Minimum: −4

**27.** Amplitude: 6
Period: $\pi$

Phase shift: $-\dfrac{\pi}{4}$

Maximum: 6
Minimum: −6

**29.** Amplitude: 2

Period: $\dfrac{2\pi}{3}$

Phase shift: $\frac{2}{3}$
Maximum: 7
Minimum: 3

**31.** Decreasing        **33.** Both        **35.** Increasing

**37.** $a = 1, b = \pi, c = 0$        **39.** $a = 2, b = 1, c = -\dfrac{\pi}{4}$

**41.** $a = 5, b = 2, c = -\pi$        **43.** Not an identity

**45.** Not an identity        **47.** Identity        **49.** $3 \sin\left(2x - \dfrac{2\pi}{3}\right)$

**51.** $\sqrt{2} \sin (8x - \pi)$

**53.** 0.89, 1.59        **55.** 0.045, 0.27

**57.** Amplitude: 0.002
Period: $\frac{1}{264}$

**59. a.** Period: 0.5 second
**b.** 5 centimeters

**61. a.** Amplitude: $110\sqrt{2}$
Period: $\frac{1}{60}$
**b.** $\frac{1}{160}$ second

**63. a.** 5 feet
**b.** Maximum: 105 feet at $t = 15$ seconds
Minimum: 5 feet at $t = 0$ seconds
**c.** 30 seconds

*Section 5* ■ *page 527*

**1.** g        **3.** b        **5.** f        **7.** e

**9.**

Period: $\dfrac{\pi}{3}$

Asymptotes: $\dfrac{\pi}{6} + \dfrac{\pi k}{3}$, $k$ an integer

**11.**

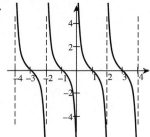

Period: 2
Asymptotes: $2k$, $k$ an integer

**13.**

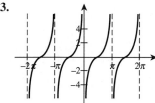

Period: $\pi$
Asymptotes: $\pi k$, $k$ an integer

**15.**

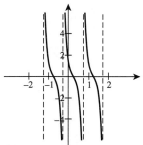

Period: 1
Asymptotes: $-\frac{1}{4} + k$, $k$ an integer

**17.**

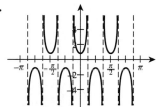

Period: $\frac{\pi}{2}$

Asymptotes: $\frac{\pi}{8} + \frac{\pi k}{4}$, $k$ an integer

**19.**

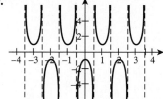

Period: 2
Asymptotes: $\frac{1}{2} + k$, $k$ an integer

**21.**

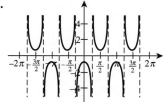

Period: $\pi$

Asymptotes: $\frac{\pi}{4} + \frac{\pi k}{2}$, $k$ an integer

**23.** 1.05        **25.** No zeros

**27. a.** $h = 5\tan\theta$

  **b.** $0 \le \theta < \dfrac{\pi}{2}$

  **c.**

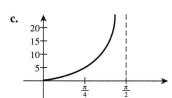

  **d.** $h$ approaches infinity.

### Section 6  ■  *page 539*

**1.** $\dfrac{\pi}{2}$    **3.** $\dfrac{2\pi}{3}$    **5.** $\dfrac{\pi}{4}$    **7.** $\dfrac{\pi}{3}$    **9.** $-\dfrac{\pi}{4}$    **11.** $\pi$

**13.** 0        **15.** Undefined        **17.** $x \approx 0.3398$, $x \approx 2.8018$

**19.** $x \approx 1.6208$, $x \approx 4.7623$        **21.** $x \approx 0.7954$, $x \approx 5.4878$

**23.** No solution        **25.** $\dfrac{\sqrt{1 - x^2}}{x}$        **27.** $\dfrac{x}{\sqrt{x^2 - 4}}$

**29.** $\dfrac{2}{x}$    **31.** $\frac{1}{3}$    **33.** $-\dfrac{2\pi}{5}$    **35.** $-7$    **37.** $\dfrac{3\pi}{4}$

**39.** $-\dfrac{\pi}{2}$    **41.** $\frac{4}{5}$    **43.** $-8$    **45.** $\frac{1}{2}$    **47.** $\dfrac{\sqrt{7}}{7}$

**49.**

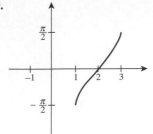

**51.**

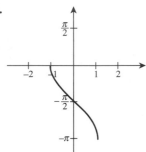

**53.**

**55.** $x \approx 0.9973$     **57.** $x \approx 0.6506$, $x \approx -1.3962$
**59.** Not an identity     **61.** Identity
**63. a.** 48 feet
  **b.** Approximately 25.64°

  **c.** $\theta = \tan^{-1}\left(\dfrac{6t}{100}\right)$

**65. a.** Approximately 59.04°

  **b.** $\theta = \tan^{-1}\left(\dfrac{2000}{x}\right)$

  **c.** Approximately 1154.7 feet

**67. a.**

| $x$ (feet)       | 20     | 50     | 100    |
|------------------|--------|--------|--------|
| $\theta$ (radians) | 1.6307 | 1.9401 | 1.7985 |

  The angle increases and then decreases.
  **b.** $x \approx 61.8$ feet

## Chapter Review ■ *page 543*

**1.** d     **3.** b     **5.** 420° and −300°, for example

**7.** $\dfrac{\pi}{2}$ and $-\dfrac{7\pi}{2}$, for example   **9.** $\dfrac{\pi}{3}$   **11.** $-\dfrac{7\pi}{6}$

**13.** Approximately 0.3194   **15.** −150°   **17.** 22.5°
**19.** Approximately 55.3846°

**21.** $\sin\theta = \frac{4}{5}$     $\cos\theta = \frac{3}{5}$
  $\tan\theta = \frac{4}{3}$     $\cot\theta = \frac{3}{4}$
  $\sec\theta = \frac{5}{3}$     $\csc\theta = \frac{5}{4}$

**23.** $\sin\theta = \dfrac{2}{3}$     $\cos\theta = \dfrac{\sqrt{5}}{3}$

  $\tan\theta = \dfrac{2\sqrt{5}}{5}$     $\cot\theta = \dfrac{\sqrt{5}}{2}$

  $\sec\theta = \dfrac{3\sqrt{5}}{5}$     $\csc\theta = \dfrac{3}{2}$

**25.** $\sin\theta = \frac{8}{17}$
  $\tan\theta = \frac{8}{15}$     $\cot\theta = \frac{15}{8}$
  $\sec\theta = \frac{17}{15}$     $\csc\theta = \frac{17}{8}$

**27.** $\sin\theta = \dfrac{1}{2}$     $\cos\theta = \dfrac{\sqrt{3}}{2}$

  $\tan\theta = \dfrac{\sqrt{3}}{3}$     $\cot\theta = \sqrt{3}$

  $\sec\theta = \dfrac{2\sqrt{3}}{3}$

**29.**

**31.**

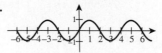

**33.** 30°   **35.** $\dfrac{\pi}{4}$   **37.** $\dfrac{\sqrt{3}}{2}$   **39.** $-\dfrac{2\sqrt{3}}{3}$   **41.** 1

**43.** 0.3420   **45.** −1.7013   **47.** −0.7047

**49.** 0.8944   **51.** $-\dfrac{\sqrt{5}}{2}$

**53. a.** 74.4759°   **b.** 1.2998 radians
**55. a.** 18.6629°   **b.** 0.3257 radian
**57.** f     **59.** g     **61.** a     **63.** h
**65.** Period: $4\pi$; amplitude: 3
**67.** Period: 2; amplitude: 2.25

**69.** Period: $\dfrac{2\pi}{3}$; amplitude: 2

**71.** Period: $2\pi$; amplitude: 2.25

**73.**

Amplitude: 1; period: 4; phase shift: 0

**75.**

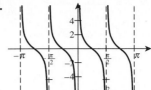

Period: $\dfrac{\pi}{2}$; asymptotes: $\dfrac{\pi k}{2}$, $k$ an integer

**77.**

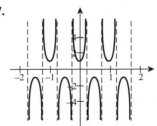

Period: 1; asymptotes: $\dfrac{1}{4} + \dfrac{k}{2}$, $k$ an integer

**79.**

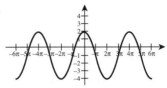

Amplitude: 3; period: $4\pi$

**81.**

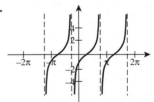

Period: $\pi$; asymptotes: $-\dfrac{\pi}{4} + \pi k$, $k$ an integer

**83.**

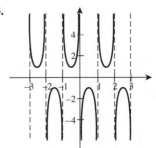

Period: 2; asymptotes: $k$, $k$ an integer

**85.**

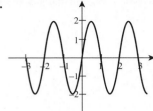

Amplitude: 2; period: 2; phase shift: $-\dfrac{1}{2}$

**87.** $a = 1, b = 2, c = 0$       **89.** $a = 3, b = 1, c = \dfrac{\pi}{4}$

**91.** 0       **93.** $\dfrac{\pi}{3}$       **95.** $-\dfrac{\pi}{6}$

**97.** $x \approx 1.8235$, $x \approx 4.4597$       **99.** $x \approx 0.6155$, $x \approx 2.5261$

**101.** $\dfrac{1}{\sqrt{1 - x^2}}$       **103.** $\dfrac{3}{\sqrt{x^2 + 9}}$       **105.** $\tfrac{4}{5}$       **107.** $\dfrac{3\sqrt{2}}{4}$

**109.** 16       **111.** 0

**113.**

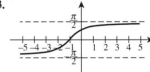

**115.**

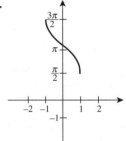

**117.** 120 radians $\approx$ 19.1 revolutions
**119.** Approximately 0.049 mile per hour
**121.** Approximately 171.38 feet
**123.** **a.** 12 months; the dieter's weight depends only on the date, not on the year.
   **b.** 10 pounds
   **c.** 210 pounds; December
   **d.** 190 pounds; June

***Chapter Test*** ■ *page 548*

**1.** False       **2.** True       **3.** False       **4.** False
**5.** True       **6.** False       **7.** True       **8.** False
**9.** True       **10.** True       **11.** $f(x) = 3 \sin 4x$, for example
**12.** $\dfrac{5\pi}{4}$, for example       **13.** $\cot x$ or $\csc x$

**14.** 1 and 0.8391, for example (the ratio of the smaller side to the larger side should be 0.8391)

**15.** $\dfrac{25\pi}{36}$

**16.** $\sin\theta = \dfrac{2\sqrt{5}}{5}$     $\cos\theta = \dfrac{\sqrt{5}}{5}$

$\tan\theta = 2$     $\cot\theta = \dfrac{1}{2}$

$\sec\theta = \sqrt{5}$     $\csc\theta = \dfrac{\sqrt{5}}{2}$

**17.** Approximately 186.6 feet     **18.** $-\dfrac{\sqrt{3}}{2}$

**19.**

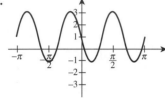

Period: $\dfrac{2\pi}{3}$; amplitude: 2, phase shift: $\dfrac{\pi}{3}$

**20.** Period: 4; amplitude: $5\frac{1}{3}$

**21.**

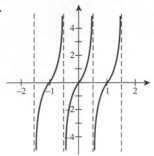

Period: 1; asymptotes: $x = \frac{1}{2} + k$, $k$ an integer

**22.** $\dfrac{2\pi}{5}$     **23.** $x \approx 0.3398$ or $x \approx 2.8018$

# CHAPTER 8

## Section 1 ■ page 558

**1.** Identity     **3.** Contradiction     **5.** Conditional equation
**7.** Contradiction     **9.** Identity     **11.** Conditional equation
**13.** Identity     **15.** Contradiction
**17.** Conditional equation; 2.65     **19.** Identity
**21.** Conditional equation; 3.14     **59.** $\sec x$     **61.** $\csc x$

**63.** $\sin x$     **65.** $\tan x = \dfrac{1}{\cot x}$     **67.** $\cos x = \pm\sqrt{1 - \sin^2 x}$

**69.** $\cot x = \dfrac{\cos x}{\pm\sqrt{1 - \cos^2 x}}$     **71.** $\sec x = \dfrac{1}{\pm\sqrt{1 - \sin^2 x}}$

**73. a.** $v = \sqrt{32r\tan\theta}$
   **b.** Approximately 82 feet per second or 56 miles per hour

## Section 2 ■ page 569

**1. a.** $\frac{63}{65}$     **b.** $-\frac{33}{65}$     **c.** $-\frac{16}{65}$     **d.** $\frac{56}{65}$
**3. a.** $-\frac{24}{25}$     **b.** 0     **c.** $-\frac{7}{25}$     **d.** $-1$
**5. a.** $-\frac{56}{65}$     **b.** $\frac{16}{65}$     **c.** $\frac{33}{65}$     **d.** $-\frac{63}{65}$

**7.** $\frac{1}{2}$     **9.** $-\dfrac{\sqrt{2}}{2}$     **11.** $\sin\dfrac{4\pi}{15}$     **13.** $\cos 3y$

**15.** $\tan 2a$     **17.** $\dfrac{\sqrt{6} - \sqrt{2}}{4}$     **19.** $\sqrt{3} - 2$

**21.** $-\cos x$     **23.** $\cos x$     **25.** $\dfrac{1}{2}\cos x + \dfrac{\sqrt{3}}{2}\sin x$

**27.** $-\dfrac{1}{2}\cos x - \dfrac{\sqrt{3}}{2}\sin x$     **29.** $\dfrac{\tan x - 1}{1 + \tan x}$     **31.** 1

**33.** $-\frac{1}{49}$

**51.** $\cos(x + y + z) = \cos x\cos y\cos z - \cos x\sin y\sin z$
   $- \cos y\sin x\sin z - \cos z\sin x\sin y$
   $\sin(x + y + z) = \cos x\cos y\sin z + \cos x\cos z\sin y$
   $+ \cos y\cos z\sin x - \sin x\sin y\sin z$

## Section 3 ■ page 572

**1. a.** $-\dfrac{24}{25}$     **b.** $\dfrac{7}{25}$     **c.** $-\dfrac{24}{7}$

**3. a.** $\dfrac{24}{25}$     **b.** $\dfrac{7}{25}$     **c.** $\dfrac{24}{7}$

**5. a.** $\dfrac{240}{289}$     **b.** $\dfrac{161}{289}$     **c.** $\dfrac{240}{161}$

**7. a.** $-\dfrac{120}{169}$     **b.** $\dfrac{119}{169}$     **c.** $-\dfrac{120}{119}$

**9. a.** $\dfrac{3}{5}$     **b.** $\dfrac{4}{5}$     **c.** $\dfrac{3}{4}$

**11. a.** $\dfrac{\sqrt{5}}{3}$    **b.** $\dfrac{2}{3}$    **c.** $\dfrac{\sqrt{5}}{2}$

**13. a.** $\dfrac{1}{3}$    **b.** $-\dfrac{2\sqrt{2}}{3}$    **c.** $-\dfrac{\sqrt{2}}{4}$

**15. a.** $\dfrac{1}{4}$    **b.** $-\dfrac{\sqrt{15}}{4}$    **c.** $-\dfrac{\sqrt{15}}{15}$

**17.** $\dfrac{1}{2}\sqrt{2-\sqrt{3}}$    **19.** $\dfrac{2}{\sqrt{2+\sqrt{2}}}$

**37. a.** $A = \dfrac{d^2}{2}\sin 2\theta$

   **b.** Approximately 359 square inches

**Section 4  ■  *page 590***

**1.** True    **3.** False    **5.** False    **7.** $\dfrac{\pi}{2}, \dfrac{3\pi}{2}$

**9.** $x \approx 1.1071, x \approx 4.2487$    **11.** $\dfrac{4\pi}{3}, \dfrac{5\pi}{3}$

**13.** $\dfrac{\pi}{6}, \dfrac{\pi}{3}, \dfrac{5\pi}{6}, \dfrac{5\pi}{3}$    **15.** $0, \dfrac{\pi}{2}, \pi, \dfrac{3\pi}{2}$

**17.** $\pi k$, $k$ an integer    **19.** $\dfrac{\pi}{3} + 2\pi k, \dfrac{5\pi}{3} + 2\pi k$, $k$ an integer

**21.** $\dfrac{\pi}{4} + \pi k, \dfrac{3\pi}{2} + 2\pi k$, $k$ an integer

**23.** $\dfrac{\pi}{2} + \pi k, \dfrac{\pi}{6} + 2\pi k, \dfrac{5\pi}{6} + 2\pi k$, $k$ an integer

**25.** $\pi k$, $k$ an integer    **27.** $\pi k, \dfrac{\pi}{4} + \dfrac{\pi}{2}k$, $k$ an integer

**29.** $\dfrac{\pi}{3} + 2\pi k, \dfrac{5\pi}{3} + 2\pi k, \pi + 2\pi k$, $k$ an integer    **31.** $\dfrac{5\pi}{4}, \dfrac{7\pi}{4}$

**33.** $\dfrac{\pi}{4}, \dfrac{5\pi}{4}$    **35.** $0, \pi$    **37.** No solution

**39.** $\dfrac{\pi}{6}, \dfrac{\pi}{2}, \dfrac{5\pi}{6}, \dfrac{7\pi}{6}, \dfrac{3\pi}{2}$

**41.** $x \approx 0.8128, x \approx 2.3288, x \approx 4.1139, x \approx 5.3108$

**43.** No solution

**45.** $x = \dfrac{3\pi}{4}, x = \dfrac{7\pi}{4}, x \approx 1.1071, x \approx 4.2487$

**47.** No solution    **49.** $\dfrac{\pi}{12}, \dfrac{5\pi}{12}, \dfrac{13\pi}{12}, \dfrac{17\pi}{12}$    **51.** $0, \pi$

**53.** $\dfrac{7\pi}{12}, \dfrac{11\pi}{12}, \dfrac{19\pi}{12}, \dfrac{23\pi}{12}$    **55.** $\dfrac{3\pi}{2}$

**57.** $x \approx 0.3547, x \approx 2.7869$    **59.** $x \approx 0.7391$

**61.** $\dfrac{\pi}{4}, \dfrac{3\pi}{4}, \dfrac{5\pi}{4}, \dfrac{7\pi}{4}$

**63. a.** 10 and 20 seconds    **b.** 30 seconds

**65.** Approximately 0.58 and 1 second
**67.** Approximately 32.08°
**69.** 0° or approximately 75.96°

**Chapter Review  ■  *page 593***

**1.** Conditional equation; approximately 0.6435
**3.** Contradiction    **5.** Identity
**17.** $\cos x = \pm\sqrt{1 - \sin^2 x}$
**19. a.** $\dfrac{77}{85}$    **b.** $-\dfrac{13}{85}$    **c.** $-\dfrac{36}{65}$    **d.** $\dfrac{84}{85}$
**21.** $\cos 60° = \dfrac{1}{2}$    **23.** $\sin 3x$    **25.** $\cos x$
**27.** $\dfrac{\sqrt{2}}{2}(\cos x + \sin x)$

**31. a.** $\dfrac{24}{25}$    **b.** $\dfrac{7}{25}$    **c.** $\dfrac{24}{7}$

**33. a.** $\dfrac{\sqrt{10}}{10}$    **b.** $-\dfrac{3\sqrt{10}}{10}$    **c.** $-\dfrac{1}{3}$

**35.** $\dfrac{\sqrt{2+\sqrt{3}}}{2}$    **43.** $\pi$    **45.** $\dfrac{\pi}{6}, \dfrac{5\pi}{6}, \dfrac{\pi}{3}, \dfrac{4\pi}{3}$

**47.** $0, \pi, \dfrac{\pi}{3}, \dfrac{5\pi}{3}$    **49.** $0, \dfrac{\pi}{3}, \dfrac{5\pi}{3}$    **51.** $0, \dfrac{2\pi}{3}, \pi, \dfrac{4\pi}{3}$

**53.** $x \approx 0.6531, x \approx 2.4885, x \approx 3.4194, x \approx 6.0053$

**55.** $\dfrac{\pi}{3}, \pi, \dfrac{5\pi}{3}$    **57.** $\dfrac{\pi}{3} + 2\pi k, \dfrac{5\pi}{3} + 2\pi k$, $k$ an integer

**59.** $\dfrac{\pi}{3} + 2\pi k, \dfrac{2\pi}{3} + 2\pi k, \dfrac{\pi}{2} + \pi k$, $k$ an integer

**61.** $\dfrac{\pi}{2} + \pi k$, $k$ an integer    **63.** $\dfrac{\pi}{4} + \pi k$

**65.** $x = 0, x \approx 0.4514, x \approx 3.6992, x \approx 4.5017$
**67.** Approximately 31.37° or 58.63°

**69.** $\dfrac{\pi}{90} + \dfrac{\pi k}{30}, \dfrac{\pi k}{30}$, $k$ an integer

**Chapter Test  ■  *page 595***

**1.** False    **2.** False    **3.** True    **4.** False
**5.** True    **6.** False    **7.** False    **8.** True
**9.** $a = 2, b = 1$, or $a = 0, b = 2$
**10.** $a = 1, b = 1$, for example    **11.** $a = 2, b = 2$, for example

**15.** $\dfrac{\tan x - 1}{1 + \tan x}$    **16.** $\dfrac{\sqrt{2-\sqrt{3}}}{2}$

**17.** Approximately 1.2490, 4.3906

**18.** $\dfrac{\pi}{6} + 2\pi k, \dfrac{5\pi}{6} + 2\pi k$, $k$ an integer    **19.** Contradiction

**20.** $\dfrac{2}{3}$ second

## CHAPTER 9

### Section 1 ■ page 606

1. $a \approx 11.79$, $b \approx 9.35$, $B = 50°$
3. $a \approx 3.21$, $A \approx 16.15°$, $C \approx 43.85°$
5. $b \approx 9.25$, $c \approx 5.49$, $A = 70$
7. $a \approx 8.2$, $b \approx 10.46$, $C = 120°$
9. $a \approx 5.28$, $b \approx 17.58$, $B = 77°$
11. $a \approx 17.32$, $A = 60°$, $B = 90°$
13. $b \approx 6.04$, $c \approx 38.49$, $A = 122°$
15. $b \approx 10.09$, $B \approx 21.85°$, $C \approx 36.15$
17. No triangle possible
19. $b \approx 18.66$, $A \approx 48.58°$, $B \approx 91.42°$;
    $b \approx 2.79$, $A \approx 131.42$, $B \approx 8.58$
21. No triangle possible
23. $a \approx 68.20$, $c \approx 61.68$, $B = 65°$
25. $b \approx 391.94$, $c \approx 150.45$, $C = 20°$
27. $a \approx 0.97$, $A \approx 121.46°$, $B \approx 6.54°$
29. 142.65    31. 3,528,060    33. 5142.13
35. Approximately 200.2 feet    37. Approximately 990 feet
39. Approximately 8.4 miles from the southern station and 6.9 miles from the northern station; approximately 5.7 miles from shore
41. Approximately 242,340 square feet; approximately $11,130

### Section 2 ■ page 616

1. $c \approx 10.67$, $A \approx 74.67°$, $B \approx 46.33°$
3. $A \approx 17.61°$, $B \approx 28.96°$, $C \approx 133.43°$
5. $a \approx 22.84$, $B \approx 139.24°$, $C \approx 5.77°$
7. No triangle possible
9. $a \approx 269.63$, $B \approx 24.68$, $C \approx 35.32$
11. $A \approx 36.18°$, $B \approx 43.53°$, $C \approx 100.29°$
13. $b \approx 22.05$, $c \approx 16.90$, $A = 71°$
15. $c \approx 16.78$, $B \approx 29.21°$, $C \approx 35.79°$
17. $c \approx 1198.32$, $A \approx 22.26°$, $B \approx 130.74°$
19. $A \approx 53.11°$, $B \approx 35.68°$, $C \approx 91.21°$
21. $a \approx 21.29$, $A \approx 134.43°$, $C \approx 17.57°$
23. No triangle possible    25. No triangle possible
27. $a \approx 1.12$, $c \approx 1.17$, $A = 37°$
29. $b \approx 124.40$, $c \approx 138.79$, $A = 64°$
31. $p \approx 1181.15$, $r \approx 602.46$, $Q = 27°$
33. $m \approx 0.0266$, $n \approx 0.0266$, $N = 20°$
35. $U \approx 22.33°$, $V \approx 46.95°$, $W \approx 110.72°$
37. No triangle possible
39. $f \approx 5503.14$, $D \approx 16.16°$, $E \approx 33.84°$
41. 82.65    43. 105.5    45. 0.5402
47. Approximately 50.6 miles
49. A horizontal distance of approximately 3647 feet (toward Elizabeth) and a vertical distance of approximately 4036 feet
51. Approximately 86.6 square feet; $\angle A \approx 60°$, $\angle B \approx 50.3°$, $\angle C \approx 69.7°$

53. **a.** Approximately 42.4 miles
    **b.** $p(t) = 80t$, $f(t) = 100 - 90t$
    **c.** $d \approx \sqrt{23{,}756.1t^2 - 28{,}284.6t + 10{,}000}$
    **d.** Approximately 0.6 hours

### Section 3 ■ page 626

1. 5    3. $\sqrt{10}$    5. 2    7. 15    9. 1

11.

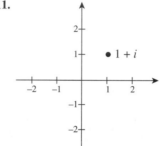

$$\sqrt{2}\left(\cos\frac{\pi}{4} + i\sin\frac{\pi}{4}\right)$$

13.

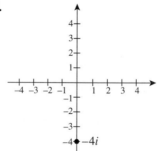

$$4\left(\cos\frac{3\pi}{2} + i\sin\frac{3\pi}{2}\right)$$

15.

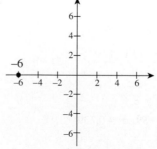

$6(\cos\pi + i\sin\pi)$

**17.**

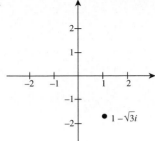

$$2\left(\cos\frac{5\pi}{3} + i\sin\frac{5\pi}{3}\right)$$

**25.**

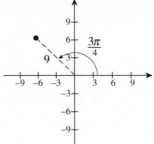

$$-\frac{9\sqrt{2}}{2} + \frac{9\sqrt{2}}{2}\,i$$

**19.**

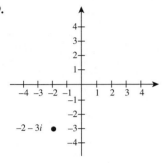

Approximately $\sqrt{13}(\cos 4.1244 + i\sin 4.1244)$

**27.**

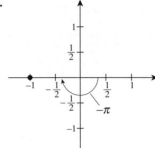

$-1$

**21.**

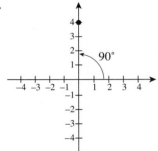

$4i$

**29.**

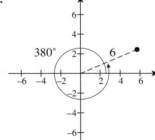

Approximately $5.6382 + 2.0521i$

**23.**

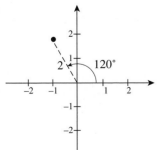

$-1 + \sqrt{3}\,i$

**31.**

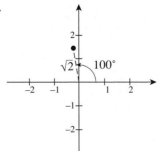

Approximately $-0.2456 + 1.3927i$

**33.**

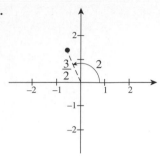

Approximately $-0.6242 + 1.3639i$

**35.** $6(\cos 70° + i \sin 70°)$    **37.** $1 - 3i$

**39.** $\dfrac{1}{2}\left[\cos\left(-\dfrac{\pi}{12}\right) + i\sin\left(-\dfrac{\pi}{12}\right)\right]$    **41.** $-2 + 8i$

**43.** $-\dfrac{7}{10} + \dfrac{11}{10}i$    **45.** $15\cos\left(\dfrac{5\pi}{4} + i\sin\dfrac{5\pi}{4}\right)$

**47.** $7\left(\cos\dfrac{11\pi}{4} + i\sin\dfrac{11\pi}{4}\right)$    **49.** $12i$    **51.** $-4$

**53.** $4\sqrt{3} - 4i$    **55.** $-\dfrac{6}{25} + \dfrac{8}{25}i$

## Section 4 ■ page 635

**1.** $32(\cos 150° + i\sin 150°)$    **3.** $16(\cos 80° + i\sin 80°)$

**5.** $256$    **7.** $64\sqrt{3} - 64i$    **9.** $117 + 44i$

**11.** $\dfrac{5\sqrt{3}}{2} + \dfrac{5}{2}i, -\dfrac{5\sqrt{3}}{2} - \dfrac{5}{2}i$

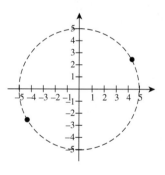

**13.** $\dfrac{\sqrt{3}}{4} + \dfrac{1}{4}i, -\dfrac{1}{4} + \dfrac{\sqrt{3}}{4}i, -\dfrac{\sqrt{3}}{4} - \dfrac{1}{4}i, \dfrac{1}{4} - \dfrac{\sqrt{3}}{4}i$

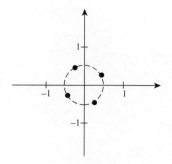

**15.** $3\sqrt{2} + 3\sqrt{2}i, -3\sqrt{2} - 3\sqrt{2}i$

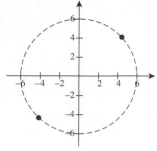

**17.** $3, -\dfrac{3}{2} + \dfrac{3\sqrt{3}}{2}i, -\dfrac{3}{2} - \dfrac{3\sqrt{3}}{2}i$

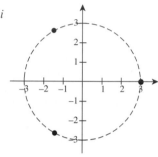

**19.** $1, \dfrac{1}{2} + \dfrac{\sqrt{3}}{2}i, -\dfrac{1}{2} + \dfrac{\sqrt{3}}{2}i, -1, -\dfrac{1}{2} - \dfrac{\sqrt{3}}{2}i, \dfrac{1}{2} - \dfrac{\sqrt{3}}{2}i$

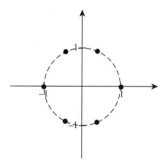

**21.** $0.3536 + 0.3536i, 0.1294 - 0.4830i, -0.4830 + 0.1294i$

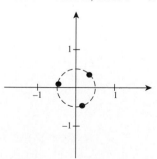

**23.** $x = \dfrac{\sqrt{2}}{2} + \dfrac{\sqrt{2}}{2} i,\ x = -\dfrac{\sqrt{2}}{2} + \dfrac{\sqrt{2}}{2} i,$

$x = -\dfrac{\sqrt{2}}{2} - \dfrac{\sqrt{2}}{2} i,\ x = \dfrac{\sqrt{2}}{2} - \dfrac{\sqrt{2}}{2} i$

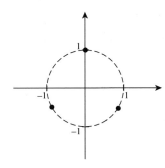

**25.** $x = i,\ x = -\dfrac{\sqrt{3}}{2} - \dfrac{1}{2} i,\ x = \dfrac{\sqrt{3}}{2} - \dfrac{1}{2} i$

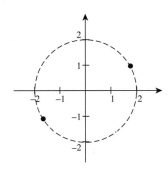

**27.** $x = \sqrt{3} + i,\ x = -\sqrt{3} - i$

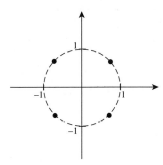

**29.** $(x + \sqrt{3} + i)(x + \sqrt{3} - i)(x - \sqrt{3} + i)(x - \sqrt{3} - i)$
$(x + 2i)(x - 2i)$

**31.** $0.9239 + 0.3827i,\ -0.9239 - 0.3827i,\ 0.3827 - 0.9239i,$
$-0.3827 + 0.9239i$

**Section 5 ■ page 646**

**1.**

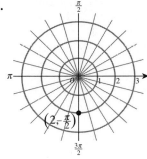

$(0, -2)$

**3.**

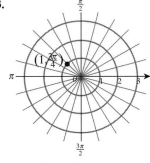

$\left( -\dfrac{\sqrt{2}}{2}, \dfrac{\sqrt{2}}{2} \right)$

**5.**

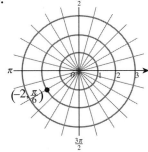

$(-\sqrt{3}, -1)$

**7.**

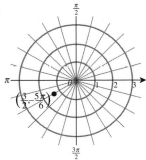

$\left( -\dfrac{3\sqrt{3}}{4}, -\dfrac{3}{4} \right)$

**9.**

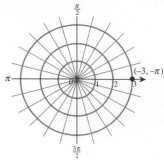

(3, 0)

**11.**

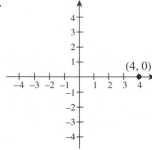

(4, 0), (−4, π)

**13.**

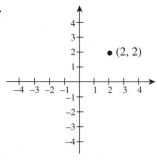

$\left(2\sqrt{2}, \dfrac{\pi}{4}\right)\left(-2\sqrt{2}, \dfrac{5\pi}{4}\right)$

**15.**

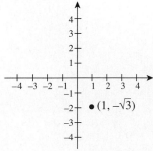

$\left(2, \dfrac{5\pi}{3}\right), \left(-2, \dfrac{2\pi}{3}\right)$

**17.**

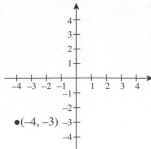

Approximately (5, 3.7851), (−5, 0.6435)

**19.**

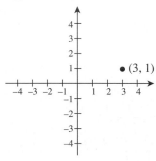

Approximately $(\sqrt{10}, 0.3218)$, $(-\sqrt{10}, 3.4633)$

**21.**

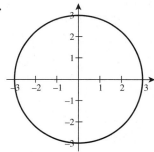

**23.**

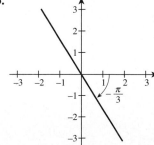

**25.**

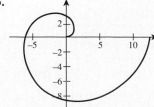

**27.**

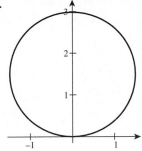

**29.**

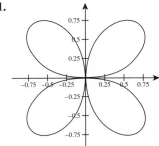

**31.**

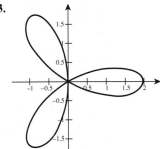

**33.**

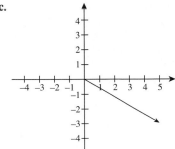

**35.** $r = 3 \csc \theta$    **37.** $\theta = \dfrac{3\pi}{4}$    **39.** $r = 3$

**41.** $\tan \theta = e$    **43.** $x^2 + y^2 = 49$    **45.** $x = 1$
**47.** $x^2 + y^2 = 2y$    **49.** $(x^2 + y^2)^3 = (x^2 - y^2)^2$
**51.** Approximately $(2, 0.52)$ and $(2, 2.62)$
**53.** Approximately $(0, 0)$ and $(1.6, -0.64)$
**55.** Approximately $(0.87, 1.05)$, $(0.87, 2.09)$, and $(0, 0)$

*Section 6* ■ *page 658*

**1. a**

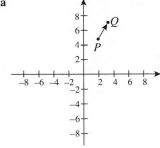

**b.** $\langle 1, 2 \rangle$
**c.**

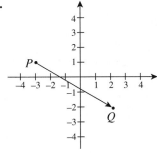

**3. a.**

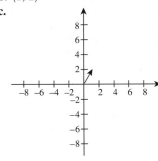

**b.** $\langle 5, -3 \rangle$
**c.**

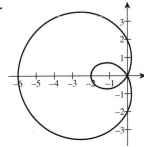

**5. a.**

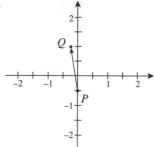

**b.** $\left\langle -\frac{1}{4}, \frac{3}{2} \right\rangle$

**c.**

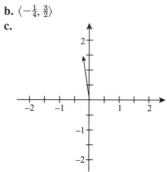

**7.** 4        **9.** 13        **11.** $\sqrt{34}$        **13.** $2\sqrt{3}$

**15.**

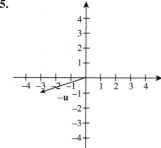

**17.**

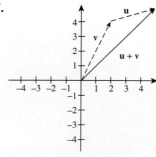

**19.**

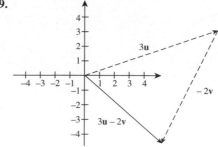

**21.**

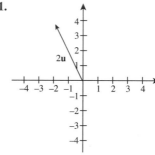

**23.**

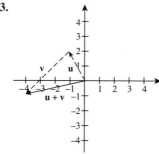

**25.**

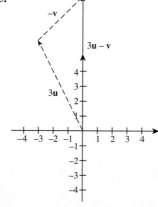

**27. a.** $\langle -1, -3 \rangle$      **b.** $\langle 3, 8 \rangle$      **c.** $\langle 0, -1 \rangle$

**29. a.** $3\sqrt{2}\,\mathbf{i} - \sqrt{2}\,\mathbf{j}$      **b.** $-7\mathbf{i} - \mathbf{j}$      **c.** $31\mathbf{i} - 2\mathbf{j}$

**31. a.** $\mathbf{i} - \mathbf{j}$      **b.** $2\mathbf{i} - 3\mathbf{j}$      **c.** $\frac{1}{5}\mathbf{i}$

**33.** **i**  **35.** $\langle -\frac{3}{5}, \frac{4}{5} \rangle$  **37.** $-\dfrac{2\sqrt{13}}{13}\,\mathbf{i} - \dfrac{3\sqrt{13}}{13}\,\mathbf{j}$

**39.** $\left\langle \dfrac{1}{2}, \dfrac{\sqrt{3}}{2} \right\rangle$  **41.** $\left\langle -\dfrac{1}{4}, -\dfrac{\sqrt{3}}{4} \right\rangle$  **43.** $\langle 0, -5 \rangle$

**45.** Approximately $\langle -0.6840, 1.8794 \rangle$

**47.** $7[(\cos 0)\mathbf{i} + (\sin 0)\mathbf{j}]$

**49.** $4\left[\left(\cos \dfrac{11\pi}{6}\right)\mathbf{i} + \left(\sin \dfrac{11\pi}{6}\right)\mathbf{j}\right]$

**51.** $\dfrac{\sqrt{2}}{8}\left[\left(\cos \dfrac{5\pi}{4}\right)\mathbf{i} + \left(\sin \dfrac{5\pi}{4}\right)\mathbf{j}\right]$

**61.** Approximately 217.5 miles per hour in the direction 350.65° (clockwise from north)

**63.** Approximately 15.9 miles per hour on a course of 33.7° clockwise from north

**65.** Approximately 217 pounds and 10.9° off course

**67.** Approximately 500 pounds

*Chapter Review* ■ *page 662*

**1.** $a \approx 11.14$, $b \approx 15.04$, $B = 81°$

**3.** $c \approx 1.49$, $A \approx 53.41°$, $C \approx 84.59°$;
  $c \approx 0.296$, $A \approx 126.59°$, $C \approx 11.41°$

**5.** $a \approx 4.97$, $B \approx 138.36°$, $C \approx 31.64°$

**7.** $A \approx 33.56°$, $B \approx 62.18°$, $C \approx 84.26°$

**9.** $A \approx 28.41°$, $B \approx 21.67°$, $C \approx 129.91°$

**11.** No triangle possible

**13.**

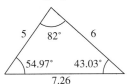

**15.**

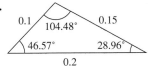

**17.** Approximately 67.62  **19.** Approximately 2993.33

**21.** 5  **23.** $\sqrt{82}$

**25.**

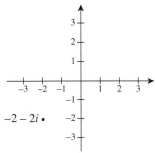

$2\sqrt{2}\left(\cos \dfrac{5\pi}{4} + i \sin \dfrac{5\pi}{4}\right)$

**27.**

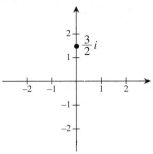

$\dfrac{3}{2}\left(\cos \dfrac{\pi}{2} + i \sin \dfrac{\pi}{2}\right)$

**29.**

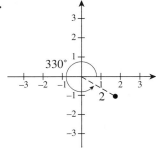

$\sqrt{3} - i$

**31.**

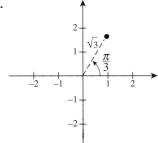

$\dfrac{\sqrt{3}}{2} + \dfrac{3}{2} i$

**33.** $10\left(\cos \dfrac{7\pi}{12} + i \sin \dfrac{7\pi}{12}\right)$  **35.** $3(\cos 200° + i \sin 200°)$

**37.** $625(\cos 40° + i \sin 40°)$  **39.** $16 + 16\sqrt{3}i$

**41.** $1 + \sqrt{3}i$, $-1 - \sqrt{3}i$

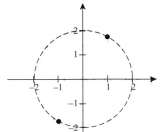

**43.** $3, -3, 3i, -3i$

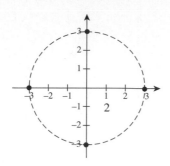

**45.** $-1, \dfrac{1}{2} + \dfrac{\sqrt{3}}{2}\,i, \dfrac{1}{2} - \dfrac{\sqrt{3}}{2}\,i$

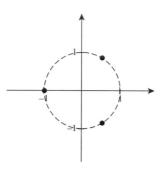

**47.**

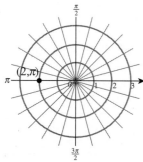

$(-2, 0)$

**49.**

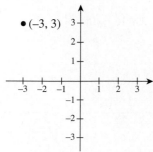

$\left(3\sqrt{2}, \dfrac{3\pi}{4}\right), \left(-3\sqrt{2}, \dfrac{7\pi}{4}\right)$

**51.**

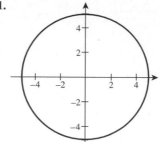

**53.**

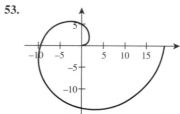

**55.**

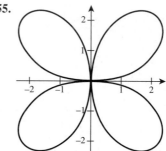

**57.** $\tan\theta = 2$     **59.** $r = 1$

**61.** $x^2 + y^2 = 1$     **63.** $x^2 + y^2 = 3y$

**65.** Approximately $(3, 0.72)$ and $(3, -0.72)$

**67. a.**

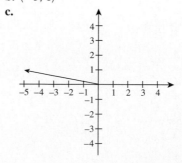

**b.** $\langle -5, 1\rangle$

**c.**

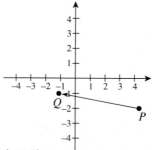

**69.** 10

**71.**

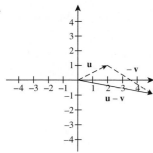

**73. a.** $\langle 8, -4 \rangle$    **b.** $\langle 0, -4 \rangle$    **c.** $\langle -3, \frac{5}{2} \rangle$

**75.** $\frac{12}{13}\mathbf{i} + \frac{5}{13}\mathbf{j}$    **77.** $\langle \sqrt{3}, 1 \rangle$

**79.** $4\left[ \left( \cos \frac{\pi}{2} \right)\mathbf{i} + \left( \sin \frac{\pi}{2} \right)\mathbf{j} \right]$

**81.** Approximately 15 minutes and 49 seconds

**83.** Approximately 7.48 miles per hour

*Chapter Test* ■ *page 665*

**1.** False    **2.** True    **3.** False    **4.** True

**5.** False    **6.** False    **7.** False    **8.** True

**9.** True    **10.** False    **11.** Answers will vary

**12.** $-i$, for example    **13.** $\mathbf{u} = \langle 2, 0 \rangle$, $\mathbf{v} = \langle -2, 0 \rangle$, for example

**14.** $-1 + i$, for example

**15.** $a \approx 9.89$, $B \approx 83.52°$, $C \approx 41.48°$

**16.** $b \approx 4.41$, $A \approx 41.47°$, $B \approx 76.53°$

**17. a.** $\sqrt{2}\left( \cos \frac{5\pi}{4} + i \sin \frac{5\pi}{4} \right)$

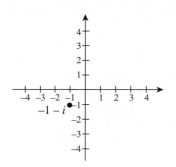

**b.** $-8i$

**c.** $\sqrt[4]{2}\left( \cos \frac{5\pi}{8} + i \sin \frac{5\pi}{8} \right)$,

$\sqrt[4]{2}\left( \cos \frac{13\pi}{8} + i \sin \frac{13\pi}{8} \right)$

**18. a.** $\mathbf{v} = \mathbf{i} + \mathbf{j}$    **b.** $\langle 1, 5 \rangle$

**19.**

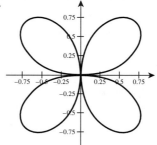

**20.** Approximately 29.2 miles

---

**CHAPTER 10**

*Section 1* ■ *page 679*

**1.** $\begin{bmatrix} 2 & 0 \\ 6 & 10 \end{bmatrix}$    **3.** $\begin{bmatrix} -2 & 11 \\ 6 & 6 \end{bmatrix}$    **5.** $\begin{bmatrix} -1 & 2 & 3 \\ -2 & 7 & -4 \end{bmatrix}$

**7.** $\begin{bmatrix} -4 & 12 \\ 3 & 13 \end{bmatrix}$    **9.** $\begin{bmatrix} 1 & 7 & 4 \\ 7 & 9 & 18 \\ 7 & 2 & -3 \end{bmatrix}$

**11.** $\begin{bmatrix} (2-a) & 0 & (1-a) \\ -a & (1-a) & 2 \end{bmatrix}$

**13.** $\begin{bmatrix} 3 & 3 \\ 7 & 14 \end{bmatrix}$    **15.** $\begin{bmatrix} -8 & -1 \\ -14 & -35 \end{bmatrix}$

**17.** $X = \begin{bmatrix} 2 & 3 \\ 4 & 1 \end{bmatrix}$    **19.** $X = \begin{bmatrix} -\frac{1}{2} & -1 & -\frac{3}{2} \\ -3 & -\frac{5}{2} & -2 \end{bmatrix}$

**21.** Not defined    **23.** Defined; $2 \times 3$    **25.** Not defined

**27.** $\begin{bmatrix} -6 \\ 6 \end{bmatrix}$    **29.** $\begin{bmatrix} 2 & 7 \\ 4 & 14 \end{bmatrix}$    **31.** $\begin{bmatrix} 3 & 9 \\ -15 & -33 \end{bmatrix}$

**33.** $\begin{bmatrix} 6 & 10 \\ 7 & 33 \\ 7 & 17 \end{bmatrix}$    **35.** $\begin{bmatrix} p & q \\ r & s \\ t & u \end{bmatrix}$    **37.** $\begin{bmatrix} 39.12 \\ 33.94 \\ 16.35 \end{bmatrix}$

**39.** $\begin{bmatrix} a & b \\ (ax + c) & (bx + d) \end{bmatrix}$    **41.** $[ai + bj + ck]$

**43.** $\begin{bmatrix} a_{31} & a_{32} & a_{33} \\ a_{21} & a_{22} & a_{23} \\ a_{11} & a_{12} & a_{13} \end{bmatrix}$

**45.** $A = [1 \ 0]$, $B = \begin{bmatrix} 0 \\ 1 \end{bmatrix}$ for example

**47.** $\begin{bmatrix} 1 & 4 \\ 0 & 1 \end{bmatrix}$    **49.** $\begin{bmatrix} 1 & 0 \\ 0 & 0 \end{bmatrix}$

**51.** $\begin{bmatrix} 1 & 4 & 14 \\ 0 & 1 & 8 \\ 0 & 0 & 1 \end{bmatrix}$    **53.** $\begin{bmatrix} 1 & 64 & 4064 \\ 0 & 1 & 128 \\ 0 & 0 & 1 \end{bmatrix}$

**59.** Hourly: $\begin{bmatrix} 230{,}000 & 320{,}000 \\ 280{,}000 & 300{,}000 \end{bmatrix}$

Salaried: $\begin{bmatrix} 125{,}000 & 250{,}000 \\ 400{,}000 & 500{,}000 \end{bmatrix}$

Sum: $\begin{bmatrix} 355{,}000 & 570{,}000 \\ 680{,}000 & 800{,}000 \end{bmatrix}$

Total major medical claims were \$355,000 for workers and \$570,000 for dependents, while total comprehensive claims were \$680,000 for workers and \$800,000 for dependents.

**61.** $S = \begin{bmatrix} 5 & 40 & 18 \\ 4 & 37 & 26 \end{bmatrix}$; $T = \begin{bmatrix} 3 \\ 2 \\ 1 \end{bmatrix}$; $ST = \begin{bmatrix} 113 \\ 112 \end{bmatrix}$

So the final score is Hornets 113, Magic 112.

**Section 2 ■ page 689**

**1.** Linear    **3.** Not linear    **5.** Not linear

**7.** $\left[\begin{array}{cc|c} 3 & 4 & 10 \\ 2 & -7 & 4 \end{array}\right]$    **9.** $\left[\begin{array}{cc|c} 1 & 0 & 3 \\ 0 & 1 & 4 \end{array}\right]$    **11.** $\left[\begin{array}{ccc|c} 1 & -1 & 2 & 1 \\ 0 & 2 & 6 & 3 \end{array}\right]$

**13.** 0    **15.** Infinitely many    **17.** 1    **19.** $x = 2, y = 1$

**21.** $x = 3, y = 7, z = 1$    **23.** $q = -1, r = 3, s = 4, t = -2$

**25.** $x = 2, y = 3$    **27.** $x = -6, y = 10, z = -4$

**29.** $x = 1, y = -2, z = -1$

**31.** Salary: \$40,000

Income from municipal bonds: \$10,000

Income from U.S. treasury bonds: \$8000

**33.** Approximately 1.5 bananas, 2 servings of milk, and 3.9 slices of wheat toast

**Section 3 ■ page 700**

**1.** $\left[\begin{array}{cc|c} 1 & 2 & 3 \\ 0 & 1 & 4 \end{array}\right]$    **3.** $\left[\begin{array}{ccc|c} 1 & 0 & \frac{1}{2} & 0 \\ 0 & 1 & 3 & 1 \\ 0 & 0 & 1 & 0 \end{array}\right]$    **5.** $\left[\begin{array}{ccc|c} 1 & 2 & 1 & 2 \\ 0 & 1 & 1 & 0 \end{array}\right]$

**7.** $\left[\begin{array}{cccc|c} 1 & 0 & 1 & 2 & 1 \\ 0 & 1 & 0 & 1 & 0 \\ 0 & 0 & 1 & 1 & 0 \\ 0 & 0 & 0 & 1 & 2 \end{array}\right]$    **9.** $x = 2, y = 1$    **11.** $p = \frac{1}{2}, q = \frac{2}{3}$

**13.** $x = 1, y = 0, z = 3$    **15.** $p = 0, q = 1, r = 2$

**17.** $w = \frac{3}{2}, x = -\frac{1}{2}, y = 0, z = 0$    **19.** $\left[\begin{array}{cc|c} 1 & 0 & -1 \\ 0 & 1 & 2 \end{array}\right]$

**21.** $\left[\begin{array}{cc|c} 1 & 0 & 2 \\ 0 & 1 & -3 \end{array}\right]$    **23.** $\left[\begin{array}{ccc|c} 1 & 0 & 0 & 2 \\ 0 & 1 & 0 & 1 \\ 0 & 0 & 1 & -1 \end{array}\right]$

**25.** $x = 3, y = -2$    **27.** $x = \frac{1}{2}, y = -\frac{1}{3}$

**29.** $x = 4, y = -1, z = 3$    **31.** $x = 0, y = 3, z = -4$

**33.** $\left[\begin{array}{cc|c} 1 & \frac{1}{2} & \frac{5}{2} \\ 0 & 0 & 0 \end{array}\right]$; $(\frac{5}{2} - \frac{1}{2}y, y)$

**35.** $\left[\begin{array}{ccc|c} 1 & 2 & 1 & 5 \\ 0 & 1 & -\frac{2}{3} & -\frac{1}{3} \\ 0 & 0 & 0 & 0 \end{array}\right]$; $(\frac{17}{3} - \frac{7}{3}z, -\frac{1}{3} + \frac{2}{3}z, z)$

**37.** $y = 2x^2 - 4x + 7$

**39.** 8 nickels, 5 dimes, and 3 quarters

**41.** Hinduism in 1500 B.C., Buddhism in 525 B.C., and Islam in A.D. 622

**43.** 4 schools of the first type, 3 schools of the second type, and one school of the third type

**Section 4 ■ page 711**

**1.** $\begin{bmatrix} 1 & 0 \\ -2 & 1 \end{bmatrix}$    **3.** $\begin{bmatrix} 0 & \frac{1}{2} \\ 3 & -\frac{3}{4} \end{bmatrix}$    **5.** Does not exist

**7.** $\begin{bmatrix} 0 & \frac{1}{b} \\ \frac{1}{a} & 0 \end{bmatrix}$    **9.** $\begin{bmatrix} \frac{1-x}{2x} & \frac{x+1}{2x} \\ \frac{1}{2} & -\frac{1}{2} \end{bmatrix}$    **11.** $\begin{bmatrix} 1 & -2 & 8 \\ 0 & 1 & -4 \\ 0 & 0 & 1 \end{bmatrix}$

**13.** $\begin{bmatrix} 1 & -2 & 0 \\ \frac{3}{4} & -1 & \frac{1}{4} \\ \frac{3}{2} & -3 & \frac{1}{2} \end{bmatrix}$    **15.** $\begin{bmatrix} 1 & 2 & 3 \\ 3 & 2 & 1 \\ 2 & 3 & 1 \end{bmatrix}$    **17.** $\begin{bmatrix} 1 & -x & x^2 \\ 0 & 1 & -x \\ 0 & 0 & 1 \end{bmatrix}$

**19.** $\frac{1}{40} \cdot \begin{bmatrix} -9 & 1 & 1 & 11 \\ 1 & 1 & 11 & -9 \\ 1 & 11 & -9 & 1 \\ 11 & -9 & 1 & 1 \end{bmatrix}$    **21.** $\begin{bmatrix} 1 & 0 & 0 & 0 & 0 & 0 & 0 \\ 1 & 1 & 0 & 0 & 0 & 0 & 0 \\ 1 & 1 & 1 & 0 & 0 & 0 & 0 \\ 1 & 1 & 1 & 1 & 0 & 0 & 0 \\ 1 & 1 & 1 & 1 & 1 & 0 & 0 \\ 1 & 1 & 1 & 1 & 1 & 1 & 0 \\ 1 & 1 & 1 & 1 & 1 & 1 & 1 \end{bmatrix}$

**23. a.** $x = \frac{10}{3}, y = \frac{13}{3}$

**b.** $x = -3, y = 1$

**c.** $x = \frac{1}{2}, y = -\frac{1}{2}$

**25. a.** $x = 5, y = 5$

**b.** $x = 4, y = 1$

**c.** $x = -2, y = 3$

**27. a.** $x = 2, y = 3, z = 4$

**b.** $x = -2, y = 0, z = 1$

**c.** $x = 5, y = 5, z = 5$

**33.** When $abcd \neq 0$; $A^{-1} = \begin{bmatrix} \frac{1}{a} & 0 & 0 & 0 \\ 0 & \frac{1}{b} & 0 & 0 \\ 0 & 0 & \frac{1}{c} & 0 \\ 0 & 0 & 0 & \frac{1}{d} \end{bmatrix}$

**35. a.** $-2$    **b.** 18    **c.** $na$

**37.** $I$

**39. a.** Approximately 35.1 minutes walking, 40.2 minutes jogging, and 44.7 minutes running

**b.** Approximately 38.6 minutes walking, 39.7 minutes jogging, and 11.7 minutes running

**41.** $f(x) = 2x^3 - 3x^2 + 5x + 1$

**Section 5 ■ page 723**

**1.** $-2$    **3.** 0    **5.** $-25$    **7.** $x$    **9.** $-10$

**11.** $-1$    **13.** $-51.406$    **15.** $a$

**17.** $(a_2b_3 - a_3b_2)i + (a_3b_1 - a_1b_3)j + (a_1b_2 - a_2b_1)k$
**19.** $-561$     **21.** 1     **23.** Invertible     **25.** Not invertible
**27.** Not invertible     **29.** Invertible     **31.** $-\frac{1}{2}$
**33.** $3, -3$     **35.** Invertible for all real $x$
**37.** $x = 9, y = 5$     **39.** $x = 1, y = -3, z = 4$
**41.** $x = \frac{87}{23}, y = -\frac{25}{23}, z = \frac{106}{23}$
**43.** $x = 1, y = 0, z = 0, w = 0$
**45.** $x = \dfrac{ce - bf}{ae - bd}, y = \dfrac{af - cd}{ae - bd}$
**51.** Approximately \$9722
**53.** Approximately 185.6 grams of milk

**19.**
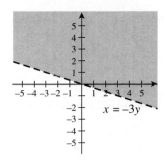

## Section 6 ■ *page 735*

**1.** $(0, 0)$: No; $(4, 1)$: No; $(-1, 3)$: Yes
**3.** $(1, 4)$: Yes; $(3, 0)$: No; $(-2, -1)$: No
**5.** $(0, 2)$: Yes; $(-1, 3)$: Yes; $(5, -1)$: No
**7.** d.     **9.** e     **11.** f

**21.**
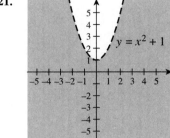

**13.**
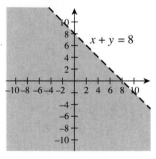

**23.**
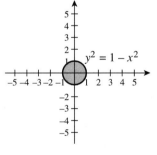

**15.**
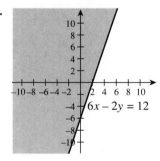

**17.**
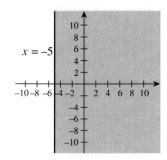

**25.**
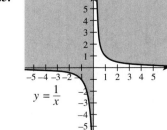

**27.**

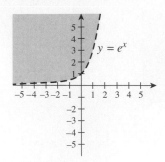

**29.** c      **31.** d

**33.**

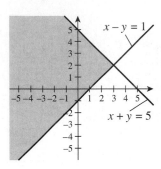

**35.**

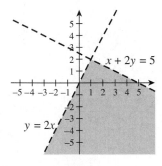

**37.**

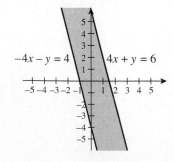

**39.**

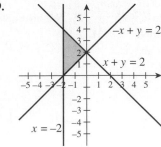

**41.**

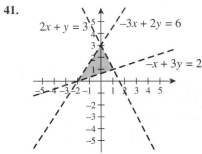

**43.**

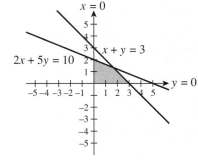

**45.**

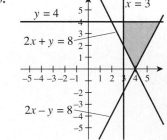

**47.**

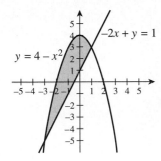

$y = 4 - x^2$
$-2x + y = 1$

**49.**

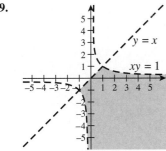

$y = x$
$xy = 1$

**51.**

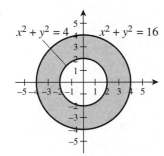

$x^2 + y^2 = 4$ $x^2 + y^2 = 16$

**53.**

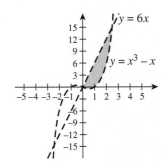

$y = 6x$
$y = x^3 - x$

**55.** $y \leq \frac{3}{4}x - 3$

**57.** $y \geq x + 2$
$\quad y \leq -\frac{2}{5}x + 2$

**59.** $x \geq 0$
$\quad y \leq -\frac{2}{3}x + 4$
$\quad y \geq \frac{2}{3}x$

**61.** $y \geq \frac{2}{3}x - \frac{4}{3}$
$\quad y \leq -\frac{3}{2}x + \frac{19}{2}$
$\quad y \geq -\frac{3}{2}x + 3$
$\quad y \leq \frac{2}{3}x + 3$

**63.** $x^2 + y^2 \leq 9$
$\quad x \geq 0$

**65.** $x \geq 6000$
$\quad y \geq 6000$
$\quad y \leq \dfrac{x}{2}$
$\quad x + y \leq 24{,}000$

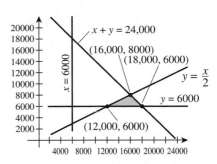

$x + y = 24{,}000$
$(16{,}000, 8000)$
$(18{,}000, 6000)$
$y = \dfrac{x}{2}$
$y = 6000$
$(12{,}000, 6000)$

**67.** $x + y \leq 1000$
$\quad 100x + 80y \leq 88{,}000$
$\quad 2x + y \leq 1600$
$\quad x \geq 200$

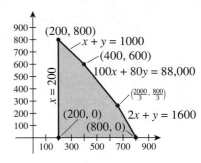

$(200, 800)$
$x + y = 1000$
$(400, 600)$
$100x + 80y = 88{,}000$
$\left(\frac{2000}{3}, \frac{800}{3}\right)$
$(200, 0)$
$2x + y = 1600$
$(800, 0)$

**Section 7** ■ *page 750*

**1.** Maximum of 76 at $(7, 2)$; minimum of 44 at $(3, 2)$
**3.** Maximum of 50 at $(5, 6)$; minimum of 0 at $(0, 0)$
**5.** Maximum of 28 at $(7, 0)$; minimum of $-32$ at $(0, 8)$
**7.** Maximum of 85 at $(4, 3)$
**9.** Minimum of 150 at $(10, 0)$
**11.** Maximum of 144 at $(12, 2)$
**13.** Minimum of 108 at $(3, 2)$
**15.** Maximum of 88 at $(4, 8)$
**17.** 22 of type A; 3 of type B
**19.** 4 days for refinery 1; 3 days for refinery 2
**21.** Steve should travel 100 miles by car and 200 miles by truck. This will cost $16.00.
**23.** 18 minutes for the senator, 12 minutes for the governor
**25.** 4 ounces of Swiss steak, $5\frac{1}{2}$ ounces of peas

**Chapter Review** ■ *page 754*

**1.** $\begin{bmatrix} 6 & -3 \\ -15 & 9 \end{bmatrix}$ **3.** $\begin{bmatrix} 3 & -1 & 2 \\ 3 & 5 & 5 \end{bmatrix}$ **5.** $\begin{bmatrix} 5 & -2 \\ 6 & 9 \end{bmatrix}$

**7.** Not defined; the number of columns of $C$ is not equal to the number of rows of $D$.

**9.** $\begin{bmatrix} 9 & 1 \\ -8 & 8 \\ 10 & 0 \end{bmatrix}$

**11.** Not defined; the number of rows of $C$ is not equal to the number of columns of $C$ ($C$ isn't square).

**13.** $\begin{bmatrix} -9 \\ -1 \end{bmatrix}$    **15.** $[2]$    **17.** $\begin{bmatrix} -12 & 4 \\ -8 & -1 \\ 6 & 9 \end{bmatrix}$

**19.** $\begin{bmatrix} b - c + 2a \\ 3b - a \\ 4b + 2c \end{bmatrix}$    **21.** $\begin{bmatrix} 5 & 1 & | & 3 \\ 1 & -3 & | & 4 \end{bmatrix}$

**23.**    $x - 4y + z = 3$
              $y - 2z = 5$
                    $z = -2$
    $x = 9, y = 1, z = -2$

**25.** $\begin{bmatrix} 1 & -3 & | & 2 \\ 0 & 1 & | & -\frac{7}{3} \end{bmatrix}$    **27.** $\begin{bmatrix} 1 & 2 & -\frac{3}{2} & | & \frac{1}{2} \\ 0 & 1 & -\frac{3}{2} & | & \frac{1}{2} \end{bmatrix}$

**29.** $x = 3, y = 5$    **31.** $x = -5, y = -4, z = -3$

**33.** $x = 0, y = -4, z = -5$    **35.** $\begin{bmatrix} 1 & 0 & | & -2 \\ 0 & 1 & | & 3 \end{bmatrix}$

**37.** $x = 5, y = -4$    **39.** $x = 2, y = 5, z = 4$

**41.** $\begin{bmatrix} -1 & -1 \\ 3 & 2 \end{bmatrix}$    **43.** $\dfrac{1}{a^2 - b^2} \cdot \begin{bmatrix} a & -b \\ -b & a \end{bmatrix}$

**45.** $\begin{bmatrix} 1 & 2 & 1 \\ 0 & 1 & 2 \\ 0 & 0 & 1 \end{bmatrix}$    **47.** No inverse

**49. a.** $x = -7, y = 10$;    **b.** $x = -2, y = 4$

**51. a.** $x = \frac{39}{4}, y = -\frac{5}{4}$    **b.** $x = \frac{9}{2}, y = -\frac{3}{2}$

**53.** $-10$    **55.** $1$    **57.** $x^3 - x$    **59.** $35$

**61.** $x = \frac{1}{4}, y = -\frac{3}{4}$    **63.** $x = 1, y = -1, z = 5$

**65.**

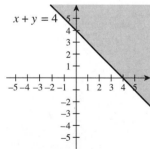

**67.**

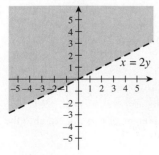

**69.**

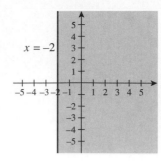

**71.**

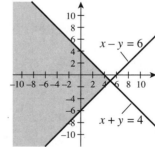

**73.**

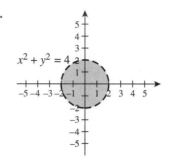

**75.**

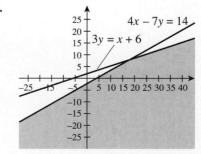

**77.**

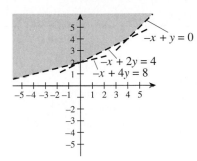

**79.**

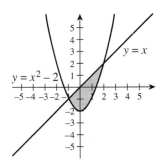

**81.**

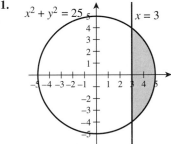

**83.** Minimum 46 at (6, 2)
Maximum 70 at (6, 5)
**85.** Maximum 68 at (4, 2)
**87.** Minimum 60 at (5, 2)
**89.** $40,000 money market, $35,000 mutual fund, $25,000 savings account
**91.** $\frac{13}{2}$
**93.** To maximize either revenue or profit plant 327.3 acres of corn and 90.9 acres of hay
**95.** $6,666.67 in bonds, $3,333.33 in mutual funds

**Chapter Test** ■ *page 757*

**1.** True      **2.** False      **3.** False      **4.** False      **5.** True
**6.** False     **7.** True       **8.** False

**9.** $\begin{bmatrix} 1 & 2 & 3 \\ 4 & 5 & 6 \end{bmatrix}$, for example      **10.** $\begin{bmatrix} 0 & 0 \\ 0 & 0 \end{bmatrix}$, for example

**11.** $\left[\begin{array}{cc|c} 1 & 1 & 2 \\ 0 & 1 & 3 \end{array}\right]$, for example

**12.** $xy = 1$
$x - y = 2$, for example
**13.** Switching two rows, for example

**14. a.** $\begin{bmatrix} -6 & -7 & 9 \\ 16 & 6 & 21 \\ -14 & -9 & -4 \end{bmatrix}$

**b.** $\begin{bmatrix} 4 & -4 & 5 \\ 12 & 11 & -3 \\ 4 & -6 & 1 \end{bmatrix}$

**c.** $\begin{bmatrix} 2 & 1 & -5 \\ -3 & 1 & -3 \\ -4 & -3 & 2 \end{bmatrix}$

**15.** $\begin{bmatrix} 3 & -5 & 1 \\ 2 & -8 & 0 \\ 3 & 0 & -1 \end{bmatrix} \begin{bmatrix} x \\ y \\ z \end{bmatrix} = \begin{bmatrix} 2 \\ -2 \\ 0 \end{bmatrix}$;
$x = \frac{13}{19}, y = \frac{8}{19}, z = \frac{39}{19}$

**16. a.** $2x - 2y + 2z = 0$
       $y - 3z = -2$
            $4z = 8$
$x = 2, y = 4, z = 2$

**b.** $\left[\begin{array}{ccc|c} 1 & 0 & 0 & 2 \\ 0 & 1 & 0 & 4 \\ 0 & 0 & 1 & 2 \end{array}\right]$

**17.** $x = 2, y = 3, z = -5$      **18.** $a = -2, b = 11, c = 4$
**19.** Approximately 71.0 million tons of waste paper, 33.2 million tons of yard waste, and 75.4 million tons of miscellaneous waste
**20.** Maximum: 0 at (0, 0)

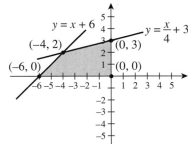

**21.** $y > \frac{3}{2}x$
$y < -x + 5$
$y > -6x$

## CHAPTER 11

### Section 1 ■ page 768

**1.** 4, 7, 10, 13     **3.** $-4, 16, -64, 256$     **5.** $1, \frac{1}{4}, \frac{1}{9}, \frac{1}{16}$

**7.** 0, 2, 0, 2     **9.** $1, -\frac{1}{2}, \frac{1}{3}, -\frac{1}{4}$     **11.** $\sqrt{2}, \sqrt{6}, 2\sqrt{3}, 2\sqrt{5}$

**13.** $2, \frac{9}{4}, \frac{64}{27}, \frac{625}{256}$     **15.** 1, 3, 6, 10     **17.** $a_n = 2n$

**19.** $a_n = (-\frac{1}{2})^n$     **21.** $a_n = \dfrac{n}{n + 1}$     **23.** $a_n = 1 - \dfrac{1}{n}$

**25.** arithmetic; $d = 3$     **27.** neither

**29.** arithmetic; $d = -4$     **31.** neither

**33.** geometric; $r = \frac{1}{10}$     **35.** $a_n = 6n + 1$

**37.** $a_n = 115 - 15n$     **39.** $a_n = \dfrac{5n - 13}{4}$     **41.** $a_{10} = 75$

**43.** $a_{15} = -60$     **45.** $a_{31} = 176$     **47.** $a_n = 3^{n-1}$

**49.** $a_n = 32(-\frac{1}{2})^{n-1}$     **51.** $a_n = 4(\frac{5}{4})^{n-1}$     **53.** $a_8 = 4374$

**55.** $a_7 = \frac{1}{81}$     **57.** $2, -3, 7, -13, 27$     **59.** 1, 2, 6, 24, 120

**61.** $2, \sqrt{2}, \sqrt[4]{2}, \sqrt[8]{2}, \sqrt[16]{2}$     **63.** 1, 2, 2, 4, 8

**65.** \$29,355; \$30,591; \$31,827; $d = 1236$; \$36,771     **67.** 74

**69.** $a_n = 1000(1.055)^{n-1}$; \$1619.09; 5.5%

**71.**

| Month | Pairs at the beginning of the month | Pairs born during the month | Pairs at the end of the month |
|---|---|---|---|
| 1 | 1 | 0 | 1 |
| 2 | 1 | 1 | 2 |
| 3 | 2 | 1 | 3 |
| 4 | 3 | 2 | 5 |
| 5 | 5 | 3 | 8 |
| 6 | 8 | 5 | 13 |
| ⋮ | ⋮ | ⋮ | ⋮ |

### Section 2 ■ page 779

**1.** 75     **3.** $\frac{71}{20}$     **5.** 51     **7.** $-15$     **9.** 60

**11.** $\frac{19}{20}$     **13.** $\frac{155}{16}$     **15.** $\displaystyle\sum_{k=1}^{6} \frac{1}{k^3}$     **17.** $\displaystyle\sum_{k=1}^{12} 5k + 2$

**19.** $\displaystyle\sum_{k=1}^{7} 3^k$     **21.** $\displaystyle\sum_{k=1}^{16} \frac{(-1)^k}{k^2}$     **23.** $\displaystyle\sum_{k=1}^{6} \frac{2^k - 1}{2^k}$     **25.** 1640

**27.** 975     **29.** $-240$     **31.** 3725     **33.** 3410

**35.** 724     **37.** 147,620     **39.** Approximately 4.5

**41.** 3.6363636     **43.** 262,140     **45.** Approximately 1.7143

**47.** Approximately $-0.0012$     **49.** 500,500     **51.** 10,000

**53.** 294     **55.** 900     **57.** Approximately \$1028.5 billion

**59.** \$5323.58     **61.** \$29,344.42     **63.** \$150

**65.** Approximately \$50.8 million

### Section 3 ■ page 791

**1.**

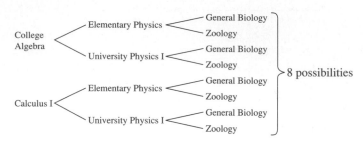

**3.**

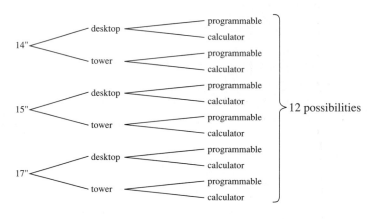

**5. a.** 125     **b.** 60

**7.** 175,760,000     **9.** 720     **11.** 720     **13.** 30

**15.** 120     **17.** $_8P_3 = 336$

**19. a.** $_{13}P_9 = 259,459,200$     **b.** $_2P_2 \cdot {}_{11}P_7 = 3,326,400$

**21.** 15     **23.** 8     **25.** 210     **27.** 56

**29. a.** $_{12}C_6 = 924$     **b.** $_5C_2 \cdot {}_7C_4 = 350$

**31.** $_{10}P_4 = 5040$     **33.** $_{40}C_5 = 658,008$

**35. a.** $_{40}P_3 = 59,280$     **b.** $40^3 = 64,000$

**37.** $_6C_4 = 15$; $_6C_4 \cdot {}_{15}C_{10} = 1260$

**39.** $2 \cdot {}_{30}C_2 = 870$ seconds; $2 \cdot {}_{60}C_2 = 3540$ seconds

**41.** $4 \cdot {}_{13}C_5 = 5148$

### Section 4 ■ page 803

**1.** $S = \{1, 2, 3, 4\}$; $n(S) = 4$

**3.** $S = \{HHH, HHT, HTH, HTT, THH, THT, TTH, TTT\}$; $n(S) = 8$

**5.** $\frac{1}{6}$     **7.** 1     **9.** $\frac{1}{4}$     **11.** $\frac{7}{12}$     **13.** $\frac{1}{52}$     **15.** $\frac{1}{13}$

**17.** $\frac{3}{26}$     **19.** $\frac{1}{4}$     **21.** $\frac{3}{8}$     **23.** $\frac{5}{18}$     **25.** $\frac{1}{6}$     **27.** $\frac{1}{12}$

**29.** $\dfrac{1}{54,145} \approx 1.85 \times 10^{-5}$    **31.** $\dfrac{33}{16,660} \approx 0.002$    **33.** $\frac{1}{16}$

**35.** $\frac{1}{8}$    **37.** $\frac{5}{16}$

**39.** $P(E) = \frac{1}{2}$; $P(F) = \frac{3}{13}$; $E \cap F$ is the event that a red face card is drawn; $P(E \cap F) = \frac{3}{26}$; $E \cup F$ is the event that a red card or a face card is drawn; $P(E \cup F) = \frac{8}{13}$; $E'$ is the event that a black card is drawn; $P(E') = \frac{1}{2}$; $F'$ is the event that a non-face card is drawn; $P(F') = \frac{10}{13}$.

**41.** $P(E) = \frac{1}{2}$; $P(F) = \frac{7}{12}$; $E \cap F$ is the event that the sum is even and less than 8; $P(E \cap F) = \frac{1}{4}$; $E \cup F$ is the event that the sum is even or less than 8; $P(E \cup F) = \frac{5}{8}$; $E'$ is the event that the sum is odd; $P(E') = \frac{1}{2}$; $F'$ is the event that the sum is greater than or equal to 8; $P(F') = \frac{5}{12}$.

**43.** 0.35

**45.** $\dfrac{1}{658,008} \approx 1.52 \times 10^{-6}$; $\dfrac{5}{164,502} \approx 3.04 \times 10^{-5}$; 329,004

**47. a.** 0.19    **b.** 0.77

**49.** $\frac{9}{28} \approx 0.32$

**51. a.** 0.179    **b.** 0.031    **c.** 0.88    **d.** 0.089    **e.** 0.059

### Section 5 ■ page 813

**1.** 4    **3.** 10    **5.** 6    **7.** 54    **9.** 6    **11.** 10

**13.** $x^5 + 5x^4y + 10x^3y^2 + 10x^2y^3 + 5xy^4 + y^5$

**15.** $16x^4 - 32x^3y + 24x^2y^2 - 8xy^3 + y^4$

**17.** $27x^3 + 135x^2y + 225xy^2 + 125y^3$

**19.** $x^5 + 5x^4 + 10x^3 + 10x^2 + 5x + 1$

**21.** $z^6 - 12z^5 + 60z^4 - 160z^3 + 240z^2 - 192z + 64$

**23.** $7 + 5\sqrt{2}$    **25.** $49 + 20\sqrt{6}$    **27.** $x^6 - 3x^4 + 3x^2 - 1$

**29.** $x^5 + 5x^3 + 10x + \dfrac{10}{x} + \dfrac{5}{x^3} + \dfrac{1}{x^5}$    **31.** $-2 + 2i$

**33.** $-64$    **35.** $i$

### Chapter Review ■ page 822

**1.** $-4, -9, -14, -19, -24$; arithmetic; $d = -5$

**3.** $e^{-1}, 2e^{-2}, 3e^{-3}, 4e^{-4}, 5e^{-5}$    **5.** $-\frac{1}{2}, \frac{3}{2}, -\frac{1}{2}, \frac{3}{2}, -\frac{1}{2}$

**7.** 2, 6, 18, 54, 162; geometric; $r = 3$

**9.** $a_n = 2(5)^{n-1}$; geometric; $r = 5$    **11.** $a_n = \dfrac{1}{n}$

**13.** $a_n = \frac{7}{2} - \frac{5}{2}n$; arithmetic; $d = -\frac{5}{2}$

**15.** $a_n = (-1)^{n-1}(n)^3$    **17.** 40    **19.** Approximately 0.4914

**21.** $\displaystyle\sum_{k=1}^{11} 5k - 2$    **23.** $\displaystyle\sum_{k=2}^{15} \dfrac{(-1)^k}{k}$    **25.** 290

**27.** Approximately 12.5    **29.** Approximately 2.1428

**31.** 7532    **33.** 36    **35.** 6720    **37.** 10    **39.** 120

**41.** $_{20}P_4 = 116,280$

**43. a.** $_{10}C_5 = 252$    **b.** $_6C_3 \cdot _4C_2 = 120$

**45.** 1024    **47.** $\frac{1}{4}$    **49.** $\frac{11}{26}$    **51.** $\frac{1}{4}$    **53.** $\frac{1}{9}$    **55.** $\frac{1}{8}$

**57.** 0.359    **59.** 0.379    **61.** 56    **63.** $-1000$

**65.** $x^4 - 4x^3y + 6x^2y^2 - 4xy^3 + y^4$    **67.** $r^3 + \dfrac{3r^2}{s} + \dfrac{3r}{s^2} + \dfrac{1}{s^3}$

**69.** $z^2\sqrt{z} + 10z^2 + 40z\sqrt{z} + 80z + 80\sqrt{z} + 32$

**77.** 210,000 in 1996; 220,500 in 1997; 231,525 in 1998; Approximately 359,171 in 2007

**79.** Approximately \$342,700.67

**81. a.** $_{80}C_{20}$    **b.** $(_{20}C_5)^4$

**83.** $\frac{1}{120}$

### Chapter Test ■ page 824

**1.** False    **2.** True    **3.** True    **4.** True    **5.** False

**6.** True    **7.** False    **8.** False

**9.** $a_n = -4n$, for example    **10.** $a_n = (\frac{1}{3})^n$, for exmaple

**11.** $\displaystyle\sum_{k=1}^{5} \dfrac{1}{k}$, for example

**12.** Tossing three coins, for example

**13.** A red ball is drawn from a bag containing one red ball, one white ball, and one blue ball, for example.

**14.** 1, 4, 6, 4, 1, for example    **15.** 4, 13, 28, 49, 76

**16.** $a_n = \dfrac{n+1}{n}$    **17.** 126    **18.** $\displaystyle\sum_{k=1}^{6} 5(\frac{1}{2})^k \approx 4.9219$

**19.** 256; $\frac{1}{256}$

**20. a.** $_{15}C_9 = 5005$    **b.** $_7C_5 \cdot _8C_4 = 1470$
The combinations would be replaced with permutations.

**21.** $\frac{1}{3}$    **22.** $\frac{8}{13}$    **23.** 393,750

**24.** $16x^4 - 96x^3y + 216x^2y^2 - 216xy^3 + 81y^4$

# INDEX

# GEOMETRIC FORMULAS

### Right triangle

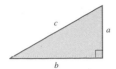

Pythagorean Theorem: $a^2 + b^2 = c^2$

### Triangle

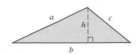

Area $= \frac{1}{2}bh$      Perimeter $= a + b + c$

### Equilateral triangle

$h = \frac{\sqrt{3}}{2}s$      Area $= \frac{\sqrt{3}}{4}s^2$

### Rectangle

Area $= lw$      Perimeter $= 2l + 2w$

### Trapezoid

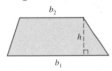

Area $= \frac{1}{2}(b_1 + b_2)h$

### Circle

Area $= \pi r^2$      Circumference $= 2\pi r$

### Rectangular box

Volume $= lwh$      Surface area $= 2lw + 2wh + 2lh$

### Right circular cylinder

Volume $= \pi r^2 h$      Lateral surface area $= 2\pi rh$

### Sphere

Volume $= \frac{4}{3}\pi r^3$      Surface area $= 4\pi r^2$

### Right square pyramid

Volume $= \frac{1}{3}b^2 h$      Lateral surface area $= b\sqrt{4h^2 + b^2}$

### Right circular cone

Volume $= \frac{1}{3}\pi r^2 h$      Lateral surface area $= \pi r\sqrt{r^2 + h^2}$

# TRIGONOMETRY

## Right Triangle Definitions

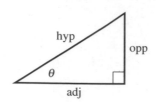

$$\sin\theta = \frac{\text{opp}}{\text{hyp}} \qquad \cos\theta = \frac{\text{adj}}{\text{hyp}} \qquad \tan\theta = \frac{\text{opp}}{\text{adj}}$$

$$\csc\theta = \frac{\text{hyp}}{\text{opp}} \qquad \sec\theta = \frac{\text{hyp}}{\text{adj}} \qquad \cot\theta = \frac{\text{adj}}{\text{opp}}$$

## Unit Circle Definitions

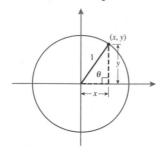

$$\sin\theta = y \qquad \cos\theta = x \qquad \tan\theta = \frac{y}{x}, x \neq 0$$

$$\csc\theta = \frac{1}{y}, y \neq 0 \qquad \sec\theta = \frac{1}{x}, x \neq 0 \quad \cot\theta = \frac{x}{y}, y \neq 0$$

## Special Angles

| $\theta$ | $\theta$ (radians) | $\sin\theta$ | $\cos\theta$ | $\tan\theta$ |
|---|---|---|---|---|
| 0° | 0 | 0 | 1 | 0 |
| 30° | $\frac{\pi}{6}$ | $\frac{1}{2}$ | $\frac{\sqrt{3}}{2}$ | $\frac{\sqrt{3}}{3}$ |
| 45° | $\frac{\pi}{4}$ | $\frac{\sqrt{2}}{2}$ | $\frac{\sqrt{2}}{2}$ | 1 |
| 60° | $\frac{\pi}{3}$ | $\frac{\sqrt{3}}{2}$ | $\frac{1}{2}$ | $\sqrt{3}$ |
| 90° | $\frac{\pi}{2}$ | 1 | 0 | undefined |
| 180° | $\pi$ | 0 | −1 | 0 |
| 270° | $\frac{3\pi}{2}$ | −1 | 0 | undefined |
| 360° | $2\pi$ | 0 | 1 | 0 |

## Ratio Identities

$$\sin\theta = \frac{1}{\csc\theta} \qquad \cos\theta = \frac{1}{\sec\theta} \qquad \tan\theta = \frac{1}{\cot\theta}$$

$$\csc\theta = \frac{1}{\sin\theta} \qquad \sec\theta = \frac{1}{\cos\theta} \qquad \cot\theta = \frac{1}{\tan\theta}$$

$$\tan\theta = \frac{\sin\theta}{\cos\theta} \qquad \cot\theta = \frac{\cos\theta}{\sin\theta}$$

## Pythagorean Identities

$$\sin^2\theta + \cos^2\theta = 1 \qquad 1 + \tan^2\theta = \sec^2\theta \qquad 1 + \cot^2\theta = \csc^2\theta$$